T0264060

Solid State Physics

Solid State Physics
Second Edition

Giuseppe Grosso
Department of Physics, University of Pisa and
NEST, Institute of Nanoscience-CNR, Pisa, Italy

Giuseppe Pastori Parravicini
Department of Physics, University of Pavia, Italy

ELSEVIER

AMSTERDAM • BOSTON • HEIDELBERG • LONDON • NEW YORK
OXFORD • PARIS • SAN DIEGO • SAN FRANCISCO • SINGAPORE
SYDNEY • TOKYO
Academic Press is an imprint of Elsevier

Academic Press is an imprint of Elsevier
The Boulevard, Langford Lane, Kidlington, Oxford OX5 1GB, UK
Radarweg 29, PO Box 211, 1000 AE Amsterdam, The Netherlands
525 B Street, Suite 1900, San Diego, CA 92101-4495, USA
225 Wyman Street, Waltham, MA 02451, USA

Second edition 2014

British Library Cataloguing in Publication Data
A catalogue record for this book is available from the British Library

Library of Congress Cataloging-in-Publication Data
A catalog record for this book is available from the Library of Congress

For information on all Academic Press publications
visit our web site at store.elsevier.com

Printed and bound by CPI Group (UK) Ltd, Croydon, CR0 4YY

ISBN: 978-0-12-385030-0

Working together
to grow libraries in
developing countries

www.elsevier.com • www.bookaid.org

Contents

CONTENTS SYNOPSIS

Chapters 1, 2, 3	Introductory information
Chapters 4, 5, 6, 7	Electronic structure of crystals
Chapters 8, 9	Adiabatic principle and lattice vibrations
Chapters 10, 11, 12, 13, 14	Scattering; optical and transport properties
Chapters 15, 16, 17	Magnetic field effects and magnetism
Chapter 18	Superconductivity

Preface to the second edition

These last years have witnessed a continuous progress in the traditional areas of solid state physics, as well as the emergence of new areas of research, rich of results and promises. The countless novelties appeared in the literature with the beginning of this century are one of the main motivations of the second edition of "Solid State Physics." Besides the inclusion of new material, the content has been greatly rationalized and the manuscript appears now in a widely renewed dress. Furthermore, this second edition is accompanied and enriched by the presence of numerous Appendices, containing better in-depth treatment of specific topics, or containing significant solved problems, for the readers willing to increase the proper technical abilities.

The chapters of "Solid State Physics" are in large part self-contained, and can be chosen with great flexibility and split in two semesters to cover an annual course for the degree in physics, or also for the degree in material science and electrical engineering. The volume is also suited for the graduate students, who have not followed similar in-depth courses in their curricula. From a didactical point of view, it is a steady characteristics of this book to enhance gradually the level of difficulty starting from the initial models and leading as near as possible to the frontier science. The purpose is to provide within the reasonable length of a textbook the general cultural baggage needed for all researchers, whose activity is oriented, or is going to be oriented, in the fascinating field of solid state physics.

We wish to express our deepest gratitude to our colleagues for their valuable suggestions, and to our students for their challenging questions; in their own way, they have made the most relevant contribution to improve and inspire this textbook. Many warm thanks are due to Donna de Weerd-Wilson for her encouragement in pursuing this project, and to Paula Callaghan, to Anita Koch, to Sharmila Vadivelan, and to Jessica Vaughan for their assistance and professionalism in the preparation of the manuscript.

Pavia and Pisa, May 2013
Giuseppe Grosso and Giuseppe Pastori Parravicini

Preface to the Second Edition

Preface to the first edition

This textbook has developed from the experience of the authors in teaching the course of Solid State Physics to students in physics at the Universities of Pavia and Pisa. The book is addressed to students at the graduate and advanced level, both oriented toward theoretical and experimental activity. No particular prerequisite is required, except for the ordinary working knowledge of wave mechanics. The degree of difficulty increases somewhat as the book progresses; however, the contents develop always in very gradual steps.

The material presented in the book has been assembled to make as economical as possible the didactical task of teaching, or learning, the various subjects. The general organization in chapters, and groups of chapters, is summarized in the synoptic table of contents. The first three chapters have a propaedeutic nature to the main entries of the book. Chapter IV starts with the analysis of the electronic structure of crystals, one of the most traditional subjects in solid state physics; Chapters V and VI concern the band theory of solids and a number of specific applications; the concepts of excitons and plasmons are given in Chapter VII. Then, in Chapters VIII and IX, the adiabatic principle and the interdependence of electronic states and lattice dynamics are studied. Having established the electronic and vibrational structure of crystals, the successive chapters from X to XIV describe several investigative techniques of crystalline properties; these include scattering of particles, optical spectroscopy, and transport measurements. Chapters XV, XVI, and XVII concern the electronic magnetism of tendentially delocalized or localized electronic systems, and cooperative magnetic effects. The final chapter is an introduction to the world of superconductivity.

From a didactical point of view, an effort is made to remain as rigorous as possible, while keeping the presentation at an accessible level. In this book, on one hand we aim to give a clear presentation of the basic physical facts, on the other hand we wish to describe them by rigorous theoretical and mathematical tools. The technical side is given the due attention and is never considered optional; in fact, a clear supporting theoretical formalism (without being pedantic) is essential to establish the limits of the physical models, and is basic enough to allow the reader eventually to move on his or her own legs, this being the ultimate purpose of a useful book.

The various chapters are organized in a self-contained way for the contents, appendixes (if any), references, and have their own progressive numeration for tables, figures, and formulae (the chapter number is added when addressing items in chapters differing from the running one). With regard to the references, these are intended only as indicative, since it is impossible to mention, let alone to comment on, all the relevant contributions of the wide literature. Since the chapters are presented in a (reasonably) self-contained way, the lecturers and readers are not compelled to follow the order in which the various subjects are discussed; the chapters, or group of chapters, can be taken up with great flexibility, selecting those topics that best fit personal tastes or needs. We will be very interested and very pleased to receive (either directly or by correspondence) comments and suggestions from lecturers and readers.

The preparation of a textbook, although general in nature, requires also a great deal of specialized information. We consider ourselves fortunate for the generous help of the colleagues and friends at the Physics Departments of the University of Pavia, University of Pisa, and Scuola Normale Superiore of Pisa; they have contributed to making this textbook far better by sharing their expertise with us. Very special thanks are due to Emilio Doni, who survived the task of reading and commenting on the whole preview-manuscript. Several other colleagues helped with their critical reading of specific chapters; we are particularly grateful to Lucio Claudio Andreani, Antonio Barone, Pietro Carretta, Alberto Di Lieto, Giorgio Guizzetti, Franco Marabelli, Liana Martinelli, Attilio Rigamonti. Many heartfelt thanks are due to Saverio Moroni for his right-on-target comments.

Before closing, we wish to thank all contributors and publishers, who gave us permission to reproduce their illustrations; their names and references are indicated adjacent each figure. We wish also to express our gratitude to Gioia Ghezzi for her encouragement, to Serena Bureau, Manjula Goonawardena, Cordelia Sealy, and Bridget Shine for their assistance in the preparation of the manuscript. Last, but always first, we thank our families for their never ending trust that the manuscript would eventually be completed and would be useful to somebody.

Pavia and Pisa, May 1999
Giuseppe Grosso and Giuseppe Pastori Parravicini

1 Electrons in One-Dimensional Periodic Potentials

Low dimensionality offers a unique opportunity to introduce some relevant concepts of solid state physics, keeping the treatment at a reasonably simple level. In the early days of solid state physics, just this desire for technical simplicity was the motivation for the studies on one-dimensional periodic potentials. More recently, with the restless developments in the world of nanoscience and nanotechnology, low-dimensional models have had a renaissance for the understanding of a number of realistic situations. At the same time a note of warning is necessary: the features of one-dimensional systems that have any relevance beyond dimensionality must be assessed situation by situation; cavalier extensions of one-dimensional results to actual three-dimensional crystals may be misleading or even completely unreliable.

The material of this chapter is organized and presented so that it can also be embodied in standard courses of "Structure of Matter" or "Quantum Mechanics"; in fact no previous knowledge is needed other than elementary ideas about wave mechanics.

This chapter begins with the presentation of the Bloch theorem for one-dimensional periodic lattices. A peculiar aspect of the energy spectrum of an electron in a periodic potential is the presence of allowed and forbidden energy regions. One-dimensional approaches are particularly suited to show from different points of view (weak binding, tight-binding, quantum tunneling, continued fractions) the mechanism of formation of energy bands in solids. The semiclassical dynamics of electrons in energy bands is then

Solid State Physics, Second Edition. http://dx.doi.org/10.1016/B978-0-12-385030-0.00001-3

considered; together with the Pauli exclusion principle for occupation of states, it gives a qualitative distinction between metals, semiconductors, and insulators.

Before beginning our trip among the one-dimensional models of primary interest in the field of solid state, we wish to notice the frequent "trespassings" that have occurred, in the course of the years, among areas quite different from the original context. For instance the Kronig-Penney model, originally suggested to justify qualitatively the formation of electronic energy bands in periodic materials, is often encountered as a precious tool in the study of electronic states in artificial superlattices, or of the propagation of electromagnetic waves in photonic crystals. Similarly, the concept of Bloch oscillations of electrons in periodic potentials and external electric fields, experimentally rather elusive even for semiconducting superlattices, has been well verified for atomic condensates in optical lattices, where gravity takes the role played by electric fields in semiconductors. Bloch oscillations of light waves have also been observed in dielectric superlattices, where a refractive index gradient plays the optical analog of the external force. Although the focus of this chapter is on one-dimensional periodic potentials, with traditional contributions going back to the early times of the foundation of quantum mechanics, it is worthwhile to be aware of the cross-fertilization of concepts and techniques between this historical area of solid state and other much younger and very active areas of research, which include for instance artificial structures and superlattices, organic crystals, photonic crystals, atomic condensates.

1.1   The Bloch Theorem for One-Dimensional Periodicity

Consider an electron in a one-dimensional potential energy $V(x)$ and the corresponding Schrödinger equation

$$-\frac{\hbar^2}{2m}\frac{d^2\psi(x)}{dx^2} + V(x)\psi(x) = E\psi(x). \tag{1.1}$$

The solutions of Eq. (1.1) for several typical forms of $V(x)$ are well known; familiar models include the free-electron case: $V(x) = 0$, the harmonic oscillator: $V(x) = (1/2)Kx^2$, the case of a uniform electric field: $V(x) = eFx$, quantum wells, and others. We focus here on the general properties of Eq. (1.1) in the case $V(x)$ is the periodic potential of a one-dimensional crystal of lattice constant a.

A potential $V(x)$, of period a, satisfies the relation

$$V(x) = V(x + ma), \tag{1.2}$$

with m arbitrary integer. The Fourier transform of a periodic potential $V(x)$ includes only plane waves of wavenumbers $h_n = n2\pi/a$, and $V(x)$ can be expressed in the form

$$V(x) = \sum_{n=-\infty}^{+\infty} V(h_n)e^{ih_n x}. \tag{1.3}$$

In general, if $V(x)$ is not periodic, it can still have a continuous Fourier transform $V(q)$ such that

$$V(x) = \int_{-\infty}^{+\infty} V(q)e^{iqx}\,dq. \tag{1.4}$$

We wish to analyze the implications on the eigenfunctions and eigenvalues of Eq. (1.1) brought about by the fact that the potential $V(x)$ is periodic, and hence its Fourier spectrum is discrete, according to Eq. (1.3).

Let us start considering Eq. (1.1) in the particular case that the periodic potential $V(x)$ vanishes (empty lattice). In the free-electron case, the wavefunctions are simply plane waves and can be written in the form

$$W_k(x) = \frac{1}{\sqrt{L}}e^{ikx}. \tag{1.5}$$

The normalization constant has been chosen such that $W_k(x)$ is normalized to 1 in the interval $0 \leq x \leq L$ (and the length L of the crystal is understood in the limit $L \rightarrow \infty$ whenever necessary). The wavenumbers k are real and the eigenvalues are $E(k) = \hbar^2 k^2/2m$. The plane waves (1.5) constitute a complete set of orthonormal functions, that can be conveniently used as an expansion set.

Let us now consider the eigenvalue problem (1.1), when the potential $V(x)$ is periodic and thus satisfies Eq. (1.3). If we apply the operator $H = (p^2/2m) + V(x)$ to the plane wave $W_k(x)$, we see that $H|W_k(x)\rangle$ belongs to the subspace S_k of plane waves of wavenumbers $k + h_n$:

$$S_k \equiv \{W_k(x), W_{k+h_1}(x), W_{k-h_1}(x), W_{k+h_2}(x), W_{k-h_2}(x)\ldots\}.$$

We also notice that the subspace S_k is *closed* under the application of the operator H to any of its elements; thus the diagonalization of the Hamiltonian operator within the subspace S_k provides eigenfunctions of H that can be labeled as $\psi_k(x)$. Notice that two subspaces S_k and $S_{k'}$ are different if k and k' are *not* related by integer multiples (positive, negative or zero) of $2\pi/a$; on the contrary, if $k \equiv k' + n2\pi/a$, then the two subspaces S_k and $S_{k'}$ coincide. This allows us to define a fundamental region of k-space, limited by $-\pi/a < k \leq \pi/a$, which includes all the different k labels giving independent S_k subspaces; this fundamental region, of length $2\pi/a$, is named *first Brillouin zone* (or simply *Brillouin zone*).

Any generic wavefunction $\psi_k(x)$, obtained by diagonalization of H within the subspace S_k, can be expressed as an appropriate linear combination of the type

$$\psi_k(x) = \sum_n c(k+h_n)\frac{1}{\sqrt{L}}e^{i(k+h_n)x}. \tag{1.6}$$

It is convenient to denote by $u_k(x)$ the function

$$u_k(x) = \sum_n c(k+h_n)\frac{1}{\sqrt{L}}e^{ih_n x} = \sum_n c(k+h_n)\frac{1}{\sqrt{L}}e^{in(2\pi/a)x}.$$

It is evident that $u_k(x)$ is a function with the same periodicity, a, as $V(x)$. Equation (1.6) then takes the form

$$\boxed{\psi_k(x) = e^{ikx}u_k(x)} \quad \text{with} \quad \boxed{u_k(x+a) = u_k(x)}. \tag{1.7}$$

This expresses the Bloch theorem: *any physically acceptable solution of the Schrödinger equation in a periodic potential takes the form of a traveling plane wave modulated on the microscopic scale by an appropriate function with the lattice periodicity.*

The Bloch theorem, summarized by Eq. (1.7), can also be written in the equivalent form

$$\boxed{\psi_k(x+t_n) = e^{ikt_n}\psi_k(x)}, \tag{1.8}$$

where $t_n = na$ is any translation in the direct lattice. It is easy to verify that Eq. (1.7) implies Eq. (1.8), and vice versa [to demonstrate the latter case, multiply both members of Eq. (1.8) by $\exp(-ikx - ikt_n)$ and denote by $u_k(x)$ the resulting periodic function]. The Bloch theorem, in the form of Eq. (1.8), shows that the values of the wavefunction $\psi_k(x)$ in any two points of the real space differing by a translation t_n are related by the phase $\exp(ikt_n)$.

We can finally notice that, in the case of a generic potential of type (1.4), the discretized expansion of Eq. (1.6) does not hold any more; in general the expansion of $\psi(x)$ includes all plane waves in the form

$$\psi(x) = \int_{-\infty}^{+\infty} c(q)e^{iqx}dq. \tag{1.9}$$

Nothing specific and general can be inferred from Eq. (1.9) about the properties of the wavefunctions and the energy spectrum of H: in fact, for aperiodic potentials, it is possible to find localized wavefunctions for the whole spectrum, itinerant ones, both types (separated by mobility edges), or even solutions belonging to fractal regions of the spectrum.

The Bloch theorem plays a central role in the physics of periodic systems; not only it characterizes the *itinerant form of the wavefunctions* summarized by Eqs. (1.7) and (1.8), but also entails the fact that the energy spectrum consists, in general, of *allowed energy regions separated by energy gaps* (as discussed in the forthcoming sections). The eigenvalues $E = E(k)$ of the Schrödinger equation (1.1), when plotted as a function of k within the first Brillouin zone, describe the band structure of the crystal; notice that $E(k) = E(-k)$, as can be seen from direct inspection of Eq. (1.1) under complex conjugate operation (or in a more formal way by use of group theory analysis of time reversal symmetry of the electronic Hamiltonian).

A feature of one-dimensional periodic potentials is that the allowed energy bands *cannot cross* each other; in fact for any allowed energy E there are only two linearly independent solutions of the differential Eq. (1.1), and the degeneracy $E(k) = E(-k)$ rules out any degeneracy at a given k. Thus, for any allowed energy band the dispersion relation $E(k)$ is a monotonic function of k for $0 \leq k \leq \pi/a$; the extremal energies occur only at $k = 0$ and π/a, where dE/dk in general vanishes. (The non-crossing of one-dimensional bands is confirmed also by group theory analysis; in one dimension the group of symmetry operations (neglecting spin) is too small to imply degeneracy of bands; on the contrary, in two- and three-dimensional crystals crossing of energy bands is possible at high symmetry points or lines in the Brillouin zone.)

So far we have considered Eq. (1.1) in the infinite interval $-\infty \leq x \leq \infty$; it is essentially equivalent from a physical point of view to consider Eq. (1.1) in the macroscopic region $0 \leq x \leq L \equiv Na$ where N is a very large but finite number (N is the number of unit cells of the crystal and is of order 10^8 for $L = 1$ cm). The reason to consider a very large macroscopic region, rather than an infinite one, is simply a matter of convenience, mainly for counting states and distributing electrons in the energy bands. In order not to affect the physics by boundary effects we use cyclic or Born-von Karman boundary conditions for the wavefunctions. This consists in the requirement

$$\psi(x + Na) \equiv \psi(x), \tag{1.10a}$$

i.e. the points x and $x + Na$ are considered as physically equivalent.

The wavefunction $\psi(x)$ must be a Bloch function of wavenumber k; then the boundary condition (1.10a) restricts the acceptable values of k to the ones that satisfy

$$e^{ikNa} = 1 \quad \Longrightarrow \quad k = \frac{2\pi}{Na}n \quad (n = 0, \pm 1, \pm 2, \dots). \tag{1.10b}$$

The density-of-states in k space is equal to $L/2\pi$, and is proportional to the length of the crystal. When the macroscopic length $L = Na$ is very large, the variable k must be thought of as a dense (although discrete) variable; notice that the first Brillouin zone contains *a number of uniformly distributed k points, equal to the number N of cells of the lattice.*

1.2 Energy Levels of a Single Quantum Well and of a Periodic Array of Quantum Wells

One of the most elementary problems in quantum mechanics is the study of the energy levels of a particle in a single quantum well. Similarly, one of the most elementary applications of the Bloch theorem is the study of the energy bands of a particle moving in a periodic array of quantum wells. The periodically repeated quantum well model was introduced by Kronig and Penney in 1931 to replace the actual crystal potential with a much more manageable piecewise constant potential; in this way, in each well (or in each barrier) the linearly independent solutions of the Schrödinger equation are simple trigonometric (or exponential) functions. Standard boundary conditions of continuity of wavefunctions and currents, combined with Bloch conditions required by periodicity, easily lead to an analytic compatibility equation for the eigenvalues of the crystal Hamiltonian. Successively, the Kronig-Penney model, with appropriate generalizations, has found a revival in the very active field of research of the electronic states of layered structures and superlattices [superlattices are artificial materials, obtained by growing on a substrate a controlled number of layers of two or more chemically similar crystals, in an appropriate periodic sequence]. Applications also include other models of field propagation, such as mechanical waves in phononic crystals and electromagnetic waves in photonic crystals.

Energy Levels of a Single Quantum Well

Before considering the one-dimensional motion of a particle in a periodic sequence of potential wells, we begin with the model of a single one-dimensional well of finite height V_0 and width w. The energy potential profile $V(x)$ is indicated in Figure 1.1; the potential is zero for $-w/2 < x < +w/2$ and equals a positive constant V_0 elsewhere. The potential has inversion symmetry around the center of the well; thus the eigenfunctions of Eq. (1.1) must be either even or odd under spatial inversion.

The general *even* solution $\psi(x) = \psi(-x)$ for bound states ($E < V_0$) takes the form

$$\psi(x) = \begin{cases} A\cos qx & \text{if } |x| < \frac{w}{2}; \quad q(E) = \sqrt{\dfrac{2mE}{\hbar^2}}; \qquad q_0 = \sqrt{\dfrac{2mV_0}{\hbar^2}}, \\[3mm] Be^{-\beta|x|} & \text{if } |x| > \frac{w}{2}; \quad \beta(E) = \sqrt{\dfrac{2m(V_0-E)}{\hbar^2}}; \quad \beta^2 + q^2 = q_0^2, \end{cases}$$

$$(1.11)$$

where $q(E)$ is the propagation wavenumber in the well and $\beta(E)$ is the damping of the wavefunction in the barrier; A and B are two arbitrary coefficients.

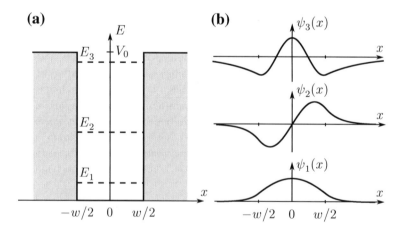

Figure 1.1 (a) Potential energy profile of a symmetric finite quantum well. In the case $V_0 = 1$ eV and $w = 14$ Å ($\approx 26\,a_B$) there are three bound states of energy $E_1 = 0.120$ eV, $E_2 = 0.464$ eV, and $E_3 = 0.936$ eV. Notice that the lowest energy levels of a quantum well in an infinite barrier $V_0 \to \infty$ of the same width are $E_1^\infty = 0.2$ eV, $E_2^\infty = 0.8$ eV, and $E_3^\infty = 1.8$ eV. (b) Normalized wavefunctions corresponding to the bound states of the finite quantum well under attention.

The standard boundary conditions requiring continuity of the wavefunction (1.11) and its derivative at $x = w/2$ (or $x = -w/2$) give

$$\begin{cases} A\cos\dfrac{qw}{2} = B\exp\left(-\dfrac{\beta w}{2}\right), \\ Aq\sin\dfrac{qw}{2} = B\beta\exp\left(-\dfrac{\beta w}{2}\right). \end{cases}$$

We thus obtain the transcendental equation

$$\boxed{\tan\dfrac{qw}{2} = \dfrac{\beta}{q}} \quad \text{(even solutions)}, \tag{1.12a}$$

which expresses the compatibility relation for even parity states.

A similar procedure can be done for the odd solutions $\psi(x) = -\psi(-x)$. The general *odd* solution, corresponding to a bound state, has the form $\psi(x) = A\sin qx$ in the well, and $\psi(x) = \text{sign}(x)Be^{-\beta|x|}$ outside. The standard boundary conditions requiring continuity of the wavefunction and its derivative at $x = w/2$ (or $x = -w/2$) give

$$\boxed{\tan\dfrac{qw}{2} = -\dfrac{q}{\beta}} \quad \text{(odd solutions)}. \tag{1.12b}$$

Using the trigonometric relation $\tan 2x = 2\tan x/(1 - \tan^2 x)$ we see that both conditions (1.12a) and (1.12b) are equivalent to the condition

$$\boxed{\tan qw = \dfrac{2q\beta}{q^2 - \beta^2}} \quad \text{(even and odd solutions)}. \tag{1.12c}$$

We will provide a graphical picture of the expressions above, after discussion of the infinite barrier case.

The case of infinite well is obtained in the limit $V_0 \to \infty$ and hence $\beta \to \infty$. Equations (1.12) are satisfied when $qw = n\pi$ ($n = 1, 2, \ldots$), and the energy spectrum becomes

$$E_n^\infty = \frac{\hbar^2}{2m}\left(\frac{n\pi}{w}\right)^2 = n^2\frac{\hbar^2}{2m}\frac{\pi^2}{w^2} \quad (n = 1, 2, \ldots), \tag{1.13}$$

where odd integers correspond to even solutions and even integers correspond to odd solutions. Equation (1.13) can also be written in the form

$$E_n^\infty = n^2\frac{\hbar^2}{2ma_B^2}\frac{\pi^2}{(w/a_B)^2} \quad\Longrightarrow\quad n^2\frac{9.87}{(w/a_B)^2}\ \text{Ryd} \quad\Longrightarrow\quad n^2\frac{134.25}{(w/a_B)^2}\ \text{eV},$$

where $a_B = 0.529$ Å is the Bohr radius and $\hbar^2/2ma_B^2 = 1$ Ryd $= 13.606$ eV. For example, for an infinite well of $w = 14$ Å ($\approx 26\,a_B$) we obtain $E_1^\infty = 0.2$ eV, $E_2^\infty = 0.8$ eV, and $E_3^\infty = 1.8$ eV.

In the case of finite height well, we have to consider the transcendental equations (1.12a) and (1.12b) here rewritten respectively in the forms

$$\tan \frac{qw}{2} = \frac{\sqrt{q_0^2 - q^2}}{q} \, (>0); \quad \cot \frac{qw}{2} = -\frac{\sqrt{q_0^2 - q^2}}{q} \, (<0). \tag{1.14}$$

We provide here a simple graphical solution, which shows how the states depend on the parameters V_0 and w.

It is convenient to introduce the dimensionless variable $x = qw/2$, and define $x_0 = q_0 w/2$. In the upper part of Figure 1.2, we plot the function $y = x \tan x$ in the domain where it is positive, and the function $y = x \cot x$ in the domain where it is negative. The intersections with the functions $\pm\sqrt{x_0^2 - x^2}$ determine the solutions of the compatibility equation (1.14). In Figure 1.2, as an exemplification, we have chosen x_0 in the interval $[\pi/2, \pi]$, in which case two solutions are possible, one even and one odd. From Figure 1.2 it is evident that in the limiting case of infinite well $x_0 \to \infty$, we recover the standard result $qw/2 = n\pi/2$, as in Eq. (1.13). In the case of finite height one-dimensional well, it is easily verified from Figure 1.2 that the number of bound states equals the number, increased by one, of confined states of the infinite well (1.13) with energy $E_n^\infty < V_0$. Thus we can notice the presence of at least one bound state, whatever values are given to w and V_0.

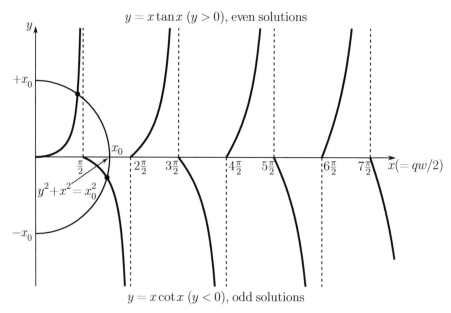

Figure 1.2 Graphical solution of the Eqs. (1.14) for a finite well. In the upper part of Figure 1.2, we plot the function $y = x \tan x$ in the domain where y is positive, in the lower part the function $y = x \cot x$ is plotted in the domain where it is negative. The intersections with the circle $y^2 + x^2 = x_0^2$ determine the solutions of the compatibility equation. Notice also that at least one bound state always occurs.

It is of interest to comment more on the presence of at least one bound state. Consider the situation in which $w \to 0$ and $V_0 \to \infty$ in such a way that the "strength" of the potential of the well, $C \equiv wV_0$, remains finite. From Eq. (1.12a), we can approximate, in this limit, the tangent function with its argument, and the energy dependent wavevector with the value corresponding to V_0, $q(E) \approx q(V_0)$. It follows:

$$\beta = \frac{1}{2}q_0^2 w, \quad q_0 = \sqrt{\frac{2mV_0}{\hbar^2}}.$$

Denoting by $E_b = V_0 - E > 0$ the binding energy, we have

$$\sqrt{\frac{2mE_b}{\hbar^2}} = \frac{1}{2}\frac{2mV_0}{\hbar^2}w \implies E_b = \frac{1}{2}\frac{m}{\hbar^2}C^2; \tag{1.15}$$

this relation gives the binding energy of an attractive one-dimensional δ-like potential of strength C, as shown also in Problem 1.

Energy Levels of Periodically Repeated Quantum Wells

Consider now the case of two (or more) potential wells; *if one decides to proceed with the same technique adopted for the single well enforcing the boundary conditions* (continuity of function and derivative at the border of the wells) one sees that the number of arbitrary coefficients and corresponding homogeneous equations increases linearly with N, where N is the number of potential wells. It becomes soon apparent that such a procedure, from the point of view of achieving analytic results and interpretations, is not particularly encouraging, even if the number of wells is as small as two or three. However, in the case the number of periodically repeated wells is infinite, we can again solve exactly the eigenvalue matching problem, by virtue of the Bloch theorem, as shown below.

Consider an infinite periodic sequence of rectangular wells, as shown in Figure 1.3; the lattice constant of the periodic array is $a = w + b$. Within the unit cell $-w < x < b$,

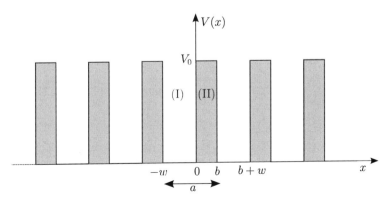

Figure 1.3 One-dimensional potential formed by a periodic array of quantum wells.

the well region $-w < x < 0$ and the barrier region $0 < x < b$ are denoted as I and II, respectively. In regions I and II, the general solution of the Schrödinger equation (1.1) for energies $0 < E < V_0$ has the form

$$\begin{cases} \psi_I(x) = Ae^{iqx} + Be^{-iqx} & -w < x < 0 & q(E) = \sqrt{2mE/\hbar^2}, \\ \psi_{II}(x) = Ce^{\beta x} + De^{-\beta x} & 0 < x < b & \beta(E) = \sqrt{2m(V_0 - E)/\hbar^2}, \end{cases}$$

(1.16)

where $q(E)$ is the propagation wavenumber in the well, $\beta(E)$ refers to the exponential damping (or growing) in the barrier, and A, B, C, D are arbitrary coefficients (we consider here positive energies E smaller than V_0; energies E larger than V_0 could be dealt with in a similar way).

The four arbitrary coefficients A, B, C, D must be chosen in such a way to satisfy the following four boundary conditions:

$$\psi_I(0) = \psi_{II}(0), \quad \left(\frac{d\psi_I}{dx}\right)_{x=0} = \left(\frac{d\psi_{II}}{dx}\right)_{x=0}, \tag{1.17a}$$

$$\psi_{II}(b) = e^{ika}\psi_I(-w), \quad \left(\frac{d\psi_{II}}{dx}\right)_{x=b} = e^{ika}\left(\frac{d\psi_I}{dx}\right)_{x=-w}. \tag{1.17b}$$

The two conditions (1.17a) impose the continuity of the wavefunction and its derivative at $x = 0$. The two conditions (1.17b) connect the wavefunction and its derivative at $x = -w$ and $x = b$ via the phase factor $\exp(ika)$, as required by the Bloch theorem.

The conditions (1.17) lead to the following linear homogeneous equations for the unknown coefficients A, B, C, D:

$$\begin{cases} A + B = C + D \\ Aiq - Biq = C\beta - D\beta \\ Ce^{\beta b} + De^{-\beta b} = e^{ika}[Ae^{-iqw} + Be^{+iqw}] \\ C\beta e^{\beta b} - D\beta e^{-\beta b} = e^{ika}[Aiqe^{-iqw} - Biqe^{+iqw}]. \end{cases}$$

The above four equations have a non-trivial solution only if the determinant of the coefficients that multiply A, B, C, D vanishes:

$$\begin{vmatrix} 1 & 1 & -1 & -1 \\ iq & -iq & -\beta & \beta \\ -e^{ika-iqw} & -e^{ika+iqw} & e^{\beta b} & e^{-\beta b} \\ -iqe^{ika-iqw} & iqe^{ika+iqw} & \beta e^{\beta b} & -\beta e^{-\beta b} \end{vmatrix} = 0.$$

The determinant can be evaluated, for instance, by expanding it with respect to the first row and by direct evaluation of the four 3×3 minors. With easy calculations and collection of terms we obtain

$$\boxed{\frac{\beta^2 - q^2}{2q\beta} \sinh \beta b \sin qw + \cosh \beta b \cos qw = \cos ka}. \tag{1.18}$$

It is possible to solve graphically the compatibility equation (1.18). Although not necessary, it is however a standard practice to adopt the further simplification to let the width of the potential barriers to approach zero and simultaneously the height of the barriers to approach infinity, conserving finite the area underneath. We thus arrive at a model potential consisting of a periodic sequence of δ-like potential barriers. In Eq. (1.18) we consider $b \to 0$ under the constraint V_0b constant; we obtain the simplified compatibility equation

$$\boxed{P\frac{\sin qa}{qa} + \cos qa = \cos ka} \quad \text{with} \quad P = \frac{mV_0ba}{\hbar^2}; \tag{1.19}$$

the dimensionless parameter P is proportional to the area V_0b of the barrier. For typical finite barriers width $a \approx b \approx 10\,a_B$ and $V_0 \approx 1$ eV, we have $P \approx 4$; then we choose $P = 3\pi/2$, as in the historical Kronig-Penney model.

The graphical solution of Eq. (1.19) is obtained plotting $F(x) = (P/x)\sin x + \cos x$ versus the dimensionless variable $x = qa$ (see Figure 1.4); the regions where $|F(x)| \le 1$, provide the energy levels $E = \hbar^2 q^2/2m$ that are allowed. From Figure 1.4, it can be seen by inspection that the energy levels are grouped into allowed energy bands separated by forbidden energy regions.

It is instructive to consider the compatibility equation (1.19) in the limiting cases of P very small or very large. In the case $P = 0$ we obviously recover the free-electron result. In the case $P \to \infty$, Eq. (1.19) is meaningful for $\sin qa \to 0$; thus the allowed energy bands become extremely narrow and the spectrum is composed by lines at energies E such that $q(E)a = n\pi (n = 1, 2, \dots)$. The energy levels are

$$E_n = n^2\frac{\hbar^2 \pi^2}{2m\, a^2} \quad (n = 1, 2, \dots),$$

and coincide with the energy levels of an infinite quantum well, of width a.

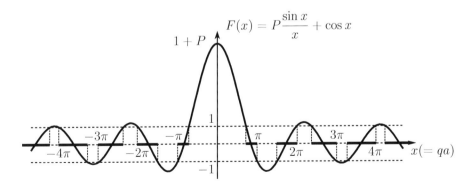

Figure 1.4 Graphical solution of the compatibility equation $(P/qa)\sin qa + \cos qa = \cos ka$ of the Kronig-Penney model. The dimensionless parameter P has been set equal to $3\pi/2$; the $qa \equiv x$ regions where the compatibility equation is satisfied have been enhanced for convenience.

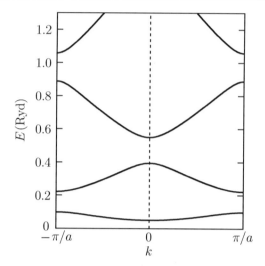

Figure 1.5 Energy bands of lowest energy in the Kronig-Penney model for $P = 3\pi/2$. The energy E (in Rydberg) is plotted as a function of k in the first Brillouin zone; the value of the lattice parameter considered is $a = 10\,a_B$ (a_B Bohr radius).

The allowed energy bands in the Kronig-Penney model can be calculated for any chosen value of P, and in Figure 1.5 we give the typical energy band structure. We can see in particular that the width of the allowed energy bands increases as E increases while the opposite occurs for the forbidden bands. We also notice that the extrema of E versus k appear at the center and at the border of the Brillouin zone, as required for any one-dimensional potential and discussed in Section 1.1, while the discontinuities of E versus k appear at the border of the Brillouin zone.

The fact that at $k = 0$ and $k = \pi/a$ we have $dE/dk = 0$ can be understood also on the basis of the following considerations. Because of time reversal symmetry we have $E(k) = E(-k)$ and the energy band is an even function of k with zero derivative at the center of the Brillouin zone. The energy $E(k)$ is an even function of k also around $k = \pi/a$, because of time reversal symmetry and invariance of $E(k)$ under any reciprocal lattice translation (in particular, the translation $h = -2\pi/a$ makes equivalent the points $\pi/a - \Delta k$ and $-\pi/a - \Delta k$, while time reversal makes equivalent the points $-\pi/a - \Delta k$ and $\pi/a + \Delta k$; then $E(\pi/a - \Delta k) = E(\pi/a + \Delta k)$).

1.3 Transfer Matrix, Resonant Tunneling, and Energy Bands

In this section we consider the electron propagation in one-dimensional crystals from the point of view of the *electron optics in solid state*, via the determination of the reflected and transmitted components of a wave impinging on a given potential. We present first some general properties of transfer matrices for the description of the elastic propagation through a potential of arbitrary shape. Then the conditions of resonant electron tunneling through a double barrier structure formed by two identical barriers is examined. Finally in the case of a crystal pictured as a periodic structure formed by identical barriers, we illustrate the formation of the energy bands from the point of view of quantum tunneling. In all three subsections, the general concepts are applied to rectangular barriers, where the simple analytic results allow a better insight and understanding of the physical aspects.

1.3.1 Transmission and Reflection of Electrons from an Arbitrary Potential

Consider an electron in a one-dimensional potential $V(x)$ of arbitrary profile within the finite region $x_L \leq x \leq x_R$, and connecting two semi-infinite regions *at the same constant potential*, taken to be zero for simplicity; the two semi-infinite regions are often addressed as *leads* and the finite region therein as *central device* or *scatterer* (see Figure 1.6). The corresponding Schrödinger equation reads

$$-\frac{\hbar^2}{2m}\frac{d^2\psi(x)}{dx^2} + V(x)\psi(x) = E\psi(x) \qquad (1.20)$$

with

$$V(x) \equiv 0 \quad \text{if} \quad x < x_L, x > x_R;$$

only in the interval $x_L < x < x_R$ the potential $V(x)$ can be different from zero.

The general solution of the Schrödinger equation for a positive energy E, in the left and right leads, can be written as

$$\begin{cases} \psi_L(x) = A_L e^{iqx} + B_L e^{-iqx} & x \leq x_L \\ \psi_R(x) = A_R e^{iqx} + B_R e^{-iqx} & x \geq x_R \end{cases} \quad q(E) = \sqrt{2mE/\hbar^2} \quad E > 0. \qquad (1.21)$$

In one dimension, at any energy E, the second order differential Schrödinger equation (1.20) has two linearly independent solutions. This entails two linear homogeneous

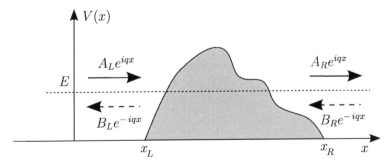

Figure 1.6 Schematic representation of a scatterer, made of a potential $V(x)$ of finite range and arbitrary shape, connected to two leads at zero potential. Arrows indicate the wavevectors of the propagating plane waves of energy E, wavenumber $q(E) = (2mE)^{1/2}/\hbar$ and amplitudes A_L, B_L, A_R, B_R.

equations for the four quantities A_L, B_L, A_R, B_R. One can choose to express any two amplitudes in terms of the other two. A frequently adopted convention expresses the amplitudes on the right (left) lead in terms of those on the left (right) lead by means of the *forward* (*backward*) *transfer matrix*. The forward transfer matrix $M(E)$ relates the amplitudes (A_R, B_R) of the incoming and outgoing waves in the right lead with the amplitudes (A_L, B_L) of the incoming and outgoing waves in the left lead, and is defined by

$$\begin{pmatrix} A_R \\ B_R \end{pmatrix} = M(E) \begin{pmatrix} A_L \\ B_L \end{pmatrix} = \begin{bmatrix} m_{11}(E) & m_{12}(E) \\ m_{21}(E) & m_{22}(E) \end{bmatrix} \begin{pmatrix} A_L \\ B_L \end{pmatrix}; \tag{1.22a}$$

(the backward transfer matrix is the inverse of the forward one). Equation (1.22a) can be written in extended form

$$\begin{cases} A_R = m_{11} A_L + m_{12} B_L; \\ B_R = m_{21} A_L + m_{22} B_L; \end{cases} \tag{1.22b}$$

for simplicity of notations, at times the energy dependence of variables is not indicated explicitly.

Alternatively, and equivalently, it is also frequent the use of the *scattering matrix* $S(E)$ that relates the outgoing wave amplitudes (B_L, A_R) from the scatterer with the incoming wave amplitudes (A_L, B_R) on the scatterer; namely:

$$\begin{pmatrix} B_L \\ A_R \end{pmatrix} \equiv S(E) \begin{pmatrix} A_L \\ B_R \end{pmatrix} \Longrightarrow S(E) = \begin{bmatrix} s_{11} & s_{12} \\ s_{21} & s_{22} \end{bmatrix} = \frac{1}{m_{22}} \begin{bmatrix} -m_{21} & 1 \\ \det M & m_{12} \end{bmatrix}; \tag{1.23}$$

the last passage in Eq. (1.23) is a straight elaboration of Eq. (1.22b). With respect to the transfer matrix, the scattering matrix approach may present advantages of numerical stability, in some situations that need large scale numerical calculations or involve strongly space dependent wavefunctions. In view of the rather simple models of our concern, we will continue our elaborations with the forward propagation, summarized by Eqs. (1.22), transferring wavefunctions from the left lead to the right one.

The transfer matrix M, corresponding to a given potential $V(x)$ and given energy E, can be obtained by (numerical or analytic) integration of the Schrödinger equation; before considering specific examples, we discuss the general properties of the transfer matrices. First of all we notice that the complex conjugate of any solution of the Schrödinger equation (1.20) is on its own right a solution at the same energy (this property reflects the time reversal symmetry of the Hamiltonian operator under attention). It follows:

$$m_{11} = m_{22}^* \quad \text{and} \quad m_{12} = m_{21}^*; \tag{1.24}$$

by virtue of time reversal symmetry the independent matrix elements of the transfer matrix are limited to two.

We can now show that the matrix M is unimodular, i.e. its determinant equals unity. This property follows from probability current conservation, in conjunction with the assumption that the constant potentials in the left and right leads are equal. (In the case the constant potentials in the left and right leads are different, some of the equations below need simple modifications.) In one dimension the probability current associated to an eigenfunction is given by

$$J(x) = \frac{1}{2m} \left[\psi^* p \psi - \psi p \psi^* \right] = -\frac{i\hbar}{2m} \left[\psi^* \frac{d\psi}{dx} - \psi \frac{d\psi^*}{dx} \right]. \tag{1.25}$$

Expression (1.25) is expected to be independent of x. In fact its differentiation gives

$$\frac{dJ(x)}{dx} = -\frac{i\hbar}{2m} \left[\psi^* \frac{d^2\psi}{dx^2} - \psi \frac{d^2\psi^*}{dx^2} \right] \equiv 0,$$

where the last equality occurs by virtue of the Schrödinger equation (1.20) and its complex conjugate. Using Eqs. (1.21) and (1.25), it is easily established that

$$J_L(x) = -\frac{i\hbar}{2m} \left[\psi_L^* \frac{d\psi_L}{dx} - \psi_L \frac{d\psi_L^*}{dx} \right] \equiv \frac{\hbar q}{m} \left[|A_L|^2 - |B_L|^2 \right]$$

and similarly for $J_R(x)$. Conservation of current probability through a structure, whose leads have the same constant potential (so that $q_L = q_R = q$), entails

$$|A_L|^2 - |B_L|^2 \equiv |A_R|^2 - |B_R|^2. \tag{1.26}$$

Using Eqs. (1.22) and (1.24) we have

$$|A_R|^2 - |B_R|^2 = |m_{11}A_L + m_{12}B_L|^2 - |m_{21}A_L + m_{22}B_L|^2$$
$$= [|m_{11}|^2 - |m_{21}|^2][|A_L|^2 - |B_L|^2] = \det M \left[|A_L|^2 - |B_L|^2 \right].$$

Comparison with Eq. (1.26) shows that *the transfer matrix between two leads at the same potential is unimodular.* We now determine the physical meaning of the independent matrix elements of the transfer matrix, and its unimodularity, in terms of reflection and transmission amplitudes and coefficients.

Reflection and Transmission Amplitudes and Coefficients

The transfer matrix provides an intuitive description of the electron tunneling through a given potential region. Consider in fact the stationary solution of the Schrödinger equation in the form of an impinging wave from the left lead, partially reflected and partially transmitted through the potential region (scattering region). In this specific case we have $B_R \equiv 0$, and Eqs. (1.22b) read

$$\begin{cases} A_R = m_{11}A_L + m_{12}B_L; \\ 0 = m_{21}A_L + m_{22}B_L; \end{cases}$$

from these two equations, we can easily obtain the reflection and transmission amplitudes and coefficients as follows.

The *reflection amplitude for a particle incoming from the left lead* is defined as $r = B_L/A_L$, and is given by

$$\boxed{r = -\frac{m_{21}}{m_{22}}}; \tag{1.27a}$$

the *reflection coefficient* is

$$R = rr^* = |r|^2 = \left| \frac{m_{21}}{m_{22}} \right|^2. \tag{1.27b}$$

The *transmission amplitude to the right lead for a particle incoming from the left lead* is $t = A_R/A_L$, and is given by

$$\boxed{t = \frac{1}{m_{22}}}, \tag{1.27c}$$

where the unimodularity of the transfer matrix has been taken into account. The *transmission coefficient* is

$$T = tt^* = |t|^2 = \left| \frac{1}{m_{22}} \right|^2. \tag{1.27d}$$

We can easily verify the general property $R + T = 1$. In terms of transmission and reflection amplitudes t and r for a particle incoming from the left lead, the transfer matrix can be written in the form

$$M = \begin{bmatrix} m_{11} & m_{12} \\ m_{21} & m_{22} \end{bmatrix} \equiv \begin{bmatrix} 1/t^* & -r^*/t^* \\ -r/t & 1/t \end{bmatrix} \tag{1.28}$$

with a clear physical meaning of its matrix elements, and its unimodularity.

A similar elaboration for the scattering matrix expressed by Eq. (1.23) gives

$$S(E) = \begin{bmatrix} s_{11} & s_{12} \\ s_{21} & s_{22} \end{bmatrix} = \begin{bmatrix} r & t' \\ t & r' \end{bmatrix},$$

where r and t are the reflection and transmission amplitudes for a particle incoming from the left, and r' and t' refer to the same quantities for a particle incoming from the right.

Given a potential $V(x)$ (localized in a given region) and the corresponding transfer matrix M, a particularly interesting situation occurs for those energies (if any) for which the transfer matrix is diagonal. This happens when

$$|m_{11}(E)| = 1 \quad \text{or equivalently} \quad m_{12}(E) = 0,$$

because of the unimodularity of the transfer matrix. These energies are called *resonance energies*. The transmission coefficient equals unity at the resonance energies; the incoming wave $\exp(iqx)$ is just transmitted with a change of phase equal to $\arg[m_{11}(E)]$, and the potential $V(x)$ is transparent to the particle flux at these special energies.

In the following we need to connect the transfer matrix M, corresponding to a given potential $V(x)$, with the transfer matrix $M(d)$ corresponding to the rigidly shifted potential $V(x - d)$. It is seen by straight elaboration of the defining Eqs. (1.21) and (1.22) that

$$\begin{pmatrix} A_R e^{+iqd} \\ B_R e^{-iqd} \end{pmatrix} \equiv M \begin{pmatrix} A_L e^{+iqd} \\ B_L e^{-iqd} \end{pmatrix} \implies \begin{pmatrix} A_R \\ B_R \end{pmatrix} \equiv M(d) \begin{pmatrix} A_L \\ B_L \end{pmatrix}$$

with

$$M(d) = \begin{bmatrix} e^{-iqd} & 0 \\ 0 & e^{+iqd} \end{bmatrix} \begin{bmatrix} m_{11} & m_{12} \\ m_{21} & m_{22} \end{bmatrix} \begin{bmatrix} e^{+iqd} & 0 \\ 0 & e^{-iqd} \end{bmatrix} \equiv \begin{bmatrix} m_{11} & m_{12}e^{-2iqd} \\ m_{21}e^{2iqd} & m_{22} \end{bmatrix};$$

(1.29)

thus for a scatterer shifted along the x-axis by the displacement d, the diagonal matrix elements of the transfer matrix do not change, while the off-diagonal elements acquire the phase $\exp(\pm 2iqd)$.

A nice property of the transfer matrix is that the total transfer matrix for a sequence of non-overlapping scatterers, described by the potentials $V_1(x), V_2(x), \ldots, V_n(x)$ ordered from left to right along the x-axis, is given by the product of transfer matrices

$$M^{(\text{tot})} = M_n \cdots M_2 M_1,$$

(1.30)

in the indicated ordered sequence. The simplicity of this composition is the key point of the transfer matrix procedure (while the composition of the scattering matrices is somewhat more elaborate, as can be seen by inspection).

As an application of the concepts explained so far, we consider here the case of tunneling through a single rectangular barrier. In the next subsection we examine the case of two barriers of equal shape and focus on the resonant tunneling effects. Finally, we will consider the case of periodically repeated rectangular barriers (i.e. the Kronig-Penney model) and the formation of energy bands within the transfer matrix framework.

Transfer Matrix for a Rectangular Barrier and for a Piecewise Potential

As an application of the concepts explained so far, we determine explicitly the transfer matrix for a rectangular barrier; such a matrix also constitutes the basic ingredient to build up the transfer matrix of any piecewise potential.

Consider the potential barrier of height V_0 in the interval $0 \le x \le b$, shown in Figure 1.7. In the whole x axis, we can distinguish three regions: the left lead, the intermediate barrier region, and the right lead. In the three regions, the general solution

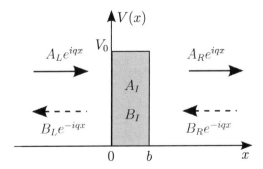

Figure 1.7 Rectangular potential barrier of height V_0 and width b connecting two leads at zero potential.

of the Schrödinger equation (1.20) for energies $0 < E < V_0$ has the form

$$\begin{cases} \psi_L(x) = A_L e^{iqx} + B_L e^{-iqx} & x < 0, \\ \psi_I(x) = A_I e^{\beta x} + B_I e^{-\beta x} & 0 < x < b, \\ \psi_R(x) = A_R e^{iqx} + B_R e^{-iqx} & x > b, \end{cases}$$

where $q^2(E) = 2mE/\hbar^2$ and $\beta^2(E) = 2m(V_0 - E)/\hbar^2$.

The standard boundary conditions of continuity of the wavefunction and its derivative at $x = 0$ give

$$\begin{cases} A_L + B_L = A_I + B_I, \\ A_L iq - B_L iq = A_I \beta - B_I \beta. \end{cases}$$

From the above equations, we can express (A_I, B_I) in terms of (A_L, B_L) in the matrix form

$$\begin{pmatrix} A_I \\ B_I \end{pmatrix} = \frac{1}{2\beta} \begin{bmatrix} iq + \beta & -iq + \beta \\ -iq + \beta & iq + \beta \end{bmatrix} \begin{pmatrix} A_L \\ B_L \end{pmatrix}. \tag{1.31a}$$

Similarly, we can consider the boundary conditions at $x = b$ and we can express (A_R, B_R) in terms of (A_I, B_I) in the form

$$\begin{pmatrix} A_R \\ B_R \end{pmatrix} = \frac{1}{2iq} \begin{bmatrix} (iq + \beta)e^{(-iq+\beta)b} & (iq - \beta)e^{(-iq-\beta)b} \\ (iq - \beta)e^{(+iq+\beta)b} & (iq + \beta)e^{(+iq-\beta)b} \end{bmatrix} \begin{pmatrix} A_I \\ B_I \end{pmatrix}. \tag{1.31b}$$

The direct multiplication of the transfer matrices in Eqs. (1.31a) and (1.31b) provides the transfer matrix M for the rectangular barrier; with straightforward calculations we obtain

$$\begin{array}{ll} m_{11} = e^{-iqb} \left[\cosh \beta b + i \dfrac{q^2 - \beta^2}{2q\beta} \sinh \beta b \right] & m_{22} = m_{11}^* \\[3mm] m_{12} = e^{-iqb}(-i)\dfrac{q^2 + \beta^2}{2q\beta} \sinh \beta b & m_{21} = m_{12}^* \end{array} \tag{1.32}$$

Expressions (1.32) provide the transfer matrix of a barrier of height V_0 in the interval $[0, b]$; for barriers rigidly shifted by a displacement d, the appropriate phase factors pointed out in Eq. (1.29) are to be included. The transfer matrix for any arbitrary piecewise potential is obtained simply multiplying in the appropriate order the matrices corresponding to each component barrier.

With the transfer matrix given in Eq. (1.32), we can analyze explicitly the electron tunneling through a rectangular potential barrier. From the relation $m_{11} = 1/t^*$, the expression of m_{11} given in Eq. (1.32), and the exponential form $t = |t| \exp(i\phi_t)$ for the transmission amplitude, we obtain

$$\frac{1}{|t|} e^{i[\phi_t + qb]} \equiv \cosh \beta b + i \frac{q^2 - \beta^2}{2q\beta} \sinh \beta b,$$

where $|t(E)|$ and $\phi_t(E)$ denote the modulus and the phase of the transmission amplitude. The above equality for the phases and for the moduli of the two members implies

$$\tan[\phi_t + qb] = \frac{q^2 - \beta^2}{2q\beta} \tanh \beta b \tag{1.33}$$

and

$$\frac{1}{|t|^2} = \cosh^2 \beta b + \frac{(q^2 - \beta^2)^2}{4q^2\beta^2} \sinh^2 \beta b \equiv 1 + \frac{(q^2 + \beta^2)^2}{4q^2\beta^2} \sinh^2 \beta b.$$

The transmission coefficient of the barrier $T(E) = |t(E)|^2$ is then

$$T(E) = \frac{1}{1 + \dfrac{V_0^2}{4E(V_0 - E)} \sinh^2 \sqrt{\dfrac{2m(V_0 - E)b^2}{\hbar^2}}} \qquad 0 \le E \le V_0. \tag{1.34a}$$

For $E = V_0$ we have

$$T(V_0) = \frac{1}{1 + \beta_0^2 b^2/4} \quad \text{with} \quad \frac{\hbar^2 \beta_0^2}{2m} = V_0. \tag{1.34b}$$

For $E \ge V_0$ similar calculations can be performed and the transmission coefficient becomes

$$T(E) = \frac{1}{1 + \dfrac{V_0^2}{4E(E - V_0)} \sin^2 \sqrt{\dfrac{2m(E - V_0)b^2}{\hbar^2}}} \qquad E \ge V_0. \tag{1.34c}$$

The typical behavior of the transmission coefficient (1.34) of a rectangular barrier is reported in Figure 1.8. For $0 \le E \le V_0$, the transmission varies monotonically

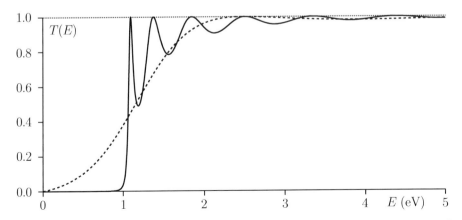

Figure 1.8 Transmission coefficient through a barrier of height $V_0 = 1$ eV and length $b = 5$ Å (dashed line), and through a barrier of height $V_0 = 1$ eV and length $b = 20$ Å (solid line).

from zero to $T(V_0)$; for a reasonably thick barrier characterized by $\beta_0 b \gg 1$, the transmission is close to zero. In such a situation, the unity can be neglected in the denominator of Eq. (1.34a) and, apart from prefactors of the order of unity, Eq. (1.34a) takes the approximate form

$$T(E) \approx \exp\left[-2\sqrt{\frac{2m(V_0 - E)}{\hbar^2}}\, b\right] \quad 0 < E < V_0; \tag{1.34d}$$

the transmission coefficient decreases exponentially, with an exponent which is linear in the width of the barrier and goes as the square root of its height. The approximate expression of Eq. (1.34d) is what expected by the semiclassical theory, considering the propagation of a wavefunction of the form $\exp(-\beta x)$ with $\beta^2 = 2m(V_0 - E)/\hbar^2$ for the length b of the barrier (and keeping the squared modulus).

For $E > V_0$ the transmission coefficient has an oscillatory behavior and approaches asymptotically to 1. In particular $T(E)$ is exactly equal to 1 at energies such that the argument of the trigonometric function in Eq. (1.34c) equals an integer multiple of π; this occurs for

$$E_n - V_0 = n^2 \frac{\hbar^2}{2m}\frac{\pi^2}{b^2} \quad n = 1, 2, \ldots$$

Comparison with the familiar energy spectrum of the infinite well shows that complete transmission $T(E) = 1$ occurs just at those energies that allow appropriate adjustment of wavefunctions at the border points of the interval $[0, b]$.

As summarized by Eq. (1.34d), the transmission coefficient for low energy electrons through a barrier is extremely sensitive to the height and width of the barrier. This high sensitivity is the key working principle of the scanning tunneling microscope [G. Binning and H. Rohrer "Scanning tunneling microscopy—from birth to adolescence" Rev. Mod. Phys. 59, 615 (1987)]. In the experiments, the sharp tip of the microscope is separated by a small gap of a few angstroms from the conducting surface of the material under investigation. The map of the tunneling current flowing between tip and sample, provides information on the surface landscape with an accuracy that may reach atomic resolution.

1.3.2   Double Barrier and Resonant Tunneling

As shown in the previous section, perfect transmission from a single rectangular barrier occurs at selected energies above the threshold V_0; however, these resonance energies occur in a region where the medium is basically transparent and the transmission coefficient, apart from oscillations, is approaching to unity. A much more interesting situation occurs in a symmetric double barrier structure where ideal transparency may occur also at energies lower than the barriers height, i.e. in regions where each individual barrier could be quite opaque; this highly selective filter effect has the largest interest in the physics of devices and is called *resonant tunneling*.

Consider a double barrier structure, in which two identical barriers of width b are displaced by a distance $d = b + w$, as indicated in Figure 1.9. The transfer matrix

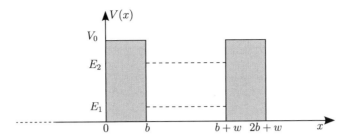

Figure 1.9 Energy potential profile of a symmetric double barrier. In the figure we also report the two bound states ($E_1 = 0.189$ eV and $E_2 = 0.711$ eV) of a well with zero potential in the region $[b, b + w]$ with $b = 5$ Å, $w = 10$ Å, and $V_0 = 1$ eV in the regions $[-\infty, b]$ and $[b + w, +\infty]$.

for the first barrier of Figure 1.9 is given by expression (1.32). The second barrier has exactly the same shape as the first barrier, except for a rigid shift in space by $b + w$, so that the transfer matrix is obtained applying Eq. (1.29). The total transfer matrix for the symmetric double barrier model is the product of the transfer matrices of the two barrier scatterers, namely:

$$M^{(tot)}(E) = \begin{pmatrix} m_{11} & m_{12}e^{-2iq(b+w)} \\ m_{21}e^{2iq(b+w)} & m_{22} \end{pmatrix} \begin{pmatrix} m_{11} & m_{12} \\ m_{21} & m_{22} \end{pmatrix}. \qquad (1.35)$$

The upper diagonal element of the transfer matrix (1.35) reads

$$m_{11}^{(tot)} = m_{11}^2 + |m_{12}|^2 e^{-2iq(b+w)} = \frac{1}{T}e^{2i\phi_t} + \frac{R}{T}e^{-2iq(b+w)},$$

where $R(E), T(E)$, and $\phi_t(E)$ are the reflection coefficient, the transmission coefficient and the phase of the transmission amplitude of a single barrier provided by Eqs. (1.27). The transmission coefficient of the double barrier is then

$$T^{(tot)} = \frac{1}{|m_{11}^{(tot)}|^2} \implies \boxed{T^{(tot)}(E) = \frac{T^2(E)}{\left|1 + R(E)e^{-2i(\phi_t + qb + qw)}\right|^2}}. \qquad (1.36)$$

From the above relation, it is evident that the transmission coefficient of the double barrier equals unity (perfect transparency) when the transmission coefficient T of the single well equals unity (and $R = 0$); resonance energies of the single barrier are resonance energies for the double barrier, too.

More importantly, it is seen by inspection that perfect transparency also occurs when

$$e^{-2i(\phi_t + qb + qw)} \equiv -1 \quad \text{or equivalently} \quad \cos(\phi_t + qb + qw) \equiv 0. \qquad (1.37)$$

The condition (1.37) can be cast in the form

$$\cos(\phi_t + qb)\cos(qw) - \sin(\phi_t + qb)\sin(qw) \equiv 0 \quad \Rightarrow \quad \tan qw = \frac{1}{\tan(\phi_t + qb)}.$$

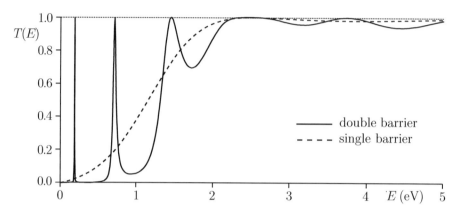

Figure 1.10 Transmission coefficient through a symmetric double rectangular barrier; each of the two component barriers has height $V_0 = 1$ eV and width $b = 5$ Å; their separation is $w = 10$ Å. The two sharp structures at energies smaller than V_0 illustrate the resonant tunneling effect. The transmission coefficient through a single rectangular barrier of height $V_0 = 1$ eV and width $b = 5$ Å is also given for comparison (dashed line).

Using Eq. (1.33) it is seen that the above condition is fully equivalent to the requirement

$$\tan qw = \frac{2q\beta}{q^2 - \beta^2} \frac{1}{\tanh \beta b}. \tag{1.38a}$$

For thick barriers such that $\beta b \gg 1$ and $\tanh \beta b \approx 1$, we have that the resonant condition (1.38a) becomes

$$\tan qw = \frac{2q\beta}{q^2 - \beta^2}. \tag{1.38b}$$

The resonant condition (1.38b) valid for thick barriers, coincides with Eq. (1.12c), which gives the condition for the eigenvalues of a single well of thickness w and depth V_0. *Thus we see that resonant tunneling is sustained by the eigenvalues of the well itself.* In Figure 1.10 we show as an example the transmission coefficient of a symmetric double barrier, with illustration of the resonant tunneling effect.

Consider now Eq. (1.36) in an energy region in which the single composing barrier is rather opaque ($T \ll 1$) and the energy is reasonably far from the condition (1.37). In the off-resonance condition, the denominator of Eq. (1.36) remains of the order of unity; the total transmission of the double barrier is of the order of T^2 and is thus strongly suppressed.

1.3.3   Electron Tunneling through a Periodic Potential

Consider a one-dimensional potential $V(x)$, of periodicity a, and let $V_0(x)$ denote the potential within the unit cell $0 \leq x \leq a$ (see Figure 1.11). For simplicity, the minima of $V(x)$ are assumed to be zero and located at the positions $t_n = na$; this is in general possible by an appropriate choice of the origin for the energy axis and for the

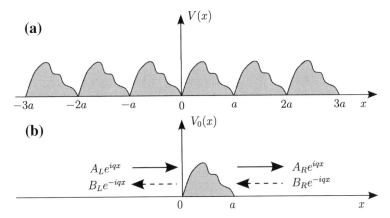

Figure 1.11 (a) Schematic representation of the periodic potential $V(x)$ of a one-dimensional crystal. (b) Potential $V_0(x)$ within the unit cell.

x-axis. With the help of the Bloch theorem and the knowledge of the transfer matrix corresponding to the potential $V_0(x)$ in the unit cell, we can express the condition of propagation of Bloch-type wavefunctions throughout the system.

As indicated in Figure 1.11b, we consider the electron tunneling through the potential $V_0(x)$, between two leads at zero potential extending in the regions $x \leq 0$ and $x \geq a$. We specify Eqs. (1.21) to the present geometry and we have

$$\begin{cases} \psi_L(x) = A_L e^{iqx} + B_L e^{-iqx} & x \leq 0, \\ \psi_R(x) = A_R e^{iqx} + B_R e^{-iqx} & x \geq a, \end{cases} \tag{1.39}$$

where $q^2(E) = 2mE/\hbar^2$, the energy E is positive, and (A_R, B_R) can be expressed in terms of (A_L, B_L) via the transfer matrix M of the potential $V_0(x)$.

We look for solutions of Eqs. (1.39) that satisfy, at the points $x = 0$ and $x = a$, the boundary conditions required by the Bloch theorem

$$\psi_R(a) = e^{ika}\psi_L(0) \quad \text{and} \quad \left(\frac{d\psi_R}{dx}\right)_{x=a} = e^{ika}\left(\frac{d\psi_L}{dx}\right)_{x=0}. \tag{1.40}$$

The Eqs. (1.40) provide two linear homogeneous equations for the arbitrary amplitudes (A_L, B_L, A_R, B_R) namely:

$$\begin{cases} A_R e^{iqa} + B_R e^{-iqa} = e^{ika} A_L + e^{ika} B_L \\ A_R e^{iqa} - B_R e^{-iqa} = e^{ika} A_L - e^{ika} B_L \end{cases} \Longrightarrow \begin{cases} A_R e^{+iqa} = e^{ika} A_L, \\ B_R e^{-iqa} = e^{ika} B_L, \end{cases} \tag{1.41}$$

where the last passage on the right has been done by summing and subtracting the two equations on the left side.

Expressing (A_R, B_R) in terms of (A_L, B_L) exploiting the transfer matrix M, we obtain from Eqs. (1.41) two linear homogeneous equations for the arbitrary coefficients

A_L and B_L

$$\begin{pmatrix} m_{11}e^{iqa} & m_{12}e^{iqa} \\ m_{21}e^{-iqa} & m_{22}e^{-iqa} \end{pmatrix} \begin{pmatrix} A_L \\ B_L \end{pmatrix} = e^{ika} \begin{pmatrix} A_L \\ B_L \end{pmatrix}. \tag{1.42}$$

The condition of Bloch propagation and existence of allowed energy bands requires the vanishing of the determinant

$$\begin{Vmatrix} m_{11}e^{iqa} - e^{ika} & m_{12}e^{iqa} \\ m_{21}e^{-iqa} & m_{22}e^{-iqa} - e^{ika} \end{Vmatrix} = 0.$$

Taking into account that the transfer matrix is unimodular, one obtains

$$\boxed{m_{11}e^{iqa} + m_{22}e^{-iqa} = 2\cos ka} \tag{1.43a}$$

or equivalently

$$\boxed{\mathrm{Re}\left[m_{11}e^{iqa}\right] = \cos ka}. \tag{1.43b}$$

The above compatibility equation could also be obtained by direct inspection of Eq. (1.42). In fact the 2×2 matrix in Eq. (1.42) has eigenvalue $\exp(ika)$; since the matrix is unimodular, it has also the eigenvalue $\exp(-ika)$ and trace $2\cos ka$.

It is convenient to express the quantities in Eqs. (1.43) in terms of the transmission amplitude $t = |t|\exp(i\phi_t)$; we have

$$m_{11}(E) = \frac{1}{t^*(E)} = \frac{1}{|t(E)|}e^{i\phi_t(E)}.$$

The compatibility condition (1.43) for the tunneling of Bloch-type wavefunctions takes the compact and significant form

$$\boxed{\frac{1}{|t(E)|}\cos[\phi_t(E) + q(E)a] = \cos ka}, \tag{1.44}$$

where $\phi_t(E)$ and $|t(E)|$ are the phase and the modulus of the transmission amplitude $t(E)$ through the potential $V_0(x)$ of the unit cell. Notice that Eq. (1.44) can be satisfied only for those values of E for which the function $(1/|t|)\cos(\phi_t + qa)$ is in magnitude lower than (or equal to) unity; since $|t(E)| \leq 1$, we can infer that the compatibility equation (1.44) leads quite generally to a sequence of allowed and forbidden energy regions.

It is instructive to analyze the band structure, produced by the periodic array of barriers of Figure 1.3, from the point of view of the electron tunneling framework. The transfer matrix M for a single barrier is given by Eq. (1.32). The energy bands are determined by Eq. (1.43b), which reads in the present case

$$\mathrm{Re}\left\{\left[\cosh\beta b + i\frac{q^2 - \beta^2}{2q\beta}\sinh\beta b\right]e^{iq(a-b)}\right\} = \cos ka,$$

where Re denotes the real part of its argument. With $a - b = w$ (see Figure 1.3), the above equation becomes

$$\cosh \beta b \cos qw - \frac{q^2 - \beta^2}{2q\beta} \sinh \beta b \sin qw = \cos ka. \tag{1.45}$$

We thus see that the compatibility equation (1.45) for the tunneling of Bloch-type wavefunctions through periodically repeated barriers, exactly coincides with Eq. (1.18), previously obtained using the ordinary boundary conditions procedure.

1.4 The Tight-Binding Model

1.4.1 Expansion in Localized Orbitals

In the previous section we have considered the origin of the energy bands within the framework of electron tunneling through a periodic potential. In this section we adopt another point of view; we imagine construction of a crystal from a hypothetical periodic one-dimensional sequence of N equal atoms. In the case of negligible interaction among atoms, the same atomic orbitals centered in the different lattice sites would have the same energy; in the presence of interaction this N-fold degeneracy is removed and evolves into an energy band. As usual in this chapter we keep our considerations at an introductory level, referring to Chapter 5 for a deeper analysis.

Let us consider a one-dimensional crystal made up of equal atoms centered in the lattice positions $t_n = na$. For each atom we focus our attention on a given local orbital ϕ_a of energy E_a (for simplicity, the atomic orbital ϕ_a is assumed to be non-degenerate and real). We try to obtain crystal wavefunctions using as basis set the N orbital functions $\phi_a(x - t_n)$ centered in the N atomic sites t_n. A schematic picture of the model crystal considered is given in Figure 1.12.

It is convenient to represent the crystal Hamiltonian H on the localized functions $\{\phi_a(x - t_n)\}$. We do not need here to be very specific on the crystal Hamiltonian, but

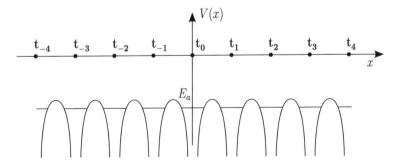

Figure 1.12 Schematic representation of the crystal potential as a superposition of atomic-like potentials, centered at the lattice sites t_n. In the tight-binding approximation, the interaction between nearby atomic-like orbitals ϕ_a of energy E_a leads to the formation of energy bands.

only exploit its translational symmetry. Because of translational symmetry, the diagonal matrix elements of H on the atomic orbitals are all equal; similarly the hopping integrals between nearest neighbor orbitals are equal. We have

$$\langle \phi_a(x - t_n)|H|\phi_a(x - t_n)\rangle = E_0 \quad \text{and} \quad \langle \phi_a(x - t_n)|H|\phi_a(x - t_{n\pm 1})\rangle = \gamma. \quad (1.46)$$

For simplicity, due to the localized nature of the atomic orbitals, the hopping integrals involving second or further apart neighbors are assumed negligible; the value of the interaction energy γ is taken to be negative (to simulate the case of s-like orbitals and attractive atomic potentials).

The localized functions $\{\phi_a(x - t_n)\}$ do not satisfy the Bloch theorem; but this can be remedied considering the linear combinations of atomic orbitals of the type

$$\Phi(k, x) = \frac{1}{\sqrt{N}} \sum_n e^{ikt_n} \phi_a(x - t_n), \quad (1.47)$$

where N is the number of unit cells of the crystal. The functions defined by Eq. (1.47) are named *Bloch sums* have delocalized character, and satisfy the Bloch theorem; in fact

$$\Phi(k, x + t_m) = e^{ikt_m} \frac{1}{\sqrt{N}} \sum_n e^{ik(t_n - t_m)} \phi_a(x - t_n + t_m) = e^{ikt_m} \Phi(k, x).$$

For simplicity we assume orthonormality of orbitals centered on different atoms; in this case, the Bloch sums (1.47) are also orthonormal.

The N orthonormal itinerant Bloch sums $\{\Phi(k, x)\}$ span the same Hilbert space as the N localized functions $\{\phi_a(x - t_n)\}$, but the great advantage is that Bloch sums of different k values cannot mix under the influence of a periodic potential. The dispersion of the energy band originated from the atomic orbitals $\{\phi_a(x - t_n)\}$ is thus given by

$$E(k) = \frac{\langle \Phi(k, x)|H|\Phi(k, x)\rangle}{\langle \Phi(k, x)|\Phi(k, x)\rangle} = \langle \Phi(k, x)|H|\Phi(k, x)\rangle.$$

The matrix elements of H between Bloch sums can be expressed via the matrix elements of H between localized orbitals and appropriate Bloch phase factors, as follows:

$$\langle \Phi(k, x)|H|\Phi(k, x)\rangle = \frac{1}{N} \sum_{mn} e^{-ik(t_m - t_n)} \langle \phi_a(x - t_m)|H|\phi_a(x - t_n)\rangle$$

$$= \sum_n e^{ikt_n} \langle \phi_a(x)|H|\phi_a(x - t_n)\rangle,$$

where the term $t_m = 0$ has been selected, and the denominator N has been dropped.

In the particular case that the matrix elements of H between atomic orbitals are given by Eq. (1.46), the above dispersion relation becomes

$$E(k) = E_0 + 2\gamma \cos ka; \quad (1.48)$$

this expression clearly shows at the most elementary level that the N-fold degenerate states of the non-interacting atoms are smeared into a continuous band of width $4|\gamma|$ (see Figure 1.13).

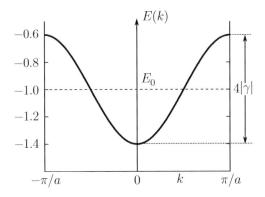

Figure 1.13 Energy band $E(k) = E_0 + 2\gamma \cos ka$ for a tight-binding model with a single orbital per site and nearest neighbor interactions; in the figure $E_0 = -1$ eV and $\gamma = -0.2$ eV.

We can expand the second member of Eq. (1.48) in powers of k for small k around the center of the Brillouin zone and retain terms up to the second order; we have

$$E(k) \approx E_0 + 2\gamma - \gamma a^2 k^2 \quad \text{for} \quad k \approx 0.$$

The quadratic term $-\gamma a^2 k^2$ can be written in the form $\hbar^2 k^2/2m^*$, similar to the one pertaining to the free-electron case, provided the effective mass m^* is defined as

$$m^* = \frac{\hbar^2}{2|\gamma|a^2}. \tag{1.49}$$

The effective mass is small if the hopping parameter γ is large, and vice versa. Notice also that the effective mass, at the border of the Brillouin zone, is negative and is the opposite of expression (1.49); this can be seen performing a series development of the second member of Eq. (1.48) around the zone edge $k = \pi/a$.

1.4.2 Tridiagonal Matrices and Continued Fractions

A matrix, whose non-vanishing elements occur only on the principal diagonal and the next upper and lower ones, is said to be tridiagonal. The crystal Hamiltonian H, with matrix elements given by Eq. (1.46), represented on a basis of localized orbitals $|f_n\rangle = |\phi_a(x - t_n)\rangle$ has the form

$$H = E_0 \sum_n |f_n\rangle\langle f_n| + \gamma \sum_n [|f_n\rangle\langle f_{n+1}| + |f_{n+1}\rangle\langle f_n|]. \tag{1.50}$$

The one-dimensional tight-binding Hamiltonian (1.50), with one orbital per site and nearest neighbor interactions, is the simplest example of problems described by a tridiagonal matrix.

The most general tridiagonal form for the Hamiltonian is

$$H = \sum_n a_n |f_n\rangle\langle f_n| + \sum_n b_{n+1} [|f_n\rangle\langle f_{n+1}| + |f_{n+1}\rangle\langle f_n|] \tag{1.51}$$

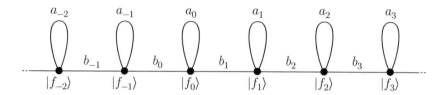

Figure 1.14 Graphical representation of the most general tridiagonal Hamiltonian. The diagonal matrix elements are denoted by a_n; the off-diagonal hopping integrals are denoted by b_n.

with appropriate values for diagonal and off-diagonal matrix elements a_n and b_n (since H is hermitian, all a_n are real; also b_n can be brought to real form, without loss of generality, just embodying appropriate phases in the basis functions). The tridiagonal operator (1.51) can be represented graphically in a chain form as indicated in Figure 1.14.

Numerous physical problems can be described by a tridiagonal Hamiltonian of the type (1.51); among them we mention the Harper model for incommensurate systems and the Anderson model for disordered systems. In the Harper model the diagonal energies are taken as $a_n = V_0 \cos (2\pi n\alpha)$ with α irrational number, while the hopping integrals are constant (taken equal to unity); it can be shown that if $0 \le V_0 < 2$ all the states are delocalized, but when $V_0 > 2$ all the states become localized. In the Anderson model the diagonal energies a_n are randomly distributed in an energy interval, while the hopping integrals are constant; in the Anderson model all states are localized, for any degree of disorder. We cannot pursue here the interesting problems posed by localization, and the wide and restless research on Hamiltonians of type (1.51).

Much more generally, *the importance of tridiagonal matrices is due to the fact that, by the Lanczos method* (see Chapter 5), *it is possible to map any quantum mechanical problem into a semi-infinite chain represented by a tridiagonal matrix.* This is the reason why tridiagonal operators have a special relevance in quantum mechanics [see for instance D. W. Bullett, R. Haydock, V. Heine and M. J. Kelly, Solid State Physics Vol.35 (Academic Press, New York 1980)].

Tridiagonal matrices have simple and elegant mathematical properties, that allow to focus on physical aspects. From a physical point of view, the most remarkable aspect is that the inversion of tridiagonal matrices is straightforward and thus the Green's function of tridiagonal operators is easily accessible.

A Note on the Inversion of Tridiagonal Matrices

Consider a tridiagonal matrix M of the form

$$M = \begin{pmatrix} \alpha_0 & \beta_1 & 0 & 0 & \cdot \\ \beta_1 & \alpha_1 & \beta_2 & 0 & \cdot \\ 0 & \beta_2 & \alpha_2 & \beta_3 & \cdot \\ 0 & 0 & \beta_3 & \alpha_3 & \cdot \\ \cdot & \cdot & \cdot & \cdot & \cdot \end{pmatrix} \qquad (1.52)$$

(for simplicity the rank of the matrix M is supposed to be finite, although arbitrary large). We consider the inverse matrix M^{-1}, and we are here primarily interested in obtaining the explicit expression of its upper left corner element $(M^{-1})_{00}$.

Let us indicate with D_0 the determinant of the matrix M; we denote by D_1 the determinant of the matrix obtained by M suppressing the first column and the first row; similarly D_2 denotes the determinant of the matrix obtained from M suppressing also the successive row and column, and so on. Development of the determinant D_0 along the elements of the first row gives

$$D_0 = \alpha_0 D_1 - \beta_1^2 D_2.$$

From the above result and considering that $(1/M)_{00} = D_1/D_0$ we obtain

$$\left(\frac{1}{M}\right)_{00} \equiv \frac{1}{D_0/D_1} = \frac{1}{\alpha_0 - \beta_1^2 \dfrac{1}{D_1/D_2}}.$$

Iterating the procedure for D_1/D_2 we obtain the continued fraction expansion

$$\left(\frac{1}{M}\right)_{00} = \cfrac{1}{\alpha_0 - \cfrac{\beta_1^2}{\alpha_1 - \cfrac{\beta_2^2}{\cdots}}}. \tag{1.53}$$

The continued fraction expansion (1.53) is truncated at the nth step if β_n vanishes.

In the case the tridiagonal matrix M extends both to positive and negative n (as for instance in Figure 1.14), we obtain

$$\left(\frac{1}{M}\right)_{00} = \cfrac{1}{\alpha_0 - \cfrac{\beta_1^2}{\alpha_1 - \cfrac{\beta_2^2}{\cdots}} - \cfrac{\beta_0^2}{\alpha_{-1} - \cfrac{\beta_{-1}^2}{\cdots}}}. \tag{1.54}$$

In the case $\beta_0 = 0$ we recover Eq. (1.53). It is also evident that we can obtain the expression of any diagonal matrix element $(1/M)_{nn}$ (this simply requires an appropriate relabeling of states). Also off-diagonal matrix elements of $(1/M)$ could be expressed, with appropriate elaboration, in terms of continued fractions.

Density-of-States and Green's Functions

Consider a system described by a Hamiltonian H, and let us indicate by ψ_m and E_m its normalized eigenfunctions and eigenvalues (supposed to be countable, for simplicity). The *total density-of-states* of the system is defined as

$$D(E) = \sum_m \delta(E - E_m). \tag{1.55}$$

It is evident that the integral $\int D(E)dE$ in any energy interval $[E_1, E_2]$ provides the number of states of the system therein.

The *density-of-states projected* on any arbitrarily chosen state of interest $|f_0\rangle$ (normalized to unity) is defined as

$$n_0(E) = \sum_m |\langle f_0 | \psi_m \rangle|^2 \delta(E - E_m). \tag{1.56}$$

Differently from $D(E)$, the projected density-of-states $n_0(E)$ (also called *local density-of-states*) gives information uniquely on the spectral region investigated by the orbital $|f_0\rangle$; we also see by inspection that $\int n_0(E)dE \equiv 1$ (due to the normalization of $|f_0\rangle$). Furthermore, the total density-of-states is the sum of the projected density-of-states on any complete orthonormal set $\{f_n\}$.

In the following we consider a single-particle time-independent Hamiltonian H represented on a set of orthonormal basis functions $\{\phi_i\}$ with $i = 1, 2, \ldots, N$ where N is supposed to be finite, although it could be arbitrarily large. In the stated conditions, although very restrictive from the point of view of many-body formalism, we can nevertheless begin to introduce some concepts of Green's functions and density-of-states at an acceptable degree of technicality, compatible with the introductory nature of this chapter.

The retarded Green's function (or resolvent) of an operator H in matrix form is defined as the inverse of the matrix $E + i\varepsilon - H$, i.e.

$$G(E) = \frac{1}{E + i\varepsilon - H}, \tag{1.57a}$$

where the real energy E is accompanied by an (infinitesimal) positive imaginary part (when a scalar quantity such as $E + i\varepsilon$ is added to an operator such as H in matrix form, it is implicitly understood that the scalar quantity is preliminarily multiplied by the identity operator). In terms of matrix elements Eq. (1.57a) can be written in the equivalent form

$$\sum_\alpha (E + i\varepsilon - H)_{m\alpha} G_{\alpha n}(E) = \delta_{mn}. \tag{1.57b}$$

Equations (1.57) apply to a single-particle operator expressed in matrix form. In the case of an Hamiltonian expressed as a differential operator, the generalization of Eqs. (1.57) will be considered elsewhere (see for instance Section 5.7). Besides the retarded Green's function, defined by Eqs. (1.57), one can similarly define the advanced Green's function with the energy E accompanied by a negative infinitesimal imaginary part. In the following, the term Green's function stands automatically for *retarded* Green's function, unless stated otherwise. [Incidentally, we mention that basic tools for the treatment of many-body Hamiltonians, or nonequilibrium systems, include time-ordered or anti-time-ordered Green's functions, lesser or greater Green's functions, besides the appropriately defined retarded or advanced Green's functions. For several aspects concerning equilibrium properties of systems described by time-independent one-electron (or at least mean field) Hamiltonians, the elementary considerations reported below on the properties of retarded Green's function are sufficient for the description of significant properties.]

A most useful property of the Green's function is its connection with the density-of-states of the system. Consider in fact a diagonal matrix element of the Green's function, for instance

$$G_{00}(E) = \langle f_0 | \frac{1}{E + i\varepsilon - H} | f_0 \rangle. \tag{1.58a}$$

It is possible to put in evidence some properties of the resolvent inserting into Eq. (1.58a) the unit operator $1 = \sum |\psi_m\rangle\langle\psi_m|$, where ψ_m are the eigenfunctions of H. We have

$$G_{00}(E) = \langle f_0 | \sum_m |\psi_m\rangle\langle\psi_m| \frac{1}{E + i\varepsilon - H} | f_0 \rangle$$

$$= \sum_m |\langle f_0 | \psi_m \rangle|^2 \frac{1}{E + i\varepsilon - E_m} = \sum_m |\langle f_0 | \psi_m \rangle|^2 \frac{E - E_m - i\varepsilon}{(E - E_m)^2 + \varepsilon^2}. \tag{1.58b}$$

From Eq. (1.58b), we see that for any $\varepsilon > 0$, $G_{00}(E)$ is analytic and its imaginary part is negative (Herglotz property). On the real energy axis, the real part of $G_{00}(E)$ exhibits poles in correspondence of the *discrete* eigenvalues of H, while the imaginary part exhibits δ-like singularities; this can be seen keeping the limit $\varepsilon \to 0^+$ in Eq. (1.58b) and using the result

$$\lim_{\varepsilon \to 0^+} \frac{1}{\pi} \frac{\varepsilon}{(E - E_m)^2 + \varepsilon^2} = \delta(E - E_m).$$

From Eqs. (1.56) and (1.58b), we obtain the standard spectral theorem

$$\boxed{n_0(E) = -\frac{1}{\pi} \lim_{\varepsilon \to 0^+} \operatorname{Im} G_{00}(E + i\varepsilon)}. \tag{1.59a}$$

The total density-of-states of the system can be expressed as the trace of the Green's function on any chosen complete orthonormal set

$$\boxed{D(E) = -\frac{1}{\pi} \lim_{\varepsilon \to 0^+} \operatorname{Im} \operatorname{Tr} G(E + i\varepsilon)}. \tag{1.59b}$$

Relations (1.59) hold regardless of the fact that the energy spectrum of H is discrete or continuous.

In general the calculation of the Green's function of an operator H requires the preliminary diagonalization of H, or appropriate elaborations of equations of motion or diagrammatic procedures. However for tridiagonal operators the calculation of the diagonal matrix elements of the resolvent is straightforward. In fact, if the operator H has the tridiagonal form of type (1.51), the operator $E - H$ is also in tridiagonal form. Then from the inversion properties of tridiagonal matrices we have for the Green's function diagonal matrix element

$$G_{00}(E) = \cfrac{1}{E - a_0 - \cfrac{b_1^2}{E - a_1 - \cfrac{b_2^2}{E - a_2 - \cfrac{b_3^2}{\cdots}}} - \cfrac{b_0^2}{E - a_{-1} - \cfrac{b_{-1}^2}{E - a_{-2} - \cfrac{b_{-2}^2}{\cdots}}}}.$$

$$\tag{1.60}$$

Notice that the infinitesimal imaginary part that always accompanies the energy is sometimes left implicit.

Application to the Tight-Binding Case

Let us consider the Hamiltonian (1.50) for the one-dimensional tight-binding crystal with one orbital per site. In this case the diagonal elements a_n are constant and equal to E_0 (without loss of generality we take $E_0 = 0$); the off-diagonal elements b_n are constant and equal to γ. The Green's function (1.60) for the infinite linear chain becomes

$$G_{00}(E) = \langle f_0 | \frac{1}{E - H} | f_0 \rangle = \cfrac{1}{E - \cfrac{2\gamma^2}{E - \cfrac{\gamma^2}{E - \cfrac{\gamma^2}{\dots}}}}. \tag{1.61a}$$

To perform the sum of the continued fraction let $t(E)$ denote the quantity

$$t(E) = \cfrac{\gamma^2}{E - \cfrac{\gamma^2}{E - \cfrac{\gamma^2}{\dots}}} = \frac{\gamma^2}{E - t(E)} \implies t(E) = \frac{E \pm \sqrt{E^2 - 4\gamma^2}}{2}. \tag{1.61b}$$

The continued fraction in Eq. (1.61a) can be summed exactly to the value

$$G_{00}(E) = \frac{1}{E - 2t(E)} = \frac{1}{\pm\sqrt{E^2 - 4\gamma^2}}; \tag{1.61c}$$

the sign of the square root must be chosen so that $\text{Im}\, G_{00}(E + i\varepsilon) < 0$ for $\varepsilon \to 0^+$. In particular from the spectral theorem the local density-of-states becomes

$$n_0(E) = \frac{1}{\pi} \frac{1}{\sqrt{4\gamma^2 - E^2}} \qquad |E| < 2|\gamma|, \tag{1.61d}$$

and $n_0(E) = 0$ for $|E| > 2|\gamma|$. Thus, the formation of allowed energy bands emerges also from the continued fraction approach to the electronic states in periodic systems.

The result (1.61d) can also be obtained starting directly from the dispersion relation $E(k) = 2\gamma \cos ka$; in fact, with simple elaborations, we have for the total density-of-states

$$D(E) = \sum_k \delta(E - 2\gamma \cos ka) = \frac{L}{2\pi} \int_{-\pi/a}^{+\pi/a} \delta(E - 2\gamma \cos ka)\, dk$$

$$= \frac{L}{2\pi} 2 \frac{1}{|2\gamma a \sin k_0 a|} = N \frac{1}{\pi} \frac{1}{\sqrt{4\gamma^2 - E^2}} \qquad |E| < 2|\gamma|.$$

In the above expression $L = Na$ denotes the length of the crystal; k_0 is one of the two zeroes of the expression $f(k) = E - 2\gamma \cos ka = 0$ when $|E| < 2|\gamma|$; furthermore, use has been made of the property $\delta[f(x)] \equiv \sum \delta(x - x_0)/|f'(x_0)|$, where the sum is over any simple zero x_0 of the function $f(x)$.

It is instructive to calculate the Green's function just on the initial site (surface site) of a semi-infinite linear chain. The Green's function at the surface orbital is

$$G_{00}^{(s)}(E) = \cfrac{1}{E - \cfrac{\gamma^2}{E - \cfrac{\gamma^2}{\cdots}}} = \frac{E \mp \sqrt{E^2 - 4\gamma^2}}{2\gamma^2}; \tag{1.62a}$$

the sign of the square root must be chosen so that Im $G_{00}^{(s)}(E + i\varepsilon) < 0$ for $\varepsilon \to 0^+$. The projected density-of-states on the surface orbital is thus

$$n_0^{(s)}(E) = -\frac{1}{\pi} \lim_{\varepsilon \to 0^+} \text{Im } G_{00}^{(s)}(E + i\varepsilon) = \frac{\sqrt{4\gamma^2 - E^2}}{2\pi\gamma^2} \quad |E| < 2\gamma. \tag{1.62b}$$

Similarly one could calculate the Green's function $G_{11}^{(s)}(E)$, $G_{22}^{(s)}(E)$, etc. on the next sites of the semi-infinite constant chain, and obtain the corresponding local density-of-states $n_1^{(s)}(E)$, $n_2^{(s)}(E)$, etc. For example:

$$G_{11}^{(s)}(E) = \cfrac{1}{E - \cfrac{\gamma^2}{E} - t(E)}; \quad G_{22}^{(s)}(E) = \cfrac{1}{E - \cfrac{\gamma^2}{E - \cfrac{\gamma^2}{E}} - t(E)}, \tag{1.62c}$$

where $t(E)$, given by Eq. (1.61b), represents the contribution of the semi-infinite part of the chain at the right side of the site under attention, while the other fractions represent the contribution of one or two sites at the left of the site under attention. Working out the imaginary part of expressions (1.62c), a simple calculation gives

$$n_1^{(s)}(E) = n_0^{(s)}(E)\frac{E^2}{\gamma^2}, \quad n_2^{(s)}(E) = n_0^{(s)}(E)\left[\frac{E^2}{\gamma^2} - 1\right]^2. \tag{1.62d}$$

The local density-of-states in the bulk of a constant infinite chain, and at the surface orbital and at the next two ones of a semi-infinite chain is reported in Figure 1.15.

We do not pursue further the properties of the Green's function of general tight-binding Hamiltonians of type (1.51); we only mention that several significant properties such as localization effects in incommensurate systems, dimerized systems, disordered systems etc. can be inferred with a systematic study of the resolvents of the corresponding Hamiltonians [see for instance J. B. Sokoloff, Phys. Rep. *126*, 190 (1985); R. Farchioni, G. Grosso and G. Pastori Parravicini, Phys. Rev. B *45*, 6383 (1992); B *53*, 4294 (1996); F. M. Izrailev, A. A. Krokhin and N. M. Makarov, Physics Reports *512*, 125 (2012) and references quoted therein].

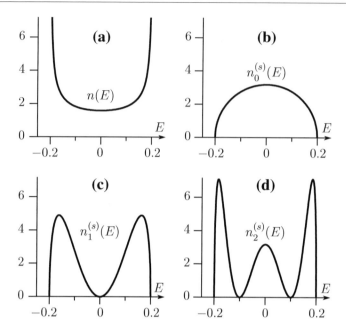

Figure 1.15 Local density-of-states on the bulk orbital of a constant infinite linear chain (a), at the surface orbital of a semi-infinite chain (b), and on the next two orbitals (c) and (d). The chosen value of γ is $\gamma = -0.1$ eV. Energies are in eV, and the density-of-states is in $(eV)^{-1}$.

1.5 Plane Waves and Nearly Free-Electron Model

1.5.1 Expansion in Plane Waves

In this section we consider the origin of the energy bands starting from the "empty lattice" situation and then switching on a weak periodic potential. For this purpose, we consider first the case of vanishing periodic potential, where the energy dispersion $E(k) = \hbar^2 k^2 / 2m$ is a continuous function of k, without gaps; then we study how a periodic potential modifies the free-electron behavior, introducing energy gaps. In Chapter 5 we will see that plane wave expansion of crystal wavefunctions, together with the concept of pseudopotential, is indeed one of the most effective methods of band structure calculation.

In Figure 1.16 we report the free-electron energy dispersion $E(k) = \hbar^2 k^2 / 2m$ as k varies in the reciprocal lattice; we also show the free-electron parabola folded within the first Brillouin zone. Folding is easily performed by means of appropriate reciprocal lattice translations by $n2\pi/a$ of arches of parabola; the convenience of folding stands in the fact that, according to the Bloch theorem, *only states vertical in the first Brillouin zone may interact* under the influence of a potential with the lattice periodicity.

We wish now to obtain the eigenvalues and eigenvectors of the crystal Hamiltonian

$$H = -\frac{\hbar^2}{2m}\frac{d^2}{dx^2} + V(x) \tag{1.63a}$$

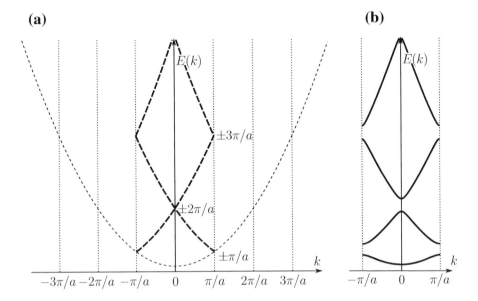

Figure 1.16 (a) Empty lattice states (free-electron parabola) and folding in the first Brillouin zone. At the center and at the border of the first Brillouin zone we have also indicated the wavenumbers of the degenerate plane waves. In the presence of a periodic potential, gaps open at the center and at the border of the Brillouin zone, as shown qualitatively by continuous lines in (b).

starting from the empty lattice eigenvalues and eigenfunctions. At a given k value, the crystal wavefunctions $\psi_k(x)$ can be expanded as a linear combination of plane waves of the type

$$W_{k_n}(x) = \frac{1}{\sqrt{L}} e^{i(k+h_n)x},$$
(1.63b)

where L is the length of the crystal and $h_n = n\,2\pi/a$. The matrix elements of H between the basis functions (1.63b) take the form

$$\langle W_{k_m} | H | W_{k_n} \rangle = \frac{\hbar^2 (k+h_n)^2}{2m} \delta_{mn} + \frac{1}{L} \int_0^L e^{-i(h_m - h_n)x} V(x)dx$$

$$= \frac{\hbar^2 (k+h_n)^2}{2m} \delta_{mn} + V(h_m - h_n),$$
(1.63c)

where $V(h)$ denotes the Fourier transform of $V(x)$.

Diagonalization of H on the basis set of plane waves then leads to the secular equation for the energy eigenvalues

$$\left\| \left[\frac{\hbar^2 (k+h_n)^2}{2m} - E \right] \delta_{mn} + V(h_m - h_n) \right\| = 0.$$
(1.64)

We consider now some significant particular cases of Eq. (1.64) in order to make more transparent the origin of allowed and forbidden energy bands in periodic potentials.

Nearly Free-Electron Approximation

Consider the general secular equation (1.64) at a given k value, for instance at $k = \pi/a$. From Figure 1.16 we see that the interacting plane waves are characterized by the wavenumbers $\pm\pi/a$ with energy $E_0 = (\hbar^2/2m)(\pi^2/a^2)$, by the wavenumbers $\pm 3\pi/a$ with energy $9E_0$, etc. For small strength of the periodic potential, we may confine our attention to the two basis functions (degenerate in the empty lattice analysis)

$$\psi_1(x) = \frac{1}{\sqrt{L}}\exp\left(i\frac{\pi}{a}x\right), \quad \psi_2(x) = \frac{1}{\sqrt{L}}\exp\left(-i\frac{\pi}{a}x\right). \tag{1.65a}$$

The diagonalization of the crystal Hamiltonian (1.63a) on the two wavefunctions (1.65a) leads to the 2×2 secular equation

$$\left\| \begin{array}{cc} E_0 - E & V_1 \\ V_1^* & E_0 - E \end{array} \right\| = 0, \tag{1.65b}$$

where $V_1 = V(2\pi/a)$ is the Fourier transform of the crystal potential corresponding to the lowest reciprocal lattice wavenumber $h_1 = 2\pi/a$. Thus, the twofold degenerate empty lattice states of energy E_0 are split by the periodic potential in the form

$$E = E_0 \pm |V_1|. \tag{1.65c}$$

The same reasoning holds for the empty lattice states degenerate at the point $k = 0$, and we can understand qualitatively, in the nearly free-electron approximation, the origin of the energy gaps of the one-dimensional crystal as due to the splitting of the twofold degeneracy of the empty lattice produced by the periodic potential.

We can also determine analytically the behavior of the energy bands near the boundary of the first Brillouin zone. Suppose that the wavenumber k is very near to the boundary, but not exactly at $\pm\pi/a$. Instead of the basis functions (1.65a), consider now the basis functions

$$\begin{cases} \psi_1(x) = \dfrac{1}{\sqrt{L}}e^{i(\pi/a-\Delta k)x} & \text{with energy} \quad E_1 = \dfrac{\hbar^2}{2m}\left(\dfrac{\pi}{a} - \Delta k\right)^2, \\[2mm] \psi_2(x) = \dfrac{1}{\sqrt{L}}e^{i(-\pi/a-\Delta k)x} & \text{with energy} \quad E_2 = \dfrac{\hbar^2}{2m}\left(\dfrac{\pi}{a} + \Delta k\right)^2. \end{cases} \tag{1.66}$$

The 2×2 determinantal equation

$$\left\| \begin{array}{cc} E_1 - E & V_1 \\ V_1^* & E_2 - E \end{array} \right\| = 0,$$

produces the two eigenvalues

$$E_\pm = \frac{1}{2}\left[E_1 + E_2 \pm \sqrt{(E_1 - E_2)^2 + 4|V_1|^2} \right].$$

Inserting in the above expression the values of E_1 and E_2 given in Eq. (1.66), we obtain the energy dispersion curves near the boundary of the Brillouin zone

$$E_\pm(\Delta k) = E_0 + \frac{\hbar^2(\Delta k)^2}{2m} \pm \sqrt{4E_0\frac{\hbar^2(\Delta k)^2}{2m} + |V_1|^2} \quad \text{with} \quad E_0 = \frac{\hbar^2}{2m}\left(\frac{\pi}{a}\right)^2.$$

When Δk is small, a series development of the square root term gives

$$E_\pm(\Delta k) = E_0 + \frac{\hbar^2(\Delta k)^2}{2m} \pm |V_1|\left[1 + \frac{2E_0}{|V_1|^2}\frac{\hbar^2(\Delta k)^2}{2m}\right] + \cdots ;$$

the electronic effective masses for the upper and lower energy bands are

$$\frac{1}{m^*} = \frac{1}{m}\left(1 \pm \frac{2E_0}{|V_1|}\right) \implies m^* = \pm m\frac{|V_1|}{2E_0} \quad \text{for} \quad |V_1| \ll E_0. \quad (1.67)$$

In the case of small energy gap ($|V_1| \ll E_0$) also the effective masses are expected to be small (at parity of other conditions). Qualitatively this is the trend actually observed in some materials; semiconductors with small energy gaps are often characterized by small effective masses (and high mobility) of carriers.

1.5.2 The Mathieu Potential and the Continued Fraction Solution

We consider now the energy bands of a crystal in the case the periodic potential is a simple cosine potential. For convenience, we choose the potential in the form

$$V(x) = 2V_1[1 - \cos(2\pi x/a)] \quad \text{with} \quad V_1 > 0,$$

so that the potential energy is zero at the minima, and positive otherwise (see Figure 1.17a). The Hamiltonian operator becomes

$$H = \frac{p^2}{2m} + 2V_1 - V_1[e^{i(2\pi/a)x} + e^{-i(2\pi/a)x}]. \quad (1.68)$$

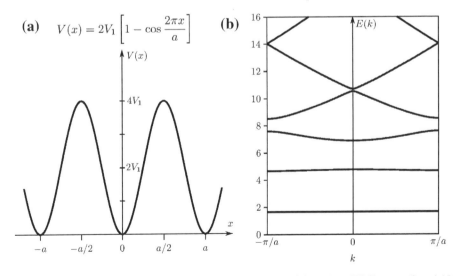

Figure 1.17 (a) Representation of the periodic Mathieu potential $V(x) = 2V_1[1 - \cos(2\pi x/a)]$ with $V_1 = 5E_0 = 5(\hbar^2/2m)(\pi/a)^2$ and $a = 5\,a_B$ (the value of V_1 is ≈ 2 Rydberg). (b) Energy bands (in Rydberg) of the described Mathieu potential.

The matrix elements of the Mathieu Hamiltonian (1.68) between plane waves have the form

$$\langle W_{k_m} | H | W_{k_n} \rangle = \left[\frac{\hbar^2 (k + h_n)^2}{2m} + 2V_1 \right] \delta_{mn} - V_1 \delta_{m,n+1} - V_1 \delta_{m,n-1}. \quad (1.69)$$

The Mathieu Hamiltonian is another important case of problems with tridiagonal form, and it is thus candidate to be solved by means of the continued fraction apparatus.

The tridiagonal form (1.69) is well suited for the evaluation of the Green's function. The diagonal matrix element $G_{00}(E)$, in particular, is given by the expression

$$G_{00}(E) = \cfrac{1}{E - a_0 - \cfrac{V_1^2}{E - a_1 - \cfrac{V_1^2}{E - a_2 - \cfrac{V_1^2}{\cdots}}} - \cfrac{V_1^2}{E - a_{-1} - \cfrac{V_1^2}{E - a_{-2} - \cfrac{V_1^2}{\cdots}}}},$$
$$(1.70)$$

where $a_n(k) = 2V_1 + (\hbar^2/2m)(k + n\,2\pi/a)^2$.

It is now easy to plot the expression (1.70) as a function of E and from the poles of $G_{00}(E)$ we can obtain the eigenvalues of the Mathieu problem. In general the number of steps to be effectively included in the continued fraction before truncation depends on the relative strength of V_1 with respect to $E_0 = (\hbar^2/2m)(\pi/a)^2$; if this ratio is of order of unity a few tens of steps in the continued fraction are found sufficient. In Figure 1.17b we report the energy bands corresponding to the Mathieu potential in the case $V_1 = 5E_0$. Notice that all the degeneracies of the empty lattice are fully removed (even if the direct coupling between degenerate wavefunctions of the empty lattice occurs only for $k = \pm\pi/a$) [for further aspects of the Mathieu problem see J. C. Slater, Phys. Rev. 87, 807 (1952)].

The Mathieu potential has been originally introduced as a model to study and mimic some features of the physics of electronic states of crystalline materials; however, the ideal experimental realization of systems described by the Mathieu model has occurred several decades later in the physics of ultracold atoms in optical lattices, where atoms of mass M (for instance sodium or rubidium) are trapped in a standing wave of laser light with wavelength $\lambda \approx 10^4$ Å. Akin phenomena occur, in two only apparently different areas of research, at scales of energy differing by nine or ten orders of magnitude; actually, the energy scale changes from $(\hbar^2/2m)(\pi/a)^2 \approx 10$ eV (m electron mass, $a \approx 1$ Å) for ordinary crystals, to $(\hbar^2/2M)(\pi/\lambda)^2 \approx 10^{-10} - 10^{-11}$ eV for ultracold atoms in optical lattices.

1.6   Some Dynamical Aspects of Electrons in Band Theory

Velocity, Quasi-Momentum, and Effective Mass of an Electron in a Band

It is useful to introduce the concept of velocity, quasi-momentum, and effective mass for an electron in a band, within a semiclassical picture. In order to introduce this argument we consider first the free-electron situation, with eigenfunctions and eigenvalues given by $W(k, x) = (1/\sqrt{L}) \exp(ikx)$ and $E(k) = \hbar^2 k^2 / 2m$. The plane waves are eigenfunctions of the operator $p = -i\hbar d/dx$ with eigenvalue $\hbar k$, which represents the momentum of the free electron.

We consider now an electron in a periodic potential. We focus on a given energy band and we indicate with $E(k)$ and $\psi(k, x)$ energies and wavefunctions as k varies in the Brillouin zone. We notice, first of all, that a Bloch function $\psi(k, x) = \exp(ikx)u(k, x)$ is not (in general) an eigenfunction of the momentum operator; in fact we have

$$p\,\psi(k, x) = -i\hbar \frac{d}{dx}[e^{ikx}u(k, x)] = \hbar k \psi(k, x) + e^{ikx}(-i)\hbar \frac{d}{dx}u(k, x);$$

this relation clearly shows that the Bloch wavefunctions $\psi(k, x)$ of the crystal Hamiltonian are not, in general, eigenfunctions of the p operator (except for the trivial empty lattice situation), and thus $\hbar k$ cannot be considered the momentum of the electron in the state $\psi(k, x)$.

However, $\psi(k, x)$ can always be expressed as a linear combination of plane waves of wavenumbers $k, k \pm 2\pi/a, k \pm 4\pi/a, \ldots$ and the *only* possible values of a measure of the observable p on the Bloch function $\psi(k, x)$ are thus $\hbar k, \hbar k \pm \hbar(2\pi/a), \hbar k \pm \hbar(4\pi/a), \ldots$ For this reason, the quantity $\hbar k$ (with k usually, but not necessarily, taken within the first Brillouin zone) is called *quasi-momentum* of the electron in the crystal, or also *crystal momentum*. For brevity, and with due caution, $\hbar k$ is often addressed simply as the *momentum of the electron*, keeping in mind that the wavenumber k and the momentum $\hbar k$ of an electron in a crystal are defined within arbitrary integer values of $2\pi/a$ and $\hbar 2\pi/a$, respectively.

It is of interest to consider the expectation value of the momentum operator in the state $\psi(k, x)$, and the closely related semiclassical electron velocity or *group velocity* $v(k)$ defined as

$$v(k) = \langle \psi(k, x)| \frac{p}{m} |\psi(k, x)\rangle. \tag{1.71}$$

For this purpose, we start from the relation

$$\langle \psi(k, x)| \frac{p^2}{2m} + V(x)|\psi(k, x)\rangle = E(k),$$

and express the crystal wavefunction in the Bloch form $\psi(k, x) = \exp(ikx)u(k, x)$; it follows:

$$\langle u(k, x)| \frac{1}{2m}(p + \hbar k)^2 + V(x)|u(k, x)\rangle = E(k). \tag{1.72}$$

We now derive both members of Eq. (1.72) with respect to the parameter k. From the derivative of the first member of Eq. (1.72) we obtain three contributions, two of them coming from the derivative of $u^*(k, x)$ and $u(k, x)$, and the other from the derivative of the operator; the two terms coming from the derivative of u^* and u equal $E(k)(d/dk)\langle u(k, x)|u(k, x)\rangle$ and give zero because the wavefunctions are normalized to one. Thus we have

$$\langle u(k, x)| \frac{\hbar}{m}(p + \hbar k)|u(k, x)\rangle = \frac{dE(k)}{dk};$$

by expressing $u(k, x)$ as $\exp(-ikx)\psi(k, x)$, we obtain the very important result

$$v(k) = \langle \psi(k, x)| \frac{p}{m} |\psi(k, x)\rangle = \frac{1}{\hbar} \frac{dE(k)}{dk} \; ; \tag{1.73}$$

in particular, the electron velocity $v(k)$ vanishes at the extrema of the band dispersion curve $E(k)$. According to Eq. (1.73), in a state k of a chosen band the *electron velocity is provided by the gradient of the dispersion curve* of the energy band under attention. It is shown below that the *electron effective mass is provided by the curvature of the dispersion curve*.

We now discuss the rate of change of the crystal momentum under external fields. To be specific, consider the effect of an external electric field F (assumed uniform and directed in the positive x-direction) on the dynamics of an electron in a given energy band. When the field acts on the electron for a small time dt, the carrier gains the energy

$$dE = (-e)Fv\,dt = (-e)F\frac{1}{\hbar}\frac{dE}{dk}dt; \quad \Longrightarrow \quad dk = (-e)F\frac{1}{\hbar}dt.$$

From the above equation, we have that the rate of change of the crystal momentum is controlled by the external force via the relation

$$\frac{d(\hbar k)}{dt} = -eF \; . \tag{1.74}$$

The above intuitive equation, known as *acceleration theorem*, has been derived on the basis of semiclassical arguments. Because of the conceptual importance of the acceleration theorem, and its peculiar implications, it is instructive to consider its general quantum mechanical elaboration.

The total Hamiltonian of an electron in a periodic potential $V(x)$ and in the presence of a uniform electric field F (in the positive x-direction) reads

$$H = \frac{p^2}{2m} + V(x) + eFx,$$

where e is the absolute value of the electronic charge. Suppose that at the time $t = t_0 \equiv 0$ an electron is prepared in the Bloch state $\psi(k_0, x)$ of wavenumber k_0; the time evolved state at time t is given by

$$\psi(x, t; F) = \exp\left\{\frac{-i}{\hbar}\left[\frac{p^2}{2m} + V(x) + eFx\right]t\right\}\psi(k_0, x). \tag{1.75}$$

Replacing the x variable with the translated variable $x + a$ in both members of the above relation, one obtains

$$\psi(x + a, t; F) = \exp\left\{\frac{-i}{\hbar}\left[\frac{p^2}{2m} + V(x) + eF(x + a)\right]t\right\}\psi(k_0, x + a)$$

$$= \exp\left\{\frac{-i}{\hbar}\left[\frac{p^2}{2m} + V(x) + eFx\right]t\right\}e^{-(i/\hbar)eFat}e^{ik_0 a}\psi(k_0, x).$$

It follows:

$$\psi(x + a, t; F) = e^{ik(t)a}\psi(x, t; F),$$

where

$$k(t) = -\frac{1}{\hbar}eFt + k_0. \tag{1.76}$$

Equation (1.76) shows that the time evolved wavefunction $\psi(x, t; F)$ is a Bloch-type wavefunction, whose wave-number $k(t)$ changes linearly in time; this exact quantum mechanical result can be evidently recast in the compact form of Eq. (1.74). Thus, for what concerns the effect of external forces (at least for applied electric fields), it is indeed justified to identify $\hbar k$ as the quasi-momentum of the electron in the crystal.

The time evolution of band states in the presence of an external field F is described by Eq. (1.75); it shows that the wavefunction $\psi(x, t; F)$ can be expressed in general as a linear combination of the crystal eigenstates (without field) of wavenumber $k(t)$ *with contributions from all the bands of the crystal*, even if the initially prepared state $\psi(k_0, x)$ belongs to a specific band. In summary: *the acceleration theorem does not entail that the band index is conserved, in general.*

Actually the tunneling rate of an electron from a given occupied band to a higher energy band due to an external electric field was first calculated by Zener, using semi-classical approaches [C. Zener "A theory of electrical breakdown of solid dielectrics" Proc. R. Soc. London, A *145*, 523 (1934)]. However, in most ordinary situations of electric field strengths and energy gaps, interband mixing induced by the electric field F can be neglected. For instance for a gap $E_G = 1$ eV and $F = 10^4$ V/cm, interband mixing would require tunneling of the electron through a triangular barrier of height 1 eV and width $d = 10^{-4}$ cm $= 10^4$ Å ($eFd = E_G$), and it is thus negligible. *In the absence of interband tunneling, an electron belonging to a given empty band, under the influence of an applied electric field explores the same band to which it initially belongs.*

When the electron explores a unique band, one can obtain a semiclassical expression for its acceleration, defined as the time derivative of the velocity of Eq. (1.73); in fact, using Eq. (1.74) we obtain

$$\frac{dv(k)}{dt} = \frac{d}{dt}\frac{1}{\hbar}\frac{dE(k)}{dk} = \frac{1}{\hbar}\frac{d^2E(k)}{dk^2}\frac{dk}{dt} = \frac{1}{\hbar^2}\frac{d^2E(k)}{dk^2}(-e)F.$$

Thus we have reduced the equation of motion for the electron in a Newton-like form, where only *external forces* appear, provided the electron effective mass is defined in the form

$$\boxed{\frac{1}{m^*} = \frac{1}{\hbar^2}\frac{d^2E(k)}{dk^2}}. \tag{1.77}$$

We can also say that the lattice periodic potential has modified the inertia of the electron according to Eq. (1.77).

The effective mass of an electron is related to the *local curvature* of the energy band at the specified k vector. The concept of effective mass is particularly useful in the neighborhood of regions in k space, where the energy dispersion curve $E(k)$ has a parabolic energy-momentum relationship. In these regions the "effective mass" is constant and, in some situations, the carriers of the crystal respond to sufficiently small and smooth external perturbations, as if the main effect of the periodic lattice is the modification of the inertia of the electron according to Eq. (1.77); this is called *effective mass approximation*. Since the effective mass concept is related to the curvature of the energy band dispersion curve $E(k)$, one can encounter carriers whose effective mass is much different from the bare electron mass, and with negative effective mass.

Bloch Oscillations and Stark-Wannier Ladders

The Bloch oscillator was suggested theoretically in Z. Phys. *52*, 555 (1928) from the semiclassical study of the motion of an electron in a crystal, in the presence of an applied electric field. When the electric field is sufficiently small, or the band gaps are sufficiently large, the interband Zener tunneling can be neglected; in this situation, *an electron in a given empty energy band of a perfect periodic crystal responds to a steady electric field F in an oscillatory way with Bloch frequency* $\hbar\omega_B = eFa$, where a is the lattice parameter.

Consider an electron of given wavenumber k in an energy band $E(k)$ under the influence of a steady electric field F; in the absence of scattering (ballistic regime), the crystal momentum changes linearly in time in agreement with Eq. (1.74), or the equivalent Eq. (1.76); the collision-free motion of the electron is described by the equations

$$k(t) = k_0 - \frac{1}{\hbar}eFt \quad \text{and} \quad v(t) = \frac{1}{\hbar}\left[\frac{dE(k)}{dk}\right]_{k=k(t)}, \tag{1.78}$$

where k_0 is the electron wavenumber at the initial time $t = 0$. In the presence of scattering processes (due to impurities or any other break of periodicity), the electron wavenumber is expected to be reset at or near some initial value within a relaxation time, which in ordinary situations may be of the order of picosecond or so.

It is interesting to analyze closely the implications of Eqs. (1.78) in the free-electron case and for an electron belonging to an actual energy band. In the free-electron case, the momentum, the velocity, and the kinetic energy of the electron increase indefinitely in time, as schematically shown in Figure 1.18a and in the essentially equivalent Figure 1.18b.

The situation becomes drastically different for an electron belonging to an energy band. With reference to Figure 1.18c, let us start for instance with an electron at $k = 0$. Under the influence of an electric field the electron is first accelerated, and acquires energy and velocity. At the top of the energy band at $k = -\pi/a$ the velocity vanishes; in the absence of interband tunneling, the electron continues its path on the same band from $k = +\pi/a$ and it begins to lose energy, until it reaches the initial state at $k = 0$. Differently from the free-electron evolution, the motion is now periodic in reciprocal

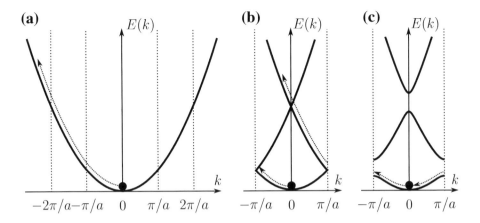

Figure 1.18 Schematic representation in k space of the motion of an electron under the influence of a constant electric field F in x direction. In the free-electron case (Figure 1.18a), the momentum, velocity and energy increase indefinitely in time. Figure 1.18b is the same as Figure 1.18a, with the free-electron parabola folded in the first Brillouin zone for convenience. In the presence of the electric field and of a periodic potential, the electron velocity oscillates and the electron undertakes a very fast oscillating motion (Figure 1.18c).

space and in real space, where the electron oscillates between its initial position and an end point.

The time T_B to complete a cycle can be obtained from the first of Eqs. (1.78) and is given by

$$\frac{2\pi}{a} = \frac{1}{\hbar} e F T_B;$$

the angular frequency ω_B of the "Bloch oscillator" is

$$\omega_B = \frac{2\pi}{T_B} = \frac{e F a}{\hbar} \implies \hbar \omega_B = e F a. \tag{1.79}$$

For instance, with a field of intensity $F = 10^3$ V/cm, we have $eF = 10^{-5}$ eV/Å; taking $\hbar = 6.5822 \cdot 10^{-16}$ eV · s, one obtains

$$\begin{cases} \hbar\omega_B = 0.01 \text{ mev} & \omega_B = 1.5192 \times 10^{10} \text{ rad/s} & \text{for } a = 1 \text{ Å,} \\ \hbar\omega_B = 1 \text{ mev} & \omega_B = 1.5192 \times 10^{12} \text{ rad/s} & \text{for } a = 100 \text{ Å,} \end{cases} \tag{1.80}$$

where typical parameters of lattices and superlattices have been considered. We can also add that Bloch oscillations of ultracold atoms in optical lattices under the effect of a uniform force (due to gravity, for instance) occur in a frequency range of the order of the kilohertz, which differs by several orders of magnitudes with respect to the values of Bloch oscillation frequencies encountered in ordinary semiconductors, as hinted from the difference of energy scales estimated in Section 1.5.2.

As an exemplification, let us consider the case that the energy band $E(k)$ is given by the tight-binding expression $E(k) = E_0 + 2\gamma \cos ka$ (see Eq. 1.48); the corresponding semiclassical electron velocity becomes

$$v(k) = \frac{1}{\hbar}\frac{dE}{dk} \implies v(t) = -\frac{2\gamma a}{\hbar} \sin\left[\left(k_0 - \frac{eFt}{\hbar}\right)a\right]; \qquad (1.81a)$$

the semiclassical position $x(t)$ of the oscillator is then

$$x(t) = \int v(t)dt \implies x(t) = x_0 - \frac{2\gamma}{eF}\cos\left[\left(k_0 - \frac{eFt}{\hbar}\right)a\right]. \qquad (1.81b)$$

Thus the electron performs a harmonic motion of angular frequency ω_B and amplitude $A = 2|\gamma|/eF$ around some equilibrium position x_0; notice that $2eFA$ represents the maximum energy change of the semiclassical oscillator at $x = x_0 \pm A$; as expected, $2eFA$ equals the band width $4|\gamma|$ of the energy band in consideration.

Until now we have considered the semiclassical motion of the electron in the highly idealized situation of a collisionless regime (disregarding Zener tunneling). Unavoidable deviations from ideal periodicity of any realistic system must be taken into account; phenomenologically we assume that the collisionless motion is limited by a time τ, that somehow represents the average time between two successive collisions. In order to achieve the actual realization of a Bloch oscillator of frequency ω_B, we must require $T_B \ll \tau$, or equivalently

$$\omega_B \tau \gg 1 \implies \hbar\omega_B \equiv eFa \gg \frac{\hbar}{\tau}, \qquad (1.82)$$

so that the electron may complete several Bloch oscillations before scattering events take place. In standard lattices, where the lattice parameter is of the order of the Angstrom and $\omega_B \approx 10^{11}$ rad/s, the condition (1.82) is hardly satisfied. However in man-made superlattices, both the lattice parameter ($a \approx 50 - 100$ Å) and the relaxation time are significantly higher than in ordinary materials, and $\omega_B \approx 10^{12}$ rad/s. This makes it possible applications of very clean artificial superlattices as sources of terahertz electromagnetic radiation.

The periodic semiclassical motion is a hint that quantization of the whole band occurs when a static electric field F is present. Indeed, strictly speaking, the electric field breaks translational symmetry, and the band structure scheme can at most be a pictorial initial approximation. To try to understand more on the subtle nature of electronic states in a potential, which is the sum of a periodic potential $V(x)$ and a linearly varying potential eFx, consider the stationary eigenvalue equation

$$\left[\frac{p^2}{2m} + V(x) + eFx\right]\psi(x) = E\psi(x), \qquad (1.83)$$

and the following heuristic considerations due to Wannier. Suppose that $\psi(x)$ is an eigenfunction of energy E of the eigenvalue equation (1.83); by making the translation $x \rightarrow x - t_n$ we have that $\psi(x - t_n)$ is also an eigenfunction of Eq. (1.83) with eigenvalue $E + eFt_n$; thus the solutions of the Schrödinger equation (1.83) are organized in what are called the Stark-Wannier ladders, with rungs separated by $\Delta W = eFa$. The role of an electric field in replacing itinerant crystalline states by a ladder of evenly spaced bound levels was pioneered by Wannier, who analyzed the spectrum of a crystal Hamiltonian in the presence of a uniform electric field in the one-band approximation [G. H. Wannier, Rev. Mod. Phys. *34*, 465 (1962)]. Later, for actual many-band Hamiltonians with sufficiently large energy gaps, it was shown that Zener tunneling does not wash out the Stark ladders, although it leads to their broadening into equidistant Stark resonances of finite width. For further aspects of Bloch electrons in the presence of a homogeneous electric field see for instance G. Bastard, J. A. Brum and R. Ferreira in Solid State Physics, Vol. *44*, 289 (1991) and references quoted therein.

From this brief digression on the Bloch oscillations and Stark-Wannier ladder, it is evident that the conditions for their observation are very demanding, but not impossible; actually, much interest have received the experimental observations of Bloch oscillations in semiconductor superlattices, in atomic condensates, in photonic superlattices, and we refer to the specialized literature for further aspects.

Before closing this subject, consider again the Bloch oscillations $\hbar\omega_B = eFa$ produced by a uniform electric field in a ideal single-band collision-free crystal. Denoting by $V = Fa$ the potential drop in the unit cell, the Bloch relation can be cast in the form $\hbar\omega = eV$, which is formally akin to the frequency of the Josephson effect (with double electronic charge of the Cooper pairs). Thus, the application of a constant external field to an ideal crystal or to coupled superconductors produces a similar frequency dependent response from the formal side. In the latter case, the spectacular accuracy of the Josephson constant $K_J = 2e/\hbar = 483597.9$ GHz/V has become a standard reference value in metrology. (Superconductivity and Josephson effects, are discussed in Chapter 18).

Velocity, Quasi-Momentum and Effective Mass of a Hole in a Band

Until now we have considered *a single electron in an empty band*; here we wish to consider the dynamical aspects of a *missing electron in a completely filled band* (the missing electron is called *hole*). We notice that the current associated to a band fully filled with electrons vanishes; in fact we have

$$I = 2\sum_k (-e)\frac{v(k)}{L} = 2\frac{-e}{L}\frac{1}{\hbar}\sum_k \frac{dE(k)}{dk} \equiv 0,$$

where $L = Na$ is the length of the crystal, $L/v(k)$ is the time required by the electron to cross the crystal, and the factor 2 takes into account spin degeneracy; the current is zero because of the general property that $E(k) = E(-k)$ is an even function of k. If an electron (of given spin) is missing in a state of wavenumber k_h, we have for the current

$$I_h = 2\sum_k (-e)\frac{v(k)}{L} - (-e)\frac{v(k_h)}{L} \equiv +e\frac{v(k_h)}{L}. \qquad (1.84)$$

From Eq. (1.84) we see that the effective current due to the presence of a hole is the opposite to the current carried by an electron in that state; thus "holes" and "electrons" have opposite charges. A similar reasoning can be done for the variation of the crystal momentum; in fact we have for a hole

$$\frac{d(\hbar k)}{dt} = +eF.$$

Finally, if we are interested on holes at the top of a fully occupied band, the negative local curvature of the energy band can be interpreted as a positive effective mass of the hole.

Conductors, Semiconductors, and Insulators

A proper distinction between conductors, semiconductors, and insulators, can be done only in the three-dimensional case. However it is of interest to show how the band structure theory can explain the huge difference of conductivity in materials considering that the energy bands of a crystal are occupied by the available electrons according to the Pauli principle and Fermi-Dirac statistics.

We first of all notice that a fully occupied band is completely ineffective for electron conductivity. In fact the electrons subjected to electric fields cannot absorb energy, unless interband transitions are involved. An insulator is composed by fully occupied bands and fully unoccupied bands and is thus ineffective to the conduction. Conversely, a typical model of a conductor is constituted by a partially filled band. In this case the energy gap between occupied and unoccupied states is zero, and it is possible to have a response to a steady electric field; the conductivity of a metal usually decreases with increasing temperature, due to the reduction of the relaxation time elapsing between two successive collisions of the free carriers (with the lattice vibrations, for instance). In semiconductors we have fully occupied and fully empty bands at $T = 0$, but the energy gap between occupied and unoccupied states is small (typically less than 1–2 eV). Varying the temperature a number of electrons occupy the conduction band and a number of holes are left in the valence bands. These thermally excited carriers depend strongly on temperature, and give rise to a conductivity, that in general highly increases with temperature.

Electronic Current Carried by a One-Dimensional Channel in the Diffusive and in the Ballistic Regime

We can conclude these introductory notes on the peculiar effects of periodicity on the properties of electrons with a few considerations on the electric current through a one-dimensional conductor.

Consider first a conductor with a single partially occupied band, and assume n carriers per unit length, characterized by an effective mass $m^\star$ and relaxation time τ between collisions. In the ordinary diffusive regime, the carriers are subjected to several scattering events before crossing the crystal of length L; then one can estimate classically the current produced by a bias voltage V applied to the device. When an electron is accelerated by the electric field $F = V/L$, it acquires a drift velocity $v_{\text{drift}} = (-e)F\tau/m^\star$, where τ is the average time since the last collision. The current through the device becomes

$$I = n(-e)v_{\text{drift}} = \frac{ne^2\tau}{m^\star}F = \frac{ne^2\tau}{m^\star}\frac{V}{L},$$

and the well-known Drude transport model is recovered (the conditions of applicability of this classical elaboration are discussed with details in Chapter 11). The above linear current-voltage relationship gives for the resistance R, and the conductance G, of the device

$$\boxed{R = \frac{m^\star}{ne^2\tau}L} \quad \text{and} \quad \boxed{G = \frac{ne^2\tau}{m^\star}\frac{1}{L}} \quad \text{(diffusive regime)}. \qquad (1.85)$$

In the diffusive regime, it is evident that the conductance depends on several parameters: the electron density, the scattering time, the features of the band (via the effective mass, or other aspects of the dispersion curve), the geometrical length of the device.

The situation changes dramatically when one considers charge transport through an ideal one-dimensional conductor, under the assumption of ballistic dynamics of electrons; in this situation the conductance of the channel is quantized to the universal value $2e^2/h$.

In the ballistic regime, the one-dimensional system under attention can be schematized as a crystal in contact with two particle reservoirs at the left and right sides of the device (see Figure 1.19). The left and right reservoirs have chemical potentials μ_L and μ_R, respectively (and zero temperature for simplicity). The electronic states of electrons moving from left to right are occupied for energies up to the chemical potential μ_L. Similarly the left moving carrier states are occupied for energies up to the chemical potential μ_R.

The occupancy of the Bloch states of the channel is illustrated in Figure 1.19b. The current carried by the ballistic device is

$$I = 2(-e)\frac{1}{L}\sum_{k \text{ occ.}} v(k) \quad \text{with} \quad v(k) = \frac{1}{\hbar}\frac{dE(k)}{dk},$$

where L is the length of the crystal, $v(k)$ is the velocity, and the factor 2 takes into account spin degeneracy. Replacing the discrete sum over k with $L/2\pi$ times the integral over k, we have

$$I = \frac{2(-e)}{2\pi}\int_{-k_R}^{+k_L}\frac{1}{\hbar}\frac{dE(k)}{dk}dk = \frac{2(-e)}{h}\int_{\mu_R}^{\mu_L}dE = \frac{2(-e)}{h}(\mu_L - \mu_R).$$

Since the applied voltage V is related to the difference of chemical potentials by the equality $(-e)V = \mu_L - \mu_R$, one obtains

$$I = \frac{2e^2}{h}V \quad \Longrightarrow \quad \boxed{R = \frac{h}{2e^2}} \quad \text{and} \quad \boxed{G = \frac{2e^2}{h}} \quad \text{(ballistic regime)}. \qquad (1.86)$$

The conductance, that relates the total current to the voltage drop takes thus the universal quantized value $2e^2/h$ for any ballistic channel, intersected by the Fermi level.

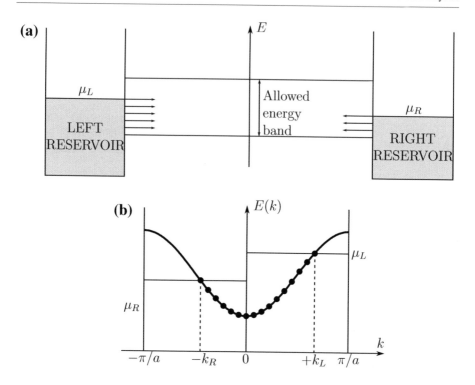

Figure 1.19 (a) Schematic representation of a ballistic device for the calculation of the conductance of an ideal one-dimensional crystal. The allowed energy band of the crystal, and the chemical potentials of the left and right reservoirs are indicated; arrows denote the carriers injected by the reservoirs. (b) Occupancy of the band states for left-moving and right-moving electrons.

Experiments performed on nanoscale constrictions (quantum point contacts) in two-dimensional electron gases showed that the conductance decreases in step-like fashion as the section of the constriction decreases, in contrast to macroscopic conductors where conductance decreases monotonically as the cross-section decreases (see, for instance, the papers by B. J. van Wees et al., and by D. A. Wharam et al. in the references). Also notice that in the ballistic regime the voltage drop occurs conceptually at the contacts of the sample with the left and right reservoirs, or the parts of the circuit that take the role of left or right particle reservoirs [see R. Landauer "Conductance determined by transmission: probes and quantized constriction resistance," J. Phys.: Condens Matter *1*, 8099 (1989)].

As in the case of the Bloch oscillator, the conditions for the realization of ballistic conductance are quite demanding, but experimentally achievable in the world of nanostructures (see, for instance, the textbook of Ferry et al. cited in the references). The conditions of ballistic transport are met in experiments involving quantum point contacts, quantum wires, quantum dots, and the spectacular quantum Hall effect. In the latter case, the von Klitzing constant $R_K = h/e^2 = 25812.806 \ \Omega$ has become the reference value of resistance in metrology, because of its unrivaled accuracy (see Chapter 15).

Finally it is worthwhile to revisit Eqs. (1.85) and (1.86) in the case the effective charge of carriers is $-e^*$ (instead of $-e$). It is seen by inspection that

$$G = \frac{ne^{*2}\tau}{m^\star}\frac{1}{L} \text{ (diffusive regime)}, \quad G = \frac{2ee^*}{h} \text{ (ballistic regime)}; \qquad (1.87)$$

the conductance of a channel is proportional to the square of e^* in the diffusive regime, while it is linear in e^* in the ballistic regime.

Appendix A.   Solved Problems and Complements

Problem 1. Bound state of an electron in an extremely narrow one-dimensional quantum well of arbitrary strength.

Problem 2. Transfer matrix approach for a periodic tight-binding Hamiltonian with one orbital per unit cell.

Problem 3. Band structure of a tight-binding Hamiltonian with two orbitals of different energies in the unit cell.

Problem 4. Band structure of a tight-binding Hamiltonian with two alternating bond strengths in the unit cell.

Problem 5. Photon energy bands in one-dimensional periodic structure with the Kronig-Penney model.

Problem 6. Electron tunneling with the semiclassical approximation.

Problem 7. Zener tunneling in a two band model semiconductor.

Problem 1. Bound State of an Electron in an Extremely Narrow One-Dimensional Quantum Well of Arbitrary Strength

Consider a one-dimensional free electron subjected to an attractive δ-like potential of arbitrary strength. Show that there is always one and only one bound state, and determine its binding energy.

The Hamiltonian under attention is

$$H = \frac{p^2}{2m} + C\delta(x) \quad \text{with} \quad C < 0.$$

We consider the eigenvalue equation $H\psi = E\psi$ and the plane waves expansion of the wavefunction

$$\psi(x) = \sum_q A(q)W(q,x) \quad \text{where} \quad W(q,x) = \frac{1}{\sqrt{L}}e^{iqx};$$

the plane waves are normalized to one in the segment of macroscopic length L. We have

$$\left[\frac{p^2}{2m} + C\delta(x)\right]\sum_q A(q)W(q,x) = E\sum_q A(q)W(q,x).$$

Taking the scalar product with $\langle W(k, x)|$ we obtain

$$\frac{\hbar^2 k^2}{2m} A(k) + C\frac{1}{L}\sum_q A(q) = E A(k).$$

For negative energies and for attractive potential (i.e. $E < 0$ and $C < 0$) we have

$$A(k) = C\frac{1}{E - \hbar^2 k^2/2m}\frac{1}{L}\sum_q A(q)$$

since the denominator cannot vanish in the stated conditions.

Summing up with respect to k, we obtain the compatibility equation

$$\frac{1}{L}\sum_k \frac{1}{E - (\hbar^2 k^2/2m)} \equiv \frac{1}{C}, \tag{A.1}$$

whose graphical solution (not reported here) shows that there is always one and only one bound state.

The determination of the binding energy can be carried out replacing in Eq. (A.1) the discrete sum over k with $L/2\pi$ times the integral in dk. We have

$$\int_{-\infty}^{+\infty} \frac{1}{|E| + \hbar^2 k^2/2m}dk = \frac{2\pi}{|C|}. \tag{A.2}$$

With the help of the indefinite integral

$$\int \frac{1}{|E| + \hbar^2 k^2/2m}dk = \sqrt{\frac{2m}{|E|\hbar^2}}\arctan\left(k\sqrt{\frac{\hbar^2}{2m|E|}}\right),$$

Equation (A.2) gives

$$\sqrt{\frac{2m}{|E|\hbar^2}} = \frac{2}{|C|}.$$

Thus the binding energy $|E| = E_b$ becomes

$$E_b = \frac{1}{2}\frac{m}{\hbar^2}|C|^2.$$

This result coincides with Eq. (1.15) of Section 1.2, and shows from a different perspective the origin of (at least) one bound state in any one-dimensional attractive well.

Problem 2. Transfer Matrix Approach for a Periodic Tight-Binding Hamiltonian with One Orbital per Unit Cell

Consider the tight-binding Hamiltonian of a one-dimensional periodic crystal, and assume for simplicity one orbital per site and nearest neighbors interactions. As in

Eq. (1.50), the tight-binding model Hamiltonian can be represented in the bra-ket notations as

$$H = E_0 \sum_n |f_n\rangle\langle f_n| + \gamma \sum_n [|f_n\rangle\langle f_{n+1}| + |f_{n+1}\rangle\langle f_n|],$$

where E_0 is the on-site energy, γ is the nearest neighbors interaction, and the basis orbitals f_n localized on the corresponding sites $t_n = na$ are assumed orthonormal. Using the transfer matrix approach, show that the band structure is given by the expression

$$E(k) = E_0 + 2\gamma \cos ka. \tag{A.3}$$

This dispersion curve was previously obtained in Section 1.4.1 using the Bloch sum procedure.

Consider the eigenvalue equation $H\psi = E\psi$, and the expansion of the wavefunctions in localized orbitals of the type $\psi = \sum c_m |f_m\rangle$; it holds

$$\left[E_0 \sum_n |f_n\rangle\langle f_n| + \gamma \sum_n |f_n\rangle\langle f_{n+1}| + \gamma \sum_n |f_{n+1}\rangle\langle f_n| \right] \sum_m c_m |f_m\rangle = E \sum_m c_m |f_m\rangle.$$

Due to the assumed orthonormality of the adopted basis functions, one has

$$E_0 \sum_n c_n |f_n\rangle + \gamma \sum_n c_{n+1} |f_n\rangle + \gamma \sum_n c_{n-1} |f_n\rangle = E \sum_n c_n |f_n\rangle.$$

Taking the scalar product with $\langle f_m|$ one obtains the set of coupled linear homogeneous equations

$$(E_0 - E)c_m + \gamma c_{m+1} + \gamma c_{m-1} = 0.$$

The above set of equations can be written in the compact matrix form

$$\begin{pmatrix} c_{m+1} \\ c_m \end{pmatrix} = \begin{bmatrix} (E - E_0)/\gamma & -1 \\ 1 & 0 \end{bmatrix} \begin{pmatrix} c_m \\ c_{m-1} \end{pmatrix} \equiv M(E) \begin{pmatrix} c_m \\ c_{m-1} \end{pmatrix}, \tag{A.4}$$

where $M(E)$ defines the transfer matrix for electron propagation in the forward direction. Using the Bloch theorem, we can connect the amplitudes c_{m+1} and c_m on any two adjacent sites with the phase factors $\exp(ika)$, and write

$$\begin{pmatrix} e^{ika} c_m \\ e^{ika} c_{m-1} \end{pmatrix} = \begin{bmatrix} (E - E_0)/\gamma & -1 \\ 1 & 0 \end{bmatrix} \begin{pmatrix} c_m \\ c_{m-1} \end{pmatrix}.$$

The above matrix equation entails the determinant compatibility condition

$$\left\| \begin{matrix} (E - E_0)/\gamma - e^{ika} & -1 \\ 1 & -e^{ika} \end{matrix} \right\| = 0; \tag{A.5}$$

the explicit elaboration leads to the band structure of Eq. (A.3).

Even more simply, one can enforce the compatibility relation between the transfer matrix of a periodic system and the Bloch form of wavefunctions; one has

$$\operatorname{Tr} M(E) = 2 \cos ka \quad \Longrightarrow \quad \frac{E - E_0}{\gamma} = 2 \cos ka, \tag{A.6}$$

and Eq. (A.3) is recovered also through this path.

Problem 3. Band Structure of a Tight-Binding Hamiltonian with Two Orbitals of Different Energies in the Unit Cell

Consider a periodic one-dimensional chain with two orbitals per unit cell, of energies E_1 and E_2, and nearest neighbor interactions. Determine the allowed energy bands with three different methods: (a) Bloch sums, (b) transfer matrix, (c) continued fractions.

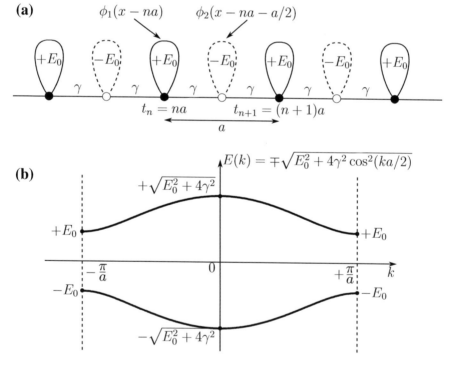

Figure 1.20 (a) Schematic representation of the tight-binding Hamiltonian of a one-dimensional chain of lattice parameter a, nearest neighbor interaction γ, and two orbitals per unit cell with alternating diagonal energies $\pm E_0$. (b) Band structure of the described one-dimensional crystal.

The periodic crystal under attention is schematically indicated in Figure 1.20, where the parameters of relevance are also reported. Without loss of generality, the alternating on-site energies are taken as $+E_0$ and $-E_0$ (having set $|E_1 - E_2| = 2E_0$, and appropriately shifted the reference zero energy).

Band structure using a basis of Bloch sums. The system under attention contains two non-equivalent orbitals in the unit cell, $\phi_1(x - t_n)$ and $\phi_2(x - t_n - a/2)$ with orbital energies $+E_0$ and $-E_0$, respectively. At every k wavenumber in the Brillouin zone, we consider the two Bloch sums

$$\Phi_1(k, x) = \frac{1}{\sqrt{N}} \sum_n e^{ikt_n} \phi_1(x - t_n) \quad \text{and}$$

$$\Phi_2(k, x) = \frac{1}{\sqrt{N}} \sum_n e^{ik(t_n + a/2)} \phi_2(x - t_n - a/2),$$

where N is the number of unit cells of the crystal. The diagonalization of the crystal Hamiltonian on the two Bloch sums gives

$$\begin{Vmatrix} +E_0 - E & 2\gamma \cos(ka/2) \\ 2\gamma \cos(ka/2) & -E_0 - E \end{Vmatrix} = 0.$$

From the above secular determinant one obtains the band structure

$$E(k) = \mp\sqrt{E_0^2 + 4\gamma^2 \cos^2 \frac{ka}{2}}, \tag{A.7}$$

which is schematically indicated in Figure 1.20b. Near the extrema at $k = 0$ and $k = \pi/a$, a series development of Eq. (A.7) gives for the effective masses

$$m^\star = \mp\frac{\hbar^2 \sqrt{E_0^2 + 4\gamma^2}}{\gamma^2 a^2} \quad \text{at } k \approx 0; \quad m^\star = \mp\frac{\hbar^2 E_0}{\gamma^2 a^2} \quad \text{at } k \approx \frac{\pi}{a}. \tag{A.8}$$

In the particular case $E_0 = 0$, the energy dispersion and the effective masses of Section 1.4.1 are evidently recovered, after performing an appropriate folding of the Brillouin zone (and some trivial adjustments of notations).

Band structure with the transfer matrix. According to Eq. (A.4), the transfer matrix for a single forward scatterer in a chain with a single orbital per unit cell is

$$M(E) = \begin{bmatrix} (E - E_0)/\gamma & -1 \\ 1 & 0 \end{bmatrix}.$$

In the present case we have to consider two steps in the unit cell, with diagonal energies $+E_0$ and $-E_0$; the product of the two corresponding transfer matrices gives

$$M^{(\text{tot})}(E) = \begin{bmatrix} \dfrac{E - E_0}{\gamma} & -1 \\ 1 & 0 \end{bmatrix} \begin{bmatrix} \dfrac{E + E_0}{\gamma} & -1 \\ 1 & 0 \end{bmatrix} = \begin{bmatrix} \dfrac{E^2 - E_0^2}{\gamma^2} - 1 & -\dfrac{E - E_0}{\gamma} \\ \dfrac{E + E_0}{\gamma} & -1 \end{bmatrix}.$$

Compatibility between the above transfer matrix and the Bloch theorem for periodic systems enforces the relation

$$\operatorname{Tr} M^{(\text{tot})}(E) = 2 \cos ka.$$

It follows:

$$\frac{E^2 - E_0^2}{\gamma^2} - 2 = 2 \cos ka \quad \Longrightarrow \quad E(k) = \mp \sqrt{E_0^2 + 2\gamma^2(1 + \cos ka)}$$

which is equivalent to Eq. (A.7).

Allowed energy bands with continued fractions. From Figure 1.20a, the Green's function matrix element on a site, for instance the site of energy E_0, is given by the continued fraction

$$G(E) = \cfrac{1}{E - E_0 - \cfrac{2\gamma^2}{E + E_0 - \cfrac{\gamma^2}{E - E_0 - \cdots}}} = \frac{1}{E - E_0 - 2t(E)}, \tag{A.9}$$

where

$$t(E) = \cfrac{\gamma^2}{E + E_0 - \cfrac{\gamma^2}{E - E_0 - t(E)}}.$$

A straight algebraic calculation gives the following expression for the tail

$$t(E) = \frac{1}{2(E + E_0)} \left[E^2 - E_0^2 \mp \sqrt{(E^2 - E_0^2)^2 - 4\gamma^2(E^2 - E_0^2)} \right].$$

The expression under square root can be written as

$$\Delta(E) = (E + E_0)(E - E_0) \left(E + \sqrt{E_0^2 + 4\gamma^2} \right) \left(E - \sqrt{E_0^2 + 4\gamma^2} \right). \tag{A.10}$$

The Green's function (A.9) becomes

$$G(E) = \frac{1}{E - E_0 - \dfrac{1}{E + E_0} \left[E^2 - E_0^2 \mp \sqrt{\Delta(E)} \right]} = \frac{E + E_0}{\mp\sqrt{\Delta(E)}}; \tag{A.11}$$

the sign of the square root must be chosen in such a way that the imaginary part of the retarded Green's function is negative. It is evident that allowed energy states are possible if the Green's function has an imaginary part, i.e. if $\Delta(E) < 0$. From Eq. (A.10), the energy regions where $\Delta(E)$ is negative and energy states are allowed are

$$-\sqrt{E_0^2 + 4\gamma^2} < E < -E_0 \quad \text{and} \quad E_0 < E < +\sqrt{E_0^2 + 4\gamma^2}, \tag{A.12}$$

and this is in agreement with the band structure of Figure 1.20b. The density-of-states can be evaluated from the spectral theorem.

Problem 4. Band Structure of a Tight-Binding Hamiltonian with Two Alternating Bond Strengths in the Unit Cell

Consider a periodic one-dimensional chain with two alternating nearest neighbor interactions γ_1 and γ_2 between adjacent sites along the chain. Determine the allowed energy bands with three different methods: (a) Bloch sums, (b) transfer matrix, and (c) continued fractions.

The periodic crystal under attention is schematically indicated in Figure 1.21a, where the parameters of relevance are also reported. The alternating interaction energies are denoted as γ_1 and γ_2 (and supposed to be both negative). The on-site energies are assumed all equal, and set to zero. The model could mimic for instance a chain of hydrogen molecules or a chain of dimers.

Band structure using a basis of Bloch sums. The system under attention contains two non-equivalent bonds and orbitals in the unit cell, $\phi_1(x - t_n)$ and $\phi_2(x - t_n - d)$. The diagonalization of the crystal Hamiltonian on the two corresponding Bloch sums gives

$$\left\| \begin{matrix} -E & \gamma_1 e^{ikd} + \gamma_2 e^{-ik(a-d)} \\ \gamma_1 e^{-ikd} + \gamma_2 e^{+ik(a-d)} & -E \end{matrix} \right\| = 0.$$

From the above secular determinant one obtains the band structure

$$E(k) = \mp \left| \gamma_1 + \gamma_2 e^{-ika} \right| = \mp \sqrt{\gamma_1^2 + \gamma_2^2 + 2\gamma_1\gamma_2 \cos ka}, \qquad \text{(A.13)}$$

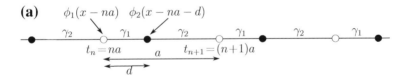

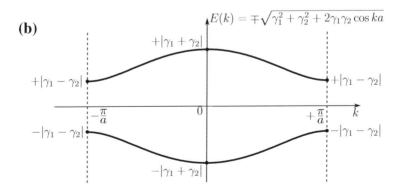

Figure 1.21 (a) Schematic representation of the tight binding Hamiltonian of a one-dimensional chain of lattice parameter a, with alternating bonds of strength γ_1 and γ_2. The orbital energies are assumed equal and taken as the zero reference energy. (b) Band structure of the described one-dimensional crystal.

which is schematically indicated in Figure 1.21b. Near the extrema at $k = 0$ and $k = \pi/a$, the effective masses are

$$m^{\star} = \mp \frac{\hbar^2}{a^2} \frac{|\gamma_1 + \gamma_2|}{\gamma_1 \gamma_2} \quad \text{at} \quad k \approx 0; \quad m^{\star} = \mp \frac{\hbar^2}{a^2} \frac{|\gamma_1 - \gamma_2|}{\gamma_1 \gamma_2} \quad \text{at} \quad k \approx \frac{\pi}{a}.$$

The effective masses become infinite in the case either γ_1 or γ_2 vanish. In the particular case $\gamma_1 = \gamma_2 = \gamma$ it is seen that the band structure coincides with the model of Problem 2, after appropriate folding of the Brillouin zone and trivial adjustment of notations.

The problem addressed above can be exploited to describe a one-dimensional dimerized chain composed of evenly spaced identical single orbital atoms, with site displacements $+u$ and $-u$, alternatively. Expanding nearest neighbor interactions to the leading order in the displacements, we can set

$$\gamma_1 = \gamma + 2\alpha u, \quad \gamma_2 = \gamma - 2\alpha u,$$

where γ is the nearest neighbor interaction in the absence of distortion, and α is an appropriate expansion coefficient. The dispersion relation (A.13) in terms of the parameters (γ, u) reads

$$E(k) = \mp \sqrt{4\gamma^2 \cos^2(ka/2) + 16\alpha^2 u^2 \sin^2(ka/2)}. \tag{A.14}$$

It is evident that the energies of the bottom (top) energy band are all decreased (increased) because of the effect of dimerization induced by the site displacements (Peierls distortion), except at the center of the Brillouin zone. In particular at $k = \pi/a$ an energy gap opens, $E_g = 8\alpha u$, proportional to the amplitude of the lattice distortion. Depending on electronic occupancy, change of lattice energy, possible many-body effects, dimerization, or other lattice distortions may occur in materials.

Band structure with the transfer matrix. The transfer matrix for a single scatterer in a linear chain, with alternating nearest neighbor interactions γ_1 and γ_2, reads

$$M(E) = \begin{bmatrix} (E - E_0)/\gamma_1 & -\gamma_2/\gamma_1 \\ 1 & 0 \end{bmatrix},$$

with exchange of γ_1 and γ_2 for the next scatterer. This expression can be worked out with a procedure similar to Eq. (A.4), to which it reduces in the particular case $\gamma_1 = \gamma_2 = \gamma$.

The product of the two transfer matrices corresponding to the two scatterers in the unit cell (setting $E_0 = 0$ in the present case) gives

$$M^{(\text{tot})}(E) = \begin{bmatrix} \dfrac{E}{\gamma_1} & -\dfrac{\gamma_2}{\gamma_1} \\ 1 & 0 \end{bmatrix} \begin{bmatrix} \dfrac{E}{\gamma_2} & -\dfrac{\gamma_1}{\gamma_2} \\ 1 & 0 \end{bmatrix} = \begin{bmatrix} \dfrac{E^2}{\gamma_1 \gamma_2} - \dfrac{\gamma_2}{\gamma_1} & -\dfrac{E}{\gamma_2} \\ \dfrac{E}{\gamma_2} & -\dfrac{\gamma_1}{\gamma_2} \end{bmatrix}.$$

Compatibility between the above transfer matrix and the Bloch theorem for periodic systems enforces the relation

$$\text{Tr}\, M^{(\text{tot})}(E) = 2\cos ka \quad \Longrightarrow \quad E(k) = \mp\sqrt{\gamma_1^2 + \gamma_2^2 + 2\gamma_1\gamma_2\cos ka}$$

in agreement with Eq. (A.13).

Allowed energy bands with continued fractions. From Figure 1.21a, the Green's function matrix element on a site, is given by the continued fraction

$$G(E) = \cfrac{1}{E - \cfrac{\gamma_1^2}{E - \cfrac{\gamma_2^2}{E - \cdots}} - \cfrac{\gamma_2^2}{E - \cfrac{\gamma_1^2}{E - \cdots}}} = \frac{1}{E - t_1(E) - t_2(E)}, \qquad (A.15)$$

where

$$t_1(E) = \cfrac{\gamma_1^2}{E - \cfrac{\gamma_2^2}{E - t_1(E)}} \quad \text{and} \quad t_2(E) = \cfrac{\gamma_2^2}{E - \cfrac{\gamma_1^2}{E - t_2(E)}}.$$

A straight algebraic calculation gives the following expression for the tails

$$t_1(E) = \frac{1}{2E}\left[E^2 + \gamma_1^2 - \gamma_2^2 \mp \sqrt{\Delta(E)}\right],$$

$$t_2(E) = \frac{1}{2E}\left[E^2 + \gamma_2^2 - \gamma_1^2 \mp \sqrt{\Delta(E)}\right],$$

where

$$\Delta(E) = \left[E^2 - (\gamma_1 + \gamma_2)^2\right]\left[E^2 - (\gamma_1 - \gamma_2)^2\right]. \qquad (A.16)$$

The Green's function (A.15) becomes

$$G(E) = \frac{E}{\mp\sqrt{\Delta}}$$

It is evident that allowed energy states are possible if the Green's function has an imaginary part, i.e. if $\Delta(E) < 0$. From Eq. (A.16), the energy regions where $\Delta(E)$ is negative and energy states are allowed are

$$-|\gamma_1 + \gamma_2| < E < -|\gamma_1 - \gamma_2| \quad \text{and} \quad +|\gamma_1 - \gamma_2| < E < +|\gamma_1 + \gamma_2|,$$

and this is in agreement with the band structure of Figure 1.21b.

Problem 5. Photon Energy Bands in One-Dimensional Periodic Structure with the Kronig-Penney Model

Consider a one-dimensional photonic crystal in which the refractive index $n(x)$ varies along the x-direction in a periodic manner, while it is invariant in the y and z directions.

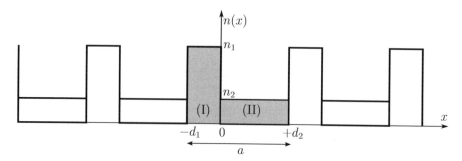

Figure 1.22 Schematic representation of the refractive index $n(x)$ of a one-dimensional photonic crystal. The dielectric is segmented into a periodically alternating regions of index n_1 and length d_1, and index n_2 and length d_2.

For simplicity, the dielectric is assumed to be made of a sequence of alternating layers of refractive index n_1 and n_2 and length d_1 and d_2, respectively (see Figure 1.22). Determine the photonic band structure for the electromagnetic modes propagating in direction perpendicular to the layers.

Photonic crystals are structured materials in which the dielectric constant is periodically modulated. In close formal analogy with the quantum theory of electrons in crystals, it is expected that the dispersion curves of electromagnetic modes are described by photonic energy bands, where gaps may occur. In the present problem, we consider the simplest model made up of alternating layers of transparent materials with different dielectric constants; this paradigmatic sample is studied adapting the original Kronig-Penney procedure to the treatment of propagating modes controlled by the Maxwell's equations.

In the one-dimensional photonic crystal, illustrated in Figure 1.22, the region I and the region II within the unit cell are occupied by materials of refractive index n_1 and n_2, respectively. In regions I and II, consider the general solution of the Maxwell's equations in the frequency domain for an electromagnetic wave propagating in the x-direction, perpendicular to the interfaces of the layered medium, and electric field parallel to them (in this situation, transverse-electric modes and transverse-magnetic modes coincide). For the x-dependence of the electric field component it holds

$$\begin{cases} E_I(x) = Ae^{iq_1x} + Be^{-iq_1x} & -d_1 < x < 0 \quad q_1(\omega) = n_1\omega/c, \\ E_{II}(x) = Ce^{iq_2x} + De^{-iq_2x} & 0 < x < +d_2 \quad q_2(\omega) = n_2\omega/c, \end{cases} \tag{A.17}$$

where $q_1(\omega)$, $q_2(\omega)$ are the propagation wavenumbers in the two media, and A, B, C, D are arbitrary coefficients.

The four arbitrary coefficients A, B, C, D must be chosen so as to satisfy the following four boundary conditions:

$$E_I(0) = E_{II}(0), \quad \left(\frac{dE_I}{dx}\right)_{x=0} = \left(\frac{dE_{II}}{dx}\right)_{x=0}, \tag{A.18}$$

$$E_{II}(d_2) = e^{ika} E_I(-d_1), \quad \left(\frac{dE_{II}}{dx}\right)_{x=d_2} = e^{ika} \left(\frac{dE_I}{dx}\right)_{x=-d_1}. \tag{A.19}$$

The two conditions (A.18) impose the continuity of the electric field and its derivative (connected to the magnetic field) at $x = 0$. The two conditions (A.19) connect electric field and its derivative at the opposite boundaries of the unit cell via the phase factor, as required by the Bloch theorem, which keeps its validity for any translational invariant system.

The conditions (A.18) and (A.19) lead to the following linear homogeneous equations for the coefficients A, B, C, D:

$$\begin{cases} A + B = C + D \\ Aq_1 - Bq_1 = Cq_2 - Dq_2 \\ Ce^{iq_2d_2} + De^{-iq_2d_2} = e^{ika}[Ae^{-iq_1d_1} + Be^{+iq_1d_1}] \\ Cq_2e^{iq_2d_2} - Dq_2e^{-iq_2d_2} = e^{ika}[Aq_1e^{-iq_1d_1} - Bq_1e^{+iq_1d_1}]. \end{cases}$$

The above four equations have a solution only if the determinant of the coefficients which multiply A, B, C, D vanishes:

$$\begin{Vmatrix} 1 & 1 & -1 & -1 \\ q_1 & -q_1 & -q_2 & q_2 \\ -e^{ika-iq_1d_1} & -e^{ika+iq_1d_1} & e^{iq_2d_2} & e^{-iq_2d_2} \\ -q_1e^{ika-iq_1d_1} & q_1e^{ika+iq_1d_1} & q_2e^{iq_2d_2} & -q_2e^{-iq_2d_2} \end{Vmatrix} = 0.$$

The determinant can be evaluated, for instance, by expanding it with respect to the first row and by direct evaluation of the four 3×3 minors. With easy calculations and collection of terms one obtains

$$\boxed{-\frac{q_1^2 + q_2^2}{2q_1q_2} \sin(q_1d_1) \sin(q_2d_2) + \cos(q_1d_1) \cos(q_2d_2) = \cos ka}. \tag{A.20}$$

As indicated by the photonic band equation (A.20), the allowed modes occur when the first member lies between $+1$ and -1.

For the graphical solution of Eq. (A.20), it is convenient to write the arguments of the trigonometric functions in the form

$$q_1d_1 = \frac{n_1}{c}\omega d_1 = 2\pi n_1 \frac{d_1}{a} \frac{\hbar\omega}{\hbar\omega_0} \quad \text{with} \quad \hbar\omega_0 = \hbar c \frac{2\pi}{a},$$

and similarly for q_2d_2. The quantity $\hbar\omega_0$ expresses the energy of the photon whose wavelength in free space equals the lattice parameter, and represents the natural unit of energy of the model under investigation. Equation (A.20) can be cast in the dimensionless form

$$F(x) = \cos ka \quad \text{with} \quad x = \frac{\hbar\omega}{\hbar\omega_0}$$

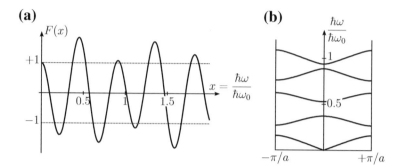

Figure 1.23 (a) Graphical solution of the compatibility equation $F(x) = \cos ka$ of the Kronig-Penney model of one-dimensional photonic crystals. The function $F(x)$ is plotted with the following values of the parameters: $n_1 = 3.42, n_2 = 1, d_1 = d_2 = 0.5\, a$. (b) Lowest photonic energy bands of the described photonic crystal.

and

$$F(x) = -\frac{1}{2}\left(\frac{n_1}{n_2} + \frac{n_2}{n_1}\right)\sin\left(2\pi n_1 \frac{d_1}{a}x\right)\sin\left(2\pi n_2 \frac{d_2}{a}x\right)$$

$$+ \cos\left(2\pi n_1 \frac{d_1}{a}x\right)\cos\left(2\pi n_2 \frac{d_2}{a}x\right). \tag{A.21}$$

As an example, let us take $n_1 = 3.42$ (refractive index of silicon), $n_2 = 1$ (refractive index of vacuum) and $d_1 = d_2 = 0.5\, a$. The function $F(x)$ is plotted versus the dimensionless parameter x in Figure 1.23a; the regions where $|F(x)| \leq 1$ provide the energy values $\hbar\omega = x\hbar\omega_0$ which are allowed. The photonic band structure, for electromagnetic modes propagating along the x-direction, are reported in Figure 1.23b.

From Figure 1.23, it can be seen by inspection that the energy levels are grouped into allowed energy bands separated by forbidden energy regions. The band extrema appear at the center and at the border of the Brillouin zone, similarly to the electronic band structure discussed in Section 1.2.

Problem 6. Electron Tunneling with the Semiclassical Approximation

Consider with the semiclassical Wentzel-Kramers-Brillouin (WKB) theory the tunneling of an electron through barriers of rectangular and triangular shapes.

In some mesoscopic devices, the transport of electronic charge includes the evaluation of tunneling probability through barrier potentials of appropriate shape. For approximate estimation of tunneling probability the semiclassical WKB theory may be sufficient in some circumstances. In the present problem we consider the application of the WKB approximation to tunneling through barriers with rectangular or triangular shapes (see Figure 1.24).

Consider first the rectangular barrier, of height V_b and thickness b, schematically illustrated in Figure 1.24a. The wavefunction of an electron, injected from the left lead

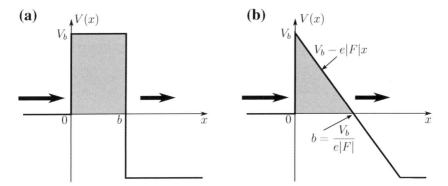

Figure 1.24 Schematic illustration of electron tunneling through a one-dimensional rectangular barrier (a) and a one-dimensional triangular barrier (b). Arrows indicate injected and transmitted particles.

with energy very small with respect to the barrier height, is subject to an exponential damping, in the barrier region, of the type

$$\psi(x) \approx e^{-\beta x} \quad \text{with} \quad \frac{\hbar^2 \beta^2}{2m} = V_b - E \approx V_b \quad \text{for} \quad E \ll V_b.$$

For barriers of low transparency, the tunneling probability (in the absence of scattering events) can be estimated as

$$T = \exp\left[-2 \int_0^b \beta \, dx\right], \quad \beta = \sqrt{\frac{2mV_b}{\hbar^2}}, \quad 0 \le E \ll V_b,$$

where the factor 2 in the exponential is due to the squared modulus of the ratio of the wavefunction at the right side and at the left side of the barrier. According to the above WKB approximation, the tunneling probability through a rectangular barrier of height V_b (measured with respect to the injected energy of the carriers) and thickness b becomes

$$T = \exp\left[-2\sqrt{\frac{2mV_b}{\hbar^2}}\, b\right] \quad \text{rectangular barrier,} \tag{A.22}$$

where m is the mass of the electron (or the effective mass of the electron-like carrier). Qualitatively it is seen that the tunneling probability increases for narrow barriers, small potential barriers, and small effective mass, as also discussed in Eqs. (1.34).

Consider now the case of a triangular barrier, of height V_b and thickness $b = V_b/e|F|$, typically encountered in the presence of electric fields, and schematically illustrated in Figure 1.24b. The wavefunction of an electron, injected from the left lead with energy $E \ll V_b$, in the barrier region is subject to an exponential damping which is now x-dependent and given by

$$\frac{\hbar^2 \beta^2(x)}{2m} = V_b - e|F|x, \quad \beta(x) = \frac{1}{\hbar}\sqrt{2m(V_b - e|F|x)},$$

where e is the absolute value of the electronic charge and F is the electric field. For barriers of low transparency, the tunneling probability (in the absence of scattering events) can be estimated with the WKB approximation, and becomes

$$T = \exp\left[-2\int_0^{V_b/e|F|} \beta(x)dx\right] = \exp\left[-2\frac{\sqrt{2m}}{\hbar}\int_0^{V_b/e|F|}\sqrt{V_b - e|F|x}\; dx\right].$$

Carrying out the integral, the tunneling through a triangular barrier in the WKB approximation takes the expression

$$T = \exp\left[-\frac{4}{3}\frac{1}{\hbar e|F|}\sqrt{2mV_b^3}\right] \quad \text{triangular barrier,} \tag{A.23}$$

where m is the mass of the electron (or the effective mass of the electron-like carrier), V_b is the height of the triangular barrier (measured with respect to the injected energy of the carriers). The above expression has the same structure as Eq. (A.22), taking into account that V_b/eF represents the width of the triangular barrier. Qualitatively it is seen that the tunneling probability increases for small effective mass, for small potential barriers and for large electric fields.

Problem 7. Zener Tunneling in a Two Band Model Semiconductor

In a two band model semiconductor, consider with the semiclassical theory the interband tunneling of a valence electron into the conduction band due to the presence of a uniform electric field.

Consider a typical two band model semiconductor, for instance the one illustrated in Figure 1.20. According to Eqs. (A.7) and (A.8) the expression of the dispersion curves of the two bands, the energy gap at the Brillouin zone border $k = \pi/a$, and the effective masses of the valence and conduction bands around the gap are

$$E(k) = \mp\sqrt{E_0^2 + 4\gamma^2\cos^2\frac{ka}{2}}; \quad E_G = 2E_0; \quad m^\star = \mp\frac{\hbar^2 E_G}{2\gamma^2 a^2}, \tag{A.24}$$

where a is the lattice parameter, γ is the nearest neighbor interaction, and $\pm E_0$ are the energies of the two orbitals in the unit cell.

To better investigate the energy region around the gap occurring at the Brillouin zone border, we can put for convenience $q = k - \pi/a$, and for small values of q, we have

$$E(q) = \mp\sqrt{E_0^2 + \gamma^2 a^2 q^2}.$$

Inverting the above expression one obtains

$$q(E) = \begin{cases} \mp\dfrac{1}{|\gamma|a}\sqrt{E^2 - E_0^2} & |E| > E_0, \\[3mm] \mp\dfrac{i}{|\gamma|a}\sqrt{E_0^2 - E^2} & |E| < E_0. \end{cases} \tag{A.25}$$

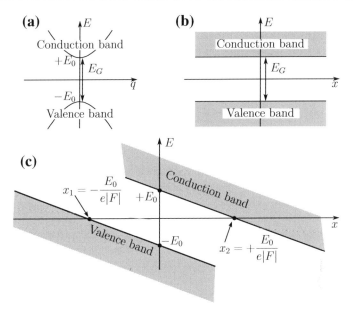

Figure 1.25 (a) Schematic representation of the dispersion curves of a two model semiconductor around the energy gap. The allowed energy bands in the absence and in the presence of a uniform electric field are indicated in figures (b) and (c), respectively.

In the case $|E| > E_0$, i.e. E belongs to the valence or the conduction band, the quantity $q(E)$ is real, and the Bloch wavefunctions contain an oscillating itinerant part $\exp(\pm iqx)$, besides the periodic modulation. On the contrary, in the case $|E| < E_0$, i.e. E belongs to the forbidden energy gap, the quantity $q(E)$ is imaginary, and there is an exponential enhancement or damping $\exp(\pm |q|x)$ of the Bloch-Floquet part of the states satisfying the Schrödinger equation (but not the boundary conditions at $\pm\infty$).

In Figure 1.25 the dispersion curves of the energy bands around the energy gap are reported, together with the behavior of the allowed energy bands in the absence and in the presence of an electric field. With reference to Figure 1.25c, it is seen that in the interval $x_1 < x < x_2$ the propagation wavenumber is purely imaginary and is given by the bottom Eq. (A.25) with $E = e|F|x$. The tunneling probability in the WKB approximation is given by the quantity

$$T = \exp\left[-2\int_{-E_0/e|F|}^{+E_0/e|F|} \frac{1}{|\gamma|a}\sqrt{E_0^2 - e^2 F^2 x^2}\,dx\right]$$

$$= \exp\left[-\frac{2}{|\gamma||e|F|a}\int_{-E_0}^{+E_0}\sqrt{E_0^2 - t^2}\,dt\right],$$

where in the last passage the change of variable $e|F|x = t$ has been performed. The integration in the exponent can be carried out exploiting the indefinite integral

$$\int \sqrt{a^2 - x^2}\,dx = \frac{1}{2}x\sqrt{a^2 - x^2} + \frac{1}{2}a^2 \arcsin\frac{x}{a} \quad a > 0,$$

and the corresponding definite integral

$$\int_{-a}^{+a} \sqrt{a^2 - x^2}\, dx = \frac{\pi}{2} a^2.$$

The tunneling probability becomes

$$T = \exp\left[-\frac{\pi E_0^2}{|\gamma||e||F|a}\right] = \exp\left[-\frac{\pi}{4}\frac{1}{\hbar e|F|}\sqrt{2m^* E_G^3}\right]; \tag{A.26}$$

in the last passage instead of the two parameters γ and E_0, use has been made of the two corresponding parameters, effective mass and energy gap, given in Eq. (A.24). The physical meaning of Eq. (A.26) emerges with evidence from the comparison with Eq. (A.23) and formal coincidence (apart almost equal numerical factors).

The number of oscillations per second in the valence band is $h\nu = e|F|a$; then the rate R at which an electron escapes from the valence band to the conduction band becomes

$$R = \frac{\omega_B}{2\pi} \cdot T = \frac{e|F|a}{h}\exp\left[-\frac{\pi}{4}\frac{1}{\hbar e|F|}\sqrt{2m^* E_G^3}\right]. \tag{A.27}$$

We have that the Zener escaping rate remains safely small (or negligible) for electric fields two (or more) orders of magnitude smaller than the electric field value F^* such that $eF^*a = E_G$. For typical values of the energy gap $E_G = 2$ eV and the lattice constant $a = 2$ Å $= 2 \cdot 10^{-8}$cm, we have $F^* = 10^8$ V/cm, and dielectric breakdown in a number of semiconductors actually requires fields of the order of 10^6 V/cm.

Further Reading

Bastard, G. (1988). *Wave mechanics applied to semiconductor heterostructures*. Les Ulis: Les Editions de Physique.

Berggren, K.-F., & Pepper, M. (2009). Electrons in one-dimension. *Philosophical Transactions of the Royal Society A, 368,* 1141.

Datta, S. (2005). *Quantum transport: Atom to transistor*. Cambridge: Cambridge University Press.

Davies, J. H. (1998). *The physics of low-dimensional semiconductors*. Cambridge: Cambridge University Press.

Economou, E. N. (2006). *Green's function in quantum physics*. Berlin: Springer.

Ferry, D. K., Goodnick, S. M., & Bird, J. (2009). *Transport in nanostructures*. Cambridge: Cambridge University Press.

Fromhold, A. T., Jr. (1981). *Quantum mechanics for applied physics and engineering*. New York: Academic Press.

Giamarchi, T. (2004). *Quantum physics in one dimension*. Oxford: Clarendon Press.

Gilmore, R. (2004). *Elementary quantum mechanics in one dimension*. Baltimore: The Johns Hopkins University Press.

Griffiths, D. J., & Steinke, C. A. (2001). Waves in locally periodic media. *American Journal of Physics, 69,* 137.

Joannopoulos, J. D., Johnson, S. G., Winn, J. N., & Meade, R. D. (2008). *Photonic crystals. Molding the flow of light*. Princeton: Princeton University Press.

Kelly, M. J. (1995). *Low-dimensional semiconductors*. Oxford: Clarendon Press.

Merzbacher, E. (1970). *Quantum mechanics*. New York: Wiley.

Mishra, S., & Satpathy, S. (2003). One-dimensional photonic crystal: The Kronig-Penney model. *Physical Review B, 68*, 045121.

Olsen, R. J., & Vignale, G. (2010). The quantum mechanics of electric conduction in crystals. *American Journal of Physics, 78*, 954.

Peierls, R. E. (1955). *Quantum theory of solids*. London: Oxford University Press.

Schönhammer, K. (2013). Physics in one dimension: Theoretical concepts for quantum many-body systems. *Journal of Physics: Condensed Matter, 25*, 014001.

van Wees, B. J., van Heuten, H., Beenakker, C. W., Williamson, J. G., Kouwenhoven, L. P., van der Marel, D., & Foxon, C. T. (1988). Quantized conductance of point contacts in a two dimensional electron gas. *Physical Review Letters, 60*, 848.

Walker, J. S., & Gathright, J. (1994). Exploring one-dimensional quantum mechanics with transfer matrices. *American Journal of Physics, 62*, 408.

Wannier, G. H. (1959). *Elements of solid state theory*. Cambridge: Cambridge University Press.

Weisbuch, C., & Vinter, B. (1991). *Quantum semiconductor structures*. London: Academic Press.

Wharam, D. A., Thornton, T. J., Newbury, R., Pepper, M., Ahmed, H., Frost, J. E. F., Asko, D. G., Peacock, D. C., Ritchie, D. A., & Jones, G. A. C. (1988). One dimensional transport and quantization of the ballistic resistance. *Journal of Physics C, 21*, L209.

Yariv, A., & Yeh, P. (2003). *Optical waves in crystals*. Hoboken, New Jersey: Wiley.

2 Geometrical Description of Crystals: Direct and Reciprocal Lattices

Crystalline solids are constituted by a regular array of identical units, periodically repeated in space. In this chapter we consider some geometrical aspects of periodic systems, and describe a few crystal structures of wide interest. The description of translational symmetry in the ordinary direct space would not be complete without the complementary notion of the reciprocal space; the Brillouin zone concept in reciprocal space, together with the Bloch theorem, is the fundamental tool for the classification of states in quantum mechanical problems with translational symmetry. Other subjects, best studied with the help of direct and reciprocal spaces, include indexing of lattice planes and directions, choice of special points in the Brillouin zones for averaging of physical quantities, some general properties of itinerant wavefunctions, energy dispersion curves, and density-of-states.

2.1 Simple Lattices and Composite Lattices

A crystal is characterized by a regular array of atoms, which repeat periodically in the space. A crystalline solid is the most familiar example of a system with long-range order; the knowledge of physical properties (for instance electron charge density, crystalline

Solid State Physics, Second Edition. http://dx.doi.org/10.1016/B978-0-12-385030-0.00002-5

field, etc.) within any arbitrary unit cell in ideal crystals implies their knowledge within any other unit cell, even macroscopically far removed from the reference one. In amorphous solids, or in liquids, the long-range order is destroyed, although some sort of short-range order may survive on a local microscopic scale. In this section we describe some general properties entailed by periodicity, and focus on the basic distinction between simple lattices and composite lattices.

2.1.1 Periodicity and Bravais Lattices

A *Bravais lattice* is defined as a regular periodic arrangement of points in space, all of them connected by translation vectors

$$\mathbf{t}_n = n_1 \mathbf{t}_1 + n_2 \mathbf{t}_2 + n_3 \mathbf{t}_3; \tag{2.1}$$

the non-coplanar vectors $\mathbf{t}_1, \mathbf{t}_2, \mathbf{t}_3$ are called *primitive or fundamental translation vectors* and n_1, n_2, n_3 are any tern of *integer* numbers (negative, positive, or zero). The parallelepiped formed by $\mathbf{t}_1, \mathbf{t}_2, \mathbf{t}_3$ is called *primitive unit cell* (or primitive cell); its volume is

$$\Omega = \mathbf{t}_1 \cdot (\mathbf{t}_2 \times \mathbf{t}_3) \tag{2.2}$$

(it is understood that the order of the primitive translation vectors is chosen in such a way to form a right-handed system). The primitive cell contains just one lattice point. The parallelepiped primitive cell is usually specified providing the lengths a, b, c of the edges and the angles α, β, γ between each pair of them.

It is apparent that the choice of the primitive translation vectors, and hence also the shape of the primitive cell, is not unique (only the volume of the unit cell is invariant). This arbitrariness can be used to select the parallelepiped primitive cell with the highest possible symmetry.

An elementary example of the mentioned arbitrariness in the choice of the primitive translation vectors is illustrated in Figure 2.1. The two-dimensional rectangular Bravais lattice generated by the fundamental vectors

$$\mathbf{t}_1 = (a, 0, 0) \quad \text{and} \quad \mathbf{t}_2 = (0, b, 0), \tag{2.3a}$$

can be described equally well, for instance, by the primitive vectors

$$\tilde{\mathbf{t}}_1 = (a, 0, 0) = \mathbf{t}_1 \quad \text{and} \quad \tilde{\mathbf{t}}_2 = (na, b, 0) = n\mathbf{t}_1 + \mathbf{t}_2 \tag{2.3b}$$

with n arbitrary integral number. It is also evident that the choice of primitive vectors and rectangular unit cell of Eq. (2.3a) appears preferable (although equivalent) to the choices of Eq. (2.3b), where the shape of the unit cell is an oblique parallelogram. In any case the area of the two-dimensional unit cell takes the invariant value ab.

From the definition of Bravais lattices, it is evidently possible to describe the whole set of lattice points (2.1) using non-primitive cells, (called *conventional unit cells*), containing an integer number of primitive cells and a corresponding number of extra

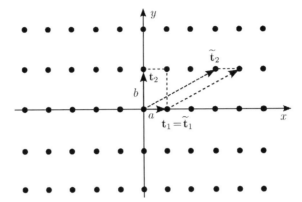

Figure 2.1 Illustration of a two-dimensional rectangular Bravais lattice. The lattice can be described by the primitive translation vectors $\mathbf{t}_1 = a(1, 0, 0)$ and $\mathbf{t}_2 = b(0, 1, 0)$. Although this is the most natural choice, infinite other choices are possible; the figure illustrates the choice $\widetilde{\mathbf{t}}_1 = (a, 0, 0)$ and $\widetilde{\mathbf{t}}_2 = (na, b, 0)$ with $n = 3$. Whatever choice is done, the area of the primitive unit cell is invariant and equals ab.

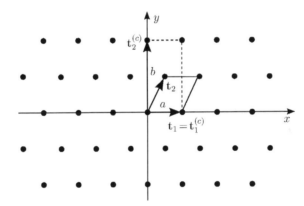

Figure 2.2 Illustration of the primitive unit cell and conventional unit cell for a two-dimensional centered rectangular Bravais lattice.

lattice points. An elementary example of the opportunity to describe some Bravais lattices with a conventional unit cell, is illustrated in Figure 2.2, that reports the two-dimensional centered rectangular Bravais lattice. The primitive lattice vectors of this periodic structure are

$$\mathbf{t}_1 = (a, 0, 0) \quad \text{and} \quad \mathbf{t}_2 = \left(\frac{a}{2}, \frac{b}{2}, 0\right), \tag{2.4a}$$

and the primitive unit cell is a parallelogram of area $ab/2$. Instead of the pictured primitive unit cell (whose shape is an oblique parallelogram), it is sometimes preferred (although equivalent) to describe the lattice by a *conventional unit cell of rectangular shape, double area, and an extra lattice point at the center of the rectangle.* The fundamental vectors of the conventional unit cell are

$$\mathbf{t}_1^{(c)} = (a, 0, 0) \quad \text{and} \quad \mathbf{t}_2^{(c)} = (0, b, 0), \tag{2.4b}$$

the conventional cell defined by $\mathbf{t}_1^{(c)}$ and $\mathbf{t}_2^{(c)}$ contains two primitive cells.

Notice that the conventional cell is not the repeating block of smallest volume; thus in the study of the effects related to the translational symmetry (for instance classification of states with wavevector quantum number, Brillouin zone concept, and in general in any problem directly or indirectly related to the *full* translational symmetry) we must *consider only the primitive translation vectors and the primitive unit cell.* Conventional cells are however of help for a pictorial description of a number of crystals.

The number of different Bravais lattices is determined by symmetry considerations. In two dimensions, the different Bravais lattices are five: oblique, rectangular (primitive and centered), square, hexagonal.

In three dimensions, symmetry considerations lead to fourteen different Bravais lattices grouped into seven crystal systems: triclinic, monoclinic (primitive and base-centered), orthorhombic (primitive, base-centered, body-centered, and face-centered), trigonal, tetragonal (primitive and base-centered), hexagonal, cubic (primitive, body-centered, and face-centered). This classification is based on the equality or inequality of the lengths a, b, c of the edges of the conventional cell, on the angles α, β, γ between each pair of them, and on possible occurrence of additional lattice points at the center of opposite faces or at the center of the cell. The seven crystal systems, in order of increasing symmetry (triclinic, monoclinic, orthorhombic, trigonal, tetragonal, hexagonal, cubic) and the fourteen Bravais lattices are reported in Figure 2.3.

Before closing, we remark that the symmetry properties of a crystal are not confined to the translational invariance, but include in general also appropriate point group operations. Throughout this chapter, we shall focus essentially on the translational symmetry and its entailed consequences; for an account of the full crystal symmetry, we refer to textbooks on group theory.

2.1.2 Simple and Composite Crystal Structures

The geometrical description of a crystal requires the specification of the *primitive translation vectors* $\mathbf{t}_1, \mathbf{t}_2, \mathbf{t}_3$ of the underlying Bravais lattice, as well as the specification of the atoms (or ions) in the primitive cell. The contents of the unit cell is described by means of an appropriate set of *basis vectors* $\mathbf{d}_1, \mathbf{d}_2, \ldots, \mathbf{d}_\nu$, which individuate the equilibrium positions of the nuclei of all the atoms (or ions) in the unit cell; for consistency, the positions of the atoms in the basis *cannot* be related by translation vectors. The above considerations can be summarized as follows:

$$\text{Crystal structure} \Longrightarrow \begin{cases} \mathbf{t}_1, \mathbf{t}_2, \mathbf{t}_3 & \text{(primitive translation vectors)} \\ \mathbf{d}_1, \mathbf{d}_2, \ldots, \mathbf{d}_\nu & \text{(basis of equal or different atoms)} \end{cases}.$$

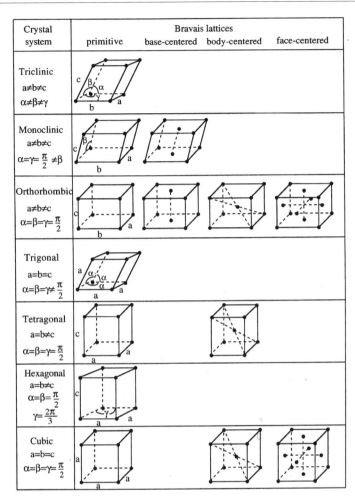

Crystal system	Bravais lattices			
	primitive	base-centered	body-centered	face-centered
Triclinic $a{\neq}b{\neq}c$ $\alpha{\neq}\beta{\neq}\gamma$				
Monoclinic $a{\neq}b{\neq}c$ $\alpha{=}\gamma{=}\frac{\pi}{2}{\neq}\beta$				
Orthorhombic $a{\neq}b{\neq}c$ $\alpha{=}\beta{=}\gamma{=}\frac{\pi}{2}$				
Trigonal $a{=}b{=}c$ $\alpha{=}\beta{=}\gamma{\neq}\frac{\pi}{2}$				
Tetragonal $a{=}b{\neq}c$ $\alpha{=}\beta{=}\gamma{=}\frac{\pi}{2}$				
Hexagonal $a{=}b{\neq}c$ $\alpha{=}\beta{=}\frac{\pi}{2}$ $\gamma{=}\frac{2\pi}{3}$				
Cubic $a{=}b{=}c$ $\alpha{=}\beta{=}\gamma{=}\frac{\pi}{2}$				

Figure 2.3 Classification of the 14 Bravais lattices into the seven crystal systems. The conventional parallelepiped cells are shown.

The geometrical description of the crystal structure is thus specified by the *triad of primitive translation vectors* of the Bravais lattice and by *the vectors forming the basis*.

A crystal with a single atom in the primitive unit cell is called *simple crystal* (or *simple lattice*); in this case, the basis contains a single vector, which can be taken to be zero (with appropriate choice of the origin in the crystal). In a simple lattice the (equilibrium) atomic positions $\mathbf{R}_n$ coincide with the Bravais translation vectors $\mathbf{t}_n$, and we have

$$\mathbf{R}_n = n_1\mathbf{t}_1 + n_2\mathbf{t}_2 + n_3\mathbf{t}_3. \tag{2.5a}$$

A crystal with two or more atoms (or ions) in the primitive unit cell is called a *composite crystal* (or a *composite lattice*). The equilibrium atomic positions are at the points

$$\mathbf{R}_n^{(1)} = \mathbf{d}_1 + n_1\mathbf{t}_1 + n_2\mathbf{t}_2 + n_3\mathbf{t}_3,$$
$$\mathbf{R}_n^{(2)} = \mathbf{d}_2 + n_1\mathbf{t}_1 + n_2\mathbf{t}_2 + n_3\mathbf{t}_3,$$

$$\dots \tag{2.5b}$$

$$\mathbf{R}_n^{(\nu)} = \mathbf{d}_\nu + n_1\mathbf{t}_1 + n_2\mathbf{t}_2 + n_3\mathbf{t}_3.$$

A composite lattice can be thought of as composed by a number of interpenetrating simple lattices (called *sublattices*) equal to the number of vectors of the basis. All the points of a given sublattice are related by translation vectors and must thus be occupied by atoms of the same type; different sublattices are occupied by atoms of the same or different type.

2.2 Geometrical Description of Some Crystal Structures

Crystal Structure of Rare-Gas Solids (Face-Centered Cubic Lattice)

We begin our brief description of some crystals of interest with the case of rare-gas solids (Ne, Ar, Kr, Xe). This crystal structure (illustrated in Figure 2.4) is obtained repeating periodically in space a *face-centered cube* (fcc), i.e. a conventional cubic cell (of edge a) with atoms at the corners and at the center of the faces. The primitive translation vectors of the fcc Bravais lattice are

$$\boxed{\mathbf{t}_1 = \frac{a}{2}(0, 1, 1), \quad \mathbf{t}_2 = \frac{a}{2}(1, 0, 1), \quad \mathbf{t}_3 = \frac{a}{2}(1, 1, 0)}. \tag{2.6}$$

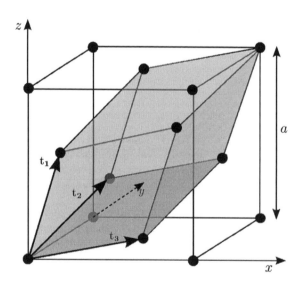

Figure 2.4 Face-centered cubic lattice and primitive translation vectors $\mathbf{t}_1, \mathbf{t}_2, \mathbf{t}_3$ given by Eqs. (2.6) in the text; the primitive cell is reported in gray.

Notice that $-\mathbf{t}_1 + \mathbf{t}_2 + \mathbf{t}_3 = a(1, 0, 0)$, $\mathbf{t}_1 - \mathbf{t}_2 + \mathbf{t}_3 = a(0, 1, 0)$, and $\mathbf{t}_1 + \mathbf{t}_2 - \mathbf{t}_3 = a(0, 0, 1)$. The volume of the primitive cell is $\Omega = \mathbf{t}_1 \cdot (\mathbf{t}_2 \times \mathbf{t}_3) = a^3/4$; thus the primitive cell, which expresses the building block of minimum volume of the crystal, is four times smaller than the volume a^3 of the conventional cubic cell.

In the fcc structure, each atom has 12 nearest neighbors (the number of nearest neighbors is called *coordination number*). The atom at the origin has 12 nearest neighbors in the positions $(a/2)(0, \pm1, \pm1)$ (and cyclic permutations) at distance $(a/2)\sqrt{2}$. It has 6 second nearest neighbors in the positions $(a/2)(\pm2, 0, 0)$ (and cyclic permutations) at distance a; 24 third nearest neighbors $(a/2)(\pm2, \pm1, \pm1)$ (and cyclic permutations) at distance $(a/2)\sqrt{6}$, etc.

The lattice constants of rare-gas solids are: $a = 4.43\,\text{Å}$ for Ne; $a = 5.26\,\text{Å}$ for Ar; $a = 5.72\,\text{Å}$ for Kr; and $a = 6.20\,\text{Å}$ for Xe. Besides rare-gas solids, there is a number of elements with monoatomic face-centered cubic structure; among them, we may mention several metals (Ag, Al, Au, Cu, Pd, Pt, and others) and some rare earth elements.

Crystal Structure of Alkali Metals (Body-Centered Cubic Lattice)

The crystal structure of alkali metals (see Figure 2.5) is obtained repeating periodically in space a body-centered cube (bcc), i.e. a conventional cubic cell with atoms at the corners and at the center of a cube of edge a. The primitive translation vectors of the bcc Bravais lattice are

$$\mathbf{t}_1 = \frac{a}{2}(-1, 1, 1), \quad \mathbf{t}_2 = \frac{a}{2}(1, -1, 1), \quad \mathbf{t}_3 = \frac{a}{2}(1, 1, -1). \tag{2.7}$$

Notice that $\mathbf{t}_1 + \mathbf{t}_2 + \mathbf{t}_3 = (a/2)(1, 1, 1)$, and $\mathbf{t}_2 + \mathbf{t}_3 = a(1, 0, 0)$, $\mathbf{t}_1 + \mathbf{t}_3 = a(0, 1, 0)$, and $\mathbf{t}_1 + \mathbf{t}_2 = a(0, 0, 1)$. The atom at the origin has 8 nearest neighbors in the positions

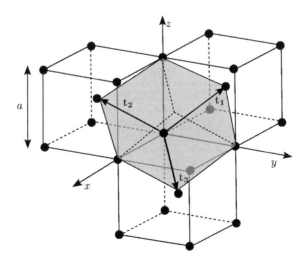

Figure 2.5 Body-centered cubic lattice and primitive translation vectors $\mathbf{t}_1, \mathbf{t}_2, \mathbf{t}_3$ given by Eqs. (2.7) in the text; the primitive cell is reported in gray.

$(a/2)(\pm 1, \pm 1, \pm 1)$ at distance $(a/2)\sqrt{3}$; there are 6 second nearest neighbors in the positions $(a/2)(\pm 2, 0, 0)$ (and cyclic permutations) at distance a, etc. The volume of the primitive unit cell is $\Omega = \mathbf{t}_1 \cdot (\mathbf{t}_2 \times \mathbf{t}_3) = a^3/2$; thus the conventional cubic cell is twice as big as the primitive unit cell.

The lattice constants of alkali metals are: $a = 3.49\,\text{Å}$ for Li; $a = 4.23\,\text{Å}$ for Na; $a = 5.23\,\text{Å}$ for K; $a = 5.59\,\text{Å}$ for Rb; and $a = 6.05\,\text{Å}$ for Cs. Other elements which crystallize in the monoatomic body-centered cubic structure include Cr ($a = 2.88\,\text{Å}$), Mo ($a = 3.15\,\text{Å}$), W ($a = 3.16\,\text{Å}$), Fe ($a = 2.87\,\text{Å}$), and Ba ($a = 5.03\,\text{Å}$).

Sodium Chloride Structure

The sodium chloride structure is shown in Figure 2.6. The crystal structure can be described as two interpenetrating fcc lattices displaced by $(a/2)(1, 1, 1)$ along the body diagonal of the conventional cube; one of the two fcc sublattices is composed by cations (Na^+) and the other by anions (Cl^-). The primitive translation vectors and the two vectors of the basis of the NaCl crystal structure are given by

$$
\begin{aligned}
&\mathbf{t}_1 = \frac{a}{2}(0, 1, 1), \quad \mathbf{t}_2 = \frac{a}{2}(1, 0, 1), \quad \mathbf{t}_3 = \frac{a}{2}(1, 1, 0) \\
&\mathbf{d}_1 = 0, \quad \mathbf{d}_2 = \frac{a}{2}(1, 1, 1)
\end{aligned}
\tag{2.8}
$$

In the NaCl structure, each anion in a given sublattice has 6 nearest neighboring cations (on the other sublattice) at distance $a/2$, 12 second nearest neighbors of the same type at distance $(a/2)\sqrt{2}$, etc. Some lattice constants of crystals with the NaCl arrangement are: $a = 5.63\,\text{Å}$ for NaCl; $a = 4.02\,\text{Å}$ for LiF; $a = 5.77\,\text{Å}$ for AgBr; $a = 6.99\,\text{Å}$ for BaTe; $a = 6.02\,\text{Å}$ for SnSe; and $a = 4.61\,\text{Å}$ for ZrN.

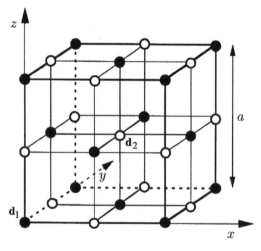

Figure 2.6 Crystal structure of sodium chloride; the end-points of the basis vectors $\mathbf{d}_1$ and $\mathbf{d}_2$, given in Eqs. (2.8) of the text, are also indicated.

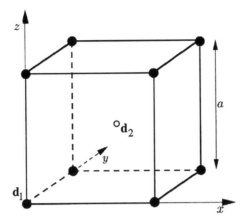

Figure 2.7 The cesium chloride structure; the end-points of the basis vectors $\mathbf{d}_1$ and $\mathbf{d}_2$, given in Eqs. (2.9) of the text, are also indicated.

Cesium Chloride Structure

The cesium chloride structure is shown in Figure 2.7. The crystal structure of CsCl can be described as two interpenetrating simple cubic lattices displaced by $(a/2)(1, 1, 1)$ along the body diagonal of the cubic cell. The underlying Bravais lattice is simple cubic, and there are two atoms Cs and Cl in the primitive unit cell; the primitive translation vectors and the two vectors of the basis are

$$\boxed{\begin{array}{l} \mathbf{t}_1 = a(1, 0, 0), \quad \mathbf{t}_2 = a(0, 1, 0), \quad \mathbf{t}_3 = a(0, 0, 1) \\ \mathbf{d}_1 = 0, \quad \mathbf{d}_2 = \dfrac{a}{2}(1, 1, 1) \end{array}} \tag{2.9}$$

In the CsCl structure, each anion in a given sublattice has 8 nearest neighboring cations (on the other sublattice) at distance $(a/2)\sqrt{3}$; there are 6 second nearest neighbors of the same type at distance a, etc. Some lattice constants of crystals with the CsCl arrangement are: $a = 4.12$ Å for CsCl; $a = 3.97$ Å for TlBr; $a = 3.33$ Å for AgCd; and $a = 2.95$ Å for CuZn.

Cubic Perovskite Structure

The cubic perovskite structure is shown in Figure 2.8 for $BaTiO_3$; this crystal remains cubic in the temperature interval from $201\,°C$ (lattice constant $a = 4.01$ Å) up to $1372\,°C$ ($a = 4.08$ Å). The crystal structure of barium titanate presents one molecule per unit cell and can be described as five interpenetrating simple cubic lattices.

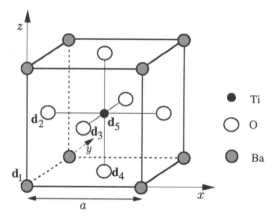

Figure 2.8 Cubic structure of barium titanate; the end-points of the basis vectors $\mathbf{d}_1$, $\mathbf{d}_2$, $\mathbf{d}_3$, $\mathbf{d}_4$, and $\mathbf{d}_5$ of Eqs. (2.10) of the text are also indicated.

The primitive translation vectors and the five vectors of the basis are

$$
\begin{aligned}
&\mathbf{t}_1 = a(1, 0, 0), \quad \mathbf{t}_2 = a(0, 1, 0), \quad \mathbf{t}_3 = a(0, 0, 1) \\
&\mathbf{d}_1 = 0, \quad \mathbf{d}_2 = \frac{a}{2}(0, 1, 1), \quad \mathbf{d}_3 = \frac{a}{2}(1, 0, 1), \quad \mathbf{d}_4 = \frac{a}{2}(1, 1, 0), \\
&\mathbf{d}_5 = \frac{a}{2}(1, 1, 1)
\end{aligned} \qquad (2.10)
$$

In Figure 2.8 barium is at the origin $\mathbf{d}_1$, oxygens are in $\mathbf{d}_2$, $\mathbf{d}_3$, $\mathbf{d}_4$, and titanium is at $\mathbf{d}_5$. Each titanium is octahedrally coordinated to six oxygens; each oxygen is coordinated with two titanium and four barium sites; each barium is surrounded by twelve oxygen sites.

Cubic Diamond Structure and Cubic Zincblende Structure

The cubic diamond structure (or simply "diamond structure") can be described as two interpenetrating fcc lattices displaced by $(a/4)(1, 1, 1)$ along the body diagonal of the conventional cube (see Figure 2.9). The underlying Bravais lattice is fcc with two carbon atoms forming the basis; the primitive translation vectors and the two vectors of the basis are

$$
\begin{aligned}
&\mathbf{t}_1 = \frac{a}{2}(0, 1, 1), \quad \mathbf{t}_2 = \frac{a}{2}(1, 0, 1), \quad \mathbf{t}_3 = \frac{a}{2}(1, 1, 0) \\
&\mathbf{d}_1 = 0, \quad \mathbf{d}_2 = \frac{a}{4}(1, 1, 1)
\end{aligned} \qquad (2.11)
$$

The coordination number is 4; each atom in a given sublattice is surrounded by four atoms of the other sublattice, at distance $(a/4)\sqrt{3}$, in a tetrahedral configuration as in Figure 2.9; there are 12 second neighbors at distance $(a/2)\sqrt{2}$. For instance, the carbon

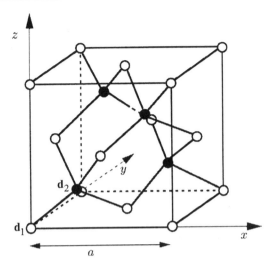

Figure 2.9 Crystal structure of diamond and zincblende; the end-points of the vectors $\mathbf{d}_1$ and $\mathbf{d}_2$ of the basis are also indicated. In the diamond structure the two sublattices are occupied by atoms of the same type; in the zincblende structure the two sublattices are occupied by different types of atoms.

atom at the origin $(\mathbf{t}_n = 0, \mathbf{d}_1 = 0)$ is surrounded by four atoms in the positions

$$\mathbf{d}_2 = \frac{a}{4}(1, 1, 1), \qquad\qquad \mathbf{d}_2 - \mathbf{t}_1 = \frac{a}{4}(1, -1, -1),$$

$$\mathbf{d}_2 - \mathbf{t}_2 = \frac{a}{4}(-1, 1, -1), \qquad \mathbf{d}_2 - \mathbf{t}_3 = \frac{a}{4}(-1, -1, 1).$$

The diamond structure is also typical of silicon, germanium, and gray tin. The lattice constants are: $a = 3.57\,\text{Å}$ for C (diamond); $a = 5.43\,\text{Å}$ for Si; $a = 5.65\,\text{Å}$ for Ge; and $a = 6.49\,\text{Å}$ for α-Sn (gray tin).

The cubic zincblende (ZnS) structure is similar to that of diamond; now the two sublattices are occupied by two different types of atoms or ions. Among crystals with this structure we mention: cubic ZnS ($a = 5.41\,\text{Å}$), AgI ($a = 6.47\,\text{Å}$), cubic BN ($a = 3.62\,\text{Å}$), and CuCl ($a = 5.41\,\text{Å}$). This structure is also typical of several III–V and II–VI groups semiconducting compounds; for instance: GaAs ($a = 5.65\,\text{Å}$), AlAs ($a = 5.66\,\text{Å}$), InAs ($a = 6.04\,\text{Å}$), GaSb ($a = 6.12\,\text{Å}$), CdTe ($a = 6.48\,\text{Å}$), HgTe ($a = 6.48\,\text{Å}$), and ZnTe ($a = 6.09\,\text{Å}$).

Two-Dimensional Graphene and Three-Dimensional Graphite

The two-dimensional honeycomb structure of graphene, a monolayer of hexagons of carbon atoms, is indicated in Figure 2.10. With the choice of axes as indicated in

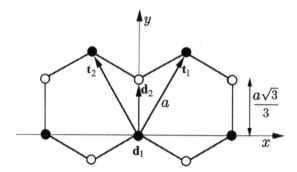

Figure 2.10 Layer structure of two-dimensional graphene. The primitive translation vectors $\mathbf{t}_1$ and $\mathbf{t}_2$, and the basis vectors $\mathbf{d}_1$ and $\mathbf{d}_2$, given by Eqs. (2.12) of the text, are also indicated.

Figure 2.10, the primitive translation vectors and the two vectors of the basis are

$$\mathbf{t}_1 = a\left(\frac{1}{2}, \frac{\sqrt{3}}{2}, 0\right), \quad \mathbf{t}_2 = a\left(-\frac{1}{2}, \frac{\sqrt{3}}{2}, 0\right)$$
$$\mathbf{d}_1 = 0, \quad \mathbf{d}_2 = a\left(0, \frac{\sqrt{3}}{3}, 0\right)$$

(2.12)

with $a = 2.46\,\text{Å}$. In the choice of Eqs. (2.12), the vectors $\mathbf{t}_1$ and $\mathbf{t}_2$ are positioned symmetrically around the y-axis, and form an angle of $2\pi/6$. Another frequent (and equivalent) choice is to take the vector $\tilde{\mathbf{t}}_1 = a(1, 0, 0)$ along the x-axis, and the vector $\tilde{\mathbf{t}}_2 = a(1/2, \sqrt{3}/2, 0)$ at angle of $2\pi/6$ with the x-axis. In the following, we continue with the choice made in Eqs. (2.12).

It can be easily noticed, by inspection, that it would be impossible to define a lattice translation vector joining an atom with any of its first neighbors; in fact, the opposite of such a vector points at the center of the hexagons, where no atom exists. The two-dimensional graphene structure is composite and consists of two sublattices of carbon atoms, corresponding to the basis vectors $\mathbf{d}_1 = 0$ and $\mathbf{d}_2 = a(0, \sqrt{3}/3, 0)$. If we replace one sublattice with boron atoms and the other with nitrogen atoms, we obtain the two-dimensional structure of hexagonal boron nitride.

The three-dimensional structure of graphite is indicated in Figure 2.11. The value of the lattice parameter in the z-direction is $c = 6.71\,\text{Å}$. The fundamental unit cell of graphite contains four carbon atoms. The primitive translation vectors and the basis for the three-dimensional graphite are given by

$$\mathbf{t}_1 = a\left(\frac{1}{2}, \frac{\sqrt{3}}{2}, 0\right), \quad \mathbf{t}_2 = a\left(-\frac{1}{2}, \frac{\sqrt{3}}{2}, 0\right), \quad \mathbf{t}_3 = c(0, 0, 1)$$
$$\mathbf{d}_1 = 0, \quad \mathbf{d}_2 = a\left(0, \frac{\sqrt{3}}{3}, 0\right), \quad \mathbf{d}_3 = c\left(0, 0, \frac{1}{2}\right), \quad \mathbf{d}_4 = \left(0, a\frac{2\sqrt{3}}{3}, \frac{c}{2}\right)$$

(2.13)

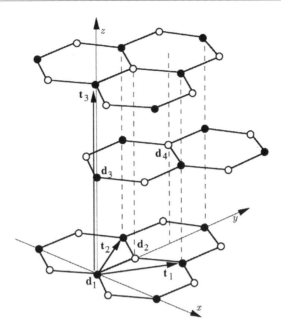

Figure 2.11 Three-dimensional crystal structure of graphite. The primitive translation vectors t_1, t_2, t_3 and the end-points of the basis vectors d_1, d_2, d_3, and d_4, given in Eqs. (2.13) of the text, are also indicated.

In the case of graphite two consecutive layers, stacked along the z-axis, are rotated by $2\pi/6$. As a consequence the atoms d_1 and d_3 have two neighbors on different adjacent planes at distance $c/2$.

Besides diamond, graphene, and graphite, other interesting crystalline forms of elemental carbon include fullerene and carbon nanotubes. The structure of carbon nanotubes [S. Ijima, Nature *354*, 56 (1991)] can be pictured as sheets of graphene appropriately wound to cylinders. The fullerene carbon allotrope is constituted by molecules C_{60} with the structure of a soccer ball, i.e. a truncated icosahedron, originally proposed by H. W. Kroto, J. R. Heath, S. C. O'Brien, R. F. Curl, and R. E. Smalley, Nature *318*, 162 (1985). The structure of C_{60} molecule is shown in Figure 2.12; the centers of the molecules are arranged in an fcc lattice, and the primitive cell of the solid contains a soccer ball molecule of 60 carbon atoms. This form of carbon has been given the name of *fullerene*, as a tribute to the geodesic studies of the architect R. Buckminster Fuller.

Hexagonal Close-Packed Structure and Ideal Hexagonal Closed-Packed Structure

Several elements crystallize in the hexagonal close-packed structure (hcp). To describe this structure consider a two-dimensional array of equilateral triangles (or regular hexagons *including* their centers) of edge a, as indicated in Figure 2.13. Then one

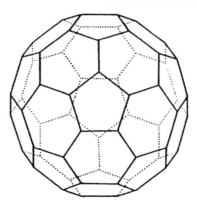

Figure 2.12 The structure of C_{60} molecule in its regular truncated icosahedron geometry. The polygon has 60 vertices and 32 faces, 12 of which are pentagonal and 20 hexagonal. The bond lengths forming pentagons are 1.47 Å; the bond lengths common to two hexagons are 1.41 Å [from P. Milani, Rivista del Nuovo Cimento *19*, N.11 (1996); with kind permission of Società Italiana di Fisica].

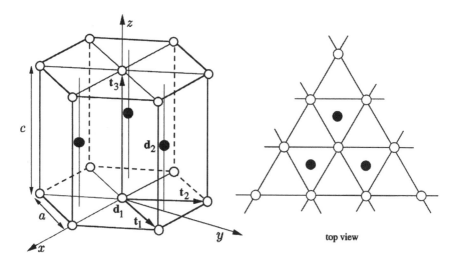

Figure 2.13 Hexagonal closed-packed structure. The primitive translation vectors t_1, t_2, t_3 and the end-points of the basis vectors d_1 and d_2, given in Eqs. (2.14) of the text, are also indicated. The top view of the structure is also shown for convenience.

stacks along the z-axis a second plane of equilateral triangles, above the centers of the other triangles, as indicated in Figure 2.13. The sequence of planes ABAB AB ... is then repeated: the third plane lies directly over the first one, the fourth over the second, and so on.

The hexagonal close-packed structure can be pictured as formed by two interpenetrating simple hexagonal Bravais lattices; the primitive translation vectors and the two basis vectors of the hcp structure are

$$
\mathbf{t}_1 = a\left(\frac{1}{2}, \frac{\sqrt{3}}{2}, 0\right), \quad \mathbf{t}_2 = a\left(-\frac{1}{2}, \frac{\sqrt{3}}{2}, 0\right), \quad \mathbf{t}_3 = c(0, 0, 1)
$$
$$
\mathbf{d}_1 = 0, \quad \mathbf{d}_2 = \left(0, a\frac{\sqrt{3}}{3}, \frac{c}{2}\right)
$$
; (2.14)

the two sublattices specified by $\mathbf{d}_1$ and $\mathbf{d}_2$ are occupied by atoms of the same type. There is no necessary link between the lattice parameters a and c, and the ratio c/a is in general different in different materials. Several metals have the hcp structure, among them Be ($a = 2.29$ Å; $c = 3.58$ Å; $c/a = 1.56$), Cd ($a = 2.98$ Å; $c = 5.62$ Å; $c/a = 1.88$), Mg ($a = 3.21$ Å; $c = 5.21$ Å; $c/a = 1.62$), Ti ($a = 2.95$ Å; $c = 4.69$ Å; $c/a = 1.59$), and Zn ($a = 2.66$ Å; $c = 4.95$ Å; $c/a = 1.86$).

The ideal hexagonal closed-packed structure occurs when the four points ($\mathbf{t}_1, \mathbf{t}_2$, $\mathbf{d}_1, \mathbf{d}_2$), specified in Eqs. (2.14) and illustrated in Figure 2.13, constitute a regular tetrahedron; the requirement that the lengths of the four edges are all equal, give the condition

$$
|\mathbf{d}_2| = |\mathbf{t}_1| \implies \frac{a^2}{3} + \frac{c^2}{4} = a^2.
$$

The *ideal ratio* c/a becomes

$$
\frac{c}{a} = \sqrt{\frac{8}{3}} = 1.633;
$$

this ratio is only approximately verified in actual hcp crystals. The ideal closed-packed hexagonal structure occurs in the case atoms behave as hard sphere touching each other.

Finally it is worth to mention that another ideal close-packing arrangement is provided by the fcc structure; the sequence of hexagonal planes is in this case ABC, ABC, etc. where C is rotated by $2\pi/6$ with respect to the B plane. These closed-packed hexagonal planes are arranged orthogonally to the body diagonal of the conventional fcc cube. To see this, we remember that in the fcc structure the atom at the origin has 12 neighbors in the positions $(a/2)(0, \pm1, \pm1)$ and cyclic permutations. The three points $P_1 = (a/2)(0, -1, -1)$, $P_2 = (a/2)(-1, 0, -1)$, and $P_3 = (a/2)(-1, -1, 0)$, describe a regular triangle (of edge $a\sqrt{2}/2$) orthogonal to the body diagonal unit vector $\mathbf{n} = (1/\sqrt{3})(1, 1, 1)$ and belong to the plane A, distant $a/\sqrt{3}$ from the origin. The six points given by $(a/2)(0, 1, -1)$ and $(a/2)(0, -1, 1)$ and cyclic permutations, describe a regular hexagon (orthogonal to $\mathbf{n}$) around the atom at the origin, and belong to plane B; finally, the three points given by $(a/2)(0, 1, 1)$ and cyclic permutations, belong to plane C. The ratio between $2a/\sqrt{3}$ and $a\sqrt{2}/2$ equals the ideal value $\sqrt{8/3}$. From this point of view, the fcc Bravais lattice can also be regarded as the cubic realization of the ideal-closed-packed arrangement of atomic spheres.

Hexagonal Wurtzite Structure

The hexagonal wurtzite structure (or simply "wurtzite structure") can be considered as formed by two interpenetrating hexagonal closed-packed lattices; in the unit cell there are four atoms of two different types, forming two molecules (see Figure 2.14). The primitive translation vectors and the basis vectors are

$$\mathbf{t}_1 = a\left(\frac{1}{2}, \frac{\sqrt{3}}{2}, 0\right), \quad \mathbf{t}_2 = a\left(-\frac{1}{2}, \frac{\sqrt{3}}{2}, 0\right), \quad \mathbf{t}_3 = c(0, 0, 1)$$

$$\mathbf{d}_1 = 0, \quad \mathbf{d}_2 = (0, 0, uc), \quad \mathbf{d}_3 = \left(0, a\frac{\sqrt{3}}{3}, \frac{c}{2}\right), \quad \mathbf{d}_4 = \left(0, a\frac{\sqrt{3}}{3}, \frac{c}{2} + uc\right)$$

$$(2.15)$$

where a and c are the lattice constants, and u is dimensionless; $\mathbf{d}_1$ and $\mathbf{d}_3$ are occupied by the same type of atom, $\mathbf{d}_2$ and $\mathbf{d}_4$ are occupied by the other type of atom. Each atom is surrounded by an (almost regular) tetrahedron of atoms of opposite sort; the tetrahedral blocks are regular if the axis ratio has the ideal value $c/a = 1.633$ and the parameter $u = 3/8 = 0.375$. The atom in $\mathbf{d}_3$, for instance, is surrounded by four

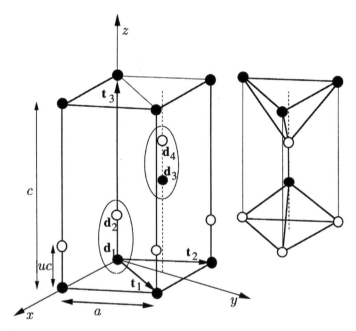

Figure 2.14 Hexagonal wurtzite structure. The primitive translation vectors $\mathbf{t}_1, \mathbf{t}_2, \mathbf{t}_3$ and the end-points of the basis vectors $\mathbf{d}_1, \mathbf{d}_2, \mathbf{d}_3$, and $\mathbf{d}_4$, given by Eqs. (2.15) of the text, are indicated; the two molecules of the unit cell are encircled for convenience. The orientations of adjacent tetrahedra surrounding $\mathbf{d}_3$ and $\mathbf{d}_4$ are also shown.

nearest neighbors of opposite type in the positions $(\mathbf{d}_2, \mathbf{d}_2 + \mathbf{t}_1, \mathbf{d}_2 + \mathbf{t}_2, \mathbf{d}_4)$. These four positions describe a regular tetrahedron if $c^2/a^2 = 8/3$. The atom in $\mathbf{d}_3$ is at the center of the regular tetrahedron if

$$|\mathbf{d}_3 - \mathbf{d}_2| \equiv |\mathbf{d}_3 - \mathbf{d}_4| \implies \frac{a^2}{3} + \left(\frac{1}{2} - u\right)^2 = u^2 c^2 \implies u = \frac{1}{4} + \frac{a^2}{3c^2} = \frac{3}{8}.$$

The adjacent tetrahedral blocks are stacked along the z-axis with the relative orientation shown in Figure 2.14 as an example. [Notice that in the cubic zincblende structure, the adjacent tetrahedral blocks are stacked along the body diagonal of the conventional cubic cell, and are rotated by $2\pi/6$ with respect to each other.] The lattice parameters for ZnO (or zincite) are: $a = 3.25\,\text{Å}$, $c = 5.21\,\text{Å}$, $u = 0.345$. For the hexagonal form of CdS we have: $a = 4.14\,\text{Å}$, $c = 6.75\,\text{Å}$, $c/a = 1.63$; and for ZnS, or wurtzite, we have: $a = 3.81\,\text{Å}$, $c = 6.23\,\text{Å}$, $c/a = 1.64$. Replacing both types of atoms of the wurtzite structure with carbon atoms gives the hexagonal diamond, which is a (slight) modification of the cubic diamond structure found in nature in some meteorites.

We close here our brief description of crystal structures, and refer to the literature for further information [see for instance R. W. G. Wyckoff "Crystal Structures" vols. 1–5 (Wiley, New York, 1963–1968)]. Beyond specific descriptions, it should be clear the main message of this section: a crystal structure is specified by the knowledge of the *primitive translation vectors* and the *basis vectors* locating the atomic positions in the unit cell; from this knowledge, any geometrical property of interest (neighbors, distances and bond lengths, angles, planes, directions, also symmetry operations) can be easily worked out.

2.3   Wigner-Seitz Primitive Cells

In describing crystal structures, we have found convenient to take the fundamental primitive cell as a parallelepiped defined by the vectors $\mathbf{t}_1, \mathbf{t}_2, \mathbf{t}_3$. There are infinite possible choices for the fundamental cell (one can vary the origin, or the primitive vectors keeping the volume constant, or both). An important and very useful choice for the primitive cell has been suggested by Wigner and Seitz.

Consider first a *simple lattice*; the translation vectors individuate all the atoms of the crystal. *The Wigner-Seitz cell about a (reference) lattice point is defined by the property that any point of the cell is closer to that lattice point than to any other.* The Wigner-Seitz cell can be operatively obtained by bisecting with perpendicular planes the vectors joining one atom with the nearest neighbors, second nearest neighbors, and so on, and considering the smallest volume enclosed.

If we have *composite lattices*, we focus on the underlying Bravais lattice; the Wigner-Seitz cell is defined, as before, by the property that any point of the cell is closer to that lattice point than any other. However, in this case, we can obtain *subcells* by bisecting with perpendicular planes the vectors joining one atom with nearest neighbors, second nearest neighbors, etc. and considering the smallest volume enclosed. For composite lattices, the Wigner-Seitz cell is constituted by a number of subcells equal to the number of atoms of the basis. Each subcell has an atom at its center. The Wigner-Seitz cells are particularly convenient in those methods of electronic state calculations that exploit appropriately the "local spherical symmetry" of the periodic crystal potential.

2.4   Reciprocal Lattices

2.4.1   Definitions and Basic Properties

For the study of crystals, besides *the direct lattice in the ordinary space*, it is important to consider also the *reciprocal lattice in the dual (or reciprocal) space*.

Given a crystal with primitive translation vectors $\mathbf{t}_1, \mathbf{t}_2, \mathbf{t}_3$ in the direct space, we consider the three primitive vectors $\mathbf{g}_1, \mathbf{g}_2, \mathbf{g}_3$ in the reciprocal space, defined by the relations

$$\boxed{\mathbf{t}_i \cdot \mathbf{g}_j = 2\pi \delta_{ij}} \tag{2.16a}$$

(the numerical factor 2π is introduced as a matter of convenience to simplify some expressions later on). If $\mathbf{t}_1, \mathbf{t}_2, \mathbf{t}_3$ are non-coplanar vectors and form a right-handed system, also $\mathbf{g}_1, \mathbf{g}_2, \mathbf{g}_3$ are non-coplanar vectors and form a right-handed system. If a *crystal rotation of* $\mathbf{t}_1, \mathbf{t}_2, \mathbf{t}_3$ *is performed in the direct space, the same rotation of* $\mathbf{g}_1, \mathbf{g}_2, \mathbf{g}_3$ *occurs in the reciprocal space.* Notice that the propagation wavevector $\mathbf{k}$ of a general plane wave $\exp(i\mathbf{k} \cdot \mathbf{r})$ has "reciprocal length" dimension, and can be conveniently represented in the reciprocal space.

All the points defined by the vectors of the type

$$\mathbf{g}_m = m_1\mathbf{g}_1 + m_2\mathbf{g}_2 + m_3\mathbf{g}_3 \tag{2.16b}$$

(with m_1, m_2, m_3 integer numbers, negative, zero, or positive) constitute the *reciprocal lattice*. Notice that the reciprocal lattice is related only to the translational properties of the crystal and not to the basis. Crystals with the same translational symmetry, but completely different basis, have the same reciprocal lattice. In Chapter 10, in the study of elastic diffraction, we will see that the geometry of the diffracted beams determines *reciprocal lattice vectors*, while the *intensity* of the diffracted beams provides information on *the basis*.

We can solve Eq. (2.16a) explicitly. For instance, $\mathbf{g}_1$ must be orthogonal both to $\mathbf{t}_2$ and $\mathbf{t}_3$ and is thus parallel to $\mathbf{t}_2 \times \mathbf{t}_3$. The condition $\mathbf{g}_1 \cdot \mathbf{t}_1 = 2\pi$ fully determines $\mathbf{g}_1$. The other vectors are similarly obtained with cyclic permutations. We have

$$\mathbf{g}_1 = \frac{2\pi}{\Omega}\mathbf{t}_2 \times \mathbf{t}_3, \quad \mathbf{g}_2 = \frac{2\pi}{\Omega}\mathbf{t}_3 \times \mathbf{t}_1, \quad \mathbf{g}_3 = \frac{2\pi}{\Omega}\mathbf{t}_1 \times \mathbf{t}_2, \quad \text{and} \quad \Omega = \mathbf{t}_1 \cdot (\mathbf{t}_2 \times \mathbf{t}_3),$$
$$\tag{2.17}$$

where the volume of the primitive cell in the direct lattice is denoted by Ω. From Eq. (2.16a) we see that the reciprocal of the reciprocal lattice is the original direct lattice.

The direct and reciprocal lattices obey some simple useful properties. First of all, we begin to observe that *the volume Ω_k of the unit cell in the reciprocal space is $(2\pi)^3$*

times the reciprocal of the volume of the unit cell in the direct lattice. In fact

$$\Omega_k = \mathbf{g}_1 \cdot (\mathbf{g}_2 \times \mathbf{g}_3) = \frac{(2\pi)^3}{\Omega^3} (\mathbf{t}_2 \times \mathbf{t}_3) \cdot [(\mathbf{t}_3 \times \mathbf{t}_1) \times (\mathbf{t}_1 \times \mathbf{t}_2)] = \frac{(2\pi)^3}{\Omega}.$$

To perform the calculation of the quantity within square brackets, we have used the following relation for the vector product among any three vectors:

$$\mathbf{v}_1 \times (\mathbf{v}_2 \times \mathbf{v}_3) \equiv \mathbf{v}_2(\mathbf{v}_1 \cdot \mathbf{v}_3) - \mathbf{v}_3(\mathbf{v}_1 \cdot \mathbf{v}_2).$$

From Eqs. (2.16), it is evident that the scalar product of any reciprocal lattice vector with any translation vector is an integer number of 2π. We can thus write

$$\mathbf{g}_m \cdot \mathbf{t}_n = \text{integer} \cdot 2\pi \tag{2.18a}$$

for any translation vector $\mathbf{t}_n$. Furthermore, if a vector $\mathbf{q}$ satisfies the relation

$$\mathbf{q} \cdot \mathbf{t}_n = \text{integer} \cdot 2\pi \tag{2.18b}$$

for any $\mathbf{t}_n$, then $\mathbf{q}$ must be a reciprocal lattice vector. To show this, we notice that it is always possible to express $\mathbf{q}$ in terms of the non-collinear vectors $\mathbf{g}_1, \mathbf{g}_2, \mathbf{g}_3$; we thus can write $\mathbf{q} = c_1\mathbf{g}_1 + c_2\mathbf{g}_2 + c_3\mathbf{g}_3$ with c_i real numbers. To satisfy Eq. (2.18b) for any $\mathbf{t}_n$ and in particular for $\mathbf{t}_1, \mathbf{t}_2, \mathbf{t}_3$ it is necessary that c_1, c_2, c_3 are integer numbers, and this means that $\mathbf{q}$ is a reciprocal lattice vector. Equations (2.18) say that all and only the reciprocal lattice vectors have scalar product with any lattice translation vector equal to an integer number of 2π.

An immediate consequence of Eqs. (2.18) is that a *plane wave* $\exp(i\mathbf{k} \cdot \mathbf{r})$ *has the lattice periodicity if and only if the wavevector* $\mathbf{k}$ *equals a reciprocal lattice vector.* In fact the function

$$W(\mathbf{r}) = e^{i\mathbf{g}_m \cdot \mathbf{r}} \tag{2.19a}$$

remains unchanged if we replace $\mathbf{r} \rightarrow \mathbf{r} + \mathbf{t}_n$ (the opposite follows from Eq. (2.18b)). Thus a *function* $f(\mathbf{r})$ *periodic in the direct lattice*, can always be expanded in the form

$$f(\mathbf{r}) = \sum_{\mathbf{g}_m} f_m e^{i\mathbf{g}_m \cdot \mathbf{r}}, \tag{2.19b}$$

where the sum is over the reciprocal lattice vectors. Similarly, a function $F(\mathbf{k})$ *periodic in the reciprocal lattice*, can always be expanded in the form

$$F(\mathbf{k}) = \sum_{\mathbf{t}_m} F_m e^{i\mathbf{k} \cdot \mathbf{t}_m}, \tag{2.19c}$$

where the sum is over the translation lattice vectors.

2.4.2   Planes and Directions in Bravais Lattices

Consider a Bravais lattice, its dual space, and a vector $\mathbf{g}_m = m_1\mathbf{g}_1 + m_2\mathbf{g}_2 + m_3\mathbf{g}_3$ of the reciprocal lattice. From Eqs. (2.16) it is evident that the scalar products $\mathbf{g}_m \cdot \mathbf{t}_n$ of a given *fixed vector* $\mathbf{g}_m$ with *all translation vectors* $\mathbf{t}_n$ constitute a ladder of the type $0, \pm 2\pi \nu, \pm 4\pi \nu, \pm 6\pi \nu, \dots$ (where ν is an integer number, not necessarily equal to 1). If $\nu \neq 1$ then the vector $\mathbf{g}_m/\nu$ is also a reciprocal lattice vector (in fact the scalar products of $\mathbf{g}_m/\nu$ with any translation vector are integer multiples of 2π); this means that the integer values m_1, m_2, m_3 of $\mathbf{g}_m$ have the integer ν as common divisor.

In this section, we confine our attention to reciprocal lattice vectors $\mathbf{g}_m$ with the *restriction that the tern of integers* m_1, m_2, m_3 *has no common divisor* (otherwise we divide by it); this means that we are considering the reciprocal vector of minimum length among all its possible multiples (in the direction of $\mathbf{g}_m$). The scalar product of $\mathbf{g}_m$ with any translation vector $\mathbf{t}_n$ gives any *integer multiples of* 2π; namely, we have

$$\mathbf{g}_m \cdot \mathbf{t}_n = 0, \pm 2\pi, \pm 4\pi, \pm 6\pi, \dots, \tag{2.20a}$$

where $\mathbf{t}_n$ is any translation vector, and m_1, m_2, m_3 are integers without common integer divisor.

We can interpret the result of Eq. (2.20a) in a geometrical form. Consider the family of planes in the direct space defined by the equations

$$\mathbf{g}_m \cdot \mathbf{r} = 0, \pm 2\pi, \pm 4\pi, \pm 6\pi, \dots \tag{2.20b}$$

The family of planes is represented in Figure 2.15. Because of Eq. (2.20a) we see that *all translation vectors* belong to the family of planes described by Eq. (2.20b); it is also apparent that the distance between two consecutive planes is

$$d = \frac{2\pi}{|\mathbf{g}_m|}. \tag{2.20c}$$

In conclusion: *Every reciprocal lattice vector* $\mathbf{g}_m$ *is normal to a family of parallel and equidistant planes containing all the direct lattice points*; the distance between two

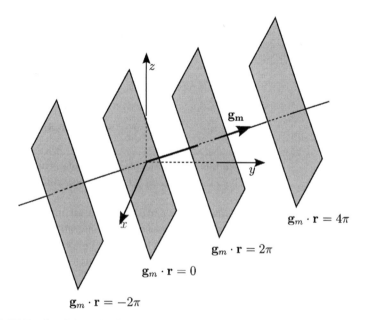

Figure 2.15 Family of planes in direct space defined by the equations $\mathbf{g}_m \cdot \mathbf{r} = 0, \pm 2\pi, \pm 4\pi, \dots$; all the translation vectors of the Bravais lattice belong to the family of planes.

successive planes is $d = 2\pi/g_m$. This property is the basis for the interpretation of the Bragg diffraction (presented in Section 10.2).

From the above properties it turns out natural to label a family of parallel lattice planes, in which we can resolve a Bravais lattice, by means of the tern of integers m_1, m_2, m_3 (with no common integer divisor) of the $\mathbf{g}_m$ vector normal to them. The three indices, conventionally enclosed by round brackets (m_1, m_2, m_3), are used to denote the orientation of the family of planes. The family of all the planes in direct space, equivalent to (m_1, m_2, m_3) by point symmetry, is indicated by curly brackets $\{m_1, m_2, m_3\}$.

It is interesting to analyze the equivalence of the present indexing scheme, based on the properties of the dual spaces, with the more traditional one, introduced by Miller in crystallography.

According to the prescription introduced by Miller, a lattice plane (i.e. a plane through three non-collinear lattice points) can be described starting from the intercepts with the primitive axes (expressed in units of $\mathbf{t}_1, \mathbf{t}_2, \mathbf{t}_3$). The intercepts themselves are not used (to avoid the occurrence of ∞ if the plane is parallel to a primitive translation vector). The reciprocals of the intercepts, multiplied by the smallest factor to convert them into integer numbers, are called *Miller indices*.

As a working illustration of the Miller indexing, consider for example in Figure 2.16 the lattice plane that intersects the axes of the primitive translation vectors at the points

$$\mathbf{R}_1 = 2\mathbf{t}_1, \quad \mathbf{R}_2 = 3\mathbf{t}_2, \quad \mathbf{R}_3 = \mathbf{t}_3. \tag{2.21}$$

The intercepts are 2, 3, 1; the corresponding reciprocals are 1/2, 1/3, 1 and the Miller indices $(3, 2, 6)$. We can immediately observe that the reciprocal lattice vector

$$\mathbf{g}_m = 3\mathbf{g}_1 + 2\mathbf{g}_2 + 6\mathbf{g}_3$$

is perpendicular to the plane passing through $\mathbf{R}_1, \mathbf{R}_2, \mathbf{R}_3$. In fact we have (by construction of the Miller indices)

$$\mathbf{g}_m \cdot \mathbf{R}_1 \equiv \mathbf{g}_m \cdot \mathbf{R}_2 \equiv \mathbf{g}_m \cdot \mathbf{R}_3.$$

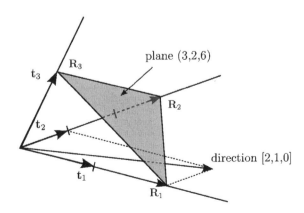

Figure 2.16 Miller indices of planes and directions in a Bravais lattice.

Thus $\mathbf{g}_m$ is perpendicular to $\mathbf{R}_1 - \mathbf{R}_2$ and to $\mathbf{R}_2 - \mathbf{R}_3$ and thus to their plane. It is evident, in general, the full equivalence of the Miller indexing scheme and the reciprocal lattice indexing of families of planes.

Before concluding, we give also the generally adopted convention to label a crystallographic direction joining two lattice points of a Bravais lattice. The relative vector, connecting the two lattice points, can always be expressed in the form

$$\mathbf{t}_n = n_1 \mathbf{t}_1 + n_2 \mathbf{t}_2 + n_3 \mathbf{t}_3,$$

where n_1, n_2, n_3 are integer numbers. Dividing the numbers n_1, n_2, n_3 by their highest common factor, we can reduce them to a non-divisible tern l_1, l_2, l_3. The direction is specified by enclosing this tern in square brackets $[l_1, l_2, l_3]$. All the directions equivalent by symmetry to $[l_1, l_2, l_3]$ are indicated by angular brackets $\langle l_1, l_2, l_3 \rangle$.

In the particular case of the primitive cubic Bravais lattice, where the axes have the same length and are mutually orthogonal, we see that the $[m_1, m_2, m_3]$ direction is always orthogonal to the (m_1, m_2, m_3) family of planes. This is not generally true in any other crystal system. For fcc and bcc lattices, it is customary to indicate directions and planes with respect to the underlying non-primitive simple cubic lattice.

2.5 Brillouin Zones

The *first Brillouin zone* (or simply the *Brillouin zone*) of the reciprocal lattice has the same definition as the Wigner-Seitz cell in the direct lattice: it has *the property that any point of the cell is closer to the chosen lattice point* (say $\mathbf{g} \equiv 0$) *than to any other*. The first Brillouin zone can be obtained by bisecting with perpendicular planes nearest neighbors reciprocal lattice vectors, second nearest neighbors (and other orders of neighbors if necessary), and considering the smallest volume enclosed. Similarly, the second Brillouin zone is obtained continuing the bisecting operations and delimiting the second volume enclosed (with exclusion of the first zone), etc. The shape of the Brillouin zone is connected to the geometry of the direct Bravais lattice, irrespectively of the content of the basis. We illustrate the construction of the first Brillouin zone with some examples.

Brillouin Zone for the Simple Cubic Lattice

The fundamental vectors in direct space of a simple cubic lattice are

$$\mathbf{t}_1 = a(1, 0, 0), \quad \mathbf{t}_2 = a(0, 1, 0), \quad \mathbf{t}_3 = a(0, 0, 1). \tag{2.22a}$$

From Eqs. (2.17) we have

$$\mathbf{g}_1 = (2\pi/a)(1, 0, 0), \quad \mathbf{g}_2 = (2\pi/a)(0, 1, 0), \quad \mathbf{g}_3 = (2\pi/a)(0, 0, 1). \tag{2.22b}$$

Thus the reciprocal lattice is still a simple cube, with edge $2\pi/a$.

The first Brillouin zone is indicated in Figure 2.17. Points of high symmetry in the cubic Brillouin zone are indicated by conventional letters: Γ denotes the origin of the Brillouin zone; X is the center of a square face at the boundaries; M is the center of a cube edge; and R is the vertex of the cube.

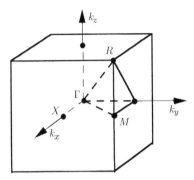

Figure 2.17 Brillouin zone for the simple cubic lattice. Some high symmetry points are indicated: $\Gamma = 0$; $X = (2\pi/a)(1/2, 0, 0)$; $M = (2\pi/a)(1/2, 1/2, 0)$; $R = (2\pi/a)(1/2, 1/2, 1/2)$.

Brillouin Zone for the Face-Centered Cubic Lattice

The fundamental vectors for an fcc Bravais lattice are

$$\mathbf{t}_1 = \frac{a}{2}(0, 1, 1), \quad \mathbf{t}_2 = \frac{a}{2}(1, 0, 1), \quad \mathbf{t}_3 = \frac{a}{2}(1, 1, 0). \tag{2.23a}$$

By applying Eqs. (2.17), we have for the fundamental vectors of the reciprocal lattice

$$\mathbf{g}_1 = \frac{2\pi}{a}(-1, 1, 1), \quad \mathbf{g}_2 = \frac{2\pi}{a}(1, -1, 1), \quad \mathbf{g}_3 = \frac{2\pi}{a}(1, 1, -1). \tag{2.23b}$$

Thus the reciprocal lattice of an fcc lattice is a bcc lattice. The Brillouin zone is the truncated octahedron shown in Figure 2.18.

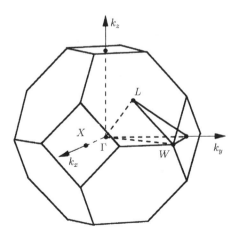

Figure 2.18 Brillouin zone for the face-centered cubic lattice (truncated octahedron). Some high symmetry points are: $\Gamma = 0$; $X = (2\pi/a)(1, 0, 0)$; $L = (2\pi/a)(1/2, 1/2, 1/2)$; $W = (2\pi/a)(1/2, 1, 0)$.

Brillouin Zone for the Body-Centered Cubic Lattice

The fundamental vectors for a bcc lattice in direct and reciprocal space are

$$\mathbf{t}_1 = \frac{a}{2}(-1, 1, 1), \quad \mathbf{t}_2 = \frac{a}{2}(1, -1, 1), \quad \mathbf{t}_3 = \frac{a}{2}(1, 1, -1) \tag{2.24a}$$

and

$$\mathbf{g}_1 = \frac{2\pi}{a}(0, 1, 1), \quad \mathbf{g}_2 = \frac{2\pi}{a}(1, 0, 1), \quad \mathbf{g}_3 = \frac{2\pi}{a}(1, 1, 0). \tag{2.24b}$$

The Brillouin zone is the regular rhombic dodecahedron shown in Figure 2.19.

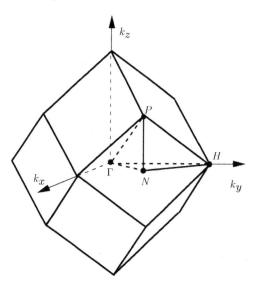

Figure 2.19 Brillouin zone for the body-centered cubic lattice (rhombic dodecahedron). Some high symmetry points are also indicated: $\Gamma = 0$; $N = (2\pi/a)(1/2, 1/2, 0)$; $P = (2\pi/a)(1/2, 1/2, 1/2)$; $H = (2\pi/a)(0, 1, 0)$.

Brillouin Zone for the Hexagonal Lattice

The fundamental translation vectors of the Bravais hexagonal lattice are

$$\mathbf{t}_1 = a\left(\frac{1}{2}, \frac{\sqrt{3}}{2}, 0\right), \quad \mathbf{t}_2 = a\left(-\frac{1}{2}, \frac{\sqrt{3}}{2}, 0\right), \quad \mathbf{t}_3 = c(0, 0, 1), \tag{2.25a}$$

and the fundamental vectors of the reciprocal lattice are

$$\mathbf{g}_1 = \frac{2\pi}{a}\left(1, \frac{\sqrt{3}}{3}, 0\right), \quad \mathbf{g}_2 = \frac{2\pi}{a}\left(-1, \frac{\sqrt{3}}{3}, 0\right), \quad \mathbf{g}_3 = \frac{2\pi}{c}(0, 0, 1). \tag{2.25b}$$

Thus the reciprocal of the hexagonal lattice is still an hexagonal lattice. The Brillouin zone is shown in Figure 2.20.

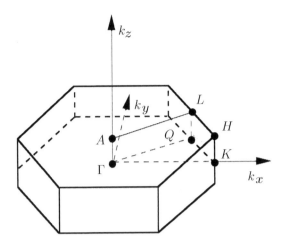

Figure 2.20 Brillouin zone for the hexagonal Bravais lattice. Some high symmetry points are also indicated: $\Gamma = 0$; $K = (2\pi/a)(2/3, 0, 0)$; $Q = (\pi/a)(1, 1/\sqrt{3}, 0)$; $A = (\pi/c)(0, 0, 1)$.

2.6 Translational Symmetry and Quantum Mechanical Aspects

2.6.1 Translational Symmetry and Bloch Wavefunctions

In Section 1.1 we have analyzed the important consequences of the periodicity on the quantum mechanical propagation of electrons in one-dimensional crystals. In this section, we consider the effect of periodicity in actual three-dimensional crystals; for this purpose we have already prepared the needed tools, in particular the notion of Brillouin zone and reciprocal lattice. Similarly to the one-dimensional case, also for three-dimensional crystals we can demonstrate the Bloch theorem: *the wavefunctions of the crystal Hamiltonian can be written as the product of a plane wave of wavevector* **k** *within the first Brillouin zone, times an appropriate periodic function.* The Bloch theorem is a general consequence of translation symmetry of the crystals; additional symmetry properties of the crystals may add further information on the wavefunctions, but cannot remove their Bloch form.

Consider the Schrödinger equation

$$\left[\frac{\mathbf{p}^2}{2m} + V(\mathbf{r})\right]\psi(\mathbf{r}) = E\psi(\mathbf{r}) \quad \text{with} \quad V(\mathbf{r}) = V(\mathbf{r} + \mathbf{t}_n) \tag{2.26}$$

for an electron in a periodic potential. Since $V(\mathbf{r})$ has the lattice periodicity, we can expand it in plane waves characterized by reciprocal lattice vectors in the form

$$V(\mathbf{r}) = \sum_{\mathbf{g}_m} V(\mathbf{g}_m)e^{i\mathbf{g}_m \cdot \mathbf{r}}. \tag{2.27a}$$

The matrix elements of a periodic potential between (normalized) plane waves of vectors **k** and **k**′ are different from zero only if the two wavevectors differ by a reciprocal lattice vector. From expansion (2.27a) we have in fact

$$\langle \frac{1}{\sqrt{V}}e^{i\mathbf{k}'\cdot\mathbf{r}}|V(\mathbf{r})|\frac{1}{\sqrt{V}}e^{i\mathbf{k}\cdot\mathbf{r}}\rangle = \sum_{\mathbf{g}_m} V(\mathbf{g}_m)\int \frac{1}{V}e^{i(-\mathbf{k}'+\mathbf{k}+\mathbf{g}_m)\cdot\mathbf{r}}\,d\mathbf{r}$$

$$= \begin{cases} 0 & \text{if } \mathbf{k}' \neq \mathbf{k}+\mathbf{g}_m \\ V(\mathbf{g}_m) & \text{if } \mathbf{k}' = \mathbf{k}+\mathbf{g}_m \end{cases} \qquad (2.27b)$$

($d\mathbf{r}$ denotes the volume element in real space).

Let us denote by $\mathbf{k}$ and $\mathbf{k}'$ two different wavevectors chosen within the first Brillouin zone (thus $\mathbf{k}' \neq \mathbf{k}+\mathbf{g}_m$ for whatever $\mathbf{g}_m$). According to Eq. (2.27b), the periodic potential $V(\mathbf{r})$ can never mix plane waves of vectors $\mathbf{k}+\mathbf{g}_n$ with plane waves of vectors $\mathbf{k}'+\mathbf{g}'_n$. Thus the eigenfunctions of the crystal Hamiltonian (2.26) can be labeled as $\psi(\mathbf{k}, \mathbf{r})$, with the wavevector $\mathbf{k}$ defined within the first Brillouin zone.

A crystal wavefunction $\psi(\mathbf{k}, \mathbf{r})$, of vector $\mathbf{k}$, must be an appropriate linear combination of plane waves of vectors $\mathbf{k}+\mathbf{g}_n$ (this being the only kind of mixing allowed by a periodic potential, and hence by the crystal Hamiltonian):

$$\psi(\mathbf{k}, \mathbf{r}) = \sum_{\mathbf{g}_n} a_n(\mathbf{k})e^{i(\mathbf{k}+\mathbf{g}_n)\cdot\mathbf{r}}. \qquad (2.28)$$

Since the functions $\exp(i\mathbf{g}_n \cdot \mathbf{r})$ have the lattice periodicity, Eq. (2.28) can be written in the form

$$\boxed{\psi(\mathbf{k}, \mathbf{r}) = e^{i\mathbf{k}\cdot\mathbf{r}}u(\mathbf{k}, \mathbf{r})}, \qquad (2.29a)$$

where $u(\mathbf{k}, \mathbf{r})$ denotes a *periodic function*. Thus the eigenfunctions of the Schrödinger equation (2.26) must take the form of the product of a plane wave, modulated on atomic scale by a periodic function (*Bloch wavefunctions*). The Bloch wavefunctions extend on the whole crystals and have thus a "delocalized" (or "itinerant") nature.

There is another equivalent way of expressing the Bloch theorem, summarized by Eq. (2.29a). We notice that

$$\psi(\mathbf{k}, \mathbf{r} + \mathbf{t}_n) = e^{i\mathbf{k}\cdot(\mathbf{r}+\mathbf{t}_n)}u(\mathbf{k}, \mathbf{r} + \mathbf{t}_n) = e^{i\mathbf{k}\cdot\mathbf{t}_n}\psi(\mathbf{k}, \mathbf{r}); \qquad (2.29b)$$

thus, the values of the wavefunction of vector $\mathbf{k}$ in any two points connected by a lattice translation $\mathbf{t}_n$ are related by the phase factor $\exp(i\mathbf{k}\cdot\mathbf{t}_n)$. Notice that Eq. (2.29b) implies Eq. (2.29a); in fact if Eq. (2.29b) is multiplied by $\exp[-i\mathbf{k}\cdot(\mathbf{r}+\mathbf{t}_n)]$, we see that the function $\exp(-i\mathbf{k}\cdot\mathbf{r})\psi(\mathbf{k}, \mathbf{r})$ is periodic, which is the same information given by Eq. (2.29a).

From the present discussion on translational symmetry, we see that the eigenfunctions of a crystal Hamiltonian with periodic potential, are characterized by a wavevector $\mathbf{k}$ defined within the first Brillouin zone. For any given $\mathbf{k}$ vector, we have in general a countably infinite set of solutions; different solutions can be labeled with an index, say n, following for instance the increasing value of energy. Suppose to vary continuously $\mathbf{k}$ in the first Brillouin zone and to follow the crystal eigenvalues $E_n(\mathbf{k})$ and the crystal wavefunctions $\psi_n(\mathbf{k}, \mathbf{r})$; we arrive thus at the concept of the *energy band structure* of electrons in crystals.

Similarly to what is already discussed in Section 1.6, the quantity $\hbar\mathbf{k}$ can be interpreted as *quasi-momentum* or *crystal momentum* of the electron in the crystal. For brevity, $\hbar\mathbf{k}$ is also addressed as the *momentum of the electron*, keeping always in mind that the *momentum of an electron in the crystal is to be understood within reciprocal lattice vectors times* $\hbar$.

The methods and technique for the calculation of the band structure of specific materials are presented in Chapter 5; here we wish to discuss some general properties of the band structure of materials, that are essentially connected once again to the periodicity (rather than the actual details of the crystal potential).

2.6.2 The Parametric k · p Hamiltonian

The general form of the Bloch wavefunctions, as the product of a plane wave and a periodic part, allows further general elaborations. We insert the Bloch form of the wavefunctions into the Schrödinger equation (2.26) and obtain

$$\left[\frac{\mathbf{p}^2}{2m} + V(\mathbf{r})\right] e^{i\mathbf{k}\cdot\mathbf{r}} u_n(\mathbf{k}, \mathbf{r}) = E_n(\mathbf{k}) e^{i\mathbf{k}\cdot\mathbf{r}} u_n(\mathbf{k}, \mathbf{r});$$

thus, the periodic parts $u_n(\mathbf{k}, \mathbf{r})$ of the electronic wavefunctions satisfy the modified Schrödinger equation

$$\left[\frac{1}{2m}(\mathbf{p} + \hbar\mathbf{k})^2 + V(\mathbf{r})\right] u_n(\mathbf{k}, \mathbf{r}) = E_n(\mathbf{k}) u_n(\mathbf{k}, \mathbf{r}). \tag{2.30}$$

The eigenvalue equation (2.30) can also be written in the form

$$H(\mathbf{k}) u_n(\mathbf{k}, \mathbf{r}) = E_n(\mathbf{k}) u_n(\mathbf{k}, \mathbf{r}),$$

where

$$H(\mathbf{k}) = \frac{1}{2m}(\mathbf{p} + \hbar\mathbf{k})^2 + V(\mathbf{r}) = \frac{\mathbf{p}^2}{2m} + V(\mathbf{r}) + \frac{\hbar}{m}\mathbf{k}\cdot\mathbf{p} + \frac{\hbar^2 k^2}{2m}.$$

The operator $H(\mathbf{k})$, besides the crystal Hamiltonian and the scalar quantity $\hbar^2 k^2/2m$, contains the operator $(\hbar/m)\mathbf{k}\cdot\mathbf{p}$, which lends the name of $\mathbf{k}\cdot\mathbf{p}$ to the whole procedure.

The operator $H(\mathbf{k})$ contains $\mathbf{k}$ as a parameter, and is one of the simplest examples of parameter dependent operators (see Section 8.4 for other examples and applications). Following a Feynman procedure, we derive both members of Eq. (2.30) with respect to the parameter $\mathbf{k}$ and obtain

$$\frac{\hbar}{m}(\mathbf{p} + \hbar\mathbf{k}) u_n(\mathbf{k}, \mathbf{r}) + H(\mathbf{k})\frac{\partial u_n(\mathbf{k}, \mathbf{r})}{\partial \mathbf{k}} = \frac{\partial E_n(\mathbf{k})}{\partial \mathbf{k}} u_n(\mathbf{k}, \mathbf{r}) + E_n(\mathbf{k})\frac{\partial u_n(\mathbf{k}, \mathbf{r})}{\partial \mathbf{k}}. \tag{2.31}$$

The projection of Eq. (2.31) on $\langle u_m(\mathbf{k}, \mathbf{r})|$ with $m \equiv n$ gives

$$\langle u_n(\mathbf{k}, \mathbf{r})|\frac{\hbar}{m}(\mathbf{p} + \hbar\mathbf{k})|u_n(\mathbf{k}, \mathbf{r})\rangle = \frac{\partial E_n(\mathbf{k})}{\partial \mathbf{k}};$$

this relation can be expressed in the form

$$\langle \psi_n(\mathbf{k}, \mathbf{r})| \frac{\mathbf{p}}{m} |\psi_n(\mathbf{k}, \mathbf{r})\rangle = \frac{1}{\hbar} \frac{\partial E_n(\mathbf{k})}{\partial \mathbf{k}}, \tag{2.32}$$

which is the trivial generalization of Eq. (1.73) to the three-dimensional crystal. The projection of Eq. (2.31) on $\langle u_m(\mathbf{k}, \mathbf{r})|$ with $m \neq n$ gives

$$\langle \psi_m(\mathbf{k}, \mathbf{r})| \frac{\hbar}{m} \mathbf{p} |\psi_n(\mathbf{k}, \mathbf{r})\rangle = -[E_m(\mathbf{k}) - E_n(\mathbf{k})]\langle u_m(\mathbf{k}, \mathbf{r})| \frac{\partial}{\partial \mathbf{k}} u_n(\mathbf{k}, \mathbf{r})\rangle. \tag{2.33}$$

The expressions (2.32) and (2.33) of the diagonal and off-diagonal matrix elements of the operator $\mathbf{p}$ between band wavefunctions are quite general and extremely useful. In particular, Eq. (2.32) allows to link the band structure $E_n(\mathbf{k})$ with the electron velocity $\mathbf{v}_n(\mathbf{k}) = (1/\hbar)\partial E_n(\mathbf{k})/\partial \mathbf{k}$ in semiclassical transport theories.

In many situations it is convenient, instead of the momentum operator $\mathbf{p}$, to evaluate the matrix elements of the coordinate operator $\mathbf{r}$. From standard commutation rules it holds

$$[H, \mathbf{r}] = \left[\frac{\mathbf{p}^2}{2m} + V(\mathbf{r}), \mathbf{r} \right] = -i \frac{\hbar}{m} \mathbf{p}.$$

Taking the matrix elements of both members of the above equation between two crystal eigenfunctions one obtains

$$\langle \psi_m(\mathbf{k}, \mathbf{r})| - i \frac{\hbar}{m} \mathbf{p} |\psi_n(\mathbf{k}, \mathbf{r})\rangle = \langle \psi_m(\mathbf{k}, \mathbf{r})|[H, \mathbf{r}]|\psi_n(\mathbf{k}, \mathbf{r})\rangle$$
$$= [E_m(\mathbf{k}) - E_n(\mathbf{k})]\langle \psi_m(\mathbf{k}, \mathbf{r})|\mathbf{r}|\psi_n(\mathbf{k}, \mathbf{r})\rangle. \tag{2.34}$$

Comparison of Eqs. (2.34) and (2.33) gives the relation

$$\langle u_m(\mathbf{k}, \mathbf{r})|\mathbf{r}|u_n(\mathbf{k}, \mathbf{r})\rangle = i \langle u_m(\mathbf{k}, \mathbf{r})| \frac{\partial}{\partial \mathbf{k}} u_n(\mathbf{k}, \mathbf{r})\rangle \quad \text{for } m \neq n. \tag{2.35}$$

This is a useful property of the off-diagonal matrix elements of the operator $\mathbf{r}$ and the parameter operator $i\partial/\partial \mathbf{k}$, on the set of periodic functions $\{u_j(\mathbf{k}, \mathbf{r})\}$ at a given selected $\mathbf{k}$ point.

In periodic materials, thanks to the Bloch theorem, the knowledge of a wavefunction $\psi_n(\mathbf{k}, \mathbf{r})$ within the unit cell entails its knowledge in the whole real space. We now show that the knowledge of the set of wavefunctions $\{\psi_n(\mathbf{k}_0, \mathbf{r})\}$ and band energies $E_n(\mathbf{k}_0)$ just at a *single* $\mathbf{k}_0$ *vector* implies the knowledge (at least in principle) of the whole band structure (wavefunctions and energies) on the entire Brillouin zone.

From the Schrödinger equation (2.30), and the general properties of an eigenvalue equation, we have that the periodic wavefunctions at any chosen specific wavevector $\mathbf{k}_0$ constitute a complete set for the expansion of the periodic wavefunctions at any other $\mathbf{k}$ wavevector. For sake of simplicity (and without loss of generality) we select $\mathbf{k}_0$

at the center of the Brillouin zone (the arguments below are readily applicable to any other selected point in the Brillouin zone).

A crystal wavefunction $\psi(\mathbf{k}, \mathbf{r})$ of vector $\mathbf{k}$ can always be expanded in the form

$$\psi(\mathbf{k}, \mathbf{r}) = \sum_n c_n(\mathbf{k}) e^{i\mathbf{k}\cdot\mathbf{r}} \psi_{n0}(\mathbf{r}) \quad \text{where} \quad \psi_{n0}(\mathbf{r}) = \psi_n(\mathbf{k}_0 = 0, \mathbf{r}). \tag{2.36a}$$

The matrix elements of the crystal Hamiltonian among the functions $\{\exp(i\mathbf{k}\cdot\mathbf{r})\psi_{n0}(\mathbf{r})\}$ are

$$M_{nn'}(\mathbf{k}) = \langle e^{i\mathbf{k}\cdot\mathbf{r}}\psi_{n0}(\mathbf{r}) | \frac{\mathbf{p}^2}{2m} + V(\mathbf{r}) | e^{i\mathbf{k}\cdot\mathbf{r}}\psi_{n'0}(\mathbf{r})\rangle$$

$$= \langle \psi_{n0}(\mathbf{r}) | \frac{\mathbf{p}^2}{2m} + V(\mathbf{r}) + \frac{\hbar}{m}\mathbf{k}\cdot\mathbf{p} + \frac{\hbar^2 k^2}{2m} | \psi_{n'0}(\mathbf{r})\rangle$$

$$= \left[E_{n0} + \frac{\hbar^2 k^2}{2m} \right] \delta_{nn'} + \frac{\hbar}{m} \langle \psi_{n0}(\mathbf{r}) | \mathbf{k}\cdot\mathbf{p} | \psi_{n'0}(\mathbf{r})\rangle. \tag{2.36b}$$

The matrix elements $M_{nn'}$ essentially require the knowledge of the band energies $E_n(\mathbf{k}_0) = E_{n0}$ and the matrix elements of the momentum operator $\mathbf{p}$ between the band wavefunctions at the selected wavevector $\mathbf{k}_0 = 0$.

From expansion (2.36a), matrix elements (2.36b), and standard variational methods in the expansion coefficients, we have that band energies and band wavefunctions at any wavevector $\mathbf{k}$ are obtained by the eigenvalues and eigenvectors of the secular equation

$$\left\| \left(E_{n0} + \frac{\hbar^2 k^2}{2m} - E \right) \delta_{nn'} + \frac{\hbar}{m}\mathbf{k}\cdot\langle\psi_{n0}|\mathbf{p}|\psi_{n'0}\rangle \right\| = 0 . \tag{2.37}$$

The secular equation (2.37) is exact and can be used in principle (and at times it is convenient also in practice) to extend the band structure, known at a particular point of the Brillouin zone (the $\mathbf{k} = 0$ point in the present case) to the *whole* Brillouin zone.

A most significant use of Eq. (2.37) is to confine $\mathbf{k}$ to a small region of the Brillouin zone, around the point $\mathbf{k} = 0$. In this case the off-diagonal matrix elements in Eq. (2.37) are small, and thus can be safely treated with standard perturbation theory. Suppose, for example, that the band of interest (say the n-th one) is non-degenerate at $\mathbf{k} = 0$ and here exhibits an extremum, so that $\langle\psi_{n0}|\mathbf{p}|\psi_{n0}\rangle$ vanishes because of Eq. (2.32). The perturbation theory for non-degenerate states, up to second order in the wavevector, gives for the band energy $E_n(\mathbf{k})$ the expression

$$E_n(\mathbf{k}) = E_{n0} + \frac{\hbar^2 k^2}{2m} + \frac{\hbar^2}{m^2} \sum_{n'(\neq n)} \frac{|\langle\psi_{n'0}|\mathbf{k}\cdot\mathbf{p}|\psi_{n0}\rangle|^2}{E_{n0} - E_{n'0}}. \tag{2.38}$$

The right-hand side of Eq. (2.38) is a quadratic function of the wavevector, and can be cast in the form

$$E_n(\mathbf{k}) = E_{n0} + \sum_{\alpha\beta} \frac{\hbar^2}{2m} \left(\frac{m}{m^*}\right)_{\alpha\beta} k_\alpha k_\beta ,$$

where the effective mass tensor reads

$$\left(\frac{m}{m^*}\right)_{\alpha\beta} = \delta_{\alpha\beta} + \frac{2}{m} \sum_{n'(\neq n)} \frac{\langle \psi_{n0}|\mathbf{p}_\alpha|\psi_{n'0}\rangle \langle \psi_{n'0}|\mathbf{p}_\beta|\psi_{n0}\rangle}{E_{n0} - E_{n'0}} \tag{2.39}$$

with $\alpha, \beta = x, y, z$.

Expression (2.39) has been often used to evaluate effective masses in crystals. For an isotropic two-band model semiconductor, with energy gap E_G and matrix element $\mathbf{p}_{cv}$ between the valence and the conduction wavefunctions at the band edges, expression (2.39) gives for the effective masses of the two bands

$$\frac{m}{m^*} \approx 1 \pm \frac{2}{m} \frac{|\mathbf{p}_{cv}|^2}{E_G};$$

this relation shows that small effective masses for electrons and holes are to be expected in small energy gap semiconductors, a qualitative trend confirmed by experience. The $k \cdot p$ approach has been profitably generalized and extended to the study of degenerate bands, also in the presence of spin-orbit coupling.

2.6.3   Cyclic Boundary Conditions

In the previous sections we have considered ideal infinite crystals. Suppose now to have a macroscopic (but finite) crystal, of volume V, in the form of a parallelepiped with edges $N_1\mathbf{t}_1$, $N_2\mathbf{t}_2$, and $N_3\mathbf{t}_3$, where N_1, N_2, N_3 are large (but finite) integer numbers. The Born-von Karman cyclic boundary conditions on the electronic wavefunctions require

$$\psi(\mathbf{r}) = \psi(\mathbf{r} + N_1\mathbf{t}_1) = \psi(\mathbf{r} + N_2\mathbf{t}_2) = \psi(\mathbf{r} + N_3\mathbf{t}_3).$$

Since $\psi(\mathbf{r})$ must be a Bloch function of vector $\mathbf{k}$, we have

$$\exp[i\mathbf{k} \cdot N_1\mathbf{t}_1] = \exp[i\mathbf{k} \cdot N_2\mathbf{t}_2] = \exp[i\mathbf{k} \cdot N_3\mathbf{t}_3] = 1. \tag{2.40}$$

The possible $\mathbf{k}$ vectors, compatible with cyclic boundary conditions (2.40), are thus

$$\mathbf{k} = \frac{m_1}{N_1}\mathbf{g}_1 + \frac{m_2}{N_2}\mathbf{g}_2 + \frac{m_3}{N_3}\mathbf{g}_3 \tag{2.41}$$

with m_i integer numbers.

The allowed $\mathbf{k}$ vectors (2.41) within the primitive unit cell of edges $\mathbf{g}_1, \mathbf{g}_2, \mathbf{g}_3$ are obtained choosing $0 \leq m_i < N_i$ $(i = 1, 2, 3)$; their number is $N = N_1 N_2 N_3$. The first Brillouin zone has the same volume as the primitive unit cell in the reciprocal lattice. *Thus the number of allowed $\mathbf{k}$ vectors in the first Brillouin zone equals the number of primitive unit cells of the crystal.*

The density of allowed $\mathbf{k}$ vectors (2.41) in the reciprocal space is uniform and is given by

$$W(\mathbf{k}) = \frac{N_1 N_2 N_3}{\Omega_k} = \frac{V}{(2\pi)^3}, \tag{2.42}$$

where V is the volume of the crystal. In some problems, we have to perform sums in reciprocal space of a given function of $\mathbf{k}$. According to Eq. (2.42), the discrete sum can be replaced by an integral as follows

$$\boxed{\sum_{\mathbf{k}} f(\mathbf{k}) \Longrightarrow \frac{V}{(2\pi)^3} \int f(\mathbf{k})d\mathbf{k}}, \tag{2.43}$$

where $d\mathbf{k}$ is the volume element in reciprocal space.

Notice also that the $\mathbf{k}$ vectors (2.41) within the first Brillouin zone satisfy the relation

$$\sum_{\mathbf{k}}^{B.Z.} e^{i\mathbf{k}\cdot\mathbf{t}_n} = \begin{cases} N_1 N_2 N_3 & \text{if } \mathbf{t}_n = 0, \\ 0 & \text{if } \mathbf{t}_n \neq 0, \end{cases} \tag{2.44a}$$

where $\mathbf{t}_n$ is any translation of the crystal such that $0 \leq n_i < N_i$ $(i = 1, 2, 3)$. The proof is simply based on the elementary algebraic property that the sum of the N_i roots of the unity equals zero. Similarly we have

$$\sum_{\mathbf{t}_n}^{N_1 N_2 N_3} e^{i\mathbf{k}\cdot\mathbf{t}_n} = \begin{cases} N_1 N_2 N_3 & \text{if } \mathbf{k} = \mathbf{g}_m, \\ 0 & \text{if } \mathbf{k} \neq \mathbf{g}_m, \end{cases} \tag{2.44b}$$

where $\mathbf{k}$ is any vector (2.41) in the reciprocal space.

2.6.4 Special k Points for Averaging Over the Brillouin Zone

In several studies of crystals (for instance in the calculation of charge density, total energies and other properties) we have to perform the average of appropriate $\mathbf{k}$-dependent functions throughout the Brillouin zone. The average value of a function $F(\mathbf{k})$ is given by

$$\overline{F} = \frac{1}{N} \sum_{\mathbf{k}}^{B.Z.} F(\mathbf{k}), \tag{2.45}$$

where N is the (large) number of allowed $\mathbf{k}$ vectors in the first Brillouin zone. The calculation of $F(\mathbf{k})$ at a given $\mathbf{k}$ vector may require laborious numerical work; then it would be highly desirable to find sets of "special $\mathbf{k}$ points" (composed by a reasonably modest number of $\mathbf{k}$ vectors) to determine efficiently the average $\overline{F}$.

To see the basic principle for this achievement we proceed as follows. Consider a function $F(\mathbf{k})$ with $\mathbf{k}$ defined in the first Brillouin zone, or, equivalently, consider a function $F(\mathbf{k})$ with the periodicity of the reciprocal lattice; its most general expansion, according to Eq. (2.19c) is

$$F(\mathbf{k}) = F_0 + \sum_{\mathbf{t}_n \neq 0} F_n e^{i\mathbf{k}\cdot\mathbf{t}_n}, \tag{2.46}$$

where $\mathbf{t}_n$ are translation vectors of the direct lattice, F_n are the appropriate Fourier coefficients, and for convenience the $\mathbf{t}_n = 0$ term has been separated from all the other terms. If we sum both members of Eq. (2.46) over all the allowed $\mathbf{k}$ vectors in the first Brillouin zone, use Eq. (2.44a) and divide by N, we recover expression (2.45) with

$\overline{F} = F_0$. Thus, in general, only the sum over a very large number of $\mathbf{k}$ vectors allows to obtain the exact average value of the function $F(\mathbf{k})$.

Consider now a function $G(\mathbf{k})$ represented in the form

$$G(\mathbf{k}) = G_0 + \sum_{\mathbf{t}_I} G_I \, e^{i\mathbf{k}\cdot\mathbf{t}_I}, \tag{2.47a}$$

where $\mathbf{t}_I$ denotes translation vectors belonging to the first shell of neighbors (all other Fourier coefficients are assumed to be zero). Let $\{\mathbf{k}_s\}$ indicate an appropriately selected set of n_s points of the Brillouin zone, chosen in such a way that

$$\sum_{\mathbf{k}\in\{\mathbf{k}_s\}} e^{i\mathbf{k}\cdot\mathbf{t}_I} = 0 \quad \text{for any } \mathbf{t}_I \in \text{first shell.} \tag{2.47b}$$

Then the average value G_0 of the function $G(\mathbf{k})$ is rigorously expressed by

$$G_0 = \frac{1}{n_s} \sum_{\mathbf{k}\in\{\mathbf{k}_s\}} G(\mathbf{k}). \tag{2.47c}$$

As an example, consider a simple cubic lattice; the six translation vectors $\mathbf{t}_I$ of the first shell are: $(\pm a, 0, 0)$, $(0, \pm a, 0)$, $(0, 0, \pm a)$. It is seen by inspection that the set of just two vectors $\mathbf{k}_1$ and $\mathbf{k}_2$, given by $\mathbf{k}_1 = -\mathbf{k}_2 = (\pi/a)(1/2, 1/2, 1/2)$, satisfies Eq. (2.47b) for any of the six vectors $\mathbf{t}_I$. The procedure can be extended to find systematically sets of wavevectors $\{\mathbf{k}_s\}$ for which $\sum_s \exp(i\mathbf{k}_s \cdot \mathbf{t}_n) = 0$ for all the translation vectors $\mathbf{t}_n$ belonging to the first shell, second shell, and so on, up to the outermost shell of interest, and we refer to the literature for details. When the number of shells to account for becomes very large, the set of special points approaches the uniform mesh of points (2.41) confined within the primitive reciprocal cell.

In several situations of physical interest, the Fourier coefficients in expression (2.46) depend only on the modulus $|\mathbf{t}_n|$; in these situations, expansion (2.46) can be recast to the form

$$F(\mathbf{k}) = F_0 + F_1 \sum_{\mathbf{t}_I} e^{i\mathbf{k}\cdot\mathbf{t}_I} + F_2 \sum_{\mathbf{t}_{II}} e^{i\mathbf{k}\cdot\mathbf{t}_{II}} + \cdots, \tag{2.48}$$

where $\mathbf{t}_I$ denotes the first shell of neighbor translation vectors, $\mathbf{t}_{II}$ denotes the second shell, etc. In this case, the determination of sets of special points is particularly instructive.

As an example let us consider a simple cubic lattice. The six translation vectors in the first shell are given by $(\pm a, 0, 0)$, $(0, \pm a, 0)$, $(0, 0, \pm a)$. The 12 vectors in the second shell are given by $(\pm a, \pm a, 0)$, $(\pm a, 0, \pm a)$, $(0, \pm a, \pm a)$. The eight vectors in the third shell are given by $(\pm a, \pm a, \pm a)$. For the simple cubic lattice, the expansion (2.48) reads

$$\begin{aligned}
F(\mathbf{k}) = F_0 &+ 2F_1(\cos k_x a + \cos k_y a + \cos k_z a) \\
&+ 4F_2(\cos k_x a \cos k_y a + \cos k_x a \cos k_z a + \cos k_y a \cos k_z a) \\
&+ 8F_3 \cos k_x a \cos k_y a \cos k_z a + \cdots
\end{aligned}$$

In the above expression, it is seen by inspection that a single point (the Baldereschi point) reduces to zero the contributions from the first, second, and third shells; this mean value point is just $\mathbf{k} = (\pi/a)(1/2, 1/2, 1/2)$. Similar procedures can be performed with the other Bravais lattices. The mean value procedure, and the generated extensions, are particularly useful for instance in performing self-consistent calculations, keeping a relative manageable number of k points into account.

2.7 Density-of-States and Critical Points

In several problems the primary interest is not in the detailed wavevector dependence of the crystal band structure but only on the density-of-states in a given energy range. Consider for simplicity a non-degenerate band $E(\mathbf{k})$; the corresponding density-of-states for this band is given by

$$D(E) = 2 \sum_{\mathbf{k}} \delta(E(\mathbf{k}) - E) = 2 \int_{\text{B.Z.}} \frac{V}{(2\pi)^3} \delta(E(\mathbf{k}) - E) \, d\mathbf{k}, \qquad (2.49)$$

where $d\mathbf{k}$ is the volume element of the reciprocal space and the factor 2 accounts for the spin degeneracy (we suppose this degeneracy is not removed). Equation (2.49) shows that contributions to the density-of-states $D(E)$ at energy E occur from the band states (if any) such that $E(\mathbf{k}) = E$; the factor $V/(2\pi)^3$ gives the uniform density of allowed $\mathbf{k}$ vectors in $\mathbf{k}$-space, as discussed in Eq. (2.42).

In Figure 2.21 we indicate schematically the two isoenergetic surfaces $E(\mathbf{k}) = E$ and $E(\mathbf{k}) = E + dE$. The distance dk between the two isoenergetic surfaces is obtained by observing that $dE = \nabla_{\mathbf{k}} E(\mathbf{k}) \cdot d\mathbf{k} = |\nabla_{\mathbf{k}} E(\mathbf{k})| dk$. Counting the volume in $\mathbf{k}$-space enclosed between the two surfaces, the expression (2.49) for the density-of-states can be transformed into the integral on the constant energy surface $E(\mathbf{k}) = E$ in the form

$$D(E) = 2 \int_{E(\mathbf{k}) \equiv E} \frac{V}{(2\pi)^3} \frac{dS}{|\nabla_{\mathbf{k}} E(\mathbf{k})|}. \qquad (2.50)$$

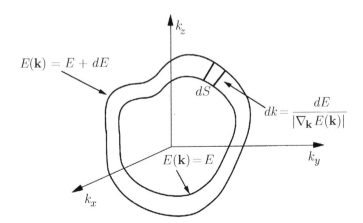

Figure 2.21 Schematic representation of two isoenergetic surfaces in the $\mathbf{k}$ space.

Expression (2.50) clearly shows that singularities in the density-of-states are expected at the *critical points*, defined as those points in **k**-space for which

$$\nabla_{\mathbf{k}} E(\mathbf{k}) = 0;$$ (2.51)

for these points the density-of-states is expected to exhibit anomalies as a function of energy.

At the critical points, Eq. (2.49) or Eq. (2.50) can be integrated analytically. In more general situations the density-of-states can be obtained numerically by appropriate sampling of the **k** vectors in the Brillouin zone.

Near a critical point, where Eq. (2.51) holds, we can expand the energy as a function of the wavevector, in quadratic form. Indicating with k_x, k_y, k_z the principal axes of the quadratic form and taking the origin at the critical point itself, we have

$$E(\mathbf{k}) = E_c \pm \frac{\hbar^2}{2m_x} k_x^2 \pm \frac{\hbar^2}{2m_y} k_y^2 \pm \frac{\hbar^2}{2m_z} k_z^2$$ (2.52)

(where we choose $m_x, m_y, m_z > 0$, while the occurrence of $\pm$ sign specifies the kind of critical point). The critical point M_0 denotes zero negative signs in expression (2.52) and is thus a minimum of $E(\mathbf{k})$; the critical point M_3 denotes three negative signs in expression (2.52) and is thus a maximum; M_1 and M_2 denote one and two negative signs, respectively, and are thus two saddle points. Thus in three dimensions we have four types of critical points. Similarly, in two-dimensional crystals we have three types of critical points M_0, M_1, and M_2: a minimum, a saddle point, and a maximum, respectively. In one-dimensional crystals we have two types of critical points: a minimum M_0 and a maximum M_1.

Near a three-dimensional critical point, the expression (2.49) for the density-of-states becomes

$$D(E) = \frac{2V}{(2\pi)^3} \int \delta\left(E_c \pm \frac{\hbar^2 k_x^2}{2m_x} \pm \frac{\hbar^2 k_y^2}{2m_y} \pm \frac{\hbar^2 k_z^2}{2m_z} - E \right) dk_x dk_y dk_z.$$ (2.53a)

For the two-dimensional case we have

$$D(E) = \frac{2L_x L_y}{(2\pi)^2} \int \delta\left(E_c \pm \frac{\hbar^2 k_x^2}{2m_x} \pm \frac{\hbar^2 k_y^2}{2m_y} - E \right) dk_x dk_y.$$ (2.53b)

For the one-dimensional case we have

$$D(E) = \frac{2L_x}{2\pi} \int \delta\left(E_c \pm \frac{\hbar^2 k_x^2}{2m_x} - E \right) dk_x.$$ (2.53c)

All types of integrals (2.53) can be easily calculated analytically with the help of the following property of the delta function:

$$\delta[f(x)] = \sum_n \frac{\delta(x - x_n)}{|f'(x_n)|},$$ (2.54)

where x_n are the simple zeroes of the function $f(x)$.

Critical Points in One-Dimensional Crystals

In one-dimensional crystals, for the critical point of type M_0 at energy E_0, we consider the integral

$$D(E) = \frac{2L_x}{2\pi} \int \delta\left(E_0 + \frac{\hbar^2 k_x^2}{2m_x} - E\right) dk_x = L_x \frac{\sqrt{2m_x}}{\pi \hbar} \int \delta(E_0 + q_x^2 - E) \, dq_x,$$

(2.55)

where the change of variable $q_x = (\hbar/\sqrt{2m_x})k_x$ has been performed. Consider the function $f(q_x) = E_0 + q_x^2 - E$; the zeroes of this function occur for $q_{x0} = \pm\sqrt{E - E_0}$ and $f'(q_{x0}) = 2q_{x0} = \pm 2\sqrt{E - E_0}$. The integral (2.55) then becomes

$$D(E) = L_x \frac{\sqrt{2m_x}}{\pi \hbar} \frac{1}{\sqrt{E - E_0}}, \quad E > E_0.$$

(2.56)

Similarly for a critical point of type M_1 at energy E_1 we have

$$D(E) = L_x \frac{\sqrt{2m_x}}{\pi \hbar} \frac{1}{\sqrt{E_1 - E}}, \quad E < E_1.$$

(2.57)

In Figure 2.22 we give the behavior of the density-of-states for the two possible critical points in one-dimensional crystals. It can be noticed that the singularities appearing in Figure 1.15a, which gives the bulk density-of-states for a linear chain evaluated with the Green's function technique, are (as expected) of the type illustrated in Figure 2.22.

Critical Points in Two-Dimensional Crystals

In two-dimensional crystals, for a critical point of type M_0 at energy E_0, we have to calculate the integral

$$D(E) = \frac{2L_x L_y}{(2\pi)^2} \int \delta\left(E_0 + \frac{\hbar^2 k_x^2}{2m_x} + \frac{\hbar^2 k_y^2}{2m_y} - E\right) dk_x dk_y.$$

(2.58)

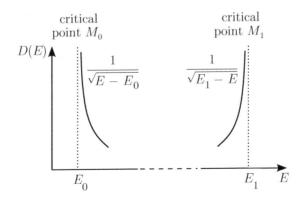

Figure 2.22 Density-of-states at the critical points M_0 (minimum) and M_1 (maximum) of a one-dimensional crystal.

It is convenient to introduce the variable $q_x = (\hbar/\sqrt{2m_x})k_x$ and $q_y = (\hbar/\sqrt{2m_y})k_y$ in Eq. (2.58), and denote by $S = L_x L_y$ the surface of the two-dimensional crystal; we have

$$D(E) = S\frac{\sqrt{m_x m_y}}{\pi^2 \hbar^2} \int \delta(E_0 + q_x^2 + q_y^2 - E)\, dq_x\, dq_y.$$

We pass to cylindrical coordinates and easily obtain

$$D(E) = S\frac{\sqrt{m_x m_y}}{\pi \hbar^2} \quad \text{for } E > E_0. \tag{2.59}$$

A similar expression holds for the critical point which is a maximum, for $E < E_2$. Thus the density-of-states is step-like at the points M_0 and M_2 of a two-dimensional crystal.

For the saddle critical point M_1 we have to evaluate the expression

$$D(E) = S\frac{\sqrt{m_x m_y}}{\pi^2 \hbar^2} \int \delta(E_1 + q_x^2 - q_y^2 - E)\, dq_x\, dq_y. \tag{2.60}$$

We suppose momentarily $E > E_1$. Consider the function $f(q_x) = E_1 + q_x^2 - q_y^2 - E$; the zeroes of this function occur for $q_{x_0} = \pm\sqrt{E - E_1 + q_y^2}$ and $f'(q_{x_0}) = 2q_{x_0}$. The integral (2.60) thus becomes

$$D(E) = S\frac{\sqrt{m_x m_y}}{\pi^2 \hbar^2} \int_{-q_c}^{q_c} \frac{1}{\sqrt{E - E_1 + q_y^2}}\, dq_y,$$

where the integral has been confined to a cutoff where the series expansion is supposed to hold. With the help of the indefinite integral

$$\int \frac{dx}{\sqrt{a^2 + x^2}} = \ln\left(x + \sqrt{a^2 + x^2}\right),$$

one has

$$D(E) = S\frac{\sqrt{m_x m_y}}{\pi^2 \hbar^2} \ln \frac{q_c + \sqrt{E - E_1 + q_c^2}}{-q_c + \sqrt{E - E_1 + q_c^2}}. \tag{2.61}$$

For $E \approx E_1$, with appropriate series development in the second member of Eq. (2.61), we obtain

$$D(E) = S\frac{\sqrt{m_x m_y}}{\pi^2 \hbar^2} \ln \frac{4q_c^2}{|E - E_1|}, \tag{2.62}$$

and we see that a logarithmic divergence occurs at the critical point (with the absolute value $|E - E_1|$, Eq. (2.62) holds also for $E < E_1$). In Figure 2.23 the behavior of the density-of-states at the two-dimensional critical points is reported.

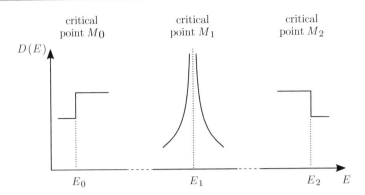

Figure 2.23 Density-of-states at the critical points of type M_0 (minimum), M_1 (saddle point), and M_2 (maximum) of a two-dimensional crystal.

Critical Points in Three-Dimensional Crystals

For three-dimensional crystals there are four types of critical points: M_0 (minimum), M_1 and M_2 (saddle points), M_3 (maximum). The analytic behavior at the three-dimensional critical points can be obtained with similar procedures:

$$\text{critical point } M_0 \quad D(E) = C_0 + V\frac{\sqrt{2m_x m_y m_z}}{\pi^2 \hbar^3}\sqrt{E - E_0} \quad \text{for } E > E_0 \quad (2.63a)$$

$$\text{critical point } M_1 \quad D(E) = C_1 - V\frac{\sqrt{2m_x m_y m_z}}{\pi^2 \hbar^3}\sqrt{E_1 - E} \quad \text{for } E < E_1 \quad (2.63b)$$

$$\text{critical point } M_2 \quad D(E) = C_2 - V\frac{\sqrt{2m_x m_y m_z}}{\pi^2 \hbar^3}\sqrt{E - E_2} \quad \text{for } E > E_2 \quad (2.63c)$$

$$\text{critical point } M_3 \quad D(E) = C_3 + V\frac{\sqrt{2m_x m_y m_z}}{\pi^2 \hbar^3}\sqrt{E_3 - E} \quad \text{for } E < E_3. \quad (2.63d)$$

In the above expressions C_i indicate either a constant (including zero) or a smoothly energy dependent quantity, while the terms with the square root are present only when the argument is positive. The results (2.63) are schematically indicated in Figure 2.24.

As an example of three-dimensional density-of-states consider the case of the cubium, a simple cubic crystal with a single s-like orbital per site, and nearest neighbor interactions only. With a straightforward generalization of the procedures leading to Eq. (1.48), we have that the band energy $E(\mathbf{k})$ of the cubium is given by the expression

$$E(\mathbf{k}) = \alpha + 2\gamma(\cos k_x a + \cos k_y a + \cos k_z a) \quad (2.64)$$

(for simplicity we take $\alpha = 0$; the hopping parameter γ is supposed to be negative). The Brillouin zone of the cubium is indicated in Figure 2.17, together with the symmetry points Γ, X, M, and R. It is seen by inspection of Eq. (2.64) that Γ is a critical point of type M_0 and energy $E_0 = -6|\gamma|$. Similarly, X and M are saddle points of type M_1 and M_2, and energy $E_1 = -2|\gamma|$ and $E_2 = 2|\gamma|$, respectively. The point R is a critical point of type M_3 and energy $E_3 = 6|\gamma|$. The density-of-states corresponding

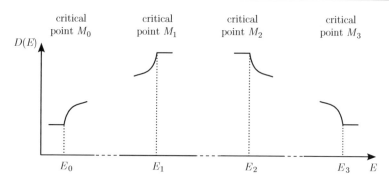

Figure 2.24 Density-of-states at the critical points of type M_0 (minimum), M_1 and M_2 (saddle points), and M_3 (maximum) of a three-dimensional crystal.

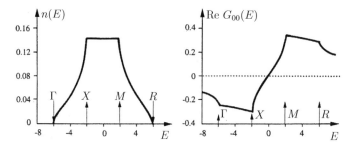

Figure 2.25 Density-of-states of the cubium, to illustrate the three-dimensional critical points. The energy E is in units of $|\gamma|$, and $n(E)$ is normalized to one. Besides $n(E) = -(1/\pi)\mathrm{Im}\, G_{00}(E)$, also the real part of the Green's function diagonal matrix element $G_{00}(E)$, on a localized orbital of the cubium, is reported for convenience [reprinted with permission from D. J. Lohrmann, L. Resca, G. Pastori Parravicini and R. D. Graft, Phys. Rev. B *40*, 8404 (1989); copyright 1989 by the American Physical Society].

to the energy band (2.64) can be computed numerically, for instance via the general definition (2.49) and appropriate sampling of the Brillouin zone, or by means of the Green's function technique and Lanczos procedure (see Section 5.8.1). The computed density-of-states of cubium is reported in Figure 2.25, and the presence of the critical points of the type and energy discussed above can be clearly noticed.

Further Reading

Altmann, S. L. (1994). *Band theory of solids: An introduction from the point of view of symmetry.* Oxford: Clarendon Press.

Baldereschi, A. (1973). Mean-value point in the Brillouin zone. *Physical Review B, 7,* 5212. For further aspects on special points see for instance: Chadi, D. J., & Cohen, M. L. (1973). Special points in the Brillouin zone. *Physical Review B, 8,* 5747; Hama, J., & Watanabe, M.

(1992). General formulae for the special points and their weighting factors in k-space integration. *Journal of Physics C, 4*, 4583.

Bassani, F., & Pastori Parravicini, G. (1975). *Electronic states and optical transitions in solids.* Oxford: Pergamon Press.

Giacovazzo, C., Monaco, H. L., Artioli, G., Viterbo, D., Ferraris, G., Gilli, G., Zanotti, G., & Catti, M. (2002). In C. Giacovazzo (Ed.), *Fundamentals of crystallography.* Oxford: Oxford University Press.

Gilat, G. (1976). Methods of Brillouin zone integration. In *Methods in computational physics* (Vol. 15, p. 317). New York: Academic Press. This article contains a survey of k-space integrations. For further aspects see for instance: Lehmann, G., & Taut, M. (1972). On the numerical calculation of the density of states and related properties. *Physica Status Solidi (b), 54*, 469; Lambin, P., & Vigneron, J. P.(1984). Computation of crystal Green's function in the complex energy plane with the use of the analytical tetrahedron method. *Physical Review B, 29*, 3430; Pickard C. J., & Payne M. C. (1999). Extrapolative approaches to Brillouin-zone integration. *Physical Review B, 59*, 4685.

Jones, H. (1975). *The theory of Brillouin zones and electronic states in crystals.* Amsterdam: North-Holland.

Koster, G. F., Dimmock, J. O., Wheeler, G., & Statz, H., *Properties of the thirty-two points groups.* Cambridge, Massachusetts: MIT Press. This book provides the conventionally used notations for symmetry points, symmetry lines, and irreducible representations.

Monkhorst H. J., & Pack J. P. (1976). Special points for Brillouin-zone integrations. *Physical Review B, 13*, 5188.

Schwarzenbach, D. (1996). *Crystallography.* Chichester: Wiley.

Sternberg, S. (1994). *Group theory and physics.* Cambridge: Cambridge University Press.

Wyckoff, R. W. G. (1963–1968). *Crystal structures* (Vols. 1–5). New York: Wiley. These volumes contain a most comprehensive description of crystal structures. See also Burns, G., & Glazer, A. M. (1981). *Space groups for solid state scientists.* New York: Academic Press; Vainshtein, B. K. (1981). *Modern crystallography.* Berlin: Springer.

3 The Sommerfeld Free-Electron Theory of Metals

Chapter Outline head

In this chapter we discuss the free-electron theory of metals, originally developed by Sommerfeld and others. The free-electron model, with its parabolic energy-wavevector dispersion curve, provides a reasonable description for conduction electrons in simple metals; it also may give useful guidelines for metals with more complicated conduction bands. Because of the simplicity of the model and its density-of-states, we can work out explicitly thermodynamic properties, and in particular the specific heat and the thermionic emission. In the Appendices we summarize for a general quantum system the thermodynamic functions of more frequent use in statistical physics. The Fermi-Dirac and Bose-Einstein distribution functions for independent fermions and bosons are discussed; finally we obtain the modified Fermi-Dirac distribution function, in a model case of correlation among electrons in localized states.

3.1 Quantum Theory of the Free-Electron Gas

An electron in a crystal feels the potential energy due to all the nuclei and all the other electrons. The determination of the crystalline potential in specific materials is a rather demanding problem (see Chapters 4 and 5). However in several metals (the so-called simple metals), it turns out reasonable to assume that the conduction

Solid State Physics, Second Edition. http://dx.doi.org/10.1016/B978-0-12-385030-0.00003-7

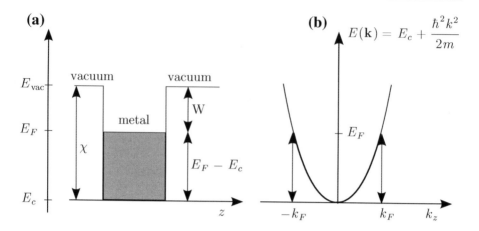

Figure 3.1 (a) The Sommerfeld model for a metal. The energy E_c denotes the bottom of the conduction band; E_{vac} denotes the energy of an electron at rest in the vacuum; the electron affinity is $\chi = E_{vac} - E_c$. The Fermi energy is denoted by E_F, and the work function W equals $\chi - E_F$. (b) Free-electron energy dispersion curve along a direction, say k_z, in the reciprocal space. At $T = 0$ all the states with $k < k_F$ (pictured in bold) are occupied by two electrons of either spin.

electrons feel an "effective" potential which is constant (or nearly constant) throughout the sample; this model, suggested by Sommerfeld in 1928, is still of value for an indicative understanding of a number of properties of some common metals.

In the free-electron Sommerfeld model, it is assumed that the dispersion energy curve for conduction electrons inside the metal takes the parabolic form

$$E(\mathbf{k}) = E_c + \frac{\hbar^2 k^2}{2m},\tag{3.1}$$

where E_c denotes an appropriate constant specific of the metal under investigation (see Figure 3.1). The number of "free electrons" of the metal is assumed to correspond to the number of electrons in the most external shell of the composing atoms. For instance in alkali metals, we expect one free electron per atom, since an alkali atom has just one electron in the most external shell. In aluminum (atomic configuration $1s^2, 2s^2 2p^6, 3s^2 3p^1$) we expect three free electrons per atom.

At first sight it could appear surprising that the conduction electrons in actual metals (or in any type of solid) feel an approximately constant potential, as strong spatial variations of the crystalline potential occur near the nuclei. However, the conduction electrons are hardly sensitive to the region close to the nuclei, because of the orthogonalization effects due to the core electron states; thus the "effective potential" or "pseudopotential" (see Section 5.4) for conduction electrons may indeed become a smoothly varying quantity, eventually approximated with a constant; these are the underlying reasons for the success of the free-electron model (or of the nearly-free-electron implementations) in a number of actual metals.

For simplicity, throughout this chapter, we set the zero of energy at the bottom E_c of the conduction band, and thus take $E_c = 0$; whenever necessary, we can reinstate the actual value of E_c by an appropriate shift in the energy scale. [In some situations, for instance in the discussion of photoelectron emission, it can be more convenient to set the zero of energy at the vacuum level, i.e. $E_{vac} = 0$; in other problems, for instance in the discussion of many-body effects, the zero of energy is generally taken at the Fermi level, i.e. $E_F = 0$; of course any choice is lawful, provided one keeps in mind which choice has been done. For instance, if the Fermi temperature T_F is defined as $k_B T_F = E_F$, it is evident that the zero of energy has been set at the bottom of the conduction band, i.e. $E_c = 0$; otherwise one should define $k_B T_F = E_F - E_c$.]

Consider now an electron gas with N free electrons in a volume V, and electron density $n = N/V$. To specify the electron density of a metal it is customary to consider the dimensionless parameter r_s connected to n by the relation

$$\frac{4}{3}\pi r_s^3 a_B^3 = \frac{1}{n} = \frac{V}{N}, \tag{3.2}$$

where $a_B = 0.529$ Å is the Bohr radius; $r_s a_B$ represents the radius of the sphere that contains in average one electron.

We evaluate r_s for alkali metals (for instance); Li, Na, K, Rb have bcc structure with cube edge a equal to 3.49 Å, 4.23 Å, 5.23 Å, and 5.59 Å, respectively. In the volume a^3 we have two atoms and two conduction electrons. From Eq. (3.2) we obtain

$$\frac{4}{3}\pi r_s^3 a_B^3 = \frac{a^3}{2} \implies r_s = \frac{a}{a_B}\frac{1}{2}\left(\frac{3}{\pi}\right)^{1/3} = 0.492\frac{a}{a_B};$$

the values for r_s are 3.25, 3.94, 4.87, and 5.20 for Li, Na, K, and Rb, respectively. A similar reasoning can be done for other crystals. For instance, the crystal structure of Al is fcc, with lattice constant $a = 4.05$ Å. In the volume a^3 we have four atoms and twelve conduction electrons; from Eq. (3.2) we have $(4/3)\pi r_s^3 a_B^3 = a^3/12$ and $r_s = 2.07$. Metallic densities of conduction electrons occur mostly in the range $2 < r_s < 6$.

The ground state of the electron system at $T = 0$ is obtained accommodating the N available electrons into the N lowest available energy levels up to the energy E_F, called the *Fermi energy*; each state of wavevector $\mathbf{k}$ and energy $E(\mathbf{k}) < E_F$ accommodates two electrons of either spin. In the $\mathbf{k}$-space, the *Fermi surface* $E(\mathbf{k}) = E_F$ separates the occupied band states of energy $E(\mathbf{k}) < E_F$ from the unoccupied band states of energy $E(\mathbf{k}) > E_F$. For the free-electron system at $T = 0$, the occupied states fill a Fermi sphere of radius k_F, as schematically indicated in Figure 3.2.

To determine k_F, we require that the total number of electrons within the Fermi sphere of wavevector k_F equals the number N of electrons available in the crystal; we must have

$$N = \sum_{\mathbf{k}}^{k<k_F} 2 = 2\frac{V}{(2\pi)^3}\frac{4}{3}\pi k_F^3 = \frac{V}{3\pi^2}k_F^3 \implies k_F^3 = 3\pi^2 n, \tag{3.3}$$

where use has been made of the standard prescription of Eq. (2.43) for converting the sum over $\mathbf{k}$ into the corresponding integral times $V/(2\pi)^3$. From Eqs. (3.2) and (3.3),

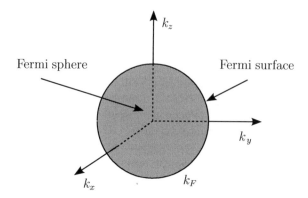

Figure 3.2 Schematic representation in the **k**-space of the Fermi sphere of occupied states and of the Fermi surface of the free-electron gas. At $T = 0$ each state of wavevector **k**, with $k < k_F$, is occupied by two electrons of either spin.

the Fermi wavevector becomes

$$k_F = \left(\frac{9\pi}{4}\right)^{1/3} \frac{1}{r_s a_B} = \frac{1.92}{r_s a_B}. \tag{3.4}$$

The *Fermi velocity* is given by

$$v_F = \frac{\hbar}{m} k_F. \tag{3.5a}$$

It is convenient to consider the dimensionless ratio v_F/c, where c is the velocity of the light. We divide both members of Eq. (3.5a) by c, and remember that the Bohr radius is $a_B = \hbar^2/me^2$ and that the fine structure constant is $1/\alpha \equiv \hbar c/e^2 = 137.036$; we obtain

$$\frac{v_F}{c} = \frac{\hbar}{mc} k_F = \alpha a_B k_F = \frac{1}{137.036} \frac{1.92}{r_s}. \tag{3.5b}$$

In the ordinary range of metallic densities, the Fermi velocity is about one hundredth of the velocity of light.

The *Fermi energy* at $T = 0$ is given by

$$E_F = \frac{\hbar^2 k_F^2}{2m} = \frac{\hbar^2}{2ma_B^2} \frac{1.92^2}{r_s^2}.$$

The Fermi energy can be expressed as

$$E_F = \frac{3.683}{r_s^2} \text{ in Rydberg} = \frac{\hbar^2}{2ma_B^2} = 13.606 \text{ eV}. \tag{3.6}$$

Typical values of E_F for metals are in the range 1–10 eV. Since 1 eV$/k_B = 11605$ K, the *Fermi temperature* T_F, defined as $k_B T_F = E_F$, is of the order of 10^4–10^5 K.

The ground-state energy E_0 at $T = 0$ of the free-electron gas is given by

$$E_0 = 2 \sum_{\mathbf{k}}^{k<k_F} \frac{\hbar^2 k^2}{2m}.$$

Converting as usual the sum over $\mathbf{k}$ into an integral times $V/(2\pi)^3$, one has

$$E_0 = 2\frac{V}{(2\pi)^3} \int_0^{k_F} 4\pi k^2 \frac{\hbar^2 k^2}{2m} dk = \frac{\hbar^2 k_F^2}{2m} \frac{V}{5\pi^2} k_F^3.$$

Using Eq. (3.3), the above expression becomes

$$\frac{E_0}{N} = \frac{3}{5} E_F = \frac{2.21}{r_s^2} \quad \text{(in Rydberg)}. \tag{3.7}$$

The average energy of each electron at zero temperature is $(3/5)E_F$.

We can here anticipate that the ground-state energy, given by Eq. (3.7), is only the leading term in the high density limit ($r_s \ll 1$) of the exact expression of the ground-state energy of the homogeneous electron gas (see for further aspects and discussion Sections 4.7 and 4.8). In spite of this and other limitations, the Sommerfeld picture is nevertheless of help to describe indicatively and preliminarily a number of properties of simple metals.

From the parabolic dispersion relation of the free-electron gas, we can pass to the density-of-states (see also Section 2.7):

$$D(E) = 2\frac{V}{(2\pi)^3} \int \delta\left(E - \frac{\hbar^2 k^2}{2m}\right) d\mathbf{k} = \frac{V}{4\pi^3} \int_0^\infty 4\pi k^2 \delta\left(E - \frac{\hbar^2 k^2}{2m}\right) dk.$$

The change of variable $\hbar^2 k^2/2m = x$ gives

$$D(E) = \frac{V}{2\pi^2} \int_0^\infty \left(\frac{2m}{\hbar^2}\right)^{3/2} x^{1/2} \delta(E - x)\, dx = \frac{V}{2\pi^2}\left(\frac{2m}{\hbar^2}\right)^{3/2} E^{1/2}, \quad E > 0. \tag{3.8a}$$

Since the integrated density-of-states up to the Fermi energy provides the total number of electrons of the sample, the density-of-states Eq. (3.8a) can be written in the form

$$D(E) = \frac{3}{2}\frac{N}{E_F}\left(\frac{E}{E_F}\right)^{1/2}, \quad E > 0. \tag{3.8b}$$

In particular for $E = E_F$ it holds

$$D(E_F) = \frac{3}{2}\frac{N}{E_F}. \tag{3.9}$$

The plot of $D(E)$ is shown in Figure 3.3.

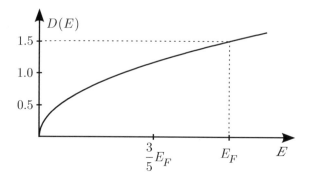

Figure 3.3 Density-of-states, in units of N/E_F, for a free-electron gas; the average electron energy $(3/5)\,E_F$ is also indicated.

3.2 Fermi-Dirac Distribution Function and Chemical Potential

Fermi-Dirac Distribution Function

Consider an assembly of identical particles at thermal equilibrium moving independently in a given volume. If the particles obey the Pauli exclusion principle, the occupation probability of a one-particle quantum state of energy E is given by the Fermi-Dirac function

$$f(E, \mu, T) = \frac{1}{e^{(E-\mu)/k_B T} + 1}, \tag{3.10}$$

where μ is the chemical potential, k_B is the Boltzmann constant, and T is the absolute temperature (see Appendix B1). The chemical potential μ for the Fermi-Dirac distribution is also addressed as the Fermi energy or Fermi level E_F; following a common use, the terms "chemical potential" and "Fermi energy," as well as the symbols μ and E_F, are used interchangeably.

The behavior of the Fermi-Dirac function is shown in Figure 3.4. If $\mu - E \gg k_B T$ the distribution function approaches unity; if $E - \mu \gg k_B T$, $f(E)$ falls exponentially to zero with a Boltzmann tail. At $T = 0$ the Fermi-Dirac distribution function becomes the step function $\Theta(\mu_0 - E)$, where μ_0 denotes the Fermi energy at $T = 0$; at zero temperature, all the states with energy lower than μ_0 are occupied and all the states with energy higher than μ_0 are empty. At finite temperature T, $f(E)$ deviates from the step function only in the thermal energy range of order $k_B T$ around $\mu(T)$.

We consider now the properties of the function $(-\partial f/\partial E)$. At $T = 0$ we have $(-\partial f/\partial E) = \delta(E - \mu_0)$. At finite temperature $T \ll T_F$, it holds approximately that $(-\partial f/\partial E) \approx \delta(E - \mu)$ (see Figure 3.4). In fact, at finite temperature the function $(-\partial f/\partial E)$ is very steep with its maximum at $E = \mu$ and differs significantly from zero in the energy range of the order of $k_B T$ around μ. It is easy to verify that $(-\partial f/\partial E)$ is

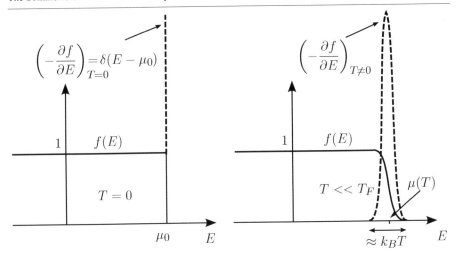

Figure 3.4 The Fermi-Dirac distribution function $f(E)$ and energy derivative $(-\partial f/\partial E)$ at $T = 0$ and at a finite temperature T, with $T \ll T_F$; μ_0 denotes the chemical potential $\mu(T)$ at $T = 0$.

an even function of E around μ and vanishes exponentially for $|E - \mu| \gg k_B T$. We also have

$$\int_{-\infty}^{+\infty} \left(-\frac{\partial f}{\partial E} \right) dE = -[f(\infty) - f(-\infty)] = 1, \tag{3.11}$$

which shows that the total area below the curve $(-\partial f/\partial E)$ is always normalized to unity.

Sommerfeld Expansion

Consider an integral of the type

$$I = \int_{-\infty}^{+\infty} G(E) \left(-\frac{\partial f}{\partial E} \right) dE, \tag{3.12}$$

where $G(E)$ is a function regular and (infinitely) differentiable around μ. We can demonstrate the following relation:

$$\boxed{\begin{aligned} \int_{-\infty}^{+\infty} G(E) \left(-\frac{\partial f}{\partial E} \right) dE = {}& G(\mu) + \frac{\pi^2}{6} (k_B T)^2 G''(\mu) \\ & + \frac{7\pi^4}{360} (k_B T)^4 G''''(\mu) + \cdots \end{aligned}} \tag{3.13}$$

where energy differentiation order is indicated by the corresponding number of apices.

To prove Eq. (3.13), consider the series expansion of $G(E)$ in powers of E around the chemical potential

$$G(E) = G(\mu) + (E - \mu)\left(\frac{dG}{dE}\right)_{E=\mu} + \frac{1}{2}(E - \mu)^2 \left(\frac{d^2G}{dE^2}\right)_{E=\mu} + \cdots$$

Inserting this expression into Eq. (3.12) one obtains

$$\int_{-\infty}^{+\infty} G(E)\left(-\frac{\partial f}{\partial E}\right) dE = G(\mu) + \frac{1}{2}G''(\mu)\int_{-\infty}^{+\infty} (E - \mu)^2 \left(\frac{\partial f}{\partial E}\right) dE + \cdots$$

Odd powers of $(E - \mu)$ in the expansion have been omitted because they give zero contribution.

The coefficient of $G''(\mu)$ in the above expression can be elaborated as follows:

$$\frac{1}{2}\int_{-\infty}^{+\infty} (E - \mu)^2 \left(-\frac{\partial f}{\partial E}\right) dE$$

$$= \int_{\mu}^{+\infty} (E - \mu)^2 \left(-\frac{\partial f}{\partial E}\right) dE \quad \text{[integrating by parts]}$$

$$= 2\int_{\mu}^{+\infty} (E - \mu) f(E)\, dE$$

$$= 2\int_{\mu}^{+\infty} \frac{E - \mu}{e^{(E-\mu)/k_B T} + 1}\, dE \quad \left[\text{set } \frac{E - \mu}{k_B T} = x\right]$$

$$= 2(k_B T)^2 \int_0^{+\infty} \frac{x}{e^x + 1}\, dx = \frac{\pi^2}{6}k_B^2 T^2.$$

In the last passage, use has been made of the definite integral

$$\int_0^{+\infty} \frac{x}{e^x + 1}\, dx = \frac{1}{2}\Gamma(2)\,\zeta(2) = \frac{\pi^2}{12} \quad \text{with} \quad \Gamma(2) = 1 \quad \text{and} \quad \zeta(2) = \frac{\pi^2}{6}$$

[see I. S. Gradshteyn and I. M. Ryzhik, Table of Integrals, Series, and Products, Academic Press, New York, 1980, p. 325]. The above elaboration explains the numerical coefficient in front of $G''(\mu)$ in expression (3.13). Successive terms in expansion (3.13) can be calculated in a similar way.

Equation (3.13) is known as the *Sommerfeld expansion*. To appreciate the fact that it is a rapidly convergent expansion for $T \ll T_F$, consider the case in which $G(E)$ has a power dependence on E of the type $G(E) \approx E^p$; then it is easy to verify that the terms of the expansion (3.13) are of the type of $(k_B T/\mu)^2$, $(k_B T/\mu)^4$, or equivalently $(T/T_F)^2$, $(T/T_F)^4$, etc.

The Sommerfeld expansion (3.13) can be rewritten also in a slightly different form. Most often the function $G(E)$ vanishes for energies E below some threshold energy and, furthermore, $G(E)$ behaves reasonably well at infinity so that $G(E)f(E) \to 0$ for $E \to \infty$. When these conditions are verified, it is convenient to perform an integration by parts in the first member of Eq. (3.13); one obtains

$$\int_{-\infty}^{+\infty} \frac{dG(E)}{dE} f(E)\, dE \equiv G(\mu) + \frac{\pi^2}{6}(k_B T)^2 G''(\mu) + \cdots \qquad (3.14)$$

Let $\Gamma(E)$ denote the derivative $G'(E)$, or equivalently

$$G(E) = \int_{-\infty}^{E} \Gamma(E')\, dE'.$$

Equation (3.14) becomes

$$\boxed{\int_{-\infty}^{+\infty} \Gamma(E) f(E)\, dE = \int_{-\infty}^{\mu} \Gamma(E)\, dE + \frac{\pi^2}{6}(k_B T)^2 \left(\frac{d\Gamma}{dE}\right)_{E=\mu} + O(T^4)}, \qquad (3.15)$$

which is a useful expression for the given function $\Gamma(E)$.

Temperature Dependence of the Chemical Potential

The chemical potential μ depends (slightly) on the temperature; in the study of several transport properties, the temperature dependence of the chemical potential (whatever small it might appear at first sight) has quite important consequences, and now we work out explicitly the behavior of $\mu(T)$.

Let $D(E)$ be the (single-particle) density-of-states for both spin directions for the metallic sample of volume V; no particular assumption on $D(E)$ or on the conduction band energy $E(\mathbf{k})$ is done at this stage. Let N be the total number of conduction electrons of the sample. At any temperature T, the chemical potential $\mu(T)$ is determined (implicitly) enforcing the equality

$$N = \int_{-\infty}^{+\infty} D(E) f(E, \mu, T)\, dE;$$

the integration is automatically restricted to the energy interval where the density-of-states is different from zero.

The above integral has in general a rather complex structure. However if $T \ll T_F$ and if $D(E)$ is reasonably smooth at $E \approx \mu$ (both conditions are in practice quite well satisfied in ordinary situations), we can perform the integral using the Sommerfeld expansion (3.15), where the function $\Gamma(E)$ stands now for $D(E)$. We have

$$N = \int_{-\infty}^{\mu} D(E)\, dE + \frac{\pi^2}{6} k_B^2 T^2 D'(\mu) + O(T^4), \qquad (3.16)$$

where the prime indicates differentiation with respect to the energy, and terms of the order of $(T/T_F)^4$ can be safely omitted since $T \ll T_F$ at the ordinary temperatures of interest. We now differentiate both members of Eq. (3.16) with respect to the temperature, taking into account that $\mu = \mu(T)$. We obtain

$$0 = D(\mu)\frac{d\mu}{dT} + \frac{\pi^2}{3} k_B^2 T D'(\mu) \qquad (3.17)$$

[we have neglected the term $(\pi^2/6)k_B^2 T^2 D''(\mu)(d\mu/dT)$ because of order $(T/T_F)^2$ with respect to $D(\mu)(d\mu/dT)$]. From Eq. (3.17) it follows:

$$\frac{d\mu}{dT} = -\frac{\pi^2}{3}k_B^2 T \frac{D'(\mu)}{D(\mu)}. \tag{3.18}$$

This equation, together with the obvious condition $\mu(T) = \mu_0$ for $T = 0$, allows us to obtain $\mu(T)$ at any temperature T, provided $T \ll T_F$. We notice that:

(i) If the density-of-states $D(E)$ for $E \approx \mu_0$ increases with energy, then the chemical potential μ decreases with temperature (and vice versa). In the case $D(E)$ is constant for $E \approx \mu_0$, the chemical potential μ is temperature independent.

(ii) For a free-electron gas $D(E) \propto E^{1/2}$. Then $D'(\mu)/D(\mu) = 1/2\mu$. The integration of Eq. (3.18) is straightforward and gives $\mu^2 - \mu_0^2 = -(\pi^2/6)k_B^2 T^2$. Then

$$\mu(T) = \mu_0 \left[1 - \frac{\pi^2}{12}\left(\frac{k_B T}{\mu_0}\right)^2 \right] = \mu_0 \left[1 - \frac{\pi^2}{12}\left(\frac{T}{T_F}\right)^2 \right]. \tag{3.19}$$

This relation gives μ as a function of T and shows that $\mu(T)$ decreases slowly with increasing temperature for the free-electron gas.

3.3 Electronic Specific Heat in Metals and Thermodynamic Functions

The *heat capacity at constant volume* of a sample is defined by

$$C_V = \left(\frac{\delta Q}{dT}\right)_V, \tag{3.20}$$

where δQ is the amount of heat transferred from the external world to the system, and dT is the corresponding change in temperature of the system, kept at constant volume V. The heat capacity is an extensive quantity, i.e. a quantity proportional to the volume of the sample; for this reason, depending on the nature of the system under investigation, it may become preferable to introduce the *specific heat* per mole, or the specific heat per unit volume, or per unit cell, or per composing atoms or electrons.

We can express the heat capacity (3.20) in convenient alternative forms. From the first law of thermodynamics, we know that the change dU of the internal energy of a system, in a transformation in which an infinitesimal quantity of heat δQ is received by the system and δL is the work done on the system by external forces, is given by

$$dU = \delta Q + \delta L.$$

In the case the external forces are only mechanical forces, exerting a pressure p on the system, we have $\delta L = -p\, dV$; if the volume V of the system is kept constant during the transformation then $\delta L = 0$; it follows $dU = \delta Q$ and

$$C_V = \left(\frac{dU}{dT}\right)_V. \tag{3.21a}$$

In the case of reversible transformations, the second law of thermodynamics states that $\delta Q = T \, dS$, where S is the entropy of the system; Eq. (3.20) gives

$$C_V = T \left(\frac{dS}{dT} \right)_V. \tag{3.21b}$$

The internal energy of the Fermi gas is

$$U(T) = \int_{-\infty}^{+\infty} E \, D(E) f(E, \mu, T) \, dE$$

$$= \int_{-\infty}^{\mu} E \, D(E) \, dE + \frac{\pi^2}{6} k_B^2 T^2 [D(\mu) + \mu D'(\mu)] + O(T^4), \tag{3.22}$$

where the Sommerfeld expansion (3.15) has been used. By differentiating with respect to the temperature both members of Eq. (3.22), and keeping only the most relevant among the terms containing $(d\mu/dT)$, we have

$$C_V = \left(\frac{dU}{dT} \right)_V = \mu D(\mu) \frac{d\mu}{dT} + \frac{\pi^2}{3} k_B^2 T [D(\mu) + \mu D'(\mu)].$$

Using Eq. (3.17), and replacing $D(\mu)$ with $D(\mu_0)$, we obtain

$$\boxed{C_V(T) = \frac{\pi^2}{3} k_B^2 T D(\mu_0)}. \tag{3.23}$$

From Eq. (3.21b), it can be noticed that expression (3.23) represents also the entropy of the electron system (see Figure 3.5).

The specific heat per unit volume $c_V = C_V/V$ becomes

$$c_V(T) = \frac{\pi^2}{3} k_B^2 T \frac{D(\mu_0)}{V} = \gamma T, \tag{3.24a}$$

where

$$\gamma = \frac{\pi^2}{3} k_B^2 \frac{D(\mu_0)}{V}. \tag{3.24b}$$

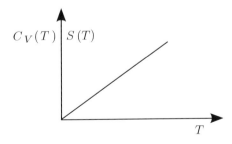

Figure 3.5 Electronic contribution $C_V(T)$ to the heat capacity of a metal at constant volume, as a function of temperature; the same expression holds for the electronic contribution to the entropy.

The contribution to the specific heat from conduction electrons is proportional to T for all temperatures of interest, since T is always much less than T_F. The electronic contribution becomes the leading one at sufficiently low temperatures, where it prevails over the T^3 Debye contribution originated by the lattice vibrations (see Section 9.5). We also notice that the knowledge of the density-of-states $D(\mu_0)$ at the Fermi energy μ_0 completely determines C_V; this gives a first insight into the importance of the electronic states at or near the Fermi surface in metals. In discussing transport phenomena, impurity screening, etc., we will further see the basic role played by the electronic states lying in the thermal interval of the order $k_B T$ around the Fermi energy.

It is interesting to specify Eq. (3.23) for the case of the free-electron gas, where $D(\mu_0) = (3/2)N/\mu_0$ according to Eq. (3.9). We have

$$C_V(T) = \frac{\pi^2}{2} k_B N \frac{T}{T_F}. \tag{3.25}$$

For comparison, we recall that the classical statistical mechanics would give the expression $C_V = (3/2)k_B N$ for the heat capacity of a gas of N noninteracting particles. The correct result (3.25) can be interpreted noticing that only the electrons in the thermal interval $k_B T$ around the Fermi energy $\mu_0 = k_B T_F$ can vary their energy, and thus the effective number of electrons with "classical behavior" is not N but rather the fraction T/T_F of N.

3.4 Thermionic Emission from Metals

We can apply the Fermi-Dirac statistics to study under very simplified conditions the thermionic emission from metals, i.e. the emission of electrons from a metal in the vacuum because of the effect of a finite temperature. For our semi-quantitative considerations, we do not consider in detail possible reflection of electrons impinging at the surface; we simply assume that all the electrons that arrive at the surface with an energy sufficient to overcome the surface barrier are transferred to the vacuum (and swept away by some small applied electric field without accumulation of space charge). The model electronic structure of the metal, with electron affinity χ and work function W, is illustrated in Figure 3.1; we wish to obtain the number of electrons which escape from the metal, kept at temperature T.

In the metal, the electrons are distributed in energy according to the Fermi-Dirac statistics. Let us indicate with z the direction normal to the surface; the current density of escaping electrons is given by

$$J_s = (-e) \int_{\sqrt{2m\chi/\hbar^2}}^{+\infty} dk_z \int_{-\infty}^{+\infty} dk_x \int_{-\infty}^{+\infty} dk_y \frac{2}{(2\pi)^3} \frac{1}{e^{[E(\mathbf{k})-\mu]/k_B T} + 1} v_z, \tag{3.26}$$

where $v_z = \hbar k_z/m$. Notice that expression (3.26), in the case v_z is just replaced by a drift velocity v independent of $\mathbf{k}$ (and the integration over k_z extends from $-\infty$ to $+\infty$), would give the standard current density $J_s = (-e)nv$. The limitation in Eq. (3.26) for the integration on k_z is just to make sure that the escaping electrons have enough kinetic energy $\hbar^2 k_z^2 / 2m \geq \chi$ in the z-direction to leave the metal.

In order to perform the integrals in equation (3.26), notice that

$$E(\mathbf{k}) - \mu = \frac{\hbar^2 k_x^2}{2m} + \frac{\hbar^2 k_y^2}{2m} + \frac{\hbar^2 k_z^2}{2m} - \mu \geq \chi - \mu \equiv W;$$

since in general the work function $W \gg k_B T$, we can safely neglect the unity in the Fermi distribution function in Eq. (3.26). Then

$$J_s = \frac{(-e)}{4\pi^3} \frac{\hbar}{m} \int_{\sqrt{2m\chi/\hbar^2}}^{+\infty} k_z \, dk_z \int_{-\infty}^{+\infty} dk_x \int_{-\infty}^{+\infty} dk_y \exp\left[\frac{\mu}{k_B T} - \frac{\hbar^2 (k_x^2 + k_y^2 + k_z^2)}{2m k_B T} \right].$$

We now use the result for Gaussian functions

$$\int_{-\infty}^{+\infty} dk_x \exp\left[-\frac{\hbar^2 k_x^2}{2m k_B T} \right] = \frac{\sqrt{2\pi m k_B T}}{\hbar}$$

and also

$$\int_{\sqrt{2m\chi/\hbar^2}}^{+\infty} k_z \, dk_z \exp\left[-\frac{\hbar^2 k_z^2}{2m k_B T} \right] = \frac{m k_B T}{\hbar^2} e^{-\chi/k_B T}.$$

The current density of escaping electrons is thus described by the Richardson expression

$$J_s = \frac{(-e)m k_B^2}{2\pi^2 \hbar^3} T^2 e^{-W/k_B T}. \tag{3.27}$$

The absolute value of the numerical factor in the Richardson law is

$$\frac{e m k_B^2}{2\pi^2 \hbar^3} = 120.4 \text{ A cm}^{-2} \text{ K}^{-2}; \tag{3.28}$$

in many metals, measured values are indeed in the range 50–120 A cm^{-2} K^{-2}.

A quantitative treatment of the electron emission from metals requires a number of refinements of the simplified model here considered. In the model of Figure 3.1, the metal-vacuum boundary is represented by an abrupt discontinuity in the potential. In reality an electron outside the metal feels an attractive image potential (to be determined in principle quantum mechanically) with consequences on the reflection coefficient of escaping electrons. The image potential also leads to a reduction of the apparent work function in the presence of applied electric fields (Schottky effect). The work function is also sensitive to various effects, for instance, surface impurities and charge modifications at the surface. Observed values of the work function for some metals are reported in Table 3.1; for a more complete list see for instance H. B. Michaelson, "Handbook of Chemistry and Physics" edited by R. C. Weast (CRC Press, Cleveland, 1962); D. E. Eastman, Phys. Rev. B **2**, 1 (1970) and references quoted therein.

Table 3.1 Experimental values of the work function W for some metals.

Metal	W (eV)	Metal	W (eV)
Li	2.49	Al	4.2
Na	2.28	Ga	4.1
K	2.24	Sn	4.4
Cs	1.81	Pb	4.0
Ag	4.3	Pt	5.6
Au	5.2	W	4.5

Appendix A. Outline of Statistical Physics and Thermodynamic Relations

A.1 Microcanonical Ensemble and Thermodynamic Quantities

The basic principles of statistical physics can be found in standard textbooks; the purpose of this Appendix is simply confined to summarize the "recipes" that link the microscopic properties of a system with its macroscopic thermodynamic quantities.

Consider a physical system composed by N identical particles confined within a volume V. Quantum mechanics provides for this confined system discretized energy levels E_m, that we label with an integer number m in increasing energy order. In the context of statistical mechanics, the distinct quantum eigenstates of the system are often referred to as *microstates* or also *accessible states* of the system.

For macroscopic systems with a large number of particles, the spacings of the different levels become extremely small (if not zero), and the total energy E of the system can be regarded as an almost continuous variable. Instead of focusing on each individual microstate, it is convenient to consider the quantity $W(E, V, N)$ which gives the total number of distinct microstates lying in the interval $(E - \Delta/2, E + \Delta/2)$, where $\Delta \ll E$.

Let (E, V, N) denote the parameters that define a particular macrostate of the given system. In general, for a given macrostate (E, V, N), there is a large number $W(E, V, N)$ of corresponding microstates lying in a small energy interval Δ around E; according to the standard concepts of the *microcanonical ensemble*, it is assumed that the macrostate (E, V, N) is equally likely to be in any of the distinct microstates $W(E, V, N)$.

The bridge between the number of microstates of a system and its thermodynamic functions is given by the celebrated Boltzmann expression for the entropy

$$\boxed{S = k_B \ln W(E, V, N)}.$$
(A.1)

From the basic expression (A.1) of the entropy, all the other thermodynamics macroscopic quantities can be obtained, with the following procedure.

Consider quasistatic transformations with (infinitesimally) slow variations of the macroscopic parameters E, V, N so as to guarantee that the transformations involve only thermodynamic equilibrium states. From Eq. (A.1), the entropy change for an infinitesimal transformation can be written as

$$dS = \left(\frac{\partial S}{\partial E}\right)_{V,N} dE + \left(\frac{\partial S}{\partial V}\right)_{E,N} dV + \left(\frac{\partial S}{\partial N}\right)_{E,V} dN. \tag{A.2}$$

From the first and second principles of thermodynamics, we can also write for any infinitesimal reversible transformation

$$dE = T\,dS - p\,dV + \mu\,dN \implies dS = \frac{1}{T}dE + \frac{p}{T}dV - \frac{\mu}{T}dN. \tag{A.3}$$

Comparison of Eqs. (A.2) and (A.3) provides the three basic relations:

$$\frac{1}{T} = +\left(\frac{\partial S}{\partial E}\right)_{V,N}, \tag{A.4}$$

$$\frac{p}{T} = +\left(\frac{\partial S}{\partial V}\right)_{E,N}, \tag{A.5}$$

$$\frac{\mu}{T} = -\left(\frac{\partial S}{\partial N}\right)_{E,V}. \tag{A.6}$$

Relation (A.4) provides the *absolute temperature T*. Relation (A.5) provides the *equation of state*, i.e. a relation among p, V, T. Relation (A.6) provides the *chemical potential* μ.

Formulae (A.4–A.6) follow from the basic expression (A.1) of entropy and from the basic laws of thermodynamics; then any other desired thermodynamic function can be obtained. For instance, the *Helmholtz free energy* is given by $F = E - TS$; the *Gibbs free energy* is $G = E - TS + pV$; the *enthalpy* is given by $H = E + pV = G + TS$.

A.2 Canonical Ensemble and Thermodynamic Quantities

Consider a physical system containing N identical particles confined within the volume V; as before we label with an integer number m all the distinct eigenstates of the system in increasing order of energy E_m.

We are now interested in the statistical description of the system in equilibrium with a surrounding thermal bath at temperature T, and thus free to exchange energy with it. The equilibrium properties of the system are calculated by averaging over all the accessible states of energy E_m, assigning to each state a weight P_m proportional to the Boltzmann factor $\exp(-E_m/k_B T)$. By requiring the normalization of the weights to unity, we obtain the *Gibbs or canonical distribution probability*

$$P_m = \frac{e^{-\beta E_m}}{\sum_m e^{-\beta E_m}} \quad \text{with} \quad \beta = \frac{1}{k_B T} \quad \text{and} \quad \sum_m P_m = 1. \tag{A.7}$$

The sum over all microscopic states of the system, appearing in the denominator of Eq. (A.7), is known as *canonical partition function*:

$$\boxed{Z(T, V, N) = \sum_m e^{-\beta E_m}}. \tag{A.8}$$

The knowledge of the canonical partition function is the passkey to the knowledge of any other thermodynamic function of the system, as apparent from the procedure summarized below.

Besides the canonical partition function, it is customary to introduce the closely related *canonical potential* $\Omega = \Omega(T, V, N)$ so defined

$$e^{-\beta\Omega} = \sum_m e^{-\beta E_m} \equiv Z(T, V, N) \qquad \Longrightarrow \qquad \Omega = -k_B T \ln Z. \qquad (A.9)$$

The probability distribution (A.7) then takes the form

$$P_m = \frac{e^{-\beta E_m}}{Z} = e^{-\beta(E_m - \Omega)}. \qquad (A.10)$$

We now show that the Helmholtz free energy F is just given by the canonical thermodynamic potential Ω defined in Eq. (A.9). All the other thermodynamic functions can then be readily worked out.

Consider an infinitesimal reversible transformation in which the temperature T is changed (while V and N are kept constant). From the first and second law of thermodynamics, for reversible transformations with constant volume and constant number of particles, we have

$$dU = T\,dS - p\,dV + \mu\,dN \qquad \Longrightarrow \qquad dS = \frac{1}{T}\,dU \quad \text{for } N \text{ and } V \text{ constant.} \quad (A.11)$$

Consider now the differential of Eq. (A.9) when the parameter T is changed (while V and N are kept constant). We have

$$e^{-\beta\Omega}\,d(\beta\Omega) = \sum_m e^{-\beta E_m} E_m\,d\beta \Longrightarrow d(\beta\Omega) = \sum_m e^{-\beta(E_m - \Omega)} E_m\,d\beta = U\,d\beta,$$

$$(A.12)$$

where U is the internal energy of the system in the canonical ensemble. The last equation, with the replacement $\beta = 1/k_B T$, can be rewritten as

$$d\frac{\Omega}{T} = U\,d\frac{1}{T} \equiv d\frac{U}{T} - \frac{1}{T}\,dU \qquad \Longrightarrow \qquad \frac{1}{T}\,dU = d\frac{U - \Omega}{T};$$

by virtue of Eq. (A.11) it follows:

$$dS = d\,\frac{U - \Omega}{T}.$$

Integrating the above expression from zero temperature to an arbitrary temperature T, and using the third principle of thermodynamics to set to zero the arbitrary integration constant, we have $S = (U - \Omega)/T$ and $\Omega = U - TS \equiv F$. This shows that the Helmholtz free energy equals the canonical potential (A.9) and is related to the partition function by the expression

$$\boxed{F \equiv \Omega = -k_B T \ln Z(T, V, N) = -k_B T \ln \sum_m e^{-\beta E_m}.} \qquad (A.13)$$

Other thermodynamic relationships can now be readily established. By differentiating the free energy, we have

$$F = U - TS \implies dF = dU - T\, dS - S\, dT.$$

On the other hand, from the first and second principles of thermodynamics, we can also write for any infinitesimal reversible transformation

$$dU = T\, dS - p\, dV + \mu\, dN.$$

Hence

$$dF = -S\, dT - p\, dV + \mu\, dN$$

and we obtain the relations:

$$S = -\left(\frac{\partial F}{\partial T}\right)_{V,N}, \tag{A.14}$$

$$p = -\left(\frac{\partial F}{\partial V}\right)_{T,N}, \tag{A.15}$$

$$\mu = +\left(\frac{\partial F}{\partial N}\right)_{T,V}. \tag{A.16}$$

The physical contents of the above relations become more transparent with some exemplification.

Consider, for instance, expression (A.14) for the entropy together with the expression (A.13) of the free energy; one obtains

$$S = -\left(\frac{\partial F}{\partial T}\right)_{V,N} = \frac{\partial}{\partial T}\left[k_B T \ln Z\right] = k_B \ln Z + k_B T \frac{1}{Z}\frac{\partial Z}{\partial T}$$

$$= k_B \ln Z + k_B \frac{1}{Z}\sum_m e^{-\beta E_m}\beta E_m = -k_B \ln\frac{1}{Z} - k_B\frac{1}{Z}\sum_m e^{-\beta E_m}\ln e^{-\beta E_m}.$$

Using Eq. (A.10) the above expression can be cast in the form

$$S = -k_B \sum_m P_m \ln P_m; \tag{A.17}$$

the Gibbs expression (A.17) shows that the entropy is determined by the probability values P_m.

Similarly the expression (A.15) for the pressure can be elaborated in the form

$$p = k_B T \left(\frac{\partial \ln Z}{\partial V}\right)_{T,N} = k_B T \frac{1}{Z}\frac{\partial}{\partial V}\sum_m e^{-\beta E_m} = -\sum_m \left(\frac{\partial E_m}{\partial V}\right)_{T,N} P_m. \tag{A.18}$$

The quantity $-\partial E_m/\partial V$ can be identified as the pressure acting on the physical system in the pure quantum mechanical state of energy E_m. Then the weighted average in Eq. (A.18) can be identified with the pressure acting on the physical system in thermodynamic equilibrium with a thermal bath of temperature T. In particular, at zero temperature we have

$$p = -\frac{\partial E_0}{\partial V};$$
(A.19)

at $T = 0$, the pressure is just the derivative of the ground-state energy of the system with respect to the volume.

A.3 Grand Canonical Ensemble and Thermodynamic Quantities

Consider a physical system containing a variable number N of identical particles confined within the volume V. We are now interested in the description of the system in thermodynamic equilibrium with a thermal reservoir at temperature T and with a particle reservoir with chemical potential μ. We label with the integer numbers (m, N) all the distinct eigenstates of the system with N particles and energy $E_{m,N}$. The equilibrium properties of the system are obtained by averaging over all its accessible states of energy $E_{m,N}$ assigning to each state a probability proportional to

$$P_{m,N} \propto e^{-\beta(E_{m,N}-\mu N)}.$$

With normalization of the weights to unity, we obtain the *grand canonical distribution probability*

$$P_{m,N} = \frac{e^{-\beta(E_{m,N}-\mu N)}}{\sum_{m,N} e^{-\beta(E_{m,N}-\mu N)}}.$$
(A.20)

The denominator in (A.20) is known as the *grand canonical partition function*

$$\boxed{Z_G(T, V, \mu) = \sum_{m,N} e^{-\beta(E_{m,N}-\mu N)}}.$$
(A.21)

In close analogy with the findings of the canonical ensemble, it can be shown that *the knowledge of the grand canonical partition function provides the passkey for the knowledge of any other thermodynamic function of the system.*

Besides the grand canonical partition function, it is customary to introduce the closely related *grand canonical potential* $\Omega_G = \Omega_G(T, V, \mu)$ so defined

$$e^{-\beta\Omega_G} = \sum_{m,N} e^{-\beta(E_m-\mu N)} \equiv Z_G(T, V, \mu) \implies \Omega_G = -k_B T \ln Z_G.$$
(A.22)

Following "mutatis mutandis" the reasoning done in Appendix A2, we can readily establish that the grand canonical potential equals

$$\boxed{\Omega_G = U - TS - \overline{N}\mu = -k_B T \ln Z_G(T, V, \mu)},$$
(A.23)

where U is the mean internal energy and $\overline{N}$ is the mean particle number in the grand canonical distribution. Relation (A.23) is the basic result of the grand canonical apparatus; all the other thermodynamic relationships follow from it.

From the first and second principles of thermodynamics we have

$$dU = T\,dS - p\,dV + \mu\,d\overline{N}.$$

By differentiating Ω_G and using the above expression, we obtain

$$d\Omega_G = dU - T\,dS - S\,dT - \overline{N}\,d\mu - \mu\,d\overline{N} = -S\,dT - p\,dV - \overline{N}\,d\mu.$$

It follows:

$$S = -\left(\frac{\partial \Omega_G}{\partial T}\right)_{V,\mu}, \tag{A.24}$$

$$p = -\left(\frac{\partial \Omega_G}{\partial V}\right)_{T,\mu}, \tag{A.25}$$

$$\overline{N} = -\left(\frac{\partial \Omega_G}{\partial \mu}\right)_{T,V}. \tag{A.26}$$

Relation (A.24) provides the entropy of the system. Relation (A.25) provides the equation of state. Relation (A.26) provides the mean number of particles. All other thermodynamic functions can be readily obtained from the equations so far established.

Appendix B. Fermi-Dirac and Bose-Einstein Statistics for Independent Particles

We consider a physical system composed by N identical particles confined within a volume V. The particles of the system are regarded as *noninteracting* (except for an arbitrary small interaction, to ensure thermodynamic equilibrium). We thus discuss the energy levels of the many-body system in terms of independent *one-particle states*.

Quantum mechanics provides for a single particle confined within the volume V discretized energy levels; we label with an integer number i all the distinct eigenstates, of energies ε_i, of the single-particle quantum problem. Since the particles are noninteracting, the total energy of the many-body system is the sum of the energies of the individual particles

$$E = \sum_i \varepsilon_i n_i, \tag{B.1}$$

where n_i denotes the number of particles with energy ε_i, and the total number of particles is given by

$$N = \sum_i n_i. \tag{B.2}$$

We now apply the general apparatus of the grand canonical ensemble in the case of noninteracting particles, taking into account the global symmetric or antisymmetric properties of the quantum states of identical particles.

B.1   Fermi-Dirac Statistics of Noninteracting Fermions

We now consider specifically a system of identical Fermi particles; as a consequence of the Pauli principle, the possible values of the occupation numbers n_i are either 0 or 1. The most general accessible state for the system of indistinguishable particles is defined by a set of numbers $\{n_i\}$ (i.e. any sequence of integer numbers equal to 0 or 1). The grand partition function (A.21) for a quantum system of noninteracting fermions becomes

$$Z_G(T, V, \mu) = \sum_{\{n_i\}} \exp\left[-\beta(\Sigma n_i \varepsilon_i - \mu \Sigma n_i)\right] \quad \text{with} \quad n_i = 0, 1$$

$$= \sum_{\{n_i\}} \exp\left[-\beta n_1(\varepsilon_1 - \mu) - \beta n_2(\varepsilon_2 - \mu) - \cdots\right]. \tag{B.3}$$

The sum over configurations in Eq. (B.3) can be carried out exactly and gives

$$Z_G(T, V, \mu) = \prod_i \left[1 + e^{-\beta(\varepsilon_i - \mu)}\right]. \tag{B.4}$$

The expression of the grand canonical potential is then

$$\Omega_G = -k_B T \ln Z_G = -k_B T \sum_i \ln\left[1 + e^{-\beta(\varepsilon_i - \mu)}\right]. \tag{B.5}$$

The grand canonical partition function (B.4), or the closely related grand canonical potential (B.5), provide the bridge to any desired thermodynamic quantity.

For instance, the average occupation number $f(\varepsilon_i)$ of a given state ε_i becomes

$$f(\varepsilon_i) = \langle n_i \rangle = \frac{1}{Z_G} e^{-\beta(\varepsilon_i - \mu)} \prod_{j(\neq i)} \left[1 + e^{-\beta(\varepsilon_j - \mu)}\right] = \frac{e^{-\beta(\varepsilon_i - \mu)}}{1 + e^{-\beta(\varepsilon_i - \mu)}};$$

the average occupancy of the one-particle states is thus described by the Fermi-Dirac distribution function

$$\boxed{f(\varepsilon_i) = \frac{1}{e^{\beta(\varepsilon_i - \mu)} + 1}}. \tag{B.6}$$

Using Eq. (A.23), the expression of the free energy of a system of noninteracting fermions reads

$$F = U - TS = \overline{N}\mu - k_B T \sum_i \ln\left[1 + e^{-\beta(\varepsilon_i - \mu)}\right]. \tag{B.7}$$

Similarly, starting from Eq. (A.24), the entropy of a system of noninteracting fermions becomes

$$S = -\left(\frac{\partial \Omega_G}{\partial T}\right)_{V,\mu} = \frac{\partial}{\partial T}\left\{k_B T \sum_i \ln\left[1 + e^{-\beta(\varepsilon_i - \mu)}\right]\right\}$$

$$= k_B \sum_i \ln\left[1 + e^{-\beta(\varepsilon_i - \mu)}\right] + k_B T \sum_i \frac{e^{-\beta(\varepsilon_i - \mu)}}{1 + e^{-\beta(\varepsilon_i - \mu)}} \frac{\varepsilon_i - \mu}{k_B T^2}.$$

We use the two identities

$$1 + e^{-\beta(\varepsilon_i - \mu)} \equiv \frac{1}{1 - f_i} \quad \text{and} \quad \frac{\varepsilon_i - \mu}{k_B T} \equiv \ln \frac{1 - f_i}{f_i},$$

where f_i denotes by brevity $f(\varepsilon_i)$. We obtain

$$S = k_B \sum_i \left[-\ln(1 - f_i) + f_i \ln \frac{1 - f_i}{f_i}\right],$$

and finally

$$\boxed{S = -k_B \sum_i \left[f_i \ln f_i + (1 - f_i) \ln(1 - f_i)\right]}, \tag{B.8}$$

which is the desired expression for the entropy of noninteracting fermions.

From Eq. (A.26) and the grand canonical potential (B.5) for fermions, we recover for the total number of particles the self-explanatory expression

$$\overline{N} = -\left(\frac{\partial \Omega_G}{\partial \mu}\right)_{T,V} = k_B T \sum_i \frac{\partial}{\partial \mu} \ln[1 + e^{-\beta(\varepsilon_i - \mu)}]$$

$$= \sum_i \frac{e^{-\beta(\varepsilon_i - \mu)}}{1 + e^{-\beta(\varepsilon_i - \mu)}} = \sum_i f_i.$$

Finally from the equation of state (A.25) and the grand canonical potential (B.5) we obtain

$$p = -\left(\frac{\partial \Omega_G}{\partial V}\right)_{T,\mu} = k_B T \sum_i \frac{e^{-\beta(\varepsilon_i - \mu)}}{1 + e^{-\beta(\varepsilon_i - \mu)}}(-\beta)\frac{\partial \varepsilon_i}{\partial V} = -\sum_i f_i \frac{\partial \varepsilon_i}{\partial V}. \tag{B.9}$$

The pressure of a system of independent fermions is given by the weighted average with the Fermi distribution function of the quantities $-\partial \varepsilon_i / \partial V$.

Consider now the particular case of a system of noninteracting particles, freely moving within a cubic box of volume $V = L^3$. The energies of the one-particle states are

$$\varepsilon = \frac{\hbar^2}{2m}\left(\frac{\pi}{L}\right)^2 \left(n_x^2 + n_y^2 + n_z^2\right), \quad n_x, n_y, n_z = 1, 2, 3, \ldots$$

The allowed energy levels are proportional to L^{-2}, or equivalently are proportional to $V^{-2/3}$. It follows

$$\frac{\partial \varepsilon}{\partial V} = -\frac{2}{3} \frac{\varepsilon}{V}.$$

Expression (B.9) thus becomes

$$p = \frac{2}{3} \frac{1}{V} \sum_i f_i \varepsilon_i \implies pV = \frac{2}{3} U, \tag{B.10}$$

which is the equation of state for the gas of independent fermions.

B.2 Bose-Einstein Statistics for Noninteracting Bosons

Consider now a system of identical noninteracting bosons confined within a volume V (for instance a dilute gas of boson atoms). We denote with ε_i the energies of the one-particle quantum states, and with $E_j = \sum n_i \varepsilon_i$ the energy levels of the whole sample; for bosons the occupancy numbers n_i can assume any integer number from zero to infinity. The grand partition function for the system of noninteracting bosons is given by expression

$$Z_G(T, V, \mu) = \sum_{\{n_i\}} \exp\left[-\beta n_1(\varepsilon_1 - \mu) - \beta n_2(\varepsilon_2 - \mu) - \cdots\right],$$

$$n_i = 0, 1, 2, \ldots \tag{B.11}$$

Notice that n_i can now take any integer value from zero to infinity, while in the case of fermions, described by Eq. (B.3), only the occupancy zero and one is possible.

The sum in Eq. (B.11) can be carried out exactly and gives for bosons

$$Z_G(T, V, \mu) = \prod_i \frac{1}{1 - e^{-\beta(\varepsilon_i - \mu)}} \quad \text{with} \quad \mu < \varepsilon_{\text{ground}} \tag{B.12}$$

with the implicit condition that the chemical potential is smaller than the energy of any quantum state of the system, and then it is smaller than the energy of the ground quantum state. The passage from Eq. (B.11) to Eq. (B.12) is done considering the sum of the geometric series

$$\sum_{n_i=0}^{\infty} e^{-\beta n_i(\varepsilon_i - \mu)} = \sum_{n_i=0}^{\infty} \left[e^{-\beta(\varepsilon_i - \mu)}\right]^{n_i} = \frac{1}{1 - e^{-\beta(\varepsilon_i - \mu)}};$$

the condition on μ in Eq. (B.12) obviously follows from the standard convergence condition of the geometrical series.

The grand canonical potential for a system of independent bosons is

$$\Omega_G = -k_B T \ln Z_G = k_B T \sum_i \ln\left[1 - e^{-\beta(\varepsilon_i - \mu)}\right]. \tag{B.13}$$

We can now obtain all the thermodynamic quantities of interest for a system of noninteracting bosons. Following "mutatis mutandis" the same procedure used in Appendix B1, it is found that the average occupancy of a given one-particle state is given by the Bose-Einstein distribution function

$$
f(\varepsilon_i) = \frac{1}{e^{\beta(\varepsilon_i - \mu)} - 1}. \tag{B.14}
$$

Similarly, the entropy of a system of noninteracting bosons is given by

$$
S = -k_B \sum_i [f_i \ln f_i - (1 + f_i) \ln (1 + f_i)]. \tag{B.15}
$$

A prediction of the Bose-Einstein statistics is that, at very low but finite temperatures, a large fraction of the total boson particles of the system collapse into the ground quantum state. This phenomenon is known as the Bose-Einstein condensation, and the fraction of particles in the ground state is called Bose condensate. [Notice that in the case of fermions, the ground one-particle state, as well as any other state, can accommodate at most one fermion, and carry at most the infinitesimal fraction $1/N$ of the total number N of the particles of the system.]

In the following, just for convenience and without loss of generality, we take the zero of energy at the lowest ground state of the noninteracting boson gas. The expression of the chemical potential is implicitly determined by enforcing the equality

$$
N = \sum_i \frac{1}{e^{\beta(\varepsilon_i - \mu)} - 1} \quad \text{with} \quad \mu < 0, \tag{B.16}
$$

where N is the total number of bosons of the system; the condition on the chemical potential comes from the fact that μ must be smaller than the ground-state energy, chosen to be zero.

We consider now the simple model of free bosonic atoms of mass M confined in a volume V (and ignore internal degrees of freedom, if any). The energy dispersion and the corresponding density-of-states are given by

$$
\varepsilon(\mathbf{k}) = \frac{\hbar^2 k^2}{2M} \quad \Longrightarrow \quad D(\varepsilon) = \frac{V}{4\pi^2} \left(\frac{2M}{\hbar^2} \right)^{3/2} \varepsilon^{1/2}, \quad \varepsilon \geq 0.
$$

[The procedure to obtain the density-of-states is of course the same adopted to obtain Eq. (3.8a), and need not be repeated here.]

In order to enforce correctly the condition (B.16), it is convenient to separate the sum on the one-particle quantum states ε_i into two terms, namely the ground quantum state (say $\varepsilon_0 = 0$) and all the other quantum levels. We have

$$
N = N_0 + N_1 \quad \text{with} \quad N_0 = \frac{1}{e^{-\beta\mu} - 1} \quad \text{and} \quad N_1 = \sum_{i(\neq 0)} \frac{1}{e^{\beta(\varepsilon_i - \mu)} - 1}. \tag{B.17}
$$

The first term N_0 can become an arbitrary large positive number for $\mu \to 0^-$. The other term N_1 can be estimated replacing the discrete sum with an integral involving the density-of-states. It holds

$$N_1 = \int_0^\infty \frac{1}{e^{\beta(\varepsilon-\mu)} - 1} D(\varepsilon)\, d\varepsilon. \tag{B.18}$$

The maximum value N_1 of bosons, that can be accommodated in the excited states, occurs for $\mu = 0$ and defines a critical value N_c given by

$$N_c = \int_0^\infty \frac{1}{e^{\beta\varepsilon} - 1} D(\varepsilon)\, d\varepsilon = \frac{V}{4\pi^2} \left(\frac{2M}{\hbar^2}\right)^{3/2} \int_0^\infty \frac{\varepsilon^{1/2}}{e^{\beta\varepsilon} - 1}\, d\varepsilon \quad [\text{set } \beta\varepsilon = x]$$

$$= \frac{V}{4\pi^2} \left(\frac{2Mk_BT}{\hbar^2}\right)^{3/2} \int_0^\infty \frac{x^{1/2}}{e^x - 1}\, dx.$$

With the use of the definite integral

$$\int_0^\infty \frac{x^{1/2}}{e^x - 1}\, dx = \Gamma(3/2)\zeta(3/2) \quad \text{with} \quad \Gamma(3/2) = \frac{\sqrt{\pi}}{2} \quad \text{and} \quad \zeta(3/2) = 2.612$$

[see I. S. Gradshteyn and I. M. Ryzhik, "Table of Integrals, Series, and Products" (Academic Press, New York, 1980), p. 325], one obtains

$$\frac{N_c}{V} = 2.612 \left(\frac{Mk_BT}{2\pi\hbar^2}\right)^{3/2}. \tag{B.19}$$

Suppose now to increase the number N of bosons, at fixed temperature T and volume V. It is apparent that the maximum number of bosons that can be accommodated in the excited states is N_c. If N becomes larger than N_c, the excess of particles $N - N_c$ must be accommodated in the ground state. The critical threshold appears in a regime where the interparticle spacing becomes comparable with the atomic de Broglie wavelength λ_T. In fact from Eq. (B.19) it is seen that

$$\frac{N_c}{V} \cdot \lambda_T^3 \approx 1 \quad \text{where} \quad \lambda_T = \left(\frac{\hbar^2}{2Mk_BT}\right)^{1/2}.$$

Quantum degeneracy and Bose-Einstein condensation occur when the de Broglie wavelengths of the individual atomic bosons begin to overlap.

From Eq. (B.19), it is seen that for a uniform three-dimensional Bose gas of noninteracting particles of density N/V, the critical temperature is given by

$$T_c = \left(\frac{1}{2.612} \frac{N}{V}\right)^{2/3} \frac{2\pi\hbar^2}{Mk_B}. \tag{B.20}$$

For extremely dilute gases of mass number ≈ 100, at densities $N/V \approx 10^{-15}$ cm^{-3}, we have $k_B T_c \approx 10^{-10}$ eV and the temperature T_c is in the microkelvin region.

The Bose-Einstein condensation has been traditionally studied as the mechanism underlying the superfluidity in liquid helium, where however the interaction between particles is rather strong, and the interpretation of the phenomenological aspects is more demanding. Since long time, the Bose-Einstein condensation has also been envisaged as the basic mechanism underlying superconductivity (where bosons, formed by Cooper pairing of two fermions, can undergo a Bose-Einstein condensation). More direct and almost ideal examples of Bose-Einstein condensations have been achieved in extremely dilute atomic gases; to avoid transitions to a liquid or a solid, it is required a very large particle separation, and a corresponding very low critical temperature in the microkelvin region or lower. For further information, and historical background, on the spectacular measurements and theoretical treatments of atomic condensates, we refer for instance to the lectures of Cornell, Ketterle, and Wieman, and to the review articles of Giorgini et al. cited in the bibliography.

Appendix C.   Modified Fermi-Dirac Statistics in a Model of Correlation Effects

In the previous Appendix B1 we have considered a physical system composed by N indistinguishable noninteracting fermions, confined within a volume V. Suppose that the one-electron Hamiltonian does not remove the spin degeneracy; in the independent particle approximation, the occupation probability of the *space orbital* (of energy ε_i), *regardless of the spin degeneracy*, is then given by

$$f(\varepsilon_i) = 2\frac{1}{e^{\beta(\varepsilon_i - \mu)} + 1}, \tag{C.1}$$

where the factor 2 accounts for the spin degeneracy of the orbital level.

In Chapter 13, in the study of doped semiconductors, we need to know not only the occupation probability of valence and conduction states (described by standard delocalized Bloch wavefunctions), but also the occupation probability of impurity localized states in the energy gap. In several situations (for instance for donor levels) the Coulomb repulsion between electrons may *prevent double occupation of a given localized orbital*. We now discuss how this effect (which is the simplest example of correlation beyond the one-electron approximation) modifies the Fermi-Dirac distribution function.

For sake of clarity, consider a band state in an allowed energy region of the crystal and described by a Bloch wavefunction; a band level can be empty, or occupied by one electron of either spin, or by two electrons of opposite spin; the four possibilities are illustrated in Figure 3.6a. Consider now an impurity state within the energy gap and described by a localized wavefunction; a donor level, for instance, can be empty, or occupied by one electron of either spin, but not by two electrons of opposite spin, because of the penalty in the electrostatic repulsion energy; the situation is illustrated in Figure 3.6b.

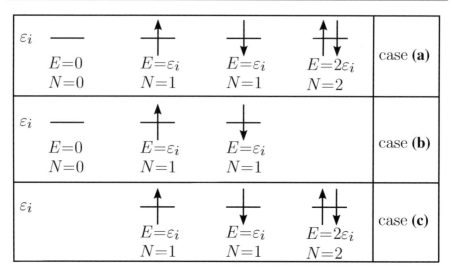

Figure 3.6 Schematic illustration of possible occupation of a given spatial orbital of energy ε_i. (a) The given level can be empty, or accept one electron of either spin, or can have two electrons with opposite spin (this condition is typical for band levels in crystals). (b) The given level can be empty, or accept one electron of either spin (this situation is typical for donor impurity levels in semiconductors). (c) The given level can accommodate one electron of either spin or two electrons with opposite spin (this condition is typical for acceptor impurity levels in semiconductors).

We calculate for both situations the average number of electrons in the state ε_i (regardless of the spin direction); the average number is given by

$$\langle n_i \rangle = \frac{\sum_{m,N} N e^{-\beta(E_{m,N} - \mu N)}}{\sum_{m,N} e^{-\beta(E_{m,N} - \mu N)}}. \tag{C.2}$$

In the case of Figure 3.6a, Eq. (C.2) gives

$$\langle n_i \rangle = \frac{e^{-\beta(\varepsilon_i - \mu)} + e^{-\beta(\varepsilon_i - \mu)} + 2e^{-\beta(2\varepsilon_i - 2\mu)}}{1 + e^{-\beta(\varepsilon_i - \mu)} + e^{-\beta(\varepsilon_i - \mu)} + e^{-\beta(2\varepsilon_i - 2\mu)}}$$

$$= \frac{2e^{-\beta(\varepsilon_i - \mu)}[1 + e^{-\beta(\varepsilon_i - \mu)}]}{[1 + e^{-\beta(\varepsilon_i - \mu)}][1 + e^{-\beta(\varepsilon_i - \mu)}]} = 2\frac{1}{e^{\beta(\varepsilon_i - \mu)} + 1}; \tag{C.3}$$

as expected, the result (C.1) is recovered.

In the case of Figure 3.6b, we have instead

$$\langle n_i \rangle = \frac{e^{-\beta(\varepsilon_i - \mu)} + e^{-\beta(\varepsilon_i - \mu)}}{1 + e^{-\beta(\varepsilon_i - \mu)} + e^{-\beta(\varepsilon_i - \mu)}} = \frac{1}{\frac{1}{2} e^{\beta(\varepsilon_i - \mu)} + 1}. \tag{C.4}$$

The occurrence of the factor 1/2 in Eq. (C.4) can be easily understood qualitatively in the limiting case of a Boltzmann tail (see Figure 3.7).

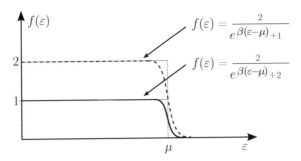

Figure 3.7 Average occupation number (dashed line) for an orbital of energy ε that can accept up to two electrons of either spin (standard Fermi-Dirac statistics). The average occupation number (solid line) for an orbital of energy ε that can accept only one electron of either spin is also reported. For $\varepsilon - \mu \gg k_B T$ the two curves coincide.

For sake of completeness, we consider also the statistics of the acceptor levels, whose electronic structure is studied in Chapter 13. An acceptor level can be empty, or occupied by one hole of either spin, but not by two holes of opposite spin because of the penalty in the electrostatic repulsion energy; this situation can be stated in terms of electrons in the same form as in Figure 3.6c. The application of Eq. (C.2) to the three possibilities illustrated in Figure 3.6c gives

$$\langle n_i \rangle = \frac{e^{-\beta(\varepsilon_i - \mu)} + e^{-\beta(\varepsilon_i - \mu)} + 2e^{-\beta(2\varepsilon_i - 2\mu)}}{e^{-\beta(\varepsilon_i - \mu)} + e^{-\beta(\varepsilon_i - \mu)} + e^{-\beta(2\varepsilon_i - 2\mu)}} = 2 - \frac{1}{\frac{1}{2} e^{-\beta(\varepsilon_i - \mu)} + 1}. \qquad (C.5)$$

If we indicate by $\langle p_i \rangle = 2 - \langle n_i \rangle$ the mean number of holes, we obtain

$$\langle p_i \rangle = \frac{1}{\frac{1}{2} e^{\beta(\mu - \varepsilon_i)} + 1}, \qquad (C.6)$$

which is just the distribution function (C.4) with the energy axis inverted for holes.

Further Reading

Ashcroft, N. W., & Mermin, N. D. (1976). *Solid state physics*. New York: Holt, Rinehart and Winston.

Blundell, S. J., & Blundell, K. M. (2010). *Concepts in thermal physics*. Oxford: Oxford University Press.

Callen, H. B. (1985). *Thermodynamics and an introduction to thermostatistics*. New York: Wiley.

Cornell, E. A., & Wieman, C. E. (2002). Bose-Einstein condensation in a dilute gas, the first 70 years and some recent experiments. *Reviews of Modern Physics, 74*, 875.

Dalfovo, F., Giorgini, S., Pitaevskii, L. P., & Stringari, S. (1999). Theory of Bose-Einstein condensation in trapped gases. *Reviews of Modern Physics, 71*, 463.

Giorgini, S., Pitaevskii, L. P., & Stringari, S. (2008). Theory of ultracold atomic Fermi gases. *Reviews of Modern Physics, 80*, 1215.

Huang, K. (1987). *Statistical mechanics*. New York: Wiley.

Ketterle, W. (2002). When atoms behave as waves: Bose-Einstein condensation and the atom laser. *Reviews of Modern Physics, 74*, 1131.

Kittel, C. (1958). *Elementary statistical physics*. New York: Wiley (Dover Publications, New York, 2004).

Kubo, R. (1988). *Statistical mechanics*. Amsterdam: North-Holland (Elsevier, Amsterdam, 2004).

Pathria, R. K., & Beale, P. D. (2011). *Statistical mechanics*. Amsterdam: Elsevier.

Reichl, L. E. (2009). *A modern course in statistical physics*. Leipzig: Wiley-VCH.

Wilson, A. H. (1953). *The theory of metals*. Cambridge: Cambridge University Press.

4 The One-Electron Approximation and Beyond

Chapter Outline head

In this chapter on the quantum mechanics of electrons in solids, we hit the heart of one of the most traditional subjects of solid state physics. A solid is constituted by a large number of electrons and nuclei in mutual interaction, and the dynamics of these particles, in general, cannot be considered separately. Throughout this chapter, however, the nuclei are imagined as *clamped in a given configuration* (usually the equilibrium configuration), so that to focus on the electronic problem. Among the approaches to the many-electron system, we consider the historic (but always actual) Hartree-Fock theory, because of its central role in the general framework of the many-body theory. We will then discuss aspects beyond the one-electron approximation, in particular the density functional theory, which has been so successful in the description of the ground-state properties of matter.

From the point of view of the organization of this book, Chapter 4 is the backbone and paves the way for the sequence of logically linked subjects, presented in Chapters from 5 to 10. In Chapter 5 we consider in detail the methods of electronic band structure calculations within the one-electron approximation. In Chapter 6 some applications to selected classes of solids are provided. Chapter 7 concerns aspects beyond the one-electron approximation, with the concepts of excitons and plasmons

Solid State Physics, Second Edition. http://dx.doi.org/10.1016/B978-0-12-385030-0.00004-9

quasiparticles. In all the mentioned chapters, the focus is on the electronic problem (within the one-electron approximation or beyond it), with the nuclei pictured as a fixed source of a static potential. Later, in Chapters 8–10 we discuss the interdependence of electronic and nuclear dynamics, the elastic field and the vibrational structure of solids, the electronic and lattice effects on the scattering of particles from crystals.

4.1 Introductory Remarks on the Many-Electron Problem

Atoms, molecules, clusters, or solids are systems composed by mutually interacting electrons and nuclei. The total non-relativistic Hamiltonian of a system of electrons (of coordinates $\mathbf{r}_i$, momenta $\mathbf{p}_i$, and charge $-e$) and nuclei (of coordinates $\mathbf{R}_I$, momenta $\mathbf{P}_I$, and charge $+z_I e$) in mutual interaction via Coulomb forces, can be written as

$$H_{\text{tot}} = \sum_i \frac{\mathbf{p}_i^2}{2m} + \sum_I \frac{\mathbf{P}_I^2}{2M_I} + \sum_i V_{\text{nucl}}(\mathbf{r}_i) + \frac{1}{2} \sum_{i \neq j} \frac{e^2}{|\mathbf{r}_i - \mathbf{r}_j|} + \frac{1}{2} \sum_{I \neq J} \frac{z_I z_J e^2}{|\mathbf{R}_I - \mathbf{R}_J|}$$

$$(4.1)$$

with

$$V_{\text{nucl}}(\mathbf{r}) = -\sum_I \frac{z_I e^2}{|\mathbf{r} - \mathbf{R}_I|}.$$

$$(4.2)$$

The terms appearing in Eq. (4.1) represent the kinetic energy of the electrons, the kinetic energy of the nuclei, the electron-nucleus attractive potential energy, the electron-electron, and nucleus-nucleus repulsive potential energies. The expression containing the electron-electron interaction involves a sum on the indices i and j, with the obvious exclusion of the self-interaction terms $i = j$; similarly in the nucleus-nucleus interaction the sum runs over I and J, with the exclusion of the self-interaction terms $I = J$. For simplicity, in the Hamiltonian (4.1) the coupling between charged particles is given by the instantaneous and spin-independent Coulomb interactions. [Coupling of particles with internal or external electromagnetic fields and in particular retardation effects due to the finite speed of light, terms of relativistic origin related to the spin and magnetic moment of the particles, and other corrections, will be included only when necessary].

The first and obvious hint to attack the Hamiltonian (4.1) is suggested by the large difference between electron and nuclear masses. The simplest approximation (although a drastic one) is just to drop the kinetic energy of the nuclei in Eq. (4.1) [a more gentle and fruitful approach within the adiabatic principle is the subject of Chapter 8]. When the approximation to zero-order in the electron-to-nucleus mass ratio is anyway adopted, the nuclear positions $\mathbf{R}_I$ become classical variables (or parameters) and the nuclei can be considered as fixed in a given selected configuration; most often the fixed configuration is the equilibrium one.

Consistently with our neglecting the nuclear kinetic energy, we gain the further bonus that the nuclear-nuclear repulsion interaction V_{NN}, represented by the last term in Eq. (4.1), is a constant for any fixed configuration; its value increases with the sample volume (and diverges if not combined with appropriate attractive terms due to the overall charge neutrality of the system) and is important to determine the total energy of the system or its cohesive energy. However for what concerns the study of the *electronic states in the rigid lattice approximation* (i.e. *without relaxation of the nuclear positions*), we can momentarily disregard the V_{NN} interaction, which is just a trivial constant for the nuclei clamped in a given configuration, and take it into account later when needed.

The many-body Hamiltonian for a system of N interacting electrons in the presence of nuclei fixed in some selected configuration can thus be written in the form

$$H_e = \sum_i^N h(\mathbf{r}_i) + \frac{1}{2} \sum_{i \neq j}^N \frac{e^2}{r_{ij}}, \tag{4.3}$$

where

$$h(\mathbf{r}) = \frac{\mathbf{p}^2}{2m} + V_{\text{nucl}}(\mathbf{r}) \tag{4.4}$$

denotes for each electron the kinetic energy operator plus the potential energy due to the nuclei. The purpose of this chapter is to present various approximations toward the solution of the many-body eigenvalue problem

$$H_e \Psi(\mathbf{r}_1\sigma_1, \mathbf{r}_2\sigma_2, \dots, \mathbf{r}_N\sigma_N) = E \Psi(\mathbf{r}_1\sigma_1, \mathbf{r}_2\sigma_2, \dots, \mathbf{r}_N\sigma_N), \tag{4.5}$$

where $\mathbf{r}_i\sigma_i$ are the space and spin variables of the ith electron. Although the Hamiltonian H_e does not include spin-dependent energy terms, the electron spin plays a fundamental role since the correct many-electron wavefunctions must be antisymmetric for interchange of the spatial and spin coordinates of any two electrons.

The pioneering approach to solve Eq. (4.5) is the *Hartree theory*, where the total N-electron ground-state wavefunction is represented by the best *simple product of N one-electron spin-orbitals*. The next major improvement, that correctly embodies the Pauli exclusion principle for identical fermions, is the *Hartree-Fock theory*, where the total N-electron ground-state wavefunction is represented by the best *antisymmetrized product of N one-electron spin-orbitals*. We also consider some aspects beyond the one-electron approximation; in this wide field, we focus on the particular role of the *density functional theory* and Kohn-Sham orbitals for the ground-state electronic density.

4.2 The Hartree Equations

To represent many-electron wavefunctions, it is convenient to consider preliminarily a complete set of orthonormal one-electron orbitals $\{\phi_i(\mathbf{r})\}$. From this, we can form a complete set of orthonormal *spin-orbitals* $\{\psi_i(\mathbf{r}\sigma)\}$, where $\psi_i(\mathbf{r}\sigma) = \phi_i(\mathbf{r})\chi_i(\sigma)$ denotes the product of the *spatial orbital* $\phi_i(\mathbf{r})$ and the *spin function* $\chi_i(\sigma)$ (either "spin-up" function α, with eigenvalue $\hbar/2$, or "spin-down" function β, with eigenvalue $-\hbar/2$). In short-hand notations, space, and spin variables are (frequently) omitted. In order to avoid possible source of confusion, throughout this chapter, we use the notation ψ_i to denote the i-th spin-orbital of interest of a given set, and the notation ϕ_i and χ_i to denote the orbital part and the spin part of the spin-orbital ψ_i. Thus the arguments of ψ_i, ϕ_i, and χ_i, when not indicated explicitly, are understood to be $\psi_i(\mathbf{r}\sigma)$, $\phi_i(\mathbf{r})$, and $\chi_i(\sigma)$, respectively.

In the Hartree theory the ground-state wavefunction of the many-body system is expressed as simple product of orthonormalized one-electron spin-orbitals in the form

$$\Psi_0(\mathbf{r}_1\sigma_1, \mathbf{r}_2\sigma_2, \ldots, \mathbf{r}_N\sigma_N) = \psi_1(\mathbf{r}_1\sigma_1)\psi_2(\mathbf{r}_2\sigma_2)\cdots\psi_N(\mathbf{r}_N\sigma_N). \tag{4.6}$$

In Eq. (4.6), the many-body wavefunction Ψ_0 is constructed by assigning any given electron to some given spin-orbital (for instance the electron of coordinates $\mathbf{r}_1\sigma_1$ is assigned to the spin-orbital ψ_1, the electron $\mathbf{r}_2\sigma_2$ is assigned to the spin-orbital ψ_2, and so on). It is apparent that the simple Hartree product (4.6) does not have the correct antisymmetry character for interchange of space and spin coordinates of any two electrons; the Pauli principle is taken into account ad hoc, avoiding multiple occupancy of any given spin-orbital. Furthermore, a simple product of type (4.6) completely neglects any correlation in the position of the electrons. Because of these serious drawbacks, we mention the Hartree theory mostly for its historical role, and confine our discussion to some heuristic considerations.

The electronic charge density $\rho(\mathbf{r})$ corresponding to the Hartree wavefunction (4.6) is given by

$$\rho(\mathbf{r}) = (-e) \sum_j^{(\text{occ})} \phi_j^*(\mathbf{r})\phi_j(\mathbf{r}), \tag{4.7}$$

where the sum runs over all occupied spin-orbitals, entering the ground-state Ψ_0. The Coulomb potential (also called the *Hartree potential*) corresponding to the electronic charge density (4.7) is

$$V_{\text{coul}}(\mathbf{r}) = \sum_j^{(\text{occ})} \int \phi_j^*(\mathbf{r}') \frac{e^2}{|\mathbf{r} - \mathbf{r}'|} \phi_j(\mathbf{r}')\, d\mathbf{r}'. \tag{4.8}$$

In the Hartree self-consistent approximation, it is assumed that each electron moves in the effective field corresponding to the Coulomb potential generated by the charge distribution of all the other $N - 1$ electrons; for simplicity, to avoid the determination of as many (slightly) different effective fields as the number of orbitals, it is assumed that the effective field for any electron is given by the Hartree potential (4.8). We thus obtain that the spin-orbitals entering in the product wavefunction (4.6) satisfy the Hartree equations

$$\left[\frac{\mathbf{p}^2}{2m} + V_{\text{nucl}}(\mathbf{r}) + V_{\text{coul}}(\mathbf{r})\right]\psi_i = \varepsilon_i\psi_i. \tag{4.9}$$

The Hartree potential is defined in terms of the occupied orbitals ψ_i and must thus be determined in a self-consistent way (usually with appropriate iterations). From an initial reasonable guess of the functions $\psi_1, \psi_2, \dots, \psi_N$ one evaluates the space charge distribution (4.7) and the corresponding Hartree potential (4.8). A new set of improved wavefunctions is then obtained from the solution of the Hartree equations (4.9). The corresponding Hartree potential (or, more often, an appropriate weighted average of input and output ones) is used to start a new cycle. Cycles are repeated up to self-consistency of input and output functions and potentials.

4.3   Identical Particles and Determinantal Wavefunctions

In this section we summarize some basic results on the antisymmetry principle for fermions, and in particular the possibility to write the basic functions of a many-body system in the form of determinantal states. It is well known that a many-electron wavefunction must be antisymmetric for interchange of the coordinates of any two electrons. Consider an N-electron system and a set $\{\psi_i\}$ ($i = 1, 2, \dots, N$) of *orthonormal one-particle spin-orbitals*. A proper antisymmetric N-electron wavefunction takes the form

$$\Psi_0(\mathbf{r}_1\sigma_1, \mathbf{r}_2\sigma_2, \dots, \mathbf{r}_N\sigma_N) = \mathcal{A}\{\psi_1(\mathbf{r}_1\sigma_1)\psi_2(\mathbf{r}_2\sigma_2)\cdots\psi_N(\mathbf{r}_N\sigma_N)\}, \quad (4.10a)$$

where $\mathcal{A}$ denotes the antisymmetrizer operator

$$\mathcal{A} = \frac{1}{\sqrt{N!}} \sum_{i=1}^{N!} (-1)^{p_i} P_i, \quad (4.10b)$$

the sum extends to all the $N!$ permutations P_i of the electronic coordinates, and $(-1)^{p_i}$ equals $+1$ or -1 for permutations of even or odd class with respect to the fundamental one. Notice that Ψ_0 is normalized to one, if the composing spin-orbitals ψ_i are orthonormal.

Expressions (4.10) can be conveniently written in the determinantal form suggested by Slater:

$$\Psi_0(\mathbf{r}_1\sigma_1, \mathbf{r}_2\sigma_2, \dots, \mathbf{r}_N\sigma_N)$$
$$= \frac{1}{\sqrt{N!}} \begin{vmatrix} \psi_1(\mathbf{r}_1\sigma_1) & \psi_1(\mathbf{r}_2\sigma_2) & \cdots & \psi_1(\mathbf{r}_N\sigma_N) \\ \psi_2(\mathbf{r}_1\sigma_1) & \psi_2(\mathbf{r}_2\sigma_2) & \cdots & \psi_2(\mathbf{r}_N\sigma_N) \\ \cdots & \cdots & \cdots & \cdots \\ \psi_N(\mathbf{r}_1\sigma_1) & \psi_N(\mathbf{r}_2\sigma_2) & \cdots & \psi_N(\mathbf{r}_N\sigma_N) \end{vmatrix}. \quad (4.11)$$

It is evident that the interchange of two columns changes the sign of the determinant consistently with the antisymmetry property of the wavefunction; moreover, occupancy of the same spin-orbital by two electrons gives two equal rows and thus the determinant equals zero.

An important property automatically embodied in determinantal wavefunctions is that electrons with parallel spin are (correctly) kept apart. To better realize this point, consider for simplicity the determinantal state (4.11) in the particular case in which the spin-orbitals have *all spin parallel* (for instance spin up). The determinantal state

(4.11) then keeps the form:

$$\Psi_0(\mathbf{r}_1\sigma_1, \mathbf{r}_2\sigma_2, \ldots, \mathbf{r}_N\sigma_N)$$

$$= \frac{1}{\sqrt{N!}} \begin{vmatrix} \phi_1(\mathbf{r}_1) & \phi_1(\mathbf{r}_2) & \cdots & \phi_1(\mathbf{r}_N) \\ \phi_2(\mathbf{r}_1) & \phi_2(\mathbf{r}_2) & \cdots & \phi_2(\mathbf{r}_N) \\ \cdots & \cdots & \cdots & \cdots \\ \phi_N(\mathbf{r}_1) & \phi_N(\mathbf{r}_2) & \cdots & \phi_N(\mathbf{r}_N) \end{vmatrix} \alpha(1)\alpha(2)\cdots\alpha(N). \qquad (4.12)$$

It is evident that *nodes* of Ψ_0 occur whenever $\mathbf{r}_i \equiv \mathbf{r}_j$; thus any two electrons cannot be in the same spatial position (with the same spin).

The many-body wavefunctions can be written in a more compact form, leaving implicit the space and spin coordinates of the electrons. Two frequently used shorthand notations for the wavefunction Ψ_0 are

$$\Psi_0 = \mathcal{A}\{\psi_1\psi_2\cdots\psi_N\} \quad \text{or} \quad \Psi_0 = \frac{1}{\sqrt{N!}}\det\{\psi_1\psi_2\cdots\psi_N\};$$

both notations clearly list the spin-orbitals entering the antisymmetrization operator or forming the Slater determinant.

4.4 Matrix Elements Between Determinantal States

The purpose of this section is to provide the explicit expression of the matrix elements of one-electron operators and two-electron operators between determinantal states. These expressions constitute a technical (but nevertheless essential) tool for handling the Hartree-Fock theory, as well as the second quantization formalism of the many-body problem. The readers more interested in the structure and physical meaning of the Hartree-Fock equations can skip this section, using their own knowledge when meeting matrix elements between determinantal states.

Expectation Values of One-particle and Two-Particle Operators on Determinantal States

The many-body electron Hamiltonian (4.3) contains two types of operators. One is the sum of one-particle operators of the form

$$G_1 = \sum_i^N h(\mathbf{r}_i); \qquad (4.13a)$$

the other type is the sum of two-particle operators of the form

$$G_2 = \frac{1}{2} \sum_{i \neq j}^N \frac{e^2}{|\mathbf{r}_i - \mathbf{r}_j|}, \qquad (4.13b)$$

which describes electron-electron interactions. We wish now to express the matrix elements of one-particle and two-particle operators between Slater determinants in terms of monoelectronic and bielectronic integrals, respectively. The elaboration is based on the elementary properties of determinants and the invariance of G_1 and G_2 under permutations of the electronic coordinates.

We begin with the evaluation of the expectation value of G_1 on a given determinantal state Ψ_0; we have

$$\langle\Psi_0|G_1|\Psi_0\rangle = \frac{1}{N!}\langle\det\{\psi_1\psi_2\cdots\psi_N\}|G_1|\det\{\psi_1\psi_2\cdots\psi_N\}\rangle. \tag{4.14a}$$

In the above expression, we can just select the identical permutation in the determinant in the left part, since all the other $N! - 1$ permutations would give the same result as the selected one. We can thus cancel $N!$ at the denominator of Eq. (4.14a), and write

$$\langle\Psi_0|G_1|\Psi_0\rangle = \langle\psi_1\psi_2\cdots\psi_N|G_1|\det\{\psi_1\psi_2\cdots\psi_N\}\rangle. \tag{4.14b}$$

Because of the orthonormality of spin-orbitals it is also evident that only the identical permutation survives in the determinant on the right part of the above matrix element. It follows

$$\langle\Psi_0|G_1|\Psi_0\rangle = \langle\psi_1\psi_2\cdots\psi_N|G_1|\psi_1\psi_2\cdots\psi_N\rangle. \tag{4.14c}$$

Comparison of Eq. (4.14a) and Eq. (4.14c) shows that, for what concerns the matrix elements of the operator G_1, it is irrelevant to use antisymmetrized products of spin-orbitals or simple (Hartree) products of the same spin-orbitals.

Inserting the expression (4.13a) of G_1 into expression (4.14c), one obtains

$$\boxed{\langle\Psi_0|G_1|\Psi_0\rangle = \sum_i\langle\psi_i|h|\psi_i\rangle}. \tag{4.15}$$

Thus, the expectation value on a determinantal state of a sum of one-electron operators equals the sum of the expectation values of the one-electron operator on the spin-orbitals, which enter in the determinantal state.

We evaluate now the expectation value of G_2 on a given determinantal state; we have

$$\langle\Psi_0|G_2|\Psi_0\rangle = \frac{1}{N!}\langle\det\{\psi_1\psi_2\cdots\psi_N\}|G_2|\det\{\psi_1\psi_2\cdots\psi_N\}\rangle$$
$$= \langle\psi_1\psi_2\cdots\psi_N|G_2|\det\{\psi_1\psi_2\cdots\psi_N\}\rangle. \tag{4.16a}$$

Consider for instance the contribution of the term e^2/r_{12}. We restore for clarity the space and spin electronic coordinates, and we easily verify that

$$\langle\psi_1(\mathbf{r}_1\sigma_1)\psi_2(\mathbf{r}_2\sigma_2)\cdots\psi_N(\mathbf{r}_N\sigma_N)|\frac{e^2}{r_{12}}|\det\{\psi_1(\mathbf{r}_1\sigma_1)\psi_2(\mathbf{r}_2\sigma_2)\cdots\psi_N(\mathbf{r}_N\sigma_N)\}\rangle$$
$$= \langle\psi_1\psi_2|\frac{e^2}{r_{12}}|\psi_1\psi_2\rangle - \langle\psi_1\psi_2|\frac{e^2}{r_{12}}|\psi_2\psi_1\rangle; \tag{4.16b}$$

in fact only the identical permutation and the permutation (of odd class) that interchanges coordinates $\mathbf{r}_2\sigma_2$ with coordinates $\mathbf{r}_1\sigma_1$ survive. The definition and elementary properties of the Coulomb and exchange bielectronic integrals appearing in the right-hand side of Eq. (4.16b) are discussed in Appendix A. In particular we remind that both bielectronic Coulomb integrals and bielectronic exchange integrals are definite positive quantities; the latter are different from zero only for parallel spins of the spin-orbitals entering in their definition.

The expectation value of G_2 on a given determinantal state thus becomes

$$
\langle \Psi_0 | G_2 | \Psi_0 \rangle = \frac{1}{2} \sum_{ij} \left[\langle \psi_i \psi_j | \frac{e^2}{r_{12}} | \psi_i \psi_j \rangle - \langle \psi_i \psi_j | \frac{e^2}{r_{12}} | \psi_j \psi_i \rangle \right]. \tag{4.17}
$$

In Eq. (4.17) we can write indifferently $\sum_{ij}$ or $\sum_{i \neq j}$, since Coulomb and exchange contributions exactly cancel each other for $i \equiv j$.

It is important to notice explicitly that the *electron-electron repulsive energy* $\langle \Psi_0 | G_2 | \Psi_0 \rangle$ *on the antisymmetrized product of spin-orbitals is always lower than the electron-electron repulsive energy on the simple Hartree product of the same spin-orbitals*; the difference is just represented by the second term in the right-hand side of Eq. (4.17), without the self-interaction terms $i = j$. Since exchange bielectronic integrals are positive quantities, different from zero only for spin-orbitals with parallel spins, we link the decrease in energy with the physical fact that electrons with parallel spin are kept apart in real space in Slater determinantal states (as already discussed in Eq. 4.12).

In the following we report other matrix elements of interest of one-electron and two-electron operators among determinantal states; however we omit details, leaving proofs as useful exercises to the interested reader, who can follow the same guidelines leading to Eqs. (4.15) and (4.17).

Determinantal States with Replacement of One Spin-Orbital

Consider two N-electron determinantal states which differ by one of the one-particle functions, say

$$
\begin{cases}
\Psi_0 = \mathcal{A}\{\psi_1 \psi_2 \cdots \psi_m \cdots \psi_N\}, \\
\Psi_{\mu,m} = \mathcal{A}\{\psi_1 \psi_2 \cdots \psi_\mu \cdots \psi_N\}.
\end{cases} \tag{4.18}
$$

The notation $\Psi_{\mu,m}$ for the singly substituted state helps to memorize that the spin-orbital ψ_m has been replaced by the spin-orbital ψ_μ, all other spin-orbitals being equal and in the same sequence.

It can be easily verified that

$$
\langle \Psi_{\mu,m} | G_1 | \Psi_0 \rangle = \langle \psi_\mu | h | \psi_m \rangle. \tag{4.19}
$$

Thus, the matrix elements of G_1 connecting two determinantal states, which differ by one of the one-particle functions, equal the matrix elements of the one-particle operator between the two (replacing and replaced) spin-orbitals. For the operator G_2 we have

$$\langle \Psi_{\mu,m} | G_2 | \Psi_0 \rangle = \sum_j \left[\langle \psi_\mu \psi_j | \frac{e^2}{r_{12}} | \psi_m \psi_j \rangle - \langle \psi_\mu \psi_j | \frac{e^2}{r_{12}} | \psi_j \psi_m \rangle \right]. \tag{4.20}$$

In Eq. (4.20) we can write indifferently $\sum_j$ or $\sum_{j(\neq m)}$, since the two terms in square brackets exactly cancel each other for $j = m$.

Determinantal States with Replacement of Two Spin-Orbitals

Consider two determinantal states which differ by two of the one-particle functions, say

$$\begin{cases} \Psi_0 = \mathcal{A}\{\psi_1 \psi_2 \cdots \psi_m \cdots \psi_n \cdots \psi_N\}, \\ \Psi_{\mu\nu,mn} = \mathcal{A}\{\psi_1 \psi_2 \cdots \psi_\mu \cdots \psi_\nu \cdots \psi_N\}. \end{cases}$$

The notation $\Psi_{\mu\nu,mn}$ for the doubly substituted state helps to memorize that the spin-orbitals ψ_m and ψ_n have been replaced by the spin-orbitals ψ_μ and ψ_ν, respectively, all other spin-orbitals being equal and in the same sequence.

A sum of one-electron operators, as G_1, has vanishing matrix elements between determinantal states which differ by two spin-orbitals. The matrix elements of G_2 are given by

$$\langle \Psi_{\mu\nu,mn} | G_2 | \Psi_0 \rangle = \langle \psi_\mu \psi_\nu | \frac{e^2}{r_{12}} | \psi_m \psi_n \rangle - \langle \psi_\mu \psi_\nu | \frac{e^2}{r_{12}} | \psi_n \psi_m \rangle. \tag{4.21}$$

Consider now two determinantal states which differ by two different pairs of the one-particle functions, namely

$$\begin{cases} \Psi_{\mu,m} = \mathcal{A}\{\psi_1 \psi_2 \cdots \psi_\mu \cdots \psi_n \cdots \psi_N\}, \\ \Psi_{\nu,n} = \mathcal{A}\{\psi_1 \psi_2 \cdots \psi_m \cdots \psi_\nu \cdots \psi_N\}. \end{cases}$$

The off-diagonal matrix elements of G_2 between the above determinantal states are

$$\langle \Psi_{\nu,n} | G_2 | \Psi_{\mu,m} \rangle = \langle \psi_m \psi_\nu | \frac{e^2}{r_{12}} | \psi_\mu \psi_n \rangle - \langle \psi_m \psi_\nu | \frac{e^2}{r_{12}} | \psi_n \psi_\mu \rangle. \tag{4.22}$$

Finally, the off-diagonal matrix elements of the operators G_1 and G_2 between determinantal states which differ by *more than two spin-orbitals are zero*.

In this section we have provided useful expressions of the matrix elements of the operator G_1 and G_2 between determinantal states. With respect to the matrix elements involving simple product wavefunctions, it can be observed that the antisymmetrization does not alter the matrix elements of the one-body operator G_1, while, for the two-body operator G_2, exchange bielectronic integrals are to be appropriately included. A more convenient formal and conceptual point of view to treat Slater determinants and related matrix elements is the use of the second quantization formalism, which is outlined in Appendix B.

4.5 The Hartree-Fock Equations

4.5.1 *Variational Approach and Hartree-Fock Equations*

Consider a system of N interacting electrons described by the Hamiltonian H_e given in Eq. (4.3). We attempt to describe *the ground-state of the many-electron system with a single determinantal state*; the best possible choice of spin-orbitals entering this determinantal state is obtained minimizing the electron energy, using the variational principle. We find that the "optimized" orbitals are solutions of the self-consistent integro-differential equations known as Hartree-Fock equations.

We approximate the ground-state of the system with a single determinantal state in the form

$$\Psi_0 = \mathcal{A}\{\psi_1 \psi_2 \cdots \psi_N\}.$$

A determinantal state Ψ_0, constituted by doubly occupied spatial orbitals, is called *spin-restricted Slater determinant*; this spin configuration is the appropriate choice for *closed-shell systems*, where electrons with spin-up and spin-down balance for each spatial orbital. However, to discuss also *open-shell systems*, we do not put at this stage any restriction on the spin part of the spin-orbitals entering the ground-state Ψ_0, which is thus in general a *spin-unrestricted Slater determinant*. The variational procedure concerns the optimized determination of the orbital parts of the spin-orbitals (for a given chosen spin configuration).

The expectation value E_0 of the electronic Hamiltonian H_e on the (normalized) state Ψ_0 can be calculated by means of Eqs. (4.15) and (4.17); it reads

$$\boxed{E_0 = \sum_i \langle \psi_i | h | \psi_i \rangle + \frac{1}{2} \sum_{ij} \left[\langle \psi_i \psi_j | \frac{e^2}{r_{12}} | \psi_i \psi_j \rangle - \langle \psi_i \psi_j | \frac{e^2}{r_{12}} | \psi_j \psi_i \rangle \right]}$$

$$(i, j = 1, \ldots, N). \qquad (4.23)$$

According to the variational principle, the N contributing spin-orbitals $\{\psi_i\}$ are varied so that the energy $E_0 \equiv E_0(\{\psi_i\})$ achieves its minimum value. This minimum value is higher (or at most equal) than the correct ground-state energy. The variation of the functional $E_0(\{\psi_i\})$ must be done under the constraint that the N spin-orbitals are normalized. With the standard technique of Lagrange multipliers, we minimize without constraints the functional

$$G(\{\psi_i\}) = \sum_i \langle \psi_i | h | \psi_i \rangle + \frac{1}{2} \sum_{ij} \left[\langle \psi_i \psi_j | \frac{e^2}{r_{12}} | \psi_i \psi_j \rangle - \langle \psi_i \psi_j | \frac{e^2}{r_{12}} | \psi_j \psi_i \rangle \right]$$

$$- \sum_i \varepsilon_i \langle \psi_i | \psi_i \rangle,$$

where ε_i are at this stage N undetermined Lagrange multipliers. [We can skip the introduction of other Lagrange multipliers to enforce orthogonality, since the optimized spin-orbitals turn out to be eigenfunctions of an hermitian operator and orthogonal anyway.]

We can now calculate the variation δG when the orbital part of the spin-orbitals ψ_i are changed by an infinitesimal amount $\psi_i + \delta \psi_i$. Consider, for instance, the variation of the term $\langle \psi_i | h | \psi_i \rangle$; we have to first order in the variations

$$\delta \langle \psi_i | h | \psi_i \rangle = \langle \delta \psi_i | h | \psi_i \rangle + \langle \psi_i | h | \delta \psi_i \rangle.$$

Similar considerations can be performed for any term appearing in the expression of $G(\{\psi_i\})$, and the variation δG can be calculated.

We can consider the variations $\delta \psi_i$ and $\delta \psi_i^*$ completely independent; for instance taking the variations $\delta \psi_i^*$ (or $\delta \psi_j^*$) arbitrary and the variations $\delta \psi_i \equiv 0$ (or $\delta \psi_j \equiv 0$) we obtain:

$$\delta G = \sum_i \langle \delta \psi_i | h | \psi_i \rangle + \sum_{ij} \left[\langle \delta \psi_i \psi_j | \frac{e^2}{r_{12}} | \psi_i \psi_j \rangle - \langle \delta \psi_i \psi_j | \frac{e^2}{r_{12}} | \psi_j \psi_i \rangle \right]$$

$$- \sum_i \varepsilon_i \langle \delta \psi_i | \psi_i \rangle. \tag{4.24}$$

In stationary conditions $\delta G = 0$; since the variations $\delta \psi_i^*$ are arbitrary, the quantity multiplying any chosen $\delta \psi_i^*$ in Eq. (4.24) must be zero. This means that the best spin-orbitals composing the ground determinantal state satisfy the set of non-linear integro-differential Hartree-Fock equations

$$\left[\frac{\mathbf{p}^2}{2m} + V_{\text{nucl}}(\mathbf{r}) + V_{\text{coul}}(\mathbf{r}) + V_{\text{exch}} \right] \psi_i = \varepsilon_i \psi_i \tag{4.25a}$$

with the Coulomb and exchange operators defined as

$$V_{\text{coul}} \psi_i (\mathbf{r}\sigma) = \sum_j^{(\text{occ})} \psi_i (\mathbf{r}\sigma) \int \psi_j^* (\mathbf{r}'\sigma') \frac{e^2}{|\mathbf{r} - \mathbf{r}'|} \psi_j (\mathbf{r}'\sigma') \, d(\mathbf{r}'\sigma'), \tag{4.25b}$$

$$V_{\text{exch}} \psi_i (\mathbf{r}\sigma) = - \sum_j^{(\text{occ})} \psi_j (\mathbf{r}\sigma) \int \psi_j^* (\mathbf{r}'\sigma') \frac{e^2}{|\mathbf{r} - \mathbf{r}'|} \psi_i (\mathbf{r}'\sigma') \, d(\mathbf{r}'\sigma') \tag{4.25c}$$

(as frequently used, in the above expressions we have relabeled $\mathbf{r}_1 \sigma_1 \to \mathbf{r}\sigma$ the space and spin variables of the electron under attention, and $\mathbf{r}_2 \sigma_2 \to \mathbf{r}'\sigma'$ the space and spin variables that are integrated and traced out; the sum over $j = 1, 2, \ldots, N$ has been relabeled as sum over the occupied spin-orbitals).

If we perform explicitly the integration over the spin variables in Eq. (4.25b), we see that the Coulomb operator is the Hartree potential $V_{\text{coul}}(\mathbf{r})$, defined by Eq. (4.8). The exchange operator of Eq. (4.25c) (also called "exchange potential" although it is not a multiplicative potential) is an integral operator, since the wavefunction on which it operates enters under integral sign. If the integration over the spin variables is performed in Eq. (4.25c), it is seen that the occupied spin-orbitals contributing to $V_{\text{exch}} \psi_i (\mathbf{r}\sigma)$ are only the *spin-orbitals with spin parallel to* $\psi_i (\mathbf{r}\sigma)$.

The Hartree-Fock equations can be written in the more compact and perfectly equivalent form

$$F \psi_i = \varepsilon_i \psi_i, \tag{4.26a}$$

where F denotes the Fock operator

$$F = h(\mathbf{r}) + V_{\text{coul}}(\mathbf{r}) + V_{\text{exch}} \quad \text{with} \quad h(\mathbf{r}) = \frac{\mathbf{p}^2}{2m} + V_{\text{nucl}}(\mathbf{r}). \tag{4.26b}$$

The Fock operator is defined in terms of the occupied spin-orbitals ψ_i and must thus be determined in a self-consistent way (usually with appropriate iterations). The Fock operator reduces the many-electron problem to a one-particle problem (within the stated approximations); although this is a major formal and practical achievement, the highest caution must be applied for a correct interpretation of the meaning of eigenfunctions and eigenvalues of the Fock operator in connection with the original many-electron problem under investigation; we thus analyze further some general properties of the Hartree-Fock equations.

4.5.2   Ground-State Energy, Ionization Energies, and Transition Energies

Consider the Hartree-Fock approximation for a system of N interacting electrons and suppose we know (at least in principle) the exact self-consistent Fock operator of the system, its eigenfunctions ψ_i, and the spin-orbital energies ε_i. The N eigenfunctions of lower energy of the Fock operator represent the occupied spin-orbitals, used to construct the Fock operator itself; all the other eigenfunctions of the operator F are unoccupied, do not enter in the construction of F itself, and are called *virtual spin-orbitals*. In Figure 4.1 we represent schematically occupied and virtual spin-orbitals, in increasing order of energy, for an interacting system of N electrons; ionized states and excited states of the N-electron system are also schematically represented.

Ground-State Electronic Energy in the Hartree-Fock Approximation

In the Hartree-Fock approximation, the ground-state electronic energy $E_0^{(\text{HF})}$ is given by expression (4.23), where the ψ_i are the self-consistent orbitals. Multiplying both members of the Fock equation (4.26a) by $\langle \psi_i |$ on the left, we obtain for the Hartree-Fock spin-orbital energies the following expression

$$\varepsilon_i \equiv \langle \psi_i | F | \psi_i \rangle = \langle \psi_i | h | \psi_i \rangle + \sum_j^{(\text{occ})} \left[\langle \psi_i \psi_j | \frac{e^2}{r_{12}} | \psi_i \psi_j \rangle - \langle \psi_i \psi_j | \frac{e^2}{r_{12}} | \psi_j \psi_i \rangle \right].$$

$$\tag{4.27}$$

Direct comparison of Eq. (4.23) with Eq. (4.27) gives for the Hartree-Fock ground-state energy the expression

$$E_0^{(\text{HF})} = \sum_i^{(\text{occ})} \varepsilon_i - \frac{1}{2} \sum_{ij}^{(\text{occ})} \left[\langle \psi_i \psi_j | \frac{e^2}{r_{12}} | \psi_i \psi_j \rangle - \langle \psi_i \psi_j | \frac{e^2}{r_{12}} | \psi_j \psi_i \rangle \right]. \tag{4.28}$$

Equation (4.28) shows that the ground-state electronic energy *is not equal to the sum of the Hartree-Fock spin-orbital energies of the occupied states*; this can be understood intuitively from the fact that otherwise the Coulomb and exchange interactions would be counted twice. The total energy of the electronic-nuclear system in the Hartree-Fock

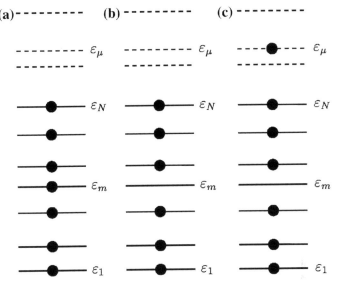

Figure 4.1 (a) Schematic representation of occupied spin-orbitals (solid lines), in increasing order of energy, for the ground-state of an N-electron system in the Hartree-Fock approximation; virtual spin-orbitals are represented with dashed lines. (b) Schematic representation of an ionized state of the N-electron system, with an electron removed from the occupied spin-orbital ψ_m (and transferred to infinity). (c) Schematic representation of an excited state of the N-electron system, with an electron removed from the occupied spin-orbital ψ_m and transferred to the virtual spin-orbital ψ_μ.

approximation is obtained adding to $E_0^{(\mathrm{HF})}$ the nucleus-nucleus Coulomb repulsion, given by the last term of the Hamiltonian (4.1).

It should be noticed that, in agreement with the variational principle, the Hartree-Fock energy $E_0^{(\mathrm{HF})}$ is *higher* than the exact ground-state energy $E_0^{(\mathrm{exact})}$ of the many-electron system, and the difference is called *correlation energy*. In spite of the importance and achievements of the Hartree-Fock approximation, corrections beyond it are often to be considered; this is due to the fact that a single determinantal state, even when built up with the best possible orbitals, remains in general a rather poor representation of the complicated ground-state wavefunction of a many-particle system.

Ionization Energy and Koopmans' Theorem

Consider an (ideal) ionization process in which an electron in the occupied state ψ_m is removed from the system and transferred to infinity (i.e. far enough from the electronic system) with zero kinetic energy; it is assumed that no modification (or "relaxation") of the other electronic orbitals occurs in the ionization process.

In the Hartree-Fock approximation, the energy $E_0^{(\mathrm{HF})}(N)$ of the ground-state with N electrons is given by Eq. (4.23); the energy $E_0^{(\mathrm{HF})}(N-1)$ of the ionized states is given by the same expression (4.23), provided the spin-orbital ψ_m is excluded from the sums. We have for the energy difference

$$
E_0^{(\mathrm{HF})}(N) - E_0^{(\mathrm{HF})}(N-1) = \langle \psi_m | h | \psi_m \rangle
$$
$$
+ \sum_j^{(\mathrm{occ})} \left[\langle \psi_m \psi_j | \frac{e^2}{r_{12}} | \psi_m \psi_j \rangle - \langle \psi_m \psi_j | \frac{e^2}{r_{12}} | \psi_j \psi_m \rangle \right] \equiv \varepsilon_m .
$$

We thus obtain the following important result, known as *Koopmans' theorem: the energy required to remove (without relaxation) an electron from the spin-orbital ψ_m is simply the opposite of the Hartree-Fock eigenvalue ε_m.* The parameters ε_m appearing in the Hartree-Fock equations (4.25) as purely formal Lagrange multipliers are thus rescued to physical interpretation by the Koopmans' theorem.

Excited States and Transition Energies

Consider now the ground-state and the singly substituted excited state

$$
\Psi_0 = \mathcal{A}\{\psi_1 \psi_2 \cdots \psi_m \cdots \psi_N\},
$$
$$
\Psi_{\mu,m} = \mathcal{A}\{\psi_1 \psi_2 \cdots \psi_\mu \cdots \psi_N\},
$$

where an electron has been removed from the occupied spin-orbital ψ_m and transferred in the previously unoccupied spin-orbital ψ_μ (we assume again that all the other occupied orbitals are unchanged). We want to determine the transition energy

$$
\Delta E = \langle \Psi_{\mu,m} | H_e | \Psi_{\mu,m} \rangle - \langle \Psi_0 | H_e | \Psi_0 \rangle .
$$

The last term in the above relation is given by expression (4.23), and the preceding one can be obtained with straight manipulations of Eq. (4.23). The transition energy ΔE then becomes

$$
\Delta E = \langle \psi_\mu | h | \psi_\mu \rangle - \langle \psi_m | h | \psi_m \rangle
$$
$$
+ \sum_j^{(\mathrm{occ})} \left[\langle \psi_\mu \psi_j | \frac{e^2}{r_{12}} | \psi_\mu \psi_j \rangle - \langle \psi_\mu \psi_j | \frac{e^2}{r_{12}} | \psi_j \psi_\mu \rangle \right]
$$
$$
- \left[\langle \psi_\mu \psi_m | \frac{e^2}{r_{12}} | \psi_\mu \psi_m \rangle - \langle \psi_\mu \psi_m | \frac{e^2}{r_{12}} | \psi_m \psi_\mu \rangle \right]
$$
$$
- \sum_j^{(\mathrm{occ})} \left[\langle \psi_j \psi_m | \frac{e^2}{r_{12}} | \psi_j \psi_m \rangle - \langle \psi_j \psi_m | \frac{e^2}{r_{12}} | \psi_m \psi_j \rangle \right] .
$$

Using Eq. (4.27) for the spin-orbital energies, we obtain

$$
\Delta E = \varepsilon_\mu - \varepsilon_m - \left[\langle \psi_\mu \psi_m | \frac{e^2}{r_{12}} | \psi_\mu \psi_m \rangle - \langle \psi_\mu \psi_m | \frac{e^2}{r_{12}} | \psi_m \psi_\mu \rangle \right] . \tag{4.29}
$$

The physical meaning of Eq. (4.29) is transparent indeed: to promote an electron from the occupied spin-orbital ψ_m to the unoccupied spin-orbital ψ_μ we need an energy equal to the difference $\varepsilon_\mu - \varepsilon_m$ of the spin-orbital energies, corrected by the *Coulomb* and *exchange* energy between the "extra electron" in the spin-orbital ψ_μ and the "missing electron" (hole) left in the initial spin-orbital ψ_m.

In crystals, ψ_μ and ψ_m are normalized Bloch functions, and thus the bielectronic integrals are of the order of $O(1/N)$ (where N is the number of unit cells of the crystal). This appears to provide a good justification for neglecting the "electron-hole interaction," expressed by the bielectronic integrals in square brackets in Eq. (4.29); thus when the wavefunctions are itinerant, we expect $\Delta E \approx \varepsilon_\mu - \varepsilon_m$ for the transition energies. Although for Bloch functions, bielectronic integrals are of the order of $O(1/N)$, we must remember that also the allowed **k** vectors (and hence wavefunctions for each band) in the Brillouin zone equal the number N of unit cells of the crystal; thus it is not a priori justified to neglect N contributions of order $O(1/N)$, and the exciton theory will discuss how to include electron-hole corrections, whenever necessary (see Section 7.1).

A final property we wish to highlight is that *the matrix elements of H_e between the Hartree-Fock ground-state and any singly substituted excited state are zero*; in fact, using Eqs. (4.19) and (4.20) it is seen that

$$\langle \Psi_{\mu,m} | H_e | \Psi_0 \rangle = \langle \psi_\mu | h | \psi_m \rangle + \sum_j^{(occ)} \left[\langle \psi_\mu \psi_j | \frac{e^2}{r_{12}} | \psi_\mu \psi_j \rangle - \langle \psi_\mu \psi_j | \frac{e^2}{r_{12}} | \psi_j \psi_\mu \rangle \right]$$

$$= \langle \psi_\mu | F | \psi_m \rangle \equiv 0 \quad \text{for} \quad \mu \neq m,$$

where F denotes the Fock operator of Eqs. (4.26). This result is known as *Brillouin theorem*. An obvious consequence of the Brillouin theorem is that energy lowering in the Hartree-Fock context requires the mixing of the ground-state with at least doubly substituted excitations of the Hartree-Fock scheme.

4.5.3 Hartree-Fock Equations and Transition Energies in Closed-Shell Systems

In the Hartree-Fock theory developed until now, no restriction has been considered on the spin part of the spin-orbitals entering the single Slater determinant, adopted to approximate the ground-state wavefunction of the electronic system. For closed-shell systems, the trial Slater determinant can be written in the more specific form

$$\Psi_0 = A\{\phi_1\alpha\,\phi_1\beta \cdots \phi_m\alpha\,\phi_m\beta \cdots \phi_{N/2}\alpha\,\phi_{N/2}\beta\}, \tag{4.30}$$

where $\phi_i(\mathbf{r})$ $(i = 1, 2, \ldots, N/2)$ indicate orthonormal orbitals. Notice that the state Ψ_0 has total spin component S_z equal to zero and total spin S equal to zero.

The general Hartree-Fock equations (4.25), carrying out explicitly the summation over spin variables, take the form

$$\left[\frac{\mathbf{p}^2}{2m} + V_{\text{nucl}}(\mathbf{r}) + V_{\text{coul}}(\mathbf{r}) + V_{\text{exch}} \right] \phi_i(\mathbf{r}) = \varepsilon_i \phi_i(\mathbf{r}) \tag{4.31a}$$

with the Coulomb and exchange operators given by

$$V_{\text{coul}}(\mathbf{r}) = 2 \sum_{j}^{N/2} \langle \phi_j(\mathbf{r}')| \frac{e^2}{|\mathbf{r} - \mathbf{r}'|} |\phi_j(\mathbf{r}')\rangle, \tag{4.31b}$$

$$V_{\text{exch}} \phi_i(\mathbf{r}) = - \sum_{j}^{N/2} \langle \phi_j(\mathbf{r}')| \frac{e^2}{|\mathbf{r} - \mathbf{r}'|} |\phi_i(\mathbf{r}')\rangle \phi_j(\mathbf{r}). \tag{4.31c}$$

We have already seen that ε_i can be interpreted as ionization energy in the Koopmans'
approximation.

We wish to discuss now the transition energies in closed-shell systems. When in the
determinantal state (4.30), an occupied orbital $\phi_m(\mathbf{r})$ is replaced by a virtual orbital
$\phi_\mu(\mathbf{r})$, we have the following four possible singly substituted excited states depending
on the initial and final spin functions:

$$\Psi_1 = \Psi_{\mu\alpha, m\beta}, \quad \Psi_2 = \Psi_{\mu\beta, m\beta}, \quad \Psi_3 = \Psi_{\mu\alpha, m\alpha}, \quad \Psi_4 = \Psi_{\mu\beta, m\alpha}. \tag{4.32}$$

The ground-state Ψ_0 and the four trial excited states Ψ_1, Ψ_2, Ψ_3, Ψ_4, where an electron
is created in ϕ_μ and a hole is left in ϕ_m, are shown schematically in Figure 4.2.

We can now diagonalize the many-electron Hamiltonian H_e on the four determinantal states (4.32) and construct the linear combinations with definite values of the
total spin component S_z and total spin multiplicity S. For this purpose, we calculate the
matrix M, whose elements are

$$M_{ij} = \langle \Psi_i | H_e | \Psi_j \rangle \quad (i, j = 1, 2, 3, 4). \tag{4.33a}$$

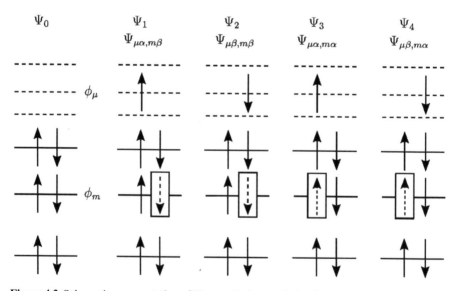

Figure 4.2 Schematic representation of Hartree-Fock occupied and virtual states (here supposed
to be discrete) of a closed-shell electronic system. In the four trial excited states Ψ_1, Ψ_2, Ψ_3,
and Ψ_4, an electron is promoted from the occupied orbital $\phi_m(\mathbf{r})$ to the virtual orbital $\phi_\mu(\mathbf{r})$.

Notice that the ground-state does not interact with the singly excited states (4.32), in agreement with the Brillouin theorem.

The explicit expression of the matrix elements M_{ij} is rather trivial. Consider for example the first diagonal element. For its evaluation we can use Eq. (4.29) (with appropriate relabeling of orbitals); it holds

$$
\begin{aligned}
M_{11} &= \langle \Psi_1 | H_e | \Psi_1 \rangle = \langle \Psi_{\mu\alpha,m\beta} | H_e | \Psi_{\mu\alpha,m\beta} \rangle \\
&= E_0 + \varepsilon_\mu - \varepsilon_m - \left[\langle \phi_{\mu\alpha}\phi_{m\beta} | \frac{e^2}{r_{12}} | \phi_{\mu\alpha}\phi_{m\beta} \rangle - \langle \phi_{\mu\alpha}\phi_{m\beta} | \frac{e^2}{r_{12}} | \phi_{m\beta}\phi_{\mu\alpha} \rangle \right] \\
&= E_0 + \varepsilon_\mu - \varepsilon_m - \langle \phi_\mu\phi_m | \frac{e^2}{r_{12}} | \phi_\mu\phi_m \rangle = E_0 + \varepsilon_\mu - \varepsilon_m - Q_{\mu m},
\end{aligned}
$$

where E_0 is the ground-state electronic energy in the Hartree-Fock approximation, ε_μ and ε_m are the Hartree-Fock spin-orbital energies, and $Q_{\mu m}$ and $J_{\mu m}$ denote the Coulomb bielectronic integral and the exchange bielectronic integral given by

$$
Q_{\mu m} = \langle \phi_\mu(\mathbf{r}_1)\phi_m(\mathbf{r}_2) | \frac{e^2}{r_{12}} | \phi_\mu(\mathbf{r}_1)\phi_m(\mathbf{r}_2) \rangle,
$$

$$
J_{\mu m} = \langle \phi_\mu(\mathbf{r}_1)\phi_m(\mathbf{r}_2) | \frac{e^2}{r_{12}} | \phi_m(\mathbf{r}_1)\phi_\mu(\mathbf{r}_2) \rangle.
$$

With similar procedures all the other diagonal matrix elements can be obtained. For what concerns off-diagonal matrix elements, since the operator H_e conserves the total component of the spin, we have immediately that the off-diagonal matrix elements M_{ij} ($i \neq j$) are all zero, except for $M_{23} = M_{32} = J_{\mu m}$. In summary, we have that the matrix M is given by the expression

$$
M = E_0 + \varepsilon_\mu - \varepsilon_m + \begin{pmatrix} -Q_{\mu m} & 0 & 0 & 0 \\ 0 & -Q_{\mu m} + J_{\mu m} & J_{\mu m} & 0 \\ 0 & J_{\mu m} & -Q_{\mu m} + J_{\mu m} & 0 \\ 0 & 0 & 0 & -Q_{\mu m} \end{pmatrix}
$$

(4.33b)

(it is implicitly understood that the scalar quantity $E_0 + \varepsilon_\mu - \varepsilon_m$ is multiplied by the 4×4 identity matrix before addition is performed).

The diagonalization of the matrix (4.33b) gives one triplet state with energy

$$
E_{\text{triplet}} = E_0 + \varepsilon_\mu - \varepsilon_m - Q_{\mu m} \tag{4.34a}
$$

and one singlet state with energy

$$
E_{\text{singlet}} = E_0 + \varepsilon_\mu - \varepsilon_m - Q_{\mu m} + 2J_{\mu m}. \tag{4.34b}
$$

For *triplet states* ($S = 1$; $S_z = 1, 0, -1$), the appropriate combinations of the basis functions (4.32) with definite spin component are

$$
\Psi_{\mu\alpha,m\beta}, \quad \frac{1}{\sqrt{2}}(\Psi_{\mu\alpha,m\alpha} - \Psi_{\mu\beta,m\beta}), \quad \Psi_{\mu\beta,m\alpha}.
$$

For the *singlet state* $(S = 0; \ S_z = 0)$ the appropriate combination is

$$\frac{1}{\sqrt{2}}(\Psi_{\mu\alpha,m\alpha} + \Psi_{\mu\beta,m\beta}).$$

For brevity, we indicate trial singly substituted excited states of a given spin multiplicity by $\Psi_{\mu,m}^{(S)}$ with $S = 0$ for the singlet state and $S = 1$ for the triplet state (for the triplet, we select arbitrary one of the three possible partners); equations (4.34) can be summarized in the compact form

$$\langle \Psi_{\mu,m}^{(S)} | H_e | \Psi_{\mu,m}^{(S)} \rangle = E_0 + \varepsilon_\mu - \varepsilon_m - \langle \phi_\mu \phi_m | \frac{e^2}{r_{12}} | \phi_\mu \phi_m \rangle$$
$$+ 2\delta_{S,0} \langle \phi_\mu \phi_m | \frac{e^2}{r_{12}} | \phi_m \phi_\mu \rangle \tag{4.35}$$

Since exchange integrals are definite positive quantities, we see that the triplet (magnetic) state is lower in energy than the singlet (non-magnetic) state, in the model we are considering.

Before concluding we also report the off-diagonal matrix elements of the operator H_e among singly substituted excited states for closed-shell systems. With the help of Eq. (4.22), and with a procedure completely similar to the one used to prove Eq. (4.35), we obtain

$$\langle \Psi_{\mu,m}^{(S)} | H_e | \Psi_{\nu,n}^{(S)} \rangle = -\langle \phi_\mu \phi_n | \frac{e^2}{r_{12}} | \phi_\nu \phi_m \rangle + 2\delta_{S,0} \langle \phi_\mu \phi_n | \frac{e^2}{r_{12}} | \phi_m \phi_\nu \rangle. \tag{4.36}$$

We will need these results in the theory of excitons in solids of Section 7.1.

4.5.4 Hartree-Fock-Slater and Hartree-Fock-Roothaan Approximations

The major obstacle to the solution of the Hartree-Fock equations is the integro-differential nature of the Fock operator. Various physical and technical elaborations have been devised for it in the literature; among them we here mention, in their general lines, the ideas of the Slater approximation and the Roothaan approximation, because of their basic role in solid state physics and in atomic and molecular physics, respectively.

Hartree-Fock-Slater Approximation

The Slater approximation starts from the fact that the exchange integral operator in the homogeneous free-electron gas can be well approximated with a local potential (as shown in Section 4.7). Then Slater proposes to replace, also for non-homogeneous systems, the exchange integral operator with the ordinary local potential, corresponding to the free-electron gas of the same local density $n(\mathbf{r})$; in this way the exchange operator takes the form

$$V_{\text{exch}}^{(\text{Slater})}(\mathbf{r}) = -\frac{3}{2}\frac{e^2}{\pi}[3\pi^2 n(\mathbf{r})]^{1/3}.$$

With the above approximation, the Fock operator (4.26) becomes an ordinary differential operator with a local potential to be determined self-consistently. The method (with appropriate implementations) has been of wide use mainly in solid state physics.

The Hartree-Fock-Slater approximation becomes particularly agile when applied to atoms, because of the spherical symmetry of the (average) atomic electron density $n(r)$. The effective atomic potential in the Hartree-Fock-Slater approximation is related to $n(r)$ in the form

$$V^{(\mathrm{HFS})}(r) = -\frac{Ze^2}{r} + \frac{e^2}{r} \int_0^r 4\pi r'^2 n(r')\, dr'$$
$$+ e^2 \int_r^\infty 4\pi r' n(r')\, dr' - \frac{3}{2}\frac{e^2}{\pi}[3\pi^2 n(r)]^{1/3}. \tag{4.37}$$

The first term in the right-hand side of Eq. (4.37) is the nuclear potential, the second and the third are the Hartree potential of the spherically symmetric charge distribution $(-e)n(r)$; the last is the Slater local exchange potential [to be precise the potential (4.37) should be appropriately corrected to preserve the asymptotic behavior $-e^2/r$, when acting on occupied wavefunctions]. Usually the Numerov method is adopted to solve the radial Schrödinger equation; self-consistency of wavefunctions and potential can be achieved with reasonable modest computational labor. Complete calculations (program code included) for all atoms is reported, for instance, by F. Herman and S. Skillman "Atomic Structure Calculations" (Prentice Hall, Englewood Cliffs, New Jersey, 1963).

The Hartree-Fock-Roothaan Method

Because of its importance in theoretical chemical physics, it is worthwhile to mention the method of Roothaan for the solution of the Hartree-Fock equations. This method introduces a finite number of localized basis spin-orbitals, on which to expand each Hartree-Fock orbital; also the integro-differential Hartree-Fock operator is represented on the chosen basis set, and the Hartree-Fock equations are thus converted into a set of self-consistent algebraic equations in the expansion coefficients, to be solved by matrix techniques. A nice feature of the Hartree-Fock-Roothaan procedure is that the quality of the results can be tested by enlarging the basis set; a limitation of the method is that numerical labor rapidly increases with the number of electrons; in practice, besides atoms, only molecules or clusters with small to moderate number of electrons (up to a few tens or so) can be handled (unless further simplifying assumptions are inserted).

As a final point we should remember the basis functions used in most calculations. The localized atomic functions are often expressed in terms of Slater-type functions; the normalized Slater-type functions are defined as

$$\phi_{nlm}(\mathbf{r}) = \frac{(2\alpha)^{n+(1/2)}}{\sqrt{(2n)!}} r^n e^{-\alpha r} Y_{lm}(\mathbf{r}),$$

where $Y_{lm}(\mathbf{r})$ denote normalized harmonic functions. Optimized coefficients for atoms are reported for instance by E. Clementi and C. Roetti "Roothaan-Hartree-Fock Atomic Wavefunctions" Atomic Data Nuclear Data Tables *14*, 177 (1974).

Another convenient set is constituted by Gaussian-type functions; normalized Gaussian orbitals are defined as

$$g_{nlm}(\mathbf{r}) = \left(\frac{2}{\pi}\right)^{1/4} \frac{2^{n+(1/2)}}{\sqrt{(2n-1)!!}} \alpha^{(2n+1)/4} r^{n-1} e^{-\alpha r^2} Y_{lm}(\mathbf{r}).$$

This choice makes it possible to perform many integrals analytically [see for instance I. Shavitt in "Methods in Computational Physics" (Eds. B. Adler, S. Fernbach and M. Rotenberg, Academic Press, New York, 1963)]. On the other hand a large number of Gaussians are in general needed to represent atomic-like wavefunctions [for further aspects see for instance S. Huzinaga, J. Andzelm, M. Klobukowski, E. Radzio-Andzelm, Y. Sakai and H. Takewaki (Eds.) "Gaussian Basis Sets for Molecular Calculation" (Elsevier, Amsterdam, 1984); R. Poirier, R. Kari and I. Csizmadia "Handbook of Gaussian Basis Sets" (Elsevier, Oxford, 1985)].

4.6 Overview of Approaches Beyond the One-Electron Approximation

In the one-electron approximation, electrons with parallel spin are correctly taken apart in real space by the use of antisymmetrized wavefunctions (see the discussion on Eq. 4.12). The depletion in real space of parallel spin particles around a reference electron occurs as a consequence of Fermi statistics, and is pictured as a sort of *exchange hole or Fermi hole* accompanying the motion of every electron. Thus, the basic inadequacy of the Hartree-Fock framework stands in its neglecting correlation between the motions of electrons with antiparallel spin; attempts beyond the Hartree-Fock approximation are mainly motivated to overcome this major fault.

Despite this and other limitations, the Hartree-Fock approximation is a milestone in the many-body electron problem from several points of view. From a *conceptual point of view*, it allows to separate exchange effects from correlation effects; from a *technical point of view*, it introduces the technicalities to face the demanding self-consistent equations that have found applications also in many other fields; from a *practical point of view*, a number of problems can be understood (at least qualitatively) within this approximation.

In most many-body systems, however, the Hartree-Fock method is insufficient for the desired accuracy; in these cases the Hartree-Fock approximation is usually considered as a starting point before more sophisticated approaches are implemented. It is outside the scope of this book to present a thorough discussion of the main lines of improvements and their physical or heuristic motivations; thus we refrain from any quantitative discussion and simply mention some developments.

One of the most important ways to go beyond the Hartree-Fock theory is offered by the quantum chemistry approaches, which consider the Hartree-Fock method as a starting point to generate a convenient set of one-electron spin-orbitals, and then insert excited configurations. These methods have different names such as configuration interaction approach, multi-configuration self-consistent field, many-body perturbation theory, depending on the adopted implementations. In essence, besides the ground Hartree-Fock state, one considers single excited states (in which one of the occupied spin-orbitals is replaced by a virtual spin-orbital), double excited states, etc. and diagonalizes the many-body Hamiltonian among a judiciously chosen set of (trial) excited states. This procedure has been widely used in theoretical quantum chemistry for atoms, molecules, and clusters, to obtain a better description of the ground-state and the lowest lying excitations of the many-electron system. In solid state physics, *excitons* in insulators and semiconductors are often studied in this spirit (see Chapter 7).

For open-shell systems, the situation may become rather complicated, when several spin-configurations are expected to produce levels within narrow energy intervals. In some cases, it is possible to describe the low-lying states by approximate *equivalent Hamiltonians* (for instance of the type of the Heisenberg spin Hamiltonian in magnetism, see Chapter 17).

A very important line of progress, is given by theoretical developments of the Green's function many-body formalism. In essence the main novelties with respect to the one-electron formalism, is the screening of the exchange potential and the introduction of the Coulomb hole (COHSEX: Coulomb hole and screened exchange approximation). This line (though not routine at all) has led a long way toward a better quantitative description of correlation effects in solids [see for instance L. Hedin and S. Lundqvist, Solid State Physics *23*, 1 (1969) and references quoted therein].

A special importance has been acquired by *the density functional theory*, in the study of the ground-state properties of a many-body system. This theory focuses on the ground-state electron density (rather than eigenfunctions) and provides a number of highly accurate results for the ground-state properties. We discuss the main aspects of the density functional theory later in this chapter. Finally we mention the important role assumed by the quantum Monte Carlo method in many-body calculations: it gives practically "exact" results for small systems and for the homogeneous electron gas. The Monte Carlo method has been used also as a precious test of the approximations introduced within the density functional theory in atoms or molecules with small number of electrons.

4.7   Electronic Properties and Phase Diagram of the Homogeneous Electron Gas

In this section we discuss some relevant properties of the *homogeneous electron gas or fluid*; this many-body system is constituted by *interacting electrons embedded in a uniform neutralizing background of positive charges (jellium model)*. The homogeneous electron gas is of major interest from several points of view. At the most elementary level, it provides a semiquantitative description of the conduction electrons in simple metals (Sommerfeld model of Chapter 3). It is the prototype system, where it is possible to illustrate analytically salient aspects and serious faults of the Hartree-Fock approximation. The electronic correlations of the uniform electron gas have long represented the natural test of the many-body diagrammatic developments, as well as of the Monte Carlo numerical applications. A thorough understanding of the variety of properties of the electron gas model represents an invaluable tool also for the investigation of real materials (as in the local density approximation to the density functional theory, discussed in Section 4.8). Last, but not least, with lowering density the uniform electron gas displays peculiar phase transitions to magnetic ordering and to Wigner crystallization (not to mention other possible lower symmetry phases). In this section, we touch some of the mentioned problems, mostly at a qualitative level.

Normal Ground State of the Homogeneous Electron Gas in the Hartree-Fock Approximation

It is instructive to analyze the jellium model within the Hartree-Fock theory, even if it is well known that this theory is completely inadequate for the description of the electron gas. Consider a system of N electrons, in the volume V, embedded in a uniform neutralizing background of positive charges. We indicate with $n = N/V$ the electron density and with r_s the dimensionless parameter related to the electronic density n by the equality $(4\pi/3)r_s^3 a_B^3 = 1/n$, where a_B is the Bohr radius; $r_s a_B$ represents the radius of the sphere containing (in average) one electron. The case of the non-interacting electron gas, i.e. the case that the electron-electron interaction can be neglected, constitutes the free-electron Sommerfeld model, characterized by a parabolic energy dispersion curve $E(\mathbf{k}) = \hbar^2 k^2/2m$ (schematically indicated in Figure 4.3). We discuss now within the Hartree-Fock framework the interacting electron gas.

The Hartree-Fock equations of the homogeneous (interacting) electron gas can be solved exactly using plane waves. The normal ground-state of the interacting gas is approximated with a single Slater determinant, formed by doubly occupied plane waves with wavevectors lying within the Fermi sea:

$$\Psi_0 = \mathcal{A}\{W_{\mathbf{k}_1}\alpha\ W_{\mathbf{k}_1}\beta \cdots W_{\mathbf{k}_{N/2}}\alpha\ W_{\mathbf{k}_{N/2}}\beta\} \quad \text{with} \quad W(\mathbf{k}_i, \mathbf{r}) = \frac{1}{\sqrt{V}}e^{i\mathbf{k}_i\cdot\mathbf{r}}. \quad (4.38)$$

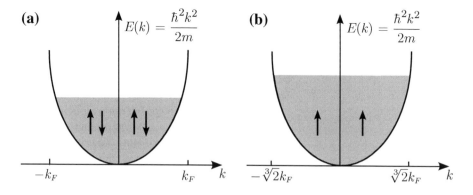

Figure 4.3 (a) Schematic representation of the *normal state* of the free-electron gas of Fermi wavevector k_F. (b) Schematic representation of the *fully polarized state* of the free-electron gas with the same density, and hence Fermi wavevector $2^{1/3}k_F$.

In the jellium model the nuclei are replaced by a uniform positively charged background; so $V_{nucl}(\mathbf{r})$ and $V_{coul}(\mathbf{r})$ exactly cancel, and the Fock operator (4.31) becomes

$$F = \frac{\mathbf{p}^2}{2m} + V_{exch}, \tag{4.39}$$

where the exchange integral operator, given by Eq. (4.31c), is built with the occupied plane waves. In the case the Hartree approximation is considered, the exchange operator is omitted in Eq. (4.39) and one recovers the free-electron Sommerfeld model.

It is well known that the kinetic energy operator $\mathbf{p}^2/2m$ is diagonal in a plane wave representation. The same property holds for the exchange integral operator built up with plane waves; in fact

$$
\begin{aligned}
V_{exch} \frac{e^{i\mathbf{k}\cdot\mathbf{r}}}{\sqrt{V}} &= -\sum_{\mathbf{q}}^{(occ)} \frac{1}{\sqrt{V}} e^{i\mathbf{q}\cdot\mathbf{r}} \int \frac{1}{\sqrt{V}} e^{-i\mathbf{q}\cdot\mathbf{r}'} \frac{e^2}{|\mathbf{r}-\mathbf{r}'|} \frac{1}{\sqrt{V}} e^{i\mathbf{k}\cdot\mathbf{r}'} \, d\mathbf{r}' \\
&= -\frac{e^{i\mathbf{k}\cdot\mathbf{r}}}{\sqrt{V}} \sum_{\mathbf{q}}^{(occ)} \int \frac{1}{V} e^{-i(\mathbf{k}-\mathbf{q})\cdot(\mathbf{r}-\mathbf{r}')} \frac{e^2}{|\mathbf{r}-\mathbf{r}'|} \, d\mathbf{r}' \\
&= -\frac{e^{i\mathbf{k}\cdot\mathbf{r}}}{\sqrt{V}} \frac{1}{V} \sum_{q<k_F} \frac{4\pi e^2}{|\mathbf{k}-\mathbf{q}|^2}.
\end{aligned}
\tag{4.40a}
$$

The last passage in Eq. (4.40a) uses the Fourier transform of the Yukawa-like potential

$$\int \frac{e^{-k_s r}}{r} e^{i\mathbf{k}\cdot\mathbf{r}} \, d\mathbf{r} = \frac{4\pi}{k^2 + k_s^2} \tag{4.40b}$$

in the particular case that the screening parameter k_s vanishes.

From Eq. (4.40a) and Eqs. (C.3) and (C.4), we see that plane waves are eigenfunctions of V_{exch} and satisfy the eigenvalue equation

$$V_{exch} W(\mathbf{k}, \mathbf{r}) = -\frac{2e^2 k_F}{\pi} F\left(\frac{k}{k_F}\right) W(\mathbf{k}, \mathbf{r}), \tag{4.41a}$$

where k_F is the Fermi wavevector, and

$$F(x) = \frac{1}{2} + \frac{1-x^2}{4x} \ln \left| \frac{1+x}{1-x} \right|. \tag{4.41b}$$

The plane waves are thus also eigenfunctions of the Fock operator (4.39) with eigenvalues

$$\varepsilon(k) = \frac{\hbar^2 k^2}{2m} - \frac{2e^2 k_F}{\pi} F\left(\frac{k}{k_F}\right). \tag{4.42}$$

Notice that the exchange contribution gradually vanishes as the electron wavevector k overcomes the Fermi wavevector k_F. The behavior of $F(x)$ and $\varepsilon(k)$ are given in Figure 4.4.

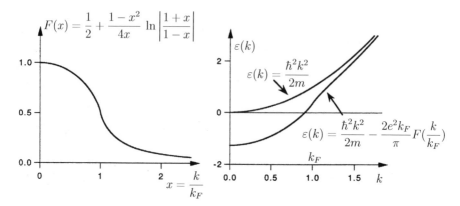

Figure 4.4 (a) Schematic plot of the function $F(x)$. (b) Kinetic energy and Hartree-Fock orbital energy as a function of the wavevector k for the homogeneous electron gas. Energies are in Rydbergs, k is in units of a_B^{-1} (inverse Bohr radius), and we have taken $k_F = 1/a_B$.

From Eq. (4.42), we can see that the Hartree-Fock treatment of the homogeneous electron gas presents quite serious drawbacks. For instance, we notice that $d\varepsilon(k)/dk$ presents an (unphysical) logarithmic divergence at $k = k_F$; this divergence would make the density-of-states vanish for $\varepsilon \approx \varepsilon_F$; on the contrary, experimental evidence shows that the density-of-states in simple metals varies rather regularly throughout the whole conduction band, Fermi energy region included. The Hartree-Fock results also produce unphysical large widths of conduction bands in metals (and unphysical large band gaps in insulators up to a factor 2 or so). From a critical reflection on Eq. (4.40), one can envisage that the unphysical singularity of the function $F(x)$ could be cured with an appropriate screening of the long-range Coulomb interaction in the exchange integrals, in line with the general findings and diagrammatic expansions of the many-body theory.

In spite of these and other inadequacies of the Hartree-Fock theory, it is interesting to evaluate the Hartree-Fock ground-state energy, as this constitutes an *upper bound* of the exact ground-state energy of the many-body system, independently on how good or how bad the Hartree-Fock approximation actually might be. The total Hartree-Fock ground-state energy can be obtained summing up the orbital energies given by Eq. (4.42) and introducing the factor 1/2 in the exchange electron-electron contribution (to avoid its double counting); we have

$$E_0^{(HF)} = 2 \sum_{k<k_F} \frac{\hbar^2 k^2}{2m} - 2 \sum_{k<k_F} \frac{1}{2} \frac{2e^2 k_F}{\pi} F\left(\frac{k}{k_F}\right), \tag{4.43}$$

where the factor 2 in front of the sums takes into account spin degeneracy. In Eq. (4.43), we can perform explicitly the sum over k for the kinetic term, while for the exchange term we can replace F with its average value $F_{av} = 3/4$ (see Eq. (C.5)); we obtain

$$E_0^{(HF)} = N \left[\frac{3}{5} \frac{\hbar^2 k_F^2}{2m} - \frac{3}{4} \frac{e^2 k_F}{\pi} \right]. \tag{4.44a}$$

Using the dimensionless parameter r_s and the relationship $k_F a_B = 1.92/r_s$, the ground-state energy per particle (in Rydberg units) becomes

$$\frac{E_0^{(HF)}}{N} = \frac{2.21}{r_s^2} - \frac{0.916}{r_s} \quad \text{Rydberg} = \frac{\hbar^2}{2ma_B^2} = \frac{e^2}{2a_B} = 13.606 \text{ eV}. \qquad (4.44b)$$

Since $2 < r_s < 6$ in ordinary metals, the exchange energy represents an important contribution to the binding energy of metals. In the high density limit (i.e. $r_s < 1$) the exact leading terms in the ground-state energy are given by Eq. (4.44b).

Another interesting remark can be made on Eqs. (4.41). From Eq. (4.41b) we see that $F(x)$ varies smoothly (from one to one half) as k goes from zero to k_F. In the hypothetical case that the function $F(x)$ is replaced by its average value 3/4 (see Eq. (C5)), the exchange operator in Eq. (4.41a) becomes a constant equal to

$$V_{\text{exch}}^{(\text{Slater})} = -\frac{3}{2}\frac{e^2}{\pi}k_F = -\frac{3}{2}\frac{e^2}{\pi}(3\pi^2 n)^{1/3}. \qquad (4.45a)$$

Slater suggested that expression (4.45a) could be adopted also for non-homogeneous electronic systems of *local density* $n(\mathbf{r})$, so obtaining the famous "$n^{1/3}$ local approximation" to the exchange operator

$$V_{\text{exch}}^{(\text{Slater})}(\mathbf{r}) = -\frac{3}{2}\frac{e^2}{\pi}[3\pi^2 n(\mathbf{r})]^{1/3}. \qquad (4.45b)$$

It is worthwhile to mention that the Slater local exchange approximation has had historically an important role in making many properties of actual materials accessible to the theory and to the interpretation; the reason is that the Hartree-Fock equations become ordinary differential equations, when the exchange operator is approximated with a local potential. In the literature, many attempts have been done to correct and improve the Slater expression (4.45b); among them we mention the once popular "X_α local approximation", where a semi-empirical parameter α multiplies the original Slater expression of Eq. (4.45b). All these attempts have come to an end with the advances of the density functional theory, mainly in the local density approximation (see Section 4.8).

Fully Spin-Polarized Ground State in the Hartree-Fock Approximation

In the Hartree-Fock approximation, the normal ground-state of the electron gas has been pictured as a Slater determinant constituted by doubly occupied plane waves of wavevectors confined within the Fermi sphere. Besides the normal ground-state, it is of interest to consider the extreme situation in which all electrons have parallel spin; we wish to compare the ground-state energy of the "normal phase" (also called the "paramagnetic phase") with the energy of the "fully polarized phase" (also called "ferromagnetic phase").

In the free-electron model, it is evident that the normal phase is *always* more stable than the polarized phase (see Figure 4.3); this is no more true in the Hartree-Fock approach. Consider in fact the Hartree-Fock ground-state energy of the normal phase given by Eq. (4.44a); the ground-state energy for the fully magnetized state, is obtained by Eq. (4.44a) with the replacement $k_F \rightarrow 2^{1/3}k_F$ (to account for the new occupied volume in reciprocal space). Thus the "fully polarized phase" is lower in energy than the "normal phase" when the condition

$$\frac{3}{5}\frac{\hbar^2}{2m}(2^{1/3}k_F)^2 - \frac{3e^2}{4\pi}(2^{1/3}k_F) < \frac{3}{5}\frac{\hbar^2}{2m}k_F^2 - \frac{3e^2}{4\pi}k_F$$

is verified. The above inequality is satisfied for

$$k_F < \frac{5}{2\pi}\frac{1}{2^{1/3}+1}\frac{1}{a_B} \quad \text{and thus} \quad r_s > 5.45 \tag{4.46}$$

($a_B = \hbar^2/me^2 =$ Bohr radius; $k_F a_B = 1.92/r_s$).

The Hartree-Fock approximation predicts that, at sufficiently low density, the spin-aligned state of the electrons is more stable than the normal unpolarized state. However, the critical value for the transition, estimated at $r_s \approx 5.45$ (a value still in the range of ordinary metallic densities), is completely unreliable; this is another example of the inadequacy of the Hartree-Fock method for treating quantitatively the electron gas. When electron correlation effects are taken into account in the jellium model, it is seen that the region of stability of the normal phase extends up to $r_s \approx 75$, as discussed below with Monte Carlo methods.

Considerations on Wigner Crystallization

At high density (small r_s), the electron jellium becomes an (almost) ideal Fermi gas with kinetic energy dominant with respect to electrostatic repulsion of electrons; the ground-state is represented by double occupied plane waves filling the Fermi sphere, in order to minimize the electron kinetic energy. At intermediate densities, a fully polarized fluid with all the spins aligned has energy lower than the normal fluid in the approximate range $75 < r_s < 100$. With further increase of the average interparticle spacing $r_s a_B$, each electron becomes trapped in the potential cage created by the Coulomb repulsion of its neighbors, and the Mott insulator phase occurs. It is generally accepted that, in three dimensions, electrons localize in a body-centered cubic lattice. The crystallization into a regular lattice of a low density electron gas was predicted by E. P. Wigner, Phys. Rev. *46*, 1002 (1934); Trans. Faraday Soc. *34*, 678 (1938).

The origin of the low density Wigner crystallization can be intuitively argued as follows. In the ground-state of the Wigner lattice, electrostatic repulsion of electrons is dominant on kinetic energy and the electrons perform zero-point oscillations around their lattice sites. The region in which each electron vibrates can be estimated by constructing a sphere of radius $r_s a_B$ surrounding the lattice site. Within such a sphere, the electron feels the restoring potential generated by the uniform positive background charge within the sphere (see Figure 4.5); contributions from other spheres are higher order multi-pole terms and are here neglected in these qualitative considerations. From classical electrostatic, the potential energy of an electron, at position r from the center

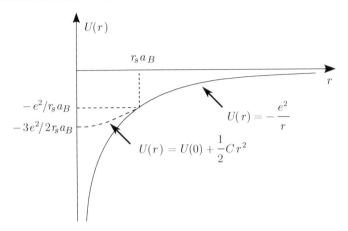

Figure 4.5 Potential energy of an electron in the Coulomb field due to a uniformly charged sphere of radius $r_s a_B$ and total charge +e. For $0 \leq r \leq r_s a_B$ the potential is quadratic, with "elastic constant" $C = e^2/r_s^3 a_B^3$.

of a sphere of radius $r_s a_B$ with a uniform charge +e is

$$U(r) = U(0) + \frac{1}{2}\frac{e^2}{r_s^3 a_B^3}r^2 \quad 0 < r < r_s a_B \tag{4.47a}$$

with $U(0) = -3e^2/2r_s a_B$. This potential is just that of an harmonic oscillator.

From the elementary properties of the harmonic oscillator, we know that the ground-state wavefunction of a particle of mass m bound to a spring of constant $C = e^2/r_s^3 a_B^3$ is a Gaussian function given by

$$\phi(r) = \left(\frac{\alpha}{\pi}\right)^{3/4} \exp\left(-\frac{1}{2}\alpha r^2\right), \tag{4.47b}$$

where

$$\alpha^2 = \frac{mC}{\hbar^2} = \frac{me^2}{\hbar^2 r_s^3 a_B^3} = \frac{1}{a_B^4 r_s^3} \tag{4.47c}$$

and a_B is the Bohr radius. The description of each electron in each cell with the localized wavefunction $\phi(r)$ is valid provided $\phi(r_s a_B) \ll \phi(0)$ (say at least one order of magnitude or so). To be conservative, we require $\alpha r_s^2 a_B^2 > 10$ and hence $\alpha^2 r_s^4 a_B^4 > 100$; using Eq. (4.47c) the condition becomes $r_s > 100$. While this numerical estimation of the critical density for Wigner crystallization is only indicative, it nevertheless hints that as r_s increases, the electrons tend to localize around appropriate lattice sites.

The Phase Diagram of the Homogeneous Electron Gas by Monte Carlo Methods

Among the numerous theoretical and numerical investigations of the uniform electron gas beyond the one-electron approximation, the quantum Monte Carlo methods have been particularly successful in providing exact constraints and (almost) exact

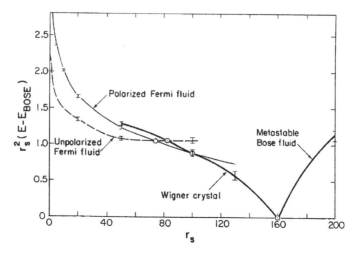

Figure 4.6 Energy (times r_s^2) in Rydberg versus r_s for several phases of the homogeneous electron gas. The figure shows that the normal Fermi fluid is stable below $r_s \approx 75$, the polarized (ferromagnetic) Fermi fluid is stable between $r_s \approx 75$ and $r_s \approx 100$, the Wigner crystal above $r_s \approx 100$. In the figure the reference energy E_{BOSE} is that of the lowest boson phase, i.e. the Bose fluid for $r_s < 160$, and the Wigner crystal for $r_s > 160$ [reprinted with permission from D. Ceperley and B. J. Alder, Phys. Rev. Lett. *45*, 566 (1980); copyright 1980 by the American Physical Society].

results, for the understanding of the properties and phase diagram of this surprising rich paradigmatic system. We briefly discuss some achievements in this direction, as illustrated in Figure 4.6, reported from the work of Ceperley and Alder (1980).

In the Hartree-Fock approach to the homogeneous electron gas, we have seen the serious drawbacks produced by the representation of the electronic system with a single trial Slater determinant. The simplest generalization of a Slater determinant is the many-body Slater-Jastrow wavefunction of the form

$$\Psi_T(\{\mathbf{r}_i \sigma_i\}) = D(N) \exp\left[-\sum_{i<j} u(r_{ij}) \right] \quad r_{ij} = |\mathbf{r}_i - \mathbf{r}_j|, \tag{4.48}$$

where $D(N)$ is a Slater determinant on N independent orbitals and $u(r)$ is a pair correlation factor. [Jastrow-type correlation factors are encountered in several other subjects of the physics, for instance in the Laughlin many-body wavefunction of the fractional Hall effect discussed in Chapter 15].

The trial wavefunction Ψ_T is the product of two-body correlation factors (whose tendency is to keep apart two electrons approaching each other) times a Slater determinant $D(N)$ of single-particle orbitals. For the normal fluid, $D(N)$ is a Slater determinant of plane waves with **k**-vectors within the Fermi sphere of radius k_F and both spin directions. For the fully polarized fluid, there is only one spin direction for each plane wave within the Fermi sphere of radius increased to $2^{1/3}k_F$, to assure the same electronic density. For the fully polarized crystal phase, $D(N)$ is a Slater determinant of Gaussian orbitals centered around the sites of a body-centered cubic lattice.

The basic strategy of the variational Monte Carlo method consists in the direct evaluation of the multi-dimensional integrals involved in the calculation of the trial energy

$$E_T = \frac{\langle \Psi_T | H_e | \Psi_T \rangle}{\langle \Psi_T | \Psi_T \rangle};$$

this is obtained expressing the integrand on a representative random sampling of points. The variational Monte Carlo method is used to minimize the variational energy with respect to the parameters of the correlation factor. For high-accuracy energy calculations, the diffusion Monte Carlo method is used to remove the variational bias; for this purpose, the trial wave-function is allowed to evolve in imaginary time, so to filter out from Ψ_T the ground-state (with the same symmetry).

In Figure 4.6 the energy of several phases of interest of the electron gas are reported. It can be seen that the normal fluid is stable for $r_s < 75$; the fully polarized fluid is stable in the region approximately given by $75 < r_s < 100$ and the Wigner crystal is stable for higher values of r_s. For sake of comparison, Figure 4.6 also considers a fictitious system of "charged bosons," with the same charge, mass, and density as the electrons. While a fermion wavefunction is totally antisymmetric under particle exchange, a boson wavefunction is totally symmetric. For the boson fluid the trial wavefunction is the Jastrow function $\Psi_T = \exp\left[-\sum_{i<j} u(r_{ij})\right]$; for the Wigner crystal of bosons, the trial wavefunction is constructed by the product of Gaussian orbitals centered at bcc sites and the Jastrow factor. Notice that the Wigner crystals of bosons and fermions have practically the same energy for the large values of r_s (> 50) of interest in the Figure 4.6 (actually, the product of Gaussian orbitals needs not to be symmetrized for large r_s).

Studies of electron gas by quantal simulation techniques have also been extended to the exact determination of static response functions, thus providing invaluable checks of theories and approximations, as discussed for instance in the works of Moroni et al. cited in the bibliography. The Monte Carlo calculations have been used also to provide precious tests of the approximations inherent the Hartree-Fock and the density functional theory in atoms, molecules, and clusters with small number of electrons. For further aspects and applications of the quantum Monte Carlo methods to inhomogeneous systems we refer for instance the review article of Needs et al. and references quoted therein.

4.8 The Density Functional Theory and the Kohn-Sham Equations

In the previous sections, the many-electron problem has been attacked by approximating as better as possible the exact ground-state many-electron wavefunction $\Psi_G(\mathbf{r}_1\sigma_1, \ldots, \mathbf{r}_N\sigma_N)$. In the density functional theory the emphasis shifts from the ground-state wavefunction to the much more manageable ground-state *one-body electron density* $n(\mathbf{r})$. The density functional theory shows that the ground-state energy of a many-particle system can be expressed as a functional of the one-body density; minimization of this functional allows in principle the determination of the actual ground-state density and ground-state properties. The success of the theory stands on its rigorous formulation and on the concomitant possibility to provide reasonably simple and accurate approximations of the functional to be minimized. The peculiarity of the density functional approach to the many-body theory is to attain a one-electron Schrödinger equation with a local effective potential for the study of the ground-state electronic density of the many-electron systems.

The Hohenberg-Kohn Theorem

Consider a system of N electrons, described by the standard many-electron Hamiltonian H_e (see Eq. 4.3). For our reasoning, it is convenient to decompose H_e into the sum of an "internal" part (kinetic energy of the electrons plus electron-electron Coulomb interactions) and "external" part (in our specific case the external part is given by the electronic-nuclear interactions). With obvious rearrangement of notations, we have:

$$H_e = H_{int} + V_{ext}, \tag{4.49a}$$

$$H_{int} = T + V_{ee} = \sum_i \frac{\mathbf{p}_i^2}{2m} + \frac{1}{2} \sum_{i \neq j} \frac{e^2}{|\mathbf{r}_i - \mathbf{r}_j|}, \tag{4.49b}$$

$$V_{ext} = \sum_i v_{ext}(\mathbf{r}_i) \quad \text{with} \quad v_{ext}(\mathbf{r}) = -\sum_I \frac{z_I e^2}{|\mathbf{r} - \mathbf{R}_I|} \equiv V_{nucl}(\mathbf{r}). \tag{4.49c}$$

For simplicity, we suppose that the ground-state $|\Psi_G\rangle$ is non-degenerate. [In principle, any degeneracy can be removed by an arbitrary small perturbation that appropriately lowers the symmetry of the system. Anyhow, the restriction of non-degeneracy, as well as other assumptions made below just for sake of simplicity, can be relaxed when required].

Let us consider as only variable of the many-electron problem the external potential $v_{ext}(\mathbf{r})$; the mass of the electrons, their charge, their number N, the form of the internal interactions are $vice\ versa$ supposed to be fixed. The Hohenberg-Kohn theorem states that $there\ is\ a\ one\text{-}to\text{-}one\ correspondence\ between\ the\ ground\text{-}state\ density\ of\ a$ $N\text{-}electron\ system\ and\ the\ external\ potential\ acting\ on\ it$; in this sense the ground-state electron density becomes the variable of interest.

The first part of the theorem is almost trivial. Suppose, in fact, to know $v_{ext}(\mathbf{r})$ and hence the total Hamiltonian H_e of the system. By solving the Schrödinger equation, one knows exactly (at least in principle) the eigenfunctions and eigenvalues of the electronic system; in particular one knows the ground-state $\Psi_G(\mathbf{r}_1, \mathbf{r}_2, \ldots, \mathbf{r}_N)$. [In the following, for simplicity, we confine our attention to systems without spin polarization, and do not indicate explicitly the spin-variables, since this is not essential to the "abstract" reasoning we are going to present]. The ground-state Ψ_G depends on the chosen external potential $v_{ext}(\mathbf{r})$ and is also denoted with the short-hand functional notation $\Psi_G[v_{ext}]$. From the knowledge of Ψ_G we obtain the one-body ground-state density $n(\mathbf{r})$ defined as

$$n(\mathbf{r}) = \langle \Psi_G(\mathbf{r}_1, \mathbf{r}_2, \ldots, \mathbf{r}_N) | \sum_i \delta(\mathbf{r} - \mathbf{r}_i) | \Psi_G(\mathbf{r}_1, \mathbf{r}_2, \ldots, \mathbf{r}_N) \rangle. \tag{4.50}$$

In passing, we notice the property

$$\langle \Psi_G | V_{ext} | \Psi_G \rangle = \int n(\mathbf{r}) v_{ext}(\mathbf{r}) \, d\mathbf{r}, \tag{4.51}$$

which is obtained multiplying both members of Eq. (4.50) by $v_{ext}(\mathbf{r})$ and integrating over the space variable.

We can summarize the contents of the previous reasoning by the following sequence:

$$v_{ext}(\mathbf{r}) \Longrightarrow \Psi_G[v_{ext}] \Longrightarrow n(\mathbf{r}).$$

Thus the knowledge of $v_{ext}(\mathbf{r})$ entails the knowledge of $\Psi_G[v_{ext}]$ and hence the knowledge of $n(\mathbf{r})$. In other words there exists a functional that links $n(\mathbf{r})$ and $v_{ext}(\mathbf{r})$, and we write

$$n(\mathbf{r}) = F[v_{ext}(\mathbf{r})]. \tag{4.52}$$

It is also evident that two external potentials, which differ by a constant in the whole space, lead to the same $n(\mathbf{r})$ (and can be considered essentially the same, upon appropriate shift by an additive constant). The conceptual scheme of this reasoning is indicated in Figure 4.7, focusing on $v_{ext}(\mathbf{r})$ and following the arrows from left toward right. The real novelty of the theorem is that the functional relation (4.52) can be inverted in the form

$$\boxed{v_{ext}(\mathbf{r}) = G[n(\mathbf{r})]}, \tag{4.53}$$

which means that from the knowledge of the ground-state density $n(\mathbf{r})$ we can determine uniquely the external potential (to within a non-essential additive constant) and thus the Hamiltonian of the system and, in principle, any other property of the system.

In order to prove Eq. (4.53), we should establish that for any given pair of external potentials $v_{ext}(\mathbf{r})$ and $\bar{v}_{ext}(\mathbf{r})$ such that $v_{ext}(\mathbf{r}) \neq \bar{v}_{ext}(\mathbf{r})$ we have $n(\mathbf{r}) \neq \bar{n}(\mathbf{r})$. This follows from the minimum property of the ground-state energy. Consider in fact the two Hamiltonians

$$H = H_{int} + V_{ext} \quad \text{and} \quad \overline{H} = H_{int} + \overline{V}_{ext} \tag{4.54}$$

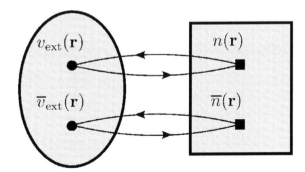

Figure 4.7 Schematic representation of the Hohenberg-Kohn theorem. *Two different* ground-state density functions must correspond to *two different* potentials, and *vice versa*.

and their ground-states $|\Psi_G\rangle$ and $|\overline{\Psi}_G\rangle$ with eigenvalues E_G and $\overline{E}_G$, respectively. We have

$$
\begin{aligned}
\langle\overline{\Psi}_G|H|\overline{\Psi}_G\rangle &= \langle\overline{\Psi}_G|H_{\text{int}} + V_{\text{ext}} + \overline{V}_{\text{ext}} - \overline{V}_{\text{ext}}|\overline{\Psi}_G\rangle \\
&= \overline{E}_G + \langle\overline{\Psi}_G|V_{\text{ext}} - \overline{V}_{\text{ext}}|\overline{\Psi}_G\rangle \\
&= \overline{E}_G + \int \overline{n}(\mathbf{r})[v_{\text{ext}}(\mathbf{r}) - \overline{v}_{\text{ext}}(\mathbf{r})]\, d\mathbf{r},
\end{aligned}
$$

where the last passage follows from Eq. (4.51). Since the ground eigenvalue of H is strictly lower than the mean value of the Hamiltonian on any other state, we have

$$
E_G < \overline{E}_G + \int \overline{n}(\mathbf{r})[v_{\text{ext}}(\mathbf{r}) - \overline{v}_{\text{ext}}(\mathbf{r})]\, d\mathbf{r}. \tag{4.55a}
$$

From a similar reasoning on $\langle\Psi_G|\overline{H}|\Psi_G\rangle$, one obtains the above expression with barred and unbarred quantities exchanged

$$
\overline{E}_G < E_G + \int n(\mathbf{r})[\overline{v}_{\text{ext}}(\mathbf{r}) - v_{\text{ext}}(\mathbf{r})]\, d\mathbf{r}. \tag{4.55b}
$$

Equations (4.55a) and (4.55b) are obviously incompatible in the hypothesis $n(\mathbf{r}) = \overline{n}(\mathbf{r})$. Therefore for $v_{\text{ext}}(\mathbf{r}) \neq \overline{v}_{\text{ext}}(\mathbf{r})$ it must be $n(\mathbf{r}) \neq \overline{n}(\mathbf{r})$, which completes the proof of the Hohenberg-Kohn theorem (schematically represented in Figure 4.7).

The Hohenberg-Kohn theorem has far reaching consequences, that require simple (although subtle) elaborations. First of all, it is convenient to define other functionals of the ground-state density of an electron system. We start from the following (obvious) functionals of the external potential $v_{\text{ext}}(\mathbf{r})$:

$$
v_{\text{ext}}(\mathbf{r}) \Longrightarrow \Psi_G[v_{\text{ext}}]
\begin{cases}
\Longrightarrow E[v_{\text{ext}}], \\
\Longrightarrow T[v_{\text{ext}}], \\
\Longrightarrow V_{\text{ee}}[v_{\text{ext}}].
\end{cases} \tag{4.56}
$$

The above logical sequence allows one to state that the ground-state energy is a functional $E[v_{\text{ext}}]$ of the external potential $v_{\text{ext}}(\mathbf{r})$; the expectation value of the electron kinetic energy on the ground-state wavefunction is a functional $T[v_{\text{ext}}]$, and the expectation value of the electron- electron interaction on the ground-state wavefunction is a functional $V_{\text{ee}}[v_{\text{ext}}]$.

Because of the Hohenberg-Kohn theorem, $v_{\text{ext}}(\mathbf{r})$ and $n(\mathbf{r})$ are in one-to-one correspondence. We thus have that the ground-state energy is a functional $E[n]$ of the ground-state density, that the expectation value of the electronic kinetic energy is a functional $T[n]$ of the ground-state density and so is also the expectation value $V_{\text{ee}}[n]$ of the electron-electron interaction.

A most important consequence of the Hohenberg-Kohn theorem is the formulation of a variational principle concerning the ground-state density of a system. Consider a system of interacting electrons in a given nuclear external potential, denoted by $v_{\text{ext}}(\mathbf{r})$. Following Hohenberg and Kohn, we construct the functional

$$
E^{(\text{HK})}[n(\mathbf{r}); v_{\text{ext}}(\mathbf{r})] \equiv \langle\Psi_G[n]|T + V_{\text{ee}} + V_{\text{ext}}|\Psi_G[n]\rangle, \tag{4.57}
$$

where $v_{ext}(\mathbf{r})$ is taken fixed, $n(\mathbf{r})$ is allowed to vary, and $\Psi_G[n]$ is the ground-state of the system with ground-state density $n(\mathbf{r})$. It is evident that the absolute minimum of the energy functional (4.57) occurs when $\Psi_G[n]$ is the ground-state of the operator $H_e = T + V_{ee} + V_{ext}$, i.e. when $n(\mathbf{r})$ is the exact electron density of the system.

The properties of the Hohenberg-Kohn energy functional (4.57), here rewritten in the form

$$E^{(HK)}[n(\mathbf{r}); v_{ext}(\mathbf{r})] = T[n(\mathbf{r})] + V_{ee}[n(\mathbf{r})] + \int v_{ext}(\mathbf{r})n(\mathbf{r})\, d\mathbf{r}, \qquad (4.58)$$

can be summarized as follows. The energy functional $E^{(HK)}[n(\mathbf{r}); v_{ext}(\mathbf{r})]$, which exists and is unique, is *minimal at the exact ground-state density*, and its minimum gives the *exact ground-state energy* of the many-body electron system. We also notice that the functional $F[n] = T[n] + V_{ee}[n]$ is *universal*, i.e. it does not depend on v_{ext}; this good news, however, is no compensation for the fact that the functional $F[n]$ is not known explicitly, and must be appropriately approximated.

In view of the foregoing minimization procedure of the Hohenberg-Kohn functional, it may be useful to recall briefly the elementary notion of functional derivative (avoiding subtleties, for simplicity). Consider a functional $G[f(\mathbf{r})]$ which associates a value G to any given function $f(\mathbf{r})$ belonging to an appropriate manifold of functions. When the function $f(\mathbf{r})$ undergoes an arbitrary small change to $f(\mathbf{r}) + \delta f(\mathbf{r})$, the functional G undergoes a corresponding change $\delta G[f] \to G[f + \delta f] - G[f]$. The functional derivative $\delta G[f]/\delta f(\mathbf{r})$ is defined through the implicit relation

$$\delta G[f] \equiv \int \frac{\delta G[f]}{\delta f(\mathbf{r})} \delta f(\mathbf{r})\, d\mathbf{r};$$

the functional derivative is thus the kernel of the integral linear relation which, acting on an arbitrary test variation $\delta f(\mathbf{r})$ of the function $f(\mathbf{r})$, produces the corresponding change δG of the functional $G[f]$. Notice that the functional derivative, once a given function f is selected, represents an ordinary (local) function of the space variable $\mathbf{r}$.

The Kohn-Sham Equations

The Kohn-Sham equations are obtained by minimizing the functional (4.58) with respect to $n(\mathbf{r})$. The key point of the Kohn-Sham method to the density functional theory is the assumption that, *for each non-uniform ground-state density $n(\mathbf{r})$ of an interacting electron system there exist a non-interacting electron system with the same non-uniform ground-state density*. This auxiliary non-interacting system (called the Kohn-Sham system) is assumed to exist, at least for non-pathologic situations (we do not enter here in the ground-state v-representability question). When the Kohn-Sham assumption is fulfilled, an important consequence follows: the ground-state density $n(\mathbf{r})$ of any interacting electron system with N electrons can be decomposed *exactly* into the sum of N independent orbital contributions of the form

$$n(\mathbf{r}) = \sum_i \phi_i^*(\mathbf{r})\phi_i(\mathbf{r}), \qquad (4.59)$$

where $\{\phi_i(\mathbf{r})\}$ $(i = 1, 2, \ldots, N)$ are orthonormal orbitals. In fact the decomposition (4.59) holds trivially for the ground-state density of any non-interacting electronic system and, by virtue of the Kohn-Sham ansatz, extends to the interacting case, too. It is worthwhile to stress explicitly the basic *conceptual* difference between the trial representation (4.59) of the ground-state density, which is *conceptually exact*, and the trial representation with a single determinant of the Hartree-Fock ground-state wavefunction, which is only *approximate*. In the former case, the minimization process provides the exact ground-state density and the exact ground-state energy of the electronic system; on the contrary, in the latter case, the minimization process provides an optimized, and yet approximate, ground-state wavefunction and an approximate upper bound of the ground-state energy.

Before progressing with the variational procedure on the functional (4.58), it is convenient to extract from it the inter-electron Coulomb interaction $V_H[n]$ (called Hartree potential)

$$V_H[n] = \frac{1}{2} \int n(\mathbf{r}) \frac{e^2}{|\mathbf{r} - \mathbf{r}'|} n(\mathbf{r}') \, d\mathbf{r} \, d\mathbf{r}' \equiv \frac{1}{2} \sum_{ij} \langle \phi_i \phi_j | \frac{e^2}{r_{12}} | \phi_i \phi_j \rangle;$$

similarly, it is convenient to extract from the functional (4.58) the kinetic energy $T_0[n]$ of the system of non-interacting electrons with the same density given by

$$T_0[n] = \sum_i \langle \phi_i | - \frac{\hbar^2 \nabla^2}{2m} | \phi_i \rangle.$$

The Hohenberg-Kohn functional (4.58) can be written as

$$E^{(HK)}[n(\mathbf{r}); v_{ext}(\mathbf{r})] = T_0[n] + V_H[n] + \int v_{ext}(\mathbf{r}) n(\mathbf{r}) \, d\mathbf{r} + E_{xc}[n], \qquad (4.60a)$$

where the *exchange-correlation functional* $E_{xc}[n]$ is defined as

$$E_{xc}[n] = T[n] - T_0[n] + V_{ee}[n] - V_H[n]. \qquad (4.60b)$$

The functional (4.60a) can be recast in the form

$$\begin{aligned} E^{(HK)}[n(\mathbf{r}); v_{ext}(\mathbf{r})] &= \sum_i \langle \phi_i | - \frac{\hbar^2 \nabla^2}{2m} + v_{ext} | \phi_i \rangle \\ &+ \frac{1}{2} \sum_{ij} \langle \phi_i \phi_j | \frac{e^2}{r_{12}} | \phi_i \phi_j \rangle + E_{xc}[n] \end{aligned} \qquad (4.61)$$

According to the standard variational procedure, we vary the N contributing orbitals $\phi_1, \phi_2, \ldots, \phi_N$ so to make stationary the energy functional (4.61), under the constraint of normalization of the wavefunctions $\{\phi_i\}$. The variational expression for the first two terms in the right-hand side of Eq. (4.61) has been already considered in Section 4.5.1, and need not to be repeated here. The only novelty from a technical point of view is thus the calculation of the variation $\delta E_{xc}[n]$. Using the notion of functional derivative, the variation $\delta E_{xc}[n]$ reads

$$\delta E_{xc}[n] = \int V_{xc}(\mathbf{r})\delta n(\mathbf{r})d\mathbf{r} = \int V_{xc}(\mathbf{r})\,\delta\sum_i \phi_i^*(\mathbf{r})\phi_i(\mathbf{r})\,d\mathbf{r}, \tag{4.62}$$

where

$$V_{xc}(\mathbf{r}) \equiv \frac{\delta E_{xc}[n]}{\delta n(\mathbf{r})} \tag{4.63}$$

denotes the derivative of the exchange-correlation functional.

A straightforward variational calculation of the functional (4.61), keeping in mind Eqs. (4.62) and (4.63), leads to the Kohn-Sham equations

$$\left[-\frac{\hbar^2 \nabla^2}{2m} + V_{nucl}(\mathbf{r}) + V_{coul}(\mathbf{r}) + V_{xc}(\mathbf{r}) \right] \phi_i(\mathbf{r}) = \varepsilon_i \phi_i(\mathbf{r}), \tag{4.64}$$

where $V_{coul}(\mathbf{r})$ denotes the Hartree potential, $V_{xc}(\mathbf{r})$ is the functional derivative of $E_{xc}[n]$, and $V_{nucl}(\mathbf{r})$ is the external potential in consideration. Once the Kohn-Sham orbitals and energies have been determined, the *exact* total ground-state energy (4.61) of the electronic system can be expressed as:

$$E_0 = \sum_i \varepsilon_i - \frac{1}{2}\sum_{ij}\langle\phi_i\phi_j|\frac{e^2}{r_{12}}|\phi_i\phi_j\rangle + E_{xc}[n] - \int V_{xc}(\mathbf{r})n(\mathbf{r})\,d\mathbf{r}. \tag{4.65}$$

We notice that the Kohn-Sham equations are standard differential equations with a rigorously local effective potential $V_{eff}(\mathbf{r}) = V_{nucl}(\mathbf{r}) + V_{coul}(\mathbf{r}) + V_{xc}(\mathbf{r})$; any difficulty in the procedure has been confined to a reasonable guess of the exchange-correlation functional $E_{xc}[n]$ (which is known only in principle). Conceptually, the Kohn-Sham equations determine exactly the electron density and the electronic energy of the ground-state; however, the orbital energies ε_i appearing in Eq. (4.64) are and remain purely formal Lagrange multipliers (not rescued for physical interpretation by any Koopmans'-like theorem); any identification of ε_i with (occupied or non-occupied) one-particle energies is to be justified (and often heuristically corrected) situation by situation. Experience shows that the calculations, performed within one version or another of the density functional theory, tend to underestimate the energy band gap in semiconductor and insulators; however, the general trend of the dispersion curves of the valence and the conduction bands is often represented to reasonably accuracy.

Local Density Approximation for the Exchange-Correlation Functional

The formal definition of the exchange-correlation functional $E_{xc}[n]$ of Eq. (4.60b) would be of no practical use, unless workable approximations can be given. Among them, the most popular is the *local density approximation*, which is particularly justified in systems with reasonably slowly varying spatial density $n(\mathbf{r})$. In the local density approximation (LDA), the exchange-correlation functional (4.60b) is approximated in the form

$$E_{xc}^{(LDA)}[n(\mathbf{r})] = \int \varepsilon_{xc}(n(\mathbf{r}))n(\mathbf{r})\, d\mathbf{r}, \tag{4.66a}$$

where $\varepsilon_{xc}(n(\mathbf{r}))$ is the many-body exchange-correlation energy per electron *of a uniform gas of interacting electrons of density* $n(\mathbf{r})$. The exchange-correlation potential produced by the functional (4.66a) becomes

$$V_{xc}^{(LDA)}(\mathbf{r}) \equiv \frac{\delta E_{xc}^{(LDA)}[n]}{\delta n(\mathbf{r})} = \varepsilon_{xc}(n(\mathbf{r})) + n(\mathbf{r})\frac{d\varepsilon_{xc}(n(\mathbf{r}))}{dn(\mathbf{r})}. \tag{4.66b}$$

In the local density approximation, the total ground-state energy (4.65) takes the expression

$$E_0^{(LDA)} = \sum_i \varepsilon_i - \frac{1}{2}\int n(\mathbf{r})\frac{e^2}{|\mathbf{r}-\mathbf{r'}|}n(\mathbf{r'})\, d\mathbf{r}\, d\mathbf{r'} - \int n(\mathbf{r})\left[\frac{d\varepsilon_{xc}}{dn}\right]_{n=n(\mathbf{r})} n(\mathbf{r})\, d\mathbf{r}.$$

$$\tag{4.66c}$$

Among the various forms, adopted for ε_{xc}, we report here the expression given by J. P. Perdew and A. Zunger, Phys. Rev. B *23*, 5048 (1981), who have parametrized the numerical results obtained by Ceperley and Alder with Monte Carlo calculations. For the unpolarized homogeneous electron gas, we have:

$$\varepsilon_{xc}(r_s) = \varepsilon_x(r_s) + \varepsilon_c(r_s) \quad \text{where} \quad \varepsilon_x(r_s) = -\frac{0.4582}{r_s},$$

$$\varepsilon_c(r_s) = \begin{cases} -0.1423/(1 + 1.0529\sqrt{r_s} + 0.3334\, r_s) & \text{for } r_s \geq 1, \\ -0.0480 + 0.0311\ln r_s - 0.0116\, r_s + 0.0020\, r_s \ln r_s & \text{for } r_s \leq 1. \end{cases}$$

In the above expressions, energies are in Hartrees (1 Hartree = 2 Rydberg); r_s is the usual dimensionless parameter defined by $(4\pi/3)(r_s a_B)^3 = 1/n$; the distinction between the exchange term and the correlation term is just for comparison with the Hartree-Fock approach.

The density functional theory, in conjunction with the local density approximation or with other more sophisticated elaborations, has become a most popular and successful tool for the investigation of the chemical and physical properties of electronic systems. The general concepts of the density functional formalism have been extended in several directions (including spin-polarized materials, multi-component systems, time-dependent phenomena, etc.) and have permeated countless areas of active research of quantum chemistry and molecular-dynamics, opening unexpected horizons in the understanding and predicting capability of the electronic and structural properties of matter.

Appendix A. Bielectronic Integrals among Spin Orbitals

Let us consider a complete set of orthonormal *spin-orbitals* of the form

$$\psi_i(\mathbf{r}\sigma) = \phi_i(\mathbf{r})\chi_i(\sigma),$$

where $\mathbf{r}$ and σ denote space and spin coordinates, $\phi_i(\mathbf{r})$ are spatial orbitals, and $\chi_i(\sigma)$ equals α or β for spin-up or spin-down electrons, respectively.

Bielectronic integrals among any four spin-orbitals are defined as follows

$$\langle \psi_i \psi_j | \frac{e^2}{r_{12}} | \psi_m \psi_n \rangle \equiv \langle \psi_i(1)\psi_j(2) | \frac{e^2}{r_{12}} | \psi_m(1)\psi_n(2) \rangle$$

$$\equiv \int \psi_i^*(\mathbf{r}_1\sigma_1)\psi_j^*(\mathbf{r}_2\sigma_2) \frac{e^2}{|\mathbf{r}_1 - \mathbf{r}_2|} \psi_m(\mathbf{r}_1\sigma_1)\psi_n(\mathbf{r}_2\sigma_2)\, d(\mathbf{r}_1\sigma_1)\, d(\mathbf{r}_2\sigma_2), \tag{A.1}$$

where integration on the space variables and summation on spin components is understood. *Notice carefully the order by which the arguments* 1 *and* 2 *appear in the definition* (A.1), since conventions different from the present one are also encountered in the literature. An obvious property of the definition (A.1) is its invariance if the order of ψ_i and ψ_j, and the order of ψ_m and ψ_n, are simultaneously interchanged. Performing the integration over the spin variables in Eq. (A.1), we have

$$\langle \psi_i \psi_j | \frac{e^2}{r_{12}} | \psi_m \psi_n \rangle = \delta_{\chi_i, \chi_m} \delta_{\chi_j, \chi_n} \int \phi_i^*(\mathbf{r}_1)\phi_j^*(\mathbf{r}_2) \frac{e^2}{r_{12}} \phi_m(\mathbf{r}_1)\phi_n(\mathbf{r}_2)\, d\mathbf{r}_1\, d\mathbf{r}_2. \tag{A.2}$$

The bielectronic integrals of Eq. (A.1) can be different from zero only if the spins of ψ_i and ψ_m are parallel to each other, and so are also the spins of ψ_j and ψ_n.

In several situations, we encounter bielectronic integrals involving two different spin-orbitals. In this case, we can have *Coulomb bielectronic integrals* (also called *direct* bielectronic integrals) defined as

$$Q_{mn} \equiv \langle \psi_m \psi_n | \frac{e^2}{r_{12}} | \psi_m \psi_n \rangle \tag{A.3}$$

and the *exchange bielectronic integrals*, defined as

$$J_{mn} \equiv \langle \psi_m \psi_n | \frac{e^2}{r_{12}} | \psi_n \psi_m \rangle. \tag{A.4}$$

The physical meaning and the nature of Coulomb and exchange bielectronic integrals is almost obvious. Consider in fact the Coulomb term defined by Eq. (A.3); it is evident that this term *does not depend on the spin part of the spin-orbitals* and it holds

$$Q_{mn} \equiv \langle \psi_m \psi_n | \frac{e^2}{r_{12}} | \psi_m \psi_n \rangle = \int \phi_m^*(\mathbf{r}_1)\phi_n^*(\mathbf{r}_2) \frac{e^2}{r_{12}} \phi_m(\mathbf{r}_1)\phi_n(\mathbf{r}_2)\, d\mathbf{r}_1\, d\mathbf{r}_2. \tag{A.5}$$

The Coulomb bielectronic integral Q_{mn} represents the interaction between two charge distributions given by $(-e)|\phi_m(\mathbf{r}_1)|^2$ and $(-e)|\phi_n(\mathbf{r}_2)|^2$, respectively; this term is of long-range nature and behaves as $1/R$ for orbitals localized at distance R.

For what concerns the exchange term, we have

$$J_{mn} \equiv \langle \psi_m \psi_n | \frac{e^2}{r_{12}} | \psi_n \psi_m \rangle = \delta_{\chi_m, \chi_n} \int \phi_m^*(\mathbf{r}_1) \phi_n^*(\mathbf{r}_2) \frac{e^2}{r_{12}} \phi_n(\mathbf{r}_1) \phi_m(\mathbf{r}_2) \, d\mathbf{r}_1 \, d\mathbf{r}_2.$$

(A.6)

Thus exchange bielectronic integrals may be different from zero only for spin-orbitals with parallel spin. Furthermore exchange bielectronic integrals are of short-range nature; in fact they go rapidly to zero if ϕ_m and ϕ_n are localized far from each other.

The exchange bielectronic integral of Eq. (A.6) can be formally interpreted as the classical electrostatic interaction between the exchange charge $(-e)\phi_m^*(\mathbf{r})\phi_n(\mathbf{r})$ at $\mathbf{r}_1$ and the complex conjugate charge $(-e)\phi_m(\mathbf{r})\phi_n^*(\mathbf{r})$ at $\mathbf{r}_2$. Such "exchange charge" is not necessarily real; even when the orbital wavefunctions are real, we see that the exchange charge is somewhere positive in space and somewhere negative; in any case, *the total exchange charge* (if so we decide to address the quantity $(-e) \int \phi_m^*(\mathbf{r})\phi_n(\mathbf{r})d\mathbf{r}$) is *exactly zero*, because of orthogonality of different spatial orbitals.

It is important to note that the *exchange bielectronic integrals, as well as the Coulomb ones, are definite-positive quantities.* That the Coulomb integrals are definite positive quantities is obvious. It is easy to show that this is true also for any exchange integral of type (A.6) (unless identically zero when the spins of the spin-orbitals are antiparallel). In fact, expanding e^2/r_{12} in plane waves

$$\frac{1}{|\mathbf{r}_1 - \mathbf{r}_2|} = \frac{1}{(2\pi)^3} \int \frac{4\pi}{k^2} e^{i\mathbf{k}\cdot(\mathbf{r}_1 - \mathbf{r}_2)} \, d\mathbf{k},$$

(A.7)

and replacing expression (A.7) into (A.6), it is evident that the integral in Eq. (A.6) is a definite-positive quantity.

From the obvious inequality

$$\int |\Psi(\mathbf{r}_1, \mathbf{r}_2)|^2 \frac{e^2}{r_{12}} d\mathbf{r}_1 \, d\mathbf{r}_2 > 0 \quad \text{with} \quad \Psi(\mathbf{r}_1, \mathbf{r}_2) = \phi_m(\mathbf{r}_1)\phi_n(\mathbf{r}_2) - \phi_m(\mathbf{r}_2)\phi_n(\mathbf{r}_1)$$

it is easily proved that $Q_{mn} \geq J_{mn} \geq 0$; this means that the exchange bielectronic integrals not only are positive but also are smaller that the corresponding Coulomb integrals.

Appendix B.   Outline of Second Quantization Formalism for Identical Fermions

In this appendix we briefly summarize some elements of second quantization for the description of a system of identical fermions. The method of second quantization is closely linked to the Slater determinants, introduced in Section 4.3, but exhibits several advantages. In fact, second quantization allows to rewrite in a more compact notation the Slater determinantal states; the Slater rules for matrix elements of operators between determinantal states are automatically embodied within the anticommutation rules of creation and annihilation operators. The second quantization formalism is of particular value for the treatment of systems where the number of particles can vary, or to discuss and handle contributions beyond the one-electron approximation.

Consider a complete set of orthonormal spin-orbitals $\{\psi_i\}$. The set of all different (normalized) Slater determinants formed selecting N (different) spin-orbitals from $\{\psi_i\}$, constitutes a complete set for the description of the N-particle system. Let us indicate with $|\Psi_0(N)\rangle$ a given Slater determinantal state with contributing spin-orbitals $\psi_{i_1}, \psi_{i_2}, \ldots, \psi_{i_N}$; we have

$$|\Psi_0(N)\rangle = \mathcal{A}\{\psi_{i_1}\psi_{i_2}\cdots\psi_{i_N}\}. \tag{B.1}$$

Next, consider a Slater determinant for $N + 1$ particles, obtained by adding an extra spin-orbital ψ_m to the contributing spin-orbitals $\psi_{i_1}, \psi_{i_2}, \cdots, \psi_{i_N}$; the change can be represented by defining a *creation operator* $c_m^\dagger$ that applied to the state $|\Psi_0(N)\rangle$ gives

$$c_m^\dagger|\Psi_0(N)\rangle = \mathcal{A}\{\psi_m\psi_{i_1}\psi_{i_2}\cdots\psi_{i_N}\}. \tag{B.2}$$

In the case ψ_m belongs to the occupied set $\psi_{i_1}, \psi_{i_2}, \ldots, \psi_{i_N}$, the new determinant (B.2) has two identical rows and thus vanishes. In a similar way, we can denote by $c_n^\dagger c_m^\dagger|\Psi_0(N)\rangle$ the operation of adding two extra spin-orbitals ψ_n and ψ_m to $|\Psi_0(N)\rangle$ and define

$$c_n^\dagger c_m^\dagger|\Psi_0(N)\rangle = \mathcal{A}\{\psi_n\psi_m\psi_{i_1}\psi_{i_2}\cdots\psi_{i_N}\}. \tag{B.3}$$

In the case ψ_n and ψ_m are equal (or belong to the occupied set $\psi_{i_1}, \psi_{i_2}, \ldots, \psi_{i_N}$) the determinant (B.3) vanishes; this means $c_n^{\dagger 2} = c_m^{\dagger 2} = 0$.

It is evident, from the fact that Slater determinants change sign for exchange of the order of any two spin-orbitals, that we have

$$c_n^\dagger c_m^\dagger|\Psi_0(N)\rangle = -c_m^\dagger c_n^\dagger|\Psi_0(N)\rangle. \tag{B.4}$$

Equation (B.4) can be written in the form

$$\left(c_n^\dagger c_m^\dagger + c_m^\dagger c_n^\dagger\right)|\Psi_0(N)\rangle \equiv 0. \tag{B.5}$$

What is important to notice is that Eq. (B.5) holds *whatever is the state* $|\Psi_0(N)\rangle$ (i.e. for any choice of the number N and for any choice of the contributing spin-orbitals); thus equation (B.5) is satisfied if and only if the anticommutation of creation operators equals zero, i.e.

$$\left\{c_n^\dagger, c_m^\dagger\right\} \equiv c_n^\dagger c_m^\dagger + c_m^\dagger c_n^\dagger = 0. \tag{B.6}$$

A determinantal state $|\Psi_0(N)\rangle$ of type (B.1) is denoted listing the contributing spin-orbitals $\psi_{i_1}, \psi_{i_2}, \ldots, \psi_{i_N}$; from the properties of the creation operators we can equally well denote the state $|\Psi_0(N)\rangle$ listing the corresponding creation operators applied to the vacuum state $|0\rangle$ (defined as the state where no electron is present). We have

$$|\Psi_0(N)\rangle = c_{i_1}^{\dagger} c_{i_2}^{\dagger} \cdots c_{i_N}^{\dagger} |0\rangle.$$

We can now define the *annihilation operator* c_m as the operator that removes a spin-orbital from any given Slater determinant as follows

$$c_m \mathcal{A}\left\{\psi_m \psi_{i_1} \psi_{i_2} \cdots \psi_{i_N}\right\} = \mathcal{A}\left\{\psi_{i_1} \psi_{i_2} \cdots \psi_{i_N}\right\}. \tag{B.7}$$

If there is no orbital ψ_m in the given Slater determinant, the action of the operator c_m is then defined to give zero. If the spin-orbital ψ_m is not in the first row of the determinant (as defined in Eq. (B.7)), it must be brought in first position; according to the number of necessary transpositions this may generate a minus sign in front of the state. In a rather similar way, as done for creation operators, we can show the anticommutation property of annihilation operators

$$\{c_n, c_m\} \equiv c_n c_m + c_m c_n = 0. \tag{B.8}$$

Finally if the anticommutation between creation and annihilation operators is examined, we can verify that creation and annihilation operators corresponding to different spin-orbitals anticommute; in the case of the same spin-orbital the anticommutation gives the identity operator. In formulae we have

$$\left\{c_n, c_m^{\dagger}\right\} \equiv c_n c_m^{\dagger} + c_m^{\dagger} c_n = \delta_{mn}. \tag{B.9}$$

It is also easy to verify that the annihilation operator c_m is the adjoint of the creation operator $c_m^{\dagger}$, i.e. $c_m = \left(c_m^{\dagger}\right)^{\dagger}$.

Operators in Second Quantization Form

The Hamiltonian of a many-electron system (see for instance Eq. (4.3)) usually contains one-electron operators and two-electron operators. It is convenient to express not only determinantal states, but also one- and two-electron operators in second quantization form.

Consider first a one-electron operator of the type

$$G_1 = \sum_i h(\mathbf{r}_i). \tag{B.10}$$

Let us indicate with $\{\psi_i\}$ a complete set of orthonormal spin-orbitals, and with $\langle\psi_m|h|\psi_n\rangle$ the matrix element of the monoelectronic operator $h(\mathbf{r})$ between any two spin-orbitals. The expression of G_1 in second quantization form is given by

$$G_1 = \sum_i h(\mathbf{r}_i) \equiv \sum_{mn} \langle\psi_m|h|\psi_n\rangle c_m^{\dagger} c_n. \tag{B.11}$$

The above equivalence means that the matrix elements of the operator G_1 calculated among Slater determinantal states with the Slater rules, and the matrix elements of the operator on the right-hand side, calculated using the anticommutation rules of creation and annihilation operators are perfectly equal on any basis set of determinantal states.

As an example, consider the determinantal state $|\Psi_0\rangle = \mathcal{A}\{\psi_1, \psi_2, \ldots, \psi_N\}$. From the Slater rules we have seen that

$$\langle\Psi_0|G_1|\Psi_0\rangle = \sum_i \langle\psi_i|h|\psi_i\rangle.$$

Operating with creation and annihilation operators we observe that

$$\langle\Psi_0|c_m^\dagger c_n|\Psi_0\rangle = \begin{cases} 1 & \text{if } m = n = \text{occupied spin-orbital,} \\ 0 & \text{otherwise} \end{cases}$$

and the equivalence is thus proved for the diagonal matrix elements of G_1; similar considerations show that the equivalence holds also for off-diagonal matrix elements.

Notice that in the particular case in which the operator h is the unit operator, we obtain just the operator that counts the number of particles

$$N_{\text{op}} = \sum_m c_m^\dagger c_m.$$

In a rather similar way, we can see that a two-body operator can be expressed in second quantization form as follows

$$G_2 = \frac{1}{2} \sum_{i \neq j} \frac{e^2}{r_{ij}} \equiv \frac{1}{2} \sum_{klmn} \langle\psi_k\psi_l| \frac{e^2}{r_{12}} |\psi_m\psi_n\rangle c_k^\dagger c_l^\dagger c_n c_m. \tag{B.12}$$

Appendix C. An Integral on the Fermi Sphere

Consider the integral of the type

$$I(k) = \frac{1}{V} \sum_{q<k_F} \frac{4\pi e^2}{|\mathbf{k}-\mathbf{q}|^2} = \frac{4\pi e^2}{(2\pi)^3} \int_{q<k_F} \frac{1}{|\mathbf{k}-\mathbf{q}|^2} d\mathbf{q}. \tag{C.1}$$

The fixed vector $\mathbf{k}$ can be taken to lie on the z direction, without loss of generality (we also suppose momentarily that k is larger than k_F, although later we remove this restriction).

To calculate the integral, let us introduce for $\mathbf{q}$ polar coordinates with polar axis along z; the integral (C.1) becomes

$$I(k) = \frac{4\pi e^2}{(2\pi)^3} \int_{q<k_F} \frac{1}{q^2 - 2kq\cos\theta + k^2} q^2 \sin\theta \, d\theta \, d\phi \, dq. \tag{C.2}$$

The integration in $d\phi$ gives 2π. Also the angular integration in $d\theta$ is easily performed with the change of variable $t = \cos\theta$ and $dt = -\sin\theta \, d\theta$. The integral $I(k)$ becomes

$$I(k) = \frac{e^2}{\pi} \frac{1}{k} \int_0^{k_F} q \ln\frac{k+q}{k-q} dq = \frac{e^2}{\pi} \frac{1}{k} \left[kq - \frac{1}{2}(k^2 - q^2) \ln\frac{k+q}{k-q} \right]_0^{k_F}.$$

We thus obtain

$$\boxed{I(k) = \frac{2e^2 k_F}{\pi} F\left(\frac{k}{k_F}\right)},$$ (C.3)

where the function $F(x)$ is given by

$$F(x) = \frac{1}{2} + \frac{1 - x^2}{4x} \ln\left|\frac{1 + x}{1 - x}\right|$$ (C.4)

(the absolute value is introduced because this makes the expression hold also for $k < k_F$).

It is of interest to calculate the average of the function $F(k/k_F)$ when k runs uniformly within the Fermi sphere. Let x indicate the dimensionless quantity $x = k/k_F$. In the interval $0 \le x \le 1$ the function $F(x)$ varies from 1 to 1/2; it is easily seen that its average value is just 3/4; in fact

$$F_{av} = \int_0^1 x^2 F(x)\, dx \bigg/ \int_0^1 x^2\, dx = 3 \int_0^1 x^2 F(x)\, dx = \frac{3}{4}.$$ (C.5)

The indefinite integral

$$\int x(1 - x^2) \ln\frac{1 + x}{1 - x}\, dx = \frac{1}{2}x - \frac{1}{6}x^3 - \frac{1}{4}(1 - x^2)^2 \ln\frac{1 + x}{1 - x}$$

has been used to obtain Eq. (C.5).

Further Reading

Atkins, P. W., & Friedman, R. S. (2010). *Molecular quantum mechanics*. New York: Oxford University Press.

Botti, S., Schindlmayr, A., Del Sole, R., & Reining, L. (2007). Time-dependent density-functional theory for extended systems. *Reports on Progress in Physics, 70*, 357.

Ceperley, D. M., & Alder, B. J. (1980). Ground state of the electron gas by a stochastic method. *Physical Review Letters, 45*, 566.

Cini, M. (2010). *Topics and methods in condensed matter theory*. Berlin: Springer.

Dreizler, R. M., & Gross, E. K. U. (1990). *Density functional theory*. Berlin: Springer.

Engel, E., & Dreizler, R. M. (2011). *Density functional theory. An advanced course*. Berlin: Springer.

Foulkes, W. M. C., Mitas, L., Needs, R. J., & Rajagopal, C. (2001). Quantum Monte Carlo simulations of solids. *Reviews of Modern Physics, 73*, 33.

Hohenberg, P., & Kohn, W. (1964). Inhomogeneous electron gas. *Physical Review, 136*, B 864.

Kohn, W. (1999). Electronic structure of matter - Wave functions and density functionals. *Reviews of Modern Physics, 71*, 1253.

Kohn, W., & Sham, L. J. (1965). Self-consistent equations including exchange and correlation effects. *Physical Review, 140*, A 1133.

Lundqvist, S., & March, N. H. (Eds.), (1983). *Theory of the inhomogeneous electron gas.* New York: Plenum Press.

Marder, M. P. (2000). *Condensed matter physics.* New York: Wiley.

Martin, R. M. (2004). *Electronic structure: Basic theory and practical methods.* Cambridge: Cambridge University Press.

McWeeny, R. & Sutcliffe, B. T. (1969). *Methods of molecular quantum mechanics.* New York: Academic Press.

Moroni, S., Ceperley, D. M., & Senatore, G. (1992). Static response from quantum Monte Carlo calculations. *Physical Review Letters, 69,* 1837; (1995). Static response and local field factor of the electron gas. *Physical Review Letters, 75,* 689.

Needs, R. J., Towler, M. D., Drummond, N. D., & López Ríos, P. (2010). Continuum variational and diffusion quantum Monte Carlo calculations. *Journal of Physics: Condensed Matter, 22,* 023201.

Pisani, C., Dovesi, R., & Roetti, C. (1988). *Hartree-Fock ab initio treatment of crystalline systems.* New York: Springer.

Pople, J. A. (1999). Quantum chemical models. *Reviews of Modern Physics, 71,* 1267.

Runge, E., & Gross, E. K. U. (1984). Density-functional theory for time-dependent systems. *Physical Review Letters, 52,* 997.

Seitz, F. (1940). *The modern theory of solids.* New York: McGraw-Hill.

Slater, J. C. (1960). *Quantum theory of atomic structure* (Vols. I and II). New York: McGraw-Hill; (1974). *The self-consistent field for molecules and solids* (Vol. IV). New York: McGraw-Hill.

Szabo, A., & Ostlund, N. S. (1989). *Modern quantum chemistry.* New York: McGraw-Hill.

Ullrich, C. A. (2012). *Time-dependent density functional theory: Concepts and applications.* Oxford: Oxford University Press.

5 Band Theory of Crystals

Chapter Outline head

The band structure of crystals is one of the most traditional subjects of solid state physics. In this chapter we overview the wealth of ideas and ingenious developments in the field of electron states in periodic potentials. The Bloch theorem is the unifying tool that flows through all the various approaches. A first line of attack consists in expanding the crystal states in appropriate sets of energy-independent Bloch functions: Bloch sums made of atomic orbitals in the tight-binding method, plane waves orthogonalized to core states in the orthogonalized plane wave method, plane waves in the pseudopotential method. A second line of attack, which includes the cellular method, the augmented plane wave method and the Green's function method, focuses on a single reference cell, adopts energy-dependent basis functions, solutions of the Schrödinger equation in the primitive cell, and enforces appropriate boundary conditions required by the Bloch theorem; the modality of imposing the boundary conditions characterizes the different cellular methods. The main approaches with strengths and weaknesses in the description of the electronic properties of classes of crystals are discussed.

Solid State Physics, Second Edition. http://dx.doi.org/10.1016/B978-0-12-385030-0.00005-0

5.1 Basic Assumptions of the Band Theory

In this chapter we describe the basic methods for calculating electronic states in crystals; we focus mainly on physical motivations, without entering into all the subtleties of the art of band structure calculations. Our purpose is to illustrate to a reader, already familiar with the Schrödinger equation for one-dimensional periodic potentials (Chapter 1), the physical concepts and the mathematical techniques which have been developed for solving the Schrödinger equation in actual three-dimensional crystals.

In the band theory of crystals, one considers the one-electron Schrödinger equation

$$\left[\frac{\mathbf{p}^2}{2m} + V(\mathbf{r})\right] \psi(\mathbf{r}) = E\psi(\mathbf{r}), \tag{5.1a}$$

where

$$V(\mathbf{r} + \mathbf{t}_n) = V(\mathbf{r}) \tag{5.1b}$$

is the periodic crystalline potential, $\mathbf{t}_n$ are translation vectors, and the eigenfunctions $\psi(\mathbf{r})$ must be of Bloch type. The eigenvalue equation (5.1) tacitly contains a number of assumptions, which have been discussed in Chapter 4, and are here summarized for convenience.

(i) *Rigid lattice approximation.* The nuclei are taken as fixed at their equilibrium positions; the large difference between the masses of the electrons and of the nuclei is the basic justification for this (nuclear vibrations and related effects are considered in Chapter 8 and following).

(ii) *One-electron approximation in local form.* In essence, the complicated many-body electron problem is simplified to a single one-electron problem with an appropriate local potential. A conceptual scheme for this is provided by the Hartree-Fock theory, with a local approximation to the exchange potential, or by the density functional theory.

(iii) *Relativistic effects are neglected.* Whenever necessary, one should replace the Schrödinger equation (5.1) with the Dirac equation, all the techniques developed for the former equation being generalizable to the latter. Often one includes relativistic terms of interest by perturbation theory. This is the case in particular of the spin-orbit coupling given by the expression

$$H_{SO} = \frac{\hbar}{4m^2c^2} \left[\boldsymbol{\sigma} \times \nabla V(\mathbf{r})\right] \cdot \mathbf{p},$$

where m is the free-electron mass, $\boldsymbol{\sigma} = (\sigma_x, \sigma_y, \sigma_z)$ are the Pauli matrices, and $V(\mathbf{r})$ is the full crystal potential (within the one-electron approximation). The main effect of the spin-orbit coupling is to mix band and spin states, producing a splitting of the non-relativistic levels.

Treatments of the electronic problem beyond the above approximations (when necessary) are numerous and routine in the literature.

To maintain presentation and technicalities at a reasonably simple level, we prefer in general to describe the methods of band structure calculations with reference to the one-electron Schrödinger equation (5.1), with a spatially periodic potential $V(\mathbf{r})$. The various procedures are normally illustrated for *simple lattices* with one atom per primitive cell; the appropriate generalizations necessary for *composite lattices*, with a basis of two or more atoms in the primitive cell, are usually obvious (or almost obvious). Furthermore we do not enter in the details of the construction of the self-consistent crystal potential; conceptually, this can be done by solving the Schrödinger equation (5.1) with a trial periodic potential and then iteratively updating the potential itself and the wavefunctions up to self-consistency; Bloch wavefunctions, and corresponding electronic charge density and potential, are usually calculated at a restricted set of "special $\mathbf{k}$ vectors" (as discussed in Section 2.6.4). For simple introductory considerations, it may be reasonable to approximate the crystalline potential as sum of spherically symmetric atomic-like potentials centered at the atomic positions.

To solve the Schrödinger equation with a periodic potential we can distinguish two main points of view. A first line of approach consists in expanding the crystal states in a convenient set of *energy-independent Bloch functions*: Bloch sums made of atomic orbitals in the tight-binding method, plane waves orthogonalized to core states in the orthogonalized plane wave method, plane waves in the pseudopotential method. A second line of approach (which includes the cellular method, the augmented plane wave method and the Green's function method) is based on the single cell formulation and adopts *energy-dependent basis functions* solutions of the Schrödinger equation in the primitive cell. In the cellular formulation one considers the Wigner-Seitz cell (or subcells), an appropriate spherically symmetric potential within it, and the energy-dependent solutions of the radial Schrödinger equation regular at the origin; these solutions constitute the basis set for the expansion of the crystal wavefunctions, which must satisfy appropriate boundary conditions at the cell boundaries required by the Bloch theorem; the modality of imposing the boundary conditions characterizes the different cellular methods.

The use of energy-independent or energy-dependent basis functions, suggested by physical motivations that may favor the choice of one procedure or the other in specific classes of materials, entails a quite distinct structure of the corresponding secular equations. *For energy-independent basis function methods, the matrix elements of the secular equation are linear in energy for any energy*, i.e. they have the canonical form $X_{ij}(E) = A_{ij} - E B_{ij}$ where the matrices A and B are energy independent (the matrix B is called overlap matrix, and must be non-singular and positive definite). For the solution of linear secular equations, highly efficient numerical techniques are available. *For energy-dependent basis function methods, the matrix elements of the secular equations have a more complicated non-linear form*, and in general the eigenvalues are worked out one-by-one searching for the zeroes of the secular determinant. When a large number of points of the Brillouin zone are processed (for ensemble average purpose, for instance), this may become a serious disadvantage, and the quest arises of appropriate linearization procedures. To reach this purpose, energy-dependent basis functions can be conveniently approximated with Taylor expansion to first order around some fixed reference energy: for all the cellular methods, ingenious "augmentation" procedures have eventually produced linear secular equations which cover a finite (but significant) energy range around the reference energy.

Table 5.1 Schematic summary of basic methods of band structure calculations. The methods either work with fixed basis functions or with energy-dependent functions. In the latter case, linearization of secular equations can be achieved when energy-dependent functions are appropriately enriched ("augmented"), for instance by first-order derivatives around some fixed reference energy.

Point of view	Methods	Energy-independent basis functions
Whole crystal formulation	Tight-binding method	Bloch sums
	Orthogonalized plane wave method	Bloch sums and plane waves
	Pseudopotential method	Plane waves

Point of view	Methods	Energy-dependent basis functions
Single cell formulation	Cellular method	Spherical waves
	Augmented plane wave method	Spherical and plane waves
	Green's function method	Spherical waves

In Table 5.1 we summarize in a schematic form the basic methods of band structure calculations and the basis functions used to describe the crystal states. Of course this scheme has to be taken only as indicative, being impossible to summarize in a concise way all the interesting variations, refinements or interplay among the various lines of approaches, and the wealth of this traditional subject of solid state physics.

5.2 The Tight-Binding Method (LCAO Method)

5.2.1 Description of the Tight-Binding Method for Simple Lattices

The tight-binding method, suggested by Bloch in 1928, consists in expanding the crystal states in linear combinations of atomic orbitals (LCAO) of the composing atoms. This method (when not applied in oversimplified forms) provides a reasonable description of occupied states in any type of crystal (metals, semiconductors, and insulators) and often also of the lowest lying conduction states.

For sake of simplicity and for keeping notations at the minimum, *we begin to consider a simple crystal, with one atom per primitive cell*. We indicate by $\phi_i(\mathbf{r})$ an atomic orbital of quantum numbers i and energy E_i of the atom centered in the reference primitive cell; similarly, $\phi_i(\mathbf{r} - \mathbf{t}_m)$ denotes the same orbital for the atom in the primitive cell individuated by $\mathbf{t}_m$. In correspondence to the atomic orbital ϕ_i, the *Bloch sum* of $\mathbf{k}$ wavevector is defined as

$$\Phi_i(\mathbf{k}, \mathbf{r}) = \frac{1}{\sqrt{N}} \sum_{\mathbf{t}_m} e^{i\mathbf{k}\cdot\mathbf{t}_m} \phi_i(\mathbf{r} - \mathbf{t}_m), \tag{5.2}$$

where N is the number of primitive cells of the crystal. In the tight-binding method, a (judicious) number of Bloch sums of vector $\mathbf{k}$ are used for expanding the crystal wavefunctions of vector $\mathbf{k}$ in the form

$$\psi(\mathbf{k}, \mathbf{r}) = \sum_i c_i(\mathbf{k})\Phi_i(\mathbf{k}, \mathbf{r}), \tag{5.3}$$

where the coefficients $c_i(\mathbf{k})$ are to be determined with standard variational methods.

From the eigenvalue equation (5.1), expansion (5.3), and application of the variational principle, we have that crystal eigenvalues and eigenfunctions are obtained from the determinantal compatibility equation

$$\boxed{\|M_{ij}(\mathbf{k}) - E\, S_{ij}(\mathbf{k})\| = 0}, \tag{5.4}$$

where $M_{ij}(\mathbf{k})$ are the matrix elements of the crystal Hamiltonian between Bloch sums, and $S_{ij}(\mathbf{k})$ are the overlap matrix elements; namely

$$M_{ij}(\mathbf{k}) = \langle \Phi_i(\mathbf{k}, \mathbf{r})|H|\Phi_j(\mathbf{k}, \mathbf{r})\rangle \tag{5.5a}$$
$$S_{ij}(\mathbf{k}) = \langle \Phi_i(\mathbf{k}, \mathbf{r})|\Phi_j(\mathbf{k}, \mathbf{r})\rangle. \tag{5.5b}$$

The interaction (5.5a) and overlap (5.5b) matrix elements can be evaluated numerically. Frequently, however, the tight-binding method is used in a semi-empirical way; here we wish to illustrate this aspect, because of its capability to give a vivid picture of some features of the band structure of several solids, with a modest amount of computational labor.

Semi-Empirical Tight-Binding Model

The semi-empirical tight-binding model adopts some rather drastic (and yet meaningful) assumptions on the overlap and Hamiltonian matrix elements (5.5). We begin to notice that, in the case of well-localized atomic orbitals, the overlap between atomic-like functions centered on different sites is reasonably small. This justifies the assumption that the localized atomic orbitals are taken as orthonormal, and so are the corresponding Bloch sums; in this approximation, the overlap matrix $S_{ij}(\mathbf{k})$ in Eq. (5.5b) equals the unit matrix δ_{ij}.

We are thus left to estimate the Hamiltonian matrix elements of Eq. (5.5a). We have

$$M_{ij}(\mathbf{k}) = \frac{1}{N}\sum_{\mathbf{t}_m\,\mathbf{t}_n} e^{i\mathbf{k}\cdot(\mathbf{t}_n-\mathbf{t}_m)}\langle \phi_i(\mathbf{r} - \mathbf{t}_m)|H|\phi_j(\mathbf{r} - \mathbf{t}_n)\rangle;$$

because of the translational invariance of H we can choose $\mathbf{t}_m = 0$ in the above expression, drop the sum over $\mathbf{t}_m$ and the factor $1/N$, and write

$$M_{ij}(\mathbf{k}) = \sum_{\mathbf{t}_n} e^{i\mathbf{k}\cdot\mathbf{t}_n}\langle \phi_i(\mathbf{r})|H|\phi_j(\mathbf{r} - \mathbf{t}_n)\rangle. \tag{5.6}$$

For semi-empirical evaluations, it is convenient to express the crystal potential as a sum of spherically symmetric atomic-like potentials $V_a(\mathbf{r} - \mathbf{t}_n)$, centered at the lattice positions. We thus approximate the crystal Hamiltonian H in the form

$$H = \frac{\mathbf{p}^2}{2m} + \sum_{\mathbf{t}_n} V_a(\mathbf{r} - \mathbf{t}_n). \tag{5.7}$$

From Eqs. (5.6) and (5.7), one obtains

$$M_{ij}(\mathbf{k}) = \sum_{\mathbf{t}_n} e^{i\mathbf{k}\cdot\mathbf{t}_n} \int \phi_i^*(\mathbf{r}) \left[\frac{\mathbf{p}^2}{2m} + V_a(\mathbf{r}) + V'(\mathbf{r}) \right] \phi_j(\mathbf{r} - \mathbf{t}_n)\, d\mathbf{r}, \qquad (5.8)$$

where $d\mathbf{r}$ is the volume element in direct space, and $V'(\mathbf{r})$ denotes the sum of all the atomic potentials of the crystal, except the contribution $V_a(\mathbf{r})$ of the atom at the origin.

We use the property that the atomic orbital $\phi_i(\mathbf{r})$ is eigenfunction of the atomic Hamiltonian with energy E_i; this fact, together with the assumption of orthonormality of localized atomic orbitals, allows us to write Eq. (5.8) in the form

$$M_{ij}(\mathbf{k}) = E_i \delta_{ij} + \sum_{\mathbf{t}_n} e^{i\mathbf{k}\cdot\mathbf{t}_n} \int \phi_i^*(\mathbf{r}) V'(\mathbf{r}) \phi_j(\mathbf{r} - \mathbf{t}_n)\, d\mathbf{r}. \qquad (5.9)$$

In Eq. (5.9) the term $\mathbf{t}_n = 0$ gives the so called *crystal field* integrals

$$I_{ij} = \int \phi_i^*(\mathbf{r}) V'(\mathbf{r}) \phi_j(\mathbf{r})\, d\mathbf{r}.$$

If the tails of the neighboring atomic-like potentials are almost constant in the region where the wavefunctions $\phi_i(\mathbf{r})$ extend, the matrix with elements I_{ij} becomes a constant diagonal matrix (which produces a rigid shift of the whole band structure, but does not influence the dispersion curves); this justifies the fact that the term $\mathbf{t}_n = 0$ in Eq. (5.9) is often reabsorbed in the diagonal energies E_i.

For what concerns the terms with $\mathbf{t}_n \neq 0$ in Eq. (5.9), we invoke again the localized nature of the atomic orbitals, so that we can limit the sum in Eq. (5.9) to a small number of neighbors. For instance, we can include first neighbor contributions, second neighbor contributions, etc. For easy qualitative "do it yourself" band structure calculations, besides assuming orthonormality of local orbitals, one considers nearest neighbor interactions adopting the *two-center approximation* (integrals involving *three* different centers are considered negligible). The Hamiltonian matrix elements (5.9) thus simplify in the form

$$M_{ij}(\mathbf{k}) = E_i \delta_{ij} + \sum_{\mathbf{t}_l} e^{i\mathbf{k}\cdot\mathbf{t}_l} \int \phi_i^*(\mathbf{r}) V_a(\mathbf{r} - \mathbf{t}_l) \phi_j(\mathbf{r} - \mathbf{t}_l)\, d\mathbf{r}, \qquad (5.10)$$

where the sum over $\mathbf{t}_l$ indicates the sum over first neighbors.

Following the pioneer work of J. C. Slater and G. F. Koster, Phys. Rev. *94*, 1498 (1954), the two-center integrals are conveniently expressed in terms of a small number of independent parameters, which are evaluated either analytically, or numerically, or semi-empirically. In the case of atomic orbitals only of type s and p, the independent integrals appearing in Eq. (5.10) are four and are labeled $V(ss\sigma)$, $V(sp\sigma)$, $V(pp\sigma)$, and $V(pp\pi)$. The convention for notations is that s or p specify the angular momentum of the orbitals, while σ, π, δ, etc. denote that the angular part with respect to the axis of quantization (along the two centers) is characterized by $\exp(im\phi)$ with $m = 0, \pm 1, \pm 2, \ldots$ A pictorial representation of the space arrangement of the orbitals involved in $V(ss\sigma)$, $V(sp\sigma)$, $V(pp\sigma)$, and $V(pp\pi)$, in the case in which the centers of the two atoms lie on the same Cartesian axis, is given in Figure 5.1. In the case in which the two atoms do not lie on the same Cartesian axis, the two-center integrals

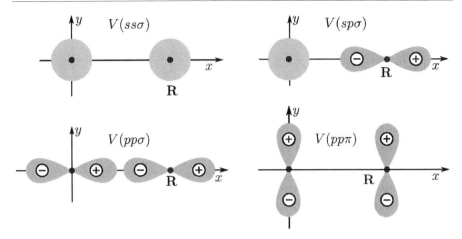

Figure 5.1 Pictorial representation of the independent two-center integrals concerning s and p atomic wavefunctions. One center is taken at the origin and the other center $\mathbf{R}$ is taken in the x-direction.

Table 5.2 Expression of two-center interaction integrals involving atomic orbitals $\phi_i(\mathbf{r})$ and $\psi_j(\mathbf{r} - \mathbf{R})$ of the type s, p_x, p_y, p_z; $V_a(\mathbf{r} - \mathbf{R}) = V_a(|\mathbf{r} - \mathbf{R}|)$ denotes a spherically symmetric potential. The director cosines of the two-center distance vector $\mathbf{R}$ are defined as $\mathbf{R} = (l_x, l_y, l_z)|\mathbf{R}|$. The other expressions of two-center integrals are obtained by cyclic permutations of the indices x, y, z.

$\int \phi_s^*(\mathbf{r}) V_a(\mathbf{r} - \mathbf{R}) \psi_s(\mathbf{r} - \mathbf{R}) \, d\mathbf{r} = V(ss\sigma)$

$\int \phi_s^*(\mathbf{r}) V_a(\mathbf{r} - \mathbf{R}) \psi_x(\mathbf{r} - \mathbf{R}) \, d\mathbf{r} = l_x V(sp\sigma)$

$\int \phi_x^*(\mathbf{r}) V_a(\mathbf{r} - \mathbf{R}) \psi_x(\mathbf{r} - \mathbf{R}) \, d\mathbf{r} = l_x^2 V(pp\sigma) + (1 - l_x^2) V(pp\pi)$

$\int \phi_x^*(\mathbf{r}) V_a(\mathbf{r} - \mathbf{R}) \psi_y(\mathbf{r} - \mathbf{R}) \, d\mathbf{r} = l_x l_y [V(pp\sigma) - V(pp\pi)]$

$\int \phi_x^*(\mathbf{r}) V_a(\mathbf{r} - \mathbf{R}) \psi_z(\mathbf{r} - \mathbf{R}) \, d\mathbf{r} = l_x l_z [V(pp\sigma) - V(pp\pi)]$

(5.10) involving atomic orbitals of type s and p depend on $V(ss\sigma)$, $V(sp\sigma)$, $V(pp\sigma)$, $V(pp\pi)$, and on the direction of the vector distance $\mathbf{R}$ between the two centers (see Table 5.2).

5.2.2 Description of the Tight-Binding Method for Composite Lattices

The extension of the tight-binding method to composite lattices formally needs only some obvious generalization, that we briefly outline. Consider a crystal with a basis of atoms in the positions $\mathbf{d}_1, \mathbf{d}_2, \ldots, \mathbf{d}_{v_b}$ in the primitive cell. Let $\phi_{\mu i}(\mathbf{r} - \mathbf{d}_\mu)$ indicate an atomic orbital of quantum numbers i and energy $E_{\mu i}$ for the atom in position $\mathbf{d}_\mu$. In terms of the atomic orbitals $\phi_{\mu i}(\mathbf{r} - \mathbf{d}_\mu - \mathbf{t}_m)$ we construct the Bloch sum

$$\Phi_{\mu i}(\mathbf{k}, \mathbf{r}) = \frac{1}{\sqrt{N}} \sum_{\mathbf{t}_m} e^{i\mathbf{k} \cdot \mathbf{t}_m} \phi_{\mu i}(\mathbf{r} - \mathbf{d}_\mu - \mathbf{t}_m). \tag{5.11a}$$

A number of Bloch sums (corresponding to atomic orbitals with the same energy or reasonably near in energy) are used as basis functions to describe the crystal wavefunctions in the chosen energy region of interest.

For semi-empirical evaluation, it is convenient to approximate the general one-electron crystalline Hamiltonian in the form

$$H = \frac{\mathbf{p}^2}{2m} + \sum_{\mathbf{t}_n\,\mathbf{d}_\nu} V_{av}(\mathbf{r} - \mathbf{d}_\nu - \mathbf{t}_n), \qquad (5.11b)$$

where $V_{av}(\mathbf{r} - \mathbf{d}_\nu - \mathbf{t}_n)$ is the atomic-like spherically symmetric potential for the atom centered in the position $\mathbf{d}_\nu$ of the lattice cell $\mathbf{t}_n$. The matrix elements of the Hamiltonian (5.11b) between the Bloch sums (5.11a) can be evaluated with the same techniques presented in the previous subsection, and the Hamiltonian matrix is then diagonalized. The mixing of Bloch sums, with different local symmetry or belonging to different sublattices, describes often intuitively hybridization effects among atomic orbitals; in particular we mention here the sp^3s^* nearest neighbor hybridization scheme for semiconductors with diamond or zincblende structures [P. Vogl, H. P. Hjalmarson and J. D. Dow, J. Phys. Chem. Solids **44**, 365 (1983)] and the most used $sp^3d^5s^*$ parametrizations [J.-M. Jancu, R. Scholz, F. Beltram and F. Bassani, Phys. Rev. B **57**, 6493 (1998); T. B. Boykin, G. Klimeck and F. Oyafuso, Phys. Rev. B **69**, 115201 (2004); Y. M. Niquet, D. Rideau, C. Tavernier and H. Jaouen, Phys. Rev. B **79**, 245201 (2009), and references quoted therein].

It can also be noticed that in the tight-binding model the spin-orbit interaction can be included, whenever necessary, in the Hamiltonian matrix, with manageable on-site spin-orbit terms of the composing atoms.

The atomistic approach of the tight-binding method allows to describe, on a rather intuitive basis, trends in energy bands and bonds in a variety of periodic systems. However, one of the greatest merits of the formalism is the possibility to describe electronic states also in non-periodic systems; within the tight-binding framework, significant models have been developed to describe electronic states in incommensurate systems, in disordered systems (in this context, we just mention the Anderson model for diagonal disorder), or even to embody effects beyond the independent electron approximation (as for instance the Hubbard model of one-site correlation effects). Further bonuses are the convenience to describe electron transport properties on a localized basis, and the possibility to carry out molecular dynamics simulations with reasonable computational effort. These are among the reasons of the continuous renaissance of the tight-binding method, too often declared in the literature as obsolete. We do not dwell on these and other aspects of the tight-binding formalism, and confine our next considerations to some elementary illustrative examples.

5.2.3   Illustrative Applications of the Tight-Binding Scheme

Energy Dispersion of a s-like Band in Face-Centered Cubic Crystals

As a first illustrative example we consider a face-centered cubic crystal with one atom per primitive cell, and a single s-like atomic orbital ϕ_s. From the localized orbitals

$\phi_s(\mathbf{r} - \mathbf{t}_m)$ we form the itinerant Bloch sum of $\mathbf{k}$ wavevector

$$\Phi_s(\mathbf{k}, \mathbf{r}) = \frac{1}{\sqrt{N}} \sum_{\mathbf{t}_m} e^{i\mathbf{k}\cdot\mathbf{t}_m} \phi_s(\mathbf{r} - \mathbf{t}_m). \tag{5.12a}$$

The energy of the s-like band is

$$E(\mathbf{k}) = \frac{\langle \Phi_s(\mathbf{k}, \mathbf{r}) | H | \Phi_s(\mathbf{k}, \mathbf{r}) \rangle}{\langle \Phi_s(\mathbf{k}, \mathbf{r}) | \Phi_s(\mathbf{k}, \mathbf{r}) \rangle}. \tag{5.12b}$$

In the empirical tight-binding method, we assume orthonormality of localized orbitals and thus the denominator in Eq. (5.12b) is unit; the numerator is evaluated in the two-center approximation, retaining only nearest neighbor interactions (see Eq. 5.10). We obtain

$$E(\mathbf{k}) = E_s + \sum_{\mathbf{t}_l} e^{i\mathbf{k}\cdot\mathbf{t}_l} \int \phi_s^*(\mathbf{r}) V_a(\mathbf{r} - \mathbf{t}_l) \phi_s(\mathbf{r} - \mathbf{t}_l) \, d\mathbf{r}, \tag{5.12c}$$

where E_s is made of the atomic eigenvalue and the crystal field term, $V_a(\mathbf{r})$ is the atomic-like potential, and the 12 nearest neighbor vectors $\mathbf{t}_l$ of the fcc structure are $(a/2)(0, \pm 1, \pm 1); (a/2)(\pm 1, 0, \pm 1); (a/2)(\pm 1, \pm 1, 0)$.

The value $V(ss\sigma)$ of the two-center potential integral appearing in Eq. (5.12c) is independent of $\mathbf{t}_l$, and can thus be factorized out of the sum. We denote by $F(\mathbf{k})$ the sum

$$F(\mathbf{k}) = \sum_{\mathbf{t}_l} e^{i\mathbf{k}\cdot\mathbf{t}_l} = 4 \left(\cos\frac{k_x a}{2} \cos\frac{k_y a}{2} + \cos\frac{k_y a}{2} \cos\frac{k_z a}{2} + \cos\frac{k_z a}{2} \cos\frac{k_x a}{2} \right).$$

Equation (5.12c) thus becomes

$$E(\mathbf{k}) = E_s + V(ss\sigma)F(\mathbf{k}), \tag{5.12d}$$

and the behavior of the dispersion curve is reported in Figure 5.2. It is evident that s-like bands *bend upward*, as $\mathbf{k}$ increases from the center of the Brillouin zone toward the border, since $V(ss\sigma)$ is negative and $F(\mathbf{k})$ has its maximum value at $\mathbf{k} = 0$.

Energy Dispersion of p-like Bands in Face-Centered Cubic Lattices

As a second illustrative example we consider the band structure corresponding to p-like atomic orbitals in face-centered cubic crystals. We have now the three Bloch sums

$$\Phi_i(\mathbf{k}, \mathbf{r}) = \frac{1}{\sqrt{N}} \sum_{\mathbf{t}_m} e^{i\mathbf{k}\cdot\mathbf{t}_m} \phi_i(\mathbf{r} - \mathbf{t}_m) \quad (i = x, y, z). \tag{5.13a}$$

The matrix elements of the crystal Hamiltonian on the basis functions (5.13a), in the two-center nearest neighbor approximation and using Table 5.2 for the expression of independent integrals, are given by

$$M_{xx}(\mathbf{k}) = E_p + 2\cos\frac{k_x a}{2} \left(\cos\frac{k_y a}{2} + \cos\frac{k_z a}{2} \right) [V(pp\sigma) + V(pp\pi)]$$
$$+ 4\cos\frac{k_y a}{2} \cos\frac{k_z a}{2} V(pp\pi)$$

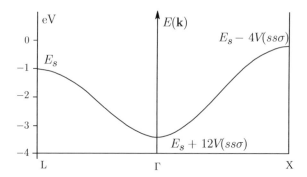

Figure 5.2 Schematic behavior of an s-like band in a fcc crystal lattice with the tight-binding method; E_s is the energy of the s-like orbital; $V(ss\sigma) < 0$ is the two-center potential integral between nearest neighbor s-like atomic orbitals. The values of the wavevector $\mathbf{k}$ at the points Γ, X, and L are $\mathbf{k} = 0$, $\mathbf{k} = (2\pi/a)(1, 0, 0)$, and $\mathbf{k} = (2\pi/a)(1/2, 1/2, 1/2)$, respectively. In the figure, we have taken $E_s = -1$ eV and $V(ss\sigma) = -0.2$ eV.

$$M_{xy}(\mathbf{k}) = -2\sin\frac{k_x a}{2}\sin\frac{k_y a}{2}\left[V(pp\sigma) - V(pp\pi)\right]$$

and cyclic permutations. The 3×3 secular determinant has the form

$$\begin{Vmatrix} M_{xx}(\mathbf{k}) - E & M_{xy}(\mathbf{k}) & M_{xz}(\mathbf{k}) \\ M_{xy}^*(\mathbf{k}) & M_{yy}(\mathbf{k}) - E & M_{yz}(\mathbf{k}) \\ M_{xz}^*(\mathbf{k}) & M_{yz}^*(\mathbf{k}) & M_{zz}(\mathbf{k}) - E \end{Vmatrix} = 0. \tag{5.13b}$$

The diagonalization of the matrix (5.13b) is straightforward at the symmetry points and lines of the Brillouin zone. At the point $\Gamma(\mathbf{k} = 0)$ the threefold degenerate eigenvalue is given by

$$E(\Gamma) = E_p + 4V(pp\sigma) + 8V(pp\pi). \tag{5.14a}$$

At the point X, $\mathbf{k} = (2\pi/a)(1, 0, 0)$, the non-degenerate level $E_1(X)$ and the twofold degenerate level $E_2(X)$ are

$$E_1(X) = E_p - 4V(pp\sigma), \quad E_2(X) = E_p - 4V(pp\pi). \tag{5.14b}$$

At the point L, $\mathbf{k} = (2\pi/a)(1/2, 1/2, 1/2)$, the non-degenerate level $E_1(L)$ and the twofold degenerate level $E_2(L)$ are

$$\begin{aligned} E_1(L) &= E_p - 4V(pp\sigma) + 4V(pp\pi), \\ E_2(L) &= E_p + 2V(pp\sigma) - 2V(pp\pi). \end{aligned} \tag{5.14c}$$

At a general $\mathbf{k}$ point of the Brillouin zone, the energy bands are given by solution of the secular equation (5.13b).

In Figure 5.3 we report the behavior of the p-like energy bands. In general, from the graphical representation of the independent integrals $V(pp\sigma)$ and $V(pp\pi)$ of

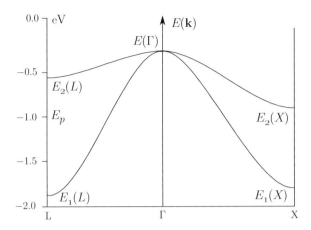

Figure 5.3 Schematic behavior of p-like bands in a fcc crystal lattice with the tight-binding method; E_p is the energy of the p-like orbitals. The values of $E(\Gamma)$, $E_1(X)$, $E_2(X)$, $E_1(L)$, $E_2(L)$ are given by Eqs. (5.14) of the text. The figure refers to the case $E_p = -1$ eV, $V(pp\sigma) = 0.2$ eV, and $V(pp\pi) = -0.02$ eV.

Figure 5.1, we expect that the former is positive and the latter is negative, and furthermore $|V(pp\sigma)| > |V(pp\pi)|$ (in the strong binding limit). Thus we see that p-bands *bend downward* as k varies from the center of the Brillouin zone toward its boundaries.

5.3   The Orthogonalized Plane Wave (OPW) Method

Considerations on Plane Waves Expansion

The choice of plane waves as basis functions to expand crystal wavefunctions (Sommerfeld and Bethe, 1933) appears to be very attractive, because of the formal simplicity of the matrix elements of the Hamiltonian. In fact the kinetic energy operator is diagonal in the plane wave representation and the matrix elements of the potential simply involve the Fourier transforms of the atomic potential (and appropriate phase factors in the case of several atoms in the primitive cell).

Let us consider a crystal, simple, or composite, with one or more atoms in the positions $\mathbf{d}_1, \mathbf{d}_2, \ldots, \mathbf{d}_{v_b}$ in the primitive cell. For the sake of simplicity, we assume that the crystal potential is approximated as sum of atomic-like potentials centered at the atomic positions; the crystal Hamiltonian thus becomes

$$H = -\nabla^2 + V(\mathbf{r}) = -\nabla^2 + \sum_{\mathbf{t}_n \mathbf{d}_v} V_v(\mathbf{r} - \mathbf{d}_v - \mathbf{t}_n), \tag{5.15}$$

where $V_v(\mathbf{r} - \mathbf{d}_v - \mathbf{t}_n)$ denote the potentials of the atoms in the v-th sublattice. For convenience, Rydberg atomic units are being used from now on (see Table 5.3), so that the kinetic energy operator $-(\hbar^2/2m)\nabla^2 \equiv -(\hbar^2/2ma_B^2)a_B^2\nabla^2$ becomes $-\nabla^2$.

Table 5.3 Expression of some fundamental constants (Bohr radius, Rydberg energy, Hartree energy, and fine structure constant). The Hartree atomic units and the Rydberg atomic units are summarized.

Expression of some fundamental constants

$$a_B = \frac{\hbar^2}{me^2} = \text{Bohr radius} = 0.529 \,\text{Å} \quad 1 \text{ Rydberg} = \frac{\hbar^2}{2ma_B^2} = \frac{e^2}{2a_B} = 13.606 \text{ eV}$$

$$\alpha^{-1} = \frac{\hbar c}{e^2} = 137.036 \quad 1 \text{ Hartree} = \frac{e^2}{a_B} = 27.21 \text{ eV}$$

Hartree atomic units: $\hbar = 1$ $m = 1$ $e = 1$

unit of length $= a_B$ unit of energy $= 1$ Hartree   velocity of light $= \alpha^{-1}$

Rydberg atomic units: $\hbar = 1$ $m = 1/2$ $e^2 = 2$

unit of length $= a_B$ unit of energy $= 1$ Rydberg   velocity of light $= 2 \cdot \alpha^{-1}$

Chosen a vector $\mathbf{k}$ in the Brillouin zone, consider the set of normalized plane waves of vector $\mathbf{k}_i = \mathbf{k} + \mathbf{h}_i$, given by

$$W(\mathbf{k}_i, \mathbf{r}) = \frac{1}{\sqrt{N\Omega}} e^{i(\mathbf{k}+\mathbf{h}_i)\cdot\mathbf{r}}, \qquad (5.16)$$

where Ω is the volume of the primitive cell, and $\mathbf{h}_i$ are reciprocal lattice vectors [we suppose that the plane waves of vector $\mathbf{k}_i = \mathbf{k} + \mathbf{h}_i$ have been ordered according to increasing values of the moduli $|\mathbf{k}_i|$ and hence of the kinetic energy k_i^2]. The plane waves $\{W(\mathbf{k}+\mathbf{h}_i, \mathbf{r})\}$ constitute a complete orthonormal set for the expansion of Bloch functions of vector $\mathbf{k}$.

The matrix elements of the crystal Hamiltonian (5.15) between plane waves are:

$$\langle W_{\mathbf{k}_i} | -\nabla^2 + V(\mathbf{r}) | W_{\mathbf{k}_j} \rangle$$

$$= k_i^2 \delta_{ij} + \frac{1}{N\Omega} \sum_{\mathbf{t}_n \mathbf{d}_\nu} \int e^{-i(\mathbf{k}+\mathbf{h}_i)\cdot\mathbf{r}} V_\nu(\mathbf{r} - \mathbf{d}_\nu - \mathbf{t}_n) e^{i(\mathbf{k}+\mathbf{h}_j)\cdot\mathbf{r}} \, d\mathbf{r}$$

$$= k_i^2 \delta_{ij} + \sum_{\mathbf{d}_\nu} e^{-i(\mathbf{h}_i - \mathbf{h}_j)\cdot\mathbf{d}_\nu} V_\nu(\mathbf{h}_i - \mathbf{h}_j), \qquad (5.17a)$$

where

$$V_\nu(\mathbf{h}_i - \mathbf{h}_j) = V_\nu(|\mathbf{h}_i - \mathbf{h}_j|) = \frac{1}{\Omega} \int e^{-i(\mathbf{h}_i - \mathbf{h}_j)\cdot\mathbf{r}} V_\nu(\mathbf{r}) \, d\mathbf{r}. \qquad (5.17b)$$

The phase $\exp[-i(\mathbf{h}_i - \mathbf{h}_j) \cdot \mathbf{d}_\nu]$ is called *structure phase factor* for the atom in the position $\mathbf{d}_\nu$ in the primitive cell, while $V_\nu(\mathbf{h}_i - \mathbf{h}_j)$ is called *form factor* for the atom. We can now expand a crystal wavefunction of vector $\mathbf{k}$ in the form

$$\psi(\mathbf{k}, \mathbf{r}) = \sum_i c_i(\mathbf{k}) \, W(\mathbf{k}_i, \mathbf{r});$$

standard variational methods on the expansion coefficients lead to the determinantal compatibility equation

$$\left\| (k_i^2 - E)\delta_{ij} + \sum_{\mathbf{d}_\nu} e^{-i(\mathbf{h}_i - \mathbf{h}_j)\cdot\mathbf{d}_\nu} V_\nu(\mathbf{h}_i - \mathbf{h}_j) \right\| = 0. \tag{5.18}$$

The secular equation (5.18) in the plane wave representation is built with elementary ingredients, which are the *empty lattice* kinetic energies k_i^2, and the *form and structure phase factors* of the atoms in the primitive cell.

Following a simple consideration due to Herring, it is easily seen that a *pure expansion of crystal states into plane waves is seriously flawed by the so called "variational collapse" problem*. Consider in fact the secular equation (5.18) and imagine to truncate and solve it including a finite number n of plane waves (with modulus $|\mathbf{k}_i|$ up to some cut-off value k_{max}). As $n \to \infty$, the lowest roots of the secular equation (5.18) precipitate toward the lowest core states of the crystal. Since core states are strongly localized in real space, their accurate description with plane waves requires an impossibly large number of them. Thus the secular equation (5.18) is completely unreliable for the crystal states of lowest energy, and is thus unreliable for any other higher energy state too.

To understand better the nature of the "variational collapse" problem, consider for instance the case of silicon ($Z = 14$), whose structure is fcc with lattice constant $a = 5.43$ Å $= 10.26\,a_B$, and primitive cell volume $\Omega = a^3/4$. The radius a_{1s} of the core state $1s$ of silicon atom can be estimated to be $a_{1s} \approx a_B/Z$, and the cut-off value $k_{max} \approx 2\pi/a_{1s} = 2\pi Z/a_B$. The number n of reciprocal lattice vectors inside a sphere of radius k_{max} is given by $(4/3)\pi k_{max}^3 = n\Omega_k = n(2\pi)^3/\Omega$; we get $n \approx Z^3(a/a_B)^3 \approx 10^6$. Thus, the secular matrix (5.18) should be of the order of $10^6 \times 10^6$ to reproduce the $1s$ state of silicon, and even larger for heavier elements.

The Principle of Orthogonalization

The variational collapse problem can be overcome with the orthogonalization procedure, introduced by C. Herring, Phys. Rev. 57, 1169 (1940). We illustrate first the essential idea, and later the development of the orthogonalized plane wave method. Consider an operator H and a complete orthonormal set $\{\phi_i\}$. The eigenvalues and eigenfunctions of H are obtained diagonalizing H within this set, i.e. solving the standard secular equation

$$\| \langle \phi_i | H | \phi_j \rangle - E\delta_{ij} \| = 0. \tag{5.19}$$

Suppose that the n_c *lowest energy eigenvalues and eigenfunctions of the operator H are already known* (this is the situation of practical interest). We denote these states and eigenvalues by ψ_c and E_c, respectively; they satisfy the eigenvalue equation

$$H\psi_c = E_c\psi_c \quad c = 1, 2, \ldots, n_c. \tag{5.20}$$

We wish to exploit this knowledge to simplify the problem of obtaining all the other eigenvalues and eigenfunctions of the operator H.

A possible way to proceed is to consider for any basis state $|\phi_i\rangle$ the modified state $|\widetilde{\phi}_i\rangle$, orthogonalized to all the known $|\psi_c\rangle$ eigenstates, given by

$$|\widetilde{\phi}_i\rangle = |\phi_i\rangle - \sum_c |\psi_c\rangle\langle\psi_c|\phi_i\rangle. \tag{5.21}$$

An eigenfunction $|\psi_i\rangle$ of H, orthogonal to the states ψ_c, can be expanded on the set of states $\{\widetilde{\phi}_i\}$ in the form

$$|\psi_i\rangle = \sum_j c_{ij} |\widetilde{\phi}_j\rangle ; \tag{5.22}$$

with standard variational methods, the energies E_i and the expansion coefficients c_{ij} are determined by solving the secular equation

$$\boxed{\| \langle\widetilde{\phi}_i|H|\widetilde{\phi}_j\rangle - E\langle\widetilde{\phi}_i|\widetilde{\phi}_j\rangle \| = 0}. \tag{5.23}$$

The secular equation (5.23) *produces only the eigenstates of H orthogonal to the already known states* $\{\psi_c\}$, while the original secular equation (5.19) obviously produces all the eigenstates of H.

We discuss now in more detail some features of the modified secular equation (5.23). We begin to notice that the modified states (5.21) are not orthonormal; in fact the overlap can be expressed in the form

$$\widetilde{S}_{ij} = \langle\widetilde{\phi}_i|\widetilde{\phi}_j\rangle = \delta_{ij} - \langle\phi_i| \left[\sum_c |\psi_c\rangle\langle\psi_c| \right] |\phi_j\rangle. \tag{5.24a}$$

Thus, a preliminary check is to verify that the overlap matrix $\widetilde{S}$ (truncated at a given finite rank of interest) is not singular and that its determinant is actually positive; this assures that the orthogonalized basis functions $\{\widetilde{\phi}_i\}$ under attention are still linearly independent and the set is not "overcomplete" [experience shows that the matrix $\widetilde{S}$ tends to become singular as its rank increases; in this case the treatment consists in diagonalizing the matrix $\widetilde{S}$ and dropping the redundant linear combination(s)].

Using Eqs. (5.20) and (5.21), it is immediately seen that the matrix elements $\widetilde{M}_{ij}$ of the secular equation (5.23) can be expressed as follows

$$\widetilde{M}_{ij} = \langle\widetilde{\phi}_i|H - E|\widetilde{\phi}_j\rangle = \langle\phi_i| \left[H - E + \sum_c (E - E_c)|\psi_c\rangle\langle\psi_c| \right] |\phi_j\rangle, \quad (5.24b)$$

where $E > E_c$. Thus the secular equation (5.23) can be written in the alternative and perfectly equivalent form

$$\boxed{\| \langle\phi_i|H + V^{(\text{rep})}|\phi_j\rangle - E\delta_{ij} \| = 0} \quad \text{with} \quad \boxed{V^{(\text{rep})} = \sum_c (E - E_c)|\psi_c\rangle\langle\psi_c|},$$

$$\tag{5.25}$$

where the energy-dependent operator $V^{(\text{rep})}$ defined above accounts exactly for the effect of orthogonalization.

We have supposed that $\{\psi_c\}$ are the lowest eigenfunctions of H; this fact gives a very simple interpretation to the additional operator $V^{(\text{rep})}$. For any $E > E_c$ $(c = 1, 2, \ldots, n_c)$, *the expectation value of* $V^{(\text{rep})}$ *on any state can never become negative*; this justifies the interpretation of $V^{(\text{rep})}$ as a kind of repulsive potential, even if its actual formal definition is quite different from an ordinary local potential.

In summary: compare the initial secular equation (5.19) on a given basis set and the secular equation (5.25) on the corresponding orthogonalized basis set. The former secular equation provides all the eigenvalues of the Hamiltonian (and is thus subject to the variational collapse problem previously discussed), the latter provides eigenvalues higher than the core ones (and is free of the variational collapse problem, if a judicious separation of localized core states and higher energy states is possible). It is remarkable that this important achievement is reached at low cost. In fact, the matrix elements of the secular equations (5.19) and (5.25) have both the canonical linear form $X_{ij}(E) = A_{ij} - E B_{ij}$, where the matrices A and B are energy-independent; for linear secular equations highly efficient numerical techniques are available.

Expansion of the Electronic Crystalline States in Orthogonalized Plane Waves

The basic difficulty of a simple plane waves expansion consists in its impossibility to describe (with a reasonable number of plane waves) the strongly localized core states; on the other hand these states are accurately described by the tight-binding method and Herring suggested how to use from the very beginning this information. There is a large number of crystals whose electronic states can be sharply separated into two classes: (i) *inner (core) states*, which are very localized spatially and very deep in energy; (ii) *outer (valence and/or conduction) states* which are spread out spatially and at higher energy. Herring proposed to describe the former by Bloch sums built from localized orbitals and the latter by plane waves orthogonalized to the core states. Orthogonalized plane waves appear a convenient set to describe itinerant states, since they are atomic-like near the nuclei where the crystal potential is atomic-like and plane waves in the interstitial regions, where the crystal potential is smooth.

Consider a plane wave $W(\mathbf{k}_j, \mathbf{r})$ of vector $\mathbf{k}_j = \mathbf{k} + \mathbf{h}_j$; we orthogonalize it to all the core Bloch sums $\Phi_c(\mathbf{k}, \mathbf{r})$ and obtain the OPW function in the form

$$|\widetilde{W}(\mathbf{k}_j, \mathbf{r})\rangle = |W(\mathbf{k}_j, \mathbf{r})\rangle - \sum_{\text{core}} |\Phi_c(\mathbf{k}, \mathbf{r})\rangle\langle\Phi_c(\mathbf{k}, \mathbf{r})|W(\mathbf{k}_j, \mathbf{r})\rangle. \tag{5.26}$$

We now expand *outer crystal states* of vector $\mathbf{k}$ into orthogonalized plane waves of vectors $\mathbf{k}_j = \mathbf{k} + \mathbf{h}_j$; standard variational procedures on the coefficients of the expansion lead to the determinantal compatibility equation

$$\|\langle\widetilde{W}_{\mathbf{k}_i}|H|\widetilde{W}_{\mathbf{k}_j}\rangle - E\langle\widetilde{W}_{\mathbf{k}_i}|\widetilde{W}_{\mathbf{k}_j}\rangle\| = 0. \tag{5.27a}$$

The lowest eigenvalue of (5.27a) corresponds to the lowest (valence or conduction) crystal eigenstate. The secular equation (5.27a), using the expression (5.26), can be written in the alternative and perfectly equivalent form

$$\boxed{\|\langle W_{\mathbf{k}_i}| - \nabla^2 + V(\mathbf{r}) + V^{(\text{rep})}|W_{\mathbf{k}_j}\rangle - E\delta_{ij}\| = 0}, \tag{5.27b}$$

where $H = -\nabla^2 + V(\mathbf{r})$ is the crystal Hamiltonian and the repulsive operator $V^{(\text{rep})}$ is defined as

$$V^{(\text{rep})} = \sum_{\text{core}} (E - E_c)|\Phi_{c\mathbf{k}}\rangle\langle\Phi_{c\mathbf{k}}|. \tag{5.27c}$$

The operator $V^{(\text{rep})}$ defined above is evidently a "non-orthodox" operator, which is energy-dependent and non-local. The operator $V^{(\text{rep})}$ has several good features: (i) it is linear in energy, and standard algebraic methods for the secular equation can be applied; (ii) the matrix elements of $V^{(\text{rep})}$ among plane waves are rather simple, since they involve Fourier transforms of core states, often represented on a set of Slater or Gaussian type orbitals; (iii) *qualitatively*, $V^{(\text{rep})}$ can be interpreted as a repulsive potential produced by the presence of core states. This is the peculiar feature that will eventually flow from the OPW method to the full maturity concepts of pseudopotentials.

In the orthogonalized plane wave method, one sets up the secular equation (5.27a), or the equivalent secular equation (5.27b), using a finite number n of OPWs (up to a given $k_{\max}$); then the stability of the eigenvalues in the energy range of interest is checked as the order of the secular equation is increased. In general a reasonable rapid convergence is obtained: a number of 10–100 OPWs being sufficient in most cases. In the OPW method, a check must be always performed to assure that the overlap matrix $\tilde{S}_{ij} = \langle\tilde{W}_{\mathbf{k}_i}|\tilde{W}_{\mathbf{k}_j}\rangle$ is not singular (otherwise the corresponding redundant linear combinations must be suppressed). From a mathematical point of view, the rapid convergence of the method is related to the *cancellation effect* between orthogonalization terms and Fourier transforms of the crystal potential when the wavevector transfers $|\mathbf{h}_i - \mathbf{h}_j|$ are large.

Evaluation of the Matrix Elements of the OPW Method

The ingredients to set up the secular equation (5.27b) of the OPW method are the matrix elements of the crystal Hamiltonian and of the repulsive operator between (ordinary) plane waves $W(\mathbf{k}_j, \mathbf{r})$. The former had already been discussed in Eqs. (5.17); the latter are now discussed beginning with the case of a simple lattice. Consider the Bloch sum

$$\Phi_c(\mathbf{k}, \mathbf{r}) = \frac{1}{\sqrt{N}} \sum_{\mathbf{t}_n} e^{i\mathbf{k}\cdot\mathbf{t}_n} \phi_c(\mathbf{r} - \mathbf{t}_n)$$

corresponding to the atomic core function $\phi_c(\mathbf{r})$; the overlap between $\Phi_c(\mathbf{k}, \mathbf{r})$ and the plane wave $W(\mathbf{k}_j, \mathbf{r})$ is

$$\langle\Phi_{c\mathbf{k}}|W_{\mathbf{k}_j}\rangle = \frac{1}{\sqrt{N}} \sum_{\mathbf{t}_n} \int e^{-i\mathbf{k}\cdot\mathbf{t}_n} \phi_c^*(\mathbf{r} - \mathbf{t}_n) \frac{e^{i(\mathbf{k}+\mathbf{h}_j)\cdot\mathbf{r}}}{\sqrt{N\Omega}} d\mathbf{r}$$

$$= \sum_{\mathbf{t}_n} \int \frac{e^{i(\mathbf{k}+\mathbf{h}_j)\cdot(\mathbf{r}-\mathbf{t}_n)}}{N\sqrt{\Omega}} \phi_c^*(\mathbf{r} - \mathbf{t}_n) d\mathbf{r} = \frac{1}{\sqrt{\Omega}} \int e^{i(\mathbf{k}+\mathbf{h}_j)\cdot\mathbf{r}} \phi_c^*(\mathbf{r}) d\mathbf{r}. \tag{5.28}$$

The familiar expansion of plane waves in spherical harmonics reads

$$e^{i\mathbf{k}_j\cdot\mathbf{r}} = 4\pi \sum_{lm} i^l j_l(k_j r) Y_{lm}^*(\mathbf{k}_j) Y_{lm}(\mathbf{r}), \tag{5.29}$$

where $Y_{lm}(\mathbf{r})$ stands for $Y_{lm}(\theta_\mathbf{r}, \phi_\mathbf{r})$, $Y_{lm}(\mathbf{k})$ stands for $Y_{lm}(\theta_\mathbf{k}, \phi_\mathbf{k})$, and $j_l(x)$ is the spherical Bessel function of order l. We write the atomic core state as product of a radial part and an angular part in the usual form $\phi_c(\mathbf{r}) = R_{nl}(r) Y_{lm}(\mathbf{r})$. The orthogonalization coefficient (5.28), relabeled as $\langle \Phi_{nlm\mathbf{k}} | W_{\mathbf{k}_j} \rangle$, becomes

$$\langle \Phi_{nlm\mathbf{k}} | W_{\mathbf{k}_j} \rangle = \frac{4\pi}{\sqrt{\Omega}} i^l \, Y_{lm}^*(\mathbf{k}_j) \int_0^\infty R_{nl}(r) j_l(k_j r) r^2 \, dr.$$

The matrix elements of the operator $V^{(\text{rep})}$ between plane waves is

$$\langle W_{\mathbf{k}_i} | V^{(\text{rep})} | W_{\mathbf{k}_j} \rangle = \sum_{nlm}^{(\text{core})} (E - E_{nl}) \langle W_{\mathbf{k}_i} | \Phi_{nlm\mathbf{k}} \rangle \langle \Phi_{nlm\mathbf{k}} | W_{\mathbf{k}_j} \rangle.$$

The sum over m ($m = -l, -l+1, \ldots, +l$) can be performed using the addition theorem for spherical harmonics

$$P_l(\hat{\mathbf{k}}_i \cdot \hat{\mathbf{k}}_j) = \frac{4\pi}{2l+1} \sum_{m=-l}^{+l} Y_{lm}^*(\mathbf{k}_i) Y_{lm}(\mathbf{k}_j), \tag{5.30}$$

where $\hat{\mathbf{k}}_i$ and $\hat{\mathbf{k}}_j$ denote the unit vectors of $\mathbf{k}_i$ and $\mathbf{k}_j$, and P_l is the Legendre polynomial of order l. After defining for every core state the *orthogonalization coefficients*

$$A_{nl}(k_i) = \left[\frac{4\pi(2l+1)}{\Omega} \right]^{1/2} \int_0^\infty R_{nl}(r) j_l(k_i r) r^2 \, dr,$$

we obtain the expression

$$\langle W_{\mathbf{k}_i} | V^{(\text{rep})} | W_{\mathbf{k}_j} \rangle = \sum_{nl}^{(\text{core})} (E - E_{nl}) P_l(\hat{\mathbf{k}}_i \cdot \hat{\mathbf{k}}_j) A_{nl}(k_i) A_{nl}(k_j). \tag{5.31}$$

The ingredients for the matrix elements of the OPW method are thus the Fourier transforms of the atomic potential at reciprocal lattice vectors, and the orthogonalization coefficients.

The matrix elements of the OPW method for *composite lattices*, with atoms in the positions $\mathbf{d}_1, \mathbf{d}_2, \ldots, \mathbf{d}_{v_b}$ in the primitive cell, are obtained in a rather similar procedure; each atom contributes with an appropriate number of core states, and the corresponding orthogonalization terms include the appropriate structure phase factors, due to the atomic positions in the primitive cell.

We conclude this discussion recalling some properties of the special functions used in the OPW method. The Legendre polynomials are obtained from the recursion relation

$$(l+1)P_{l+1}(x) - (2l+1)x P_l(x) + l P_{l-1}(x) = 0 \quad \text{with}$$
$$P_0(x) = 1 \quad \text{and} \quad P_1(x) = x.$$

The spherical Bessel functions are the solutions, regular at the origin, of the radial wave equation (with zero potential and unit energy)

$$\frac{d^2 R}{dr^2} + \frac{2}{r}\frac{dR}{dr} + \left[1 - \frac{l(l+1)}{r^2} \right] R = 0.$$

The first few Bessel functions are

$$j_0(x) = \frac{\sin x}{x}; \quad j_1(x) = \frac{\sin x}{x^2} - \frac{\cos x}{x};$$

$$j_2(x) = \left(\frac{3}{x^3} - \frac{1}{x} \right) \sin x - \frac{3}{x^2} \cos x. \tag{5.32a}$$

Any Bessel function of interest can be obtained from the recurrence relation

$$j_{l+1}(x) = \frac{2l+1}{x} j_l(x) - j_{l-1}(x) \quad (l \geq 1). \tag{5.32b}$$

For small values of x it holds

$$j_l(x) \rightarrow \frac{x^l}{1 \cdot 3 \cdot \ldots \cdot (2l+1)} \quad \text{as } x \rightarrow 0. \tag{5.32c}$$

In the following, we need also the Neumann functions, solutions of the radial wave equation (with zero potential and unit energy) that are singular at the origin. The first few Neumann functions are

$$n_0(x) = -\frac{\cos x}{x}; \quad n_1(x) = -\frac{\cos x}{x^2} - \frac{\sin x}{x};$$

$$n_2(x) = -\left(\frac{3}{x^3} - \frac{1}{x} \right) \cos x - \frac{3}{x^2} \sin x.$$

Any Neumann function of interest can be obtained from the recurrence relations corresponding to Eq. (5.32b) where j_l is substituted by n_l. For small values of x it holds

$$n_l(x) \rightarrow -\frac{1 \cdot 3 \cdot \ldots \cdot (2l-1)}{x^{l+1}} \quad \text{as } x \rightarrow 0.$$

For a thorough review of Bessel and Neumann functions see, for instance, M. Abramowitz and I. A. Stegun "Handbook of Mathematical Functions" (Dover Publications, New York, 1972).

Illustrative Example: Valence and Conduction Bands of Lithium Hydride

As an application of the orthogonalized plane wave method, we consider the band structure of lithium hydride. The crystal structure of LiH is fcc, with lattice parameter $a = 7.72 \, a_B$; there are two ions in the primitive cell: H^- in the position $\mathbf{d}_1 = 0$ and Li^+ in the position $\mathbf{d}_2 = (a/2)(1, 0, 0)$. There are only four electrons in the primitive cell, and this simplicity adds to the interest in this prototype material.

The Li^+ ion has the closed shell electronic configuration $1s^2$. We have only a core state and thus the application of the OPW method is particularly simple and significant. To describe valence and conduction bands, we orthogonalize the plane waves to the Bloch sum formed with the $1s$ wavefunctions of Li^+. The matrix elements appearing in the OPW method, including the exchange potential in its non-local form, can be evaluated analytically. The band structure of lithium hydride is reported in Figure 5.4. The core band at ≈ -60 eV originates from the $1s$ core orbitals of the Li^+ ions.

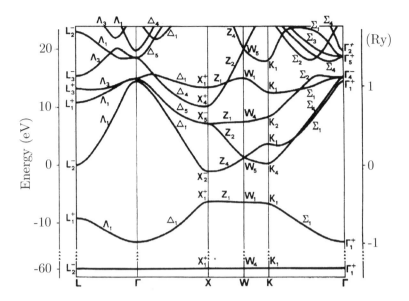

Figure 5.4 Band structure of solid lithium hydride with the orthogonalized plane wave method [reprinted with permission from S. Baroni, G. Pastori Parravicini and G. Pezzica, Phys. Rev. B *32*, 4077 (1985); copyright 1985 by the American Physical Society].

The valence band is *s*-like, with its minimum at Γ and its maximum at the point X of the Brillouin zone; the valence band is mostly related to the $1s$ orbitals of the H^- ions. The conduction band has its minimum at X; the material is an insulator with a direct energy gap of about 5 eV.

It is of interest to report the empty lattice energy states in face-centered cubic lattices at some relevant symmetry points (Figure 5.5). The general trend of valence and conduction bands of Figure 5.4 can be inferred from the empty lattice states of Figure 5.5. In fact the accidental degeneracies of the empty lattice are removed by the introduction of the crystal potential and repulsive potential (due to orthogonalization) in agreement with the qualitative rules of perturbation theory; this suggests that the "effective potential" $V(\mathbf{r}) + V^{(\text{rep})}$ is reasonably smooth, so that valence and conduction band wavefunctions can be well approximated by a relatively small number of orthogonalized plane waves.

5.4 The Pseudopotential Method

Introduction and Empirical Pseudopotentials for Crystals

In the study of the electronic states in solids, we can often separate the chemically inert core states of the atoms from the chemically active valence states. The former do not change significantly from free atoms to solids, and do not influence the chemical

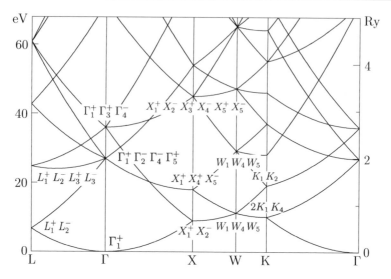

Figure 5.5 Empty lattice energy states for face-centered cubic lattices at the symmetry points L, Γ, X, W, K; the lattice parameter is taken to be $a = 7.72\ a_B$, as in LiH. The symmetry notations are those of G. F. Koster, J. O. Dimmock, R. C. Wheeler and H. Statz "Properties of the Thirty-Two Point Groups" (MIT, Cambridge, Massachusetts, 1963).

properties of materials; the latter, with self-consistent readjustment of wavefunctions and charge densities to different environment situations, determine the actual properties of materials.

We have seen that *the basic strategy of the* OPW *method is to exploit the presence of the core states* for an efficient determination of the valence and conduction energy bands, as well as of the crystalline wavefunctions in the whole sample, i.e. *both in the core regions and outside them.* As we shall see, *the basic strategy of the pseudopotential method is to get rid of the core states, by replacing the strong crystalline potential with a weak pseudopotential,* capable of an efficient determination of valence and conduction energy bands; for what concerns the corresponding pseudowavefunctions, these are required to represent the genuine crystalline wavefunctions only *outside the core regions,* without worrying of the smoothing occurring in the chemically inert core regions. The link and evolution between the OPW method and the pseudopotential method can be understood with the following arguments.

In the orthogonalized plane wave method, the core states are taken into account by enforcing the orthogonalization requirement of valence and conduction states to them. According to the OPW method, the valence and conduction states at the energies E are obtained solving the secular equation corresponding to the effective operator

$$H^{(\text{eff})} = -\nabla^2 + V(\mathbf{r}) + \sum_{\text{core}} (E - E_c)|\Phi_{c\mathbf{k}}\rangle\langle\Phi_{c\mathbf{k}}|, \tag{5.33}$$

where $-\nabla^2 + V(\mathbf{r})$ is the ordinary crystal Hamiltonian; the additional operator is an energy-dependent non-local operator, whose origin is related to the presence of core states. At first sight, the *formal aspect* of the orthogonalization-related terms appearing in Eq. (5.33) (with their non-local and energy-dependent nature) does not seem particularly encouraging for simplifications (in the OPW these terms are evaluated just as they stand); nevertheless the physical aspect is rather appealing. In essence, the effect of the orthogonalization terms is to cancel (to a large extent) the true all-electron crystal potential $V(\mathbf{r})$ just in the core region, where $V(\mathbf{r})$ is particularly strong. From this, the hope arises that the "effective potential" (true potential plus orthogonalization terms) appearing in Eq. (5.33), can be somehow mimicked by an appropriate weak *pseudopotential* $V^{(\text{pseudo})}$, smoother than the true potential $V(\mathbf{r})$ *within* the core region, and equal to it *outside*.

To corroborate this point of view with a simple example, consider for instance the case of Na atom (electronic configuration: $1s^2$; $2s^2 2p^6$; $3s^1$). The atomic potential $V_a(r)$ behaves as $-11e^2/r$ for small r and as $-e^2/r$ for large r. We consider $1s$, $2s$, and $2p$ shells as atomic core states, and we try to infer the atomic pseudopotential acting on the $3s$ optical electron; from the fact that the orbital energy of the $3s$ state is ≈ -5.14 eV, we can safely conclude that the effective potential in the core region is much softer than the hydrogen potential $-e^2/r$ (whose ground energy is -13.6 eV); thus the true potential $V_a(r)$ has been largely canceled in the core region. Much more general plausibility and formal arguments are available in the literature to support this point of view; in particular, pseudopotentials have appeared in solid state physics following the pioneering works of J. C. Phillips and L. Kleinman, Phys. Rev. *116*, 287 (1959); F. Bassani and V. Celli, J. Phys. Chem. Solids *20*, 64 (1961).

The above guidelines have been combined with physical ingenuity to mimic as better as possible the exact operator $V(\mathbf{r}) + \sum_c (E - E_c)|\Phi_{c\mathbf{k}}\rangle\langle\Phi_{c\mathbf{k}}|$ of Eq. (5.33) with a smooth pseudopotential (possibly local and energy-independent). The simplest *empirical* way to introduce a crystal pseudopotential is to express it as sum of local spherically symmetric atomic pseudopotentials centered in the atomic positions. In this case, in analogy to Eq. (5.18), the secular equation for *valence and conduction* states becomes

$$\left\| (k_i^2 - E)\delta_{ij} + \sum_{\mathbf{d}_\nu} e^{-i(\mathbf{h}_i - \mathbf{h}_j)\cdot\mathbf{d}_\nu} V_\nu^{(\text{pseudo})}(\mathbf{h}_i - \mathbf{h}_j) \right\| = 0 . \tag{5.34}$$

It is worthwhile to stress that Eq. (5.34) is formally similar to Eq. (5.18), but strongly different from it in the meaning. Eq. (5.18) is built with the true atomic potentials; it thus describes all crystal states (core, valence, and conduction) and is subject to the "variational collapse" problem, previously discussed. Equation (5.34) is built with atomic pseudopotentials; it thus describes only valence and conduction states, and is mathematically free of the variational collapse problem.

The matrix elements of the secular equation (5.34) can be expressed in terms of a few Fourier components of the atomic pseudopotentials at the reciprocal lattice vectors; often these components are considered as disposable parameters. Their number can be significantly limited by requiring that these parameters are zero when the transfer wavevector $q = |\mathbf{h}_i - \mathbf{h}_j|$ is sufficiently large (so that the cancellation expected for small r is at work). The small number of disposable parameters is determined in such a way to fit basic properties of the crystal (for instance energy gap, some relevant transition energies, etc.). With a little amount of computational labor and with a limited knowledge of the experimental data of the solid, one can derive the whole band structure and explain a number of properties (such as optical constants, photoemission spectra, structural properties); this kind of approach is referred to in the literature as *empirical pseudopotential method*.

As an illustrative application, we briefly consider the semiconductors with the diamond or zincblende structure; this structure is fcc with two atoms in the primitive cell in the positions $\mathbf{d}_1 = 0$ and $\mathbf{d}_2 = (a/4)(1, 1, 1)$. The secular equation (5.34) of the empirical pseudopotential method becomes:

$$\left\| (k_i^2 - E)\delta_{ij} + V_1^{(\text{pseudo})}(\mathbf{h}_i - \mathbf{h}_j) + e^{-i(\mathbf{h}_i - \mathbf{h}_j)\cdot\mathbf{d}_2} V_2^{(\text{pseudo})}(\mathbf{h}_i - \mathbf{h}_j) \right\| = 0$$

$$(5.35)$$

[in actual calculations, it is generally preferred to choose the origin of coordinates halfway between the two atoms of the primitive cell, and break up the crystal pseudopotential into a symmetric and an antisymmetric part; for details we refer to the classic work of M. L. Cohen and T. K. Bergstresser, Phys. Rev. *141*, 789 (1966)].

The first five shells of the fcc reciprocal lattice vectors are: $(2\pi/a)(0, 0, 0)$ [1 vector]; $(2\pi/a)(\pm 1, \pm 1, \pm 1)$ [8 vectors]; $(2\pi/a)(\pm 2, 0, 0)$ and permutations [6 vectors]; $(2\pi/a)(\pm 2, \pm 2, 0)$ and permutations [12 vectors]; $(2\pi/a)(\pm 3, \pm 1, \pm 1)$ and permutations [24 vectors]. The magnitude of the vectors of the first five shells are 0, $\sqrt{3}$, $\sqrt{4}$, $\sqrt{8}$, $\sqrt{11}$ times $(2\pi/a)$ (for the lowest ones, see Figure 5.5 at Γ), and only these magnitudes are allowed to have a non-zero pseudopotential in the paper of Cohen and Bergstresser. The value of the crystal pseudopotential for $\mathbf{h} = 0$ is irrelevant, since it merely adds a constant to all energy levels. We notice that $\exp(-i\mathbf{h}\cdot\mathbf{d}_2) = -1$ for any of the six vectors of the type $(2\pi/a)(\pm 2, 0, 0)$, and $\exp(-i\mathbf{h}\cdot\mathbf{d}_2) = +1$ for any of the 12 vectors of the type $(2\pi/a)(\pm 2, \pm 2, 0)$. Thus, the secular Eq. (5.35) for homopolar substances contains only three independent parameters: $V(\sqrt{3})$, $V(\sqrt{8})$, and $V(\sqrt{11})$. For instance, for silicon, the fitted pseudopotential form factors are found to be: $V(\sqrt{3}) = -0.21$ Ry, $V(\sqrt{8}) = 0.04$ Ry, $V(\sqrt{11}) = 0.08$ Ry, while for germanium: $V(\sqrt{3}) = -0.23$ Ry, $V(\sqrt{8}) = 0.01$ Ry, and $V(\sqrt{11}) = 0.06$ Ry. For heteropolar compounds, the secular equation (5.35) contains six independent parameters, namely: $V_1(\sqrt{3})$, $V_2(\sqrt{3})$, $V_1(\sqrt{4}) - V_2(\sqrt{4})$, $V_1(\sqrt{8}) + V_2(\sqrt{8})$, $V_1(\sqrt{11})$, $V_2(\sqrt{11})$. A so reasonably small number of independent parameters is appealing, and explains why the empirical pseudopotential method has been extremely useful for understanding band structure and trends in several crystals.

As an example, we report in Figure 5.6 the energy bands of GaAs, a direct gap material of great importance in optoelectronics. The valence bands of GaAs have the maximum at $\mathbf{k} = 0$; the symmetry of the top valence state is Γ_{15} (a threefold degenerate level). The minimum of the conduction bands which also occurs at $\mathbf{k} = 0$, is the total-symmetric state Γ_1; the energy band gap is $E_G = E(\Gamma_1) - E(\Gamma_{15}) \approx 1.4$ eV. Notice that the minima of the conduction bands at L_1 and X_1 are not far above Γ_1, and this is at the origin of the Gunn effect in the transport properties (see Section 13.4.1).

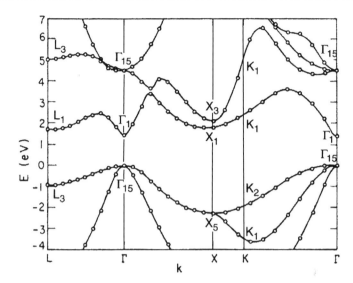

Figure 5.6 Energy bands of GaAs by the empirical pseudopotential method [reprinted with permission from M. L. Cohen and T. K. Bergstresser, Phys. Rev. *141*, 789 (1966); copyright 1966 by the American Physical Society].

Atomic Pseudopotentials

An appealing evolution of the pseudopotential framework appears in the literature in the 1970s, and consists in a skillful use of atomic properties to fix the pseudopotentials of the atoms composing the solid. There is not a unique recipe for the elaboration of atomic pseudopotentials, and this has sometimes cast (unjustified) doubts on the reliability of the procedure. In spite of some degree of arbitrariness, the essential guidelines apply on whatever pseudopotential model and assure, in general, a reasonable standard of quality.

The first step is an *all-electron* self-consistent atomic calculation, usually in the local density approximation. The atomic potential $V_a(r)$ and the atomic wavefunctions are solutions of the radial Schrödinger equation

$$\left[-\frac{d^2}{dr^2} + \frac{l(l+1)}{r^2} + V_a(r)\right] r R_{nl}(r) = E_{nl}\, r R_{nl}(r). \tag{5.36}$$

Once the all-electron potential $V_a(r)$ is known, one proceeds toward the determination of a *pseudopotential for the pseudoatom free from the core states*, and yet capable to describe the physics and chemistry of the external valence electrons. Most often, the pseudopotential is taken in a form of the type

$$V_a^{(\text{pseudo})}(r) = \begin{cases} V(r; \{\lambda_i\}) & \text{for } r < r_c \\ V_a(r) & \text{for } r > r_c \end{cases}. \tag{5.37}$$

Inside the core region, the pseudopotential (5.37) is approximated by some reasonable functional form, containing one or more adjustable parameters $\{\lambda_i\}$; outside the core region, the pseudopotential (5.37) coincides with $V_a(r)$. The radius r_c (although it carries some degree of arbitrariness from model to model) is however indicative of the extension of the core states of the atom. The one or more adjustable parameters $\{\lambda_i\}$ are then determined by fitting to a number of experimental inputs (for instance atomic spectroscopic data for outer electrons, electron-atom cross-section, etc.).

In considering the needed steps to elaborate an all-electron atomic calculation into a useful atomic pseudopotential, we illustrate the specific case of the Na atom. We suppose to have performed the all-electron calculation and to know the atomic potential $V_{Na}(r)$, the eigenvalues and the radial wavefunctions of the occupied states. We consider $1s$, $2s$, and $2p$ states as core states, and we outline possible procedures to construct an atomic pseudopotential free from the core states, and suitable to describe the properties of the external $3s$ optical electron.

The simplest form of parametrization of type (5.37), as suggested by Ashcroft, and Heine and Abarenkov, is

$$
V_{Na}^{(pseudo)}(r) = \begin{cases} A & \text{for } r < r_c \\ -\dfrac{e^2}{r} & \text{for } r > r_c \end{cases}
\tag{5.38}
$$

[for other forms see for instance F. Nogueira, C. Fiolhais and J. P. Perdew, Phys. Rev. B 59, 2570 (1999) and references quoted therein]. In Eq. (5.38), the pseudopotential is constant for $r < r_c$, and correctly equals $-e^2/r$ for $r > r_c$. The adjustable parameters A and r_c are chosen so that the ground-state energy of the pseudoatom equals the orbital energy $E_{3s} = -5.14$ eV of the Na atom (this quantity is taken either from theoretical calculations or from experimental spectroscopic measurements). If the flexibility of the pseudopotential allows it, other properties (for instance higher eigenvalues) of the optical electron are enforced.

Ab initio Norm-Conserving Pseudopotentials

A major breakthrough in the theory of pseudopotentials has occurred in 1979 with the concept of norm-conserving pseudopotentials of Hamann, Schlüter, and Chiang. This has further improved the quality of the last generation pseudopotentials, allowing major developments in the field of electronic state calculations. The concept starts from the technical observation *that in the region outside the core, where the true potential and the pseudopotential coincide, the atomic radial wavefunction $R_a(r)$ and the corresponding atomic pseudowavefunction $R_a^{(pseudo)}(r)$ are proportional to each other, but in general are not rigorously equal* (as they should be). In fact integration of the radial Schrödinger equation (at the same energy) from infinity obviously produces the same wavefunctions within a normalization factor as far as the pseudopotential and true potential coincide (the two normalizations are in general different, as they depend on the integration up to the origin, the very region where the true potential and the pseudopotential are different). The basic principle of the *norm-conserving pseudopotential is just to enforce the condition $R_a^{(pseudo)}(r) \equiv R_a(r)$ for $r > r_c$*; this assures that pseudo-charge density and true charge density outside the core region are perfectly equal.

We present now some considerations for the generation of norm-conserving pseudopotentials for atoms; for simplicity, we make reference to some aspects of the procedure introduced by G. P. Kerker, J. Phys. C *13*, L189 (1980) (the wealth of implementations and variants cannot be summarized here; in the "art" of pseudopotentials there is wide space for special requirements and ingenuity). In agreement to the general concepts of norm-conserving pseudopotentials, the Kerker procedure *does not parametrize the pseudopotential* (as done for instance in Eq. 5.38); *rather it parametrizes the pseudowavefunction* in the form

$$R_a^{(pseudo)}(r) = \begin{cases} r^l e^{p(r)} & \text{for } r < r_c \\ R_a(r) & \text{for } r > r_c \end{cases}.$$

(5.39a)

In Eq. (5.39a), l is the angular momentum of the radial wavefunction $R_a(r)$ of lowest energy (and not included in the core states); $p(r)$ is a polynomial of degree 4 of the type

$$p(r) = \lambda_0 + \sum_{i=2}^{n} \lambda_i r^i \quad \Longrightarrow \quad p(r) = \lambda_0 + \lambda_2 r^2 + \lambda_3 r^3 + \lambda_4 r^4$$

(5.39b)

and $\lambda_0, \lambda_2, \lambda_3, \lambda_4$ are four disposable parameters; the coefficient λ_1 is taken as zero so that $p'(r) \div r$ for $r \to 0$, and $V_a^{(pseudo)}(r)$ at $r = 0$ is not singular, as can be seen from Eq. (5.39c) below. It is evident that also for $r < r_c$ the radial pseudowavefunction $R_a^{(pseudo)}(r)$ is nodeless (this requirement *excludes* that undesired pseudowavefunctions of lower energy might exist). The four coefficients of the polynomial $p(r)$ are determined by the following four conditions: normalization to one of the wavefunction (this means that charge conservation within the core region is guaranteed), continuity of function, first and second derivative at r_c.

Once the coefficients $\lambda_0, \lambda_2, \lambda_3, \lambda_4$ and hence $p(r)$ are determined, $R_a^{(pseudo)}(r)$ is replaced into the radial Schrödinger equation (5.36), and one obtains for the norm-conserving pseudopotential the analytic expression

$$V_a^{(pseudo)}(r) = \begin{cases} E_a + 2\dfrac{l+1}{r} p'(r) + p''(r) + [p'(r)]^2 & \text{for } r < r_c \\ V_a(r) & \text{for } r > r_c \end{cases},$$

(5.39c)

where E_a is the energy of the atomic radial wavefunction $R_a(r)$ under consideration.

In the construction of pseudopotentials one arrives in general to a *pseudopotential different for each angular momentum*. The matrix elements of such angular momentum dependent pseudopotential between plane waves are still manageable enough to be explicitly calculated, using the standard spherical waves expansion of plane waves. Another significant aspect of the art of generating norm-conserving pseudopotentials concerns their transferability to different environments, this being a prerequisite for accurate first principle calculations of properties of solids. As an example, we consider the silicon atom and in Figure 5.7a we report the all-electron wavefunctions $3s$, $3p$, and $3d$, and the corresponding nodeless pseudowavefunctions; the latter coincide with the true wavefunctions for large r, and appropriately extrapolate to zero for small r.

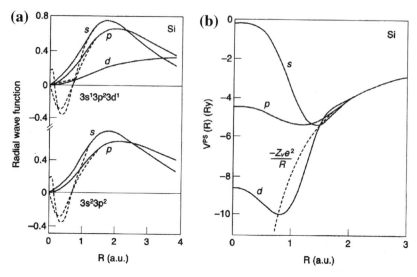

Figure 5.7 (a) Comparison of the pseudowavefunction (solid lines) and the corresponding all-electron wavefunctions (dashed lines) for the configurations $3s^2 3p^2$ and $3s^1 3p^2 3d^1$ of silicon. (b) Non-local pseudopotential of Si for angular momentum $l = 0, 1, 2$. The dashed line denotes the Coulomb potential of a point-like atomic core [reprinted with permission from M. T. Yin and M. L. Cohen, Phys. Rev. B **26**, 5668 (1982); copyright 1982 by the American Physical Society].

The non-local pseudopotentials of Si for $l = 0, 1, 2$ are reported in Figure 5.7b, and are shown to be rather smooth as compared with the Coulomb potential of a (fictitious) point-like atomic core.

The Kerker method can be generalized by increasing the order n of the polynomial $p(r)$ in Eq. (5.39b), and the additional variational freedom can be exploited to improve the smoothness properties of pseudopotentials, as achieved for instance by Troullier and Martins (1991). Ultra-soft pseudopotentials have been elaborated by Vanderbilt (1990), who introduces an additional variational freedom constituted by a non-local overlap operator, accompanying the non-local pseudopotential operator. This scheme is particularly useful for systems such as first-row elements and transition-metal elements, characterized by a significant degree of valence-electron localization, since the radial functions $2p$ and $3d$ are nodeless.

5.5   The Cellular Method

The methods of electronic state calculations so far described (tight-binding, orthogo-nalized plane waves, pseudopotentials) involve expansion sets of *energy-independent basis functions* (plane waves, atomic-like orbitals, etc.). The novelty of the cellu-lar method, which is the prototype of a family of related methods such as the aug-mented plane wave method and the Green's function method, is to adopt expansion sets made of *energy-dependent basis functions*. The cellular methods, by virtue of the flexibility offered by the energy-dependent basis functions, present in general the advantage of good convergence with a relatively small number of basis functions. The disadvantage is a more complicated form of the matrix element, that are now energy-dependent, and are no more linear in energy [i.e. they do not have the "canonical" form $X_{ij}(E) = A_{ij} - E B_{ij}$, where the matrices A and B are energy-independent]. The eigenvalues of a non-linear secular equation must be traced back one-by-one, searching for the zeroes of the determinant: this disadvantage may become particularly serious if the energy bands and crystal wavefunctions are to be evaluated in many points of the Brillouin zone, for averaging purposes, self-consistent calculations, or other reasons. In order to cope with these situations, the cellular methods have evolved into full maturity with the formulation of the corresponding linearized versions, which maintain the original framework, and yet restore "canonical" secular equations, which cover an appropriate energy range. In this and in the next two sections, we present the most relevant concepts and techniques of the cellular methods.

The cellular method was originally introduced by Wigner and Seitz (1933–1934) in the study of the s-like conduction band in alkali metals, and explicitly formulated in a more general context by Slater (1934). In the cellular method attention is focused to the Wigner-Seitz primitive cell; within it, *the crystal potential is assumed to be spherically symmetric*, the radial Schrödinger equation is solved, and appropriate *boundary conditions at the surface of the primitive cell* are applied. The cellular method was widely used for about three decades since its original formulation; then with the progress of the more satisfactory augmented plane wave and Green's function methods, the cellular approach has gradually lost its importance for practical calculations. Despite this, the cellular method well deserves illustration, because of its important concepts and its historical relevance.

We consider first crystals with one atom per primitive cell. As primitive cell we choose the Wigner-Seitz polyhedron, and we make the basic assumption that the actual crystal potential within the polyhedron has spherical symmetry; in many practical cases this assumption is well justified because of the dominant contribution of the atomic potential at the center of the Wigner-Seitz cell.

A crystal state $\psi(\mathbf{k}, \mathbf{r})$ of energy $E(\mathbf{k})$ can be expanded within the Wigner-Seitz cell in the form

$$\psi(\mathbf{k}, \mathbf{r}) = \sum_{lm} c_{lm}(\mathbf{k}) Y_{lm}(\mathbf{r}) R_l(E, r) \qquad \mathbf{r} \in \Omega_{WS}, \qquad (5.40)$$

where $l = 0, 1, 2, \ldots$ and $m = -l, -l + 1, \ldots, +l$; $Y_{lm}(\mathbf{r})$ are spherical harmonic functions [$Y_{lm}(\mathbf{r})$ stands for $Y_{lm}(\theta, \phi)$ where θ, ϕ are the polar coordinates of $\mathbf{r}$ and the origin is at the center of the Wigner-Seitz cell Ω_{WS}]; c_{lm} are appropriate coefficients not yet specified. The function $R_l(E, r)$ is the solution regular at the origin of the radial wave equation

$$\frac{d^2(r R_l)}{dr^2} = \left[V(r) + \frac{l(l + 1)}{r^2} - E \right] (r R_l)$$

and $V(r)$ is the spherically symmetric cellular potential; $R_l(E, r)$ can be obtained by integrating numerically, outward from the origin, the above radial wave equation.

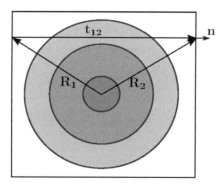

Figure 5.8 Schematic visualization of the Wigner-Seitz cell, of the spherically symmetric cellular potential, and of a pair of surface conjugate points R_1 and R_2 for imposing boundary conditions.

At the surface of the Wigner-Seitz polyhedron, a crystal state of vector $\mathbf{k}$ must satisfy the boundary conditions required by the Bloch theorem

$$\psi(\mathbf{k}, \mathbf{R}_2) = e^{i\mathbf{k} \cdot \mathbf{t}_{12}} \psi(\mathbf{k}, \mathbf{R}_1) \tag{5.41a}$$

$$\mathbf{n} \cdot \nabla \psi(\mathbf{k}, \mathbf{R}_2) = e^{i\mathbf{k} \cdot \mathbf{t}_{12}} \mathbf{n} \cdot \nabla \psi(\mathbf{k}, \mathbf{R}_1) \tag{5.41b}$$

for every pair of points $\mathbf{R}_1$ and $\mathbf{R}_2$ on opposite faces of the Wigner-Seitz cell, connected by a lattice translational vector $\mathbf{t}_{12}$; $\mathbf{n}$ is a unit vector normal to the pair of faces to which $\mathbf{R}_1$ and $\mathbf{R}_2$ belong (see Figure 5.8).

In principle the arbitrary coefficients $c_{lm}(\mathbf{k})$ and the crystal energies $E(\mathbf{k})$ are determined by inserting the expansion of Eq. (5.40) into Eqs. (5.41), considering the resulting homogeneous equations in the coefficients c_{lm} and requiring that the corresponding determinant vanishes. In practice one restricts the sum in Eq. (5.40) to a finite number of l values (for instance if one wishes to reproduce s, p, and d character of crystal states, $l_{max} = 2$ and there are nine disposable coefficients c_{lm}) and selects a corresponding finite number of boundary conditions. [In the case the selected boundary conditions are redundant with respect to the number of disposable parameters, variational procedures for minimization of boundary mismatches can be applied.] Practical recipes for judicious selections of surface points have been provided, and often tested for the case $V(r) = 0$, for which the exact eigenfunctions are plane waves and the exact eigenvalues are $(\mathbf{k} + \mathbf{h}_n)^2$ (*Shockley empty lattice test*). However the unavoidable arbitrariness in the choice of surface points constitutes the basic limitation of the cellular method.

At this point it is interesting to mention the original calculation of Wigner and Seitz for the bottom of the conduction band (which occurs at $\mathbf{k} = 0$) for alkali metals. Wigner and Seitz limited the expansion in Eq. (5.40) to the term $R_0(E, r)$ and replaced the Wigner-Seitz polyhedron with a sphere of radius r_s of equal volume. For $\mathbf{k} = 0$, the boundary condition of Eq. (5.41a) is automatically satisfied, while Eq. (5.41b) implies $dR_0(E, r)/dr = 0$ for $r = r_s$. This condition for the solid should be contrasted with the standard condition of regularity at infinity for the radial wavefunctions of the isolated atoms; this difference in boundary conditions explains in part the cohesive energy of alkali metals.

The cellular method has been also extended to composite crystals. In this case the primitive cell is divided into a number of subcells equal to the number of atoms in the primitive cell. Within each Wigner-Seitz subcell, the crystal potential is assumed to have spherical symmetry, polar coordinates with respect to the center are introduced, and expansion in spherical waves considered. Besides boundary conditions at the surface of the cell, one has now to consider standard continuity conditions at the subcell surfaces.

We do not dwell in further aspects of the cellular method, also because of its presently marginal importance for actual calculations. The major merit of the cellular approach consists in the introduction of energy-dependent basis functions, the key idea better developed in the augmented plane wave method and Green's function method, described in the next two sections.

5.6   The Augmented Plane Wave (APW) Method

Description of the Method

The augmented plane wave (APW) method was originally proposed by Slater in 1937 in order to overcome the problem of boundary conditions of the cellular method. The APW method finds its motivation in the fact that, in many solids, the crystal potential energy can be well approximated by a *muffin-tin potential*: so is called a potential spherically symmetric within non-overlapping spheres centered about each atom, and constant elsewhere (see Figure 5.9); the shape of the potential is reminiscent of the shape

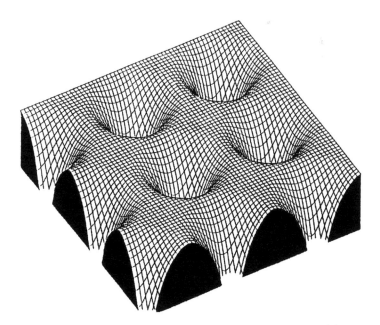

Figure 5.9 Schematic representation of the muffin-tin potential.

of a popular kitchen mold for muffins. The essential idea of the APW method consists in representing the crystal states by cellular type functions within the muffin-tin spheres of radius r_{MT} and plane waves outside them; a much easier matching problem at the sphere surface provides the band states more satisfactorily than in the cellular method.

We begin to illustrate the method for the case of simple crystals with one atom per primitive cell. We assume the crystal potential in the muffin-tin form; the constant value of the muffin-tin potential outside the non-overlapping spheres of radius r_s can be taken as zero, upon appropriate shift of the energy scale.

Within the primitive cell we consider the Schrödinger equation (in atomic units)

$$\left[-\nabla^2 + V(r) \right] \psi(\mathbf{k}, \mathbf{r}) = E \psi(\mathbf{k}, \mathbf{r}),$$

where $V(r)$ is the spherically symmetric muffin-tin potential ($V(r) = 0$ for $r > r_s$). As in the cellular method, we indicate by $R_l(E, r)$ the radial wavefunction which is regular at the origin and satisfies the radial wave equation

$$\frac{d^2(r R_l)}{dr^2} = \left[V(r) + \frac{l(l+1)}{r^2} - E \right] (r R_l). \tag{5.42}$$

An augmented plane wave $A(\mathbf{k} + \mathbf{h}_i, \mathbf{r}, E)$ is defined as a plane wave of vector $\mathbf{k} + \mathbf{h}_i$ outside the sphere r_s, which joins continuously with a cellular-type function inside it. By using the familiar expansion (5.29) of a plane wave in spherical harmonics, we find the following expression for an augmented plane wave:

$$A(\mathbf{k}_i, \mathbf{r}, E) = \begin{cases} \dfrac{4\pi}{\sqrt{\Omega}} \displaystyle\sum_{lm} i^l \dfrac{j_l(k_i r_s)}{R_l(E, r_s)} Y^*_{lm}(\mathbf{k}_i) Y_{lm}(\mathbf{r}) R_l(E, r) & \mathbf{r} \in \Omega_s \\[12pt] \dfrac{e^{i\mathbf{k}_i \cdot \mathbf{r}}}{\sqrt{\Omega}} \equiv \dfrac{4\pi}{\sqrt{\Omega}} \displaystyle\sum_{lm} i^l j_l(k_i r) Y^*_{lm}(\mathbf{k}_i) Y_{lm}(\mathbf{r}) & \mathbf{r} \in \Omega_{WS} - \Omega_s \end{cases} \tag{5.43}$$

where Ω is the volume of the primitive cell, j_l is the spherical Bessel functions of order l, Ω_{WS}, and Ω_s indicate the Wigner-Seitz cell and the inscribed sphere of radius r_s, respectively. Notice that an augmented plane wave is a continuous function, but in general a discontinuity in its slope at $r = r_s$ remains.

A crystal eigenfunction $\psi(\mathbf{k}, \mathbf{r})$ with energy $E(\mathbf{k})$ can be expanded within the Wigner-Seitz cell, in the form

$$\psi(\mathbf{k}, \mathbf{r}) = \sum_j a_j(\mathbf{k}) A(\mathbf{k} + \mathbf{h}_j, \mathbf{r}, E),$$

where $\mathbf{h}_j$ are the vectors of the reciprocal space. The function $\psi(\mathbf{k}, \mathbf{r})$ and its gradient are well behaved everywhere in the primitive cell, though the expansion functions have discontinuous first derivative at the surface of the sphere of radius r_s. By substituting the above expansion in the Schrödinger equation and using the variational procedure with respect to the coefficients a_j, the following secular equation is obtained

$$\boxed{\| \langle A(\mathbf{k}_i, \mathbf{r}, E) | H - E | A(\mathbf{k}_j, \mathbf{r}, E) \rangle \| = 0}. \tag{5.44}$$

The matrix elements $M_{ij}(\mathbf{k}, E)$ of the secular equation (5.44) are worked out in detail in Appendix A, and Eq. (A.5) is here reported for convenience

$$M_{ij}(\mathbf{k}, E) = \left(k_j^2 - E\right)\delta_{ij} - \frac{4\pi r_s^2}{\Omega}(\mathbf{k}_i \cdot \mathbf{k}_j - E)\frac{j_1(|\mathbf{k}_i - \mathbf{k}_j|r_s)}{|\mathbf{k}_i - \mathbf{k}_j|}$$
$$+ \frac{4\pi r_s^2}{\Omega}\sum_{l=0}^{\infty}(2l+1)P_l\left(\widehat{\mathbf{k}}_i \cdot \widehat{\mathbf{k}}_j\right)j_l(k_i r_s)j_l(k_j r_s)\frac{R_l'(E, r_s)}{R_l(E, r_s)}, \qquad (5.45)$$

where $\mathbf{k}_i$ is a short hand notation for $\mathbf{k} + \mathbf{h}_i$, the unit vector in the direction of $\mathbf{k}_i$ is denoted by $\widehat{\mathbf{k}}_i$ and $P_l(z)$ is the Legendre polynomial of order l; we have also indicated

$$R_l'(E, r_s) = \left[\frac{dR_l(E, r)}{dr}\right]_{r=r_s}.$$

Contrary to first sight impression, the energy-dependent structure of the matrix elements of the APW method embodies highly rewarding features (at the cost of little complication), as will be soon apparent.

The generalization to composite lattice is straightforward. Instead of a single atom per primitive cell we have a number ν_b of atoms, centered at $\mathbf{d}_1, \mathbf{d}_2, \ldots, \mathbf{d}_{\nu_b}$, each of them surrounded by non-overlapping spheres of radii $r_{s1}, r_{s2}, \ldots, r_{s\nu_b}$. The matrix elements for composite lattices are obtained putting an appropriate label on the second and third terms in the right-hand side of Eq. (5.45), multiplying by the structure phase factor $\exp[-i(\mathbf{h}_i - \mathbf{h}_j) \cdot \mathbf{d}_\nu](\nu = 1, 2, \ldots, \nu_b)$ and summing over ν.

Some Remarks on the APW Matrix Elements

The matrix elements $M_{ij}(\mathbf{k}, E)$, given by Eq. (5.45), contain E both explicitly and implicitly through the logarithmic derivative R_l'/R_l. For practical calculations one may proceed in this way: (i) an appropriate truncation in $|\mathbf{k}_i|_{\max}$ (i.e. in the order of the determinant) and $l_{\max}$ is chosen; (ii) the matrix elements $M_{ij}(\mathbf{k}, E)$ are computed at regular intervals of E (say 0.001 Ry); the determinant $\|M_{ij}(\mathbf{k}, E)\|$ is plotted as a function of E and eigenvalues are obtained as the zeroes of the curve (when needed, the corresponding eigenvectors can be determined and crystal wavefunctions explicitly found); (iii) the convergence and stability of the method can be tested by increasing both $|\mathbf{k}_i|_{\max}$ and $l_{\max}$. The favorable aspect of the method lies in its very rapid convergence, a number of 10–30 augmented plane waves (and $l_{\max} \approx 5$) being sufficient in most practical cases; this well rewards the amount of computational labor involved.

Another aspect which makes the APW method such a powerful one is the fact that the degree of localization of crystal states is irrelevant. Remember that the tight-binding method is applicable for reasonably localized states; on the other hand, the orthogonalized-plane-wave method or its pseudopotential version requires an arbitrary (and sometimes questionable) separation of crystal states in well localized core states and spread-out valence or conduction states. The APW method describes in a natural way strong localization, weak localization, or any possible intermediate situation.

This can be intuitively understood considering the two limiting cases of free (or quasi free) and strongly localized electron states. For the former case, it can be observed that the APW method automatically satisfies the empty lattice test: in the limit of vanishing muffin-tin potential, augmented plane waves and ordinary plane waves coincide. Also in the case of resonances, i.e. $R_l(E, r_s) \to 0$ or equivalently $R'_l(E, r_s)/R_l(E, r_s) \to \infty$ it is apparent that small changes of E around the resonance can counterbalance large change of **k** in the Brillouin zone, so describing tendentially flat bands.

Another merit of the APW method is the possibility (whenever necessary) to include in the formalism also the part of the crystalline potential beyond the muffin-tin form; this requires the computation of the corrections, produced by the difference between the true crystalline potential and the approximate muffin-tin form, on the basis of augmented plane waves initially adopted.

From a computational point of view, the most demanding part of the APW matrix elements (5.45) is the evaluation of the logarithmic derivatives $R'_l(E, r)/R_l(E, r)$ at the sphere radius r_s. The wavefunction $R_l(E, r)$ is obtained by integrating numerically (usually with the very simple Numerov method) the radial wave equation (5.42), from the origin outward. The functions $R_l(E, r)$ are thus determined, except for an inessential multiplicative constant, which is irrelevant for the logarithmic derivatives $R'_l(E, r)/R_l(E, r)$.

We notice that the differential equation (5.42) implies the identity

$$\int_0^{r_s} R_l^2(E, r) r^2 dr \equiv -R_l^2(E, r_s) r_s^2 \frac{\partial}{\partial E} \frac{R'_l(E, r_s)}{R_l(E, r_s)}, \tag{5.46}$$

as discussed in Eq. (B.2) of Problem 1. In Eq. (5.46) the first member is always positive; it follows that the logarithmic derivatives $R'_l(E, r)/R_l(E, r)$ *are always monotonically decreasing functions of energy* and exhibit poles at the resonance energies, which are defined as the energies for which $R_l(E, r_s) = 0$.

As a final comment, we observe that the matrix elements of the APW method given by Eq. (5.45) depend obviously on r_s. Notice that r_s is to some extent arbitrary and must satisfy the relation $r_{MT} \leq r_s \leq r_{max}$, where r_{MT} is the radius of the assumed muffin-tin potential and r_{max} is the maximum radius consistent with non-overlapping spheres. In practical situations one usually chooses $r_{MT} \equiv r_s \equiv r_{max}$. However, in principle, one can have $r_{MT} < r_{max}$; in this situation, r_s can be chosen arbitrarily in the interval $[r_{MT}, r_{max}]$, the matrix elements M_{ij} do depend on r_s, but the eigenvalues of Eq. (5.45) do not.

It would be more satisfactory, from a pure formal point of view, a method whose matrix elements are independent of r_s. This would occur if matrix elements do not contain the *logarithmic derivatives of the radial wavefunctions, but rather the phase shifts of the muffin-tin potential*; for $r > r_{MT}$ the logarithmic derivatives depend on r, while the phase shifts are independent of it (see also Problem 2). This last step is achieved in the Green's function method, which is the most far-reaching development of the cellular and APW methods.

Considerations on the Linearized APW Method

In the APW method the crystal wavefunctions are expanded in *energy-dependent* spherical waves within the muffin-tin sphere. The energy dependence of the basis functions is reflected in the complicate non-linear energy dependence of the matrix elements of the secular equations. For repeated calculations of energy bands toward self-consistency, or optical properties and other integrated properties through the whole Brillouin zone, this may become a real disadvantage and a limitation.

The possibility and concepts of linearization were introduced by O. K. Andersen at the beginning of 1970s. The basic idea of linearization consists in augmenting the crystal wavefunctions with the combined use both of the radial functions $R_l(E, r)$ and of their derivatives $\dot{R}_l(E, r)$ with respect to the energy, specified at a particular but arbitrary value E_0 of the energy. The key point that determines the success of the method lies in the fact that the radial wavefunctions within the atomic sphere vary smoothly with energy, in an appreciable energy range. With wavefunctions correct to first order in the difference $E - E_0$ between the actual energy E and the chosen linearization energy E_0, it is expected that the energies are correct at order $(E - E_0)^2$ (at least), in an appropriate range around the reference energy. The combined use of the wavefunctions $R_l(E_0, r)$ and $\dot{R}_l(E_0, r)$ leads to secular equations which are *linear* in energy, and the search of eigenvalues and eigenvectors can be handled as a usual algebraic problem.

In the augmented plane wave method we have seen that a plane wave in the interstitial region can be matched continuously with the spherical waves $R_l(E, r)Y_{lm}(\mathbf{r})$, but discontinuities in derivatives cannot be eliminated (see Eq. 5.43). If inside the muffin-tin sphere we use not only the spherical waves $R_l(E_0, r)Y_{lm}(\mathbf{r})$ but also $\dot{R}_l(E_0, r)Y_{lm}(\mathbf{r})$, it is possible to match a plane wave in value and first derivative. The use of these linearized augmented plane waves, continuous in value and in derivative, and fixed at a chosen reference energy, leads to the linearized augmented plane wave method (LAPW). Around the reference energy, the linearized formalism enjoys the "physical advantages" (accuracy and convergence) of the original formulation, together with the "computational advantages" offered by fixed energy-independent basis functions, and we refer to the specific literature for further details.

Before closing this section, it is worthwhile to mention the significant development introduced by Blöchl in 1994 and known as the Projector Augmented-Wave method. The PAW method constitutes an innovative and ingenious cross-fertilization of the concepts, procedures, and codes of the LAPW method and of the norm-conserving and ultra-soft pseudopotential procedures. The formalism exploits the versatility of the augmentation principle and the formal simplicity of the pseudopotential princi-ple, the key point being the possibility to reconstruct the all-electron wavefunctions from the pseudowavefunctions of the PAW approximation. For details and the state-of-the-art description of the approach see for instance the topical review by Enkovaara et al. (2010) and reference quoted therein.

5.7   The Green's Function Method (KKR Method)

The Green's function method was proposed by Korringa (1947), Kohn and Rostoker (1954) (and therefore called also KKR method), and Morse (1956) in different though equivalent forms. In the KKR method the Schrödinger equation is transformed into an equivalent integral equation and, once a muffin-tin potential is assumed, the problem of boundary conditions is elegantly solved through the fulfillment of appropriate surface integrals, related to the Green's theorem. A nice feature of the method is the sharp separation of the procedure into two parts: (i) structural aspects of the lattice; (ii) phase shifts of the spherical muffin-tin potential.

From the point of view of electronic state calculation for periodic crystals, it is generally agreed that KKR is even more rapidly convergent than the APW method; but this advantage is to some extent counterbalanced by the somewhat more complicated expressions of the KKR structural coefficients. However the real peculiar novelty of the KKR methodology (i.e. sharp separation of *geometry* from *potential*) has made the method generalizable to the treatment of impurities, clusters, homogeneously disordered alloys, and photon energy bands in metallic ceramics. Probably these are the most important modern aspects of the KKR method; however, we keep in line with the purpose of this chapter and we limit ourselves to the study of electrons in fully periodic materials.

5.7.1   Scattering Integral Equation for a Generic Potential

In this section we briefly summarize the Green's function technique to transform into integral form the Schrödinger equation, here rewritten for convenience as

$$(E + \nabla^2)\psi(\mathbf{r}) = V(\mathbf{r})\psi(\mathbf{r}). \tag{5.47}$$

The free-particle Green's function $g(\mathbf{r}, \mathbf{r}_0, E)$ for a δ-like source at $\mathbf{r}_0$ is defined by means of the equation

$$(E + \nabla^2)\, g(\mathbf{r}, \mathbf{r}_0, E) = \delta(\mathbf{r} - \mathbf{r}_0), \tag{5.48}$$

with the requirement that $g(\mathbf{r}, \mathbf{r}_0, E)$ obeys the same boundary conditions as $\psi(\mathbf{r})$. It is straightforward to verify that the differential eigenvalue equation (5.47) is equivalent to the integral eigenvalue equation

$$\psi(\mathbf{r}) = \phi(\mathbf{r}) + \int g(\mathbf{r}, \mathbf{r}_0, E) V(\mathbf{r}_0)\psi(\mathbf{r}_0)\, d\mathbf{r}_0, \tag{5.49}$$

where $\phi(\mathbf{r})$ denotes a solution at energy E (if any) of the homogeneous differential equation $(E + \nabla^2)\phi(\mathbf{r}) = 0$. The equivalence of Eq. (5.49) with Eq. (5.47) follows by direct application of the operator $(E + \nabla^2)$ to both members of Eq. (5.49).

It is convenient to express the Green's function of the free-particle Hamiltonian $H_0 = -\nabla^2$ without sticking to the $\mathbf{r}$-representation. The free-particle retarded Green's function $g(E)$ can be expressed as

$$g(E) = \frac{1}{E + i\varepsilon - H_0} = \frac{1}{E + i\varepsilon + \nabla^2} \qquad \varepsilon \to 0^+; \tag{5.50}$$

it is understood that the real energy E is always accompanied by an infinitesimal positive imaginary part (sometimes, for brevity of notations, the imaginary part is not explicitly indicated; however, the imaginary part $i\varepsilon$ must never be overlooked, and must be reinstated whenever necessary). The integral eigenvalue equation (5.49) can now be written in the form

$$|\psi\rangle = |\phi\rangle + \frac{1}{E + i\varepsilon - H_0} V |\psi\rangle. \tag{5.51}$$

It is convenient to obtain a spectral representation of the Green's function $g(E)$ in terms of the eigenvalues and eigenfunctions of the operator H_0 itself; these are plane waves $W(\mathbf{k}, \mathbf{r}) = (1/\sqrt{V}) \exp(i\mathbf{k} \cdot \mathbf{r})$ of energy k^2. The Green's function (5.50) can be evidently written as

$$g(E) = \frac{1}{E + i\varepsilon - H_0} \sum_{\mathbf{k}} |W_{\mathbf{k}}\rangle\langle W_{\mathbf{k}}| = \sum_{\mathbf{k}} \frac{1}{E + i\varepsilon - k^2} |W_{\mathbf{k}}\rangle\langle W_{\mathbf{k}}|. \tag{5.52a}$$

In the $\mathbf{r}$-representation we have

$$g(\mathbf{r}, \mathbf{r}_0, E) = \frac{1}{V} \sum_{\mathbf{k}} \frac{e^{i\mathbf{k}\cdot\mathbf{r}} e^{-i\mathbf{k}\cdot\mathbf{r}_0}}{E + i\varepsilon - k^2}. \tag{5.52b}$$

Notice that the free-particle Green's function depends on the variables $\mathbf{r}$ and $\mathbf{r}_0$ through their difference.

The discrete sum over $\mathbf{k}$ in Eq. (5.52b) can be replaced (as usual) with $V/(2\pi)^3$ times the integral on the $\mathbf{k}$ variable; then

$$g(\mathbf{r} - \mathbf{r}_0, E) = \frac{1}{(2\pi)^3} \int \frac{1}{E + i\varepsilon - k^2} e^{i\mathbf{k}\cdot(\mathbf{r}-\mathbf{r}_0)} d\mathbf{k}, \tag{5.53a}$$

where $d\mathbf{k}$ denotes the volume element in the reciprocal space. Performing the integration one obtains

$$\boxed{g(\mathbf{r} - \mathbf{r}_0, E) = -\frac{1}{4\pi} \frac{e^{i\alpha|\mathbf{r}-\mathbf{r}_0|}}{|\mathbf{r} - \mathbf{r}_0|},} \tag{5.53b}$$

where $\alpha^2 = E$ or, more specifically,

$$\alpha(E) = \begin{cases} \sqrt{E} & \text{for } E > 0 \\ i\sqrt{|E|} & \text{for } E < 0 \end{cases}. \tag{5.53c}$$

The free-particle Green's is exponentially damped for negative energies, while it has an oscillating long-range behavior for positive energies.

The integration in Eq. (5.53a) has been carried out with the following procedure. We introduce polar coordinates in $\mathbf{k}$ space with $\mathbf{r} - \mathbf{r}_0$ as k_z direction, and obtain

$$g(\mathbf{r} - \mathbf{r}_0, E) = \frac{1}{(2\pi)^3} 2\pi \int_0^\infty \frac{k^2}{E + i\varepsilon - k^2} dk \int_0^\pi e^{ik|\mathbf{r}-\mathbf{r}_0|\cos\theta} \sin\theta \, d\theta.$$

The integral in the angular variable θ gives

$$g(\mathbf{r} - \mathbf{r}_0, E) = \frac{1}{4\pi^2 i |\mathbf{r} - \mathbf{r}_0|} \int_{-\infty}^\infty \frac{k}{E + i\varepsilon - k^2} e^{ik|\mathbf{r}-\mathbf{r}_0|} dk. \tag{5.54}$$

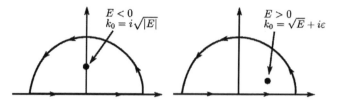

Figure 5.10 Integration contour for evaluating the Green's function of the free-electron problem.

Consider first the integration in Eq. (5.54) in the case $E < 0$. The integral can be evaluated closing the integration contour in the upper part of the complex plane, as indicated in Figure 5.10. We have to consider one simple pole at $k = k_0 \equiv i\sqrt{|E|}$ and the residue of the integrand function is $(-1/2)\exp(ik_0|\mathbf{r} - \mathbf{r}_0|)$. For negative energies Eq. (5.53b) is obtained. A similar procedure proves that Eq. (5.53b) holds also for positive energies.

5.7.2 Scattering Integral Equation for a Periodic Muffin-Tin Potential

In the case of a periodic crystal potential, we can specify the integral Eq. (5.49) taking into account that the wavefunctions must be of Bloch type; we have

$$\psi(\mathbf{k}, \mathbf{r}) = \int_{\text{crystal}} g(\mathbf{r} - \mathbf{r}_0, E)V(\mathbf{r}_0)\psi(\mathbf{k}, \mathbf{r}_0)\,d\mathbf{r}_0, \tag{5.55a}$$

where the integration extends on the crystal volume; the solution $\phi(\mathbf{r})$ of the homogeneous equation is set equal to zero, since in general there is no incident plane wave in the infinite crystal.

The wavefunction $\psi(\mathbf{k}, \mathbf{r})$, at any two points $\mathbf{r}$ and $\mathbf{r} + \mathbf{t}_n$ related by a translation vector, takes values related by the Bloch phase factor $\exp(i\mathbf{k} \cdot \mathbf{t}_n)$. This allows to restrict the integration in Eq. (5.55a) to a single primitive cell (for convenience the Wigner-Seitz cell centered at $\mathbf{t}_n = 0$). One obtains the following integral equation

$$\psi(\mathbf{k}, \mathbf{r}) = \int_{\Omega_{\text{WS}}} g(\mathbf{k}, \mathbf{r} - \mathbf{r}_0, E)V(\mathbf{r}_0)\psi(\mathbf{k}, \mathbf{r}_0)\,d\mathbf{r}_0 \,, \tag{5.55b}$$

where the kernel of the integral equation has the expression

$$g(\mathbf{k}, \mathbf{r} - \mathbf{r}_0, E) = \sum_{\mathbf{t}_n} e^{i\mathbf{k}\cdot\mathbf{t}_n}\, g(\mathbf{r} - \mathbf{r}_0 - \mathbf{t}_n, E) \tag{5.55c}$$

and is called *the greenian of vector* $\mathbf{k}$. Using Eq. (5.53b) the greenian of vector $\mathbf{k}$ becomes

$$g(\mathbf{k}, \mathbf{r} - \mathbf{r}_0, E) = -\frac{1}{4\pi}\sum_{\mathbf{t}_n} e^{i\mathbf{k}\cdot\mathbf{t}_n}\frac{e^{i\alpha|\mathbf{r}-\mathbf{r}_0-\mathbf{t}_n|}}{|\mathbf{r} - \mathbf{r}_0 - \mathbf{t}_n|}, \tag{5.56}$$

where the sum is over the translation vectors of the direct lattice.

We now assume the muffin-tin form of the potential; the integration region in Eq. (5.55b) can then be reduced to the sphere Ω_s of radius $r_s (\geq r_{MT})$ inscribed in the Wigner-Seitz polyhedron (we limit our attention to simple crystals, since the extension to composite crystals only requires some straightforward generalizations). We have thus that the crystal eigenfunctions of vector $\mathbf{k}$ satisfy the integral equation

$$\psi(\mathbf{k}, \mathbf{r}) = \int_{\Omega_s} g(\mathbf{k}, \mathbf{r} - \mathbf{r}_0, E) V(\mathbf{r}_0) \psi(\mathbf{k}, \mathbf{r}_0) \, d\mathbf{r}_0.$$

Because of Eq. (5.47) it holds

$$\int_{\Omega_s} g(\mathbf{k}, \mathbf{r} - \mathbf{r}_0, E)(E + \nabla_0^2) \psi(\mathbf{k}, \mathbf{r}_0) \, d\mathbf{r}_0 = \psi(\mathbf{k}, \mathbf{r}). \tag{5.57}$$

We use the identity

$$g \nabla^2 \psi = (\nabla^2 g) \psi + \nabla \cdot [g \nabla \psi - \psi \nabla g],$$

where g and ψ are any two continuous and differentiable functions. We insert this identity into Eq. (5.57) and use the Green's theorem to transform the volume integral of the divergence of $(g \nabla \psi - \psi \nabla g)$ into a more convenient surface integral; we arrive at the basic "boundary condition" equation

$$\boxed{\int_{S_0} \left[g(\mathbf{k}, \mathbf{r} - \mathbf{r}_0, E) \frac{\partial \psi(\mathbf{k}, \mathbf{r}_0)}{\partial r_0} - \psi(\mathbf{k}, \mathbf{r}_0) \frac{\partial g(\mathbf{k}, \mathbf{r} - \mathbf{r}_0, E)}{\partial r_0} \right]_{r_0 = r_s} dS_0 = 0},$$

$$\tag{5.58}$$

where S_0 is the surface of the sphere Ω_s and $\mathbf{r}$ is any point within it. In a crystal, with muffin-tin form of the potential, any crystal wavefunction of vector $\mathbf{k}$ and energy E is such that the surface integral, defined in the left-hand side of Eq. (5.58), identically vanishes for any $\mathbf{r} \in \Omega_s$. In a crystal, with *no* muffin-tin form of the potential, a similar result holds, but the great complication occurs that the integral involves the surface of the primitive cell rather than the quite convenient spherical surface S_0.

In order to perform explicitly the surface integrals in Eq. (5.58) we need an expression of the greenian in terms of appropriate (structure) coefficients, multiplied by factorized functions of $\mathbf{r}$ and $\mathbf{r}_0$ separately. The desired and exact expression of the greenian, in terms of spherical Bessel and Neumann functions, is worked out in Appendix C and summarized by Eq. (C.14), here repeated for convenience

$$g(\mathbf{k}, \mathbf{r} - \mathbf{r}_0, E) = \alpha \sum_{lm} j_l(\alpha r) Y_{lm}(\mathbf{r}) n_l(\alpha r_0) Y_{lm}^*(\mathbf{r}_0)$$

$$+ \sum_{lm\,l'm'} j_l(\alpha r) Y_{lm}(\mathbf{r}) \Gamma_{lm,l'm'}(\mathbf{k}, E) j_{l'}(\alpha r_0) Y_{l'm'}^*(\mathbf{r}_0), \tag{5.59}$$

where $\alpha = \sqrt{E}$, $\mathbf{r}$ and $\mathbf{r}_0$ are within the Wigner-Seitz cell, and $r < r_0$. [In the case $r > r_0$ we have to exchange r and r_0 in the first term in the right-hand side of Eq. (5.59).]

The coefficients $\Gamma_{lm,l'm'}(\mathbf{k}, E)$ are called structure coefficients: for any chosen point in the Brillouin zone and for any chosen energy they depend exclusively on the crystal structure and not on the crystal potential.

As in the cellular method, also in the Green's function method the crystal wavefunctions are expanded in spherical waves which must satisfy the Schrödinger equation with the muffin-tin potential

$$\psi(\mathbf{k}, \mathbf{r}_0) = \sum_{l''m''} c_{l''m''}(\mathbf{k}) R_{l''}(E, r_0) Y_{l''m''}(\mathbf{r}_0), \tag{5.60}$$

but now the unknown expansion coefficients and the crystal eigenvalues are determined in a natural way through the fulfillment of the boundary conditions summarized by Eq. (5.58). As usual the radial function $R_l(E, r)$ indicates the solution, regular at the origin, of the corresponding radial Schrödinger equation.

The radial function for $r > r_{MT}$ can be expressed as a linear combination of the two linearly independent solutions of the Schrödinger equation when the potential is constant and equal to zero. For convenience, the normalization of $R_l(E, r)$ is chosen so as to satisfy the following asymptotic behavior

$$R_l(E, r) = n_l(\alpha r) - \cotg \eta_l(E) \cdot j_l(\alpha r) \quad \text{for} \quad r \geq r_{MT}, \tag{5.61}$$

where j_l and n_l are the spherical Bessel and Neumann functions, and $\eta_l(E)$ are the *phase shifts* due to the muffin-tin potential at the energy E and for the wave of angular momentum l. It is useful to recall the Wronskian relations for Bessel and Neumann functions

$$[j_l(x), n_l(x)] \equiv j_l(x) n_l'(x) - j_l'(x) n_l(x) = \frac{1}{x^2}.$$

Then, the following two Wronskian relations hold

$$[n_l, R_l] = \cotg \eta_l(E) \frac{1}{\alpha r_s^2}; \quad [j_l, R_l] = \frac{1}{\alpha r_s^2} \quad [r = r_s \geq r_{MT}]$$

[the derivative entering in the Wronskian relations are intended with respect to the space variable].

We now insert expression (5.59) for the greenian and expansion (5.60) for the trial wavefunction into Eq. (5.58); for any arbitrary $\mathbf{r} \in \Omega_s$ we obtain

$$\alpha \cotg \eta_l(E) c_{lm} + \sum_{l'm'} \Gamma_{lm,l'm'}(\mathbf{k}, E) c_{l'm'} = 0.$$

The above system of linear homogeneous equations for the arbitrary coefficients c_{lm} leads to the compatibility determinantal equation

$$\left\| \Gamma_{lm,l'm'}(\mathbf{k}, E) + \sqrt{E} \cotg \eta_l(E) \delta_{ll'} \delta_{mm'} \right\| = 0 . \tag{5.62}$$

The determinant in Eq. (5.62) is plotted versus the energy E and its zeroes provide the crystal eigenvalues at the chosen $\mathbf{k}$ vector.

A remarkable feature of the basic secular equation (5.62) is the net separation of the information concerning the crystal structure, completely embodied in the structure coefficients $\Gamma_{lm,l'm'}(\mathbf{k}, E)$ and the effect of the potential summarized by the phase shifts $\eta_l(E)$. Furthermore, the dimension of the secular equation needed in actual calculations is expected to be reasonably small; for instance, to reproduce the s, p, and d character of the bands of interest, it is sufficient to consider the angular momentum values $l = 0, 1, 2$ and the secular equation becomes a 9×9 matrix. From a formal point of view the KKR method is the most satisfactory tool to solve the problem of boundary conditions, when the muffin-tin form for the potential is assumed.

A rather demanding aspect of the procedure is the evaluation of the structure coefficients entering the secular equation of the KKR method. We can notice that for negative values of the energy, the free Green's function is exponentially damped since $\alpha = i\sqrt{|E|}$, and the direct-lattice sum for the greenian of Eq. (5.56) converges exponentially fast. However, for positive values of the energy, the free-electron Green's function has a long-range oscillatory behavior, and the direct-lattice sum in Eq. (5.56) becomes slowly conditionally convergent.

The standard technique for achieving a satisfactory accuracy in the evaluation of the structure constants at any desired energy, is the application of the Ewald algorithm, which takes advantage and appropriately balances summations both in direct and reciprocal space. The detailed procedure, remarkable both for the mathematic aspects and for the inspiring underlying physical ideas is reported in Appendix C.

Among the most important developments, flowing from the Green's function scattering formalism, a particular role have assumed the breakthrough concepts of muffin-tin orbitals and linearization, introduced by O. K. Andersen at the beginning of seventies. Without entering the physical and mathematical aspects of linearization, we simply mention that harmonic functions of the Neumann type in the interstitial region can be appropriately "augmented" with spherical waves, inside non-overlapping muffin-tin spheres, to produce muffin-tin orbitals that can be efficiently used as local basis functions for electronic states calculations. Furthermore, augmentation of Neumann functions in the interstitial region with radial wavefunctions and their energy derivatives, at some chosen reference energy, has led to the linearized muffin-tin orbital method (LMTO), which has emerged as one of the most precious methods for electronic state calculations.

5.8 Iterative Methods in Electronic Structure Calculations

5.8.1 The Lanczos or Recursion Method

The Lanczos method (1950), or the closely-related recursion method (Haydock et al., 1972), is a very convenient approach for the determination of the eigensolutions of matrices, especially those of very large rank and sparse (i.e. with many matrix elements equal to zero). In the field of electronic state calculations, the recursion method has been originally introduced in connection with a local basis representation of the electronic states in solids; successively it has been used also in connection with other basis sets (i.e. plane waves, muffin-tin orbitals). The method is very useful for periodic systems; it becomes really essential in aperiodic materials, when a very large number of orbitals must be taken into account and when translational symmetry is (partially or completely) absent, so that the concepts and simplifications implied by the Bloch theorem are not at work.

The essential principle of the Lanczos recursion method is very simple and very general at the same time. Consider a quantum system, an operator H, and a number N (arbitrary large) of orthonormal basis states $\{|\phi_i\rangle\}(i = 1, 2, \ldots, N)$. Starting from any given state $|f_0\rangle$ belonging to the space spanned by $\{|\phi_i\rangle\}$, and operating with H, *the recursion method provides a one-dimensional chain representation of the original quantum system.*

To illustrate the procedure, consider an operator H (usually the Hamiltonian of the system) and an initial normalized state $|f_0\rangle$, arbitrarily chosen; the starting state $|f_0\rangle$ is often addressed as the *seed state* of the procedure. We apply the operator H to the initial state and subtract from $H|f_0\rangle$ its projection on the initial state, so to obtain a new state $|F_1\rangle$, orthogonal to $|f_0\rangle$. The operation of orthogonalization can be conveniently performed by means of the projection operator $P_0 = |f_0\rangle\langle f_0|$, and the state $|F_1\rangle$ can be expressed as

$$|F_1\rangle = (1 - P_0)H|f_0\rangle = H|f_0\rangle - a_0|f_0\rangle, \tag{5.63a}$$

where

$$a_0 = \langle f_0|H|f_0\rangle. \tag{5.63b}$$

Let us indicate with b_1 the normalization of $|F_1\rangle$, and with $|f_1\rangle$ the corresponding normalized state, namely

$$b_1^2 = \langle F_1|F_1\rangle, \quad |f_1\rangle = \frac{1}{b_1}|F_1\rangle.$$

In a rather similar way we proceed now by applying the operator H to the state $|f_1\rangle$; then $H|f_1\rangle$ is orthogonalized to both $|f_1\rangle$ and $|f_0\rangle$, obtaining the state $|F_2\rangle$. We have

$$|F_2\rangle = (1 - P_1)(1 - P_0)H|f_1\rangle = (1 - P_1 - P_0)H|f_1\rangle,$$

where $P_1 = |f_1\rangle\langle f_1|$ denotes the projection operator on the state $|f_1\rangle$. We thus obtain

$$|F_2\rangle = H|f_1\rangle - a_1|f_1\rangle - b_1|f_0\rangle, \tag{5.64a}$$

where

$$a_1 = \langle f_1|H|f_1\rangle, \quad b_2^2 = \langle F_2|F_2\rangle \quad \text{and} \quad |f_2\rangle = \frac{1}{b_2}|F_2\rangle. \tag{5.64b}$$

The procedure is indicated schematically in Figure 5.11.

Let us now proceed to the next logical step of the sequence, and here we arrive at the key point of the ingenious, and yet so simple, Lanczos procedure. Consider in fact the state $|F_3\rangle$ obtained by orthogonalization of $H|f_2\rangle$ to the previous states f_0, f_1, f_2; we have

$$|F_3\rangle = (1 - P_2)(1 - P_1)(1 - P_0)H|f_2\rangle = (1 - P_2 - P_1 - P_0)H|f_2\rangle.$$

We notice that $P_0 H|f_2\rangle \equiv 0$ since $\langle f_0|H|f_2\rangle \equiv 0$; thus *the iteration procedure leads to a three-term relation.* The most remarkable feature of the method is that the orthogonalization procedure never includes more than three terms.

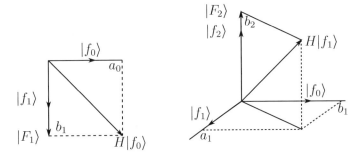

Figure 5.11 Schematic representation of the Lanczos recursion procedure.

Iterating n times the procedure, orthogonalizing the new generated state $H|f_n\rangle$ only to the two predecessors, $|f_n\rangle$ and $|f_{n-1}\rangle$ one obtains

$$|F_{n+1}\rangle = H|f_n\rangle - a_n|f_n\rangle - b_n|f_{n-1}\rangle. \tag{5.65a}$$

The next pairs of coefficients are given by

$$b_{n+1}^2 = \langle F_{n+1}|F_{n+1}\rangle \quad \text{and} \quad a_{n+1} = \langle f_{n+1}|H|f_{n+1}\rangle. \tag{5.65b}$$

The three-term relation (5.65a) is all we need to generate the orthonormal basis set $\{|f_n\rangle\}$ and the set of coefficients $\{a_n\}$ and $\{b_n\}$. This is a remarkable advantage, both conceptual and technical, in the specific calculations carried out by a computer code. On the basis set $\{|f_n\rangle\}$, the operator H is represented by the tridiagonal matrix

$$H = \begin{pmatrix} a_0 & b_1 & & & \\ b_1 & a_1 & b_2 & & \\ & b_2 & a_2 & b_3 & \\ & & \cdot & \cdot & \cdot \\ & & & \cdot & \cdot \end{pmatrix}; \tag{5.66a}$$

it can also be written in the form

$$H = \sum_{n=0}^{\infty} a_n|f_n\rangle\langle f_n| + \sum_{n=0}^{\infty} b_{n+1}\left[|f_n\rangle\langle f_{n+1}| + |f_{n+1}\rangle\langle f_n|\right] \tag{5.66b}$$

and represented diagrammatically as in Figure 5.12.

A most important feature of the tridiagonal matrix (5.66a) is the possibility to obtain very easily the Green's function; in particular the matrix element $G_{00}(E)$ of the Green's function is given by the continued fraction expansion (see Section 1.4.2)

$$G_{00}(E) = \langle f_0| \frac{1}{E - H} |f_0\rangle = \cfrac{1}{E - a_0 - \cfrac{b_1^2}{E - a_1 - \cfrac{b_2^2}{E - a_2 - \cdots}}} \tag{5.67a}$$

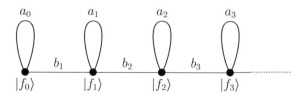

Figure 5.12 Schematic representation of the linear chain generated by the Lanczos procedure.

According to Eq. (1.59a), the local density-of-states $n_0(E)$, projected on $|f_0\rangle$, of the Hamiltonian operator H is given by

$$n_0(E) = -\frac{1}{\pi} \lim_{\varepsilon \to 0^+} \operatorname{Im} G_{00}(E + i\varepsilon), \tag{5.67b}$$

where ε is an infinitesimal positive quantity.

The recursion method allows (at least in principle) the one-dimensional chain representation of any quantum mechanical system, with an arbitrary number N of states. No explicit diagonalization of the original matrix H is required for performing such a transformation. The one-dimensional chain representation allows to describe economically the system through the parameters a_n and b_n (i.e. $2N - 1$ parameters rather than $N(N + 1)/2$ matrix elements $\langle \phi_i | H | \phi_j \rangle$; for $N = 1000$ for instance, we have to compare approximately two thousand with one million!).

A peculiar feature of the Lanczos procedure is that the number of steps to be carried out in specific problems may be reasonably small (say 10 up to 100 or so). Once the original matrix H is put into tridiagonal form, successive diagonalizations of small order tridiagonal matrices allow to infer the eigenvalues of H, especially those whose eigenfunctions have large overlap with the seed state.

In some problems, it is even possible to reasonably infer the asymptotic behavior of expansion (5.67a) and reproduce the analytic properties of the Green's function through the analytic theory of continued fractions. The asymptotic behavior of the continued fraction coefficients a_n and b_n is connected with the compactness of the spectrum and with the presence of critical points. We cannot dwell on this and other interesting aspects of the recursion method, and on its interplay with other formalisms, such as the method of moments, the memory function approach, and the equation of motion method [see for instance the review articles by G. Grosso and G. Pastori Parravicini, Adv. Chem. Phys. *63*, 81 (1985); *63*, 133 (1985) and references quoted therein]; we rather prefer to clarify the procedure with an example.

Illustrative Example of the Recursion Method

To illustrate the recursion method, we consider for instance the case of the "quadratum", which is a simple square lattice with a single orbital per site and nearest neighbor interactions. The Hamiltonian is given by

$$H = E_0 \sum_{ij} |\phi_{ij}\rangle\langle\phi_{ij}|$$

$$+ t \sum_{ij} \left[|\phi_{ij}\rangle\langle\phi_{i+1,j}| + |\phi_{ij}\rangle\langle\phi_{i-1,j}| + |\phi_{i+1,j}\rangle\langle\phi_{ij}| + |\phi_{i-1,j}\rangle\langle\phi_{ij}| \right]$$

$$+ t \sum_{ij} \left[|\phi_{ij}\rangle\langle\phi_{i,j+1}| + |\phi_{ij}\rangle\langle\phi_{i,j-1}| + |\phi_{i,j+1}\rangle\langle\phi_{ij}| + |\phi_{i,j-1}\rangle\langle\phi_{ij}| \right]. \quad (5.68)$$

We take for convenience $E_0 = 0$ and $t = 1$. The Hamiltonian operator (5.68) is represented by a sparse matrix (non-vanishing interactions are confined to nearest neighbor orbitals), and is indicated graphically in Figure 5.13.

Before starting the Lanczos procedure, we need to establish explicitly the effect of the operator H on a generic state of the form

$$|u\rangle = \sum_{ij} c_{ij} |\phi_{ij}\rangle. \quad (5.69a)$$

From Eq. (5.68), we see that

$$H|u\rangle = \sum_{ij} \gamma_{ij} |\phi_{ij}\rangle \quad \text{with} \quad \gamma_{ij} = c_{i-1,j} + c_{i+1,j} + c_{i,j-1} + c_{i,j+1}. \quad (5.69b)$$

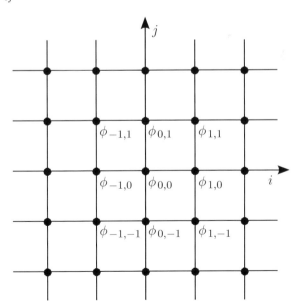

Figure 5.13 Schematic graphical representation of a simple square lattice, with a single orbital per site, and nearest neighbor interactions; ϕ_{ij} indicates the orbital localized in the position (i, j) (the edge of the quadratum is taken as unit).

With the help of Eqs. (5.69), one can easily carry out the Lanczos procedure; choosing as seed state the orbital at the origin, i.e. $|f_0\rangle = |\phi_{00}\rangle$, one obtains

$$|F_1\rangle = |\phi_{0,1}\rangle + |\phi_{0,-1}\rangle + |\phi_{1,0}\rangle + |\phi_{-1,0}\rangle;$$

the normalization of $|F_1\rangle$ gives $b_1^2 = \langle F_1|F_1\rangle = 4$. At the second step, one obtains

$$|F_2\rangle = |\phi_{1,1}\rangle + |\phi_{-1,1}\rangle + |\phi_{-1,-1}\rangle + |\phi_{1,-1}\rangle + \frac{1}{2}\left[|\phi_{2,0}\rangle + |\phi_{0,2}\rangle + |\phi_{-2,0}\rangle + |\phi_{0,-2}\rangle\right].$$

The normalization of $|F_2\rangle$ gives $b_2^2 = \langle F_2|F_2\rangle = 5$. The procedure can be continued so to obtain all the desired continued fraction parameters (with the help of a straightforward computer code). The density-of-states can then be computed from Eq. (5.67b).

As a second illustrative example, the reader could consider the case of the "cubium", which is a simple cubic lattice with a single orbital per site and nearest neighbor interactions. Choosing as seed state an atomic orbital, the continued fraction expression of the Green's function can be established; the real part of the Green's function, as well as the density-of-states of the cubium calculated by means of Eq. (5.67b), have been reported in Figure 2.25.

5.8.2 Modified Lanczos Method for Excited States

In the Lanczos method, the states $\{|f_n\rangle\}$ generated by the three term relations (5.65) are in principle orthonormal. However, it is well known that finite precision arithmetic induces rounding errors in the procedure, which may give rise to loss of precision in the results (loss of orthogonality among the generated states and appearance of spurious "ghost" states). In a number of situations where high accuracy is needed, there is little benefit to pursue further steps of the Lanczos approach, and alternative procedures must be found.

There is a vast literature that concerns the analysis and solution of the numerical problems that can be met in the Lanczos method; in particular, for the ground state of quantum systems, we refer to the paper of E. Dagotto, Rev. Mod. Phys. *66*, 763 (1994) and references quoted therein. Here, we briefly mention an intuitive extension of the Lanczos procedure for the evaluation of the *excited states* of quantum systems in a desired energy range [G. Grosso, L. Martinelli and G. Pastori Parravicini, Nuovo Cimento D *15*, 269 (1993); Phys. Rev. B *51*, 13033 (1995) and references quoted therein].

To obtain excited states of H, we consider a trial energy E_t and the auxiliary operator $A = (H - E_t)^2$ (other forms are suitable as well, but we focus on this form due to its simplicity). The ground state of the operator A is obviously the excited state of H nearest in energy to the chosen energy E_t. We can thus start the Lanczos iterations with the operator A and a seed state $|f_0\rangle$; however, to avoid any possible numerical problem, we perform only one Lanczos step and obtain the state $|f_1\rangle$. At this stage we diagonalize A on the two states $|f_0\rangle, |f_1\rangle$, and select the lowest eigenvalue of the 2×2 matrix and the corresponding eigenfunction $|f_g\rangle$. We can now use $|f_g\rangle$ as seed state and iterate the whole procedure; we obtain thus an iterative approach that systematically converges toward the exact eigenvalue and eigenfunction of the ground state of the operator A

(which is also the excited state of H nearest in energy to the chosen energy E_t). The convergence can be conveniently accelerated with appropriate procedures, and we refer to the literature for a more detailed description of the required technical aspects and applications.

5.8.3 Renormalization Method for Electronic Systems

The basic idea of the renormalization method is to infer the properties of systems, whose description requires a large (or infinite) number of basis functions, by reducing progressively the dimension of the space of preserved basis functions. In its ultimate motivation, the origin of this strategy can be traced back to the concept of renormalization introduced in the theory of phase transitions by Wilson (see Section 17.6).

We begin our treatment of the renormalization procedure for electronic state calculations by considering the Dyson equation for the Green's function of a quantum system. Consider an operator H, arbitrarily split into two parts

$$H = H_0 + W. \tag{5.70a}$$

If we indicate with $g(E)$ the Green's function of the operator H_0, and with $G(E)$ the Green's function of the operator H, we have that the two Green's functions are related via the Dyson equation

$$G(E) = g(E) + g(E)WG(E). \tag{5.70b}$$

The proof is simple indeed. We start from the identity

$$E - H_0 = (E - H_0 - W) + W,$$

and multiply both members of the above equation by $1/(E - H_0)$ (multiplication is performed, say, on the left); then we multiply by $1/(E - H_0 - W)$ (on the right) and obtain

$$\frac{1}{E - H_0 - W} = \frac{1}{E - H_0} + \frac{1}{E - H_0} W \frac{1}{E - H_0 - W},$$

which proves the Dyson equation (5.70b).

Consider now an operator H represented in a given orthonormal set by a $N \times N$ matrix (in specific applications N can be very large, say from thousand to a million). We separate the representative space S of dimension N into a subspace S_A of dimension N_A and a subspace S_B of dimension $N_B = N - N_A$ (in some situations the subspace S_A has dimension $N - 1$, and the subspace S_B is composed just by a single state). Whatever arbitrary separation has been performed, the operator H can be written, and then split, in the form

$$H = \begin{pmatrix} H_{AA} & H_{AB} \\ H_{BA} & H_{BB} \end{pmatrix} \quad \text{and} \quad H = H_0 + W = \begin{pmatrix} H_{AA} & 0 \\ 0 & H_{BB} \end{pmatrix} + \begin{pmatrix} 0 & H_{AB} \\ H_{BA} & 0 \end{pmatrix}.$$

Application of the Dyson equation (5.70b) gives

$$\begin{pmatrix} G_{AA} & G_{AB} \\ G_{BA} & G_{BB} \end{pmatrix} = \begin{pmatrix} g_{AA} & 0 \\ 0 & g_{BB} \end{pmatrix} + \begin{pmatrix} g_{AA} & 0 \\ 0 & g_{BB} \end{pmatrix} \begin{pmatrix} 0 & H_{AB} \\ H_{BA} & 0 \end{pmatrix} \begin{pmatrix} G_{AA} & G_{AB} \\ G_{BA} & G_{BB} \end{pmatrix},$$

where $g_{AA}(E) = (E - H_{AA})^{-1}$ and $g_{BB}(E) = (E - H_{BB})^{-1}$. Performing the algebraic matrix operations indicated in the above expression, it follows

$$G_{AA}(E) = g_{AA}(E) + g_{AA}(E)H_{AB}G_{BA}(E)$$
$$G_{BA}(E) = g_{BB}(E)H_{BA}G_{AA}(E)$$

(and similar expressions for exchange of the subscripts A and B). With straightforward properties of matrix multiplication we have

$$G_{AA}(E) = \cfrac{1}{E - H_{AA} - H_{AB}\cfrac{1}{E - H_{BB}}H_{BA}}. \tag{5.71}$$

Until now we have performed exact algebraic transformations. The interpretation of Eq. (5.71) is immediate. The Green's function in the subspace S_A is determined by a "renormalized" or "effective" Hamiltonian which is given by

$$H^{(\mathrm{eff})}(E) = H_{AA} + H_{AB}\frac{1}{E - H_{BB}}H_{BA}.$$

The physical meaning of the renormalized Hamiltonian is self-explanatory: besides H_{AA}, it contains the effect of an excursion from A to B, a propagation within B, and an excursion back to A.

Suppose that the splitting of the original space spanned by H is such that in the subspace S_B the Green's function is known. We can *eliminate* (or as it is commonly said in the literature *decimate*) all the states of the subspace S_B, considering within the subspace S_A the effective Hamiltonian

$$H^{(\mathrm{eff})}(E) = H_{AA} + \Sigma_{AA}(E) \tag{5.72a}$$

with

$$\Sigma_{AA}(E) = H_{AB}\frac{1}{E - H_{BB}}H_{BA}. \tag{5.72b}$$

The operator $\Sigma_{AA}(E)$, whose origin is linked to the elimination of the subspace S_B, is called *self-energy operator*; the self-energy operator $\Sigma_{AA}(E)$, added to H_{AA}, gives the effective Hamiltonian on the preserved subspace S_A, now formally decoupled from the subspace S_B.

The decimation procedure can always be performed, but it is of practical help when the advantages outnumber the disadvantages. Let us see in general when this can occur. The advantage of the decimation procedure stands in having to handle a subspace S_A of dimension smaller than the original one. The price for this is the necessity to evaluate the appropriate self-energy operator, whose matrix elements are energy-dependent. In general situations it is not convenient to proceed toward a partial elimination of the states of the preserved subspace S_A, because of the complicated structure of the self-energy operator $\Sigma_{AA}(E)$. However in particular but very important cases (Bethe lattices, one-dimensional lattices, multi-layer structures) it is possible to find peculiar decimations (the analog of the "renormalization group transformations" of phase transitions), which can be applied "ad libitum" to the preserved subspace itself.

In particular, multi-layer structures, superlattices, and quantum wells, as well as very interesting incommensurate systems have been studied with the iterative renormalization procedure. Here we limit our attention to two particular significant examples, which are exactly soluble with the renormalization method, and are encountered very often in the forthcoming chapters.

Example 1. Discrete State Interacting with a Manifold (Discrete or Continuous) of States

The first example we consider is that of a discrete state interacting with a continuum of states; for simplicity, the continuum is mimicked with an (arbitrarily) large number N of discrete states. The Hamiltonian can be written as

$$H = H_0 + W, \tag{5.73a}$$

$$H_0 = E_0 |\phi_0\rangle\langle\phi_0| + \sum_{i=1}^{N} E_i |\phi_i\rangle\langle\phi_i|, \tag{5.73b}$$

$$W = \frac{1}{\sqrt{N}} \sum_{i=1}^{N} \left[\gamma_i |\phi_0\rangle\langle\phi_i| + \gamma_i^* |\phi_i\rangle\langle\phi_0| \right]. \tag{5.73c}$$

The energies E_i of the quasi-continuum manifold of states $|\phi_i\rangle (i = 1, 2, \ldots, N)$ extend from $E_1 = E_{\min}$ to $E_N = E_{\max}$; the energy E_0 of the discrete state can be either degenerate with the manifold (we shall consider this case in discussing the Fano model in Section 12.6) or at lower energy (we shall reduce to this case in discussing the Kondo effect in Section 16.6). We discuss here explicitly this latter case, schematically indicated in Figure 5.14.

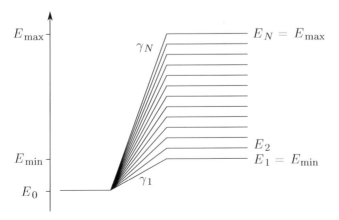

Figure 5.14 Schematic representation of a system consisting of a discrete state of energy E_0, interacting with a manifold of states in the energy range $E_1 = E_{\min}$ and $E_N = E_{\max}$.

Using the renormalization procedure, we can *eliminate* one-by-one all the states $|\phi_i\rangle (i = 1, 2, \ldots, N)$ and we obtain

$$G_{00}(E) = \cfrac{1}{E - E_0 - \cfrac{1}{N} \sum_i \cfrac{|\gamma_i|^2}{E - E_i}}.$$

The poles of this Green's function are given by the equation

$$\boxed{E - E_0 = \frac{1}{N} \sum_i \frac{|\gamma_i|^2}{E - E_i}}.$$

(5.74)

The above equation is easily solved graphically, as indicated in Figure 5.15.

The solutions are given by the intercepts of the straight line through E_0 and slope one, with the monotonically decreasing curve $(1/N) \sum |\gamma_i|^2/(E - E_i)$. Besides the solutions reminiscent of the individual states of the quasi-continuum, Figure 5.15 shows that there is always formation of a "collective state", with energy less than E_0; the adjective "collective" is used to indicate that this state is jointly determined by the discrete state and the states of the quasi-continuum (at least those of lower energy).

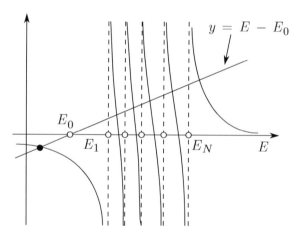

Figure 5.15 Graphical solution of Eq. (5.74) of the text. The system consists of a discrete state E_0 interacting with a quasi-continuum of states extending from $E_{\min} = E_1$ to $E_{\max} = E_N$. We illustrate the case $E_0 \lesssim E_{\min}$. Notice that a "collective state" always appears with energy smaller than E_0.

Example 2. Eigenvalues of an Operator in the Presence of a Constant Coupling Between Basis Functions

Consider an Hamiltonian of the form

$$H = H_0 + W$$

(5.75a)

with

$$H_0 = \sum_i E_i |\phi_i\rangle\langle\phi_i| \quad \text{and} \quad W = \frac{w_0}{N} \sum_{ij} |\phi_i\rangle\langle\phi_j|. \tag{5.75b}$$

The operator H_0 is diagonal with energies E_i ($i = 1, 2, \ldots, N$); the operator W has all the diagonal and off-diagonal matrix elements equal to a constant, denoted as w_0/N. We can obtain the eigenvalues of $H = H_0 + W$ with the following procedure.

Consider the (normalized) state

$$|u_0\rangle = \frac{1}{\sqrt{N}} \sum_i |\phi_i\rangle.$$

The diagonal matrix element on the state $|u_0\rangle$ of the Dyson equation (5.70b) reads

$$\langle u_0|G(E)|u_0\rangle = \langle u_0|g(E)|u_0\rangle + \langle u_0|g(E)WG(E)|u_0\rangle;$$

since W can be expressed as $w_0|u_0\rangle\langle u_0|$, we immediately obtain

$$\langle u_0|G(E)|u_0\rangle = \frac{\langle u_0|g(E)|u_0\rangle}{1 - w_0\langle u_0|g(E)|u_0\rangle}. \tag{5.76}$$

We notice that

$$g_{00}(E) = \langle u_0|g(E)|u_0\rangle = \frac{1}{N} \sum_i \frac{1}{E - E_i};$$

we thus have that the poles of $G(E)$, given by the zeroes of the denominator of Eq. (5.76), are determined by the condition

$$\boxed{\frac{1}{w_0} = \frac{1}{N} \sum_i \frac{1}{E - E_i}}. \tag{5.77}$$

The graphical solution of Eq. (5.77) is indicated in Figure 5.16. The graphical solution is obtained by the intercepts of the horizontal line $1/w_0$ with the curve $g_{00}(E)$. The curve $g_{00}(E)$ is regular and monotonically decreasing for $-\infty < E < E_1$ and $E_N < E < \infty$; furthermore we have:

$$\lim_{E \to E_1} g_{00}(E) = -\infty \quad \text{for} \quad E < E_1,$$
$$\lim_{E \to E_N} g_{00}(E) = +\infty \quad \text{for} \quad E > E_N. \tag{5.78}$$

In the case the N states E_i are discrete (as in Figure 5.16), we have always the formation of a *split-off state* (or "collective state"), with energy outside the interval $E_1 \leq E \leq E_N$; the split-off state has energy smaller than E_1 if $w_0 < 0$, and energy larger than E_N if $w_0 > 0$.

It is interesting to consider the limiting case of a genuine continuum, i.e. the case $N \to \infty$. In this case $g_{00}(E)$ is again regular and monotonically decreasing in the

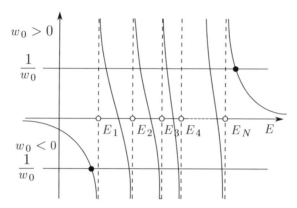

Figure 5.16 Graphical solution of Eq. (5.77) of the text and formation of split-off states.

intervals $-\infty < E < E_1$ and $E_N < E < \infty$, but the limits (5.78) can be either finite or infinite. In the case the limits (5.78) are finite, a critical strength $|w_0|$ is necessary to form split-off states. We shall encounter in other parts of this book examples of either type (for instance, in the case of Cooper pairs in metals, their formation occurs whatever small the interaction is; in the theory of deep impurity states, a threshold is in general required for their formation).

Appendix A.   Matrix Elements of the Augmented Plane Wave Method

This Appendix provides some technical aspects concerning the matrix elements of the APW method. An augmented plane wave $A(\mathbf{k} + \mathbf{h}_i, \mathbf{r}, E)$ is defined as a plane wave of vector $\mathbf{k}_i = \mathbf{k} + \mathbf{h}_i$ outside the sphere r_s, which joins continuously with a cellular type function inside it. Its expression is

$$A(\mathbf{k}_i, \mathbf{r}, E) = \begin{cases} A_{\text{in}}(\mathbf{k}_i, \mathbf{r}, E) = \dfrac{4\pi}{\sqrt{\Omega}} \displaystyle\sum_{lm} i^l \dfrac{j_l(k_i r_s)}{R_l(E, r_s)} Y_{lm}^*(\mathbf{k}_i) Y_{lm}(\mathbf{r}) R_l(E, r) & \mathbf{r} \in \Omega_s \\[4mm] A_{\text{out}}(\mathbf{k}_i, \mathbf{r}) = \dfrac{e^{i\mathbf{k}_i \cdot \mathbf{r}}}{\sqrt{\Omega}} \equiv \dfrac{4\pi}{\sqrt{\Omega}} \displaystyle\sum_{lm} i^l j_l(k_i r) Y_{lm}^*(\mathbf{k}_i) Y_{lm}(\mathbf{r}) & \mathbf{r} \in \Omega_{\text{WS}} - \Omega_s \end{cases}$$

where j_l is the order l spherical Bessel function, Ω is the volume of the primitive cell, Ω_{WS} and Ω_s indicate the Wigner-Seitz cell and the inscribed sphere of radius r_s, respectively. Notice that an augmented plane wave is a continuous function, but in general a discontinuity in its slope at $r = r_s$ remains. Expansion of crystal states in terms of augmented plane waves, and standard variational procedures, lead to the secular equation

$$\| \langle A(\mathbf{k}_i, \mathbf{r}, E) | H - E | A(\mathbf{k}_j, \mathbf{r}, E) \rangle \| = 0, \tag{A.1}$$

where H is the crystal Hamiltonian with a crystalline potential in the muffin-tin form. The explicit expression of the matrix elements requires some care because of the

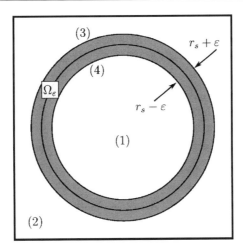

Figure 5.17 Schematic representation of the four contributions (two volume contributions and two surface contributions) for the calculation of the matrix elements of the crystal Hamiltonian between augmented plane waves.

discontinuity in the slope at $r = r_s$ of the basis wavefunctions; thus, we begin to show how to handle the discontinuity.

Let $u(\mathbf{r})$ and $v(\mathbf{r})$ be *continuous functions with a kink, i.e. with a finite discontinuity in the gradient*, at $r = r_s$. Let $u_{\text{in}}(\mathbf{r})$ and $u_{\text{out}}(\mathbf{r})$ indicate the function $u(\mathbf{r})$ inside and outside the sphere of radius r_s. We consider the integral of the kinetic energy operator $-\nabla^2$ between $v(\mathbf{r})$ and $u(\mathbf{r})$ through an arbitrary small volume Ω_ε containing the spherical surface S of radius $r = r_s$ (the region of integration is shown in Figure 5.17. We have

$$\int_{\Omega_\varepsilon} v(\mathbf{r})[-\nabla^2 u(\mathbf{r})]\, d\mathbf{r} = -\int_{\Omega_\varepsilon} \nabla \cdot [v(\mathbf{r})\nabla u(\mathbf{r})]\, d\mathbf{r} + \int_{\Omega_\varepsilon} \nabla v(\mathbf{r}) \cdot \nabla u(\mathbf{r})\, d\mathbf{r}$$

(as can be seen by inspection of the integrands). The first integral in the right-hand side of the above equation can be performed using the divergence theorem, while the second integral in the right-hand can be neglected for $\Omega_\varepsilon \to 0$. It follows

$$\int_{\Omega_\varepsilon} v(\mathbf{r})[-\nabla^2 u(\mathbf{r})]\, d\mathbf{r} = \int_S v(\mathbf{r}) \left[\frac{\partial}{\partial r} u_{\text{in}}(\mathbf{r}) - \frac{\partial}{\partial r} u_{\text{out}}(\mathbf{r}) \right] dS\,; \qquad \text{(A.2)}$$

the two surface contributions in the right-hand side of Eq. (A.2) do not cancel because of the discontinuity of the gradient of the wavefunction $u(\mathbf{r})$ at $r = r_s$.

We have now all the ingredients for the calculation of the matrix elements $M_{ij}(\mathbf{k}, E)$ of the crystal Hamiltonian operator $H - E$ between any two augmented plane waves;

their expression reads

$$M_{ij}(\mathbf{k}, E) = \int_{\Omega_{WS}} A^*(\mathbf{k}_i, \mathbf{r}, E)[-\nabla^2 + V(r) - E]A(\mathbf{k}_j, \mathbf{r}, E)\, d\mathbf{r}. \tag{A.3}$$

The matrix elements $M_{ij}(\mathbf{k}, E)$ can be split into four contributions

$$M_{ij} = M_{ij}^{(1)} + M_{ij}^{(2)} + M_{ij}^{(3)} + M_{ij}^{(4)},$$

i.e. two volume integrals corresponding to the two regions in which the primitive cell is partitioned, and two surface integrals at the spheres of radius $r_s \pm \varepsilon$ (see Figure 5.17). The surface contributions appear due to the kink of the augmented plane waves at $r = r_s$.

Let us consider separately these four contributions. For the first volume integral we have

$$M_{ij}^{(1)} = \int_{\Omega_s} A_{in}^*(\mathbf{k}_i, \mathbf{r}, E) \left[-\nabla^2 + V(r) - E \right] A_{in}(\mathbf{k}_j, \mathbf{r}, E)\, d\mathbf{r} \equiv 0.$$

This term is exactly zero because an APW satisfies the Schrödinger equation inside the sphere Ω_s.

The second volume integral is

$$M_{ij}^{(2)} = \int_{\Omega_{WS}-\Omega_s} A_{out}^*(\mathbf{k}_i, \mathbf{r}, E) \left[-\nabla^2 - E \right] A_{out}(\mathbf{k}_j, \mathbf{r}, E)\, d\mathbf{r}.$$

In the interstitial region $\Omega_{WS} - \Omega_s$, the APW is simply a plane wave; then we have

$$M_{ij}^{(2)} = \frac{k_j^2 - E}{\Omega} \left[\int_{\Omega_{WS}} e^{i(\mathbf{k}_j - \mathbf{k}_i)\cdot\mathbf{r}}\, d\mathbf{r} - \int_{\Omega_s} e^{i(\mathbf{k}_j - \mathbf{k}_i)\cdot\mathbf{r}}\, d\mathbf{r} \right]$$

$$= (k_j^2 - E)\delta_{ij} - \left(k_j^2 - E \right) \frac{4\pi r_s^2}{\Omega} \frac{j_1(|\mathbf{k}_i - \mathbf{k}_j|r_s)}{|\mathbf{k}_i - \mathbf{k}_j|}.$$

We pass now to consider the surface integrals, originated by the kink in the basis functions.

The surface integral $M_{ij}^{(3)}$ on the external side of the spherical surface, according to Eq. (A.2), becomes

$$M_{ij}^{(3)} = -\int_S A_{out}^*(\mathbf{k}_i, \mathbf{r}) \frac{\partial}{\partial r} A_{out}(\mathbf{k}_j, \mathbf{r})\, dS$$

$$= -\frac{16\pi^2}{\Omega} \sum_{lm\, l'm'} \int_S (-i)^l j_l(k_i r_s) Y_{lm}(\mathbf{k}_i) Y_{lm}^*(\mathbf{r}) i^{l'} j_{l'}'(k_j r_s) Y_{l'm'}^*(\mathbf{k}_j) Y_{l'm'}(\mathbf{r})\, dS$$

where

$$j_{l'}'(k_j r) = \left[\frac{d}{dr} j_{l'}(k_j r) \right]_{r=r_s}.$$

Using the standard orthonormality relations of the spherical harmonics, and the addition theorem of Eq. (5.30) for spherical harmonics, it follows

$$M_{ij}^{(3)} = -\frac{4\pi r_s^2}{\Omega} \sum_l (2l+1) j_l(k_i r_s) j_l(k_j r_s) P_l(\widehat{\mathbf{k}}_i \cdot \widehat{\mathbf{k}}_j) \frac{j_l'(k_j r_s)}{j_l(k_j r_s)}.$$

The surface integral $M_{ij}^{(4)}$ on the internal side of the spherical surface can be obtained in a similar way and is given by

$$M_{ij}^{(4)} = \frac{4\pi r_s^2}{\Omega} \sum_l (2l+1) j_l(k_i r_s) j_l(k_j r_s) P_l(\widehat{\mathbf{k}}_i \cdot \widehat{\mathbf{k}}_j) \frac{R_l'(E, r_s)}{R_l(E, r_s)}.$$

Summing up the previous results we have

$$M_{ij}(\mathbf{k}, E) = \left(k_j^2 - E\right) \delta_{ij} - \frac{4\pi r_s^2}{\Omega} \left(k_j^2 - E\right) \frac{j_1(|\mathbf{k}_i - \mathbf{k}_j| r_s)}{|\mathbf{k}_i - \mathbf{k}_j|}$$
$$+ \frac{4\pi r_s^2}{\Omega} \sum_l (2l+1) j_l(k_i r_s) j_l(k_j r_s) P_l(\widehat{\mathbf{k}}_i \cdot \widehat{\mathbf{k}}_j) \left[\frac{R_l'(E, r_s)}{R_l(E, r_s)} - \frac{j_l'(k_j r_s)}{j_l(k_j r_s)}\right]. \quad (A.4)$$

Equation (A.4) does not show, at first sight, the condition $M_{ij} = M_{ji}$; to put in evidence explicitly the hermiticity of the (real) matrix M, we perform a last elaboration. Consider the identity

$$j_0(|\mathbf{k}_i - \mathbf{k}_j| r_s) = \sum_l (2l+1) P_l(\widehat{\mathbf{k}}_i \cdot \widehat{\mathbf{k}}_j) \, j_l(k_i r_s) j_l(k_j r_s),$$

and derive both members with respect to k_j. In performing the derivation, we remember the standard relation $j_0'(x) = -j_1(x)$ and

$$\frac{\partial}{\partial k_j} |\mathbf{k}_i - \mathbf{k}_j| = \frac{\partial}{\partial k_j} \sqrt{k_i^2 + k_j^2 - 2 k_i k_j \cos \Theta_{ij}} = \frac{1}{|\mathbf{k}_i - \mathbf{k}_j|} \frac{1}{k_j} \left(k_j^2 - \mathbf{k}_i \cdot \mathbf{k}_j\right).$$

We obtain

$$\left(\mathbf{k}_i \cdot \mathbf{k}_j - k_j^2\right) \frac{j_1(|\mathbf{k}_i - \mathbf{k}_j| r_s)}{|\mathbf{k}_i - \mathbf{k}_j|} \equiv \sum_l (2l+1) P_l(\widehat{\mathbf{k}}_i \cdot \widehat{\mathbf{k}}_j) \, j_l(k_i r_s) j_l'(k_j r_s).$$

Using this identity we can transform the matrix elements of Eq. (A.4) in the form

$$M_{ij}(\mathbf{k}, E) = \left(k_j^2 - E\right) \delta_{ij} - \frac{4\pi r_s^2}{\Omega} (\mathbf{k}_i \cdot \mathbf{k}_j - E) \frac{j_1(|\mathbf{k}_i - \mathbf{k}_j| r_s)}{|\mathbf{k}_i - \mathbf{k}_j|}$$
$$+ \frac{4\pi r_s^2}{\Omega} \sum_{l=0}^{\infty} (2l+1) P_l(\widehat{\mathbf{k}}_i \cdot \widehat{\mathbf{k}}_j) \, j_l(k_i r_s) j_l(k_j r_s) \frac{R_l'(E, r_s)}{R_l(E, r_s)}, \quad (A.5)$$

where the symmetric property of the matrix is now evident.

Appendix B.   Solved Problems and Complements

Problem 1. Logarithmic derivatives as function of energy.
Problem 2. Phase shifts and logarithmic derivatives.
Problem 3. Error function and free-electron Green's function.

Problem 1. Logarithmic Derivatives as Function of Energy

Show that the logarithmic derivatives $R'_l(E, r)/R_l(E, r)$ of any radial wavefunction are monotonically decreasing functions of energy.

Consider a radial wavefunction $R_l(E, r)$, which is regular at the origin and satisfies the radial wave equation

$$\frac{d^2(r R_l)}{dr^2} = \left[V(r) + \frac{l(l+1)}{r^2} - E \right] (r R_l) \tag{B.1}$$

for a spherically symmetric potential $V(r)$. At two arbitrary chosen energies E_1 and E_2, consider Eq. (B.1) for $U_l(E_1, r) \equiv r R_l(E_1, r)$ and for $U_l(E_2, r) \equiv r R_l(E_2, r)$; namely:

$$\frac{d^2 U_l(E_1, r)}{dr^2} = \left[V(r) + \frac{l(l+1)}{r^2} - E_1 \right] U_l(E_1, r),$$

$$\frac{d^2 U_l(E_2, r)}{dr^2} = \left[V(r) + \frac{l(l+1)}{r^2} - E_2 \right] U_l(E_2, r).$$

We now multiply the radial equation for $U_l(E_1, r)$ by $U_l(E_2, r)$ and vice versa, subtract, integrate over the radial coordinate and perform an integration by parts, and finally let $E_1 \to E_2$. We obtain the identity

$$\int_0^{r_s} R_l^2(E, r) r^2 dr \equiv -R_l^2(E, r_s) r_s^2 \frac{\partial}{\partial E} \frac{R'_l(E, r_s)}{R_l(E, r_s)}. \tag{B.2}$$

The first member of Eq. (B.2) is always positive; it follows that the logarithmic derivatives $R'_l(E, r)/R_l(E, r)$ *are always monotonically decreasing functions of energy* and exhibit poles at the resonance energies, which are defined as the energies for which $R_l(E, r_s) = 0$.

Problem 2. Phase Shifts and Logarithmic Derivatives

Consider a spherically symmetric potential, which vanishes outside the muffin-tin sphere, and the radial function $R_l(E, r)$, solution of Eq. (B.1), regular at the origin. Express the phase shifts $\eta_l(E)$ in term of the logarithmic derivative of the radial wavefunctions outside the muffin-tin sphere.

Consider a radial wavefunction $R_l(E, r)$, which is regular at the origin and satisfies the radial wave equation

$$\frac{d^2(r R_l)}{dr^2} = \left[V(r) + \frac{l(l+1)}{r^2} - E \right] (r R_l)$$

for a spherically symmetric potential $V(r)$. Suppose that $V(r) \equiv 0$ for $r > r_{MT}$. The radial function for $r > r_{MT}$ can be expressed as a linear combination of the two linearly independent solutions of the Schrödinger equation when the potential is constant and equal to zero. The asymptotic form of the radial function $R_l(E, r)$ can be expressed as

$$R_l(E, r) = n_l(\alpha r) - \cotg \eta_l(E) \cdot j_l(\alpha r) \quad \text{for} \quad r \geq r_{MT}, \tag{B.3}$$

where j_l and n_l are the spherical Bessel and Neumann functions, and $\eta_l(E)$ are the *phase shifts* produced by the muffin-tin potential for the wave of angular momentum l and energy E.

From Eq. (B.3) and its derivative with respect to r, for any $r > r_{MT}$ we have

$$\frac{R_l'(E, r)}{R_l(E, r)} = \frac{n_l'(\alpha r) - \cotg \eta_l(E) \cdot j_l'(\alpha r)}{n_l(\alpha r) - \cotg \eta_l(E) \cdot j_l(\alpha r)} \qquad r > r_{MT}.$$

It follows

$$\cotg \eta_l(E) = \frac{n_l(\alpha r)\left[R_l'(E, r)/R_l(E, r)\right] - n_l'(\alpha r)}{j_l(\alpha r)\left[R_l'(E, r)/R_l(E, r)\right] - j_l'(\alpha r)}, \tag{B.4}$$

where n_l, j_l, R_l, n_l', j_l', R_l' are all calculated at any arbitrary point outside the muffin-tin sphere. A nice feature of Eq. (B.4) is the irrelevance of the normalization chosen for R_l, since the logarithmic derivative R_l'/R_l is obviously independent of normalization.

Problem 3. Error Function and Free-Electron Green's Function

Exploiting the properties of the error function, provide an integral representation of the free-electron Green's function.

The error function of real argument is defined as

$$\text{erf}(x) = \Phi(x) = \frac{2}{\sqrt{\pi}} \int_0^x e^{-t^2} dt = \frac{2}{\sqrt{\pi}} \sum_{n=0}^{\infty} \frac{(-1)^n x^{2n+1}}{n!\,(2n+1)}. \tag{B.5}$$

An obvious property of the error function is

$$\frac{d\Phi(x)}{dx} = \frac{2}{\sqrt{\pi}} e^{-x^2} = \frac{2}{\sqrt{\pi}} \sum_{n=0}^{\infty} \frac{(-1)^n x^{2n}}{n!}.$$

The error function increases monotonically from the value $\Phi(0) = 0$ to $\Phi(\infty) = 1$, as x increases in the real positive axis. Also notice that $\Phi(-x) = -\Phi(x)$ (odd function), and $\Phi'(-x) = \Phi'(x)$ (even function). Using the series development of the error function and its derivative, it can be easily seen that

$$2x\Phi(x) + \Phi'(x) = -\frac{2}{\sqrt{\pi}} \sum_{n=0}^{\infty} \frac{(-1)^n x^{2n}}{n!\,(2n-1)} \tag{B.6}$$

a relation useful in the following.

The error function of complex argument is defined as an extension in the complex plane of Eq. (B.5), namely;

$$\text{erf}(z) = \Phi(z) = \frac{2}{\sqrt{\pi}} \int_0^z e^{-t^2} dt = \frac{2}{\sqrt{\pi}} \sum_{n=0}^{\infty} \frac{(-1)^n z^{2n+1}}{n! (2n+1)}. \tag{B.7}$$

In the case $z \to \infty$, the path of integration is subject to the restriction $|\arg t| < \pi/4$ for $t \to \infty$ along the path; this condition guarantees the convergence of the integral, and we have erf $z \to 1$ for $z \to \infty$ if $|\arg z| < \pi/4$.

The error function is useful in order to express analytically countless indefinite and definite integrals, as thorough discussed for instance by M. Abramowitz and I. A. Stegun "Handbook of Mathematical Functions" (Dover Publications, New York, 1972). In the following we need the integral

$$\frac{2}{\sqrt{\pi}} \int \exp\left[-R^2 x^2 + \frac{E}{4x^2}\right] dx = \frac{e^{i\alpha R}}{2R} \Phi\left(Rx + \frac{i\alpha}{2x}\right) + \frac{e^{-i\alpha R}}{2R} \Phi\left(Rx - \frac{i\alpha}{2x}\right),$$

$$\tag{B.8}$$

where $R > 0$, and $\alpha = \sqrt{E}$. The square root of the energy for positive or negative values is

$$\alpha = \begin{cases} \sqrt{E} & \text{for } E > 0 \\ \sqrt{E} = i\sqrt{|E|} & \text{for } E < 0 \end{cases}.$$

The demonstration of Eq. (B.8) is easily verified by observing that the derivative of the second member of Eq. (B.8) equals the integrand function at the first member.

With the help of Eq. (B.8), the definite integral in the interval $[0, +\infty]$ becomes

$$\frac{2}{\sqrt{\pi}} \int_0^\infty \exp\left[-R^2 x^2 + \frac{E}{4x^2}\right] dx = \frac{e^{i\alpha R}}{R}. \tag{B.9}$$

Equation (B.9) provides the well-known *integral representation of the Hankel function of first kind*. [For positive E, the integration path near the origin must be chosen in such a way that the argument of complex variable z is smaller than $-\pi/4$; in fact $\arg z < -\pi/4$ guarantees that $\exp(E/4z^2) \to 0$ for $z \to 0$. No path restriction occurs for negative values of E.]

Using Eq. (B.9) we obtain for the free-particle Green's function $g(\mathbf{r} - \mathbf{r}_0, E)$ the integral representation

$$g(\mathbf{r} - \mathbf{r}_0, E) = -\frac{1}{4\pi} \frac{e^{i\alpha|\mathbf{r}-\mathbf{r}_0|}}{|\mathbf{r} - \mathbf{r}_0|} = -\frac{1}{2\pi^{3/2}} \int_0^\infty \exp\left[-|\mathbf{r} - \mathbf{r}_0|^2 x^2 + \frac{E}{4x^2}\right] dx,$$

$$\tag{B.10}$$

a relation widely used in the Green's function formalism.

With the help of Eq. (B.8), another definite integral of interest can be obtained, covering the region $[\sqrt{\eta}/2, +\infty]$ (with η real positive arbitrary quantity). It holds.

$$
\frac{2}{\sqrt{\pi}} \int_{\sqrt{\eta}/2}^{\infty} \exp\left[-R^2 x^2 + \frac{E}{4x^2}\right] dx
$$

$$
= \frac{\cos(\alpha R)}{R} - \frac{e^{i\alpha R}}{2R} \Phi\left(R\frac{\sqrt{\eta}}{2} + \frac{i\alpha}{\sqrt{\eta}}\right) - \frac{e^{-i\alpha R}}{2R} \Phi\left(R\frac{\sqrt{\eta}}{2} - \frac{i\alpha}{\sqrt{\eta}}\right).
$$

From the above equation, taking the limit $R \to 0$ of the last two terms, we have

$$
\frac{2}{\sqrt{\pi}} \int_{\sqrt{\eta}/2}^{\infty} \exp\left[-R^2 x^2 + \frac{E}{4x^2}\right] dx = \frac{\cos(\alpha R)}{R} + \frac{\sqrt{\eta}}{\sqrt{\pi}} \sum_{n=0}^{\infty} \frac{(E/\eta)^n}{n!(2n-1)} \quad R \to 0,
$$

$$\tag{B.11}$$

a result easily established considering the series development of the exponential functions and error functions, for small R, and taking the $R \to 0$ limit.

Appendix C. Evaluation of the Structure Coefficients of the KKR Method with the Ewald Procedure

This Appendix provides the technical aspects for an efficient expression of the structure coefficients of the KKR method. The free-particle Green's function is given by the expression

$$
g(\mathbf{r} - \mathbf{r}_0, E) = -\frac{1}{4\pi} \frac{e^{i\alpha|\mathbf{r}-\mathbf{r}_0|}}{|\mathbf{r} - \mathbf{r}_0|} \quad \text{with} \quad \alpha(E) = \begin{cases} \sqrt{E} & E > 0 \\ i\sqrt{|E|} & E < 0 \end{cases}.
$$

Two basic tools are needed for its efficient elaboration: (i) the integral representation of $g(\mathbf{r} - \mathbf{r}_0, E)$, already discussed in Problem 3 (ii) the general Neumann expansion, which reads

$$
-\frac{1}{4\pi} \frac{e^{i\alpha|\mathbf{r}-\mathbf{r}_0|}}{|\mathbf{r} - \mathbf{r}_0|} = -\frac{1}{4\pi} \frac{\cos(\alpha|\mathbf{r}-\mathbf{r}_0|)}{|\mathbf{r} - \mathbf{r}_0|} - \frac{i}{4\pi} \frac{\sin(\alpha|\mathbf{r}-\mathbf{r}_0|)}{|\mathbf{r} - \mathbf{r}_0|}
$$

$$
= \frac{\alpha}{4\pi}[n_0(\alpha|\mathbf{r}-\mathbf{r}_0|) - ij_0(\alpha|\mathbf{r}-\mathbf{r}_0|)]
$$

$$
= \alpha \sum_{lm} j_l(\alpha r)Y_{lm}(\mathbf{r})[n_l(\alpha r_0) - ij_l(\alpha r_0)]Y_{lm}^*(\mathbf{r}_0), \tag{C.1}
$$

where $r < r_0$ and j_l and n_l are spherical Bessel and Neumann functions, respectively. [In the case $r > r_0$ we have to exchange $\mathbf{r}$ and $\mathbf{r}_0$ in the right-hand side of Eq. (C.1)]. In Eq. (C.1) the spatial variables $\mathbf{r}$ and $\mathbf{r}_0$ span the whole crystal space and are not confined to some primitive cell.

The general expression of the greenian of vector $\mathbf{k}$ is

$$g(\mathbf{k}, \mathbf{R}, E) = -\frac{1}{4\pi} \sum_{t_n} e^{i\mathbf{k}\cdot t_n} \frac{e^{i\alpha|\mathbf{R}-t_n|}}{|\mathbf{R}-t_n|} \quad \text{with} \quad \mathbf{R} = \mathbf{r} - \mathbf{r}_0.$$

From the Neumann expansion, for $|\mathbf{R}| < t_n \neq 0$ (i.e. under the restriction that R is smaller than any non-zero lattice translation) we can cast the greenian in the form

$$g(\mathbf{k}, \mathbf{R}, E) = -\frac{1}{4\pi} \frac{\cos(\alpha R)}{R} + \sum_{lm} D_{lm}(\mathbf{k}, E) j_l(\alpha R) Y_{lm}(\mathbf{R}) \quad |\mathbf{R}| < t_n \neq 0,$$

(C.2)

where we have separated the singular term, expressed by means of the Neumann function $n_0(\alpha R)$, from the regular ones, expressed by means of the Bessel functions of any order. *The quantities $D_{lm}(\mathbf{k}, E)$ in the regular part of the greenian are called the reduced structure coefficients.* The direct calculation of $D_{lm}(\mathbf{k}, E)$ using the expression in direct space coming out from the Neumann expansion is not particularly useful, because is flawed by the slow convergence (at least for positive energies). We will thus use the Ewald algorithm, which takes advantage of an appropriate balance summation both in direct space and reciprocal space.

The alternative expression of the greenian of vector $\mathbf{k}$, using the integral representation given in Eq. (B.10), reads.

$$\begin{aligned} g(\mathbf{k}, \mathbf{R}, E) &= -\frac{1}{4\pi} \sum_{t_n} e^{i\mathbf{k}\cdot t_n} \frac{e^{i\alpha|\mathbf{R}-t_n|}}{|\mathbf{R}-t_n|} \\ &= -\frac{1}{2\pi^{3/2}} \sum_{t_n} e^{i\mathbf{k}\cdot t_n} \int_0^\infty \exp\left[-|\mathbf{R}-t_n|^2 x^2 + \frac{E}{4x^2}\right] dx. \end{aligned}$$

(C.3)

The determination of the reduced structure coefficients is performed comparing Eqs. (C.3) and (C.2) and enforcing their identity in the particularly convenient limit $R \to 0$.

Following the Ewald recipe, it is convenient to split the integration in Eq. (C.3) into two parts: an integration from the origin to an arbitrary point $\sqrt{\eta}/2$ (with η positive and arbitrary quantity), and the integration from $\sqrt{\eta}/2$ to infinity. The greenian (C.3) can be written as

$$g(\mathbf{k}, \mathbf{R}, E) = g^{(1)}(\mathbf{k}, \mathbf{R}, E) + g^{(2)}(\mathbf{k}, \mathbf{R}, E) + g^{(3)}(R, E),$$

where

$$g^{(1)}(\mathbf{k}, \mathbf{R}, E) = -\frac{1}{2\pi^{3/2}} \sum_{t_n} e^{i\mathbf{k}\cdot t_n} \int_0^{\sqrt{\eta}/2} \exp\left[-(\mathbf{R}-t_n)^2 x^2 + \frac{E}{4x^2}\right] dx, \quad \text{(C.4)}$$

$$g^{(2)}(\mathbf{k}, \mathbf{R}, E) = -\frac{1}{2\pi^{3/2}} \sum_{\mathbf{t}_n \neq 0} e^{i\mathbf{k}\cdot\mathbf{t}_n} \int_{\sqrt{\eta}/2}^{\infty} \exp\left[-(\mathbf{R} - \mathbf{t}_n)^2 x^2 + \frac{E}{4x^2}\right] dx, \quad (C.5)$$

$$g^{(3)}(R, E) = -\frac{1}{2\pi^{3/2}} \int_{\sqrt{\eta}/2}^{\infty} \exp\left[-R^2 x^2 + \frac{E}{4x^2}\right] dx. \quad (C.6)$$

Notice that the sum in Eq. (C.5) omits the term $\mathbf{t}_n = 0$; the omitted term $\mathbf{t}_n = 0$ is just given by Eq. (C.6).

Consider this last term, which is the simplest. Using Eq. (B.11) it holds

$$g^{(3)}(R, E) = -\frac{1}{4\pi} \frac{\cos(\alpha R)}{R} - \frac{1}{4\pi} \frac{\sqrt{\eta}}{\sqrt{\pi}} \sum_{n=0}^{\infty} \frac{(E/\eta)^n}{n!(2n-1)}.$$

Enforcing the identity

$$-\frac{1}{4\pi} \frac{\sqrt{\eta}}{\sqrt{\pi}} \sum_{n=0}^{\infty} \frac{(E/\eta)^n}{n!(2n-1)} = D_{00}^{(3)}(E) j_0(\alpha R) Y_{00}(\mathbf{R}) \qquad \text{for} \quad R \to 0$$

one obtains

$$D_{00}^{(3)}(E) = -\frac{\sqrt{\eta}}{2\pi} \sum_{n=0}^{\infty} \frac{(E/\eta)^n}{n!(2n-1)}. \quad (C.7)$$

We continue now with the elaboration of the contribution $g^{(1)}(\mathbf{k}, \mathbf{R}, E)$. First of all we notice the so called *Ewald identity* given by

$$\boxed{\sum_{\mathbf{t}_n} e^{i\mathbf{k}\cdot\mathbf{t}_n} \exp\left[-(\mathbf{R} - \mathbf{t}_n)^2 x^2\right] = \frac{1}{\Omega} \frac{\pi^{3/2}}{x^3} \sum_{\mathbf{h}_j} e^{i(\mathbf{k}+\mathbf{h}_j)\cdot\mathbf{R}} \exp\left[-\frac{k_j^2}{4x^2}\right]}, \quad (C.8)$$

where $\mathbf{k}_j$ stands for $\mathbf{k} + \mathbf{h}_j$ with $\mathbf{h}_j$ generic vector of the reciprocal lattice. In essence, the Ewald identity expresses a sum of (sharp/broad) Gaussians in the direct lattice as an appropriate sum of (broad/sharp) Gaussians in the reciprocal lattice. The Ewald identity can be easily demonstrated by inspection. In fact, consider the second member of Eq. (C.8), multiply by $\exp[-i(\mathbf{k} + \mathbf{h}_n) \cdot \mathbf{R}]$ and integrate in $d\mathbf{R}$ in the volume V of the crystal; one obtains $(V/\Omega)(\pi^{3/2}/x^3)\exp(-k_n^2/4x^2)$ where $V/\Omega = N$ is the number of primitive cells of volume Ω in the crystal. The same result is obtained if one considers the first member of Eq. (C.8), multiplies by $\exp[-i(\mathbf{k} + \mathbf{h}_n) \cdot \mathbf{R}]$ and integrates in $d\mathbf{R}$ using the relation

$$\int e^{-R^2 x^2} e^{-i\mathbf{k}_n \cdot \mathbf{R}} d\mathbf{R} = \frac{\pi^{3/2}}{x^3} \exp\left[-\frac{k_n^2}{4x^2}\right]$$

which expresses the well-known property that the Fourier transform of a Gaussian is still a Gaussian.

Using the Ewald identity, Eq. (C.4) takes the form.

$$g^{(1)}(\mathbf{k}, \mathbf{R}, E) = -\frac{1}{2\Omega} \sum_{\mathbf{k}_n} e^{i\mathbf{k}_n \cdot \mathbf{R}} \int_0^{\sqrt{\eta}/2} \frac{1}{x^3} \exp\left[\frac{E - k_n^2}{4x^2}\right] dx$$

$$= \frac{1}{\Omega} \sum_{\mathbf{k}_n} e^{i\mathbf{k}_n \cdot \mathbf{R}} \frac{1}{E - k_n^2} \exp\left[\frac{E - k_n^2}{\eta}\right], \tag{C.9}$$

where the integral in dx has been performed with the substitution $1/x^2 = t$ (and taking into account the path restriction in the integral representation of the Hankel function). With the standard expansion of plane waves $\exp(i\mathbf{k}_n \cdot \mathbf{R})$ into spherical harmonics and Bessel functions one obtains

$$g^{(1)}(\mathbf{k}, \mathbf{R}, E) = \frac{4\pi}{\Omega} \sum_{\mathbf{k}_n} \sum_{lm} i^l j_l(k_n R) Y_{lm}^*(\mathbf{k}_n) Y_{lm}(\mathbf{R}) \frac{\exp\left[(E - k_n^2)/\eta\right]}{(E - k_n^2)}$$

$$\equiv \sum_{lm} D_{lm}^{(1)}(\mathbf{k}, E) j_l(\alpha R) Y_{lm}(\mathbf{R}) \qquad |\mathbf{R}| < t_n \neq 0. \tag{C.10}$$

In the last line we have enforced the identity satisfied by the regular part of the greenian for any point in the indicated domain. Using the property $j_l(x) \to x^l/(2l + 1)!!$ for $x \to 0$, for small R it holds

$$\frac{j_l(k_n R)}{j_l(\alpha R)} \to \frac{k_n^l}{\alpha^l} = \frac{k_n^l}{\sqrt{E}^l}.$$

We thus obtain the expression

$$D_{lm}^{(1)}(\mathbf{k}, E) = \frac{4\pi}{\Omega} \frac{i^l}{\sqrt{E}^l} \sum_{\mathbf{k}_n} \frac{k_n^l}{E - k_n^2} Y_{lm}^*(\mathbf{k}_n) \exp\left[(E - k_n^2)/\eta\right]. \tag{C.11}$$

The elaboration of the contribution $g^{(2)}(\mathbf{k}, \mathbf{R}, E)$ and the corresponding coefficient $D_{lm}^{(2)}(\mathbf{k}, E)$ is rather straight. We notice that

$$e^{-(\mathbf{R}-\mathbf{t}_n)^2 x^2} = e^{-R^2 x^2 - t_n^2 x^2} e^{2x^2 \mathbf{t}_n \cdot \mathbf{R}}$$

$$\equiv e^{-R^2 x^2 - t_n^2 x^2} 4\pi \sum_{lm} i^l j_l(-i2x^2 R t_n) Y_{lm}^*(\mathbf{t}_n) Y_{lm}(\mathbf{R}),$$

where the usual expansion in spherical harmonics of $\exp[i\mathbf{k} \cdot \mathbf{R}]$ has been considered, with $\mathbf{k} = -i2x^2 \mathbf{t}_n$. The coefficients $D_{lm}^{(2)}(\mathbf{k}, E)$ are then obtained.

In summary, the reduced structure coefficients are given by the following expression

$$D_{lm}(\mathbf{k}, E) = D_{lm}^{(1)}(\mathbf{k}, E) + D_{lm}^{(2)}(\mathbf{k}, E) + D_{00}^{(3)}(E) \delta_{l0} \delta_{m0}, \tag{C.12a}$$

where

$$D_{lm}^{(1)}(\mathbf{k}, E) = \frac{4\pi}{\Omega} \frac{i^l}{\sqrt{E}^l} \sum_{\mathbf{k}_n} \frac{k_n^l}{E - k_n^2} Y_{lm}^*(\mathbf{k}_n) \exp\left[(E - k_n^2)/\eta\right] \tag{C.12b}$$

$$D_{lm}^{(2)}(\mathbf{k}, E) = \frac{-1}{\sqrt{\pi}} \frac{2^{l+1}}{\sqrt{E}^l} \sum_{\mathbf{t}_n \neq 0} e^{i\mathbf{k}\cdot\mathbf{t}_n} |\mathbf{t}_n|^l Y_{lm}^*(\mathbf{t}_n) \int_{\sqrt{\eta}/2}^{\infty} x^{2l} \exp\left[-x^2 t_n^2 + \frac{E}{4x^2}\right] dx$$

$$\text{(C.12c)}$$

$$D_{00}^{(3)}(E) = -\frac{\sqrt{\eta}}{2\pi} \sum_{n=0}^{\infty} \frac{(E/\eta)^n}{n!(2n-1)}. \tag{C.12d}$$

The arbitrary positive parameter η is chosen to optimize simultaneously the convergence of the series in direct and reciprocal spaces. The outcome of the Ewald procedure is the expression of the structure coefficients by means of absolutely convergent series in both real and reciprocal space.

Now that an efficient expression of the reduced structure coefficients has been achieved, we write Eq. (C.2) in extended form putting explicitly $\mathbf{R} = \mathbf{r} - \mathbf{r}_0$; for $|\mathbf{r} - \mathbf{r}_0| < t_n(\neq 0)$, one obtains

$$g(\mathbf{k}, \mathbf{r} - \mathbf{r}_0, E) = -\frac{1}{4\pi} \frac{\cos\alpha|\mathbf{r} - \mathbf{r}_0|}{|\mathbf{r} - \mathbf{r}_0|} + \sum_{lm} D_{lm}(\mathbf{k}, E) j_l(\alpha|\mathbf{r} - \mathbf{r}_0|) Y_{lm}(\mathbf{r} - \mathbf{r}_0). \tag{C.13}$$

For analytic applications, *it is needed to factorize the greenian into a sum of terms depending separately on $\mathbf{r}$ and $\mathbf{r}_0$*. For the singular part, this purpose is already achieved and contained in Eq. (C.1). For the regular part we avail of the identity

$$i^L j_L(\alpha|\mathbf{r}-\mathbf{r}_0|) Y_{LM}(\mathbf{r}-\mathbf{r}_0) = 4\pi \sum_{lm\,l'm'} i^{l-l'} C_{LM;lm,l'm'} j_l(\alpha r) j_{l'}(\alpha r_0) Y_{lm}(\mathbf{r}) Y_{l'm'}^*(\mathbf{r}_0),$$

where

$$C_{LM;lm,l'm'} = \int Y_{LM}(\mathbf{k}) Y_{lm}^*(\mathbf{k}) Y_{l'm'}(\mathbf{k}) dS_{\mathbf{k}}$$

are the Gaunt coefficients. The identity is proved by expanding $\exp[i\mathbf{k} \cdot (\mathbf{r} - \mathbf{r}_0)]$ in spherical harmonics and then comparing with the product of the expansion in spherical harmonics of $\exp(i\mathbf{k} \cdot \mathbf{r})$ and $\exp(-i\mathbf{k} \cdot \mathbf{r}_0)$ separately, and putting $|\mathbf{k}| = \alpha$.

The greenian thus takes the factorized form

$$g(\mathbf{k}, \mathbf{r} - \mathbf{r}_0, E) = \alpha \sum_{lm} j_l(\alpha r) Y_{lm}(\mathbf{r}) n_l(\alpha r_0) Y_{lm}^*(\mathbf{r}_0)$$

$$+ \sum_{lm\,l'm'} j_l(\alpha r) Y_{lm}(\mathbf{r}) \Gamma_{lm,l'm'}(\mathbf{k}, E) j_{l'}(\alpha r_0) Y_{l'm'}^*(\mathbf{r}_0), \tag{C.14}$$

where $r < r_0 < t_n(\neq 0)$, and

$$\Gamma_{lm,l'm'}(\mathbf{k}, E) = 4\pi i^{l-l'} \sum_{LM} i^{-L} D_{LM}(\mathbf{k}, E) C_{LM;lm,l'm'}$$

are the *structure coefficients* entering the secular equation of the KKR method.

Further Reading

Altmann, S. L. (1994). *Band theory of solids.* Oxford: Clarendon Press.

Bachelet, G. B., Hamann, D. R., & Schlüter, M. (1982). Pseudopotentials that work: From H to Pu. *Physical Review B, 26,* 4199.

Bassani, F., & Pastori Parravicini, G. (1975). *Electronic states and optical transitions in solids.* Oxford: Pergamon Press.

Blöchl, P. E. (1994). Projector augmented-wave method. *Physical Review B, 50,* 17953.

Callaway, J. (1991). *Quantum theory of the solid state.* New York: Academic Press.

Economou, E. N. (2006). *Green's functions in quantum physics.* Berlin: Springer.

Enkovaara, J., et al. (2010). Electronic structure calculations with GPAW: a real-space implementation of the projector augmented-wave method. *Journal of Physics: Condensed Matter, 22,* 253202.

Eschrig, H. (1989). *Optimized LCAO method and the electronic structure of extended systems.* Berlin: Springer.

Fletcher, G. C. (1971). *Electron band theory of solids.* Amsterdam: North-Holland.

Giannozzi, P., Grosso, G., & Pastori Parravicini, G. (1990). Theory of electronic states in lattices and superlattices. *Rivista del Nuovo Cimento, 13,* N.3.

Goringe, C. M., Bowler, D. R., & Hernández, E. (1997). Tight-binding modelling of materials. *Reports on Progress in Physics, 60,* 1447.

Hamann, D. R., Schlüter, M., & Chiang, C. (1979). Norm-conserving pseudopotential. *Physical Review Letters, 43,* 1494.

Harrison, W. A. (1980). *Electronic structure and the properties of solids.* San Francisco: Freeman.

Heine, V. (1980). Electronic structure from the point of view of the local atomic environment; Bullett, D. W. The renaissance and quantitative development of the tight-binding method; Haydock, R. The recursive solution of the Schrödinger equation; Kelly, M. J. Applications of the recursion method to the electronic structure from an atomic point of view. In *Solid state physics* (Vol. 35). New York: Academic Press.

Kaxiras, E. (2003). *Atomic and electronic structure of solids.* Cambridge: Cambridge University Press.

Kerker, G. P. (1980). Non-singular atomic pseudopotentials for solid state applications. *Journal of Physics C: Solid State Physics, 13,* L189.

Lanczos, C. (1988). *Applied analysis.* New York: Dover Publications.

Loucks, T. L. (1967). *Augmented plane wave method: A guide to perform electronic structure calculations.* New York: Benjamin.

Martin, R. M. (2004). *Electronic structure. Basic theory and practical methods.* Cambridge: Cambridge University Press.

Nemoshkalenko, V. V., & Antonov, V. N. (1998). *Computational methods in solid state physics.* Amsterdam: Gordon and Breach.

Papaconstantopoulos, D. (2013). *Handbook of the band structure of elemental solids.* New York: Springer.

Singh, D. J., & Nordström, L. (2006). *Planewaves, pseudopotentials, and the LAPW method.* New York: Springer.

Singleton, J. (2001). *Band theory and electronic properties of solids.* Oxford: Oxford University Press.

Skriever, H. L. (1984). *The LMTO method.* Berlin: Springer.

Slater, J. C. (1974). *Quantum theory of molecules and solids* (Vols. 1–4). New York: McGraw-Hill.

Sutton, A. (1993). *Electronic structure of materials.* Oxford: Clarendon Press.

Troullier, N., & Martins, J. L. (1991). Efficient pseudopotentials for plane-wave calculations. *Physical Review B, 43,* 1993.

Vanderbilt, D. (1990). Soft self-consistent pseudopotentials in a generalized eigenvalue formalism. *Physical Review B, 41,* 7892.

6 Electronic Properties of Selected Crystals

Chapter Outline head

In this chapter we give a brief survey of the electronic structure of some crystals of particular interest from a fundamental or technological point of view. The methods of electronic band structure calculations, in conjunction with the advances in the density functional formalism and many-body techniques, have made it possible in principle (and often also in practice) to describe the band structure, charge density, and cohesive energy of crystals, starting from the very knowledge of their chemical composition. However, simple heuristic schemes are still of great value, because of their capability in describing physical properties on a more intuitive basis.

In this chapter we discuss the electronic band structure and ground-state properties of some selected materials; the focus is not on the properties of single specific crystals, but rather on the trends in similar compounds. We begin with the description of rare-gas solids, which are large gap insulators formed by weakly interacting closed-shell neutral atoms. We then describe ionic crystals, constituted by strongly interacting closed-shell ions. In the discussion of crystals made up by open-shell units, we consider typical examples of covalent semiconductors and the metallic bond. The last subject concerns some electronic features of carbon allotropes, with particular attention on the Dirac points in the band structure of graphene. In spite of the narrowness of the topics, included mostly at a semi-quantitative level, this chapter should nevertheless convey some of the ideas underlying the material science investigations in solids.

Solid State Physics, Second Edition. http://dx.doi.org/10.1016/B978-0-12-385030-0.00006-2

6.1   Band Structure and Cohesive Energy of Rare-Gas Solids

6.1.1   General Features of Band Structure of Rare-Gas Solids

Rare-gas solids (Ne, Ar, Kr, Xe) are the simplest crystals made up of closed-shell atoms (helium is a quantum crystal), and constitute a class of large gap insulators. Neon atom has the closed-shell electronic configuration $1s^2 2s^2 2p^6$. The electronic configuration of argon is [Ne] $3s^2 3p^6$; the configuration of krypton is [Ar] $3d^{10} 4s^2 4p^6$; that of xenon is [Kr] $4d^{10} 5s^2 5p^6$. Rare-gas solids crystallize in the face-centered cubic lattice (as described in Chapter 2), with one atom per primitive cell.

The band structure of rare-gas solids is rather simple, at least in its essential lines. The inner core states give rise to (almost) dispersionless fully occupied bands. The occupied p-states of higher energy give rise to fully occupied valence bands of p-like nature; these bands present the maximum at the point Γ of the Brillouin zone, and bend downward as $\mathbf{k}$ increases out of the center. This general trend is expected qualitatively from a tight-binding description of localized p-states in fcc Bravais lattices (as discussed in Section 5.2.3). The conduction bands are almost free-electron like; in fact electrons can move almost freely in the crystal, with exclusion of the (small) fraction of volume, where core and valence orbitals extend. Solid neon with an energy gap of ≈ 21.5 eV is the crystal with the highest energy gap.

We consider, as an exemplification, the band structure of solid argon. In the ground state of the argon atom, the orbital energies are $E_{1s} = -3205.9$ eV, $E_{2s} = -326.3$ eV, $E_{2p} = -249.34$ eV, $E_{3s} = -29.24$ eV, and $E_{3p} = -15.82$ eV. The inner core states $1s, 2s, 2p_x, 2p_y, 2p_z$ of the argon atom give rise to (fully occupied) crystal core bands. The valence bands, originated from the $3s$ and $3p$ orbitals, and the conduction bands of solid argon can be described, for instance, by the orthogonalized plane wave method, where plane waves are orthogonalized to the five Bloch sums formed with $1s, 2s, 2p_x, 2p_y, 2p_z$ orbitals of argon. Using a Gaussian representation of atomic orbitals, the matrix elements appearing in the OPW method can be evaluated analytically, and accurate calculations can be performed. In the calculations reported in Figure 6.1 the maximum number of 259 OPWs has been used; at Γ this corresponds to consider 16 shells of reciprocal lattice vectors, up to the waves of type $(2\pi/a)(4, 4, 2)$ and $(2\pi/a)(6, 0, 0)$.

The general trend of the valence bands of Figure 6.1 can be understood within the tight-binding framework, because of the strong localization of the occupied orbitals on the scale of the lattice parameter. The general trend of the conduction bands can be understood starting from the empty lattice analysis for fcc structures (Figure 5.5); the removal of the empty lattice degeneracies is in qualitative agreement with what expected from the presence of a smooth repulsive pseudopotential.

From the correlated energy bands of Figure 6.1, we see that the maximum of the valence bands of argon is the threefold degenerate state $E(\Gamma_4^-) = -13.7$ eV. The width of the valence bands is $E(\Gamma_4^-) - E(L_2^-) = 1.7$ eV. The minimum of the conduction bands is the total symmetric state Γ_1^+ of energy $E(\Gamma_1^+) = 0.9$ eV. The calculated energy gap is $E_G = E(\Gamma_1^+) - E(\Gamma_4^-) = 14.6$ eV. All the above calculated values are in rather good agreement with the experimental band structure parameters of Ar reported in Table 6.1. The table also reports the relevant band structure parameters of

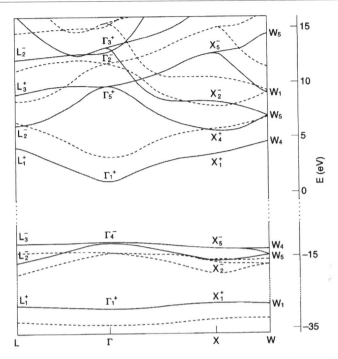

Figure 6.1 Energy bands of solid argon with inclusion of correlation effects (solid line). For comparison the energy bands in the Hartree-Fock approximation are also reported (dashed lines); correlation effects shift (almost rigidly) the Hartree-Fock conduction bands to lower energies, and the Hartree-Fock valence bands to higher energies [reprinted with permission from S. Baroni, G. Grosso and G. Pastori Parravicini, Phys. Rev. B *29*, 2891 (1984); copyright 1984 by the American Physical Society].

Table 6.1 Relevant parameters of the band structure of rare-gas solids as obtained from experiments. The energies are in eV [from N. Schwentner, F.-J. Himpsel, V. Saile, M. Skibowski, W. Steinmann, and E. E. Koch, Phys. Rev. Lett. *34*, 528 (1975); copyright 1975 by the American Physical Society].

	Ne	**Ar**	**Kr**	**Xe**
Band-gap energy E_G	21.7	14.2	11.6	9.3
Top valence band $E(\Gamma_4^-)$	−20.3	−13.8	−11.9	−9.8
Bottom conduction band $E(\Gamma_1^+)$	1.4	0.4	−0.3	−0.5
Valence band width	1.3	1.7	2.3	3.0
$\Delta_{SO}^{(gas)}$	0.14	0.22	0.67	1.31
$\Delta_{SO}^{(solid)}$	≈0.1	0.2	0.64	1.3

Ne, Kr, and Xe; for completeness the spin-orbit splittings of the more external occupied atomic p-states and of the corresponding valence bands at $\mathbf{k} = 0$ are also given.

From the data of Table 6.1, it can be noticed that there is a systematic trend of the relevant band structure parameters passing from neon to xenon. The trend can be

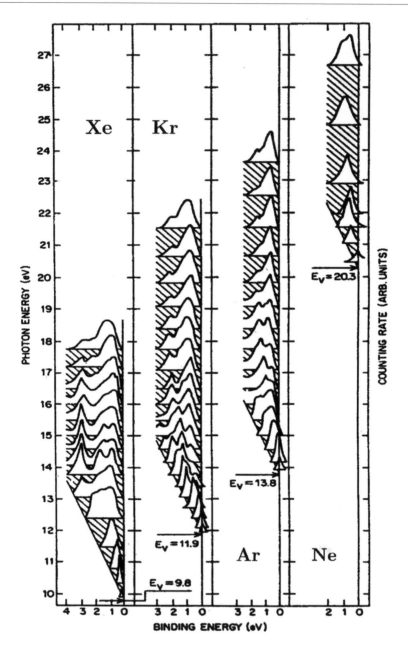

Figure 6.2 Energy distribution curves of emitted photoelectrons for solid Ne, Ar, Kr, and Xe, at various photon energies. The zero-count line for each individual curve is shifted upwards proportional to the exciting photon energy. E_V represents the vacuum energy measured from the top of the valence band [reprinted with permission from N. Schwentner, F.-J. Himpsel, V. Saile, M. Skibowski, W. Steinmann, and E. E. Koch, *Phys. Rev. Lett.* *34*, 528 (1975); copyright 1975 by the American Physical Society].

best noticed also from Figure 6.2, which reports the experimental energy distribution curves for the photoelectrons emitted from the valence bands of solid Ne, Ar, Kr, and Xe. Photoemission measurements basically consist in shining monochromatic photons of energy $\hbar\omega$ on a target crystal and measuring the kinetic energy of the electrons escaped from the surface.

The analysis of the energy distribution curves of the emitted photoelectrons permits us to infer relevant information on the band structure. In fact the energy distribution curves are expected to be reminiscent of the major features and edges of the actual density-of-states of occupied bands, shifted by $\hbar\omega$. This is shown for rare-gas solids in Figure 6.2 for different energies of the incident photons; when the energy of the incident photon is high enough to excite electrons from the bottom of the valence band, the widths of the energy distribution curves are independent of $\hbar\omega$ and provide the total width of the valence bands; furthermore, with appropriate extrapolation, the position of the vacuum level E_V with respect to the top of the valence band can be obtained with high accuracy.

6.1.2   Cohesive Energy of Rare-Gas Solids

The study of the ground-state energy of solids provides several properties and in particular the equilibrium lattice constant, cohesive energy, compressibility, and bulk modulus. In principle, the density functional theory (appropriately corrected to embody van der Waals interactions, when needed) can be applied to investigate the electron charge density and other ground-state properties of materials [see for instance the special issue on van der Waals interactions in advanced materials, J. Phys.: Condens. Matter *24*, 420201 (2012) edited by P. Hyldgaard and T. S. Rahman]. However, we wish here to confine ourselves to intuitive semi-quantitative considerations.

In rare-gas solids, the total energy can be expressed reasonably well in terms of pair interaction (or two-body interaction). For the present discussion we make the standard approach to consider only an effective two-body interaction (although the importance of three or multi-body interactions is well established, for instance in the description of the stability of the fcc phase with respect to the hcp phase).

The interaction potential $U(R)$ between two isolated atoms can be calculated, in principle, starting from the quantum mechanical approach. It is interesting to notice that a pure Hartree-Fock calculation of two isolated rare-gas atoms at distance R predicts repulsion between the two atoms at any R. Thus rare-gas solids would not be bound in the Hartree-Fock approximation.

A widely used semi-empirical description approximates the interatomic potential between two neutral closed-shell atoms, at distance R, with the *Lennard-Jones potential* of the type

$$U(R) = 4\varepsilon \left[\left(\frac{\sigma_0}{R} \right)^{12} - \left(\frac{\sigma_0}{R} \right)^6 \right].$$
(6.1)

The first term is a crude way to represent classically the fact that filled atomic shells constitute an almost impenetrable core. The inverse 12 power law has been chosen for analytic simplicity (and for historical reasons); a high exponent in the repulsive term simulates its short-range nature. The second term in Eq. (6.1) corresponds to the fluctuating dipole-dipole interaction, which is attractive; the inverse sixth power law is the leading term for the interaction of a dipole with a polarizable neutral atom (van der Waals interaction). The two parameters $-\varepsilon$ and σ_0 represent, respectively, the minimum value of $U(R)$ and the value at which $U(R)$ vanishes.

The Lennard-Jones potential (6.1) for a pair of rare-gas atoms at distance R can also be re-cast in the equivalent form

$$U(R) = \varepsilon \left[\left(\frac{\sigma}{R} \right)^{12} - 2 \left(\frac{\sigma}{R} \right)^{6} \right] \quad \text{with} \quad \sigma = 2^{1/6} \sigma_0; \tag{6.2}$$

the parameters ε and σ represent the depth and the position of the potential well. The expression (6.2) of the Lennard-Jones potential is slightly more convenient when comparing the equilibrium distance of the atoms in the solid with the equilibrium distance σ of the atoms composing the pairs, and for this reason it is adopted in the following. The quantities ε and σ are often considered as two phenomenological parameters, to be determined via some appropriate elaboration of experimental data. As an example, in Table 6.2 we report a set of parameters for rare-gas atoms, inferred from molecular dynamics simulation of their vibrational and structural properties.

Once the pair interaction is known (either semi-empirically or from first principles), we can obtain the total energy of the solid summing up the interactions for all pairs of atoms composing the crystal (three-body corrections are assumed negligible; zero point

Table 6.2 Parameters of the Lennard-Jones potential $U(R) = \varepsilon[(\sigma/R)^{12} - 2(\sigma/R)^6]$ for a pair of rare-gas atoms, as provided by S. Gonçalves and H. Bonadeo, Phys. Rev. B *46*, 10738 (1992). In the table, we also report the static properties of rare-gas solids (nearest neighbor distance, cohesive energy, and bulk modulus) calculated from the given set of parameters ε and σ. The experimental values of the nearest neighbor distance are taken from R. W. G. Wyckoff "Crystal Structures" (Interscience, New York, 1963). The experimental binding energies are quoted by E. R. Dobbs and G. O. Jones, Rep. Progr. Phys. *20*, 516 (1957). The experimental values of the bulk modulus are quoted by P. Korpium and E. Lüscher in "Rare Gas Solids" (edited by M. L. Klein and J. A. Venables) vol. II, p. 729 (Academic Press, London, 1977).

Pair of rare-gas atoms				Rare-gas solids			
		Nearest neighbor distance (Å)		Binding energy (eV/atom)		Bulk modulus (kbar)	
σ(Å)	ε(eV)	Calc. Eq. (6.6)	Exp.	Calc. Eq. (6.7)	Exp.	Calc. Eq. (6.10)	Exp.
Ne	3.25	0.0024	3.16	3.13	0.021	0.020	11.9	11.1
Ar	3.87	0.0098	3.76	3.72	0.084	0.080	28.8	28.6
Kr	4.11	0.0135	3.99	4.05	0.116	0.112	33.1	34.1
Xe	4.46	0.0185	4.33	4.38	0.159	0.166	35.5	37.9

vibrational energy is also neglected). We have for the total energy $U_S(R)$ of the solid

$$U_S(R) = \frac{1}{2} N\varepsilon \left[\sum_{j(\neq i)} \left(\frac{\sigma}{R_{ij}} \right)^{12} - 2 \sum_{j(\neq i)} \left(\frac{\sigma}{R_{ij}} \right)^{6} \right],$$ (6.3)

where N is the total number of atoms, R_{ij} is the distance between two atoms i and j, and the factor 1/2 appears in order not to include twice each pair of atoms. Because of the translational symmetry, the sum over j in the right-hand side of Eq. (6.3) is independent of the chosen reference atom with label i; the i-th atom can thus be taken at the origin.

Let us express R_{ij} in the product form $p_{ij}R$, where R is the distance between nearest neighbors and p_{ij} is dimensionless. For the fcc structure, we have 12 first neighbors at distance R ($p_{ij} = 1$), 6 second neighbors at distance $R\sqrt{2}$ ($p_{ij} = \sqrt{2}$), 24 third neighbors at distance $R\sqrt{3}$ ($p_{ij} = \sqrt{3}$), etc. We consider thus the lattice sums

$$A_{12} = \sum_{j(\neq i)} \left(\frac{1}{p_{ij}} \right)^{12} = 12 + \frac{6}{\sqrt{2}^{12}} + \frac{24}{\sqrt{3}^{12}} + \cdots = 12.132,$$ (6.4a)

$$A_6 = \sum_{j(\neq i)} \left(\frac{1}{p_{ij}} \right)^{6} = 12 + \frac{6}{\sqrt{2}^{6}} + \frac{24}{\sqrt{3}^{6}} + \cdots = 14.454.$$ (6.4b)

The lattice sums (6.4) are rapidly convergent and can be obtained by direct summation in real space; however, whenever desired or needed, the acceleration of the convergence of the lattice sums can be worked out with the Ewald method (see Problem 4). Using the coefficients A_{12} and A_6, we can express the crystal energy (6.3) in the form

$$U_S(R) = \frac{1}{2} N\varepsilon \left[A_{12} \left(\frac{\sigma}{R} \right)^{12} - 2A_6 \left(\frac{\sigma}{R} \right)^{6} \right].$$ (6.5)

The behavior of $U_S(R)$ is given in Figure 6.3. From Eq. (6.5), we can now easily obtain the equilibrium lattice constant, the cohesive energy, and the bulk modulus.

The *equilibrium nearest neighbor constant* R_0 is determined by the condition that the derivative of $U_S(R)$ vanishes; we have

$$\boxed{R_0 = \left(\frac{A_{12}}{A_6} \right)^{1/6} \cdot \sigma = 0.971\,\sigma}.$$ (6.6)

The *equilibrium cohesive energy* (disregarding vibrational effects) is given by

$$\boxed{U_S(R_0) = -\frac{1}{2} N\varepsilon \frac{A_6^2}{A_{12}} = -8.61\, N\varepsilon}.$$ (6.7)

Equation (6.6) shows that the ratio R_0/σ is smaller than 1 (since $A_{12} < A_6$) and is independent of the material (at least within the assumed model). Equation (6.7) shows that the cohesive energy per atom is proportional to the binding energy ε, and again the proportionality constant is independent of the material.

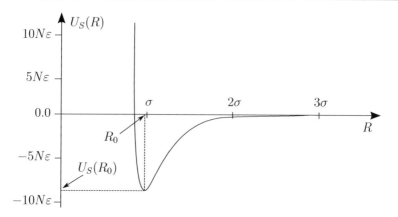

Figure 6.3 Behavior of the crystal energy $U_S(R)$ for a rare-gas solid as a function of the nearest neighbor distance R. The minimum of the curve provides the cohesive energy $U_S(R_0)$ and the equilibrium position R_0. The slope is related to the applied pressure, and the curvature near R_0 is related to the bulk modulus.

We pass now to consider the *compressibility and bulk modulus* in our simple model. At $T = 0$ the pressure p is given by $p = -dU_S/dV$. The isothermal compressibility κ_T is defined as $\kappa_T = -(1/V)(dV/dp)$. The inverse of the compressibility is called bulk modulus B. From the definition of pressure and compressibility, we have for the bulk modulus at zero temperature

$$B_0 = \left(-V\frac{dp}{dV}\right)_{V=V_0} = \left(V\frac{d^2U_S}{dV^2}\right)_{V=V_0}, \tag{6.8}$$

where V_0 is the equilibrium volume of the crystal. In essence, the bulk modulus is a measure of the stiffness (or hardness) of the crystal.

In face-centered cubic structures, the relation between the volume V of the solid and the nearest neighbor distance R is

$$V = \frac{1}{\sqrt{2}}NR^3 \implies \frac{dV}{dR} = \frac{3}{\sqrt{2}}NR^2; \quad \frac{dR}{dV} = \frac{\sqrt{2}}{3NR^2}$$

(in fact the volume V of a fcc structure with N atoms is $V = Na^3/4$, and the nearest neighbor distance is $R = a\sqrt{2}/2$). The derivative of the total energy with respect to the volume becomes

$$\frac{dU_S}{dV} = \frac{dU_S}{dR}\frac{dR}{dV}; \quad \frac{d^2U_S}{dV^2} = \frac{d^2U_S}{dR^2}\left(\frac{dR}{dV}\right)^2 + \frac{dU_S}{dR}\frac{d^2R}{dV^2}.$$

At equilibrium $dU_S/dR = 0$, and the bulk modulus (6.8) becomes

$$B_0 = V_0\left(\frac{dR}{dV}\right)^2_{V=V_0}\left(\frac{d^2U_S}{dR^2}\right)_{R=R_0}. \tag{6.9}$$

Using the given relation for dR/dV, and Eq. (6.5) one has

$$B_0 = \frac{\sqrt{2}}{9} \frac{1}{N R_0} \left(\frac{d^2 U_S}{dR^2} \right)_{R=R_0} = \frac{2\sqrt{2}}{3} \frac{\varepsilon}{\sigma^3} \frac{\sigma^9}{R_0^9} \left[13 A_{12} \frac{\sigma^6}{R_0^6} - 7 A_6 \right];$$

with the help of Eq. (6.6) it follows

$$\boxed{B_0 = 4\sqrt{2} \frac{\varepsilon}{\sigma^3} \left(\frac{A_6}{A_{12}} \right)^{3/2} \cdot A_6 = 106.3 \frac{\varepsilon}{\sigma^3}.} \tag{6.10}$$

For the evaluation of the bulk modulus (6.10) with the data of Table 6.2, where ε is expressed in eV and σ in Å, it is useful to remember the conversion factors

$$\frac{eV}{Å^3} = \frac{1.60219 \cdot 10^{-19}}{10^{-30}} \frac{J}{m^3} = 1.60219 \cdot 10^{11} \frac{N}{m^2} = 1.60219 \text{ Mbar.}$$

The unit of pressure *Newton per square meter* is referred to as *Pascal*. The unit of pressure *bar* equals exactly 100 kPa [one *bar* corresponds approximately to the atmospheric pressure at sea level]. In the following we will adopt frequently megabar (Mbar) or kilobar (kbar) units, depending on the class of materials under investigation.

The values of the equilibrium distance, cohesive energy, and bulk modulus for rare-gas solids, calculated using Eqs. (6.6), (6.7), and (6.10) are also reported in Table 6.2; the reasonable description of these and other properties with a simple two parameter model, explains the popularity of the Lennard-Jones scheme in the investigation of weakly bound van der Waals crystals.

6.2 Electronic Properties of Ionic Crystals

6.2.1 Introductory Remarks and Madelung Constant

Crystals that can be pictured as formed by positive and negative ions are referred to as *ionic crystals*. Some properties of these solids can be understood starting from the extreme approximation of a lattice made of point wise charges (*point-ion lattice*); the influence of the actual internal constitution of ions can then be introduced in the model, and overall trends in electronic properties can often be understood at a reasonable elementary level.

An important quantity in the study of ionic crystals is the Coulomb interaction energy (Madelung energy) between ions of the crystals in the point-ion approximation. As we shall see, this quantity accounts for most of (i) cohesive energy of the crystal, (ii) stability of the structure, (iii) energy gap of the material. We thus begin the study of ionic crystals with the evaluation of the potential energy of a given reference ion (say a negative ion taken at the origin) due to the Coulomb interaction with all the other positive and negative ions of the crystal. It is usual to express this potential energy in the form

$$V(0) = -\alpha_M \frac{e^2}{R}, \tag{6.11}$$

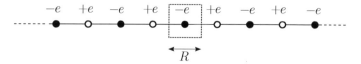

Figure 6.4 Schematic representation of a one-dimensional ionic crystal in the point-ion approximation, for the calculation of the Madelung constant. The nearest neighbor distance is denoted by R. The ion at the origin has been surrounded by a box.

where R is the nearest neighbor distance, α_M is the Madelung constant discussed below, $\pm e$ are the point charges at the lattice sites. The Madelung constant is a dimensionless constant, characteristic of the geometrical crystal structure under consideration.

The evaluation of the Madelung constant requires a special care, because of the long-range character of the Coulomb interaction. To give an idea of the slow convergence of the series that are encountered in the direct evaluation of the Madelung constant, consider the one-dimensional model of ionic crystal indicated in Figure 6.4. In this case, the potential energy of the ion at the origin reads

$$V(0) = -\frac{e^2}{R} 2 \left[\frac{1}{1} - \frac{1}{2} + \frac{1}{3} - \frac{1}{4} + \frac{1}{5} - \cdots \right] = -\frac{e^2}{R} 2 \ln 2.$$

The series in the square brackets is a slowly, conditionally convergent, series; in this case, however, the sum can be performed analytically and gives $\ln(1 + x)$ with $x = 1$. The Madelung constant for a strictly one-dimensional model is thus $\alpha_M = 2 \ln 2$.

In actual three-dimensional crystals, the evaluation of the Madelung constant is faced with the problem of summing up numerically series, which are slowly and conditionally convergent. Let us consider for instance the NaCl structure. Suppose we take a Cl^- anion at the origin; we find 6 ions Na^+ at distance R (where R denotes the nearest neighbor distance), 12 ions Cl^- at $R\sqrt{2}$, 8 ions Na^+ at $R\sqrt{3}$, 6 ions Cl^- at distance $R\sqrt{4}$, etc.; the ions up to the distance $R\sqrt{12}$ are listed in Table 6.3. The Coulomb energy of this ion in the field of all the other ions is thus

$$V(0) = -\frac{e^2}{R} \left[\frac{6}{\sqrt{1}} - \frac{12}{\sqrt{2}} + \frac{8}{\sqrt{3}} - \frac{6}{\sqrt{4}} + \frac{24}{\sqrt{5}} - \frac{24}{\sqrt{6}} - \frac{12}{\sqrt{8}} + \cdots \right]$$

$$= -\frac{e^2}{R} [6.000 - 8.486 + 4.619 - 3.000 + 10.733 - 9.798 - 4.243 + \cdots].$$

$$(6.12)$$

By inspection of the terms indicated in the square brackets in Eq. (6.12), it is not possible to infer when the direct summation converges, even approximately, to the value $\alpha_M = 1.7476$ of the NaCl structure (see Table 6.4). The most used procedures to evaluate lattice sums of the type shown in Eq. (6.12) are the Evjen and the Ewald methods.

The Evjen Method for the Evaluation of the Madelung Constant

The Evjen method [H. M. Evjen, Phys. Rev. *39*, 675 (1932)] for the evaluation of the Madelung constant considers the crystal as formed by neutral (or approximately

Table 6.3 List of neighbors in NaCl structure with Cl^- at the origin; all ions up to the distance $R\sqrt{12}$ are indicated. In the last column, for each shell of neighbors, we indicate the charge *inside* the cube of edge $4R$ surrounding the reference ion; including the ion at the origin, this cube contains 32 positive and 32 negative electronic charges.

Reference atom at the origin Cl^-			
Number and type of ions	*Representative position*	*Distance from the origin*	*Charge within the cube of edge $4R$*
$6Na^+$	$R(1, 0, 0)$	$R\sqrt{1}$	$+6e$
$12Cl^-$	$R(1, 1, 0)$	$R\sqrt{2}$	$-12e$
$8Na^+$	$R(1, 1, 1)$	$R\sqrt{3}$	$+8e$
$6Cl^-$	$R(2, 0, 0)$	$R\sqrt{4}$	$-3e$
$24Na^+$	$R(2, 1, 0)$	$R\sqrt{5}$	$+12e$
$24Cl^-$	$R(2, 1, 1)$	$R\sqrt{6}$	$-12e$
$12Cl^-$	$R(2, 2, 0)$	$R\sqrt{8}$	$-3e$
$24Na^+$	$R(2, 2, 1)$	$R\sqrt{9}$	$+6e$
$6Na^+$	$R(3, 0, 0)$	$R\sqrt{9}$	0
$24Cl^-$	$R(3, 1, 0)$	$R\sqrt{10}$	0
$24Na^+$	$R(3, 1, 1)$	$R\sqrt{11}$	0
$8Cl^-$	$R(2, 2, 2)$	$R\sqrt{12}$	$-e$

Table 6.4 Madelung constant for some cubic structures.

Crystal structure	Madelung constant
Cesium chloride	$\alpha_M = 1.7627$
Sodium chloride	$\alpha_M = 1.7476$
Zincblende	$\alpha_M = 1.6381$

neutral) groups of ions (the Evjen cell). The electrostatic energy of a finite cell is evaluated by direct summation. As Evjen cells of increasing size are considered, one expects a rapidly convergent series, due to the short-range nature of the potential of groups of neutral atoms. A similar reasoning has been adopted by Harrison for the direct evaluation of the Madelung sums within a sphere of large radius [W. A. Harrison, Phys. Rev. B *73*, 212103 (2006)].

To give an exemplification of the Evjen method, consider the NaCl structure, a reference ion (for instance Cl^- taken at the origin), and a cube of edge $4R$ surrounding the reference ion. In the Evjen method, all the ions *inside* the cube are considered as they stand and all the ions *outside* the cube are neglected. Ions at the *border* of the cube are assigned an appropriate fraction; this fraction is 1/2 for an ion at the surface, 1/4 for an ion at the edge, and 1/8 for an ion at the corner of the cube. In the last column of Table 6.3, the charge within the cube of edge $4R$, contributed by each shell of neighbors, is reported; from it, we see that the potential energy of the point ion at the origin is given by

$$V(0) = -\frac{e^2}{R}\left[\frac{6}{1} - \frac{12}{\sqrt{2}} + \frac{8}{\sqrt{3}} - \frac{3}{\sqrt{4}} + \frac{12}{\sqrt{5}} - \frac{12}{\sqrt{6}} - \frac{3}{\sqrt{8}} + \frac{6}{\sqrt{9}} - \frac{1}{\sqrt{12}}\right]$$

$$= -1.751\frac{e^2}{R},$$

a result already very near to the exact one $-\alpha_M e^2/R$ with $\alpha_M = 1.7476$.

The Ewald Method for the Evaluation of the Madelung Constant

The Ewald technique [P. P. Ewald, Ann. Physik *64*, 253 (1921)] provides a very elegant procedure to transform a slow conditionally convergent lattice series into the sum of two fast absolutely convergent series in real and reciprocal space. The Ewald method essentially consists in using and balancing appropriately both real space and reciprocal space summations; this is achieved by splitting the function to be summed into a short-range part, to be computed as it stands in real space, and a long-range part to be represented and computed in reciprocal space. For the evaluation of the Madelung constant, the basic idea of Ewald is to replace point charges with Gaussian charge distributions, most conveniently handled in reciprocal space; the difference between the *initial point charges* and the *dress of Gaussian distributions* can be recognized as a system of *completely screened point charges*, which produces a short-range potential conveniently evaluated in real space. To implement the method, we need a few elementary facts concerning Gaussian functions, and the intimately related error function.

We remind the usual definition of the *error function*

$$\text{erf}(x) = \frac{2}{\sqrt{\pi}} \int_0^x e^{-t^2} \, dt. \tag{6.13}$$

The error function increases monotonically from 0 to 1 as x varies from zero to infinity. We notice that

$$\text{erf}(x) = \frac{2}{\sqrt{\pi}}\left(x - \frac{x^3}{3} + \cdots\right) \quad \text{for} \quad x \ll 1;$$

the derivative of the error function is the Gaussian function

$$\frac{d}{dx}\text{erf}(x) = \frac{2}{\sqrt{\pi}}e^{-x^2}.$$

It is also useful to consider the complementary error function $\text{erfc}(x)$ defined from

$$\boxed{\text{erf}(x) + \text{erfc}(x) = 1}. \tag{6.14}$$

The reason for introducing the error function is the following. Consider a Gaussian charge distribution $\rho(\mathbf{r})$, normalized to $+e$, and spherically symmetric around the origin in real space

$$\rho(\mathbf{r}) \equiv (+e)\left(\frac{\eta}{\pi}\right)^{3/2} e^{-\eta r^2}. \tag{6.15a}$$

The electrostatic potential $\phi(\mathbf{r})$, corresponding to the Gaussian charge density of Eq. (6.15a) is simply

$$\phi(\mathbf{r}) = +\frac{e}{r}\text{erf}(\sqrt{\eta}\, r). \tag{6.15b}$$

In fact the operator $-\nabla^2$ applied to the electrostatic potential (6.15b) gives

$$-\nabla^2\left[\frac{e}{r}\text{erf}(\sqrt{\eta}\, r)\right] = -\frac{e}{r^2}\frac{\partial}{\partial r}\left[r^2\frac{\partial}{\partial r}\frac{\text{erf}(\sqrt{\eta}\, r)}{r}\right] = 4\pi e\left(\frac{\eta}{\pi}\right)^{3/2}e^{-\eta r^2}. \tag{6.15c}$$

The first passage in Eq. (6.15c) is just the action of the operator ∇^2 on a spherically symmetric function; the second passage is obtained performing explicitly the derivatives with respect to r. Equation (6.15c) shows that the charge density (6.15a) and the electrostatic potential (6.15b) indeed satisfy the Poisson equation $-\nabla^2\phi(\mathbf{r}) = 4\pi\rho(\mathbf{r})$.

Notice that in the particular case of $\eta \to \infty$, the Gaussian charge density (6.15a) represents the point charge $(+e)\delta(\mathbf{r})$, and the potential (6.15b) gives just the Coulomb potential e/r, which is singular at the origin. For a Gaussian charge density, the potential $\phi(r) = (e/r)\text{erf}(\sqrt{\eta}\, r)$ is regular at the origin, where it takes the value

$$\phi(0) = 2e\left(\frac{\eta}{\pi}\right)^{1/2}. \tag{6.16}$$

We now remember that the Fourier transform of a Gaussian function is still a Gaussian function; specifically, we have the relation

$$\left(\frac{\eta}{\pi}\right)^{3/2}\int_{-\infty}^{+\infty}e^{-\eta r^2}\,e^{i\mathbf{k}\cdot\mathbf{r}}\,d\mathbf{r} = e^{-k^2/4\eta}, \tag{6.17a}$$

where the integral in $d\mathbf{r} \equiv dx\,dy\,dz$ extends in the whole real space. Another useful relation is the following Fourier transform

$$\int_{-\infty}^{+\infty}\frac{e^2}{r}\text{erf}(\sqrt{\eta}\, r)e^{i\mathbf{k}\cdot\mathbf{r}}\,d\mathbf{r} = \frac{4\pi e^2}{k^2}e^{-k^2/4\eta} \quad (\mathbf{k} \neq 0). \tag{6.17b}$$

It is seen by inspection that the derivation with respect to η of both members of Eq. (6.17b) gives back Eq. (6.17a). In the case the variable $\mathbf{r}$ is shifted to $\mathbf{r} - \mathbf{R}$, Eq. (6.17b) gives

$$\left\langle\frac{e^2}{|\mathbf{r} - \mathbf{R}|}\text{erf}(\sqrt{\eta}\,|\mathbf{r} - \mathbf{R}|)\,\Big|\,e^{i\mathbf{k}\cdot\mathbf{r}}\right\rangle = e^{i\mathbf{k}\cdot\mathbf{R}}\frac{4\pi e^2}{k^2}e^{-k^2/4\eta} \quad (\mathbf{k} \neq 0), \tag{6.17c}$$

where the bra-ket notation has been used for later convenience.

In the following we use the simple trick (suggested by Eq. (6.14)) to write a point charge potential $-e^2/r$ (or similarly $+e^2/r$) in the form

$$-\frac{e^2}{r} \equiv -\frac{e^2}{r}\text{erf}(\sqrt{\eta}\, r) - \frac{e^2}{r}\text{erfc}(\sqrt{\eta}\, r) \tag{6.18}$$

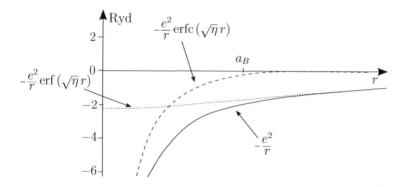

Figure 6.5 Decomposition of the potential $-e^2/r$ (singular at the origin and of long-range nature) into a contribution $-(e^2/r)\text{erf}(\sqrt{\eta}\,r)$ (regular at the origin and of long-range nature) and a contribution $-(e^2/r)\text{erfc}(\sqrt{\eta}\,r)$ (singular at the origin and of short-range nature). Energies are expressed in Rydbergs, length in Bohr radius units; we have chosen $\sqrt{\eta} = a_B^{-1}$.

(see Figure 6.5). The first term on the right-hand side of Eq. (6.18) is the potential generated by a Gaussian charge density; it is of *long-range nature and regular at the origin* (in lattice sums, it is conveniently handled in reciprocal space). The second term on the right-hand side of Eq. (6.18) is the potential generated by a point charge located at the origin and dressed by a neutralizing Gaussian charge density; this contribution is of *short-range nature and singular at the origin* (in lattice sums with origin excluded, it is conveniently handled in real space).

Madelung Constant with the Ewald Method for a Two-Sublattice Structure

We consider now a crystal constituted by two sublattices, occupied by anions and cations of point charges $(-e)$ and $(+e)$, respectively; the two sublattices are described by the positions $\mathbf{d}_1 + \mathbf{t}_n$ and $\mathbf{d}_2 + \mathbf{t}_n$ (with $\mathbf{d}_1 = 0$, $\mathbf{d}_2 \neq 0$, and $\mathbf{t}_n$ translation vectors). We take, for instance, a negative ion position as a reference site, as in Figure 6.6.

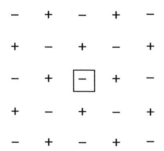

Figure 6.6 Schematic representation of a two-sublattice ionic crystal in the point-ion approximation, for the calculation of the Madelung constant. The ion at the origin has been surrounded by a box.

The potential energy $V(\mathbf{r})$ felt by a particle of charge $(-e)$ because of the field of all the ions of the crystal (except the reference one) is

$$V(\mathbf{r}) = \sum_{\mathbf{t}_n(\neq 0)} \frac{e^2}{|\mathbf{r}-\mathbf{t}_n|} - \sum_{\mathbf{t}_n} \frac{e^2}{|\mathbf{r}-\mathbf{t}_n-\mathbf{d}_2|}. \tag{6.19}$$

The first term in the right-hand side of Eq. (6.19) is the repulsive contribution due to the sublattice of negative ions, while the second term is the attractive contribution due to the sublattice of positive ions.

Taking into account Eq. (6.18), we can write Eq. (6.19) in the form

$$V(\mathbf{r}) = V_1(\mathbf{r}) + V_2(\mathbf{r}),$$

where

$$V_1(\mathbf{r}) = \sum_{\mathbf{t}_n(\neq 0)} \frac{e^2}{|\mathbf{r}-\mathbf{t}_n|} \operatorname{erfc}\left(\sqrt{\eta}\,|\mathbf{r}-\mathbf{t}_n|\right) - \sum_{\mathbf{t}_n} \frac{e^2}{|\mathbf{r}-\mathbf{t}_n-\mathbf{d}_2|} \operatorname{erfc}\left(\sqrt{\eta}\,|\mathbf{r}-\mathbf{t}_n-\mathbf{d}_2|\right) \tag{6.20a}$$

and

$$V_2(\mathbf{r}) = \sum_{\mathbf{t}_n(\neq 0)} \frac{e^2}{|\mathbf{r}-\mathbf{t}_n|} \operatorname{erf}\left(\sqrt{\eta}\,|\mathbf{r}-\mathbf{t}_n|\right) - \sum_{\mathbf{t}_n} \frac{e^2}{|\mathbf{r}-\mathbf{t}_n-\mathbf{d}_2|} \operatorname{erf}\left(\sqrt{\eta}\,|\mathbf{r}-\mathbf{t}_n-\mathbf{d}_2|\right). \tag{6.20b}$$

In particular, the term $V_1(0)$ is given by

$$V_1(0) = \sum_{\mathbf{t}_n(\neq 0)} \frac{e^2}{t_n} \operatorname{erfc}\left(\sqrt{\eta}\,t_n\right) - \sum_{\mathbf{t}_n} \frac{e^2}{|\mathbf{t}_n+\mathbf{d}_2|} \operatorname{erfc}\left(\sqrt{\eta}\,|\mathbf{t}_n+\mathbf{d}_2|\right). \tag{6.21}$$

Because of the short-range nature of the complementary error function, the term $V_1(0)$ is conveniently calculated as it stands, i.e. by direct summation in real space.

We now elaborate the term $V_2(\mathbf{r})$ of Eq. (6.20b); this term is more conveniently transformed into a reciprocal space sum, with the following procedure. We write $V_2(\mathbf{r})$ in the form

$$V_2(\mathbf{r}) = U(\mathbf{r}) - \frac{e^2}{r} \operatorname{erf}\left(\sqrt{\eta}\,r\right), \tag{6.22a}$$

where $U(\mathbf{r})$ is given by

$$U(\mathbf{r}) = \sum_{\mathbf{t}_n} \frac{e^2}{|\mathbf{r}-\mathbf{t}_n|} \operatorname{erf}\left(\sqrt{\eta}\,|\mathbf{r}-\mathbf{t}_n|\right) - \sum_{\mathbf{t}_n} \frac{e^2}{|\mathbf{r}-\mathbf{t}_n-\mathbf{d}_2|} \operatorname{erf}\left(\sqrt{\eta}\,|\mathbf{r}-\mathbf{t}_n-\mathbf{d}_2|\right); \tag{6.22b}$$

notice that $U(\mathbf{r})$ is a periodic function.

From the complete set of periodic functions $\left\{(1/\sqrt{N\Omega}) \exp\left(i\mathbf{h}_m \cdot \mathbf{r}\right)\right\}$, normalized to one in the volume $V = N\Omega$ of the crystal, we form the unit operator for projection of *periodic* functions

$$\frac{1}{N\Omega} \sum_{\mathbf{h}_m} |e^{i\mathbf{h}_m \cdot \mathbf{r}} \rangle \langle e^{i\mathbf{h}_m \cdot \mathbf{r}}| \equiv 1,$$

where the sum extends to all the vectors $\mathbf{h}_m$ of the reciprocal lattice. By applying the above identity to the periodic function $U(\mathbf{r})$, we have

$$U(\mathbf{r}) = \frac{1}{N\Omega} \sum_{\mathbf{h}_m} e^{i\mathbf{h}_m \cdot \mathbf{r}} \langle e^{i\mathbf{h}_m \cdot \mathbf{r}} | U(\mathbf{r}) \rangle. \tag{6.23a}$$

The scalar product appearing in Eq. (6.23a) can be evaluated explicitly, and one obtains

$$\left\langle e^{i\mathbf{h}_m \cdot \mathbf{r}} | U(\mathbf{r}) \right\rangle = N \left(1 - e^{-i\mathbf{h}_m \cdot \mathbf{d}_2}\right) \frac{4\pi e^2}{h_m^2} e^{-h_m^2/4\eta} \quad (h_m \neq 0), \tag{6.23b}$$

where use has been made of Eqs. (6.22b) and (6.17c). When $\mathbf{h}_m \equiv 0$, the scalar product (6.23b) represents the average of $U(\mathbf{r})$ on the crystal volume; such an average is evidently equal to zero, because of the overall neutrality of the system of positive and negative (identical) charge distributions.

From Eq. (6.22a) and Eqs. (6.23) we obtain

$$V_2(0) = U(0) - 2e^2 \sqrt{\frac{\eta}{\pi}} = \frac{4\pi e^2}{\Omega} \sum_{\mathbf{h}_m(\neq 0)} \left(1 - e^{-i\mathbf{h}_m \cdot \mathbf{d}_2}\right) \frac{1}{h_m^2} e^{-h_m^2/4\eta} - 2e^2 \sqrt{\frac{\eta}{\pi}}. \tag{6.24}$$

Collecting Eqs. (6.21) and (6.24), we obtain for the Madelung energy the final expression

$$V(0) = \sum_{\mathbf{t}_n(\neq 0)} \frac{e^2}{t_n} \text{erfc}\left(\sqrt{\eta}\, t_n\right) - \sum_{\mathbf{t}_n} \frac{e^2}{|\mathbf{t}_n + \mathbf{d}_2|} \text{erfc}\left(\sqrt{\eta}\, |\mathbf{t}_n + \mathbf{d}_2|\right)$$

$$+ \frac{4\pi e^2}{\Omega} \sum_{\mathbf{h}_m(\neq 0)} \left(1 - e^{-i\mathbf{h}_m \cdot \mathbf{d}_2}\right) \frac{1}{h_m^2} e^{-h_m^2/4\eta} - 2e^2 \sqrt{\frac{\eta}{\pi}}. \tag{6.25}$$

The arbitrary parameter η controls the convergence of the two summations, and is chosen to optimize simultaneously the convergence of the series in direct and reciprocal spaces.

Many ionic crystals occur in cubic structures: the cesium chloride structure (coordination number 8), sodium chloride structure (coordination number 6), zincblende structure (coordination number 4). The Madelung constant is (slightly) higher for higher coordination number, as can be seen from Table 6.4.

6.2.2 Considerations on Bands and Bonds in Ionic Crystals

In the class of ionic crystals, alkali halides (formed with one element of column I and one element of column VII of the periodic table) are the most typical compounds. Also

II–VI crystals can be often pictured as formed by (doubly ionized) ions, although the partial covalent character of the atomic interactions may become significant. In III–V compounds the binding is mostly covalent in nature, while IV–IV materials are the typical covalent compounds. We confine here our attention to some aspects of the band structure of alkali halides crystals.

Alkali halides are in general large gap insulators; the electrons completely fill the core and valence bands, while the conduction bands are empty and well separated in energy from the valence states. To understand in essential lines the origin of the gap and the role of the Madelung energy, we consider as an exemplification the case of NaCl. In the ionic picture, the crystal is described as a system of interacting Cl^- and Na^+ ions; the ions have the closed-shell electronic configuration: Cl^- ($1s^2, 2s^2 2p^6, 3s^2 3p^6$) and Na^+ ($1s^2, 2s^2 2p^6$).

Some features of the band structure of NaCl can be envisaged with the following considerations. The orbital energy $E_{3p}(Cl^-)$ is given by the (negative) of the ionization energy of the Cl^- ion; this in turns equals the electron affinity of the Cl atom (≈ 3.7 eV); we thus estimate $E_{3p}(Cl^-) = -3.7$ eV. In the crystal, an electron in the reasonably localized $3p$ orbital of Cl^- feels the Madelung energy $-\alpha_M e^2/R_0 \approx -9.0$ eV ($R_0 = 5.27\, a_B$ for NaCl), due to all the surrounding ions in the point-ion approximation. We thus estimate that the topmost valence bands (in the point-ion approximation) are located around $E_v = E_{3p}(Cl^-) - \alpha_M e^2/R_0 = -12.7$ eV. Rather in a similar way, we estimate the orbital energy $E_{3s}(Na^+) = -5.1$ eV as the (negative) of the ionization energy of the Na atom (≈ 5.1 eV). We can reasonably take the energy of the bottom of the conduction band as $E_c = E_{3s}(Na^+) + \alpha_M e^2/R_0 = +3.9$ eV. The estimated energy gap, in the point-ion approximation, is $E_G \approx E_{3s}(Na^+) - E_{3p}(Cl^-) + 2\alpha_M e^2/R_0 = 16.6$ eV, and is mostly determined by the Madelung energy.

From the above considerations, we expect that the topmost valence bands are p-like and originated from the $3p$ orbitals of the anion, while the bottom conduction band is s-like and originated from the $3s$ orbital of the cation. These speculations, as well as the role of the Madelung energy in making NaCl a large gap insulator, are corroborated by detailed band structure calculations (Figure 6.7). The estimates done above within the point-ion approximation for the band edges compare reasonably well with the Hartree-Fock calculations of NaCl. In Figure 6.7 we also report the correlated energy bands, which properly take into account the internal structure and the polarizability of the Cl^- and Na^+ ions. With respect to the Hartree-Fock calculations, the correlation effects lower the conduction bands and increase the valence bands, similarly to the situation described in Figure 6.1. In NaCl, the inclusion of correlation effects leads to a calculated band gap of ≈ 10 eV and a valence band width of ≈ 3 eV. Optical absorption measurements and photoemission measurements provide a fundamental band gap for NaCl of about 9 eV, and a valence band width of about 3 eV.

Similar considerations can be made for other ionic crystals; since the Madelung energy $\alpha_M e^2/R_0$ is ordinarily in the range 5–10 eV, this is also the expected order of magnitude of the energy gap. In particular, because of its small value of R_0, LiF is expected to have the largest gap among alkali halides; this is the origin of the utilization of LiF as a transparent window for near ultraviolet spectroscopy (up to the cut-off energy of ≈ 11.8 eV).

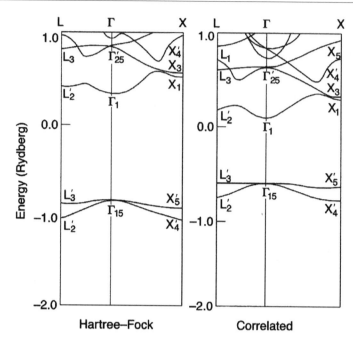

Figure 6.7 Hartree-Fock and correlated energy bands of NaCl. The state Γ_{15} is the top of the valence bands and the state Γ_1 is the bottom of the conduction bands [reprinted with permission from A. B. Kunz, Phys. Rev. B **26**, 2056 (1982); copyright 1982 by the American Physical Society].

Considerations on Cohesive Energy in Ionic Crystals

We wish to mention here the semi-empirical Born model, to supply evidence for some features of bonds and cohesive energy in ionic crystals. Consider a typical ionic crystal, such as alkali halides, composed by N positive and N negative ions of site charge $\pm e$. The total energy of the crystal is taken as the sum of pairwise interactions, containing a repulsive part of short-range nature and an attractive part originated by the long-range Coulomb interactions.

The total energy of the solid, relative to the free ions, can be expressed in the semi-empirical form

$$U_S(R) = N \left(\frac{\lambda}{R^n} - \alpha_M \frac{e^2}{R} \right), \tag{6.26}$$

where R is the nearest neighbor distance, α_M is the Madelung constant, λ and n are semi-empirical parameters. The term $-\alpha_M e^2/R$ represents the attractive energy of an ion due to the presence of all the other ions, in the strictly point-ion approximation. The term λ/R^n represents the repulsive energy of an ion (of finite size) due to the presence of all the other ions (of finite size) of the lattice; it has been expressed by a power law, following an assumption by Born. In other frequently adopted models, the repulsive energy is assumed to vary exponentially with the distance, keeping in better line with the quantum mechanical results; however, we do not enter in details, since the qualitative features we are going to work out are to a large extent independent whether the inverse power-law representation or the exponential representation of the short-range repulsive energy is adopted

Using Eq. (6.26), we can now proceed to the evaluation of the equilibrium lattice constant, the cohesive energy and the bulk modulus (similarly to what previously done in rare-gas solids in Section 6.1.2). As an application, we will consider here the case of NaCl crystal for which the Madelung constant is $\alpha_M = 1.7476$. The nearest neighbor distance is $R_0 = 2.789$ Å $= 5.27\,a_B$ and the Madelung energy $\alpha_M e^2/R_0 = 9.02$ eV. The experimental cohesive energy per NaCl molecule is $E_{coh} = -8.01$ eV and the experimental bulk modulus is $B_0 = 0.266$ Mbar. [The experimental values are quoted from the paper by N. Karasawa and W. A. Goddard III, J. Phys. Chem. 93, 7320 (1989).]

At the equilibrium distance, we have

$$\left(\frac{dU_S}{dR}\right)_{R=R_0} = 0 \quad \text{hence} \quad \lambda = \frac{\alpha_M e^2}{n} R_0^{n-1}.$$

The cohesive energy $U_S(R_0)$ becomes

$$U_S(R_0) = -N\alpha_M \frac{e^2}{R_0}\left(1 - \frac{1}{n}\right). \tag{6.27}$$

In the case of NaCl, using the reported experimental values for the nearest neighbor distance and for the cohesive energy, one obtains $n \approx 9$. Similarly to NaCl, typical values of n for other alkali halides are of the order of ≈ 10 or so; then, from Eq. (6.27), it can be inferred that *the cohesive energy of an ionic crystal is mostly determined by the Madelung energy of the crystal.*

We can now evaluate the bulk modulus given by Eqs. (6.8) and (6.9)

$$B_0 = V_0\left(\frac{d^2U_S}{dV^2}\right)_{V=V_0} = V_0\left(\frac{dR}{dV}\right)^2_{V=V_0}\left(\frac{d^2U_S}{dR^2}\right)_{R=R_0}.$$

For the NaCl structure, the relation between volume and nearest neighbor distance is $V = 2NR^3$. It follows

$$B_0 = \frac{1}{18NR_0}\left(\frac{d^2U_S}{dR^2}\right)_{R=R_0} = (n-1)\frac{e^2\alpha_M}{18R_0^4}. \tag{6.28}$$

In the case of NaCl, the bulk modulus estimated from Eq. (6.28) is $B_0 = 0.292$ Mbar, reasonably near to the quoted experimental value ≈ 0.266 Mbar. It should be noticed that the bulk modulus of alkali halides is much higher than the values of rare-gas solids, as expected from the stronger nature of interatomic interactions.

From the above considerations, in spite of the model simplifications, the relevance of the electrostatic contribution in the cohesive energy of ionic crystals emerges with evidence. From this, we can also attempt an empirical criterion to estimate the relative stability of the cesium chloride structure, compared with sodium chloride structure and zincblende structure. Because of the higher Madelung constant (see Table 6.4), the CsCl structure should be more stable than the other two, at parity of nearest neighbor distance.

In ionic crystals, the ions have a closed-shell configuration, and this makes it possible to define (theoretically or experimentally) ionic radii, which are (approximately) transferable quantities. Thus we can analyze (at least at a preliminary level) the stability of different structures, requiring that positive and negative nearest neighbor ions touch in the solid.

In the following reasoning, we assume ions as hard spheres of radii R_A and R_C (we suppose that R_A is the larger of the two radii). Consider the cesium chloride structure and let a denote the edge of the conventional cube; the nearest neighbor distance and the second nearest neighbor distance are given, respectively, by

$$d_I = \frac{a\sqrt{3}}{2} \quad \text{and} \quad d_{II} = a \quad \text{(CsCl structure)}. \tag{6.29a}$$

The condition that positive and negative ions touch each other gives one equality and one inequality

$$\begin{cases} R_A + R_C = a\sqrt{3}/2 \\ 2R_A \leq a \end{cases}.$$

Solving for a, we have

$$2R_A \leq \frac{2}{\sqrt{3}}(R_A + R_C) \quad \text{and} \quad \frac{R_A}{R_C} \leq \frac{1}{\sqrt{3}-1} = 1.37 \quad \text{(CsCl structure)}. \tag{6.29b}$$

Thus the cesium chloride structure allows touching of ions if the ratio of the ionic radii satisfies the condition (6.29b).

A similar reasoning can be done for the NaCl structure. In this structure, indicating by a the edge of the conventional cube, we have that the nearest neighbor distance and the second nearest neighbor distance are given by

$$d_I = \frac{a}{2} \quad \text{and} \quad d_{II} = a\frac{\sqrt{2}}{2} \quad \text{(NaCl structure)}. \tag{6.30a}$$

The condition that positive and negative ions can touch each other gives one equality and one inequality

$$\begin{cases} R_A + R_C = a/2 \\ 2R_A \leq a\sqrt{2}/2 \end{cases}.$$

Solving for a, we have

$$\frac{R_A}{R_C} \leq \frac{1}{\sqrt{2}-1} = 2.44 \quad \text{(NaCl structure)}. \tag{6.30b}$$

Thus the sodium chloride structure allows touching of ions if the ratio of the ionic radii satisfies the condition (6.30b).

Finally we can consider the zincblende structure. In this structure, indicating as usual by a the edge of the conventional cube, we have that nearest and second nearest neighbor distances are

$$d_I = \frac{a\sqrt{3}}{4} \quad \text{and} \quad d_{II} = a\frac{\sqrt{2}}{2}. \tag{6.31a}$$

The condition that positive and negative ions can touch each other leads to

$$R_A/R_C \leq 2 + \sqrt{6} = 4.45 \quad \text{(zincblende structure)}. \tag{6.31b}$$

The above arguments suggest that ions with almost equal radii tend to crystallize in CsCl structure, while ions with very different radii prefer sodium chloride or zincblende structure, a trend that is actually observed in ionic crystals. Of course these considerations are far from complete since the actual ionic structure of anions and cations has been ignored. For in-depth understanding of the electronic properties of ionic crystals, full *ab initio* calculations are required, usually elaborated within the density functional theory.

6.3 Covalent Crystals with Diamond Structure

Rare-gas crystals and ionic crystals are examples of crystals composed by closed-shell units; closed-shell units have no (or little) tendency to produce overlap of electron clouds. In crystals composed by open-shell units, there may be a tendency to a redistribution of the electronic charge; the electronic charge, that accumulates midway two nearest neighbor nuclei, enjoys attractive interaction from both nuclei and is responsible of the covalent bond.

Among the numerous investigations in the literature on the group IV covalent solids, we briefly mention here some results obtained with the pseudopotential method; in this approach the effective Hamiltonian (sum of kinetic energy and a weak pseudopotential) is conveniently expressed and diagonalized using the basis set of plane waves. In Figure 6.8 we give in sequence the band structure of silicon, germanium, and gray-tin. For the heavier elements, the spin-orbit effects are included.

The group IV elements, diamond, silicon, germanium, and gray-tin, crystallize in the diamond structure, which is fcc with two atoms per unit cell. In the free atoms, there are four electrons in the most external shell: C $(2s^2, 2p^2)$, Si $(3s^2, 3p^2)$, Ge $(4s^2, 4p^2)$, Sn $(5s^2, 5p^2)$; inner shells are completely filled. The most external s and p_x, p_y, p_z atomic orbitals of the two atoms in the unit cell form bonding (and antibonding) combinations, which generate the highest valence bands (and the bottom conduction bands) of the crystal. For what concerns the energy gap, diamond is a strong insulator with indirect energy gap of 5.48 eV, silicon and germanium are semiconductors with indirect energy gap 1.17 eV and 0.74 eV, respectively (at $T = 0$), gray-tin is a semi-metal and lead is a metal. In spite of this quite different behavior, from band theory and symmetry considerations it is possible to understand the differences (and the similarities) in the electronic structure of these materials.

Of special interest is the trend of the energy bands of diamond, silicon, and germanium around the fundamental gap. In diamond the position of the minimum of the conduction band along the ΓX direction occurs at the point $\mathbf{k} \approx (2\pi/a)(0.78, 0, 0)$; there are thus six equivalent minima in the Brillouin zone. In silicon, in the ΓX direction,

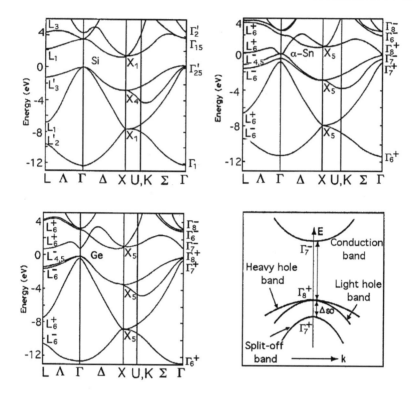

Figure 6.8 Band structure for silicon, germanium, and gray-tin [reprinted with permission from J. R. Chelikowsky and M. L. Cohen, Phys. Rev. B *14*, 556 (1976); copyright 1976 by the American Physical Society]. The typical band structure of cubic semi-conductors near the gap at **k** = 0 is also schematically reported; Δ_{SO} denotes the spin-orbit splitting.

the minimum of the conduction band occurs at the point $\mathbf{k} \approx (2\pi/a)(0.85, 0, 0)$; there are again six equivalent minima, nearer to the boundary of the Brillouin zone. In germanium, the minimum of the conduction band occurs at the boundary point $\mathbf{k} = (2\pi/a)(1/2, 1/2, 1/2)$ and there are thus four equivalent minima in the Brillouin zone (since $\mathbf{k}$ and $-\mathbf{k}$ are related by a reciprocal lattice vector). The top of the valence bands is at Γ'_{25}, in diamond, germanium, and silicon. The state Γ'_{25} arises from the three p_x, p_y, p_z bonding orbitals, which are degenerate at the center of the Brillouin zone; considering the spin of the electrons, Γ'_{25} is sixfold degenerate. Spin-orbit interaction splits the topmost valence bands into two groups: $p_{3/2}$ bands (heavy-hole and light-hole bands) with degeneracy 4 at Γ point (Γ_8^+ state), and $p_{1/2}$ bands (split-off band) with degeneracy 2 (Γ_7^+ state). The spin-orbit separation of the valence bands at Γ is negligible in diamond, it is 0.044 eV for Si, 0.29 eV for Ge, and 0.80 eV for gray-tin. A special feature occurs for gray-tin, where the energy of the s-like antibonding state decreases and becomes intermediate between the spin-orbit split Γ_8^+ and Γ_7^+ states; gray-tin becomes thus a semi-metal with energy gap zero at the point Γ of the Brillouin zone.

We wish now to mention a classic investigation on the static structural properties, crystal stability and phase transformation of silicon and germanium, based on

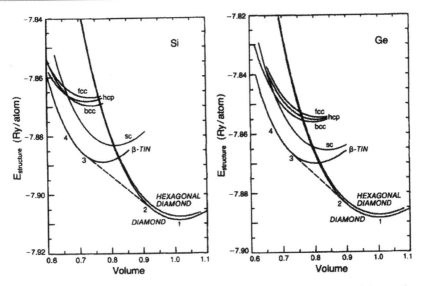

Figure 6.9 Total energy curves for seven structures of Si and Ge, as a function of the atomic volume (normalized to the experimental volume). Dashed line is the common tangent of the energy curves for the diamond phase and the β-tin phase [reprinted with permission from M. T. Yin and M. L. Cohen, Phys. Rev. B 26, 5668 (1982); copyright 1982 by the American Physical Society].

norm-conserving pseudopotential approach within the local-density-functional approximation. The total energy curves are reported in Figure 6.9 for seven crystal structures: face-centered cubic lattice, body-centered cubic lattice, hexagonal close-packed, simple cubic, cubic diamond, hexagonal diamond (wurtzite), and β-tin. From Figure 6.9 it is seen that the diamond structure lies lowest in energy. The minimum of the total energy curve, and its curvature, determine the lattice constant, the cohesive energy and the bulk modulus (see Table 6.5). It can also be noticed that the β-tin structure is lower in energy than the diamond structure at small volumes; hence a solid-solid phase

Table 6.5 Calculated and measured equilibrium properties of silicon and germanium [from M. T. Yin and M. L. Cohen, Phys. Rev. B 26, 5668 (1982); copyright 1982 by the American Physical Society].

	Lattice constant (Å)	Nearest neighbor distance (Å)	Binding energy (eV/atom)	Bulk modulus (Mbar)
Si				
Calculation	5.451	2.360	4.84	0.98
Experiment	5.429	2.351	4.63	0.99
Ge				
Calculation	5.655	2.449	4.26	0.73
Experiment	5.652	2.447	3.85	0.77

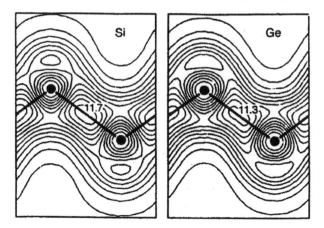

Figure 6.10 Calculated pseudovalence charge density in the (1 1 0) plane for Si and Ge. The contours are in units of electrons per atomic volume, with a contour step of 1. Atomic positions are denoted by black dots; atomic chains are denoted by straight lines [reprinted with permission from M. T. Yin and M. L. Cohen, Phys. Rev. B *26*, 5668 (1982); copyright 1982 by the American Physical Society].

transition should occur at the estimated transition pressure of $\approx$ 100 kbar, given by the slope of the dashed line in Figure 6.9.

In Table 6.5 we report calculated and measured static properties of silicon and germanium, and notice the satisfactory agreement with experimental data. It is instructive to make a comparison of the order of magnitude of the static properties of rare-gas solids (Table 6.2) with those of the typical covalent crystals (Table 6.5); it can be seen that covalent crystals have smaller nearest neighbor distance, and stronger cohesive energy and bulk modulus. For diamond the cohesive energy is 7.37 eV/atom, and the bulk modulus is as high as 4.423 Mbar; the bulk modulus decreases, but remains in the Mbar range, for silicon and germanium (0.99 and 0.77 Mbar, respectively).

Before concluding this section, we show the pseudovalence charge densities for Si and Ge (see Figure 6.10) [outside the core region pseudovalence charge distribution and true valence charge distribution coincide, because of the norm-conserving property of the pseudopotential used in the calculations]. The maximum value of the valence charge density occurs midway nearest neighbor atomic positions; the pile-up of the covalent bonding charge, elongated along the atomic chains, is clearly visualized in Figure 6.10.

6.4 Band Structures and Fermi Surfaces of Some Metals

The metallic state is the most common situation for crystals of the elements of the periodic table: out of the more than 100 elements, about 70 are metals. A most important quantity of metals is the *Fermi energy* E_F, defined (at $T = 0$) by the requirement that the total number of band states with energy smaller than E_F exactly equals the total number of electrons available in the crystal. In metals, the Fermi level occurs within a band or intersects a number of bands; the surface in **k**-space that connects all the **k**-points, corresponding to crystal states of energy E_F, is called *Fermi surface*. The electron transport properties of metals are determined by the properties of the Fermi surface and of the density-of-states in the small thermal shell of energy $k_B T$ around it. The very rich phenomenology of metals is related to the variety of peculiar shapes of the Fermi surface, determined by the actual crystal potential.

In general, the study of metals requires the evaluation of the energy band structure $E_n(\mathbf{k})$ at a rather high number of $\mathbf{k}$-points in the first Brillouin zone (say several thousands or so). This allows a reliable determination of the Fermi energy, of the topology of the Fermi surface, and the density-of-states curve. The points $\mathbf{k}$ belonging to the Fermi surface are given by the implicit equation

$$\boxed{E_n(\mathbf{k}) \equiv E_F}. \tag{6.32}$$

The equation can be solved graphically plotting any given branch $E = E_n(\mathbf{k})$ versus $\mathbf{k}$ in various directions in the first Brillouin zone; from the intersection of the curve with the line $E = E_F$ the points $\mathbf{k}_F$ on the Fermi surface can be obtained, and the Fermi surface drawn. We illustrate the above considerations with a few simple examples of band structures and Fermi surfaces.

The *alkali metals* (Li, Na, K, Rb, Cs) are the typical *simple metals*; so are addressed those metals with an almost free-electron picture for the conduction band. An alkali atom has one "optical electron" in the state ns^1, with $n = 2, 3, 4, 5, 6$ for lithium, sodium, potassium, rubidium, and cesium, respectively, and fully occupied inner shells. The alkali metals crystallize in the bcc structure, with one atom per unit cell. The lowest conduction band is s-like in nature, corresponding to the partially occupied ns^1 external atomic orbital; the Fermi surface lies within the Brillouin zone quite far from zone boundaries. As an example, the band structure and the Fermi level of sodium is reported in Figure 6.11.

Although the Fermi surface in alkali metals is almost spherical and well inside the first Brillouin zone, there are nevertheless significant deviations, with influence on the conductivity, magnetoresistance, thermoelectric power, etc. These deviations are relatively high in lithium and almost absent in sodium, that can be considered the simplest of the simple metals. For instance, at the bottom of the conduction band, the effective mass m_c^* (in units of the free-electron mass m) is $m_c^* = 1.33, 0.97, 0.86, 0.78$ and 0.73 for Li, Na, K, Rb, and Cs, respectively. This trend of the effective mass in the sequence of alkali metals can be interpreted with the help of the $\mathbf{k} \cdot \mathbf{p}$ method (described in Section 2.6.2). At the bottom of the conduction band, taking into account the cubic symmetry of the materials, we can rewrite Eq. (2.39) in the form

$$\frac{m}{m_c^*} = 1 + \frac{2}{m} \sum_{n(\neq c)} \frac{|\langle \psi_n(\mathbf{k}_0, \mathbf{r})| p_x |\psi_c(\mathbf{k}_0, \mathbf{r})\rangle|^2}{E_c(\mathbf{k}_0) - E_n(\mathbf{k}_0)}, \tag{6.33}$$

where $E_n(\mathbf{k}_0)$ and $\psi_n(\mathbf{k}_0, \mathbf{r})$ are eigenvalues and eigenfunctions at the center of the Brillouin zone, and the subscript c refers to the partially occupied conduction band. Equation (6.33) easily accounts why Li has an effective mass larger than the free-electron mass m (the only possible interactions are with higher energy bands); sodium has effective mass almost equal to m (there is an approximate balance in the interactions with lower and higher energy bands); finally the other alkali metals K, Rb, and Cs have a conduction effective mass smaller than the free-electron mass.

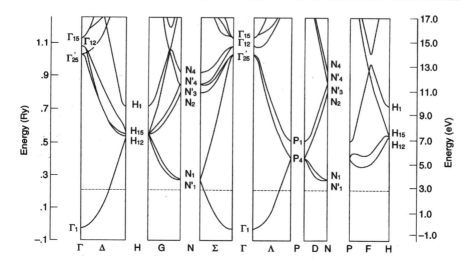

Figure 6.11 Band structure of sodium. The Fermi energy is indicated by a dotted line [from D. A. Papaconstantopoulos "Handbook of the Band Structure of Elemental Solids" (Plenum, New York, 1986); with kind permission from Springer Science and Business Media B. V.].

The *divalent metals* (Be, Mg, Ca, Sr, Ba and Zn, Cd, Hg) are formed by atoms with two electrons in the most external ns^2 orbital, and fully occupied inner shells. The lowest conduction bands are mostly s-like in nature, and closely related to the occupied ns^2 external atomic orbital. These crystals, having two optical electrons per atom, would give completely filled and completely empty bands, in the absence of energy overlap of different bands. This is not the case, as shown for instance in Figure 6.12, where we report the conduction band structure of calcium (fcc Bravais lattice). The overlap of bands makes calcium a typical example of metal, with an equal number of electrons and holes: the carriers are constituted by the electrons occupying pockets in the second conduction band, and an equal number of empty levels (holes) to the full occupancy of the first conduction band.

We have seen that the conduction bands in alkali metals are essentially free-electron like (also referred to as "itinerant"), with moderate modifications brought about by the crystalline potential. The situation can change dramatically in the presence of hybridization effects between tendentially itinerant bands (originating from orbitals of type s or p) and tendentially localized bands (originating from orbitals of type d or f). A very interesting case is constituted by noble metals. The *noble metals* (Cu, Ag, Au) are monovalent and crystallize in fcc Bravais lattice. The s and d mixing has a profound effect on the band structure and Fermi surface topology of these materials (in the absence of s-d hybridization, these materials would be very much like the alkali metals).

The electronic configuration of Cu is [Ar] $3d^{10}4s^1$. Similarly for Ag we have [Kr] $4d^{10}5s^1$ and for Au the electronic configuration is [Xe] $5d^{10}6s^1$. We consider for instance the band structure of copper. The bands arising from $1s$, $2s$, $2p$, $3s$, and $3p$ orbitals of Cu have very low energy and can be considered as core states flat in k-space, fully occupied. Thus we have to consider only the states arising from $3d$ and $4s$ atomic

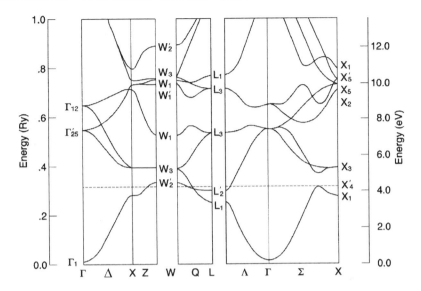

Figure 6.12 Band structure of calcium. The Fermi energy is indicated by a dotted line [from D. A. Papaconstantopoulos "Handbook of the Band Structure of Elemental Solids" (Plenum, New York, 1986); with kind permission from Springer Science and Business Media B. V.].

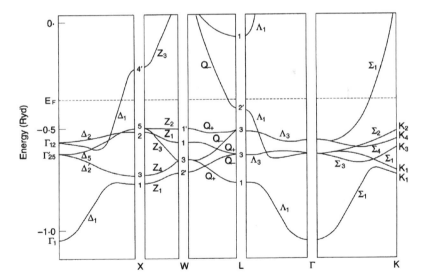

Figure 6.13 Band structure of copper [reprinted with permission from G. A. Burdick, Phys. Rev. *129*, 138 (1963); copyright 1963 by the American Physical Society].

orbitals. In Figure 6.13 we report the band structure of Cu; its overall aspect is just that of a rather wide s-like band, that hybridizes with rather narrow d-like bands.

From Figure 6.13, we can now reconstruct the Fermi surface of copper, which is given schematically in Figure 6.14. For this aim, we apply the procedure summarized in

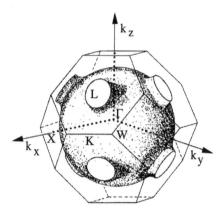

Figure 6.14 Brillouin zone for the face-centered cubic lattice and Fermi surface of copper.

Eq. (6.32), along various directions in the Brillouin zone. Let us consider the direction ΓX, for instance; we see from Figure 6.13 that there is one intersection of the band structure with the curve $E = E_F$; a similar situation occurs for instance along the ΓK direction. If we consider the ΓL direction, no intersection is found. If we consider the LW direction on the hexagonal face of the Brillouin zone, we find an intersection point. Putting these (and other similar) facts together, we can easily arrive at the Fermi surface schematically indicated in Figure 6.14 and originally proposed by Pippard. The Fermi surface bulges out in $\langle 111 \rangle$ direction, makes contact with the hexagonal faces of the Brillouin zone, and is thus an *open Fermi surface*. As a typical consequence of this topology, profound effects on the transport properties (in magnetoresistance, for instance) may occur.

In noble metals, the mostly s-like band at the Fermi surface is separated in energy from the fully occupied d states, although their presence influences the energy dispersion of the s-like electrons, with the peculiar effects in the Fermi surface shape described above. In transition and rare-earth metals, the electronic states at and near the Fermi surface are determined by the mixing of localized inner d or f electrons with external s or p electrons; a variety of situations can occur in transition metals (partially filled $3d$, $4d$, or $5d$ orbitals), and rare-earth metals (partially filled $4f$ and $5f$ orbitals). An additional complication is that the ground state of several metals is magnetic, and that in general no simple model can include easily magnetic effects; we thus have to refer to the literature for investigations and details.

We wish to conclude these simple remarks on metals with a few considerations on the nature of the metallic bond. A preliminary, although *crude* estimation, can be inferred from the "idealized" jellium model of a metal, which assumes that the electrons of the Fermi sea are embedded in a uniform background of neutralizing positive charges. In the jellium model (see Section 4.7), the ground-state energy per electron in the Hartree-Fock approximation is given by

$$E_0 = \left[\frac{2.21}{r_s^2} - \frac{0.916}{r_s} \right] \text{ (in Rydberg).} \tag{6.34}$$

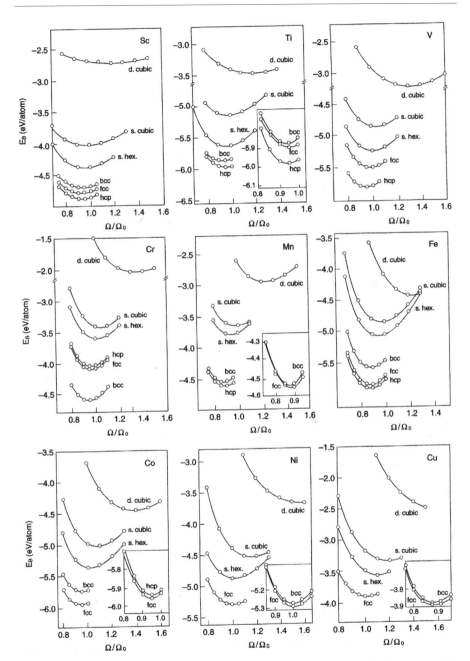

Figure 6.15 Structural energy-volume curves in the $3d$ transition metals: cohesive energy (eV) versus atomic volume normalized to the experimental atomic volume Ω_0. Structures are fcc, bcc, hcp, simple hexagonal, simple cubic, and diamond cubic; curves in insets are shown with expanded scales [reprinted with permission from A. T. Paxton, M. Methfessel, and H. M. Polatoglou, Phys. Rev. B *41*, 8127 (1990); copyright 1990 by the American Physical Society].

The minimum of $E_0(r_s)$ occurs when the dimensionless parameter r_s equals $r_{s0} = 4.825$; the corresponding cohesive energy is $E(r_{s0}) = -1.29$ eV.

In actual metals the correlation energy and the internal structure of ions are of major importance in determining the metallic bond. With the developments of self-consistent band structure calculations and implementation of the density functional formalism, much progress has been done for an accurate account for the total energy of metals [see for instance Y. Mishin, D. Farkas, M. J. Mehl, and D. A. Papaconstantopoulos, Phys. Rev. B 59, 3393 (1999) and references quoted therein].

As an exemplification we report in Figure 6.15 a study on the total energy of the first-row transition metals, for various structures; the study is based on the linearized muffin-tin orbital method and the local-density-functional approximation. The first-row transition elements from Sc to Cu have the core states $1s^2$, $2s^2 2p^6$, $3s^2 3p^6$ fully occupied; the electronic configuration of the external $3d$ and $4s$ shells is listed for convenience, together with the crystal structure: Sc ($3d^1 4s^2$; hcp); Ti ($3d^2 4s^2$; hcp); V ($3d^3 4s^2$; bcc); Cr ($3d^5 4s^1$; bcc); Mn ($3d^5 4s^2$; nearly hcp); Fe ($3d^6 4s^2$; bcc); Co ($3d^7 4s^2$; hcp); Ni ($3d^8 4s^2$; fcc); Cu ($3d^{10} 4s^1$; fcc). In Figure 6.15 we report for several structures the energy-volume curves. It must be noticed that the calculations do not include magnetic effects; thus the magnetic ground state of transition metals near the end of the row are not expected to be determined very accurately; in principle, magnetic effects could be introduced by spin-polarized calculations. From Figure 6.15, we see that the crystal structure of all non-magnetic crystals are correctly reproduced. We also see that the typical binding energy is of the order of 4–5 eV/atom. Also the value of the equilibrium lattice constant is predicted with good accuracy. The bulk modulus elaborated from the structural energy-volume curves is of the order of the Mbar for the various metals.

6.5 Carbon-Based Materials and Electronic Structure of Graphene

Allotropes of Elemental Carbon

In Section 6.3, in the study of features and trends of covalent crystals made of group IV elements, we have seen some electronic properties of diamond; due to its typical tetrahedral coordination structure and sp^3 hybridization of carbon orbitals, this large gap insulator is characterized by strong covalent bonds and strong bulk modulus.

Graphite is the other most traditional allotrope of elemental carbon. As discussed in Chapter 2, the three-dimensional graphite is characterized by a planar structure of hexagonal rings of carbon atoms, with strong covalent bonds corresponding to sp^2 hybridization within the stacking planes, and weak van der Waals forces between different planes. Graphite represents the paradigmatic example of layered materials, which have always attracted much attention for the peculiar effects induced by the structural anisotropy on their electronic, optical, and transport properties.

Diamond and graphite, the fcc and the hexagonal packing forms of elemental carbon atoms, have been enriched in the arena of elemental carbon allotropes by fullerenes in 1985 [H. W. Kroto, J. R. Heath, S. C. O'Brien, R. F. Curl, and R. E. Smalley, Nature

318, 162 (1985)], carbon nanotubes in 1991 [S. Ijima, Nature *354*, 56 (1991)], and graphene in 2004 [K. S. Novoselov, A. K. Geim, S. V. Morozov, D. Jiang, Y. Zhang, S. V. Dubonos, I.V. Grigorieva, and A. A. Firsov, Science *306*, 666 (2004)].

The materials of the fullerene family, in particular built with the C_{60} spherical molecule, have received great attention for their chemical and physical properties, superconducting behavior when doped by metals, perspectives of technological applications in organic electronics. Similarly, quasi one-dimensional carbon nanotubes and two-dimensional graphene have opened novel perspectives in nanoscience and nanotechnology. The advances in the area of the new carbon allotropes and the interdisciplinary context of their realizations have been recognized in the scientific community by the 1996 Nobel Prize in Chemistry awarded to Kroto, Curl, and Smalley for the discovery of fullerenes, and the 2010 Nobel Prize in Physics awarded to Geim and Novoselov for the discovery of graphene.

In particular graphene, with its peculiar band structure, provides the basic elements for the comprehension of the electronic structure of carbon allotropes. Most of the features of graphene are essentially governed by the special linear energy-momentum dispersion relation at the corners of the Brillouin zone. The importance of the two-dimensional p_z electron gas is that it mimics the behavior of relativistic massless Dirac fermions, with striking consequences on electronic states, transport properties, optical properties, and magnetic field effects. Some of the basic properties can be understood within a simple tight-binding model, by virtue of the underlying symmetry properties of the honeycomb lattice, as shown below.

Energy Dispersions of σ and of π Bands in Graphene

The two-dimensional monolayer structure of graphene is described by the typical honeycomb lattice shown in Figure 6.16. The primitive translation vectors, and the two carbon positions in the unit cell are given by

$$\mathbf{t}_1 = a\left(\frac{1}{2}, \frac{\sqrt{3}}{2}\right), \quad \mathbf{t}_2 = a\left(-\frac{1}{2}, \frac{\sqrt{3}}{2}\right), \quad \mathbf{d}_1 = 0, \quad \mathbf{d}_2 = a\left(0, \frac{\sqrt{3}}{3}\right);$$

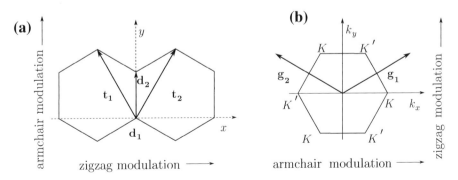

Figure 6.16 (a) Primitive translation vectors and basis vectors of graphene. (b) Brillouin zone for the honeycomb lattice. Zigzag and armchair modulation directions of repeated cells in the real and in the reciprocal lattices are indicated.

the lattice constant is $a = 2.46$ Å $= 4.65\,a_B$, and the nearest neighbor distance is $a_0 = 1.42$ Å $= 2.68\,a_B$. The primitive reciprocal vectors and the high symmetry points of the hexagonal Brillouin zone are

$$\mathbf{g}_1, \mathbf{g}_2 = \frac{2\pi}{a}\left(\pm 1, \frac{\sqrt{3}}{3}\right),$$

$$\Gamma = 0, \qquad K, K' = \frac{2\pi}{a}\left(\pm\frac{2}{3}, 0\right), \qquad Q = \frac{2\pi}{a}\left(\frac{1}{2}, \frac{\sqrt{3}}{6}\right).$$

The relevant features of the direct and reciprocal lattice are reported in Figure 6.16.

The electronic configuration of the carbon atom is $1s^2$, $2s^2 2p^2$, and the orbital energies are approximately $E_{1s} = -21.37$, $E_{2s} = -1.29$, and $E_{2p} = -0.66$ (Rydberg). The $1s$ core atomic orbitals are strongly localized near the nuclei, and give rise to dispersionless crystal core bands. In order to study the valence and the (lowest) conduction bands, we diagonalize the crystal Hamiltonian on the basis set of the eight Bloch sums, formed with $2s$, $2p_x$, $2p_y$, and $2p_z$ orbitals for each of the two carbon atoms in the unit cell. Since there are eight valence electrons per unit cell, we expect four completely occupied bands (if the four lowest lying bands do not overlap in energy with upper lying four energy bands).

The band wavefunctions originated from s, p_x, and p_y orbitals (σ bands) are even under reflection in the plane of graphite; they do not mix with band wavefunctions originated from p_z orbitals (π bands), which are odd under reflection in the plane of graphene. Thus σ-bands and π-bands can be studied separately.

For σ-bands, we start from the six basis Bloch sums built from the s, p_x, p_y orbitals of the two carbon atoms in the unit cell

$$\Phi_{1\alpha}(\mathbf{k}, \mathbf{r}) = \frac{1}{\sqrt{N}} \sum_{\mathbf{t}_m} e^{i\mathbf{k}\cdot\mathbf{t}_m} \phi_\alpha(\mathbf{r} - \mathbf{d}_1 - \mathbf{t}_m) \quad \alpha = s, p_x, p_y, \tag{6.35a}$$

$$\Phi_{2\alpha}(\mathbf{k}, \mathbf{r}) = \frac{1}{\sqrt{N}} \sum_{\mathbf{t}_m} e^{i\mathbf{k}\cdot\mathbf{t}_m} \phi_\alpha(\mathbf{r} - \mathbf{d}_2 - \mathbf{t}_m) \quad \alpha = s, p_x, p_y. \tag{6.35b}$$

Within the framework of the tight-binding scheme, we can set up the appropriate 6×6 secular determinant, whose diagonalization provides the σ-bands. The three σ-bands of bonding character are well separated in energy from the three antibonding σ-bands by an energy gap of ≈ 5.6 eV, as can be seen in Figure 6.17. It is worthwhile to notice that the hybridization of s, p_x, and p_y orbitals of carbon atoms forming the graphene sheet, generates filled energy bands of electrons in the two-dimensional layer; these bands are inert from the point of view of charge transport, but the strong σ bonds in the plane guarantee the exceptional mechanical stability of the material. Indeed, from the observation that the $C-C$ length is 1.42 Å in graphene (compared with 1.54 Å in diamond) it is expected that the layer sp^2 covalent bonding is quite strong.

We come then to the crucial role of p_z orbitals. Within the wide forbidden gap separating the sp^2 bonding and antibonding states, the p_z orbitals support a *two-dimensional electron gas of massless relativistic-like particles* which is topologically quite different

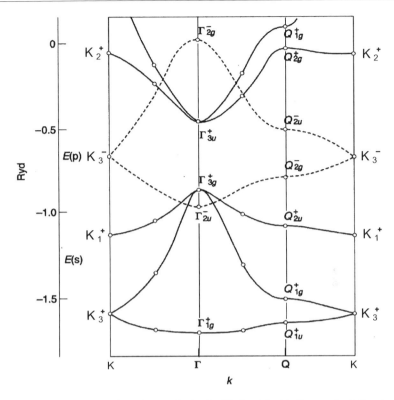

Figure 6.17 Band structure of graphene, obtained with the tight-binding method. Bands which are even and odd under reflection in the plane of graphene are indicated with continuous and broken lines respectively. The energies of the valence states of the carbon atom are also indicated. The values of **k** at the points Γ, K, and Q, are $\mathbf{k} = 0$, $\mathbf{k} = (2\pi/a)(2/3, 0)$, and $\mathbf{k} = (2\pi/a)(1/2, \sqrt{3}/6)$, respectively [from F. Bassani and G. Pastori Parravicini, Nuovo Cimento *50* B, 95 (1967); with kind permission from Società Italiana di Fisica].

with respect to the *traditional two-dimensional electron gas of Schrödinger particles* formed at semiconductor heterojunctions. The p_z orbitals of carbon atoms on the two sublattices of the honeycomb lattice give rise to the highest valence band and to the lowest conduction band, which become degenerate at the corners of the Brillouin zone for symmetry reasons.

For π-bands, we start from the two basis Bloch sums built from the p_z orbitals of the two carbon atoms in the primitive cell

$$\Phi_{1,2}(\mathbf{k}, \mathbf{r}) = \frac{1}{\sqrt{N}} \sum_{\mathbf{t}_m} e^{i\mathbf{k}\cdot\mathbf{t}_m} \phi_z(\mathbf{r} - \mathbf{d}_{1,2} - \mathbf{t}_m), \tag{6.36}$$

The essential features of the π-bands dispersion curves can be easily captured within the semi-empirical tight-binding method (Section 5.2). Assuming nearest neighbor interactions, with estimated parameter $t \approx -3\,\text{eV}$, the representation of the crystal

Hamiltonian on the basis functions (6.36) leads to the 2×2 determinantal equation

$$\begin{Vmatrix} E_p - E & t F(\mathbf{k}) \\ t F(\mathbf{k})^* & E_p - E \end{Vmatrix} = 0, \tag{6.37}$$

where E_p is the orbital energy of the p_z function (including crystal field effects; often the orbital energy is taken as the reference energy and set equal to zero). The geometrical form factor $F(\mathbf{k})$ denotes the sum

$$F(\mathbf{k}) = \sum_{\mathbf{t}_l} e^{i\mathbf{k}\cdot\mathbf{t}_l} = 1 + 2\cos\frac{k_x a}{2}\exp\left(-i\frac{\sqrt{3}k_y a}{2}\right); \tag{6.38}$$

the sum runs over the vectors $0, -\mathbf{t}_1, -\mathbf{t}_2$, since the atom at $\mathbf{d}_1$ has three nearest neighbors located at $\mathbf{d}_2, \mathbf{d}_2 - \mathbf{t}_1$, and $\mathbf{d}_2 - \mathbf{t}_2$.

From the secular equation (6.37) and expression (6.38), we can easily understand the overall behavior of the π-bands of graphene (reported in Figure 6.17). We notice that $F(\mathbf{k})$ takes its maximum value at $\Gamma(\mathbf{k} = 0)$; here the separation between π-bands is maximum. The value of $F(\mathbf{k})$ decreases to one at $Q[\mathbf{k} = (2\pi/a)(1/2, \sqrt{3}/6)]$, which turns out to be a saddle point (see Section 2.7). The value of $F(\mathbf{k})$ vanishes at the corner point $K[\mathbf{k} = (2\pi/a)(2/3, 0)]$ of the two-dimensional Brillouin zone; the π-bands are degenerate at the point K, and the material is said to be a semi-metal.

We consider now the electronic structure of graphene near the neutrality point and we focus on some general aspects related to the underlying bipartite hexagonal lattice. Near the degeneracy points K and K', the geometrical form factor can be expanded in Taylor series to first order in the wavevector measured from the K and K' points. The resulting linearized expressions of the form factors are

$$F_1(k) = -\frac{\sqrt{3}}{2}a(k_x - ik_y) \quad K \text{ point}; \quad F_2(k) = +\frac{\sqrt{3}}{2}a(k_x + ik_y) \quad K' \text{ point};$$

The matrix Hamiltonian near the degeneracy points K or K' can be recast in the form

$$H_K = \begin{vmatrix} 0 & v_F[\hbar k_x - i\hbar k_y] \\ v_F[\hbar k_x + i\hbar k_y] & 0 \end{vmatrix}; \quad H_{K'} = -H_K^\star; \quad v_F = \frac{\sqrt{3}}{2}\frac{a|t|}{\hbar}, \tag{6.39}$$

[the minus sign between $H_{K'}$ and $H_K^\star$ is irrelevant and could be omitted, if necessary, with an appropriate re-orientation of local axes in reciprocal space]. The eigenvalues of H_K and $H_{K'}$ have the typical conical shape

$$E(\mathbf{k}) = \pm v_F \hbar k \quad \text{with} \quad k = \sqrt{k_x^2 + k_y^2}, \tag{6.40}$$

where v_F represents the Fermi velocity. Graphene is a material with zero energy gap, and charge carriers behave around the neutrality point like relativistic particles of rest mass equal to zero. This dispersion relation is at the heart of the striking relativistic quantum Hall effect in graphene, of occurrence of Klein tunneling, and other quantum electrodynamics effects.

Before concluding these remarks on the band structure of graphene, we wish to mention that besides massless Dirac electrons also massive Dirac particles can occur in systems with honeycomb topology, under appropriate perturbation. A way to transform the massless Dirac particles of graphene into massive Dirac particles, consists in growing graphene on a substrate of the iso-electronic and iso-structural boron nitride lattice. The orbital energy of the two sublattices of carbon atoms is influenced in a different way by the crystalline fields I_N and I_B of the underlying nitrogen or boron atoms. With respect to Eq. (6.39), the nearest neighbor tight-binding Hamiltonian is modified in the form

$$H_K = \begin{vmatrix} \Delta & v_F[\hbar k_x - i\hbar k_y] \\ v_F[\hbar k_x + i\hbar k_y] & -\Delta \end{vmatrix}, \tag{6.41}$$

where Δ is half of the difference of the diagonal energy on the two sublattices (and the energy zero has been shifted at the center of the energy gap). Instead of Eq. (6.40) we have now the dispersion curves

$$E(\mathbf{k}) = \mp \sqrt{\Delta^2 + v_F^2 \hbar^2 k^2}. \tag{6.42}$$

It is seen that the above dispersion curves are formally the same produced by the Dirac equation, with the light velocity c replaced by the Fermi velocity v_F and rest mass $m_0 = \Delta/v_F^2$. Other "perturbations" capable to give a mass to the graphene carriers could include bi-layer perturbation, passivation of edges, formation of ribbons and nanomeshes, and other manipulations, to arrive at gapped graphene materials for electronic applications.

Appendix A.   Solved Problems and Complements

Problem 1. Van der Waals forces between two neutral and spherically symmetric atoms or molecules.

Problem 2. Cohesive energy in ionic crystals with an exponential model for the repulsive energy.

Problem 3. Expression of the Madelung energy and partition of the direct and reciprocal lattice sums.

Problem 4. Ewald acceleration method for inverse power-law interactions.

Problem 5. Dirac equation for two-dimensional electrons and bands of graphene.

Problem 1. Van der Waals Forces Between Two Neutral and Spherically Symmetric Atoms or Molecules

Discuss the origin of the van der Waals forces between two neutral, spherically symmetric, and non-overlapping quantum systems. Show that these dispersion forces are attractive and the leading term goes as the inverse six power of the distance.

Consider two neutral and spherically symmetric atoms, or molecules, sufficiently far apart that their electronic clouds do not overlap. If the electrons are considered

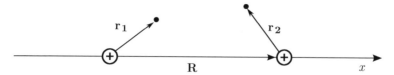

Figure 6.18 Schematic description of two hydrogen atoms with the nuclei fixed at the origin and at distance $|\mathbf{R}|$, while $\mathbf{r}_1$ and $\mathbf{r}_2$ denote the positions of the electrons with respect to the first and second nuclei, respectively; for simplicity we choose the x-axis along $\mathbf{R}$.

classically as a static distribution of charge, there would be no interaction between the two electronic systems. In reality the electrons move even in the lowest stationary state, and although the average dipole moment is zero, each atom is momentarily a dipole, inducing in the other an additional dipole, and leading to an attractive interaction with the inverse six power law of distance.

To infer the essential features and the origin of the mutual interactions, we begin for simplicity to consider two hydrogen atoms, very far apart so that their electron clouds do not overlap. The nuclei are supposed to be fixed at distance $|\mathbf{R}|$ (with $\mathbf{R}$ taken along the x-direction), with one electron near the first nucleus in the relative position $\mathbf{r}_1$ and another electron near the second nucleus at the relative position $\mathbf{r}_2$, as schematically indicated in Figure 6.18.

The total Hamiltonian of the quantum system of Figure 6.18, besides the sum of the Hamiltonians of the two isolated atoms, contains the additional perturbation term given by the following expression

$$V = \frac{e^2}{R} - \frac{e^2}{|\mathbf{R} - \mathbf{r}_1|} - \frac{e^2}{|\mathbf{R} + \mathbf{r}_2|} + \frac{e^2}{|\mathbf{R} - \mathbf{r}_1 + \mathbf{r}_2|}. \tag{A.1}$$

It is convenient to expand the above energy potential in Taylor series assuming that the relative coordinates of all electrons are much smaller than the nuclear distance R.

For this aim, consider the expression

$$\frac{1}{|\mathbf{R} + \mathbf{u}|} = \frac{1}{\sqrt{(R + u_x)^2 + u_y^2 + u_z^2}} = \frac{1}{R} \frac{1}{\sqrt{1 + 2u_x/R + u^2/R^2}}$$

where $\mathbf{u}$ denotes a generic vector. In the case $u \ll R$, we have the series expansion up to quadratic terms

$$\frac{1}{|\mathbf{R} + \mathbf{u}|} = \frac{1}{R} - \frac{u_x}{R^2} + \frac{1}{2R^3} \left[2u_x^2 - u_y^2 - u_z^2 \right] + \cdots \tag{A.2}$$

Using Eq. (A.2) we have in particular

$$-\frac{1}{|\mathbf{R} - \mathbf{r}_1|} = -\frac{1}{R} - \frac{x_1}{R^2} - \frac{1}{2R^3} \left[2x_1^2 - y_1^2 - z_1^2 \right] + \cdots$$

$$-\frac{1}{|\mathbf{R} + \mathbf{r}_2|} = -\frac{1}{R} + \frac{x_2}{R^2} - \frac{1}{2R^3} \left[2x_2^2 - y_2^2 - z_2^2 \right] + \cdots$$

$$\frac{1}{|\mathbf{R} - \mathbf{r}_1 + \mathbf{r}_2|} = \frac{1}{R} + \frac{x_1 - x_2}{R^2} + \frac{1}{2R^3} \left[2(x_1 - x_2)^2 - (y_1 - y_2)^2 - (z_1 - z_2)^2 \right] + \cdots .$$

Inserting these results into Eq. (A.1) provides for the perturbation up to second order in the electronic coordinates

$$V = -\frac{e^2}{R^3} \left[2x_1 x_2 - y_1 y_2 - z_1 z_2 \right]. \tag{A.3}$$

The expression represents the interaction energy of two electric dipoles in the instantaneous configuration of the two electrons. [Higher order terms in Eq. (A.3) like $1/R^4$ or $1/R^5$, etc. represent dipole-quadrupole energy, quadrupole-quadrupole energy, etc. In the case of multi-electron atoms, expression (A.3) must be appropriately summed over all pairs of electrons, one electron near the first nucleus and the other electron near the second nucleus.]

Assuming that the total system is in the ground state, the energy change of the ground state to second order in the electronic coordinates becomes

$$W(R) = \sum_{n(\neq 0)} \frac{|V_{0n}|^2}{E_0 - E_n} \propto -\frac{1}{R^6}, \tag{A.4}$$

where the label n denotes all the states of the system, and the sum must be performed on all states but the ground one $n = 0$. It is evident that the interaction W, which goes as $1/R^6$, is attractive since $E_0 < E_n$ and the numerators are definite positive quantities. Due to the role played by dipoles, the van der Waals forces are also referred to as dispersion forces, for analogy with optical dispersion properties.

Problem 2. Cohesive Energy in Ionic Crystals with an Exponential Model for the Repulsive Energy

Discuss the cohesive energy of ionic crystals assuming an exponential behavior of the repulsive energy between pairs of closed-shell nearest neighbor ions.

In Section 6.2.2 we have considered the total energy of alkali halides crystals using an inverse power-law model for the repulsive energy (see Eq. (6.26)). Here it is assumed that the repulsive energy varies exponentially with the distance. The total energy of the solid, relative to the free ions, is then expressed in the semi-empirical form

$$U_S(R) = N \left(\lambda e^{-R/\rho} - \alpha_M \frac{e^2}{R} \right), \tag{A.5}$$

where N is the number of anions (or cations) of the crystal, R is the nearest neighbor distance, α_M is the Madelung constant, λ and ρ are semi-empirical parameters. The term $-\alpha_M e^2/R$ represents the attractive energy of an ion due to the presence of all the other ions, in the strictly point-ion approximation.

Using Eq. (A.5), we can now proceed to the evaluation of the equilibrium lattice constant, the cohesive energy, and the bulk modulus. As an application, we consider again the case of NaCl crystal characterized with the following parameters: nearest neighbor distance $R_0 = 2.789$ Å $= 5.27 \, a_B$, Madelung energy $\alpha_M e^2/R_0 = 9.02$ eV, experimental cohesive energy per NaCl molecule $E_{\text{coh}} = -8.01$ eV, experimental bulk modulus $B_0 = 0.266$ Mbar.

At the equilibrium distance, we have

$$\left(\frac{dU_S}{dR}\right)_{R=R_0} = 0 \implies \frac{\lambda}{\rho} e^{-R_0/\rho} - \alpha_M \frac{e^2}{R_0^2} = 0.$$

The cohesive energy $U_S(R_0)$ becomes

$$U_S(R_0) = -N\alpha_M \frac{e^2}{R_0}\left(1 - \frac{\rho}{R_0}\right). \tag{A.6}$$

In the case of NaCl, using the reported experimental values for the nearest neighbor distance and for the cohesive energy, one obtains

$$\frac{\rho}{R_0} = 0.112 \implies \rho = 0.312 \text{ Å}.$$

In the NaCl structure, the relation between volume and nearest neighbor distance R is $V = 2NR^3$; from this, the bulk modulus becomes

$$B_0 = V_0\left(\frac{d^2 U_S}{dV^2}\right)_{V=V_0} = V_0 \left(\frac{dR}{dV}\right)_{V=V_0}^2 \left(\frac{d^2 U_S}{dR^2}\right)_{R=R_0} = \frac{\alpha_M e^2}{18 R_0^4}\left(\frac{R_0}{\rho} - 2\right). \tag{A.7}$$

In the case of NaCl, the bulk modulus estimated from Eq. (A.7) is $B_0 = 0.256$ Mbar, again reasonably near to the quoted experimental value ≈ 0.266 Mbar. The semi-empirical estimation of the bulk modulus is to a large extent independent whether the inverse power-law representation or the exponential representation of the short-range repulsive energy is adopted, since both are prepared to fit the short-range interaction, at least around the range of distances of interest.

Problem 3. Expression of the Madelung Energy and Partition of the Direct and Reciprocal Lattice Sums

Consider the expression of the Madelung energy, given as sum on direct and reciprocal lattices. Show explicitly that the splitting parameter η is a dummy variable.

We have seen from Eq. (6.25) that the Madelung energy for a crystal with two sublattices reads (using atomic units for convenience and setting $e^2 = 1$)

$$V(0) = \sum_{t_n(\neq 0)} \frac{1}{t_n} \text{erfc}\left(\sqrt{\eta}\, t_n\right) - \sum_{t_n} \frac{1}{|t_n + d_2|} \text{erfc}\left(\sqrt{\eta}\, |t_n + d_2|\right)$$

$$+ \frac{4\pi}{\Omega} \sum_{h_m(\neq 0)} \left(1 - e^{-i h_m \cdot d_2}\right) \frac{1}{h_m^2} e^{-h_m^2/4\eta} - 2\sqrt{\frac{\eta}{\pi}}, \tag{A.8}$$

where the arbitrary parameter η controls the convergence of the two summations. The expression $V(0) \equiv f(\eta)$ can be split in the form

$$f(\eta) = f_1(\eta) - f_2(\eta),$$

where

$$f_1(\eta) = \sum_{\mathbf{t}_n(\neq 0)} \frac{1}{t_n} \operatorname{erfc}\left(\sqrt{\eta}\, t_n\right) + \frac{4\pi}{\Omega} \sum_{\mathbf{h}_m(\neq 0)} \frac{1}{h_m^2} e^{-h_m^2/4\eta} - 2\sqrt{\frac{\eta}{\pi}}$$

$$f_2(\eta) = \sum_{\mathbf{t}_n} \frac{e^2}{|\mathbf{t}_n + \mathbf{d}_2|} \operatorname{erfc}\left(\sqrt{\eta}\, |\mathbf{t}_n + \mathbf{d}_2|\right) + \frac{4\pi}{\Omega} \sum_{\mathbf{h}_m(\neq 0)} e^{-i\mathbf{h}_m \cdot \mathbf{d}_2} \frac{1}{h_m^2} e^{-h_m^2/4\eta}.$$

Performing the derivative of the first contribution we obtain

$$\frac{df_1(\eta)}{d\eta} = -\frac{1}{\sqrt{\pi \eta}} \sum_{\mathbf{t}_n(\neq 0)} e^{-\eta t_n^2} + \frac{\pi}{\Omega} \frac{1}{\eta^2} \sum_{\mathbf{h}_m(\neq 0)} e^{-h_m^2/4\eta} - \frac{1}{\sqrt{\pi \eta}}$$

$$= -\frac{1}{\sqrt{\pi \eta}} \sum_{\mathbf{t}_n} e^{-\eta t_n^2} + \frac{\pi}{\Omega} \frac{1}{\eta^2} \sum_{\mathbf{h}_m} e^{-h_m^2/4\eta} - \frac{\pi}{\Omega \eta^2} \equiv -\frac{\pi}{\Omega \eta^2}.$$

The reason of the great simplification in the above equation is due to the identity

$$\sum_{\mathbf{t}_n} e^{-\eta t_n^2} = \frac{1}{\Omega} \left(\frac{\pi}{\eta}\right)^{3/2} \sum_{\mathbf{h}_m} e^{-h_m^2/4\eta}, \tag{A.9}$$

which is just the general Ewald identity of Eq. (5.C8), here re-written for the particular case $\mathbf{k} = 0$, $\mathbf{R} = 0$, $x^2 = \eta$. A quite similar calculation shows that the same result is achieved with the derivative $df_2(\eta)/d\eta$. It follows that $f(\eta)$ is a constant, and the splitting parameter a dummy variable.

Problem 4. Ewald Acceleration Method for Inverse Power-Law Interactions

Determine the electrostatic potential for a periodic array of unit point charges, and then discuss the close formal analogy with the calculation of the free-electron Green's function for a periodic array of unit excitations, and other inverse power-law interactions.

In Section 6.2.1 we have considered the superposition of the Coulomb potentials created by a periodic array of point-like charges; we have used the concepts of the Ewald procedure, but we have somewhat simplified the technical part exploiting physical properties of electrostatic potentials. Here we re-visit the general Ewald procedure, because it can handle quite different and demanding situations within a unique algorithm.

Consider the energy potential of a periodic array of point charges

$$V(\mathbf{r}) = \sum_{\mathbf{t}_n} \frac{e^2}{|\mathbf{r} - \mathbf{t}_n|}, \tag{A.10}$$

(in the following we set $e^2 = 1$ in atomic units).

The Ewald procedure starts with the well-known integral representation of the $1/R$ interaction in terms of Gaussian functions:

$$\boxed{\frac{1}{R} = \frac{2}{\sqrt{\pi}} \int_0^\infty e^{-R^2 s^2}\, ds} \quad (R > 0);$$

(A.11)

the above representation is easily verified performing the change of variable $Rs = t$ in the integral.

Following the Ewald recipe, it is convenient to split the integration in Eq. (A.11) into two parts: an integration from the origin to an arbitrary point $\sqrt{\eta}$ (where η is a positive, but otherwise arbitrary quantity), and the integration from $\sqrt{\eta}$ to infinity. We have

$$\frac{1}{R} = \frac{2}{\sqrt{\pi}} \int_0^{\sqrt{\eta}} e^{-R^2 s^2}\, ds + \frac{2}{\sqrt{\pi}} \int_{\sqrt{\eta}}^\infty e^{-R^2 s^2}\, ds \equiv \frac{\mathrm{erf}(\sqrt{\eta} R)}{R} + \frac{\mathrm{erfc}(\sqrt{\eta} R)}{R},$$

where the last passage is obtained with the substitution $Rs = t$ in the integration variable. Using the above result, Eq. (A.10) can be split in the form

$$V(\mathbf{r}) = V_1(\mathbf{r}) + V_2(\mathbf{r}),$$

(A.12)

where

$$V_1(\mathbf{r}) = \frac{2}{\sqrt{\pi}} \sum_{\mathbf{t}_n} \int_0^{\sqrt{\eta}} e^{-(\mathbf{r}-\mathbf{t}_n)^2 s^2}\, ds \equiv \sum_{\mathbf{t}_n} \frac{1}{|\mathbf{r} - \mathbf{t}_n|} \mathrm{erf}(\sqrt{\eta}\, |\mathbf{r} - \mathbf{t}_n|),$$

(A.13)

and

$$V_2(\mathbf{r}) = \frac{2}{\sqrt{\pi}} \sum_{\mathbf{t}_n} \int_{\sqrt{\eta}}^\infty e^{-(\mathbf{r}-\mathbf{t}_n)^2 s^2}\, ds \equiv \sum_{\mathbf{t}_n} \frac{1}{|\mathbf{r} - \mathbf{t}_n|} \mathrm{erfc}(\sqrt{\eta}\, |\mathbf{r} - \mathbf{t}_n|).$$

(A.14)

The term $V_2(\mathbf{r})$ is conveniently calculated in direct space, due to the short-range nature of the complementary error function.

The long-range term $V_1(\mathbf{r})$ is conveniently calculated in reciprocal space using the identity

$$\sum_{\mathbf{t}_n} \exp\left[-(\mathbf{r} - \mathbf{t}_n)^2 s^2\right] = \frac{1}{\Omega} \frac{\pi^{3/2}}{s^3} \sum_{\mathbf{h}_m} e^{i\mathbf{h}_m \cdot \mathbf{r}} \exp\left[-\frac{h_m^2}{4s^2}\right],$$

(A.15)

which is just the general Ewald identity of Eq. (5.C8), here reported for the particular case $\mathbf{k} = 0$ (and some trivial changes of notation). Using the identity (A.15), we can cast Eq. (A.13) in the form

$$V_1(\mathbf{r}) = \frac{2\pi}{\Omega} \sum_{\mathbf{h}_m} e^{i\mathbf{h}_m \cdot \mathbf{r}} \int_0^{\sqrt{\eta}} \frac{1}{s^3} \exp\left[\frac{-h_m^2}{4s^2}\right]\, ds \equiv \frac{4\pi}{\Omega} \sum_{\mathbf{h}_m} e^{i\mathbf{k}_m \cdot \mathbf{r}} \frac{1}{h_m^2} \exp\left[\frac{-h_m^2}{4\eta}\right];$$

(A.16)

in the last passage, use has been made of the indefinite integral

$$\int \frac{1}{x^3} \exp[-a^2/4x^2]\,dx = \frac{1}{2a^2} \exp[-a^2/4x^2].$$

With proper account of the vacant ion (in order to exclude the self-interacting term), one re-obtains the Madelung constant of Eq. (A.8).

The Ewald procedure, carried out along the lines illustrated above, is suitable for handling many other demanding situations. For instance, for the calculation of the free-electron Green's function for a periodic array of unit excitations, instead of the sum of Coulomb contributions as in Eq. (A.10), one considers the sum of free-electron Green's functions; then, instead of the integral representation (A.11) one considers the integral representation of the free-electron Green's function

$$\frac{e^{i\alpha R}}{R} = \frac{2}{\sqrt{\pi}} \int_0^\infty \exp\left[-R^2 s^2 + E/4s^2\right]\,ds \quad (R > 0\,;\, \alpha = \sqrt{E}). \tag{A.17}$$

The procedure has been already embodied in the discussion following Eq. (5.B10), and need not be repeated here.

Similarly, for the inverse power-law interactions, the starting point of the Ewald procedure is the Gaussian representation

$$\frac{1}{R^n} = \frac{1}{\Gamma(n/2)} \int_0^\infty s^{n-1} \exp\left[-R^2 s^2\right]\,ds \quad (R > 0), \tag{A.18}$$

which reduces to Eq. (A.11) for $n = 1$. Following the same pattern, the Ewald acceleration method for inverse power-laws interactions can be worked out, and we refer for details to the paper of Mazars cited in the bibliography.

Problem 5. Dirac Equation for Two-dimensional Electrons and Band Structure of Graphene

Consider the Dirac equation for a massless and massive two-dimensional free electron, and show the formal analogy with the band structure of planar graphene near the neutrality point and gapped graphene-like materials.

For an electron of rest mass m_0 traveling in the three-dimensional free space, the 4×4 Dirac operator reads

$$H_{\text{Dirac}} = c\,\boldsymbol{\alpha} \cdot \mathbf{p} + m_0 c^2 \beta, \tag{A.19}$$

where $\mathbf{p} = -i\hbar\nabla$, and $\alpha_x, \alpha_y, \alpha_z$, and β are the 4×4 Dirac matrices. The expressions of the Dirac matrices in terms of the 2×2 Pauli matrices $\sigma_x, \sigma_y, \sigma_z$ and 2×2 unit matrix $\mathbf{1}$ are

$$\alpha_x = \begin{pmatrix} 0 & \sigma_x \\ \sigma_x & 0 \end{pmatrix} = \begin{pmatrix} 0 & 0 & 0 & 1 \\ 0 & 0 & 1 & 0 \\ 0 & 1 & 0 & 0 \\ 1 & 0 & 0 & 0 \end{pmatrix}; \quad \alpha_y = \begin{pmatrix} 0 & \sigma_y \\ \sigma_y & 0 \end{pmatrix} = \begin{pmatrix} 0 & 0 & 0 & -i \\ 0 & 0 & i & 0 \\ 0 & -i & 0 & 0 \\ i & 0 & 0 & 0 \end{pmatrix},$$

$$\alpha_z = \begin{pmatrix} 0 & \sigma_z \\ \sigma_z & 0 \end{pmatrix} = \begin{pmatrix} 0 & 0 & 1 & 0 \\ 0 & 0 & 0 & -1 \\ 1 & 0 & 0 & 0 \\ 0 & -1 & 0 & 0 \end{pmatrix}; \ \beta = \begin{pmatrix} \mathbf{1} & 0 \\ 0 & -\mathbf{1} \end{pmatrix} = \begin{pmatrix} 1 & 0 & 0 & 0 \\ 0 & 1 & 0 & 0 \\ 0 & 0 & -1 & 0 \\ 0 & 0 & 0 & -1 \end{pmatrix}.$$

When considering plane waves of propagation wavevector $\mathbf{k} = (k_x, k_y, k_z)$, the Dirac operator (A19) reads

$$H(\mathbf{k}) = c\hbar \boldsymbol{\alpha} \cdot \mathbf{k} + m_0 c^2 \beta, \tag{A.20}$$

or in extended form

$$H(\mathbf{k}) = \begin{bmatrix} m_0 c^2 & 0 & c\hbar k_z & c\hbar k_x - ic\hbar k_y \\ 0 & m_0 c^2 & c\hbar k_x + ic\hbar k_y & -c\hbar k_z \\ c\hbar k_z & c\hbar k_x - ic\hbar k_y & -m_0 c^2 & 0 \\ c\hbar k_x + ic\hbar k_y & -c\hbar k_z & 0 & -m_0 c^2 \end{bmatrix}.$$

It is easily verified that the above Hamiltonian squared is a constant. Then, the two twice-degenerate eigenvalues of the above Hamiltonian are

$$W = \mp\sqrt{(m_0 c^2)^2 + c^2 \hbar^2 k^2}. \tag{A.21}$$

Limiting our considerations to plane waves with zero k_z, we obtain

$$H(k_x, k_y) = \begin{bmatrix} m_0 c^2 & 0 & 0 & c\hbar k_x - ic\hbar k_y \\ 0 & m_0 c^2 & c\hbar k_x + ic\hbar k_y & 0 \\ 0 & c\hbar k_x - ic\hbar k_y & -m_0 c^2 & 0 \\ c\hbar k_x + ic\hbar k_y & 0 & 0 & -m_0 c^2 \end{bmatrix}.$$

The above determinant is split in two independent 2×2 determinants, whose formal equivalence with Eqs. (6.39) and (6.41) can be seen by inspection, replacing the light velocity c with the Fermi velocity v_F, and the electron mass m_0 with the effective mass $m^* = \Delta / v_F^2$.

Further Reading

Coulson, C. A. (1961). *Valence*. Oxford: Oxford University Press.

Dresselhaus, M. S. (1998). The wonderful world of carbon. In S. Yoshimura, & R. P. H. Chang (Eds.), *Supercarbon, synthesis, properties and applications*. Berlin: Springer.

Flowers, B. H., & Mendoza E. (1970). *Properties of matter*. London: Wiley.

Geim, A. K. (2011). Random walk to graphene. *Reviews of Modern Physics, 83*, 851.

Harrison, W. A. (1980). *Electronic structure and properties of solids: The physics of the chemical bond*. San Francisco: Freeman; (2004). *Elementary electronic structure*. Singapore: World Scientific.

Klein, M. L., & Venables J. A. (Eds.), (1976/1977). *Rare gas solids* (Vols. I and II). London: Academic Press.

Mazars, M. (2010). Ewald methods for inverse power-law interactions in tridimensional and quasi-two-dimensional systems. *Journal of Physics A: Mathematical and Theoretical, 43*, 425002.

Mooser, E. (1986). Bonds and bands in semiconductors. In P. N. Butcher, N. March, & M. P. Tosi (Eds.), *Crystalline semiconducting materials and devices*. New York: Plenum Press.

Novoselov, K. S. (2011). Graphene: materials in the flatland. *Reviews of Modern Physics, 83*, 837.

Papaconstantopoulos, D. A. (1986). *Handbook of the band structure of elemental solids*. New York: Plenum Press; (2013). New York: Springer [This book contains useful discussions of trends of band structures in elemental crystals and tight-binding parametrization].

Pauling, L. (1960). *The nature of the chemical bond*. New York: Cornell University Press.

Phillips, J. C., & Lukovsky, G. (2009). *Bonds and bands in semiconductors*. New York: Momentum Press.

Rohrer, G. S. (2001). *Structure and bonding in crystalline materials*. Cambridge: Cambridge University Press.

Sutton, A. (1993). *Electronic structure of materials*. Oxford: Clarendon Press.

Tosi, M. P. (1964). Cohesion of ionic solids in the Born model. In F. Seitz, & D. Turnbull (Eds.), *Solid state physics* (Vol. 16, pp. 1). New York: Academic Press.

Tyagi, S. (2005). New series representation for the Madelung constant. *Progress of Theoretical Physics, 114*, 517.

7 Excitons, Plasmons, and Dielectric Screening in Crystals

Chapter Outline head

In the previous two chapters we have considered the band theory of crystals and provided some specific applications. The whole treatment was based on the one-electron approximation; in spite of its great merits, it is often necessary to proceed beyond the independent particle approximation and investigate the many-body effects on the band structure of crystals.

Within the one-electron formalism, the elementary electronic excitations of a crystal are constituted by individual electron-hole pairs. In general, many-body effects not only influence the continuum of single-particle excitations (where resonances may be introduced), but more important, may lead to new excitations in previously forbidden energy regions; the new features, referred to as collective excitations, are regarded as the fingerprint of many-body effects. In this chapter, we focus on excitons and plasmons, which are typical collective effects of a system of interacting electrons; they are present in any type of crystal (metals, semiconductors, and insulators) and are commonly detected by optical measurements and electron energy-loss measurements.

The linear response formalism is adopted to study in homogeneous media the quantum expression of the longitudinal dielectric function, from which bulk collective

Solid State Physics, Second Edition. http://dx.doi.org/10.1016/B978-0-12-385030-0.00007-4

plasmon excitations can be inferred. The dielectric screening is analyzed in a number of relatively simple though significant models that can be used as guidelines for more complicated situations. The chapter concludes with the description of surface plasmon modes and surface polaritons.

7.1 Exciton States in Crystals

Introductory Remarks

Excitons are excited states of the crystals, whose description lies beyond the one-electron approximation and the band theory approach. We begin the discussion of many-body effects in crystals with the study of exciton states in semiconductors or insulators (excitonic effects on the optical properties are described in Chapter 12).

In the one-electron approximation, the electrons are considered to move independently in an appropriate self-consistent periodic potential. In semiconductors and insulators, the fully occupied valence bands are separated from the fully empty conduction bands by an energy gap E_G, which represents the threshold energy for band-to-band electronic transitions (in the independent particle approximation). Many-body effects modify the physical picture: the extra electron in the conduction band and the hole left behind in the valence band can be visualized as interacting via a Coulomb-like field. Thus, the possibility opens of bound "electron-hole" states (*excitons*) with excitation energies lower than the energy gap, as suggested in the original works of Frenkel and Peierls.

Exciton effects are particularly spectacular near the fundamental energy gap of insulators and semiconductors (*valence excitons*), and at the onset of transitions from deep energy bands (*core excitons*); they influence also the continuum of band-to-band transitions. Exciton binding energies vary drastically depending on the kind of material. For instance in neon, a large gap insulator, the exciton binding energy is about 4 eV (see Figure 7.1). In the small gap semiconductor GaAs, the binding energy of valence excitons is a few meV (see Figure 7.2). In metals, excitons from core states are believed to have vanishingly small binding energies, due to metallic screening of the electron-hole interaction; however, exciton effects may strongly enhance the optical absorption from core states to the continuum above the Fermi level, and actually produce the *exciton edge singularities* experimentally observed in many metals in the X-ray region.

Exciton States in the Two-Band Model

We consider now the principles for the quantum description of the exciton states in solids. The basic Hamiltonian of the many-electron system is given by

$$H_e = \sum_i \frac{\mathbf{p}_i^2}{2m} - \sum_{i,I} \frac{Z_I e^2}{|\mathbf{r}_i - \mathbf{R}_{I0}|} + \frac{1}{2} \sum_{i \neq j} \frac{e^2}{|\mathbf{r}_i - \mathbf{r}_j|}, \tag{7.1}$$

where the terms appearing in Eq. (7.1) represent the kinetic energy of the electrons, the electronic-nuclear and the electronic-electronic interaction energy, respectively. The nuclei are supposed fixed at the equilibrium configuration $\{\mathbf{R}_{I0}\}$ and the constant

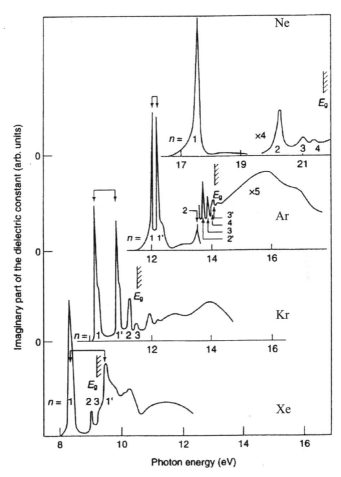

Figure 7.1 Imaginary part of the dielectric constant of rare-gas solids [from G. Zimmerer, in "Excited-State Spectroscopy in Solids", Proc. Intern. School of Physics "Enrico Fermi", edited by U. M. Grassano and N. Terzi (North-Holland, Amsterdam, 1987); with kind permission of Società Italiana di Fisica]. The pronounced structures in the optical absorption spectrum of Ne represent the exciton series $n = 1, 2, 3, \ldots$ converging to the energy gap $E_G = 21.69$ eV of the material. In the heavier rare-gas solids Ar, Kr, Xe the atomic spin-orbit splitting (and hence the splitting in the valence bands) increases; besides the exciton series $n = 1, 2, 3, \ldots$ converging to the energy gap, some spin-orbit partners $n' = 1', 2', 3', \ldots$ are resolved in the experiments.

nucleus-nucleus Coulomb repulsion becomes irrelevant in the study of the energy difference of the electronic states. Spin-orbit terms and other relativistic corrections are omitted for simplicity. In Chapter 4, we have discussed at length how to approximate the many-electron Hamiltonian (7.1) with an appropriate one-electron operator. Now, we face the task to overcome the limitations inherent in the independent particle picture.

To keep formalism at the essential, we consider here the simplest possible model of a semiconductor or insulator, i.e. a *two-band model*, with just one valence band (fully

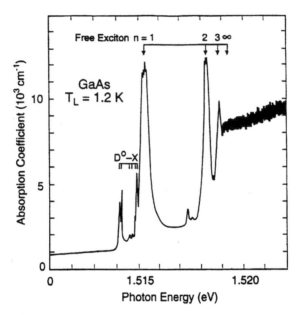

Figure 7.2 Absorption spectrum at 1.2 K of a pure sample of GaAs containing residual donors. The $n = 1, 2, 3$ exciton peaks and the extrapolated energy gap are indicated by arrows. The peak $D^0 - X$ corresponds to the creation of an exciton (X) bound to a neutral donor (D^0) [from R. G. Ulbrich and C. Weisbuch, unpublished; C. Weisbuch, Thesis, Université Paris VII, 1977. For further details see C. Weisbuch, H. Benisty and R. Houdré, J. Luminescence 85, 271 (2000)].

occupied by electrons of either spin) and a fully empty conduction band; a direct gap at the center of the Brillouin zone is assumed (see Figure 7.3). All the other bands of the crystal are not accounted for in detail, but are assumed to influence some physical property (such as the dielectric constant ε of the medium), to be phenomenologically embodied in the two-band model.

Within the one-electron approximation, the ground state of the two-band model is given by the Slater determinant, formed with double occupied valence wavefunctions $\phi_{v\mathbf{k}_i}$ of allowed wavevector $\mathbf{k}_i$ within the first Brillouin zone; we have

$$\Psi_0 = \mathcal{A}\{\phi_{v\mathbf{k}_1}\alpha \, \phi_{v\mathbf{k}_1}\beta \ldots \phi_{v\mathbf{k}_N}\alpha \, \phi_{v\mathbf{k}_N}\beta\}, \tag{7.2}$$

where N is the number of unit cells of the crystal, of volume $V = N\Omega$; the valence (and conduction) wavefunctions are normalized to one in the volume V of the crystal.

The ground state Ψ_0 has total spin equal to zero and total wavevector equal to zero. In the one-electron approximation, the elementary excitations of given total wavevector $\mathbf{k}_{ex}$ are constituted by electron-hole pairs in which an electron from the orbital $\phi_{v\mathbf{k}}$ is transferred to the orbital $\phi_{c\mathbf{k}+\mathbf{k}_{ex}}$. The spin of the hole and the spin of the electron forming the pair can be either up or down, and it is thus possible to construct four different trial excited states. Similarly to what discussed in Section 4.5.3, it is convenient to consider single-particle excitations $\Phi^{(S)}_{c\mathbf{k}+\mathbf{k}_{ex},v\mathbf{k}}$ of definite spin multiplicity S (equal to 0 or to 1).

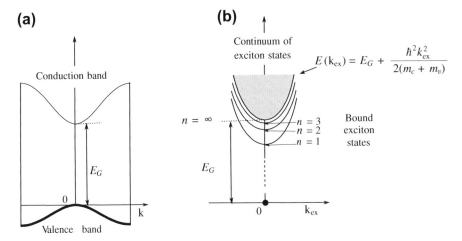

Figure 7.3 (a) Schematic representation of a direct two-band model semiconductor. The lowest conduction band and the highest valence band are shown, assuming band extrema at $\mathbf{k} = 0$ and effective masses m_c and m_v, respectively. The valence band region has been reported in bold to remind its full occupancy by electrons. (b) The electronic ground state of the crystal E_0 (with fully occupied valence band and total wavevector $k_{ex} = 0$) is chosen as zero of energy. The excitonic energy spectrum at $k_{ex} = 0$ is made of discrete levels below E_G and a continuous part above E_G. The curve $E(\mathbf{k}_{ex}) = E_G + \hbar^2 k_{ex}^2/2(m_c + m_v)$ separates the region of discrete exciton levels from the continuum.

The orthonormal basis functions $\{\Phi_{c\mathbf{k}+\mathbf{k}_{ex}, v\mathbf{k}}^{(S)}\}$, when $\mathbf{k}$ varies throughout the first Brillouin zone, can be conveniently used to describe (at least approximately) the *excited states* (or *excitons*) of the many-electron Hamiltonian H_e. We thus expand the exciton wavefunctions of total wavevector $\mathbf{k}_{ex}$ and total spin S in the form

$$\Psi_{ex} = \sum_{\mathbf{k}'} A(\mathbf{k}') \Phi_{c\mathbf{k}'+\mathbf{k}_{ex}, v\mathbf{k}'}^{(S)}. \tag{7.3a}$$

We insert this expansion into the Schrödinger equation $H_e \Psi_{ex} = E \Psi_{ex}$, project on $\langle \Phi_{c\mathbf{k}+\mathbf{k}_{ex}, v\mathbf{k}}^{(S)}|$, and obtain for the expansion coefficients the set of linear homogeneous equations

$$\sum_{\mathbf{k}'} \langle \Phi_{c\mathbf{k}+\mathbf{k}_{ex}, v\mathbf{k}}^{(S)} | H_e | \Phi_{c\mathbf{k}'+\mathbf{k}_{ex}, v\mathbf{k}'}^{(S)} \rangle A(\mathbf{k}') = E A(\mathbf{k}). \tag{7.3b}$$

Using Eqs. (4.35) and (4.36), the matrix elements of H_e in Eq. (7.3b) can be expressed in terms of bielectronic integrals involving the band wavefunctions. Thus the coefficients $A(\mathbf{k})$ satisfy the integral equation

$$\left[E_c(\mathbf{k} + \mathbf{k}_{ex}) - E_v(\mathbf{k}) - E \right] A(\mathbf{k}) + \sum_{\mathbf{k}'} U(\mathbf{k}, \mathbf{k}'; \mathbf{k}_{ex}) A(\mathbf{k}') = 0, \tag{7.4}$$

where the kernel of the integral equation reads

$$U(\mathbf{k}, \mathbf{k}'; \mathbf{k}_{ex}) = U_1(\mathbf{k}, \mathbf{k}'; \mathbf{k}_{ex}) + U_2(\mathbf{k}, \mathbf{k}'; \mathbf{k}_{ex}) \tag{7.5}$$

with

$$U_1(\mathbf{k}, \mathbf{k}'; \mathbf{k}_{\mathrm{ex}}) = -\langle \phi_{c\mathbf{k}+\mathbf{k}_{\mathrm{ex}}} \phi_{v\mathbf{k}'} | \frac{e^2}{r_{12}} | \phi_{c\mathbf{k}'+\mathbf{k}_{\mathrm{ex}}} \phi_{v\mathbf{k}} \rangle ,$$

$$U_2(\mathbf{k}, \mathbf{k}'; \mathbf{k}_{\mathrm{ex}}) = 2\delta_{S,0} \langle \phi_{c\mathbf{k}+\mathbf{k}_{\mathrm{ex}}} \phi_{v\mathbf{k}'} | \frac{e^2}{r_{12}} | \phi_{v\mathbf{k}} \phi_{c\mathbf{k}'+\mathbf{k}_{\mathrm{ex}}} \rangle .$$

The general properties of the bielectronic integrals are discussed in Appendix 4.A. In Eq. (7.5) the kernel is split into two parts which have a different role in determining the excitonic states (as elaborated in detail below). The first term is spin-independent, is basically peaked in **k**-space and controls to a large extent the Rydberg-like excitonic series. The second term is spin-dependent, is tendentially flat in **k**-space and is responsible of the singlet-triplet separation and other energy shifts, especially of the lowest members of the excitonic series.

Approximate Solution of the Exciton Integral Equation (Two-Band Model)

The integral Eq. (7.4), with the kernel (7.5), is the basic eigenvalue equation for excitons in the two-band model. Even in this highly idealized approximation (that neglects, for instance, the multi-band structure of valence or conduction bands, spin-orbit interactions, many-body effects of the other bands, and multi-configuration interactions) the accurate solution of the exciton problem may be rather demanding, as can be inferred from the fact that the Eq. (7.4) is an integral equation (rather than a more familiar differential type equation) and the kernel involves bielectronic integrals (a typical fingerprint of many-body effects). Without entering too much in technical aspects, nevertheless it is instructive to show that the integral Eq. (7.4) can be brought (approximately) to the form of the Schrödinger equation of a hydrogen-like system, with appropriate effective mass and screened Coulomb interaction, and possible short-range corrections.

The kernel (7.5) is expressed in terms of bielectronic integrals among two pairs of wavefunctions, and this expression is so complicated that it calls for some simplification. Indeed, a reasonable simplification can be found in periodic structures, where the orbitals $\phi_{c\mathbf{k}}$ and $\phi_{v\mathbf{k}}$ entering in Eq. (7.5) are of Bloch form and thus can be expressed as the product of plane waves and periodic parts (normalized to unity in the volume V of the crystal). Due to the different spatial character of the periodic parts (periodic and rapidly changing in the unit cell) compared with the spatial character of plane waves (non-periodic but smoothly varying on the unit cell, at least for small vectors in the Brillouin zone) we can factorize (approximately) the bielectronic integrals in a contribution involving the smooth part of the wavefunction and a contribution involving the rapidly changing part. For example, the first term of the kernel (7.5) can be elaborated as follows

$$U_1(\mathbf{k}, \mathbf{k}'; \mathbf{k}_{\mathrm{ex}}) = -\langle \phi_{c\mathbf{k}+\mathbf{k}_{\mathrm{ex}}} \phi_{v\mathbf{k}'} | \frac{e^2}{r_{12}} | \phi_{c\mathbf{k}'+\mathbf{k}_{\mathrm{ex}}} \phi_{v\mathbf{k}} \rangle$$

$$= -\int u^*_{c\mathbf{k}+\mathbf{k}_{\mathrm{ex}}}(\mathbf{r}_1) u^*_{v\mathbf{k}'}(\mathbf{r}_2) u_{c\mathbf{k}'+\mathbf{k}_{\mathrm{ex}}}(\mathbf{r}_1) u_{v\mathbf{k}}(\mathbf{r}_2) e^{-i(\mathbf{k}-\mathbf{k}')\cdot(\mathbf{r}_1-\mathbf{r}_2)} \frac{e^2}{r_{12}} \, d\mathbf{r}_1 \, d\mathbf{r}_2$$

$$\approx -\langle u_{c\mathbf{k}+\mathbf{k}_{\mathrm{ex}}} | u_{c\mathbf{k}'+\mathbf{k}_{\mathrm{ex}}} \rangle \langle u_{v\mathbf{k}'} | u_{v\mathbf{k}} \rangle \frac{1}{V^2} \int e^{-i(\mathbf{k}-\mathbf{k}')\cdot(\mathbf{r}_1-\mathbf{r}_2)} \frac{e^2}{r_{12}} \, d\mathbf{r}_1 \, d\mathbf{r}_2 . \quad (7.6)$$

The last step has been done replacing the product of any two periodic functions having the same argument with the average value on the volume of the crystal; for instance:

$$u^*_{c\mathbf{k}+\mathbf{k}_{ex}}(\mathbf{r}_1)u_{c\mathbf{k}'+\mathbf{k}_{ex}}(\mathbf{r}_1) \longrightarrow \frac{1}{V}\left\langle u_{c\mathbf{k}+\mathbf{k}_{ex}}|u_{c\mathbf{k}'+\mathbf{k}_{ex}}\right\rangle.$$

With the further approximation

$$\left\langle u_{c\mathbf{k}+\mathbf{k}_{ex}}|u_{c\mathbf{k}'+\mathbf{k}_{ex}}\right\rangle \approx 1 \quad\text{and}\quad \left\langle u_{v\mathbf{k}'}|u_{v\mathbf{k}}\right\rangle \approx 1 \quad\text{for}\quad \mathbf{k}\approx\mathbf{k}',$$

and setting $\mathbf{r} = \mathbf{r}_1 - \mathbf{r}_2$, the expression (7.6) becomes

$$U_1(\mathbf{k},\mathbf{k}';\mathbf{k}_{ex}) = -\frac{1}{V}\int e^{-i(\mathbf{k}-\mathbf{k}')\cdot\mathbf{r}}\frac{e^2}{r}\,d\mathbf{r} = -\frac{1}{V}\frac{4\pi e^2}{|\mathbf{k}-\mathbf{k}'|^2} \quad\text{for}\quad \mathbf{k}\approx\mathbf{k}'. \quad (7.7a)$$

Within the stated approximations, it is seen that the kernel U_1 for the exciton states is the Fourier transform of the (unscreened) Coulomb interaction $-e^2/r$.

In the following, we take the liberty to screen the Coulomb interaction with an appropriate background dielectric constant ε of the medium (for simplicity, ε is assumed to be frequency and momentum independent). A full many-body treatment, which decouples the given pair of bands from all the others, justifies the screening of the electron-hole interaction $-e^2/r$ and suggests more sophisticated screening procedures [see for instance G. D. Mahan "Many-Particle Physics" (Kluwer Academic, New York, 2000)].

For singlet excitons ($S = 0$), a similar elaboration can be carried out for the term U_2 in the kernel (7.5). In this case we can approximate

$$u^*_{c\mathbf{k}+\mathbf{k}_{ex}}(\mathbf{r}_1)u_{v\mathbf{k}}(\mathbf{r}_1) \approx \frac{1}{V}\left\langle u_{c\mathbf{k}+\mathbf{k}_{ex}}|u_{v\mathbf{k}}\right\rangle.$$

We are particularly interested in excitons with $\mathbf{k}_{ex}$ small with respect to the Brillouin zone, because of their role in optical transitions (see Chapter 12). In the long wavelength limit ($\mathbf{k}_{ex} \to 0$) of the exciton wavevector, we can expand the conduction band wavefunction to first order in $\mathbf{k}_{ex}$. Then we obtain

$$\langle u_{c\mathbf{k}+\mathbf{k}_{ex}}|u_{v\mathbf{k}}\rangle = \langle u_{c\mathbf{k}} + \mathbf{k}_{ex}\cdot\frac{\partial u_{c\mathbf{k}}}{\partial\mathbf{k}}|u_{v\mathbf{k}}\rangle$$

$$= \mathbf{k}_{ex}\cdot\langle\frac{\partial}{\partial\mathbf{k}}u_{c\mathbf{k}}|u_{v\mathbf{k}}\rangle = i\mathbf{k}_{ex}\cdot\langle u_{c\mathbf{k}}|\mathbf{r}|u_{v\mathbf{k}}\rangle = i\mathbf{k}_{ex}\cdot\mathbf{r}_{cv},$$

where the last step has been done using Eq. (2.35). We thus obtain for the U_2 term the expression

$$U_2(\mathbf{k},\mathbf{k}';\mathbf{k}_{ex}) = \frac{1}{V}2\delta_{S,0}\frac{4\pi e^2}{k_{ex}^2}(\mathbf{k}_{ex}\cdot\mathbf{r}_{cv})(\mathbf{k}_{ex}\cdot\mathbf{r}^*_{cv}) \quad (7.7b)$$

which is essentially the classical dipole-dipole interaction energy of a static system of dipoles of electric moment $\mathbf{d}_{cv}/N = e\mathbf{r}_{cv}/N$, localized on each of the N unit cells of the crystal. Also the dipole-dipole interaction should be screened by the background dielectric constant ε of the medium (for simplicity). Differently from U_1, which is strongly peaked for $\mathbf{k} \approx \mathbf{k'}$, the term U_2 turns out to be independent (or possibly weakly dependent) on $\mathbf{k}$ or $\mathbf{k'}$, and represents thus a short-range interaction in real space. This term has in general a minor (or negligible) effect on the exciton binding energies, especially when the radius of the exciton is sufficiently large.

We now specify the integral equation (7.4), with the kernels $U_1 + U_2$ appropriately screened, in the case the valence and conduction bands are both parabolic in the $\mathbf{k}$-region of interest around the center of the Brillouin zone (see Figure 7.3). For parabolic bands we have

$$E_c(\mathbf{k}) = E_G + \frac{\hbar^2 k^2}{2m_c} \quad \text{and} \quad E_v(\mathbf{k}) = -\frac{\hbar^2 k^2}{2m_v} \quad (m_c > 0, \ m_v > 0).$$

It follows

$$\begin{aligned}
E_c(\mathbf{k} + \mathbf{k}_{ex}) - E_v(\mathbf{k}) &= E_G + \frac{\hbar^2}{2m_c}(\mathbf{k} + \mathbf{k}_{ex})^2 + \frac{\hbar^2 k^2}{2m_v} \\
&\equiv E_G + \frac{\hbar^2}{2\mu_{ex}}\left(\mathbf{k} + \frac{\mu_{ex}}{m_c}\mathbf{k}_{ex}\right)^2 + \frac{\hbar^2 k_{ex}^2}{2(m_c + m_v)}, \quad (7.8a)
\end{aligned}$$

where $\mu_{ex}^{-1} = m_c^{-1} + m_v^{-1}$; μ_{ex} is called the reduced effective mass of the exciton. From Eq. (7.8a) it is seen that the minimum value of single-particle excitations of given wavevector $\mathbf{k}_{ex}$ is

$$\text{Min}\left[E_c(\mathbf{k} + \mathbf{k}_{ex}) - E_v(\mathbf{k})\right] = E_G + \frac{\hbar^2 k_{ex}^2}{2(m_c + m_v)}. \quad (7.8b)$$

The quantity (7.8b) represents the threshold energy of the continuum of individual electron-hole excitations (and is plotted in Figure 7.3b); it is the sum of the energy gap and the kinetic energy associated to the center-of-mass motion of the electron-hole pair.

The integral Eq. (7.4), for excitons with $\mathbf{k}_{ex} \approx 0$, takes the form

$$\begin{aligned}
&\left[E_G + \frac{\hbar^2 k^2}{2\mu_{ex}} - E\right] A(\mathbf{k}) \\
&\quad - \frac{1}{V}\sum_{\mathbf{k'}} \frac{4\pi e^2}{\varepsilon|\mathbf{k} - \mathbf{k'}|^2} A(\mathbf{k'}) + \delta_{S,0}\frac{8\pi}{\varepsilon}|\widehat{\mathbf{k}}_{ex} \cdot \mathbf{d}_{cv}|^2 \frac{1}{V}\sum_{\mathbf{k'}} A(\mathbf{k'}) = 0, \quad (7.9)
\end{aligned}$$

where $\widehat{\mathbf{k}}_{ex} = \mathbf{k}_{ex}/|\mathbf{k}_{ex}|$. In order to solve Eq. (7.9) it is convenient to introduce the so-called *envelope function, normalized to unity*, and defined as

$$F(\mathbf{r}) = \frac{1}{\sqrt{V}}\sum_{\mathbf{k'}} A(\mathbf{k'}) e^{i\mathbf{k'} \cdot \mathbf{r}}. \quad (7.10)$$

We can transform the eigenvalue integral equation (7.9) for the $A(\mathbf{k})$ function into a more familiar differential eigenvalue equation for the envelope function $F(\mathbf{r})$. For this purpose we remark the following relations:

$$A(\mathbf{k}) = \frac{1}{\sqrt{V}} \int F(\mathbf{r}) e^{-i\mathbf{k}\cdot\mathbf{r}} \, d\mathbf{r} \, ; \quad k^2 A(\mathbf{k}) = \frac{1}{\sqrt{V}} \int [-\nabla^2 F(\mathbf{r})] e^{-i\mathbf{k}\cdot\mathbf{r}} \, d\mathbf{r} \, ;$$

$$-\frac{1}{\sqrt{V}} \sum_{\mathbf{k}'} \frac{4\pi e^2}{\varepsilon |\mathbf{k} - \mathbf{k}'|^2} A(\mathbf{k}') = -\int \frac{e^2}{\varepsilon r} F(\mathbf{r}) e^{-i\mathbf{k}\cdot\mathbf{r}} \, d\mathbf{r} \, ;$$

$$\frac{1}{\sqrt{V}} \sum_{\mathbf{k}'} A(\mathbf{k}') = \int \delta(\mathbf{r}) F(\mathbf{r}) e^{-i\mathbf{k}\cdot\mathbf{r}} \, d\mathbf{r} \, ;$$

[the above four relations are easily proved putting in their second members the expression (7.10) for the envelope function]. We insert the above relations into Eq. (7.9), and obtain the equation for the envelope function of the form

$$\left[-\frac{\hbar^2 \nabla^2}{2\mu_{\text{ex}}} - \frac{e^2}{\varepsilon r} + \delta_{S,0} \frac{8\pi}{\varepsilon} |\hat{\mathbf{k}}_{\text{ex}} \cdot \mathbf{d}_{\text{cv}}|^2 \delta(\mathbf{r}) \right] F(\mathbf{r}) = (E - E_G) F(\mathbf{r}) . \tag{7.11}$$

Approximations of the kernel more accurate than the adopted Eqs. (7.7) would introduce corrective terms in the Schrödinger Eq. (7.11), without changing however its basic structure.

Equation (7.11), omitting momentarily the short-range spin-dependent term, provides a simple description of the exciton states in terms of a hydrogen-like atom in a polarizable medium (Wannier model). The effective Rydberg of the exciton problem (i.e. the binding energy $E_b^{(\text{ex})}$ of the lowest exciton state) can be estimated from the relation

$$E_b^{(\text{ex})} \approx 13.6 \, \frac{\mu_{\text{ex}}}{m} \frac{1}{\varepsilon^2} \quad (\text{in eV}) \qquad (1 \text{ Ryd} = 13.606 \text{ eV}).$$

Typically in semiconductors $\varepsilon \approx 10$ and $\mu_{\text{ex}} \approx 0.1 \, m$ and thus the binding energy is of the order of a few millielectronvolts. For strong insulators (such as neon for instance) we have $\varepsilon \approx 2$ and $\mu_{\text{ex}} \approx m$ and the binding energy is of the order of a few electronvolts. In any case, exciton levels below the gap tend to group in well-defined hydrogenic-like series (as seen in Figures 7.1 and 7.2). The effective radius of the ground exciton state can be estimated from the relation

$$a_{\text{ex}} = a_B \frac{m}{\mu_{\text{ex}}} \varepsilon \qquad (a_B = 0.529 \text{ Å}).$$

With the above estimated values of μ_{ex} and ε, we see that the wavefunction of electrons and holes bound together may extend over a few unit cells in large gap insulators, and over several thousand unit cells in weakly bound excitons. The Schrödinger equation of type (7.11) was originally introduced in the literature to describe shallow excitons; appropriately implemented by a more accurate description of the electron-hole interaction at small distances, it can be used to describe also strongly bound valence or core excitons [see for instance S. Baroni, G. Grosso, L. Martinelli, and G. Pastori Parravicini, Phys. Rev. B 20, 1713 (1979); Phys. Rev. B 22, 6440 (1980), and references quoted therein].

An exciton can propagate in the crystal and transport energy, without transport of net charge. The treatment for excitons with $\mathbf{k}_{ex} \neq 0$, only requires the trivial inclusion of the center-of-mass kinetic energy $\hbar^2 k_{ex}^2/2(m_c + m_v)$, as can be seen considering Eq. (7.4) and Eqs. (7.8), and keeping $\mathbf{k}_{ex} \neq 0$.

We add a few considerations on the short-range term in Eq. (7.11). This term is present only for singlet excitons ($S = 0$) and is responsible of the energy difference between transverse excitons ($\mathbf{k}_{ex} \perp \mathbf{d}_{cv}$) and longitudinal excitons ($\mathbf{k}_{ex} \| \mathbf{d}_{cv}$). The expectation value of the short-range term on $F(\mathbf{r})$ is proportional to $|F(0)|^2$, which can be interpreted as the probability that the electron and the hole of the exciton are on the same position. Thus short-range corrections are expected to have a role especially for the lowest members of the s-like exciton series; for higher members and for shallow excitons $|F(0)|^2$ decreases and becomes rapidly negligible.

We do not enter in deeper analysis of screening and local field effects, or consider extensions to multi-band models, or discuss further the adopted simplifications; in spite of these limits, the two-band model presented here allows a reasonably intuitive picture of the excitonic effects and a feeling of how to construct more general approaches.

7.2   Plasmon Excitations in Crystals

From the discussion of the previous section, excitons can be viewed as a coherent combination of initially independent electron-hole excitations, coordinated by the long-range Coulomb field. In this section, we focus on the concept of plasma oscillations, or plasmons, which are longitudinal electron-density oscillations, brought about by the Coulomb interaction between the electrons. Plasmon excitations can be viewed as *collective* electronic charge-density oscillations (or fluctuations) formed by *individual* electron-hole pairs fluctuations, organized by the long-range Coulomb field.

Plasmon excitations can be intuitively described with the following classical picture. Consider a free-electron gas of average density n, embedded in a uniform and fixed background of neutralizing positive charges. Suppose that the electrons originally at $\mathbf{r}$ undergo a time-dependent longitudinal displacement $u(\mathbf{r}, t)$ of the form

$$\mathbf{u}(\mathbf{r}, t) = \mathbf{u}_0 e^{i(\mathbf{q}\cdot\mathbf{r}-\omega t)} + \text{c.c.} \qquad \mathbf{u}_0 \| \mathbf{q}, \qquad (7.12a)$$

where c.c. indicates complex conjugate of the previous term. The polarization of the system due to the electronic displacement is $\mathbf{P}(\mathbf{r}, t) = n(-e)\mathbf{u}(\mathbf{r}, t)$; the associated microscopic charge is

$$\rho_{micr}(\mathbf{r}, t) = -\text{div}\,\mathbf{P}(\mathbf{r}, t) = ine\,\mathbf{q} \cdot \mathbf{u}_0\,e^{i(\mathbf{q}\cdot\mathbf{r}-\omega t)} + \text{c.c.} \qquad \mathbf{u}_0 \| \mathbf{q}.$$

For longitudinal displacements $\mathbf{u}_0$ and $\mathbf{q}$ are parallel, and $\rho_{micr} \neq 0$; the longitudinal electric field satisfying $\text{div}\,\mathbf{E} = 4\pi\rho_{micr}$ is given by

$$\mathbf{E}(\mathbf{r}, t) = -4\pi\mathbf{P}(\mathbf{r}, t) = 4\pi n e\,\mathbf{u}(\mathbf{r}, t). \qquad (7.12b)$$

We can thus write the classical equation of motion for each electron of the gas

$$m\ddot{\mathbf{u}}(\mathbf{r}, t) = -e\mathbf{E}(\mathbf{r}, t) = -4\pi n e^2\,\mathbf{u}(\mathbf{r}, t). \qquad (7.12c)$$

This equation shows that the longitudinal electron oscillations occur with the frequency ω_p (*plasma frequency*) given by

$$\boxed{\omega_p^2 = \frac{4\pi n e^2}{m}}.$$ (7.13)

Notice that this classic model fails to predict a wavevector dependence of the frequency of the longitudinal charge fluctuations.

The range of values of the plasmon quanta can be obtained inserting into Eq. (7.13) the dimensionless parameter r_s related to the electron density by the expression $(4/3)\pi r_s^3 a_B^3 = 1/n$. One obtains

$$\hbar^2 \omega_p^2 = \frac{12}{r_s^3} \frac{\hbar^2}{2ma_B^2} \frac{e^2}{2a_B} \longrightarrow \hbar\omega_p = \sqrt{\frac{12}{r_s^3}} \text{ Rydberg;}$$

it also holds

$$\frac{\hbar\omega_p}{E_F} = 0.94 \sqrt{r_s}.$$

For ordinary metals with $2 < r_s < 6$, typical values of $\hbar\omega_p$ are in the range 3–17 eV.

The bulk plasmons in isotropic and homogeneous media are purely longitudinal waves, and cannot couple with transverse electromagnetic waves. The experimental detection of plasma excitations is often inferred from energy-loss measurements of monoenergetic electron beams, transmitted or reflected by a sample. In Figure 7.4

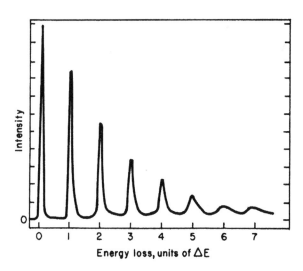

Figure 7.4 Energy-loss spectrum (at very small angles) for a beam of 20 keV primary electrons passing through an aluminum film of thickness $\approx$ 2580 Å. The unit of energy-loss ΔE ($\approx$15 eV) is the plasmon excitation energy for Al. Zero-loss, first-loss, second-loss, and other plural-loss peaks are clearly detected [reprinted with permission from L. Marton, J. A. Simpson, H. A. Fowler and N. Swanson, Phys. Rev. *126*, 182 (1962); copyright 1962 by the American Physical Society].

we report, as an example, the energy-loss spectrogram of a beam of 20 keV primary electrons, transmitted through an Al thin film; the sharp peaks in inelastic scatterings correspond to plural plasmon excitations.

Coherent electron density oscillations are possible not only in the bulk of media, but also on the surface or at the interfaces between different materials, or in nanoparticles. In contrast with volume plasmons, that involve pure longitudinal electric field waves, surface plasmons involve appropriate longitudinal, and transverse electric and magnetic field, and coupling to photons becomes possible. Considerations on the classical description of surface plasmons modes are the subject of the final section of this chapter.

In the other sections, we consider some relevant aspects of dielectric screening and volume plasmons from a quantistic point of view. For this purpose, we establish within the linear response theory the microscopic quantum expression of the longitudinal dielectric function of crystals; plasmons are then inferred from the identification of self-sustaining collective motions of electron particles. Plasmons can be considered (ideally) as elementary excitations of the interacting electron system, and are a typical many-body longitudinal-wave effect.

7.3 Static Dielectric Screening in Metals within the Thomas-Fermi Model

Consider a medium and an extra point charge Ze embedded in it. In a dielectric (semiconductor or insulator) it is well known that the test charge Ze is surrounded by a screening charge equal to $-(1 - 1/\varepsilon_s)Ze$, where ε_s is the static dielectric constant of the medium. In a metal instead the screening is complete; the *bare Coulomb potential* (born from the external charge Ze) is modified into a (total) *screened Coulomb potential* (born from the external charge Ze and the screening charge induced in the metal), with the following peculiar properties: (i) The screened Coulomb field is cut off at a characteristic distance of the order of k_F^{-1}; the electrons are very effective in screening external charges. (ii) Weakly decaying long-range oscillations of electron density occur (Friedel oscillations); Knight shift in nuclear magnetic resonance of the nuclei near impurities in metals confirms these oscillations. (iii) The change of electron density must be finite at the origin (where the point charge Ze is located). In fact the lifetime for positron annihilation in metals (which is proportional to the electron density) is finite. Also the Knight shift of the nuclei of impurities is finite.

To understand qualitatively the highly effective shielding in metals, we consider first the Thomas-Fermi model. This model explains quite well the exponential screening at intermediate distances, but fails in predicting finite induced charge at the origin ($r \approx 0$) and long-range oscillations ($r \approx \infty$). [Further aspects of dielectric screening in metals are discussed in Appendix A.]

Suppose that the electrons of a metal, at the point $\mathbf{r}$, are subjected to a total static potential

$$U(\mathbf{r}) = U(\mathbf{q})\, e^{i\mathbf{q}\cdot\mathbf{r}} + \text{c.c.} \,;$$

the corresponding electric field is of longitudinal type and is given by

$$\mathbf{E}(\mathbf{r}) = -\nabla \frac{U(\mathbf{r})}{(-e)} = \frac{1}{e} i\, \mathbf{q}\, U(\mathbf{q}) e^{i\mathbf{q}\cdot\mathbf{r}} + \text{c.c.} \qquad \mathbf{E}\|\mathbf{q}.$$

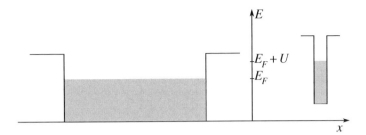

Figure 7.5 Schematic representation of the linearized Thomas-Fermi approximation. On the left we have represented a metal with Fermi energy E_F, and density-of-states $n(E_F)$ (per unit volume and both spin directions). On the right we have represented an (isolated) piece of the same metal, with all band structure rigidly shifted by a superimposed energy U. When the contact is established between the reservoir on the left and the (small) sample on the right, the Fermi energy of the sample equalizes E_F; as a consequence, the induced electron density in the sample is $n_{ind} = -n(E_F)U$.

In order to determine the static dielectric function $\varepsilon(\mathbf{q}, \omega = 0) = \varepsilon(\mathbf{q})$ of the metal, we have to establish the induced electron density $n_{ind}(\mathbf{r})$ related to the perturbation potential $U(\mathbf{r})$. A very simple answer is provided by the Thomas-Fermi approximation (Figure 7.5), that requires that the Fermi level of the material remains constant throughout its volume.

With reference to Figure 7.5, the induced change of electron density is then

$$n_{ind}(\mathbf{r}) = -n(E_F)U(\mathbf{r}) = -n(E_F)U(\mathbf{q})e^{i\mathbf{q}\cdot\mathbf{r}} + \text{c.c.},$$

where $n(E_F)$ is the density-of-states of the metal (per unit volume and both spin directions) at the Fermi energy (we have also assumed that $U(\mathbf{r})$ is small enough that the density-of-states of the metal can be taken as constant). The induced change of electron charge density is $\rho_{ind}(\mathbf{r}) = -e\, n_{ind}(\mathbf{r})$. The macroscopic polarization field becomes

$$\text{div } \mathbf{P} = -\rho_{ind} \quad \Longrightarrow \quad \mathbf{P}(\mathbf{r}) = \frac{e\,n(E_F)}{q^2} i\,\mathbf{q}\,U(\mathbf{q})e^{i\mathbf{q}\cdot\mathbf{r}} + \text{c.c.}.$$

The static dielectric function of the medium becomes

$$\varepsilon(q) = 1 + 4\pi\frac{\mathbf{P}}{\mathbf{E}} = 1 + \frac{4\pi e^2 n(E_F)}{q^2},$$

which is known as the Thomas-Fermi dielectric function.

The above result can also be written as

$$\boxed{\varepsilon_{TF}(q) = 1 + \frac{k_{TF}^2}{q^2}} \quad k_{TF}^2 = 4\pi e^2 n(E_F). \tag{7.14}$$

For a free-electron gas, one has

$$k_{TF}^2 = 4\pi e^2 \frac{3}{2}\frac{n}{E_F} = 6\pi e^2 \frac{k_F^3}{3\pi^2}\frac{2m}{\hbar^2 k_F^2} = \frac{4}{\pi}\frac{k_F}{a_B},$$

where $a_B = \hbar^2/me^2$ is the Bohr radius. With $k_F = 1.92/r_s a_B$, and $2 < r_s < 6$ at ordinary metallic densities, the screening wavenumber k_{TF} turns out to be of the order of a_B^{-1}.

To better illustrate the meaning of k_{TF}, consider an extra point charge Ze, at the origin in the medium. The bare Coulomb potential energy is

$$U_{ext}(\mathbf{r}) = -\frac{Ze^2}{r} \equiv -\frac{1}{(2\pi)^3} \int \frac{4\pi Ze^2}{q^2} e^{i\mathbf{q}\cdot\mathbf{r}} d\mathbf{q} \tag{7.15a}$$

(the last passage uses the standard Fourier representation of $1/r$). The total potential energy $U(\mathbf{r})$ is obtained screening with $\varepsilon(q)$ the integrand in Eq. (7.15a)

$$U(\mathbf{r}) = -\frac{1}{(2\pi)^3} \int \frac{1}{\varepsilon(q)} \frac{4\pi Ze^2}{q^2} e^{i\mathbf{q}\cdot\mathbf{r}} d\mathbf{q}. \tag{7.15b}$$

Using the Thomas-Fermi dielectric function, one obtains

$$U_{TF}(\mathbf{r}) = -\frac{1}{(2\pi)^3} \int \frac{4\pi Ze^2}{q^2 + k_{TF}^2} e^{i\mathbf{q}\cdot\mathbf{r}} d\mathbf{q} = -\frac{Ze^2}{r} \exp(-rk_{TF}).$$

Thus, the bare long-range Coulomb interaction Ze^2/r is transformed into an exponentially damped interaction with screening length equal to $1/k_{TF}$.

We consider now the induced charge density around the point impurity Ze (inserted at the origin). Using the Poisson equation, we have

$$-\nabla^2 \frac{U(\mathbf{r}) - U_{ext}(\mathbf{r})}{(-e)} = 4\pi \rho_{ind}(\mathbf{r}).$$

Inserting Eqs. (7.15) in the above expression, one finds

$$\rho_{ind}(\mathbf{r}) = Ze \frac{1}{(2\pi)^3} \int \left[\frac{1}{\varepsilon(q)} - 1\right] e^{i\mathbf{q}\cdot\mathbf{r}} d\mathbf{q}; \tag{7.16}$$

the total induced screening charge Q_S is

$$Q_S = \int \rho_{ind}(\mathbf{r}) d\mathbf{r} = Ze \left[\frac{1}{\varepsilon(0)} - 1\right].$$

This equation shows that in metals, where $\varepsilon(0) \to \infty$, the total displaced charge is $Q_S = -Ze$, and the screening is complete.

In the case of the Thomas-Fermi dielectric function, Eq. (7.16) gives

$$\rho_{ind}^{TF}(\mathbf{r}) = -\frac{Ze}{(2\pi)^3} \int \frac{k_{TF}^2}{q^2 + k_{TF}^2} e^{i\mathbf{q}\cdot\mathbf{r}} d\mathbf{q} = -e\, n(E_F) \frac{Ze^2}{r} \exp(-rk_{TF});$$

thus $\rho_{ind}(0)$ is singular at the origin (contrary to experimental evidence, such as finite lifetime of positrons in metals and the finite Knight shift of impurity atoms). In order to have a finite value of $\rho_{ind}(0)$ in Eq. (7.16), it is necessary that $\varepsilon(q) - 1$ for large q decreases more rapidly than q^{-2}. The behavior of $\rho_{ind}^{TF}(\mathbf{r})$ at large r furthermore does not contain the Friedel oscillations of the induced charge. Thus the Thomas-Fermi dielectric screening needs improvements to provide the correct behavior of $\rho_{ind}(\mathbf{r})$ at small and large distances.

7.4 The Longitudinal Dielectric Function within the Linear Response Theory

In this section we consider some general features of the longitudinal response function of a quantum system of electrons to driving perturbations, and the connected longitudinal dielectric function. We begin with a few general assumptions on the sample and the field not yet in interaction. The sample (in the absence of perturbations) is assumed to be homogeneous (or nearly so). The "bare" or "external" perturbation (in the absence of the sample) is assumed to be arbitrarily small, and periodic in space and time with wavevector $\mathbf{q}$ and angular frequency ω. The external perturbation energy felt by an electron is thus taken in the form

$$U_{ext}(\mathbf{r}, t) = U_{ext}(\mathbf{q}, \omega) \, e^{i(\mathbf{q}\cdot\mathbf{r}-\omega t)} + \text{c.c.}, \tag{7.17}$$

where $U_{ext}(\mathbf{q}, \omega)$ is the (infinitesimal) amplitude of the external perturbation, and c.c. indicates the complex conjugate of the previous term. When carrying out calculations, it is always assumed (implicitly or explicitly) that the potential (7.17) is turned on adiabatically at $t = -\infty$, i.e. it is multiplied by $\exp(\eta t/\hbar)$ with $\eta \to 0^+$ (η is an infinitesimal positive energy).

We notice that the electrostatic potential $\phi_{ext}(\mathbf{r}, t)$ corresponding to the energy potential $U_{ext}(\mathbf{r}, t)$ is given by $\phi_{ext} = U_{ext}/(-e)$, where e is the modulus of the electronic charge. The electric field becomes

$$\mathbf{E}_{ext}(\mathbf{r}, t) = -\nabla\phi_{ext}(\mathbf{r}, t) = \frac{1}{e} U_{ext}(\mathbf{q}, \omega) i \, \mathbf{q} \, e^{i(\mathbf{q}\cdot\mathbf{r}-\omega t)} + \text{c.c.} \tag{7.18}$$

The electric field (7.18) is of *longitudinal type*, i.e. it is parallel to the propagation wavevector $\mathbf{q}$; it is also irrotational (curl $\mathbf{E}_{ext} \equiv 0$), consistently with the fact that it has been derived by a scalar potential.

When the sample is exposed to the external perturbation (7.17), a modulation of the electron wavefunctions, and thus a modulation of the electronic charge density, is forced within the sample. In *homogeneous systems* and in the *linear response regime*, the *total self-consistent potential* $U(\mathbf{r}, t)$ (born both from *external and internal* sources) is expected to depend on space and time with the same wavevector $\mathbf{q}$ and the same frequency ω as the external potential (7.17). One can thus write for the total driving perturbation

$$U(\mathbf{r}, t) = U(\mathbf{q}, \omega) \, e^{i(\mathbf{q}\cdot\mathbf{r}-\omega t)} + \text{c.c.} \tag{7.19}$$

In the linear response regime, the amplitudes $U_{ext}(\mathbf{q}, \omega)$ and $U(\mathbf{q}, \omega)$ are proportional to each other, and the macroscopic dielectric function $\varepsilon(\mathbf{q}, \omega)$ can then be defined as

$$U_{ext}(\mathbf{q}, \omega) = \varepsilon(\mathbf{q}, \omega) U(\mathbf{q}, \omega); \tag{7.20}$$

relation (7.20) is the analog of the standard relation $\mathbf{D} = \varepsilon\mathbf{E}$ between the displacement field $\mathbf{D}$ (due to external sources only) and the total electric field $\mathbf{E}$ in a medium (due to external and internal sources).

The dielectric function $\varepsilon(\mathbf{q}, \omega)$ for a homogeneous system satisfies some general properties. From reality condition of both members of Eqs. (7.17) and (7.19), it is inferred $\varepsilon(\mathbf{q}, \omega) = \varepsilon^*(-\mathbf{q}, -\omega)$. Since the system is homogeneous and isotropic, perturbations of wavevector $\mathbf{q}$ or $-\mathbf{q}$ produce similar effects and we thus expect $\varepsilon(\mathbf{q}, \omega) = \varepsilon(-\mathbf{q}, \omega)$. The two previous properties imply $\varepsilon(\mathbf{q}, \omega) = \varepsilon^*(\mathbf{q}, -\omega)$; this shows that *the real part* $\varepsilon_1(\mathbf{q}, \omega)$ *of the dielectric function is an even function of* ω, *while the imaginary part* $\varepsilon_2(\mathbf{q}, \omega)$ *is an odd function of* ω. The physical requirement that the response is causal, linear, and finite, ensures that $\varepsilon(\mathbf{q}, \omega)$ is analytical in the upper half of the complex ω plane and obeys the Kramers-Kronig relations

$$\varepsilon_1(\mathbf{q}, \omega) = 1 + \frac{1}{\pi} P \int_{-\infty}^{\infty} \frac{\varepsilon_2(\mathbf{q}, \omega')}{\omega' - \omega} d\omega';$$

$$\varepsilon_2(\mathbf{q}, \omega) = -\frac{1}{\pi} P \int_{-\infty}^{\infty} \frac{\varepsilon_1(\mathbf{q}, \omega') - 1}{\omega' - \omega} d\omega', \quad (7.21)$$

where P denotes the principal part of the integral. The above relations can also equivalently be expressed as integrals on the positive values of the frequencies in the form

$$\varepsilon_1(\mathbf{q}, \omega) = 1 + \frac{2}{\pi} P \int_{0}^{\infty} \omega' \frac{\varepsilon_2(\mathbf{q}, \omega')}{\omega'^2 - \omega^2} d\omega';$$

$$\varepsilon_2(\mathbf{q}, \omega) = -\frac{2\omega}{\pi} P \int_{0}^{\infty} \frac{\varepsilon_1(\mathbf{q}, \omega') - 1}{\omega'^2 - \omega^2} d\omega'. \quad (7.22)$$

From Eq. (7.20), in the case $\varepsilon(\mathbf{q}, \omega) = 0$, we can have $U(\mathbf{q}, \omega) \neq 0$ even when $U_{ext}(\mathbf{q}, \omega)$ vanishes, i.e. there is the possibility of electron-density oscillations in the system even in the absence of any external perturbation. Thus *the zeroes of the longitudinal dielectric function identify the plasmon-type collective excitations* of the system.

Before considering specific models and applications to crystals for the longitudinal dielectric function, it is important a comment on the assumption of *homogeneity*. Crystals, strictly speaking, are not homogeneous materials; in particular the crystal electron density is invariant under lattice translations, but not under arbitrary displacements. In a periodic medium, the induced electron density, besides a wavevector $\mathbf{q}$ component as the applied field, also may include other wavevector components $\mathbf{q} + \mathbf{g}_m$, with $\mathbf{g}_m$ reciprocal lattice vectors. In general "local field effects" due to inhomogeneity of the crystal within the unit cell are rather complicated to account for; we ignore them in the following and we refer to the literature for their discussion [see for instance the review article by W. Hanke (1978) and references quoted therein].

After these preliminary remarks, we provide the *quantum expression of the longitudinal dielectric function for homogeneous (or nearly homogeneous) materials, within the linear response theory.* Our elaborations are specific for quantum systems described within the one-electron approximation and are based on the application of the standard Fermi golden rule. In the literature there is a variety of derivations for the dielectric functions of electronic systems at different degrees of sophistication (self-consistent field approach, random-phase approximation, equation of motion for the density-matrix operator, etc.). In the following our attention is confined to media supposed isotropic, homogeneous and linear. In spite of these restrictions, the treatment is far from trivial and already contains relevant concepts and procedures that are encountered (and possibly generalized) in the sophisticated approaches pursued to describe the response of many-electron systems to external fields. The reader can find the instructive derivation of the quantum expression of the dielectric function in linear and homogeneous media in Appendix B. Here we summarize the results without proof so that the physical aspects and concepts of the linear response theory can be followed more easily.

Consider an arbitrary (periodic or aperiodic) electron system of volume V, and suppose that it can be described by the one-electron Hamiltonian

$$H_0 = -\frac{\hbar^2 \nabla^2}{2m} + V(\mathbf{r}). \tag{7.23}$$

We indicate by $\{\psi_\alpha(\mathbf{r})\}$ and E_α the wavefunctions and the eigenvalues of H_0 (the wavefunctions are normalized to 1 within the volume V of the sample). The occupancy of the states is determined by the Fermi-Dirac function $f(E_\alpha)$ (ordinarily considered at zero temperature). Spin degeneracy is taken into account by appropriate inclusion of a factor two, where required. The general results here obtained are later specified in the case of periodic systems, where band wavefunctions are of Bloch type.

The microscopic quantum expression of the longitudinal dielectric function, given in Appendix B, reads

$$\varepsilon(\mathbf{q}, \omega) = 1 + \frac{8\pi e^2}{q^2} \frac{1}{V} \sum_{\alpha\beta} \frac{|\langle \psi_\beta | e^{i\mathbf{q}\cdot\mathbf{r}} | \psi_\alpha \rangle|^2}{E_\beta - E_\alpha - \hbar\omega - i\eta} [f(E_\alpha) - f(E_\beta)] \tag{7.24}$$

with η positive infinitesimal quantity. The explicit separation of the real and imaginary parts of the dielectric function in Eq. (7.24), can be obtained either directly by inspection, or using the Dirac identity (or its complex conjugate)

$$\lim_{\eta \to 0^+} \frac{1}{x - i\eta} = P\frac{1}{x} + i\pi \delta(x), \tag{7.25}$$

where x is real, and P denotes the Cauchy principal value integral (i.e. any integration involving the product of $1/x$ by a function of x must be intended in principal part). The frequently used relation (7.25) follows from the algebraic equality

$$\frac{1}{x - i\eta} = \frac{x}{x^2 + \eta^2} + i\pi \frac{1}{\pi} \frac{\eta}{x^2 + \eta^2} \quad \text{for} \quad \eta \to 0^+.$$

Equation (7.24) gives the longitudinal dielectric function of a generic system under the assumption of linear response, homogeneity, and negligible local field effects. The linear response framework is quite general in nature, and its explicit specification to periodic systems will be done at a later stage.

The next sections of this chapter are devoted to make explicit the wealth of information contained in the quantum expression (7.24) of the dielectric response function, with particular attention to the static impurity screening, to the dynamical effects and plasmon quasiparticles, and to the electron energy-loss function and measurements.

7.5 Dielectric Screening within the Lindhard Model

7.5.1 Static Dielectric Screening in Simple Metals with the Lindhard Model

We consider here the simplest model of a metal, the free-electron gas, whose band wavefunctions are plane waves (normalized to one in the volume V of the crystal) and the energies are $E(\mathbf{k}) = \hbar^2 k^2/2m$. The occupancy of the states is determined by the Fermi-Dirac distribution function $f(E(\mathbf{k}))$, or for brevity $f(\mathbf{k})$, ordinarily considered at the temperature $T = 0$.

The static dielectric function $\varepsilon(\mathbf{q}, 0) \equiv \varepsilon(q)$ can be obtained specifying the general results of the linear response theory, summarized by Eq. (7.24), to the case of a single band with plane waves eigenfunctions, and taking the limit $\omega \to 0$; one finds

$$\varepsilon(q) = 1 + \frac{8\pi e^2}{q^2} \frac{1}{V} \sum_{\mathbf{k}} \frac{f(\mathbf{k}) - f(\mathbf{k} + \mathbf{q})}{E(\mathbf{k} + \mathbf{q}) - E(\mathbf{k}) - i\eta} \qquad (7.26)$$

an expression which is known as the *static Lindhard dielectric function* for the free-electron gas.

It is possible to evaluate Eq. (7.26) in analytic form, as shown in Appendix C. From Eq. (C.9) we have for its real part

$$\varepsilon_1(q, 0) = \varepsilon_L(q) = 1 + \frac{1}{2} \frac{k_{TF}^2}{q^2} + \frac{1}{2} \frac{k_{TF}^2}{q^2} \frac{k_F}{q} \left(1 - \frac{q^2}{4k_F^2}\right) \ln \left|\frac{2k_F + q}{2k_F - q}\right|. \quad (7.27a)$$

Introducing the dimensionless quantity $x = q/2k_F$, Eq. (7.27a) can be recast in the form

$$\varepsilon_L(q) = 1 + \frac{k_{TF}^2}{q^2} F(x), \qquad (7.27b)$$

where the function $F(x)$ is given by

$$F(x) = \frac{1}{2} + \frac{1 - x^2}{4x} \ln \left|\frac{1 + x}{1 - x}\right|. \qquad (7.27c)$$

The function $F(x)$ has been already discussed in Section 4.7, although in a different context. The behavior of $\varepsilon_L(q)$ is illustrated in Figure 7.6.

In spite of the extreme simplicity of the model, the Lindhard function (7.27a) already embodies some key features, that are common to more sophisticated many-body treatments. Consider in fact the behavior of $\varepsilon_L(q)$ in the long wavelength limit, in the short wavelength limit, and in the intermediate region $q \approx 2k_F$.

For *small* q ($q \ll 2k_F$) the Lindhard function gives the same result as the linearized Thomas-Fermi theory; in fact for small q, we have $x \ll 1$ and $F(x) \approx 1$.

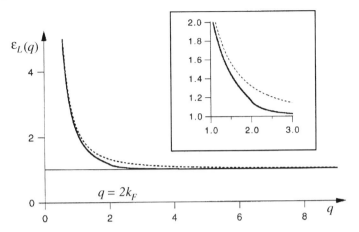

Figure 7.6 Behavior of the static Lindhard (solid line) and Thomas-Fermi (dashed line) dielectric function of a free-electron gas. The inset shows the behavior of $\varepsilon_L(q)$ near $q = 2k_F$ (we have chosen $k_F = 1\, a_B^{-1}$ or equivalently $r_s = 1.92$; q is in units of a_B^{-1}).

For *large* q ($q \gg 2k_F$), we have $x \gg 1$ and $F(x) \approx 1/3x^2$; the Lindhard function has the asymptotic behavior

$$\varepsilon_L(q) \to 1 + \frac{k_{TF}^2}{q^2}\frac{1}{3(q/2k_F)^2} = 1 + \frac{4}{3}\frac{k_{TF}^2 k_F^2}{q^4} \qquad q \gg 2k_F.$$

A decrease of $\varepsilon_L(q) - 1$ as q^{-4} assures a well-behaved screening charge density at the origin (see Eq. (7.16)).

For *intermediate* q ($q \approx 2k_F$) the Lindhard dielectric function is continuous for $q = 2k_F$, but with a logarithmic singularity in the derivative. Consider in fact Eq. (7.27a) for $q \approx 2k_F$; the singular part $\widetilde{\varepsilon}_L$ of the dielectric function is proportional to

$$\widetilde{\varepsilon}_L(q) \approx (q - 2k_F)\ln|q - 2k_F| \qquad q \approx 2k_F.$$

For the first and second derivative we have

$$\frac{d\widetilde{\varepsilon}_L(q)}{dq} \approx \ln|q - 2k_F| \quad \text{and} \quad \frac{d^2\widetilde{\varepsilon}_L(q)}{dq^2} \approx \frac{1}{q - 2k_F} \quad \text{for} \quad q \approx 2k_F. \qquad (7.28)$$

We can easily understand the origin of the discontinuity in the slope of $\varepsilon_L(q)$ for $q \approx 2k_F$ considering the energy denominators appearing in the sum over $\mathbf{k}$ in Eq. (7.26). For $q < 2k_F$ contributions with $E(\mathbf{k} + \mathbf{q}) \approx E(\mathbf{k})$ are possible, but this is no more possible for $q > 2k_F$ (as illustrated in Figure 7.7).

The singularity (7.28) in reciprocal space generates oscillations of the screening charge in real space (see Figure 7.8). To illustrate this effect, consider the screening charge induced by a point charge $+Ze$ inserted in the metal (at the origin). From Eq. (7.16) and performing the angular integration, we have for the screening charge in the Lindhard approximation

$$\rho_{ind}^L(\mathbf{r}) = Ze\frac{1}{(2\pi)^3}\int\left[\frac{1}{\varepsilon_L(q)} - 1\right]e^{i\mathbf{q}\cdot\mathbf{r}}\,d\mathbf{q} = -Ze\frac{1}{r}\int_0^\infty g_L(q)\sin qr\,dq, \qquad (7.29a)$$

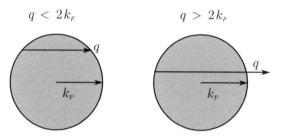

$$q < 2k_F \qquad\qquad q > 2k_F$$

Figure 7.7 Fermi sphere and excitations with wavevector transfer $q < 2k_F$ and $q > 2k_F$; $E(\mathbf{k} + \mathbf{q}) \approx E(\mathbf{k})$ is possible in the former case, but not in the latter.

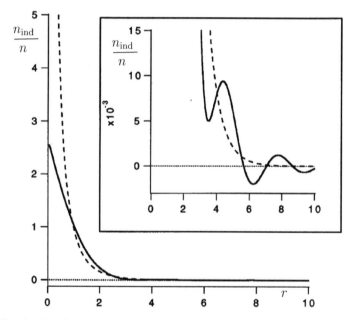

Figure 7.8 Behavior of the induced electron density $n_{\text{ind}}(r)$, relative to the unperturbed density n, when a point charge $+e$ is inserted in a metal at the origin. The Thomas-Fermi approximation is indicated with a dashed line, the Lindhard approximation with a continuous line (r is in units of the Bohr radius a_B; $k_F = 1a_B^{-1}$; $r_s = 1.92$). The inset shows n_{ind}/n at large r with a change of scale; Friedel oscillations are evident in the Lindhard approximation.

where the expression of $g_L(q)$ is

$$g_L(q) = \frac{q}{2\pi^2} \frac{\varepsilon_L(q) - 1}{\varepsilon_L(q)}. \tag{7.29b}$$

From the asymptotic behavior of $\varepsilon_L(q)$ for small and large wavenumbers, we have that $g_L(q)$ vanishes both for large and small values of q. The derivative $g'_L(q)$ vanishes for $q = \infty$ and is finite for $q = 0$. It is also seen by inspection that $g''_L(q)$ presents the same singularity found in Eq. (7.28) for $\widetilde{\varepsilon}''_L(q)$.

We now perform two successive integrations by parts in Eq. (7.29a) and obtain

$$\rho_{\text{ind}}^L(r) = Ze\frac{1}{r^3}\int_0^\infty g_L''(q)\sin qr\, dq. \tag{7.30a}$$

We look for the behavior of $\rho_{\text{ind}}^L(r)$ at very large r. Because of the rapid oscillations of $\sin qr$ as $r \to \infty$, the most important contribution in the integration (7.30a) comes from a small region $(2k_F - \Delta < q < 2k_F + \Delta)$ around $2k_F$, where $g_L''(q)$ is singular. Keeping only the dominant term in Eq. (7.30a), we have

$$\rho_{\text{ind}}^L(r) \propto Ze\frac{1}{r^3}\int_{2k_F-\Delta}^{2k_F+\Delta}\frac{1}{q-2k_F}\sin qr\, dq \qquad r \to \infty. \tag{7.30b}$$

Inserting the equality

$$\sin qr = \sin[(q-2k_F)r]\cos 2k_F r + \cos[(q-2k_F)r]\sin 2k_F r$$

into Eq. (7.30b), and omitting the term that vanishes for parity considerations, one has

$$\rho_{\text{ind}}^L(r) \propto Ze\frac{\cos 2k_F r}{r^3}\int_{2k_F-\Delta}^{2k_F+\Delta}\frac{\sin(q-2k_F)r}{q-2k_F}dq = Ze\frac{\cos 2k_F r}{r^3}\int_{-r\Delta}^{r\Delta}\frac{\sin x}{x}dx. \tag{7.30c}$$

The integration limits in Eq. (7.30c) can be extended from $-\infty$ to $+\infty$ (and the integral is π); this yields the asymptotic expression

$$\rho_{\text{ind}}^L(r) \propto Ze\frac{\cos 2k_F r}{r^3} \qquad r \to \infty. \tag{7.30d}$$

The oscillations of $\rho_{\text{ind}}^L(r)$, at large r, are known as Friedel oscillations. The singularity in the dielectric function is responsible also for the Kohn anomalies in the vibrational spectrum of metals for $q \approx 2k_F$ [W. Kohn, Phys. Rev. Lett. 2, 393 (1959)].

7.5.2 Dynamic Dielectric Screening in Simple Metals and Plasmon Modes

The quantum expression of the Lindhard dynamical dielectric function $\varepsilon(\mathbf{q}, \omega)$ of the free-electron gas can be obtained inserting plane waves into the general Eq. (7.24) for the dielectric function of a single band metal model; one finds

$$\varepsilon(\mathbf{q}, \omega) = 1 + \frac{8\pi e^2}{q^2}\frac{1}{V}\sum_{\mathbf{k}}\frac{f(\mathbf{k}) - f(\mathbf{k}+\mathbf{q})}{E(\mathbf{k}+\mathbf{q}) - E(\mathbf{k}) - \hbar\omega - i\eta}. \tag{7.31}$$

In the previous subsection, we have discussed the most significant aspects embodied in the static expression $\varepsilon(\mathbf{q}, 0)$, evaluated along the zero frequency axis. In the present subsection we focus on the most significant dynamical aspects embodied in $\varepsilon(0, \omega)$ along the zero momentum transfer axis.

In Appendix C it is shown how to evaluate the Lindhard function in analytic form in the whole (q, ω) plane, which includes of course the long wavelength limit $\mathbf{q} \approx 0$ (see

Eq. (C.10)). In this limit, however, we prefer a more direct and intuitive manipulation of Eq. (7.31).

It is useful to rewrite expression (7.31) in a slightly different form, by performing the replacement $\mathbf{k} + \mathbf{q} \rightarrow -\mathbf{k}'$ in the term containing $f(\mathbf{k} + \mathbf{q})$ (then $\mathbf{k}'$ is relabeled as $\mathbf{k}$); we obtain

$$\varepsilon(\mathbf{q}, \omega) = 1 + \frac{16\pi e^2}{q^2} \frac{1}{V} \sum_{\mathbf{k}} f(\mathbf{k}) \frac{E(\mathbf{k} + \mathbf{q}) - E(\mathbf{k})}{[E(\mathbf{k} + \mathbf{q}) - E(\mathbf{k})]^2 - (\hbar\omega + i\eta)^2}. \quad (7.32)$$

Expression (7.32) is somewhat more convenient for analytic investigations, since the sum in the reciprocal space involves the wave vectors confined in the Fermi sphere, centered at the origin of the $\mathbf{k}$-space.

The energy difference in the denominator of Eq. (7.32) reads

$$E(\mathbf{k} + \mathbf{q}) - E(\mathbf{k}) = \frac{\hbar^2}{2m}(2\mathbf{k} \cdot \mathbf{q} + q^2) = \frac{\hbar^2}{2m}(2qk_z + q^2),$$

where the vector $\mathbf{q}$ is taken in the z-direction (without loss of generality). For $\mathbf{q} \rightarrow 0$, Eq. (7.32) can be simplified by performing a series development in powers of the momentum transfer q. The expansion, taking into account that odd powers of k_z give zero contribution to the sum in the reciprocal lattice, becomes

$$\varepsilon(\mathbf{q} \rightarrow 0, \omega) = 1 - \frac{16\pi e^2}{q^2} \frac{1}{V} \sum_{\mathbf{k}} f(\mathbf{k}) \left[\frac{E(\mathbf{k} + \mathbf{q}) - E(\mathbf{k})}{(\hbar\omega + i\eta)^2} + \frac{[E(\mathbf{k} + \mathbf{q}) - E(\mathbf{k})]^3}{(\hbar\omega + i\eta)^4} + \cdots \right]$$

$$= 1 - \frac{16\pi e^2}{q^2} \frac{1}{V} \sum_{\mathbf{k}} f(\mathbf{k}) \left[\frac{1}{(\hbar\omega + i\eta)^2} \frac{\hbar^2 q^2}{2m} + \frac{1}{(\hbar\omega + i\eta)^4} \frac{\hbar^6}{8m^3} 12 k_z^2 q^4 + \cdots \right]$$

$$= 1 - \frac{8\pi e^2}{m(\omega + i\eta)^2} \frac{1}{V} \sum_{\mathbf{k}} f(\mathbf{k}) - \frac{8\pi e^2}{m(\omega + i\eta)^4} q^2 \frac{1}{V} \sum_{\mathbf{k}} f(\mathbf{k}) \left(\frac{\hbar k}{m} \right)^2 - \cdots,$$

$$(7.33a)$$

where η denotes a positive infinitesimal quantity. We notice that

$$\frac{1}{V} \sum_{\mathbf{k}} f(\mathbf{k}) = \frac{n}{2} \quad \text{and} \quad \frac{1}{V} \sum_{\mathbf{k}} \left(\frac{\hbar k}{m} \right)^2 f(\mathbf{k}) = \frac{3}{5} v_F^2 \frac{n}{2} \qquad \left(v_F = \frac{\hbar k_F}{m} \right).$$

Using the above expressions, Eq. (7.33a) becomes

$$\varepsilon(q \rightarrow 0, \omega) = 1 - \frac{\omega_p^2}{(\omega + i\eta)^2} - \frac{3}{5} \frac{\omega_p^2}{(\omega + i\eta)^4} v_F^2 q^2 - \cdots \qquad \omega_p^2 = \frac{4\pi n e^2}{m},$$

$$(7.33b)$$

where ω_p is the standard expression of the plasma frequency.

From Eq. (7.33b), the Lindhard dynamic dielectric function for $q = 0$ becomes

$$\varepsilon(0, \omega) = 1 - \frac{\omega_p^2}{\omega^2 + i\omega/\tau}, \quad (7.34)$$

where we have done the replacement $2\eta = 1/\tau$; in the Lindhard treatment the broadening parameter η is an infinitesimal positive quantity and τ is thus an arbitrary large quantity.

We now take the liberty to consider τ as a sort of finite phenomenological lifetime for the electrons moving in the crystal. Within this assumption, Eq. (7.34) can be recognized as the Drude dielectric response of a system of free-electron-like carriers with relaxation time τ. For the forthcoming qualitative discussion of plasmons in crystals, instead of the Lindhard model (7.34) with $\tau \to \infty$ (which gives sharp resonances), it is more convenient to adopt the Drude model, again expressed by Eq. (7.34), with finite phenomenological values of τ. [Other aspects of the Drude model in connection with intraband optical properties of metals are discussed in Section 11.2; we remind that in the long wavelength limit under attention, transverse and longitudinal dielectric functions are numerically equal.]

The knowledge of the longitudinal dielectric function $\varepsilon = \varepsilon_1 + i\varepsilon_2$ of a material allows to calculate the energy-loss function $-\mathrm{Im}\,1/\varepsilon = \varepsilon_2/(\varepsilon_1^2 + \varepsilon_2^2)$ and to infer important information on plasmon excitations [the function $-\mathrm{Im}\,1/\varepsilon$ determines the characteristic energy losses suffered by fast charged particles traversing the material, as can be inferred from Eq. (B.15)]. The reciprocal of the Drude formula (7.34) gives the so-called "inverted" Drude-Sellmeier formula

$$\frac{1}{\varepsilon(0, \omega)} = 1 - \frac{\omega_p^2}{\omega_p^2 - \omega^2 - i\omega/\tau};$$

from the above expression we obtain

$$-\mathrm{Im}\,\frac{1}{\varepsilon(0, \omega)} = \frac{\omega}{\tau}\frac{\omega_p^2}{(\omega^2 - \omega_p^2)^2 + \omega^2/\tau^2}. \tag{7.35}$$

The energy-loss function (7.35) has its maximum around $\hbar\omega \approx \hbar\omega_p$ and has approximately a Lorenzian shape of energy half-width $\Delta = \hbar/\tau$. When $-\mathrm{Im}\,1/\varepsilon$ is sharply peaked near $\omega = \omega_p$, the mechanism of energy loss of fast electrons mainly occurs via excitations of longitudinal currents at the plasma frequency.

In actual metals, the $-\mathrm{Im}\,1/\varepsilon$ plasma resonances have typical widths Δ in the range $[0.05$–$0.25]\hbar\omega_p$; narrow-line resonances ($\Delta/\hbar\omega_p \approx 0.05$) occur, for instance, in aluminum, magnesium and tin, while beryllium presents a rather broad-line resonance ($\Delta/\hbar\omega_p \approx 0.26$). As an example, we report in Figure 7.9 observed values of $\varepsilon_1(0, \omega)$, $\varepsilon_2(0, \omega)$, and $-\mathrm{Im}\,1/\varepsilon(0, \omega)$ in beryllium around the plasma resonance. The energy-loss function has its maximum at $\hbar\omega_p = 18.4$ eV, a value corresponding to the plasma frequency of an ideal free-electron gas with 2.0 electrons per atom; the half-width Δ is measured as $\Delta = 4.7$ eV, corresponding to a lifetime $\tau = \hbar/\Delta = 1.4 \times 10^{-16}$ s. Such a short lifetime of plasmons may entail several decay channels (background of interband single-particle transitions, collisional effects, or other mechanisms) and we refer to the literature for further analysis [see for instance N. Swanson, J. Opt. Soc. Am. 54, 1130 (1964); H. A. Fowler, and J. J. Filliben, J. Appl. Phys. 52, 6701 (1981) and references quoted therein].

From the dynamical dielectric function of a medium, one can obtain the dispersion curve of plasmons from the locus of points in $(\mathbf{q}, \omega)$ space where $\varepsilon(\mathbf{q}, \omega) \equiv 0$. In the limit of $\mathbf{q} \to 0$ and $\hbar\omega > E(\mathbf{k} + \mathbf{q}) - E(\mathbf{k})$, we have $\varepsilon_2(\mathbf{q}, \omega) = 0$; the zeroes of

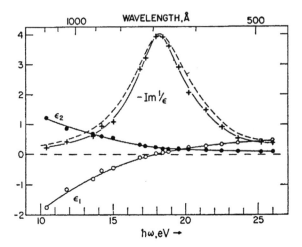

Figure 7.9 Dielectric function $\varepsilon_1(0, \omega)$, $\varepsilon_2(0, \omega)$ and $-\mathrm{Im}\, 1/\varepsilon(0, \omega)$ for beryllium, obtained from optical measurements. The dashed curve for $-\mathrm{Im}\, 1/\varepsilon(0, \omega)$ is elaborated from electron-scattering experiments [reprinted with permission from J. Toots, H. A. Fowler, and L. Marton, Phys. Rev. *172*, 670 (1968); copyright 1968 by the American Physical Society].

Eq. (7.33b) occur for

$$1 - \frac{\omega_p^2}{\omega^2} - \frac{3}{5}\frac{\omega_p^2}{\omega^4}v_F^2 q^2 = 0. \tag{7.36a}$$

The solution of Eq. (7.36a) provides the dispersion curve of plasmon modes for small q wavevectors

$$\omega_{\mathrm{plasmon}}(q) = \omega_p \left[1 + \frac{3}{10}\frac{v_F^2 q^2}{\omega_p^2} + \cdots \right]. \tag{7.36b}$$

Notice that the dispersion curve (7.36b) is rather flat (and this may justify simplified models that neglect dispersion altogether).

With increasing values of q, the plasmon modes are obtained from the (numerical) solution of the equation $\varepsilon_1(q, \omega) = 0$ in the region where no electron-hole excitations are possible (region I of Figure 7.14). With further increase of q, the plasmon band merges into the continuum of single-particle excitations (see again Figure 7.14); this establishes a cut-off q_c of the maximum wavevector of plasmon modes representing electronic charge fluctuations correlated by long-range Coulomb interactions. From Figure 7.14, we can estimate the cut-off momentum q_c from the equality

$$\hbar\omega_p \approx \frac{\hbar^2}{2m}\left(q_c^2 + 2q_c k_F \right) \approx \frac{\hbar^2}{m}q_c k_F \quad \rightarrow \quad q_c = \frac{\omega_p}{v_F} = \frac{\omega_p}{c}\frac{c}{v_F}.$$

With typical values $\hbar\omega_p \approx 10$ eV, and Fermi velocity one hundredth of the speed of light, we can estimate $\lambda_c \approx 10$ Å.

It is instructive to consider the occurrence of plasma oscillations looking at the zeroes of expression (7.32) with a graphic approach. For this purpose, we rewrite Eq. (7.32) in the equivalent form (obtained by adding a part that vanishes identically)

$$\varepsilon(\mathbf{q}, \omega) = 1 + \frac{16\pi e^2}{q^2} \frac{1}{V} \sum_{\mathbf{k}} f(\mathbf{k})[1 - f(\mathbf{k}+\mathbf{q})] \frac{E(\mathbf{k}+\mathbf{q}) - E(\mathbf{k})}{[E(\mathbf{k}+\mathbf{q}) - E(\mathbf{k})]^2 - (\hbar\omega + i\eta)^2}. \quad (7.37a)$$

We notice that $f(\mathbf{k})[1 - f(\mathbf{k}+\mathbf{q})]$ is different from zero (at $T = 0$) only between the energy of the occupied states $\mathbf{k}$ below the Fermi level and the energy of the empty states $\mathbf{k}+\mathbf{q}$ above the Fermi level; then $E(\mathbf{k}+\mathbf{q}) - E(\mathbf{k}) > 0$ and Eq. (7.37a) takes the general structure

$$\varepsilon(\mathbf{q}, \omega) = 1 + \omega_p^2 \sum_n \frac{C_n(\mathbf{q})}{\omega_n^2(\mathbf{q}) - (\omega + i\eta)^2} \quad \text{with} \quad C_n > 0 \quad \text{and} \quad \sum_n C_n \equiv 1,$$

$$(7.37b)$$

where ω_n are the *excitation frequencies of the system in the independent particle approximation* and C_n are *definite-positive quantities that can be interpreted as the oscillator strengths of the resonant frequencies* ω_n. The sum of the oscillator strengths saturates to unity, so that the universal asymptotic behavior for frequencies much larger of the resonant frequencies is satisfied (see Eq. (B.19)).

For simplicity, and just for qualitative remarks, consider the system with a finite (although very large) number N of resonant frequencies ω_n in the interval $[\omega_1, \omega_N]$ such that ω_N is sufficiently smaller than ω_p. The real part $\varepsilon_1(\mathbf{q}, \omega)$ of Eq. (7.37b) (in the limit $\eta \to 0^+$) is schematically indicated in Figure 7.10. The zeroes of $\varepsilon_1(\mathbf{q}, \omega)$ occur in correspondence with the (nearly continuum) of single-particle excitations. In addition, there is one discrete root; this collective state represents electron-hole charge fluctuations correlated by the long-range Coulomb interactions.

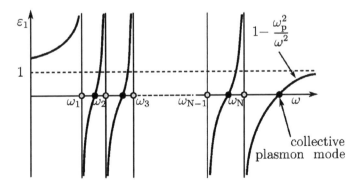

Figure 7.10 Qualitative behavior of the real part of the longitudinal dielectric function of a system with proper frequencies ω_n (indicated by empty dots) in the interval $[\omega_1, \omega_N]$, with $\omega_N \ll \omega_p$. Since $\varepsilon_1(\mathbf{q}, \omega)$ is an even function of ω, only the part $\omega > 0$ is indicated; the zeroes of $\varepsilon_1(\mathbf{q}, \omega)$ are indicated by full dots; notice the split-off collective state.

Before closing, we can notice that the popularity of the Lindhard model stands in the fact that the dielectric function can be evaluated in a closed analytic form for any given frequency and wavevector (as shown in Appendix C). The availability of closed analytic results is useful to establish guidelines or extrapolations in the study of realistic materials, where laborious numerical calculations become unavoidable. The Lindhard function has been also modified for simple (but still significant) descriptions of the dielectric screening in model semiconductors [see for instance E. Tosatti and G. Pastori Parravicini, J. Phys. Chem. Solids *32*, 623 (1971); Z. H. Levine and S. G. Louie, Phys. Rev. B *25*, 6310 (1982) and references quoted therein].

7.6 Quantum Expression of the Longitudinal Dielectric Function in Crystals

We consider now the quantum expression of the longitudinal dielectric function in periodic structures, where the electronic wavefunctions are Bloch functions. We indicate by $\psi_{m\mathbf{k}}(\mathbf{r})$ the band wavefunctions (normalized to one in the volume V of the crystal) and by $E_{m\mathbf{k}}$ the corresponding band energies; the occupancy of the states is determined by the Fermi-Dirac function $f(E_{m\mathbf{k}})$ (ordinarily considered at zero temperature). Spin degeneracy is included with appropriate insertion of a factor two, where required.

The general expression of the dielectric function $\varepsilon(\mathbf{q}, \omega)$ in terms of the wavevector $\mathbf{q}$ and frequency ω can be obtained from Eq. (7.24), considering the non-vanishing matrix elements of the perturbing wave $\exp(i\mathbf{q} \cdot \mathbf{r})$ between Bloch functions. We have

$$\varepsilon(\mathbf{q}, \omega) = 1 + \frac{8\pi e^2}{q^2} \frac{1}{V} \sum_{mn\mathbf{k}} \frac{\left|\langle \psi_{n\,\mathbf{k}+\mathbf{q}} \left| e^{i\mathbf{q}\cdot\mathbf{r}} \right| \psi_{m\mathbf{k}} \rangle\right|^2 [f(E_{m\mathbf{k}}) - f(E_{n\,\mathbf{k}+\mathbf{q}})]}{E_{n\,\mathbf{k}+\mathbf{q}} - E_{m\mathbf{k}} - \hbar\omega - i\eta}. \quad (7.38)$$

Expression (7.38) holds for any type of crystal (metal, semiconductor, or insulator); from it, the Lindhard expression (7.31) is recovered when Bloch functions are simply plane waves and band energies are free-electron like.

In the case of semiconductors and insulators (at $T = 0$) we do not need the explicit presence of the Fermi-Dirac distribution function. For dielectrics from Eq. (7.38) we obtain for the imaginary part of the dielectric function for positive frequencies

$$\varepsilon_2(\mathbf{q}, \omega) = \frac{8\pi^2 e^2}{q^2} \frac{1}{V} \sum_{vc\mathbf{k}} |\langle \psi_{c\,\mathbf{k}+\mathbf{q}} | e^{i\mathbf{q}\cdot\mathbf{r}} | \psi_{v\mathbf{k}} \rangle|^2 \delta(E_{c\,\mathbf{k}+\mathbf{q}} - E_{v\mathbf{k}} - \hbar\omega), \quad (7.39a)$$

where v labels the valence bands (fully occupied) and c labels the conduction bands (fully empty). The resulting expression of the real part

$$\varepsilon_1(\mathbf{q}, \omega) = 1 + \frac{16\pi e^2}{q^2} \frac{1}{V} \sum_{vc\mathbf{k}} \frac{\left|\langle \psi_{c\,\mathbf{k}+\mathbf{q}} \left| e^{i\mathbf{q}\cdot\mathbf{r}} \right| \psi_{v\mathbf{k}} \rangle\right|^2 (E_{c\,\mathbf{k}+\mathbf{q}} - E_{v\mathbf{k}})}{(E_{c\,\mathbf{k}+\mathbf{q}} - E_{v\mathbf{k}})^2 - \hbar^2\omega^2}, \quad (7.39b)$$

follows directly from the appropriate Kramers-Kronig dispersion relation of Eqs. (7.22).

In the case of a semiconductor or insulator, we can obtain a model of $\varepsilon(\mathbf{q} \approx 0, \omega)$ by replacing the energy differences $(E_{c\,\mathbf{k}+\mathbf{q}} - E_{v\mathbf{k}})$ in Eqs. (7.39) with some finite average excitation energy $E_{av} = \hbar\omega_{av}$. Then exploiting the sum rules (B.18) and the asymptotic behavior (B.19), we obtain

$$\varepsilon_1(0, \omega) = 1 + \frac{\omega_p^2}{\omega_{av}^2 - \omega^2}, \quad \varepsilon_2(0, \omega) = \frac{\pi}{2} \frac{\omega_p^2}{\omega_{av}} \delta(\omega - \omega_{av}) \quad \text{for} \quad \omega > 0. \quad (7.40)$$

Notice that $\varepsilon_1(0, \omega)$ vanishes for $\omega \equiv \omega_{pl} = (\omega_p^2 + \omega_{av}^2)^{1/2}$; there is thus a blue shift of the plasma frequency ω_{pl} in semiconductors and (especially) in insulators due to the average energy gap, a fact confirmed by more rigorous treatments. The model (7.40) also predicts a finite static dielectric constant $\varepsilon_s = 1 + \omega_p^2/\omega_{av}^2$ (often used to estimate semi-empirically ω_{av}).

Before closing, we add a few comments on the static dielectric function of semiconductors and insulators. In the static limit $\omega \to 0$, the imaginary part $\varepsilon_2(\mathbf{q}, 0)$ of Eq. (7.39a) vanishes. For the real part $\varepsilon_1(\mathbf{q}, 0)$, Eq. (7.39b) gives

$$\varepsilon_1(\mathbf{q}, 0) = 1 + \frac{16\pi e^2}{q^2} \sum_{vc} \int_{B.Z.} \frac{d\mathbf{k}}{(2\pi)^3} \frac{|\langle \psi_{c\,\mathbf{k}+\mathbf{q}} | e^{i\mathbf{q}\cdot\mathbf{r}} | \psi_{v\mathbf{k}}\rangle|^2}{E_{c\,\mathbf{k}+\mathbf{q}} - E_{v\mathbf{k}}},$$

where the discrete sum over $\mathbf{k}$ has been transformed as usual into an integral over the first Brillouin zone, times $V/(2\pi)^3$. The presence of the energy gap makes the denominator in the above expression always finite. In the long wavelength limit, $\varepsilon_1(\mathbf{q}, 0)$ tends to a finite value $\varepsilon_1(0)$, which represents the electronic contribution to the static dielectric constant of the material. For values of $\mathbf{q}$ larger than the smallest reciprocal lattice vector, $\varepsilon_1(\mathbf{q}, 0)$ drops rapidly to zero. As an example, in Figure 7.11 we report the static dielectric function of silicon and germanium, calculated numerically; in the same figure different simplified models are also reported.

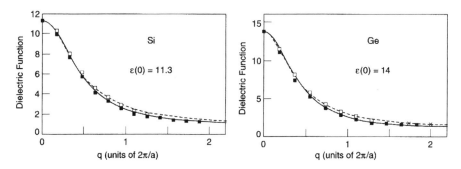

Figure 7.11 Static longitudinal dielectric function $\varepsilon_1(\mathbf{q}, 0)$ for silicon and germanium. Closed boxes [$\mathbf{q}$ along the $(1, 1, 1)$ direction] and open boxes [$\mathbf{q}$ along the $(1, 0, 0)$ direction] are from numerical calculations of J. P. Walter and M. L. Cohen, Phys. Rev. B **2**, 1831 (1970). The solid line is the model dielectric function of G. Cappellini, R. Del Sole, L. Reining, and F. Bechstedt, Phys. Rev. B **47**, 9892 (1993) (copyright 1993 by the American Physical Society). The dashed curve represents the model of Z. H. Levine and S. G. Louie, Phys. Rev. B **25**, 6310 (1982).

7.7   Surface Plasmons and Surface Polaritons

The bulk plasmons in homogeneous materials, so far considered, are longitudinal charge density waves, characterized by a non-vanishing longitudinal electric field and vanishing magnetic field components. Collective electronic charge density fluctuations are also possible at the surface or at the interface between two materials, or in films, layered systems, nanoparticles, and other mesoscopic nanostructures. Differently from bulk plasmon modes, surface plasmon modes are in general characterized by appropriate longitudinal and transverse electric and magnetic field components, and coupling with photons is thus possible. Among the attractive aspects of surface plasmons, of primary interest is the possibility to concentrate light in subwavelength regions, overcoming the diffraction limits of traditional optics. It is not our intent to enter deeply in the fascinating area of plasmonics and its many variegated aspects; rather, we confine our discussion to the classical treatment of an ideal metal-dielectric interface, because this model, although very simple, already contains some features of relevance.

The model system, here considered, consists of two semi-infinite media (typically a metal and a dielectric) separated by a planar interface taken in the x, y plane, as shown in Figure 7.12. For simplicity, the dielectric function $\varepsilon_1(\omega)$ of the metal is supposed to be local (i.e. independent on the momentum transfer), and the function ε_2 of the dielectric is supposed to be constant in the frequency-momentum region of practical interest.

We wish now to establish under which conditions an electromagnetic excitation may propagate along the surface $z = 0$ with the fields damped away both in the positive and negative z-direction. A surface plasmon propagating along the interface in the x, y

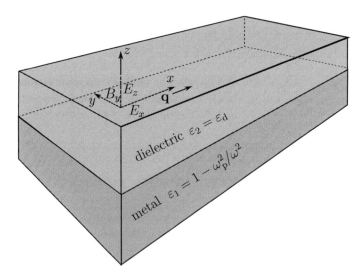

Figure 7.12 Schematic representation of two semi-infinite media for the study of surface plasmon modes. In typical metal-dielectric surfaces, the dielectric function of the metal is taken $\varepsilon_1(\omega) = 1 - \omega_p^2/\omega^2$, while the dielectric material is characterized by the frequency-independent positive constant ε_d (in the frequency range of practical interest).

plane, must be accompanied by a component of the electric field normal to the surface, because of the surface polarization charges. Choosing the x-axis along the propagation direction (see Figure 7.12), the electric and magnetic fields can be written in the form

$$
\begin{aligned}
\mathbf{E}_1(x, z, t) &= (E_{1x}, 0, E_{1z}) e^{i(q_1 x - \omega t)} e^{+k_1 z} & z < 0, \\
\mathbf{B}_1(x, z, t) &= (0, B_{1y}, 0)\, e^{i(q_1 x - \omega t)} e^{+k_1 z} & z < 0, \\
\mathbf{E}_2(x, z, t) &= (E_{2x}, 0, E_{2z}) e^{i(q_2 x - \omega t)} e^{-k_2 z} & z > 0, \\
\mathbf{B}_2(x, z, t) &= (0, B_{2y}, 0)\, e^{i(q_2 x - \omega t)} e^{-k_2 z} & z > 0,
\end{aligned}
\tag{7.41}
$$

where the subscripts 1 and 2 refer to the two semi-infinite media, q_1 and q_2 are real propagation wavenumbers, and k_1 and k_2 are real positive quantities representing the exponential decay of the fields with increasing distance from the interface $z = 0$. The fields in Eq. (7.41) must satisfy the Maxwell's equations and the appropriate boundary conditions at the interface between the two materials.

The Maxwell's equations in the absence of external sources and for non-magnetic materials ($\mathbf{B} = \mathbf{H}$) read

$$
\operatorname{div}(\varepsilon_i \mathbf{E}_i) = 0; \quad \operatorname{div} \mathbf{B}_i = 0,
\tag{7.42}
$$

$$
\operatorname{curl} \mathbf{E}_i = -\frac{1}{c} \frac{\partial \mathbf{B}_i}{\partial t}; \quad \operatorname{curl} \mathbf{B}_i = \varepsilon_i \frac{1}{c} \frac{\partial \mathbf{E}_i}{\partial t},
\tag{7.43}
$$

where the index $i = 1$ refers to the medium with $z < 0$ (the metal), while the index $i = 2$ refers to the medium with $z > 0$ (the dielectric).

We begin to simplify the parameters entering Eq. (7.41), taking proper account of the boundary conditions. The requirement that the components of the electric and magnetic fields parallel to the surface are continuous give:

$$
E_{1x} = E_{2x}; \quad B_{1y} = B_{2y}; \quad q_1 = q_2.
$$

Using the above equalities, we can recast Eq. (7.41) in the form

$$
\begin{aligned}
\mathbf{E}_1(x, z, t) &= (E_x, 0, E_{1z}) e^{i(qx - \omega t)} e^{+k_1 z} & z < 0, \\
\mathbf{B}_1(x, z, t) &= (0, B_y, 0)\, e^{i(qx - \omega t)} e^{+k_1 z} & z < 0, \\
\mathbf{E}_2(x, z, t) &= (E_x, 0, E_{2z}) e^{i(qx - \omega t)} e^{-k_2 z} & z > 0, \\
\mathbf{B}_2(x, z, t) &= (0, B_y, 0)\, e^{i(qx - \omega t)} e^{-k_2 z} & z > 0,
\end{aligned}
\tag{7.44}
$$

where the labels 1 or 2 to E_x, B_y, and q have been omitted because superfluous.

The continuity of the normal component of the displacement field gives

$$
\varepsilon_1 E_{1z} = \varepsilon_2 E_{2z}.
\tag{7.45a}
$$

From the first of the Maxwell's equation (7.42) applied to the electric fields in Eq. (7.44), one obtains for the two media

$$
iq E_x + k_1 E_{1z} = 0 \quad \text{and} \quad iq E_x - k_2 E_{2z} = 0.
\tag{7.45b}
$$

The three Eqs. (7.45) give the compatibility relation (surface plasmon condition)

$$\boxed{\frac{\varepsilon_1}{k_1} + \frac{\varepsilon_2}{k_2} = 0}.$$ (7.46)

So far, we have satisfied the boundary conditions and the vanishing divergence of the fields. We have now to consider the Maxwell's equations (7.43) with the curl operators.

Momentarily, however, we consider the *condition of occurrence of surface plasmons when retardation effects are ignored*, i.e. when the speed of light is allowed to become infinitely large, and the curl of the fields is set to zero. With this assumption, it is seen that curl $\mathbf{E}_1 = 0$ implies $iq\,E_{1z} - k_1 E_x = 0$; thus using Eqs. (7.45b) we have $k_1 = q$, and similarly $k_2 = q$. Therefore, Eq. (7.46) provides the *non-retarded surface-plasmon condition*

$$\boxed{\varepsilon_1 + \varepsilon_2 \equiv 0}.$$ (7.47)

Consider for instance the interface between a metal and a dielectric, with dielectric functions

$$\varepsilon_1 = 1 - \frac{\omega_p^2}{\omega^2} \quad \text{and} \quad \varepsilon_2 = \varepsilon_d$$ (7.48)

($\varepsilon_d = 1$ in the case of the vacuum). From the condition (7.47), the non-retarded surface-plasmon mode is possible for

$$1 - \frac{\omega_p^2}{\omega^2} + \varepsilon_d \equiv 0 \implies \omega_s = \frac{\omega_p}{\sqrt{1 + \varepsilon_d}}$$

(in the case of the vacuum, the surface plasmon resonance is $\omega_s = \omega_p/\sqrt{2}$).

We pass now to the proper account of the retardation effects due to the finiteness of the speed of light. Consider the first of Eqs. (7.43), take the curl of both members, and use the second of Eqs. (7.43) to eliminate the magnetic field. We obtain

$$\text{curl curl } \mathbf{E}_i = -\frac{\varepsilon_i}{c^2}\ddot{\mathbf{E}}_i \implies \nabla^2 \mathbf{E}_i = \frac{\varepsilon_i}{c^2}\ddot{\mathbf{E}}_i$$

[the last step follows from the vectorial identity: curl curl $\equiv$ grad div $- \nabla^2$]. The last equation, applied to the electric fields in Eqs. (7.44) gives

$$q^2 = k_1^2 + \frac{\varepsilon_1}{c^2}\omega^2; \quad q^2 = k_2^2 + \frac{\varepsilon_2}{c^2}\omega^2.$$ (7.49)

Multiplying the first of the above equations by $1/\varepsilon_1^2$, the second one by $1/\varepsilon_2^2$, subtracting and taking into account Eq. (7.46), we arrive at the basic compatibility relation for surface plasmons

$$\boxed{q^2 = \frac{\omega^2}{c^2}\frac{\varepsilon_1\varepsilon_2}{\varepsilon_1 + \varepsilon_2}}.$$ (7.50)

In the case of metal-dielectric interface, we have that real values of $q(\omega)$ are possible for frequencies in the interval $[0 < \omega < \omega_s]$ (both numerator and denominator are negative) and in the interval $\omega > \omega_p$ (both numerator and denominator are positive).

Using Eqs. (7.50) and (7.49), the damping parameters controlling the spatial extent of the electromagnetic field in the two media take the expressions

$$k_1^2 = \frac{\omega^2}{c^2} \frac{-\varepsilon_1^2}{\varepsilon_1 + \varepsilon_2} \quad \text{and} \quad k_2^2 = \frac{\omega^2}{c^2} \frac{-\varepsilon_2^2}{\varepsilon_1 + \varepsilon_2}; \tag{7.51}$$

thus k_1 and k_2 are real and positive parameters under the condition $\varepsilon_1 + \varepsilon_2 < 0$, and become arbitrary large for $\varepsilon_1 + \varepsilon_2 = 0$, meaning that the electromagnetic fields are confined to a $\delta(z)$-like sheet at $z = 0$ for the surface plasmon mode of frequency ω_s. Around this energy a strong concentration of electromagnetic field near the interface is possible.

In the particular case of the surface between a metal and a dielectric, with dielectric functions (7.48), Eq. (7.50) becomes

$$q^2 = \frac{\omega^2}{c^2} \varepsilon_d \frac{\omega^2 - \omega_p^2}{(\varepsilon_d + 1)\omega^2 - \omega_p^2}.$$

The same equation, solved in terms of q^2, gives for the dispersion curve of the surface plasmon localized around the interface the expression

$$\omega^2 = \frac{1}{2\varepsilon_d} \left[\varepsilon_d \omega_p^2 + (\varepsilon_d + 1)c^2 q^2 \right.$$
$$\left. - \sqrt{\varepsilon_d^2 \omega_p^4 + 2\varepsilon_d(\varepsilon_d - 1)\omega_p^2 c^2 q^2 + (\varepsilon_d + 1)^2 c^4 q^4} \right]. \tag{7.52}$$

This dispersion curve is represented in Figure 7.13; by straight series expansion it is seen that the curve is comprised between the light line in the dielectric and the frequency of the non-retarded surface plasmon mode.

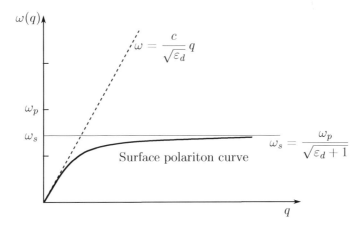

Figure 7.13 Dispersion curve of the surface plasmon polariton at the interface between a metal with plasma frequency $\hbar\omega_p = 10\,\text{eV}$ and a dielectric medium with $\varepsilon_d = 2$. The frequency of the non-retarded surface plasmon mode is $\omega_s = \omega_p/\sqrt{\varepsilon_d + 1}$. The dashed line denotes the dielectric light line $\omega = (c/\sqrt{\varepsilon_d})q$. [No propagation is possible for $\omega_s \leq \omega \leq \omega_p$. No localization around the surface is possible for $\omega > \omega_p$.]

Surface plasmons and surface polaritons formed at metallic-dielectric surfaces are quasiparticles originated by collective oscillations of conduction electrons coupled with the electromagnetic field of light. Collective electronic excitations at metallic surfaces, and more generally on patterned materials, metallic nanoparticles, and other mesoscopic systems, open the possibility to concentrate and manipulate light, with interesting applications in surface characterization and sensors, and in nanoscale photonic devices. The range of possible applications of plasmonics also extends to a number of interdisciplinary areas, including light manipulation for information technology, photocatalytic processes for environment protection, efficient photoenergetic processes in the context of the energy problem, biomedical applications [see for instance N. A. Garcia "Surface plasmons in metallic nanoparticles: fundamentals and applications" J. Phys. D: Appl. Phys. **44**, 283001 (2011), and references quoted therein].

Appendix A.   Friedel Sum Rule and Fumi Theorem

We can examine some features of the screening in metals, representing the metal as a *large spherical box of radius R*, in which the charge of positive ions is distributed uniformly and the conduction electrons can move freely. When no impurity is present at the origin of the large box, the free-electron wavefunctions are of the form $\psi_{klm}(\mathbf{r}) = A j_l(kr) Y_{lm}(\mathbf{r})$, where j_l is the spherical Bessel function of order $l (l = 0, 1, 2, \ldots)$, $Y_{lm}(\mathbf{r})(m = l, l-1, \ldots, -l)$ are spherical harmonics, and A is a normalization factor. For large values of the argument, the asymptotic behaviors of the Bessel functions are $j_l(x) \to (1/x) \sin (x - l\pi/2)$. We can thus write for the normalized wavefunctions within the sphere (of large radius R) the following expression

$$\psi_{klm}(\mathbf{r}) = \sqrt{\frac{2}{R}} \frac{1}{r} \sin \left(kr - \frac{1}{2} l\pi \right) Y_{lm}(\mathbf{r}) \qquad r \to \infty. \tag{A.1}$$

The boundary condition that the unperturbed wavefunctions vanish at $r = R$ gives

$$kR - \frac{1}{2} l\pi = n\pi \quad (n = 1, 2, \ldots). \tag{A.2}$$

The number dn of allowed states between k and $k + dk$ is obtained by differentiating both members of Eq. (A.2); we have $R dk = \pi dn$. Thus the (unperturbed) density-of-states (for any chosen value of l) is

$$D_0(k) = \frac{dn}{dk} = \frac{R}{\pi}. \tag{A.3}$$

Consider now the stationary solutions of the Schrödinger equation in the presence of the total spherical potential $U(r)$ (born by the external charge Ze located at the origin, and the accompanying screening charge). The stationary solutions are now

$$\widetilde{\psi}_{klm}(\mathbf{r}) = \sqrt{\frac{2}{R}} \frac{1}{r} \sin \left[kr + \delta_l(k) - \frac{1}{2} l\pi \right] Y_{lm}(\mathbf{r}) \qquad r \to \infty, \tag{A.4}$$

where $\delta_l(k)$ is the phase shift of the l-th wave due to the screened potential. The boundary condition that the perturbed wavefunctions vanish at $r = R$ gives

$$kR + \delta_l(k) - \frac{1}{2}l\pi = n\pi \quad (n = 1, 2, \ldots). \tag{A.5}$$

By differentiating both members of Eq. (A.5), we obtain $Rdk + \delta'_l(k)dk = \pi dn$. The perturbed density-of-states $D(k) = dn/dk$ is given by

$$D(k) = \frac{R}{\pi} + \frac{1}{\pi}\frac{d\delta_l}{dk} = D_0(k) + \frac{1}{\pi}\frac{d\delta_l}{dk}. \tag{A.6}$$

From Eq. (A.6), considering that the degeneracy of an l-orbital is $2(2l + 1)$ (spin degeneracy included), we have that the total number of displaced states up to the Fermi wavevector k_F is

$$\Delta n = \sum_l 2(2l + 1) \int_0^{k_F} \frac{1}{\pi}\frac{d\delta_l(k)}{dk}dk = \frac{2}{\pi}\sum_l (2l + 1)[\delta_l(k_F) - \delta_l(0)]. \tag{A.7}$$

If the number of bound states of $U(r)$ is zero, then it is also zero the sum involving the phase shifts $\delta_l(0)$ (Levinson theorem). From the physical requirement of perfect shielding, we have

$$\boxed{\frac{2}{\pi}\sum_l (2l + 1)\delta_l(k_F) \equiv Z}. \tag{A.8}$$

Relation (A.8) is the Friedel sum rule [J. Friedel, Phil. Mag. *43*, 153 (1952); Adv. Phys. *3*, 446 (1954)]; it expresses the fact that the total displaced electron charge is equal in magnitude, and opposite in sign, to the external point charge inserted in the metal.

From Eq. (A.4) we notice that the electron wavefunctions undergo phase shifts in order to screen the impurity potential; this fact entails also a *change of the kinetic energy of the free electrons* [F. G. Fumi, Phil. Mag. *46*, 1007 (1955)]. From Eq. (A.5) and Eq. (A.2), we see that the change Δk of the wavenumber of an l electron is $\Delta k \cdot R = -\delta_l(k)$, and the change in energy is $\hbar^2 k\Delta k/m$. Thus the change of kinetic energy of the free electrons because of the presence of the impurity is

$$\boxed{\Delta E_{\mathrm{el}} = -\sum_l 2(2l + 1) \int_0^{k_F} \frac{\hbar^2 k\delta_l(k)}{mR}\frac{R}{\pi} dk}, \tag{A.9}$$

where R/π is the density-of-states for l electrons; the integral extends from the bottom of the conduction band up to the Fermi energy assuming no bound state. Far from the impurity and far from its perfectly screened charge, the potential energy is zero and the kinetic term is the relevant part of the Hamiltonian.

The Friedel and the Fumi theorems have been of major importance for understanding the electronic structure of impurities in metals, and remain useful guidelines also after the advent of first-principle density functional treatments.

Appendix B.   Quantum Expression of the Longitudinal Dielectric Function in Materials with the Linear Response Theory

In this Appendix we provide the *quantum expression of the longitudinal dielectric function for homogeneous (or nearly homogeneous) materials, within the linear response theory*. The elaborations we present are specific for quantum systems of independent electrons, and are based on the application of the standard Fermi golden rule.

Consider an arbitrary (periodic or aperiodic) electron system of volume V, and suppose that it can be described by the one-electron Hamiltonian

$$H_0 = -\frac{\hbar^2 \nabla^2}{2m} + V(\mathbf{r}), \tag{B.1}$$

where $V(\mathbf{r})$ is the energy potential for an electron in the sample. We indicate by $\{\psi_\alpha(\mathbf{r})\}$ and E_α the wavefunctions and the eigenvalues of H_0 (the wavefunctions are normalized to 1 within the volume V of the sample). The occupancy of the states is determined by the Fermi-Dirac function $f(E_\alpha)$ (ordinarily considered at zero temperature). Spin degeneracy is taken into account by appropriate inclusion of a factor two, where required. The Hamiltonian H_0 is invariant by time-reversal symmetry, so that its wavefunctions can be thought of as real valued, when useful.

We examine the response of the electron system to a time-dependent energy perturbation of wavevector $\mathbf{q}$ and frequency ω, of the type

$$U(\mathbf{r}, t) = U_0 \, e^{i(\mathbf{q} \cdot \mathbf{r} - \omega t)} + \text{c.c.}, \tag{B.2}$$

where $U_0 = U_0(\mathbf{q}, \omega)$ is the (infinitesimal) amplitude of the (total) driving perturbation, and c.c. indicates the complex conjugate of the previous term. The electrostatic potential corresponding to the perturbation (B.2) is $\phi(\mathbf{r}, t) = U(\mathbf{r}, t)/(-e)$; the electric field $\mathbf{E} = -\nabla \phi$ is a longitudinal field parallel to $\mathbf{q}$ and given by

$$\mathbf{E}(\mathbf{r}, t) = E_0 \widehat{\mathbf{e}} e^{i(\mathbf{q} \cdot \mathbf{r} - \omega t)} + \text{c.c.} \quad \text{with} \quad E_0 = \frac{iq U_0}{e} \quad \text{and} \quad \widehat{\mathbf{e}} \parallel \mathbf{E} \parallel \mathbf{q}, \tag{B.3}$$

where $\widehat{\mathbf{e}}$ denotes the unit vector in the direction of the electric field, and is parallel to the propagation wavevector $\mathbf{q}$.

In the presence of the perturbation (B.2), the Hamiltonian of the system becomes

$$H = H_0 + U_0 \, e^{i(\mathbf{q} \cdot \mathbf{r} - \omega t)} + U_0^* \, e^{-i(\mathbf{q} \cdot \mathbf{r} - \omega t)}. \tag{B.4}$$

The two time-dependent terms in Eq. (B.4) induce transitions among the states of H_0, with absorption and emission of energy $\hbar\omega$, respectively. The probability per unit time that an electron initially in the state $|\psi_\alpha\rangle$ is transferred to the final state $|\psi_\beta\rangle$ because of the first time-dependent term in Eq. (B.4), is given by the Fermi golden rule

$$P_{\beta \leftarrow \alpha} = \frac{2\pi}{\hbar} |\langle \psi_\beta | U_0 e^{i\mathbf{q} \cdot \mathbf{r}} | \psi_\alpha \rangle|^2 \, \delta(E_\beta - E_\alpha - \hbar\omega). \tag{B.5}$$

Quite similarly, the transition probability that an electron initially in the state $|\psi_\beta\rangle$ is transferred to the final state $|\psi_\alpha\rangle$ is

$$P_{\alpha \leftarrow \beta} = \frac{2\pi}{\hbar} |\langle \psi_\alpha | U_0 e^{-i\mathbf{q} \cdot \mathbf{r}} | \psi_\beta \rangle|^2 \, \delta(E_\alpha - E_\beta + \hbar\omega). \tag{B.6}$$

We suppose that the electronic states are occupied in agreement to the Fermi-Dirac distribution function $f(E)$. We multiply Eq. (B.5) by $f(E_\alpha)(1 - f(E_\beta))$ and Eq. (B.6) by $f(E_\beta)(1 - f(E_\alpha))$, subtract the two expressions and sum up over all initial and final states; we obtain that the net number of transitions per unit time involving absorption or emission processes of energy $\hbar\omega$ is given by the expression

$$W(\mathbf{q}, \omega) = \frac{2\pi}{\hbar} 2 \sum_{\alpha\beta} |\langle \psi_\beta | U_0 e^{i\mathbf{q} \cdot \mathbf{r}} | \psi_\alpha \rangle|^2 \delta(E_\beta - E_\alpha - \hbar\omega) \, [f(E_\alpha) - f(E_\beta)], \tag{B.7}$$

where the factor 2 in front of the summation takes into account the spin degeneracy.

The energy per unit time, i.e. the power, dissipated in the system of volume V is given by

$$\mathcal{P}(\mathbf{q}, \omega) = \hbar\omega W(\mathbf{q}, \omega)$$
$$= 4\pi\omega |U_0|^2 \sum_{\alpha\beta} |\langle \psi_\beta | e^{i\mathbf{q} \cdot \mathbf{r}} | \psi_\alpha \rangle|^2 \delta(E_\beta - E_\alpha - \hbar\omega)[f(E_\alpha) - f(E_\beta)]. \tag{B.8}$$

Since $f(E_\alpha) > f(E_\beta)$ for any $E_\alpha < E_\beta$, it is evident that $\mathcal{P}(\mathbf{q}, \omega) \geq 0$ for any value (positive or negative) of the frequency ω.

In an isotropic and homogeneous medium, in the presence of the applied electric field (B.3) of wavevector $\mathbf{q}$ and frequency ω, it is reasonable to expect an induced current density parallel to the electric field, proportional to it for small field strength, and with the same time and space dependence; we can thus write

$$\mathbf{J}(\mathbf{r}, t) = \sigma(\mathbf{q}, \omega) E_0 \, \hat{\mathbf{e}} \, e^{i(\mathbf{q} \cdot \mathbf{r} - \omega t)} + \text{c.c.} \, ,$$

where $\sigma(\mathbf{q}, \omega)$ defines the *longitudinal conductivity* function. When a current density $\mathbf{J}$ flows in a medium in the presence of an electric field $\mathbf{E}$, the energy per unit time dissipated in the system of volume V is

$$\int_V \mathbf{J} \cdot \mathbf{E} \, d\mathbf{r} = \int_V \left[\sigma(\mathbf{q}, \omega) E_0 \, e^{i(\mathbf{q} \cdot \mathbf{r} - \omega t)} + \text{c.c.} \right] \left[E_0 \, e^{i(\mathbf{q} \cdot \mathbf{r} - \omega t)} + \text{c.c.} \right] d\mathbf{r}$$
$$= \sigma(\mathbf{q}, \omega) |E_0|^2 V + \text{c.c.} = 2\sigma_1(\mathbf{q}, \omega) \frac{q^2}{e^2} |U_0|^2 V.$$

This classical expression of the power dissipated in the system, can be identified with the quantum expression $\hbar\omega W(\mathbf{q}, \omega)$; we obtain

$$\sigma_1(\mathbf{q}, \omega) = \frac{1}{2} \frac{e^2}{q^2} \frac{\hbar\omega W(\mathbf{q}, \omega)}{|U_0|^2 V}, \tag{B.9}$$

which constitutes the basic relationship linking the real part of the conductivity to the microscopic quantum expression (B.7).

It is well known that the dielectric function is related to the conductivity function by the expression $\varepsilon = 1 + 4\pi i\sigma/\omega$; this relationship follows from the requirement

$$\frac{\partial \mathbf{D}}{\partial t} = \frac{\partial \mathbf{E}}{\partial t} + 4\pi \mathbf{J} \quad \text{with} \quad \mathbf{D} = \varepsilon \mathbf{E} \quad \text{and} \quad \mathbf{J} = \sigma \mathbf{E},$$

and the form $\exp(-i\omega t)$ for the time dependence of the fields. We thus have

$$\varepsilon_2(\mathbf{q}, \omega) = \frac{4\pi}{\omega}\sigma_1(\mathbf{q}, \omega) = \frac{2\pi \hbar e^2}{q^2}\frac{1}{V}\frac{W(\mathbf{q}, \omega)}{|U_0|^2}. \tag{B.10}$$

Using Eq. (B.7), the quantum expression of the imaginary part of the longitudinal dielectric function becomes

$$\varepsilon_2(\mathbf{q}, \omega) = \frac{8\pi^2 e^2}{q^2}\frac{1}{V}\sum_{\alpha\beta}|\langle\psi_\beta|e^{i\mathbf{q}\cdot\mathbf{r}}|\psi_\alpha\rangle|^2 \delta(E_\beta - E_\alpha - \hbar\omega)\,[f(E_\alpha) - f(E_\beta)]\,. \tag{B.11}$$

The real part of the dielectric function is obtained inserting Eq. (B.11) into the Kramers-Kronig relation

$$\varepsilon_1(\mathbf{q}, \omega) = 1 + \frac{1}{\pi}P\int_{-\infty}^{+\infty}\frac{\varepsilon_2(\mathbf{q}, \omega')}{\omega' - \omega}\,d\omega';$$

we have

$$\varepsilon_1(\mathbf{q}, \omega) = 1 + \frac{8\pi e^2}{q^2}\frac{1}{V}\sum_{\alpha\beta}|\langle\psi_\beta|e^{i\mathbf{q}\cdot\mathbf{r}}\psi_\alpha\rangle|^2\frac{f(E_\alpha) - f(E_\beta)}{E_\beta - E_\alpha - \hbar\omega}\,. \tag{B.12}$$

Equations (B.11) and (B.12) give the longitudinal dielectric function of a generic system, with eigenfunctions $|\psi_\alpha\rangle$ and eigenvalues E_α, under the assumption of linear response, homogeneity, and negligible local field effects. Eqs. (B.11) and (B.12) can be rewritten in the unique formula

$$\varepsilon(\mathbf{q}, \omega) = 1 + \frac{8\pi e^2}{q^2}\frac{1}{V}\sum_{\alpha\beta}\frac{|\langle\psi_\beta|e^{i\mathbf{q}\cdot\mathbf{r}}|\psi_\alpha\rangle|^2}{E_\beta - E_\alpha - \hbar\omega - i\eta}[f(E_\alpha) - f(E_\beta)]\,. \tag{B.13}$$

The separation of the real and imaginary parts of the dielectric function in Eq. (B.13), and the equivalence with Eqs. (B.11) and (B.12), can be verified using the well-known identity

$$\lim_{\eta\to 0^+}\frac{1}{x - i\eta} = P\frac{1}{x} + i\pi\delta(x),$$

where x is real and P denotes the principal part.

From Eqs. (B.8) and (B.11), we see that the power dissipated in the medium, due to the driving perturbation (B.2), can be written as

$$P(\mathbf{q}, \omega) = V \frac{\omega q^2}{2\pi e^2} \varepsilon_2(\mathbf{q}, \omega) |U_0(\mathbf{q}, \omega)|^2. \tag{B.14}$$

Equation (B.14) shows that the power dissipated in the medium, due to a driving perturbation of wavevector $\mathbf{q}$, frequency ω, and amplitude $U_0(\mathbf{q}, \omega)$, is essentially determined by the *imaginary part* $\varepsilon_2(\mathbf{q}, \omega)$ *of the dielectric function*.

In analogy with the relation $\mathbf{D} = \varepsilon \mathbf{E}$ between the induction vector $\mathbf{D}$ (due to external sources only) and the total electric field $\mathbf{E}$ (due to external and internal sources), we write $U_{\text{ext}} = \varepsilon(\mathbf{q}, \omega) U_0$, which connects the amplitude of the external perturbation to the amplitude of the total perturbation driving the system. Eq. (B.14) can thus be written also in the equivalent form

$$P(\mathbf{q}, \omega) = V \frac{\omega q^2}{2\pi e^2} \frac{\varepsilon_2(\mathbf{q}, \omega)}{|\varepsilon(\mathbf{q}, \omega)|^2} |U_{\text{ext}}(\mathbf{q}, \omega)|^2. \tag{B.15}$$

Equation (B.15) shows that the power dissipated in the medium, due to an external perturbation of wavevector $\mathbf{q}$, frequency ω, and amplitude $U_{\text{ext}}(\mathbf{q}, \omega)$, is essentially determined by the quantity

$$-\text{Im} \frac{1}{\varepsilon(\mathbf{q}, \omega)} = \frac{\varepsilon_2(\mathbf{q}, \omega)}{\varepsilon_1^2(\mathbf{q}, \omega) + \varepsilon_2^2(\mathbf{q}, \omega)}. \tag{B.16}$$

which is called *energy-loss function*.

Sum Rules for the Dielectric Function

We can now discuss other general properties of the dielectric function with the following argument. The Hamiltonian operator of the form (B.1) satisfies the general commutation relation

$$[H_0, e^{i\mathbf{q}\cdot\mathbf{r}}] = e^{i\mathbf{q}\cdot\mathbf{r}} \left(\frac{\hbar^2 q^2}{2m} - \frac{\hbar^2}{m} i\mathbf{q} \cdot \nabla \right).$$

For the double commutator we have

$$\left[e^{-i\mathbf{q}\cdot\mathbf{r}}, \left[H_0, e^{i\mathbf{q}\cdot\mathbf{r}} \right] \right] = \frac{\hbar^2 q^2}{m}.$$

Taking the expectation value of the double commutator on any chosen state ψ_α, and inserting appropriately the unit operator $1 = \sum |\psi_\beta\rangle\langle\psi_\beta|$ (and observing that the Hamiltonian H_0 is invariant for time-reversal symmetry) we obtain

$$\sum_\beta |\langle\psi_\beta|e^{i\mathbf{q}\cdot\mathbf{r}}|\psi_\alpha\rangle|^2 (E_\beta - E_\alpha) = \frac{\hbar^2 q^2}{2m}. \tag{B.17}$$

We sum up the above relation over the occupied orbitals ψ_α, and divide by the volume V of the system; we obtain the useful relation

$$\frac{1}{V} \sum_{\alpha\beta} f(E_\alpha) |\langle \psi_\beta | e^{i\mathbf{q}\cdot\mathbf{r}} | \psi_\alpha \rangle|^2 (E_\beta - E_\alpha) = \frac{\hbar^2 q^2}{2m} \frac{n}{2}, \tag{B.18}$$

where n is the number of electrons of both spin directions per unit volume, and $n/2$ is the number of occupied orbitals per unit volume.

The exact result (B.18) allows us to establish some general and significant properties of the real and imaginary parts of the dielectric function. From Eqs. (B.11) and (B.18), we obtain the sum rule

$$\int_0^\infty \omega\, \varepsilon_2(\mathbf{q}, \omega)\, d\omega = \frac{\pi}{2} \omega_p^2 \qquad \omega_p^2 = \frac{4\pi n e^2}{m}, \tag{B.19}$$

where ω_p is the plasma frequency. From Eq. (B.19), and the Kramers-Kronig relation

$$\varepsilon_1(\mathbf{q}, \omega) = 1 + \frac{2}{\pi} P \int_0^{+\infty} \frac{\omega' \varepsilon_2(\mathbf{q}, \omega')}{\omega'^2 - \omega^2}\, d\omega',$$

in the limit of $\hbar\omega$ much larger than any resonant frequency of the medium, we have the exact asymptotic behavior for the real part of the dielectric function

$$\varepsilon_1(\mathbf{q}, \omega) \to 1 - \frac{\omega_p^2}{\omega^2} \quad \text{for} \quad \omega \to \infty; \tag{B.20}$$

this relation shows that in the high frequency limit the response of any medium is essentially free-electron like.

Dielectric Function in the Long Wavelength Limit

We now add a comment on the dielectric function in the long wavelength limit. In the limit $\mathbf{q} \to 0$, the matrix elements appearing in Eq. (B.13) can be written in the form

$$\langle \psi_\beta | e^{i\mathbf{q}\cdot\mathbf{r}} | \psi_\alpha \rangle = i\mathbf{q} \cdot \langle \psi_\beta | \mathbf{r} | \psi_\alpha \rangle = \frac{\hbar}{m} \frac{i\mathbf{q} \cdot \langle \psi_\beta | \mathbf{p} | \psi_\alpha \rangle}{E_\beta - E_\alpha} \quad \text{with} \quad E_\alpha \neq E_\beta. \tag{B.21}$$

The first passage in Eq. (B.21) is obtained with the series expansion of the exponential to first order in q; the substitution of the matrix elements of $\mathbf{r}$ with matrix elements of $\mathbf{p}$ is done exploiting the commutation rule $[H_0, \mathbf{r}] = -i(\hbar/m)\mathbf{p}$ of the Hamiltonian operator (B.1).

Let us indicate by $\hat{\mathbf{e}}$ the unit vector in the direction $\mathbf{q}$ of the applied perturbing longitudinal field. Using Eq. (B.21), we can recast Eq. (B.13) into the form

$$\varepsilon(0, \omega) = 1 + \frac{8\pi e^2}{m^2} \frac{1}{V} \sum_{\alpha\beta} \frac{|\langle \psi_\beta | \hat{\mathbf{e}} \cdot \mathbf{p} | \psi_\alpha \rangle|^2}{[(E_\beta - E_\alpha)/\hbar]^2} \frac{f(E_\alpha) - f(E_\beta)}{E_\beta - E_\alpha - \hbar\omega - i\eta}. \tag{B.22}$$

Equation (B.22) gives an alternative expression of the longitudinal dielectric function of linear and homogeneous materials, valid in the long wavelength limit.

Appendix C.   Lindhard Dielectric Function for the Free-Electron Gas

The purpose of this Appendix is to evaluate analytically the Lindhard longitudinal dielectric function of the free-electron gas, given by Eq. (7.31) of the text

$$\varepsilon(\mathbf{q}, \omega) = 1 + \frac{8\pi e^2}{q^2} \frac{1}{V} \sum_{\mathbf{k}} \frac{f(\mathbf{k}) - f(\mathbf{k} + \mathbf{q})}{E(\mathbf{k} + \mathbf{q}) - E(\mathbf{k}) - \hbar\omega - i\eta}, \tag{C.1}$$

where $E(\mathbf{k}) = \hbar^2 k^2 / 2m$ is the free-electron energy, $f(\mathbf{k})$ is the Fermi-Dirac distribution function, spin degeneracy is already included, and the limit $\eta \to 0^+$ is understood.

It is convenient to rewrite expression (C.1) in a slightly different form, by performing the replacement $\mathbf{k} + \mathbf{q} \to -\mathbf{k}'$ in the term containing $f(\mathbf{k} + \mathbf{q})$ (then $\mathbf{k}'$ is relabeled as $\mathbf{k}$); we have

$$\varepsilon(\mathbf{q}, \omega) = 1 + \frac{8\pi e^2}{q^2 V} \sum_{\mathbf{k}} \left[\frac{f(\mathbf{k})}{E(\mathbf{k} + \mathbf{q}) - E(\mathbf{k}) - \hbar\omega - i\eta} + \frac{f(\mathbf{k})}{E(\mathbf{k} + \mathbf{q}) - E(\mathbf{k}) + \hbar\omega + i\eta} \right]. \tag{C.2}$$

For brevity, the two sums over $\mathbf{k}$ appearing in the right-hand side of Eq. (C.2) are denoted as S_1 and S_2, respectively.

We consider the limit of zero temperature, and as usual we convert the sum over $\mathbf{k}$ into the integral in $d\mathbf{k}$ times $V/(2\pi)^3$. The expression of S_1 becomes

$$S_1 = \frac{e^2}{\pi^2 q^2} \int_{\substack{\text{Fermi} \\ \text{sphere}}} \frac{1}{E(\mathbf{k} + \mathbf{q}) - E(\mathbf{k}) - \hbar\omega - i\eta} \, d\mathbf{k}. \tag{C.3}$$

It is useful to rewrite the denominator in the form

$$\frac{\hbar^2 q^2}{2m} + \frac{\hbar^2}{m} \mathbf{k}\cdot\mathbf{q} - \hbar\omega - i\eta = \frac{\hbar^2 q}{m}(z_1 k_F + k\cos\theta), \quad z_1 = \frac{q}{2k_F} - \frac{\omega}{(\hbar k_F/m)\, q} - i\eta,$$

where z_1 is a convenient dimensionless complex quantity, $\hbar k_F/m$ is the Fermi velocity, and, without loss of generality, the vector $\mathbf{q}$ is assumed to lie along the polar axis k_z of the reciprocal space. The expression of S_1 becomes

$$S_1 = \frac{me^2}{\pi^2 \hbar^2 q^3} \cdot 2\pi \int_0^{k_F} k^2 dk \int_0^\pi \frac{1}{z_1 k_F + k\cos\theta} \sin\theta \, d\theta. \tag{C.4}$$

The two integrals in Eq. (C.4) are both easily performed. The first integration gives

$$\int_0^\pi \frac{1}{z_1 k_F + k\cos\theta} \sin\theta \, d\theta = \frac{1}{k} \ln \frac{z_1 k_F + k}{z_1 k_F - k}.$$

The second integration in Eq. (C.4) can again be performed analytically using the following indefinite integral

$$\int x \ln \frac{a+x}{a-x} \, dx = ax + \frac{1}{2}(x^2 - a^2) \ln \frac{a+x}{a-x}.$$

We obtain

$$\int_0^{k_F} k \ln \frac{z_1 k_F + k}{z_1 k_F - k} \, dk = k_F^2 \left[z_1 + \frac{1 - z_1^2}{2} \ln \frac{z_1 + 1}{z_1 - 1} \right].$$

In summary

$$S_1 = \frac{2me^2}{\pi \hbar^2 k_F} \cdot \frac{k_F^3}{q^3} \left[z_1 + \frac{1 - z_1^2}{2} \ln \frac{z_1 + 1}{z_1 - 1} \right], \tag{C.5}$$

where all factors are dimensionless.

The first dimensionless factor in Eq. (C.5) can be expressed in terms of the Thomas-Fermi screening wave number previously encountered and so defined

$$k_{TF}^2 = 4\pi e^2 n(E_F) \equiv \frac{4me^2 k_F}{\pi \hbar^2} \implies \boxed{\frac{2me^2}{\pi \hbar^2 k_F} = \frac{1}{2} \frac{k_{TF}^2}{k_F^2}} \tag{C.6}$$

$$\left[\text{use: } n(E_F) = \frac{3}{2} \frac{n}{E_F}, \; n = \frac{k_F^3}{3\pi^2}, \; E_F = \frac{\hbar^2 k_F^2}{2m} \right].$$

The expression of S_1 can be rewritten in the form

$$S_1 = \frac{1}{2} \frac{k_{TF}^2}{q^2} \frac{k_F}{q} \left[z_1 + \frac{1 - z_1^2}{2} \ln \frac{z_1 + 1}{z_1 - 1} \right] \qquad z_1 = \frac{q}{2k_F} - \frac{m\omega}{\hbar q k_F} - i\eta.$$

A quite similar expression holds for the second sum that reads

$$2 = \frac{1}{2} \frac{k_{TF}^2}{q^2} \frac{k_F}{q} \left[z_2 + \frac{1 - z_2^2}{2} \ln \frac{z_2 + 1}{z_2 - 1} \right] \qquad z_2 = \frac{q}{2k_F} + \frac{m\omega}{\hbar q k_F} + i\eta.$$

The dielectric function (C.1) becomes

$$\boxed{\varepsilon(q, \omega) = 1 + \frac{1}{2} \frac{k_{TF}^2}{q^2} + \frac{1}{4} \frac{k_{TF}^2}{q^2} \frac{k_F}{q} \left[(1 - z_1^2) \ln \frac{z_1 + 1}{z_1 - 1} + (1 - z_2^2) \ln \frac{z_2 + 1}{z_2 - 1} \right]} \tag{C.7}$$

$$z_1 = x_1 - i\eta, \quad x_1 = \frac{q}{2k_F} - \frac{m\omega}{\hbar q k_F}; \quad z_2 = x_2 + i\eta, \quad x_2 = \frac{q}{2k_F} + \frac{m\omega}{\hbar q k_F},$$

where x_1, x_2 denote the real parts of z_1, z_2, respectively.

In Eq. (C.7) we separate real and imaginary parts taking the limit $\eta \to 0^+$. The logarithmic function of complex argument is defined with the cut from 0 to $-\infty$; we have $\ln z = \ln |z| + i \arg z$ with $-\pi < \arg z < +\pi$. The real part of the dielectric function (C.7) becomes

$$\boxed{\varepsilon_1(q, \omega) = 1 + \frac{1}{2} \frac{k_{TF}^2}{q^2} + \frac{1}{4} \frac{k_{TF}^2}{q^2} \frac{k_F}{q} \left[(1 - x_1^2) \ln \left| \frac{x_1 + 1}{x_1 - 1} \right| + (1 - x_2^2) \ln \left| \frac{x_2 + 1}{x_2 - 1} \right| \right]}.$$

$$\tag{C.8}$$

In particular for the static dielectric function we have $x_1 = x_2 = q/2k_F$, and Eq. (C.8) gives

$$\varepsilon_1(q,0) = 1 + \frac{1}{2}\frac{k_{TF}^2}{q^2} + \frac{1}{2}\frac{k_{TF}^2}{q^2}\frac{k_F}{q}\left(1 - \frac{q^2}{4k_F^2}\right)\ln\left|\frac{2k_F+q}{2k_F-q}\right|. \tag{C.9}$$

For the dynamic dielectric function at sufficiently high frequencies, we obtain the simple expression

$$\boxed{\varepsilon_1(q,\omega) = 1 - \frac{\omega_p^2}{\omega^2} - \frac{3}{5}\frac{\omega_p^2}{\omega^4}v_F^2 q^2 - \cdots, \quad \hbar\omega >> \frac{\hbar^2}{2m}(2qk_F - q^2)\,};} \tag{C.10}$$

this result can be established by performing appropriate series developments in Eq. (C.8), and is left as an exercise to the reader.

It is also straightforward to obtain the imaginary part of the dielectric function (C.7). In fact we have

$$\text{Im}\ln\frac{z_1+1}{z_1-1} = \text{Im}\ln[x_1+1-i\eta] - \text{Im}\ln[x_1-1-i\eta]$$

$$= \begin{cases} 0 & \text{if } x_1 < -1 \text{ or } x_1 > +1 \\ \pi & \text{if } -1 < x_1 < +1 \end{cases}.$$

With similar elaborations, the imaginary part of the dielectric function (C.7) can be cast in the form

$$\boxed{\varepsilon_2(q,\omega) = \frac{1}{4}\frac{k_{TF}^2}{q^2}\frac{k_F}{q}\pi\left[\left(1-x_1^2\right)\Theta\left(1-x_1^2\right) - \left(1-x_2^2\right)\Theta\left(1-x_2^2\right)\right],} \tag{C.11}$$

where Θ is the step function $[\Theta(x) = 0$ for $x < 0$, and $\Theta(x) = 1$ for $x > 0]$.

We can give a more explicit expression of the function (C.11), considering in the (q,ω) plane the four parabolas defined by $x_{1,2} = \mp 1$; namely:

$$x_1 = \mp 1 \Rightarrow \hbar\omega = \frac{\hbar^2}{2m}(q^2 \pm 2qk_F); \quad x_2 = \mp 1 \Rightarrow \hbar\omega = \frac{\hbar^2}{2m}(-q^2 \mp 2qk_F). \tag{C.12}$$

These parabolas divide the first quadrant of the (q,ω) plane (with positive q and ω) into four regions, as indicated in Figure 7.14. In each region, a straight evaluation of Eq. (C.11) gives

$$\varepsilon_2(q,\omega) = \begin{cases} \dfrac{\pi}{4}\dfrac{k_{TF}^2}{q^2}\dfrac{k_F}{q}\dfrac{\hbar\omega}{E_F} & \text{in region II} \\[3mm] \dfrac{\pi}{4}\dfrac{k_{TF}^2}{q^2}\dfrac{k_F}{q}\left[1 - \left(\dfrac{q}{2k_F} - \dfrac{m\omega}{\hbar qk_F}\right)^2\right] & \text{in region III} \\[3mm] 0 & \text{in the other cases} \end{cases}. \tag{C.13}$$

From Eq. (C.13) and Figure 7.14, we see that $\varepsilon_2(q,\omega)$ vanishes in regions I and IV; only in regions II and III is $\varepsilon_2(q,\omega)$ different from zero, and depends linearly on ω in the region II and quadratically in the region III.

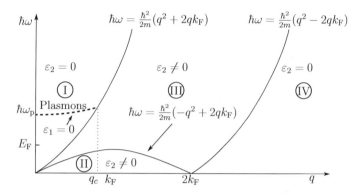

Figure 7.14 Imaginary part of the Lindhard dielectric function. The first quadrant in the (q, ω) plane is divided into four regions by the parabolas defined by Eq. (C.12). The dashed line, where simultaneously $\varepsilon_1(q, \omega)$ and $\varepsilon_2(q, \omega)$ vanish, gives the dispersion curve of plasmon modes. The plasmon cut-off wavevector q_c is also indicated.

We note that in the regions I and IV electron-hole single-particle excitations of vector **q** are not possible (and thus ε_2 vanishes there). Consider in fact the energy difference

$$\Delta(\mathbf{k} + \mathbf{q}, \mathbf{k}) = E(\mathbf{k} + \mathbf{q}) - E(\mathbf{k}) = \frac{\hbar^2}{2m}(q^2 + 2\mathbf{q} \cdot \mathbf{k}), \qquad (C.14)$$

with $f(\mathbf{k}) \equiv 1$ and $f(\mathbf{k} + \mathbf{q}) \equiv 0$. Expression (C.14) represents the energy required to excite an electron within the Fermi sphere to an empty state outside it with wavevector transfer **q**. It is evident that its maximum value, $\Delta_{\max}$, is obtained when **k** lies in the direction of **q** and $|\mathbf{k}| = k_F$; it holds

$$\Delta_{\max} = \frac{\hbar^2}{2m}(q^2 + 2qk_F).$$

If $\hbar\omega > \Delta_{\max}$ (region I), no electron-hole excitation of wavevector transfer q is possible. Similarly for $q > 2k_F$ the minimum value of expression (C.14) occurs for **k** antiparallel to **q** and $|\mathbf{k}| = k_F$; it holds

$$\Delta_{\min} = \frac{\hbar^2}{2m}(q^2 - 2qk_F).$$

Thus no electron-hole excitation is possible in region IV either.

Appendix D.   Quantum Expression of the Transverse Dielectric Function in Materials with the Linear Response Theory

In this Chapter we have so far considered the *longitudinal dielectric function* of linear and homogeneous materials, and the entailed phenomenological aspects, such as energy-loss measurements of impinging electron beams.

In this Appendix, we consider the quantum expression of the *transverse dielectric function*, which is needed for the interpretation and understanding of optical properties. For what concerns experimental aspects, there is an extreme differentiation of techniques using beams of charged particles to explore longitudinal response functions and beams of photons to explore transverse response functions; on the contrary from a formal point of view the generality of the linear response formalism makes the theoretical description of longitudinal and transverse response functions rather similar (actually, in the long wavelength limit, the transverse, and longitudinal dielectric functions are just equal). For this reason we anticipate here some theoretical results concerning the optical properties of materials, that will be discussed in great detail in Chapters 11 and 12.

Thus, the purpose of this Appendix is to provide the *quantum expression of the transverse dielectric function for homogeneous (or nearly homogeneous) materials, within the linear response theory*. The quantum expression of the longitudinal dielectric function has been considered in Appendix B. In the present Appendix, within the same approximations of linearity and homogeneity, we consider the transverse dielectric function of an electron system. The formal procedure for the longitudinal and transverse functions is so similar, that little has to be changed in the former treatment to obtain the latter one.

Consider a general electronic system, described by the Hamiltonian

$$H_0 = \frac{\mathbf{p}^2}{2m} + V(\mathbf{r}), \tag{D.1}$$

where $V(\mathbf{r})$ is the energy potential for an electron in the sample. We indicate by $\psi_n(\mathbf{r})$ and E_n the eigenfunctions and eigenvalues of H_0 (the eigenfunctions are normalized to unity in the volume V of the crystal); spin degeneracy is taken into account by including a factor 2, whenever needed. The occupancy of the states is determined by the Fermi-Dirac function $f(E_n)$ (in most situations, the temperature $T = 0$ is considered).

In the presence of an electromagnetic field, the Hamiltonian H of the system is obtained from the unperturbed Hamiltonian H_0 replacing the momentum operator $\mathbf{p}$ with the generalized momentum $\mathbf{p} + (e/c)\mathbf{A}(\mathbf{r}, t)$, where $\mathbf{A}$ is the vector potential of the electromagnetic field and e is the modulus of the electronic charge. Without loss of generality, we take the scalar potential as zero and we adopt the Coulomb gauge $\text{div}\mathbf{A} = 0$, so that the terms $\mathbf{A} \cdot \mathbf{p}$ and $\mathbf{p} \cdot \mathbf{A}$ coincide. Neglecting the term in $\mathbf{A}^2$, the Hamiltonian of the system in the presence of a radiation field thus becomes

$$H = H_0 + \frac{e}{mc}\mathbf{A}(\mathbf{r}, t) \cdot \mathbf{p}. \tag{D.2}$$

We consider a *transverse electromagnetic plane wave* of frequency ω, wavevector $\mathbf{q}$ and polarization vector $\widehat{\mathbf{e}}$, described by the vector potential

$$\mathbf{A}(\mathbf{r}, t) = A_0 \widehat{\mathbf{e}} e^{i(\mathbf{q}\cdot\mathbf{r}-\omega t)} + \text{c.c.} \qquad \widehat{\mathbf{e}} \perp \mathbf{q}, \tag{D.3}$$

where c.c. indicates the complex conjugate of the previous term, and $A_0 = A_0(\mathbf{q}, \omega)$ is the (infinitesimal) amplitude of the transverse field. Under the stated approximations the Hamiltonian H becomes

$$H = H_0 + \frac{eA_0}{mc} e^{i(\mathbf{q}\cdot\mathbf{r}-\omega t)}\widehat{\mathbf{e}} \cdot \mathbf{p} + \frac{eA_0^\star}{mc} e^{-i(\mathbf{q}\cdot\mathbf{r}-\omega t)}\widehat{\mathbf{e}} \cdot \mathbf{p}. \tag{D.4}$$

We also notice that the electric field in the medium associated to the vector potential $\mathbf{A}(\mathbf{r}, t)$, given by Eq. (D.3), is

$$\mathbf{E}(\mathbf{r}, t) = -\frac{1}{c} \frac{\partial \mathbf{A}}{\partial t} = E_0 \hat{\mathbf{e}} e^{i(\mathbf{q} \cdot \mathbf{r} - \omega t)} + \text{c.c.} \quad \text{with} \quad E_0 = i \omega \frac{A_0}{c}. \quad (D.5)$$

In close analogy to the treatment of Appendix B, we observe that the time-dependent terms in the right-hand side of Eq. (D.4) induce transitions among the states of H_0, and can be treated according to standard time-dependent perturbation theory. The first time-dependent term in Eq. (D.4) gives rise to absorption of radiation, while the second term gives rise to emission. Taking into account first order transition probabilities and occupation of states, we obtain that the net number of transitions per unit time involving the energy $\hbar \omega$ is given by the expression

$$W(\mathbf{q}, \omega) = \frac{2\pi}{\hbar} \left| \frac{e A_0}{mc} \right|^2 2 \sum_{ij} |\langle \psi_j | e^{i\mathbf{q} \cdot \mathbf{r}} \hat{\mathbf{e}} \cdot \mathbf{p} | \psi_i \rangle|^2 \delta(E_j - E_i - \hbar \omega)[f(E_i) - f(E_j)], \quad (D.6)$$

where the factor 2 in front of the summation takes into account the spin degeneracy. Similarly to Eq. (B.10), the imaginary part of the dielectric function is related to $W(\mathbf{q}, \omega)$ by the expression

$$\varepsilon_2(\mathbf{q}, \omega) = \frac{2\pi \hbar c^2}{\omega^2} \frac{1}{V} \frac{W(\mathbf{q}, \omega)}{|A_0|^2}. \quad (D.7)$$

At this stage there is no need to repeat further a procedure already outlined in Appendix B, and we arrive straight at the result.

The quantum expression of the transverse dielectric function is

$$\varepsilon(\mathbf{q}, \omega) = 1 + \frac{8\pi e^2}{m^2} \frac{1}{V} \sum_{ij} \frac{\left| \langle \psi_j | e^{i\mathbf{q} \cdot \mathbf{r}} \hat{\mathbf{e}} \cdot \mathbf{p} | \psi_i \rangle \right|^2}{(E_j - E_i)^2 / \hbar^2} \frac{[f(E_i) - f(E_j)]}{E_j - E_i - \hbar \omega - i\eta}, \quad (D.8)$$

where $\eta \to 0^+$. Equation (D.8) is the basic expression of the *transverse dielectric function* of an electronic system of volume V, with one-electron wavefunctions $\psi_i(\mathbf{r})$ and eigenvalues E_i.

It is worthwhile to notice that Eq. (D.8) coincides with Eq. (B.22) in the particular case $\mathbf{q} = 0$. As expected, the transverse and longitudinal dielectric functions of isotropic media are equal in the long wavelength limit $\mathbf{q} \to 0$, while they are in general different for $\mathbf{q} \neq 0$.

It is useful to give also the expression of the transverse conductivity, linked to the transverse dielectric function by the relation $\varepsilon = 1 + 4\pi i \sigma / \omega$. We have

$$\sigma(\mathbf{q}, \omega) = \frac{2e^2}{m^2} \frac{1}{V} \sum_{ij} \frac{\left| \langle \psi_j | e^{i\mathbf{q} \cdot \mathbf{r}} \hat{\mathbf{e}} \cdot \mathbf{p} | \psi_i \rangle \right|^2}{(E_j - E_i)/\hbar} \frac{(-i)[f(E_i) - f(E_j)]}{E_j - E_i - \hbar \omega - i\eta}. \quad (D.9)$$

Equation (D.9) is the quantum mechanical expression of the transverse conductivity function of homogeneous materials; in crystals, it treats on the same footing intraband and interband contributions.

Further Reading

Agranovich, V. M., & Ginzburg, V. L. (1966). *Spatial dispersion in crystal optics and the theory of excitons.* New York: Interscience Publishing.

Andreani, L. C. (1995). Optical transitions, excitons, and polaritons in bulk and low-dimensional semiconductor structures. In E. Burnstein, & C. Weishbuch (Eds.), *Confined electrons and photons.* New York: Plenum Press.

Andreani, L. C. (2003). Exciton-polaritons in confined systems. In B. Deveaud, A. Quattropani, & P. Schwendimann (Eds.), *Proceedings of the international school of physics "Enrico Fermi" course CL.* Amsterdam: IOS Press.

Bassani, F., & Pastori Parravicini, G. (1975). *Electronic states and optical transitions in solids.* Oxford: Pergamon Press.

Cho, K. (Ed.). (1979). *Excitons.* Berlin: Springer.

Giuliani, G., & Vignale, G. (2005). *Quantum theory of the electron liquid.* Cambridge: Cambridge University Press.

Hanke, W. (1978). Dielectric theory of elementary excitations in crystals. *Advances in Physics, 27,* 287.

Harrison, W. A. (1979). *Solid state theory.* New York: Dover Publication.

Kittel, C. (1987). *Quantum theory of solids.* New York: Wiley.

Knox, R. (1963). *Theory of excitons.* New York: Academic Press.

Maier, S. A. (2007). *Plasmonics: Fundamentals and applications.* Berlin: Springer.

Martin, P. A., & Rothen, F. (2004). *Many-body problems and quantum field theory.* Berlin: Springer.

Nakajima, S., Toyozawa, Y., & Abe, R. (1980). *The physics of elementary excitations.* Berlin: Springer.

Pines, D. (1963). *Elementary excitations in solids.* New York: Benjamin.

Pitarke, J. M., Silkin, V. M., Chulkov, E. V., & Echenique, P. M. (2007). Theory of surface plasmons and surface-plasmon polaritons. *Reports on Progress in Physics, 70,* 1.

Raether, H. (1980). *Excitation of plasmons and interband transitions by electrons.* Berlin: Springer.

Rashba, E. I., & Sturge, M. D. (Eds.). (1982). *Excitons.* Amsterdam: North-Holland.

Schnatterly, S. E. (1979). Inelastic electron scattering spectroscopy. In H. Ehrenreich, F. Seitz, & D. Turnbull (Eds.), *Solid state physics* (Vol. 34, pp. 275). New York: Academic Press.

Wallis, R. F., & Balkanski, M. (1986). *Many-body aspects of solid state spectroscopy.* Amsterdam: North-Holland.

Zayats, A. V., Smolyaninov, I. I., & Maradudin, A. A. (2005). Nano-optics of surface plasmon polaritons. *Physics Reports, 408,* 131.

Ziman, J. M. (1972). *Principles of the theory of solids.* Cambridge: Cambridge University Press.

8 Interacting Electronic-Nuclear Systems and the Adiabatic Principle

In the study of the electronic structure of crystals we have so far considered the nuclei fixed in a given spatial configuration, usually the equilibrium one. In this chapter we analyze the consequences related to the fact that the nuclei have indeed a mass much larger than the electron mass, but yet finite. Thus the solution of the electronic problem with the nuclei frozen in a given configuration can only be considered as a preliminary step, from which to start a more realistic investigation.

In this chapter we develop the quantum theory for systems of electrons and nuclei in interaction. We focus on the concepts embodied in the adiabatic approximation and on some aspects beyond it; these ideas are the natural link between the electronic properties of crystals and the lattice dynamics, the two most traditional subjects of solid state physics. The parametric dependence of the crystal Hamiltonian from the nuclear coordinates in the adiabatic approximation offers the occasion to present some general properties of parameter dependent operators, and in particular the Hellmann-Feynman theorem, and the concept of geometric Berry phase. The general principles of this chapter are further elaborated and put at work in several other parts of the book; thus, the reader can choose to give initially an overall view to this chapter and later re-visit in depth its topics, as these are encountered in specific contexts.

Solid State Physics, Second Edition. http://dx.doi.org/10.1016/B978-0-12-385030-0.00008-6

8.1 Interacting Electronic-Nuclear Systems and Adiabatic Potential-Energy Surfaces

A crystal is composed by a large number of interacting particles and consequently its theoretical treatment cannot avoid appropriate approximations. The starting one is suggested by the large difference in the masses of the nuclei and of the electrons; in most situations, this allows one to decouple the dynamics of the fast variables (the electrons) from the dynamics of the slow variables (the nuclei). The decoupling of electron and nuclear dynamics (when it is possible at all) is conceptually achieved by means of the so-called "adiabatic" procedure, whose basic principles can be summarized as follows.

In a first stage the *nuclei are supposed fixed in selected spatial configurations*, and attention is focused on the electronic eigenvalues as a function of the chosen nuclear coordinates; these eigenvalues describe the so-called *adiabatic potential-energy surfaces*, also simply denoted as *potential surfaces* or *adiabatic surfaces* (or *sheets*). Most often, the ground or a few lowest energy adiabatic surfaces are of major interest.

The potential surfaces of electronic-nuclear systems may be non-degenerate or may exhibit degeneracy points (see Figure 8.1). In some systems, the ground potential surface under attention is non-degenerate in the whole domain of nuclear coordinates of interest (*ordinary Born-Oppenheimer systems*). In other systems *two or more potential surfaces are degenerate* (or nearly degenerate) at some point in the nuclear coordinate space. Degeneracy includes a variety of situations: Jahn-Teller systems, characterized by non-vanishing spatial gradients of the potential surfaces at the degeneracy point, Renner-Teller systems in linear molecules with vanishing gradients at the degeneracy point, systems in Kramers degenerate electronic states.

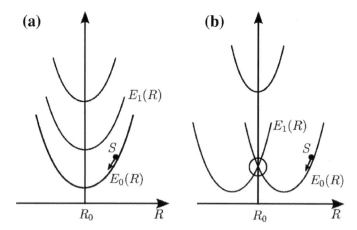

Figure 8.1 Schematic representation of adiabatic potential-energy surfaces $E_i(R)$ (the multidimensional nuclear variable R is represented as a one-dimensional parameter). (a) The adiabatic surface of interest is supposed non-degenerate, and the nuclear dynamics of the system S is mostly determined by the single non-degenerate potential surface under consideration. (b) Two (or more) adiabatic surfaces are degenerate at R_0, and the nuclear dynamics of the system S is determined by the whole set of degenerate adiabatic surfaces.

Once the potential-energy surfaces $E_i(R)$ are known as functions of the multidimensional nuclear coordinates (collectively indicated by R), the nuclear dynamics can be studied. From a classical point of view, the forces acting on the nuclei are just given by the negative gradient of the potential-energy surface to which the system belongs. If the potential surface under consideration is well separated from all the others, it seems reasonable to assume that the nuclei move (classically or quantistically) on the selected potential surface itself (Figure 8.1a). In the case two (or more) adiabatic surfaces are degenerate at some configuration, the nuclear dynamics is determined jointly by the multiple potential-energy surfaces (as schematically indicated in Figure 8.1b). In either cases, the limitation to one or a number of potential-energy surfaces is only a useful starting approximation and (whenever necessary) the influence of the other adiabatic surfaces should be taken into account.

The key point at the foundation of the quantum description of any polynuclear system (molecules, clusters, solids) is the large difference of the masses of electrons and nuclei. A general electronic-nuclear system can be pictured as composed by a massive slowly moving part (the nuclear subsystem) and a fast moving part (the electronic subsystem), that quickly adjusts to the nuclear configuration. It has been shown by Born and Oppenheimer how to describe the interdependence between the electronic energies and the nuclear dynamics [M. Born and R. Oppenheimer, Ann. Physique *84*, 457 (1927)].

A crystal is composed by mutually interacting electrons and nuclei, and the total Hamiltonian includes the kinetic energy of the nuclei (T_N), the kinetic energy of the electrons (T_e), and all possible electron-electron (V_{ee}), electron-nucleus (V_{eN}), nucleus-nucleus (V_{NN}) interactions. If we consider only Coulomb forces, the expression of the Hamiltonian in the non-relativistic limit (and disregarding coupling with electromagnetic fields and other interactions) becomes

$$H_{tot} = T_N + T_e + V_{ee} + V_{eN} + V_{NN}$$
$$= -\sum_I \frac{\hbar^2 \nabla_I^2}{2M_I} - \sum_i \frac{\hbar^2 \nabla_i^2}{2m} + \frac{1}{2} \sum_{i \neq j} \frac{e^2}{|\mathbf{r}_i - \mathbf{r}_j|} - \sum_{iI} \frac{z_I e^2}{|\mathbf{r}_i - \mathbf{R}_I|} + \frac{1}{2} \sum_{I \neq J} \frac{z_I z_J e^2}{|\mathbf{R}_I - \mathbf{R}_J|},$$

$$(8.1)$$

where the indices i, j refer to the electrons, and the indices I, J to the nuclei (of charge $z_I e$ and $z_J e$, respectively).

It is convenient to partition the total Hamiltonian (8.1) in the form

$$H_{tot} = T_N(R) + T_e(r) + V(r, R), \qquad (8.2)$$

where T_N is the nuclear kinetic operator, T_e is the electronic kinetic operator, and $V(r, R)$ denotes *the sum of all the electrostatic interactions between particles (i.e. between electrons V_{ee}, between nuclei V_{NN}, and between nuclei and electrons V_{eN}).* The multidimensional nuclear variable R is a shorthand notation for all the $3N$ cartesian coordinates of the nuclei of the system; when required by clarity, in place of R we use the notations $\{R_I\}(I = 1, 2, \ldots, 3N)$ or also $\mathbf{R}_I(I = 1, 2, \ldots, N)$. The nuclear kinetic energy is indicated in shorthand notation as $-(\hbar^2/2M)(\partial^2 \ldots / \partial R^2)$. Similarly r is a shorthand notation for all space and spin electron coordinates of the system. In the following, whenever necessary, we replace shorthand notations with the extended notations specific for the system under investigation. [In actual calculations, it is often convenient to break the crystal into a collection of ions (nuclei plus core electrons) and valence electrons, with appropriate pseudopotentials taking care of the disregarded electronic core wavefunctions; the present discussion also applies to these descriptions.]

The Schrödinger equation to be solved for the system of electrons and nuclei in interaction is

$$[T_N(R) + T_e(r) + V(r, R)]\Psi(r, R) = W\Psi(r, R)] . \tag{8.3}$$

The eigenvalues W and the eigenfunctions $\Psi(r, R)$ of the combined electronic-nuclear system are called "vibronic energies" and "vibronic wavefunctions", respectively.

In order to attack the eigenvalue equation (8.3), we begin to observe what happens of the total Hamiltonian H_{tot} if the nuclear masses M_I, much larger than the electron mass m, would actually be treated as infinite. *In this case the nuclear kinetic operator T_N could be dropped and the nuclei could be thought as "fixed" in some assigned configuration R.* In the realistic situation of large (and yet finite) nuclear masses M_I, we can estimate qualitatively the order of magnitude of the error in the kinetic energy, introduced neglecting T_N in Eq. (8.3), with the following argument. Consider the case of a vibronic wavefunction $\Psi(r, R)$ made up of functions of the type $\phi(\mathbf{r} - \mathbf{R})$, representing an electron $\mathbf{r}$ bound to a nucleus localized at $\mathbf{R}$. The expectation values of the operators $\nabla_{\mathbf{R}}^2$ and $\nabla_{\mathbf{r}}^2$ on the wavefunctions of the type $\phi(\mathbf{r} - \mathbf{R})$ are obviously equal; thus the mean kinetic energy of the nuclei is of the order of $m/M \approx 10^{-3}$ smaller than the mean kinetic energy of the electrons.

In the "fixed" lattice approximation, obtained by ignoring altogether the nuclear kinetic operator in the total Hamiltonian (8.2), we are left with the so-called electronic adiabatic Hamiltonian $H_e(r; R)$ given by

$$H_e(r; R) = T_e + V(r, R). \tag{8.4}$$

In this Hamiltonian the *variables R appear simply as parameters* (instead of quantum dynamical observables); thus $H_e(r; R)$ belongs to the class of parameter dependent operators; a semicolon is used to denote this parametric dependence. The eigenvalue equation for the electronic Hamiltonian $H_e(r; R)$ is

$$H_e(r; R)\psi_n(r; R) = E_n(R)\psi_n(r; R) . \tag{8.5}$$

The electronic wavefunctions $\psi_n(r; R)$, as well as the eigenvalues $E_n(R)$, depend on the parameters R; the suffix n summarizes the electronic quantum numbers.

In Chapter 4 we have seen (at least in principle) how to determine the eigenvalues and eigenfunctions of a many-electron system with the nuclei fixed in a given spatial configuration R (usually, but non-necessarily, the equilibrium configuration R_0). As we vary the parameters R, the eigenvalues $E_n(R)$ define the adiabatic potential-energy surfaces, and $\psi_n(r; R)$ define the set of adiabatic electronic wavefunctions. With the help of the adiabatic surfaces and adiabatic wavefunctions, we can now proceed to the study of the *nuclear dynamics (or lattice dynamics)* of the atoms or ions composing the crystal; for this purpose, we consider first the case that the nuclear dynamics is controlled by a single non-degenerate adiabatic surface (Section 8.2) and then the case of multiple adiabatic surfaces (Section 8.3).

8.2   Non-Degenerate Adiabatic Surface and Nuclear Dynamics

8.2.1   Classical Nuclear Dynamics of Born-Oppenheimer Systems

Electronic-nuclear systems, whose ground adiabatic sheet $E_0(R)$ is non-degenerate and well separated in energy from all the other adiabatic surfaces, are referred to as Born-Oppenheimer systems. In these systems, it seems reasonable (at least in a first approach) to ignore all the other adiabatic surfaces.

From a classical point of view, the nuclei moving on the ground non-degenerate adiabatic sheet can be pictured as a set of material points of mass M_I, subjected to forces given by the negative gradient of the potential energy:

$$\boxed{M_I \ddot{R}_I = -\frac{\partial E_0(\{R_J\})}{\partial R_I}}.$$ (8.6)

The determination of the trajectories obeying Eq. (8.6) (with appropriate initial conditions) is the classical subject of *Molecular Dynamics*.

Often one is interested in the behavior of the ground adiabatic surface near its absolute minimum at R_0. In crystals we expect that the nuclear displacements $u_I = R_I - R_{I0}$ from the equilibrium positions $\{R_{I0}\}$ are small with respect to the lattice constant. The potential function $E_0(R)$ can then be expanded in Taylor series around the equilibrium positions

$$E_0(R) = E_0(R_0) + \frac{1}{2} \sum_{IJ} \left(\frac{\partial^2 E_0}{\partial R_I \partial R_J} \right)_0 u_I u_J + \text{higher order terms.}$$ (8.7)

Notice that terms linear in u_I do not appear, because all the derivatives $\partial E_0 / \partial R_I$ vanish at the equilibrium configuration. If the Taylor series is truncated up to second order and inserted into Eq. (8.6), we obtain in the harmonic approximation a very simple picture of the lattice dynamics as an ensemble of point masses interacting with harmonic springs, with force constants given by the second derivatives of the ground state energy with respect to the nuclear positions:

$$M_I \ddot{\mathbf{u}}_I = -\sum_J \left(\frac{\partial^2 E_0}{\partial \mathbf{u}_I \partial \mathbf{u}_J} \right)_0 \mathbf{u}_J.$$ (8.8)

The vibrational modes of frequency ω are obtained assuming a periodic displacement for each nucleus

$$\mathbf{u}_I(t) = \mathbf{A}_I e^{-i\omega t}$$

where $\mathbf{A}_I$ denote the amplitudes of the nuclear displacements. This leads to the following eigenvalue equation

$$-\omega^2 M_I \mathbf{A}_I = -\sum_J \left(\frac{\partial^2 E_0}{\partial \mathbf{u}_I \partial \mathbf{u}_J} \right)_0 \mathbf{A}_J.$$ (8.9)

These equations for the "small oscillations" of the nuclei around their equilibrium positions will be extensively studied in the next chapter, in dealing with lattice vibrations and phonons.

Most significant properties of solids (such as the equilibrium configuration, bulk modulus, force constants) are determined from the knowledge of the ground adiabatic potential-energy surface $E_0(\{R_I\})$. The ground-state total energy of a given electronic-nuclear system can often be obtained with various degrees of sophistication, ranging from semi-empirical to fully ab initio approaches. Among the semi-empirical approaches, the Lennard-Jones potential or the Born effective interionic potential, discussed in Chapter 6, are perhaps the most elementary examples. With the development of the density functional formalism, the determination of the ground-state total energy has been put on a firm theoretical basis. Furthermore, the advent of the Car-Parrinello method has made possible to reach with unprecedented efficiency the equilibrium configurations of polynuclear systems, by simultaneous relaxation of nuclear coordinates and electronic wavefunctions [R. Car and M. Parrinello "Unified approach for molecular dynamics and density functional theory" Phys. Rev. Lett. *55*, 2471 (1985)]. The Car-Parrinello method has opened new perspectives particularly in the physics of complicated systems; major applications concern ab initio studies of liquids, clusters, and solids, also in the presence of impurities and complex surface reconstructions.

8.2.2   Quantum Nuclear Dynamics of Born-Oppenheimer Systems

We describe now some relevant aspects of the quantum theory of the lattice dynamics. At first sight one could expect that the quantum treatment eventually ends up with the quantum transcript of the classical equations of motion of Molecular Dynamics; however, this is not the whole story, and a detailed analysis is needed to determine the circumstances that make this transcription possible.

Consider ordinary Born-Oppenheimer systems with ground potential energy $E_0(R)$ well separated in energy from all the other adiabatic surfaces, and non-degenerate adiabatic electronic wavefunctions $\psi_0(r; R)$. In these systems, it is reasonable to approximate the vibronic wavefunctions of the electronic-nuclear system in the Born-Oppenheimer product form

$$\Psi_{\text{trial}}(r, R) \approx \chi(R)\psi_0(r; R), \tag{8.10}$$

where $\chi(R)$ depends only on the nuclear coordinates and has to be determined following standard variational principles. The *product wavefunction* $\chi(R)\psi_0(r; R)$ *assumes that the electronic system is strictly confined in the ground adiabatic surface* and that the effect of all the other adiabatic sheets is negligible.

Before using the adiabatic wavefunctions $\psi_0(r; R)$, or any other adiabatic wavefunction $\psi_m(r; R)$, as basis functions for vibronic states, it is understood that $\psi_0(r; R)$ (thought as function of R) is *continuous and single-valued in the parameter space R* (as it is routinely required for the dependence on the electronic space coordinates r). The requirements of continuity and single-valuedness do not determine univocally the adiabatic electronic wavefunctions $\psi_0(r; R)$; in fact any change of phases of the type

$$\widetilde{\psi}_0(r; R) = e^{i\alpha(R)}\psi_0(r; R); \tag{8.11}$$

where $\alpha(R)$ is any real, continuous, and single-valued function, defines a new set of adiabatic wavefunctions essentially interchangeable with the old set; *this arbitrariness in the choice of phases is called gauge arbitrariness* and any transformation of type (8.11) is called *gauge transformation* (see Section 8.5 for further considerations).

We can now apply the variational principle to obtain the equation satisfied by the optimized choice of $\chi(R)$ (supposed to be normalized). Equivalently, in a shorthand way, we may substitute Eq. (8.10) into the Schödinger equation (8.3), multiply at the left by $\psi_0^*(r; R)$ and integrate with respect to r; this produces the eigenvalue equation

$$\left[-\frac{\hbar^2}{2M} \frac{\partial^2}{\partial R^2} + E_0(R) \right] \chi(R) + \Lambda_{00}(R)\chi(R) = W\chi(R), \tag{8.12}$$

where the operator

$$\Lambda_{00}(R) = -\frac{\hbar^2}{2M} \langle \psi_0(r; R) | \frac{\partial^2 \psi_0(r; R)}{\partial R^2} \rangle - \frac{\hbar^2}{M} \langle \psi_0(r; R) | \frac{\partial \psi_0(r; R)}{\partial R} \rangle \cdot \frac{\partial}{\partial R} \tag{8.13}$$

results from the application of T_N to the electronic wavefunction $\psi_0(r; R)$. The Schrödinger equation (8.12) shows that the effective total potential that determines the nuclear dynamics is $E_0(R) + \Lambda_{00}(R)$. Thus *besides the easily predictable "adiabatic potential"* $E_0(R)$ we have the *additional operator* $\Lambda_{00}(R)$ (sometimes called "non-adiabatic operator"); we notice that the former is a gauge independent potential, while the latter is a gauge dependent operator.

The presence of the operator $\Lambda_{00}(R)$ should not be perceived as a complication. Rather such an operator is essential to guarantee that the vibronic eigenfunctions $\chi(R)\psi_0(r; R)$ and the vibronic eigenvalues W are not affected by gauge transformations of type (8.11). Within the general framework and concepts of the geometrical Berry phase, the non-trivial role of the $\Lambda_{00}(R)$ operator in specific situations has been clearly established (for an application see for instance Section 8.3.2). In particular the quantity $\langle \psi_0 | \partial \psi_0 / \partial R \rangle$ (times i) can be recognized as the *geometric Berry connection vector*, whose far reaching and surprising properties will be presented in Section 8.5.

We can better understand the structure and physical meaning of the eigenvalue equation (8.12) with the following arguments. We notice that at any particular nuclear configuration it is possible to choose the adiabatic electronic wavefunction $\psi_0(r; R)$ to be *real-valued*. In fact, if $\psi_0(r; R)$ is an eigenfunction of the Schrödinger equation (8.5) with energy $E_0(R)$ also its complex conjugate wavefunction is eigenfunction with the same eigenvalue; since $E_0(R)$ is assumed to be non-degenerate $\psi_0(r; R)$ and $\psi_0^*(r; R)$ must be linearly dependent and can be put in real form. [More generally we notice that the electronic Hamiltonian $H_e(r; R)$ is invariant under time-reversal symmetry. When the electron spin can be disregarded, the time-reversal operator essentially becomes the complex conjugation operation, and the eigenfunctions (whether degenerate or not) can be taken as real. When the electron spin cannot be disregarded, we refer for properties of the time-reversal operator to textbooks on group theory, for instance, F. Bassani and G. Pastori Parravicini "Electronic States and Optical Transitions in Solids" (Pergamon Press, Oxford 1975).]

In general, the choice of reality of wavefunctions at all nuclear configurations is of no particular value, unless the choice still preserves continuity with respect to R. In the particular case of a discrete non-degenerate adiabatic surface, in the presence of time-reversal symmetry in spinless systems, the constraints of reality, continuity and single-valuedness can all be satisfied simultaneously (the proof of this key issue is postponed to Section 8.5, where discussing the Berry phase concept). We thus examine Eqs. (8.12) and (8.13) in the *"preferential gauge"*, in which the *adiabatic wavefunctions* $\psi_0(r; R)$ *are real-valued, besides being continuous and single valued.*

The first term in the right-hand side of Eq. (8.13) is expected to be a small correction (of order m/M) of the electronic kinetic energy [the estimate holds under assumption that the wavefunction $\psi_0(r; R)$ is made up of functions of the type $\phi(r - R)$, which depend on the relative coordinates of electronic and nuclear positions]. The electronic kinetic energy enters automatically into the adiabatic potential $E_0(R)$ and it is thus often not essential to include the above-mentioned term. The second term appearing in the right-hand side of Eq. (8.13) identically vanishes for real and normalized wavefunctions; in fact we have

$$
\langle \psi_0(r; R)| \frac{\partial}{\partial R} \psi_0(r; R) \rangle = \int \psi_0^*(r; R) \frac{\partial}{\partial R} \psi_0(r; R) \, dr
$$

$$
= \int \psi_0(r; R) \frac{\partial}{\partial R} \psi_0(r; R) \, dr = \frac{1}{2} \frac{\partial}{\partial R} \int \psi_0(r; R) \psi_0(r; R) \, dr \equiv 0
$$

(the derivation of the normalization constant is trivially zero).

When the preferential gauge of real, continuous, and single-valued adiabatic wavefunctions, exists, and is adopted, Eq. (8.12) becomes

$$
\boxed{\left[-\frac{\hbar^2}{2M} \frac{\partial^2}{\partial R^2} + E_0(R) \right] \chi(R) = W \chi(R)} . \tag{8.14}
$$

According to the Born-Oppenheimer adiabatic approximation, summarized by Eq. (8.14), the nuclear dynamics is described by a Schrödinger equation with an effective potential $E_0(R)$ given by the adiabatic potential-energy surface. Equation (8.14) has the great merit to accomplish a full decoupling between nuclear and electronic degrees of freedom. The electronic coordinates and momenta do not enter directly into Eq. (8.14), but only indirectly via the electronic energy $E_0(R)$, thought of as a function of the nuclear coordinates.

Before concluding, we notice that the Born-Oppenheimer "single-product vibronic wavefunctions" of the form $\chi(R)\psi_0(r; R)$, with $\chi(R)$ satisfying Eq. (8.12) or Eq. (8.14), are the "best wavefunctions" of H_{tot} in a variational sense, but are not genuine wavefunctions of the combined electronic-nuclear system. The exact vibronic wavefunctions of the total Hamiltonian H_{tot} are in fact appropriate linear combinations of Born-Oppenheimer products of type (8.10), associated to different adiabatic surfaces; in other words, *when the nuclei are allowed to move, the electronic-nuclear system cannot be strictly confined to a given adiabatic surface.* As long as the mixing of different adiabatic surfaces can be considered as a small perturbation, the separation of nuclear, and electronic wavefunctions is justified; otherwise the mixing of different adiabatic surfaces must be appropriately taken into account. In general, first principle account of the part of the electronic-nuclear interaction that causes the mixing (*electron-phonon interaction*) is rather demanding, and is often estimated on the basis of simplified semi-empirical models.

Case of a Single One-Dimensional Vibrational Mode

As an illustration of our discussion, we consider briefly the particular case where the multi-dimensional variables $\{R_I\}$ reduce to a single one-dimensional variable. This happens for instance for a diatomic molecule, where the internuclear distance is the parameter of interest; also in the study of more complicated systems, such as localized impurities in solids, one may have to handle one or a small number of collective variables.

We consider a dimer, with the two nuclei at the distance R (in a fixed direction), and we indicate with $\psi_g(r; R)$ and $E_g(R)$ the electronic wavefunction and the energy of the ground state (assumed non-degenerate). Near the (absolute) minimum R_0 we expand $E_g(R)$ up to second order in the displacements $R - R_0$ in the form

$$E_g(R) = E_g(R_0) + \frac{1}{2}C(R - R_0)^2. \tag{8.15a}$$

Within the adiabatic approximation for the ground adiabatic surface, the vibronic wavefunctions have the "product form"

$$\Psi_{gm}(r, R) = \chi_m(R)\psi_g(r; R); \tag{8.15b}$$

the functions $\chi_m(R)$ are the Hermite solutions of the harmonic oscillator equation

$$\left[-\frac{\hbar^2}{2M^*} \frac{\partial^2}{\partial R^2} + E_g(R_0) + \frac{1}{2}C(R - R_0)^2 \right] \chi_m(R) = W_m \chi_m(R), \tag{8.15c}$$

where M^* is the reduced mass of the two nuclei. The vibronic energies are given by

$$W_{gm} = E_g(R_0) + \left(m + \frac{1}{2} \right) \hbar\omega \quad \text{with} \quad \omega = \sqrt{C/M^*} \quad (m = 0, 1, 2, \dots). \tag{8.15d}$$

The model is schematically illustrated in Figure 8.2.

We consider now the lowest excited energy adiabatic sheet $E_e(R)$ (assumed non-degenerate) and the corresponding electronic eigenfunctions $\psi_e(r; R)$. We suppose that $E_e(R)$ has the minimum at the same value R_0 and with the same curvature C as the ground-energy sheet $E_g(R)$. Within the adiabatic approximation for the excited adiabatic sheet, the vibronic wavefunctions have the product form

$$\Psi_{en}(r, R) = \chi_n(R)\psi_e(r; R), \tag{8.16a}$$

where $\chi_n(R)$ are the solutions of the harmonic oscillator of frequency ω. The quantized energies of the nuclear motion associated with the excited adiabatic sheet are given by

$$W_{en} = E_e(R_0) + \left(n + \frac{1}{2} \right) \hbar\omega \quad (n = 0, 1, 2, \dots) \tag{8.16b}$$

and are schematically shown in Figure 8.2a. The separate quantization of the different adiabatic sheets is only a (useful) starting approximation; in fact, it is relaxed by a proper account of the electron-phonon operator.

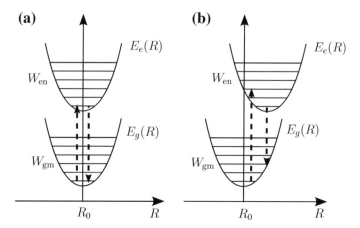

Figure 8.2 Schematic behavior of the adiabatic electronic energies $E_g(R)$ and $E_e(R)$ of the ground-state and lowest excited state of a model dimer system (R is assumed to be a one-dimensional variable). In case (a) the minima of the adiabatic potential sheets occur at the same configuration. In case (b) the minima are in different configurations (Franck-Condon model). The quantized vibronic energies are also indicated.

In Figure 8.2a, the ground adiabatic sheet and the higher one are non-degenerate with minima at the same configuration R_0. The structure of Figure 8.2a can be considered as indicative of the electronic-nuclear systems, whose electron wavefunctions are reasonably delocalized (as it happens for valence and conduction states in ordinary crystals). In this case, the electronic charge density of the system is much the same in the ground-state or in the (lowest lying) excited states, and so are the forces $-\partial E_m(R)/\partial R$ [in fact the Hellmann-Feynman theorem of Section 8.4 shows that the forces equal the negative gradients of the classical electrostatic potential due to the nuclei and electronic charge density].

In Figure 8.2b the ground adiabatic sheet and the higher one are non-degenerate, and their minima occur at different configurations. Figure 8.2b is known as *Seitz model*, or *Franck-Condon model*; it is indicative for the adiabatic structure of impurities in crystals with well-localized electronic wavefunctions; in fact excitations from localized impurity orbitals may be accompanied by significant local change of electron density and hence of local forces. These situations are of particular interest in optical transitions, and may lead to substantial energy shifts in absorption and luminescence processes, as we shall see in detail in the study of the optical properties of impurities (see Section 12.7).

8.3 Degenerate Adiabatic Surfaces and Jahn-Teller Systems

8.3.1 *Degenerate Adiabatic Surfaces and Nuclear Dynamics*

In the previous section, we have assumed that the adiabatic potential-energy surface under attention is non-degenerate. We consider now electronic-nuclear systems in

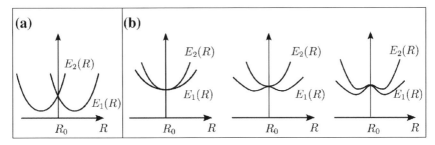

Figure 8.3 Schematic representation of possible topologies of two adiabatic potential-energy surfaces near a degeneracy point R_0 in nuclear coordinate space. In (a) we report the typical Jahn-Teller "conical intersection", with the presence of linear terms in $R - R_0$. In (b) we report Renner-Teller "glancing intersections", with the absence of linear terms in $R - R_0$, and positive or negative second derivatives. In general, terms linear in certain symmetry-breaking distortions are always possible for any type of degeneracy (except for Kramers degeneracy and some levels in linear molecules).

which two or more electronic wavefunctions $\{\psi_1(r; R_0), \psi_2(r; R_0), \ldots, \psi_\nu(r; R_0)\}$ are degenerate (for symmetry reasons) at the high symmetry nuclear configuration R_0, and are thus quasi-degenerate for values of R near to R_0. The behavior (or the so-called *topology*) of the potential-energy surfaces $E_n(R)(n = 1, 2, \ldots, \nu)$ near the intersection point R_0 can be distinguished essentially into *two classes, depending whether or not R_0 is a stationary point for all the potential-energy surfaces $E_n(R)(n = 1, 2, \ldots, \nu)$*; in other words, the distinction concerns whether or not the gradients $\nabla_R E_n(R) \equiv 0$ for $R \to R_0$. By far the most common situation is that in which the gradients of the potential-energy surfaces are different from zero (at least in some directions in the nuclear coordinate space) and R_0 is not a stationary point: these systems are called *Jahn-Teller systems*.

To help to visualize different topologies, in Figure 8.3 we indicate schematically possible behavior of two Born-Oppenheimer potential-energy surfaces near the degeneracy point R_0 in nuclear coordinate space. In Figure 8.3a, we report the typical Jahn-Teller intersection with the presence of linear terms in $R - R_0$; the local topology of the potential-energy surfaces is that of a "double cone" and the name "conical intersection" has been coined for it. In Figure 8.3b we show possible kinds of stable or unstable "glancing intersections" where R_0 is a stationary point (Renner-Teller intersections). The basic difference between Jahn-Teller and Renner-Teller intersections, and the consequent implications, can be clarified at the light of the geometric Berry phase, discussed in Section 8.5, and in the solved problems of Appendix B.

The topology of the adiabatic potential surfaces near a degeneracy point is regulated by the Jahn-Teller theorem [H. A. Jahn and E. Teller, Proc. Roy. Soc. A *161*, 220 (1937)]. The theorem states that *any electronically degenerate system can lower its energy (and is thus intrinsically unstable) under certain asymmetric distortions of the nuclear framework*. The theorem has an obvious corollary: the initially electronically degenerate system undergoes symmetry-breaking distortions and eventually reaches an equilibrium position, in which the symmetry of the nuclear framework is low enough to completely remove the electronic degeneracy. [The only exceptions of the Jahn-Teller theorem concern some levels in linear molecules and the twofold Kramers degeneracy. The former case is of limited interest in solids; the latter case is due to time-reversal symmetry and cannot be lifted reducing the symmetry of the nuclear framework. As a matter of fact, the discovery of exceptions was established preliminarily to the rule; for the historical background, see E. Teller "The Jahn–Teller effect. Its history and applicability" Physica *114* A, 14 (1982).]

In order to clarify the physical arguments underlying the Jahn-Teller theorem, we need to describe the behavior of the adiabatic energy surfaces near the degeneracy point R_0. For this purpose we consider the matrix elements of the electron Hamiltonian $H_e(r; R) = T_e + V(r, R)$ within the degenerate manifold of electronic wavefunctions $\{\psi_n(r; R_0)\}(n = 1, 2, \ldots, \nu)$. For configurations R reasonably near to the high symmetry configuration R_0, the adiabatic energy surfaces $E_n(R)$ and the adiabatic electronic wavefunctions $\{\psi_n(r; R)\}(n = 1, 2, \ldots, \nu)$ are determined from the eigenvalues and eigenvectors of the secular equation

$$\|U_{mn}(R) - E\delta_{mn}\| = 0, \tag{8.17}$$

where

$$U_{mn}(R) = \langle \psi_m(r; R_0)|H_e(r; R)|\psi_n(r; R_0)\rangle \quad (m, n = 1, 2, \ldots, \nu).$$

The above matrix elements can be recast in the form

$$
\begin{aligned}
U_{mn}(R) &= \langle \psi_m(r; R_0)|H_e(r; R_0) + H_e(r; R) - H_e(r; R_0)|\psi_n(r; R_0)\rangle \\
&= E(R_0)\delta_{mn} + \langle \psi_m(r; R_0)|H_e(r; R) - H_e(r; R_0)|\psi_n(r; R_0)\rangle \\
&= E(R_0)\delta_{mn} + \langle \psi_m(r; R_0)|V(r, R) - V(r, R_0)|\psi_n(r; R_0)\rangle, \tag{8.18}
\end{aligned}
$$

where $E(R_0)$ denotes the energy of the adiabatic surfaces at the degeneracy point.

The *topology of the potential sheets* obtained by solving the secular equation (8.17), and specifically the assessment *whether or not R_0 is a stationary point*, can be established by direct inspection of the matrix elements $U_{mn}(R)$ for $R \approx R_0$ to first order in the displacements $R - R_0$. The topology of the potential sheets is dictated by the presence or absence of terms linear in the displacements in Eq. (8.18); thus, we have to examine whether the matrix elements

$$M_{mn}(R_0) = \langle \psi_m(r; R_0)| \left[\frac{\partial V(r, R)}{\partial R} \right]_{R=R_0} |\psi_n(r; R_0)\rangle \tag{8.19}$$

can be different from zero or not.

From group theory considerations it has been shown by Jahn and Teller that in general the matrix elements (8.19) can be different from zero, except the situation of linear molecules, and Kramers degeneracy. The proof is done listing one-by-one every irreducible representation, that classifies the basis functions $\psi_n(r; R_0)(n = 1, 2, \ldots, \nu)$. Then one considers the representation according to which the product functions $\psi_m^*(r; R_0)\psi_n(r; R_0)$ transform. In general, it is realized by inspection that there are (non-total-symmetric) linear combinations of nuclear coordinates transforming as one (or more) of the irreducible representations contained in the representation of the product functions (Jahn-Teller active modes). Thus, in general, no symmetry reason exists why the matrix elements (8.19) should vanish (apart the exceptions mentioned above).

Jahn-Teller systems may present different manifestations depending on the strength of the coupling between electronic and nuclear operators. In Jahn-Teller systems, the adiabatic potential surfaces near the confluence point R_0 vary linearly in certain directions of asymmetric nuclear displacements; a number of equivalent minima are thus expected away from the high symmetry point R_0. For strong coupling, the minima of the adiabatic sheets are distant in R space and a large amount of energy is required for the system to migrate from one minimum to another. In this case we have a "permanent" lowering of the local symmetry and the *Jahn-Teller effect is said to be static*. If the coupling is weak, the system may tunnel from a configuration to the others and is thus delocalized on all the equivalent configurations. In this case, the *Jahn-Teller effect is said to be dynamic*; if one considers a reasonable period of time, there is in average no permanent distortion and no lowering of the symmetry. Of course this (sharp) distinction is only qualitative; indeed it is even possible that the same system shows a static or a dynamic Jahn-Teller effect depending for instance from the characteristic time of the specific experiment (such as spin resonance, nuclear magnetic resonance), from the sample temperature, or other conditions.

After this discussion of the topology of the adiabatic potential surfaces around the degeneracy points, we pass now to the description of the nuclear dynamics in Jahn-Teller systems, or more generally, in any electronic-nuclear system where a number of adiabatic surfaces are degenerate (or nearly degenerate in the so called pseudo-Jahn-Teller systems) in the parameter domain of interest, and thus deserve a treatment on the same footing. The electronic-nuclear wavefunctions can be expressed in the form

$$\Psi(r, R) = \sum_{n=1}^{\nu} \chi_n(R)\psi_n(r; R_0),$$ (8.20)

where $\chi_n(R)$ are suitable expansion coefficients on the subset of the electronic wavefunctions $\psi_n(r; R_0)$ degenerate (or close in energy) at R_0. A vibronic wavefunction of type (8.20) permits the appropriate mixing of the different potential surfaces. To determine the equations obeyed by the vibrational functions $\chi_n(R)$, we replace Eq. (8.20) into Eq. (8.3), multiply on the left by $\psi_m^*(r; R_0)$ and integrate over the electronic coordinates. We obtain for the nuclear motion the following system of coupled differential equations

$$-\frac{\hbar^2}{2M}\frac{\partial^2}{\partial R^2}\chi_m(R) + \sum_{n=1}^{\nu} U_{mn}(R)\chi_n(R) = W\chi_m(R)$$ (8.21)

with

$$U_{mn}(R) = E_m(R_0)\delta_{mn} + \langle \psi_m(r; R_0)|V(r, R) - V(r, R_0)|\psi_n(r; R_0)\rangle.$$

In most practical situations, it is sufficient to expand the above matrix elements to include linear and quadratic terms in $R - R_0$. The non-diagonal elements clearly show that near degeneracy points the nuclear dynamics is determined by all *the adiabatic sheets on the same footing, and not by each of them individually.*

8.3.2 The Jahn-Teller Effect for Doubly Degenerate Electronic States

We now illustrate the key aspects of the Jahn-Teller effect in a few prototype systems, that can be worked out analytically to a large extent. Among possible examples, we select the regular triangular and the regular octahedral molecules; the interest is not confined to molecules but extends, for instance, to the study of certain defects and complexes in crystals in the "quasi-molecular" approximation. Since degeneracy is a consequence of symmetry, a systematic study of the vibronic systems requires the knowledge of group theory; thus most often a newcomer, not yet familiar with group theory, does not approach the fascinating subject of the Jahn-Teller effect and its manifestations. In order to make this field accessible also to readers with little knowledge of the theory of symmetries, we address the subject avoiding the formal apparatus of the abstract group theory, and rather we choose to adopt simple microscopic models and intuitive arguments.

The $E \otimes \varepsilon$ Jahn-Teller System

Probably one of the most studied Jahn-Teller system is the so-called $E \otimes \varepsilon$ system, in which an *electronic doublet* (E) *interacts linearly with a doublet of vibrational degenerate states* (ε) [it is customary to denote the symmetry of the electronic states by upper case Latin letters, and the symmetry of vibrational states by lower case Greek letters]. A possible realization of the $E \otimes \varepsilon$ Jahn-Teller system is given by two degenerate electronic wavefunctions, at the center of a regular triangular molecule, interacting with the doublet of vibrational states of the triangular complex (see Figure 8.4).

The geometry of the regular triangular molecule is indicated in Figure 8.4. The motion of the nuclei is described in terms of *normal coordinates*; these are combination of Cartesian coordinates determined by the symmetry properties of the system under consideration. [For their definition in the case of triangular molecules we refer for instance to the didactic article of A. Nussbaum, Am. J. Phys. *36*, 529 (1968); for

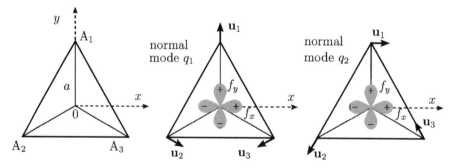

Figure 8.4 Geometry of the regular triangular molecule and nuclear displacements of the degenerate vibrational mode ε. The doublet of electronic states (f_x, f_y) centered at the origin is also indicated.

Table 8.1 Equilibrium positions and normal displacements of the vibrational mode ε in the regular triangular molecule.

Equilibrium position	Normal mode q_1	Normal mode q_2
$\mathbf{d}_1 = a(0, 1)$	$\mathbf{u}_1(q_1) = q_1 \left(0, \dfrac{\sqrt{3}}{3} \right)$	$\mathbf{u}_1(q_2) = q_2 \left(\dfrac{\sqrt{3}}{3}, 0 \right)$
$\mathbf{d}_2 = a \left(\dfrac{-\sqrt{3}}{2}, \dfrac{-1}{2} \right)$	$\mathbf{u}_2(q_1) = q_1 \left(\dfrac{1}{2}, \dfrac{-\sqrt{3}}{6} \right)$	$\mathbf{u}_2(q_2) = q_2 \left(\dfrac{-\sqrt{3}}{6}, \dfrac{-1}{2} \right)$
$\mathbf{d}_3 = a \left(\dfrac{\sqrt{3}}{2}, \dfrac{-1}{2} \right)$	$\mathbf{u}_3(q_1) = q_1 \left(\dfrac{-1}{2}, \dfrac{-\sqrt{3}}{6} \right)$	$\mathbf{u}_3(q_2) = q_2 \left(\dfrac{-\sqrt{3}}{6}, \dfrac{1}{2} \right)$

other molecules we refer to standard books such as J. Hertzberg "Molecular Spectra and Molecular Structure" Vol. 2 (Van Nostrand, New York 1966).] The normal vibrations of the equilateral triangle molecule consist of a totally symmetric mode, which does not alter the symmetry of the structure, and a doublet of degenerate modes (called ε); the nuclear displacements corresponding to the two partner components q_1 and q_2 of the doubly degenerate mode ε are indicated in Figure 8.4. In Table 8.1 the equilibrium positions of the atomic sites and the normal displacements of the vibrational mode ε are reported.

We discuss now the structure of the interaction matrix for the $E \otimes \varepsilon$ vibronic system with the following arguments. Consider two degenerate electronic wavefunctions (f_x, f_y), respectively with p_x-like and p_y-like symmetry, localized at the center O of the molecule; we assume that the environment potential felt by the electrons in O can be considered as the sum of atomic-like spherically symmetric potentials $V_a(\mathbf{r})$ centered in the positions $\mathbf{R}_i = \mathbf{d}_i + \mathbf{u}_i(q_1) + \mathbf{u}_i(q_2)$, $(i = 1, 2, 3)$ of the triangular cage. We consider then the 2×2 matrix of the type

$$U_{\alpha\beta}(q_1, q_2) = \langle f_\alpha | V_a(\mathbf{r} - \mathbf{R}_1) + V_a(\mathbf{r} - \mathbf{R}_2) + V_a(\mathbf{r} - \mathbf{R}_3) | f_\beta \rangle \quad (\alpha, \beta = x, y).$$
$$(8.22)$$

The matrix elements (8.22) can be expressed in terms of independent parameters, following the same techniques introduced by Slater and Koster, and explained for the tight-binding method (see Section 5.2.1). In the adopted intuitive model, the electron-lattice coupling matrix (8.22) is controlled by the symmetry of the triangular cage of atoms $\mathbf{R}_1, \mathbf{R}_2, \mathbf{R}_3$, through their director cosines and their distance from the origin. The elementary elaborations are explicitly reported in Appendix A for several prototype Jahn-Teller systems.

The structure of the linear electron-lattice coupling for a regular triangular molecule, provided in Eq. (A.4), reads

$$U(q_1, q_2) = \gamma \begin{pmatrix} -q_1 & q_2 \\ q_2 & q_1 \end{pmatrix},$$
$$(8.23)$$

where γ is a real parameter that depends on the specific system under attention, while the matrix structure is universal for any $E \otimes \varepsilon$ Jahn-Teller system. Higher order terms (when necessary) are also reported in Eq. (A.4). From Eq. (8.23) it is seen that the mode q_1 splits the degeneracy of the electronic states (f_x, f_y) without mixing them, while the mode q_2 also acts to mix them.

The matrix (8.23) of the $E \otimes \varepsilon$ Jahn-Teller system belongs to the class of low rank matrices of particular significance often encountered in physics. In this class we can mention the 2×2 Pauli matrices at the basis of the quantum description of the electron spin, the rank two matrices at the basis of the symmetry between protons and neutrons, the 4×4 Dirac matrices at the basis of the quantum equation compatible with the special theory of relativity. Similarly, the 2×2 parametric matrix (8.23) emerges as a most significant paradigmatic model of electron-phonon coupling. Last, but not least, the matrix (8.23) is isomorphic (within a similarity transformation) to the matrix Hamiltonian describing the neutrino particles, as well as the electronic low energy excitations in the honeycomb topology of graphene.

We consider now the full Hamiltonian of the coupled vibronic system $E \otimes \varepsilon$. The Hamiltonian of the uncoupled system must be modified introducing explicitly the interaction matrix between the doublet E of electronic states and the doublet ε of the normal modes. The coupled differential equations (8.21) describe the nuclear dynamics controlled by degenerate adiabatic surfaces; for the $E \otimes \varepsilon$ vibronic system, the corresponding Hamiltonian, expressed in matrix form, reads

$$H = -\frac{\hbar^2}{2M} \frac{\partial^2}{\partial q_1^2} - \frac{\hbar^2}{2M} \frac{\partial^2}{\partial q_2^2} + \gamma \begin{pmatrix} -q_1 & q_2 \\ q_2 & q_1 \end{pmatrix} + \frac{1}{2} C(q_1^2 + q_2^2) \qquad (8.24)$$

(when a scalar operator or a function is added to a matrix, it is implicitly understood that the scalar is multiplied by the identity matrix before the algebraic summation is performed). The terms in Eq. (8.24) that survive when $\gamma \equiv 0$ represent the kinetic energy and the potential energy of two harmonic oscillators with coordinates q_1 and q_2. The matrix in Eq. (8.24) represents the linear coupling between the pair of degenerate vibrational modes (q_1, q_2) and the pair of degenerate electronic states (f_x, f_y). Equation (8.24) expresses the famous vibronic Hamiltonian of the $E \otimes \varepsilon$ system, which constitutes the simplest example of (non-trivial) Jahn-Teller system.

Adiabatic Surfaces of the $E \otimes \varepsilon$ Vibronic System

The solution of the $E \otimes \varepsilon$ vibronic system, described by the Hamiltonian (8.24), has engaged several authors and different mathematical techniques, including the analytic theory of continued fractions (the continued fraction solution is presented in Section 12.7, while studying the manifestations of the Jahn-Teller effect on optical spectra). Here we wish to discuss some aspects of the adiabatic surfaces of this simple and yet surprisingly rich model.

If the nuclei are regarded as fixed and their kinetic energy operator is ignored in Eq. (8.24), the potential-energy matrix becomes

$$V(q_1, q_2) = \gamma \begin{pmatrix} -q_1 & q_2 \\ q_2 & q_1 \end{pmatrix} + \frac{1}{2} C(q_1^2 + q_2^2). \qquad (8.25a)$$

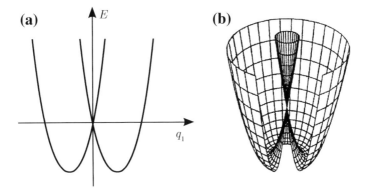

Figure 8.5 Adiabatic potential-energy surfaces of the $E \otimes \varepsilon$ Jahn-Teller system. (a) The section of the two branches with the $q_2 = 0$ plane. (b) When rotation by 2π through the energy axis is performed, the lower branch takes the shape of a "Mexican Hat" and the upper branch takes the shape of a conical "Wizard Hat" [from Lai-Shengh Wang, B. Niu, Y. T. Lee, D. A. Shirley, E. Ghelichkhani and E. R. Grant, J. Chem. Phys. *93*, 6318 (1990); copyright 1990 by the American Physical Society].

We can diagonalize the above 2×2 matrix and introduce the polar coordinates (q, θ) in the $\mathbf{q} \equiv (q_1, q_2)$ plane. The two adiabatic potential surfaces, or branches, have the expressions

$$E_1(q) = -\gamma q + \frac{1}{2}Cq^2, \qquad E_2(q) = \gamma q + \frac{1}{2}Cq^2. \tag{8.25b}$$

(from now on we assume $\gamma > 0$; a change of sign of γ simply would interchange the upper and lower branches). A section of the two branches along the q_1 axis is plotted in Figure 8.5, together with the rotation figure generated by it and known as the "Mexican Hat". The minimum of $E_1(q)$ occurs for $q_0 = \gamma/C$ and its value is $E_0 = -\gamma^2/2C$; the depth $|E_0|$ of the minimum is called *Jahn-Teller energy*, E_{JT}, of the system.

The adiabatic electronic eigenfunctions have been studied in Problem 1 of the Appendix B. Exploiting Eq. (B.7), we can express the eigenfunctions and eigenvalues of the potential-energy matrix (8.25a) in the form

$$\begin{cases} |\psi_1\rangle = e^{-i\theta/2} \cos(\theta/2) |f_x\rangle - e^{-i\theta/2} \sin(\theta/2) |f_y\rangle & E_1(q) = -\gamma q + \frac{1}{2}Cq^2 \\ |\psi_2\rangle = e^{+i\theta/2} \sin(\theta/2) |f_x\rangle + e^{+i\theta/2} \cos(\theta/2) |f_y\rangle & E_2(q) = +\gamma q + \frac{1}{2}Cq^2 \end{cases}.$$

$$\tag{8.26}$$

Notice that a *"preferential gauge" in which the adiabatic wavefunctions are real and, at the same time, continuous, and single-valued does not exist for the $E \otimes \varepsilon$ vibronic system.* [A preferential gauge does not exist even for the pseudo-Jahn-Teller $E \otimes \varepsilon$ system, where no degeneracy of the adiabatic surfaces occur, as thoroughly studied in Problems 1 and 2].

The dynamic problem described by the Hamiltonian (8.24) can be dealt exactly from a mathematical point of view for any value of the coupling constant γ (see Section 12.7).

Here we focus specifically on large values of γ, and discuss the $E \otimes \varepsilon$ model with an instructive "ad hoc" procedure, which holds in the strong-coupling limit. In this case the circular trough of the "Mexican Hat" is deep (Figure 8.5) and the nuclear motion is essentially confined there. For low-lying vibronic levels, far from the conical intersection, the nuclei never move on the upper potential-energy surface; thus the vibronic wavefunction can be approximated with the Born-Oppenheimer product of the type

$$\Psi_{\text{trial}} = \chi(q, \theta)\psi_1(r; q, \theta),$$

where both ψ_1 and χ are *single-valued functions* of the vibrational coordinates (q, θ).

The best vibrational functions $\chi(q, \theta)$ satisfy the general differential equation (8.12), which in the present context takes the form

$$-\frac{\hbar^2}{2M}\left[\frac{\partial^2}{\partial q^2} + \frac{1}{q}\frac{\partial}{\partial q} + \frac{1}{q^2}\frac{\partial^2}{\partial\theta^2}\right]\chi(q, \theta) + E_1(q)\chi(q, \theta) + \Lambda_{11}\chi(q, \theta) = W\chi(q, \theta).$$

$$(8.27)$$

The first operator in Eq. (8.27) is the nuclear kinetic energy in polar coordinates; $E_1(q)$ is the adiabatic potential-energy surface of the lowest branch (the Mexican Hat); the remaining operator is the non-adiabatic operator defined by Eq. (8.13), which in the present context reads

$$\Lambda_{11}(q, \theta) = -\frac{\hbar^2}{2M}\langle\psi_1|\nabla_{\mathbf{q}}^2|\psi_1\rangle - \frac{\hbar^2}{M}\langle\psi_1|\nabla_{\mathbf{q}}|\psi_1\rangle \cdot \nabla_{\mathbf{q}}$$

The gradient operator and the kinetic energy operator in polar coordinates are given by

$$\nabla_{\mathbf{q}} = \mathbf{e}_q\frac{\partial}{\partial q} + \mathbf{e}_\theta\frac{1}{q}\frac{\partial}{\partial\theta} \quad \text{and} \quad \nabla_{\mathbf{q}}^2 = \frac{\partial^2}{\partial q^2} + \frac{1}{q}\frac{\partial}{\partial q} + \frac{1}{q^2}\frac{\partial^2}{\partial\theta^2}.$$

Their matrix elements on the wavefunction ψ_1 become

$$\langle\psi_1|\nabla_{\mathbf{q}}^2|\psi_1\rangle = -\frac{1}{2q^2} \quad \text{and} \quad \langle\psi_1|\frac{1}{q}\frac{\partial}{\partial\theta}|\psi_1\rangle\frac{1}{q}\frac{\partial}{\partial\theta} = -\frac{i}{2q^2}\frac{\partial}{\partial\theta};$$

then

$$\Lambda_{11}(q, \theta) = \frac{\hbar^2}{4Mq^2} + \frac{\hbar^2}{2Mq^2}i\frac{\partial}{\partial\theta}$$

is the explicit expression of the non-adiabatic operator to be inserted in Eq. (8.27).

The solutions of Eq. (8.27) can be factorized into the product of a θ-dependent function and a q-dependent function in the form

$$\chi(q, \theta) = e^{im\theta}\frac{g(q)}{\sqrt{q}} \quad (m = 0, \pm1, \pm2, \ldots), \quad (8.28)$$

where the function $1/\sqrt{q}$ has been introduced in view of further elaborations. Inserting Eq. (8.28) into Eq. (8.27), after straightforward elaboration, we obtain the one-dimensional Schrödinger equation

$$\left[-\frac{\hbar^2}{2M}\frac{\partial^2}{\partial q^2} + E_1(q)\right] g(q) + \frac{\hbar^2}{2Mq^2}\left(m - \frac{1}{2}\right)^2 g(q) = W g(q). \tag{8.29}$$

We notice that

$$E_1(q) = -\gamma q + \frac{1}{2}Cq^2 \equiv -\frac{\gamma^2}{2C} + \frac{1}{2}C\left(q - \frac{\gamma}{C}\right)^2.$$

In the strong-coupling limit the trough minimum $E_0 = -\gamma^2/2C$ is very deep; the wavefunctions satisfying Eq. (8.29) for low-lying energy levels are concentrated at $q \approx q_0 = \gamma/C$ and the variable q in the expression $\hbar^2/2Mq^2$ can be replaced by q_0. The eigenvalues of Eq. (8.29) then become

$$W_{nj} = E_0 + \left(n + \frac{1}{2}\right)\hbar\omega + \frac{\hbar^2}{2Mq_0^2} j^2 \quad \left(n = 0, 1, 2, \ldots; j = \pm\frac{1}{2}, \pm\frac{3}{2}, \ldots\right) \tag{8.30}$$

where $\omega = \sqrt{C/M}$, and j stands for the half odd integer $m - 1/2$. The above eigenvalues and the corresponding wavefunctions intuitively describe the combination of a radial harmonic oscillator centered around q_0 and a two-dimensional rigid rotor, moving in the lowest adiabatic sheet.

It is interesting to notice that the ground-state of the $E \otimes \varepsilon$ vibronic system is characterized by the quantum numbers $n = 0$ and $j = \pm 1/2$, and is thus doubly degenerate. The responsibility of this twofold degeneracy can be traced back to the conical intersection of the upper and lower adiabatic surfaces at the origin of the (q_1, q_2) plane. It is the conical intersection that prevents the existence of a "preferential gauge" of real-valued, continuous, and single-valued adiabatic functions; this is so regardless of the fact that the nuclei never move on the upper potential-energy surface. Notice that in the hypothetical case of existence of the "preferential gauge", the quantum number j would take the integer values $0, \pm 1, \pm 2, \ldots$ and the ground state would be non-degenerate. More detailed theoretical solutions of the dynamic vibronic problem, as well as experimental evidence, confirm that the ground state of the $E \otimes \varepsilon$ model with linear coupling is a doublet [for the more general case of the presence of quadratic coupling see I. B. Bersuker (2006) and references therein].

Before concluding this section, we notice that the Hamiltonian (8.24), obtained with reference to the triangular molecule geometry of Figure 8.4, also describes the interaction of a doublet of electronic states with a doublet of vibrational modes in other high symmetry environments. Here we wish to mention the classical case of Cu^{2+} ions in various materials. The Cu^{2+} ion has the electronic configuration d^9, and can be pictured as a single hole in a closed d shell. When the ion is embedded in an octahedral cage of negative charges, the d-states split into a triplet (at lower energy) and a doublet (at higher energy); the hole occupies the doublet and suffers a Jahn-Teller interaction with the doublet of vibrational modes of the surrounding octahedron (see Figure 8.6). It is believed that the Jahn-Teller Cu^{2+} ion plays an important role in the physical and structural properties in the high-temperature copper oxide superconductors, which typically

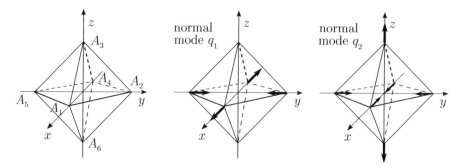

Figure 8.6 Geometry of the octahedral complex and degenerate vibrational modes q_1 and q_2, partner of the ε representation.

contain CuO_6 octahedra or CuO_5 pyramids; a wide debate in the literature concerns a possible contribution of the Jahn-Teller effect in the pairing mechanism of carriers in high-temperature superconductivity [see for instance the review by M. C. M. O'Brien et al. (1993) and the textbook of I. B. Bersuker (2006), and references quoted therein].

8.3.3   The Jahn-Teller Effect for Triply Degenerate Electronic States

The $T \otimes \varepsilon$ Jahn-Teller System

The $T \otimes \varepsilon$ Jahn-Teller system concerns an *electronic triplet interacting with a doublet of vibrational modes* in full cubic symmetry. The geometry of an octahedral complex (the molecule of SF_6 for instance) is indicated in Figure 8.6. In the same figure we also indicate the normal mode displacements q_1 and q_2 of the nuclei, corresponding to the double degenerate ε mode. In Table 8.2 the equilibrium positions and normal displacements of the vibrational mode ε in the cubic symmetry are reported.

Table 8.2 Equilibrium positions and normal displacements of the ε vibrational modes q_1 and q_2 in the cubic symmetry environment.

$\mathbf{d}_1 = a(+1, 0, 0)$	$\mathbf{u}_1(q_1) = q_1 \left(\dfrac{1}{2}, 0, 0 \right)$	$\mathbf{u}_1(q_2) = q_2 \left(\dfrac{-1}{2\sqrt{3}}, 0, 0 \right)$
$\mathbf{d}_2 = a(0, +1, 0)$	$\mathbf{u}_2(q_1) = q_1 \left(0, -\dfrac{1}{2}, 0 \right)$	$\mathbf{u}_2(q_2) = q_2 \left(0, \dfrac{-1}{2\sqrt{3}}, 0 \right)$
$\mathbf{d}_3 = a(0, 0, +1)$	$\mathbf{u}_3(q_1) = 0$	$\mathbf{u}_3(q_2) = q_2 \left(0, 0, \dfrac{1}{\sqrt{3}} \right)$
$\mathbf{d}_4 = a(-1, 0, 0)$	$\mathbf{u}_4(q_1) = q_1 \left(-\dfrac{1}{2}, 0, 0 \right)$	$\mathbf{u}_4(q_2) = q_2 \left(\dfrac{1}{2\sqrt{3}}, 0, 0 \right)$
$\mathbf{d}_5 = a(0, -1, 0)$	$\mathbf{u}_5(q_1) = q_1 \left(0, \dfrac{1}{2}, 0 \right)$	$\mathbf{u}_5(q_2) = q_2 \left(0, \dfrac{1}{2\sqrt{3}}, 0 \right)$
$\mathbf{d}_6 = a(0, 0, -1)$	$\mathbf{u}_6(q_1) = 0$	$\mathbf{u}_6(q_2) = q_2 \left(0, 0, \dfrac{-1}{\sqrt{3}} \right)$

We can easily establish the matrix for the linear interaction of the triplet electronic states with the doublet of normal modes with the simplified approach of Appendix A. The matrix potential for the $T \otimes \varepsilon$ model becomes

$$V(q_1, q_2) = \gamma \begin{pmatrix} q_1 - q_2/\sqrt{3} & 0 & 0 \\ 0 & -q_1 - q_2/\sqrt{3} & 0 \\ 0 & 0 & +2q_2/\sqrt{3} \end{pmatrix} + \frac{1}{2}C(q_1^2 + q_2^2),$$

(8.31)

where C denotes the spring constant of the oscillator modes, and the interaction matrix is provided by Eq. (A.8).

The potential matrix (8.31) is diagonal and the ε distortion fails to mix electronic states. The energies of the adiabatic sheets and the coordinates of the minima are

$$E_1(q_1, q_2) = \gamma \left(q_1 - \frac{1}{\sqrt{3}} q_2 \right) + \frac{1}{2}C(q_1^2 + q_2^2) \qquad q_1^0 = -\frac{\gamma}{C}; \quad q_2^0 = \frac{\gamma}{\sqrt{3}C}$$

$$E_2(q_1, q_2) = -\gamma \left(q_1 + \frac{1}{\sqrt{3}} q_2 \right) + \frac{1}{2}C(q_1^2 + q_2^2) \qquad q_1^0 = \frac{\gamma}{C}; \quad q_2^0 = \frac{\gamma}{\sqrt{3}C}$$

$$E_3(q_1, q_2) = \gamma \frac{2}{\sqrt{3}} q_2 + \frac{1}{2}C(q_1^2 + q_2^2) \qquad q_1^0 = 0; \quad q_2^0 = -\frac{2\gamma}{\sqrt{3}C}.$$

From the values q_1^0 and q_2^0 and Table 8.2, the Cartesian components of the displacements can be obtained; it is seen by inspection that the three minima describe a distorted octahedron, extended along one axis, and compressed along the other two.

The minimum energy is $E_{min} = -2\gamma^2/3C$ for any of the three valleys. The adiabatic surfaces in the (q_1, q_2) space are the three "disjoint" paraboloids indicated in Figure 8.7. Thus a triplet electronic state should always suffer a static Jahn-Teller distortion (but inclusion of spin-orbit coupling may act to mix the different valleys).

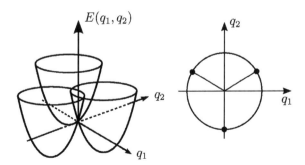

Figure 8.7 Schematic representation of the adiabatic-potential energy surfaces in the case of a triply degenerate state T interacting with the ε modes (q_1, q_2) in the cubic symmetry environment. The coordinates of the minima of the potential are also shown in a top view.

Table 8.3 Normal displacements of the τ vibrational modes q_1, q_2 and q_3 in the cubic symmetry environment.

Normal mode q_1	Normal mode q_2	Normal mode q_3
$\mathbf{u}_1(q_1) = 0$	$\mathbf{u}_1(q_2) = q_2\left(0,0,\frac{1}{2}\right)$	$\mathbf{u}_1(q_3) = q_3\left(0,\frac{1}{2},0\right)$
$\mathbf{u}_2(q_1) = q_1\left(0,0,\frac{1}{2}\right)$	$\mathbf{u}_2(q_2) = 0$	$\mathbf{u}_2(q_3) = q_3\left(\frac{1}{2},0,0\right)$
$\mathbf{u}_3(q_1) = q_1\left(0,\frac{1}{2},0\right)$	$\mathbf{u}_3(q_2) = q_2\left(\frac{1}{2},0,0\right)$	$\mathbf{u}_3(q_3) = 0$
$\mathbf{u}_4(q_1) = 0$	$\mathbf{u}_4(q_2) = q_2\left(0,0,-\frac{1}{2}\right)$	$\mathbf{u}_4(q_3) = q_3\left(0,-\frac{1}{2},0\right)$
$\mathbf{u}_5(q_1) = q_1\left(0,0,-\frac{1}{2}\right)$	$\mathbf{u}_5(q_2) = 0$	$\mathbf{u}_5(q_3) = q_3\left(-\frac{1}{2},0,0\right)$
$\mathbf{u}_6(q_1) = q_1\left(0,-\frac{1}{2},0\right)$	$\mathbf{u}_6(q_2) = q_2\left(-\frac{1}{2},0,0\right)$	$\mathbf{u}_6(q_3) = 0$

The $T \otimes \tau$ Jahn-Teller System

In the $T \otimes \tau$ Jahn-Teller system, an *electronic triplet T interacts linearly with a triplet of vibrational modes* τ in a full cubic symmetry. The triplet of vibrational modes τ (even under space inversion symmetry) is indicated in Figure 8.8, while in Table 8.3 we give the displacements corresponding to each degenerate vibrational mode.

The matrix potential describing the $T \otimes \tau$ model can be worked out with the same procedure adopted before. With the help of Eq. (A.10) one obtains

$$V(q_1, q_2, q_3) = \gamma \begin{pmatrix} 0 & q_3 & q_2 \\ q_3 & 0 & q_1 \\ q_2 & q_1 & 0 \end{pmatrix} + \frac{1}{2}C(q_1^2 + q_2^2 + q_3^2) \tag{8.32}$$

(in the following we suppose $\gamma > 0$).

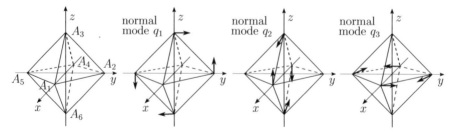

Figure 8.8 Geometry of the octahedral complex and normal coordinates q_1, q_2, q_3 of the triply degenerate vibrational mode τ (even under space inversion symmetry).

The eigenvalues of the 3×3 matrix appearing in the right-hand side of Eq. (8.32) are determined by the equation

$$E^3 - \gamma^2(q_1^2 + q_2^2 + q_3^2)E - 2\gamma^3 q_1 q_2 q_3 = 0. \tag{8.33a}$$

If $E(q_1, q_2, q_3)$ indicate the solutions of Eq. (8.33a), the adiabatic potential-energy surfaces are given by

$$E^{(ad)}(q_1, q_2, q_3) = E(q_1, q_2, q_3) + \frac{1}{2}C(q_1^2 + q_2^2 + q_3^2). \tag{8.33b}$$

In order to obtain in a more explicit way the adiabatic potential-energy surfaces, consider in the q_1, q_2, q_3 space the sphere of radius q, given by $q_1^2 + q_2^2 + q_3^2 = q^2$. We indicate with l_1, l_2, l_3 the director cosines of a point P in q-space, and introduce the dimensionless variable $\lambda = E/\gamma q$. Equation (8.33a) takes now the form

$$\lambda^3 - \lambda - 2l_1 l_2 l_3 = 0, \tag{8.34a}$$

and its graphical solution is shown in Figure 8.9.

It can then immediately be seen that the lowest root of Eq. (8.34a) is $\lambda = -2/\sqrt{3}$; it occurs in any of the four directions

$$\frac{1}{\sqrt{3}}(-1, -1, -1), \quad \frac{1}{\sqrt{3}}(-1, 1, 1), \quad \frac{1}{\sqrt{3}}(1, -1, 1), \quad \frac{1}{\sqrt{3}}(1, 1, -1). \tag{8.34b}$$

The energy of the ground adiabatic sheet for q along any of the directions (8.34b) is

$$E_0^{(ad)}(q_1, q_2, q_3) = -\frac{2\gamma}{\sqrt{3}}q + \frac{1}{2}Cq^2. \tag{8.34c}$$

Equation (8.34c) takes its minimum value $E_0 = -2\gamma^2/3C$ for $q_0 = 2\gamma/C\sqrt{3}$. The eigenvalue Eq. (8.34a) can be solved for any other point in q-space of interest and the adiabatic potential-energy branches can be worked out.

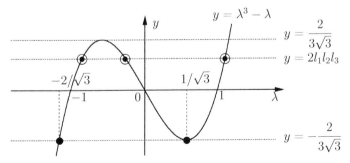

Figure 8.9 Graphical solution of the equation $\lambda^3 - \lambda = 2l_1 l_2 l_3$; the roots are obtained from the intersection of the cubic curve $y = \lambda^3 - \lambda$ with the straight line $y = 2l_1 l_2 l_3$. In particular at the point $P = (1/\sqrt{3})(-1, -1, -1)$ we have $l_1 l_2 l_3 = -1/(3\sqrt{3})$; the equation $\lambda^3 - \lambda = -2/(3\sqrt{3})$ has the single root $\lambda_1 = -2/\sqrt{3}$ and the doubly degenerate roots $\lambda_2 = \lambda_3 = 1/\sqrt{3}$.

Among the numerous problems investigated with the $T \otimes \tau$ vibronic model, we briefly mention here the so-called EL2 impurity center in GaAs. An accredited model in the literature is that this impurity is an isolated antisite defect, with As anion replacing Ga in the cation lattice [see M. Kaminska, Phys. Scripta T *19*, 551 (1987)]. From the four valley structure of the $T \otimes \tau$ ground adiabatic sheet, it is possible to infer that the sequence of vibronic states is a triplet state followed by a total symmetric singlet state; the presence of this state is indeed important to explain the non-linear behavior of splitting of the ground triplet under uniaxial stress.

We have introduced so far a few typical examples of Jahn-Teller systems. These simple models have been considered because of their own interest and because they may provide qualitative guidelines in much more complicated and demanding situations. In general the coupling between electronic and nuclear vibrational degrees of freedom are much more complicated than those here introduced, because they may involve several Jahn-Teller active modes of different frequency or symmetry. Another interesting extension is the pseudo Jahn-Teller effect that concerns almost degenerate adiabatic sheets, that nevertheless have to be considered on the same footing. For instance it occurs between the s and p levels of the relaxed F center and other color centers; the vibronic model is essential to explain the luminescence of these centers and the decay times. Further phenomenology and properties of vibronic systems can be found in the wide literature; see for instance G. Bevilacqua et al. (1999) and references quoted therein.

8.4   The Hellmann-Feynman Theorem and Electronic-Nuclear Systems

8.4.1   *General Considerations on the Hellmann-Feynman Theorem*

Parametric dependent operators present some general aspects, that we are going to examine; in this section we consider the Hellmann-Feynman theorem and, in the next section, we consider the geometric Berry phase.

Consider an operator $H(\lambda)$ which depends on one parameter, or a set of parameters, collectively indicated as λ. The electron Hamiltonian operator $H_e(r; R)$ defined by Eq. (8.4) is just an example, in which the set of parameters λ are the coordinates R of the nuclei fixed in a given configuration. Consider the eigenvalue equation

$$H(\lambda)\psi_n(r; \lambda) = E_n(\lambda)\psi_n(r; \lambda), \tag{8.35}$$

where the variable r denotes all the internal coordinates of the system. We imagine to vary continuously λ in a given region of the parameter space and follow the eigenvalues $E_n(\lambda)$ and the stationary states $\psi_n(r; \lambda)$, obtaining *parameter dependent energy sheets*; for simplicity we suppose that the sheet under attention has no degeneracy in the domain of parameters λ of interest.

The representation of the operator $H(\lambda)$ in terms of its eigenstates is

$$\langle \psi_m(r; \lambda)|H(\lambda)|\psi_n(r; \lambda)\rangle = E_n(\lambda)\delta_{mn}. \tag{8.36a}$$

We have also the orthonormality relations

$$\langle \psi_m(r; \lambda) | \psi_n(r; \lambda) \rangle = \delta_{mn}. \tag{8.36b}$$

We now derive both members of Eqs. (8.36) with respect to some component, say λ_i, of the set of parameters λ; in doing this we assume that it is possible to interchange integration on the internal coordinates r and partial differentiation with respect to the parameters λ.

The derivation of both members of Eq. (8.36b) with respect to λ_i gives

$$\langle \frac{\partial}{\partial \lambda_i} \psi_m(r; \lambda) | \psi_n(r; \lambda) \rangle = -\langle \psi_m(r; \lambda) | \frac{\partial}{\partial \lambda_i} \psi_n(r; \lambda) \rangle; \tag{8.37a}$$

the above equation is a trivial consequence of orthonormality of wavefunctions. In the particular case $m = n$, Eq. (8.37a) implies

$$\boxed{\langle \psi_m(r; \lambda) | \frac{\partial}{\partial \lambda_i} \psi_m(r; \lambda) \rangle = \text{pure imaginary quantity}}. \tag{8.37b}$$

The derivation of both members of Eq. (8.36a) with respect to λ_i gives

$$\langle \frac{\partial \psi_m}{\partial \lambda_i} | H | \psi_n \rangle + \langle \psi_m | \frac{\partial H}{\partial \lambda_i} | \psi_n \rangle + \langle \psi_m | H | \frac{\partial \psi_n}{\partial \lambda_i} \rangle = \frac{\partial E_n}{\partial \lambda_i} \delta_{mn}.$$

Since $\psi_m(r; \lambda)$ and $\psi_n(r; \lambda)$ are eigenfunctions of $H(\lambda)$ with energy $E_m(\lambda)$ and $E_n(\lambda)$, respectively, it follows

$$\langle \frac{\partial \psi_m}{\partial \lambda_i} | \psi_n \rangle E_n + \langle \psi_m | \frac{\partial H}{\partial \lambda_i} | \psi_n \rangle + E_m \langle \psi_m | \frac{\partial \psi_n}{\partial \lambda_i} \rangle = \frac{\partial E_n}{\partial \lambda_i} \delta_{mn}.$$

With the help of Eq. (8.37a), the above relation becomes

$$\langle \psi_m | \frac{\partial H}{\partial \lambda_i} | \psi_n \rangle + [E_m - E_n] \langle \psi_m | \frac{\partial \psi_n}{\partial \lambda_i} \rangle = \frac{\partial E_n}{\partial \lambda_i} \delta_{mn}. \tag{8.38}$$

The specification of Eq. (8.38) to the case $m \equiv n$, gives the Hellmann-Feynman theorem

$$\boxed{\langle \psi_m(r; \lambda) | \frac{\partial H(\lambda)}{\partial \lambda_i} | \psi_m(r; \lambda) \rangle = \frac{\partial E_m(\lambda)}{\partial \lambda_i}}. \tag{8.39}$$

This theorem is very important because it provides the gradient along which to move in the case we wish to minimize the energy on a given energy sheet; the gradient on a given energy sheet along λ_i can conveniently be obtained from the expectation value of the quantum mechanical operator $\partial H(\lambda)/\partial \lambda_i$ on the appropriate wavefunction.

The specification of Eq. (8.38) to the case $m \neq n$, gives the Epstein generalization of the Hellmann-Feynman theorem

$$\boxed{\langle \psi_m(r; \lambda) | \frac{\partial}{\partial \lambda_i} \psi_n(r; \lambda) \rangle = \frac{1}{E_n(\lambda) - E_m(\lambda)} \langle \psi_m(r; \lambda) | \frac{\partial H(\lambda)}{\partial \lambda_i} | \psi_n(r; \lambda) \rangle} \tag{8.40}$$

with $E_n(\lambda) \neq E_m(\lambda)$. The theorems (8.39) and (8.40) are completely general, and apply whatever is the physical meaning of the external parameters λ (and internal coordinates r). Notice that in Section 2.6.2 we discussed an elementary application of the Hellmann-Feynman theorem; in that case the parametric Hamiltonian was the so-called $\mathbf{k} \cdot \mathbf{p}$ Hamiltonian and the parameters λ were just the components of the wavevector $\mathbf{k}$.

8.4.2   Charge Density and Atomic Forces

We apply now the Hellmann-Feynman theorem to interacting electronic-nuclear systems, with the nuclei fixed in a given configuration; in this case the parameters λ are just the coordinates R of the nuclei in the electron Hamiltonian $H_e(r; R)$ of Eq. (8.4). The "atomic force" $\mathbf{F}_K = -\partial E / \partial \mathbf{R}_K$ on the nucleus at $\mathbf{R}_K$ is given by differentiation of the adiabatic potential-energy function $E(R)$ (supposed non-degenerate) with respect to the nuclear coordinate $\mathbf{R}_K$. With straightforward elaborations, we can now prove that the "atomic forces" are just the "electrostatic forces", born from all the other nuclear point charges and from the total one-body electronic charge density.

Consider the electronic problem with the nuclei fixed in a given configuration R; the electronic Hamiltonian $H_e(r; R)$ is here rewritten for convenience in the form

$$H_e(r; R) = -\sum_i \frac{\hbar^2 \nabla_i^2}{2m} + \frac{1}{2} \sum_{i \neq j} \frac{e^2}{|\mathbf{r}_i - \mathbf{r}_j|} - \sum_{iI} \frac{z_I e^2}{|\mathbf{r}_i - \mathbf{R}_I|} + \frac{1}{2} \sum_{I \neq J} \frac{z_I z_J e^2}{|\mathbf{R}_I - \mathbf{R}_J|}.$$
(8.41)

Let $\psi(r; R)$ denote the wavefunction of a given adiabatic surface (supposed to be non-degenerate) and $E(R)$ the corresponding energy. It holds

$$\langle \psi(r; R) | H_e(r; R) | \psi(r; R) \rangle = E(R).$$

Derivation of both members of the above equation with respect to the nuclear coordinate $\mathbf{R}_K$ of the K-th nucleus, and application of the Hellmann-Feynman theorem, gives

$$\langle \psi(r; R) | \frac{\partial}{\partial \mathbf{R}_K} H_e(r; R) | \psi(r; R) \rangle = \frac{\partial E(R)}{\partial \mathbf{R}_K}.$$
(8.42a)

The gradient of the operator $H_e(r; R)$ is given by

$$\frac{\partial}{\partial \mathbf{R}_K} H_e(r; R) = -\frac{\partial}{\partial \mathbf{R}_K} \sum_i \frac{z_K e^2}{|\mathbf{r}_i - \mathbf{R}_K|} + \frac{\partial}{\partial \mathbf{R}_K} \sum_{I(\neq K)} \frac{z_I z_K e^2}{|\mathbf{R}_I - \mathbf{R}_K|}.$$
(8.42b)

Using Eqs. (8.42), the force on the nucleus K can be cast in the form

$$\mathbf{F}_K = -\frac{\partial E}{\partial \mathbf{R}_K} = \langle \psi(r; R) | \frac{\partial}{\partial \mathbf{R}_K} \sum_i \frac{z_K e^2}{|\mathbf{r}_i - \mathbf{R}_K|} | \psi(r; R) \rangle - \frac{\partial}{\partial \mathbf{R}_K} \sum_{I(\neq K)} \frac{z_I z_K e^2}{|\mathbf{R}_I - \mathbf{R}_K|}.$$

Let $n(\mathbf{r}; R)$ denote the one-body electron density corresponding to the many-body wavefunction $\psi(r; R)$ (for the definition see for instance Section 4.8); the above expression of the forces reads

$$\boxed{\mathbf{F}_K = -\frac{\partial E}{\partial \mathbf{R}_K} = \int n(\mathbf{r}; R) \frac{\partial}{\partial \mathbf{R}_K} \frac{z_K e^2}{|\mathbf{r} - \mathbf{R}_K|} d\mathbf{r} - \frac{\partial}{\partial \mathbf{R}_K} \sum_{I(\neq K)} \frac{z_I z_K e^2}{|\mathbf{R}_I - \mathbf{R}_K|}}. \qquad (8.43)$$

Thus the force acting on a given nucleus equals the negative gradient of the classical electrostatic potential-energy, originated from all the other point charged nuclei and from the quantum mechanical electronic charge distribution. [Notice that this description, based on average electronic distribution, does not include for instance the electromagnetic interactions between fluctuating atomic dipoles, contributing to the van der Waals forces.]

Equation (8.43) is rich of physical consequences; once the one-body electron density is known, one can determine the forces on the nuclei, and then the nuclear motion itself; this is indeed the microscopic foundation of the (classical or quantum) lattice dynamics, see for instance S. Baroni et al. (2001). Equation (8.43) contains in germ the working principle of the atomic force microscope [G. Binning, C. F. Quate and Ch. Gerber, Phys. Rev. Lett. *12*, 930 (1986).] This type of atomic resolution microscope maps the force between the sample surface and a sharp tip probe, and is a most useful tool for visualization and investigation of surface profiles. In typical experimental situations, a cantilever is vibrated at a given frequency above the surface to be investigated; the surface image is then elaborated by monitoring the frequency change due to the tip-surface interactions. [See for instance the review article "Theories of scanning probe microscopes at the atomic scale" by W. A. Hofer, A. S. Foster and A. L. Shluger, Rev. Mod. Phys. 75, 1287 (2003) and references quoted therein.]

8.5 Parametric Hamiltonians and Berry Phase

Consider the most generic quantum Hamiltonian $H(\mathbf{R})$, that depends on a set of parameters $(R_1, R_2, \dots)$ collectively indicated as $\mathbf{R}$. In this section, for sake of simplicity, the parameter space $\mathbf{R}$ is assumed to span an ordinary three-dimensional space (we can thus use familiar vector relations, without need of generalization of vector algebra to a multidimensional space). We consider the eigenfunctions $\psi_m(r; \mathbf{R})$ and the eigenvalues $E_m(\mathbf{R})$ of the Schrödinger equation

$$H(\mathbf{R})\psi_m(r; \mathbf{R}) = E_m(\mathbf{R})\psi_m(r; \mathbf{R}), \qquad (8.44)$$

where all the internal variables of the system are collectively indicated as r. We discuss here some *general properties of parameter dependent eigenvalue equations*, which are independent of the specific meaning of the "internal variables r" and the "external variables (or parameters) $\mathbf{R}$".

It is evident that the eigenvalue equation (8.44) at different $\mathbf{R}$ does not uniquely define the wavefunctions $\psi_m(r; \mathbf{R})$, because an $\mathbf{R}$-dependent arbitrariness in the phases of $\psi_n(r; \mathbf{R})$ remains (gauge freedom). Suppose that $\psi_m(r; \mathbf{R})$ constitute a set of continuous and single-valued functions in the connected parameter domain D of interest;

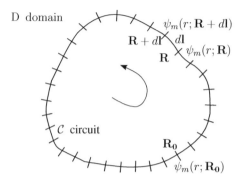

Figure 8.10 Schematic representation of the sequence of states $\psi_m(r;\ \mathbf{R})$ as $\mathbf{R}$ is changed along a circuit C drawn in the parameter space D, where no degeneracy occurs.

then any gauge transformation of the type

$$\widetilde{\psi}_m(r;\mathbf{R}) = e^{i\alpha(\mathbf{R})}\psi_m(r;\mathbf{R}), \tag{8.45}$$

where $\alpha(\mathbf{R})$ is a real, continuous and single-valued function, provides an equivalent set of continuous and single-valued wavefunctions in the domain D. However, in spite of this gauge freedom, it is possible to identify gauge invariant geometric phases, according to a procedure initiated by Berry.

In the following we focus our attention on a given discrete adiabatic surface, say the m-th one, of energy $E_m(\mathbf{R})$ and eigenfunctions $\psi_m(r;\mathbf{R})$, assumed to be non-degenerate at any $\mathbf{R}$. Most often, but not necessarily, the surface of interest is the ground adiabatic sheet.

In the parameter space $\mathbf{R}$, let us draw an arbitrary circuit C, and along it consider two contiguous points $\mathbf{R}$ and $\mathbf{R} + d\mathbf{l}$, and the corresponding wavefunctions $\psi_m(r;\mathbf{R})$ and $\psi_m(r;\mathbf{R} + d\mathbf{l})$ (see Figure 8.10). The infinitesimal phase difference $d\phi$ between the two functions is most naturally defined as

$$e^{id\phi} \equiv \frac{\langle\psi_m(r;\mathbf{R}+d\mathbf{l})|\psi_m(r;\mathbf{R})\rangle}{|\langle\psi_m(r;\mathbf{R}+d\mathbf{l})|\psi_m(r;\mathbf{R})\rangle|}. \tag{8.46}$$

The second member of Eq. (8.46) is just the ratio between the scalar product of two wavefunctions (in general a complex number) and the modulus of the scalar product itself; this defines the phase factor reported in the first member of Eq. (8.46). The real quantity $d\phi$ can be referred as the infinitesimal phase acquired by the quantum system moving from $\mathbf{R}$ to $\mathbf{R} + d\mathbf{l}$ in the given adiabatic sheet along the circuit C. However, the quantity $d\phi$ cannot have a precise physical meaning, since *it is clearly a gauge dependent quantity*, i.e. it depends on the actual phases embodied in the two wavefunctions of the scalar product.

It is possible to express $d\phi$ in a more effective form, by expanding $\psi_m(r;\mathbf{R} + d\mathbf{l})$ in Taylor series up to first order in $d\mathbf{l}$ and expanding the exponential to first order in

$d\phi$; this linearization procedure gives

$$1 + id\phi = \langle \psi_m(r; \mathbf{R}) + \frac{\partial}{\partial \mathbf{R}} \psi_m(r; \mathbf{R}) \cdot d\mathbf{l} | \psi_m(r; \mathbf{R}) \rangle$$

$$= 1 - \langle \psi_m(r; \mathbf{R}) | \frac{\partial}{\partial \mathbf{R}} \psi_m(r; \mathbf{R}) \rangle \cdot d\mathbf{l};$$

it follows

$$d\phi = i \langle \psi_m(r; \mathbf{R}) | \frac{\partial}{\partial \mathbf{R}} \psi_m(r; \mathbf{R}) \rangle \cdot d\mathbf{l}. \tag{8.47a}$$

The matrix element of the gradient operator in the above equation is purely imaginary (this is nothing more than a trivial consequence of normalization of wavefunctions). Then the real infinitesimal phase $d\phi$ can be written in the equivalent form

$$d\phi = -\text{Im} \langle \psi_m(r; \mathbf{R}) | \frac{\partial}{\partial \mathbf{R}} \psi_m(r; \mathbf{R}) \rangle \cdot d\mathbf{l}, \tag{8.47b}$$

where Im stands for "imaginary part of"; notice that Im (…) denotes a *real* quantity. Expression (8.47a) and expression (8.47b) are of course equivalent, but the latter one is often preferred because it has the advantage to exhibit explicitly that the infinitesimal phase change $d\phi$ is a real quantity.

The *total geometric phase* $\gamma_m(\mathcal{C})$, *or Berry phase*, introduced in completing a circuit $\mathcal{C}$ in the parameter space is defined as

$$\gamma_m(\mathcal{C}) = i \oint_{\mathcal{C}} \langle \psi_m(r; \mathbf{R}) | \frac{\partial}{\partial \mathbf{R}} \psi_m(r; \mathbf{R}) \rangle \cdot d\mathbf{l}. \tag{8.48}$$

The "Berry phase" is thus the phase acquired by a quantum system moving along a circuit $\mathcal{C}$ on a given adiabatic surface.

It is convenient to rewrite the expression of the Berry phase in the equivalent form

$$\boxed{\gamma_m(\mathcal{C}) = \oint_{\mathcal{C}} \mathbf{A}_m(\mathbf{R}) \cdot d\mathbf{l}}, \tag{8.49a}$$

where $\mathbf{A}_m(\mathbf{R})$ is a kind of local vector (which depends on the sheet m) defined as

$$\boxed{\mathbf{A}_m(\mathbf{R}) = i \langle \psi_m(r; \mathbf{R}) | \nabla_{\mathbf{R}} \psi_m(r; \mathbf{R}) \rangle} \tag{8.49b}$$

(the gradient $\partial \dots / \partial \mathbf{R}$ is now denoted by $\nabla_{\mathbf{R}}$ for simplicity). The real vectorial quantity $\mathbf{A}_m(\mathbf{R})$ is called *Berry vector potential*, or *geometric vector potential*, or *fictitious vector potential* (for the forthcoming analogy with the familiar vector potential of electromagnetism). The vector $\mathbf{A}_m(\mathbf{R})$ is also commonly addressed as the *Berry connection*, because it controls the "connection" between the wavefunctions $\psi_m(r; \mathbf{R})$ and $\psi_m(r; \mathbf{R} + d\mathbf{R})$ at two nearby points in the parameter space. The Berry connection is thus defined as the expectation value of the gradient operator in the parameter domain (times i) on the adiabatic wavefunction of the sheet under attention.

It is now easy to show that the geometric phase $\gamma_m(\mathcal{C})$ is *gauge invariant*, in spite of the fact that every infinitesimal contribution is *gauge dependent*. Consider in fact the gauge transformation defined by Eq. (8.45); we have

$$\widetilde{\mathbf{A}}_m(\mathbf{R}) = i\langle \widetilde{\psi}_m(r;\mathbf{R})|\nabla_{\mathbf{R}}\widetilde{\psi}_m(r;\mathbf{R})\rangle = i\langle e^{i\alpha(\mathbf{R})}\psi_m(r;\mathbf{R})|\nabla_{\mathbf{R}}\,e^{i\alpha(\mathbf{R})}\psi_m(r;\mathbf{R})\rangle$$
$$= i\langle \psi_m(r;\mathbf{R})|\nabla_{\mathbf{R}}\psi_m(r;\mathbf{R})\rangle - \nabla_{\mathbf{R}}\alpha(\mathbf{R}).$$

The relation so obtained can be written as

$$\widetilde{\mathbf{A}}_m(\mathbf{R}) = \mathbf{A}_m(\mathbf{R}) - \nabla_{\mathbf{R}}\alpha(\mathbf{R})\,; \tag{8.50}$$

the extra term, being the gradient of the regular single-valued function $\alpha(\mathbf{R})$, does not contribute to the circulation $\gamma_m(\mathcal{C})$ on a closed loop.

The expression (8.49a) of the Berry phase defines a gauge independent quantity as a sum of contributions, which taken one-by-one are gauge dependent. It would be desirable to express the Berry phase directly in terms of gauge invariant contributions; this is possible by using the Stokes theorem to transform the line integral along $\mathcal{C}$ into an integral on a surface S of contour $\mathcal{C}$. Exploiting Stokes theorem, Eq. (8.49a) can be written in the form

$$\gamma_m(\mathcal{C}) = \oint_{\mathcal{C}} \mathbf{A}_m(\mathbf{R}) \cdot d\mathbf{l} = \int_S \left[\nabla_{\mathbf{R}} \times \mathbf{A}_m(\mathbf{R})\right] \cdot d\mathbf{S}\,.$$

From Eq. (8.50), it is seen that $\nabla_{\mathbf{R}} \times \mathbf{A}_m = \nabla_{\mathbf{R}} \times \widetilde{\mathbf{A}}_m$, and the invariant character of the Berry phase to gauge transformations is thus confirmed.

It is of interest to derive an explicit expression of the quantity

$$\mathbf{B}_m(\mathbf{R}) = \nabla_{\mathbf{R}} \times \mathbf{A}_m(\mathbf{R})\,; \tag{8.51}$$

the field $\mathbf{B}_m(\mathbf{R})$ is called *Berry curvature* or also *fictitious magnetic field* (again borrowing from the electromagnetism analogy). Using elementary vector algebra, we have

$$\mathbf{B}_m(\mathbf{R}) = -\text{Im}\,\nabla_{\mathbf{R}} \times \langle \psi_m(r;\mathbf{R})|\nabla_{\mathbf{R}}\psi_m(r;\mathbf{R})\rangle$$
$$= -\text{Im}\,\langle \nabla_{\mathbf{R}}\psi_m(r;\mathbf{R})| \times |\nabla_{\mathbf{R}}\psi_m(r;\mathbf{R})\rangle$$
$$= -\text{Im}\,\sum_{n(\neq m)} \langle \nabla_{\mathbf{R}}\psi_m(r;\mathbf{R})|\psi_n(r;\mathbf{R})\rangle \times \langle \psi_n(r;\mathbf{R})|\nabla_{\mathbf{R}}\psi_m(r;\mathbf{R})\rangle,$$

where in the last passage a complete set $|\psi_n\rangle\langle\psi_n|$ has been inserted. The exclusion of the term $n = m$ in the summation is justified by the fact that it vanishes identically (it involves the vector product of two vectors, which differ only for the sign). Since for $m \neq n$ we can use the Epstein generalization of the Hellmann-Feynman theorem (see Eq. 8.40), we obtain the expression

$$\mathbf{B}_m(\mathbf{R}) = -\text{Im}\,\sum_{n(\neq m)} \frac{\langle \psi_m(r;\mathbf{R})|\nabla_{\mathbf{R}}H|\psi_n(r;\mathbf{R})\rangle \times \langle \psi_n(r;\mathbf{R})|\nabla_{\mathbf{R}}H|\psi_m(r;\mathbf{R})\rangle}{[E_n(\mathbf{R}) - E_m(\mathbf{R})]^2}$$

$$\tag{8.52}$$

and

$$\gamma_m(\mathcal{C}) = \int_S \mathbf{B}_m(\mathbf{R}) \cdot d\mathbf{S}. \tag{8.53}$$

Thus the Berry phase $\gamma_m(\mathcal{C})$ can be interpreted as the flux of $\mathbf{B}_m(\mathbf{R})$ across a surface of contour $\mathcal{C}$.

The expression (8.52) constitutes a most remarkable result: the Berry curvature $\mathbf{B}_m(\mathbf{R})$ is clearly independent of the phases embodied in the wavefunctions; actually it is even unnecessary to require that the wavefunctions are single-valued in the parameter space. The concepts of Berry connection and Berry field acquire a familiar aspect, by virtue of the electromagnetic analogy, with $\mathbf{A}_m(\mathbf{R})$ playing the role of a "fictitious vector potential" and $\mathbf{B}_m(\mathbf{R}) = \nabla_\mathbf{R} \times \mathbf{A}_m(\mathbf{R})$ the role of a "fictitious magnetic field".

Considerations on the Berry Phase for Electronic-Nuclear Systems

We consider now the geometrical Berry phase concepts for the electronic-nuclear systems, in which case the parameter dependent Hamiltonian under attention is the electron Hamiltonian $H_e(r; R)$ discussed in Section 8.1. This Hamiltonian is invariant under time-reversal symmetry, as discussed in Section 8.2.2. If the effect of spin is negligible, the adiabatic wavefunctions $\psi_m(r; R)$ can be taken in real form, and we assume this to be the case.

We discuss first an ordinary Born-Oppenheimer system, whose ground adiabatic surface is never degenerate. The Berry field on the ground adiabatic surface can be calculated with the fundamental Eq. (8.52). Since the adiabatic wavefunctions can be put in real form, it is seen that the Berry curvature of the ground adiabatic surface is zero everywhere in the parameter space, and hence the Berry connection and the Berry phases are trivially vanishing, too. The geometric vector $\langle\psi_0|\nabla_\mathbf{R}|\psi_0\rangle$ entering into the Eq. (8.13) for the non-adiabatic operator $\Lambda_{00}(R)$ vanishes. A preferential gauge of real, continuous, and single-valued adiabatic wavefunctions is possible, and the lattice dynamics is controlled by the electronic energy in agreement with Eq. (8.14).

We discuss now the case in which two or more adiabatic surfaces are degenerate at some symmetry configuration R_0 (Jahn-Teller systems or Renner-Teller systems). It is evident that Eq. (8.52) for the Berry field presents vanishingly small denominators for $R \approx R_0$, and cannot be applied as it stands to handle a degeneracy point. [We remember that the elaborations leading to Eq. (8.52) assume that the adiabatic surface under attention is nondegenerate]. To remain within the limits of validity of the presented elaborations, we can envisage to remove the "offending" degeneracy point with an infinitesimal small perturbation, that produces an energy gap among the adiabatic surfaces, confluent in R_0. Sufficiently far from R_0 the perturbation has a negligible effect on the potential energies and on the adiabatic wavefunctions, which remain real and entail vanishing Berry curvature (8.52); around R_0 the Berry curvature (8.52) is expected to become singular by virtue of the vanishing small denominators.

Having established these general guidelines, we can move to specific examples. For instance, in the case of the $E \otimes \varepsilon$ Jahn-Teller system it is seen that the fictitious magnetic field is zero everywhere except at the conical intersection, where it has a delta-like singularity; the geometric vector potential is induced by an infinitesimal narrow flux tube, with an associate phase of π. Similarly for a typical Renner-Teller system the infinitesimal flux tube has an associate phase of 2π. The main breakthrough of the far reaching Berry formalism is that some physical properties of quantum parametric systems can be revisited at the light of the gauge-invariant Berry phase and appreciated in their deepest transcendent aspects. Some significant examples are reported in the set of problems and complements reported in Appendix B. In the next section, the role of Berry phase formalism to give an elegant and workable solution of the long standing problem of macroscopic polarization of crystalline solids is analyzed.

8.6 The Berry Phase Theory of the Macroscopic Electric Polarization in Crystals

The macroscopic electric polarization of materials plays a fundamental role in the phenomenological description of dielectrics. In the literature, most often, simplified and intuitive models are adopted for the electric polarization of systems of charged particles. More recently, the progress of the methods of electronic state calculations, and in particular of the density functional method, have made possible accurate first-principle investigations of the ground-state properties of interacting electronic-nuclear systems. In this section, we discuss some aspects of the quantum theory of polarization of crystalline dielectrics and the role assumed in the theory by the geometric Berry phase.

Consider a crystal of volume $V = N\Omega$, formed by an arbitrary (large) number N of identical unit cells of volume Ω, and let $\rho_{tot}(\mathbf{r})$ denote the total electronic and nuclear (or ionic) charge density. The charge distribution in the crystal is the sum of the nuclear (or ionic) contribution and of the electronic contribution and reads

$$\rho_{tot}(\mathbf{r}) = \sum_j e z_j \delta(\mathbf{r} - \mathbf{R}_j) - e\, n_{el}(\mathbf{r}), \qquad (8.54)$$

where e is the absolute value of the electronic charge, $\mathbf{R}_j$ are the positions of the pointwise nuclei (or ions) of charge $e z_j$, and $n_{el}(\mathbf{r})$ is the electronic density.

The charge distribution $\rho_{tot}(\mathbf{r})$ averages to zero within any unit cell, as well as within the whole crystal composed by an integer number of unit cells (periodic boundary conditions are assumed). In any unit cell it holds

$$\int_{cell} \rho_{tot}(\mathbf{r})\, d\mathbf{r} \equiv 0 \quad \text{(neutrality condition)}.$$

One could be tempted to define the average polarization of the crystal (electric dipole moment per unit volume) via the relation

$$\mathbf{P} = \frac{1}{\Omega} \int_{cell} \mathbf{r}\, \rho_{tot}(\mathbf{r})\, d\mathbf{r}. \qquad (8.55)$$

However such a definition of polarization vector $\mathbf{P}$ is fruitless (or *ill defined*), since the value of $\mathbf{P}$ depends on the details (shape and location) of the unit cell chosen in the bulk crystal. In Figure 8.11 a simple illustration is given to understand why $\mathbf{P}$, if defined via Eq. (8.55), is cell dependent, and does not represent a genuine bulk property of the material.

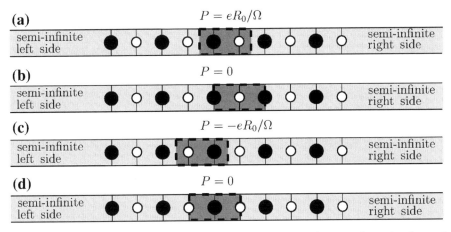

Figure 8.11 Schematic illustration of the electric polarization within a one-dimensional crystal, composed by unit cells of volume Ω. The model crystal is constituted by uniformly charged spheres of charge $+e$ and $-e$ (white and black circles, respectively) at distance R_0. As an example, four different ways to choose the unit cell are indicated. In case (a) the electric polarization is $+eR_0/\Omega$; in case (b) and (d) it is zero; in case (c) it is $-eR_0/\Omega$. Other shifts of the unit cell would produce any other value of the polarization between $\mp eR_0/\Omega$.

The problem of electric polarization remained unsolved until a Berry-phase formula was proposed in the 1990s in a series of seminal works by King-Smith, Resta, and Vanderbilt (some of their works are cited in the bibliography). The basic guideline of the novel procedure is to address the issue of *changes in the polarization of a crystalline solid*, rather than to address the ill defined problem of *what is the polarization of a crystalline solid*. It is shown below that infinitesimal (and then finite) changes of polarization are well defined; thus all the physical effects related to *changes of polarization* can be evaluated unambiguously and compared with experimental measurements. Experimental means to produce changes of the polarization of crystals include application of a stress (piezoelectric crystals), or changes of temperature (pyroelectric crystals), or spontaneous polarization reversal by electric fields (ferroelectric crystals).

To keep things at the essential, consider a dielectric composed by two sublattices (GaAs for instance), and let λ indicate a parameter that regulates with continuity the relative position of one sublattice with respect to the other. The parameter λ is arranged to take values 0 and 1 at the initial and final configuration. In the independent particle approximation, the Hamiltonian of the electronic system takes the form

$$H = \frac{\mathbf{p}^2}{2m} + V(\mathbf{r}, \lambda), \tag{8.56}$$

where $V(\mathbf{r}, \lambda)$ is the periodic crystalline potential (or a convenient pseudo-potential), continuously dependent on the parameter λ. For simplicity, it is assumed that the fundamental translation vectors remain unchanged, so that the Brillouin zone remains unchanged too, and that the material remains a semiconductor or an insulator for all the values of λ in the interval $[0,1]$. For any assigned value of λ, and for any wavevector within the first Brillouin zone, we denote with $E_{n\mathbf{k}}(\lambda)$ and $\psi_{n\mathbf{k}}(\mathbf{r}, \lambda)$ the eigenvalues and the crystalline wavefunctions (most often the Kohn-Sham orbitals). The band wave-functions are normalized to unity in the volume $V = N\Omega$ of the crystal, and have the standard form

$$\psi_{n\mathbf{k}}(\mathbf{r}, \lambda) = e^{i\mathbf{k}\cdot\mathbf{r}} u_{n\mathbf{k}}(\mathbf{r}, \lambda),$$

where $u_{n\mathbf{k}}(\mathbf{r}, \lambda)$ denote the lattice periodic part of the Bloch wavefunctions.

We focus now on the change of the ionic and electronic polarization as λ is varied to $\lambda + d\lambda$.

Consider first the change of the electric dipole of a system when a pointwise ion of charge ez_j, initially at the point $\mathbf{R}_j$, is shifted to $\mathbf{R}_j + d\mathbf{R}_j$; the dipole change is $ez_j d\mathbf{R}_j$. The change of ionic polarization is trivial, because of the classical pointwise nature of ionic charges.

For electrons, described by quantum wavefunctions, the electronic contribution to the polarization of the crystal reads

$$\mathbf{P}_{\text{el}}(\lambda) = \frac{-e}{V} \sum_{n\mathbf{k}}^{(\text{occ})} \int_V \mathbf{r}\, |\psi_{n\mathbf{k}}(\mathbf{r},\lambda)|^2\, d\mathbf{r} = -\frac{e}{V} \sum_{n\mathbf{k}}^{(\text{occ})} \langle u_{n\mathbf{k}}|\, \mathbf{r}\, |u_{n\mathbf{k}}\rangle, \tag{8.57}$$

where the sum is over the occupied bands of the insulator or semiconductor under investigation.

The quantity $\mathbf{P}_{\text{el}}$ is ill defined, because so are the diagonal matrix elements of the position operator between itinerant or periodic wavefunctions. The quantity $\mathbf{P}_{\text{el}}$, although "ill defined" from a physical point of view, is "well defined" from a mathematical point of view, since it involves integrals carried out on a finite volume $V = N\Omega$, where the wavefunctions are normalized. We can thus consider the derivative of $\mathbf{P}_{\text{el}}(\lambda)$ with respect to λ, and obtain

$$\frac{\partial \mathbf{P}_{\text{el}}(\lambda)}{\partial \lambda} = -\frac{e}{V} \sum_{n\mathbf{k}}^{(\text{occ})} 2\text{Re}\, \langle u_{n\mathbf{k}}|\, \mathbf{r}\, |\frac{\partial}{\partial \lambda} u_{n\mathbf{k}}\rangle, \tag{8.58}$$

where "Re" stands for "real part of" what follows. Notice that each term $\langle u_{n\mathbf{k}}|\mathbf{r}|u_{n\mathbf{k}}\rangle$ entering in the sum in Eq. (8.57) is independent of the phases embodied in the cell periodic functions. The same property holds, of course, for each term appearing in the sum in Eq. (8.58); for instance

$$u_{n\mathbf{k}} \rightarrow \tilde{u}_{n\mathbf{k}} \equiv u_{n\mathbf{k}} e^{i\chi(\lambda)} \implies \text{Re}\, \langle \tilde{u}_{n\mathbf{k}}|\, \mathbf{r}\, |\frac{\partial}{\partial \lambda} \tilde{u}_{n\mathbf{k}}\rangle \equiv \text{Re}\, \langle u_{n\mathbf{k}}|\, \mathbf{r}\, |\frac{\partial}{\partial \lambda} u_{n\mathbf{k}}\rangle,$$

where the phase $\chi(\lambda)$ is an arbitrary differentiable real function of the parameter λ.

We now show the basic property

$$\boxed{\text{Re}\, \langle u_{n\mathbf{k}}|\, \mathbf{r}\, |\frac{\partial}{\partial \lambda} u_{n\mathbf{k}}\rangle \quad \text{is a well-defined quantity.}}$$

For this purpose, consider the standard projector operator

$$\sum_{\alpha} |u_{\alpha\mathbf{k}}\rangle\langle u_{\alpha\mathbf{k}}| \equiv 1 \tag{8.59}$$

and write

$$\sum_{\alpha} |u_{\alpha \mathbf{k}}\rangle \langle u_{\alpha \mathbf{k}}| \frac{\partial}{\partial \lambda} u_{n\mathbf{k}}\rangle \equiv |\frac{\partial}{\partial \lambda} u_{n\mathbf{k}}\rangle.$$

It follows

$$\mathrm{Re}\, \langle u_{n\mathbf{k}}| \mathbf{r}| \frac{\partial}{\partial \lambda} u_{n\mathbf{k}}\rangle = \mathrm{Re} \sum_{\alpha} \langle u_{n\mathbf{k}}| \mathbf{r}| u_{\alpha \mathbf{k}}\rangle \langle u_{\alpha \mathbf{k}}| \frac{\partial}{\partial \lambda} u_{n\mathbf{k}}\rangle$$

$$= \mathrm{Re} \sum_{\alpha(\neq n)} \langle u_{n\mathbf{k}}| \mathbf{r}| u_{\alpha \mathbf{k}}\rangle \langle u_{\alpha \mathbf{k}}| \frac{\partial}{\partial \lambda} u_{n\mathbf{k}}\rangle ; \qquad (8.60)$$

the term with $\alpha = n$ has been excluded from the sum because it is a pure imaginary quantity.

We have thus arrived at the remarkable *conclusion that although the electronic macroscopic polarization of a crystal (8.57) is ill defined, its derivatives are well defined. The key point allowing this, is the fact that only off-diagonal matrix elements of the position operator enter into Eq. (8.60), while the initial expression (8.57) of the polarization involves the ill defined diagonal matrix elements of the position operator.* [R. Resta, Ferroelectrics *136*, 51 (1992)].

Having established this point, for the off-diagonal matrix elements of the position operator we exploit Eq. (2.35) (a by-product of the Epstein-Hellmann-Feynman theorems), here rewritten in the form

$$\langle u_{n\mathbf{k}}| \mathbf{r}| u_{\alpha \mathbf{k}}\rangle = -i \langle \frac{\partial}{\partial \mathbf{k}} u_{n\mathbf{k}}| u_{\alpha \mathbf{k}}\rangle \qquad \alpha \neq n.$$

Eq. (8.60) becomes

$$\mathrm{Re}\, \langle u_{n\mathbf{k}}| \mathbf{r}| \frac{\partial}{\partial \lambda} u_{n\mathbf{k}}\rangle = \mathrm{Im} \sum_{\alpha(\neq n)} \langle \frac{\partial}{\partial \mathbf{k}} u_{n\mathbf{k}}| u_{\alpha \mathbf{k}}\rangle \langle u_{\alpha \mathbf{k}}| \frac{\partial}{\partial \lambda} u_{n\mathbf{k}}\rangle .$$

In the above equation we can drop the restriction $\alpha(\neq n)$ (since the product of two pure imaginary quantities is a real quantity). By virtue of the closure property (8.59), one obtains

$$\boxed{\mathrm{Re}\, \langle u_{n\mathbf{k}}| \mathbf{r}| \frac{\partial}{\partial \lambda} u_{n\mathbf{k}}\rangle = \mathrm{Im} \langle \frac{\partial}{\partial \mathbf{k}} u_{n\mathbf{k}}| \frac{\partial}{\partial \lambda} u_{n\mathbf{k}}\rangle} ; \qquad (8.61)$$

this constitutes a remarkable expression, free of sums over intermediate states.

Using Eq. (8.61), the expression (8.58) of the derivative of the electronic contribution to the average crystal polarization can be cast in the form

$$\frac{\partial \mathbf{P}_{\mathrm{el}}(\lambda)}{\partial \lambda} = \frac{2(-e)}{V} \sum_{n\mathbf{k}}^{(\mathrm{occ})} 2\, \mathrm{Im} \langle \frac{\partial}{\partial \mathbf{k}} u_{n\mathbf{k}}| \frac{\partial}{\partial \lambda} u_{n\mathbf{k}}\rangle,$$

where the first factor 2 takes into account spin degeneracy (assuming spin paired bands) and the sum is over all the n occupied bands of the semiconductor or insulator under attention. The sum over $\mathbf{k}$ can be converted, in the thermodynamic limit, into the integral over the Brillouin zone times $V/(2\pi)^3$; it follows

$$\frac{\partial \mathbf{P}_{el}(\lambda)}{\partial \lambda} = -\frac{e}{4\pi^3} \sum_n^{(occ)} \int_{B.Z.} d\mathbf{k} \, 2 \, \text{Im} \, \langle \frac{\partial}{\partial \mathbf{k}} u_{n\mathbf{k}} | \frac{\partial}{\partial \lambda} u_{n\mathbf{k}} \rangle. \tag{8.62}$$

The change of electronic polarization between two configurations then reads

$$\Delta \mathbf{P}_{el} = \int_0^1 \frac{\partial \mathbf{P}_{el}(\lambda)}{\partial \lambda} \, d\lambda.$$

Manipulation with Stokes theorem of the above integrals shows that the total change of polarization between two configurations, say $\lambda = 0$ and $\lambda = 1$, is given by

$$\boxed{\Delta \mathbf{P} = \Delta \mathbf{P}_{ion} + [\mathbf{P}_{el}(1) - \mathbf{P}_{el}(0)]}, \tag{8.63}$$

where $\Delta \mathbf{P}_{ion}$ denotes the trivial change of polarization of the pointwise nuclear (or ionic) charges, and

$$\boxed{\mathbf{P}_{el}(\lambda) = \frac{e}{4\pi^3} \, \text{Im} \sum_n^{(occ)} \int_{B.Z.} d\mathbf{k} \, \langle u_n(\mathbf{k}, \mathbf{r}; \lambda) | \nabla_{\mathbf{k}} | u_n(\mathbf{k}, \mathbf{r}; \lambda) \rangle}. \tag{8.64}$$

The sum is over all occupied energy bands of the crystalline system assumed to remain insulating for any value of the parameter λ from the initial configuration to the final one. Equation (8.64) is known as the King-Smith and Vanderbilt formula for the electronic polarization of crystalline solids. The passage from Eq. (8.62) to Eq. (8.64) is reported in the last part of this section.

The integrand in Eq. (8.64) can be recognized as the Berry connection vector for each of the bands of the occupied manifold. The integration on the first Brillouin zone mimics a closed circuit: at any two wavevectors at the surface of the Brillouin zone that differ by a reciprocal lattice vector the Bloch wavefunctions satisfy

$$\psi_n(\mathbf{k}, \mathbf{r}; \lambda) \equiv \psi_n(\mathbf{k}', \mathbf{r}; \lambda) \quad \text{for} \quad \mathbf{k} \equiv \mathbf{k}' + \mathbf{G}, \tag{8.65}$$

and the cell periodic functions must be compatible with such relation. It is remarkable that the value of Eq. (8.64) is perfectly well defined whatever is the gauge chosen for the functions $u_n(\mathbf{k}, \mathbf{r}; \lambda)$, provided the wavefunctions are defined continuously between $\lambda_1(=0)$ and $\lambda_2(=1)$ for all $\mathbf{k}$ in the first Brillouin zone. [For practical calculations, in view also of unavoidable discretization procedures for the calculation of the integral, it is convenient to consider Eq. (8.64) at the end points $\lambda = 0, 1$, without enforcing continuity through the λ path. This "low cost" simplification produces an electronic polarization that is well defined within polarization quanta equal to $e\mathbf{R}/\Omega$, where $\mathbf{R}$ is a lattice vector. In general, for sufficiently small steps of λ, there is no problem to distinguish the physical result of interest from the numerical one, which may contain lattice quantum replica.]

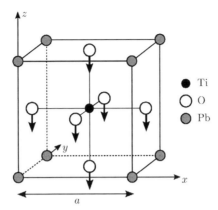

Figure 8.12 Centro-symmetrical structure and tetragonal distortion of the typical perovskite ferroelectric material PbTiO$_3$. The origin is taken at the Pb site. The arrows indicate the atomic displacements of the oxygen atoms (the much smaller displacement of Ti is not reported).

As an example we consider the application of the Berry phase theory of polarization to perovskite crystals, whose typical structure is indicated in Figure 8.12. The parameter λ controls the displacement of the oxygen atoms with respect to the cubic environment of metallic atoms. The case $\lambda = 0$ corresponds to the centro-symmetric situation, while the case $\lambda = 1$ corresponds to the final tetragonal structure. The change of polarization between the two configurations provides the spontaneous polarization of the ferroelectric material. Detailed calculations show that the spontaneous polarization of perovskite ferroelectric materials are in the range of about 0.3–0.6 C/m^2.

Before concluding this subject, we mention that the electric polarization as well as orbital magnetization of crystals can be studied exploiting itinerant Bloch wavefunctions, and also with appropriate transformation to localized wavefunctions, elaborated within the modern theory of Wannier functions (see for instance the review of Marzari et al. and references quoted therein).

Berry Phase Expression for the Crystal Polarization and Stokes Theorem

Consider Eq. (8.62) specified for simplicity to the case of a single occupied band, and to a one-dimensional crystal. The change of electronic polarization (per unit length) of a one-dimensional crystal reads

$$\frac{\partial P_{\text{el}}(\lambda)}{\partial \lambda} = -\frac{e}{\pi} \int_{-\pi/a}^{+\pi/a} dk \, 2 \, \text{Im} \, \langle \frac{\partial}{\partial k} u_n(k, x; \lambda) | \frac{\partial}{\partial \lambda} u_n(k, x; \lambda) \rangle. \tag{8.66}$$

The total change ΔP_{el} in electronic polarization is obtained by integrating in $d\lambda$ within the range $0 \leq \lambda \leq 1$. [The forthcoming elaboration is general in nature and it applies, apart details, to the three-dimensional crystals, too. In practice, the integration over the three-dimensional Brillouin zone is carried out performing integrations over one variable (say k_z), once a number of discretized "special points" are chosen for the other two variables k_x, k_y.]

The change of electronic polarization is

$$\Delta P_{el} = -\frac{e}{\pi} \int_0^1 d\lambda \int_{-\pi/a}^{+\pi/a} 2\,\text{Im}\,\langle \frac{\partial}{\partial k} u_n(k, x; \lambda) | \frac{\partial}{\partial \lambda} u_n(k, x; \lambda)\rangle dk. \qquad (8.67)$$

The integrand in Eq. (8.67), exploiting the identity $2\,\text{Im}\,z = \text{Im}\,z - \text{Im}\,z^*$, can be elaborated as follows

$$2\,\text{Im}\,\langle \frac{\partial}{\partial k} u_n(k, x; \lambda) | \frac{\partial}{\partial \lambda} u_n(k, x; \lambda)\rangle$$

$$= \text{Im}\left[\langle \frac{\partial}{\partial k} u_n(k, x; \lambda) | \frac{\partial}{\partial \lambda} u_n(k, x; \lambda)\rangle - \langle \frac{\partial}{\partial \lambda} u_n(k, x; \lambda) | \frac{\partial}{\partial k} u_n(k, x; \lambda)\rangle \right]$$

$$= \text{Im}\left[\frac{\partial}{\partial k}\langle u_n(k, x; \lambda) | \frac{\partial}{\partial \lambda} u_n(k, x; \lambda)\rangle - \frac{\partial}{\partial \lambda}\langle u_n(k, x; \lambda) | \frac{\partial}{\partial k} u_n(k, x; \lambda)\rangle \right]$$

$$= \text{Im}\,\nabla \times \langle u_n(k, x; \lambda) | \nabla u_n(k, x; \lambda)\rangle,$$

where the gradient operator is respect to the variables (k, λ) and the quantity $\nabla \times \mathbf{A}$ denotes the component $\partial A_\lambda / \partial k - \partial A_k / \partial \lambda$. Equation (8.67) then becomes

$$\Delta P_{el} = -\frac{e}{\pi}\,\text{Im} \int_0^1 d\lambda \int_{-\pi/a}^{+\pi/a} \nabla \times \langle u_n(k, x; \lambda) | \nabla u_n(k, x; \lambda)\rangle dk\,d\lambda$$

$$= -\frac{e}{\pi}\,\text{Im} \oint_C \langle u_n(k, x; \lambda) | \nabla u_n(k, x; \lambda)\rangle \cdot d\mathbf{l}, \qquad (8.68)$$

where the integration in the (k, λ) space has been reduced, by Stokes circuitation theorem, to the evaluation of a line integral along the circuit indicated in Figure 8.13.

In normal insulators, the two contributions to the line integral along the λ segments at $k = \pi/a$ and $k = -\pi/a$ give zero; in fact

$$\psi_n\left(\frac{\pi}{a}, x; \lambda\right) \equiv \psi_n\left(-\frac{\pi}{a}, x; \lambda\right) \Longrightarrow e^{i(\pi/a)x} u_n\left(\frac{\pi}{a}, x; \lambda\right)$$

$$\equiv e^{-i(\pi/a)x} u_n\left(-\frac{\pi}{a}, x; \lambda\right);$$

this shows that $u_n(\pm\pi/a, x; \lambda)$ differ simply by a phase factor independent of λ parameter. Integration of Eq. (8.68) gives the desired result

$$\Delta P_{el} = P_{el}(1) - P_{el}(0), \qquad (8.69)$$

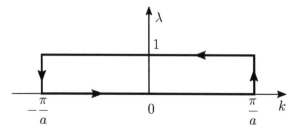

Figure 8.13 Integration circuit for the calculation of electronic polarization with the Berry phase formalism.

where

$$P_{el}(\lambda) = \frac{e}{\pi} \, \text{Im} \int_{-\pi/a}^{+\pi/a} \langle u_n(k, x; \lambda) | \nabla_k | u_n(k, x; \lambda) \rangle \, dk. \tag{8.70}$$

The same procedure used to pass from Eqs. (8.66) to (8.70) can be implemented to pass from Eqs. (8.62) to (8.64).

Appendix A. Simplified Evaluation of Typical Jahn-Teller and Renner-Teller Matrices

The purpose of this Appendix is to provide the structure of electron-phonon interaction matrices in paradigmatic electronic-nuclear systems using a simple and intuitive approach, which does not require the knowledge of the abstract group theory; the present treatment can be easily followed also by readers non yet familiar with the quantum theory of symmetries. The approach is adapted from the historical Slater-Koster technique to express two-center integrals in terms of independent parameters, routinely used in the tight-binding band theory of crystals. In conjunction with series developments appropriate for small nuclear displacements around the equilibrium situation, linear (or higher order) coupling matrices can be worked out.

Model Calculation of the Electron-Phonon Interaction Matrices within a Localized Orbital Framework

Consider a doublet of electronic orbitals (f_x, f_y) with p_x-like and p_y-like symmetry localized around some reference atom taken at the origin. Consider now the perturbation produced on the doublet by an atomic-like spherically symmetric potential $V_a(\mathbf{r} - \mathbf{R})$ centered at some atomic position $\mathbf{R}$. The matrix elements of the perturbing potential on the electronic doublet can be expressed in terms of independent parameters, following the same techniques introduced by Slater and Koster, and explained for the tight-binding method (see Section 5.2.1); it holds

$$\int f_x^*(\mathbf{r}) V_a(\mathbf{r} - \mathbf{R}) f_x(\mathbf{r}) \, d\mathbf{r} = l_x^2 I(pp\sigma, R) + (1 - l_x^2) I(pp\pi, R) \equiv l_x^2 f(R) + g(R)$$

$$\int f_x^*(\mathbf{r}) V_a(\mathbf{r} - \mathbf{R}) f_y(\mathbf{r}) \, d\mathbf{r} = l_x l_y [I(pp\sigma, R) - I(pp\pi, R)] \equiv l_x l_y f(R)$$

$$\int f_y^*(\mathbf{r}) V_a(\mathbf{r} - \mathbf{R}) f_y(\mathbf{r}) \, d\mathbf{r} = l_y^2 I(pp\sigma, R) + (1 - l_y^2) I(pp\pi, R) \equiv l_y^2 f(R) + g(R)$$

$$\tag{A.1}$$

In Eqs. (A.1), f_x and f_y are p_x-like and p_y-like (real) functions centered at the origin; $V_a(\mathbf{r} - \mathbf{R}) = V_a(|\mathbf{r} - \mathbf{R}|)$ is a localized spherically symmetric potential centered in the position individuated by the vector $\mathbf{R}$, with director cosines l_x, l_y, l_z. The only two independent integrals, for a given distance R, are $I(pp\sigma, R)$ and $I(pp\pi, R)$; their difference is indicated with $f(R)$, and $g(R)$ stands for $I(pp\pi, R)$. According to Eqs. (A.1), the perturbation produced by $V_a(\mathbf{r} - \mathbf{R})$ on the electronic doublet (omitting for simplicity the $g(R)$ term which does not produce any splitting) can be summarized in the form

$$U(\mathbf{R}) = f(R) \begin{pmatrix} l_x^2 & l_x l_y \\ l_y l_x & l_y^2 \end{pmatrix} \qquad \mathbf{R} \equiv R(l_x, l_y, l_x) . \tag{A.2}$$

For the electronic doublet in consideration, the 2×2 perturbation matrix $U(\mathbf{R})$ is real, symmetric, and essentially controlled by the director cosines of the actual atomic center.

With a similar treatment, we can consider a triplet of electronic orbitals (f_x, f_y, f_z); the perturbing potential of a spherically symmetric atomic-like potential centered at position $\mathbf{R}$ is represented by the matrix

$$U(\mathbf{R}) = f(R) \begin{pmatrix} l_x^2 & l_x l_y & l_x l_z \\ l_y l_x & l_y^2 & l_y l_z \\ l_z l_x & l_z l_y & l_z^2 \end{pmatrix} \qquad \mathbf{R} \equiv R(l_x, l_y, l_x) . \tag{A.3}$$

We can now apply the above results to obtain the structure of the perturbation produced by typical triangular or octahedral cages of atoms surrounding a doublet or a triplet of electronic states.

Evaluation of the Jahn-Teller Matrices for the $E \otimes \varepsilon$ System up to Second Order in the Normal Mode Displacements

We begin with the study of the electron-phonon coupling matrices for the widely studied $E \otimes \varepsilon$ system (see Section 8.3.2), and then the same procedure is repeated and applied to other systems of interest.

The purpose of this elaboration is to show that the structure of the Jahn-Teller electron-phonon coupling matrix for a regular triangular molecule up to quadratic terms in the normal coordinates q_1 and q_2 has the expression

$$U(q_1, q_2) = \gamma \begin{bmatrix} -q_1 & q_2 \\ q_2 & q_1 \end{bmatrix} + g \begin{bmatrix} q_1^2 - q_2^2 & 2q_1 q_2 \\ 2q_1 q_2 & -q_1^2 + q_2^2 \end{bmatrix} , \tag{A.4}$$

where γ and g are real parameters that depend on the specific system under attention, while the two matrices, linear and quadratic in the normal coordinates, are universal for any $E \otimes \varepsilon$ Jahn-Teller system. Higher order terms could also be obtained by inspection, following and extending the procedure given below.

Consider the regular triangular cage and the normal-mode distorted geometries pictured in Figure 8.4 and summarized in Table 8.1. For convenience the coordinates of the three atomic centers $\mathbf{R}_1, \mathbf{R}_2, \mathbf{R}_3$ displaced from the symmetrical positions are explicitly reported in Table 8.4, together with the director cosines and the distances from the origin, up to linear terms in the normal coordinates. Using Eq. (A.2), we can calculate the perturbative matrix on the electronic doublet produced by a spherically symmetric potential centered on each of the three atoms. The coupling matrix produced by the triangular cage is taken as the sum of equal spherically symmetric potentials centered in the positions $\mathbf{R}_1, \mathbf{R}_2, \mathbf{R}_3$.

Table 8.4 Displaced nuclear positions $\mathbf{R}_i(q_1, q_2)$ ($i = 1, 2, 3$) versus normal displacements modes q_1, q_2 of the vibrational mode ε in the equilateral triangle molecule. The director cosines of $\mathbf{R}_i$ and the distance $|\mathbf{R}_i|$ are also reported to first order in the normal coordinates.

$$\mathbf{R}_1(q_1, q_2) = \mathbf{d}_1 + \mathbf{u}_1(q_1) + \mathbf{u}_1(q_2) = \left(\frac{\sqrt{3}}{3}q_2, a + \frac{\sqrt{3}}{3}q_1 \right)$$

$$l_x^2 \approx 0; \; l_x l_y \approx \frac{\sqrt{3}}{3a}\frac{q_2}{3a}; \; l_y^2 \approx 1; \; |\mathbf{R}_1| \approx a + \frac{\sqrt{3}}{3}q_1$$

$$\mathbf{R}_2(q_1, q_2) = \mathbf{d}_2 + \mathbf{u}_2(q_1) + \mathbf{u}_2(q_2) = \left(-\frac{\sqrt{3}}{2}a + \frac{1}{2}q_1 - \frac{\sqrt{3}}{6}q_2, -\frac{1}{2}a - \frac{\sqrt{3}}{6}q_1 - \frac{1}{2}q_2 \right)$$

$$l_x^2 \approx \frac{3}{4} - \frac{\sqrt{3}q_1}{4a} - \frac{q_2}{4a} \approx 1 - l_y^2; \; l_x l_y \approx \frac{\sqrt{3}}{4} + \frac{q_1}{4a} + \frac{\sqrt{3}q_2}{12a}; \; |\mathbf{R}_2| \approx a - \frac{\sqrt{3}}{6}q_1 + \frac{1}{2}q_2$$

$$\mathbf{R}_3(q_1, q_2) = \mathbf{d}_2 + \mathbf{u}_3(q_1) + \mathbf{u}_3(q_2) = \left(\frac{\sqrt{3}}{2}a - \frac{1}{2}q_1 - \frac{\sqrt{3}}{6}q_2, -\frac{1}{2}a - \frac{\sqrt{3}}{6}q_1 + \frac{1}{2}q_2 \right)$$

$$l_x^2 \approx \frac{3}{4} - \frac{\sqrt{3}q_1}{4a} + \frac{q_2}{4a} \approx 1 - l_y^2; \; l_x l_y \approx -\frac{\sqrt{3}}{4} - \frac{q_1}{4a} + \frac{\sqrt{3}q_2}{12a}; \; |\mathbf{R}_3| \approx a - \frac{\sqrt{3}}{6}q_1 - \frac{1}{2}q_2$$

The perturbation matrix produced on the electronic doublet by a spherically symmetric potential centered on the atom at $\mathbf{R}_1$ gives

$$U(\mathbf{R}_1) = f\left(a + \frac{\sqrt{3}}{3}q_1\right) \begin{bmatrix} 0 & \dfrac{\sqrt{3}q_2}{3a} \\ \dfrac{\sqrt{3}q_2}{3a} & 1 \end{bmatrix} \tag{A.5}$$

The perturbation matrix on the doublet produced by the atom at $\mathbf{R}_2$ is

$$U(\mathbf{R}_2) = f\left(a - \frac{\sqrt{3}}{6}q_1 + \frac{1}{2}q_2\right) \begin{bmatrix} \dfrac{3}{4} - \dfrac{\sqrt{3}\,q_1}{4a} - \dfrac{q_2}{4a} & \dfrac{\sqrt{3}}{4} + \dfrac{q_1}{4a} + \dfrac{\sqrt{3}q_2}{12a} \\ \dfrac{\sqrt{3}}{4} + \dfrac{q_1}{4a} + \dfrac{\sqrt{3}q_2}{12a} & \dfrac{1}{4} + \dfrac{\sqrt{3}q_1}{4a} + \dfrac{q_2}{4a} \end{bmatrix} \tag{A.6}$$

The perturbation matrix due to the atom at $\mathbf{R}_3$ gives

$$U(\mathbf{R}_3) = f\left(a - \frac{\sqrt{3}}{6}q_1 - \frac{1}{2}q_2\right) \begin{bmatrix} \dfrac{3}{4} - \dfrac{\sqrt{3}\,q_1}{4a} + \dfrac{q_2}{4a} & -\dfrac{\sqrt{3}}{4} - \dfrac{q_1}{4a} + \dfrac{\sqrt{3}q_2}{12a} \\ -\dfrac{\sqrt{3}}{4} - \dfrac{q_1}{4a} + \dfrac{\sqrt{3}q_2}{12a} & \dfrac{1}{4} + \dfrac{\sqrt{3}q_1}{4a} - \dfrac{q_2}{4a} \end{bmatrix} \tag{A.7}$$

In Eq. (A.5) consider the series expansion

$$f\left(a + \frac{\sqrt{3}}{3}q_1\right) = f(a) + \frac{\sqrt{3}}{3}q_1 f'(a) + \cdots$$

and similarly for the other functions in Eq. (A.6) and Eq. (A.7). Summing up the perturbations provided by Eqs. (A5–A7), we obtain for the linear and quadratic Jahn-Teller matrices the expressions reported in Eq. (A.4), with

$$\gamma = \frac{\sqrt{3}}{2a} f(a) + \frac{\sqrt{3}}{4} f'(a) \qquad \text{and} \qquad g = \frac{1}{4a} f'(a).$$

The values of the above parameters are, of course, model dependent; on the contrary, the general structure of the matrices in Eq. (A.4) is only determined by symmetry, and could be worked out with formal group theory, regardless reference to any specific model.

Evaluation of the Linear Jahn-Teller Interaction Matrix for the $T \otimes \varepsilon$ System

We pass now to the elaboration of the electron-phonon coupling matrix for the $T \otimes \varepsilon$ system (see Section 8.3.3), repeating mutatis mutandis the same procedure used so far. The $T \otimes \varepsilon$ system describes an electronic triplet interacting with a doublet of vibrational modes in cubic symmetry, as pictured in Figure 8.6.

The purpose of this elaboration is to show that the structure of the Jahn-Teller matrix, linear in the normal mode coordinates q_1 and q_2, takes the expression

$$U(q_1, q_2) = \gamma \begin{pmatrix} q_1 - q_2\sqrt{3}/3 & 0 & 0 \\ 0 & -q_1 - q_2\sqrt{3}/3 & 0 \\ 0 & 0 & +q_2 2\sqrt{3}/3 \end{pmatrix}, \qquad \text{(A.8)}$$

where γ is a real parameter that depends on the specific system, while the matrix structure is characteristic of any $T \otimes \varepsilon$ Jahn-Teller system. Higher order terms, if needed, could also be obtained extending the procedure given below.

Consider the perturbation matrix produced on the electron doublet by each of the six atoms of the surrounding octahedral cage. For instance for the atom along the positive x-axis, the position vector, and the corresponding director cosines are

$$\mathbf{R}_1 = \left(a + \frac{1}{2}q_1 - \frac{\sqrt{3}}{6}q_2, 0, 0 \right)$$

$$|\mathbf{R}_1| = a + \frac{1}{2}q_1 - \frac{\sqrt{3}}{6}q_2 \qquad l_\alpha l_\beta = 0 \qquad \text{except } l_x^2 = 1.$$

Using Eq. (A.3), the perturbation matrix reads

$$U(\mathbf{R}_1) = f\left(a + \frac{1}{2}q_1 - \frac{\sqrt{3}}{6}q_2 \right) \begin{pmatrix} 1 & 0 & 0 \\ 0 & 0 & 0 \\ 0 & 0 & 0 \end{pmatrix}. \qquad \text{(A.9)}$$

After linearization of the f function and summation on all six perturbation matrices, one obtains for the linear Jahn-Teller matrix the expression reported in Eq. (A.8), with $\gamma = f'(a)$.

Evaluation of the Linear Jahn-Teller Interaction Matrix for the $T \otimes \tau$ System

The $T \otimes \tau$ vibronic system describes an electronic triplet interacting with a triplet of vibrational modes q_1, q_2, q_3 (see Figure 8.8). The purpose of this elaboration is to show that the structure of the Jahn-Teller matrix, linear in the normal modes coordinates, has the expression

$$U(q_1, q_2, q_3) = \gamma \begin{pmatrix} 0 & q_3 & q_2 \\ q_3 & 0 & q_1 \\ q_2 & q_1 & 0 \end{pmatrix}, \tag{A.10}$$

where γ is a real parameter that depends on the specific system, while the matrix structure is characteristic of any $E \otimes \tau$ Jahn-Teller system.

Consider the perturbation matrix produced on the electron triplet by each of the six atoms of the surrounding octahedral cage. For instance for the atom along the positive x-axis, the nuclear position vector, its modulus and its director cosines (at the lowest order in the normal mode coordinates) are

$$\mathbf{R}_1 = \left(a, \frac{1}{2}q_3, \frac{1}{2}q_2 \right), \quad |\mathbf{R}_1| = a + \frac{1}{2}q_2 + \frac{1}{2}q_3, \quad l_x = 1, \quad l_y = \frac{q_3}{2a}, \quad l_z = \frac{q_2}{2a}.$$

Using Eq. (A.3) the perturbation matrix, linearized in the normal mode coordinates, reads

$$U(\mathbf{R}_1) = f(a) \begin{pmatrix} 1 & q_3/2a & q_2/2a \\ q_3/2a & 0 & 0 \\ q_2/2a & 0 & 0 \end{pmatrix}. \tag{A.11}$$

After summation on all six perturbation matrices (and disregarding irrelevant scalar quantities, which do not produce splitting of states but only their rigid shift), one obtains for the linear Jahn-Teller matrix the expression reported in Eq. (A.10), with $\gamma = 2f(a)/a$.

Evaluation of the Typical Quadratic Renner-Teller Interaction Matrix in Linear Molecules

Given a linear molecule with atoms lined up in the z-direction, consider the perturbation matrix on the doublet of states (f_x, f_y) produced by an atomic center in vibration perpendicularly to the z-direction (as schematically indicated in Figure 8.14). In this geometry, the structure of the Renner-Teller interaction matrix is quadratic in the normal mode coordinates q_x, q_y, and is given by

$$U(q_x, q_y) = \gamma \begin{bmatrix} q_x^2 - q_y^2 & 2q_x q_y \\ 2q_x q_y & -q_x^2 + q_y^2 \end{bmatrix}, \tag{A.12}$$

where γ is a real parameter that depends on the specific molecule, while the matrix structure is characteristic of the Renner-Teller vibronic system involving an electronic doublet with $l = 1$ angular momentum.

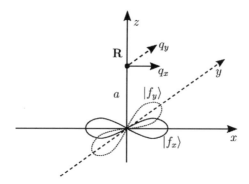

Figure 8.14 Geometry of the linear molecule and nuclear displacement of a given atom corresponding to the degenerate (q_x, q_y) vibrational mode. The doublet of electronic states (f_x, f_y) centered at the origin is also indicated.

According to Figure 8.14, the vector position and director cosines of a nucleus, along the z-axis at the equilibrium distance a from the origin, reads

$$\mathbf{R} = (q_x, q_y, a); \quad l_x^2 \approx \frac{q_x^2}{a^2}, \quad l_y^2 \approx \frac{q_y^2}{a^2}, \quad l_x l_y \approx \frac{q_x q_y}{a^2}; \quad |\mathbf{R}| \approx a + \frac{q_x^2}{2a} + \frac{q_y^2}{2a},$$

where expansion up to quadratic terms has been considered. Using Eq. (A.2), the interaction matrix quadratic in the normal coordinates becomes

$$U(\mathbf{R}) = f(a) \frac{1}{a^2} \begin{bmatrix} q_x^2 & q_x q_y \\ q_x q_y & q_y^2 \end{bmatrix}. \tag{A.13}$$

The trace of the matrix in Eq. (A.13) is $q_x^2 + q_y^2$; to obtain a zero trace matrix, we subtract $(1/2)(q_x^2 + q_y^2)$ (multiplied by the unit matrix). Disregarding the irrelevant scalar quantity (that does not produce any splitting), we recover the Renner-Teller matrix reported in Eq. (A.12), with the value $\gamma = f(a)/2a^2$ in the considered model.

The general form of Renner-Teller interaction matrices can be obtained at any order of the phonon variables and with electron orbital functions of different symmetry (p-like, d-like, f-like etc.). [See for instance G. Bevilacqua, L. Martinelli and G. Pastori Parravicini "Renner-Teller interaction matrices and Green's function formalism" in Advances in Quantum Chemistry, edited by J. R. Sabin and E. Brändas, vol. 44, p.45 (Elsevier, Amsterdam 2003)]. Also notice that the matrix in Eq. (A.12) and the second matrix in Eq. (A.4) have the same structure, essentially for symmetry reasons.

According to the Jahn-Teller theorem, any orbitally degenerate electronic state in a symmetric environment of surrounding nuclei (except the collinear case) is unstable. [We disregard here for simplicity the effect of spin, which is often insignificant for the present considerations]. Linear molecules alone constitute the exception: the exception can be easily understood with elementary angular momentum considerations. A nucleus moving off z-axis of the molecule (see Figure 8.14), taking for convenience as variables $q_x + iq_y$ and $q_x - iq_y$, produces a perturbation characterized by angular momentum

$l_z = \pm 1$. A twofold degenerate state (f_x, f_y), or equivalently $(f_x + if_y, f_x - if_y)$, constitute a doublet of states with angular momentum $\Lambda_z = \pm 1$. Non-vanishing perturbative matrix on the doublet requires a perturbation with angular momentum at least equal to 2. Thus the interaction matrix of the Renner-Teller system must be at least quadratic in the normal coordinates. This entails that the linear position of the nuclei in the molecule is not necessarily unstable for general symmetry reasons (although it may be unstable in specific situations depending on the type and strengths of the bonds formation).

It is instructive to consider the adiabatic potential-energy surfaces of the vibronic Renner-Teller system described by the matrix Hamiltonian

$$H_{RT} = \gamma \begin{bmatrix} q_x^2 - q_y^2 & 2q_x q_y \\ 2q_x q_y & -q_x^2 + q_y^2 \end{bmatrix} + \frac{1}{2} C(q_x^2 + q_y^2) \tag{A.14}$$

[as usual when a scalar quantity, such as the elastic energy $(1/2)C(q_x^2 + q_y^2)$, is added to a matrix it is automatically understood that the scalar quantity is preliminary multiplied by the unit matrix.] Diagonalization of Eq. (A.14) gives for the adiabatic energy surfaces

$$W = \left(\pm \gamma + \frac{1}{2} C \right) q^2 \qquad \text{with} \qquad q^2 = q_x^2 + q_y^2.$$

These surfaces have an extremum at the point of degeneracy, with no instability if $\gamma < C/2$ and with instability for $\gamma > C/2$.

Appendix B.   Solved Problems and Complements

Problem 1. Berry phase of the $E \otimes \varepsilon$ Jahn-Teller system.
Problem 2. Berry phase of the Hamiltonian of Dirac carriers in honeycomb materials.
Problem 3. Berry magnetic monopole and geometrical phase factors.
Problem 4. Berry phase of a typical Renner-Teller coupling matrix.
Problem 5. Berry phase in honeycomb bilayers.

Problem 1. Berry Phase of the $E \otimes \varepsilon$ Jahn-Teller System

Consider the parametric matrix M of rank two given by

$$M = \begin{bmatrix} -q_1 & q_2 \\ q_2 & q_1 \end{bmatrix} = -q_1 |f_x\rangle\langle f_x| + q_2 |f_x\rangle\langle f_y| + q_2 |f_y\rangle\langle f_x| + q_1 |f_y\rangle\langle f_y|, \tag{B.1}$$

where the parameters q_1 and q_2 are real quantities. Determine the eigenvalues, the eigenfunctions and the Berry phase for circuits enclosing the origin of the parameter space.

The matrix M represents the basic structure of the linear electron-lattice interaction in the $E \otimes \varepsilon$ Jahn-Teller system; f_x, f_y denote the two orthonormal p-like basis functions of the electronic doublet (or any other doublet with the same symmetry properties). The eigenvalues of the matrix are obtained solving the determinantal equation

$$||M - E|| = \begin{Vmatrix} -q_1 - E & q_2 \\ q_2 & q_1 - E \end{Vmatrix} = -q_1^2 + E^2 - q_2^2 = 0.$$

The eigenvalues are

$$E = \mp\sqrt{q_1^2 + q_2^2} = \mp q,$$

where q denotes the modulus of the distance of the point (q_1, q_2) from the origin. The eigenvector corresponding to the lowest eigenvalue is obtained by the matrix equation

$$\begin{bmatrix} -q_1 & q_2 \\ q_2 & q_1 \end{bmatrix} \begin{bmatrix} C_1 \\ C_2 \end{bmatrix} = -q \begin{bmatrix} C_1 \\ C_2 \end{bmatrix}.$$

Performing the standard row by column product, we obtain two (redundant) equations

$$\begin{aligned} -q_1 C_1 + q_2 C_2 &= -q\, C_1, \\ q_2 C_1 + q_1 C_2 &= -q\, C_2 \end{aligned} \tag{B.2}$$

It is slightly more convenient (although not necessary) to multiply the first of the above equations by $(-i)$ and sum up to the second. We obtain

$$i(q_1 - iq_2 - q)C_1 = (-q_1 + iq_2 - q)C_2. \tag{B.3}$$

The above elaboration is suggested by the convenience to handle quantities of the type $q_1 \pm iq_2$, rather than q_1 or q_2 separately.

Using polar coordinates in the (q_1, q_2) plane, we have

$$\frac{C_1}{C_2} = \frac{1}{i}\frac{+e^{-i\theta}+1}{-e^{-i\theta}+1}.$$

The above ratio is well behaved in all the parameter space (q_1, q_2), except at the origin where the two adiabatic surfaces are degenerate. The not yet normalized coefficients can be taken as proportional to the numerator and denominator respectively:

$$\begin{cases} C_1 \approx e^{-i\theta} + 1 \\ C_2 \approx i(-e^{-i\theta} + 1) \end{cases} \tag{B.4}$$

After enforcing the normalization, the expression of the normalized coefficients becomes

$$\begin{cases} C_1(\theta) = \dfrac{1}{2}(+e^{-i\theta} + 1) = +e^{-i\theta/2}\cos(\theta/2) \\ C_2(\theta) = \dfrac{i}{2}(-e^{-i\theta} + 1) = -e^{-i\theta/2}\sin(\theta/2) \end{cases} \tag{B.5}$$

It is important to notice that, without altering anything else, the amplitudes $C_1(\theta)$ and $C_2(\theta)$ can be simultaneously multiplied by a phase factor of modulus unity. In particular we can consider gauge transformations of the type

$$C_1(\theta), C_2(\theta) \Longrightarrow C_1(\theta)e^{im\theta}, C_2(\theta)e^{im\theta} \quad (m \text{ integer}); \tag{B.6}$$

the only restriction on the gauge freedom (B.6) is m integer number, in order to preserve the single-valuedness property of the eigenfunctions.

A similar treatment can be done for the upper sheet, simply replacing $-q$ with $+q$. We can choose the eigenfunctions of the studied matrix in the form

$$\begin{cases} |\psi_1\rangle = e^{-i\theta/2} \cos(\theta/2) |f_x\rangle - e^{-i\theta/2} \sin(\theta/2) |f_y\rangle & E_1 = -q \\ |\psi_2\rangle = e^{+i\theta/2} \sin(\theta/2) |f_x\rangle + e^{+i\theta/2} \cos(\theta/2) |f_y\rangle & E_2 = +q \end{cases}. \tag{B.7}$$

Equations (B.7) constitute a possible choice of one-valued wavefunctions among all the possible gauge transformations of type (B.6).

For the calculation of the geometrical vector potential, and then of the Berry phase, we need the matrix elements of the gradient operator

$$\mathbf{V} = \mathbf{e}_q \frac{\partial}{\partial q} + \mathbf{e}_\theta \frac{1}{q} \frac{\partial}{\partial \theta}$$

on the eigenfunctions (B.7). We have

$$\langle \psi_1 | \frac{\partial}{\partial \theta} | \psi_1 \rangle = e^{+i\theta/2} \cos \frac{\theta}{2} \cdot \frac{\partial}{\partial \theta} \left[e^{-i\theta/2} \cos \frac{\theta}{2} \right]$$
$$+ e^{+i\theta/2} \sin \frac{\theta}{2} \cdot \frac{\partial}{\partial \theta} \left[e^{-i\theta/2} \sin \frac{\theta}{2} \right] = -\frac{i}{2};$$

and similarly

$$\langle \psi_2 | \frac{\partial}{\partial \theta} | \psi_2 \rangle = +\frac{i}{2}.$$

The *fictitious vector potential* or *Berry connection field*, on the lower energy adiabatic sheet becomes

$$\mathbf{A}_1(q, \theta) = i \langle \psi_1 | \mathbf{V} | \psi_1 \rangle = i \frac{1}{q} \langle \psi_1 | \frac{\partial}{\partial \theta} | \psi_1 \rangle \mathbf{e}_\theta = +\frac{1}{2q} \mathbf{e}_\theta = +\pi \frac{1}{2\pi q} \mathbf{e}_\theta.$$

Similarly for the upper sheet one finds

$$\mathbf{A}_2(q, \theta) = i \langle \psi_2 | \mathbf{V} | \psi_2 \rangle = -\pi \frac{1}{2\pi q} \mathbf{e}_\theta.$$

The Berry phase $\gamma_1(\mathcal{C})$ and $\gamma_2(\mathcal{C})$ on a circuit that surrounds the origin in the lower or upper sheet become

$$\gamma_1(\mathcal{C}) = +\pi \quad (\text{modulus } 2\pi), \quad \gamma_2(\mathcal{C}) = -\pi \quad (\text{modulus } 2\pi). \tag{B.8}$$

[the modulus 2π uncertainty has been added to take into account the gauge freedom of type (B.6)]. Equation (B.8) shows that the Berry phase can be interpreted as originated from an infinitesimal flux tube, located at the origin of the parameter space, carrying a finite magnetic flux of value $\pm\pi$ (modulus 2π). The flux tube strength changes sign if complex conjugate operation (i.e. time-reversal symmetry) is applied to wavefunctions (B.7).

According to Eq. (B.8), the Berry phase acquired by the wavefunctions (B.7) transported around the origin is equal to $\pm\pi$, within integer multiples of 2π. To better understand this uncertainty, which is due to the presence of the degeneracy point at $q = 0$ and the entailed gauge freedom summarized in Eq. (B.6), it is instructive to remove the "offending" degeneracy point with a finite or an infinitesimally small perturbation. To be specific, besides studying the $E \otimes \varepsilon$ Jahn-Teller matrix (B.1) it is rewarding to study the pseudo-Jahn-Teller matrix

$$P = \begin{bmatrix} -q_1 & q_2 + i\Delta \\ q_2 - i\Delta & q_1, \end{bmatrix}, \tag{B.9}$$

where the real parameter Δ controls the splitting of the two adiabatic surfaces (the splitting can be produced for instance by Zeeman effect due to an external magnetic field, a perturbation that removes time-reversal symmetry). The adiabatic surfaces of the P-matrix are never degenerate, the wavefunctions can be chosen regular in the whole (q_1, q_2) plane included the origin, and the limit of vanishing Δ can be discussed. When this procedure is pursued, it becomes apparent that the degeneracy point has the effect to produce a localized infinitesimal flux tube, of phase $\pm\pi$, the only (minor) arbitrariness being the direction of the induced fictitious magnetic field (which is controlled by the sign of Δ, as $\Delta \to 0$). This is shown in detail in Problem 2.

Before passing to the next problem, it is worthwhile to point out that the treatment of a given parametric matrix can, of course, be automatically extended to any other matrix related by a unitary transformation. For example, the electron-phonon coupling matrix $M(q_1, q_2))$ of Eq. (B.1) is expressed on the set of the electron doublet (f_x, f_y). It is instructive to express the same matrix on the electronic functions of definite angular momentum given by

$$\phi_1 = \frac{1}{\sqrt{2}}(-f_x + if_y), \quad \phi_2 = \frac{1}{\sqrt{2}}(f_x + if_y). \tag{B.10}$$

The matrix M of Eq. (B.1) in the old and in the new basis functions takes the form

$$\begin{bmatrix} -q_1 & q_2 \\ q_2 & q_1 \end{bmatrix} \text{ set } (f_x, f_y) \implies \begin{bmatrix} 0 & q_1 - iq_2 \\ q_1 + iq_2 & 0 \end{bmatrix} \text{ set } (\phi_1, \phi_2). \tag{B.11}$$

The second matrix of expression (B.11) can be recognized as the matrix describing the structure of the low energy electronic excitations at the Dirac points of graphene (see Section 6.5). Thus the study of the Berry phase of the $E \otimes \varepsilon$ Jahn-Teller system and the Berry phase of graphene are essentially equivalent. A fictitious Berry flux tube of strength $\pm\pi$ occurs at the degeneracy points in both systems, and some properties of these systems, so different phenomenologically, can be better appreciated within this unifying and universal framework. A direct manifestation of the non-trivial Berry phase of π in the $E \otimes \varepsilon$ Jahn-Teller system is the twofold degeneracy of the ground vibronic levels and the entailed effects on optical properties [see Section 8.3.2; see also for instance G. Bevilacqua, L. Martinelli and G. Pastori Parravicini "Effect of a magnetic field on an $E \otimes \varepsilon$ Jahn-Teller system: Berry phase and optical properties" Phys. Rev. B 63, 132403 (2001), and references quoted therein]. In a quite different phenomenological context, a direct manifestation of the non-trivial Berry phase of π in the graphene wavefunctions is the "non-conventional" quantum Hall conductance across the neutrality region. This is just an example of how apparently different effects in different materials, are linked together by the same Berry phase of the describing Hamiltonians.

Problem 2. Berry Phase of the Hamiltonian of Massless and Massive Dirac Carriers in Graphene and Gapped Honeycomb Materials

Determine the eigenvalues, the eigenfunctions, and the Berry phase of the parametric Hamiltonian given by

$$H = \begin{bmatrix} \Delta & q_x - iq_y \\ q_x + iq_y & -\Delta \end{bmatrix}, \tag{B.12}$$

where the parameters q_x, q_y, Δ are real. The quantity Δ controls the energy gap between the two adiabatic surfaces of H. Show that in the limiting case $\Delta \to 0^+$, the Berry phases of the lower and upper adiabatic surfaces take the values $+\pi$ and $-\pi$, respectively, without modulus 2π uncertainty. For $\Delta \to 0^-$ the sign of the Berry phases is reversed.

In Section 6.5, the electronic band structures of two-dimensional graphene and boron nitride have been studied throughout the Brillouin zone. Here the attention is confined to the low energy electronic excitations and to the physical effects occurring in this energy region. In the study of massless Dirac electrons in graphene and massive Dirac particles in gapped materials with honeycomb topology we have encountered the following tight-binding Hamiltonian at the two Dirac points

$$H_1 = \begin{bmatrix} +\Delta & v_F\hbar(k_x - ik_y) \\ v_F\hbar(k_x + ik_y) & -\Delta \end{bmatrix}; \qquad H_2 = H_1^*$$

By setting $q_x = v_F\hbar k_x$ and $q_y = v_F\hbar k_y$, the matrix H_1 becomes equal to the matrix (B.12) under investigation. [The matrix H_2 equals the complex conjugate of H_1, and does not need an independent elaboration]. The matrix H describes the low energy excitations of graphene for $\Delta = 0$ and gapped graphene-like materials for $\Delta \neq 0$ (see Section 6.5). The matrix H of Eq. (B.12) is also isomorphic with the electron-lattice interaction matrix (B.9) of the $E \otimes \varepsilon$ pseudo-Jahn-Teller systems, once the matrix (B.9) is represented on the basis functions (B.10) of definite angular momentum.

In the following we consider in detail the case of finite and positive Δ, included the limit $\Delta \to 0^+$; the case of negative Δ, included the limit $\Delta \to 0^-$ can be elaborated in a similar way.

The eigenvalues of the H matrix are obtained solving the determinantal equation

$$\|H - E\| = \begin{Vmatrix} \Delta - E & q_x - iq_y \\ q_x + iq_y & -\Delta - E \end{Vmatrix} = -\Delta^2 + E^2 - (q_x^2 + q_y^2) = 0.$$

The eigenvalues are

$$E = \mp\sqrt{q^2 + \Delta^2},$$

and the adiabatic surfaces are never degenerate for finite values of Δ. The energy gap between the adiabatic surfaces is $2|\Delta|$ and occurs at $q = 0$.

The eigenfunction corresponding to the lowest eigenvalue is determined by the expression

$$
\begin{bmatrix} \Delta & q_x - iq_y \\ q_x + iq_y & -\Delta \end{bmatrix} \begin{bmatrix} C_1 \\ C_2 \end{bmatrix} = -\sqrt{q^2 + \Delta^2} \begin{bmatrix} C_1 \\ C_2 \end{bmatrix}.
$$

The standard row by column product provides two (redundant) equations; the first of them reads

$$
\Delta \cdot C_1 + (q_x - iq_y)C_2 = -\sqrt{q^2 + \Delta^2} \cdot C_1 \; ;
$$

it follows

$$
\frac{C_1}{C_2} = -\frac{q_x - iq_y}{\Delta + \sqrt{q^2 + \Delta^2}} = -\frac{qe^{-i\theta}}{\Delta + \sqrt{q^2 + \Delta^2}}.
$$

The not yet normalized coefficients can be taken as

$$
\begin{cases} C_1 \approx -qe^{-i\theta} \\ C_2 \approx \Delta + \sqrt{q^2 + \Delta^2} \end{cases}. \tag{B.13}
$$

For normalization purpose, we evaluate

$$
|C_1|^2 + |C_2|^2 = 2q^2 + 2\Delta^2 + 2\Delta\sqrt{q^2 + \Delta^2}.
$$

The expression of the normalized coefficients for the lower energy adiabatic surface, up to gauge transformations, becomes

$$
C_1(q, \theta) = \frac{1}{\sqrt{2}} \frac{-qe^{-i\theta}}{\sqrt{q^2 + \Delta^2 + \Delta\sqrt{q^2 + \Delta^2}}} \qquad C_2(q, \theta) = \frac{1}{\sqrt{2}} \frac{\Delta + \sqrt{q^2 + \Delta^2}}{\sqrt{q^2 + \Delta^2 + \Delta\sqrt{q^2 + \Delta^2}}}. \tag{B.14}
$$

For the upper adiabatic energy surface, we have just to change the sign of the eigenvalue, and up to gauge transformations, we can choose

$$
D_1(q, \theta) = \frac{1}{\sqrt{2}} \frac{-q}{\sqrt{q^2 + \Delta^2 - \Delta\sqrt{q^2 + \Delta^2}}} \qquad D_2(q, \theta) = \frac{1}{\sqrt{2}} \frac{(\Delta - \sqrt{q^2 + \Delta^2})e^{+i\theta}}{\sqrt{q^2 + \Delta^2 - \Delta\sqrt{q^2 + \Delta^2}}}. \tag{B.15}
$$

The wavefunctions and eigenvalues can be summarized as

$$
\boxed{
\begin{aligned}
|\psi_1\rangle &= C_1(q, \theta) |\phi_1\rangle + C_2(q, \theta) |\phi_2\rangle & E_1 &= -\sqrt{q^2 + \Delta^2} \\
|\psi_2\rangle &= D_1(q, \theta) |\phi_1\rangle + D_2(q, \theta) |\phi_2\rangle & E_2 &= +\sqrt{q^2 + \Delta^2}
\end{aligned}
} \tag{B.16}
$$

From Eqs. (B.14, B.15) we have for $\Delta > 0$ and $q \to 0$ the limits

$$
C_1(0, \theta) = 0, \quad C_2(0, \theta) = 1, \quad D_1(0, \theta) = -1, \quad D_2(0, \theta) = 0.
$$

Thus the wavefunctions (B.16) are regular on the whole parameter space; no gauge arbitrariness related to multiplication by $\exp(im\theta)$ (for any integer $m \neq 0$) is possible because it would destroy the continuity of the wavefunctions at $q \to 0$. [In the case

$\Delta < 0$, the same reasoning holds after multiplication of Eqs. (B.14) by $\exp(i\theta)$ and multiplication of Eqs. (B.15) by $\exp(-i\theta)$.]

We pass now to the study of the Berry phase with two equivalent procedures: (i) the contour integral of the *Berry connection using wavefunctions regular in all the parameter space* (ii) the surface integral of the fictitious magnetic field on a surface of chosen contour; with this powerful procedure the Berry phase *is completely independent of any chosen gauge, and even independent of single-valuedness of wavefunctions.* The two procedures give of course the same results, putting in evidence complementary aspects.

Berry phase with the contour integral of the Berry connection. We continue to examine explicitly the case $\Delta > 0$ (the case of negative gap parameter can be treated similarly). Consider the matrix elements of the gradient operator on the eigenfunctions (B.14–B.16), which are regular in the whole parameter space. A matrix element of interest is

$$\langle \psi_1 | \frac{\partial}{\partial \theta} | \psi_1 \rangle = C_1^*(q, \theta) \frac{\partial}{\partial \theta} C_1(q, \theta) = \frac{q^2}{q^2 + \Delta^2 + \Delta\sqrt{q^2 + \Delta^2}} \frac{-i}{2}.$$

We notice the trivial identity

$$\frac{q^2}{q^2 + \Delta^2 + \Delta\sqrt{q^2 + \Delta^2}} \equiv 1 - \frac{\Delta}{\sqrt{q^2 + \Delta^2}}.$$

It follows

$$\boxed{\langle \psi_1 | \frac{\partial}{\partial \theta} | \psi_1 \rangle = -\frac{i}{2} \left(1 - \frac{\Delta}{\sqrt{q^2 + \Delta^2}} \right)} \quad \text{and} \quad \boxed{\langle \psi_1 | \frac{\partial}{\partial q} | \psi_1 \rangle = 0} \, ;$$

the last relation vanishes because it can be cast in the form of the derivative of a constant.

The "fictitious vector potential" takes the expression

$$\mathbf{A}_1(q, \theta) = i \langle \psi_1 | \nabla | \psi_1 \rangle = \frac{1}{2q} \left[1 - \frac{\Delta}{\sqrt{q^2 + \Delta^2}} \right] \mathbf{e}_\theta.$$

The Berry phase $\gamma_1(q)$ on a circular circuit that surrounds the origin on the ground sheet, and a similar calculation of $\gamma_2(q)$ for the upper sheet, give

$$\gamma_1(q) = -\gamma_2(q) = +\pi \left[1 - \frac{\Delta}{\sqrt{q^2 + \Delta^2}} \right] \quad (\Delta > 0) \tag{B.17}$$

It can be seen that the Berry phase changes gradually from zero to $\pm\pi$ as the circuit is enlarged, for the lower and upper adiabatic surface, respectively. For $\Delta \to 0^+$ the phases become $\pm\pi$ for any circuit that encompasses the origin. The results are schematically reported in Figure 8.15; they complete and generalize the elaborations of Problem 1.

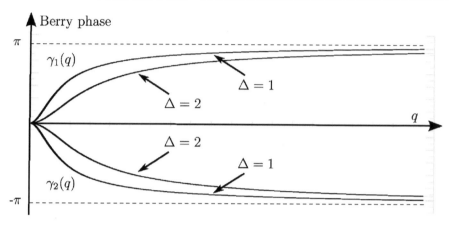

Figure 8.15 Typical behavior of the Berry phase $\gamma_1(q)$ for the ground sheet and $\gamma_2(q)$ for the upper sheet of the matrix H, that describes low energy excitations in gapped graphene or the interaction matrix in the pseudo $E \otimes \varepsilon$ Jahn-Teller system. In the figure the gap parameter Δ is taken as positive. If $\Delta < 0$, in the figure we must replace Δ with $|\Delta|$ while $\gamma_1(q)$ and $\gamma_2(q)$ are exchanged.

Berry phase with the surface integral of the fictitious magnetic field. In the main text we have seen that the fictitious magnetic field is given by Eq. (8.52). In the present case, which concerns two adiabatic surfaces (never degenerate) in a two-dimensional parameter space, the fictitious magnetic field is perpendicular to the (q_x, q_y) parameter space and its value on the lower adiabatic surface reads

$$\mathbf{B}_1(\mathbf{q}) = -\mathrm{Im}\left[\langle\psi_1| \frac{\partial H}{\partial q_x} |\psi_2\rangle\langle\psi_2| \frac{\partial H}{\partial q_y} |\psi_1\rangle - \langle\psi_1| \frac{\partial H}{\partial q_y} |\psi_2\rangle\langle\psi_2| \frac{\partial H}{\partial q_x} |\psi_1\rangle \right] \frac{1}{(E_{2\mathbf{q}} - E_{1\mathbf{q}})^2} \mathbf{e}_z.$$

$$(\mathrm{B}.18)$$

The remarkable expression (B.18) is clearly independent of the phases embodied in the wavefunctions. Furthermore it shows explicitly that

$$\mathbf{B}_1(\mathbf{q}) \equiv -\mathbf{B}_2(\mathbf{q}).$$

To evaluate the matrix elements entering into Eq. (B.18), consider for instance

$$\langle\psi_1| \frac{\partial H}{\partial q_x} |\psi_2\rangle = \begin{pmatrix} C_1^* & C_2^* \end{pmatrix} \begin{pmatrix} 0 & 1 \\ 1 & 0 \end{pmatrix} \begin{pmatrix} D_1 \\ D_2 \end{pmatrix} = C_1^* D_2 + C_2^* D_1.$$

With straight algebra we obtain

$$\langle\psi_1| \frac{\partial H}{\partial q_x} |\psi_2\rangle = \left[-\frac{\Delta}{\sqrt{q^2 + \Delta^2}} \cos\theta + i \sin\theta \right] e^{+i\theta} = \langle\psi_2| \frac{\partial H}{\partial q_x} |\psi_1\rangle^*$$

$$\langle\psi_1| \frac{\partial H}{\partial q_y} |\psi_2\rangle = \left[-\frac{\Delta}{\sqrt{q^2 + \Delta^2}} \sin\theta - i \cos\theta \right] e^{+i\theta} = \langle\psi_2| \frac{\partial H}{\partial q_y} |\psi_1\rangle^*$$

Inserting the above results in Eq. (B.18), it is seen that the fictitious magnetic field on the lower adiabatic sheet is given by

$$\mathbf{B}_1(\mathbf{q}) = \frac{1}{2} \frac{\Delta}{(q^2 + \Delta^2)^{3/2}} \, \mathbf{e}_z \,. \tag{B.19}$$

[The procedure closely follows the article by G. Bevilacqua et al., Phys. Rev. B *63*, 132403 (2001).]

Consider a circuit $\mathcal{C}$ and a surface S of contour $\mathcal{C}$ in the parameter domain. The Berry phase $\gamma_1(\mathcal{C})$ is given by the flux of $\mathbf{B}_1(\mathbf{q})$ across a surface of contour $\mathcal{C}$

$$\gamma_1(\mathcal{C}) = \int_S \mathbf{B}_1(\mathbf{q}) \cdot d\mathbf{S} \,. \tag{B.20}$$

For a circular circuit of radius q we have

$$\gamma_1(q) = \int_0^q 2\pi q \, \frac{1}{2} \frac{\Delta}{(q^2 + \Delta^2)^{3/2}} \, dq \,. \tag{B.21}$$

Using the indefinite integral

$$\int \frac{q}{(q^2 + \Delta^2)^{3/2}} \, dq = -\frac{1}{\sqrt{q^2 + \Delta^2}}$$

we obtain

$$\gamma_1(q) = \pi \Delta \left[-\frac{1}{\sqrt{q^2 + \Delta^2}} + \frac{1}{\Delta} \right] = \pi \left[1 - \frac{\Delta}{\sqrt{q^2 + \Delta^2}} \right], \tag{B.22}$$

in agreement (of course) with the results (B.17) of the previous procedure. Equation (B.22) provides the exact expression of the Berry phase, free from any modulus of 2π.

Problem 3. Berry Magnetic Monopole and Geometrical Phase Factors

Consider the parametric Hamiltonian given by

$$K = \begin{bmatrix} q_z & q_x - iq_y \\ q_x + iq_y & -q_z \end{bmatrix} \equiv q_x\sigma_x + q_y\sigma_y + q_z\sigma_z \equiv \mathbf{q} \cdot \boldsymbol{\sigma}, \tag{B.23}$$

where the parameters q_x, q_y, q_z are real quantities, and $\sigma_x, \sigma_y, \sigma_z$ are the Pauli matrices. Show that the Berry fictitious magnetic field can be described as due to a monopole centered at the origin of the parameter space.

The matrix K and the matrix H of Eq. (B.12) in Problem 2 are obviously the same once the gap parameter Δ is relabeled as q_z. In the present problem q_z is treated on the same footing as q_x and q_y. It is also evident that all the needed ingredients are already elaborated in Problem 2, so we can proceed rapidly.

From Eq. (B.19), we have seen that the fictitious magnetic field on the lower adiabatic sheet (when q_z is momentarily thought as a fixed parameter) is given by

$$\mathbf{B}_1(q_x, q_y, q_z) = \frac{1}{2} \frac{q_z}{(q_x^2 + q_y^2 + q_z^2)^{3/2}} \, \mathbf{e}_z \,.$$

The isotropic spin symmetry of Eq. (B.23), or explicit evaluation, suggest that in general the fictitious magnetic field takes the expression

$$\mathbf{B}_1(q_x, q_y, q_z) = \frac{1}{2} \frac{1}{(q_x^2 + q_y^2 + q_z^2)^{3/2}} (q_x \mathbf{e}_x + q_y \mathbf{e}_y + q_z \mathbf{e}_z) \,;$$

this Berry magnetic field can be written in the form

$$\mathbf{B}_1(\mathbf{q}) = \frac{1}{2q^3} \, \mathbf{q} \,; \tag{B.24}$$

it represents the magnetic field of a monopole of strength $1/2$ located at the degeneracy point.

Consider a circuit $\mathcal{C}$ and a surface S of contour $\mathcal{C}$ in the parameter domain. The Berry phase $\gamma_1(\mathcal{C})$ is given by the flux of $\mathbf{B}_1(\mathbf{q})$ across a surface of contour $\mathcal{C}$

$$\gamma_1(\mathcal{C}) = \int_S \mathbf{B}_1(\mathbf{q}) \cdot d\mathbf{S} \,.$$

In the case of the Berry monopole (B.24), we can write

$$\gamma_1(\mathcal{C}) = \frac{1}{2} \Omega(\mathcal{C}), \tag{B.25}$$

where $\Omega(\mathcal{C})$ denotes the *solid angle* that $\mathcal{C}$ subtends at the degeneracy point. In accordance with Eq. (B.25), notice that $\gamma_1(q)$ in Eq. (B.21) is just the product of the monopole strength $1/2$ by the solid angle subtended by a circular circuit around the q_z-axis in the $q_z \equiv \Delta$ plane.

Finally it can be noticed that for $q_z \equiv \Delta \to 0^+$ the Berry phases equal π for any circuit that encompasses the origin; this can be interpreted as originated by a semi-infinite Dirac string singularity attached to the magnetic monopole of strength $1/2$ located at the degeneracy point.

Problem 4. Berry Phase of a Typical Renner-Teller Coupling Matrix

Consider the parametric matrix M that describes a typical quadratic Renner-Teller electron-phonon coupling

$$M = \begin{bmatrix} q_x^2 - q_y^2 & 2q_x q_y \\ 2q_x q_y & -q_x^2 + q_y^2 \end{bmatrix} \quad \text{(basis electronic set } f_x, f_y), \tag{B.26}$$

where the parameters q_x and q_y are real quantities. Determine the eigenvalues, the eigenfunctions and the Berry phase for circuits enclosing the origin of the parameter space.

This problem has several points in common with Problem 1, so we can be rather schematic. The eigenvalues of the M-matrix are obtained solving the determinantal equation

$$\left\| \begin{matrix} q_x^2 - q_y^2 - E & 2q_x q_y \\ 2q_x q_y & -q_x^2 + q_y^2 - E \end{matrix} \right\| = 0 \quad \Longrightarrow \quad E = \mp q^2.$$

The eigenfunction corresponding to the lowest eigenvalue is determined by the expression

$$\begin{bmatrix} q_x^2 - q_y^2 & 2q_x q_y \\ 2q_x q_y & -q_x^2 + q_y^2 \end{bmatrix} \begin{bmatrix} C_1 \\ C_2 \end{bmatrix} = -q^2 \begin{bmatrix} C_1 \\ C_2 \end{bmatrix}.$$

Performing the standard row by column product, we obtain two (redundant) equations; the first one (for instance) reads

$$(q_x^2 - q_y^2)C_1 + 2q_x q_y C_2 = -q^2 C_1 \quad \rightarrow \quad \frac{C_1}{C_2} = -\frac{q_y}{q_x} = -\frac{\sin\theta}{\cos\theta}$$

Similarly for the upper sheet we have

$$(q_x^2 - q_y^2)D_1 + 2q_x q_y D_2 = +q^2 D_1 \quad \rightarrow \quad \frac{D_1}{D_2} = \frac{q_x}{q_y} = \frac{\cos\theta}{\sin\theta}$$

The eigenfunctions, up to gauge transformations of type $\exp(im\theta)$ with integral m, are

$$\begin{cases} |\psi_1\rangle = \sin\theta|f_x\rangle - \cos\theta|f_y\rangle \\ |\psi_2\rangle = \cos\theta|f_x\rangle + \sin\theta|f_y\rangle \end{cases} \tag{B.27}$$

The functions are real, one-valued, and continuous, except the degeneracy point $q = 0$. The Berry connection vanishes on the basis functions (B.27); thus the Berry phase is zero (modulus 2π). Because of the presence of the degeneracy point $q = 0$, we cannot say at this stage whether the Berry phase is trivially zero or not, if the degeneracy is removed by an infinitesimal small perturbation; this is discussed in the next problem where we will see that the Berry phase is $\pm 2\pi$.

Before passing to the next problem, we notice that the electron-phonon coupling matrix $M(q_x, q_y)$ of Eq. (B.26) is expressed on the set of the electron doublet (f_x, f_y). It is instructive to express the same matrix on the electronic functions of definite angular momentum chosen as follows

$$\phi_1 = \frac{1}{\sqrt{2}}(f_x + if_y), \quad \phi_2 = \frac{1}{\sqrt{2}}(f_x - if_y). \tag{B.28}$$

The matrix M of Eq. (B.26) in the old and in the new basis functions takes the form

$$\begin{bmatrix} q_x^2 - q_y^2 & 2q_x q_y \\ 2q_x q_y & -q_x^2 + q_y^2 \end{bmatrix} \quad \text{set } (f_x, f_y) \Longrightarrow$$

$$\begin{bmatrix} 0 & (q_x - iq_y)^2 \\ (q_x + iq_y)^2 & 0 \end{bmatrix} \quad \text{set } (\phi_1, \phi_2).$$

The last matrix can be recognized as the matrix describing the low energy excitations at one of the Dirac points of graphene bilayers (the other Dirac point could be treated similarly). Thus the study of the Berry phase of the Renner-Teller electron-phonon coupling and the Berry phase of graphene bilayers are essentially equivalent, since the corresponding matrices are related by a similarity transformation.

Problem 5. Berry Phase of the Hamiltonian of Low Energy Carriers in Graphene Bilayers and Gapped Honeycomb Materials

Determine the eigenvalues, the eigenfunctions, and the Berry phase of the parametric Hamiltonian

$$H = \begin{bmatrix} \Delta & (q_x - iq_y)^2 \\ (q_x + iq_y)^2 & -\Delta \end{bmatrix}. \tag{B.29}$$

where q_x, q_y, Δ are real, and Δ is the parameter that controls the energy gap.

The matrix H represents the low energy excitation of graphene bilayers for $\Delta = 0$ and gapped graphene bilayers for $\Delta \neq 0$. It also generalizes the essential structure of typical Renner-Teller systems. This problem has several points in common with Problem 2, so we can be rather schematic. As in Problem 2, we consider explicitly only the case $\Delta > 0$ (since the case $\Delta < 0$ simply needs some appropriate adjustments).

The eigenvalues of the H matrix are obtained solving the determinantal equation

$$\|H - E\| = \begin{Vmatrix} \Delta - E & (q_x - iq_y)^2 \\ (q_x + iq_y)^2 & -\Delta - E \end{Vmatrix} \rightarrow E = \mp\sqrt{q^4 + \Delta^2}$$

with $q^2 = q_x^2 + q_y^2$. The eigenfunction corresponding to the lowest eigenvalue is determined by the expression

$$\begin{bmatrix} \Delta & (q_x - iq_y)^2 \\ (q_x + iq_y)^2 & -\Delta \end{bmatrix} \begin{bmatrix} C_1 \\ C_2 \end{bmatrix} = -\sqrt{q^4 + \Delta^2} \begin{bmatrix} C_1 \\ C_2 \end{bmatrix}.$$

Performing the standard row by column product, we obtain two (redundant) equations; the first of them reads

$$\Delta \cdot C_1 + (q_x - iq_y)^2 C_2 = -\sqrt{q^4 + \Delta^2}\, C_1.$$

Using polar coordinates in the (q_x, q_y) plane, we have

$$\frac{C_1}{C_2} = \frac{-q^2 e^{-2i\theta}}{\Delta + \sqrt{q^4 + \Delta^2}}.$$

The expression of the normalized coefficients for the lower energy adiabatic surface becomes

$$C_1(q, \theta) = \frac{1}{\sqrt{2}} \frac{-q^2 e^{-2i\theta}}{\sqrt{q^4 + \Delta^2 + \Delta\sqrt{q^4 + \Delta^2}}} \qquad C_2(q, \theta) = \frac{1}{\sqrt{2}} \frac{\Delta + \sqrt{q^4 + \Delta^2}}{\sqrt{q^4 + \Delta^2 + \Delta\sqrt{q^4 + \Delta^2}}}.$$

$$\tag{B.30}$$

The steps performed in the present treatment to arrive at Eqs. (B.30) and the steps performed in Problem 2 to arrive at Eqs. (B.14) are so similar that there is no need to go on explicitly with procedures and discussions. We come directly to the results.

With the same procedures used to establish Eq. (B.19), we evaluate the fictitious magnetic field on the lower and upper adiabatic sheets, and obtain

$$\mathbf{B}_1(\mathbf{q}) = \frac{2q^2\Delta}{(q^4 + \Delta^2)^{3/2}} \, \mathbf{e}_z = -\mathbf{B}_2(\mathbf{q}).$$

For a circular circuit of radius q the Berry phase takes the values

$$\gamma_1(q) = -\gamma_2(q) = 2\pi\Delta \left[-\frac{1}{\sqrt{q^4 + \Delta^2}} + \frac{1}{\Delta} \right].$$

In the case $\Delta \to 0^+$, it is seen that the Berry phase has the non trivial value $\pm 2\pi$ free from any modulus of 2π.

Further Reading

Allen, M. P., & Tildesley, D. J. (1989). *Computer simulation of liquids*. Oxford: Clarendon.

Baroni, S., de Gironcoli, S., Dal Corso, A., & Giannozzi, P. (2001). Phonons and related crystal properties from density-functional perturbation theory. *Reviews of Modern Physics, 73*, 515. This paper contains a most significant development of the Hellmann-Feynman forces for the first principle calculations of the lattice dynamics.

Berry, M. V. (1984). Quantal phase factors accompanying adiabatic changes. *Proceedings of the Royal Society London A, 392*, 45.

Bersuker, I. B. (2006). *The Jahn-Teller effect*. Cambridge: Cambridge University Press.

Bersuker, I. B., & Polinger, V. Z. (1989). *Vibronic interactions in molecules and crystals*. New York: Springer.

Bevilacqua, G., Martinelli, L., & Terzi, N. (Eds.). (1999). *Electron-phonon dynamics and Jahn-Teller effect*. Singapore: World Scientific.

Born, M., & Huang, K. (1954). *Dynamical theory of crystal lattices*. Oxford: Oxford University Press.

Englman, R. (1972). *The Jahn-Teller effect in molecules and crystals*. New York, Wiley.

Garg, A. (2010). Berry phase near degeneracy: Beyond the simplest case. *American Journal of Physics, 78*, 661. This paper contains a review on the Berry connection and Berry curvature, with application to magnetic molecular solids.

King-Smith, R. D., & Vanderbilt, D.(1993). Theory of polarization of crystalline solids. *Physical Review, B, 47*, 1651; Electric polarization as a bulk quantity and its relation to surface charge, *Physical Review B, 48*, 4442.

Maradudin A. A., Montroll, E. W., Weiss G. H., & Ipatova I. P. (1971). *Theory of lattice dynamics in the harmonic approximation*. New York: Academic Press.

Marks, D., & Hutter, J. (2009). *Ab initio molecular dynamics. Basic theory and advanced methods*. Cambridge: Cambridge University Press.

Marzari, N., Mostofi, A. A., Yates, J. R., Souza, I., & Vanderbilt, D. (2012). Maximally localized Wannier functions: Theory and applications. *Reviews of Modern Physics, 84*, 1419.

O'Brien, M. C. M., & Chancey, C. C. (1993). The Jahn-Teller effect: An introduction and current review. *American Journal of Physics, 61*, 688.

Payne, M. C., Teter, M. P., Allan, D. C., Arias, T. A., & Joannopoulos, J. D. (1992). Iterative minimization techniques for ab-initio total-energy calculations: Molecular dynamics and conjugate gradients. *Reviews of Modern Physics, 64*, 1045.

Resta, R. (1992). Theory of the electric polarization in crystals. *Ferroelectrics, 136*, 51.

Resta, R. (1994). Macroscopic polarization in crystalline dielectrics: The geometric phase approach. *Reviews of Modern Physics, 66*, 899.

Resta, R., & Vanderbilt, D. (2007). Theory of polarization: A modern approach. In K. Rabe, C. H. Ahn, & J.-M.Triscone (Eds.), *Physics of ferroelectrics* (Vol. 105, p. 31). Berlin: Springer.

Shapere, A., & Wilczek, F. (Eds.) (1989). *Geometric phases in physics*. Singapore: World Scientific.

Yarkony, D. Y.(1996). Diabolical conical intersections. *Reviews of Modern Physics, 68*, 985.

9 Lattice Dynamics of Crystals

Chapter Outline head

In the previous chapter we have laid down the basic concepts of the quantum mechanical theory of electrons and nuclei in mutual interaction. According to the adiabatic principle, the system of light particles (the electrons) is preliminarily studied with the heavy particles (the nuclei) clamped in a given spatial configuration; the total ground-state energy of the electronic system, thought of as a function of the nuclear coordinates, becomes then the potential energy for the nuclear motion. We are interested here in small displacements around the equilibrium configuration of the ground adiabatic energy surface (supposed to be non-degenerate). Within the harmonic approximation, we describe the dispersion curves for normal mode propagation in crystals, and introduce the concept of *phonons*, as traveling quanta of vibrational energy. In the study of the lattice vibrations of ionic or partially ionic crystals (polar crystals), essentially new features appear in the long wave length limit; in polar crystals, the coupling of optical vibrational branches with the electromagnetic field leads to the concept of mixed phonon-photon quasiparticles, known as *polaritons*.

9.1 Dynamics of Monoatomic One-Dimensional Lattices

General Introductory Remarks

The conceptual framework for the description of nuclear motion and lattice vibrations is based on the Born-Oppenheimer adiabatic approximation (Section 8.2): the

Solid State Physics, Second Edition. http://dx.doi.org/10.1016/B978-0-12-385030-0.00009-8

Table 9.1 Energy units of most frequent use for phonons.

$h\nu$ $(\nu = 10^{12}$ Hz$) = 4.1357$ meV	$\hbar\omega$ $(\omega = 10^{13}$ rad/s$) = 6.5822$ meV
1 eV$/hc = 8065$ cm^{-1}	1 eV$/k_B = 11605$ K
for brevity: 1 eV $\leftrightarrow$ 8065 cm^{-1}	1 eV $\leftrightarrow$ 11605 K
1 meV $\leftrightarrow$ 8.065 cm^{-1}	1 cm^{-1} $\leftrightarrow$ 0.124 meV

preliminary step is the knowledge of the energy of the ground-state (supposed non-degenerate) of the interacting electronic-nuclear system as a function of the nuclear configurations. We have seen that the total ground-state energy becomes the "potential energy" for the nuclear motion. In view of small nuclear oscillations from equilibrium positions, the potential energy is expanded to second order in the nuclear displacements. The matrix formed with the expansion coefficients is called *dynamical matrix*, and its diagonalization produces the *normal vibrational modes* or *phonon states* of the system under investigation. A typical vibration spectrum extends in general to the infrared region, up to energies $\hbar\omega$ of several tens of meV; for convenience, we summarize in Table 9.1 the energy units for phonons, often adopted in the literature.

It is worthwhile to notice since now that the *dynamical matrix procedure*, by its own intrinsic structure, neglects the interaction of phonons with the retarded modes of the electromagnetic field; retardation effects due to the finite speed of light will be properly considered when necessary, and lead to the concept of new quasiparticles called *polaritons*. In the first sections of this chapter, we focus our attention to the relevant aspects of the dynamical matrix approach to phonon states. The necessary modifications introduced by the interaction of phonons with the retarded electromagnetic field will be considered in Section 9.7.

Dynamics of a Monoatomic One-Dimensional Lattice

To describe the lattice vibrations of crystals, we consider first linear chains of equal atoms (present section), then linear chains with a basis of different atoms (Section 9.2), and finally general three-dimensional structures (Section 9.3). This sequence of increasing sophistication is adopted because some physical concepts are better illustrated in one-dimensional situations, where notations and technicalities can be kept at the minimal level.

So, we begin by considering a one-dimensional chain, of lattice constant a, formed by a (large) number N of atoms of mass M. We indicate by u_n the longitudinal displacement of the nth atom from the equilibrium position $t_n = na$, at a particular time (see Figure 9.1). We denote by $E_0(\{u_n\})$ the total ground-state energy of the interacting electronic-nuclear system, with the *nuclei fixed in the positions* $R_n = na + u_n$. The ground state of the crystal is supposed to be non-degenerate for all configurations $\{u_n\}$ of interest.

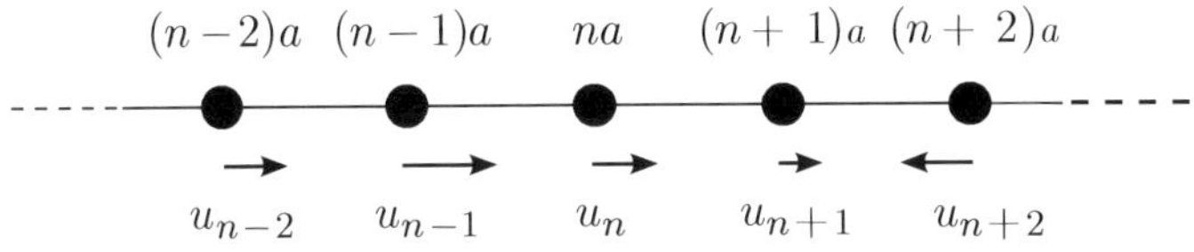

$(n-2)a$ $(n-1)a$ na $(n+1)a$ $(n+2)a$

u_{n-2} u_{n-1} u_n u_{n+1} u_{n+2}

Figure 9.1 Longitudinal displacements in a one-dimensional monoatomic lattice. The equilibrium positions $t_n = na$ are indicated by circles; the displacements u_n at a given instant are indicated by arrows.

In the study of small oscillations, it is convenient to expand the energy of the ground Born-Oppenheimer sheet in increasing powers of the displacements:

$$E_0(\{u_n\}) = E_0(0) + \frac{1}{2} \sum_{nn'} \left(\frac{\partial^2 E_0}{\partial u_n \partial u_{n'}} \right)_0 u_n u_{n'}$$

$$+ \frac{1}{3!} \sum_{nn'n''} \left(\frac{\partial^3 E_0}{\partial u_n \partial u_{n'} \partial u_{n''}} \right)_0 u_n u_{n'} u_{n''} + \cdots \tag{9.1}$$

(derivatives evaluated at the equilibrium configuration carry the subscript 0). In the expansion (9.1), the linear terms in the displacements are not present since $\partial E_0 / \partial u_n = 0$ at the equilibrium configuration. The total ground-state energy $E_0(0)$ at the equilibrium configuration is important in the discussion of the cohesive energy, but irrelevant in the discussion of lattice vibrations. The truncation to quadratic terms in the expansion (9.1) is called "harmonic approximation"; in general the anharmonic terms (cubic, quartic terms, and so on) are taken into account only after the harmonic approximation has been carried out.

In the harmonic approximation, which is appropriate for sufficiently small displacements, the total crystal energy (9.1) becomes

$$E_0^{(\mathrm{harm})}(\{u_n\}) = E_0(0) + \frac{1}{2} \sum_{nn'} D_{nn'} u_n u_{n'}, \tag{9.2}$$

where

$$D_{nn'} = \left(\frac{\partial^2 E_0}{\partial u_n \partial u_{n'}} \right)_0 \tag{9.3}$$

denote the second derivatives of the ground-state energy $E_0(\{u_n\})$ evaluated at the equilibrium configuration. The quantities $D_{nn'}$ are called *interatomic force constants*, and the matrix D formed with the interatomic force constants $D_{nn'}$ is called *dynamical matrix*. The force constants $D_{nn'}$ represent the proportionality coefficients connecting the forces acting on the nuclei with the displacements suffered by the nuclei; in the harmonic approximation, in fact, we have

$$F_n = -\frac{\partial E_0^{(\mathrm{harm})}}{\partial u_n} = -\sum_{n'} D_{nn'} u_{n'}. \tag{9.4}$$

There are some general symmetries and constraints that must be obeyed by the dynamical matrix D. From the definition (9.3), it is seen that the matrix D is real and symmetric

$$D_{nn'} = D_{n'n}. \tag{9.5a}$$

The translational symmetry of the lattice requires

$$D_{nn'} = D_{mm'} \quad \text{if} \quad t_n - t_{n'} = t_m - t_{m'}. \tag{9.5b}$$

Furthermore we have the general and important "sum rule"

$$\sum_{n'} D_{nn'} \equiv 0 \quad \text{for any} \quad n; \tag{9.5c}$$

this is a trivial consequence of the fact that the forces, given by Eq. (9.4), vanish not only when all nuclear displacements are zero, but also when all nuclear displacements are equal. Equations (9.5a) and (9.5c) show that the sum of the matrix elements of any row or any column of the force constant matrix D vanishes.

We consider now the classical equation of motion for the n-th nucleus of mass M in the position $R_n = na + u_n$ under the force F_n; we have

$$M\ddot{u}_n = -\sum_{n'} D_{nn'} u_{n'} \tag{9.6}$$

with $n = 1, 2, \ldots, N$. The set of coupled differential equations (9.6) can be solved, in general, looking for solutions periodic in time of the form $u_n(t) = A_n \exp(-i\omega t)$. In the present case, we can take advantage of the translational symmetry of the force-constant matrix D in real space; this suggests to solve Eq. (9.6) looking for solutions in the form of traveling waves, periodic in space and time, of the type

$$u_n(t) = A e^{i(qna - \omega t)}, \tag{9.7}$$

where the amplitudes A of the displacements are the same for all the sites, while the phases are controlled by the Bloch theorem. Replacing Eq. (9.7) into Eq. (9.6), we have that the phonon frequencies ω are given by the relation

$$-M\omega^2 A = -\sum_{n'} D_{nn'} e^{-iq(na - n'a)} A.$$

The above expression can be recast in the form

$$\boxed{M\omega^2(q) = D(q)}, \tag{9.8a}$$

where

$$\boxed{D(q) = \sum_{n'} D_{nn'} e^{-iq(na - n'a)}}; \tag{9.8b}$$

notice that the Fourier transform $D(q)$ of the force constant matrix elements $D_{nn'}$ does not depend on the specific value of n because of the property (9.5b).

Equation (9.8a) provides the dispersion relation $\omega = \omega(q)$ connecting the phonon frequency to the phonon wavenumber of the traveling plane wave of type (9.7). Notice that the displacements (9.7) are unaffected by changes of q by integer multiples of $2\pi/a$; the independent values of q are confined within the Brillouin zone $-\pi/a < q \le \pi/a$. When standard Born-von Karman boundary conditions are applied, i.e. one requires $u_n(t) \equiv u_{n+N}(t)$, the allowed values of q within the Brillouin zone become discretized with values $m(2\pi/Na)$ and m integer; the number of allowed wavenumbers within the Brillouin zone equals the number of unit cells of the crystal (see Section 1.1). We now apply our analysis to the specific case of a linear chain of atoms with nearest neighbor interactions only.

Linear Monoatomic Chain with Nearest Neighbor Interactions

To show the essential aspects of the lattice vibrations in the linear chain, we suppose that the only relevant interatomic interactions occur between nearest neighbor atoms; this means that the only force constants different from zero are D_{nn}, $D_{n\,n+1}$, and $D_{n-1\,n}$. From the general properties of the force constants summarized by Eqs. (9.5), it is seen that there is a unique independent parameter (denoted below by C), and it holds

$$D_{nn} = 2C, \quad D_{n\,n+1} = D_{n-1\,n} = -C. \tag{9.9}$$

The energy (9.2) of the linear chain, in the harmonic approximation and nearest neighbor interaction becomes (taking $E_0(0)$ as the reference energy)

$$E_0^{(\text{harm})} = \frac{1}{2}C \sum_n (2u_n^2 - u_n u_{n+1} - u_n u_{n-1}) \equiv \frac{1}{2}C \sum_n (u_n - u_{n+1})^2. \tag{9.10}$$

Equation (9.10) is quite intuitive and represents the elastic energy of a chain of atoms, connected to nearest neighbors with springs of constant C.

The classical equations of motion (9.6) for the nuclear vibrations are thus

$$\boxed{M\ddot{u}_n = -C(2u_n - u_{n+1} - u_{n-1})} \tag{9.11}$$

for any n integer number. The set of discrete coupled differential equations (9.11) can be solved looking for traveling waves, periodic in space and time, of the form $u_n(t) = A \exp(iqna - i\omega t)$. By direct substitution, or equivalently from Eqs. (9.8), one obtains

$$-M\omega^2 = -C(2 - e^{iqa} - e^{-iqa}) = -4C \sin^2 \frac{1}{2}qa. \tag{9.12a}$$

The dispersion relation for normal modes is thus

$$\omega = \sqrt{\frac{4C}{M}} \left| \sin \frac{1}{2}qa \right|, \tag{9.12b}$$

and is illustrated in Figure 9.2; we see that the spectrum of vibrational frequencies extends from zero to a cutoff frequency $\omega_{\max} = \sqrt{4C/M}$.

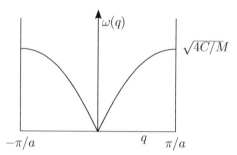

Figure 9.2 Phonon dispersion curve for a monoatomic linear lattice with nearest neighbor interactions only; the Brillouin zone is the segment between $-\pi/a$ and $+\pi/a$.

It is interesting to consider the normal modes in the *long wavelength limit qa* $\ll 1$. The dispersion relation (9.12b) takes the form

$$\omega = \sqrt{\frac{C}{M}}\, aq \equiv v_s q \quad (qa \ll 1);$$

the proportionality coefficient v_s between phonon frequency ω and phonon wavenumber q represents the velocity of the sound in the medium, and is given by

$$v_s \equiv \sqrt{\frac{C}{M}}\, a. \tag{9.12c}$$

From Eq. (9.12b), we can estimate the value of the cutoff angular frequency

$$\omega_{\max} = 2\sqrt{\frac{C}{M}} = \frac{2v_s}{a} \approx \frac{10^5 \text{ cm/s}}{10^{-8} \text{ cm}} = 10^{13} \text{ rad/s}.$$

In the long wavelength limit, we can perform a continuous approximation to the set of discretized coupled differential equations (9.11); in fact $(-2u_n + u_{n+1} + u_{n-1})/a^2$ can be considered as the finite difference expression of the second order derivative $\partial^2 u/\partial x^2$. Equation (9.11) is thus equivalent to $M\ddot{u} = Ca^2 \partial^2 u/\partial x^2$, and the propagation velocity of the elastic wave is again given by Eq. (9.12c).

9.2 Dynamics of Diatomic One-Dimensional Lattices

Consider now the dynamics of a diatomic linear chain, of lattice constant a_0, with two atoms of mass M_1 and M_2 in the unit cell; this model is the prototype of a crystal with basis. In the equilibrium configuration, we assume that the atoms of mass M_1 occupy the sublattice positions $R_n^{(1)} = na_0$, while the atoms of mass M_2 occupy the sublattice positions $R_n^{(2)} = (n + 1/2)a_0$ (see Figure 9.3).

We denote by u_n the displacements of the atoms of mass M_1 and with v_n the displacements of the atoms of mass M_2. For simplicity we assume that only nearest

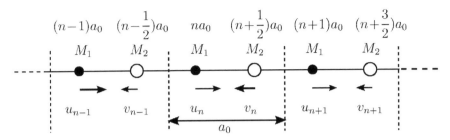

Figure 9.3 Longitudinal displacements in a one-dimensional diatomic lattice. The equilibrium positions of the two sublattices of atoms, of mass M_1 and M_2, are indicated by black and white circles, respectively; the displacements u_n and v_n at a given instant are indicated by arrows.

neighbor atoms interact with elastic forces of spring constant C. The classical equations of motion for the two types of particles are

$$
\begin{aligned}
M_1 \ddot{u}_n &= -C(2u_n - v_{n-1} - v_n) \\
M_2 \ddot{v}_n &= -C(2v_n - u_n - u_{n+1})
\end{aligned}
\tag{9.13}
$$

for any integer number n.

To solve the set of discrete coupled differential equations (9.13), we look for traveling waves, periodic in space and time, of the form

$$
u_n(t) = A_1 \, e^{i(qna_0 - \omega t)} \quad \text{and} \quad v_n(t) = A_2 \, e^{i(qna_0 + qa_0/2 - \omega t)}.
\tag{9.14}
$$

Notice that the vibrations of atoms of the same sublattice in different cells have the same amplitudes and phase relations of Bloch type.

Replacing (9.14) into (9.13) gives

$$
\begin{aligned}
-M_1\omega^2 A_1 &= -C(2A_1 - A_2 \, e^{-iqa_0/2} - A_2 \, e^{iqa_0/2}), \\
-M_2\omega^2 A_2 &= -C(2A_2 - A_1 \, e^{-iqa_0/2} - A_1 \, e^{iqa_0/2}).
\end{aligned}
$$

The two linear homogeneous equations in the two unknown amplitudes A_1 and A_2 have a non-trivial solution if the determinant of the coefficients of A_1 and A_2 is zero, namely

$$
\begin{vmatrix}
2C - M_1\omega^2 & -2C \cos (qa_0/2) \\
-2C \cos (qa_0/2) & 2C - M_2\omega^2
\end{vmatrix} = 0.
$$

The eigenvalues are given by

$$
\omega^2 = C\left(\frac{1}{M_1} + \frac{1}{M_2}\right) \pm C\sqrt{\left(\frac{1}{M_1} + \frac{1}{M_2}\right)^2 - \frac{4\sin^2 (qa_0/2)}{M_1 M_2}},
\tag{9.15a}
$$

and the corresponding amplitudes satisfy the relation

$$
\frac{A_1}{A_2} = \frac{2C \cos (qa_0/2)}{2C - M_1\omega^2}.
\tag{9.15b}
$$

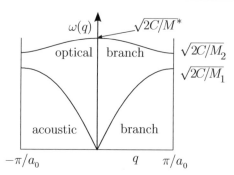

Figure 9.4 Phonon dispersion curves of a diatomic linear chain, with nearest neighbor atoms interacting with spring constant C. The masses of the atoms are M_1 and M_2 (with $M_1 > M_2$); M^* is the reduced mass.

The dispersion relations (9.15a) are illustrated in Figure 9.4. We have now two branches, the lower one called "acoustic" and the upper one called "optical," with a frequency gap between them. In the particular case that $M_1 \equiv M_2$, the gap between the acoustic branch and the optical branch disappears and we recover the result of the previous section [in the case $M_1 \equiv M_2$, Figures 9.4 and 9.2 coincide exactly, once the trivial folding due to the fact that $a_0 = 2a$ is performed].

Long Wavelength Limit of a Diatomic Linear Chain: Acoustic and Optical Modes

We consider now the normal modes of a diatomic chain in the long wavelength limit $q a_0 \ll 1$. For small $q a_0$, the two roots ω^2 and the corresponding amplitudes, given by Eqs. (9.15), are

$$\omega^2 = \frac{2C}{M_1 + M_2}\left(\frac{a_0}{2}\right)^2 q^2 + O(q^4), \qquad A_1 = A_2 \quad \text{acoustic branch,}$$

$$\omega^2 = \frac{2C}{M^*} + O(q^2), \quad \frac{1}{M^*} = \frac{1}{M_1} + \frac{1}{M_2}, \quad M_1 A_1 = -M_2 A_2 \quad \text{optical branch,}$$

where M^* denotes the reduced mass.

In the acoustic branch, in the long wavelength limit, the atoms vibrate in phase and with the same amplitude; the frequency ω is proportional to the wavenumber q, and the proportionality coefficient, the sound velocity v_s, is given by

$$v_s = \sqrt{\frac{C}{(M_1 + M_2)/2}}\, \frac{a_0}{2}.$$

The above relation is the obvious counterpart of expression (9.12c), if one notices that the average mass $(M_1 + M_2)/2$ replaces the mass M; a and $a_0/2$ denote nearest neighbor distance in the monoatomic and diatomic models, respectively; C is the spring constant.

In the upper branch, A_1 and A_2 have opposite signs and absolute values inversely proportional to the atomic masses; this means that the two atoms in the unit cell move in opposite directions, while the "center of mass" of the unit cell remains fixed. In the long wavelength limit, the frequency $\omega(q)$ of the optical branch tends to the finite value $\omega_0 = \sqrt{2C/M^*}$.

It is instructive to consider the optical modes of the diatomic chain in the continuous approximation; this is useful also to justify the origin of the name attached to such a branch. For optical modes in the long wavelength limit, we can assume that nearest neighbor atoms on the same sublattice have the same displacements. Equations (9.13) can be recast in the form

$$M_1 \ddot{u}_n = -2C(u_n - v_n), \tag{9.16a}$$

$$M_2 \ddot{v}_n = -2C(v_n - u_n). \tag{9.16b}$$

In optical modes, the atoms in the two sublattices move against each other; so it is convenient to discuss optical vibrations in terms of the *relative displacement variable* $w_n = u_n - v_n$. If we divide both members of Eq. (9.16a) by M_1, both members of Eq. (9.16b) by M_2, and subtract, we obtain

$$\ddot{w}_n = -\omega_0^2 w_n \quad \text{with} \quad \omega_0 = \sqrt{\frac{2C}{M^*}}; \tag{9.17}$$

the relative motion of the atoms M_1 and M_2 around the center of mass of the unit cell is harmonic with frequency ω_0.

The optical modes owe their name to the fact that they are expected to couple strongly with electromagnetic fields (of appropriate frequency). Suppose in fact that the diatomic crystal can be pictured as composed by ionic (or partially ionic) units, with effective net charge $\pm e^*$ (*polar crystals*). In polar crystals, the relative displacement variable w_n of the atoms in a given unit cell is accompanied by a dynamic electric dipole $d_n = e^* w_n$. Thus the mechanical motion of the lattice should be studied taking in proper account also the coupling to electromagnetic fields. In the presence of electric fields, the equations of motion (9.16) have to be modified in the form

$$M_1 \ddot{u}_n = -2C(u_n - v_n) + e^* E_{\text{loc}}, \tag{9.18a}$$

$$M_2 \ddot{v}_n = -2C(v_n - u_n) - e^* E_{\text{loc}}, \tag{9.18b}$$

where E_{loc} denotes the local electric field acting at the lattice sites. In the presence of the electric field, the equation of motion of the relative displacement w_n, or equivalently of the oscillating dipole $d_n = e^* w_n$, becomes

$$\ddot{w}_n = -\omega_0^2 w_n + \frac{e^*}{M^*} E_{\text{loc}} \quad \text{or} \quad \ddot{d}_n = -\omega_0^2 d_n + \frac{e^{*2}}{M^*} E_{\text{loc}}; \quad \omega_0 = \sqrt{\frac{2C}{M^*}}. \tag{9.19}$$

The forced-oscillation structure of Eqs. (9.19) shows that the coupling effects between the mechanical system of proper frequency ω_0 and the driving field of frequency ω are particularly significant when the frequency ω is resonant or almost resonant with the optical mode frequency ω_0. A detailed analysis of the coupling of photon and phonon modes in polar crystals, and the resulting polariton effects are discussed in Section 9.7.

9.3   Dynamics of General Three-Dimensional Crystals

Crystal Dynamical Matrix and Phonon Frequencies

In the previous two sections, we have discussed the lattice dynamics of one-dimensional crystals, with or without a basis. We complete now the subject, addressing the general problem of lattice dynamics of three-dimensional crystals, with or without a basis; for this purpose, we follow step-by-step the script of the previous two sections.

Consider a general three-dimensional crystal, with N unit cells, translation vectors $\mathbf{t}_n$, and a basis of atoms in the positions $\mathbf{d}_1, \mathbf{d}_2, \ldots, \mathbf{d}_{\nu_b}$. We label atoms with two indices $(n\nu)$, where the (Latin) index n denotes the unit cells of the crystal and the (Greek) index ν the atoms inside the unit cell. According to the general principles of the adiabatic approximation, we consider first *the nuclei fixed in the positions* $\mathbf{t}_n + \mathbf{d}_\nu + \mathbf{u}_{n\nu}$, and we denote by $E_0(\{\mathbf{u}_{n\nu}\})$ the energy of the electronic-nuclear system in the ground Born-Oppenheimer potential surface (supposed to be non-degenerate). The expansion of E_0 up to second order in the displacements from the equilibrium positions (*harmonic approximation*) gives

$$E_0^{(\text{harm})}(\{\mathbf{u}_{n\nu}\}) = E_0(0) + \frac{1}{2} \sum_{n\nu\alpha, n'\nu'\alpha'} D_{n\nu\alpha, n'\nu'\alpha'} u_{n\nu\alpha} u_{n'\nu'\alpha'}, \tag{9.20a}$$

where $\alpha, \alpha' = x, y, z$; $\nu, \nu' = 1, 2, \ldots, \nu_b$; and $n = 1, 2, \ldots, N$. In the expansion (9.20a), the linear terms in the displacements are not present since vanishing at the equilibrium configuration. The "interatomic force constants" are defined as the second derivative of $E_0(\{\mathbf{u}_{n\nu}\})$ evaluated at the equilibrium configuration

$$D_{n\nu\alpha, n'\nu'\alpha'} = \left(\frac{\partial^2 E_0}{\partial u_{n\nu\alpha} \partial u_{n'\nu'\alpha'}} \right)_0. \tag{9.20b}$$

The matrix D formed by the interatomic force constants is called *the dynamical matrix of the crystal in real space*. [Notice that in the literature, one can also encounter the essentially equivalent dynamical matrix $D/\sqrt{M_\nu M_{\nu'}}$ scaled by the nuclear masses.]

The matrix D obeys some general properties and constraints. From the definition (9.20b), it follows that the matrix D is real and symmetric:

$$D_{n\nu\alpha, n'\nu'\alpha'} = D_{n'\nu'\alpha', n\nu\alpha}. \tag{9.21a}$$

The translation symmetry of the lattice implies

$$D_{n\nu\alpha, n'\nu'\alpha'} = D_{m\nu\alpha, m'\nu'\alpha'} \quad \text{if} \quad \mathbf{t}_n - \mathbf{t}_{n'} = \mathbf{t}_m - \mathbf{t}_{m'} \tag{9.21b}$$

(the presence of point symmetry operations may produce further constraints, that can be analyzed with group theory considerations). Furthermore we have the general and important "sum rule"

$$\sum_{n'\nu'} D_{n\nu\alpha, n'\nu'\alpha'} \equiv 0, \tag{9.21c}$$

which represents a straight generalization of Eq. (9.5c).

The classical equations of motion for the dynamics of the nuclei in the harmonic approximation read

$$M_\nu \ddot{u}_{n\nu\alpha} = -\sum_{n'\nu'\alpha'} D_{n\nu\alpha,n'\nu'\alpha'}\, u_{n'\nu'\alpha'}\,, \tag{9.22}$$

where $n, n' = 1, 2, \ldots, N;\ \nu, \nu' = 1, 2, \ldots, \nu_b;\ \alpha, \alpha' = x, y, z$. The presence of translational symmetry suggests to solve the set of coupled differential equations (9.22) looking for solutions in the form of traveling waves of the type

$$\mathbf{u}_{n\nu}(t) = \mathbf{A}_\nu(\mathbf{q}, \omega)e^{i(\mathbf{q}\cdot\mathbf{t}_n - \omega t)}. \tag{9.23a}$$

The *polarization vectors* $A_{\nu\alpha}(\mathbf{q}, \omega)$ $(\nu = 1, 2, \ldots, \nu_b;\ \alpha = x, y, z)$ of the vibrations of the nuclei within the primitive cell are left unspecified and are determined below by the solution of an appropriate secular equation. Replacing (9.23a) into (9.22), we have

$$-M_\nu\, \omega^2 A_{\nu\alpha} = -\sum_{n'\nu'\alpha'} D_{n\nu\alpha,n'\nu'\alpha'} e^{-i\mathbf{q}\cdot(\mathbf{t}_n - \mathbf{t}_{n'})} A_{\nu'\alpha'}. \tag{9.23b}$$

Non-trivial solutions of Eq. (9.23b) are obtained by solving the determinantal equation

$$\left\| D_{\nu\alpha,\nu'\alpha'}(\mathbf{q}) - M_\nu\, \omega^2 \delta_{\alpha\alpha'}\delta_{\nu,\nu'} \right\| = 0\,, \tag{9.23c}$$

where

$$D_{\nu\alpha,\nu'\alpha'}(\mathbf{q}) = \sum_{n'} D_{n\nu\alpha,n'\nu'\alpha'} e^{-i\mathbf{q}\cdot(\mathbf{t}_n - \mathbf{t}_{n'})}. \tag{9.23d}$$

The above matrix $D(\mathbf{q})$ is called *the dynamical matrix of the crystal in reciprocal space.*

Equation (9.23c) is the fundamental eigenvalue equation for the normal modes of a crystal. The dynamical matrix $D(\mathbf{q})$ has dimension $3\nu_b$, where ν_b is the number of atoms forming the basis of the unit cell. The secular equation (9.23c) produces $3\nu_b$ eigenvalues (called phonons or normal modes); at every vector $\mathbf{q}$ we have thus $3\nu_b$ normal modes, giving rise to $3\nu_b$ phonon branches as $\mathbf{q}$ is varied within the first Brillouin zone. For a crystal with N unit cells, the total number of normal modes equals $3\nu_b N$, i.e. three times the total number of atoms. Let $\omega(\mathbf{q}, p)$ $(p = 1, 2, \ldots, 3\nu_b)$ denote the frequency of the p-th normal mode of wavevector $\mathbf{q}$, and $\mathbf{A}_\nu(\mathbf{q}, p)$ $(\nu = 1, 2, \ldots, \nu_b)$ the corresponding polarization vectors. A mode $\omega(\mathbf{q}, p)$ is called "longitudinal" (or "transverse") in the case the polarization vectors $\mathbf{A}_\nu(\mathbf{q}, p)$ are parallel to $\mathbf{q}$ (or perpendicular to it). Modes that involve oscillating electric dipoles are called "optically active" since they can couple directly with electromagnetic fields.

From the standard algebraic properties of matrix eigenvalue equations, the eigenvectors of the secular equation (9.23c) satisfy the orthogonality relations

$$\sum_{\nu\alpha} M_\nu A^*_{\nu\alpha}(\mathbf{q}, p) A_{\nu\alpha}(\mathbf{q}, p') = \delta_{p,p'}.$$

The above relation can be written in the extended form

$$M_1 \mathbf{A}_1^*(\mathbf{q}, p) \cdot \mathbf{A}_1(\mathbf{q}, p') + \cdots + M_{v_b} \mathbf{A}_{v_b}^*(\mathbf{q}, p) \cdot \mathbf{A}_{v_b}(\mathbf{q}, p') = \delta_{p,p'}. \tag{9.24}$$

In the case of a monoatomic three-dimensional Bravais lattice, $v_b = 1$ in Eq. (9.24), and for the three branches it holds $\mathbf{A}^*(\mathbf{q}, p) \cdot \mathbf{A}(\mathbf{q}, p') = \delta_{p,p'}$.

It is interesting to consider the eigenvalue equation (9.23c) in the long wavelength limit $\mathbf{q} \to 0$. From the sum rule (9.21c) and Eq. (9.23d) it follows

$$\sum_{v'} D_{v\alpha, v'\alpha'}(\mathbf{q} = 0) \equiv 0;$$

in particular we have

$$\sum_{v'\alpha'} D_{v\alpha, v'\alpha'}(\mathbf{q} = 0) A_{\alpha'} \equiv 0. \tag{9.25a}$$

The above equation shows that vibrations with amplitudes $A_{v'\alpha'} \equiv A_{\alpha'}$ equal for all the atoms of the basis satisfy the secular equation (9.23c) with frequency $\omega = 0$. Thus *any three-dimensional crystal presents three acoustic branches* with $\omega(\mathbf{q}) \to 0$ as $\mathbf{q} \to 0$, with all atoms in the unit cell vibrating in phase and with the same amplitude. The remaining $3v_b - 3$ "optical" modes vibrate in such a way that the motion of the center of mass of the cell is unaltered; in fact the orthogonality relation (9.24) to the acoustic modes, with polarization vectors $\mathbf{A}^*$ independent of the atomic index, gives

$$M_1 \mathbf{A}_1 + M_2 \mathbf{A}_2 + \cdots + M_{v_b} \mathbf{A}_{v_b} = 0 \tag{9.25b}$$

for any of the optical modes.

Phonon Dispersion Curves with the Crystal Dynamical Matrix and Short-Range or Long-Range Nature of Force Constants

We consider now a few illustrative examples of phonon dispersion curves, obtained within the dynamical matrix framework and compared with scattering measurements. Interatomic force constants and phonon dispersion curves have long been evaluated on the basis of empirical assumptions, physical insight, and ingenious models of interactions. The assumption of "two-body" central forces, acting between pairs of atoms composing the lattice, is the most common one; however in many cases this approximation is too drastic, and angular forces and torsional forces are to be included appropriately. The interatomic force constants of the dynamical matrix are often considered as disposable or semiempirical parameters.

Successively, first principle methods based on the density-functional theory have permitted the accurate calculation of the total ground-state energy of several materials. In particular the density-functional perturbation theory (see S. Baroni et al.), based on the knowledge of the ground-state *electronic charge density* and *its linear response* to nuclear displacements perturbation, has emerged as a most successful *ab initio* method for the evaluation of the interatomic force constants. This has made it possible the ambitious project of predicting the vibrational structure of a wide variety of materials, and has opened new perspectives in the field.

Once a specific approach of interatomic forces has been chosen, the dynamical matrix in reciprocal space can be set up, and diagonalization at several points of the Brillouin zone provides the phonon dispersion curves of the crystal. Some crystals can be intuitively described as consisting of neutral atoms interacting with *short-range forces*; the interatomic force constants extend to a reasonably small number of shells (say nearest neighbors, second nearest neighbors, and possibly a few more up to 10 shells or so), and become safely negligible afterwards. Examples of crystals with short-range nature of interatomic forces include simple crystals with Bravais lattice structures (vibrations concern neutral atoms), metals (the electron gas is very effective in screening Coulomb interactions), homopolar elemental semiconductors (such as silicon and germanium).

In Figure 9.5 we show the phonon dispersion curves of aluminum, together with the experimental measurements obtained with coherent inelastic scattering of slow neutrons. Since aluminum crystallizes in a simple fcc Bravais lattice, the phonon dispersion curves consist of three acoustic branches. The acoustic modes are degenerate at $\mathbf{q} = 0$; along the high symmetry directions ΓL and ΓX the two transverse modes are degenerate; in directions of low or no symmetry, the three acoustic branches are non-degenerate.

In Figure 9.6 we show the phonon dispersion curves of silicon and germanium. Since these elemental semiconductors crystallize in a fcc lattice with two atoms per unit cell, we have now three acoustic branches and three optical branches. The acoustic branches as well as the optical branches are degenerate at $\mathbf{q} = 0$. Along the

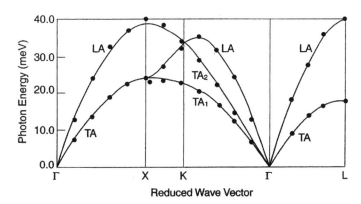

Figure 9.5 Phonon dispersion curves of aluminum along symmetry directions. The solid lines represents the calculations of A. A. Quong and B. M. Klein, Phys. Rev. B *46*, 10734 (1992) (copyright 1992 by the American Physical Society). Longitudinal and transverse acoustic branches are indicated by LA and TA (or TA₁ and TA₂), respectively. The experimental points are from the papers of G. Gilat and R. M. Nicklow, Phys. Rev. *143*, 487 (1966) and R. Stedman, S. Almqvist and G. Nilsson, Phys. Rev. *162*, 549 (1967).

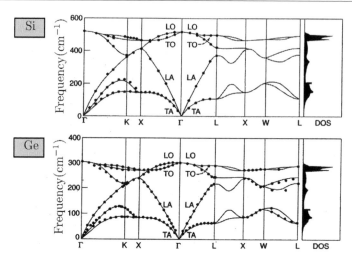

Figure 9.6 Phonon dispersion curves and density-of-states of Si and Ge calculated by P. Gian-nozzi, S. de Gironcoli, P. Pavone and S. Baroni, Phys. Rev. B *43*, 7231 (1991) (copyright 1991 by the American Physical Society). Longitudinal and transverse acoustic (or optical) modes are indicated by LA and TA (LO and TO), respectively. The experimental points are from G. Dolling, in "Inelastic Scattering of Neutrons in Solids and Liquids" edited by S. Ekland (IAEA, Vienna, 1963) Vol. II, p. 37; G. Nilsson and G. Nelin, Phys. Rev. B *3*, 364 (1971) and Phys. Rev. B *6*, 3777 (1972). Conversion to meV units can be done noting that $1 \text{ cm}^{-1} = 0.124$ meV.

high symmetry directions ΓX and ΓL the two transverse acoustic modes and the two transverse optical modes are degenerate; in directions of low or no symmetry, all modes are non-degenerate.

In polar crystals, such as ionic crystals and heteropolar semiconductors, the crystal lattice can be modeled as constituted by charged ions interacting both with *short-range forces and long-range Coulomb forces*. In the ionic picture of polar crystals, appropriate site dependent "Born effective charges" are attributed to the ions of the different sublattices to mimic long-range interactions.

The crystal dynamical matrix for polar crystals is obtained by direct summation of short-range terms and use of the Ewald method for long-range Coulomb terms. It should be noticed that the dynamical matrix of polar crystals at small q is particularly vulnerable to the long-range nature of the force constants in real space. The effects of the long-range part of the interatomic Coulomb interactions in polar crystals can be qualitatively understood assuming to expand the Coulomb interactions in multipoles, and to retain at the lowest order of approximation the dipole term. It is well known that the classical electrostatic dipole-dipole interaction energy is different for dipoles embedded in transverse or longitudinal fields in the long wavelength limit; actually, this fact is reflected on the dynamical matrix, which has a "pathological" behavior for small wavevector (called "non-analyticity"), and is responsible of the transverse-longitudinal splitting of optical phonons (see Section 9.7 for further aspects). The effects due to the long-range nature of interatomic forces are evident in the optical phonon branches of

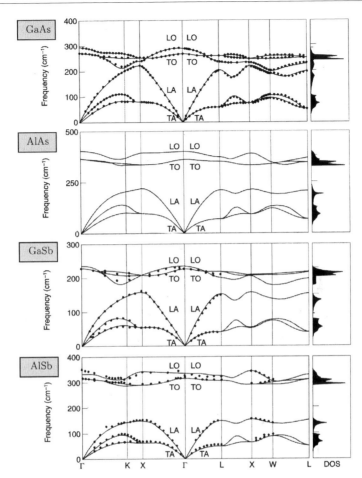

Figure 9.7 Calculated phonon dispersion curves and density-of-states for binary semiconductors GaAs, AlAs, GaSb, and AlSb [reprinted with permission from P. Giannozzi, S. de Gironcoli, P. Pavone, and S. Baroni, Phys. Rev. B *43*, 7231 (1991); copyright 1991 by the American Physical Society]. Longitudinal and transverse acoustic (or optical) modes are indicated by LA and TA (LO and TO), respectively. Conversion to meV units can be done noting that 1 cm$^{-1} = 0.124$ meV.

polar semiconductors (see Figure 9.7), and are even more important in typical ionic crystals (see Figure 9.8).

In Figure 9.7 we show the phonon dispersion curves for heteropolar semiconductors GaAs, AlAs, GaSb, and AlSb. In these crystals the interatomic forces include long-range Coulomb interaction, because of the partial ionic nature of the chemical bond. Since these heteropolar semiconductors crystallize in the fcc lattice with two atoms per unit cell, the phonon curves present three acoustic, and three optical branches; as expected, the acoustic and optical branches are well separated in crystals where the mass difference of the two atoms in the unit cell is large. A most important feature of

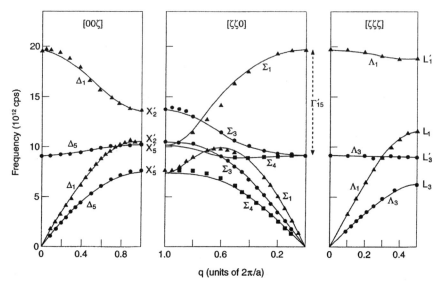

Figure 9.8 Measured phonon dispersion curves along three directions of high symmetry in LiF; the solid curves are a best least-squares fit of a parameter model [reprinted with permission from G. Dolling, H. G. Smith, R. M. Nicklow, P. R. Vijayaraghavan and M. K. Wilkinson, Phys. Rev. *168*, 970 (1968); copyright 1968 by the American Physical Society]. Notice that 10^{12} Hz $= 4.137$ meV.

Table 9.2 Frequencies ω_{LO} and ω_{TO} in cm^{-1} ($= 0.124$ meV) of longitudinal optical and transverse optical phonons for six semiconductors. The calculations are taken from P. Giannozzi, S. de Gironcoli, P. Pavone and S. Baroni, Phys. Rev. B *43*, 7231 (1991), to which we refer for further details; experimental data are in parentheses. The static and the high-frequency dielectric constants are also given; notice that the ratio $\omega_{LO}^2/\omega_{TO}^2$ equals (within experimental error) the ratio $\varepsilon_s/\varepsilon_\infty$.

	Si	Ge	GaAs	AlAs	GaSb	AlSb
ω_{LO}	517 (517)	306 (304)	291 (291)	400 (402)	237 (233)	334 (344)
ω_{TO}	517 (517)	306 (304)	271 (271)	363 (361)	230 (224)	316 (323)
$\omega_{LO}^2/\omega_{TO}^2$	1	1	1.15 (1.17)	1.22 (1.24)	1.06 (1.08)	1.12 (1.14)
ε_s	12.1	16.5	12.40	10.06	15.69	12.04
ε_∞	12.1	16.5	10.60	8.16	14.44	10.24
$\varepsilon_s/\varepsilon_\infty$	1	1	1.17	1.23	1.09	1.17

Figure 9.7 is the longitudinal-transverse splitting of optical modes at $\mathbf{q} \approx 0$; this splitting is the fingerprint of the long-range nature of interatomic forces, and is connected with the break of cubic symmetry, due to the induced dipoles accompanying the vibrational modes. A simplified model for the study of the long wavelength optical phonons

is given in Section 9.7. In Table 9.2 we report for convenience the frequencies of the optical phonons at the center of the Brillouin zone for the elemental semiconductors Si and Ge, and for the polar semiconductors GaAs, AlAs, GaSb, AlSb.

As a final example, we report in Figure 9.8 the phonon branches of LiF. Lithium fluoride is a typical ionic material with the NaCl structure; there are two ions in the unit cell, and there are thus three acoustic and three optical branches. The long-range nature of interionic forces produces a strong longitudinal-transverse splitting of optical modes at $\mathbf{q} \approx 0$. In LiF the ratio of the low-frequency dielectric constant ($\varepsilon_s = 8.9$) and high-frequency dielectric constant ($\varepsilon_\infty = 1.9$) is relatively large, and so is the squared ratio of the measured LO and TO mode frequencies.

9.4 Quantum Theory of the Harmonic Crystal

9.4.1 Canonical Transformation to Independent Oscillators

In the previous sections, we have discussed the lattice vibrations by means of the classical equations of motion. We can revisit the problem from a quantum mechanical point of view, and show that the classical and quantum treatments are completely equivalent as far as dispersion curves are concerned; on the other hand, the quantum treatment of the elastic field shows that energies are discretized into quanta, called phonons.

In Section 9.1, we have considered the classical dynamics of a monoatomic linear chain; we consider now the quantum mechanical counterpart of the same problem. In the harmonic approximation, and nearest neighbor interactions (see Eq. (9.10)), the Hamiltonian of the linear chain becomes

$$H = \sum_n \frac{1}{2M} p_n^2 + \frac{1}{2} C \sum_n \left(2u_n^2 - u_n u_{n+1} - u_n u_{n-1} \right), \tag{9.26}$$

where u_n and p_n are the coordinate and conjugate moment of the nucleus at the n-th site; these observables obey the commutation rules

$$[u_n, p_{n'}] = i\hbar \delta_{n,n'}; \quad [u_n, u_{n'}] = [p_n, p_{n'}] = 0. \tag{9.27}$$

Instead of the dynamical variables u_n and p_n, it is convenient to perform a canonical transformation, with the final aim to put in diagonal form the Hamiltonian (9.26). We pass to the collective variables so defined

$$p_q = \frac{1}{\sqrt{N}} \sum_{t_n} e^{+iqt_n} p_n, \qquad u_q = \frac{1}{\sqrt{N}} \sum_{t_n} e^{-iqt_n} u_n. \tag{9.28a}$$

The collective operators are defined as linear combinations of the dynamical variables of all the nuclei, with appropriate phase factors $\exp(\pm iqt_n)$ as suggested by the Bloch form. The inverse transformations are

$$p_n = \frac{1}{\sqrt{N}} \sum_q e^{-iqt_n} p_q, \qquad u_n = \frac{1}{\sqrt{N}} \sum_q e^{+iqt_n} u_q. \tag{9.28b}$$

The passage between direct and inverse transformations exploits the standard properties

$$\frac{1}{N}\sum_{t_n} e^{-i(q-q')t_n} = \delta_{q,q'} \quad \text{and} \quad \frac{1}{N}\sum_{q} e^{-iq(t_n - t_{n'})} = \delta_{n,n'}. \tag{9.29}$$

It is easily seen that the transformations (9.28) from the set of operators u_n, p_n, to the set of operators u_q, p_q are canonical, i.e. the commutation rules are preserved:

$$[u_q, p_{q'}] = i\hbar\delta_{qq'}; \quad [u_q, u_{q'}] = [p_q, p_{q'}] = 0.$$

Notice also the following relationships $u_q^\dagger = u_{-q}$ and $p_q^\dagger = p_{-q}$; this means that the hermitian conjugates of collective operators of wavenumber q are the linear combinations of wavenumber $-q$.

We can now express the original Hamiltonian in terms of the new collective displacements and conjugate momenta. We have

$$\sum_n u_n^2 = \frac{1}{N}\sum_n \sum_{qq'} e^{i(q+q')t_n} u_q u_{q'} = \sum_q u_q u_{-q},$$

and

$$\sum_n p_n^2 = \sum_n p_q p_{-q}.$$

Similarly we have

$$\sum_n u_n u_{n+1} = \frac{1}{N}\sum_n \sum_{qq'} e^{i(q+q')t_n + iq'a} u_q u_{q'} = \sum_q u_q u_{-q} e^{-iqa}.$$

The Hamiltonian (9.26) thus becomes

$$H = \sum_q \frac{1}{2M} p_q p_{-q} + \frac{1}{2}C \sum_q u_q u_{-q}(2 - e^{-iqa} - e^{iqa}).$$

We thus obtain the following relation:

$$\boxed{H = \sum_q \left[\frac{1}{2M} p_q p_{-q} + \frac{1}{2} M\omega_q^2 u_q u_{-q} \right]}, \tag{9.30}$$

where

$$\omega^2(q) = \frac{C}{M}(2 - e^{-iqa} - e^{iqa}) = \frac{4C}{M}\sin^2\frac{qa}{2}.$$

This relation shows that the linear chain of N coupled harmonic oscillators is equivalent to N uncoupled normal modes of the appropriate frequency specified above, a result in agreement with Eq. (9.12b).

9.4.2   Quantization of the Elastic Field and Phonons

It is useful to perform another canonical transformation to appropriate creation and annihilation operators. In analogy with the quantum theory of the harmonic oscillator (see Appendix A) we define

$$a_q = \sqrt{\frac{M\omega_q}{2\hbar}}\, u_q + i\sqrt{\frac{1}{2M\hbar\omega_q}}\, p_{-q}, \tag{9.31a}$$

$$a_q^\dagger = \sqrt{\frac{M\omega_q}{2\hbar}}\, u_{-q} - i\sqrt{\frac{1}{2M\hbar\omega_q}}\, p_q. \tag{9.31b}$$

The commutation rules are preserved by the transformations (9.31):

$$\left[a_q, a_{q'}^\dagger\right] = \delta_{qq'}; \quad \left[a_q, a_{q'}\right] = \left[a_q^\dagger, a_{q'}^\dagger\right] = 0.$$

The inverse transformations of Eqs. (9.31) are

$$u_q = \sqrt{\frac{\hbar}{2M\omega_q}}\left(a_q + a_{-q}^\dagger\right) \quad \text{and} \quad p_q = (-i)\sqrt{\frac{M\hbar\omega_q}{2}}\left(a_{-q} - a_q^\dagger\right),$$

as can be easily verified by inspection.

Using Eq. (9.28b), the expression of the local atomic displacements and of the local conjugate momenta can be cast in the form

$$u_n = \frac{1}{\sqrt{N}}\sum_q e^{iq t_n}\sqrt{\frac{\hbar}{2M\omega_q}}\left(a_q + a_{-q}^\dagger\right), \tag{9.32a}$$

$$p_n = \frac{1}{\sqrt{N}}\sum_q e^{-iq t_n}(-i)\sqrt{\frac{M\hbar\omega_q}{2}}\left(a_{-q} - a_q^\dagger\right). \tag{9.32b}$$

We now re-write the Hamiltonian (9.30) of the elastic field in second quantization form. We have

$$\frac{1}{2M} p_q p_{-q} = \frac{1}{2M}(-i)^2 \frac{M\hbar\omega_q}{2}\left(a_{-q} - a_q^\dagger\right)\left(a_q - a_{-q}^\dagger\right)$$

$$= -\frac{\hbar\omega_q}{4}\left[a_{-q}a_q - a_{-q}a_{-q}^\dagger - a_q^\dagger a_q + a_q^\dagger a_{-q}^\dagger\right].$$

Similarly it holds

$$\frac{1}{2}M\omega_q^2 u_q u_{-q} = \frac{1}{2}M\omega_q^2 \frac{\hbar}{2M\omega_q}(a_q + a_{-q}^\dagger)(a_{-q} + a_q^\dagger)$$

$$= \frac{\hbar\omega_q}{4}\left[a_q a_{-q} + a_q a_q^\dagger + a_{-q}^\dagger a_{-q} + a_{-q}^\dagger a_q^\dagger\right].$$

It follows

$$\frac{1}{2M} p_q p_{-q} + \frac{1}{2}M^2\omega_q^2 u_q u_{-q} = \frac{\hbar\omega_q}{2}\left[a_q^\dagger a_q + a_{-q}^\dagger a_{-q} + 1\right].$$

The Hamiltonian (9.30) of the linear chain then becomes

$$
H = \sum_q \hbar\omega(q) \left(a_q^\dagger a_q + \frac{1}{2} \right),
\tag{9.33}
$$

which is the sum of the Hamiltonians of N independent linear harmonic oscillators of frequency $\omega(q)$.

Completely similar analysis and conclusions could be performed for the diatomic linear chain of Section 9.2, and the general three-dimensional crystal of Section 9.3; the dispersion relations provided by the quantum mechanical treatment, and by the classical treatment are the same, since the unitary transformation from localized variables to itinerant (or collective) variables are the same in the classical and quantum treatment. The quantum theory thus recovers the same $\omega = \omega(\mathbf{q}, p)$ dispersion curves of the classical theory; however the quantum theory leads to the quantization of the elastic field in terms of phonons, which can be considered as traveling quasiparticles of energy $\hbar\omega(\mathbf{q}, p)$, wavevector $\mathbf{q}$, and branch index p ($p = 1, 2, \ldots, 3v_b$).

9.5   Lattice Heat Capacity. Einstein and Debye Models

Consider a crystal composed by N unit cells and a basis of v_b atoms in the unit cell; the crystal volume is $V = N\Omega$, and the total number of atoms is $N_a = N v_b$. In the harmonic approximation, the system of N_a vibrating atoms is equivalent to a system of $3N_a$ independent (one-dimensional) oscillators of frequency $\omega = \omega(\mathbf{q}, p)$, where $\mathbf{q}$ assumes N allowed values in the first Brillouin zone, and p runs over the $3v_b$, branches of the phonon dispersion curves. The average vibrational energy of the harmonic crystal is the sum of independent phonon contributions; according to the Bose-Einstein statistics, we have

$$
U_{\text{vibr}}(T) = \sum_{\mathbf{q}p} \left[\frac{\hbar\omega(\mathbf{q}, p)}{e^{\hbar\omega(\mathbf{q},p)/k_B T} - 1} + \frac{1}{2}\hbar\omega(\mathbf{q}, p) \right].
$$

The *lattice heat capacity at constant volume*, using Eq. (3.21a), is given by

$$
C_V(T) = \frac{\partial U_{\text{vibr}}}{\partial T} = \frac{\partial}{\partial T} \sum_{\mathbf{q}p} \frac{\hbar\omega(\mathbf{q}, p)}{e^{\hbar\omega(\mathbf{q},p)/k_B T} - 1}.
\tag{9.34}
$$

The above expression allows the numerical calculation of the lattice heat capacity of crystals, once the phonon dispersion curves $\omega(\mathbf{q}, p)$ are known. In the following we consider the application of Eq. (9.34) to simple models of dispersion curves, that can be worked out analytically.

Einstein Model

In the Einstein model, the actual frequencies of the normal modes are replaced by a unique (average) frequency ω_e (Einstein frequency). If N_a is the total number of atoms,

Eq. (9.34) for the heat capacity at constant volume becomes

$$C_V(T) = 3N_a \frac{\partial}{\partial T} \frac{\hbar\omega_e}{e^{\hbar\omega_e/k_B T} - 1} = 3N_a k_B \left(\frac{\hbar\omega_e}{k_B T}\right)^2 \frac{e^{\hbar\omega_e/k_B T}}{(e^{\hbar\omega_e/k_B T} - 1)^2}.$$

The behavior of $C_V(T)$ in the low and high temperature limits is

$$C_V \to e^{-\hbar\omega_e/k_B T} \quad \text{for} \quad k_B T \ll \hbar\omega_e$$

and

$$C_V \to 3N_a k_B \quad \text{for} \quad k_B T \gg \hbar\omega_e.$$

In the high temperature limit the Einstein model recovers the Dulong and Petit value $3N_a k_B$. In the low temperature limit, the Einstein model predicts for $C_V(T)$ an exponentially vanishing behavior, contrary to the T^3 experimental law. The origin of this discrepancy is the presence in crystals of the phonon acoustic branches, which cannot be mimicked by a unique Einstein frequency, and actually need a more realistic description.

Debye Model

Any three-dimensional crystal, with or without a basis, presents three acoustic branches with linear dispersion $\omega = v_s q$ for small q. For simplicity we assume the same sound velocity v_s for each of the three acoustic branches and extend the linear dispersion relation to the whole Brillouin zone. To avoid inessential details, we approximate the Brillouin zone with a sphere (Debye sphere) of equal volume (in order to preserve the total number of allowed wavevectors); we indicate with q_D the radius of the Debye sphere and define $\omega_D = v_s q_D$ as the *cutoff Debye frequency*. We notice that $(4/3)\pi q_D^3 = (2\pi)^3/\Omega$, where Ω is the volume of the unit cell in the direct space.

The density of phonon states corresponding to a branch with linear dispersion relation $\omega = v_s q$ is easily obtained. In fact the number of states $D(\omega)\,d\omega$ with frequency in the interval $[\omega, \omega + d\omega]$ equals the number of states in the reciprocal space with wavevector between $[q, q + dq]$; namely:

$$D(\omega)\,d\omega = \frac{V}{(2\pi)^3} 4\pi q^2 dq = \frac{V}{(2\pi)^3} 4\pi \frac{\omega^2}{v_s^2} d\frac{\omega}{v_s}.$$

It follows

$$D(\omega) = \frac{V}{(2\pi)^3} 4\pi \frac{\omega^2}{v_s^3} = \frac{N\Omega}{(2\pi)^3} \frac{4\pi q_D^3}{3} \frac{3\omega^2}{\omega_D^3} = N\frac{3\omega^2}{\omega_D^3}, \quad 0 \leq \omega \leq \omega_D, \qquad (9.35)$$

where N is the number of unit cells of the crystal.

The contribution of the three acoustic branches to the average vibrational energy (apart the constant zero point energy) is

$$U_{vibr}^{(acoustic)}(T) = 3 \int_0^{\omega_D} N\frac{3\omega^2}{\omega_D^3} \frac{\hbar\omega}{e^{\hbar\omega/k_B T} - 1}\,d\omega. \qquad (9.36)$$

It is convenient to perform the change of variables $x = \hbar\omega/k_B T$ and define $x_D = \hbar\omega_D/k_B T = T_D/T$, where $T_D = \hbar\omega_D/k_B$ is called Debye temperature. The expression (9.36) becomes

$$U_{\text{vibr}}^{(\text{acoustic})}(T) = 9Nk_B T \left(\frac{T}{T_D}\right)^3 \int_0^{x_D} \frac{x^3}{e^x - 1} \, dx. \tag{9.37}$$

In the high temperature limit $T_D \ll T$, $x_D \ll 1$ and $e^x - 1 \approx x$. The integral in Eq. (9.37) then gives $x_D^3/3$ and hence $U = 3Nk_B T$; thus for $T \gg T_D$, the heat capacity of the three acoustic branches approaches the value $3Nk_B$ of the Dulong and Petit law.

In the low temperature limit $T \ll T_D$ we can replace $x_D = \infty$, and the integral in Eq. (9.37) equals $\pi^4/15$ [I. S. Gradshteyn and M. Ryzhik, "Table of Integrals, Series, and Products" (Academic, New York, 1980, p. 325)]. Equation (9.37) thus gives

$$U_{\text{vibr}}^{(\text{acoustic})}(T) = \frac{3}{5}\pi^4 Nk_B \frac{T^4}{T_D^3}, \quad T \ll T_D.$$

Correspondingly, in the low temperature region the heat capacity becomes

$$C_V(T) = \frac{12}{5}\pi^4 Nk_B \frac{T^3}{T_D^3}, \quad T \ll T_D$$

and the correct experimental T^3 behavior is reproduced.

The Debye model can be refined in several ways. For instance the three acoustic branches could be treated with different sound velocities. In the case of crystals with a basis, one could use the Debye model for the acoustic modes and the Einstein model for the optical modes. We notice that, in the high temperature limit, anharmonic effects are of increasing importance, and corrections to the Dulong and Petit value are likely to be of significance. In metals, besides the vibrational contribution to the internal energy, we have to consider the electronic contribution; the electronic contribution to the heat capacity is proportional to T at any temperature and may become the dominant term at very low temperatures (see Section 3.3). We notice finally that the T^3 law depends on the crystal dimensionality. In a two-dimensional crystal, instead of Eq. (9.35), the density-of-states $D(\omega)$ is proportional to ω and the low temperature lattice heat capacity is characterized by a T^2 power law. Similarly, in an ideal one-dimensional crystal, one would obtain a lattice heat capacity linear in the temperature.

9.6   Considerations on Anharmonic Effects and Melting of Solids

So far we have confined our attention to the harmonic approximation for the lattice vibrations; in this approximation, phonons are elementary excitations of the elastic field, which do not decay, and do not interact with each other. The anharmonic terms, which correspond to cubic, quartic, and successive terms in the series expansion of the crystal potential energy, have quite important consequences; for instance, cubic terms make possible three-phonon processes in which one-phonon decays into two phonons or two phonons merge into one. Among the physical effects of anharmonicity, we mention the thermal expansion of solids, the change of normal mode frequencies with temperature (or other parameters), the thermal resistivity, the broadening of one-phonon peaks in neutron scattering experiments, the solid-liquid transition. It is not our intention to discuss the wealth of problems related to anharmonicity; here we simply provide some intuitive remarks concerning the amplitude of localized motions of the atoms and the Lindemann criterion of melting.

The mean quadratic displacement of a given atom about its equilibrium position is an important quantity, which influences X-ray scattering, neutron scattering, Mössbauer effect, and determines the solid-liquid transition. Thus it is useful to estimate the mean quadratic displacement of an atom around its equilibrium position as a function of temperature.

For simplicity, consider a three-dimensional crystal with N unit cells and one atom per unit cell. In this case the normal modes consist of three acoustic branches, of frequency $\omega(\mathbf{q}, p)$, and polarization vectors $\mathbf{A}(\mathbf{q}, p)$ ($p = 1, 2, 3$). To avoid unessential details, the three acoustic branches are assumed degenerate, so that $\omega(\mathbf{q}, p)$ does not depend on the branch index p. Furthermore, for any wavevector, the three polarization vectors $\mathbf{A}(\mathbf{q}, p)$ form an orthonormal tern, which we intend to orient parallel to some fixed reference frame. The displacement $\mathbf{u}_n$ of the atom at $\mathbf{t}_n$ can be expanded in normal modes; in analogy to Eq. (9.32a), the component of $\mathbf{u}_n$ along a given z-direction reads

$$u_{nz} = \sum_{\mathbf{q}} \sqrt{\frac{\hbar}{2NM\omega(\mathbf{q})}} \, e^{i\mathbf{q}\cdot\mathbf{t}_n} \left[a_{\mathbf{q}} + a_{-\mathbf{q}}^{\dagger} \right],$$

where the sum extends on phonons polarized along the chosen z-direction (one phonon for each given wavevector).

We can now calculate the ensemble average $\langle u_{nz}^2 \rangle$ of the quadratic displacement u_{nz}^2. Without loss of generality, we take $\mathbf{t}_n = 0$ and use the results

$$\langle a_{\mathbf{q}}^{\dagger} a_{\mathbf{q}} \rangle = \frac{1}{\exp(\hbar\omega_{\mathbf{q}}/k_B T) - 1}, \qquad \langle a_{\mathbf{q}} a_{\mathbf{q}}^{\dagger} \rangle = \langle a_{\mathbf{q}}^{\dagger} a_{\mathbf{q}} \rangle + 1,$$

where $\langle a_{\mathbf{q}}^{\dagger} a_{\mathbf{q}} \rangle$ is the well-known Bose population factor (see Appendix A). The average square displacement $\langle u_z^2 \rangle$ of each atom thus becomes

$$\langle u_z^2 \rangle = \sum_{\mathbf{q}} \frac{\hbar}{2NM\omega(\mathbf{q})} \left[\frac{2}{\exp(\hbar\omega_{\mathbf{q}}/k_B T) - 1} + 1 \right]. \tag{9.38}$$

It is instructive to calculate the average square displacement for the Debye model of the phonon spectrum. We have already seen that the density-of-states for any of the three branches is given by Eq. (9.35). Equation (9.38) thus becomes

$$\langle u_z^2 \rangle = \int_0^{\omega_D} \frac{\hbar}{2NM\omega} \left(\frac{2}{e^{\hbar\omega/k_B T} - 1} + 1 \right) N \frac{3\omega^2}{\omega_D^3} \, d\omega.$$

If we introduce the dimensionless variable $x = \hbar\omega/k_B T$ and define $\hbar\omega_D = k_B T_D$, we have

$$\langle u_z^2 \rangle = 3\frac{\hbar^2 T^2}{M k_B T_D^3} \int_0^{T_D/T} \left(\frac{1}{e^x - 1} + \frac{1}{2} \right) x\, dx. \tag{9.39}$$

The integral can be easily performed in the limits of very high temperatures ($T \gg T_D$) using a series development of the exponential, and of very low temperatures ($T \ll T_D$). We have respectively

$$\langle u_z^2 \rangle = \begin{cases} \dfrac{3}{4}\dfrac{\hbar^2}{M k_B T_D} & \text{for } T \ll T_D, \\[2mm] 3\dfrac{\hbar^2 T}{M k_B T_D^2} & \text{for } T \gg T_D. \end{cases} \tag{9.40}$$

From Eqs. (9.40), it is seen that the value of $\langle u_z^2 \rangle$ at zero temperature is only somewhat smaller than the value of $\langle u_z^2 \rangle$ at the Debye temperature; in fact $\langle u_z^2 \rangle_{T=T_D} \approx 4\langle u_z^2 \rangle_{T=0}$.

We can now establish a simple qualitative criterion for melting. Let r_0 be the mean radius of the unit cell, and consider the ratio

$$f = \frac{\sqrt{\langle u_x^2 \rangle + \langle u_y^2 \rangle + \langle u_z^2 \rangle}}{r_0} = \sqrt{\frac{9\hbar^2 T}{M k_B T_D^2 r_0^2}}, \qquad T \gg T_D. \tag{9.41a}$$

When the ratio f reaches a phenomenological value f_c (almost independent of the specific solid in consideration), melting is expected to occur. The melting temperature is thus given by the Lindemann formula

$$T_m = \frac{f_c^2}{9\hbar^2} M k_B T_D^2 r_0^2; \tag{9.41b}$$

the critical value f_c turns out to be of the order of 0.2–0.3 in many solids.

Another interesting speculation concerns the stability of ideal and infinite one-dimensional and two-dimensional crystals. In these cases, the calculation of the mean quadratic displacement in the plane or in the chain (using the appropriate density of phonon states) leads to a divergent value at any finite temperature. Thus one-dimensional and two-dimensional crystals are unstable in the harmonic approximation; some three-dimensional interaction (whatever small with respect to intralayer or intrachain interaction) is necessary to stabilize low-dimensional structures.

The lattice instability of ideal and infinite low-dimensional systems has been inferred from the presence of arbitrary small frequency modes in the long wavelength limit. It is evident that finite size effects (or other deviations from ideal periodicity) may invalidate assumptions (and conclusions); thus much caution is needed to extend results elaborated for ideal situations to realistic ones. Typical is the situation of graphene, a monolayer of carbon atoms studied in Section 6.5, which can cover relatively large free-standing areas in spite of its two-dimensional structure.

9.7 Optical Phonons and Polaritons in Polar Crystals

9.7.1 General Considerations

In the previous sections we have studied the crystal lattice vibrations by means of the *dynamical matrix formalism* (or "interatomic force constant method"), which constitutes a workable and invaluable tool for the treatment of systems of electronic-nuclear particles interacting through Coulomb potentials. The tenet of this approach, as summarized in the introduction of Section 9.1, is based on the adiabatic principle and the ground Born-Oppenheimer energy surface, supposed non-degenerate. [In the case the lowest adiabatic surface is degenerate or quasi-degenerate with other potential surfaces, we have to handle a Jahn-Teller or a pseudo Jahn-Teller system; in this regard, we only mention that a simple paradigmatic Jahn-Teller system has been discussed in the previous chapter.] The expansion coefficients of the non-degenerate ground potential energy to second order in the nuclear displacements, represent the interatomic force constant, and the diagonalization of the corresponding matrix gives the normal vibrational modes or phonons of the crystal. This dynamical matrix procedure is a valuable starting point for further elaborations (such as phonon-phonon interactions due to anharmonic terms, or electron-phonon interactions due to the phonon-induced perturbation of the electronic potential). A more fundamental intrinsic limit of the dynamical matrix approach is its neglecting the interaction of phonons with the retarded modes of the electromagnetic field. The focus of this section is the study of retardation effects due to the finite speed of light, and the concept of new quasiparticles called *polaritons*.

In Section 9.3, in the discussion of the phonon dispersion curves of some materials, we have noticed that in polar crystals (such as ionic crystals and heteropolar semiconductors), optical phonons are electric-dipole active. In these materials it is thus expected that the mechanical waves (phonons) and the electromagnetic waves (photons) can be mutually coupled to each other, giving rise to mixed phonon-photon modes, which are the genuine internal states of the system.

As a preliminary to further considerations, let us briefly summarize the distinctive features of vibrational curves of homopolar and heteropolar cubic crystals with two atoms per unit cell, as given by *the interatomic force constant method* (for specific examples see Figure 9.6 for homopolar silicon and germanium, and Figure 9.7 for heteropolar semiconductors). From the general discussion of Section 9.3, we have seen that the dispersion curves of a diatomic crystal consist of three acoustic branches and three optical branches. In cubic crystals, for vectors $\mathbf{q}$ along high symmetry directions, there are two degenerate transverse acoustic modes (TA) and one longitudinal acoustic mode (LA). Similarly, there are two degenerate transverse optical modes (TO) and one longitudinal optical mode (LO), whose frequencies go to finite values ω_{TO} and ω_{LO} in the long wavelength limit. In nonpolar diatomic cubic crystals, the optical modes for small wavelengths are degenerate, i.e. $\omega_{LO} \equiv \omega_{TO}$ (Figure 9.9a). On the contrary in polar crystals, when retardation effects are neglected, the frequencies ω_{LO} and ω_{TO} of longitudinal and transverse optical phonons are different, in spite of the cubic symmetry of the crystal (Figure 9.9b); this symmetry-breaking effect is due to the long-range nature and anisotropy of the electrostatic dipole-dipole forces.

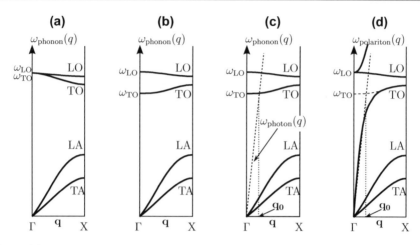

Figure 9.9 (a) Schematic behavior of phonon dispersion curves of a cubic *homopolar* semiconductor (or insulator) with two equal atoms per unit cell, along a high symmetry direction. (b) Schematic behavior of phonon dispersion curves of a cubic *heteropolar* semiconductor (or insulator) with two different atoms per unit cell, when retardation effects are ignored. (c) Superposition of the phonon dispersion curves of the previous diagram and photon dispersion curve $\omega(\mathbf{q}) = cq/\sqrt{\varepsilon_\infty}$, assuming momentarily no coupling between electromagnetic waves and lattice vibrations. In the figure, the slope of the photon dispersion curve (of the order of the velocity of light) and the slope of the acoustic modes (of the order of sound velocity) could not be drawn in scale. (d) Schematic picture of *polariton effects*. Phonons and photons with nearly equal wavevectors and energies interact and determine the polariton dispersion curves, indicated by solid lines (dashed lines repeat the non-ineracting situation, for a direct comparison).

The longitudinal-transverse splitting has implications on the infrared dielectric properties of polar semiconductors and insulators. These materials are strongly reflecting in the frequency region $\omega_{TO} < \omega < \omega_{LO}$; repeated reflexions have long been used to select a band of wavelengths of infrared radiation, which is known as Reststrahlen (residual rays) radiation. Furthermore, the frequencies ω_{LO} and ω_{TO} satisfy the Lyddane-Sachs-Teller relation

$$\boxed{\frac{\omega_{LO}^2}{\omega_{TO}^2} = \frac{\varepsilon_s}{\varepsilon_\infty}}, \tag{9.42}$$

where ε_s is the static dielectric constant and ε_∞ is the high-frequency dielectric constant. By ε_∞ we mean the infrared dielectric constant at frequencies much higher than a typical phonon frequency, $\approx 10^{13}$ s^{-1} (so that ionic displacement contribution can be neglected) and much smaller than any electronic transition frequency, $\approx 10^{15}$ s^{-1}; ε_∞ is determined by the electronic contribution to the static dielectric constant. If ω_{LO} and ω_{TO} are significantly different, the same occurs for ε_s and ε_∞ (and vice versa); in some materials (such as ferroelectric ionic crystals), ω_{TO} is anomalously small and ε_s anomalously large.

In order to obtain a preliminary qualitative picture of the region in the Brillouin zone, where retardation effects are important, it is convenient to superimpose the phonon dispersion curve of Figure 9.9b with the photon dispersion curve $\omega(\mathbf{q}) = cq/\sqrt{\varepsilon_\infty}$, assuming momentarily no coupling between electromagnetic waves and lattice vibrations. In Figure 9.9c, $q_0 \approx \sqrt{\varepsilon_\infty}\omega_{TO}/c$ denotes the wavevector of the photon of energy ω_{TO}. Photons and phonons strongly interact near the crossing point q_0 of the corresponding dispersion curves, which are modified into polariton dispersion curves (Figure 9.9d). Polariton effects extend from $q = 0$ to approximately q_0, which is a fraction of the order of $v_s/c \approx 10^{-5}$ of the Brillouin zone dimension (this region is reported in an expanded scale in Figures 9.10 and 9.11). Notice that polariton effects restore the threefold degeneracy of the state of frequency ω_{LO} at $q = 0$.

We pass now to interpret the phenomenological properties of polar crystals mentioned above, with a suitable continuous model.

9.7.2 Lattice Vibrations in Polar Crystals and Polaritons

The Continuous Approximation for Optical Vibrational Modes in Isotropic Materials

We can establish a reasonable simple model for polaritons, combining a continuous approximation for the description of the mechanical waves of the optical modes and the Maxwell's equations for the description of the electromagnetic waves [J. J. Hopfield and D. G. Thomas, Phys. Rev. *132*, 563 (1963)]. We confine our attention here to polar

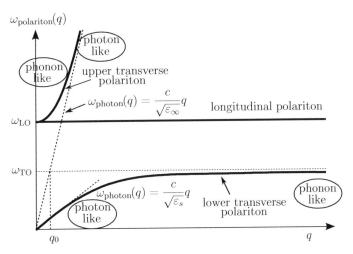

Figure 9.10 Schematic description of polaritons. The dispersion curves of uncoupled transverse phonons and photons are shown by dashed lines; q_0 is the crossing point $q_0 = \omega_{TO}/(c/\sqrt{\varepsilon_\infty})$; dispersion curves of the longitudinal phonon, and transverse polaritons are shown by solid lines. In the lower polariton branch, the character of the dispersion curve changes from photon-like for $q < q_0$ to phonon-like for $q > q_0$; in the upper branch, the character changes from phonon-like to photon-like as q increases.

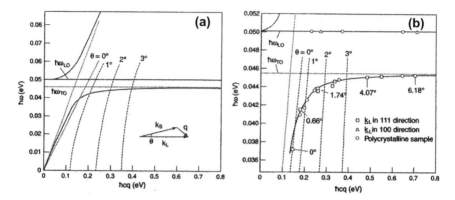

Figure 9.11 Polariton dispersion curves in GaP. (a) The vector diagram of Raman spectroscopy measurements is also indicated; $\mathbf{k}_L$, $\mathbf{k}_S$, and $\mathbf{q}$ are the wavevectors of the incident laser photon, scattered Stokes photon, and polariton; θ is the scattering angle. Values of energies and wavevectors which are kinematically possible at angle θ are shown by long dashed lines. (b) The plot of the observed energies and wavectors of polaritons and LO phonons are given. The figures are reprinted with permission from C. H. Henry and J. J. Hopfield, Phys. Rev. Lett. *15*, 964 (1965); copyright 1965 by the American Physical Society.

cubic crystals with two atoms (cation and anion) in the unit cell, of effective charge $\pm e^*$, mass M_1 and M_2 (and reduced mass M^*). In optical vibrational modes, cations and anions move against each other; in analogy with Eq. (9.19), in isotropic cubic crystals, the relative displacement $\mathbf{w}_n$ of the ions in the n-th unit cell, or the corresponding electric dipole $\mathbf{d}_n = e^* \mathbf{w}_n$, obey the equation of motion of the form

$$\ddot{\mathbf{w}}_n = -\omega_0^2 \mathbf{w}_n + \frac{e^*}{M^*}\mathbf{E} \implies \ddot{\mathbf{d}}_n = -\omega_0^2 \mathbf{d}_n + \frac{e^{*2}}{M^*}\mathbf{E};\tag{9.43}$$

for simplicity, in this section, the local electric field acting on the ions is taken equal to the average electric field entering the Maxwell's equations.

The electric polarization field due to the collection of oscillating dipoles in the crystal can be written as

$$\mathbf{P}_{\text{ion}}(\mathbf{r},t) = \frac{N}{V}\,\mathbf{d}(\mathbf{r},t),$$

where the discrete index n in Eq. (9.43), labeling unit cells, has been replaced by the continuous space variable; as usual N indicates the number of unit cells and V the volume of the crystal. The equation of motion of the polarization is then

$$\ddot{\mathbf{P}}_{\text{ion}} = -\omega_0^2 \mathbf{P}_{\text{ion}} + \frac{N}{V}\frac{e^{*2}}{M^*}\mathbf{E}.$$

The last equation is more significantly written in the form

$$\boxed{\ddot{\mathbf{P}}_{\text{ion}} = -\omega_0^2 \mathbf{P}_{\text{ion}} + \omega_0^2 \frac{\varepsilon_s - \varepsilon_\infty}{4\pi}\mathbf{E}},\tag{9.44}$$

where the two phenomenological dielectric constants ε_s and ε_∞ at low-and high-frequency replace the two microscopic constants e^* and M^*. In fact, assuming separation of the lattice and the electronic contributions to the polarization, in static conditions we have

$$\varepsilon_s = 1 + 4\pi \frac{P_{el}}{E} + 4\pi \frac{P_{ion}}{E} = \varepsilon_\infty + 4\pi \frac{P_{ion}}{E} \implies \frac{P_{ion}}{E} = \frac{\varepsilon_s - \varepsilon_\infty}{4\pi};$$

this result is automatically contained in Eq. (9.44), when considered in the static limit. Equation (9.44) is the very useful "constitutive" equation of polar crystals; it couples the electric polarization, produced by the vibrating lattice of ions, to the electric field in the crystal; the phenomenological coupling constant is given by $\omega_0^2(\varepsilon_s - \varepsilon_\infty)/4\pi$.

We now consider the above constitutive equation together with the Maxwell's equations for the electromagnetic fields. In the absence of external charges and currents, the Maxwell's equations read

$$\begin{cases} \text{div } \mathbf{D} = 0 \qquad \text{div } \mathbf{B} = 0, \\ \text{curl } \mathbf{E} = -\dfrac{1}{c}\dfrac{\partial \mathbf{B}}{\partial t} \qquad \text{curl } \mathbf{H} = \dfrac{1}{c}\dfrac{\partial \mathbf{D}}{\partial t}. \end{cases} \qquad (9.45)$$

In non-magnetic materials we also assume $\mathbf{B} = \mathbf{H}$. The electric-magnetic-mechanical modes of propagations controlled by Eqs. (9.44) and (9.45) are called polaritons. We consider first longitudinal polaritons modes and then the dispersion relations for the transverse polaritons modes.

Longitudinal Polariton Modes

We consider first the propagation in the medium of longitudinal optical vibrations, in which case the polarization field and the electric field are also expected to be of longitudinal type with the form

$$\mathbf{P}_{ion}(\mathbf{r}, t) = \mathbf{P}_0\, e^{i(\mathbf{q}\cdot\mathbf{r}-\omega t)} \quad \text{with } \mathbf{P}_0 \parallel \mathbf{q}, \qquad (9.46a)$$

$$\mathbf{E}(\mathbf{r}, t) = \mathbf{E}_0\, e^{i(\mathbf{q}\cdot\mathbf{r}-\omega t)} \quad \text{with } \mathbf{E}_0 \parallel \mathbf{q}. \qquad (9.46b)$$

Since the propagation modes are longitudinal, the Maxwell's equations containing the curl-operator are automatically satisfied, and the only equations to care are those involving the divergence. The magnetic field is then zero (since its divergence and curl are zero), while for the electric field we have to consider the constitutive equation and the first of the four Maxwell's equations.

The electric induction vector is defined as

$$\mathbf{D} = \mathbf{E} + 4\pi \mathbf{P}_{el} + 4\pi \mathbf{P}_{ion} = \varepsilon_\infty \mathbf{E} + 4\pi \mathbf{P}_{ion}.$$

For longitudinal modes we have thus the following equations:

$$\begin{cases} \ddot{\mathbf{P}}_{ion} = -\omega_0^2\, \mathbf{P}_{ion} + \omega_0^2 \dfrac{\varepsilon_s - \varepsilon_\infty}{4\pi} \mathbf{E}, \\ \text{div}[\varepsilon_\infty \mathbf{E} + 4\pi \mathbf{P}_{ion}] = 0. \end{cases}$$

The exponential time dependence of Eqs. (9.46), and the projection on the direction of the field give

$$\begin{cases} -\omega^2 P_0 = -\omega_0^2 P_0 + \omega_0^2 \dfrac{\varepsilon_s - \varepsilon_\infty}{4\pi} E_0 \,, \\ \varepsilon_\infty E_0 + 4\pi P_0 = 0 \,. \end{cases}$$

The second of the two linear equations does not contain any frequency dependence, and both are independent of light velocity and $\mathbf{q}$-wavevector; thus, for the longitudinal waves one obtains the dispersion less relation

$$longitudinal\ modes: \quad \omega^2 = \omega_0^2 \frac{\varepsilon_s}{\varepsilon_\infty} \equiv \omega_{LO}^2 \quad with \quad E_0 = -\frac{4\pi}{\varepsilon_\infty} P_0 \,. \qquad (9.47)$$

The effect of the long-range Coulomb field on the longitudinal optical vibrations does not introduce any dispersion; it however does increase the "restoring forces" and does increase the oscillation frequency from ω_0 (the value neglecting long-range contributions) to ω_{LO}. Since for longitudinal phonons in the bulk isotropic materials the electromagnetic field is purely electrostatic, the shift of the longitudinal mode with respect to the frequency ω_0 (which neglects long-range dipole-dipole interactions), is contained in the dynamical matrix formalism, and is here recovered from a different point of view. The real novelty of the present treatment does not concern longitudinal modes, where retardation effects are absent, but the forthcoming study of the propagation of transverse fields.

Transverse Polariton Modes

Consider now the propagation in the medium of *transverse optical vibrations*, in which case the polarization field and the electric field are of the type

$$\mathbf{P}_{\text{ion}}(\mathbf{r}, t) = \mathbf{P}_0 \, e^{i(\mathbf{q}\cdot\mathbf{r} - \omega t)} \quad with \quad \mathbf{P}_0 \perp \mathbf{q}\,, \qquad (9.48a)$$

$$\mathbf{E}(\mathbf{r}, t) = \mathbf{E}_0 \, e^{i(\mathbf{q}\cdot\mathbf{r} - \omega t)} \quad with \quad \mathbf{E}_0 \perp \mathbf{q}\,. \qquad (9.48b)$$

In this case the two Maxwell's equations (9.45) with the divergence are automatically satisfied and we have to consider the other two involving curl-operators:

$$\operatorname{curl} \mathbf{E} = -\frac{1}{c} \frac{\partial \mathbf{B}}{\partial t}\,, \qquad (9.49a)$$

$$\operatorname{curl} \mathbf{H} = \frac{1}{c} \frac{\partial \mathbf{D}}{\partial t} = \frac{\varepsilon_\infty}{c} \frac{\partial \mathbf{E}}{\partial t} + \frac{4\pi}{c} \frac{\partial \mathbf{P}_{\text{ion}}}{\partial t}\,. \qquad (9.49b)$$

Instead of two first order differential equations in the magnetic and electric field, it is convenient and equivalent to consider a single second order differential equation in the electric field; taking the curl-operator of both members of the above equations, and assuming $\mathbf{B} = \mathbf{H}$ we have

$$\operatorname{curl}\operatorname{curl} \mathbf{E} = -\frac{\varepsilon_\infty}{c^2} \ddot{\mathbf{E}} - \frac{4\pi}{c^2} \ddot{\mathbf{P}}_{\text{ion}}\,.$$

We now use the well-known relation

$$\text{curl curl} \ldots = \text{grad div} \ldots - \nabla^2 \ldots$$

We have thus to consider the two equations

$$\begin{cases} \ddot{\mathbf{P}}_{ion} = -\omega_0^2 \mathbf{P}_{ion} + \omega_0^2 \dfrac{\varepsilon_s - \varepsilon_\infty}{4\pi} \mathbf{E}, \\ -\nabla^2 \mathbf{E} = -\dfrac{\varepsilon_\infty}{c^2} \ddot{\mathbf{E}} - \dfrac{4\pi}{c^2} \ddot{\mathbf{P}}_{ion}. \end{cases} \tag{9.50}$$

The exponential time dependence of Eqs. (9.48), and the projection in the direction of the field give

$$\begin{cases} -\omega^2 P_0 = -\omega_0^2 P_0 + \omega_0^2 \dfrac{\varepsilon_s - \varepsilon_\infty}{4\pi} E_0 \\ q^2 E_0 = \dfrac{\varepsilon_\infty}{c^2} \omega^2 E_0 + \dfrac{4\pi}{c^2} \omega^2 P_0 \end{cases} \implies \begin{vmatrix} \omega_0^2 - \omega^2 & -\omega_0^2 \dfrac{\varepsilon_s - \varepsilon_\infty}{4\pi} \\ -4\pi\omega^2 & c^2 q^2 - \varepsilon_\infty \omega^2 \end{vmatrix} = 0.$$

The determinant vanishes for

$$\varepsilon_\infty \omega^4 - \left(\omega_0^2 \varepsilon_s + c^2 q^2 \right) \omega^2 + \omega_0^2 c^2 q^2 = 0. \tag{9.51}$$

We can solve for ω^2 and obtain the dispersion relation for polaritons

$$\omega^2 = \frac{1}{2\varepsilon_\infty} \left[\omega_0^2 \varepsilon_s + c^2 q^2 \pm \sqrt{(\omega_0^2 \varepsilon_s + c^2 q^2)^2 - 4\omega_0^2 c^2 q^2 \varepsilon_\infty} \right]. \tag{9.52}$$

The dispersion curves for polaritons, given by Eq. (9.52), are schematically shown in Figure 9.10.

It is interesting to examine the lower and upper polariton branches in the limit of $q < q_0$ and $q > q_0$, where q_0 is the point for which $(c/\sqrt{\varepsilon_\infty})q_0 \equiv \omega_{TO}$. For $q \gg q_0$, the two solutions of Eq. (9.52), and the corresponding amplitudes E_0 and P_0 are

$$\begin{cases} \omega^2 = \omega_0^2 = \omega_{TO}^2 & E_0 = 0 \quad P_0 \neq 0 \\ \omega^2 = \dfrac{c^2}{\varepsilon_\infty} q^2 & P_0 = 0 \quad E_0 \neq 0 \end{cases} \qquad q \gg q_0 = \frac{\omega_{TO}}{c/\sqrt{\varepsilon_\infty}};$$

this shows that the lower branch is a pure mechanical wave (with $E_0 = 0$) and the upper branch is a pure electromagnetic wave (with $P_0 = 0$). In summary: for $q \gg q_0$ the retardation effects are negligible.

For $q \ll q_0$, the two solutions of Eq. (9.52), and the corresponding amplitudes E_0 and P_0 are

$$\begin{cases} \omega^2 = \omega_0^2 \dfrac{\varepsilon_s}{\varepsilon_\infty} = \omega_{LO}^2 & E_0 = -\dfrac{4\pi}{\varepsilon_\infty} P_0 \\ \omega^2 = \dfrac{c^2}{\varepsilon_s} q^2 & P_0 = \dfrac{\varepsilon_s - \varepsilon_\infty}{4\pi} E_0 \end{cases} \qquad q \leq q_0 = \frac{\omega_{TO}}{c/\sqrt{\varepsilon_\infty}}.$$

For $q < q_0$ the interaction of the phonon with the retarded electromagnetic field must be taken into account. At $q = 0$ this interaction produces a shift by an amount exactly equal to the longitudinal-transverse splitting. For q smaller or near q_0 the polariton modes have the character of coupled mechanical-electromagnetic waves, with E_0 and P_0 *simultaneously* different from zero. Notice in particular that for $\omega = \omega_{LO}$ we have $E_0 = -(4\pi/\varepsilon_\infty)P_0$ both for transverse and longitudinal waves (see Eq. (9.47)), and the degeneracy of transverse and longitudinal modes is restored at $\mathbf{q} = 0$.

As an illustrative example of the concepts developed so far, we report in Figure 9.11 the polariton dispersion curves in GaP. The dispersion curves for optical phonons of long wavelength, in the absence of coupling to photons, are horizontal straight lines; transverse optical phonons and photons in the small wavevector limit are strongly coupled by the phonon-photon interaction and lead to the polariton dispersion curves.

We have seen that dipole-active phonons and photons with nearly the same energy and wavevector are strongly coupled; we notice that quite similar coupling effects occur also for photons and other dipole-active quasiparticles encountered in the physics of crystals. In general, the term "polariton" is used to denote mixed states produced from the coupling of dipole-carrying quasiparticles (phonons, excitons, plasmons) with the retarded modes of the electromagnetic field. In particular, the mixed exciton-photon states are called exciton-polaritons and their dispersion curves have a behavior qualitatively similar to the polariton curves so far discussed (see for instance the review articles of L. C. Andreani, and references quoted therein). An example of dispersion curves of exciton-polaritons in CuCl is reported in Figure 9.12.

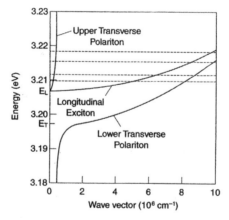

Figure 9.12 Dispersion curves of the transverse exciton-polaritons of CuCl in the anomalous dispersive region. The dashed lines refer to the laser energies used to study luminescence line shape due the decay from bound pairs of excitons [reprinted with permission from J. K. Pribram, G. L. Koos, F. Bassani and J. P. Wolfe, Phys. Rev. B **28**, 1048 (1983); copyright 1983 by the American Physical Society].

Infrared Dielectric Properties of Polar Crystals

We can now discuss the infrared dielectric properties of polar crystals exploiting the "constitutive" equation (9.44), which couples the ionic polarization to the electric field in the medium, here re-written for convenience in the form

$$\ddot{\mathbf{P}}_{\text{ion}} = -\omega_0^2 \, \mathbf{P}_{\text{ion}} - \eta \dot{\mathbf{P}}_{\text{ion}} + \omega_0^2 \frac{\varepsilon_s - \varepsilon_\infty}{4\pi} \, \mathbf{E} \,, \tag{9.53}$$

where we have taken the liberty to add to the equation of motion a frictional term controlled by the parameter $\eta > 0$ (which can be considered finite or eventually as an infinitesimal positive quantity). Consider the response of the system to a time dependent driving electric field, periodic in space and time, of the form

$$\mathbf{E}(\mathbf{r}, t) = \mathbf{E}_0 \, e^{i(\mathbf{q} \cdot \mathbf{r} - \omega t)}. \tag{9.54a}$$

By analogy with Eq. (9.54a), we assume for the ionic polarization the expression

$$\mathbf{P}_{\text{ion}}(\mathbf{r}, t) = \mathbf{P}_0 \, e^{i(\mathbf{q} \cdot \mathbf{r} - \omega t)}. \tag{9.54b}$$

Replacing Eqs. (9.54) into Eq. (9.53), one obtains

$$\mathbf{P}_0 = \frac{\omega_0^2}{\omega_0^2 - \omega^2 - i\eta\,\omega} \frac{\varepsilon_s - \varepsilon_\infty}{4\pi} \, \mathbf{E}_0 \tag{9.55}$$

(for the present model under consideration, it is irrelevant whether $\mathbf{E}_0$ and $\mathbf{P}_0$ are parallel or orthogonal to the vector $\mathbf{q}$).

The dielectric function is given by $\varepsilon(\omega) = \varepsilon_\infty + 4\pi P_0/E_0$, where as before ε_∞ denotes the dielectric constant due to the electronic polarizability (at frequencies well-below any electronic transition resonance). Using Eq. (9.55), the dielectric function of the polar crystal becomes

$$\boxed{\varepsilon(\omega) = \varepsilon_\infty + \frac{\omega_0^2}{\omega_0^2 - \omega^2 - i\eta\omega}(\varepsilon_s - \varepsilon_\infty)}. \tag{9.56}$$

The real and imaginary parts of the dielectric function are schematically indicated in Figure 9.13 (for clarity of the figure η is taken small, but finite).

In the limit $\eta \to 0^+$ the real part of Eq. (9.56) reads

$$\varepsilon_1(\omega) = \varepsilon_\infty + \frac{\omega_0^2}{\omega_0^2 - \omega^2}(\varepsilon_s - \varepsilon_\infty) = \frac{\varepsilon_s \, \omega_0^2 - \varepsilon_\infty \, \omega^2}{\omega_0^2 - \omega^2}. \tag{9.57a}$$

In the same limit $\eta \to 0^+$ the imaginary part of Eq. (9.56) becomes

$$\varepsilon_2(\omega) = \omega_0^2(\varepsilon_s - \varepsilon_\infty) \, \text{Im} \, \frac{1}{\omega_0^2 - \omega^2 - i\eta\omega} = \omega_0^2(\varepsilon_s - \varepsilon_\infty) \, \pi \, \text{sign}(\omega) \, \delta(\omega_0^2 - \omega^2)$$

and

$$\varepsilon_2(\omega) = \frac{\pi \omega_0(\varepsilon_s - \varepsilon_\infty)}{2} \left[\delta(\omega - \omega_0) - \delta(\omega + \omega_0) \right]. \tag{9.57b}$$

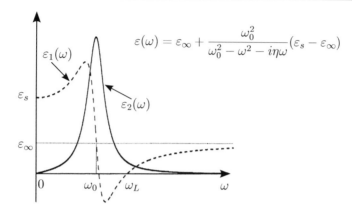

$$\varepsilon(\omega) = \varepsilon_\infty + \frac{\omega_0^2}{\omega_0^2 - \omega^2 - i\eta\omega}(\varepsilon_s - \varepsilon_\infty)$$

Figure 9.13 Schematic behavior of the real and imaginary part of the dielectric function $\varepsilon(\omega)$ of a polar crystal in the infrared region. Since $\varepsilon_1(\omega)$ and $\varepsilon_2(\omega)$ are even and odd functions of ω, respectively, only the part $\omega > 0$ is indicated. For $\eta \to 0^+$, the real part $\varepsilon_1(\omega)$ presents a pole at $\omega = \omega_0$, while the imaginary part $\varepsilon_2(\omega)$ presents a δ-like singularity at $\omega = \omega_0$.

At positive frequencies, $\varepsilon_1(\omega)$ exhibits a pole for $\omega = \omega_0 = \omega_{TO}$ (transverse phonon frequency) and has a zero for $\omega = \omega_0\sqrt{\varepsilon_s/\varepsilon_\infty} = \omega_{LO}$ (longitudinal phonon frequency); the values ω_{TO} and ω_{LO} satisfy the Lyddane-Sachs-Teller relation. The dielectric function $\varepsilon_1(\omega)$ is negative for $\omega_{TO} < \omega < \omega_{LO}$. In this region the reflectivity equals one, and the electromagnetic propagation in the crystal is forbidden. Outside the interval $[\omega_{TO}, \omega_{LO}]$ the dielectric function $\varepsilon_1(\omega)$ is positive, and polariton modes are possible.

The knowledge of the dielectric response of the system, together with the standard Maxwell's equations, allow one to infer the elementary excitations of the system. For longitudinal modes, the zero divergence of the induction electric field entails

$$\varepsilon(\omega) = 0 \quad \Longrightarrow \quad \omega^2 = \frac{\varepsilon_s}{\varepsilon_\infty}\omega_0^2 \quad \textit{longitudinal modes.}$$

For transverse waves, curl-operations on the fields of the Maxwell's equations (see Eqs. (9.50) and similar) entail

$$q^2 = \frac{\varepsilon(\omega)}{c^2}\omega^2 \quad \textit{transverse polaritons.}$$

Using for $\varepsilon_1(\omega)$ the expression (9.57a), one obtains

$$\omega^2 \frac{\varepsilon_s \omega_0^2 - \varepsilon_\infty \omega^2}{\omega_0^2 - \omega^2} = c^2 q^2. \tag{9.58}$$

It can be immediately seen that Eq. (9.58) coincides exactly with Eq. (9.51), and thus defines the dispersion curves of the polaritons in the crystal.

9.7.3   Local Field Effects on Polaritons

The Internal Field According to Lorentz

In the discussion of Section 9.7.2 we have assumed that the macroscopically averaged electric field $\mathbf{E}$ and the local field $\mathbf{E}_{loc}$ are the same. In solids, however, there can be significant differences between the two fields, and a central (and not easy) problem in the theory of dielectrics is the calculation of the electric field at the position of a given atom or molecule. Without entering in all the subtleties of this problem, we here briefly discuss the internal electric field according to Lorentz.

Consider an isotropic dielectric crystal with the shape of a bar, very long in the z-direction (see Figure 9.14). Imagine that aligned microscopic electric dipoles are set up at the lattice points so to give rise to a uniform polarization $\mathbf{P}$, e.g. in the z-direction. We notice that since $\mathbf{P}$ is uniform, we have $\rho_{micr} = -\text{div} \, \mathbf{P} = 0$ and no volume microscopic charge density is accompanying the polarization. We also notice that the geometry of the (thin and very long) bar is chosen so that $\mathbf{P}$ is parallel to the surface of the sample; thus no discontinuity of the normal component of $\mathbf{P}$ occurs at the surfaces, and no microscopic surface charge density occurs either. Because of the absence of any internal and external charges, we have $\text{div} \, \mathbf{E} = 0$; in stationary situations also curl $\mathbf{E} = 0$ and the electric field is thus zero.

Although the uniformly polarized specimen in the geometry of Figure 9.14 is in a *null electric field*, it is easily realized that the *local electric field acting on the microscopic dipoles*, which are the origin of the macroscopic polarization field $\mathbf{P}$, is in general different from zero. For simplicity, we confine our attention to the extreme tight-binding limit, in which the crystal can be viewed as a collection of microscopic electric dipoles, well-localized around the lattice sites; in this case, the local field can be obtained with the following arguments.

Imagine to carve a small spherical region around the site at which the local field is to be evaluated. The medium contained in this region is considered as a discretized collection of dipoles and we determine the electric field at the center of the cavity by summing up the electric fields generated by every dipole (except the one at the origin);

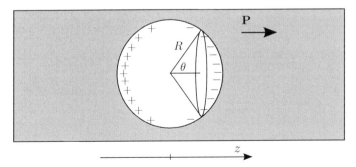

Figure 9.14 Schematic representation of the Lorentz cavity for the calculation of the local electric field. The sample (with the ideal shape of a thin and infinitely long bar) is supposed to be uniformly polarized with $\mathbf{P}$ parallel to the surface of the sample.

when certain conditions of symmetry are fulfilled, the sum may vanish. For simplicity, we focus our attention on structures with sufficiently high local symmetry (some cubic structures, for instance), so that the electric field generated by the point-like dipoles within the cavity vanishes at its center.

The uniformly polarized medium outside the Lorentz cavity is dealt with in the continuum approximation (Figure 9.14). The contribution due to the dipoles outside the ideal cavity, of radius R, can be obtained noticing that the discontinuity of the component of **P** normal to the surface implies a microscopic density of surface polarization charge given by

$$\sigma_p = P_n = -P \cos \theta.$$

The electric field in the z-direction at the center of the cavity due to the polarization charges at the cavity surface is

$$E_0 = \int_0^\pi (-P \cos \theta) \cdot R \, d\theta \cdot 2\pi R \sin \theta \cdot \frac{-1}{R^2} \cos \theta = \frac{4\pi}{3} P. \tag{9.59}$$

From Eq. (9.59), we see that in the specified isotropic materials in null electric field, the local field is $(4\pi/3)\mathbf{P}$. If the electric field **E** applied to the material is different from zero, the local field acting on the dipoles is the sum of **E** and $(4\pi/3)\mathbf{P}$. In summary, in the Lorentz model, the relationship between local field, average macroscopic field and electric polarization is given by

$$\mathbf{E}_{\text{loc}} = \mathbf{E} + \frac{4\pi}{3} \mathbf{P}. \tag{9.60a}$$

The above expression has several limitations. These may be due to the overlapping of electronic clouds, or dipolar fields far from homogeneity. At times the Lorentz field is better approximated by the generalized form

$$\mathbf{E}_{\text{loc}} = \mathbf{E} + \gamma \frac{4\pi}{3} \mathbf{P}, \tag{9.60b}$$

where γ is a semiempirical parameter. The case $\gamma = 0$ indicates no distinction between local and average field (this is the case of free electrons or essentially spread out wavefunctions), while $\gamma = 1$ is the case of strong localized dipoles in highly symmetric crystals. In even more refined models, γ may be different for different sublattices.

Internal Field, Polarizability, and Dielectric Constant of Materials

Consider a system that can be visualized as constituted by N atoms (or molecules) in the volume V, and suppose for simplicity that the interaction between different atoms (or molecules) can be neglected. We wish to express the dielectric constant ε of the material in terms of the polarizability α of the composing units.

In the presence of an applied field **E**, the average polarization due to induced dipoles of polarizability α is

$$\mathbf{P} = \frac{N}{V} \alpha \, \mathbf{E}_{\text{loc}} = \frac{N}{V} \alpha \left(\mathbf{E} + \gamma \frac{4\pi}{3} \mathbf{P} \right).$$

Hence

$$\mathbf{P} = \frac{(N/V)\alpha}{1 - \gamma(4\pi/3)(N/V)\alpha}\mathbf{E}. \tag{9.61}$$

It is interesting a brief discussion of Eq. (9.61). In the case the local field and the macroscopic field are the same ($\gamma = 0$), we have $\mathbf{P} = (N/V)\alpha\mathbf{E}$; the polarization $\mathbf{P}$ is thus finite for any finite polarizability. In the case the local field and the macroscopic field are different ($\gamma \neq 0$), the polarization tends to diverge if

$$\gamma \frac{4\pi}{3} \frac{N}{V} \alpha \to 1 ; \tag{9.62}$$

this condition is known as "polarization catastrophe." For ordinary dielectrics, the denominator in Eq. (9.61) is safely far from vanishing condition. For very special crystals, candidate to become ferroelectric, the polarization catastrophe considerations are basic for understanding physical and structural properties near phase transition. Notice that the polarization catastrophe concept is inherent to the local field theory ($\gamma \neq 0$) and is essentially a cooperative effect.

From Eq. (9.61), we obtain for the dielectric constant the expression

$$\varepsilon = 1 + 4\pi \frac{P}{E} \implies \boxed{\varepsilon = 1 + \frac{4\pi(N/V)\alpha}{1 - \gamma(4\pi/3)(N/V)\alpha}}. \tag{9.63}$$

Equation (9.63) expresses the dielectric constant in terms of the polarizability α of the composing units and of the parameter γ, which characterizes the local field. In the specific case $\gamma = 1$ (Lorentz field), Eq. (9.63) can also be written in the form

$$\frac{\varepsilon - 1}{\varepsilon + 2} = \frac{4\pi}{3} \frac{N}{V}\alpha ,$$

which is named Lorentz-Lorenz (or Clausius-Mossotti) relation.

Infrared Dielectric Function and Polaritons in Polar Crystals in the Presence of Local Field Effects

In Section 9.7.2 the study of polaritons and optical properties of polar crystals in the infrared region has been done starting from the equation of motion (9.43). In the presence of local field effects we have rather to consider the equation of motion of the type

$$\ddot{\mathbf{w}} = -\omega_0^2 \mathbf{w} - \eta \dot{\mathbf{w}} + \frac{e^*}{M^*}\mathbf{E}_{\mathrm{loc}} \tag{9.64}$$

(again, it is convenient to add a frictional term, controlled by the positive parameter η, which can be eventually considered an infinitesimal quantity). We now study the consequences brought about by the fact that the local electric field and the macroscopic electric field may be different.

Consider Eq. (9.64) when the electric field and the relative displacement are periodic in space and time with the form

$$\mathbf{E}_{\text{loc}} = \mathbf{E}_0 \, e^{i(\mathbf{q} \cdot \mathbf{r} - \omega t)} \quad \text{and} \quad \mathbf{w} = \mathbf{w}_0 \, e^{i(\mathbf{q} \cdot \mathbf{r} - \omega t)}.$$

We obtain

$$-\omega^2 \, \mathbf{w}_0 = -\omega_0^2 \, \mathbf{w}_0 + i\eta \, \omega \, \mathbf{w}_0 + \frac{e^*}{M^*} \mathbf{E}_0 \quad \Longrightarrow \quad \mathbf{w}_0 = \frac{e^*}{M^*} \frac{1}{\omega_0^2 - \omega^2 - i\eta \, \omega} \mathbf{E}_0.$$

Thus the polarizability due to the lattice ionic displacements is

$$\alpha(\omega) \equiv \frac{e^* w_0}{E_0} = \alpha_{\text{ion}} \frac{\omega_0^2}{\omega_0^2 - \omega^2 - i\eta\omega} \quad \text{with} \quad \alpha_{\text{ion}} = \frac{e^{*2}}{M^* \omega_0^2}.$$

The *ionic* polarizability has a significant frequency dependence in the infrared region.

Let us indicate with α_+ and α_- the *electronic* polarizabilities of the cation and the anion of the polar crystal, and with α_{el} their sum; in the infrared region we can neglect any frequency dependence of electronic polarizabilities. Assuming that electronic and ionic polarizabilities add up, the total polarizability (per unit cell) becomes

$$\alpha_{\text{tot}}(\omega) = \alpha_{\text{el}} + \alpha_{\text{ion}} \frac{\omega_0^2}{\omega_0^2 - \omega^2 - i\eta\omega} = \frac{(\alpha_{\text{el}} + \alpha_{\text{ion}})\omega_0^2 - \alpha_{\text{el}}(\omega^2 + i\eta\omega)}{\omega_0^2 - \omega^2 - i\eta\omega}. \quad (9.65)$$

Inserting Eq. (9.65) into Eq. (9.63), the dielectric function becomes

$$\varepsilon(\omega) = 1 + \frac{4\pi(N/V)\alpha_{\text{tot}}(\omega)}{1 - \gamma(4\pi/3)(N/V)\alpha_{\text{tot}}(\omega)}. \quad (9.66)$$

It is convenient to introduce the two quantities so defined

$$A_{\text{el}} = \frac{4\pi}{3} \frac{N}{V} \alpha_{\text{el}} \quad \text{and} \quad A_{\text{ion}} = \frac{4\pi}{3} \frac{N}{V} \alpha_{\text{ion}}.$$

With straight algebra Eq. (9.66) can be cast in the form

$$\varepsilon(\omega) = 1 + \frac{3(A_{\text{el}} + A_{\text{ion}})\omega_0^2 - 3A_{\text{el}}(\omega^2 + i\eta\omega)}{(1 - \gamma A_{\text{el}} - \gamma A_{\text{ion}})\omega_0^2 - (1 - \gamma A_{\text{el}})(\omega^2 + i\eta\omega)}. \quad (9.67)$$

With an eye to the denominator of Eq. (9.67) we define the "renormalized transverse frequency" ω_{TO} as

$$\omega_{TO}^2 = \omega_0^2 \frac{1 - \gamma A_{\text{el}} - \gamma A_{\text{ion}}}{1 - \gamma A_{\text{el}}}. \quad (9.68)$$

Notice that the renormalized transverse frequency is always less than the initial frequency ω_0; the two frequencies become equal only in the case local field effects are negligible, which means $\gamma = 0$. From Eqs. (9.67) and (9.68) we obtain

$$\varepsilon(\omega) = 1 + \frac{3(A_{\text{el}} + A_{\text{ion}})\omega_0^2 - 3A_{\text{el}}(\omega^2 + i\eta\omega)}{1 - \gamma A_{\text{el}}} \frac{1}{\omega_{TO}^2 - (\omega^2 + i\eta\omega)}. \quad (9.69)$$

The above expression, in the limiting case of static and high-frequency regions, takes the values

$$\varepsilon_s = 1 + \frac{3(A_{el} + A_{ion})\omega_0^2}{(1 - \gamma A_{el})\omega_{TO}^2} \quad \text{and} \quad \varepsilon_\infty = 1 + \frac{3A_{el}}{1 - \gamma A_{el}}.$$

Equation (9.69) thus becomes

$$\varepsilon(\omega) = 1 + \frac{(\varepsilon_s - 1)\omega_{TO}^2 - (\varepsilon_\infty - 1)(\omega^2 + i\eta\omega)}{\omega_{TO}^2 - (\omega^2 + i\eta\omega)}$$

$$\equiv \varepsilon_\infty + \frac{\omega_{TO}^2}{\omega_{TO}^2 - \omega^2 - i\eta\omega}(\varepsilon_s - \varepsilon_\infty). \qquad (9.70)$$

Comparison of Eq. (9.70) with Eq. (9.56) is self-explanatory; we see that local field effects do not change the form of $\varepsilon(\omega)$, except for the "renormalization" of the transverse and longitudinal frequencies ω_{TO} and ω_{LO}. In particular the transverse frequency (9.68) decreases with respect to the short-range value ω_0 as an effect of long-range Coulomb interaction and tends to become *soft*. In any case the renormalized transverse and longitudinal frequencies are still related by the Lyddane-Sachs-Teller relation, as this depends on the analytic structure of the response function, rather than on the details of the local fields. We do not have to discuss again the polariton dispersion curves, as the treatment can be performed following step-by-step the previous section, once transverse and longitudinal frequencies are renormalized.

An interesting implication of the local field theory is the possible occurrence of *soft phonon modes*. From Eq. (9.66) we see that ω_{TO} is reduced with respect to ω_0, since the long-range Coulomb interaction tends to counteract the short-range restoring forces. The limiting case of $\omega_{TO} \rightarrow 0$ denotes an incipient lattice instability, since the relative positions of anions and cations in the unit cell tend to displace with zero restoring forces. From Eq. (9.68) it is seen that the condition $\omega_{TO} = 0$ occurs for

$$\omega_{TO} \equiv 1 - \gamma \frac{4\pi}{3}\frac{N}{V}(\alpha_{el} + \alpha_{ion}) \equiv 0 \implies \gamma \frac{4\pi}{3}\frac{N}{V}\alpha_{tot}(0) \equiv 1;$$

this show that the "lattice instability condition" $\omega_{TO} = 0$ and the "polarization catastrophe" condition (9.62) are equivalent.

In polar materials, in the case due to anharmonicity or some other mechanism, the frequency ω_{TO} tends to zero at some temperature, from the Lyddane-Sachs-Teller relation we expect that ε_s tends to infinity. Thus a polar crystal, which exhibits a transverse-optical branch with a low-frequency mode ω_{TO}, is candidate to develop an extraordinary large polarization. Eventually the crystal might undergo a phase transition and acquire a spontaneous polarization, even in the absence of external fields. This is the typical behavior of displacive ferroelectrics such as barium titanate and lead titanate, whose transition temperatures are 381 K and 763 K, respectively. Neutron measurements and far infrared optical measurements well support the role of a soft transverse branch in some perovskite ionic crystals. [For further information on the displacive ferroelectrics, as well as other mechanisms of ferroelectricity, see for instance the monograph of R. Blinc (2011) and references quoted therein.]

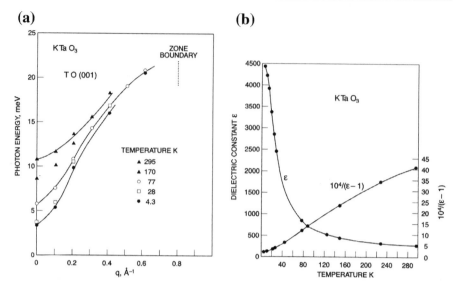

Figure 9.15 (a) Temperature dependence of the soft transverse-optical branch in KTaO$_3$ [reprinted with permission from G. Shirane, R. Nathans and V. J. Minkiewicz, Phys. Rev. *157*, 396 (1967); copyright 1967 by the American Physical Society]. (b) Dielectric constant and reciprocal susceptibility of KTaO$_3$ as a function of temperature [reprinted with permission from S. H. Wemple, Phys. Rev. *137*, A1575 (1965); copyright 1965 by the American Physical Society].

As an example, we consider the case of perovskite potassium tantalite, an incipient ferroelectric material, where the frequency of the transverse optical mode decreases with temperature but never goes to zero. We report in Figure 9.15 the temperature dependence of the soft transverse optical branch (studied by inelastic neutron scattering techniques), as well as the dielectric constant measurements. From Figure 9.15 it can be seen that the phonon energy of the soft mode at $q = 0$ is 10.7 meV at 295 K, and decreases to 3.1 meV at 4 K; correspondingly the dielectric constant passes from the value $\varepsilon = 243$ at 295 K to very large values (exceeding several thousands) at low temperatures.

Appendix A.   Quantum Theory of the Linear Harmonic Oscillator

Creation and Annihilation Operators

We summarize here some results of the quantum theory of the linear harmonic oscillator, that are preliminary and useful for the discussion of lattice vibrations of crystals.

Consider a one-dimensional harmonic oscillator, of angular frequency ω, described by the Hamiltonian

$$H = \frac{1}{2M} p_x^2 + \frac{1}{2} M \omega^2 x^2.$$

(A.1)

It is convenient to express the above Hamiltonian putting in evidence the natural unit of energy $\hbar\omega$ of the harmonic oscillator, namely:

$$H = \frac{1}{2}\hbar\omega\left[\frac{1}{M\hbar\omega}p_x^2 + \frac{M\omega}{\hbar}x^2\right].$$

The above expression can be further cast into the form

$$H = \frac{1}{2}\hbar\omega\left[\frac{l_0^2}{\hbar^2}p_x^2 + \frac{1}{l_0^2}x^2\right] \qquad \text{with} \qquad l_0^2 = \frac{\hbar}{M\omega}, \tag{A.2}$$

where l_0 is the natural unit of length of the harmonic oscillator under attention.

The structure of Eq. (A.2) suggests to introduce the lowering (or annihilation) operator and the raising (or creation) operator, defined by the following linear transformations of the observables x and p_x:

$$a = \frac{1}{\sqrt{2}}\left[\frac{1}{l_0}x + i\frac{l_0}{\hbar}p_x\right], \quad a^\dagger = \frac{1}{\sqrt{2}}\left[\frac{1}{l_0}x - i\frac{l_0}{\hbar}p_x\right]; \quad l_0 = \sqrt{\frac{\hbar}{M\omega}}. \tag{A.3}$$

The inverse transformations are

$$x = \frac{1}{\sqrt{2}}l_0(a + a^\dagger), \quad p_x = \frac{-i}{\sqrt{2}}\frac{\hbar}{l_0}(a - a^\dagger). \tag{A.4}$$

The operators a and $a^\dagger$ satisfy the commutation rule

$$[a, a^\dagger] = aa^\dagger - a^\dagger a = 1. \tag{A.5}$$

In terms of a and $a^\dagger$, the Hamiltonian (A.1), or the equivalent one (A.2), takes the form

$$H = \hbar\omega\left(a^\dagger a + \frac{1}{2}\right), \tag{A.6}$$

as can be easily verified inserting expressions (A.4) into Eq. (A.2).

In order to work out eigenvalues and eigenfunctions of the Hamiltonian (A.6), we note a few relationships from the commutation relation (A.5). We have

$$aa^\dagger = a^\dagger a + 1, \quad aa^{\dagger 2} = (a^\dagger a + 1)a^\dagger = a^{\dagger 2}a + 2a^\dagger,$$

and in general

$$aa^{\dagger n} = a^{\dagger n}a + na^{\dagger n-1}. \tag{A.7}$$

Let $|0\rangle$ denote the normalized state that satisfies the equation $a|0\rangle = 0$; and let $|n\rangle$ indicate the normalized state

$$|n\rangle = \frac{1}{\sqrt{n!}}a^{\dagger n}|0\rangle. \tag{A.8}$$

The correctness of the normalization follows from the observation that

$$\langle 0|a^n a^{\dagger n}|0\rangle = \langle 0|a^{n-1} a a^{\dagger n}|0\rangle = n\langle 0|a^{n-1} a^{\dagger n-1}|0\rangle = n!,$$

where use has been done of Eq. (A.7). With similar procedures, it follows

$$a^\dagger a|n\rangle = n|n\rangle.$$

The number operator $a^\dagger a$ indicates the number of quanta (phonons) in the state $|n\rangle$. The eigenvalues of the Hamiltonian (A.6) are thus $E_n = (n + \frac{1}{2})\hbar\omega$ with $n = 0, 1, 2, \ldots$

From the expression (A.8) of the normalized eigenstates of the harmonic oscillator, we see that the operators a and $a^\dagger$ satisfy the relations

$$a|n\rangle = \sqrt{n}|n-1\rangle, \quad a^\dagger|n\rangle = \sqrt{n+1}|n+1\rangle.$$

We also notice that

$$\langle n|a^{\dagger p} a^p|n\rangle = \langle n|a^{\dagger p}|n-p\rangle\langle n-p|a^p|n\rangle = (\sqrt{n-p+1} \cdots \cdot \sqrt{n})^2, \quad n \geq p.$$

Thus

$$\langle n|a^{\dagger p} a^p|n\rangle = \begin{cases} n!/(n-p)! & \text{if } n \geq p, \\ 0 & \text{if } n < p. \end{cases} \tag{A.9}$$

Statistical Average of Operators

At thermodynamic equilibrium, the statistical average of an operator A is defined as

$$\langle A\rangle = \sum_{n=0}^{\infty} P_n\langle n|A|n\rangle, \tag{A.10}$$

where

$$P_n = \frac{e^{-\left(n+\frac{1}{2}\right)\hbar\omega/k_B T}}{\sum_m e^{-(m+\frac{1}{2})\hbar\omega/k_B T}} = \frac{e^{-n\hbar\omega/k_B T}}{\sum_m (e^{-\hbar\omega/k_B T})^m}.$$

Summing up the geometric series in the denominator, and replacing it into Eq. (A.10), one obtains

$$\boxed{\langle A\rangle = (1-z)\sum_{n=0}^{\infty} z^n\langle n|A|n\rangle \quad \text{with} \quad z = \exp\left(-\hbar\omega/k_B T\right).} \tag{A.11}$$

Using Eq. (A.11) the thermal average of the number operator reads

$$\langle a^\dagger a\rangle = (1-z)\sum_{n=0}^{\infty} z^n\langle n|a^\dagger a|n\rangle = (1-z)\sum_{n=0}^{\infty} n\, z^n$$

$$= (1-z)z\frac{\partial}{\partial z}\sum_{n=0}^{\infty} z^n = \frac{z}{1-z} = \frac{1}{e^{\hbar\omega/k_B T} - 1}, \tag{A.12}$$

which expresses the standard Bose-Einstein statistics. It also holds

$$\langle aa^\dagger \rangle = \langle a^\dagger a \rangle + 1; \quad \langle aa \rangle = \langle a^\dagger a^\dagger \rangle = 0.$$

With a little of algebra, we can prove the following relation:

$$\boxed{\langle a^{\dagger p} a^p \rangle = p! \langle a^\dagger a \rangle^p} \quad \text{for any } p = 0, 1, 2, \ldots \tag{A.13}$$

In fact, from Eqs. (A.11) and (A.9) we have

$$\langle a^{\dagger p} a^p \rangle = (1 - z) \sum_{n(\geq p)} \frac{n!}{(n-p)!} z^n = (1 - z) \sum_{n=0}^{\infty} \frac{(n+p)!}{n!} z^{n+p}$$

$$= p! z^p (1 - z) \sum_{n=0}^{\infty} \frac{(n+p)!}{n! p!} z^n = p! z^p (1 - z) \frac{1}{(1-z)^{p+1}} = p! \frac{z^p}{(1-z)^p}.$$

This last result together with Eq. (A.12) proves Eq. (A.13).

Weyl Identity

We establish two identities (now the Weyl identity and later the Bloch identity), which are very useful in the study of the correlation functions and Debye-Waller factor in the scattering theory of the harmonic crystal (see Chapter 10).

Consider any *two linear operators A and B under the assumption that they commute with their commutator*; for the two corresponding exponentials the following multiplication rule holds:

$$\boxed{e^A e^B = e^{A+B} e^{[A,B]/2}} \quad \text{if } [A, [A, B]] \equiv [B, [A, B]] \equiv 0. \tag{A.14}$$

The proof of the above identity can be performed, for instance, following a procedure due to Glauber. We replace momentarily the operators A and B by xA and xB, respectively, where the real parameter x will be set equal to 1 at the end of the reasoning. We consider then the following two operators depending on the x parameter:

$$F_1(x) = e^{x(A+B)} e^{x^2[A,B]/2} \quad \text{and} \quad F_2(x) = e^{xA} e^{xB}. \tag{A.15}$$

It is seen by inspection that the operator functions $F_1(x)$ satisfy the differential equation

$$\frac{dF_1}{dx} = \{A + B + x[A, B]\} F_1(x). \tag{A.16}$$

Quite similarly, the differentiation of $F_2(x)$ gives

$$\frac{dF_2}{dx} = \left[A + e^{xA} B e^{-xA} \right] F_2(x) = \{A + B + x[A, B]\} F_2(x). \tag{A.17}$$

The last passage in Eq. (A.17) is a straight elaboration of the often used operator identity

$$e^{-S}Oe^{S} = \left[1 - S + \frac{1}{2!}S^2 - \frac{1}{3!}S^3 + \cdots\right]O\left[1 + S + \frac{1}{2!}S^2 + \frac{1}{3!}S^3 + \cdots\right]$$

$$= O + [O, S] + \frac{1}{2!}[[O, S], S] + \frac{1}{3!}[[[O, S], S], S] + \cdots, \qquad (A.18)$$

which holds for any operator O and S. The particular case $S = -xA$ and $O = B$ gives $\exp(xA)B\exp(-xA) = B + x[A, B]$, which is the result used to establish the last passage in Eq. (A.17). The operator functions $F_1(x)$ and $F_2(x)$ satisfy the same differential equation, with the same boundary condition $F_1(0) = F_2(0)$. This implies $F_1(x) = F_2(x)$ for any x; in particular for $x = 1$ we recover Eq. (A.14).

Bloch Identity

We now prove that any operator C, arbitrary linear combination of phonon operators a and $a^\dagger$, satisfies the Bloch identity

$$\boxed{\langle e^C \rangle = e^{\langle C^2 \rangle/2}}; \qquad (A.19)$$

this theorem states that the thermal average of the exponential of an operator, linear in a and $a^\dagger$, is just the exponential of half of the thermal average of the squared operator itself.

Consider in fact the linear combination of the phonon operators a and $a^\dagger$ of the form

$$C = c_1 a^\dagger + c_2 a$$

with c_1 and c_2 arbitrary complex numbers. We remark that

$$\langle C^2 \rangle = c_1 c_2 [2\langle a^\dagger a \rangle + 1].$$

Using the Weyl identity it follows

$$e^C = e^{c_1 a^\dagger + c_2 a} = e^{c_1 a^\dagger} e^{c_2 a} e^{c_1 c_2/2}.$$

Performing the thermal average one gets

$$\langle e^C \rangle = e^{c_1 c_2/2} \langle e^{c_1 a^\dagger} e^{c_2 a} \rangle = e^{c_1 c_2/2} \sum_{mn} \frac{c_1^m c_2^n}{m! n!} \langle a^{\dagger m} a^n \rangle.$$

In the double sum only the terms with $m = n$ survive, and using Eq. (A.13) it follows

$$\langle e^C \rangle = e^{c_1 c_2/2} \sum_m \frac{(c_1 c_2)^m}{m!} \langle a^\dagger a \rangle^m = e^{c_1 c_2 [2\langle a^\dagger a \rangle + 1]/2} = e^{\langle C^2 \rangle/2},$$

and the Bloch identity is thus proved.

From the Bloch identity and the Weyl identity, we can obtain the following important result. Let A and B indicate two operators linear in creation and annihilation operators; it holds

$$\boxed{\langle e^A e^B \rangle = \langle e^{A+B} \rangle \, e^{[A,B]/2} = e^{\langle A^2 + 2AB + B^2 \rangle / 2}}. \tag{A.20}$$

This relation will be used in Chapter 10, in the study of the dynamical structure factor for the scattering of particles from harmonic crystals.

Displaced Harmonic Oscillator

In Eq. (A.1) the one-dimensional harmonic oscillator under attention is centered at the origin of the x-axis. Consider now the same harmonic oscillator, displaced by the quantity x_0 along the x-axis; the corresponding Hamiltonian can be cast in the form

$$\widetilde{H} = \frac{1}{2M} p_x^2 + \frac{1}{2} M \omega^2 (x - x_0)^2 = \frac{1}{2} \hbar \omega \left[\frac{l_0^2}{\hbar^2} p_x^2 + \frac{1}{l_0^2} (x - x_0)^2 \right], \quad l_0^2 = \frac{\hbar}{M\omega}. \tag{A.21}$$

It is customary to characterize the spatial shift by means of the dimensionless Huang-Rhys parameter S, that expresses the elastic energy (in units $\hbar\omega$) of the oscillator in x_0 from the equilibrium position at $x = 0$; namely:

$$S\hbar\omega \equiv \frac{1}{2} M \omega^2 x_0^2 \implies x_0 = \sqrt{\frac{2\hbar S}{M\omega}} = l_0 \sqrt{2S}. \tag{A.22}$$

The phonon creation and annihilation operators of the displaced oscillator are linked to the corresponding one of the undisplaced oscillator by the following relations:

$$\begin{cases} \widetilde{a} = \dfrac{1}{\sqrt{2}} \left[\dfrac{1}{l_0} (x - x_0) + i \dfrac{l_0}{\hbar} p_x \right] = a - \sqrt{S}, \\[4mm] \widetilde{a}^\dagger = \dfrac{1}{\sqrt{2}} \left[\dfrac{1}{l_0} (x - x_0) - i \dfrac{l_0}{\hbar} p_x \right] = a^\dagger - \sqrt{S}. \end{cases} \tag{A.23}$$

In terms of annihilation and creation operators, the Hamiltonian (A.21) of the displaced harmonic oscillator reads

$$\widetilde{H} = \hbar\omega \left(\widetilde{a}^\dagger \widetilde{a} + \frac{1}{2} \right) = \hbar\omega \, a^\dagger a - \sqrt{S} \hbar\omega (a^\dagger + a) + \left(S + \frac{1}{2} \right) \hbar\omega \tag{A.24}$$

as can be easily verified.

To connect the states $|\widetilde{\phi}_n\rangle$ of the displaced oscillator to the corresponding $|\phi_n\rangle$ of the undisplaced oscillator, we exploit the general property of translation operators

$$f(x - x_0) = f(x) - x_0 f'(x) + \frac{1}{2!} x_0^2 f''(x) + \cdots = \exp\left[-x_0 \frac{\partial}{\partial x} \right] f(x).$$

Using Eqs. (A.4) and (A.22) one obtains

$$\frac{\partial}{\partial x} = \frac{1}{\sqrt{2} l_0} (a - a^\dagger) \implies -x_0 \frac{\partial}{\partial x} = \sqrt{S} (a^\dagger - a);$$

it follows

$$|\widetilde{\phi}_n\rangle = e^{\sqrt{S}(a^\dagger - a)}|\phi_n\rangle.$$

The overlap matrix elements between displaced and undisplaced harmonic oscillator wavefunctions can be obtained analytically. In particular it holds

$$\langle\widetilde{\phi}_n|\phi_0\rangle = \langle\phi_n|e^{-\sqrt{S}(a^\dagger - a)}|\phi_0\rangle = \langle\phi_n|e^{-\sqrt{S}a^\dagger}e^{\sqrt{S}a}|\phi_0\rangle e^{-[-\sqrt{S}a^\dagger,\sqrt{S}a]/2}$$

$$= \langle\phi_n|e^{-\sqrt{S}a^\dagger}|\phi_0\rangle e^{-S/2} = (-1)^n\sqrt{\frac{S^n}{n!}}\, e^{-S/2}, \qquad (A.25)$$

where use has been made of the Weyl identity and of standard series development of exponentials. The above results will be useful for the study of the optical properties of the Franck-Condon model, in Chapter 12.

Further Reading

Andreani, L. C. (1995). Optical transitions, excitons, and polaritons in bulk and low dimensional semiconductor structures. In E. Burnstein, & C. Weisbuch (Eds.), *Confined electrons and photons*. New York: Plenum Press.

Andreani, L. C. (2003). Exciton-polaritons in confined systems. In B. Deveaud, A. Quattropani, & P. Schwendimann (Eds.) *Proceedings of the international school of physics "Enrico Fermi" course CL*. Amsterdam: IOS Press.

Baroni, S., de Gironcoli, S., Dal Corso, A., & Giannozzi, P. (2001). Phonons and related crystal properties from density-functional perturbation theory. *Reviews of Modern Physics, 73*, 515.

Bilz, H., & Kress, W. (1979). *Phonon dispersion relations in insulators*. Berlin: Springer.

Blinc, R. (2011). *Advanced ferroelectricity*. Oxford: Oxford University Press.

Born, M., & Huang, K. (1954). *Dynamical theory of crystal lattices*. Oxford: Oxford University Press.

Brillouin, L. (1953). *Wave propagation in periodic structures*. New York: Dover.

Choquard, P. F. (1967). *The anharmonic crystal*. New York: Benjamin.

Di Bartolo, B., & Powell, R. C. (1976). *Phonons and resonances in solids*. New York: Wiley.

Fröhlich, H. (1958). *Theory of dielectrics*. Oxford: Clarendon Press.

Grimvall, G. (1981). *The electron-phonon interaction in metals*. Amsterdam: North-Holland.

Lines, M. E., & Glass, A. M. (1977). *Principles and applications of ferroelectrics and related materials*. Oxford: Clarendon Press.

Lorentz, H. A. (1952). *The theory of electrons*. New York: Dover.

Maradudin, A. A., Montroll, E. W., Weiss, G. H., & Ipatova, I. P. (1971). *Theory of lattice dynamics in the harmonic approximation*. New York: Academic Press.

Ridley, B. K. (2009). *Electrons and phonons in semiconductor multilayers*. Cambridge: Cambridge University Press.

Srivastava, G. P. (1990). *The physics of phonons*. Bristol: Adam Hilger.

Stroscio, M. A., & Dutta, M. (2001). *Phonons in nanostructures*. Cambridge: Cambridge University Press.

Ziman, J. M. (1960). *Electrons and phonons*. Oxford: Clarendon Press.

10 Scattering of Particles by Crystals

Elastic and inelastic scattering of particles by crystals is studied, with particular attention to the scattering of photons and neutrons. The guidelines provided by the laws of conservation of energy and momentum in the investigation of the elementary excitations of periodic systems are discussed. This analysis is then enriched with the quantum theory elaboration of single and double differential cross-sections. The effect of the lattice vibrations on scattering processes and the quantum origin of the Debye-Waller factor are also illustrated. The Debye-Waller factor is crucial in controlling the elastic and inelastic scattering of particles, as well as the recoilless and recoil emission and absorption lines in the Mössbauer effect. Scattering of particles provides a wealth of invaluable information of single-particle excitations and collective excitations of the materials under investigation.

10.1 General Considerations

Elastic and inelastic scattering of particles (such as photons, neutrons, and electrons) by a crystal may provide important information on the crystal structure, electron density, energy-wavevector dispersion laws of elementary excitations (phonons, polaritons, excitons, plasmons, magnons, etc.). Generally X-rays and neutrons are well suited for

Solid State Physics, Second Edition. http://dx.doi.org/10.1016/B978-0-12-385030-0.00010-4

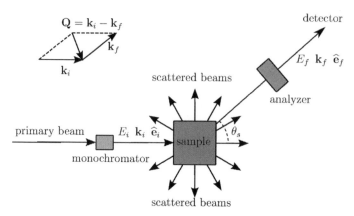

Figure 10.1 Schematic experimental set-up for scattering measurements. The momentum transfer to the target is $\hbar\mathbf{Q} = \hbar\mathbf{k}_i - \hbar\mathbf{k}_f$, and the energy transfer to the target is $E = E_i - E_f$.

analysis of bulk properties because they easily penetrate in the crystals, while electrons are used for analysis of surfaces and thin films, due to their poor penetration in the sample. Scattering of X-rays is essentially produced by the interaction of the electric field of the radiation with the electronic charge density of the crystal. In contrast with X-rays, electrons are efficiently scattered by the nuclei, as well by the electrons of the sample, because of the strong Coulomb interactions with charged particles. Neutrons interact via nuclear forces with the nuclei, and thus are sensitive to their spatial distribution and vibrations; neutrons also interact with electronic magnetic moments, and thus give useful information on magnetic materials.

A schematic representation of the experimental set-up for scattering measurements is indicated in Figure 10.1. A beam of incident particles (photons, or neutrons, or electrons) is collimated and monochromatized, so as to select particles with an initial momentum $\hbar\mathbf{k}_i$, initial energy E_i, and initial polarization state $\widehat{\mathbf{e}}_i$. The monochromatic beam impinges on the sample and is diffused (partially elastically and partially inelastically) in the space. A polarization analyzer selects particles of final energy E_f, final momentum $\hbar\mathbf{k}_f$, and final polarization state $\widehat{\mathbf{e}}_f$. From the measurements of the momentum transfer $\hbar\mathbf{Q} = \hbar\mathbf{k}_i - \hbar\mathbf{k}_f$, of the energy transfer $E = E_i - E_f$, and of the intensity of the scattered beam, information on the structural and dynamical properties of the sample can be inferred. For this purpose it is convenient to consider first some general aspects on photonic, neutronic, and electronic beams, their energy-wavelength relations, and their use as probes in scattering experiments.

Photons

The elastic scattering by X-rays is the most traditional tool to obtain information on the crystal structure and electronic charge distribution. An electromagnetic wave is scattered by electrons, which act as oscillating dipoles under the influence of the wave electric field. For photons, the energy-wavevector relation is $\hbar\omega = \hbar c k = \hbar c 2\pi/\lambda$;

with $\hbar\omega$ expressed in eV and λ in Å, one obtains

$$\text{photons:} \quad \lambda = \frac{12398.5}{\hbar\omega} \quad (\lambda \text{ in Å and } \hbar\omega \text{ in eV}). \tag{10.1a}$$

For the determination of the crystal structures, one usually works in the ≈ 10–50 keV energy range, for which λ is of the order of the interatomic spacings.

Inelastic scattering of photons is also a very important tool for the investigation of crystals. Very accurate information on phonons (or polaritons) is obtained by inelastic scattering of visible light (Brillouin and Raman scattering); however, the wavevector of visible light is very small on the scale of the Brillouin zone dimension, and thus only the $q \approx 0$ region can be explored. Information on phonons throughout the Brillouin zone can be inferred from inelastic scattering of X-rays. Typically, the energy of the incident photons is in the range ≈ 10 keV, while the energy shift due to emission or absorption of phonons is as small as ≈ 10 meV; thus high resolution experimental techniques have been developed for detecting so small energy shifts. Finally, we notice that X-rays can be scattered inelastically by electrons, through the Compton mechanism (the energy shift of photons in typical experiments is of the order of a few keV); the analysis of Compton profiles provides important information on the ground-state electron momentum density.

Neutrons

The interaction of neutrons with solids is an important area of research for the determination of crystal structures, lattice dynamics, and magnetic properties of materials. For neutrons, the energy-wavevector relationship is $E = \hbar^2 k^2 / 2M_n$, where $k = 2\pi/\lambda$ and M_n is the neutron mass; with E expressed in eV, and λ in Å, one obtains

$$\text{neutrons:} \quad \lambda = \frac{0.2862}{\sqrt{E}} \quad (\lambda \text{ in Å and } E \text{ in eV}). \tag{10.1b}$$

Neutrons with wavelength of the order of 1 Å have energy of about 80 meV, a value of the same order as $k_B T$ at room temperature ($k_B T = 23.538$ meV for $T = 273.15$ K); thus, thermal neutrons, obtained after moderation from high flux nuclear reactors or pulsed sources, have momenta and energies comparable with those of phonons; for this reason, inelastic scattering of neutrons is the most natural and accurate method for investigation of phonon dispersion curves.

Neutrons interact with nucleons through nuclear forces. The interaction between a neutron at position $\mathbf{r}$ and a nucleus located at $\mathbf{R}$ is usually described by the Fermi pseudo-potential

$$V(\mathbf{r}, \mathbf{R}) = \frac{2\pi\hbar^2}{M_n} b\,\delta(\mathbf{r} - \mathbf{R}),$$

where b is a phenomenological parameter, called Fermi length (or scattering length) of the given nucleus and M_n is the neutron mass. The scattering length b varies irregularly across the periodic table, can assume positive or negative values, and its order of magnitude is 10^{-13} cm $= 10^{-5}$ Å, the range of nuclear forces.

The scattering amplitude for X-rays is related to the number Z of electrons of the atoms; the scattering amplitude for neutrons is related to the scattering length b, which varies in an erratic way with the mass number of the nuclei; thus neutron diffraction is often used as a precious complementary tool to the X-ray diffraction experiments. For instance, light elements can be investigated in the presence of heavy ones (a difficult or impossible job with X-rays). Isotopes of the same element are equally efficient as X-ray scatterers, but may behave differently as neutron scatterers; in particular the large difference in the scattering length for hydrogen and deuterium can be exploited for the study of complicated molecular systems by appropriately replacing one isotope with the other one. We also notice that a nucleus acts as a point scatterer for thermal neutrons (the wavelength of thermal neutrons is of the order of 1 Å while the scattering length of the nuclei is of the order of 10^{-5} Å); thus the scattering amplitude for neutrons does not depend on the scattering angle; this isotropy leads to improved resolution capability in the study of diffraction beams with high values of momentum transfer, and is particularly important in the study of complex systems.

Neutrons are particles with magnetic moment and interact with matter also through electromagnetic forces. Thus neutrons also take notice of unpaired spin electrons and are an invaluable tool for the characterization of magnetic structures.

Electrons

For electrons the relationship between energy and wavevector is $E = \hbar^2 k^2 / 2m$, where $k = 2\pi/\lambda$ and m is the electron mass; we have

$$\text{electrons:} \quad \lambda = \frac{12.264}{\sqrt{E}} \quad (\lambda \text{ in Å and } E \text{ in eV}). \tag{10.1c}$$

Electrons with wavelength of the order of 1 Å have energy of about 150 eV; the energy region $10 \text{ eV} < E < 10^3 \text{ eV}$ is particularly suitable for diffraction experiments and is called low energy electron diffraction region.

Electrons are charged particles and thus they interact strongly with the nuclei and the electrons of the crystal via Coulomb forces. Electrons penetrate into the crystal for small distances and elastic electron diffraction is particularly convenient for investigation of surfaces and thin layers. Inelastic scattering of electrons with appropriate incident kinetic energy gives information on the energy spectrum of the excitations localized at the surface; energy-loss experiments with fast electrons also provide information on excitons and plasmons in the crystal.

10.2   Elastic Scattering of X-rays from Crystals and the Thomson Approximation

In this section we discuss the *elastic scattering or diffraction* of X-rays from crystals, assuming the atoms fixed at the equilibrium positions (the effect of the thermal motion of atoms and inelastic scattering processes are studied later). We shall see that the geometric distribution of the diffracted beams gives direct information on the reciprocal lattice vectors of the crystal, while the intensity of the diffracted beams provides information on the contents of the unit cell. Although we discuss here the specific case of elastic scattering of X-rays, the results that depend essentially on the wave nature of the incident particle apply as well to other incident beams.

10.2.1 Elastic Scattering of X-rays and Bragg Diffraction Condition

Consider an incident radiation beam of frequency ω, propagation vector $\mathbf{k}_i$ ($\omega = ck_i$), polarization vector $\widehat{\mathbf{e}}_i$ ($\widehat{\mathbf{e}}_i \perp \mathbf{k}_i$), and amplitude E_0; the electric field of the linearly polarized monochromatic electromagnetic wave can be expressed as

$$\mathbf{E}(\mathbf{r}, t) = \widehat{\mathbf{e}}_i \, E_0 \, e^{i(\mathbf{k}_i \cdot \mathbf{r} - \omega t)}. \tag{10.2a}$$

A free electron, located at the point $\mathbf{r}$, under the influence of the impinging (high frequency) electric field is accelerated according to the classical equation of motion

$$m \, \ddot{\mathbf{u}}(\mathbf{r}, t) = (-e)\mathbf{E}(\mathbf{r}, t) \implies \ddot{\mathbf{u}}(\mathbf{r}, t) = \frac{(-e)}{m}\widehat{\mathbf{e}}_i \, E_0 \, e^{i(\mathbf{k}_i \cdot \mathbf{r} - \omega t)}. \tag{10.2b}$$

The electron follows the periodic variations of $\mathbf{E}$, acts as an oscillating electric dipole and radiates electromagnetic waves at the same frequency as the incident one. Magnetic dipole radiation and other contributions are assumed negligible.

According to classical electrodynamics, the electric field at the point P at distance R from the dipole centered at $\mathbf{r}$ (when R is much larger than the wavelength of the incident radiation) is given by the expression

$$\begin{aligned}
\mathbf{E}_d(\mathbf{r}, \mathbf{R}, t) &= \frac{-e}{c^2 R} \left\{ \ddot{\mathbf{u}}\left(\mathbf{r}, t - R/c\right) - \left[\widehat{\mathbf{k}}_f \cdot \ddot{\mathbf{u}}\left(\mathbf{r}, t - R/c\right) \right] \widehat{\mathbf{k}}_f \right\} \\
&= \frac{1}{R} \frac{e^2}{mc^2} E_0 \, e^{i\mathbf{k}_i \cdot \mathbf{r}} e^{-i\omega(t - R/c)} \left[\widehat{\mathbf{e}}_i - (\widehat{\mathbf{k}}_f \cdot \widehat{\mathbf{e}}_i) \widehat{\mathbf{k}}_f \right],
\end{aligned} \tag{10.3}$$

where $\widehat{\mathbf{k}}_f \equiv \mathbf{k}_f / k_f$ is the unit vector in the direction of $\mathbf{k}_f$. From the above equation, it is seen that the polarization vector $\widehat{\mathbf{e}}_d$ of the dipolar scattered field $\mathbf{E}_d$ propagating along the $\mathbf{R}$ (or $\mathbf{k}_f$) direction is

$$\widehat{\mathbf{e}}_d = \frac{\widehat{\mathbf{e}}_i - (\widehat{\mathbf{k}}_f \cdot \widehat{\mathbf{e}}_i) \widehat{\mathbf{k}}_f}{|\widehat{\mathbf{e}}_i - (\widehat{\mathbf{k}}_f \cdot \widehat{\mathbf{e}}_i) \widehat{\mathbf{k}}_f|};$$

the vector $\widehat{\mathbf{e}}_d$ lies in the plane spanned by the two vectors $\widehat{\mathbf{e}}_i$ and $\widehat{\mathbf{k}}_f$, and is orthogonal to the latter. The denominator of the above expression reads

$$|\widehat{\mathbf{e}}_i - (\widehat{\mathbf{k}}_f \cdot \widehat{\mathbf{e}}_i) \widehat{\mathbf{k}}_f| \equiv \sqrt{1 - (\widehat{\mathbf{k}}_f \cdot \widehat{\mathbf{e}}_i)^2} = \sin \psi \quad \text{with} \quad \cos \psi = \widehat{\mathbf{k}}_f \cdot \widehat{\mathbf{e}}_i,$$

where ψ is the angle between the polarization vector of the incident radiation and the geometrical direction selected for the analyzer, as shown in Figure 10.2.

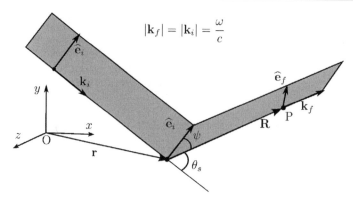

Figure 10.2 Schematic representation of an incident electromagnetic wave (of frequency ω, propagation vector $\mathbf{k}_i$, polarization unit vector $\widehat{\mathbf{e}}_i$) elastically scattered from a free electron at the point $\mathbf{r}$ and detected at the point $P \equiv \mathbf{r} + \mathbf{R}$. The scattering angle between the vectors $\mathbf{k}_i$ and $\mathbf{k}_f$ is denoted by θ_s. The angle between the polarization vector of the incident field and the propagation vector of the scattered radiation is denoted by ψ.

For elastic scattering, $\omega/c = k_f$ and we can also write Eq. (10.3) in the form

$$\mathbf{E}_d(\mathbf{r}, \mathbf{R}, t) = \frac{1}{R} \frac{e^2}{mc^2} E_0 \, e^{i\mathbf{k}_i \cdot \mathbf{r}} e^{i(\mathbf{k}_f \cdot \mathbf{R} - \omega t)} \left[\widehat{\mathbf{e}}_i - (\widehat{\mathbf{k}}_f \cdot \widehat{\mathbf{e}}_i)\widehat{\mathbf{k}}_f \right]. \tag{10.4}$$

Let $\widehat{\mathbf{e}}_f$ denote the unit vector that gives the polarization state of photons accepted by the analyzer. From the modulus squared of $\mathbf{E}_d \cdot \widehat{\mathbf{e}}_f$ (and the transverse condition $\widehat{\mathbf{e}}_f \cdot \widehat{\mathbf{k}}_f = 0$), the intensity of the scattered field monitored by the detector takes the expression

$$I_s(\mathbf{R}) = I_0 \frac{1}{R^2} \left(\frac{e^2}{mc^2} \right)^2 (\widehat{\mathbf{e}}_i \cdot \widehat{\mathbf{e}}_f)^2, \tag{10.5a}$$

where I_0 is the intensity of the incident field. From Eq. (10.5a) we see that the intensity of the scattered wave decreases as the inverse square of the distance, and the geometrical dependence is fully controlled by the scalar product of the initial and final selected polarizations. It can also be added that the mass at the denominator ensures that the scattering from nuclei can be safely ignored, with respect to the scattering from electrons.

In the description of scattering processes and in measurements, it is customary to consider the differential cross-section $d\sigma/d\Omega$, which is defined as the number of particles scattered per unit time per solid angle $d\Omega = \sin\theta \, d\theta \, d\phi$, divided by the incident flux density. Then, in the second member of Eq. (10.5a), the quantity multiplying I_0/R^2 can be recognized as the differential cross-section for elastic scattering of polarized beams interacting with a single free electron; such a quantity is known as Thomson cross-section and reads

$$\left(\frac{d\sigma}{d\Omega} \right)_{\text{Th}} = \left(\frac{e^2}{mc^2} \right)^2 (\widehat{\mathbf{e}}_i \cdot \widehat{\mathbf{e}}_f)^2 = r_0^2 (\widehat{\mathbf{e}}_i \cdot \widehat{\mathbf{e}}_f)^2, \tag{10.5b}$$

where $e^2/mc^2 = r_0 = 2.82 \times 10^{-13}$ cm is the classical radius of the electron. The Thomson cross-section is as small as ≈ 0.1 barn (1 barn $= 10^{-24}$ cm^2) and the scattering can be safely termed as weak. For scattering and detection of unpolarized photons, it is shown in Problem 1 that the scattering cross-section becomes

$$\left(\frac{d\sigma}{d\Omega}\right)^{\text{(unpol)}}_{\text{Th}} = \left(\frac{e^2}{mc^2}\right)^2 \frac{1 + \cos^2 \theta_s}{2}, \tag{10.5c}$$

where θ_s is the scattering angle. We remind that the classical treatment, followed so far, only considers the coupling of the *free* electronic charge (i.e. *no resonance in the energy range of interest*) with the electric field of the radiation, while other couplings (for example interactions involving the electron spin, resonant effects) have been disregarded; but the limit of validity of the Thomson treatment will emerge more clearly with the quantum analysis of Section 10.2.3.

Let us now consider the Thomson scattering by two electrons, one at the origin O ($\mathbf{r} = 0$) and the other located at the point P (at position $\mathbf{r} \neq 0$) (see Figure 10.3). The two electrons are both accelerated by the electric field of the impinging electromagnetic wave, and both radiate. Using Eq. (10.4), the scattered fields from the electron located at the origin and from the electron located at the point P, apart common terms, contain the phase factors:

$$\text{(electron at } \mathbf{r} = 0) \quad \mathbf{E}_d(0, \mathbf{R}_1, t) \div e^{i\mathbf{k}_f \cdot \mathbf{R}_1}, \tag{10.6a}$$

$$\text{(electron at } \mathbf{r} \neq 0) \quad \mathbf{E}_d(\mathbf{r}, \mathbf{R}_2, t) \div e^{i\mathbf{k}_i \cdot \mathbf{r}} e^{i\mathbf{k}_f \cdot \mathbf{R}_2}. \tag{10.6b}$$

From Figure 10.3, we see that $\mathbf{k}_f \cdot \mathbf{R}_2 = \mathbf{k}_f \cdot \mathbf{R}_1 - \mathbf{k}_f \cdot \mathbf{r}$; thus, the phase factors in Eqs. (10.6a,b) differ by quantity $\exp(i\mathbf{Q} \cdot \mathbf{r})$, where $\mathbf{Q} = \mathbf{k}_i - \mathbf{k}_f$ is the momentum transfer (in units $\hbar$) of the scattering process (the quantity $\mathbf{k}_i - \mathbf{k}_f$, or sometimes its opposite, is also referred to as the scattering vector). The phase factor can be traced back to the difference in optical path length from the wave scattered from P and the wave scattered

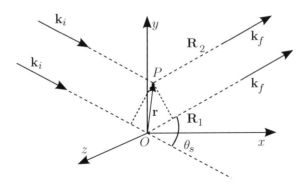

Figure 10.3 Schematic geometry of an electromagnetic field scattered from one electron at the origin O and one electron at the point P; the difference in optical path length between the wave scattered from O and the wave scattered from P is $\mathbf{k}_i \cdot \mathbf{r} - \mathbf{k}_f \cdot \mathbf{r} \equiv \mathbf{Q} \cdot \mathbf{r}$, where $\mathbf{Q} = \mathbf{k}_i - \mathbf{k}_f$ is the momentum transfer in units $\hbar$.

from O (see Figure 10.3). Then, the differential cross-section for elastic scattering of a radiation beam interacting with two free electrons, one at the origin and one at $\mathbf{r}$ is given by

$$\frac{d\sigma}{d\Omega} = \left(\frac{d\sigma}{d\Omega}\right)_{\text{Th}} \left|1 + e^{i\mathbf{Q}\cdot\mathbf{r}}\right|^2, \tag{10.7}$$

where the quantity within modulus squared takes proper account of the effect of geometrical phases on the scattered waves.

We can generalize Eq. (10.7) to a quantum system with total electron density $n_{\text{el}}(\mathbf{r})$. The differential cross-section for (nonresonant) scattering of sufficiently hard X-rays is then given by

$$\frac{d\sigma}{d\Omega} = \left(\frac{d\sigma}{d\Omega}\right)_{\text{Th}} \left|\int n_{\text{el}}(\mathbf{r})e^{i\mathbf{Q}\cdot\mathbf{r}}\,d\mathbf{r}\right|^2.$$

The quantity within modulus in the above equation is *the Fourier transform of the electronic density and is called form factor*; its modulus square is called *structure factor*. We can thus write

$$\boxed{\frac{d\sigma}{d\Omega} = \left(\frac{d\sigma}{d\Omega}\right)_{\text{Th}} |F(\mathbf{Q})|^2 = \left(\frac{d\sigma}{d\Omega}\right)_{\text{Th}} S(\mathbf{Q})}, \tag{10.8a}$$

where

$$\boxed{F(\mathbf{Q}) = \int n_{\text{el}}(\mathbf{r})e^{i\mathbf{Q}\cdot\mathbf{r}}\,d\mathbf{r}} \quad \text{and} \quad \boxed{S(\mathbf{Q}) = |F(\mathbf{Q})|^2} \tag{10.8b}$$

are the form factor and structure factor of the electronic system under investigation. We can summarize the results so far obtained by observing that the *intensity of X-ray scattering is controlled by the modulus squared of the Fourier transform of the electron density of the system.*

Condition for Elastic Scattering in Periodic Systems

In crystals $n_{\text{el}}(\mathbf{r})$ is a periodic function; thus its Fourier coefficients, i.e. the crystal form factors, can be different from zero only in correspondence to reciprocal lattice vectors; this means that occurrence of diffraction peaks requires

$$\boxed{\mathbf{Q} = \mathbf{k}_i - \mathbf{k}_f \equiv \mathbf{G}}, \tag{10.9}$$

where $\mathbf{G}$ is a reciprocal lattice vector. Thus a necessary condition for X-ray diffraction is that the difference between the incident and the scattered radiation wavevectors equals a reciprocal lattice vector. Equation (10.9) can also be written in the form

$$\hbar\mathbf{k}_i = \hbar\mathbf{k}_f + \hbar\mathbf{G};$$

thus in the scattering process the momentum is preserved within reciprocal lattice vectors times $\hbar$.

It is instructive to show that the diffraction condition (10.9), deduced by von Laue, is equivalent to the intuitive description by Bragg, who considers specular reflection of the incident radiation by a family of lattice planes; as illustrated in Figure 10.4, the geometrical condition for the coherent scattering from two successive planes (and hence from the whole sequence of parallel planes) requires

$$\boxed{2\,d\sin\theta = n\lambda}\,, \tag{10.10}$$

where n is an integer, λ is the wavelength of the scattered (and incident) wave, $\theta_s = 2\theta$ is the scattering angle, and d is the distance between adjacent planes of the family.

For convenience in Figure 10.4 we also report the geometrical construction of the vector $\mathbf{Q} = \mathbf{k}_i - \mathbf{k}_f$. Since for elastic scattering $|\mathbf{k}_i| = |\mathbf{k}_f| = 2\pi/\lambda$, we have

$$|\mathbf{Q}| = 2|\mathbf{k}_i|\sin\theta = 2\frac{2\pi}{\lambda}\sin\theta = n\frac{2\pi}{d}\,, \tag{10.11}$$

where the last equality follows from the Bragg law (10.10). From Figure 10.4 it is seen that $\mathbf{Q}$ is perpendicular to the family of lattice planes with distance d; furthermore, Eq. (10.11) shows that the magnitude $|\mathbf{Q}|$ is an integer multiple of the quantity $2\pi/d$; these two observations (together with the general properties of real and reciprocal spaces studied in Section 2.4.2) allow us to conclude that $\mathbf{Q}$ must be a reciprocal lattice vector. The Laue condition (10.9) for occurrence of "diffracted beams" is thus fully equivalent to the Bragg condition (10.10) for occurrence of "reflected beams," and the expressions diffracted beam and reflected beam in this context become synonymous.

From the Bragg condition (10.10), the possibility of elastic scattering occurs only if $\lambda < 2d$. Thus λ must be of the order of the Å or less. Furthermore λ cannot be much smaller than the interatomic distance, otherwise experimental arrangements at glancing angles are necessary to detect diffraction peaks with small momentum transfer. This restricts the ordinary frequency range of interest to the X-ray region.

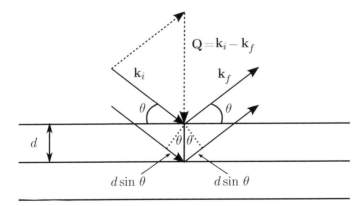

Figure 10.4 Waves reflected from successive planes reinforce if $2d\sin\theta$ equals an integer number $n\lambda$ of wavelengths. The geometrical construction of $\mathbf{Q} = \mathbf{k}_i - \mathbf{k}_f$ is also provided.

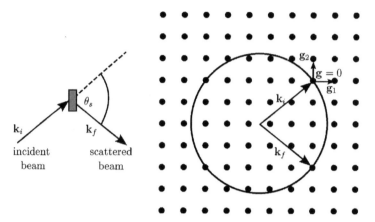

Figure 10.5 Schematic representation of the Ewald construction in the reciprocal lattice (supposed cubic for simplicity) of allowed wavevectors for elastic diffraction.

Ewald Construction

We can consider the following geometrical construction due to Ewald for an easy determination of the occurrence of (elastic) diffraction peaks. Suppose to have an incident monochromatic beam of particles of wavevector $\mathbf{k}_i$ and a diffracted beam with propagation wavevector $\mathbf{k}_f$. For elastic scattering, the conservation of energy and the conservation of momentum (within reciprocal lattice vectors $\mathbf{G}$ times $\hbar$) imply

$$|\mathbf{k}_i| = |\mathbf{k}_f| \quad \text{and} \quad \mathbf{k}_i = \mathbf{k}_f + \mathbf{G}. \tag{10.12}$$

The Ewald construction (see Figure 10.5) permits a simple geometrical interpretation of Eq. (10.12). In the reciprocal space, the vector $\mathbf{k}_i$ is drawn in such a way that its tip terminates at the reciprocal lattice vector $\mathbf{G} = 0$. With the center at the origin of the vector $\mathbf{k}_i$, the sphere of radius k_i is also drawn; if this sphere intersects, besides $\mathbf{G} = 0$, one (or more) points of the reciprocal lattice, conditions (10.12) are satisfied and Bragg diffractions are possible. It is evident that diffraction may occur only if the magnitude of k_i exceeds one half of the magnitude of the smallest (non-vanishing) $\mathbf{G}$ vector.

Methods of X-ray Scattering

In general, if we consider a *monochromatic X-ray beam and a fixed crystal orientation*, we have no possibility of elastic diffraction. Some experimental methods are thus adopted to perform X-ray diffraction measurements:

a. *The Laue method.* In the Laue method the incident radiation covers a wide band of wavelengths. Correspondingly, all Ewald spheres with radii within the appropriate range of incident wavevectors are to be drawn in the reciprocal lattice; whenever the Ewald spheres intercept points of the reciprocal lattice, elastic diffractions become possible.

b. *The Bragg method* (rotating crystal method). In the Bragg method, the incident radiation is monochromatic, but the crystal is capable of rotational motion around a given axis. The radius of the Ewald sphere is thus fixed, but the reciprocal space rotates following the crystal rotation in direct space; when points of the reciprocal lattice intercept the Ewald sphere, elastic diffractions become possible.

c. *The Debye-Scherrer method* (powder method). The incident radiation is monochromatic, but now the single (rotating) crystal is replaced by a polycrystalline specimen (powder) with random orientation of the composing crystallites.

We shall not enter into the details of the various experimental methods. We simply notice that *from each elastically diffracted beam a specific reciprocal lattice vector* $\mathbf{G} = \mathbf{k}_i - \mathbf{k}_f$ *is individuated*. A sufficiently detailed map of the reciprocal lattice vectors permits to infer the fundamental vectors $\mathbf{g}_1, \mathbf{g}_2, \mathbf{g}_3$ (and hence the fundamental translation vectors of the direct lattice, via the standard relation $\mathbf{g}_i \cdot \mathbf{t}_j = 2\pi \delta_{ij}$).

10.2.2 Elastic Scattering of X-rays and Intensity of Diffracted Beams

Until now we have focused our attention on the geometrical aspects of diffraction (i.e. on the change $\mathbf{Q} = \mathbf{k}_i - \mathbf{k}_f$ between the incident and scattered wavevectors). Very important information on the electronic charge distribution can be obtained by the measurements of the intensity of the diffracted beams, and comparison with appropriate theoretical models. In the literature, ab initio calculations of crystal ground-state electronic density $n_{\mathrm{el}}(\mathbf{r})$, and form factors $F(\mathbf{Q})$ have been performed for an increasing number of crystals. Here, however, we focus on simplified models useful for a semiquantitative understanding of Bragg reflections intensities.

Consider first an atom with electron density $n_a(\mathbf{r})$; the *atomic form factor* $f_a(\mathbf{Q})$ is defined as the Fourier transform

$$f_a(\mathbf{Q}) = \int e^{i\mathbf{Q}\cdot\mathbf{r}} n_a(\mathbf{r}) \, d\mathbf{r}. \tag{10.13a}$$

In isolated atoms, the electron cloud $n_a(\mathbf{r})$ is spherically symmetric, and Eq. (10.13a) can be simplified to an integral over the radial variable

$$f_a(\mathbf{Q}) = \int_0^\infty n_a(r) \frac{\sin Qr}{Qr} 4\pi r^2 \, dr. \tag{10.13b}$$

The atomic form factor depends only on the magnitude (but not on the orientation) of the transfer vector $\mathbf{Q}$; from Eq. (10.13b) it is evident that for an atom with Z electrons, the atomic form factor changes monotonically from Z to zero, as Q increases from zero to large values.

When atoms are assembled to build a solid, their electronic clouds are no more spherically symmetric. However, for a preliminary analysis of the crystal form factors, it is useful to approximate the crystal electron density $n_{\mathrm{el}}(\mathbf{r})$ as a *sum of spherically symmetric contributions, centered at the various atomic positions* (for brevity we call this the *local spherical model* for the crystalline electronic distribution). We write thus

$$n_{\mathrm{el}}(\mathbf{r}) = \sum_{\mathbf{t}_n} \sum_{\mathbf{d}_\nu} n_{a\nu}(\mathbf{r} - \mathbf{t}_n - \mathbf{d}_\nu), \tag{10.14}$$

where $\mathbf{t}_n$ are lattice translational vectors, $\mathbf{d}_\nu$ are the positions of the atoms in the unit cell, and $n_{av}(\mathbf{r}-\mathbf{t}_n-\mathbf{d}_\nu) \equiv n_{av}(|\mathbf{r}-\mathbf{t}_n-\mathbf{d}_\nu|)$ indicates the contribution of the atom at the site $\mathbf{t}_n-\mathbf{d}_\nu$ to the total electron density. Notice that Eq. (10.14) is a fairly accurate description of the "core electronic contribution," while for the "valence electronic contribution" some degree of non-sphericity is expected to occur (the environment of a lattice point, in fact, has the local point crystal symmetry, and not the full rotational symmetry). In particular, bonding charge tends to accumulate midway between interacting atoms and cannot be accurately described by the expression (10.14). Charge asymmetry is at the origin of subtle but interesting effects determining a (weak) intensity for some "forbidden reflections" of the local spherical model, as we shall discuss below.

The Fourier transform of the crystal electronic density (10.14) is

$$F(\mathbf{Q}) = \sum_{\mathbf{t}_n} \sum_{\mathbf{d}_\nu} \int e^{i\mathbf{Q}\cdot\mathbf{r}} \, n_{av}(\mathbf{r}-\mathbf{t}_n-\mathbf{d}_\nu) \, d\mathbf{r}. \tag{10.15}$$

Insertion in Eq. (10.15) of the unit quantity $\exp[i\mathbf{Q}\cdot(\mathbf{t}_n+\mathbf{d}_\nu)]\exp[-i\mathbf{Q}\cdot(\mathbf{t}_n+\mathbf{d}_\nu)]$, and straight manipulations give

$$F(\mathbf{Q}) = \sum_{\mathbf{t}_n} e^{i\mathbf{Q}\cdot\mathbf{t}_n} \sum_{\mathbf{d}_\nu} e^{i\mathbf{Q}\cdot\mathbf{d}_\nu} f_{av}(\mathbf{Q}), \tag{10.16a}$$

where

$$f_{av}(\mathbf{Q}) = \int e^{i\mathbf{Q}\cdot\mathbf{r}} \, n_{av}(\mathbf{r}) \, d\mathbf{r}. \tag{10.16b}$$

In Eq. (10.16a), the sum over the lattice translations is different from zero if $\mathbf{Q}$ equals a reciprocal lattice vector, in agreement with the Laue condition given in Eq. (10.9). Once the Laue condition is satisfied, the crystal form factors $F(\mathbf{G})$ become

$$\boxed{F(\mathbf{G}) = N \sum_{\mathbf{d}_\nu} e^{i\mathbf{G}\cdot\mathbf{d}_\nu} f_{av}(\mathbf{G})}. \tag{10.17}$$

We see that the crystal form factors $F(\mathbf{G})$ are expressed in terms of the atomic form factors $f_{av}(\mathbf{G})$ of the atoms in the position $\mathbf{d}_\nu$ of the unit cell and of the phase factors $\exp(i\mathbf{G}\cdot\mathbf{d}_\nu)$.

In the case of crystals with a single atom per unit cell, the crystal form factors $F(\mathbf{G})$ are proportional to the atomic form factors of the only type of atom composing the crystal; thus $|F(\mathbf{G})|^2$ are expected to decrease monotonically as $|\mathbf{G}|$ increases, this being the behavior of the atomic form factors.

A more interesting situation occurs in crystals with a basis of two or more atoms in the unit cell; in this case, systematic enhancement or weakening (or absence) of diffraction beams may occur. If all the atoms in the unit cell are equal and contribute the same spherically symmetric electron cloud around the appropriate centers, the form factors $f_{av}(\mathbf{G}) \equiv f_a(\mathbf{G})$ can be factorized out from the sum in Eq. (10.17) and we obtain

$$F(\mathbf{G}) = N f_a(\mathbf{G}) f_b(\mathbf{G}), \tag{10.18a}$$

where the *geometrical form factor* $f_b(\mathbf{G})$ *of the basis* is defined as the sum of the phase factors $e^{i\mathbf{G}\cdot\mathbf{d}_\nu}$ of the (identical) atoms composing the primitive cell

$$f_b(\mathbf{G}) = \sum_{\mathbf{d}_\nu} e^{i\mathbf{G}\cdot\mathbf{d}_\nu}; \tag{10.18b}$$

the geometrical form factor may lead to systematic absence of diffraction beams ("forbidden reflections").

As an application of Eqs. (10.18), consider the elemental semiconductors with the diamond structure. As described in Figure 2.9, the diamond structure consists of a fcc Bravais lattice with a basis of two equal atoms in the unit cell in the positions $\mathbf{d}_1 = 0$ and $\mathbf{d}_2 = (a/4)(1, 1, 1)$. The fundamental vectors of the reciprocal lattice are $\mathbf{g}_1 = (2\pi/a)(-1, 1, 1)$, $\mathbf{g}_2 = (2\pi/a)(1, -1, 1)$, and $\mathbf{g}_3 = (2\pi/a)(1, 1, -1)$. It is easily seen by inspection that the most general vector $\mathbf{G}$ of the reciprocal lattice has the form $\mathbf{G} = (2\pi/a)(h_1, h_2, h_3)$ with h_1, h_2, h_3 all odd, or all even integers.

For a preliminary analysis of the crystal form factors, we assume that the crystal electron density is made by the superposition of identical spherically symmetric contributions, centered at the various atomic positions (local spherical approximation). The geometrical factor of Eq. (10.18b) in the case of the diamond structure becomes

$$f_b(\mathbf{G}) = e^{i\mathbf{G}\cdot\mathbf{d}_1} + e^{i\mathbf{G}\cdot\mathbf{d}_2} = 1 + e^{i\pi(h_1+h_2+h_3)/2}. \tag{10.19a}$$

The possible values of $f_b(\mathbf{G})$ are the following:

$$f_b(\mathbf{G}) = \begin{cases} 1+i & \text{if } h_1, h_2, h_3 \text{ are all odd and } h_1 + h_2 + h_3 = 4n + 1, \\ 1-i & \text{if } h_1, h_2, h_3 \text{ are all odd and } h_1 + h_2 + h_3 = 4n + 3, \\ 2 & \text{if } h_1, h_2, h_3 \text{ are all even and } h_1 + h_2 + h_3 = 4n, \\ 0 & \text{if } h_1, h_2, h_3 \text{ are all even and } h_1 + h_2 + h_3 = 4n + 2. \end{cases} \tag{10.19b}$$

From Eq. (10.19b), we see that the reflections with h_1, h_2, h_3 odd numbers have the same value of $|f_b(\mathbf{G})|^2$. On the contrary, the reflections with h_1, h_2, h_3 even numbers satisfying $h_1+h_2+h_3 = 4n+2$ (n integer) *are forbidden in the local spherical approximation*. The first few "forbidden" reflections (ordered in increasing values of $|\mathbf{G}|$) correspond to reciprocal lattice vectors $(2\pi/a)(2, 0, 0)$, $(2\pi/a)(2, 2, 2)$, $(2\pi/a)(4, 2, 0)$, $(2\pi/a)(4, 4, 2)$, etc. (see Table 10.1).

Actually, the charge cloud around each site in the diamond structure has tetrahedral symmetry (and not spherical symmetry); because of the particular orientation of the tetrahedra with respect to each other, tetrahedral deformations invert at each successive atomic plane in the $[1, 1, 1]$ direction, as schematically indicated in Figure 10.6. This charge asymmetry transforms some of the "forbidden reflections" into "weakly allowed reflections" [notice that also anharmonic effects could have similar consequences, but we shall not concern ourselves with this mechanism].

A simple qualitative model that allows to mimic the tetrahedral symmetry of the covalent charge, and thus to distinguish "forbidden reflections" from "weakly allowed

Table 10.1 Theoretical and experimental X-ray form factors (absolute values) for Si, Ge, and GaP (in units of electrons per unit cell). For elemental semiconductors, reflections forbidden in the local spherical approximation are denoted by ($*$); reflections forbidden in the local tetrahedral symmetry are denoted by ($**$). The theoretical results are taken from C. S. Wang and B. M. Klein, Phys. Rev. B *24*, 3393 (1981) (copyright 1981 by the American Physical Society) and we refer to this paper for further details and comments on experiments.

Reciprocal lattice vectors (in units $2\pi/a$)	Number of vectors in the shell	Si		Ge		GaP	
		Theory	*Exp.*	*Theory*	*Exp.*	*Theory*	*Exp.*
$(0,0,0)$	1	28.00	28.00	64.00	64.00	46.00	46.00
$(1,1,1)$	8	15.11	15.19	38.83	39.42	28.84	28.83
$(2,0,0)$ $(*)(**)$	6	–	–	–	–	14.63	14.40
$(2,2,0)$	12	17.26	17.30	47.23	47.44	31.77	32.19
$(3,1,1)$	24	11.37	11.35	31.29	31.37	22.89	22.92
$(2,2,2)$ $(*)$	8	0.25	0.38	0.22	0.26	12.45	12.79
$(4,0,0)$	6	14.92	14.89	40.56	40.50	26.95	26.19
$(3,3,1)$	24	10.17	10.25	27.39	27.72	19.69	19.43
$(4,2,0)(*)(**)$	24	–	–	–	–	10.44	10.48
$(4,2,2)$	24	13.37	13.42	35.91	36.10	23.79	23.86
$(3,3,3)$	8	9.07	9.08	24.35	24.50	17.34	17.24
$(5,1,1)$	24	9.08	9.11	24.35	–	17.35	17.24
$(4,4,0)$	12	12.04	12.08	32.23	32.34	21.32	21.01

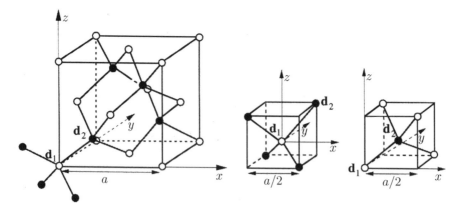

Figure 10.6 Diamond structure with the two fcc composing sublattices represented with white and shaded circles, respectively. The inversion of bond orientations in the two fcc sublattices is illustrated in the two smaller cubes, which provide the four atoms surrounding $\mathbf{d_1} = 0$ and $\mathbf{d_2} = (a/4)(1, 1, 1)$.

reflections" consists in simulating the bond charge in the unit cell as the sum of eight spherically symmetric contributions centered in the positions

$$
\begin{aligned}
\mathbf{d}_1 &= \frac{a}{4}(\gamma, \gamma, \gamma), & \mathbf{d}_5 &= \frac{a}{4}(1-\gamma, 1-\gamma, 1-\gamma), \\
\mathbf{d}_2 &= \frac{a}{4}(\gamma, -\gamma, -\gamma), & \mathbf{d}_6 &= \frac{a}{4}(1-\gamma, 1+\gamma, 1+\gamma), \\
\mathbf{d}_3 &= \frac{a}{4}(-\gamma, \gamma, -\gamma), & \mathbf{d}_7 &= \frac{a}{4}(1+\gamma, 1-\gamma, 1+\gamma), \\
\mathbf{d}_4 &= \frac{a}{4}(-\gamma, -\gamma, \gamma), & \mathbf{d}_8 &= \frac{a}{4}(1+\gamma, 1+\gamma, 1-\gamma).
\end{aligned}
\tag{10.20a}
$$

The four centers $\mathbf{d}_\nu (\nu = 1, \ldots, 4)$ surround the site $\mathbf{R} = 0$ along the four nearest neighbor directions; similarly, the four centers $\mathbf{d}_\nu$ ($\nu = 5, \ldots, 8$) surround the site $\mathbf{R} = (a/4)(1, 1, 1)$ along the four directions of formation of the covalent bond; the fraction of electrons taking part in it, or the specific value of the parameter γ ($0 < \gamma < 1/2$) are irrelevant, as far as we limit ourselves to work out the "selection rules."

We now calculate the geometrical structure factor $f_b(\mathbf{G}) = \sum \exp(i\mathbf{G} \cdot \mathbf{d}_\nu)$, where the sum runs over the eight vectors given by Eq. (10.20a), and the reciprocal lattice vectors $\mathbf{G} = (2\pi/a)(h_1, h_2, h_3)$ satisfy $h_1 + h_2 + h_3 = 4n + 2$; we obtain

$$
\begin{aligned}
f_b(\mathbf{G}) &= 2i \left[\sin\frac{\pi\gamma}{2}(h_1 + h_2 + h_3) + \sin\frac{\pi\gamma}{2}(h_1 - h_2 - h_3) \right. \\
&\quad \left. + \sin\frac{\pi\gamma}{2}(-h_1 + h_2 - h_3) + \sin\frac{\pi\gamma}{2}(-h_1 - h_2 + h_3) \right] \\
&= -8i \sin\frac{\pi\gamma h_1}{2} \sin\frac{\pi\gamma h_2}{2} \sin\frac{\pi\gamma h_3}{2}.
\end{aligned}
\tag{10.20b}
$$

From Eq. (10.20b) we see that $f_b(\mathbf{G}) \equiv 0$ if one (or more) of the integers h_1 or h_2 or h_3 is zero, while $f_b(\mathbf{G}) \neq 0$ if none of the integers h_1 or h_2 or h_3 is zero. We thus see, for instance, that the forbidden reflections $(2\pi/a)(2, 2, 2)$, $(2\pi/a)$ $(4, 4, 2)$, etc. are (weakly) allowed in tetrahedral symmetry, while the reflections $(2\pi/a)(2, 0, 0)$, $(2\pi/a)(4, 2, 0)$, etc. remain forbidden also in the adopted tetrahedral symmetry model. The study of weakly allowed reflections appears thus an invaluable tool for investigating wavefunctions, responsible of the chemical bond.

In Table 10.1, we give theoretical and experimental X-ray structure factors for silicon and germanium. We notice in particular that the structure factor for the $(2\pi/a)(2, 2, 2)$ reflection (forbidden in the local spherical approximation) is indeed different from zero (although very weak) in tetrahedral symmetry. Notice also that the lack of local spherical symmetry explains why the reflections $(2\pi/a)(3, 3, 3)$ and $(2\pi/a)(5, 1, 1)$ may have (slightly) different intensities, even if corresponding to reciprocal lattice vectors with the same modulus $(2\pi/a)\sqrt{27}$.

The study of the structure factors of elemental semiconductors is just an example of the wealth of information provided by X-ray scattering experiments. In Table 10.1, we report for a useful comparison the X-ray structure factors of GaP, a binary III–V compound with zincblende structure; the two atoms in the unit cell have rather different atomic numbers ($Z = 15$ for P and $Z = 31$ for Ga) and atomic form factors; in this case, it can be seen that the crystal form factors for

$(2\pi/a)(2, 0, 0)$, $(2\pi/a)(2, 2, 2)$, $(2\pi/a)(4, 2, 0)$ reflections are somewhat smaller, but of the same order of magnitude as the other ones.

A similarly fruitful analysis could be carried out for other classes of composite crystals. In particular for lithium hydride (with two core electrons and two valence electrons per unit cell), the ground-state density matrix and theoretical X-ray structure factors can be examined with relatively elementary techniques; comparison with experimental data can be used to test the quantum mechanical models for the valence wavefunctions associated to the s-like orbitals of the hydride ions [see for instance G. Grosso and G. Pastori Parravicini, Phys. Rev. B *17*, 3421 (1978), and references quoted therein].

10.2.3 Quantum Theory Analysis of the Thomson Cross-Section

In Section 10.2.1, on the basis of classical electrodynamics and the Thomson model, we have seen that the differential cross-section for elastic scattering of X-rays from an electron system is given by Eq. (10.8), which reads

$$\frac{d\sigma}{d\Omega} = \left(\frac{e^2}{mc^2}\right)^2 (\hat{\mathbf{e}}_i \cdot \hat{\mathbf{e}}_f)^2 |F(\mathbf{Q})|^2 \quad \text{with} \quad F(\mathbf{Q}) = \int n_{\text{el}}(\mathbf{r}) e^{i\mathbf{Q}\cdot\mathbf{r}} \, d\mathbf{r} \, ; \quad (10.21)$$

the form factor $F(\mathbf{Q})$ is the Fourier transform of the total electron density. The purpose of this subsection is to recover the Thomson cross-section with a full quantum mechanical treatment; the quantum formulation has the merit to put in better evidence the assumptions and the limits of applicability of the Thomson expression (the reader can choose to skip the technical details and pass directly to the summary of the physical facts reported at the end of this subsection).

Suppose the Hamiltonian of the electronic system is given by

$$H_0 = \frac{\mathbf{p}^2}{2m} + V(\mathbf{r}), \tag{10.22}$$

where $\mathbf{p}$ is the electron momentum and $V(\mathbf{r})$ is the (mean field) potential energy. [For introductory purposes and for sake of simplicity, it is sufficient to confine momentarily notations to systems with just a single electron (the hydrogen atom, for instance); at the end of the elaboration, the single-particle density is formally replaced by the one-body electron density of the N-particle system.] To find the coupling to the electromagnetic field, one replaces in the standard manner the momentum $\mathbf{p}$ with the generalized momentum $\mathbf{p} + (e/c)\mathbf{A}(\mathbf{r})$ (and adopts the Coulomb gauge for the vector potential).

The coupling Hamiltonian then contains two distinct terms that read

$$H_1 = \frac{e}{mc} \mathbf{A}(\mathbf{r}) \cdot \mathbf{p}, \tag{10.23a}$$

$$H_2 = \frac{e^2}{2mc^2} \mathbf{A}(\mathbf{r}) \cdot \mathbf{A}(\mathbf{r}). \tag{10.23b}$$

The first-order perturbation diagrams are shown in Figure 10.7. In first-order perturbation theory H_1 gives rise to one-photon absorption or emission, and only in the second order to scattering. On the contrary, in first-order perturbation theory H_2 gives rise

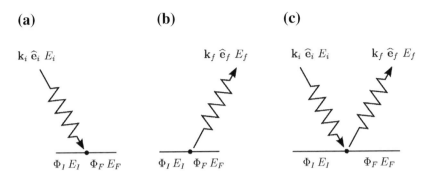

Figure 10.7 (a) and (b) Schematic diagrams of one-photon absorption and one-photon emission provided by the $\mathbf{A} \cdot \mathbf{p}$ term. (c) Schematic diagram of the scattering processes provided by the $\mathbf{A} \cdot \mathbf{A}$ term. In the figures, E_i, $\mathbf{k}_i$, $\widehat{\mathbf{e}}_i$ and E_f, $\mathbf{k}_f$, $\widehat{\mathbf{e}}_f$ denote energies, wavevectors, and polarizations of the incoming and outgoing photon. Initial and final energies and states of the electronic system are labeled by I and F, respectively.

to scattering processes. This justifies considering in the following only the term H_2 (*nonresonant X-ray scattering*): this term is expected to be more important than the scattering produced by H_1 to second order, except possibly at or near resonance conditions (*resonant X-ray scattering*). Disregarding resonant scattering, we proceed to the calculation of the nonresonant X-ray scattering cross-section within the lowest order Born approximation of the term proportional to $\mathbf{A} \cdot \mathbf{A}$.

Consider an incident photon, defined by its wavevector $\mathbf{k}_i$, polarization $\widehat{\mathbf{e}}_i$, energy $E_i = c\hbar k_i$, and the scattering into a final state of wavevector $\mathbf{k}_f$, polarization $\widehat{\mathbf{e}}_f$, energy $E_f = c\hbar k_f$. In the scattering process, the energy transfer to the target, i.e. to the electron system, is $E = E_i - E_f$, and the momentum transfer $\hbar\mathbf{Q} = \hbar\mathbf{k}_i - \hbar\mathbf{k}_f$. The energies and wavefunctions of the target before and after collision are indicated by E_I, $\Phi_I(\mathbf{r})$ and E_F, $\Phi_F(\mathbf{r})$, respectively.

The probability per unit time of a transition of the probe-target system between the indicated initial state and final state is given by the Fermi golden rule

$$P_{\mathbf{k}_f\widehat{\mathbf{e}}_f, \Phi_F \leftarrow \mathbf{k}_i\widehat{\mathbf{e}}_i \Phi_I} = \frac{2\pi}{\hbar} \left| \langle \mathbf{k}_f\widehat{\mathbf{e}}_f, \Phi_F | \frac{e^2}{2mc^2} \mathbf{A} \cdot \mathbf{A} | \mathbf{k}_i\widehat{\mathbf{e}}_i, \Phi_I \rangle \right|^2 \delta(E_f + E_F - E_i - E_I).$$

We are here interested only in elastic scattering, with the target in its ground state Φ_0 before and after the scattering events. This situation is described by the transition rate

$$P_{\mathbf{k}_f\widehat{\mathbf{e}}_f \leftarrow \mathbf{k}_i\widehat{\mathbf{e}}_i} = \frac{2\pi}{\hbar} \left(\frac{e^2}{2mc^2} \right)^2 \left| \langle \mathbf{k}_f\widehat{\mathbf{e}}_f, \Phi_0 | \mathbf{A} \cdot \mathbf{A} | \mathbf{k}_i\widehat{\mathbf{e}}_i, \Phi_0 \rangle \right|^2 \delta(E_f - E_i). \quad (10.24)$$

For the evaluation of the matrix element, it is convenient to expand the vector potential in terms of photon creation, and annihilation operators as follows:

$$\mathbf{A}(\mathbf{r}) = \frac{1}{\sqrt{V}} \sum_{\mathbf{k}\alpha} \sqrt{\frac{2\pi\hbar c}{k}} \left[\widehat{\mathbf{e}}_\alpha a_{\mathbf{k}\alpha} e^{i\mathbf{k}\cdot\mathbf{r}} + \widehat{\mathbf{e}}_\alpha a_{\mathbf{k}\alpha}^\dagger e^{-i\mathbf{k}\cdot\mathbf{r}} \right], \quad (10.25)$$

where V is the volume of the system and α indicates the photon state. Straight calculations give

$$\langle \mathbf{k}_f \widehat{\mathbf{e}}_f | \, \mathbf{A} \cdot \mathbf{A} \, | \mathbf{k}_i \widehat{\mathbf{e}}_i \rangle = \frac{1}{V} \frac{2\pi \hbar c}{\sqrt{k_i k_f}} (\widehat{\mathbf{e}}_i \cdot \widehat{\mathbf{e}}_f) \, 2 \, e^{i(\mathbf{k}_i - \mathbf{k}_f) \cdot \mathbf{r}}.$$

Inserting the above result into Eq. (10.24) gives

$$P_{\mathbf{k}_f \widehat{\mathbf{e}}_f \leftarrow \mathbf{k}_i \widehat{\mathbf{e}}_i} = \frac{2\pi}{\hbar} \left(\frac{e^2}{2mc^2} \right)^2 \frac{1}{V^2} \frac{4\pi^2 \hbar^2 c^2}{k_i k_f} (\widehat{\mathbf{e}}_i \cdot \widehat{\mathbf{e}}_f)^2$$

$$\times 4 \left| \langle \Phi_0(\mathbf{r}) | e^{i\mathbf{Q} \cdot \mathbf{r}} | \Phi_0(\mathbf{r}) \rangle \right|^2 \delta(E_f - E_i), \tag{10.26}$$

where $\mathbf{Q} = \mathbf{k}_i - \mathbf{k}_f$.

The double differential scattering cross-section, is proportional to the transition rate, apart trivial kinematic factors (see Problem 2); according to Eq. (A.3) it holds

$$\frac{d^2\sigma}{dE_f d\Omega} = \frac{V^2}{(2\pi)^3} \frac{k_f^2}{\hbar c^2} P_{\mathbf{k}_f \widehat{\mathbf{e}}_f \leftarrow \mathbf{k}_i \widehat{\mathbf{e}}_i}.$$

Inserting Eq. (10.26) in the above expression, one obtains

$$\frac{d^2\sigma}{dE_f d\Omega} = \left(\frac{e^2}{mc^2} \right)^2 (\widehat{\mathbf{e}}_i \cdot \widehat{\mathbf{e}}_f)^2 |F(\mathbf{Q})|^2 \delta(E_f - E_i), \tag{10.27}$$

where $F(\mathbf{Q})$ denotes the Fourier transform of the electron density of the ground state. The single differential cross-section, obtained integrating on the energy variable the double differential expression, can be recognized as the Thomson expression of Eq. (10.21).

The motivation of the present quantum treatment for elastic scattering of photons is to assess in straight manner the main adopted approximations. These can be summarized as follows: (i) the interaction of the electronic charges with electromagnetic field has been considered in the extreme non-relativistic limit (approximation to zero order in the ratio v/c); (ii) the direct coupling of the field with the nuclei has been neglected (approximation to zero order in the electron-to-nuclei mass ratio); (iii) magnetic terms have been neglected (approximation to zero order in the electron spin); and (iv) resonant effects have been disregarded.

Although in scattering of sufficiently hard X-rays the above approximations are often reasonably satisfied, it is evident the wealth of additional experimental information supplied whenever one or more of the above approximations are not applicable. Being impossible to enter in so many details concerning the elastic and inelastic scattering of particles, in the next two sections we choose to focus mainly on the guidelines given by the laws of conservation of energy and momentum. With these basic conservation conditions firmly established, in the remaining sections we consider the quantum theory of double differential cross-sections for neutron scattering, the effects of lattice vibrations and the key role of the Debye-Waller factor in the study of elastic and inelastic processes in general, and the Mössbauer effect in particular.

10.3   Compton Scattering and Electron Momentum Density

The Compton effect is the inelastic scattering of a photon (usually X-ray or γ-ray) by an electron; when the target electron is moving, the Compton-scattered radiation is also Doppler-broadened, and its energy distribution at a given scattering angle is called Compton profile. Measurements of Compton profiles of materials give information on the electron momentum density, projected along the scattering direction. The technique is particularly sensitive to the most external electronic wavefunctions, which describe slowly moving electrons and are responsible of the chemical bond. A closely related method which provides similar information is the study of the angular correlation in positron annihilation.

When a photon is scattered by an electron, its wavelength shift can be obtained from energy and momentum conservation laws. The scattering process is schematically represented in Figure 10.8; $\mathbf{k}_i$ and $\mathbf{k}_f$ denote the wavevector of the incident and scattered photon, respectively, and $\hbar\omega_i$ and $\hbar\omega_f$ their energy; θ_s is the scattering angle and $\mathbf{q}_f$ the final wavevector of the electron (the initial wavevector of the electron is supposed to be $\mathbf{q}_i = 0$). The conservation laws of momentum and energy require

$$\mathbf{k}_i = \mathbf{k}_f + \mathbf{q}_f \tag{10.28a}$$

and

$$\hbar\omega_i + mc^2 = \hbar\omega_f + \sqrt{m^2c^4 + \hbar^2c^2q_f^2}, \tag{10.28b}$$

where the relativistic expression of the electron energy is used.

We now eliminate from Eqs. (10.28) the final electron wavevector $\mathbf{q}_f$; for this purpose we compare the value of q_f^2 obtained from Eq. (10.28a) and the one obtained from Eq. (10.28b). From Eq. (10.28a) we have

$$q_f^2 = (\mathbf{k}_i - \mathbf{k}_f)^2 = k_i^2 - 2k_ik_f\cos\theta_s + k_f^2 = (k_i - k_f)^2 + 4k_ik_f\sin^2\frac{\theta_s}{2}. \tag{10.29a}$$

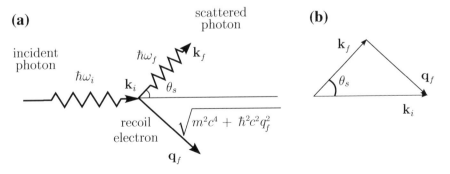

Figure 10.8 Schematic representation of the Compton scattering, through an angle θ_s, of a photon impinging on an electron initially at rest.

From Eq. (10.28b) we have

$$\hbar c(k_i - k_f) + mc^2 = \sqrt{m^2 c^4 + \hbar^2 c^2 q_f^2} \, ;$$

squaring both members of the above relation, one obtains

$$q_f^2 = (k_i - k_f)^2 + \frac{2mc}{\hbar}(k_i - k_f). \tag{10.29b}$$

Direct comparison of Eqs. (10.29a,b) gives

$$k_i - k_f = \frac{2\hbar}{mc} k_i k_f \sin^2 \frac{\theta_s}{2}.$$

With the replacement $k_i = 2\pi/\lambda_i$ and $k_f = 2\pi/\lambda_f$, we obtain for the increase of the wavelength of the scattered radiation the well-known Compton relation

$$\boxed{\Delta\lambda = \lambda_f - \lambda_i = \frac{2h}{mc}\sin^2\frac{\theta_s}{2}}, \tag{10.30}$$

where $2h/mc = 0.0485$ Å.

Using the relations $\lambda_i = 2\pi c/\omega_i$ and $\lambda_f = 2\pi c/\omega_f$, Eq. (10.30) can be recast into the equivalent form

$$\omega_f = \omega_i \frac{1}{1 + \dfrac{\hbar\omega_i}{mc^2} 2 \sin^2 \dfrac{\theta_s}{2}}. \tag{10.31a}$$

At low energies such that $\hbar\omega_i \ll mc^2$ ($= 0.511$ MeV), the above expression becomes

$$\frac{\omega_i - \omega_f}{\omega_i} = \frac{\hbar\omega_i}{mc^2} 2 \sin^2 \frac{\theta_s}{2}, \tag{10.31b}$$

and the relative Compton shift is controlled by the fraction $\hbar\omega_i/mc^2$.

Experiments are usually performed at high scattering angles (150–170°), and the photon energy shifts are quite measurable. For instance for the Kα radiation of molybdenum ($\hbar\omega_i \approx 17.6$ keV), Eq. (10.31b) gives for the energy shift of the backscattered photon the value ≈ 1.2 keV; this is also the energy transferred to the recoil electron. The numerical example considered shows that as long as the energy of X-ray is small with respect to mc^2, we have $k_i \approx k_f$ and the scattering of photons can be considered "quasi-elastic." Moreover, if the energy of the photon is in the hard X-ray region (10–50 keV) the electron recoil energy is of the order of a few keV; this energy greatly exceeds the binding energy of outer electrons and justifies the impulse approximation in the quantitative treatment of the collision between photon and electron.

Consider now the Compton scattering between an incoming photon and a moving electron. In this case, the scattered photon beam is also Doppler-broadened, and the wavelength shift can be obtained from the standard conservation laws of energy and momentum. In view of the relatively small initial and final kinetic energy of electrons (and for sake of simplicity) we use the non-relativistic expression of the electron kinetic energy and we have

$$\mathbf{k}_i + \mathbf{q}_i = \mathbf{k}_f + \mathbf{q}_f \tag{10.32a}$$

and

$$\hbar\omega_i + \frac{\hbar^2 q_i^2}{2m} = \hbar\omega_f + \frac{\hbar^2 q_f^2}{2m}. \tag{10.32b}$$

Let $E = \hbar\omega_i - \hbar\omega_f$ denote the energy difference of the incoming and outgoing photon, and $\mathbf{Q} = \mathbf{k}_i - \mathbf{k}_f$ the scattering vector. From the conservation laws (10.32) one obtains

$$\boxed{E = \frac{\hbar^2 Q^2}{2m} + \frac{\hbar^2}{m}\mathbf{q}_i \cdot \mathbf{Q}}; \tag{10.33}$$

this equality can be easily verified replacing $\mathbf{Q} = \mathbf{q}_f - \mathbf{q}_i \; [\equiv \mathbf{k}_i - \mathbf{k}_f]$ in Eq. (10.33). The first term in Eq. (10.33) is always positive and describes the Compton shift, whereas the second term expresses the Doppler broadening of the scattered radiation (see Figure 10.9).

The kinematic relation of Eq. (10.33) can be further simplified in the so called "quasi-elastic scattering approximation." In the triangle formed by the vectors $(\mathbf{k}_i, \mathbf{k}_f, \mathbf{Q})$, in several practical situations, the difference of the lengths $|\mathbf{k}_i| - |\mathbf{k}_f|$ is very small with respect to either lengths $|\mathbf{k}_i|$ or $|\mathbf{k}_f|$; as a consequence the scattering vector $\mathbf{Q}$ can be considered the basis of an isosceles triangle with length $Q \approx 2k_i \sin(\theta_s/2)$. Inserting this value of Q in Eq. (10.33), and indicating with z the direction of the scattering vector, one obtains

$$E = \frac{\hbar^2}{m}k_i^2 \, 2\sin^2\frac{\theta_s}{2} + \frac{\hbar}{m}p_z k_i \, 2\sin\frac{\theta_s}{2} \quad \text{where} \quad p_z = \hbar q_{iz}; \tag{10.34}$$

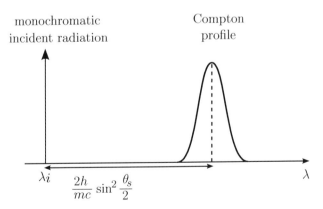

Figure 10.9 Schematic representation of the profile of the Compton-scattered radiation. The incident radiation is supposed to be monochromatic, with wavelength λ_i; the scattered radiation is shifted by $(2h/mc)\sin^2(\theta_s/2)$ and Doppler-broadened.

the quantity $p_z = \hbar q_{iz}$ is the *component of the momentum of the incoming electron along the X-ray scattering vector*. The above expression, using $k_i = \omega_i/c$ and $E = \hbar\omega_i - \hbar\omega_f$ can be cast in the form

$$\frac{\omega_i - \omega_f}{\omega_i} = \frac{\hbar\omega_i}{mc^2} 2\sin^2\frac{\theta_s}{2} + \frac{p_z}{mc} 2\sin\frac{\theta_s}{2}, \qquad (10.35)$$

which is an obvious generalization of Eq. (10.31b). As an estimation of the Doppler shift, consider for instance electrons at the Fermi surface in simple metals; we have $p_F/mc = v_F/c \approx 1/100$; thus also the Doppler shift $\Delta\omega/\omega_i$ is of the order of 1/100 and can be easily detected.

The above semiclassical analysis shows that *the Compton profile is proportional to the projection of the electron momentum along the scattering direction* (say the z-direction). Thus if $n(\mathbf{p})$ denotes the electron momentum density of the ground state of the sample, the theoretical Compton profile $J(p_z)$ is calculated as

$$J(p_z) = \int\int n(\mathbf{p}) \, dp_x \, dp_y. \qquad (10.36)$$

A more rigorous theory of Compton lineshapes, done in terms of second order time-dependent perturbation theory, gives formal support and establishes validity limits of the semi-classical analysis, summarized by expression (10.36). As an example, we report in Figure 10.10 experimental and computed isotropic Compton profile of LiH crystals; the Compton profile at small wavevectors, and in particular $J(0)$, is dominated by the behavior of the hydride ion wavefunctions; a proper account of the overlap of orbitals on different centers is necessary to correctly describe the ground-state electron momentum density of lithium hydride crystals.

The complete relativistic treatment of the Compton amplitude does depend also on the spin of the electron (in the semiclassical treatment, justified in the extreme non-relativistic limit, X-rays couple exclusively with the charge of the electrons by means of the electric field of the electromagnetic wave). In particular, in the case of circularly polarized photons, aligned spins induce an asymmetry in the Compton scattered radiation. In principle, from the change of the spectral distribution of scattered circularly polarized light when the polarization (or the magnetic field) is reversed, it is possible to infer the "magnetic Compton profiles" given by

$$I_{\text{magnetic}}(p_z) = \int\int [n_\uparrow(\mathbf{p}) - n_\downarrow(\mathbf{p})] \, dp_x \, dp_y,$$

where $n_\uparrow(\mathbf{p})$ and $n_\downarrow(\mathbf{p})$ are the electron momentum densities for spin-up and spin-down electrons, respectively. With the development of synchrotron radiation facilities, intense beams of circularly polarized X-rays have become available for the investigation of the Compton profiles of unpaired electrons in magnetic materials (magnetic Compton scattering). Notice that also the elastic scattering of X-rays may provide information on the electronic magnetic structure of a crystal, since the scattering amplitude in a relativistic treatment depends on the spins of the electrons (magnetic X-ray scattering).

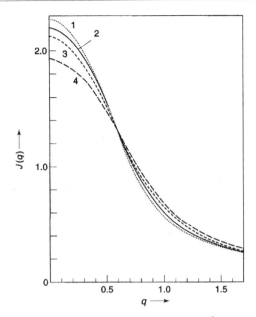

Figure 10.10 Compton profile of lithium hydride (q is in units a_B^{-1}, inverse Bohr radius). Curves 1 and 2 correspond to the free ion model with overlap and to the cluster model with overlap of Li^+ and H^- wavefunctions; the calculations are taken from G. Grosso, G. Pastori Parravicini and R. Resta, Phys. Stat. Sol. (b) *73*, 371 (1976). Curves 3 and 4 give the experimental results of W. C. Phillips and R. W. Weiss, Phys. Rev. *182*, 923 (1969) and those of J. Felsteiner, R. Fox and S. Kahane, Phys. Rev. B *6*, 4689 (1972), respectively.

10.4 Inelastic Scattering of Particles and Phonons Spectra of Crystals

In the previous sections we have considered elastic and inelastic scattering of X-rays from the electron charge distribution of the crystal, neglecting the lattice motion. In reality, even at zero absolute temperature, the nuclei move around their equilibrium positions, and the effects of lattice vibrations must be taken into account, especially in processes with small or zero energy transfer. The vibrational motion modifies the intensity of the elastically scattered waves and makes it possible inelastic processes with phonon emission or absorption.

In this section we focus on conservation laws of energy and crystal momentum for one-phonon processes, which are at the basis of the experimental determination of phonons spectra of solids. We do not need (at least at this stage) to be very specific about the coupling mechanism between the probe beam and the crystal; we only remind that for periodic structures the crystal momentum is preserved within reciprocal lattice vectors times $\hbar$; in general, processes involving $\mathbf{G} = 0$ and $\mathbf{G} \neq 0$ are referred as *normal* and *umklapp* processes, respectively. Information on phonon spectra of crystals can be obtained by various spectroscopic techniques, including scattering of X-rays,

Brillouin and Raman scattering, neutron techniques. In particular neutron spectro-
scopies, discussed here only on the basis of conservation laws, are then addressed in
the next sections by the detailed expressions of the double differential cross-sections
and by a detailed analysis of the relative role of elastic and inelastic processes, according
to the Debye-Waller factor.

Inelastic Scattering of X-rays by Phonons

In addition to elastic scattering, X-rays can be scattered inelastically with the absorption
or emission of one (or more) phonons. Away from directions satisfying the Bragg
condition, one can measure a diffuse background of radiation with a range of photon
energies, which reflect multi-phonon processes. Since phonon energies (≈ 10 meV) are
much smaller than X-ray photon energies (≈ 10 keV), in most experiments scattered
beams are not analyzed in energy; information on the phonon dispersion curves can be
disentangled from experimental data only indirectly and with limited accuracy.

With the development of synchrotron radiation sources with high photon flux in
conjunction with the improvements in crystal optics, inelastic X-ray scattering experiments with meV energy resolution have become possible [see for instance F. Sette,
G. Ruocco, M. Krisch, U. Bergmann, C. Masciovecchio, V. Mazzacurati, G. Signorelli
and R. Verbeni, Phys. Rev. Lett. *75*, 850 (1995); T. Scopigno, G. Ruocco and F. Sette,
Rev. Mod. Phys. *77*, 881 (2005) and references therein]. High resolution is achieved by
varying with millikelvin precision the temperature of the analyzer (or polarizer) crystal,
so controlling with high accuracy the distance between the family of reflecting planes
and thus the Bragg selected wavelength. Also notice that the momentum transfer is
determined only by the scattering angle θ_s, via the relation $q = 2k_i \sin(\theta_s/2)$, where
$\mathbf{k}_i$ is the wavevector of the incident photon; this is so because the fractional change of
energy of the X-ray photon is very small ($\approx 10^{-6}$), and $|\mathbf{k}_i|$ and $|\mathbf{k}_f|$ are practically equal.

The development of synchrotron radiation spectroscopy with meV resolution has
made possible the disclosure of previously unaccessible energy-momentum kinematic
regions (see Problem 4). Inelastic X-ray scattering measurements have become a precious tool for obtaining information on the phonon dynamics in solids, and on the
collective excitations in liquids.

Inelastic Scattering of Visible Light by Phonons: Brillouin and Raman Scattering

The inelastic scattering of optical radiation is a very sensitive tool for the experimental
determination of phonon energies in the small wavevector limit. Consider a monochromatic optical beam (usually a laser beam in the visible region) that propagates in a
crystal and undergoes inelastic scattering with emission of a phonon (Stokes process)
or absorption of a phonon (anti-Stokes process). In the photon frequency region of
interest, the crystal is assumed to be transparent and characterized by refractive index
n, so that the dispersion relation for photons $\omega = ck/n$ holds for both incident and
scattered radiation.

Consider a scattering process (see Figure 10.11), in which a photon propagating
in the medium with wavevector $\mathbf{k}_i$ and frequency $\omega_i = ck_i/n$ is scattered into the

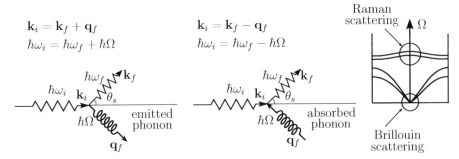

$$\mathbf{k}_i = \mathbf{k}_f + \mathbf{q}_f$$
$$\hbar\omega_i = \hbar\omega_f + \hbar\Omega$$

$$\mathbf{k}_i = \mathbf{k}_f - \mathbf{q}_f$$
$$\hbar\omega_i = \hbar\omega_f - \hbar\Omega$$

Raman scattering

emitted phonon

absorbed phonon

Brillouin scattering

Figure 10.11 Scattering of a photon through an angle θ_s with creation of a phonon (Stokes process) or annihilation of a phonon (anti-Stokes process); $\mathbf{k}_i$ and $\mathbf{k}_f$ denote the propagation wavevectors of the photon in the medium, before and after the scattering process. For photons in the visible region, a very small part of the phonon dispersion curves can be explored.

state $\mathbf{k}_f$ and $\omega_f = ck_f/n$, with emission of a phonon of wavevector $\mathbf{q}$ and frequency $\Omega(\mathbf{q}, p)$, where p labels a phonon branch (scattering processes with absorption of a phonon could be treated similarly). The laws of conservation of momentum and energy require

$$\mathbf{k}_i = \mathbf{k}_f + \mathbf{q} \quad \text{and} \quad \hbar\omega_i = \hbar\omega_f + \hbar\Omega(\mathbf{q}, p). \tag{10.37}$$

In the visible region, both $\mathbf{k}_i$ and $\mathbf{k}_f$ are much smaller than the Brillouin zone dimensions; thus the vector $\mathbf{q}$ defined by Eqs. (10.37) is already within the Brillouin zone (very near to the center) without need of adding a reciprocal lattice vector. The conservation relations (10.37) are very stringent indeed: from the determination of $\mathbf{q} = \mathbf{k}_i - \mathbf{k}_f$ and the measure of $E = \hbar\omega_i - \hbar\omega_f$ one can obtain the wavevector $\mathbf{q}$ and the energy $\hbar\Omega(\mathbf{q}, p)$ of the phonon taking part to the scattering process.

The conservation relations (10.37) can be elaborated taking into account that the phonon energies (≈ 0.01 eV) are much smaller than photon energies (≈ 1 eV) used in typical experimental situations; when the fractional change of the photon energy is very small, the photon scattering is referred as "quasi-elastic," and it holds $|\mathbf{k}_i| \approx |\mathbf{k}_f|$. In the quasi-elastic scattering process, the wavevector transfer q is controlled only by the scattering angle θ_s and is given with good approximation by

$$\boxed{q = 2k_i \sin \frac{\theta_s}{2}}. \tag{10.38a}$$

From the relation $\omega_i = ck_i/n$ it follows

$$q \approx 2\,\omega_i\, \frac{n}{c} \sin \frac{\theta_s}{2}. \tag{10.38b}$$

The light scattering with the emission or absorption of acoustic phonons is called *Brillouin scattering*. For acoustic phonons of small wavevector $\Omega(q) = v_s q$ (where v_s is the sound velocity), and Eq. (10.38b) gives

$$\boxed{\Omega \approx 2\omega_i\, n\, \frac{v_s}{c} \sin \frac{\theta_s}{2}}. \tag{10.38c}$$

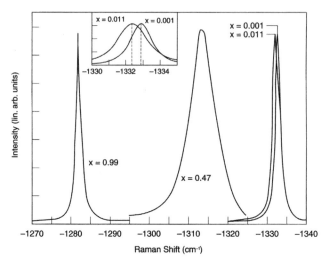

Figure 10.12 Raman spectra of diamond specimens $^{12}C_{1-x}{}^{13}C_x$ for $x = 0.001, x = 0.011$ (natural composition), $x = 0.47$ and $x = 0.99$; the change of the optical mode frequency ω_0 with the isotopic composition x follows approximately the $M^{-1/2}$ dependence, where M is the average atomic mass [reprinted with permission from R. Vogelgesang, A. K. Ramdas, S. Rodriguez, M. Grimsditch and T. R. Anthony, Phys. Rev. B *54*, 3989 (1996); copyright 1996 by the American Physical Society].

If the velocity v_s depends on the different acoustic branches, the Brillouin spectrum exhibits a characteristic structure, reflecting the different acoustic phonons. The measurements of the frequency shifts $\pm\Omega$ of the Brillouin components of the scattered light yield accurate information on the acoustic phonons with wavevector q close to the zone center.

The light scattering with the emission or the absorption of optical phonons is called *Raman scattering*. In this case, the dispersion relation of optical phonons can be neglected and the Raman spectra reveal peaks corresponding to optical phonons with wavevector q essentially at the center of the Brillouin zone. In Figure 10.12 we report as an example the Raman spectra of diamond for several isotopic compositions (the Raman shift is the opposite of the energy transferred to the system). Diamond is a non-polar crystal with the zone center optical modes triply degenerate and infrared inactive; the large value of the optical mode frequency ω_0 is produced by the strong covalent bond and the light mass of ^{12}C or ^{13}C constituent atoms. In polar crystals, the Raman scattering of laser light has become a precious tool for investigating polariton dispersion curves (see for instance Figure 9.11).

Inelastic Scattering of Neutrons by Phonons

The inelastic scattering of neutrons by phonons is a powerful tool for investigating the phonon dispersion curves over the whole Brillouin zone. Consider a neutron of wavevector $\mathbf{k}_i$ and energy E_i, which is scattered into the state $\mathbf{k}_f$ and energy E_f, with emission of a phonon of wavevector $\mathbf{q}$ and frequency $\Omega(\mathbf{q}, p)$, where p labels a phonon branch (scattering processes with absorption of a phonon could be handled similarly). The conservation laws for momentum and energy give

$$\mathbf{k}_i = \mathbf{k}_f + \mathbf{q} + \mathbf{G} \tag{10.39a}$$

and

$$\frac{\hbar^2 k_i^2}{2M_n} = \frac{\hbar^2 k_f^2}{2M_n} + \hbar\Omega(\mathbf{q}, p); \tag{10.39b}$$

a reciprocal lattice vector $\mathbf{G}$ is in general required in the second member of Eq. (10.39a).
   Equations (10.39) can be compacted into a unique equation

$$\frac{\hbar^2 k_i^2}{2M_n} = \frac{\hbar^2 k_f^2}{2M_n} + \hbar\Omega(\mathbf{k}_i - \mathbf{k}_f, p). \tag{10.40}$$

The highly restrictive aspect of the relation (10.40) can be understood imagining to fix $\mathbf{k}_i$ and also the direction of observation $\widehat{\mathbf{k}}_f = \mathbf{k}_f/k_f$; thus only k_f remains as a disposable parameter and presumably only one (or a few) discretized values of k_f can satisfy the relation (10.40). In particular, in the case $\hbar^2 k_i^2/2M_n > \hbar\Omega(\mathbf{k}_i, p)$, it is easily seen by inspection that there is (at least) a solution of Eq. (10.40) in any p-th branch and for any given direction of observation $\widehat{\mathbf{k}}_f$.

   Neutron scattering measurements are typically performed by means of a "triple-axis spectrometer" (or some appropriate implementation). A beam of collimated and polyenergetic neutrons strikes a (orientable) single-crystal; by means of Bragg reflection, neutrons of a given wavelength are selected. The resulting beam of monoenergetic neutrons strikes a single-crystal sample (capable of different orientations). The scattered neutrons are then analyzed by means of a second (orientable) Bragg scatterer. By use of conservation laws, one can determine the phonon wavevector $\mathbf{q}$, besides the measured energy change $\hbar\Omega(\mathbf{q}, p)$; thus the dispersion relations $\hbar\Omega(\mathbf{q}, p)$ can be worked out over the Brillouin zone.

10.5 Quantum Theory of Elastic and Inelastic Scattering of Neutrons

In this section we give the basic concepts and quantum theory for the microscopic expression of the differential cross-sections for scattering of particles in condensed matter. We consider in particular the study of the double differential cross-section for the diffusion of neutrons by a crystal, because the simplicity of the probe-sample interaction is reflected in the (relative) simplicity of the needed elaborations.

   The essential kinematic aspects of a neutron beam colliding on a crystal are indicated in Figure 10.13. The nuclei of the crystal can be pictured as a collection of scatterers $\{\mathbf{R}_1, \mathbf{R}_2, \ldots, \mathbf{R}_N\}$ in vibration around their equilibrium lattice positions. The

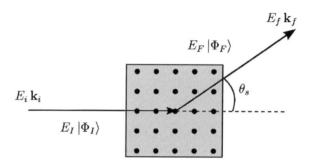

Figure 10.13 Diffusion of particles from a crystal and effect of the thermal motion of the lattice; $E_I, |\Phi_I\rangle$, E_F and $|\Phi_F\rangle$ denote energies and vibrational states of the whole crystal lattice before and after the scattering; E_i, $\mathbf{k}_i$, E_f, and $\mathbf{k}_f$ are the energies and the wavevectors of the probe particle before and after the scattering. The energy transfer to the target is $E = E_i - E_f$ and the momentum transfer to the target is $\hbar\mathbf{Q} = \hbar\mathbf{k}_i - \hbar\mathbf{k}_f$.

complete set of wavefunctions and the energies of the crystalline lattice Hamiltonian $H(\{\mathbf{R}_i\}, \{\mathbf{P}_i\})$ are denoted by $\Phi_I(\{\mathbf{R}_m\})$ and E_I, respectively.

Neutrons interact with the nucleons of a crystal via nuclear forces. For thermal or low energy neutrons, the particle-nucleus coupling can be described by the Fermi pseudo-potential

$$V_p(\mathbf{r} - \mathbf{R}) = \frac{2\pi\hbar^2}{M_n} b\,\delta(\mathbf{r} - \mathbf{R}), \tag{10.41}$$

where $\mathbf{r}$ is the coordinate of the impinging neutron, $\mathbf{R}$ is the coordinate of a given scatterer nucleus, and b is the Fermi length. To avoid inessential details, we assume that all the nuclei are of the same isotope (and have thus the same scattering length), and also neglect any other interaction except the pseudo-potential interaction (10.41). Then the probe-sample interaction can be written as

$$V_{\text{probe-sample}}(\mathbf{r}; \{\mathbf{R}_m\}) = \sum_m V_p(\mathbf{r} - \mathbf{R}_m); \tag{10.42}$$

this form, represents the sum of contributions $V_p(\mathbf{r})$ of identical scattering units centered at the various positions $\mathbf{R}_m$.

Consider an incoming neutron, described by plane wave $|W_{\mathbf{k}_i}\rangle$ of vector $\mathbf{k}_i$ and energy E_i. After collision with the target, suppose that the neutron is scattered into the plane wave $|W_{\mathbf{k}_f}\rangle$ of vector $\mathbf{k}_f$ and energy E_f. In the scattering process, the energy transfer to the target is

$$E = E_i - E_f$$

and the momentum transfer to the target is

$$\hbar\mathbf{Q} = \hbar\mathbf{k}_i - \hbar\mathbf{k}_f.$$

The energies and wavefunctions of the target before and after collision are indicated by E_I, $\Phi_I(\{\mathbf{R}_i\})$ and E_F, $\Phi_F(\{\mathbf{R}_i\})$, respectively. The forthcoming quantum treatment is based on the "kinematic approximation" (or first order Born approximation), which holds for weak interactions and negligible multiple scattering processes.

The probability per unit time of a transition of the probe-target system from an initial state $|W_{\mathbf{k}_i}\Phi_I\rangle$ to a final state $|W_{\mathbf{k}_f}\Phi_F\rangle$ is given by the Fermi golden rule

$$P_{\mathbf{k}_f\Phi_F\leftarrow\mathbf{k}_i\Phi_I} = \frac{2\pi}{\hbar}\left|\langle W_{\mathbf{k}_f}\Phi_F|\sum_m V_p(\mathbf{r}-\mathbf{R}_m)|W_{\mathbf{k}_i}\Phi_I\rangle\right|^2 \delta(E_f + E_F - E_i - E_I).$$

$$(10.43\text{a})$$

The basic step in the elaboration of the above equation exploits the fact that incident and scattered particles are plane waves, and the contribution of every probe-nucleus term only depends on the difference of the two coordinates. For any chosen term it holds

$$\langle W_{\mathbf{k}_f}|V_p(\mathbf{r}-\mathbf{R}_m)|W_{\mathbf{k}_i}\rangle = \frac{1}{V}\int e^{-i\mathbf{k}_f\cdot\mathbf{r}}V_p(\mathbf{r}-\mathbf{R}_m)e^{i\mathbf{k}_i\cdot\mathbf{r}}\,d\mathbf{r}$$

$$= \frac{1}{V}e^{i(\mathbf{k}_i-\mathbf{k}_f)\cdot\mathbf{R}_m}\int e^{i(\mathbf{k}_i-\mathbf{k}_f)\cdot\mathbf{r}}V_p(\mathbf{r})\,d\mathbf{r} \equiv \frac{1}{V}e^{i\mathbf{Q}\cdot\mathbf{R}_m}V_p(\mathbf{Q}),$$

where V is the volume over which the initial and final plane waves are normalized, and

$$V_p(\mathbf{Q}) = \int e^{i\mathbf{Q}\cdot\mathbf{r}}V_p(\mathbf{r})\,d\mathbf{r} = \int e^{i\mathbf{Q}\cdot\mathbf{r}}\frac{2\pi\hbar^2}{M_n}b\,\delta(\mathbf{r})\,d\mathbf{r} = \frac{2\pi\hbar^2}{M_n}b$$

is the Fourier transform of the interaction potential. Since all the scattering units are assumed to be equivalent, the Fourier transform can be factorized out, and Eq. (10.43a) becomes

$$P_{\mathbf{k}_f\Phi_F\leftarrow\mathbf{k}_i\Phi_I} = b^2\frac{8\pi^3\hbar^3}{M_n^2}\frac{1}{V^2}\left|\langle\Phi_F|\sum_m e^{i\mathbf{Q}\cdot\mathbf{R}_m}|\Phi_I\rangle\right|^2\delta(E_F-E_I-E), \quad (10.43\text{b})$$

where $\mathbf{Q}$ is the momentum transfer and E is the energy transfer.

We perform the sum over all the states $|\Phi_F\rangle$ of the scatterer, and obtain the probability rate for a scattering to a final state of the probe (regardless of the final state of the target). Generally the target is not in a pure state $|\Phi_I\rangle$ (typically the ground state at zero temperature), but rather it is spread over all its accessible states in thermal equilibrium, with probability $p_I = \exp[-E_I/k_BT]/\sum\exp[-E_I/k_BT]$. Then the transition rate of an incident particle $\mathbf{k}_i$ to the plane wave $\mathbf{k}_f$, produced by a scatterer at thermal equilibrium, is given by

$$P_{\mathbf{k}_f \leftarrow \mathbf{k}_i} = b^2 \frac{8\pi^3 \hbar^3}{M_n^2} \frac{1}{V^2} \sum_{F,I} p_I \left| \langle \Phi_F | \sum_m e^{i\mathbf{Q} \cdot \mathbf{R}_m} | \Phi_I \rangle \right|^2 \delta(E_F - E_I - E). \quad (10.44)$$

This expression can be re-written in the form

$$\boxed{P_{\mathbf{k}_f \leftarrow \mathbf{k}_i} = b^2 \frac{8\pi^3 \hbar^3}{M_n^2} \frac{1}{V^2} S(\mathbf{Q}, E)}, \quad (10.45a)$$

where the dynamical structure factor $S(\mathbf{Q}, E)$ is defined as

$$\boxed{S(\mathbf{Q}, E) = \sum_{F,I} p_I \left| \langle \Phi_F | \sum_m e^{i\mathbf{Q} \cdot \mathbf{R}_m} | \Phi_I \rangle \right|^2 \delta(E_F - E_I - E)}. \quad (10.45b)$$

Notice that the dynamical structure factor depends exclusively on the wavefunctions and energies of the system under investigation (i.e. without any interaction with the probe).

The measured signal in scattering experiments is determined by the double differential scattering cross-section, which is proportional to the transition rate, apart trivial kinematic factors (see Problem 2). According to Eq. (A.2), for the scattering of neutrons we have

$$\frac{d^2\sigma}{dE_f \, d\Omega} = \frac{V^2}{(2\pi)^3} \frac{M_n^2}{\hbar^3} \frac{k_f}{k_i} P_{\mathbf{k}_f \leftarrow \mathbf{k}_i}.$$

We arrive thus at the elegant and self-explanatory result

$$\boxed{\frac{d^2\sigma}{dE_f \, d\Omega} = b^2 \frac{k_f}{k_i} S(\mathbf{Q}, E)}. \quad (10.46)$$

Equation (10.46) is the basic theoretical description of neutron scattering within the linear response formalism, valid when the coupling between the probe and the sample is weak. The double differential cross-section contains the product of three terms: (i) The first term describes the intensity of the probe-sample coupling [it is constant in the case of contact interaction for neutrons, but is in general a smooth $\mathbf{Q}$-dependent quantity for other particles]. (ii) The second is a kinematic term depending on the dispersion curves of the probe particles. (iii) The third term is exclusively related to the elementary excitations, characteristic of the system under investigation.

The dynamical structure factor is the final outcome of the measurements; however from a theoretical side, Eq. (10.45b) as it stands does not seem encouraging for actual calculations. A major step to overcome this difficulty exploits a general quantum mechanical identity, that transforms the golden rule structure of Eq. (10.45b) into a conveniently and manageable autocorrelation function (see Problem 3). Using Eq. (A.5) after setting $A = \sum \exp{(i\mathbf{Q} \cdot \mathbf{R}_m)}$ therein, we have

$$S(\mathbf{Q}, E) = \frac{1}{\hbar} \int_{-\infty}^{+\infty} \frac{dt}{2\pi} e^{-iEt/\hbar} \sum_{m\,n} \langle e^{-i\mathbf{Q}\cdot\mathbf{R}_m} \, e^{i\mathbf{Q}\cdot\mathbf{R}_n(t)} \rangle , \qquad (10.47)$$

where the symbol $\langle \cdots \rangle$ indicates the thermal average and time-dependent operators are intended in the Heisenberg representation. The interpretation of Eq. (10.47) is very appealing: *the dynamical structure factor $S(\mathbf{Q}, E)$ can be viewed as the space and time Fourier transform of the autocorrelation function of the observable that couples the sample to the probe.* In fact from Eq. (10.42), it is seen that the probe-sample interaction is controlled by the observable

$$n(\mathbf{r}) = \sum_m \delta(\mathbf{r} - \mathbf{R}_m),$$

whose space Fourier transform is

$$n(\mathbf{Q}) = \sum_m \int e^{i\mathbf{Q}\cdot\mathbf{r}} \delta(\mathbf{r} - \mathbf{R}_m) \, d\mathbf{r} = \sum_m e^{i\mathbf{Q}\cdot\mathbf{R}_m};$$

the operators $n(\mathbf{r})$ and $n(\mathbf{Q})$ can be recognized as the nuclear (or atomic) density number and the corresponding density number fluctuations of the system under investigation. The expression (10.47) for the dynamical structure factor in terms of the atomic density-density autocorrelation function, besides its formal elegance, is also very convenient for the actual calculations (or for reasonable approximations) based on various many-body techniques. We can add that the basic concepts of this section, with appropriate implementations, also apply to other types of scattering (for instance magnetic scattering of neutrons in magnetic materials) or to other types of particles (X-rays and other beams). To get some familiarity on the wealth of information contained in Eq. (10.47), in the next Sections 10.6 and 10.7 we will consider a number of significant examples.

10.6 Dynamical Structure Factor for Harmonic Displacements and Debye-Waller Factor

In this section we evaluate the dynamical structure factor for harmonic displacements. We begin with the simple but significant case of a single harmonic oscillator (see Figure 10.14): this model paves the way to the discussion of the Debye-Waller factor in crystals and the γ-emission or absorption from nuclei in solids. Next, the discussion is generalized to actual crystals, where all the nuclei of the lattice are undergoing normal modes oscillations.

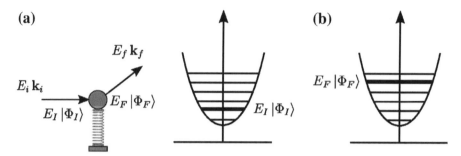

Figure 10.14 (a) Diffusion of particles by a scatterer, represented by an elastically bound nucleus of mass M and frequency $\hbar\omega_0$. (b) Schematic representation of initial and final energy levels and states of the harmonic oscillator scatterer, of given frequency ω_0.

10.6.1 Dynamical Structure Factor of a Three-Dimensional Harmonic Oscillator

It is instructive to calculate explicitly the dynamical structure factor in the case the target is described by an isotropic three-dimensional harmonic oscillator of mass M and angular frequency ω_0. In the situation under attention the structure factor of Eq. (10.47) reads

$$S(\mathbf{Q}, E) = \frac{1}{\hbar} \int_{-\infty}^{+\infty} \frac{dt}{2\pi}\, e^{-iEt/\hbar} \langle e^{-i\mathbf{Q}\cdot\mathbf{R}}\, e^{i\mathbf{Q}\cdot\mathbf{R}(t)} \rangle, \tag{10.48}$$

where $\mathbf{R}$ denotes the coordinate of the single harmonic scatterer. We notice again that the dynamical structure factor is related to the elementary excitations characteristic of the harmonic oscillator under investigation, and is fully independent of the impinging beam or the specific details of the probe-scattering coupling.

It is convenient to indicate with $\mathbf{u} = (u_x, u_y, u_z)$ the displacement of the scatterer from the equilibrium position, and use creation and annihilation operators, for a straight calculation of thermal averages. We have

$$u_x = \sqrt{\frac{\hbar}{2M\omega_0}}\left[a_x + a_x^\dagger\right], \tag{10.49a}$$

and similar expressions for u_y and u_z. From elementary properties of harmonic oscillators (see Appendix 9.A) we obtain

$$\langle u_x^2 \rangle = \frac{\hbar}{2M\omega_0}\langle (a_x + a_x^\dagger)(a_x + a_x^\dagger) \rangle = \frac{\hbar}{2M\omega_0}\left(2\langle n \rangle + 1\right),$$

where the average number of phonons $\langle n \rangle = \langle a_x^\dagger a_x \rangle$ is given by the Bose-Einstein distribution function $1/(\exp(\hbar\omega_0/k_BT) - 1)$. Then

$$\langle u_x^2 \rangle = \frac{\hbar}{2M\omega_0}\left(\frac{2}{e^{\hbar\omega_0/k_BT} - 1} + 1\right). \tag{10.49b}$$

In the Heisenberg representation we have

$$u_x(t) = \sqrt{\frac{\hbar}{2M\omega_0}} \left[a_x e^{-i\omega_0 t} + a_x^\dagger e^{i\omega_0 t} \right] \tag{10.49c}$$

and $\langle u_x^2(t) \rangle = \langle u_x^2 \rangle$.

According to Eq. (9.A20), it is convenient to remind the identity

$$\langle e^A e^B \rangle = e^{\langle A^2 + 2AB + B^2 \rangle / 2}, \tag{10.50}$$

where A and B are any two operators linear in creation and annihilation operators. We use the above expression with $A = -i\mathbf{Q} \cdot \mathbf{u}$ and $B = i\mathbf{Q} \cdot \mathbf{u}(t)$ (without loss of generality, whenever useful, we can suppose that $\mathbf{Q}$ is in the x-direction). Equation (10.48) thus takes the form

$$\boxed{S(\mathbf{Q}, E) = e^{-2W} \frac{1}{\hbar} \int_{-\infty}^{+\infty} \frac{dt}{2\pi} e^{-iEt/\hbar} e^{\langle \mathbf{Q} \cdot \mathbf{u}\, \mathbf{Q} \cdot \mathbf{u}(t) \rangle},} \tag{10.51a}$$

where

$$2W = \langle (\mathbf{Q} \cdot \mathbf{u})^2 \rangle. \tag{10.51b}$$

The quantity $\exp(-2W)$ is called the *Debye-Waller factor*; its physical consequences will be apparent soon.

Using Eq. (10.49b), the explicit expression of $2W$ reads

$$2W = \frac{\hbar^2 Q^2}{2M} \frac{1}{\hbar\omega_0} \left(\frac{2}{e^{\hbar\omega_0/k_B T} - 1} + 1 \right).$$

In the limiting case of temperatures much smaller or much higher than the reference temperature $T_0 = \hbar\omega_0/k_B$ it holds

$$2W \equiv \langle (\mathbf{Q} \cdot \mathbf{u})^2 \rangle = \begin{cases} \dfrac{\hbar^2 Q^2}{2M} \dfrac{1}{\hbar\omega_0} & \text{for } T \ll T_0, \\[2ex] \dfrac{2T}{T_0} \dfrac{\hbar^2 Q^2}{2M} \dfrac{1}{\hbar\omega_0} & \text{for } T \gg T_0. \end{cases} \tag{10.52}$$

Thus, for all the temperatures of usual interest (ranging from $T = 0$ to temperatures smaller or near T_0), we see that the important physical effect determining the quantity $2W$ is *just the ratio $E_R/\hbar\omega_0$ between the target recoil energy $E_R = \hbar^2 Q^2/2M$ and the harmonic oscillator quantum of energy $\hbar\omega_0$*.

It is convenient to represent the dynamical structure factor in Eq. (10.51a) in terms of a series of displacement-displacement correlation functions representing zero-phonon, one-phonon, and multi-phonon scattering processes

$$S(\mathbf{Q}, E) = e^{-2W} \frac{1}{\hbar} \int_{-\infty}^{+\infty} \frac{dt}{2\pi} e^{-iEt\hbar} \sum_{m=0}^{\infty} \frac{1}{m!} \left[\langle \mathbf{Q} \cdot \mathbf{u}\, \mathbf{Q} \cdot \mathbf{u}(t) \rangle \right]^m. \tag{10.53}$$

We consider now in particular the zero-phonon and the one-phonon contributions to the expansion, given by the $m = 0$ and $m = 1$ terms, respectively.

Zero-Phonon Contribution (Elastic Scattering)

The zero-phonon contribution to the dynamical structure factor describes elastic scattering processes and is given by

$$S_0(\mathbf{Q}, E) = e^{-2W} \frac{1}{\hbar} \int_{-\infty}^{+\infty} \frac{dt}{2\pi} e^{-iEt/\hbar} = e^{-2W} \delta(E). \tag{10.54}$$

The above expression shows that *the intensity for elastic scattering from an oscillating target with respect to the intensity of elastic scattering from a fixed target is reduced by the Debye-Waller factor* $\exp(-2W)$. Thus, the fraction f of elastic (or "recoil-free") scattering processes can be written as

$$f = \exp(-2W) = \exp(-Q^2 \langle u_x^2 \rangle),$$

where $\langle u_x^2 \rangle$ is the mean square vibrational amplitude of the scatterer in the direction of $\mathbf{Q}$. Notice that the recoilless fraction f equals one for a "rigid" scatterer, i.e. for a scatterer where $(\langle u_x^2 \rangle)^{1/2}$ is much smaller than the wavelength $2\pi/|\mathbf{Q}|$.

In the temperature range $0 \le T \approx T_0$, using Eq. (10.52), the recoil-free fraction can be expressed as

$$f = \exp(-E_R/\hbar\omega_0),$$

where $E_R = \hbar^2 Q^2/2M$ is the recoil energy of the scatterer. In the case $E_R \ll \hbar\omega_0$, we see that most of the scattering takes place as if *the target is held fixed*; in the case $E_R > \hbar\omega_0$ the elastic scattering is strongly reduced with respect to inelastic processes.

One-Phonon Contribution

The one-phonon contribution to the dynamical structure factor is

$$S_1(\mathbf{Q}, E) = e^{-2W} \frac{1}{\hbar} \int_{-\infty}^{+\infty} \frac{dt}{2\pi} e^{-iEt/\hbar} \langle \mathbf{Q} \cdot \mathbf{u} \, \mathbf{Q} \cdot \mathbf{u}(t) \rangle.$$

Using Eqs. (10.49) we have

$$\langle \mathbf{Q} \cdot \mathbf{u} \, \mathbf{Q} \cdot \mathbf{u}(t) \rangle = \frac{\hbar Q^2}{2M\omega_0} \langle (a_x + a_x^\dagger)(a_x e^{-i\omega_0 t} + a_x^\dagger e^{i\omega_0 t}) \rangle$$

$$= \frac{\hbar^2 Q^2}{2M} \frac{1}{\hbar\omega_0} [\langle n \rangle e^{-i\omega_0 t} + \langle n + 1 \rangle e^{i\omega_0 t}].$$

With the help of the above expression, we obtain

$$S_1(\mathbf{Q}, E) = e^{-2W} \frac{\hbar^2 Q^2}{2M} \frac{1}{\hbar\omega_0} [\langle n \rangle \delta(E + \hbar\omega_0) + \langle n + 1 \rangle \delta(E - \hbar\omega_0)], \tag{10.55}$$

where $E = E_i - E_f$. The dynamical structure factor $S_1(\mathbf{Q}, E)$ consists of sharp delta-function peaks describing one-phonon absorption and emission processes; the Bose thermal factors $\langle n \rangle$ and $\langle n \rangle + 1$ are appropriate for processes in which phonons are absorbed or emitted, respectively (notice that at $T = 0$ absorption processes are suppressed). The relative importance of zero-phonon *versus* one-phonon contributions is in essence dictated by the ratio between the recoil energy and the quantum of energy $\hbar\omega_0$. This same ratio controls the relative importance of multi-phonon scattering processes in which two or more phonons are absorbed or emitted.

10.6.2 Dynamical Structure Factor of a Three-Dimensional Harmonic Crystal

In the previous section we have analyzed the diffusion of particles by a single dynamic scatterer, schematized as an atom or a nucleus performing harmonic oscillations, and we have seen that the relative importance of *elastic* and *inelastic processes* is essentially determined by the Debye-Waller factor. We wish now to study the scattering of particles by an actual crystal with N nuclei in thermal vibration, and to assess the relative importance of elastic and inelastic processes.

The dynamic structure factor of the crystal is given by Eq. (10.47) here re-written for convenience

$$S(\mathbf{Q}, E) = \frac{1}{\hbar} \int_{-\infty}^{+\infty} \frac{dt}{2\pi} e^{-iEt/\hbar} \sum_{mn} \langle e^{-i\mathbf{Q}\cdot\mathbf{R}_m} e^{i\mathbf{Q}\cdot\mathbf{R}_n(t)} \rangle; \qquad (10.56)$$

as previously, the symbol $\langle \cdots \rangle$ indicates the thermal average. Exploiting translational symmetry for the crystal, and assuming the harmonic approximation for the lattice vibrations, we can now calculate the relevant aspects of the structure factor.

A first simplification comes from the translational symmetry of the lattice. We indicate the atomic positions in the Bravais lattice as $\mathbf{R}_n = \mathbf{t}_n + \mathbf{u}_n$, where $\mathbf{t}_n$ is a lattice translation vector and $\mathbf{u}_n$ is the displacement of the atom from its equilibrium position. We can simplify the double sum in Eq. (10.56) with N times a unique sum with fixed $\mathbf{R}_m$ (for instance $\mathbf{R}_m = 0$). We have

$$S(\mathbf{Q}, E) = N \sum_{\mathbf{t}_n} e^{i\mathbf{Q}\cdot\mathbf{t}_n} \frac{1}{\hbar} \int_{-\infty}^{+\infty} \frac{dt}{2\pi} e^{-iEt/\hbar} \langle e^{-i\mathbf{Q}\cdot\mathbf{u}_0} e^{i\mathbf{Q}\cdot\mathbf{u}_n(t)} \rangle.$$

We now use for *harmonic crystals* the identity expressed by Eq. (10.50), with $A = -i\mathbf{Q} \cdot \mathbf{u}_0$ and $B = i\mathbf{Q} \cdot \mathbf{u}_n(t)$. The dynamic structure factor takes the form

$$\boxed{S(\mathbf{Q}, E) = Ne^{-2W} \sum_{\mathbf{t}_n} e^{i\mathbf{Q}\cdot\mathbf{t}_n} \frac{1}{\hbar} \int_{-\infty}^{+\infty} \frac{dt}{2\pi} e^{-iEt/\hbar} e^{\langle \mathbf{Q}\cdot\mathbf{u}_0 \, \mathbf{Q}\cdot\mathbf{u}_n(t) \rangle},} \qquad (10.57a)$$

where

$$\boxed{2W = \langle (\mathbf{Q} \cdot \mathbf{u}_0)^2 \rangle} \qquad (10.57b)$$

and $\exp{(-2W)}$ is the Debye-Waller factor. The quantity $\langle(\mathbf{Q}\cdot\mathbf{u}_0)^2\rangle$ for a Bravais crystal lattice has been calculated in Chapter 9; using Eq. (9.40), we have

$$2W \equiv \langle(\mathbf{Q}\cdot\mathbf{u}_0)^2\rangle = \begin{cases} \dfrac{3}{2}\dfrac{\hbar^2 Q^2}{2M}\dfrac{1}{k_B T_D} & \text{for } T \ll T_D, \\[2ex] \dfrac{6T}{T_D}\dfrac{\hbar^2 Q^2}{2M}\dfrac{1}{k_B T_D} & \text{for } T \gg T_D. \end{cases} \tag{10.58}$$

Apart for some numerical factors, it is apparent the perfect analogy between Eqs. (10.58) and (10.52) of the simplified treatment of the previous section.

The meaning of Eq. (10.57a) becomes evident if the second exponential occurring under integral sign is expanded in series; we have

$$S(\mathbf{Q}, E) = \sum_{m=0}^{\infty} S_m(\mathbf{Q}, E), \tag{10.59a}$$

where

$$S_m(\mathbf{Q}, E) = N e^{-2W} \sum_{\mathbf{t}_n} e^{i\mathbf{Q}\cdot\mathbf{t}_n} \frac{1}{\hbar} \int_{-\infty}^{+\infty} \frac{dt}{2\pi} e^{-iEt/\hbar} \frac{1}{m!} [\langle\mathbf{Q}\cdot\mathbf{u}_0\,\mathbf{Q}\cdot\mathbf{u}_n(t)\rangle]^m. \tag{10.59b}$$

We see that the m-th term in the expansion gives the contribution of the processes involving m phonons (emitted or absorbed); we consider now in particular the zero-phonon and one-phonon contributions.

Zero-Phonon Contribution (Elastic Scattering)

The zero-phonon contribution to the dynamical structure factor is

$$S_0(\mathbf{Q}, E) = N e^{-2W} \sum_{\mathbf{t}_n} e^{i\mathbf{Q}\cdot\mathbf{t}_n} \delta(E). \tag{10.60}$$

The sum over $\mathbf{t}_n$ is different from zero only if $\mathbf{Q}$ equals a reciprocal lattice vector $\mathbf{G}$, in this case it gives N. We thus recover the standard result that the Bragg diffraction pattern for elastic scattering maps the reciprocal lattice vectors. Notice that $S_0(\mathbf{G}, E)$ is proportional to N^2 (and not to N), as expected from the fact that the Bragg scattering is a coherent process. We also see that the intensity of the diffracted beams is reduced by the Debye-Waller factor $\exp{(-2W(\mathbf{G}))}$, which decreases as the magnitude of momentum transfer increases.

One-Phonon Contribution

The one-phonon contribution to the dynamical structure factor is

$$S_1(\mathbf{Q}, E) = N e^{-2W} \sum_{\mathbf{t}_n} e^{i\mathbf{Q}\cdot\mathbf{t}_n} \frac{1}{\hbar} \int_{-\infty}^{+\infty} \frac{dt}{2\pi} e^{-iEt/\hbar} \langle\mathbf{Q}\cdot\mathbf{u}_0\,\mathbf{Q}\cdot\mathbf{u}_n(t)\rangle. \tag{10.61}$$

To calculate explicitly the thermal average in the above equation, we express the operators $\mathbf{u}_n(t)$ in terms of the normal modes operators. For a simple lattice, the normal modes consist of three acoustic branches; for simplicity we assume that the three acoustic branches are degenerate. The component $u_{nx}(t)$ along a given direction (say x) of the displacement of the atom in the n-th cell is given by

$$u_{nx}(t) = \frac{1}{\sqrt{N}} \sum_{\mathbf{q}} \sqrt{\frac{\hbar}{2M\omega_{\mathbf{q}}}} \left[a_{\mathbf{q}} e^{i(\mathbf{q}\cdot\mathbf{t}_n - \omega_{\mathbf{q}}t)} + a_{\mathbf{q}}^{\dagger} e^{-i(\mathbf{q}\cdot\mathbf{t}_n - \omega_{\mathbf{q}}t)} \right].$$

In particular for $t = 0$ and $\mathbf{t}_n = 0$, we have

$$u_{0x} = \frac{1}{\sqrt{N}} \sum_{\mathbf{q}} \sqrt{\frac{\hbar}{2M\omega_{\mathbf{q}}}} \left[a_{\mathbf{q}} + a_{\mathbf{q}}^{\dagger} \right].$$

Hence

$$\langle \mathbf{Q}\cdot\mathbf{u}_0 \, \mathbf{Q}\cdot\mathbf{u}_n(t) \rangle = \frac{\hbar^2 Q^2}{2M} \frac{1}{N} \sum_{\mathbf{q}} \frac{1}{\hbar\omega_{\mathbf{q}}} \left[n_{\mathbf{q}} \, e^{i(\mathbf{q}\cdot\mathbf{t}_n - \omega_{\mathbf{q}}t)} + (n_{\mathbf{q}} + 1) e^{-i(\mathbf{q}\cdot\mathbf{t}_n - \omega_{\mathbf{q}}t)} \right],$$

where $n_{\mathbf{q}} = 1/[\exp{(\hbar\omega_{\mathbf{q}}/k_B T)} - 1]$ is the Bose population factor.

Inserting the above expression into Eq. (10.61), we obtain

$$S_1(\mathbf{Q}, E) = e^{-2W} \sum_{\mathbf{q}\, \mathbf{t}_n} \frac{\hbar^2 Q^2}{2M} \frac{1}{\hbar\omega_{\mathbf{q}}} \exp[i(\mathbf{Q} + \mathbf{q})\cdot\mathbf{t}_n] n_{\mathbf{q}} \, \delta(E + \hbar\omega_{\mathbf{q}})$$

$$+ e^{-2W} \sum_{\mathbf{q}\, \mathbf{t}_n} \frac{\hbar^2 Q^2}{2M} \frac{1}{\hbar\omega_{\mathbf{q}}} \exp[i(\mathbf{Q} - \mathbf{q})\cdot\mathbf{t}_n] (n_{\mathbf{q}} + 1) \, \delta(E - \hbar\omega_{\mathbf{q}}), \quad (10.62)$$

where $\mathbf{Q} = \mathbf{k}_i - \mathbf{k}_f$ and $E = E_i - E_f$ are the momentum transfer and the energy transfer to the target. The dynamical structure factor $S_1(\mathbf{Q}, E)$ consists of sharp delta-function peaks describing one-phonon absorption and emission processes, compatible with conservation laws of energy and momentum. The thermal factors $n_{\mathbf{q}}$ and $n_{\mathbf{q}} + 1$ are appropriate for processes in which phonons are absorbed or emitted, respectively. Notice that phonon annihilation processes are proportional to the Bose population factor $n_{\mathbf{q}}$ and vanish as the temperature is decreased; on the contrary, phonon creation processes are proportional to $n_{\mathbf{q}} + 1$, and can also occur in the lattice at zero temperature.

Multi-Phonon Contributions

We can proceed with higher order terms in the exponential and describe scattering processes in which two or more phonons are absorbed or emitted. The conservation laws for these higher order processes are less selective, and multi-phonon effects in many practical situations provide a rather unstructured background that does not smear out the important information provided by zero-phonon and one-phonon contribution. The effect of masking of course increases for high $|\mathbf{Q}|$ and at high temperature, due to the Debye-Waller factor.

10.7 Mössbauer Effect

Consider the emission of a low energy γ-ray ($E_\gamma = 10$–100 keV) from a nucleus, in the hypothetical situation that the emitting nucleus is *held fixed*; a typical emission lineshape could be (approximately) expected to have a Lorentzian form

$$I(E) = \frac{1}{\pi} \frac{\Gamma}{(E - E_0)^2 + \Gamma^2}, \tag{10.63}$$

where E is the emitted energy, E_0 is the energy separation between the two nuclear states of interest, and Γ is the linewidth broadening (essentially determined by the natural radiative width). The uncertainty Γ in the energy transition corresponds to a lifetime $\tau = \hbar/\Gamma$ of the excited state of the nucleus. For instance, in the case of the isotope ^{57}Fe, the parameters of the emission spectrum are taken to be $E_0 = 14.4$ keV, $\tau = 141$ ns, $\Gamma = \hbar/\tau \approx 4.66 \times 10^{-9}$ eV. In the case of ^{191}Ir, we have $E_0 = 129$ keV, $\tau = 0.13$ ns, $\Gamma \approx 5 \times 10^{-6}$ eV.

If the nucleus is free and is initially at rest, after the emission of a γ-ray of momentum $\hbar \mathbf{k}_\gamma$ ($k_\gamma = \omega_\gamma/c$) it recoils with momentum $-\hbar \mathbf{k}_\gamma$. The recoil kinetic energy of a nucleus of mass M is

$$E_R = \frac{\hbar^2 k_\gamma^2}{2M} = \frac{\hbar^2 \omega_\gamma^2}{2Mc^2} \approx \frac{E_0^2}{2Mc^2} \tag{10.64}$$

(notice that $M_p c^2 = 938$ MeV for the proton mass M_p); thus, the emission spectrum of gammas from a *free nucleus at rest* would be centered at the energy $E_0 - E_R$. The energy shift E_R is in general orders of magnitude larger than the natural linewidth Γ of the nucleus. For instance, in the case of ^{57}Fe we have $Mc^2 = 57 M_p c^2 = 57 \cdot 938$ MeV, and the recoil energy becomes $E_R = 1.9$ meV, while $\Gamma = 4.66 \times 10^{-9}$ eV; in the case of ^{191}Ir, the recoil energy is $E_R = 46$ meV, and $\Gamma = 5 \times 10^{-6}$ eV.

If the emitting nuclei are supposed to be free and in thermal equilibrium at the temperature T, we expect for their mean quadratic velocity $(1/2)M \langle v^2 \rangle = (3/2)k_B T$; the γ-ray emission line is broadened by longitudinal Doppler effect, which gives for the percentual variation of the energy of the emitted line (to first order in v/c)

$$\frac{\Delta E}{E_0} \approx \frac{\sqrt{\langle v^2 \rangle}}{c} = \sqrt{\frac{3k_B T}{Mc^2}}. \tag{10.65}$$

From Eqs. (10.64) and (10.65), the energy broadening due to the Doppler effect is estimated $\Delta E \approx \sqrt{E_R k_B T}$, a value in general orders of magnitude higher than the natural width.

The above considerations show that the γ-ray emission spectrum from *free nuclei* exhibits a shift and a broadening much larger than the natural width. On the contrary, and in striking contrast with the free nuclei situation, we can now see that the γ-ray *emission spectrum from nuclei bound in a crystal lattice contains a sharp emission line, which is neither shifted nor broadened with respect to the natural width (Mössbauer effect)*.

The formal theory for γ-light emission intensity from a nucleus, bound elastically to a fixed center or (more realistically) bound elastically in a crystal lattice, is completely similar to what has been done in the previous two sections, except for some minor adjustments. As in the general theory of particle scattering by crystals, the fraction of elastic (i.e. zero-phonon) emission processes, which produce the recoilless Mössbauer line, is controlled by the Debye-Waller factor $\exp(-2W)$. From Eq. (10.57b), with the replacement $|\mathbf{Q}| = k_\gamma$, the recoil-free fraction of emitted gammas can be written as

$$f = \exp(-2W) = \exp\left(-k_\gamma^2 \langle u_{0x}^2 \rangle\right), \tag{10.66}$$

where $\langle u_{0x}^2 \rangle$ is the mean square vibrational amplitude of the emitting nucleus (in the direction of the emission of the gamma ray). Notice that the recoilless fraction f drops to zero in the case $\langle u_{0x}^2 \rangle$ is not bound or is large with respect to $1/k_\gamma$. The recoil fraction of emitted gammas, involving creation or annihilation of one or more phonons, is regulated by the supplemental fraction $1 - f$.

For temperatures lower than the Debye temperature, from Eqs.(10.58) and (10.66) we have

$$f = \exp\left[-\frac{3}{2}\frac{\hbar^2 k_\gamma^2}{2M}\frac{1}{k_B T_D}\right] = \exp\left[-\frac{3}{2}\frac{E_R}{k_B T_D}\right].$$

Thus if the recoil energy E_R is less (or not much higher) than $k_B T_D$, we have that a substantial part of the emission process occurs with *no broadening and no shift* of the emission line, described by Eq. (10.63). Typical values of f are 0.92 for the 14.4 keV gamma ray of ^{57}Fe ($E_R = 1.9$ meV, $T_D = 400$ K $= 34.5$ meV), and $f = 0.06$ for the 129 keV gamma ray of ^{191}Ir ($E_R = 46$ meV, $T_D = 285$ K $= 24.6$ meV). The condition $E_R \leq k_B T_D$ limits the Mössbauer effect to γ-rays up to ≈ 150 keV or so.

In addition to the sharp peak at energy E_0 and width Γ, the emission spectrum presents a broad background (of the order of tens of meV) due to one-, two- and multi-phonon processes. The presence of the so sharp central line in low energy γ-ray emission is very appealing for spectroscopists and has found numerous applications in fundamental problems of solid state physics, chemical physics, and relativity tests.

The above considerations on the photon emission can be extended to the phonon absorption from the nuclear resonances of nuclei bound in crystals. The absorption spectrum consists of a narrow recoil-free peak at the energy of the nuclear transition (controlled by the recoilless fraction f), plus a broad background due to phonon assisted transitions (controlled by the recoil fraction $1 - f$). Important information on the lattice dynamics can be obtained by the analysis of the recoil fraction of resonant nuclear absorptions.

In Figure 10.15 we report a typical example of measured absorption spectrum of X-rays from the 14.413 keV nuclear resonance of ^{57}Fe [X-rays are produced by synchrotron radiation; the intensity of nuclear absorption is measured by counting the fluorescence events, resulting from nuclear internal conversion processes]. The observed absorption spectrum consists of an elastic peak and sidebands at lower and higher energies (inset of Figure 10.15). The central peak corresponds to the nuclear absorption without recoil. The sideband above (below) the central peak corresponds to inelastic absorption accompanied by creation (annihilation) of phonons. The observed asymmetry in the spectra, as usual, is related to the fact that phonon annihilation probability is

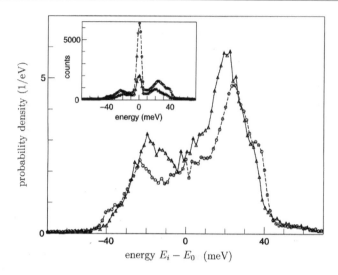

Figure 10.15 Absorption probability density for α-iron (circles, dashed line) and stainless steel (triangles, solid line), after removing the elastic peak. The inset shows the raw data. The energy of the nuclear resonance is denoted by E_0 ($E_0 = 14.413$ keV in ^{52}Fe), while E_i denotes the energy of the incident radiation [reprinted with permission from W. Sturhahn, T. S. Toellner, E. E. Alp, X. Zhang, M. Ando, Y. Yoda, S. Kikuta, M. Seto, C. W. Kimball and B. Dabrowski, Phys. Rev. Lett. *74*, 3832 (1995); copyright 1995 by the American Physical Society].

proportional to the number $\langle n \rangle$ of existing phonons, while phonon creation probability is proportional to $\langle n \rangle + 1$. The inelastic nuclear absorption by synchrotron radiation has found several successful applications in the study of lattice dynamics and in providing direct measurements of the phonon density-of-states.

Appendix A.   Solved Problems and Complements

Problem 1. Thomson scattering of unpolarized light.
Problem 2. Scattering cross-section and transition rates.
Problem 3. Fermi golden rule and van Hove correlation functions.
Problem 4. Kinematics of the scattering processes.
Problem 5. Scattering length and neutron cross-section.

Problem 1. Thomson Scattering of Unpolarized Light

Show that the differential cross-section for elastic scattering of unpolarized light by a free electron is given by the Thomson expression

$$\left(\frac{d\sigma}{d\Omega} \right)_{\text{Th}}^{(\text{unpol})} = \left(\frac{e^2}{mc^2} \right)^2 \frac{1 + \cos^2 \theta_s}{2}, \tag{A.1}$$

where θ_s is the scattering angle.

In Section 10.2.1, following classical electrodynamics, we have considered the acceleration of a free electron under the influence of an oscillating electric field of polarization vector $\widehat{\mathbf{e}}_i$ and wavevector $\mathbf{k}_i$; according to Eq. (10.3) the dipolar field is

$$\mathbf{E}_d(\mathbf{r}, \mathbf{R}, t) = \frac{1}{R} \frac{e^2}{mc^2} E_0 \, e^{i\mathbf{k}_i \cdot \mathbf{r}} e^{-i\omega(t - R/c)} \left[\widehat{\mathbf{e}}_i - (\widehat{\mathbf{k}}_f \cdot \widehat{\mathbf{e}}_i)\widehat{\mathbf{k}}_f \right],$$

where $\widehat{\mathbf{k}}_f$ is the unit vector in the direction of $\mathbf{k}_f$. From the modulus squared of $|\mathbf{E}_d|^2$, we obtain for the intensity of the scattered field

$$I_s(\mathbf{R}) = I_0 \frac{r_0^2}{R^2} \left[\widehat{\mathbf{e}}_i - (\widehat{\mathbf{k}}_f \cdot \widehat{\mathbf{e}}_i)\widehat{\mathbf{k}}_f \right]^2 = I_0 \frac{r_0^2}{R^2} \left[1 - (\widehat{\mathbf{e}}_i \cdot \widehat{\mathbf{k}}_f)^2 \right],$$

where I_0 is the intensity of the incident field and $r_0 = e^2/mc^2$ is the electron radius.

In order to perform easily the average on the incident polarization direction $\widehat{\mathbf{e}}_i$ consider the unit vector $\widehat{\mathbf{v}}$ that lies in the plane spanned by $\mathbf{k}_i$ and $\mathbf{k}_f$ and is orthogonal to $\mathbf{k}_i$. It is seen by inspection that

$$\mathbf{v} = \frac{\widehat{\mathbf{k}}_f - \widehat{\mathbf{k}}_i (\widehat{\mathbf{k}}_i \cdot \widehat{\mathbf{k}}_f)}{\sqrt{1 - (\widehat{\mathbf{k}}_i \cdot \widehat{\mathbf{k}}_f)^2}} = \frac{\widehat{\mathbf{k}}_f - \widehat{\mathbf{k}}_i \cos\theta_s}{\sin\theta_s} \qquad \text{where} \qquad \cos\theta_s = \widehat{\mathbf{k}}_i \cdot \widehat{\mathbf{k}}_f.$$

Multiplication of the above equality by $\widehat{\mathbf{e}}_i$ gives $\widehat{\mathbf{e}}_i \cdot \widehat{\mathbf{k}}_f = \widehat{\mathbf{e}}_i \cdot \widehat{\mathbf{v}} \sin\theta_s$. Then we have

$$\left[1 - (\widehat{\mathbf{e}}_i \cdot \widehat{\mathbf{k}}_f)^2 \right] = 1 - (\widehat{\mathbf{e}}_i \cdot \widehat{\mathbf{v}})^2 \sin^2\theta_s \implies 1 - \frac{1}{2}\sin^2\theta_s \equiv \frac{1 + \cos^2\theta_s}{2},$$

where we have inserted the value 1/2 for the average of $(\widehat{\mathbf{e}}_i \cdot \widehat{\mathbf{v}})^2 = \cos^2\phi$, ϕ being the polarization angle. The angular dependent factor in the right-hand side of Eq. (A.1) is thus demonstrated.

Problem 2. Scattering Cross-Section for Neutrons and Photons, in Term of the Transition Rates

Taking proper account of the kinetic of neutrons show that the double differential cross-section for scattering of neutrons in the energy interval dE_f and in the solid angle $d\Omega$, becomes

$$\boxed{\frac{d^2\sigma}{dE_f \, d\Omega} = \frac{V^2}{(2\pi)^3} \frac{M^2}{\hbar^3} \frac{k_f}{k_i} P_{\mathbf{k}_f \leftarrow \mathbf{k}_i}} \qquad \text{NEUTRONS,} \qquad (A.2)$$

where $P_{\mathbf{k}_f \leftarrow \mathbf{k}_i}$ is the transition rate for the particle scattering from the plane wave $\mathbf{k}_i$ to the plane wave $\mathbf{k}_f$. Similarly, taking proper account of the kinetics of photons show that double differential cross-section for scattering of photons in the energy interval dE_f and in the solid angle $d\Omega$, becomes

$$\boxed{\frac{d^2\sigma}{dE_f \, d\Omega} = \frac{V^2}{(2\pi)^3} \frac{k_f^2}{\hbar c^2} P_{\mathbf{k}_f \widehat{\mathbf{e}}_f \leftarrow \mathbf{k}_i \widehat{\mathbf{e}}_i}} \qquad \text{PHOTONS,} \qquad (A.3)$$

where $P_{k_f\hat{e}_f \leftarrow k_i\hat{e}_i}$ is the transition rate for the photon scattering from the plane wave $k_i\hat{e}_i$ to the plane wave $k_f\hat{e}_f$.

The double differential cross-section $d^2\sigma/dE_f d\Omega$, which is the quantity measured in scattering experiments, equals the transition rate times the density-of-states divided by the incident particle current density:

$$\frac{d^2\sigma}{dE_f d\Omega} = \frac{1}{J} \frac{d^2n}{dE_f d\Omega} P_{k_f \leftarrow k_i}. \tag{A.4}$$

For neutrons, the number of states in the volume element $d\mathbf{k}_f$ around $\mathbf{k}_f$, is given by

$$\frac{V}{(2\pi)^3}d\mathbf{k}_f = \frac{V}{(2\pi)^3}k_f^2\, dk_f\, d\Omega = \frac{V}{(2\pi)^3}k_f \frac{M}{\hbar^2}dE_f d\Omega,$$

where $V/(2\pi)^3$ represents as usual the density-of-states in $\mathbf{k}$ space; for the last passage, use has been made of the energy-wavevector relation $E_f = (\hbar^2/2M)k_f^2$ for a particle of mass M, and hence of the relation $dE_f = (\hbar^2/M)k_f\, dk_f$. We can thus write

$$\frac{d^2n}{dE_f d\Omega} = \frac{V}{(2\pi)^3}k_f \frac{M}{\hbar^2}.$$

The flux of incident particles of mass M, initial momentum $\hbar k_i$, velocity $\hbar k_i/M$, and wavefunction $\psi_i = (1/\sqrt{V})\exp(i\mathbf{k}_i \cdot \mathbf{r})$ is

$$J_i = \frac{\hbar k_i}{M} \frac{1}{V}.$$

Inserting the two above results into Eq. (A.4) provides the demonstration of Eq. (A.2). Similar considerations can be performed for the scattering of photons, taking into account that the dispersion curve for photons is $E_f = \hbar c k_f$. For the density-of-states and for the current we have, respectively:

$$\frac{d^2n}{dE_f d\Omega} = \frac{V}{(2\pi)^3}\frac{k_f^2}{\hbar c} \quad \text{and} \quad J_i = \frac{c}{V}.$$

With these results, Eq. (A.3) is also proved.

Problem 3. Fermi Golden Rule and van Hove Correlation Functions

Consider a quantum system described by a time-independent Hamiltonian H, and let $\{\Phi_I\}$ denote the complete set of eigenfunctions and E_I the corresponding eigenvalues. Let

$$A(t) = e^{iHt/\hbar}A(0)e^{-iHt/\hbar}$$

denote an arbitrary chosen operator in the Heisenberg representation. Show the following identity:

$$\sum_{IJ} p_I |\langle \Phi_J|A|\Phi_I\rangle|^2 \delta(E_J - E_I - E) \equiv \frac{1}{\hbar}\int_{-\infty}^{+\infty}\frac{dt}{2\pi}e^{-iEt/\hbar}\langle A^\dagger(0)A(t)\rangle, \tag{A.5}$$

where the first member is a typical Fermi golden rule summation for the quantum system in the canonical ensemble, and the second member is the time Fourier transform of the van Hove self-correlation function $\langle A^\dagger(0)A(t)\rangle$.

The quantity $\langle A^\dagger(0)A(t)\rangle$ denotes the thermal average of $A^\dagger(0)A(t)$; it is called auto-correlation function of the operator A, and is defined as

$$\langle A^\dagger(0)A(t)\rangle = \sum_I p_I \langle \Phi_I | A^\dagger(0)A(t)|\Phi_I\rangle \quad \text{where } p_I = \frac{\exp[-E_I/k_B T]}{\sum_I \exp[-E_I/k_B T]}.$$

(A.6)

In the following an operator whose time dependence is not explicitly indicated, is intended at $t = 0$, the instant at which the Heisenberg and the Schrödinger representations coincide.

With straight elaborations we obtain

$$\langle \Phi_I | A^\dagger(0)A(t)|\Phi_I\rangle = \langle \Phi_I | A^\dagger e^{iHt/\hbar} A e^{-iHt/\hbar}|\Phi_I\rangle$$
$$= \sum_J \langle \Phi_I | A^\dagger e^{iHt/\hbar}|\Phi_J\rangle\langle \Phi_J | A e^{-iHt/\hbar}|\Phi_I\rangle = \sum_J e^{i(E_J - E_I)t/\hbar}|\langle \Phi_J | A|\Phi_I\rangle|^2.$$

Insertion of the above result into Eq. (A.6) gives

$$\langle A^\dagger(0)A(t)\rangle \equiv \sum_{IJ} e^{i(E_J - E_I)t/\hbar}\, p_I |\langle \Phi_J | A|\Phi_I\rangle|^2.$$

Performing the time Fourier transform of both members we obtain

$$\frac{1}{\hbar}\int_{-\infty}^{+\infty} \frac{dt}{2\pi}\, e^{-iEt/\hbar}\,\langle A^\dagger(0)A(t)\rangle$$
$$\equiv \sum_{IJ} \frac{1}{\hbar}\int_{-\infty}^{+\infty} \frac{dt}{2\pi}\, e^{-iEt/\hbar}\, e^{i(E_J - E_I)t/\hbar}\, p_I |\langle \Phi_J | A|\Phi_I\rangle|^2.$$

With the help of the integral representation for the delta function

$$\delta(E_J - E_I - E) = \frac{1}{\hbar}\int_{-\infty}^{+\infty} \frac{dt}{2\pi} e^{i(E_J - E_I - E)t/\hbar}$$

the demonstration of Eq. (A.5) is completed. Notice that the demonstration is very general, and it holds regardless the number or the nature of interacting particles in the Hamiltonian under considerations.

Problem 4. Kinematics of the Scattering Processes

Consider a crystal lattice and typical acoustic branches with dispersion $E = \hbar v_s Q$ (for sufficiently small Q), where v_s is the sound speed. Show that, in inelastic X-ray scattering, there is no limitation on the detection of phonon-like excitations at a chosen momentum transfer Q. In the case of scattering by neutrons of velocity $v_n = \hbar k/M_n$

show that the elementary excitations $E = \hbar v_s Q$ are outside the accessible kinematic region if $v_s > v_n$.

The momentum and energy conservation laws impose that

$$E = E_i - E_f; \quad \mathbf{Q} = \mathbf{k}_i - \mathbf{k}_f, \tag{A.7}$$

where E and $\mathbf{Q}$ are the energy transfer and momentum transfer to the target, and E_i, $\mathbf{k}_i$ and E_f, $\mathbf{k}_f$ are the energy and momentum of the probe particle before and after scattering. The need of satisfying the energy and momentum conservation laws define the $(Q - E)$ region accessible to the probe.

We consider specifically the case $k_i > k_f$ and therefore $E_i > E_f$ (similar considerations can be carried out in the opposite case). Squaring the transfer wavevector, we have

$$Q^2 = k_i^2 + k_f^2 - 2k_i k_f \cos \theta_s = (k_i - k_f)^2 + 4k_i k_f \sin^2 \frac{\theta_s}{2}.$$

Then

$$\frac{E}{Q} = \frac{E_i - E_f}{\sqrt{(k_i - k_f)^2 + 4k_i k_f \sin^2 (\theta_s/2)}} < \frac{E_i - E_f}{k_i - k_f}. \tag{A.8}$$

In the case of photon scattering, $E(k_{i,f}) = \hbar c k_{i,f}$, and Eq. (10.8) does not entail any kinematic restriction.

In the case of neutrons, we have

$$\frac{E}{Q} < \frac{E_i - E_f}{k_i - k_f} = \frac{\hbar^2}{2M_n} \frac{k_i^2 - k_f^2}{k_i - k_f} < \frac{\hbar^2 k_i}{M_n} \quad \Longrightarrow \quad v_s < v_n.$$

In the case the velocity of sound is greater than the velocity of the incoming neutron, the phonon-like elementary excitations cannot be studied by neutron technique.

Problem 5. Scattering Length and Neutron Cross-Section

Suppose the neutron-nucleus interaction is described by the Fermi pseudo-potential

$$V_p(\mathbf{r} - \mathbf{R}) = \frac{2\pi \hbar^2}{M_n} b \, \delta(\mathbf{r} - \mathbf{R}), \tag{A.9}$$

where $\mathbf{r}$ is the coordinate of the impinging particle, $\mathbf{R}$ is the coordinate of the scattering nucleus, and b is the Fermi length. In the hypothetical situation that the target nucleus is held fixed (static approximation), show that the differential cross-section is $|b|^2$, and the total cross-section is $4\pi |b|^2$.

Let the incoming particle be described by a plane wave $|W_{\mathbf{k}_i}\rangle$ of vector $\mathbf{k}_i$ and energy E_i; we are interested in the transition probability rate of the incoming particle into a final state $|W_{\mathbf{k}_f}\rangle$ of vector $\mathbf{k}_f$ and energy E_f.

Within the static approximation, the scatterer coordinate $\mathbf{R}$ becomes a parameter, rather than a quantum observable; furthermore the scattering is elastic. At the lowest order of perturbation theory (Born approximation), the probability per unit time that the perturbation potential (A.9) induces a transition from the initial state $|W_{\mathbf{k}_i}\rangle$ to a final state $|W_{\mathbf{k}_f}\rangle$ is given by the Fermi golden rule

$$P_{\mathbf{k}_f \leftarrow \mathbf{k}_i} = \frac{2\pi}{\hbar} |\langle W_{\mathbf{k}_f}|V_p(\mathbf{r}-\mathbf{R})|W_{\mathbf{k}_i}\rangle|^2 \delta(E_f - E_i). \tag{A.10}$$

The matrix element in the above equation reads

$$\langle W_{\mathbf{k}_f}|V_p(\mathbf{r}-\mathbf{R})|W_{\mathbf{k}_i}\rangle = \frac{1}{V}\int e^{-i\mathbf{k}_f\cdot\mathbf{r}}V_p(\mathbf{r}-\mathbf{R})e^{i\mathbf{k}_i\cdot\mathbf{r}} \equiv \frac{1}{V}e^{i\mathbf{Q}\cdot\mathbf{R}}\frac{2\pi\hbar^2}{M_n}b,$$

where V is the volume over which the initial and final plane waves are normalized. The transition rate becomes

$$P_{\mathbf{k}_f \leftarrow \mathbf{k}_i} = |b|^2 \frac{8\pi^3\hbar^3}{M_n^2}\frac{1}{V^2}\delta(E_f - E_i).$$

According to Eq. (A.2), the double differential cross-section (for the elastic scattering events of our concern) becomes

$$\frac{d^2\sigma}{dE_f d\Omega} = \frac{V^2}{(2\pi)^3}\frac{M_n^2}{\hbar^3}P_{\mathbf{k}_f \leftarrow \mathbf{k}_i} \equiv |b|^2\delta(E_f - E_i). \tag{A.11}$$

Integration of the double differential cross-section on the energy variable gives $d\sigma/d\Omega = |b|^2$, whereas the total cross-section becomes $\sigma = 4\pi|b|^2$.

Further Reading

Ashcroft, N. W., & Mermin, N. D. (1976). *Solid state physics*. New York: Holt, Rinehart and Winston.

Bacon, G. E. (1975). *Neutron diffraction*. Oxford: Clarendon Press.

Birkholz, M. (2006). *Thin film analysis by X-ray scattering*. Weinheim: Wiley-VCH.

Brockhouse, B. N. (1995). Slow neutron spectroscopy and the grand atlas of the physical world. *Reviews of Modern Physics, 67*, 735.

Chen, Y.-L., & Yang, D.-P. (2007). *Mössbauer effect in lattice dynamics*. Weinheim: Wiley-VCH.

Cooper, M. J. (1985). Compton scattering and electron momentum determination. *Reports on Progress in Physics, 48*, 415.

Cowley, J. M. (1995). *Diffraction physics*. Amsterdam: North-Holland.

Hayes, W., & Loudon, R. (1978). *Scattering of light by crystals*. New York: Wiley.

Hüfner, S. (2003). *Photoelectron spectroscopy*. Berlin: Springer.

Ibach, H., & Lüth, H. (2009). *Solid-state physics*. Berlin: Springer.

Jackson, J. D. (1998). *Classical electrodynamics*. New York: Wiley.

Lovesey, S. W. (1984). *Theory of neutron scattering from condensed matter*. Oxford: Clarendon Press.

Panofsky, W. K. H., & Phillips, M. (1962). *Classical electricity and magnetism*. Reading, Mass: Addison-Wesley.

Schwarzenbach, D. (1996). *Crystallography*. Chichester: Wiley.

Shirane, G., Shapiro, S. M., & Tranquada, J. M. (2002). *Neutron scattering with a triple-axis spectrometer*. Cambridge: Cambridge University Press.

Wertheim, G. K. (1964). *Mössbauer effect: Principles and applications*. New York: Academic Press.

Willmott, P. (2011). *An introduction to synchrotron radiation: Techniques and applications*. Chichester: Wiley.

Yoshida, Y., & Langouche, G. (Eds.). (2013). *Mössbauer spectroscopy*. Berlin: Springer.

Yu, P. Y., & Cardona, M. (2010). *Fundamentals of semiconductors*. Berlin: Springer.

11 Optical and Transport Properties of Metals

Chapter Outline head

This is the first of four interrelated chapters, in which we study the optical and transport properties of crystals. Similarly to its historical role in atomic and molecular physics, optical spectroscopy has always accompanied and stimulated major progresses also in the physics of crystals. The importance of transport phenomena for the development of solid state electronics and devices is also well known. In this chapter we consider optical and transport properties of metals. The optical constants of semiconductors and insulators are discussed in Chapter 12; the transport properties of homogeneous and inhomogeneous semiconductors are presented in Chapter 13 and 14, respectively.

This chapter begins with a section, rather general in nature, that describes the phenomenological optical constants of homogeneous materials. We then consider the optical properties due to *intraband transitions* of carriers; this contribution is typical of the conduction electrons in metals, but also semiconductors (with small bandgap or strongly doped) may present a significant intraband contribution at finite temperature. The optical properties due to *interband transitions* are discussed in the next chapter; interband transitions are basic in insulators and semiconductors, where bands are either fully occupied or fully empty at zero temperature, and are of relevance also in metals at sufficiently high excitation energies.

In the study of the optical properties of free carriers, classical and semiclassical approaches maintain their usefulness, and are preliminary to full quantum mechanical

Solid State Physics, Second Edition. http://dx.doi.org/10.1016/B978-0-12-385030-0.00011-6

treatments. We thus consider the optical properties of free carriers within the classical theory of Drude, because of its simplicity and pedagogical value. Successively, we consider static and dynamic conductivity of free carriers by means of the semiclassical transport equation of Boltzmann; this treatment is then compared with the linear response formalism of the quantum mechanical theory. The transport equations in the presence of electric fields and temperature gradients are also examined, and the thermoelectric effects of metals are discussed.

11.1 Macroscopic Theory of Optical Constants in Homogeneous Materials

The propagation of electromagnetic waves in materials is described by the Maxwell's equations, complemented by appropriate constitutive equations of the specific system under attention. In this section we summarize the optical phenomenological constants, more commonly investigated in experiments. The considerations of the present section are rather general and may apply to homogeneous materials of different nature (metals, semiconductors, and insulators), while the following sections refer almost exclusively to simple metals.

We study for simplicity non-magnetic media ($\mathbf{B} = \mathbf{H}$; $\mu = 1$) in the absence of external charges and currents ($\rho_{ext} = 0$; $\mathbf{J}_{ext} = 0$). The Maxwell's equations for electromagnetic waves propagating in a medium and interacting with the internal charge density $\rho(\mathbf{r}, t)$ and internal current density $\mathbf{J}(\mathbf{r}, t)$ are (in Gauss units)

$$
\begin{aligned}
\operatorname{div} \mathbf{E} &= 4\pi\rho, \quad \operatorname{curl} \mathbf{E} = -\frac{1}{c}\frac{\partial \mathbf{B}}{\partial t} \\
\operatorname{div} \mathbf{B} &= 0, \quad \operatorname{curl} \mathbf{B} = \frac{1}{c}\frac{\partial \mathbf{E}}{\partial t} + \frac{4\pi}{c}\mathbf{J}
\end{aligned}
\qquad (11.1)
$$

In the following, several quantities and relations are elaborated in appropriate convenient dimensionless notations, so that to be independent on the actual units adopted in the starting Maxwell's equations.

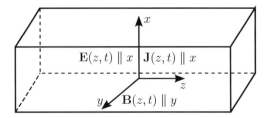

Figure 11.1 Geometry chosen for the description of transverse electromagnetic fields in isotropic materials. The electric field and the internal density current are in the x-direction, the magnetic field is in the y-direction, the wave propagation is along the z-direction.

We consider the propagation, along say the z-direction, of *transverse electromagnetic waves of given angular frequency* ω *in isotropic materials*, and specify the Maxwell's equations in the geometry of Figure 11.1. We take the electric field $\mathbf{E}$ in the x-direction, the magnetic field $\mathbf{B}$ in the y-direction, and the current density $\mathbf{J}$ parallel to the electric field; we can write

$$\mathbf{E}(\mathbf{r}, t) = E(z)e^{-i\omega t}\,\widehat{\mathbf{e}}_x \quad \Longrightarrow \quad \mathbf{E} \parallel x, \tag{11.2a}$$

$$\mathbf{B}(\mathbf{r}, t) = B(z)e^{-i\omega t}\,\widehat{\mathbf{e}}_y \quad \Longrightarrow \quad \mathbf{B} \parallel y, \tag{11.2b}$$

$$\mathbf{J}(\mathbf{r}, t) = J(z)e^{-i\omega t}\,\widehat{\mathbf{e}}_x \quad \Longrightarrow \quad \mathbf{J} \parallel x, \tag{11.2c}$$

where $\widehat{\mathbf{e}}_x$ and $\widehat{\mathbf{e}}_y$ denote the unit vectors along the indicated axes. From Eq. (11.2a) we have div $\mathbf{E} = 0$; combined with the first Maxwell's equation, it entails $\rho = 0$. From Eq. (11.2b) it is seen that the Maxwell's equation div $\mathbf{B} = 0$ is satisfied. The remaining two Maxwell's equations give

$$\frac{dE(z)}{dz} = \frac{i\omega}{c}B(z), \tag{11.3a}$$

$$-\frac{dB(z)}{dz} = -\frac{i\omega}{c}E(z) + \frac{4\pi}{c}J(z). \tag{11.3b}$$

Solving for $E(z)$ we obtain

$$\boxed{\frac{d^2E(z)}{dz^2} = -\frac{\omega^2}{c^2}E(z) - \frac{4\pi i\omega}{c^2}J(z)}. \tag{11.4}$$

This is the basic information, provided by the Maxwell's equations, on the electric field and current density; to proceed further we must establish between them a phenomenological or microscopic constitutive relationship.

Local and Non-Local Constitutive Relationship for the Conductivity

The above Eq. (11.4) can be closed by appropriate assumption and elaboration of the conductivity response function σ, connecting the internal current density $\mathbf{J}$ to the electric field $\mathbf{E}$ in the specific medium under investigation. Alternatively and equivalently, one could use, for instance, the constitutive equation linking the polarization $\mathbf{P}$ to the electric field $\mathbf{E}$ via the polarizability response function, or the dielectric response function linking the induction electric field $\mathbf{D} = \mathbf{E} + 4\pi\mathbf{P}$ to the electric field; in the following we focus on the conductivity response function.

Within the linear response approximation, we can write quite generally that the α-component of the current density is related to the β-components of the electric field by the relation

$$J_\alpha(\mathbf{r}, t) = \sum_\beta \int d\mathbf{r}' \int_{-\infty}^{t} \sigma_{\alpha\beta}(\mathbf{r}, \mathbf{r}', t, t') E_\beta(\mathbf{r}', t')\, dt' \qquad (\alpha, \beta = x, y, z),$$

$$\tag{11.5}$$

where $\sigma_{\alpha\beta}(\mathbf{r}, \mathbf{r}', t, t')$ is the conductivity tensor, which in general depends separately on the two space variables and on the two time variables.

In isotropic media, the conductivity tensor takes the form $\sigma_{\alpha\beta} = \sigma \delta_{\alpha\beta}$, and can be treated as a scalar quantity. We assume that the scalar conductivity σ is a homogeneous function both of time coordinates and spatial coordinates. [While the assumption of time translation invariance is standard, the assumption of spatial homogeneity needs at least a brief comment. We notice that even an ideally perfect crystal is not rigorously homogeneous on a microscopic unit cell scale. Thus in general, the response functions of a crystal to driving fields depend on two spatial variable separately, and not through their difference. In the following we adopt the simplification of neglecting "local field effects" induced by inhomogeneity of the sample, since this is reasonably justified in a number of relevant situations.]

With the adopted assumptions of time and space translational invariance, Eq. (11.5) becomes

$$\mathbf{J}(\mathbf{r}, t) = \int_{-\infty}^{+\infty} d\mathbf{r}' \int_{-\infty}^{+\infty} dt' \sigma(\mathbf{r} - \mathbf{r}', t - t') \mathbf{E}(\mathbf{r}', t') ; \qquad (11.6)$$

the extension to $+\infty$ of the integration in dt' is allowed by the causality principle which entails

$$\sigma(\mathbf{r} - \mathbf{r}', t - t') \equiv 0 \quad \text{for } t' > t.$$

The homogeneous approximation of the kernel in Eq. (11.6) suggests to express it in terms of the Fourier transforms of the quantities of interest. The Fourier transform of the conductivity function is defined as

$$\sigma(\mathbf{q}, \omega) = \int_{-\infty}^{+\infty} e^{i(\mathbf{q} \cdot \mathbf{r} - \omega t)} \sigma(\mathbf{r}, t) \, d\mathbf{r} \, dt ;$$

similar definitions hold for $\mathbf{J}(\mathbf{q}, \omega)$ and $\mathbf{E}(\mathbf{q}, \omega)$. Multiplication of both members of Eq. (11.6) by $\exp[(i(\mathbf{q} \cdot \mathbf{r} - \omega t)]$ and integration over $d\mathbf{r} \, dt$ gives the relation

$$\boxed{\mathbf{J}(\mathbf{q}, \omega) = \sigma(\mathbf{q}, \omega) \mathbf{E}(\mathbf{q}, \omega)} . \qquad (11.7)$$

Equation (11.6) in real space-time domain, or equivalently Eq. (11.7) in momentum-frequency domain, entail a non-local spatial relationship between current density and electric field. In fact $\mathbf{J}(\mathbf{r}, t)$ at a given point $\mathbf{r}$ is determined by the knowledge of the electric field distribution in the whole space around $\mathbf{r}$ and not simply by the value of the electric field at the chosen point.

In the approximation of local response, it is assumed that $\mathbf{J}(\mathbf{r}, t)$ depends only on the electric field $\mathbf{E}(\mathbf{r}, t')$ in the point $\mathbf{r}$, at any previous time $t' < t$. This assumption entails that the conductivity function in real space takes the form

$$\sigma(\mathbf{r} - \mathbf{r}', t - t') = \delta(\mathbf{r} - \mathbf{r}') \sigma(t - t').$$

The Fourier transform becomes independent of the wavevector and we have

$$\sigma(\mathbf{q}, \omega) = \sigma(\mathbf{q} = 0, \omega) = \sigma(\omega),$$

where $\sigma(\omega)$ is the complex *conductivity function* of the medium.

Thus, the *local-response regime corresponds to neglecting the wavevector dependence of the response function*. In this situation the constitutive equation, linking the current density to the electric field takes the form

$$\mathbf{J}(\mathbf{r}) = \sigma(\omega)\mathbf{E}(\mathbf{r}),\tag{11.8}$$

where the assumed time dependence of the fields $\mathbf{J}$ and $\mathbf{E}$ is specified through the angular frequency ω of the conductivity. According to Eq. (11.8), the current density at a given point in the material is proportional to the value of the electric field at the same point: this is the so-called *local-response regime*. A local relationship between current and field can be justified when the average distance travelled by the carriers is small with respect to the length of spatial variation of the electric field; otherwise, one should follow the trajectory of the carriers to find the effect of the spatially varying electric field: this is the so-called *non-local-response regime*.

Local Relationship for the Conductivity and Expression of the Optical Constants

In the rest of this section, we start from the local constitutive equation (11.8) between the current density and the electric field; then we work out the other frequency-dependent optical constants (or optical functions), which are commonly used for the description of the optical properties of matter. The basic equation (11.4) joined with the local assumption gives

$$\frac{d^2 E(z)}{dz^2} = -\frac{\omega^2}{c^2}\left[1 + \frac{4\pi i \sigma(\omega)}{\omega}\right]E(z).\tag{11.9}$$

The solution of this equation inside the material is a damped (or undamped) wave, which can be written in the form

$$E(z) = E_0 e^{i(\omega/c)Nz},\tag{11.10a}$$

where the *complex refractive index* $N(\omega)$ is given by

$$\boxed{N^2 = 1 + \frac{4\pi i \sigma(\omega)}{\omega}}.\tag{11.10b}$$

The magnetic field associated to the electric field $E(z)$ is given by

$$B(z) = N\,E(z),\tag{11.10c}$$

as can be seen by applying Eq. (11.3a).

It is customary to write the complex refractive index in the form $N = n + ik$, where n is the ordinary *refractive index* and k the *extinction coefficient*; Eq. (11.10a) can be re-written in the form

$$E(z) = E_0 e^{i(\omega/c)nz} e^{-(\omega/c)kz}.\tag{11.11}$$

From Eq. (11.11), we see that the velocity of the electromagnetic wave in the medium is c/n; the penetration depth (*classical skin depth*, defined as the distance at which the field amplitude drops of $1/e$) is

$$\delta(\omega) = \frac{c}{\omega k(\omega)}.$$

(11.12a)

The intensity of an electromagnetic field is proportional to $|E(z)|^2$, and from Eq. (11.11) we have $I(z) = I_0 \exp(-2\omega k z/c)$. The *attenuation coefficient* of the wave is thus

$$\alpha(\omega) = \frac{2\omega k(\omega)}{c} \equiv \frac{2}{\delta(\omega)};$$

(11.12b)

the attenuation coefficient is also referred to as *absorption coefficient*.

Equation (11.10b) can also be written in the more effective form $N^2 = \varepsilon$, where ε is the *complex dielectric function*. Indicating $\varepsilon = \varepsilon_1 + i\varepsilon_2$, we have

$$\varepsilon_1 = n^2 - k^2 \quad \text{and} \quad \varepsilon_2 = 2nk;$$

and inversely

$$n^2 = \frac{1}{2}\left(\varepsilon_1 + \sqrt{\varepsilon_1^2 + \varepsilon_2^2}\right) \quad \text{and} \quad k^2 = \frac{1}{2}\left(-\varepsilon_1 + \sqrt{\varepsilon_1^2 + \varepsilon_2^2}\right).$$

From Eq. (11.10b), and $N^2 = \varepsilon$, it is seen that the conductivity and the dielectric function satisfy the relation

$$\varepsilon(\omega) = 1 + \frac{4\pi i \sigma(\omega)}{\omega}.$$

(11.13)

Separation of the real and imaginary part gives

$$\varepsilon_1(\omega) = 1 - \frac{4\pi \sigma_2(\omega)}{\omega} \quad \text{and} \quad \varepsilon_2(\omega) = \frac{4\pi \sigma_1(\omega)}{\omega};$$

these equations link the real and imaginary part of the dielectric function to the imaginary and real part of the conductivity.

Consider now an electromagnetic wave that impinges at normal incidence on the surface of an isotropic material, where it is partially transmitted and partially reflected. In the geometry of Figure 11.2 we have for the spatial part

$$E(z) = E_t e^{i(\omega/c)Nz} \qquad\qquad z > 0,$$

(11.14a)

$$E(z) = E_i e^{i(\omega/c)z} + E_r e^{-i(\omega/c)z} \quad z < 0.$$

(11.14b)

From Eqs. (11.14) the continuity condition at $z = 0$ of the electric field $E(z)$ parallel to the surface gives $E_t = E_i + E_r$. From Eqs. (11.3a) and (11.14), the continuity

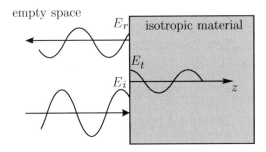

Figure 11.2 Schematic representation of incident, reflected and transmitted electromagnetic waves at the surface (the $z = 0$ plane) of an isotropic material; the normal incident situation is considered.

condition of the y-components of the magnetic field gives $N E_t = E_i - E_r$. The ratio E_r/E_i equals $(1 - N)/(1 + N)$; thus *the reflectivity R at normal incidence* is

$$R = \left| \frac{E_r}{E_i} \right|^2 = \left| \frac{1 - N}{1 + N} \right|^2 = \frac{(n - 1)^2 + k^2}{(n + 1)^2 + k^2}. \tag{11.15}$$

Another convenient measurable quantity for describing a surface is the *surface impedance Z* defined as

$$Z = \frac{4\pi}{c} \frac{E(0)}{B(0)}. \tag{11.16a}$$

From Eqs. (11.3a) and (11.10a) we have also

$$Z = \frac{4\pi i \omega}{c^2} \left(\frac{E(z)}{dE(z)/dz} \right)_{z=0} = \frac{4\pi}{cN}. \tag{11.16b}$$

The relationships between optical constants in isotropic media, in the linear and local-response regime, are summarized in Table 11.1.

Table 11.1 Relationships between optical constants in isotropic media in the linear and local-response regime.

Conductivity	$\sigma = \sigma_1 + i\sigma_2$
Dielectric constant	$\varepsilon = \varepsilon_1 + i\varepsilon_2, \quad \varepsilon = 1 + \dfrac{4\pi i \sigma}{\omega}, \quad \varepsilon_1 = 1 - \dfrac{4\pi \sigma_2}{\omega}, \quad \varepsilon_2 = \dfrac{4\pi \sigma_1}{\omega}$
Refractive index	$N = n + ik, \quad \varepsilon = N^2, \quad \varepsilon_1 = n^2 - k^2, \quad \varepsilon_2 = 2nk$
	$n^2 = \dfrac{1}{2}\left(\varepsilon_1 + \sqrt{\varepsilon_1^2 + \varepsilon_2^2} \right), \quad k^2 = \dfrac{1}{2}\left(-\varepsilon_1 + \sqrt{\varepsilon_1^2 + \varepsilon_2^2} \right)$
Absorption coefficient	$\alpha(\omega) = 2\omega k/c \equiv \omega \varepsilon_2 / nc$
Classical skin depth	$\delta = c/\omega k$
Surface impedance	$Z = 4\pi/cN$
Reflectivity	$R = \dfrac{(n - 1)^2 + k^2}{(n + 1)^2 + k^2}$ (at normal incidence)

Before closing, it is very instructive to consider the implications of the occurrence $\varepsilon_1(\omega) > 0$ and $\varepsilon_2(\omega) = 0$ for some given ω region; then $n(\omega) = \sqrt{\varepsilon_1(\omega)}$ and $k(\omega) = 0$. From Eqs. (11.2) and (11.11), it is seen that the medium can sustain *undamped transverse waves* (constituted by an electromagnetic field and an induced current density) *of frequency ω and propagation vector* **q**, with dispersion relation

$$q = \frac{\omega}{c} n, \quad \text{or equivalently} \quad \omega = \frac{c}{n} q = \frac{c}{\sqrt{\varepsilon_1(\omega)}} q;$$

in the local-response regime we are considering, $\varepsilon_1(q, \omega) = \varepsilon_1(0, \omega) = \varepsilon_1(\omega)$. The undamped transverse waves of the system can be interpreted as elementary excitations of energy $\hbar\omega$ and momentum $\hbar\mathbf{q}$. [An example of these concepts has been considered in Section 9.7.2, in the study of the infrared dielectric properties of polar crystals and *polaritons*. We should also remember that in the case of longitudinal waves and longitudinal dielectric function, *plasmon* excitations are inferred from the vanishing of both ε_1 and ε_2, as discussed in particular in Sections 7.5 and 7.6.]

11.2 The Drude Theory of the Optical Properties of Free Carriers

Consider a free-electron gas with n carriers per unit volume, each with effective mass m and charge $(-e)$; the carriers are embedded in a uniform background of neutralizing positive charge. [The internal structure and polarizability of the ions is here ignored, for simplicity.] In the Drude theory, the response of the carriers to a spatially uniform driving electric field of frequency ω and amplitude $\mathbf{E}_0$, is described via the classical equation of motion

$$m\ddot{\mathbf{u}}(t) = -\frac{m}{\tau}\dot{\mathbf{u}}(t) + (-e)\mathbf{E}_0 e^{-i\omega t}, \tag{11.17}$$

where $\mathbf{u}(t)$ is the displacement of the particle and τ is a phenomenological relaxation time. The viscous damping term $-(m/\tau)\dot{\mathbf{u}}$ takes into account semi-empirically the dissipative mechanisms, originated by the random collisions between the electron and whatever kind of impurities, phonons, and other imperfections of the crystal. [Strictly speaking, the second term in the right-hand side of Eq. (11.17) should be accompanied by its complex conjugate, so that the electric field acting on the electron is real; for simplicity, we omit the complex conjugate term, when nothing would change either in the formal treatment or in the final results.]

In Eq. (11.17), the displacement $\mathbf{u}(t)$ is expected of the type $\mathbf{u}(t) = \mathbf{u}_0 \exp(-i\omega t)$, with the same direction and frequency of the driving electric field; insertion into Eq. (11.17) gives

$$\mathbf{u}_0 = \frac{e\tau}{m} \frac{1}{\omega(i + \omega\tau)} \mathbf{E}_0.$$

The free-carrier contribution to the current density is

$$\mathbf{J}(t) = n(-e)\dot{\mathbf{u}}(t) = n(-e)(-i\omega)\mathbf{u}_0 e^{-i\omega t} = \frac{ne^2\tau}{m} \frac{1}{1 - i\omega\tau} \mathbf{E}_0 e^{-i\omega t}.$$

The Drude frequency dependent complex conductivity is thus

$$\sigma(\omega) = \sigma_0 \frac{1}{1 - i\omega\tau} \qquad \text{with} \qquad \sigma_0 = \frac{ne^2\tau}{m}, \qquad (11.18)$$

where σ_0 is the static conductivity. The dynamical conductivity (11.18) combines a wealth of relevant physical information with all the merits of simplicity, and explains why the Drude model, formulated in 1900, still remains a precious tool for the understanding of features of intraband transitions in solids.

The expression (11.18) of the conductivity contains two important parameters, the carrier density n and the relaxation time τ, that set the two energy scales of the phenomena under investigation.

The relaxation times in typical situations are of the order $\tau \approx 10^{-14}$ s, and the corresponding relaxation energies $\gamma = \hbar/\tau$ are of the order of 0.1 eV [with $\hbar = 6.852 \times 10^{16}$ eV $\cdot$ s].

The other scale of energy comes out from the relation connecting the static conductivity to the free-electron plasma frequency ω_p via the carrier density:

$$\sigma_0 = \frac{ne^2\tau}{m} = \frac{\tau}{4\pi}\omega_p^2 \qquad \text{where} \qquad \omega_p^2 = \frac{4\pi ne^2}{m}. \qquad (11.19)$$

To estimate the plasma energy, we use the standard expression $(4/3)\pi r_s^3 a_B^3 = 1/n$ that links the electron density to the dimensionless parameter r_s, and obtain

$$\hbar^2\omega_p^2 = \hbar^2 \frac{4\pi e^2}{m} \frac{1}{(4/3)\pi r_s^3 a_B^3} = \frac{12}{r_s^3} \frac{\hbar^2}{2ma_B^2} \frac{e^2}{2a_B} \implies \hbar\omega_p = \sqrt{\frac{12}{r_s^3}} \text{ Rydberg.}$$

For ordinary metals with $2 < r_s < 6$, the values of $\hbar\omega_p$ are in the range 3–17 eV. Thus in simple metals in ordinary situations, the typical energy $\hbar\omega_p \approx 10$ eV is about two orders of magnitude larger than the typical relaxation energy $\gamma = \hbar/\tau \approx 0.1$ eV.

When a current density $\mathbf{J}$ flows in a medium driven by an electric field $\mathbf{E}$, the power dissipated in the unit volume is

$$\mathbf{J} \cdot \mathbf{E} = \left[\sigma(\omega)\mathbf{E}_0 e^{-i\omega t} + \sigma^*(\omega)\mathbf{E}_0 e^{+i\omega t} \right] \cdot \left[\mathbf{E}_0 e^{-i\omega t} + \mathbf{E}_0 e^{+i\omega t} \right],$$

where the amplitude $\mathbf{E}_0$ is taken as real. Averaging on time, we have that

$$\langle \mathbf{J} \cdot \mathbf{E} \rangle = [\sigma(\omega) + \sigma^*(\omega)]|\mathbf{E}_0|^2 = \sigma_1(\omega)\langle|\mathbf{E}|^2\rangle \qquad \text{with} \qquad \sigma_1(\omega) = \sigma_0\frac{1}{1 + \omega^2\tau^2},$$

where σ_1 denotes the real part of the conductivity. Thus for $\omega\tau \ll 1$ the metal is strongly dissipative, while it becomes (almost) lossless if $\omega\tau \gg 1$.

Other properties of the metal are seen in more direct form elaborating the other optical constants. In particular from Eq. (11.13) the complex dielectric constant becomes

$$\varepsilon(\omega) = 1 - \frac{\omega_p^2}{\omega(\omega + i/\tau)}; \qquad (11.20a)$$

the real and imaginary parts of the dielectric function are

$$\varepsilon_1(\omega) = 1 - \frac{\omega_p^2 \tau^2}{1 + \omega^2 \tau^2} \qquad \text{and} \qquad \varepsilon_2(\omega) = \frac{\omega_p^2 \tau}{\omega(1 + \omega^2 \tau^2)}. \tag{11.20b}$$

Proceeding in the study of the optical properties of free carriers in metals, we can roughly distinguish three frequency regions, called low-frequency region ($\omega \ll 1/\tau$), the high-frequency region ($1/\tau \approx \omega \ll \omega_p$) and the ultraviolet region ($\omega \approx \omega_p$ and $\omega > \omega_p$).

Low-Frequency Region ($\omega\tau \ll 1 \ll \omega_p\tau$ or equivalently $\hbar\omega \ll \gamma \ll \hbar\omega_p$)

In the low-frequency region $\omega\tau$ is negligible with respect to 1, and the material is strongly dissipative. The dielectric function (11.20) becomes

$$\varepsilon_1(\omega) \approx 1 - \omega_p^2 \tau^2 \approx -\omega_p^2 \tau^2 \quad \text{and} \quad \varepsilon_2(\omega) \approx \frac{\omega_p^2 \tau}{\omega}.$$

We see that $\varepsilon_1(\omega)$ is negative and constant, while $\varepsilon_2(\omega)$ is singular for $\omega \to 0$. Notice that $|\varepsilon_2| \gg |\varepsilon_1|$.

From the relation $N^2 = \varepsilon \approx i\varepsilon_2$, it follows $N \approx \varepsilon_2(1+i)/\sqrt{2}$. The refractive index n and the extinction coefficient k are almost equal and given by

$$n(\omega) \approx k(\omega) \approx \sqrt{\frac{\varepsilon_2(\omega)}{2}} = \sqrt{\frac{\omega_p^2 \tau}{2\omega}}.$$

Thus also $n(\omega)$ and $k(\omega)$ are singular for $\omega \to 0$. Since $n(\omega) \gg 1$ the reflectivity is almost 1 and the metal is strongly reflecting: metals are excellent mirrors in this frequency region.

The penetration depth is

$$\delta(\omega) = \frac{c}{\omega k(\omega)} \approx \frac{c}{\omega_p} \sqrt{\frac{2}{\omega\tau}} \gg \frac{c}{\omega_p} \equiv \frac{\ell_0}{2\pi},$$

where the length $\ell_0 \equiv 2\pi c/\omega_p$ is typically in the range $\ell_0 \simeq 1000$ Å $= 10^{-5}$ cm for $\hbar\omega_p = 10$ eV (the length ℓ_0 is by definition the wavelength of a photon of energy $\hbar\omega_p$). To estimate the penetration depth in ordinary metals, assume as typical values $\tau \approx 10^{-14}$ s, and $\hbar\omega_p \approx 10$ eV. For $\omega \approx 100$ rad/s the penetration depth is of the order of centimeters; the penetration depth decreases to the value $\simeq \ell_0$ when the frequency of the electromagnetic field is of the order of the relaxation frequency.

High-Frequency Region ($1 < \omega\tau \ll \omega_p\tau$ or equivalently $\gamma < \hbar\omega \ll \hbar\omega_p$)

The high-frequency region, extends approximately from the relaxation frequency to the plasma frequency. When $\omega\tau$ is much larger than unity, we can neglect unity in the denominators of Eqs. (11.20b) and the dielectric function becomes

$$\varepsilon_1(\omega) \approx 1 - \frac{\omega_p^2}{\omega^2} \approx -\frac{\omega_p^2}{\omega^2} \qquad \text{and} \qquad \varepsilon_2(\omega) = \frac{\omega_p^2}{\omega^3 \tau}. \tag{11.21}$$

Thus $\varepsilon_1(\omega)$ is still negative, and the imaginary part of the dielectric function is much smaller than the real part. From $N^2 = (n + ik)^2 = \varepsilon_1 + i\varepsilon_2$, and taking into account that $\varepsilon_1 < 0$ and $|\varepsilon_2| \ll |\varepsilon_1|$, we obtain

$$k(\omega) \approx \sqrt{-\varepsilon_1} = \frac{\omega_p}{\omega} \quad \text{and} \quad n(\omega) \approx \frac{1}{2}\frac{\varepsilon_2}{\sqrt{-\varepsilon_1}} = \frac{1}{2}\frac{\omega_p}{\omega^2 \tau}.$$

The above relations show that $k(\omega) \gg n(\omega) \gg 1$; since the refractive index $n(\omega)$ is still much larger than 1, the reflectivity is almost 1 and the metal is strongly reflecting also in this frequency region. The penetration depth, is given by

$$\delta(\omega) = \frac{c}{\omega k(\omega)} \approx \frac{c}{\omega_p} \equiv \frac{\ell_0}{2\pi},$$

and is now approximately constant.

Ultraviolet Region ($\omega \approx \omega_p$ and $\omega > \omega_p$)

In this region, Eqs. (11.20) for the dielectric function read

$$\varepsilon_1(\omega) \approx 1 - \frac{\omega_p^2}{\omega^2} \quad \text{and} \quad \varepsilon_2(\omega) \approx \frac{\omega_p^2}{\omega^3 \tau} \approx 0.$$

We see that $\varepsilon_1(\omega)$ is positive for $\omega \geq \omega_p$; the reflectivity changes from (almost) one to (almost) zero, when ω increases above the plasma frequency, and the metal becomes transparent for $\omega > \omega_p$. The schematic behavior of the dielectric function of a metal in the Drude model is given in Figure 11.3.

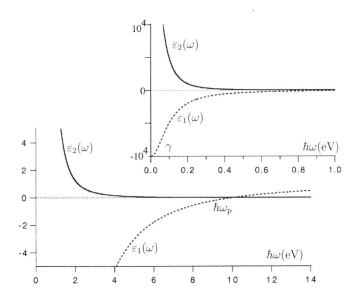

Figure 11.3 Behavior of real and imaginary part of the dielectric function of a free-electron gas with the Drude model; we have taken $\hbar\omega_p = 10$ eV and $\gamma = \hbar/\tau = 0.1$ eV.

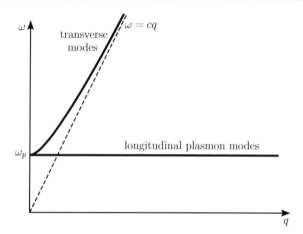

Figure 11.4 Representation of the dispersion curves of transverse electromagnetic modes and longitudinal plasmons in a bulk metallic system.

As discussed at the end of Section 11.1, the occurrence $\varepsilon_1(\omega) > 0$ and $\varepsilon_2(\omega) = 0$ entails that the medium can sustain undamped transverse waves, whose frequency and propagation vector are linked by the dispersion relation

$$\omega = \frac{c}{\sqrt{\varepsilon_1(\omega)}}\, q, \quad \varepsilon_1(\omega) = 1 - \frac{\omega_p^2}{\omega^2} \quad \Longrightarrow \quad \boxed{\omega = \sqrt{\omega_p^2 + c^2 q^2}}\ .$$

The above dispersion relation for transverse electromagnetic waves is reported in Figure 11.4. It is seen that transverse electromagnetic modes in the bulk material are possible for $\omega > \omega_p$; for high values of q the dispersion relation approaches the light line. In the figure the (almost flat) dispersion curve for longitudinal electromagnetic modes (studied in Section 7.5) is also reported for completeness.

Illustrative Examples of Application of the Drude Theory

The Drude model of intraband transitions describes an ideal system of free electrons, with spherical Fermi surface, a single relaxation time, in the local-response regime. In spite of these restrictions, the Drude theory applies reasonably well for the description of the optical properties of a number of metals in the red and infrared region, sufficiently below the threshold of interband electronic transitions. As an illustration, we consider here the optical constants of copper (Figures 11.5 and 11.6) and alkali metals (Figure 11.7); in these materials the Fermi surfaces are almost spheroid single sheets, and at sufficiently long wavelengths the optical properties are determined almost entirely by the free electrons of the sample.

In Figure 11.5, the measured imaginary part of the dielectric function of copper is reported in the wavelength range from 0.365 to 2.5 μm ($\hbar\omega$ is in the energy range from 3.4 to 0.5 eV). At the low frequencies, $\varepsilon_2(\omega)$ decreases rapidly with increasing ω; then there is a sudden increase for $\hbar\omega \approx 2.1$ eV, which marks the onset of interband

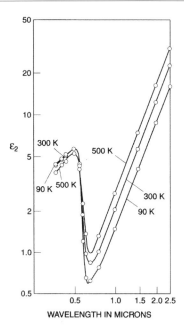

Figure 11.5 Imaginary part of the dielectric function of copper at different temperatures [reprinted with permission from S. Roberts, Phys. Rev. *118*, 1509 (1960); copyright 1960 by the American Physical Society].

electronic transitions; below this threshold, the Drude theory is very useful for an overall understanding of the optical constants.

The complex dielectric function $\varepsilon(\omega)$ of a metal, can be expressed as the sum of a Drude free-carrier contribution, given by Eq. (11.21) for $\omega\tau \gg 1$, and an interband contribution; this latter, for frequencies sufficiently below the onset of interband transitions can be considered as a constant, here denoted as $\varepsilon_{\text{inter}}$; we have thus

$$\varepsilon(\omega) = 1 - \frac{\omega_p^2}{\omega^2} + i\frac{\omega_p^2}{\omega^3\tau} + \varepsilon_{\text{inter}}.$$

With the replacement $\omega = ck = 2\pi c/\lambda$, it follows

$$\varepsilon_1(\lambda) = 1 + \varepsilon_{\text{inter}} - \frac{\omega_p^2}{4\pi^2 c^2}\lambda^2 \quad \text{and} \quad \varepsilon_2(\lambda) = \frac{\omega_p^2}{8\pi^3 c^3\tau}\lambda^3,$$

where λ is the free-space wavelength of light. From the above expressions, we expect that $\varepsilon_1(\lambda)$ varies linearly with λ^2, and the same occurs for $\varepsilon_2(\lambda)/\lambda$; we also expect that ε_2 is rather sensitive to the temperature and increases with temperature, since the relaxation time is expected to decrease with temperature. In Figure 11.6 we report the experimental data of ε_1 and ε_2/λ versus λ^2 for copper in the infrared and red region: it can be noticed that the data fall almost exactly on straight lines, and the temperature dependence of ε_2/λ is in agreement with what expected. The fact that the straight lines

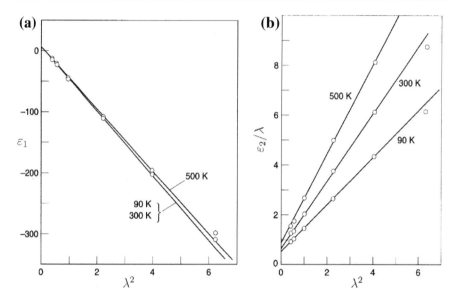

Figure 11.6 Dielectric function of copper in the infrared and red region at different temperatures. (a) ε_1 versus square of wavelength (with λ in micron); (b) ε_2/λ versus square of wavelength (with λ in micron) [reprinted with permission from S. Roberts, Phys. Rev. *118*, 1509 (1960); copyright 1960 by the American Physical Society].

representing ε_2/λ do not pass through the origin, however, needs a proper account of the anomalous skin effect, and we refer to the paper by Roberts (cited in Figure 11.6) for specific considerations on this point.

As another example, we consider the optical properties of alkali metals in the region from 0.5 eV to 4 eV. In the relaxation region $\omega\tau \gg 1$, the optical conductivity $\sigma_1(\omega)$ for intraband transitions becomes

$$\sigma_1(\omega) = \frac{\omega\varepsilon_2(\omega)}{4\pi} \approx \frac{\omega_p^2}{4\pi\tau\omega^2}.$$

The experimental values of the optical conductivity for four alkali metals are reported in Figure 11.7; it can be seen that in the infrared region the simple dependence $\sigma_1(\omega) \sim \omega^{-2}$ is closely followed. Defining a phenomenological lifetime

$$\tau = \frac{\omega_p^2}{4\pi\sigma_1(\omega)\omega^2},$$

then, if the Drude theory model is applicable, one expects that τ should be constant. Typical values of τ are of the order of 10^{-14}s; however, from a close analysis of experimental data, it is seen that τ is not exactly a constant quantity, and depends somewhat on frequency. The optical absorption of alkali metals has been interpreted considering intraband and interband transitions within the nearly free-electron model, and we refer to the paper by Smith (cited in Figure 11.7) for further considerations.

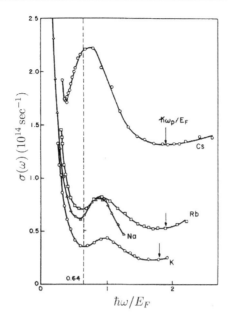

Figure 11.7 Optical conductivity of the four alkali metals Na, K, Rb, Cs plotted against the normalized frequency $\hbar\omega/E_F$, where E_F is the free-electron Fermi energy [reprinted with permission from N. V. Smith, Phys. Rev. B 2, 2840 (1970); copyright 1970 by the American Physical Society].

Range of Validity of the Drude Theory

The Drude theory gives an overall comprehension of the intraband contribution to the optical constants of metals; however seldom if ever it is sufficient for a quantitative account of experimental observations (even in the frequency regions where interband contributions can be safely ignored or worked out separately). There are several reasons for this. (i) The conduction band structure of the electrons in actual metals may be significantly different from the parabolic energy band assumption (with effective mass m), implicit in the Drude model. (ii) The relaxation time appears as an energy independent phenomenological parameter, while an analysis of scattering mechanisms implies an appropriate energy dependence. (iii) A more basic reason is the assumption that spatial dispersion can be neglected.

In the Drude theory the current density at a given point in the material is taken as proportional to the value of the electric field at that same point as summarized in Eq. (11.8). This assumption can be justified when the average distance travelled by the carriers is small with respect to the length of spatial variation of the electric field.

We can estimate the mean free path travelled by the carriers with the following arguments. Consider first the case $\omega\tau \ll 1$, corresponding to a relaxation time τ much smaller than the period $T = 2\pi/\omega$ of the electromagnetic field. The displacement of an electron with typical Fermi velocity v_F between two collisions is about

$$\Lambda_0 \approx v_F \tau \quad (\omega\tau \ll 1). \tag{11.22a}$$

In the case $\omega\tau \gg 1$, the displacement of an electron with velocity v_F during the period $T = 2\pi/\omega$ of the electromagnetic field is $\approx v_F/\omega$, and we have

$$\Lambda(\omega) = v_F \frac{1}{\omega} = \frac{v_F \tau}{\omega\tau} \quad (\omega\tau \gg 1). \tag{11.22b}$$

Equations (11.22a,b) apply to the low- and high-frequency regions, respectively; they can be more conveniently elaborated into a unique equation by defining the frequency dependent complex mean free path

$$\boxed{\Lambda(\omega) = v_F \tau \frac{1}{1 - i\omega\tau}} . \tag{11.23}$$

It is evident that $|\Lambda(\omega)|$ reproduces the estimated mean free path of Eqs. (11.22) both for $\omega\tau \ll 1$ and for $\omega\tau \gg 1$, and furthermore Eq. (11.23) constitutes a simple and reasonable interpolation for the mean free path at any frequency.

We have also to estimate the distance along which the electromagnetic field changes significantly. For this purpose consider Eq. (11.11); taking into account that $n(\omega)$ is much smaller than $k(\omega)$ (or at most comparable) in all the frequency range of interest below the plasma frequency, we can approximate the distance of significant change of the field with the penetration depth $\delta(\omega)$ of the electromagnetic wave in the metal.

We can now compare the penetration depth $\delta(\omega)$ with the estimated mean free path $|\Lambda(\omega)|$ of carriers. The region of local regime, also called of normal skin effect is characterized by the condition

$$|\Lambda(\omega)| \ll \delta(\omega). \tag{11.24a}$$

The region where the local approximation is no more valid, and rather it holds

$$|\Lambda(\omega)| \gg \delta(\omega) \tag{11.24b}$$

is referred to as extreme anomalous region.

To put to a test the local-response assumption, consider for instance a metal in the typical region of frequency around the relaxation frequency, where we have seen that the penetration depth δ (when the local-response assumption holds) is of the order of 1000 Å for $\hbar\omega_p \approx 10$ eV. In ordinary metals we have $\tau \approx 10^{-14}$ s, and $\Lambda(\omega) \approx v_F\tau \approx 10^8 \times 10^{-14} \approx 10^{-6}$ cm; then the criterium of Eq. (11.24a) appears reasonably satisfied in the whole range of frequencies (including the "vulnerable" relaxation frequency region). In a very pure specimen at helium temperature we can have $\tau \approx 10^{-11}$ s or so. The electron mean free path in very pure samples at very low temperature may become of the order $\Lambda(\omega) \approx v_F\tau \approx 10^8 \times 10^{-11} \approx 10^{-3}$ cm. When $\Lambda(\omega) \gg \delta$, the current density at a given point can no longer be determined by the electric field at the same point, and the local approximation thus far considered is no more applicable.

In very pure and good conductors at very low temperatures (copper for instance), this extreme anomalous region may extend significantly around the relaxation frequency. The determination of the optical constants when spatial dispersion has to be taken into account requires more sophisticated semiclassical approaches based on the Boltzmann transport equation which is the subject of the next section.

11.3 Transport Properties and Boltzmann Equation

In the study of intraband transport processes, we focus for simplicity on metals with a unique partially filled conduction band of energy $E = E(\mathbf{k})$ (multiband situations, where two or more bands are partially filled with electrons, would require appropriate elaborations of the one-band model we are going to describe). At thermodynamic equilibrium, the probability of occupation of an energy level $E(\mathbf{k})$ is given by the Fermi-Dirac distribution function

$$f_0(\mathbf{k}) = \frac{1}{e^{(E(\mathbf{k})-\mu)/k_BT} + 1}, \tag{11.25}$$

where T is the temperature of the sample and μ is the chemical potential (the chemical potential will be denoted by μ or by E_F, indifferently); $f_0(\mathbf{k})$ is a shorthand notation for $f_0(E(\mathbf{k}))$.

When external perturbations (electric fields, magnetic fields, temperature gradients) are applied to the sample, the electron distribution is disturbed from the equilibrium Fermi-Dirac function. In general the disturbed distribution function $f(\mathbf{r}, \mathbf{k}, t)$, in addition to $\mathbf{k}$, depends also on the real space coordinate $\mathbf{r}$, and on time t; the quantity $f(\mathbf{r}, \mathbf{k}, t)\, d\mathbf{r}\, d\mathbf{k}/4\pi^3$ gives the number of electrons at time t in the volume element $d\mathbf{r}\, d\mathbf{k}$ around the point $(\mathbf{r}, \mathbf{k})$ of the "phase space."

According to the semiclassical dynamics of carriers in a given energy band, an electron in the point $(\mathbf{r}, \mathbf{k})$ at time t evolves toward the point $(\mathbf{r} + \mathbf{v_k}\, dt, \mathbf{k} + (\mathbf{F}/\hbar)\, dt)$ at time $t + dt$, where

$$\mathbf{v_k} = \frac{1}{\hbar}\frac{\partial E(\mathbf{k})}{\partial \mathbf{k}}, \qquad \frac{d(\hbar\mathbf{k})}{dt} = \mathbf{F}, \tag{11.26}$$

and $\mathbf{F}$ denotes the external force acting on the carriers. During the motion, collision processes (due for instance to lattice vibrations, impurities, boundaries) may cause a net rate of change $[\partial f/\partial t]_{\text{coll}}$ of the number of electrons in the phase space volume $d\mathbf{r}\, d\mathbf{k}$. Using Liouville theorem (volumes in phase space are preserved by the semiclassical equations of motion) we must have for the distribution function

$$f\left(\mathbf{r} + \mathbf{v}\, dt, \mathbf{k} + \frac{\mathbf{F}}{\hbar}\, dt, t + dt\right) \equiv f(\mathbf{r}, \mathbf{k}, t) + \left[\frac{\partial f}{\partial t}\right]_{\text{coll}} dt. \tag{11.27a}$$

Equation (11.27a) expresses the detailed balance in each volume $d\mathbf{r}\, d\mathbf{k}$ of the number of carriers, when moving in the phase space under the action of external fields and in the presence of collision processes (as indicated in pictorial form in Figure 11.8).

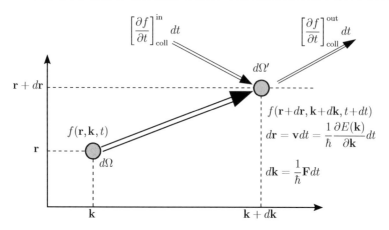

Figure 11.8 Schematic representation of the conservation of the number of electrons moving in the space phase $\mathbf{r}$, $\mathbf{k}$. The region $d\Omega$ around $\mathbf{r}$, $\mathbf{k}$ at time t evolves into a new region $d\Omega'$, whose volume is the same as $d\Omega$ (Liouville theorem). The distribution function $f(\mathbf{r}+d\mathbf{r}, \mathbf{k}+d\mathbf{k}, t+dt)$ equals $f(\mathbf{r}, \mathbf{k}, t)$ supplemented by the net change $[\partial f/\partial t]_{\text{coll}}\, dt = [\partial f/\partial t]_{\text{coll}}^{\text{in}}\, dt - [\partial f/\partial t]_{\text{coll}}^{\text{out}}\, dt$ of the number of electrons forced in and ejected out because of collision processes.

The left member of Eq. (11.27a) can be expanded in Taylor series up to first order, and we obtain the Boltzmann equation

$$
\frac{\partial f}{\partial \mathbf{r}} \cdot \mathbf{v} + \frac{\partial f}{\partial \mathbf{k}} \cdot \frac{\mathbf{F}}{\hbar} + \frac{\partial f}{\partial t} = \left[\frac{\partial f}{\partial t} \right]_{\text{coll}} , \tag{11.27b}
$$

where $\partial f/\partial \mathbf{r}$ and $\partial f/\partial \mathbf{k}$ stand for $\nabla_{\mathbf{r}} f$ and $\nabla_{\mathbf{k}} f$, respectively.

A crucial aspect in the transport theory is just the collision term, which makes the Boltzmann equation (11.27b) a formidable integro-differential equation. In general, the Boltzmann collision operator takes the form

$$
\left[\frac{\partial f}{\partial t} \right]_{\text{coll}} = I_{\text{coll}}[f(\mathbf{r}, \mathbf{k}, t)],
$$

where I_{coll} is an integral operator acting on $f(\mathbf{r}, \mathbf{k}, t)$; it describes the effect on the distribution function of the scattering processes (produced for instance by impurities, vacancies, lattices vibrations, physical surfaces, etc.). The change of the distribution function at a given $\mathbf{k}$ vector equals the number of electrons forced into $\mathbf{k}$ by collisions minus the number of electrons ejected from $\mathbf{k}$ by collisions. The expression of I_{coll} is of the type

$$
I_{\text{coll}}[f(\mathbf{r}, \mathbf{k}, t)] = \frac{1}{(2\pi)^3} \int \left[P_{\mathbf{k} \leftarrow \mathbf{k}'} f_{\mathbf{k}'}(1 - f_{\mathbf{k}}) - P_{\mathbf{k}' \leftarrow \mathbf{k}} f_{\mathbf{k}}(1 - f_{\mathbf{k}'}) \right] d\mathbf{k}',
$$

where $P_{\mathbf{k}' \leftarrow \mathbf{k}}$ denotes the probability per unit time that an electron in the $\mathbf{k}$ state is scattered into the $\mathbf{k}'$ state; the factor of the type $f(\mathbf{r}, \mathbf{k}, t)[1 - f(\mathbf{r}, \mathbf{k}', t)]$ is due to the Pauli principle and takes into account the occupancy of the initial states and the non-occupancy of the final states.

A major simplification is obtained when the deviation of f from the thermal equilibrium distribution f_0 is small: it is customary to assume that the rate of change of f due to collisions is proportional to the deviation itself, i.e.

$$\left[\frac{\partial f}{\partial t}\right]_{coll} = -\frac{f - f_0}{\tau},$$

where τ denotes the appropriate proportionality coefficient and is called *relaxation time*; in general $\tau = \tau(\mathbf{k}, \mathbf{r})$ depends on the energy $E(\mathbf{k})$ and on position, and is often considered a semi-empirical parameter. In the *relaxation time approximation*, the Boltzmann equation (11.27b) becomes the rather manageable partial differential equation

$$\boxed{\frac{\partial f}{\partial \mathbf{r}} \cdot \mathbf{v} + \frac{1}{\hbar}\frac{\partial f}{\partial \mathbf{k}} \cdot \mathbf{F} + \frac{\partial f}{\partial t} = -\frac{f - f_0}{\tau}}. \tag{11.28}$$

The role of the relaxation time can be further clarified by considering the case of sudden removal of external forces ($\mathbf{F} = 0$) at $t = 0$ in a uniform system ($\partial f/\partial \mathbf{r}) = 0$; from Eq. (11.28) one obtains

$$f(\mathbf{k}, t) = f_0 + [f(\mathbf{k}, t = 0) - f_0]e^{-t/\tau},$$

and the distribution function moves to its equilibrium value f_0 exponentially with time constant τ.

The non-equilibrium distribution function f, evaluated with the Boltzmann equation, permits the investigation of a number of transport phenomena due to intraband electronic processes. Transport coefficients can be inferred from the general expression of charge current density and the energy current density (or energy flux) given by

$$\mathbf{J} = \frac{1}{4\pi^3} \int (-e)\,\mathbf{v}\,f\,d\mathbf{k}, \tag{11.29a}$$

$$\mathbf{U} = \frac{1}{4\pi^3} \int E\,\mathbf{v}\,f\,d\mathbf{k}, \tag{11.29b}$$

where the factor $2/(2\pi)^3$ takes into account spin degeneracy and density of allowed points in $\mathbf{k}$ space per unit volume (the $\mathbf{k}$-dependence of the group velocity, band energy and distribution function is left implicit, for simplicity of notations). Transport properties, dissipation of irreversible heat, generation or subtraction of reversible heat, can be discussed using the above equations, and the principles of the thermodynamics.

In the next section we apply the transport equations to establish the electrical conductivity in constant and oscillating electric fields; in particular we take into account the spatial dispersion of the electric field and examine the extreme anomalous region. Successively, we describe the thermoelectric effects, due to the simultaneous presence of electric fields and temperature gradients.

As we shall see in the next sections, the Boltzmann transport equation describes a variety of transport phenomena in bulk crystalline solids. Whenever the scattering processes are reasonably weak and the applied fields are reasonably low, it makes sense to consider the joint effects of fields and collisions in driving the system out of equilibrium, within the perturbative approach and mean-field spirit, implicit in the Boltzmann equation. The description of transport phenomena by means of a distribution function is doomed to fail for small or ultrasmall electronic devices, whose spatial scales (micron or submicron dimensions) may become comparable or shorter than the semiclassical mean free path, or the quantum wavelength of the particles responsible of transport; for these systems quantum transport techniques must be applied from the very beginning. We do not dwell on these and other shortcomings of the Boltzmann transport equation, and refer to the literature for the achievements and progress in the theories of nonequilibrium quantum transport and the role of many-body effects. [Among the wide literature, we refer for instance to the reviews by Stroscio (1990), and by Dubi et al. (2011), and to the textbook by Ferry et al. (2009) and references quoted therein.]

11.4   Static and Dynamic Conductivity in Metals

11.4.1   Static Conductivity with the Boltzmann Equation

We consider the static conductivity of a metal by using the Boltzmann approach. For a homogeneous material in a uniform and steady electric field $\mathbf{E}$, the distribution function f depends only on $\mathbf{k}$ and the Boltzmann equation (11.28) becomes

$$\frac{1}{\hbar} \frac{\partial f}{\partial \mathbf{k}} \cdot (-e)\mathbf{E} = -\frac{f - f_0}{\tau}. \tag{11.30a}$$

For low electric fields, we can assume that $f - f_0$ is linear in the field strength, and we can thus put $f \approx f_0$ in the first member of Eq. (11.30a); we obtain

$$f = f_0 + \frac{e\tau}{\hbar} \frac{\partial f_0}{\partial \mathbf{k}} \cdot \mathbf{E}. \tag{11.30b}$$

In ordinary situations $e\tau |\mathbf{E}| \ll \hbar k_F$ (take for instance $E \approx 10^4$ V/cm $= 10^{-4}$ V/Å, $\hbar/\tau \approx 0.1$ eV and $k_F \approx 1$ Å^{-1}). Thus Eq. (11.30b), at the lowest order in $e\tau |\mathbf{E}|/\hbar k_F$, can be written as $f(\mathbf{k}) \approx f_0(\mathbf{k} + e\tau \mathbf{E}/\hbar)$. The nonequilibrium distribution function $f(\mathbf{k})$ and the equilibrium distribution function $f_0(\mathbf{k})$ are schematically indicated in Figure 11.9.

It is convenient to write $f = f_0 + f_1$, and recast Eq. (11.30b) in the form

$$f_1 = \frac{e\tau}{\hbar} \frac{\partial f_0}{\partial \mathbf{k}} \cdot \mathbf{E} = \frac{e\tau}{\hbar} \frac{\partial f_0}{\partial E} \frac{\partial E(\mathbf{k})}{\partial \mathbf{k}} \cdot \mathbf{E} = e\tau \frac{\partial f_0}{\partial E} \mathbf{v} \cdot \mathbf{E},$$

where $\mathbf{v} = (1/\hbar)\partial E(\mathbf{k})/\partial \mathbf{k}$ is the semiclassical expression of the velocity. Inserting f_1 in the expression (11.29a) for the current density (f_0 gives zero contribution), one obtains

$$\mathbf{J} = \frac{e^2}{4\pi^3} \int \tau \left(-\frac{\partial f_0}{\partial E}\right) \mathbf{v}(\mathbf{v} \cdot \mathbf{E}) \, d\mathbf{k} \quad \Longrightarrow \quad J_\alpha = \sum_\beta \sigma_{\alpha\beta} E_\beta \tag{11.31}$$

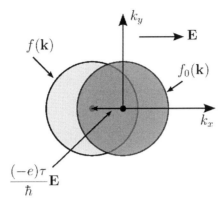

Figure 11.9 Schematic representation of the equilibrium distribution function $f_0(\mathbf{k})$ and of the nonequilibrium distribution function $f(\mathbf{k}) \approx f_0(\mathbf{k}+e\tau\hbar^{-1}\mathbf{E})$ in the presence of a static electric field $\mathbf{E}$. The effect of the field is to shift the whole Fermi sphere by $-e\tau\hbar^{-1}\mathbf{E}$ in the $\mathbf{k}$ space. For sake of clarity of the figure, the value of $-e\tau\hbar^{-1}\mathbf{E}$ has been magnified and taken of the same order as k_F.

with $\alpha, \beta = x, y, z$; the above relation defines the conductivity tensor linking the Cartesian components of the current density to the Cartesian components of the electric field.

For simplicity we assume that the material is isotropic so that $\mathbf{J}$ and $\mathbf{E}$ are parallel, and the proportionality constant is the static conductivity σ_0; indicating by $\hat{\mathbf{e}} = \mathbf{E}/|\mathbf{E}|$ the unit vector in the direction of the electric field, and projecting $\mathbf{J}$ along $\hat{\mathbf{e}}$, we obtain

$$\sigma_0 = \frac{e^2}{4\pi^3} \int \tau(\hat{\mathbf{e}} \cdot \mathbf{v})^2 \left(-\frac{\partial f_0}{\partial E}\right) d\mathbf{k} \; . \tag{11.32}$$

It is well known that the Fermi distribution function $f_0(E)$ changes sharply from unity to zero within a small interval ($\approx k_B T$) around the Fermi level, and that $(-\partial f_0/\partial E)$ is significantly different from zero in this same interval. Thus the conductivity σ_0 is basically determined from the features of the conduction bands in the thermal shell $k_B T$ around the Fermi energy E_F; this conclusion holds not only for the conductivity but also for any other transport coefficient in metals.

To estimate the conductivity given by expression (11.32), one can replace $(\hat{\mathbf{e}} \cdot \mathbf{v})^2$ with $v^2/3$, and approximate the derivative of the Fermi function with a δ-function; it follows

$$\sigma_0 = \frac{e^2}{12\pi^3} \int \tau v^2 \delta[E(\mathbf{k}) - E_F] d\mathbf{k}.$$

The integration can be performed exploiting the identity

$$\int \delta[E(\mathbf{k}) - E_0] d\mathbf{k} \equiv \int_{E(\mathbf{k})\equiv E_0} \frac{dS}{|\nabla_\mathbf{k} E(\mathbf{k})|} \; . \tag{11.33}$$

The expression of the conductivity thus reduces to an integration over the Fermi surface

$$\sigma_0 = \frac{e^2}{12\pi^3} \int_{\substack{\text{Fermi} \\ \text{surface}}} \tau v^2 \frac{dS}{|\nabla_{\mathbf{k}} E(\mathbf{k})|} = \frac{e^2}{12\pi^3 \hbar} \int_{\substack{\text{Fermi} \\ \text{surface}}} \tau v \, dS. \tag{11.34}$$

In the particular case of a parabolic conduction band with effective mass m^*, Eq. (11.34) gives

$$\sigma_0 = \frac{e^2}{12\pi^3 \hbar} \tau_F v_F 4\pi k_F^2 = \frac{ne^2}{m^*} \tau_F \tag{11.35}$$

($v_F = \hbar k_F/m^*$ and $k_F^3 = 3\pi^2 n$). Notice that the rather naive result (11.35), in which *all the electrons seem to take part to the transport*, is valid only for parabolic bands; in reality, the generally valid Eq. (11.34) shows that *in any case* (parabolic or non-parabolic bands) only *the electrons at (or near) the Fermi surface can change their state under perturbations and are relevant in the transport phenomena.*

11.4.2 Frequency and Wavevector Dependence of the Conductivity

In the previous section we have considered the conductivity in the presence of a steady and uniform electric field. We consider now the conductivity function in the presence of a transverse electric field, periodic in space and time, given by

$$\mathbf{E}(\mathbf{r}, t) = \mathbf{E}_0 e^{i(\mathbf{q} \cdot \mathbf{r} - \omega t)}, \quad \mathbf{E}_0 \perp \mathbf{q}.$$

The present section is confined to the treatment of transverse fields (in the case of longitudinal fields $\mathbf{E}_0 \parallel \mathbf{q}$, similar procedures can be applied).

We start again from the Boltzmann equation (11.28) that now becomes

$$\frac{\partial f}{\partial \mathbf{r}} \cdot \mathbf{v} + \frac{1}{\hbar} \frac{\partial f}{\partial \mathbf{k}} \cdot (-e)\mathbf{E} + \frac{\partial f}{\partial t} = -\frac{f - f_0}{\tau}. \tag{11.36a}$$

As before, we write $f = f_0 + f_1$, and assume that f_1 is linear in the applied field; for consistency, we neglect the term $(1/\hbar)(\partial f_1/\partial \mathbf{k}) \cdot (-e)\mathbf{E}$ in Eq. (11.36a) and we have

$$\frac{\partial f_1}{\partial \mathbf{r}} \cdot \mathbf{v} + \frac{1}{\hbar} \frac{\partial f_0}{\partial \mathbf{k}} \cdot (-e)\mathbf{E} + \frac{\partial f_1}{\partial t} = -\frac{f_1}{\tau}. \tag{11.36b}$$

In an isotropic material, in the linear response regime, f_1 is assumed to have the same space and time dependence as the electric field; inserting

$$f_1(\mathbf{r}, \mathbf{k}, t) = \Phi(\mathbf{k}) e^{i(\mathbf{q} \cdot \mathbf{r} - \omega t)}$$

into Eq. (11.36b), one obtains

$$i\mathbf{q} \cdot \mathbf{v}\Phi(\mathbf{k}) + \frac{1}{\hbar} \frac{\partial f_0}{\partial \mathbf{k}} (-e) \cdot \mathbf{E}_0 - i\omega\Phi(\mathbf{k}) = -\frac{\Phi(\mathbf{k})}{\tau}.$$

Thus, the function $\Phi(\mathbf{k})$ is given by

$$\Phi(\mathbf{k}) = \frac{e\tau\mathbf{v}\cdot\mathbf{E}_0}{1 - i\tau(\omega - \mathbf{q}\cdot\mathbf{v})} \frac{\partial f_0}{\partial E}.$$

The current density, evaluated with Eq. (11.29a), becomes

$$\mathbf{J} = \frac{1}{4\pi^3}\int(-e)\mathbf{v}f_1 d\mathbf{k} = \frac{e^2}{4\pi^3}\int\tau\left(-\frac{\partial f_0}{\partial E}\right)\mathbf{v}\frac{\mathbf{v}\cdot\mathbf{E}}{1 - i\tau(\omega - \mathbf{q}\cdot\mathbf{v})}d\mathbf{k}. \quad (11.37)$$

In the case of isotropy $\mathbf{J}$ and $\mathbf{E}$ are parallel, and the proportionality constant is the conductivity $\sigma(\mathbf{q}, \omega)$; indicating by $\hat{\mathbf{e}} = \mathbf{E}/|\mathbf{E}|$ the polarization vector, from Eq. (11.37) we obtain for the conductivity

$$\boxed{\sigma(\mathbf{q}, \omega) = \frac{e^2}{4\pi^3}\int\frac{\tau(\hat{\mathbf{e}}\cdot\mathbf{v})^2}{1 - i\tau(\omega - \mathbf{q}\cdot\mathbf{v})}\left(-\frac{\partial f_0}{\partial E}\right)d\mathbf{k}}. \quad (11.38)$$

In the case of isotropic medium the direction of $\hat{\mathbf{e}}$ becomes irrelevant.

Equation (11.38) for the frequency and wavevector dependent conductivity $\sigma(\mathbf{q}, \omega)$ is the natural generalization of Eq. (11.32), to which it reduces in the long wavelength limit $\mathbf{q} \to 0$ and static limit $\omega \to 0$. Furthermore, when spatial dispersion can be neglected, the expression for $\sigma(\mathbf{q} = 0, \omega) \equiv \sigma(\omega)$ given by Eq. (11.38) becomes

$$\boxed{\sigma(\omega) = \frac{\sigma_0}{1 - i\omega\tau}} \quad \text{with} \quad \boxed{\sigma_0 = \frac{e^2}{4\pi^3}\int\tau(\hat{\mathbf{e}}\cdot\mathbf{v})^2\left(-\frac{\partial f_0}{\partial E}\right)d\mathbf{k}};$$

this relation shows how to generalize the Drude theory, summarized by Eq. (11.18), and how to consider explicitly the band structure of the material. More importantly, for $\mathbf{q} \neq 0$ Eq. (11.38) allows in principle to describe the optical properties of materials, including the effects of spatial dispersion of the electromagnetic field, as discussed in the next subsection.

11.4.3 Pippard Ineffectiveness Concept and the Anomalous Skin Effect

The physical meaning and concepts encrypted in Eq. (11.38) for the transverse conductivity can be better clarified considering first the particular case of a parabolic band profile. For this prototype model, a simple and analytic expression of the conductivity can be worked out, and the model can provide an illustration of the important Pippard ineffectiveness concept and guidelines for more complicated situations.

Transverse Conductivity for a Parabolic Energy Band

We now specify Eq. (11.38) in the case of a transverse electric field with polarization vector $\hat{\mathbf{e}}$ in x-direction and propagation vector $\mathbf{q}$ in z-direction, and obtain

$$\sigma(q, \omega) = \frac{e^2}{4\pi^3}\int\frac{\tau v_x^2}{1 - i\tau(\omega - qv_z)}\left(-\frac{\partial f_0}{\partial E}\right)d\mathbf{k}. \quad (11.39)$$

In the case of a parabolic energy band, the above integral can be worked out exactly (as shown in Problem 2). The analytic expression reads

$$\sigma(q,\omega) = \frac{3}{4}\frac{\sigma_0}{1-i\omega\tau}\left[\frac{2}{s^2}+\frac{s^2-1}{s^3}\ln\frac{1+s}{1-s}\right] \qquad \text{with} \qquad s = \frac{iqv_F\tau}{1-i\omega\tau}.$$

$$(11.40)$$

This represents in the whole (q,ω) domain the frequency and momentum-transfer conductivity obtained within the Boltzmann framework, in the case of a parabolic energy band.

The physical meaning of the parameter s can be established by expressing it in terms of the frequency dependent mean free path defined in Eq. (11.23):

$$s = iq\Lambda(\omega), \quad \Lambda(\omega) = \frac{v_F\tau}{1-i\omega\tau}.$$

The values $|s| \ll 1$, or equivalently $|\Lambda(\omega)| \ll 1/q$, entail a mean free path of the carriers much smaller than the wavelength of the electromagnetic wave, and individuate the *Drude or normal* (q,ω) *region*. On the same token, the values $|s| \gg 1$ individuate the *extreme anomalous* (q,ω) *region*.

It is very instructive to consider Eq. (11.40) in the limiting case of normal and extremely anomalous regions. In the normal region $|s| \ll 1$, the expression in square brackets in Eq. (11.40) equals 4/3; in this limit one recovers for the conductivity the Drude result

$$\sigma(q,\omega) = \frac{\sigma_0}{1-i\omega\tau}, \quad |s| \ll 1. \qquad (11.41a)$$

In the extreme anomalous region $|s| \gg 1$, the expression in square brackets in Eq. (11.40) equals $i\pi/s$, and the conductivity becomes

$$\sigma(q,\omega) = \frac{3}{4}\sigma_0\frac{\pi}{v_F\tau q}, \quad |s| \gg 1, \qquad (11.41b)$$

a result which is *real, and independent both from frequency and from relaxation time*. The conductivity in the extreme anomalous region is quite different from the conductivity of the Drude theory. The most interesting aspect of expression (11.41b) is that it does not depend on the relaxation time; in fact

$$\frac{\sigma_0}{v_F\tau} = \frac{ne^2}{m^*v_F} = \frac{k_F^3}{3\pi^2}\frac{e^2}{\hbar k_F} = \frac{e^2}{3\pi^2\hbar}k_F^2.$$

In the extreme anomalous region, measurements of the surface resistance (real part of the surface impedance) provide an accurate tool for the investigation of the Fermi wavevector, and of the features related to the shape of the Fermi surface (see also Problem 3).

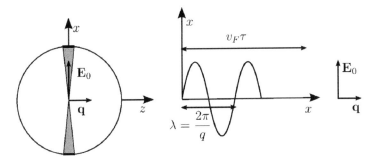

Figure 11.10 Illustration of Pippard ineffectiveness concept. Out of all the electrons within the Fermi sphere, only those traveling (almost) parallel to $\mathbf{E}_0$ can absorb energy and contribute to the conductivity, in the extreme anomalous region.

The result (11.41b) can be qualitatively understood using the Pippard ineffectiveness concept (see Figure 11.10). Consider first a static and uniform electric field $\mathbf{E}_0$ (in the x-direction); in this case the conductivity is just $\sigma_0 = ne^2\tau/m$, and all the electrons seem to participate to the transport. Consider now an oscillatory electric field with amplitude $\mathbf{E}_0$ and wavevector $\mathbf{q}$ (in the z-direction). We notice that $|s| \gg 1$ implies $qv_F\tau \gg 1$. It is evident that an electron with velocity $\mathbf{v}_F$ parallel to $\mathbf{q}$ sees an electric field that has changed its sign $\approx qv_F\tau$ times, before a collision takes place. Thus, if $qv_F\tau \gg 1$ we expect that the *effective electrons capable to absorb energy* are only the very electrons with velocity almost parallel to $\mathbf{E}_0$ such that $0 \leq \mathbf{q} \cdot \mathbf{v}_F \tau \leq 1$. These electrons are a fraction $f \approx 1/qv_F\tau$ of all the electrons and correspondingly σ_0 is reduced to $\approx \sigma_0 f$, in qualitative agreement with Eq. (11.41b).

Anomalous Skin Effect for Copper

Thus far, only spherical Fermi surfaces have been discussed. The general problem of an arbitrary Fermi surface and of Pippard ineffectiveness concept, is not much different conceptually, although technically more demanding. Consider more closely the structure of Eq. (11.38) in the case the integration (at zero temperature) is extended over a Fermi surface that has cubic symmetry, but is not spherically symmetric. In the extreme anomalous region, in the integration in Eq. (11.38), the only electrons that contribute to the conductivity are the ones in a *small belt of the Fermi surface* such that

$$E(\mathbf{k}) = E_F \quad \text{and} \quad \widehat{\mathbf{n}} \cdot \mathbf{q} \approx 0,$$

where $\widehat{\mathbf{n}} = \widehat{\mathbf{n}}(\mathbf{k})$ denotes the unit vector perpendicular to the Fermi surface at the point $\mathbf{k}$.

As one might expect, a great deal can be learned on the shapes of the Fermi surfaces from measurements of surface resistance for different crystallographic orientations. As an example, we mention the historical work by Pippard on the surface resistance of copper for different orientations, in the extreme anomalous region [A. B. Pippard, Phil. Trans. Roy. Soc. (London) A *250*, 325 (1957)]. In spite of the fact that copper is an fcc cubic crystal, the surface resistances with orientation (100) and (110) differ by about a factor 2. As pointed out by Pippard, the shape of the Fermi surface, that best fits the measurements, is an almost spheroidal single sheet, which makes contact with the Brillouin zone hexagonal faces (see Figure 6.14); the study of the anomalous skin effect has provided a direct experimental evidence of the open nature of the Fermi surface of copper.

11.5 Boltzmann Treatment and Quantum Treatment of Intraband Transitions

The Boltzmann treatment of intraband transport phenomena is based on the semiclassical equations of motion, given by Eq. (11.26), and proper account of collision effects. It is instructive to analyse what links a semiclassical treatment, which *takes into account the band energy dispersion but ignores wavefunctions*, with the quantum mechanical treatment, whose ingredients are *both energies and wavefunctions*.

To avoid inessential details, consider the simplest model for a metal, with a unique conduction band of energy dispersion $E_{\mathbf{k}}$ and Bloch wavefunctions $\psi_{\mathbf{k}}$. In this specific one-band model, the transverse conductivity $\sigma(\mathbf{q}, \omega)$ provided by the linear response theory is given by Eq. (7.D9), and reads (with trivial adjustments of notations)

$$\sigma(\mathbf{q}, \omega) = \frac{2e^2}{m^2} \int \frac{d\mathbf{k}}{(2\pi)^3} \frac{|\langle \psi_{\mathbf{k}+\mathbf{q}} | e^{i\mathbf{q}\cdot\mathbf{r}} \widehat{\mathbf{e}} \cdot \mathbf{p} | \psi_{\mathbf{k}} \rangle|^2}{(E_{\mathbf{k}+\mathbf{q}} - E_{\mathbf{k}})/\hbar} \frac{(-i)[f_0(E_{\mathbf{k}}) - f_0(E_{\mathbf{k}+\mathbf{q}})]}{E_{\mathbf{k}+\mathbf{q}} - E_{\mathbf{k}} - \hbar\omega - i\eta},$$

(11.42)

where f_0 is the Fermi-Dirac equilibrium distribution function, and the limit $\eta \to 0^+$ is understood.

We now make on Eq. (11.42) some approximations, which are justified for sufficiently small $\mathbf{q}$ vectors, i.e. under the condition $|\mathbf{q}| < k_F$. The change of the distribution function is approximated with the series development

$$f_0(E_{\mathbf{k}+\mathbf{q}}) - f_0(E_{\mathbf{k}}) \approx \frac{\partial f_0}{\partial E} (E_{\mathbf{k}+\mathbf{q}} - E_{\mathbf{k}});$$

(11.43a)

the matrix elements in Eq. (11.42) can be simplified for small radiation wavevectors so to obtain

$$\frac{1}{m} \langle \psi_{\mathbf{k}+\mathbf{q}} | e^{i\mathbf{q}\cdot\mathbf{r}} \widehat{\mathbf{e}} \cdot \mathbf{p} | \psi_{\mathbf{k}} \rangle \approx \frac{1}{m} \langle \psi_{\mathbf{k}} | \widehat{\mathbf{e}} \cdot \mathbf{p} | \psi_{\mathbf{k}} \rangle = \widehat{\mathbf{e}} \cdot \mathbf{v}.$$

(11.43b)

With the approximations (11.43), the conductivity becomes

$$\sigma(\mathbf{q}, \omega) = \frac{e^2 \hbar}{4\pi^3} \int d\mathbf{k} \, (\widehat{\mathbf{e}} \cdot \mathbf{v})^2 \frac{-i}{E_{\mathbf{k}+\mathbf{q}} - E_{\mathbf{k}} - \hbar\omega - i\eta} \left(-\frac{\partial f_0}{\partial E} \right).$$

We can expand $E_{\mathbf{k}+\mathbf{q}} - E_{\mathbf{k}} \approx \mathbf{q} \cdot (\partial E/\partial \mathbf{k}) = \mathbf{q} \cdot \mathbf{v}\hbar$, and make the replacement $\eta = \hbar/\tau$ ($\eta \to 0+$ entails $\tau \to +\infty$). The quantum expression for the conductivity becomes

$$\sigma(\mathbf{q}, \omega) = \frac{e^2}{4\pi^3} \int \frac{\tau(\widehat{\mathbf{e}} \cdot \mathbf{v})^2}{1 - i\tau(\omega - \mathbf{q} \cdot \mathbf{v})} \left(-\frac{\partial f_0}{\partial E} \right) d\mathbf{k},$$

(11.44)

where the parameter τ is an arbitrary large quantity. The quantum expression of the conductivity of Eq. (11.44) holds only in the ideal case of negligible broadening (or very large lifetime) of electronic states. On the contrary, the semiclassical Boltzmann approach of Eq. (11.38) holds also for finite lifetimes, and could be extended to more realistic materials characterized by energy-dependent broadening parameters.

The above considerations help to clarify some aspects of the semiclassical approach and of the quantum treatment. Within the adopted approximations, the Boltzmann framework appears suitable in treating intraband transitions, where relaxation times play a key role. On the other side, the higher energy band-to-band transitions are relatively less sensible to broadening effects, which often may be ignored in the quantum treatment of interband optical properties. For these reasons, in the present chapter on intraband transport we focus on the semiclassical Boltzmann approach, while in the next chapter on interband transitions the quantum approach is followed.

11.6 The Boltzmann Equation in Electric Fields and Temperature Gradients

11.6.1 The Transport Equations in General Form

In the previous sections we have studied transport effects due to the presence of electric fields in samples at uniform temperature (i.e. in *isothermal* conditions); we consider now transport equations in the presence of electric fields and temperature gradients. As usual, we consider the simplest possible electronic structure of the metal with a single conduction band of interest of energy $E(\mathbf{k})$; the influence (if any) of the temperature on the energy band structure $E(\mathbf{k})$ is assumed to be negligible.

In a crystal kept at non-uniform temperature, it is convenient to define the local equilibrium distribution function $f_0(\mathbf{k}, \mathbf{r})$ as

$$f_0(\mathbf{k}, \mathbf{r}) = \frac{1}{\exp[(E(\mathbf{k}) - \mu(\mathbf{r}))/k_B T(\mathbf{r})] + 1}; \tag{11.45}$$

the local equilibrium distribution function $f_0(\mathbf{k}, \mathbf{r})$, in addition to $\mathbf{k}$, depends implicitly on $\mathbf{r}$ since the local temperature $T = T(\mathbf{r})$ is a function of $\mathbf{r}$, and the chemical potential $\mu = \mu(T(\mathbf{r}), n(\mathbf{r})) = \mu(\mathbf{r})$ depends on $\mathbf{r}$ via the local temperature $T(\mathbf{r})$ and the local electron density $n(\mathbf{r})$. Notice that the local chemical potential $\mu(\mathbf{r})$ at the point $\mathbf{r}$, entering in Eq. (11.45), is just the *chemical potential of an ideal infinite sample at thermodynamic equilibrium, characterized by band structure $E(\mathbf{k})$, uniform temperature T equal to $T(\mathbf{r})$, and uniform electron density n equal to $n(\mathbf{r})$.*

In the following we need the gradients of f_0 with respect to $\mathbf{k}$ and $\mathbf{r}$; these are given by

$$\frac{\partial f_0}{\partial \mathbf{r}} = \frac{\partial f_0}{\partial E} k_B T \frac{\partial}{\partial \mathbf{r}} \frac{E(\mathbf{k}) - \mu}{k_B T} = \frac{\partial f_0}{\partial E} \left[-\frac{\partial \mu}{\partial \mathbf{r}} - \frac{E(\mathbf{k}) - \mu}{T} \frac{\partial T}{\partial \mathbf{r}} \right], \tag{11.46a}$$

and

$$\frac{1}{\hbar}\frac{\partial f_0}{\partial \mathbf{k}} = \frac{1}{\hbar}\frac{\partial f_0}{\partial E}\frac{\partial E(\mathbf{k})}{\partial \mathbf{k}} = \frac{\partial f_0}{\partial E}\mathbf{v}(\mathbf{k}). \tag{11.46b}$$

To simplify somewhat the notations, the $\mathbf{k}$-dependence of the group velocity and of the energy band is often left implicit.

The Boltzmann equation (11.28) for the stationary distribution $f(\mathbf{r}, \mathbf{k})$ in the presence of an electric field $\mathbf{E}$ and temperature gradient is

$$\frac{\partial f}{\partial \mathbf{r}}\cdot \mathbf{v} + \frac{1}{\hbar}\frac{\partial f}{\partial \mathbf{k}}\cdot(-e)\mathbf{E} = -\frac{f - f_0}{\tau} = -\frac{f_1}{\tau}.$$

Since the electric field and temperature gradient are usually small, we can assume that f_1 is linear in these variables; then we can put $f = f_0$ on the left-hand side of the above equation and obtain

$$f_1 = -\tau \frac{\partial f_0}{\partial \mathbf{r}}\cdot \mathbf{v} - \frac{\tau}{\hbar}\frac{\partial f_0}{\partial \mathbf{k}}\cdot(-e)\mathbf{E}.$$

Using Eqs. (11.46), the stationary nonequilibrium distribution function becomes

$$f_1 = \left(-\frac{\partial f_0}{\partial E}\right)\tau\left[-e\,\mathbf{E} - \nabla\mu - \frac{E - \mu}{T}\nabla T\right]\cdot \mathbf{v}, \tag{11.47}$$

where $\nabla = \partial/\partial \mathbf{r}$ indicates the gradient with respect to space variable $\mathbf{r}$.

We remind the general charge and energy transport equations (11.29), here re-written in a slightly different form

$$\mathbf{J} = \frac{1}{4\pi^3}\int (-e)\,\mathbf{v}\,f_1\,d\mathbf{k}; \qquad \mathbf{U} = \frac{1}{4\pi^3}\int (E - \mu)\,\mathbf{v}\,f_1\,d\mathbf{k} - \frac{\mu}{e}\mathbf{J}. \tag{11.48}$$

Furthermore (though not strictly necessary) we suppose that the system is isotropic, so that transport kinetic parameters become scalar quantities, rather than tensors. Inserting Eq. (11.47) into Eq. (11.48) one obtains

$$\begin{cases} \mathbf{J} = e\,K_0\left[e\,\mathbf{E} + \nabla\mu\right] + e\,\frac{K_1}{T}\nabla T, \\[2mm] \mathbf{U} = -K_1\left[e\,\mathbf{E} + \nabla\mu\right] - \frac{K_2}{T}\nabla T - \frac{\mu}{e}\mathbf{J}. \end{cases} \tag{11.49}$$

The expressions of the *kinetic coefficients* K_0, K_1, K_2 are

$$K_n = \frac{1}{4\pi^3}\int \tau\,(\hat{\mathbf{e}}\cdot\mathbf{v})^2\,(E - \mu)^n\left(-\frac{\partial f_0}{\partial E}\right)d\mathbf{k}, \quad n = 0, 1, 2, \tag{11.50}$$

where $\hat{\mathbf{e}}$ is the unit vector in the direction of the electric field; under the assumption of isotropy, the direction of $\hat{\mathbf{e}}$ becomes irrelevant, and can be taken, for instance, in the x-direction.

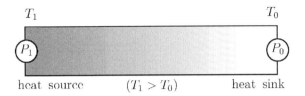

Figure 11.11 Schematic representation of a bar of homogeneous material, whose ends are kept at different temperatures.

It is convenient to rewrite the two basic transport equations (11.49) in a slightly different form, which is more suitable for the interpretation of the thermoelectric phenomena. The first of Eq. (11.49) can be cast in the form

$$\mathbf{J} = e^2 K_0 \left[\mathbf{E} + \frac{1}{e} \nabla\mu - S(T)\nabla T \right] \quad \text{with} \quad S(T) = -\frac{1}{e\,T}\frac{K_1}{K_0}, \quad (11.51)$$

where the transport coefficient S is called *absolute thermoelectric power or Seebeck coefficient*. From Eqs. (11.51) we see that the current density $\mathbf{J}$ consists of three contributions. The first term $e^2 K_0\,\mathbf{E}$ is the standard drift term $\sigma_0\,\mathbf{E}$, where σ_0 is the conductivity of the metal. The second one is due to the inhomogeneity (i.e. to the $\mathbf{r}$-dependence) of the chemical potential. The third is due to the presence of a temperature gradient. It is interesting to notice that the energy dissipated per unit time and unit volume

$$\mathcal{P} = \mathbf{E}\cdot\mathbf{J} = \frac{J^2}{\sigma_0} - \frac{1}{e}\nabla\mu\cdot\mathbf{J} + S(T)\nabla T\cdot\mathbf{J},$$

besides the essentially positive Joule term J^2/σ_0 (irreversible heat), contains two additional terms linear in J, which can be either positive or negative (reversible heat).

For what concerns the transport equation for the energy flux, it is convenient to obtain $[e\,\mathbf{E} + \nabla\mu]$ from the first of Eqs. (11.49) and replace it into the second one; this gives

$$\mathbf{U} = \left[-\frac{K_1}{e\,K_0} - \frac{\mu}{e} \right]\mathbf{J} - k_e\nabla T \quad \text{with} \quad k_e = \frac{1}{T}\left(K_2 - \frac{K_1^2}{K_0} \right), \quad (11.52)$$

where the transport parameter k_e is called *electron thermal conductivity*.

The physical meaning of k_e is easily established if one considers a metal in the presence of a uniform temperature gradient ∇T and in open circuit situation, so that $\mathbf{J} = 0$ (see Figure 11.11). In this case, Eq. (11.52) takes the form $\mathbf{U} = -k_e\nabla T$; this shows that the energy density (or heat density) flowing through the device is opposite and linear to the temperature gradient, with proportionality constant k_e.

11.6.2 Thermoelectric Phenomena

We discuss now some applications of Eqs. (11.51) and (11.52), which are the basic equations controlling transport in (isotropic) materials. After a few preliminary considerations on equilibrium conditions and isothermal conditions, we pass to study some typical thermoelectric circuits.

Drift and Diffusion Currents in Isothermal Conditions

As a first application, consider the electron current density in a metal in isothermal conditions, but with a non-uniform carrier concentration $\nabla n \neq 0$; this implies $\nabla \mu \neq 0$. Putting $\nabla T = 0$ into Eq. (11.51) we have

$$\mathbf{J} = \sigma_0 \left[\mathbf{E} + \frac{1}{e} \nabla \mu \right]. \tag{11.53}$$

We can thus distinguish a *drift current density* $\mathbf{J}_{\text{drift}} = \sigma_0 \mathbf{E}$ and a *diffusion current density* $\mathbf{J}_{\text{diff}} = \sigma_0 \nabla \mu / e$.

Consider, for example, a metal with free-electron-like conduction band. In this case $\sigma_0 = n e^2 \tau / m^*$, the chemical potential reads $\mu = (\hbar^2/2m^*)(3\pi^2 n)^{2/3}$ and $\nabla \mu / \mu = (2/3) \nabla n / n$. The current density (11.53) in the metal can thus be written as

$$\boxed{\mathbf{J} = n e \mu_e \mathbf{E} + e D \nabla n} \,, \tag{11.54}$$

where $\mu_e = e\tau/m^*$ is the *electron mobility*, and D is the *diffusion coefficient*

$$\boxed{D = \frac{2}{3} \frac{E_F}{e} \mu_e} \,, \tag{11.55a}$$

where the chemical potential has been here indicated with E_F.

In the case the free-electron gas is non-degenerate and follows the Boltzmann distribution, we have $\nabla n / n = \nabla \mu / k_B T$. The current density is again given by Eq. (11.54), but now the diffusion coefficient becomes

$$\boxed{D = \frac{k_B T}{e} \mu_e} \,. \tag{11.55b}$$

Relations (11.55a) and (11.55b) are the *Einstein relations* between mobility and diffusion coefficient for the degenerate and non-degenerate electron gas, respectively.

Consider now Eq. (11.51) in isothermal conditions $\nabla T = 0$ and in open circuit situation $\mathbf{J} = 0$. Indicating with $\phi(\mathbf{r})$ the electrostatic potential we have

$$-\nabla \phi + \frac{1}{e} \nabla \mu = 0 \quad \Longrightarrow \quad (-e)\phi + \mu \equiv \text{const.}$$

As expected the electrochemical potential $\mu + (-e)\phi$ is uniform throughout the sample in equilibrium conditions.

Seebeck Effect and Thermoelectric Power

When a temperature gradient is established in a long bar (in open circuit situation) an electric field has to set in, so to prevent any net carrier flux. Consider in fact a specimen with a cool end at temperature T_0 and a hot end at temperature T_1 (see Figure 11.11). In open circuit situation $\mathbf{J} = 0$ and the electric field can be obtained from Eq. (11.51) in the form

$$\mathbf{E} = -\frac{1}{e}\nabla\mu + S(T)\nabla T. \tag{11.56}$$

The potential difference between the end points P_0 and P_1, at temperatures T_0 and T_1, is

$$\phi_1 - \phi_0 = -\int_{P_0}^{P_1} \mathbf{E} \cdot d\mathbf{l} = \frac{1}{e}(\mu_1 - \mu_0) - \int_{T_0}^{T_1} S(T)\,dT. \tag{11.57}$$

Thus the difference of the electrochemical potentials at the ends of the bar is related to the line integral of the Seebeck coefficient.

The thermoelectric power $S(T)$ of a material can be measured by means of the standard bimetallic circuit of Figure 11.12, in which the two junctions between metal A and metal B are kept at different temperatures. Using the relation

$$-\nabla\phi = -\frac{1}{e}\nabla\mu + S(T)\nabla T,$$

it is easy to evaluate the potential difference at the extremal points P_0 and P_3 (kept at the same temperature $T_0 = T_3$ so that $\mu_{0A} \equiv \mu_{3A}$). The integrand is the scalar product of the above expression for $d\mathbf{l}$, and the integral can be carried out along any line going from P_0 to P_3 within the circuit (conveniently broken into five parts). We have

$$\phi_{1A} - \phi_{0A} = \frac{1}{e}(\mu_{1A} - \mu_{0A}) - \int_{T_0}^{T_1} S_A(T)\,dT,$$

$$\phi_{1B} - \phi_{1A} = \frac{1}{e}(\mu_{1B} - \mu_{1A}),$$

$$\phi_{2B} - \phi_{1B} = \frac{1}{e}(\mu_{2B} - \mu_{1B}) - \int_{T_1}^{T_2} S_B(T)\,dT,$$

$$\phi_{2A} - \phi_{2B} = \frac{1}{e}(\mu_{2A} - \mu_{2B}),$$

$$\phi_{3A} - \phi_{2A} = \frac{1}{e}(\mu_{3A} - \mu_{2A}) - \int_{T_2}^{T_3} S_A(T)\,dT,$$

where $T_i = T(P_i)$. Summing up the above relations we have

$$\phi_{3A} - \phi_{0A} = -\int_{T_0}^{T_1} S_A(T)\,dT - \int_{T_1}^{T_2} S_B(T)\,dT - \int_{T_2}^{T_0} S_A(T)\,dT.$$

It follows

$$\boxed{\phi_{3A} - \phi_{0A} = \int_{T_1}^{T_2} S_A(T)\,dT - \int_{T_1}^{T_2} S_B(T)\,dT}. \tag{11.58}$$

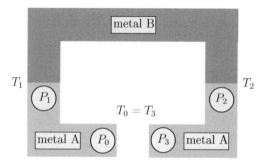

Figure 11.12 Standard bimetallic circuit to measure the thermoelectric effect. The two junctions between the metals are kept at different temperatures ($T_1 \neq T_2$); a voltage appears between points P_0 and P_3.

Thus if we choose a material B with $S_B(T)$ known (often lead is taken because its thermoelectric power is negligible) and vary T_2 with respect to T_1 we can obtain an experimental determination of $S_A(T)$, by measuring the potential difference $\phi_{3A} - \phi_{0A}$.

Thomson Effect

When an electric current flows in a given homogeneous material in the presence of a temperature gradient, heat is released or absorbed reversibly at a rate depending on the current density and on the nature of the material; if the direction of current is reversed, the Thomson effect also changes sign (contrary to the Joule heating effect).

To study the Thomson effect, we imagine that temperature gradient, electric field and density current depend on a single direction (say x), and we consider a (small) cylinder of section Σ_0 and length d_0 with its axis parallel to $\mathbf{J}$, and two sections at the temperatures T_A and T_B, respectively; for simplicity we also suppose that the temperature is kept constant and equal to T_A on the left side of the cylinder, while it is kept constant and equal to T_B on the right side of the cylinder. The geometry is schematically indicated in Figure 11.13.

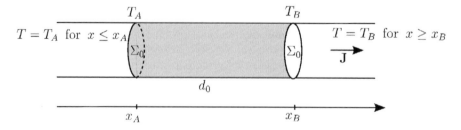

Figure 11.13 Schematic figure for the calculation of the Thomson coefficient.

When a current $\mathbf{J}$ flows from a point at temperature T_A to a point at temperature T_B reversible heat is generated in the cylinder at the rate

$$\frac{\delta Q}{dt} = -J\Sigma_0 \int_{T_A}^{T_B} K_{\text{rev}}(T)\, dT, \tag{11.59}$$

where $K_{\text{rev}}(T)$ is known as Thomson coefficient, and $J\Sigma_0 = I$ is the current through the cylinder of section Σ_0 under consideration. We now prove that the *Thomson coefficient is related to the absolute thermoelectric power through the relationship*

$$\boxed{K_{\text{rev}}(T) = T\,\frac{dS(T)}{dT}}. \tag{11.60}$$

The internal energy fluxes across the basis of the cylinder at sections A and B are respectively

$$\mathbf{U}_A = \left[-\frac{K_1(T_A)}{e K_0(T_A)} - \frac{\mu_A}{e} \right]\mathbf{J} \quad \text{and} \quad \mathbf{U}_B = \left[-\frac{K_1(T_B)}{e K_0(T_B)} - \frac{\mu_B}{e} \right]\mathbf{J},$$

as can be seen from Eq. (11.52) (taking into account that temperature gradients at the left and right sides of the cylinder are assumed to be zero).

The heat δQ generated in the time dt in the cylinder of volume $V = \Sigma_0 d_0$ is given by

$$\delta Q = dU + \delta L,$$

where dU is the energy which accumulates in the time dt because of the unbalance between energy flowing in and out of the considered cylinder, and δL is the work performed by the electric field in the time dt. We have

$$\frac{dU}{dt} = \left[-\frac{K_1(T_A)}{e K_0(T_A)} - \frac{\mu_A}{e} + \frac{K_1(T_B)}{e K_0(T_B)} + \frac{\mu_B}{e} \right] J\Sigma_0. \tag{11.61a}$$

Similarly, using Eq. (11.51) for the electric field, we have

$$\begin{aligned}
\frac{\delta L}{dt} &= \Sigma_0 \int_A^B \mathbf{J} \cdot \mathbf{E}\, dl = \Sigma_0 \int_A^B J \left[\frac{1}{\sigma_0} J - \frac{1}{e}\nabla\mu + S(T)\nabla T \right] dl \\
&= \Sigma_0 \left[\frac{1}{\sigma_0} J^2 d_0 - \frac{J}{e}(\mu_B - \mu_A) + J \int_{T_A}^{T_B} S(T)\, dT \right].
\end{aligned} \tag{11.61b}$$

From Eqs. (11.61), and disregarding the Joule heating $J^2 V/\sigma_0$, we obtain for the reversible heat generation rate

$$\frac{\delta Q}{dt} = \left[-\frac{K_1(T_A)}{eK_0(T_A)} + \frac{K_1(T_B)}{eK_0(T_B)} \right] J\Sigma_0 + J\Sigma_0 \int_{T_A}^{T_B} S(T)\, dT$$

$$= [T_A S_A(T) - T_B S_B(T)] J\Sigma_0 + J\Sigma_0 \int_{T_A}^{T_B} S(T)\, dT.$$

Performing an integration by parts, it follows

$$\frac{\delta Q}{dt} = -J\Sigma_0 \int_{T_A}^{T_B} T\, dS(T) = -J\Sigma_0 \int_{T_A}^{T_B} T\, \frac{dS}{dT}\, dT \qquad (11.62)$$

and this proves the anticipated Eq. (11.60) for the Thomson coefficient.

Peltier Effect

Heat is generated reversibly not only when current flows in a given homogeneous material in the presence of temperature gradient, but also when current flows across a junction between two contacting materials (Peltier effect). If the direction of current changes, the Peltier effect changes sign (contrary to the Joule heating effect).

For a quantitative analysis consider the standard bimetallic circuit of Figure 11.14 in isothermal conditions ($\nabla T = 0$) and with a current density J flowing throughout the circuit. Across the contact between metal A and metal B the rate (per unit time and section of area Σ_0) of reversible heat released or absorbed is

$$\frac{\delta Q}{dt} = \Pi_{AB} J\Sigma_0, \qquad (11.63)$$

where J is supposed to flow from metal A to metal B. The Peltier coefficient of a given metal is connected to the Seebeck coefficient by the simple equality

$$\boxed{\Pi(T) = T S(T)}\,, \qquad (11.64)$$

together with the relation $\Pi_{AB} = \Pi_A - \Pi_B$.

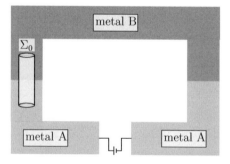

Figure 11.14 Standard bimetallic circuit for illustration of the Peltier effect; the temperature is uniform throughout the whole circuit.

To show this, we can use Eq. (11.62) (with a trivial extension of its meaning) keeping T constant, and S changing not because of temperature but because of inhomogeneity in the material. We have

$$\frac{\delta Q}{dt} = -J \Sigma_0 T [S_B(T) - S_A(T)],$$

and Eq. (11.64) is thus proved.

Considerations on Other Transport Effects

We have seen that the Boltzmann equation is very useful for the description of transport effects in metals. The transport phenomena we have investigated have been confined to the simplest situations in which the driving perturbation is a static or oscillating electric field (electrical conductivity effects), or a static electric field and a temperature gradient (thermal conductivity, Seebeck, Peltier, and Thomson effects).

The Boltzmann equation is of major help for several other transport phenomena. These include transport effects in the presence of electric and magnetic fields (Hall and magnetoresistivity effects), galvanomagnetic effects (Righi-Leduc effect, Nernst effect, Etthingshausen effect, etc.), "anomalies" or "giant effects" (in particular situations, for instance in the presence of magnetic impurities). A variety of challenging situations occur when the "phonon thermal bath," that usually ensures relaxation toward equilibrium of the electron distribution function, is itself dragged out from thermal equilibrium.

The Boltzmann equation has been widely applied to describe transport properties in semiconductors, following essentially the same semiclassical concepts given for metals, but keeping in mind some obvious differences. Among these, the fact that the distribution of conduction electrons in a semiconductor is in general non-degenerate, because of the low density of carriers. Furthermore we have to consider both electrons in the conduction band and holes in the valence band; and in general their contribution to a given transport phenomenon is not simply additive. These differences may lead to profound effects with respect to metals. For instance the electronic conductivity of a semiconductor is in general several orders of magnitude lower than that of a metal; nevertheless the thermoelectric power of a semiconductor, due to the presence of the energy gap, is in general much higher than typical thermopowers of metals. We do not dwell on other semiclassical aspects of transport properties, except for the discussion in Section 15.5 of magnetoresistivity and the Hall effect.

11.6.3 Transport Coefficients and Efficiency of Good Thermoelectric Materials

Thermoelectric Materials and Efficiency Parameter

Thermoelectric materials, for solid state devices without moving parts, have always attracted great technological interest especially for realization of refrigerators or power generators. Thermoelectric generators use the Seebeck effect to produce a voltage difference, while refrigerators make use of the Peltier effect for cooling purpose; these solid state devices are extremely reliable, although at the moment not yet competitive with conventional vapor-compressors or other mechanical systems, because of low efficiency. In this section we analyze more closely the transport properties of crystals, and, most importantly, we try to infer qualitatively what should be the features of electronic band structure, that characterize good thermoelectric materials.

So far in the study of the thermoelectric effects, we have considered several transport coefficients (electric conductivity, Seebeck coefficient, thermal conductivity of electrons, etc.). However there is an important parameter, dubbed as efficiency parameter, that is routinely used to characterize thermoelectric materials. The efficiency parameter p of thermoelectric materials (also called *figure of merit* $p = ZT$) is the dimensionless quantity defined as

$$p = \frac{T\sigma S^2}{k_e + k_l},$$ (11.65a)

where T is the temperature, σ is the conductivity, S is the Seebeck coefficient, k_e is the thermal electron conductivity, and k_l is the thermal lattice conductivity. In situations in which the lattice contribution to the thermal conductivity is (or can be made) sufficiently smaller with respect to electron contribution, the efficiency parameter can be approximated by the upper value

$$p_e = \frac{T\sigma S^2}{k_e} \quad (k_l \ll k_e),$$ (11.65b)

where the subscript to p denotes that we are here taking into account only the electronic thermal conductivity. Similarly, in situations in which $k_e \ll k_l$ (typically, but not exclusively, doped semiconductors because of the small number of carriers) the efficiency parameter can be approximated by the value

$$p_l = \frac{T\sigma S^2}{k_l} \quad (k_e \ll k_l).$$ (11.65c)

One of the best and most studied thermoelectric materials is Bi_2Te_3, which at room temperature has efficiency parameter near unity; however competitive thermoelectrics should have values significantly higher. This explains the continued search of novel thermoelectric materials with high efficiency parameters, and the importance of guidelines and qualitative criteria to design them. Thermoelectric materials are usually heavily doped semiconductors in degenerate regime and the theory of metals may be applied.

General Expression of the Kinetic Coefficients

The general expression of the kinetic coefficients for isotropic materials in the one-band approximation is given by Eq. (11.50) here re-written

$$K_n = \frac{1}{4\pi^3} \int \tau \, (\hat{\mathbf{e}} \cdot \mathbf{v})^2 \, (E - \mu)^n \left(-\frac{\partial f_0}{\partial E} \right) d\mathbf{k}, \quad n = 0, 1, 2.$$

The three-dimensional integral in $d\mathbf{k}$ throughout the Brillouin zone is conveniently broken into a two-dimensional integral on constant energy surfaces and an integration on the energy variable. For this purpose we introduce the *generalized transport distribution function* $\Sigma(E)$ so defined

$$\Sigma(E) = \frac{1}{4\pi^3} \int \tau(\hat{\mathbf{e}} \cdot \mathbf{v})^2 \delta\left[E(\mathbf{k}) - E\right] d\mathbf{k} = \frac{1}{4\pi^3} \int_{E(\mathbf{k})=E} \frac{\tau(\hat{\mathbf{e}} \cdot \mathbf{v})^2}{|\nabla E_{\mathbf{k}}|} \, dS.$$

$$(11.66)$$

For instance, for the free-electron gas and energy-independent relaxation time, the generalized transport distribution function has the power law form $\Sigma(E) = C E^p$ with $p = 3/2$; this and other models of transport distribution functions will be considered for qualitative considerations.

In terms of the generalized transport distribution function of the material under investigation, the kinetic parameters take the form

$$\boxed{K_n = \int \left(-\frac{\partial f_0}{\partial E}\right) (E - \mu)^n \, \Sigma(E) \, dE, \quad n = 0, 1, 2} \, . \tag{11.67}$$

Without entering in the details of specific materials, it is instructive to estimate the kinetic coefficients and the transport parameters in some simple but significant models for the transport distribution function $\Sigma(E)$.

Case of Smooth $\Sigma(E)$ and the Sommerfeld Expansion

Consider now the case that the generalized distribution function $\Sigma(E)$ is reasonably smooth around the chemical potential, so that we can exploit the Sommerfeld expansion for the calculation of the kinetic parameters. In the case of simple metals, for example, we have already seen that the Sommerfeld expansion is well justified.

The Sommerfeld expansion, introduced in Section 3.2, applies to any function $G(E)$ sufficiently smooth in the thermal shell $k_B T$ around the Fermi energy, and reads

$$\int \left(-\frac{\partial f_0}{\partial E}\right) G(E) \, dE = G(\mu) + \frac{\pi^2}{6} k_B^2 T^2 \left(\frac{d^2 G}{dE^2}\right)_{E=\mu} + O(T^4).$$

Setting $G(E) = (E - \mu)^n \Sigma(E)$ in the above equation, the kinetic coefficients (11.67), taking into account for simplicity only the leading terms, become

$$K_0 = \Sigma(\mu) + O(T^2), \tag{11.68a}$$

$$K_1 = \frac{\pi^2}{3} k_B^2 T^2 \Sigma'(\mu) + O(T^4), \tag{11.68b}$$

$$K_2 = \frac{\pi^2}{3} k_B^2 T^2 \Sigma(\mu) + O(T^4), \tag{11.68c}$$

where the function $\Sigma(E)$ and its derivative are calculated at the Fermi energy $E = \mu$. Using Eqs. (11.68), we can obtain the conductivity, the Seebeck coefficient, the electron thermal conductivity, and the figure of merit (for what concerns the electronic part) of a thermoelectric material. In all the cases where the Sommerfeld expansion is applicable, they have the expression

$$\sigma_0(\mu, T) \equiv e^2 K_0 = e^2 \Sigma(\mu), \tag{11.69a}$$

$$S(\mu, T) \equiv -\frac{1}{e} \frac{K_1}{T} \frac{K_1}{K_0} = -\frac{\pi^2}{3} \frac{k_B}{e} k_B T \frac{\Sigma'(\mu)}{\Sigma(\mu)}, \tag{11.69b}$$

$$k_e(\mu, T) \equiv \frac{1}{T} \left(K_2 - \frac{K_1^2}{K_0} \right) = \frac{\pi^2}{3} \frac{k_B^2}{e^2} T \sigma_0, \tag{11.69c}$$

$$p_e(\mu, T) \equiv \frac{T \sigma_0 S^2}{k_e} = \frac{\pi^2}{3} k_B^2 T^2 \left[\frac{\Sigma'(\mu)}{\Sigma(\mu)} \right]^2. \tag{11.69d}$$

A few comments on the above results are worthwhile. First, consider the thermoelectric power $S(T)$, whose natural unit of measure is

$$\frac{k_B}{e} \equiv 86.17 \ \mu V/K;$$

this value also sets the order of magnitude of interesting thermoelectric materials. The validity of Eqs. (11.69) automatically implies that the distribution function is reasonably smooth around the Fermi energy; this means $k_B T \ \Sigma'(\mu)/\Sigma(\mu) \ll 1$. The occurrence of this condition, tendentially depresses the Seebeck coefficient given by Eq. (11.69b), and this effect is even stronger on the efficiency parameter given by Eq. (11.69d).

Transport coefficients in simple metals can be well-described within the Sommerfeld expansion, because of the smooth character of the density-of-states and related generalized transport distribution function. It is instructive to estimate the transport coefficients in simple metals, noticing however that the specific transport properties of actual materials are rather sensitive to the energy dependence of the relaxation time and to the peculiarities of the Fermi surface. From Eq. (11.69b), it is seen that the thermoelectric power can be either negative or positive depending on the sign of $d\Sigma/dE$ at the Fermi energy. To evaluate the order of magnitudes, suppose that the generalized distribution function has a power law form of the type $\Sigma(E) \approx C E^p$ (for the free-electron gas $p = 3/2$). Then we can estimate

$$k_B T \frac{\Sigma'(\mu)}{\Sigma(\mu)} \approx \frac{k_B T}{\mu} = \frac{k_B T}{k_B T_F} = \frac{T}{T_F}; \tag{11.70a}$$

from Eqs. (11.69b,d) it follows

$$S(T) \approx -\frac{k_B}{e} \frac{T}{T_F} \quad \text{and} \quad p_e(T) \approx \left(\frac{T}{T_F} \right)^2. \tag{11.70b}$$

With $T \approx 300$ K and $T_F \approx 100$ T we expect a (negative) thermoelectric power of the order of $\mu V/K$ in normal metals at room temperature. It is also evident that the figure of merit of ordinary simple metals is very poor.

Another consideration concerns the relation between the electron thermal conductivity and the electrical conductivity. From the expression (11.69c) of the electron thermal conductivity k_e, we see that the ratio of thermal to electrical conductivity is proportional to T (Wiedemann-Franz law). Consider now the ratio

$$\frac{k_e}{T\sigma_0} = L \equiv \frac{\pi^2}{3}\frac{k_B^2}{e^2},$$

which is known as *Lorentz number*. The Lorentz number would actually be a universal constant (independent from the specific metal, temperature and relaxation time), provided the approximations done in the transport equations are justified. If one goes over the whole treatment, one realizes that the most vulnerable point is the relaxation time approximation of the collision term. This approximation is justified above the Debye temperature, where the electron-phonon scattering is the dominant process, and at very low temperature, where the impurity scattering is dominant. In both temperature regimes, the ratio $k_e/T\sigma_0$ is approximately the same for all metals. At intermediate temperatures, however significant deviations may occur.

Case of Peaked Generalized Transport Distribution Function $\Sigma(E)$

What can be learned from Eqs. (11.69) and the comments done so far, is that smooth distribution functions are not likely to produce good thermoelectric materials. From this matter-of-fact consideration, several investigations in the literature have focused on metals and materials with sharp transport distribution functions. Without entering in specific details, we limit our considerations to a few qualitative remarks.

To model a peaked generalized distribution function $\Sigma(E)$, we choose for simplicity a Lorentzian function of the type

$$\Sigma(E) = C\frac{\Gamma}{(E - E_0)^2 + \Gamma^2},\tag{11.71}$$

where E_0 is the resonance energy, Γ is the width, and C is taken as constant. We study the qualitative dependence of the kinetic parameters for large and small values of Γ with respect to the thermal energy k_BT.

We examine first the case $k_BT \ll \Gamma$. This assumption mimics the situation where the generalized distribution function changes smoothly on the thermal energy scale, and the Sommerfeld expansion is still applicable; then the results expressed by Eqs. (11.69) are still valid. We observe that for the Lorentzian function (11.71) it holds

$$\frac{\Sigma'(E)}{\Sigma(E)} = \frac{-2(E - E_0)}{(E - E_0)^2 + \Gamma^2} \quad \text{and} \quad -\frac{1}{\Gamma} \le \frac{-2(E - E_0)}{(E - E_0)^2 + \Gamma^2} \le \frac{1}{\Gamma}.$$

We can thus estimate

$$\frac{\Sigma'(\mu)}{\Sigma(\mu)} \approx \pm\frac{1}{\Gamma} \quad \text{for} \quad \mu \approx E_0 \mp \Gamma.\tag{11.72a}$$

In the case the chemical potential can be settled at or near the energies $E_0 \mp \Gamma$, from Eqs. (11.72a) and (11.69b,d) we obtain for the thermoelectric coefficient and the efficiency parameter

$$S(T) \approx \mp \frac{k_B}{e} \frac{k_B T}{\Gamma} \quad \text{and} \quad p_e(T) \approx \left(\frac{k_B T}{\Gamma}\right)^2. \tag{11.72b}$$

From the comparison of these results with the ones of Eq. (11.70b), it is evident the significant benefit achieved in thermoelectric power and efficiency values when the broadening width Γ is in the range $k_B T < \Gamma \ll k_B T_F$. However, the benefit cannot be extended too much, since the condition $k_B T < \Gamma$ must be satisfied, and the Wiedemann-Franz law of Eq. (11.69c) is still at work.

We discuss now the opposite case of extremely narrow resonances, such that $\Gamma \ll k_B T$. Evidently in this situation the Sommerfeld expansion cannot be applied; nevertheless, an instructive discussion can be elaborated assuming for simplicity a δ-like shape for the transport distribution function

$$\Sigma(E) = C\delta(E - E_0), \tag{11.73a}$$

where E_0 is the resonance energy, and C is an appropriate constant. Inserting Eq. (11.73a) into the expression (11.67) of the kinetic parameters gives

$$K_n = C \left(-\frac{\partial f_0}{\partial E}\right)_{E=E_0} (E_0 - \mu)^n, \quad n = 0, 1, 2. \tag{11.73b}$$

From the above results, it is worthwhile to notice that the electron thermal conductivity $k_e = (1/T)(K_2 - K_1^2/K_0)$ becomes exactly zero, and the corresponding efficiency parameter p_e becomes ideally infinity (of course in such a situation k_l cannot be neglected any more). This occurs because k_e is proportional to the variance of an appropriate distribution function, and the variance of a δ-like function is exactly zero. The physical reason of the vanishing of the electron thermal conductivity is due to the fact that no heat flow is possible without a spread of allowed energies; if the band width of allowed energies is much smaller than $k_B T$, charge current can flow without being accompanied by heat flow, and any limitation on the performance of thermoelectrics originated by the Wiedemann-Franz law is overcome.

Without taking too seriously the above results (based on the hypothetical model of δ-like transport distribution function), it is evident the general message that sharp transport distribution functions could lead to increased performance. This explains the wide attention in the literature devoted to transition-metal compounds, rare-earth compounds, multiband metals and alloys, superlattices and other low-dimensional systems, where sharp structures in the density-of-states and in the generalized transport function can occur in the operative range of interest. For what concerns the quest of materials with lattice thermal conductivity as low as possible, much attention has been focused on thermoelectric crystals with great chemical complexity, and in particular on skutterudites (such as $CoAs_3$ and $CoSb_3$) and clathrates (such as Na_8Si_{46} and $Na_{24}Si_{136}$). A prominent feature of these compounds is the presence of "cavities," where interstitially placed atoms can act as phonon-scattering centers to depress the thermal conductivity. [For further information see for instance the review by Mahan (1998), the review by Singh (2001), or the book of Nolas et al. (2001), and references quoted therein.]

Appendix A.   Solved Problems and Complements

Problem 1. Kramers-Kronig relations.
Problem 2. Transverse conductivity for a parabolic energy band.
Problem 3. Surface impedance in the extreme anomalous region.
Problem 4. Electron thermal conductivity and the Lorentz number.

Problem 1. Kramers-Kronig Relations

Consider a complex function $f(z)$ which is analytic on the upper half of the complex plane, including the real axis for simplicity. It is also supposed that the function goes to zero at infinity (otherwise we subtract its value at infinity). Then show that the real and imaginary part of the function $f(z)$ on the real axis satisfy the Kramers-Kronig relations.

Given the function $f(z)$ with the stated analytic properties, consider the function defined as

$$g(z) = \frac{f(z)}{z + i\varepsilon - x_0},$$

where x_0 is any chosen point in the real axis, and ε is an infinitesimal positive quantity. We have evidently that the integral of $g(z)$ on the real axis vanishes:

$$\int_{-\infty}^{+\infty} \frac{f(x)}{x + i\varepsilon - x_0} \, dx \equiv 0. \tag{A.1}$$

By exploiting the well-known Dirac identity (see Eq. (7.25))

$$\frac{1}{x + i\varepsilon - x_0} = P \frac{1}{x - x_0} - i\pi \delta(x - x_0),$$

we can cast Eq. (A.1) in the form

$$P \int_{-\infty}^{+\infty} \frac{f(x)}{x - x_0} \, dx - i\pi f(x_0) \equiv 0. \tag{A.2}$$

We write explicitly $f(x) = \text{Re } f(x) + i \,\text{Im } f(x)$ in Eq. (A.2), and obtain

$$\text{Re } f(x_0) = \frac{1}{\pi} P \int_{-\infty}^{+\infty} \frac{\text{Im } f(x)}{x - x_0} \, dx, \quad \text{Im } f(x_0) = -\frac{1}{\pi} P \int_{-\infty}^{+\infty} \frac{\text{Re } f(x)}{x - x_0} \, dx, \tag{A.3}$$

which are the Kramers-Kronig relations.

Problem 2. Transverse Conductivity for a Parabolic Energy Band

Using the Boltzmann equation, obtain the analytic expression of the transverse conductivity for a spherical energy band in the frequency-momentum domain.

Within the Boltzmann transport equation, the electron conductivity under attention is given by Eq. (11.39) and reads

$$\sigma(q, \omega) = \frac{e^2}{4\pi^3} \int \frac{\tau v_x^2}{1 - i\tau(\omega - qv_z)} \left(-\frac{\partial f_0}{\partial E} \right) d\mathbf{k}. \tag{A.4}$$

At zero temperature the energy derivative of the Fermi function becomes a δ-like function and we obtain

$$\sigma(q, \omega) = \frac{e^2}{4\pi^3} \int_{\substack{\text{Fermi} \\ \text{surface}}} \frac{\tau v_x^2}{1 - i\omega\tau + iqv_z\tau} \frac{dS}{\hbar v_F}, \tag{A.5}$$

where we have assumed an energy band with parabolic dispersion.

The integral over the spherical Fermi surface can be performed analytically using polar coordinates and the standard relations

$$dS = k_F^2 \sin\theta \, d\theta \, d\phi, \quad v_x = v_F \sin\theta \cos\phi, \quad v_z = v_F \cos\theta.$$

Performing the integral over ϕ (which gives π) we have

$$\sigma(q, \omega) = \frac{e^2 \tau}{4\pi^3 \hbar} v_F k_F^2 \, \pi \int_0^\pi \frac{\sin^3\theta}{1 - i\omega\tau + iqv_F\tau \cos\theta} \, d\theta.$$

With the change of variable $x = \cos\theta$ the above expression becomes

$$\sigma(q, \omega) = \frac{3}{4}\sigma_0 \int_{-1}^{+1} \frac{1 - x^2}{1 - i\omega\tau + iqv_F\tau x} dx,$$

where use has been made of the relationship $k_F^3 = 3\pi^2 n$ and $\sigma_0 = ne^2\tau/m$. It is convenient to define the parameter

$$s = \frac{iqv_F\tau}{1 - i\omega\tau},$$

and exploit the indefinite integral

$$\int \frac{1 - x^2}{1 + sx} dx = -\frac{x^2}{2s} + \frac{x}{s^2} + \frac{s^2 - 1}{s^3} \ln(1 + sx).$$

The conductivity becomes

$$\sigma(q, \omega) = \frac{3}{4} \frac{\sigma_0}{1 - i\omega\tau} \left[\frac{2}{s^2} + \frac{s^2 - 1}{s^3} \ln\frac{1 + s}{1 - s} \right]. \tag{A.6}$$

Expression (A.6) for the transverse conductivity holds on the whole (q, ω) plane for a parabolic band structure, within the Boltzmann approximation.

Problem 3. Surface Impedance in the Extreme Anomalous Region

Discuss the surface impedance for a parabolic energy band in the extreme anomalous region, and show that it is independent from relaxation processes

Consider the propagation of an electromagnetic wave in a metal in general situations, including the *extreme anomalous region*. The basic relation between current density and electric field, derived from Maxwell's equations, is given by Eq. (11.4), here re-written for convenience

$$\frac{d^2 E(z)}{dz^2} = -\frac{\omega^2}{c^2} E(z) - \frac{4\pi i \omega}{c^2} J(z). \tag{A.7}$$

For a proper account of spatial dispersion, one must complement Eq. (A.7) with the non-local relationship

$$J(q) = \sigma(q, \omega) E(q), \tag{A.8}$$

where $\sigma(q, \omega)$ is the frequency and wavevector dependent conductivity of the system, and $E(q)$ and $J(q)$ are the Fourier components of the electric field and current density, respectively.

The relation (A.7) refers to the geometry of Figure 11.2, where a metal fills half of the space $(z > 0)$. For simplicity we assume specular reflection of the electrons approaching the surface of the metal. The assumption of completely specular reflection is equivalent to consider the remaining half of space $(z < 0)$ filled with another piece of the same metal. The two pieces of metal are mirror images one of the other. The electric field is damped in both $+z$ and $-z$ directions and the boundary condition on the electric field is

$$\left(\frac{dE}{dz}\right)_{0+} = -\left(\frac{dE}{dz}\right)_{0-}. \tag{A.9}$$

A simple trick to account for the boundary condition (A.9) consists in adding to Eq. (A.7) a delta-like term in the form

$$\frac{d^2 E(z)}{dz^2} = -\frac{\omega^2}{c^2} E(z) - \frac{4\pi i \omega}{c^2} J(z) + 2\left(\frac{dE}{dz}\right)_{0+} \delta(z). \tag{A.10}$$

In fact, if we multiply both members of Eq. (A.10) by dz, integrate from $-\varepsilon$ to $+\varepsilon$, and let $\varepsilon \to 0$, we obtain a result in agreement with Eq. (A.9).

It is convenient to expand $E(z)$, $J(z)$, and $\delta(z)$ in plane waves:

$$\left\{ \begin{array}{c} E(z) \\ J(z) \\ \delta(z) \end{array} \right\} = \int_{-\infty}^{+\infty} \left\{ \begin{array}{c} E(q) \\ J(q) \\ \delta(q) \end{array} \right\} e^{iqz}\, dq. \tag{A.11}$$

Replacing the above equations into Eq. (A.10), and taking into account Eq. (A.8) gives

$$E(q)\left[-q^2 + \frac{\omega^2}{c^2} + \frac{4\pi i \omega\, \sigma(q, \omega)}{c^2} \right] = \frac{1}{\pi}\left(\frac{dE}{dz}\right)_{0+}.$$

Inserting the above result into the first of Eq. (A.11) it follows

$$E(z) = \frac{1}{\pi}\left(\frac{dE}{dz}\right)_{0+} \int_{-\infty}^{+\infty} \frac{e^{iqz}}{-q^2 + \frac{\omega^2}{c^2} + \frac{4\pi i\omega\sigma(q,\omega)}{c^2}}\, dq. \tag{A.12}$$

Consider first Eq. (A.12) in the case σ does not depend on q. In this case the denominator in Eq. (A.12) is conveniently written as $-q^2 + (\omega^2/c^2)N^2$, where N^2 denotes the quantity $1 + 4\pi i\sigma/\omega$; the integral can be performed with the method of residues and the standard result of Eq. (11.10a) is recovered.

In the extreme anomalous region, the integral in Eq. (A.12) is more complicated; to give an idea of the results, we evaluate Eq. (A.12) in the case $z = 0$ and the displacement current term $(\omega/c)^2$ is negligible. Using expression (11.41b) for $\sigma(q,\omega)$, and the general property $\sigma(q,\omega) = \sigma(-q,\omega)$, one obtains

$$E(0) = \frac{2}{\pi}\left(\frac{dE}{dz}\right)_{0+}\int_0^{+\infty}\frac{1}{-q^2 + i(A/q)}\,dq = \frac{2}{\pi}\left(\frac{dE}{dz}\right)_{0+}\frac{1}{A^{1/3}}\int_0^{+\infty}\frac{x}{-x^3 + i}\,dx \tag{A.13}$$

with

$$A = \frac{4\pi\omega}{c^2}\frac{3\pi\sigma_0}{4v_F\tau} = \frac{3\pi^2\omega\sigma_0}{c^2 v_F\tau}.$$

The integral in Eq. (A.13) can be carried out with the help of the definite integral

$$\int_0^{+\infty}\frac{x^{\mu-1}}{x^\nu + 1}\,dx = \frac{\pi}{\nu}\frac{1}{\sin(\mu\pi/\nu)}, \qquad \mathrm{Re}\,\nu \geq \mathrm{Re}\,\mu > 0$$

[see I. S. Gradshtein and I. M. Ryzhik "Table of Integrals, Series and Products" (Academic Press, New York, 1980), p. 292]. We have

$$\int_0^\infty\frac{x}{-x^3 + i}\,dx = -\int_0^\infty\frac{x^4}{x^6 + 1}\,dx - i\int_0^\infty\frac{x}{x^6 + 1}\,dx = -\frac{\pi\sqrt{3}}{9}(\sqrt{3} + i). \tag{A.14}$$

From the above results and Eq. (11.16b), one obtains for the surface impedance in the anomalous region the expression

$$Z_{an} = \frac{4\pi i\omega}{c^2}\left(\frac{E}{dE/dz}\right)_{0+} = \frac{8}{9}\left(\frac{\sqrt{3}\pi\omega^2 v_F\tau}{c^4\sigma_0}\right)^{1/3}(1 - i\sqrt{3}). \tag{A.15}$$

The most interesting aspect of expression Eq. (A.15) is that it does not depend on the relaxation time; in fact

$$\frac{\sigma_0}{v_F\tau} = \frac{ne^2}{m^* v_F} = \frac{k_F^3}{3\pi^2}\frac{e^2}{\hbar k_F} = \frac{e^2}{3\pi^2\hbar}k_F^2.$$

In the extreme anomalous region, measurements of the surface resistance (real part of the surface impedance) provide an accurate tool for the investigation of the Fermi wavevector.

Problem 4. Electron Thermal Conductivity and the Lorentz Number

Discuss the thermal electron conductivity for a parabolic energy band of free electrons, with elementary kinetic considerations.

It is of interest to give a simplified derivation of the thermal electron conductivity of a free-electron gas, and of the Lorentz number, with elementary kinetic considerations, without resort to the Boltzmann equation.

Consider a metal with free-carriers density n, and average internal energy per unit volume $u[T(x)]$, where for simplicity the temperature changes only in the x-direction; we also assume momentarily that all the carriers move in the x-direction with velocities $\pm v_F$. The energy flux at the point x is

$$J_Q = \frac{n}{2} v_F \frac{1}{n} u[T(x - v_F \tau)] - \frac{n}{2} v_F \frac{1}{n} u[T(x + v_F \tau)] \equiv v_F^2 \tau \frac{du}{dT}\left(-\frac{dT}{dx}\right).$$

If we indicate with $c_V = du/dT$ the electron specific heat per unit volume, we have $k_e = v_F^2 \tau c_V$. For a three-dimensional metal we have to include a factor $1/3$ to average on all possible orientations of v_F. We find the expression

$$\boxed{k_e = \frac{1}{3} c_V v_F^2 \tau} \tag{A.16}$$

for the thermal conductivity per particles of velocity v_F, heat capacity c_V per unit volume and relaxation time τ. Using for the free-electron gas the standard expressions of the specific heat, Fermi energy and static conductivity

$$c_V = \frac{\pi^2}{2} \frac{k_B T}{E_F} n k_B, \quad E_F = \frac{1}{2} m v_F^2, \quad \sigma = \frac{n e^2 \tau}{m}$$

one obtains

$$k_e = \frac{\pi^2}{3} \frac{n\tau}{m} k_B^2 T \implies k_e = \frac{\pi^2}{3}\left(\frac{k_B}{e}\right)^2 \sigma T;$$

in agreement with the Wiedemann-Franz law and Lorentz number.

Further Reading

Abelès, F. (Ed.). (1972). *Optical properties of solids*. Amsterdam: North-Holland.

Abrikosov, A. A. (1988). *Fundamentals of the theory of normal metals*. Amsterdam: North-Holland.

Born, M., & Wolf, E. (1999). *Principles of optics*. Cambridge: Cambridge University Press.

Dubi, Y., & Di Ventra, M. (2011). Heat flow and thermoelectricity in atomic and molecular junctions. *Reviews of Modern Physics, 83*, 131.

Ferry, D. K., Goodnick, S. M., & Bird, J. (2009). *Transport in nanostructures*. Cambridge: Cambridge University Press.

Harrison, W. A., & Webb, M. B. (Eds.). (1960). *The Fermi surface*. New York: Wiley.

Jackson, J. D. (1998). *Classical electrodynamics*. New York: Wiley.

Mahan, G. D. (1998). Good thermoelectrics. In H. Ehrenreich, & F. Spaepen (Eds.), *Solid state physics* (Vol. 51, p. 81). San Diego: Academic Press.

Nolas, G. S., Sharp, J., & Goldsmid, H. J. (2001). *Thermoelectrics*. Berlin: Springer.

Palik, E. D. (Ed.) (1985 and 1991). *Handbook of the optical constants of solids* (Vols. I and II). New York: Academic Press.

Pippard, A. B. (1965). *Dynamics of conduction electrons*. New York: Gordon and Breach.

Singh, D. J. (2001). Theoretical and computational approaches for identifying and optimizing novel thermoelectric materials. In R. K. Willardson, & E. R. Weber (Eds.), *Semiconductors and semimetals* (Vol. 70, p. 125). San Diego: Academic Press.

Smith, R. A. (1978). *Semiconductors*. Cambridge: Cambridge University Press.

Smith, A. C., Janak, J. F., & Adler, R. B. (1967). *Electronic conduction in solids*. New York: McGraw-Hill.

Stern, F. (1963). Elementary optical properties of solids. In H. Ehrenreich, F. Seitz, & D. Turnbull (Eds.), *Solid state physics* (Vol. 15, p. 299). New York: Academic Press.

Stroscio, M. A. (1990). Quantum mechanical corrections to classical transport in submicron/ultra submicron dimensions. In C. T. Wang (Ed.) *Introduction to semiconductor technology*, New York: Wiley.

Wilson, A. H. (1953). *The theory of metals*. Cambridge: Cambridge University Press.

Wooten, F. (1972). *Optical properties of solids*. New York: Academic Press.

12 Optical Properties of Semiconductors and Insulators

Chapter Outline head

In the previous chapter we have considered the role of *intraband electronic transitions* in the optical and transport properties of materials with partially occupied energy bands (metals, or also semiconductors with a sufficiently high number of free carriers); here, we consider *interband electronic transitions* in materials with fully occupied or fully empty energy bands (insulators, and semiconductors apart a possible free carrier contribution). Interband transitions are also of relevance in metals, usually at energies above the intraband contribution.

This chapter begins with a section, rather general in nature, that provides the quantum mechanical expression of the transverse optical constants in homogeneous materials within the linear response formalism (Kubo-Greenwood theory). Then, the study of the direct band-to-band optical transitions by means of first-order perturbation theory is developed within the dipole approximation, and the role of the critical points in the joint density-of-states is analyzed. In the next two sections, aspects beyond the linear response theory are also discussed, in order to interpret phonon-assisted indirect interband transitions and two-photon absorption.

The remaining sections of this chapter contain a selection of significant applications of the linear response theory analyzed with different approaches, including the Green's function technique and the continued fraction apparatus. We consider first excitonic effects on the optical transitions; excitons are particularly important in large gap insulators, where the dielectric constant is small and the effective masses are large. We describe then discrete states coupled to a continuum and the Fano absorption profiles.

Solid State Physics, Second Edition. http://dx.doi.org/10.1016/B978-0-12-385030-0.00012-8

Finally the optical properties of coupled electron-phonon systems are considered, with focus on the Franck-Condon vibronic model and on typical Jahn-Teller systems.

Transition rates for absorption or emission processes at first and at higher orders of perturbation theory are summarized for convenience in Appendix A. Some useful elaborations of the optical constants in terms of Green's functions are provided in Appendix B.

12.1 Transverse Dielectric Function and Optical Constants in Homogeneous Media

The purpose of this chapter is the study of the response functions and optical constants of crystals in the presence of transverse electromagnetic fields. In the case of longitudinal perturbations, phenomenological and theoretical aspects have been described in detail in Chapter 7 for homogeneous media, in the linear response regime. Then, in Appendix 7.D, we have also noticed formal analogies (and differences) between longitudinal and transverse response functions. In this section, within the standard approximations of homogeneity and linearity, we focus on the optical properties of materials, in particular semiconductors and insulators. The general analytic properties of the transverse response functions related to the physical requirement that the response is causal, linear, and finite are the same as the ones discussed in Chapter 7 for the longitudinal case and need not be repeated here; we simply remind that the real part of the dielectric function is an even function of the frequency, while the imaginary part is an odd function; real and imaginary parts of the dielectric function are linked by the Kramers-Kronig relations.

Consider a general (periodic or aperiodic) electronic system, described by the one-electron Hamiltonian

$$H_0 = \frac{\mathbf{p}^2}{2m} + V(\mathbf{r}), \tag{12.1}$$

where $V(\mathbf{r})$ is the energy potential for an electron in the sample. Let $\psi_n(\mathbf{r})$ and E_n denote the eigenfunctions and eigenvalues of H_0 (the eigenfunctions are normalized to unity in the volume V of the crystal); spin degeneracy is taken into account by including a factor 2, whenever needed. The Hamiltonian of the electron system in interaction with the electromagnetic field described by the vector potential $\mathbf{A}(\mathbf{r}, t)$ in the Coulomb gauge (scalar potential equal to zero and divergence of vector potential equal to zero) becomes

$$H = \frac{1}{2m} \left[\mathbf{p} + \frac{e}{c} \mathbf{A}(\mathbf{r}, t) \right]^2 + V(\mathbf{r}) = H_0 + \frac{e}{mc} \mathbf{A}(\mathbf{r}, t) \cdot \mathbf{p} + \frac{e^2}{2mc^2} \mathbf{A}^2(\mathbf{r}, t), \tag{12.2}$$

where e is the modulus of the electron charge. The last two terms in the right-hand side of Eq. (12.2) represent the quantum mechanical interaction between the radiation and the charge of the electron; for simplicity, any other interaction term, as for instance the interaction of the electromagnetic field with the electron spin, is ignored. In the following we also disregard the term in $\mathbf{A}^2$, whose role in scattering processes has been discussed in Chapter 10.

We consider a *transverse electromagnetic plane wave* of frequency ω, wavevector $\mathbf{q}$ and polarization vector $\widehat{\mathbf{e}}$, described by the vector potential

$$\mathbf{A}(\mathbf{r}, t) = A_0 \widehat{\mathbf{e}} \, e^{i(\mathbf{q}\cdot\mathbf{r}-\omega t)} + \text{c.c.} \qquad \widehat{\mathbf{e}} \perp \mathbf{q},$$

where c.c. indicates the complex conjugate of the previous term; for simplicity, the amplitude A_0 is chosen to be real, and the frequency ω is assumed positive. Under the stated approximations the Hamiltonian H of the electron system in the presence of the electromagnetic field becomes

$$H = H_0 + \frac{eA_0}{mc} e^{i(\mathbf{q}\cdot\mathbf{r}-\omega t)} \, \widehat{\mathbf{e}} \cdot \mathbf{p} + \frac{eA_0}{mc} e^{-i(\mathbf{q}\cdot\mathbf{r}-\omega t)} \, \widehat{\mathbf{e}} \cdot \mathbf{p}. \tag{12.3}$$

The time-dependent terms in the right-hand side of Eq. (12.3) induce transitions among the states of H_0, and can be treated according to standard time-dependent perturbation theory (see Appendix A). The first time-dependent term in Eq. (12.3) gives rise to absorption of radiation, while the second term gives rise to emission. The probability per unit time that an electron initially in the state $|\psi_i\rangle$ is transferred to the final state $|\psi_j\rangle$ (supposed empty) with *absorption of one-photon of energy* $\hbar\omega$ is given by the Fermi golden rule

$$P_{j\leftarrow i} = \frac{2\pi}{\hbar} \left(\frac{eA_0}{mc} \right)^2 |\langle\psi_j|e^{i\mathbf{q}\cdot\mathbf{r}} \, \widehat{\mathbf{e}} \cdot \mathbf{p}|\psi_i\rangle|^2 \delta(E_j - E_i - \hbar\omega). \tag{12.4a}$$

Quite similarly, the transition probability that an electron initially in the state $|\psi_j\rangle$ is transferred to the final state $|\psi_i\rangle$ (supposed empty) with *emission of one-photon of energy* $\hbar\omega$ is

$$P_{i\leftarrow j} = \frac{2\pi}{\hbar} \left(\frac{eA_0}{mc} \right)^2 |\langle\psi_i|e^{-i\mathbf{q}\cdot\mathbf{r}} \, \widehat{\mathbf{e}} \cdot \mathbf{p}|\psi_j\rangle|^2 \delta(E_i - E_j + \hbar\omega). \tag{12.4b}$$

The two processes of emission or absorption satisfy the principle of detailed balance, and are schematically represented in Figure 12.1.

For a system at thermal equilibrium (or very near to it), the electronic states are occupied in agreement to the standard Fermi-Dirac distribution function $f(E)$; the net number of transitions per unit time involving energy $\hbar\omega$ is given by the expression

$$W(\mathbf{q}, \omega) = \frac{2\pi}{\hbar} \left(\frac{eA_0}{mc} \right)^2 2 \sum_{ij} |\langle\psi_j|e^{i\mathbf{q}\cdot\mathbf{r}} \, \widehat{\mathbf{e}}\cdot\mathbf{p}|\psi_i\rangle|^2 \delta(E_j - E_i - \hbar\omega)[f(E_i) - f(E_j)],$$

$$\tag{12.5}$$

where the factor 2 in front of the summation takes into account the spin degeneracy.

Equation (12.5) is the basic tool for the microscopic expression of the various optical response functions. Among them, we can focus for instance on the dielectric function,

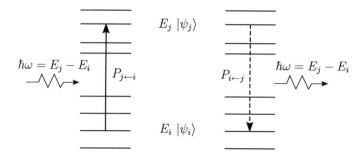

Figure 12.1 Schematic representation of the radiative transition probability between a pair of electronic states $|\psi_i\rangle$ and $|\psi_j\rangle$, of energy E_i and E_j; any transition between these two states involves the absorption, or the emission of a photon with energy $\hbar\omega = E_j - E_i$.

relating the displacement field to the electric field. We have seen (Appendix 7.D) that the link between the imaginary part of the dielectric function and the microscopic transition rate is

$$\varepsilon_2(\mathbf{q}, \omega) = \frac{2\pi\hbar c^2}{\omega^2} \frac{1}{V} \frac{W(\mathbf{q}, \omega)}{A_0^2}. \tag{12.6}$$

Using Eqs. (12.5) and (12.6), the quantum expression of the imaginary part of the transverse dielectric function becomes

$$\varepsilon_2(\mathbf{q}, \omega) = \frac{8\pi^2 e^2}{m^2 \omega^2} \frac{1}{V} \sum_{ij} |\langle\psi_j|e^{i\mathbf{q}\cdot\mathbf{r}} \, \hat{\mathbf{e}}\cdot\mathbf{p}|\psi_i\rangle|^2 \delta(E_j - E_i - \hbar\omega)[f(E_i) - f(E_j)]. \tag{12.7}$$

Notice that $\varepsilon_2(\mathbf{q}, \omega)$ obeys the general relation $\varepsilon_2(\mathbf{q}, -\omega) = -\varepsilon_2(\mathbf{q}, \omega)$. The complete expression (both real and imaginary part) of the complex dielectric function can be written in the form

$$\varepsilon(\mathbf{q}, \omega) = 1 + \frac{8\pi e^2}{m^2} \frac{1}{V} \sum_{ij} \frac{|\langle\psi_j|e^{i\mathbf{q}\cdot\mathbf{r}} \, \hat{\mathbf{e}} \cdot \mathbf{p}|\psi_i\rangle|^2}{(E_j - E_i)^2/\hbar^2} \frac{[f(E_i) - f(E_j)]}{E_j - E_i - \hbar\omega - i\eta}, \tag{12.8}$$

where η is a positive infinitesimal. It may be useful to remind that the real and imaginary parts of an expression of type (12.8) can be explicitly worked out using the Dirac identity discussed in Eq. (7.25) and given by

$$\lim_{\eta\to 0^+} \frac{1}{x - i\eta} = P\frac{1}{x} + i\pi\delta(x), \tag{12.9}$$

where x is real, P denotes the principal part and η is a positive infinitesimal quantity.

Equation (12.8) is the basic expression of the *transverse dielectric function* of an electronic system described by a one-electron Hamiltonian H_0, with wavefunctions $\psi_i(\mathbf{r})$, and eigenvalues E_i (the factor two taking account of the spin degeneracy is already included). It is worthwhile to mention, that appropriate equivalent expressions in terms of the Green's functions of the operator H_0, rather than its eigenfunctions and eigenvalues, can be elaborated (see for details Appendix B).

Other response functions can be used, whenever convenient in specific situations. For instance in the case the momentum transfer can be neglected, it might be convenient to consider the absorption coefficient, which is related the imaginary part of the dielectric function and to the refractive index via the expression

$$\alpha(\omega) = \frac{\omega}{cn(\omega)} \varepsilon_2(\omega).$$

[For simplicity, in the case the momentum transfer $\mathbf{q}$ is vanishing small, optical constants are expressed as a function of the frequency variable only.] Using Eq. (12.6), we can give for the absorption coefficient the following quantum expression

$$\boxed{\alpha(\omega) = \frac{2\pi\hbar c}{n(\omega)\omega} \frac{1}{V} \frac{W(\omega)}{A_0^2}.} \qquad (12.10)$$

The characterization of bulk materials with the absorption coefficient may be particularly convenient in energy regions where the medium is reasonably transparent and the refractive index is independent (or at least smoothly dependent) on frequency; notice that in these situations the structure and the peaks of $\varepsilon_2(\omega)$ and $\alpha(\omega)$ are rather similar.

In the following our attention is mainly devoted to the optical properties of insulators or semiconductors, characterized (at $T = 0$) by *fully occupied or fully empty* one-electron states. In this situation, the explicit presence of the Fermi-Dirac distribution function in the expression of the dielectric function can be dropped; in particular Eq. (12.7) can be cast in the form

$$\varepsilon_2(\mathbf{q}, \omega) = \frac{8\pi^2 e^2}{m^2\omega^2} \frac{1}{V} \sum_i^{(occ)} \sum_j^{(unocc)} |\langle\psi_j|e^{i\mathbf{q}\cdot\mathbf{r}} \,\widehat{\mathbf{e}}\cdot\mathbf{p}|\psi_i\rangle|^2 \delta(E_j - E_i - \hbar\omega). \qquad (12.11)$$

The above relation holds for positive frequencies, while for negative frequencies we resort to the general property $\varepsilon_2(\mathbf{q}, -\omega) = -\varepsilon_2(\mathbf{q}, \omega)$.

Before closing, we provide a useful elaboration of the dielectric function, which applies in the *particular case* that *all optical transitions originate from the ground state* $|\Phi_g\rangle$ (supposed to be non degenerate) of an electronic system. In this situation, the sum over occupied states in Eq. (12.11) is confined to $|\Phi_g\rangle$, and the imaginary part of the dielectric function at positive frequencies (and vanishing momentum transfer) can be written as

$$\varepsilon_2(\omega) = \frac{8\pi^2 e^2}{m^2\omega^2} \frac{1}{V} I(\hbar\omega) \qquad (12.12a)$$

with

$$I(\hbar\omega) = \sum_f |\langle\psi_f|T|\Phi_g\rangle|^2 \delta(E_f - E_g - \hbar\omega), \qquad (12.12b)$$

where $T \equiv \widehat{\mathbf{e}} \cdot \mathbf{p}$, is the dipole operator, and $|\psi_f\rangle$ are the eigenstates of the Hamiltonian H_0 of the system. From Eq. (12.12a), we see that the structure of $\varepsilon_2(\omega)$ is essentially determined by the *lineshape function* $I(\hbar\omega)$, defined by Eq. (12.12b).

A convenient expression of Eqs. (12.12) in terms of the Green's function can be obtained considering the *dipole carrying state* defined as

$$|\chi\rangle = T|\Phi_g\rangle, \quad \text{where} \quad T \equiv \widehat{\mathbf{e}} \cdot \mathbf{p}. \tag{12.13a}$$

The diagonal matrix element of the Green's function $G(E_g + \hbar\omega)$ on the dipole carrying state is

$$G_{\chi\chi}(E_g + \hbar\omega) = \langle\chi|\frac{1}{E_g + \hbar\omega + i\eta - H_0}|\chi\rangle. \tag{12.13b}$$

Inserting the unit operator $1 \equiv \sum |\psi_f\rangle\langle\psi_f|$ in the above expression, and using Eqs. (12.13a) and (12.9), $I(\hbar\omega)$ can be written in the convenient and compact form (the limit $\eta \to 0^+$ is implicitly understood)

$$I(\hbar\omega) = -\frac{1}{\pi}\operatorname{Im} G_{\chi\chi}(E_g + \hbar\omega).$$

When the focus is on the *lineshape* $I(E)$ (where E stands for $\hbar\omega$) and not on its actual location on the energy axis, we can choose $E_g = 0$ as the reference energy, and write

$$\boxed{I(E) = -\frac{1}{\pi}\operatorname{Im} G_{\chi\chi}(E)}. \tag{12.14}$$

Equation (12.14) describes the optical transitions originating from a unique initial ground state, and links the shape of the imaginary part of the dielectric function with the imaginary part of the diagonal matrix element of the Green's function on the dipole carrying state.

12.2 Quantum Theory of Band-to-Band Optical Transitions and Critical Points

In the previous section, we have established the general tools for the investigation of optical properties of linear and homogeneous systems, described by a one-electron Hamiltonian. We apply now the quantum theory of optical transitions to semiconductors and insulators with fully occupied valence bands and fully empty conduction bands. Consider a radiation field, of wavevector $\mathbf{q}$ and frequency ω, and let us examine the matrix elements for the electronic transitions from an initial valence state $\psi_{v\mathbf{k}_i}$, of wavevector $\mathbf{k}_i$ (and given spin) into a final conduction state $\psi_{c\mathbf{k}_j}$ of wavevector $\mathbf{k}_j$ (and same spin):

$$\langle\psi_{c\mathbf{k}_j}|e^{i\mathbf{q}\cdot\mathbf{r}}\,\widehat{\mathbf{e}}\cdot\mathbf{p}|\psi_{v\mathbf{k}_i}\rangle = \int \psi_{c\mathbf{k}_j}^*(\mathbf{r})e^{i\mathbf{q}\cdot\mathbf{r}}\,\widehat{\mathbf{e}}\cdot\mathbf{p}\,\psi_{v\mathbf{k}_i}(\mathbf{r})\,d\mathbf{r}. \tag{12.15a}$$

We notice that the derivative $\mathbf{p}\psi_{v\mathbf{k}_i} = -i\hbar\,\partial\psi_{v\mathbf{k}_i}/\partial\mathbf{r}$ of the Bloch function of vector $\mathbf{k}_i$ produces a Bloch function of wavevector $\mathbf{k}_i$; also notice that multiplication by $\exp(i\mathbf{q}\cdot\mathbf{r})$ of a Bloch function of wavevector $\mathbf{k}_i$ produces a Bloch function of wavevector $\mathbf{q}+\mathbf{k}_i$. Thus the matrix element (12.15a) can be different from zero only if

$$\mathbf{k}_j = \mathbf{q} + \mathbf{k}_i + \mathbf{g}, \tag{12.15b}$$

where $\mathbf{g}$ is a vector of the reciprocal lattice (including the null vector).

In the following, we consider experimental situations (infrared region, visible, up to near and far ultraviolet region), where the *wavelength of the incident radiation is much larger than the lattice parameter*; in these situations, the photon wavevector $\mathbf{q}$ of the incident radiation is small compared to the range of values of $\mathbf{k}$ within the first Brillouin zone; thus we may neglect $\mathbf{q}$ in Eq. (12.15b) (*dipole approximation*), and put $\mathbf{k}_j = \mathbf{k}_i$ (*direct or vertical transitions*). The *optical transitions are thus vertical* in an energy band diagram, as schematized in Figure 12.2.

We now apply Eq. (12.11) to the case of semiconductors and insulators with fully occupied valence bands and fully empty conduction bands; we denote $\varepsilon_2(\mathbf{q}=0,\omega)$ by $\varepsilon_2(\omega)$, and take into account that only vertical transitions are possible; we obtain

$$\varepsilon_2(\omega) = \frac{8\pi^2 e^2}{m^2\omega^2}\frac{1}{V}\sum_{cv}\sum_{\mathbf{k}}|\langle\psi_{c\mathbf{k}}|\widehat{\mathbf{e}}\cdot\mathbf{p}|\psi_{v\mathbf{k}}\rangle|^2\delta(E_{c\mathbf{k}}-E_{v\mathbf{k}}-\hbar\omega)$$

$$= \frac{8\pi^2 e^2}{m^2\omega^2}\sum_{cv}\int_{\text{B.Z.}}\frac{d\mathbf{k}}{(2\pi)^3}|\widehat{\mathbf{e}}\cdot\mathbf{M}_{cv}(\mathbf{k})|^2\delta(E_{c\mathbf{k}}-E_{v\mathbf{k}}-\hbar\omega); \tag{12.16}$$

in the above expression ω is positive, the sum runs over every couple of valence and conduction bands, and $\mathbf{M}_{cv}(\mathbf{k})$ denotes the dipole matrix element $\langle\psi_{c\mathbf{k}}|\mathbf{p}|\psi_{v\mathbf{k}}\rangle$.

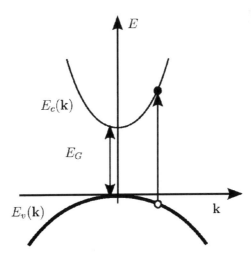

Figure 12.2 Schematic representation of the top valence band and bottom conduction band of a direct gap semiconductor or insulator. Optical transitions are vertical in the energy band diagram. [The valence band has been reported in bold to remind its full occupancy by electrons.]

It is well known that in atoms, molecules, and (small) clusters, the absorption of radiation (at least below the photoionization edge) exhibits sharp lines, which correspond to transitions to discrete excited states, allowed in the dipole approximation. Also in the expression of band-to-band transitions the absorption spectrum can have (reasonably) sharp structures, which however correspond in this case to critical points in the joint density-of-states.

Consider in fact a given couple of valence and conduction bands; in general the dipole matrix element $\widehat{\mathbf{e}} \cdot \mathbf{M}_{cv}(\mathbf{k})$ is a smooth function of $\mathbf{k}$ over the Brillouin zone (unless zero at some symmetry point for group theory considerations), and its average value can be factorized out of Eq. (12.16). Then, the contribution to the optical constants from a couple of valence and conduction bands is determined by the so called *joint density-of-states*

$$J_{cv}(\omega) = \int_{B.Z.} \frac{d\mathbf{k}}{(2\pi)^3} \delta(E_{c\mathbf{k}} - E_{v\mathbf{k}} - \hbar\omega). \tag{12.17a}$$

Critical points in the joint density-of-states are by definition those for which

$$\nabla_{\mathbf{k}}(E_{c\mathbf{k}} - E_{v\mathbf{k}}) \equiv 0. \tag{12.17b}$$

For these points the joint density-of-states exhibits rapid variation versus energy. This can be easily seen integrating analytically Eq. (12.17a) in the neighborhood of a critical point (12.17b).

The type of critical points, the singular behavior in the joint density-of-states $J_{cv}(\omega)$ and hence of the imaginary part $\varepsilon_2(\omega)$ of the dielectric function, the dependence on the dimensionality, are the same as those worked out in Section 2.7 (in the discussion of the band density-of-states), and are not repeated here. We remember that three-dimensional crystals present four types of singularities (minimum, two types of saddle points, and maximum) denoted M_0, M_1 and M_2, M_3, respectively. The singularities become sharper as the dimensionality decreases; in particular the joint density-of-states at the saddle point of a two-dimensional crystal presents a logarithmic-like singularity (for details see Section 2.7).

In order to give an idea of the role of the critical points on the optical constants in semiconductors, we discuss the classic example of the optical properties of germanium. In Figure 12.3, we report the band structure of germanium, together with important critical points for the interband transitions. The first direct transition in Ge is $\Gamma'_{25} \rightarrow \Gamma'_2$, which contributes an M_0 singularity around ≈ 1 eV. In the energy region around ≈ 2 eV, the transition $L'_3 \rightarrow L_1$ contributes an M_0 critical point, and the $\Lambda_3 \rightarrow \Lambda_1$ transitions along the Λ direction contribute an M_1 critical point; other relevant singularities at higher energies are indicated in Figure 12.3. In Figure 12.4, we report the imaginary part of the dielectric constant of Ge, and the overall satisfactory comparison with experimental results. It is apparent that the interband transition picture is of major importance for the interpretation of the optical properties of semiconductors and especially of the wealth of experimental features obtained with modulation spectroscopy. In fact, modulation techniques allow to pick out with high accuracy the discontinuities in the optical structures, and these critical points provide invaluable information on specific electronic transitions between energy bands of the crystal.

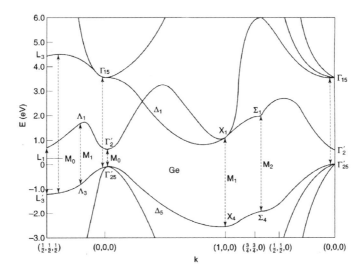

Figure 12.3 Energy bands of germanium along symmetry directions, with the empirical pseu-dopotential method. The **k** wavevector is in units of $2\pi/a$. Relevant direct interband edges are indicated by arrows [reprinted with permission from D. Brust, J. C. Phillips, and F. Bassani, Phys. Rev. Lett. *9*, 94 (1962); copyright 1962 by the American Physical Society].

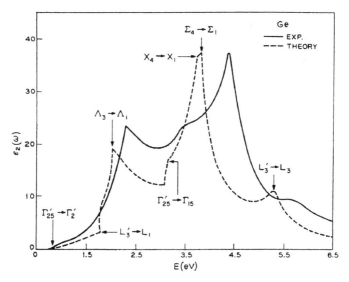

Figure 12.4 Spectral structure of $\varepsilon_2(\omega)$ (solid line) compared with the theoretical results of interband transitions for Ge (dashed line with edges emphasized to account for critical points) [reprinted with permission from D. Brust, J. C. Phillips, and F. Bassani, Phys. Rev. Lett. *9*, 94 (1962) and D. Brust, Phys. Rev. *134*, A1337 (1964); copyright 1964 by the American Physical Society].

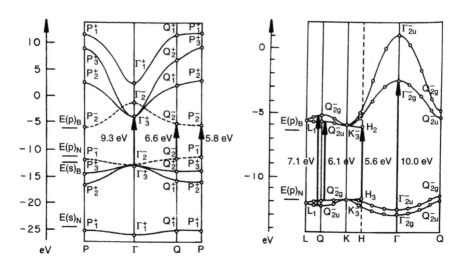

Figure 12.5 Two-dimensional (left) and three-dimensional (right) band structures of hexagonal boron nitride. Relevant optical transitions are also indicated [the calculations are due to E. Doni and G. Pastori Parravicini, Nuovo Cimento *64* B, 117 (1969); interpretation of experimental data is due to D. M. Hoffman, G. L. Doll, and P. C. Eklund, Phys. Rev. B *30*, 6051 (1984)].

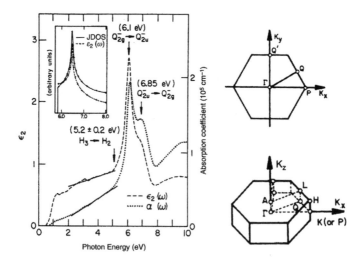

Figure 12.6 Imaginary dielectric constant $\varepsilon_2(\omega)$ and absorption coefficient $\alpha(\omega)$ due to interband electronic transitions in hexagonal boron nitride from experimental measurements [from D. M. Hoffman, G. L. Doll and P. C. Eklund, Phys. Rev. B *30*, 6051 (1984); copyright 1984 by the American Physical Society]. The inset is the joint density-of-states and $\varepsilon_2(\omega)$ calculated in the two-dimensional approximation near the saddle point singularity. The Brillouin zone of two-dimensional and three-dimensional hexagonal boron nitride are also reported for convenience.

We have already remarked that the behavior of the singularities at the critical points depends on the dimensionality; thus for two-dimensional crystals and layered materials we expect sharper structures in the optical properties. Nice examples are offered by graphene, graphite, and hexagonal boron nitride, and we consider here some theoretical and experimental data for the latter. In Figure 12.5, we report the two- and three-dimensional band structure of the hexagonal BN, with indication also of the most important transitions. In Figure 12.6 the experimental imaginary dielectric constant $\varepsilon_2(\omega)$ and the absorption coefficient $\alpha(\omega)$ are reported; it can be noticed the reasonable overall agreement between experiments and theoretical assignment of direct interband transitions.

12.3   Indirect Phonon-Assisted Transitions

We have seen that the absorption of photons in crystals entails vertical transitions in the energy band diagram, because of the small momentum carried by photons (in the ordinary visible and near ultraviolet region). In several semiconductors and insulators, the conduction band and valence band extrema occur at different points of the Brillouin zone (*indirect gap materials*). Non-vertical transitions near the energy gap may occur provided some source supplies the momentum needed for total crystal momentum conservation; phonon gas or impurity centers can accommodate any appropriate momentum and thus indirect optical transitions assisted by phonons or impurities become possible. In the following we treat explicitly only the phonon case.

We illustrate the essential aspects of indirect transitions, considering the simplest model of indirect gap semiconductor (or insulator) with the top of the valence band at the center of the Brillouin zone, and the minimum of the conduction band at the point $\mathbf{q}_0 \neq 0$ of the Brillouin zone; often $\mathbf{q}_0$ is at or near the zone boundary. [In general, a number of equivalent minima occur in the conduction band structure $E_c(\mathbf{k})$ depending on the point group symmetry of the crystal, and the various contributions must be added together.] An indirect band structure is schematically indicated in Figure 12.7.

The highest valence band and the lowest conduction band are supposed to be parabolic near their extrema, with isotropic effective masses. As shown in Figure 12.7, the band gap E_G of the semiconductor is indirect, and is smaller than the first allowed direct energy gap $E_G + \Delta$. Thus when the photon energy $\hbar\omega$ of the incident beam is approximately in the energy range $[E_G, E_G + \Delta]$, optical transitions can occur only through indirect processes.

In order to evaluate some relevant aspects of indirect phonon-assisted transitions, we use second-order perturbation theory considering both the optical perturbation $H_{\text{em}}(\mathbf{r}, t)$ due to the electromagnetic field and the phonon perturbation $H_{\text{ep}}(\mathbf{r}, t)$, due to the lattice vibrations. We have already seen that an electromagnetic field of frequency ω (in the dipole approximation) implies a time-dependent perturbation Hamiltonian of the form

$$H_{\text{em}}(\mathbf{r}, t) = \frac{eA_0}{mc}\,\widehat{\mathbf{e}} \cdot \mathbf{p}\, e^{-i\omega t} + \text{h.c.} , \qquad (12.18\text{a})$$

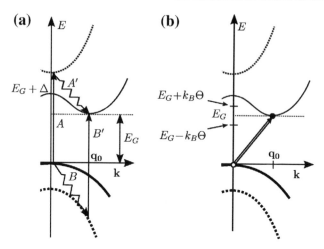

Figure 12.7 Model band structure for estimating the absorption coefficient due to indirect transitions. In (a) we indicate schematically possible second-order processes contributing to the scattering of an electron from the top of the valence band $\psi_{v\mathbf{k}_1}$ ($\mathbf{k}_1 \approx 0$) to the bottom of the conduction band $\psi_{c\mathbf{k}_2}$ ($\mathbf{k}_2 \approx \mathbf{q}_0$). The mechanism AA' consists of a direct transition from $\psi_{v\mathbf{k}_1}$ to a virtual state $\psi_{\alpha\mathbf{k}_1}$ (occupied or empty), followed by a phonon-assisted scattering to $\psi_{c\mathbf{k}_2}$. The mechanism BB' consists of a phonon-assisted scattering from $\psi_{v\mathbf{k}_1}$ to a virtual state $\psi_{\beta\mathbf{k}_2}$ (occupied or empty), followed by a direct transition to $\psi_{c\mathbf{k}_2}$ (energy conservation and Pauli exclusion principle can be ignored in the virtual states, as discussed in the text). In (b) the so evaluated second-order scattering amplitude is indicated with a double line. The energy thresholds for photon absorption, with absorption or emission of a phonon of energy $k_B\Theta$, are also indicated.

where h.c. indicates the hermitian conjugate of the previous term. The perturbation due to the phonon field can be described (in its essential lines) in the form

$$H_{\text{ep}}(\mathbf{r}, t) = \frac{1}{\sqrt{N}} \sum_{\mathbf{q}}^{\text{B.Z.}} V_p(\mathbf{q}, \mathbf{r}) e^{-i\omega_{\mathbf{q}}t} a_{\mathbf{q}} + \text{h.c.}, \tag{12.18b}$$

where N is the number of unit cells of the crystal, $V_p(\mathbf{q}, \mathbf{r})$ denotes a perturbation function of wavevector $\mathbf{q}$, while $\omega_{\mathbf{q}}$ is the frequency of a normal mode at the indicated wavevector (to avoid inessential details we consider a unique phonon branch) and $a_{\mathbf{q}}$ is the annihilation operator for a phonon. For what concerns the operator $V_p(\mathbf{q}, \mathbf{r})$ we do not need to be very specific; in fact we simply exploit the property that it may scatter electrons from an initial wavevector to a final wavevector differing by $\mathbf{q}$. Whenever needed, the explicit expression of $V_p(\mathbf{q}, \mathbf{r})$ can be inferred (at least in principle) considering the change in the crystal potential when the nuclei are displaced from their equilibrium position according to the phonon mode.

With the use of optical and phonon perturbations (12.18), it is now straightforward to apply *second-order perturbation theory* (along the lines described in Appendix A). In the presence of both photon and phonon perturbation fields, we can consider second-order processes in which a photon and a phonon are both absorbed, or both emitted, or one is absorbed and the other emitted. We consider first the processes in which a photon and a phonon are both absorbed (and thus we neglect the hermitian conjugate partners in Eqs. (12.18), because they would give rise to photon and phonon emission).

According to second-order perturbation theory, the probability per unit time that an electron, initially in the valence state $|\psi_{v\mathbf{k}_1}\rangle$ is scattered into the conduction state $|\psi_{c\mathbf{k}_2}\rangle$ *with absorption of a photon and a phonon* becomes

$$P_{c\mathbf{k}_2 \leftarrow v\mathbf{k}_1} = \frac{2\pi}{\hbar} |t_{c\mathbf{k}_2 \leftarrow v\mathbf{k}_1}|^2 \delta(E_{c\mathbf{k}_2} - E_{v\mathbf{k}_1} - \hbar\omega - \hbar\omega_{\mathbf{q}}); \tag{12.19a}$$

the matrix element for the indirect scattering process reads

$$t_{c\mathbf{k}_2 \leftarrow v\mathbf{k}_1} = \frac{eA_0}{mc} \sqrt{n_{\mathbf{q}}} \frac{1}{\sqrt{N}} \left[\sum_{\alpha} \frac{\langle \psi_{c\mathbf{k}_2} | V_p(\mathbf{q}, \mathbf{r}) | \psi_{\alpha\mathbf{k}_1} \rangle \langle \psi_{\alpha\mathbf{k}_1} | \hat{\mathbf{e}} \cdot \mathbf{p} | \psi_{v\mathbf{k}_1} \rangle}{E_v(\mathbf{k}_1) - E_\alpha(\mathbf{k}_1) + \hbar\omega} \right.$$

$$\left. + \sum_{\beta} \frac{\langle \psi_{c\mathbf{k}_2} | \hat{\mathbf{e}} \cdot \mathbf{p} | \psi_{\beta\mathbf{k}_2} \rangle \langle \psi_{\beta\mathbf{k}_2} | V_p(\mathbf{q}, \mathbf{r}) | \psi_{v\mathbf{k}_1} \rangle}{E_v(\mathbf{k}_1) - E_\beta(\mathbf{k}_2) + \hbar\omega_{\mathbf{q}}} \right], \tag{12.19b}$$

where

$$n_{\mathbf{q}} = \frac{1}{\exp(\hbar\omega_{\mathbf{q}}/k_B T) - 1} \tag{12.19c}$$

is the Bose population factor, and matrix elements involving phonon annihilation processes are proportional to $\sqrt{n_{\mathbf{q}}}$. The first sum in square brackets in Eq. (12.19b) includes all the channels in which H_{em} induces a transition from the initial band state $\psi_{v\mathbf{k}_1}$ to any crystal band states $\psi_{\alpha\mathbf{k}_1}$, regardless if occupied or empty (the states $\psi_{\alpha\mathbf{k}_1}$ are often referred to as "intermediate" or "virtual" states); the phonon perturbation completes the transition by taking the electron from $\psi_{\alpha\mathbf{k}_1}$ to the final state $\psi_{c\mathbf{k}_2}$. Alternatively, the second sum in square brackets considers all the channels, in which the first step is performed by the phonon field and is then completed by the optical perturbation. A schematic representation of the second-order perturbation channels is indicated in Figure 12.7a. [Notice that the Pauli exclusion principle in the intermediate states is not operative, essentially because the operators H_{em} and H_{ep} commute. A quite different situation will be encountered in the Kondo effect, illustrated in Section 16.5.2.].

The total number $W(\omega)$ of transitions per unit time involving a photon of energy $\hbar\omega$ is obtained summing Eq. (12.19a) over all initial states in the valence band and all final states in the conduction band. The matrix elements are commonly assumed to be practically independent of the $\mathbf{k}$ states involved. From Eq. (12.10) we also see that, in a sufficiently small energy window near the threshold of the absorption processes, the absorption coefficient is practically proportional to $W(\omega)$. The absorption coefficient for phonon absorption is thus essentially controlled by the Bose population factor and by the indirect density-of-states namely:

$$\alpha_{(\text{phonon abs})}(\omega) \div n_{\mathbf{q}} J_{(\text{phonon abs})}(\omega), \tag{12.20a}$$

where the density-of-states for phonon absorption in indirect transitions is defined as

$$J_{(\text{phonon abs})}(\omega) = \int_{\text{B.Z.}} \int_{\text{B.Z.}} \frac{1}{(2\pi)^3 (2\pi)^3} d\mathbf{k}_1 \, d\mathbf{k}_2 \, \delta(E_{c\mathbf{k}_2} - E_{v\mathbf{k}_1} - \hbar\omega - k_B\Theta) \, ;$$

$$(12.20b)$$

it is also assumed $\hbar\omega_\mathbf{q} \approx \hbar\omega_{\mathbf{q}_0} \equiv k_B\Theta$ for indirect transitions at or near the threshold.

We now estimate the joint density-of-states for indirect transitions assuming for simplicity parabolic behavior for the dispersion curves of the conduction band and of the valence band in proximity of their extrema (as indicated in Figure 12.7); then

$$J_{(\text{phonon abs})}(\omega) = \frac{(4\pi)^2}{(2\pi)^6} \int_0^\infty \int_0^\infty k_1^2 k_2^2 \, \delta(\frac{\hbar^2 k_2^2}{2m_c^*} + \frac{\hbar^2 k_1^2}{2m_v^*} + E_G - \hbar\omega - k_B\Theta) \, dk_1 \, dk_2 .$$

To perform the integral, we make the substitutions

$$\frac{\hbar^2 k_2^2}{2m_c^*} = x, \quad \frac{\hbar^2 k_1^2}{2m_v^*} = y, \quad k_2 \, dk_2 = \frac{m_c^*}{\hbar^2} \, dx, \quad k_1 \, dk_1 = \frac{m_v^*}{\hbar^2} \, dy$$

and obtain

$$J_{(\text{phonon abs})}(\omega) = \frac{(m_c^* m_v^*)^{3/2}}{2\pi^4 \hbar^6} \int_0^\infty \int_0^\infty \sqrt{x}\sqrt{y} \, \delta(x + y - b) \, dx \, dy ,$$

where $b = \hbar\omega - E_G + k_B\Theta$. The δ-function under the sign of integral can be different from zero only for positive values of b. Performing the integral over dy one gets

$$J_{(\text{phonon abs})}(\omega) = \frac{(m_c^* m_v^*)^{3/2}}{2\pi^4 \hbar^6} \int_0^b \sqrt{x}\sqrt{b - x} \, dx .$$

$$(12.21a)$$

The definite integral in Eq. (12.21a) equals $\pi b^2/8$, as can be easily verified using the following indefinite integral

$$\int \sqrt{bx - x^2} \, dx = \frac{2x - b}{4} \sqrt{bx - x^2} + \frac{b^2}{8} \arcsin \frac{2x - b}{b} \quad (b > 0).$$

The joint density-of-states for indirect transitions thus becomes

$$J_{(\text{phonon abs})}(\omega) = \frac{(m_v^* m_c^*)^{3/2}}{16\pi^3 \hbar^6} (\hbar\omega - E_G + k_B\Theta)^2 \quad \text{for} \quad \hbar\omega > E_G - k_B\Theta. \quad (12.21b)$$

A similar treatment holds for the phonon emission, with the replacement of $+k_B\Theta$ (phonon absorption) by $-k_B\Theta$ (phonon emission).

Together with the joint density-of-states for indirect transitions with phonon absorption or emission, we also take into account that the transition probabilities involving one-phonon annihilation or creation are proportional to

$$n_q = \frac{1}{\exp(\Theta/T) - 1} \quad \text{and} \quad n_q + 1 = \frac{1}{1 - \exp(-\Theta/T)},$$

respectively. We thus obtain for the absorption coefficient for indirect transitions the following behavior

$$\alpha(\omega) = C \left[\frac{(\hbar\omega - E_G + k_B\Theta)^2}{e^{\Theta/T} - 1} + \frac{(\hbar\omega - E_G - k_B\Theta)^2}{1 - e^{-\Theta/T}} \right], \quad (12.22)$$

where $C(>0)$ is approximately a constant independent of energy and temperature; it is also understood that the first term in square bracket is present only if $\hbar\omega > E_G - k_B\Theta$, and the second term is present only if $\hbar\omega > E_G + k_B\Theta$.

A few comments on Eq. (12.22) are worthwhile. The absorption coefficient for indirect transitions is in general much smaller that the one for direct transitions. The shape of the absorption spectrum for direct transitions between two parabolic bands goes as the square root of energy (M_0 critical point) due to the conservation of the initial and final electronic momentum; for indirect transitions the shape for the absorption spectrum is quadratic in energy, due to the non-conservation of the initial and final electron momentum, whose difference can be taken from, or given to, the phonons of the crystal. In the direct case, the absorption spectrum is independent (or poorly dependent) on temperature and the threshold is the direct energy gap (assuming excitonic effects negligible); the phonon-assisted transitions are strongly temperature dependent because of the phonon population factors and any allowed phonon branch gives rise to two thresholds, at energies equal to the indirect energy gap increased or decreased by the quantum of phonon energy.

The first term in brackets in Eq. (12.22) refers to photon absorption *with annihilation of one phonon* of energy $k_B\Theta$, and the threshold is just $\hbar\omega = E_G - k_B\Theta$. The second term in brackets refers to photon absorption with *creation of one phonon* of energy $k_B\Theta$; the energy threshold for this process is $\hbar\omega = E_G + k_B\Theta$. Thus when $\alpha^{1/2}(\omega)$ is plotted against $\hbar\omega$, a straight line is expected for $E_G - k_B\Theta < \hbar\omega < E_G + k_B\Theta$; and (approximately) a steeper straight line is expected for $\hbar\omega > E_G + k_B\Theta$. In Figure 12.8 we indicate the behavior of the square root of the absorption coefficients with the photon energy at various temperatures for silicon and germanium. With $\Theta = 600$ K for silicon and $\Theta = 260$ K for germanium, a reasonable agreement with the experimental results is obtained using the law (12.22) for indirect transitions [for further aspects see also G. G. Macfarlane, T. P. McLean, V. Roberts, and J. E. Quarrington, Phys. Rev. *108*, 1377 (1957); Phys. Rev. *111*, 1245 (1958)].

Before concluding, we mention the effect of dimensionality on the indirect transitions; the joint density-of-states (12.20b) presents a linear dependence on the photon energy for two-dimensional bands and steps for one-dimensional bands; this behavior should be observable in indirect transitions in quantum well structures and quantum wires, respectively.

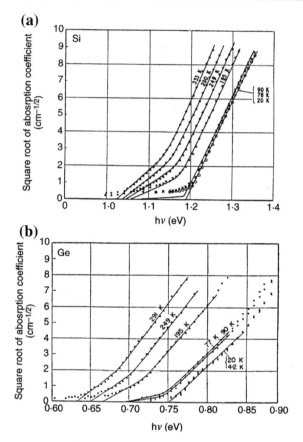

Figure 12.8 (a) Indirect transitions in silicon; dots indicate experimental measurements; the full lines are calculated using Eq. (12.22) of the text. (b) Indirect transitions in germanium [reprinted with permission from G. G. Macfarlane and V. Roberts, Phys. Rev. *98*, 1865 (1955); Phys. Rev. *97*, 1714 (1955); copyright 1955 by the American Physical Society].

12.4 Two-Photon Absorption

The advent of very intense radiation sources, and the development of sophisticated spectroscopic techniques, have made it possible to study experimentally processes in which two (or more) quanta are simultaneously absorbed to promote an electron transition. From a theoretical point of view the possibility of two photon absorption was investigated by Göppert-Mayer in 1931, about 30 years before the development of laser light made experiments possible.

We consider the optical constants of a solid in the presence of two radiation beams, of frequency ω_1 and ω_2, with N_1 and N_2 photons per unit volume. In order to avoid one-photon absorption of either beam, the frequencies ω_1 and ω_2 are chosen in such a way that $\hbar\omega_1 < E_G$ and $\hbar\omega_2 < E_G$, but $\hbar\omega_1 + \hbar\omega_2 > E_G$. Thus two-photon spectroscopy is usually adopted to investigate excited states in the approximate energy range $E_G < E < 2E_G$ (with one-photon spectroscopy one can explore in principle the whole energy range). Despite this restriction, two-photon spectroscopy is an invaluable complementary tool to the ordinary one-photon spectroscopy. The symmetry of the accessible final states, and a variety of selection rules, can provide detailed information on the electronic band structure around the fundamental energy gap. Furthermore, the relatively low value of the absorption coefficient makes easier to detect bulk effects, with a decreased importance of surface contamination.

With the advent of synchrotron radiation sources, it has also been possible to extend two-photon spectroscopy to explore edges, originated by transitions from core states to the conduction band; for this purpose one can use radiation with frequency somewhat lower than the one corresponding to core-band conduction-band threshold (where the material is reasonably transparent) and another beam (with frequency lower than the energy gap). In this way detailed information on transitions from core states in semiconductors and insulators have been obtained.

The theoretical treatment of two-photon absorption requires the application of second-order perturbation theory (along the lines presented in Appendix A). We write the vector potentials of the two incident beams in the form

$$\mathbf{A}_1(\mathbf{r}, t) = A_{01} \, \widehat{\mathbf{e}}_1 \, e^{i(\mathbf{q}_1 \cdot \mathbf{r} - \omega_1 t)} + \text{c.c.}$$

$$\mathbf{A}_2(\mathbf{r}, t) = A_{02} \, \widehat{\mathbf{e}}_2 \, e^{i(\mathbf{q}_2 \cdot \mathbf{r} - \omega_2 t)} + \text{c.c.}$$

The probability per unit time of a transition from the state $\psi_{v\mathbf{k}}$ (and given spin direction) to the state $\psi_{c\mathbf{k}}$ (and same spin orientation) considering any intermediate state $\psi_{\beta\mathbf{k}}$ (with the same spin orientation) is

$$
P_{c\mathbf{k} \leftarrow v\mathbf{k}} = \frac{2\pi}{\hbar} \left(\frac{e A_{01}}{mc} \right)^2 \left(\frac{e A_{02}}{mc} \right)^2
$$

$$
\cdot \left| \sum_\beta (1 + P_{12}) \frac{\widehat{\mathbf{e}}_2 \cdot \mathbf{M}_{c\beta}(\mathbf{k}) \, \widehat{\mathbf{e}}_1 \cdot \mathbf{M}_{\beta v}(\mathbf{k})}{E_v(\mathbf{k}) - E_\beta(\mathbf{k}) + \hbar\omega_1} \right|^2 \delta(E_{c\mathbf{k}} - E_{v\mathbf{k}} - \hbar\omega_1 - \hbar\omega_2),
$$

$$(12.23)$$

where P_{12} is the operator which exchanges $\widehat{\mathbf{e}}_1$ and $\hbar\omega_1$ with $\widehat{\mathbf{e}}_2$ and $\hbar\omega_2$. Notice that the sum over intermediate states in Eq. (12.23) extends over all band states of the crystal, and energy conservation and Pauli exclusion principle are not operative on the intermediate states. A schematic representation of two-photon absorption processes is provided in Figure 12.9.

The total number $W(\omega_1; \omega_2)$ of transitions per unit time involving the simultaneous absorption of a photon ω_1 and a photon ω_2 is obtained summing Eq. (12.23) over all $\mathbf{k}$ vectors of the first Brillouin zone, and inserting a factor 2 to account for spin degeneracy. The absorption coefficient is related to $W(\omega_1; \omega_2)$ by Eq. (12.10); thus the absorption coefficient for photons of energy $\hbar\omega_1$, in the presence of N_2 photons per

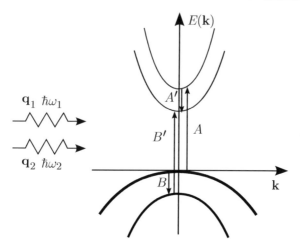

Figure 12.9 Model band structure for estimating the absorption coefficient due to two-photon absorption. We indicate schematically by AA' and BB' possible second order processes contributing to the transition of an electron from the top of the valence band to the bottom of the conduction band (energy conservation and Pauli exclusion principle in the intermediate states are not operative, similarly to the discussion of Figure 12.7).

unit volume of energy $\hbar\omega_2$, becomes

$$\alpha(\omega_1; \omega_2) = C \int_{\text{B.Z.}} \frac{d\mathbf{k}}{(2\pi)^3} \left| \sum_{\beta} (1 + P_{12}) \frac{\widehat{\mathbf{e}}_2 \cdot \mathbf{M}_{c\beta}(\mathbf{k})\widehat{\mathbf{e}}_1 \cdot \mathbf{M}_{\beta v}(\mathbf{k})}{E_v(\mathbf{k}) - E_\beta(\mathbf{k}) + \hbar\omega_1} \right|^2$$
$$\times \delta(E_{c\mathbf{k}} - E_{v\mathbf{k}} - \hbar\omega_1 - \hbar\omega_2), \tag{12.24}$$

where the quantity C (approximately constant or at least smoothly dependent on energy in the spectral region of interest) is given by

$$C = \frac{8\pi^2 e^2}{m^2 c n(\omega_1)\omega_1} \left(\frac{eA_{02}}{mc} \right)^2 = \frac{16\pi^3 \hbar e^4 N_2}{m^4 c n(\omega_1) n^2(\omega_2)\omega_1\omega_2};$$

in the last passage, for obtaining A_{02}, use has been made of the relation

$$\frac{n^2(\omega_2) A_{02}^2 \omega_2^2}{2\pi c^2} = N_2 \hbar\omega_2$$

that connects the classical energy density of the electromagnetic field with the number of photons per unit volume.

An analysis of the matrix elements appearing in Eq. (12.24) is particularly rich and instructive from the point of view of group theory. We notice that *dipole selection rules* for two-photon absorption and for one-photon absorption are in general different; for instance in crystals with inversion symmetry, accessible final states have the same parity as the initial state in two-photon spectroscopy and opposite parity in one-photon spectroscopy. Once a final state of a given symmetry is dipole allowed in two-photon transitions, the actual matrix elements depend on the direction of $\widehat{\mathbf{e}}_1$ and $\widehat{\mathbf{e}}_2$ relative to each other or relative to the crystallographic axes (*geometrical selection rules*). Furthermore, the actual matrix elements also depend on the frequencies ω_1 and ω_2, and may vanish when ω_1 is equal (or near) to ω_2 (*dynamical selection rules*); this is a kind of quantum effect related to the operator P_{12} which exchanges photon 1 with photon 2. [For further aspects see for instance E. Doni, R. Girlanda, and G. Pastori Parravicini, Phys. Stat. Sol. (b) *65*, 203 (1974); (b) *88*, 773 (1978), and references quoted therein.]

12.5   Exciton Effects on the Optical Properties

In Section 12.2, we have studied the optical properties of crystals within the one-electron approximation for the electronic states. Exciton states are beyond the independent particle approximation, and we wish now to study their effects on the optical properties of crystals. The theory of excitons in semiconductors and insulators has been considered in Section 7.1; in the simplest picture, excitons arise as a consequence of the interaction between the electron promoted in the conduction band and the hole left behind in the valence band. For sake of simplicity, we consider here a two-band model semiconductor, as schematized in Figure 12.2. As a preliminary step, we outline the optical properties of this model semiconductor in the independent particle approximation; then we consider the many-body exciton corrections following the treatment of R. J. Elliott, Phys. Rev. *108*, 1384 (1957).

Isotropic Two-Band Model Semiconductor and Band-to-Band Transitions

Consider an *isotropic two-band model semiconductor* (as schematized in Figure 12.2), with valence and conduction band extrema at the same point (for instance at the center of the Brillouin zone) and parabolic dispersion energy curves. We have

$$E_c(\mathbf{k}) = E_G + \frac{\hbar^2 k^2}{2m_c^*}; \qquad E_v(\mathbf{k}) = -\frac{\hbar^2 k^2}{2m_v^*} \qquad m_c^* > 0, \; m_v^* > 0$$

$$E_c(\mathbf{k}) - E_v(\mathbf{k}) = E_G + \frac{\hbar^2 k^2}{2\mu}; \qquad \frac{1}{\mu} = \frac{1}{m_c^*} + \frac{1}{m_v^*}.$$

The dielectric function (12.16), in the specific case of the isotropic two-band model, becomes

$$\varepsilon_2(\omega) = \frac{8\pi^2 e^2}{m^2 \omega^2} \int_{\text{B.Z.}} \frac{d\mathbf{k}}{(2\pi)^3} |\widehat{\mathbf{e}} \cdot \mathbf{M}_{cv}(\mathbf{k})|^2 \delta(E_G + \frac{\hbar^2 k^2}{2\mu} - \hbar\omega). \qquad (12.25)$$

We consider first the case of *dipole-allowed* interband transitions between the valence and the conduction band extrema (*first-class transitions*). When the dipole matrix element $\widehat{\mathbf{e}} \cdot \mathbf{M}_{cv}(\mathbf{k})$ is different from zero at the point $\mathbf{k} = 0$, it can be taken approximately as constant, say C_1, in the whereabouts; Eq. (12.25) becomes

$$\varepsilon_2(\omega) = \frac{8\pi^2 e^2}{m^2 \omega^2} |C_1|^2 \int_0^\infty \frac{1}{(2\pi)^3} 4\pi k^2 \, \delta(E_G + \frac{\hbar^2 k^2}{2\mu} - \hbar\omega) \, dk.$$

The integral can be performed easily with the substitution $x = \hbar^2 k^2 / 2\mu$ and one obtains

$$
\text{first-class transitions:} \qquad \varepsilon_2(\omega) = \frac{A_1}{\omega^2}(\hbar\omega - E_G)^{1/2} \qquad \hbar\omega > E_G \,, \qquad (12.26)
$$

where the constant is $A_1 = (2e^2/m^2)|C_1|^2(2\mu/\hbar^2)^{3/2}$. Near the threshold of allowed direct transitions, the singular part of ε_2 is of the type $(\hbar\omega - E_G)^{1/2}$.

We consider now the case of *dipole-forbidden* transitions between the valence and the conduction band extrema (*second-class transitions*). When the dipole matrix element $\mathbf{M}_{cv}(\mathbf{k})$ vanishes at $\mathbf{k} = 0$, it can be taken approximately linear in $\mathbf{k}$ in the form $\mathbf{M}_{cv}(\mathbf{k}) \approx C_2\mathbf{k}$. With straightforward elaboration of Eq. (12.25) one gets

$$
\text{second-class transitions:} \qquad \varepsilon_2(\omega) = \frac{A_2}{\omega^2}(\hbar\omega - E_G)^{3/2} \qquad \hbar\omega > E_G \,, \qquad (12.27)
$$

where the constant is $A_2 = (2e^2/3m^2)|C_2|^2(2\mu/\hbar^2)^{5/2}$. Near the threshold of forbidden direct transitions, the singular part of ε_2 is of the type $(\hbar\omega - E_G)^{3/2}$.

Exciton Effects in Isotropic Two-Band Model Semiconductor

The theory of exciton states in semiconductors and insulators has been considered in Section 7.1; the theory requires (often demanding) solutions of appropriate coupled integral equations. In several crystals, especially in semiconductors with small effective masses and high dielectric constant, the description of excitons can be (reasonably) mimicked in terms of the hydrogen atom model of effective reduced mass μ in a polarizable medium of dielectric constant ε. The envelope function $F(\mathbf{r})$, that describes the relative motion of the electron-hole pair, satisfies the Schrödinger equation

$$
\left[-\frac{\hbar^2\nabla^2}{2\mu} - \frac{e^2}{\varepsilon r} \right] F(\mathbf{r}) = (E - E_G)F(\mathbf{r}). \qquad (12.28)
$$

The effective Rydberg energy R_{ex} and the effective radius a_{ex} of the hydrogenic atom (12.28) are given by the scaling laws

$$
R_{\text{ex}} = R\frac{\mu}{m}\frac{1}{\varepsilon^2} \qquad \text{and} \qquad a_{\text{ex}} = a_B\frac{m}{\mu}\varepsilon,
$$

where $R = 13.606$ eV is the Rydberg energy and $a_B = 0.529$ Å is the Bohr radius.

As discussed in Section 7.1, the singlet exciton states with vanishing wavevector, $k_{\text{ex}} \to 0$ (the very ones of interest in optical transitions), can be expressed as a linear combination of trial excited states $\Psi_{c\mathbf{k},v\mathbf{k}}^{(S=0)}$ (of total spin $S = 0$ and total wavevector equal to zero) in the form

$$
\Psi_{\text{ex}} = \sum_{\mathbf{k}} A(\mathbf{k})\Psi_{c\mathbf{k},v\mathbf{k}}^{(S=0)} \,; \qquad (12.29a)
$$

the coefficients $A(\mathbf{k})$ are related to the normalized envelope function $F(\mathbf{r})$ by the Fourier transform

$$F(\mathbf{r}) = \frac{1}{\sqrt{V}} \sum_{\mathbf{k}} A(\mathbf{k}) e^{i\mathbf{k}\cdot\mathbf{r}}, \tag{12.29b}$$

where V is the volume of the system, and $F(\mathbf{r})$ is normalized to one. The probability per unit time of a transition from the fundamental ground-state Ψ_0 of the semiconductor to the exciton state Ψ_{ex}, induced by an electromagnetic wave of angular frequency ω and polarization unit vector $\hat{\mathbf{e}}$, is given in the dipole approximation by the expression

$$P_{\Psi_{\text{ex}} \leftarrow \Psi_0} = \frac{2\pi}{\hbar} \left(\frac{eA_0}{mc} \right)^2 2 \left| \sum_{\mathbf{k}} A(\mathbf{k}) \langle \psi_{c\mathbf{k}} | \hat{\mathbf{e}} \cdot \mathbf{p} | \psi_{v\mathbf{k}} \rangle \right|^2 \delta(E_{\text{ex}} - E_0 - \hbar\omega) \tag{12.29c}$$

(where a factor 2 takes into account spin degeneracy).

We specify Eq. (12.29c) for *first-class transitions* assuming $\langle \psi_{c\mathbf{k}} | \hat{\mathbf{e}} \cdot \mathbf{p} | \psi_{v\mathbf{k}} \rangle = C_1$ as a constant; using Eq. (12.29b) for $\mathbf{r} = 0$ one obtains

$$P_{\Psi_{\text{ex}} \leftarrow \Psi_0} = \frac{2\pi}{\hbar} \left(\frac{eA_0}{mc} \right)^2 2|C_1|^2 V |F(0)|^2 \delta(E_{\text{ex}} - E_0 - \hbar\omega). \tag{12.30}$$

It is well known that the envelope function $F(\mathbf{r})$, satisfying the hydrogenic Schrödinger equation (12.28), is different from zero at the origin only for spherically symmetric wavefunctions (i.e. for s-states); thus only s-like excitons can be excited in the case of first-class transitions. For bound s-states of principal quantum number $n = 1, 2, 3, \ldots$ we have $F(0) = 1/\sqrt{\pi a_{\text{ex}}^3 n^3}$. The absorption spectrum for energies below the energy gap consists of a series of lines with energies

$$E_n = E_G - \frac{R_{\text{ex}}}{n^2} \quad (n = 1, 2, 3, \ldots) \quad \text{and intensities} \quad I_n \div \frac{1}{\pi a_{\text{ex}}^3 n^3}. \tag{12.31a}$$

For photon energies $\hbar\omega > E_G$, we consider the envelope function $F(E, \mathbf{r})$, with $E = \hbar\omega - E_G$, describing the s-like ionized states of the hydrogenic system (12.28); as an effect of the electron-hole Coulomb interaction, the dielectric function (12.26) of interband transitions is modified by the factor $|F(E, \mathbf{r} = 0)|^2 = \pi x \exp(\pi x)/\sinh(\pi x)$, where $x = \sqrt{R_{\text{ex}}/E}$ [for the properties of the hydrogen wavefunctions for bound and unbound states see for instance N. F. Mott and H. S. W. Massey "Theory of Atomic Collisions" (Clarendon Press, Oxford, 1949)]. The dielectric function including exciton effects thus becomes

$$\varepsilon_2^{(\text{ex})}(\omega) = \varepsilon_2(\omega) \frac{\pi x e^{\pi x}}{\sinh \pi x} \qquad x = \sqrt{\frac{R_{\text{ex}}}{\hbar\omega - E_G}} \qquad \text{for} \qquad \hbar\omega > E_G, \tag{12.31b}$$

where $\varepsilon_2(\omega)$ is given by Eq. (12.26). Thus, exciton effects not only introduce discrete levels at energies below the energy gap, but also modify the optical properties in the continuum in an energy region extending for several R_{ex} above the energy gap.

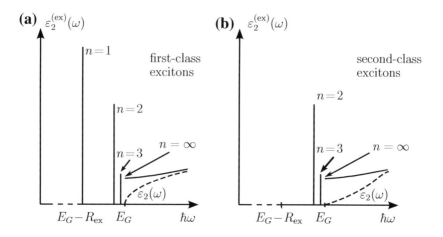

Figure 12.10 Schematic diagram of the imaginary part of the dielectric function for first-class and second-class optical transitions neglecting exciton effects (dashed lines) and including exciton effects (solid lines).

In Figure 12.10a we give a schematic diagram of the imaginary part of the dielectric function for first-class transitions. Experimental data of the absorption spectrum at the fundamental edge of rare gas solids (dipole-allowed insulators) and gallium arsenide (dipole-allowed semiconductor) are reported in Figures 7.1 and 7.2; it can be seen that the relevant experimental features of the excitonic absorption spectrum are well understood qualitatively within the simplified model so far discussed; furthermore, whenever necessary, the model can be appropriately implemented to improve quantitative agreement with experiments.

A similar analysis can be carried out for *second-class transitions*. In this case the dipole matrix element can be taken in the form $\langle \psi_{c\mathbf{k}} | \mathbf{p} | \psi_{v\mathbf{k}} \rangle = C_2 \mathbf{k}$; from Eqs. (12.29b) and (12.29c), one obtains

$$P_{\Psi_{\text{ex}} \leftarrow \Psi_0} = \frac{2\pi}{\hbar} \left(\frac{eA_0}{mc} \right)^2 2 |C_2|^2 V |\hat{\mathbf{e}} \cdot \nabla F(\mathbf{r})_{\mathbf{r}=0}|^2 \delta(E_{\text{ex}} - E_0 - \hbar\omega).$$

Notice that $\nabla F(\mathbf{r})$ can be different from zero at $\mathbf{r} = 0$ only for envelope functions with p-character. The absorption below the energy gap consists of a set of lines with energies

$$E_n = E_G - \frac{R_{\text{ex}}}{n^2} \quad (n = 2, 3, 4, \ldots) \quad \text{and intensities} \quad I_n \div \frac{n^2 - 1}{\pi \, a_{\text{ex}}^5 n^5}. \quad (12.32a)$$

Notice that the $n = 1$ line is missing in this theory and indeed, when experimentally detected, the $n = 1$ line is much weaker than the following members of the series. For energies higher than the energy gap we have

$$\varepsilon_2^{(\text{ex})}(\omega) = \varepsilon_2(\omega) \frac{\pi x (1 + x^2) e^{\pi x}}{\sinh \pi x} \qquad x = \sqrt{\frac{R}{\hbar\omega - E_G}}, \quad (12.32b)$$

where $\hbar\omega > E_G$, and $\varepsilon_2(\omega)$ is given by Eq. (12.27). A schematic diagram of the dielectric function for second-class transitions is reported in Figure 12.10b.

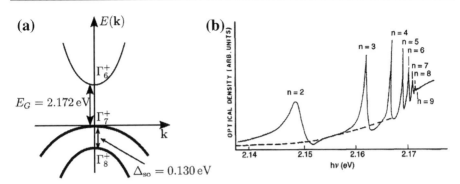

Figure 12.11 (a) Schematic band structure of Cu_2O showing the conduction band, and the valence bands, split by spin-orbit interaction. (b) Absorption spectrum of the yellow exciton series in Cu_2O at 1.8 K [from K. Shindo, T. Goto and T. Anzai, J. Phys. Soc. Japan *36*, 753 (1974)]. Since the band-to-band transition is dipole forbidden at the symmetry point $\mathbf{k} = 0$, the exciton series begins with the line $n = 2$.

As an example of second-class exciton transitions we consider here the case of excitons at the fundamental edge of Cu_2O crystals (see Figure 12.11). Cuprous oxide is a direct gap material with the minimum of the conduction band (Γ_6^+) and the maxima of the valence bands (Γ_7^+ and Γ_8^+) at the point $\mathbf{k} = 0$ of the Brillouin zone; the energy gap is about $E_G = 2.172$ eV; the splitting $\Delta_{SO} = 0.130$ eV between the valence states Γ_7^+ and Γ_8^+ is due to the spin-orbit interaction. Two exciton series have been reported, originating from the Γ_7^+ and Γ_8^+ valence bands, and called the yellow and green exciton series, respectively. Since the conduction and the valence bands have the same parity (the even parity with respect to the spatial inversion is denoted by the superscript +), the dipole matrix element between them vanishes, and the results of second-class transitions apply. Experimental data of the yellow exciton series in cuprous oxide are reported in Figure 12.11b, where transitions to exciton levels from $n = 2$ to $n = 9$ are detected. The asymmetry of the absorption lines is due to their interaction with some background continuum (see the Fano effect in next section).

We conclude these considerations on the optical properties of Cu_2O by comparing the one-photon and two-photon absorption spectra to exciton states. In one-photon spectroscopy of Cu_2O, we have seen that the accessible final states are p-like excitons; hence, following the considerations on the dipole selection rules of Section 12.4, we infer that in two-photon spectroscopy the accessible final states are s-like and d-like excitons, as nicely shown by the experimental data of Figure 12.12.

Considerations on Excitonic-Polaritons and Other Remarks

It is of interest to comment briefly on the behavior of the optical constants of a material near an extremely narrow exciton line, say the $n = 1$ exciton line of a first-class exciton series. The exciton line of interest, at energy $\hbar\omega_{ex} = E_{ex} - E_0$, is supposed to be well separated from all the other partners of the series and from the continuum. From Eq. (12.30), we notice explicitly that *the transition rate to exciton lines is proportional*

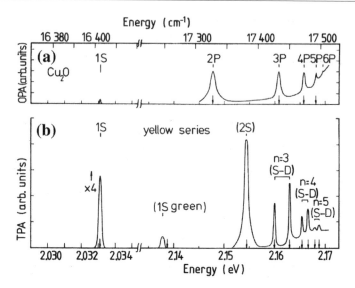

Figure 12.12 (a) One-photon absorption (OPA) and (b) two-photon absorption (TPA) spectra of Cu_2O. One-photon data at 4.2 K are from J. B. Grun and S. Nikitine, J. Phys. (Paris) *23*, 159 (1962). Two-photon data at 4.5 K are from Ch. Uihlein, D. Fröhlich and R. Kenklies, Phys. Rev. B*23*, 2731 (1981) (copyright 1981 by the American Physical Society). The arrows indicate the calculated exciton energies in an appropriate multiband model. (Notice: 1 $cm^{-1} = 0.124$ meV.)

to the volume V of the system because so is the modulus squared of the dipole matrix element [on the contrary, the transition rate to one-electron excited states is independent of the volume of the crystal because so is the dipole matrix element between two Bloch functions; for band-to-band transitions, it is the density-of-states that is proportional to the volume]. Inserting Eq. (12.30) into Eq. (12.6), we see that $\varepsilon_2(\omega)$ for $\omega = \omega_{ex}$ takes the form of a delta-like function of the type

$$\boxed{\varepsilon_2(\omega) = A\delta(\omega_{ex} - \omega)} \qquad A = \frac{8\pi^2 e^2}{m^2 \omega_{ex}^2 \hbar} |C_1|^2 |F(0)|^2, \qquad (12.33a)$$

where A is the *finite oscillator strength* of the discrete exciton line.

Using the Kramers-Kronig dispersion relations in the form of Eqs. (7.22), the real part of the dielectric function becomes

$$\varepsilon_1(\omega) = 1 + \frac{2}{\pi} P \int_0^{\omega_{ex}+\delta} \frac{\omega' \varepsilon_2(\omega')}{\omega'^2 - \omega^2} d\omega' + \frac{2}{\pi} P \int_{\omega_{ex}+\delta}^{+\infty} \frac{\omega' \varepsilon_2(\omega')}{\omega'^2 - \omega^2} d\omega'$$

$$= 1 + \frac{2}{\pi} A \frac{\omega_{ex}}{\omega_{ex}^2 - \omega^2} + \frac{2}{\pi} P \int_{\omega_{ex}+\delta}^{+\infty} \frac{\omega' \varepsilon_2(\omega')}{\omega'^2 - \omega^2} d\omega',$$

where δ is a (small) positive quantity. We suppose that the contribution of the last integral is approximately constant in the frequency region below $\omega \approx \omega_{ex}$; such a contribution, added to 1, is referred to as the background dielectric constant ε_b; in the frequency range extending from zero to about ω_{ex} we have

$$\varepsilon_1(\omega) = \varepsilon_b + \frac{2}{\pi} A \frac{\omega_{ex}}{\omega_{ex}^2 - \omega^2}. \tag{12.33b}$$

We indicate by $\varepsilon_s \equiv \varepsilon_1(0)$ the static dielectric function, and re-write Eq. (12.33b) in the form

$$\varepsilon_1(\omega) = \varepsilon_b + (\varepsilon_s - \varepsilon_b) \frac{\omega_{ex}^2}{\omega_{ex}^2 - \omega^2} = \frac{\varepsilon_s \omega_{ex}^2 - \varepsilon_b \omega^2}{\omega_{ex}^2 - \omega^2}. \tag{12.33c}$$

The dielectric function of Eqs. (12.33) (apart for some obvious change in notations) has the same form as the dielectric function of Eqs. (9.57), which directly led to the concept of polariton states in polar crystals. The analogy, of course, is not only formal but also substantial, and the mixed exciton-photon modes propagating in the crystal are called "exciton-polaritons." The experimental dispersion curves of phonon-polaritons in GaP in the infrared region have been reported in Figure 9.11; the figure can be usefully compared with the experimental dispersion curves of exciton-polaritons in CuCl in the visible region reported in Figure 9.12.

The study of the many-body effects in the optical spectra of solids gives a wealth of experimental and theoretical information. The interpretation of interband spectra is often a challenging problem, which is complicated and enriched by the presence of degenerate bands, spin-orbit interaction, anisotropy effects, autoionization effects with some background continuum. The exciton binding energies vary from a few electronvolts in large gap insulators (where the dielectric constant is relatively small and effective masses relatively large) to a few milli-electronvolts in small gap semiconductors. Core excitons in semiconductors and insulators (and in particular comparison of binding energies and oscillator strengths with valence excitons) are also of major interest. We wish also to mention that in metals the many-body exciton effects are responsible of the sharpness of the absorption coefficient from certain core states to the conduction band, at or near the Fermi level.

12.6   Fano Resonances and Absorption Lineshapes

In this section, we study the absorption lineshapes occurring when one or more discrete levels interact with a continuum of states. The problem was originally considered by Fano in 1935 and later revisited by the same author within the Green's function formalism [U. Fano, Nuovo Cimento *12*, 156 (1935); Phys. Rev. *124*, 1866 (1961)]. The most remarkable feature of the absorption spectrum is the occurrence of peculiar "asymmetric" lineshapes and "window" lineshapes, generated by the configuration interaction (mixing) between the discrete and continuum states.

Consider a discrete state $|\Phi_e\rangle$ interacting ("resonant") with a continuum set of states $|\Phi_n\rangle$ (also called "ionization channel of scattering states"). This system can be schematized as in Figure 12.13a, where also the ground state $|\Phi_g\rangle$ from which the transitions take place is indicated. Physical systems that can be reduced to such a schematization are numerous: for instance, $|\Phi_g\rangle$ may represent the ground electronic state of a

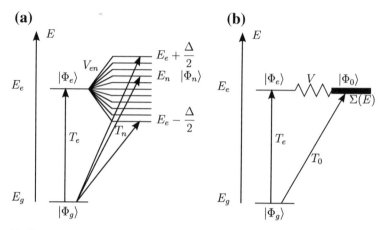

Figure 12.13 (a) Schematic representation of a discrete state $|\Phi_e\rangle$ (of energy E_e) interacting with a continuum of states $|\Phi_n\rangle$ ($n = 1, 2, \ldots, N$) with energies in the interval $[E_e - \Delta/2, E_e + \Delta/2]$; $|\Phi_g\rangle$ is the ground state from which transitions take place; T_e and T_n denote the dipole matrix elements connecting the ground state to the excited states. (b) Schematic representation of the reduced two-level Hamiltonian describing the electronic system; E_e and $\Sigma(E)$ are the site energies of $|\Phi_e\rangle$ and $|\Phi_0\rangle$, respectively; V is the interaction matrix element.

semiconductor (or insulator); $|\Phi_e\rangle$ may represent a discrete member of an exciton Rydberg series involving a core band and the conduction band of lowest energy; the states $|\Phi_n\rangle$ ($n = 1, 2, \ldots, N$ and N very large) may mimic a continuum of excitations from valence states to conduction states at energies far above the fundamental threshold.

The Hamiltonian of a system constituted by a discrete state $|\Phi_e\rangle$ of energy E_e, the continuum of states $|\Phi_n\rangle$ and their interactions, can be represented as follows

$$H = E_e|\Phi_e\rangle\langle\Phi_e| + \sum_{n=1}^{N} E_n|\Phi_n\rangle\langle\Phi_n| + \frac{1}{\sqrt{N}}\sum_{n=1}^{N}[V_{en}|\Phi_e\rangle\langle\Phi_n| + V_{en}^*|\Phi_n\rangle\langle\Phi_e|],$$

(12.34)

where all basis states are normalized to unity. In order to avoid inessential details, we assume for simplicity that the interaction matrix elements $V_{en} = V$ are independent of the index n labeling the states of the continuum. The Hamiltonian (12.34) can then be written as

$$H = E_e|\Phi_e\rangle\langle\Phi_e| + H_{\text{cont}} + V|\Phi_e\rangle\langle\Phi_0| + V^*|\Phi_0\rangle\langle\Phi_e|,$$ (12.35a)

where

$$H_{\text{cont}} = \sum_{n=1}^{N} E_n|\Phi_n\rangle\langle\Phi_n| \quad \text{and} \quad |\Phi_0\rangle = \frac{1}{\sqrt{N}}\sum_{n=1}^{N}|\Phi_n\rangle;$$ (12.35b)

in the above expressions, H_{cont} denotes the Hamiltonian corresponding to the continuum of states $|\Phi_n\rangle$ ($n = 1, 2, 3, \ldots$), and $|\Phi_0\rangle$ is a linear combination (normalized to unity) of them.

The basic obstacle in dealing with the Hamiltonian (12.35) is the arbitrary large number of states of the continuum to be handled. A possible way for a proper account of the continuum exploits the Green's function $G(E) = (E - H_{\text{cont}})^{-1}$ and specifically its diagonal matrix element

$$G_{00}(E) = \langle \Phi_0 | \frac{1}{E - H_{\text{cont}}} | \Phi_0 \rangle = \frac{1}{E - \Sigma(E)}; \qquad (12.36)$$

as usual, it is understood that the energy E is accompanied by a small imaginary part $i\eta$ and the limit $\eta \to 0^+$ is taken. The explicit calculation of the self-energy $\Sigma(E)$ can be done (for instance) with the recursion method, constructing the semi-infinite chain of states beginning with the seed state $|\Phi_0\rangle$ and applying the operator H_{cont}; also notice that in the case the dipole matrix elements T_n from the ground state to the continuum are independent of the index n, all the hierarchical states of the semi-infinite chain (but the first one) are dipole forbidden (as a consequence of their orthogonalization to any other state of the chain and to $|\Phi_0\rangle$ in particular). The Hamiltonian (12.35) can thus be mapped into the effective equivalent Hamiltonian with two states

$$H_{\text{eff}} = E_e |\Phi_e\rangle\langle\Phi_e| + \Sigma(E)|\Phi_0\rangle\langle\Phi_0| + V|\Phi_e\rangle\langle\Phi_0| + V^*|\Phi_0\rangle\langle\Phi_e|; \qquad (12.37)$$

the reduced two-level Hamitonian H_{eff}, which mimics the excited states of the electron system, is indicated schematically in Figure 12.13b.

On the basis $(|\Phi_e\rangle, |\Phi_0\rangle)$, the Hamiltonian H_{eff} and the operator $E - H_{\text{eff}}$ can be written in the matrix form

$$H_{\text{eff}} = \begin{bmatrix} 0 & V \\ V^* & \Sigma(E) \end{bmatrix}, \qquad E - H_{\text{eff}} = \begin{bmatrix} E & -V \\ -V^* & E - \Sigma(E) \end{bmatrix};$$

(for convenience, the energy E_e of the resonant state is assumed, without loss of generality, as the zero reference energy). From a straightforward inversion of the above two-by-two matrix, one obtains the Green's function

$$G_{\text{eff}}(E) = \frac{1}{E - H_{\text{eff}}} = \frac{1}{E(E - \Sigma) - VV^*} \begin{bmatrix} E - \Sigma & V \\ V^* & E \end{bmatrix}. \qquad (12.38)$$

Consider the dipole carrying state

$$|\chi\rangle = T_e|\Phi_e\rangle + T_0|\Phi_0\rangle;$$

the diagonal matrix element of the Green's function (12.38) on the dipole carrying state reads

$$\langle\chi|G_{\text{eff}}(E)|\chi\rangle = \frac{|T_e|^2(E - \Sigma) + T_e^*T_0V + T_eT_0^*V^* + |T_0|^2 E}{E(E - \Sigma) - VV^*}.$$

Multiplying numerator and denominator of the above expression by the complex conjugate of the denominator and keeping the imaginary part, one obtains

$$\text{Im}\,\langle\chi|G_{\text{eff}}(E)|\chi\rangle = \frac{|T_e|^2|V|^2 + (T_e^*T_0V + T_eT_0^*V^*)E + |T_0|^2 E^2}{|E(E - \Sigma) - VV^*|^2}\,\text{Im}\,\Sigma(E).$$

Use of Eq. (12.14) provides for the lineshape the expression

$$I(E) = \frac{|T_0 E + T_e V^*|^2}{|-E\Sigma(E) + E^2 - VV^*|^2}\left(-\frac{1}{\pi}\right)\text{Im}\,\Sigma(E).$$
(12.39)

For an overall discussion of Eq. (12.39), we do not need to be very specific about $\Sigma(E)$ and $G_{00}(E)$ in Eq. (12.36). We simply observe that in the case the continuum of states extends smoothly in an energy interval of the order of Δ, around the resonant state energy $E_e = 0$, we can take for simplicity

$$\Sigma(E) = -i\frac{\Delta}{\pi}.$$
(12.40)

In fact, with this approximation, the density-of-states of the continuum becomes

$$n(E) = -\frac{1}{\pi}\text{Im}\,G_{00}(E) = -\frac{1}{\pi}\text{Im}\,\frac{1}{E + i\Delta/\pi} = \frac{1}{\pi}\frac{\Delta/\pi}{E^2 + (\Delta/\pi)^2};$$

the shape looks like a Lorentzian curve of parameter Δ/π.

We can give a significant interpretation of Eq. (12.39) using Eq. (12.40); assuming $|E| \ll |V|$ Eq. (12.39) becomes

$$I(E) = \frac{|T_0 E + T_e V^*|^2}{E^2\Delta^2 + \pi^2|V|^4}\Delta.$$
(12.41)

It is convenient to define the *broadening parameter* Γ (and express the photon energy in units of $\Gamma/2$) and *the profile index q* as follows

$$\Gamma = 2\pi\frac{|V|^2}{\Delta}, \quad \varepsilon = \frac{E}{\Gamma/2}, \quad q = \frac{1}{\pi}\frac{T_e}{T_0}\frac{\Delta}{V}.$$

Equation (12.41) becomes

$$I(E) = |T_0|^2\frac{1}{\Delta}\frac{|\varepsilon + q|^2}{\varepsilon^2 + 1}.$$
(12.42)

In most ordinary situations the Hamiltonian of the (spinless) electronic system is invariant under time-reversal symmetry, wavefunctions can be taken as real and so also the profile index q is real. The family of curves giving the natural lineshape $I(E)$ for different real values of the index profile q is given in Figure 12.14. It is apparent that in the limit $q \to \infty$ the above equation describes a symmetric Lorentzian lineshape, whereas in the opposite limit $q \to 0$ it describes a symmetric antiresonance line.

In order to clarify better the meaning of Eq. (12.42), consider the limiting case in which only the discrete state $|\Phi_e\rangle$ is dipole connected to the ground state (i.e. $T_e \neq 0$ and $T_0 \equiv 0$). In this case the profile index $q \to \infty$ and, in this limit, Eq. (12.42) gives

$$I(E) = \frac{1}{\Delta}\frac{|qT_0|^2}{\varepsilon^2 + 1} = |T_e|^2\frac{1}{\pi}\frac{\Gamma/2}{E^2 + (\Gamma/2)^2}.$$
(12.43a)

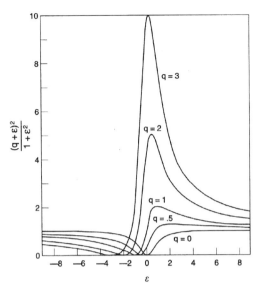

Figure 12.14 Natural lineshapes for different real values of q (reverse the scale of abscissas for negative q); asymmetry and antiresonances caused by interference between the discrete state and the background are shown [reprinted with permission from U. Fano, Phys. Rev. *124*, 1866 (1961); copyright 1961 by the American Physical Society].

Equation (12.43a) is a symmetric Lorentzian curve, with broadening parameter Γ. Consider now the opposite limiting case in which only the seed state $|\Phi_0\rangle$ is dipole connected to the ground state (i.e. $T_e \equiv 0$ and $T_0 \neq 0$). In this case the index profile q vanishes and Eq. (12.42) becomes

$$I(E) = |T_0|^2 \frac{1}{\Delta} \frac{\varepsilon^2}{\varepsilon^2 + 1} = |T_0|^2 \frac{1}{\Delta}\left(1 - \frac{1}{\varepsilon^2 + 1}\right). \qquad (12.43b)$$

The above equation shows that $I(E)$ vanishes for $\varepsilon = 0$, i.e. at the resonance energy of the discrete state; the dipole-forbidden discrete state makes its presence felt because it carves a "window" of width Γ in the background absorption $|T_0|^2/\Delta$ (*spectral repulsion principle*).

In general the dipole transitions are allowed both to the discrete and to the continuum states, and the profile index q is finite. Equation (12.42) for real q, can be written as

$$I(E) = |T_0|^2 \frac{1}{\Delta}\left[1 + \frac{q^2}{\varepsilon^2 + 1} + \frac{2q\varepsilon - 1}{\varepsilon^2 + 1}\right]. \qquad (12.43c)$$

The three terms in the right-hand side of the above equation are the background absorption, a symmetric Lorentzian peak, and a term that represents the effect of interference; note in particular that for $\varepsilon = -q$ no transition takes place (*antiresonance*).

As an illustration of Fano profiles, we report in Figures 12.15 and 12.16 the absorption spectra of neon and argon rare gas atoms and solids in the soft X-ray region;

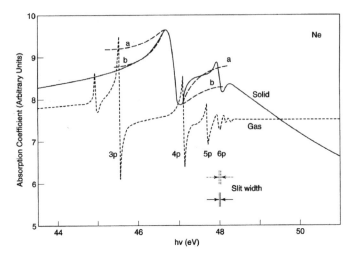

Figure 12.15 Absorption spectrum of solid Ne (solid curve) and gaseous Ne (dashed curve). The theoretical curves a and b are fitted on the absorption cross-section of Fano, with two different sets of parameters (see Table 12.1) [reprinted with permission from R. Haensel, G. Keitel, C. Kunz, and P. Schreiber, Phys. Rev. Lett. *25*, 208 (1970); copyright 1970 by the American Physical Society].

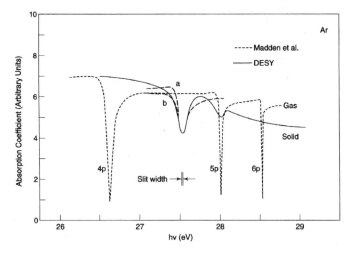

Figure 12.16 Absorption spectrum of solid Ar (solid curve) and gaseous Ar (dashed curve). The theoretical curves a and b are fitted on the absorption cross-section of Fano, with two different sets of parameters (see Table 12.1) [reprinted with permission from R. Haensel, G. Keitel, C. Kunz, and P. Schreiber, Phys. Rev. Lett. *25*, 208 (1970) (copyright 1970 by the American Physical Society); the absorption spectrum of gaseous Ar is given by R. P. Madden, D. L. Ederer, and K. Codling, Phys. Rev. *177*, 136 (1969)].

Table 12.1 Lineshape parameters of the X-ray absorption resonance in neon and argon (E_0 is the resonance energy, Γ the broadening parameter, q the lineshape parameter) [reported from R. Haensel, G. Keitel, C. Kunz, P. Schreiber, Phys. Rev. Lett. *25*, 208 (1970) (copyright 1970 by the American Physical Society)].

	E_0 (eV)	Γ (eV)	q
Ne gas	45.55	0.013	-1.6
Ne solid (fit a)	46.90	0.34	-0.76
Ne solid (fit b)	46.81	0.36	-1.31
Ar gas	26.614	0.08	-0.22
Ar solid (fit a)	27.515	0.12	-0.39
Ar solid (fit b)	27.525	0.13	-0.20

in Table 12.1 we report the lineshape parameters, that best fit the absorption cross-section of the Fano theory. Notice the asymmetric lines occurring in the absorption spectrum of neon (where q is relatively large), and the window lines occurring in the absorption spectrum of argon (where q is relatively small).

The model considered so far for a single resonant state interacting with one ionization channel of scattering states can be generalized to many resonances and complicated systems. The Fano concepts of quantum interferences have permeated countless fields of atomic physics and condensed matter, including areas of investigations of photon devices, nanoscale structures, and ultracold atom lattices. [For a review of Fano resonance in various physical systems, including an historical background, see for instance Miroshnichenko et al. (2010) and references quoted therein.]

12.7 Optical Properties of Vibronic Systems

Thus far, in the study of the optical transitions, we have tacitly assumed that the *nuclear equilibrium positions are the same, before and after the electronic excitations take place* (this is equivalent to assume that the adiabatic potential surfaces of interest have their minima at the same nuclear configuration with the same curvature). Although this is the ordinary situation, we have seen in Chapter 8 examples of vibronic systems of more general nature. When electronic transitions occur between adiabatic potential sheets with minima in different nuclear configurations (or between degenerate adiabatic sheets), the lattice cannot be disregarded and important effects on the optical properties arise. We thus complete this chapter with a survey of the basic principles of the optical constants of vibronic systems.

12.7.1 Optical Properties of the Franck-Condon Vibronic Model

The simplest vibronic model is the Franck-Condon model which is constituted by an *electronic system with only two levels for each lattice configuration and a one-dimensional vibrational mode* for the lattice (as schematized in Figure 8.2b, and reconsidered in further detail in Figure 12.17). Several physical systems can be schematized, at least as a first approach, with such a model. The simplest case is that of a diatomic molecule with two (non-degenerate) electronic states, between which optical transitions can occur. Also in the study of localized impurities in solids, the Franck-Condon model has found wide applications; it is the basis for a simple semiquantitative explanation for the lineshapes for photon absorption and for luminescence (i.e. photon emission after excitation) in the optical properties of localized centers.

Consider an electronic system with two (non-degenerate) levels, the ground state ψ_g and the excited state ψ_e, and let $E_g(q)$ and $E_e(q)$ indicate the energies of the corresponding adiabatic potential surfaces; the normal coordinate q may represent, for instance, the deviation of the nearest neighbor nuclear distance from its equilibrium value. In the ground state, the equilibrium value of the (collective) coordinate q is assumed to be zero; in the excited state, the equilibrium value is assumed to be q_0; upper and lower adiabatic potential surfaces are assumed to have the same curvature C. The schematic representation of the model is indicated in Figure 12.17. With the choice of axes and origin as in Figure 12.17, we have

$$\begin{cases} E_g(q) = \tfrac{1}{2}Cq^2, \\ E_e(q) = E_0 + \tfrac{1}{2}C(q - q_0)^2. \end{cases} \tag{12.44}$$

Notice that E_0 represents the energy difference between the minima of the ground and the excited adiabatic potentials, and is called *"zero-phonon transition energy;"* the elastic constant C is related to the frequency ω of the nuclear motion, and the mass M by the usual equation $C = M\omega^2$.

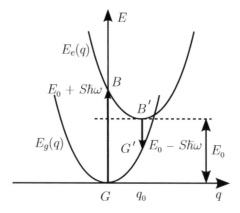

Figure 12.17 Configuration-coordinate diagram for allowed transitions between two non-degenerate electronic states (Franck-Condon model). The adiabatic potential surfaces for the ground state and the lowest excited state are indicated as a function of a single normal coordinate q.

It is useful to characterize the model by means of the dimensionless Huang-Rhys parameter S defined as

$$S\hbar\omega \equiv \frac{1}{2}Cq_0^2 = \frac{1}{2}M\omega^2 q_0^2, \tag{12.45a}$$

or equivalently

$$q_0 = \sqrt{\frac{2\hbar S}{M\omega}}. \tag{12.45b}$$

In the model it is also assumed that $E_0 > S\hbar\omega$; i.e. the energy difference between the two minima of the adiabatic surfaces is larger than the elastic shift energy. We observe that the energy difference between lower and upper levels at $q = 0$ and $q = q_0$ is given by

$$\begin{aligned}
E_e(0) - E_g(0) &= E_0 + S\hbar\omega, \\
E_e(q_0) - E_g(q_0) &= E_0 - S\hbar\omega.
\end{aligned} \tag{12.45c}$$

We now examine the features of the photon absorption spectrum and luminescence, preliminarily from the semiclassical point of view, and then with the quantum mechanical treatment.

In accordance with an intuitive picture, known as "Franck-Condon principle," the nuclear configuration of a system (molecules, clusters, impurities, and crystals) cannot change during the short time of an electron transition; in other terms, optical transitions are *vertical* in the configuration diagram, because the nuclei do not move of any significant amount during the electronic transitions. When the system is initially at the minimum of the ground adiabatic sheet of Figure 12.17, we expect that the *absorption spectrum* is peaked at the energy

$$E_{\text{abs}} = E_0 + S\hbar\omega. \tag{12.46a}$$

Similarly, when the system is initially at the minimum of the excited adiabatic sheet, we expect that the *emission spectrum* is peaked at the energy

$$E_{\text{emiss}} = E_0 - S\hbar\omega. \tag{12.46b}$$

The difference between the peaks of the absorption and emission bands is thus expected to be $2S\hbar\omega$.

We can examine more closely the photon absorption and photon emission processes. Consider the system initially at the minimum of the ground adiabatic sheet (point G in Figure 12.17). During the optical transition from G to B the nuclei remain at rest, leading to an absorption energy $E_0 + S\hbar\omega$. After the absorption, in the excited adiabatic curve, the nuclei are no more in the equilibrium position, and the system moves to the minimum B′ at energy E_0 releasing an energy equal to $S\hbar\omega$; this relaxation process occurs in a time of the order of $\approx 10^{-13}$ s, the typical time of lattice vibrations. We thus see that the dimensionless parameter S takes the meaning of the average number of phonons accompanying the optical transition; $S\hbar\omega$ is the energy transferred into vibrational energy (and then into heat). The system in the excited state B$'$ survives for a lifetime of the order of $\approx 10^{-8}$ s (in typical ordinary situations), which is about 10^5 times the period of the lattice vibrations. The emission from B$'$ to G$'$ takes place vertically leading to emission energy $E_0 - S\hbar\omega$; the system then relaxes from G$'$ to G releasing an energy $S\hbar\omega$ (transformed into heat). Thus luminescence band and absorption band are shifted by $2S\hbar\omega$, which is the energy transformed into heat in the cycle G $\rightarrow$ B $\rightarrow$ B$'$ $\rightarrow$ G$'$ $\rightarrow$ G.

Finally, we note that in the particular case that the Huang-Rhys parameter S vanishes (or equivalently $q_0 \equiv 0$), the optical absorption and the optical emission consist of a unique sharp line at the resonance energy E_0, and the nuclear lattice can be disregarded. We pass now to study quantitatively the lineshapes predicted within the Franck-Condon model when S is different from zero (or equivalently $q_0 \neq 0$).

Quantum Treatment of the Optical Properties of the Franck-Condon Vibronic Model

We consider the Franck-Condon model from a quantum mechanical point of view, taking into account the motion of the nuclei; the energy due to this motion is quantized in phonon quanta, which are in general much smaller than the transition electronic energies. For the quantum treatment we exploit phonon creation and annihilation operators, $a^\dagger$ and a, corresponding to the harmonic oscillator of potential energy $(1/2)M\omega^2 q^2$, centered at the origin $q = 0$ of the configurational coordinate q (the *oscillator at the origin* is also referred as *undisplaced oscillator*).

The eigenstates and eigenvalues of the vibronic system in the ground adiabatic surface are

$$|\psi_g, \phi_n\rangle \quad \text{and} \quad E_{gn} = E_g(0) + \left(n + \frac{1}{2}\right)\hbar\omega \qquad n = 0, 1, 2, \ldots, \qquad (12.47a)$$

where ψ_g is the ground state of the electronic system and ϕ_n are the eigenfunctions of the undisplaced harmonic oscillator. Similarly, the eigenstates and eigenvalues of the vibronic system in the excited adiabatic surface are

$$|\psi_e, \tilde{\phi}_n\rangle \quad \text{and} \quad E_{en} = E_0 + \left(n + \frac{1}{2}\right)\hbar\omega \qquad n = 0, 1, 2, \ldots, \qquad (12.47b)$$

where ψ_e is the excited state of the electronic system and $\tilde{\phi}_n$ are the eigenfunctions of the displaced harmonic oscillator.

We remind that the overlap between displaced and undisplaced harmonic oscillator wavefunctions can be calculated analytically; in particular we have

$$\langle \tilde{\phi}_n | \phi_0 \rangle = (-1)^n \sqrt{\frac{S^n}{n!}}\, e^{-S/2}, \qquad (12.47c)$$

where use has been made of Eq. (9.A25).

The absorption lineshape at zero temperature (i.e. when the system is initially in the ground electronic state and ground vibrational state) is obtained via the golden rule

$$I_{abs}(E) = \frac{2\pi}{\hbar} \left| \langle \psi_e, \tilde{\phi}_n | T | \psi_g, \phi_0 \rangle \right|^2 \delta(E_{en} - E_{g0} - E)$$

$$= \frac{2\pi}{\hbar} |T_{eg}|^2 \left| \langle \tilde{\phi}_n | \phi_0 \rangle \right|^2 \delta(E_0 + n\hbar\omega - E), \tag{12.48}$$

where T_{eg} is the dipole matrix element between the ground electronic state and the excited electronic state. Using Eq. (12.47c), the lineshape (12.48) (normalized to one) takes the form

$$\boxed{I_{abs}(E) = \frac{1}{n!} S^n e^{-S} \delta(E_0 + n\hbar\omega - E) \quad (n = 0, 1, 2, \dots).} \tag{12.49}$$

The absorption spectrum $I_{abs}(E)$ is a palisade of equally spaced lines (at energies E_0, $E_0 + \hbar\omega$, $E_0 + 2\hbar\omega \cdots$) with intensities varying according to the Poisson distribution, as schematically shown in Figure 12.18a. The factor $\exp(-S)$ gives the fractional intensity of the *zero-phonon line*, and is the optical analog of the Debye-Waller factor. The maximum of the Poisson distribution occurs for $n \approx S$, and the absorption spectrum is then peaked at energy $E_0 + S\hbar\omega$, in agreement with Eq. (12.46a).

A quite similar analysis can be carried out for the luminescence spectrum occurring when the system is initially in the excited electronic state and ground vibrational state. The emission spectrum is given by

$$\boxed{I_{emiss}(E) = \frac{1}{n!} S^n e^{-S} \delta(E_0 - n\hbar\omega - E) \quad (n = 0, 1, 2, \dots).} \tag{12.50}$$

The emission spectum $I_{emiss}(E)$ is a palisade of equally spaced lines at energies E_0, $E_0 - \hbar\omega$, $E_0 - 2\hbar\omega$ (see Figure 12.18b), and it is peaked at the energy $E_0 - S\hbar\omega$, in agreement with Eq. (12.46b).

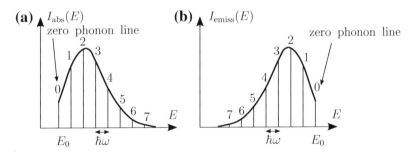

Figure 12.18 (a) Palisade spectrum of optical absorption for the two-level single-mode model (the system is initially in the ground electronic state and ground vibrational state); the zero-phonon line is indicated by an arrow. (b) Palisade spectrum of optical emission for the two-level single-mode model (the system is initially in the excited electronic state and ground vibrational state); the zero-phonon line is indicated by an arrow. In the figures we have chosen $S = 2.5$.

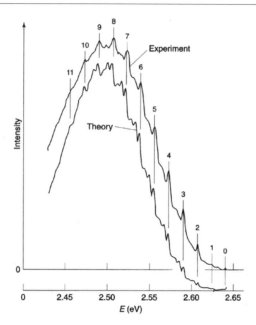

Figure 12.19 Comparison of the theoretical and experimental luminescence intensity of AgBr at 2 K containing residual iodine atoms. [Experimental results are taken from W. Czaja, J. Phys. C *16*, 3197 (1983); theoretical results and interpretation are taken from A. Testa, W. Czaja, A. Quattropani, and P. Schwendimann, J. Phys. C *20*, 1253 (1987)].

As an illustration of the above considerations, we report in Figure 12.19 the phonon-assisted luminescence of iodine impurities in AgBr. The experiments are interpreted in the framework of a conventional configuration coordinate model, in which both optical and acoustic phonons are coupled to the excited electronic states.

Solution of the Franck-Condon Model by Continued Fractions

The optical absorption spectrum (or also the emission spectrum) of the Franck-Condon model can be described by means of continued fractions. Consider the Hamiltonian $\widetilde{H}$ for the displaced oscillator in the excited adiabatic sheet

$$\widetilde{H} = \frac{p^2}{2M} + \frac{1}{2}C(q - q_0)^2 = \hbar\omega\left(\widetilde{a}^\dagger\widetilde{a} + \frac{1}{2}\right),\tag{12.51a}$$

where the tilted operators refer to the displaced harmonic oscillator. Such a Hamiltonian can be expressed in terms of annihilation and creation operators of the undisplaced harmonic oscillator. Since $\widetilde{a}^\dagger = a^\dagger - \sqrt{S}$ (see Eq. (9.A23)) we have

$$\widetilde{H} = \hbar\omega a^\dagger a - \sqrt{S}\hbar\omega(a^\dagger + a) + \left(S + \frac{1}{2}\right)\hbar\omega.\tag{12.51b}$$

We now represent $\widetilde{H}$ on the basis of the states $\{\phi_n\}$ of the oscillator at the origin:

$$\widetilde{H} = \sum_{m=0}^{\infty} \sum_{n=0}^{\infty} |\phi_m\rangle \langle \phi_m | \widetilde{H} | \phi_n \rangle \langle \phi_n | = \left(S + \frac{1}{2} \right) \hbar\omega + \sum_{n=0}^{\infty} n\hbar\omega |\phi_n\rangle \langle \phi_n |$$

$$+ \sqrt{S} \hbar\omega \sum_{n=0}^{\infty} \sqrt{n+1} \left[|\phi_n\rangle \langle \phi_{n+1}| + |\phi_{n+1}\rangle \langle \phi_n| \right]. \qquad (12.51c)$$

The above Hamiltonian has the graphical representation given in Figure 12.20.

The Hamiltonian $\widetilde{H}$ is already in tridiagonal form (see Section 1.4.2); thus the diagonal matrix element of the Green's function $G(E) = (E - \widetilde{H})^{-1}$ on the state ϕ_0 is given by

$$G_{00}(E) = \langle \phi_0 | \frac{1}{E - \widetilde{H}} | \phi_0 \rangle = \cfrac{1}{E - a_0 - \cfrac{b_1^2}{E - a_1 - \cfrac{b_2^2}{E - a_2 - \cdots}}} \qquad (12.52a)$$

with

$$a_n = \left(S + \frac{1}{2} \right) \hbar\omega + n\hbar\omega \quad (n = 0, 1, 2, \dots) \text{ and } b_n^2 = nS(\hbar\omega)^2 \quad (n = 1, 2, 3, \dots).$$
$$(12.52b)$$

The ground state $|\psi_g, \phi_0\rangle$ of the Franck-Condon model is dipole connected only to the state $|\psi_e, \phi_0\rangle$, which is thus the dipole carrying state; according to Eq. (12.14) the absorption spectrum is then given by $-(1/\pi)\text{Im} \, G_{00}(E + i\eta)$ with η positive infinitesimal quantity.

The Green's function $G_{00}(E)$ can be determined with any desired accuracy considering a sufficient number of steps of the continued fraction (12.52); the absorption spectrum can then be worked out and the results of Figure 12.18 recovered. The method of continued fractions (an optional in the case of the Franck-Condon model) becomes particularly convenient in the case of generalized vibronic models, because of its flexibility in the elaboration of the Green's function on the dipole carrying states of interest.

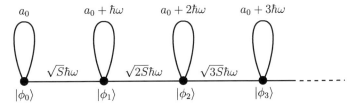

Figure 12.20 Graphic representation of the Hamiltonian of the displaced oscillator on the basis of the states $\{\phi_n\}$ of the oscillator at the origin; $a_0 = (S + 1/2)\hbar\omega$.

12.7.2   Optical Properties of Typical Jahn-Teller Systems

We consider now optical transitions involving degenerate vibronic systems and examine the consequences of the Jahn-Teller effect on the absorption spectrum. Several typical Jahn-Teller systems have been discussed in Section 8.3; here, we focus on the simplest Jahn-Teller system, the $E \otimes \varepsilon$ vibronic system, because of its own interest and because it gives a feeling of how to treat more complicated situations. We consider thus a system, consisting of a non-degenerate ground-state adiabatic sheet and an orbital doublet E of excited states, interacting with a two-dimensional vibrational mode of symmetry ε. The adiabatic potential surfaces are indicated schematically in Figure 12.21.

We have already seen in Eq. (8.24) that the full Hamiltonian of the vibronic $E \otimes \varepsilon$ system is

$$H = -\frac{\hbar^2}{2M}\left(\frac{\partial^2}{\partial q_1^2} + \frac{\partial^2}{\partial q_2^2}\right) + \gamma\begin{pmatrix} -q_1 & q_2 \\ q_2 & q_1 \end{pmatrix} + \frac{1}{2}M\omega^2(q_1^2 + q_2^2), \quad (12.53)$$

where γ is the linear coupling constant, ω the frequency of the vibrational mode, q_1 and q_2 are the vibrational normal coordinates.

Let us indicate the vibrational states by $|lm\rangle$ (the integer numbers l and m denote phonon occupation numbers), and the degenerate electronic states by $|\psi_1\rangle$ and $|\psi_2\rangle$. In the basis $|\psi_i, lm\rangle$ ($i = 1, 2; l, m = 0, 1, 2, \ldots$) the Hamiltonian (12.53) can be conveniently written as

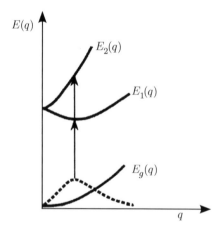

Figure 12.21 Adiabatic surfaces $E_1(q)$ and $E_2(q)$ of the vibronic $E \otimes \varepsilon$ system as a function of the radial coordinate $q = (q_1^2 + q_2^2)^{1/2}$; the non-degenerate ground adiabatic surface $E_g(q)$, from which transition begins, is also indicated. The radial density distribution for a two-dimensional oscillator is shown by a dotted line. The origin of the two principal maxima in the semiclassical band shape is indicated by two solid arrows.

$$H = H_e + H_L + H_{eL}, \tag{12.54}$$

$$H_e + H_L = \sum_{lm} \big[E_e + (l+m+1)\hbar\omega\big]\big[|\psi_1, lm\rangle\langle\psi_1, lm| + |\psi_2, lm\rangle\langle\psi_2, lm|\big]$$

$$H_{eL} = k_E\hbar\omega \sum_{lm} \big[-|\psi_1, lm\rangle\langle\psi_1, lm| + |\psi_2, lm\rangle\langle\psi_2, lm|\big](a_1 + a_1^\dagger)$$

$$+ k_E\hbar\omega \sum_{lm} \big[|\psi_1, lm\rangle\langle\psi_1, lm| + |\psi_2, lm\rangle\langle\psi_2, lm|\big](a_2 + a_2^\dagger),$$

with $k_E = \gamma/[(2M\hbar)^{1/2}\omega^{3/2}]$.

The dynamical problem can be solved for instance by applying the recursion method (see Section 5.8.1). Starting with the seed state $|f_0\rangle = |\psi_1, 00\rangle$ (or with the partner state $|\psi_2, 00\rangle$), we can easily verify by inspection that the parameters of the recursion method can be obtained *analytically*; they are

$$a_n = (n+1)\hbar\omega \ (n = 0, 1, 2, \ldots), \quad b_n^2 = 2\,\mathrm{Int}\left(\frac{n+1}{2}\right)(k_E\hbar\omega)^2 \ (n = 1, 2, 3, \ldots),$$

where $\mathrm{Int}(x)$ indicates the integer part of x. We define as usual the Huang-Rhys factor S in the form $S = k_E^2$ (or equivalently $S\hbar\omega = E_{JT}$ where E_{JT} is the Jahn-Teller energy

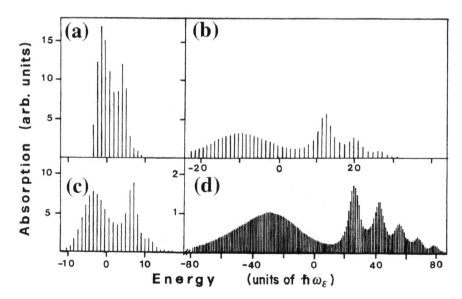

Figure 12.22 Absorption spectrum for a transition $A \to E$ for a vibronic system $E \otimes \varepsilon$ (A denotes the non-degenerate electronic ground-state and E denotes the doubly-degenerate excited electronic state). The values of the dimensionless Huang-Rhys parameter S are 4, 16, 100, 900 in (a)–(d) [reprinted with permission from L. Martinelli, M. Passaro, G. Pastori Parravicini, Phys. Rev. B *43*, 8395 (1991); copyright 1991 by the American Physical Society.]

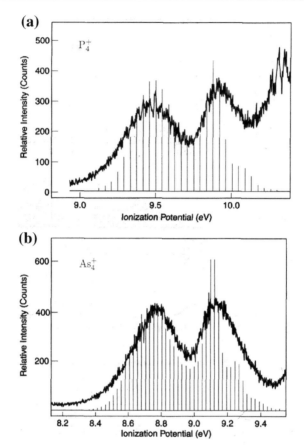

Figure 12.23 Experimental photoelectron spectrum of P_4^+ (a) and As_4^+ (b), and best fit with a linear $E \otimes \varepsilon$ Jahn-Teller model [reprinted with permission from Lai-Sheng Wang, B. Niu, Y. T. Lee, D. A. Shirley, E. Ghelichkhani, and E. R. Grant, J. Chem. Phys. *93*, 6318 (1990); copyright 1990 by the American Physical Society].

of the vibronic system); we obtain the Green's function

$$G_{00}(E) = \cfrac{1}{E - \hbar\omega - \cfrac{2S(\hbar\omega)^2}{E - 2\hbar\omega - \cfrac{2S(\hbar\omega)^2}{E - 3\hbar\omega - \cfrac{4S(\hbar\omega)^2}{E - 4\hbar\omega - \cdots}}}}. \qquad (12.55)$$

The absorption spectrum is given by $(-1/\pi)\mathrm{Im}\,G_{00}(E+i\eta)$ with $\eta \to 0^+$ in agreement with Eq. (12.14).

The lineshapes can be easily obtained numerically from the Green's function (12.55) and are shown in Figure 12.22 for several values of the dimensionless parameter S.

It can be seen that the absorption spectrum consists of two main bands, in qualitative agreement with the semiclassical considerations of Figure 12.21; the oscillations in energy appearing for large S are due to the resonances between the vibrational levels of the two adiabatic surfaces, and are known as Slonzewski resonances. Many other Jahn-Teller systems have been studied in terms of appropriate continued fractions. Also the modified Lanczos method for excited states (see Section 5.8.2) has been very fruitful to properly evaluate absorption lineshapes [see for instance G. Bevilacqua, L. Martinelli, and G. Pastori Parravicini, Phys. Rev. B *54*, 7626 (1996) and references quoted therein].

As an illustration of optical transitions involving $E \otimes \varepsilon$ Jahn-Teller systems, we report in Figure 12.23 high resolution photoelectron spectra of tetramers of group V elements. The tetramers of P and As have tetrahedral symmetry with a (non-degenerate) total symmetric ground-state; the lowest excited state of the tetrameric cations is an electronic orbital doublet, subject to Jahn-Teller distortion. In Figure 12.23, the experimental photoelectron spectra are reasonably well understood on the basis of the $E \otimes \varepsilon$ Jahn-Teller model, at least for what concerns the features of the two main bands.

Appendix A. Transitions Rates at First and Higher Orders of Perturbation Theory

In the study of optical transitions in solids we have used the time-dependent perturbation theory. It may be useful to recall here in a self-contained way some relevant aspects of the procedure, because we need transition rates at first and at higher orders of perturbation. In the literature one can find a variety of different approaches (that may appear sometimes rather formal, especially when involving two-body operators); thus some general guidelines may be useful.

General Expression for the Evolution of Quantum States

Consider a system described by a static Hamiltonian H_0, in the presence of a time-dependent perturbation operator $V(t)$. The states of the system evolve in time according to the time-dependent Schrödinger equation

$$i\hbar \frac{\partial \psi(t)}{\partial t} = [H_0 + V(t)]\psi(t). \tag{A.1}$$

In order to obtain the transition amplitudes, induced by the perturbation, between the eigenstates of the unperturbed Hamiltonian H_0, it is convenient to perform a unitary transformation (controlled by H_0) to the *interaction picture*. In the interaction picture, wavefunctions $\psi^{(I)}(t)$ and operators (for instance $V^{(I)}(t)$) at time t are related to the corresponding wavefunctions and operators at the same time t in the standard Schrödinger representation by the transformations

$$\psi^{(I)}(t) = e^{(i/\hbar)H_0 t}\psi(t), \tag{A.2}$$

and

$$V^{(I)}(t) = e^{(i/\hbar)H_0 t} V(t) e^{-(i/\hbar)H_0 t}. \tag{A.3}$$

In the interaction picture the equation of the motion for states reads

$$\boxed{i\hbar \frac{\partial \psi^{(I)}(t)}{\partial t} = V^{(I)}(t)\psi^{(I)}(t)} ; \tag{A.4}$$

in fact, it is easily verified that the expressions (A.2) and (A.3) inserted into Eq. (A.4), provide back Eq. (A.1).

The equation of motion (A.4) can be elaborated by introducing the time-evolution operator $U(t, t_0)$ defined so that

$$\psi^{(I)}(t) = U(t, t_0)\psi^{(I)}(t_0); \tag{A.5}$$

the operator $U(t, t_0)$ links any given wavefunction $\psi^{(I)}(t_0)$, considered at some initial arbitrary time t_0, with its time-evolved wavefunction $\psi^{(I)}(t)$ at time t. Substituting Eq. (A.5) into Eq. (A.4) we have that the temporal evolution operator satisfies the equation of motion

$$i\hbar \frac{\partial U(t, t_0)}{\partial t} = V^{(I)}(t)U(t, t_0); \tag{A.6}$$

this first-order differential equation in time variable, together with the obvious boundary condition $U(t_0, t_0) = 1$, determines in principle the operator U at any time.

We can integrate both sides of Eq. (A.6) from t_0 to t, embody the boundary condition at $t = t_0$, and so obtain the integral equation

$$U(t, t_0) = 1 - \frac{i}{\hbar} \int_{t_0}^{t} V^{(I)}(t_1) U(t_1, t_0) \, dt_1. \tag{A.7}$$

The integral equation (A.7) can be solved by iteration. We replace the "zero-order approximation" $U = 1$ into the right-hand side of Eq. (A.7), and obtain in the left-hand side the first-order approximation $U^{(1)}$; later we replace the first-order approximation in the right-hand side of Eq. (A.7) and obtain in the left-hand side the second-order approximation, and so on. We obtain at the successive orders of approximations

$$U(t, t_0) = U^{(0)}(t, t_0) + U^{(1)}(t, t_0) + U^{(2)}(t, t_0) + \cdots ,$$

where

$$U^{(0)}(t, t_0) = 1,$$

$$U^{(1)}(t, t_0) = -\frac{i}{\hbar} \int_{t_0}^{t} dt_1 \, V^{(I)}(t_1),$$

$$U^{(2)}(t, t_0) = \frac{(-i)^2}{\hbar^2} \int_{t_0}^{t} dt_1 \int_{t_0}^{t_1} dt_2 \, V^{(I)}(t_1) V^{(I)}(t_2) \tag{A.8}$$

and so on.

It is customary to write the operator $U(t, t_0)$ in a much more compact form suitable for important developments (especially in many-body theory), introducing the chronological ordering operator T, which orders operators so that *late* times are at the *left* (or equivalently *remote* times are at the *right*). For instance, the T-product of any two operators $A(t_1)$ and $B(t_2)$ is given by

$$T\{A(t_1)B(t_2)\} = \begin{cases} A(t_1)B(t_2) & \text{if} \quad t_1 > t_2 \\ B(t_2)A(t_1) & \text{if} \quad t_2 > t_1 \end{cases}.$$

After a few moments thought, Eq. (A.8) for $t > t_0$ can be written as

$$U(t, t_0) = 1 - \frac{i}{\hbar} \int_{t_0}^{t} dt_1 V^{(I)}(t_1) + \frac{1}{2!} \frac{(-i)^2}{\hbar^2} \int_{t_0}^{t} dt_1 \int_{t_0}^{t} dt_2 T\{V^{(I)}(t_1) V^{(I)}(t_2)\} + \cdots$$

In a more compact form, we write

$$\boxed{U(t, t_0) = T \exp\left[-\frac{i}{\hbar} \int_{t_0}^{t} V^{(I)}(t_1)\, dt_1 \right].} \tag{A.9}$$

The symbol T in front of the exponential means that one must first expand the exponential in Taylor series and then order chronologically the products of the type $V^{(I)}(t_1) V^{(I)}(t_2) \cdots$ in each term of the series expansion. The symbol T takes proper account of the fact that, in general, $V^{(I)}(t)$ and $V^{(I)}(t')$ *do not commute* for $t \neq t'$, and the *correct sequence* to place them is just the one, where remote time arguments systematically appear at the right. The operator T becomes irrelevant in the particular situations in which the perturbation operators commute at any time. Finally also notice that Eq. (A.9) although very elegant in form, is perfectly equivalent to the set of Eqs. (A.8).

Application to a Two State Hamiltonian

The evolution operator (A.9) is of utmost value (actually with the Wick theorem, it is at the basis of the many-body perturbative theory), even if in general an exact solution is not possible. However, to get some familiarity with the structure of the evolution operator, we consider an extremely simple paradigmatic two-level system, which can be solved exactly applying Eq. (A.9) (or also with alternative elementary methods).

Consider a two-level quantum system, where the two states are supposed to be degenerate and coupled by a time independent perturbation operator. The total Hamiltonian of the model under attention is of the type

$$H = H_0 + V \quad \text{with} \quad H_0 = E_0 \begin{bmatrix} 1 & 0 \\ 0 & 1 \end{bmatrix} \quad \text{and} \quad V = W \begin{bmatrix} 0 & 1 \\ 1 & 0 \end{bmatrix},$$

where E_0 is the energy of the two levels, and W is the coupling strength. Since the two-levels of the unperturbed Hamiltonian are supposed degenerate, H_0 is proportional to the unit operator, and as a consequence the perturbation operator in the interaction picture $V^{(I)} \equiv V$ remains time-independent. In applying Eq. (A.9) to the present case, the time ordered operator is irrelevant, and we have

$$U(t, 0) = \exp\left[-\frac{i}{\hbar} V t\right] = 1 - \frac{i}{\hbar} V t + \frac{1}{2!}(-\frac{i}{\hbar})^2 V^2 t^2 + \cdots$$

After noticing that V^2 is proportional to the unit matrix, one obtains for the time evolution operator U the exact result

$$U(t, 0) = \begin{bmatrix} \cos \Omega t & -i \sin \Omega t \\ -i \sin \Omega t & \cos \Omega t \end{bmatrix} \quad \text{with} \quad \hbar \Omega = W.$$

This shows that, if the system is prepared initially in one of the two chosen basis states of H_0, oscillations occur with Rabi frequency $\Omega = W/\hbar$ back and forth between the two states.

Transition Rates at Various Orders of Perturbation Theory

Consider now the time-dependent Eq. (A.1) in the case the perturbation operator has the oscillatory form

$$V(t) = V_0 e^{-i\omega t} e^{(\eta/\hbar)t} \quad \eta \to 0^+, \tag{A.10}$$

where V_0 is a static operator; the infinitesimal energy η has been added so that the perturbation is switched on gently (i.e. adiabatically) starting from an infinite remote time. [The discussion of the effect of the hermitian conjugate partner $V^\dagger(t)$ is considered later in Eq. (A.18)].

In the interaction picture, the perturbation operator becomes

$$V^{(I)}(t) \equiv e^{(i/\hbar)H_0 t} V_0 e^{-i\omega t + (\eta/\hbar)t} e^{-(i/\hbar)H_0 t}.$$

At *first-order* of perturbation theory, according to Eq. (A.8) and setting $t_0 = -\infty$, we have

$$U^{(1)}(t, -\infty) = -\frac{i}{\hbar} \int_{-\infty}^{t} dt_1 V^{(I)}(t_1).$$

The matrix elements of the first-order evolution operator between any two eigenfunctions of H_0 read

$$\begin{aligned}
\langle \psi_f | U^{(1)}(t, -\infty) | \psi_i \rangle &= -\frac{i}{\hbar} \int_{-\infty}^{t} e^{-(i/\hbar)(E_i - E_f + \hbar\omega + i\eta)t_1} dt_1 \langle \psi_f | V_0 | \psi_i \rangle \\
&= \frac{e^{-(i/\hbar)(E_i - E_f + \hbar\omega + i\eta)t}}{E_i - E_f + \hbar\omega + i\eta} \langle \psi_f | V_0 | \psi_i \rangle.
\end{aligned} \tag{A.11}$$

At *second-order* of perturbation theory, according to Eq. (A.8), we have

$$U^{(2)}(t, -\infty) = \frac{(-i)^2}{\hbar^2} \int_{-\infty}^{t} dt_1 \int_{-\infty}^{t_1} dt_2 V^{(I)}(t_1) V^{(I)}(t_2).$$

The matrix elements of the second-order evolution operator between any two eigenfunctions of H_0 read

$$\langle \psi_f | U^{(2)}(t, -\infty) | \psi_i \rangle = \frac{(-i)^2}{\hbar^2} \sum_\alpha \int_{-\infty}^t dt_1 \langle \psi_f | V^{(I)}(t_1) | \psi_\alpha \rangle \int_{-\infty}^{t_1} dt_2 \langle \psi_\alpha | V^{(I)}(t_2) | \psi_i \rangle$$

$$= \frac{-i}{\hbar} \sum_\alpha \int_{-\infty}^t dt_1 e^{-(i/\hbar)(E_i - E_f + 2\hbar\omega + 2i\eta)t_1} \frac{\langle \psi_f | V_0 | \psi_\alpha \rangle \langle \psi_\alpha | V_0 | \psi_i \rangle}{E_i - E_\alpha + \hbar\omega + i\eta}$$

$$= \frac{e^{-(i/\hbar)(E_i - E_f + 2\hbar\omega + 2i\eta)t}}{E_i - E_f + 2\hbar\omega + 2i\eta} \sum_\alpha \frac{\langle \psi_f | V_0 | \psi_\alpha \rangle \langle \psi_\alpha | V_0 | \psi_i \rangle}{E_i - E_\alpha + \hbar\omega + i\eta}. \tag{A.12}$$

Similarly, one could go on and obtain the amplitudes to any desired higher order.

The results (A.11) and (A.12) can be interpreted in a more effective way. We can define the *first-order transition probability* per unit time of a transition from an initial state ψ_i to a final state ψ_f as

$$P^{(1)}_{f \leftarrow i} = \frac{d}{dt} \left| \langle \psi_f | U^{(1)}(t, -\infty) | \psi_i \rangle \right|^2.$$

Using the relation (A.11), the transition probability per unit time becomes

$$P^{(1)}_{f \leftarrow i} = \frac{2}{\hbar} |\langle \psi_f | V_0 | \psi_i \rangle|^2 \frac{\eta e^{(2\eta/\hbar)t}}{(E_f - E_i - \hbar\omega)^2 + \eta^2}. \tag{A.13}$$

Exploiting the representation of the δ-function in the form

$$\delta(E - E_0) = \frac{1}{\pi} \lim_{\eta \to 0^+} \frac{\eta}{(E - E_0)^2 + \eta^2} \tag{A.14}$$

and performing the limit of $\eta \to 0^+$, the expression (A.13) can be recast in the form

$$\boxed{P^{(1)}_{f \leftarrow i} = \frac{2\pi}{\hbar} |\langle \psi_f | V_0 | \psi_i \rangle|^2 \delta(E_f - E_i - \hbar\omega)}; \tag{A.15}$$

this result represents the well-known *Fermi golden rule for first order transition probability rate* in steady state situation.

We now proceed to second-order; from Eq. (A.12) and similar elaborations, it follows

$$\boxed{P^{(2)}_{f \leftarrow i} = \frac{2\pi}{\hbar} \left| \sum_\alpha \frac{\langle \psi_f | V_0 | \psi_\alpha \rangle \langle \psi_\alpha | V_0 | \psi_i \rangle}{E_i + \hbar\omega - E_\alpha} \right|^2 \delta(E_f - E_i - \hbar\omega - \hbar\omega)}. \tag{A.16}$$

Similarly, the expression for the third-order transition probability is given by

$$\boxed{P^{(3)}_{f \leftarrow i} = \frac{2\pi}{\hbar} \left| \sum_{\alpha\beta} \frac{\langle \psi_f | V_0 | \psi_\beta \rangle \langle \psi_\beta | V_0 \psi_\alpha \rangle \langle \psi_\alpha | V_0 | \psi_i \rangle}{(E_i + 2\hbar\omega - E_\beta)(E_i + \hbar\omega - E_\alpha)} \right|^2 \delta(E_f - E_i - 3\hbar\omega)}. \tag{A.17}$$

A note of warning must be considered when a denominator is zero or near to zero. From a qualitative point of view these situations give rise to resonant effects; from a quantitative point of view the series development should be replaced by a more accurate analysis.

The above considerations can be extended in the case the perturbation has the form

$$V(t) = V_0\, e^{-i\omega t} e^{(\eta/\hbar)t} + V_0^\dagger e^{+i\omega t} e^{(\eta/\hbar)t}. \tag{A.18}$$

Besides the general expressions of Eqs. (A.15)–(A.17) for the absorption of one or more quanta of energy, one also obtains the general expressions for the emission of one or more quanta, or for mixed absorption-emission processes; the expressions are similar to Eqs. (A.15)–(A.17) with appropriate replacements of $+\hbar\omega$ by $-\hbar\omega$ for emission processes.

Appendix B. Optical Constants, Green's Function and Kubo-Greenwood Relation

According to Eq. (12.7) or Eq. (12.8), *the calculation of the optical constants of a linear and homogeneous system requires the knowledge of the eigenfunctions and eigenvalues of its Hamiltonian H_0*. In problems where the explicit diagonalization of H_0 poses difficult computational problems, it may be convenient to express the optical properties using the Green's function of the operator H_0. The reason is that the evaluation of Green's function matrix elements does not necessarily require the diagonalization of H_0 as an intermediate step (see for instance the discussion of the recursion method in Section 5.8); this advantage may become particularly rewarding in the treatment of disordered systems, since the Green's function formalism allows often to perform averaging procedures in a very effective way.

The *retarded*, and *advanced*, Green's functions of the one-electron Hamiltonian H_0 are defined as

$$G^R(E) = \frac{1}{E + i\eta - H_0}, \quad G^A(E) = \frac{1}{E - i\eta - H_0} \quad \text{with } \eta \to 0^+, \tag{B.1}$$

where the superscript R and A for retarded and advanced have been written explicitly; as usual, the real energy E is accompanied by a small imaginary part ($+i\eta$ for the retarded Green's function, and $-i\eta$ for the advanced one), and the limit $\eta \to 0^+$ is understood.

Besides the retarded and advanced Green's functions, it is convenient to consider the spectral Green's function defined by the expression

$$G^S(E) = \frac{i}{2\pi}\left[G^R(E) - G^A(E)\right] \quad \Longrightarrow \quad G^S(E) = \delta(E - H_0);$$

in the last passage use has been made of Eq. (A.14).

Consider, for instance, the conductivity $\sigma(0, \omega) = \sigma(\omega)$ in the long wavelength limit. The real part of the conductivity is linked to the imaginary part of the dielectric function by the relation $\sigma_1 = (\omega/4\pi)\varepsilon_2$. Using Eq. (12.7) we have

$$\sigma_1(\omega) = \frac{2\pi e^2}{m^2\omega} \frac{1}{V} \sum_{ij} |\langle\psi_j|\widehat{\mathbf{e}} \cdot \mathbf{p}|\psi_i\rangle|^2 \delta(E_j - E_i - \hbar\omega)[f(E_i) - f(E_j)]. \quad \text{(B.2)}$$

We now show that $\sigma_1(\omega)$, when expressed by means of the Green's function, takes the *Kubo-Greenwood* form

$$\sigma_1(\omega) = \frac{2\pi e^2 \hbar}{m^2} \frac{1}{V} \int_{-\infty}^{+\infty} dE \frac{f(E) - f(E+\hbar\omega)}{\hbar\omega} \text{Tr}\left\{\widehat{\mathbf{e}} \cdot \mathbf{p}\, G^S(E+\hbar\omega)\,\widehat{\mathbf{e}} \cdot \mathbf{p}\, G^S(E)\right\},$$
$$\text{(B.3)}$$

where Tr stands for the trace operation, to be performed on any chosen complete set. The two Eqs. (B.2) and (B.3) are fully equivalent, in spite of the fact that the former is built with the eigenfunctions and eigenvalues of H_0, while the latter is built in terms of the Green's function of H_0.

To prove Eq. (B.3), it is convenient to carry out the trace on the complete set of the eigenfunctions ψ_i of H_0, and exploit the closure property $\sum |\psi_j\rangle\langle\psi_j| = 1$; we have

$$\text{Tr}\left\{\widehat{\mathbf{e}} \cdot \mathbf{p}\, G^S(E+\hbar\omega)\,\widehat{\mathbf{e}} \cdot \mathbf{p}\, G^S(E)\right\}$$
$$= \sum_{ij}\langle\psi_i|\widehat{\mathbf{e}} \cdot \mathbf{p}\, G^S(E+\hbar\omega)|\psi_j\rangle\langle\psi_j|\widehat{\mathbf{e}} \cdot \mathbf{p}\, G^S(E)|\psi_i\rangle$$
$$= \sum_{ij}|\langle\psi_j|\widehat{\mathbf{e}} \cdot \mathbf{p}|\psi_i\rangle|^2 \delta(E+\hbar\omega - E_j)\delta(E - E_i);$$

insertion of the above expression into Eq. (B.3) shows its equivalence to Eq. (B.2).

As an application of Eq. (B.3), consider for instance the static conductivity of a metal. For $\omega \to 0$, the first term within the integral in Eq. (B.3) becomes $(-\partial f/\partial E)$; in a metal $(-\partial f/\partial E) \approx \delta(E - E_F)$. The integration on the energy can then be performed and the static electron conductivity becomes

$$\sigma_1(\omega = 0) = \frac{2\pi e^2 \hbar}{m^2} \frac{1}{V} \text{Tr}\left\{\widehat{\mathbf{e}} \cdot \mathbf{p}\, G^S(E_F)\,\widehat{\mathbf{e}} \cdot \mathbf{p}\, G^S(E_F)\right\}.$$

The above expression has been widely applied for the evaluation of the static electronic conductivity in disordered metals and alloys.

Further Reading

Agranovich, V. M, & Ginzburg, V. L. (1984). *Crystal optics with spatial dispersion and excitons.* Berlin: Springer.

Balkanski, M. (1994). *Optical properties of semiconductors.* Amsterdam: North-Holland.

Bassani, F., & Pastori Parravicini, G. (1975). *Electronic states and optical transitions in solids.* Oxford: Pergamon Press.

Bersuker, I. B. (2006). *The Jahn-Teller effect.* Cambridge: Cambridge University Press.

Born, M., & Wolf, E. (2002). *Principles of optics.* Cambridge: Cambridge University Press.

Cardona, M. (1969). Modulation spectroscopy. In F. Seitz, D. Turnbull, & H. Ehrenreich (Eds.), *Solid state physics, suppl.* (Vol. 11). New York: Academic Press.

Cohen, M. L., & Chelikowsky, J. R. (1989). *Electronic structure and optical properties of semiconductors.* Berlin: Springer.

Fano, U., & Cooper, J. W. (1968). Spectral distribution of atomic oscillator strength. *Reviews of Modern Physics, 40,* 441.

Greenaway, D. L., & Harbeke, G. (1968). *Optical properties and band structure of semiconductors.* Oxford: Pergamon Press.

Haug, H., & Koch, S. W. (2009). *Quantum theory of the optical and electronic properties of semiconductors.* Singapore: World Scientific.

Hodgson, J. N. (1970). *Optical absorption and dispersion in solids.* New York: Chapman and Hall.

Miroshnichenko, A. E., Flach, S., & Kivshar, Y. S. (2010). Fano resonances in nanoscale structures. *Reviews in Modern Physics, 82,* 2257.

Palik, E. D. (Ed.) (1985/1991). *Handbook of the optical constants of solids* (Vols. I and II). New York: Academic Press.

Pankove, J. I. (2010). *Optical processes in semiconductors.* New York: Dover.

Ridley, B. K. (2000). *Quantum processes in semiconductors.* New York: Oxford University Press.

Wooten, F. (1972). *Optical properties of solids.* New York: Academic Press.

Yu, P. Y., & Cardona, M. (2010). *Fundamentals of semiconductors.* Berlin: Springer.

13 Transport in Intrinsic and Homogeneously Doped Semiconductors

---

Chapter Outline head

This is the first of two partner chapters concerning electronic transport phenomena in semiconductors. In this chapter we consider transport processes in homogeneous semiconductors, which are free from impurities or uniformly doped with (donor or acceptor) impurities. In Chapter 14 we consider semiconductor structures that are inhomogeneous, either in the doping or in the chemical composition. The presence of a finite energy gap between valence and conduction states has a profound effect on carrier density and transport properties. In particular, the Fermi level within the energy gap can be easily engineered by the presence of even a small number of appropriate impurities; this fact ultimately opens the wide field of solid state electronics and device applications.

13.1   Fermi Level and Carrier Density in Intrinsic Semiconductors

A semiconductor (or insulator) at zero temperature is constituted by fully occupied valence bands and fully empty conduction bands. An energy gap $E_G = E_c - E_v$ separates the lowest energy level E_c of the conduction bands from the topmost energy level E_v of the valence bands. The fundamental energy gap E_G and the lattice constant of some semiconductors are reported in Figure 13.1.

Solid State Physics, Second Edition. http://dx.doi.org/10.1016/B978-0-12-385030-0.00013-X

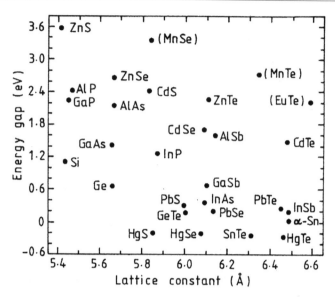

Figure 13.1 Energy gap and lattice constant of some semiconductor crystals at room temperature [from M. Jaros "Physics and Applications of Semiconductor Microstructures" (Clarendon Press, Oxford 1989); by permission of Oxford University Press].

In ideal defect-free semiconductors (*intrinsic semiconductors*), no level exists in the forbidden energy gap. On the contrary in semiconductors containing defects (*extrinsic semiconductors*), impurity levels may be introduced within the energy gap, with significant modifications of the carrier concentration and transport properties. Our purpose is to study the carrier distribution at thermal equilibrium, first in intrinsic semiconductors and then in extrinsic ones.

Consider an intrinsic semiconductor, and let $n_c(E) = D_c(E)/V$ and $n_v(E) = D_v(E)/V$ indicate the *density-of-states per unit volume in the conduction and in the valence bands*, respectively. At zero temperature, all the valence states are occupied and all the conduction states are empty, as schematically indicated in Figure 13.2. At finite

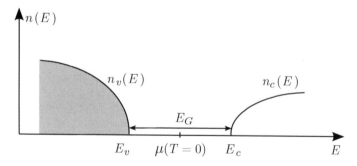

Figure 13.2 Schematic representation of the density-of-states of an intrinsic semiconductor. The top of the valence band is E_v and the bottom of the conduction band is E_c. The energy gap is $E_G = E_c - E_v$; the Fermi level at zero temperature lies at the middle of the band gap.

temperature T, a number of electrons from the valence bands are thermally excited to the conduction bands; the occupancy probability of the allowed band structure levels of energy E is given by the Fermi-Dirac distribution function

$$f(E) = \frac{1}{e^{(E-\mu)/k_B T} + 1}, \tag{13.1}$$

where μ is the chemical potential (the terms "chemical potential μ" and "Fermi level E_F" are used interchangeably).

The density of electrons at temperature T in the conduction bands is given by the expression

$$n_0(T) = \int_{E_c}^{\infty} n_c(E) f(E) \, dE = \int_{E_c}^{\infty} n_c(E) \frac{1}{e^{(E-\mu)/k_B T} + 1} \, dE, \tag{13.2a}$$

where the subscript 0 to the electron concentration is just to remind that the quantity refers to thermal equilibrium. Similarly, the density of missing electrons (or equivalently the density of holes) at temperature T in the valence bands is determined by the density-of-states in the valence bands and the complementary to 1 of the Fermi-Dirac distribution function. We have the expression

$$p_0(T) = \int_{-\infty}^{E_v} n_v(E)(1 - f(E)) \, dE = \int_{-\infty}^{E_v} n_v(E) \frac{1}{e^{(\mu-E)/k_B T} + 1} \, dE. \tag{13.2b}$$

The integrals in Eqs. (13.2) extend (in principle) to the whole energy regions where $n_c(E)$ and $n_v(E)$ are different from zero; in practice, as seen below, only the energy regions close to the band edges E_c and E_v are of relevance.

For an intrinsic semiconductor, the chemical potential $\mu(T)$ is determined by the requirement that the number of electrons in the conduction bands equals the number of holes left in the valence bands

$$n_0(T) = p_0(T). \tag{13.3}$$

Before solving explicitly the balance Eq. (13.3), we make the following considerations. Since the distribution function $f(E)$ is a step function around $E = \mu$ at zero temperature, and at the same time $n_0(T = 0) \equiv p_0(T = 0) \equiv 0$ in an intrinsic semiconductor, we have from Eqs. (13.2) that the chemical potential must *lie within the energy gap* at zero temperature, i.e. $E_v < \mu(T = 0) < E_c$. Furthermore, in the particular case that the density-of-states of the semiconductor is symmetric with respect to the middle of the energy gap, the balance of electrons and holes obviously requires that the chemical potential, at any temperature, coincides with the mid-gap energy. Even if the density-of-states in the valence and conduction bands (in the energy range of order $k_B T$ around the band edges) are not specular, it is evident that a small shift of the chemical potential of order $k_B T$ (toward the edge with lower density-of-states) is sufficient to equalize the number of electrons and the number of holes.

A semiconductor (either intrinsic or extrinsic) is said to be *non-degenerate* if the chemical potential $\mu(T)$ lies within the energy gap and is separated from the band edges by several $k_B T$ (say $\approx 5\, k_B T$ or more); the *non-degeneracy conditions* are

$$\boxed{E_v < \mu(T) < E_c, \quad \text{with} \quad E_c - \mu(T) \gg k_B T \quad \text{and} \quad \mu(T) - E_v \gg k_B T}. \quad (13.4)$$

When conditions (13.4) are satisfied, the Fermi-Dirac distribution function $f(E)$ in Eq. (13.2a), as well as the complementary distribution function $1 - f(E)$ in Eq. (13.2b), can be simplified with their corresponding Maxwell-Boltzmann exponential distributions.

We pass now to a quantitative analysis of Eq. (13.3). From the previous discussion, we expect that in general an intrinsic semiconductor is non-degenerate at any temperature of interest. For a non-degenerate semiconductor, the expression (13.2a) for the electron density in the conduction bands simplifies in the form

$$n_0(T) = N_c(T)e^{-(E_c-\mu)/k_B T}, \quad (13.5a)$$

where

$$N_c(T) \equiv \int_{E_c}^{\infty} n_c(E)e^{-(E-E_c)/k_B T}\, dE. \quad (13.5b)$$

Similarly, for a non-degenerate semiconductor, the expression for the hole density in the valence bands takes the simplified form

$$p_0(T) = N_v(T)e^{-(\mu-E_v)/k_B T}, \quad (13.5c)$$

where

$$N_v(T) \equiv \int_{-\infty}^{E_v} n_v(E)e^{-(E_v-E)/k_B T}\, dE. \quad (13.5d)$$

The quantities $N_c(T)$ and $N_v(T)$ are referred as the *effective conduction band and valence band density-of-states*, respectively. Thus a non-degenerate semiconductor can be schematized as a two-level system, where the whole conduction bands can be replaced by a single level of energy E_c and degeneracy $N_c(T)$, and the whole valence bands can be replaced by a single level of energy E_v and degeneracy $N_v(T)$.

The chemical potential μ_i in an intrinsic semiconductor is obtained by the requirement that expressions (13.5a) and (13.5c) coincide for $\mu = \mu_i$:

$$N_c(T)e^{-(E_c-\mu_i)/k_B T} = N_v(T)e^{-(\mu_i-E_v)/k_B T}.$$

Taking the logarithms of both members we have

$$\boxed{\mu_i(T) = \frac{1}{2}(E_v + E_c) + \frac{1}{2}k_B T \ln \frac{N_v(T)}{N_c(T)}} \quad \text{(intrinsic semiconductor).} \quad (13.6)$$

This equation shows that the chemical potential of the intrinsic semiconductor at zero temperature is located at the mid-gap energy; at finite temperature T, the change of μ_i is of the order of $k_B T$. Inserting Eq. (13.6) into Eqs. (13.5a) and (13.5c), we obtain for *the intrinsic concentrations* $n_i(T)$ and $p_i(T)$ of electrons and holes at equilibrium

$$n_i(T) = p_i(T) = \sqrt{N_c(T)N_v(T)}\, e^{-E_G/2k_B T}.$$

The temperature dependence of the intrinsic carrier concentration is mainly controlled by the exponential dependence on the energy band gap. In addition one has to consider the (smoother) temperature dependence of the effective density-of-states of the valence and conduction bands.

From Eqs. (13.5a) and (13.5c), we obtain for the product $n_0(T)p_0(T)$ the expression

$$\boxed{n_0(T)p_0(T) = N_c(T)N_v(T)e^{-E_G/k_BT} \equiv n_i^2(T) = p_i^2(T)},\qquad (13.7)$$

which is known as *mass-action law*. It is important to notice that the product $n_0(T)p_0(T)$ does not depend on the value of the chemical potential; thus, at fixed temperature, the product $n_0(T)p_0(T)$ keeps the same value regardless the impurity concentration in the semiconductor; while the actual values of $n_0(T)$ and $p_0(T)$ depend on the chemical potential (which is different for intrinsic or extrinsic semiconductors) their product does not. For any intrinsic or extrinsic semiconductor at thermodynamic equilibrium, the mass-action law enables to find the hole density if the electron density is known or vice versa.

It is instructive to specify the above results in the case of the isotropic and parabolic two-band model semiconductor, schematized in Figure 13.3. The density-of-states per unit volume in the conduction band for $E > E_c$ is

$$n_c(E) = \int \frac{2}{(2\pi)^3}\delta(E_c(\mathbf{k}) - E)\,d\mathbf{k} = \int_0^{\infty} \frac{2}{(2\pi)^3}\delta\left(\frac{\hbar^2k^2}{2m_c^*} + E_c - E\right)4\pi k^2 dk,$$

where the factor 2 takes into account spin degeneracy; the integral is easily carried out after performing the change of variable $x = \hbar^2k^2/2m_c^*$ and gives

$$n_c(E) = \frac{1}{2\pi^2}\left(\frac{2m_c^*}{\hbar^2}\right)^{3/2}(E - E_c)^{1/2} \quad \text{for} \quad E \geq E_c. \qquad (13.8a)$$

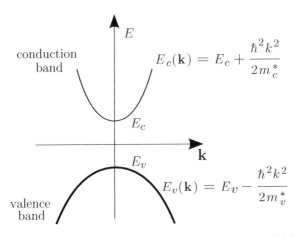

Figure 13.3 Isotropic two-band model semiconductor; the valence band profile has been reported in bold to remind its full occupancy by electrons at $T = 0$.

We insert Eq. (13.8a) into Eq. (13.5b), use the relation $\int_0^\infty \sqrt{t}\, e^{-t}\, dt = \sqrt{\pi}/2$, and obtain for the effective density-of-states in the conduction band

$$
\begin{aligned}
N_c(T) &= \frac{\sqrt{\pi}}{2} \frac{1}{2\pi^2} \left(\frac{2m_c^*}{\hbar^2} \right)^{3/2} (k_B T)^{3/2} \\
&= \frac{1}{4\pi^{3/2}} \left(\frac{m_c^*}{m} \frac{T}{T_0} \right)^{3/2} \left(\frac{k_B T_0}{\hbar^2/2ma_B^2} \right)^{3/2} \frac{1}{a_B^3} \\
&= 2.534 \left(\frac{m_c^*}{m} \frac{T}{T_0} \right)^{3/2} \cdot 10^{19} \text{ cm}^{-3},
\end{aligned}
\tag{13.8b}
$$

where $T_0 = 300$ K has been chosen as the reference temperature, $k_B T_0 \approx 0.026$ eV, $\hbar^2/2ma_B^2 = 1$ Ryd $= 13.606$ eV, and $a_B = 0.529 \times 10^{-8}$ cm.

Similar expressions can be worked out for the holes in the valence band; the effective density-of-states for the valence bands becomes

$$
N_v(T) = \frac{\sqrt{\pi}}{2} \frac{1}{2\pi^2} \left(\frac{2m_v^*}{\hbar^2} \right)^{3/2} (k_B T)^{3/2}.
\tag{13.8c}
$$

In the case of the isotropic two-band model semiconductor, Eq. (13.6) takes the form

$$
\mu_i(T) = \frac{1}{2}(E_v + E_c) + \frac{3}{4} k_B T \ln \frac{m_v^*}{m_c^*}.
\tag{13.8d}
$$

From Eq. (13.8d) we see that the intrinsic chemical potential remains close to the mid-gap energy. In particular, the intrinsic chemical potential is independent of temperature if the hole and electron effective masses are equal; otherwise it changes by a (small) amount of the order of $k_B T$ toward the band edge with lower effective mass (i.e. lower density-of-states).

The above considerations for a two-band model semiconductor are useful to estimate the order of magnitude of relevant quantities, and can be implemented to describe realistic multiband semiconductors. Some properties of germanium, silicon, and gallium arsenide at 300 K are summarized for convenience in Table 13.1; for a more complete list of data, we refer for instance to the book of Smith and to the book of Bhattacharya, quoted in the references of this chapter.

13.2   Impurity Levels in Semiconductors

Band States and Impurity Levels

In the band theory of crystals (Chapter 5), we have considered in detail the Schrödinger equation for periodic materials

$$
\left[\frac{\mathbf{p}^2}{2m} + V(\mathbf{r}) \right] \psi_n(\mathbf{k}, \mathbf{r}) = E_n(\mathbf{k}) \psi_n(\mathbf{k}, \mathbf{r});
\tag{13.9}
$$

Table 13.1 Some properties of Si, Ge, and GaAs at 300 K.

	Si	Ge	GaAs
Crystal structure and lattice constant (Å)	diamond 5.431	diamond 5.646	zincblende 5.653
Energy gap (eV) (direct or indirect)	1.12 (indirect)	0.66 (indirect)	1.43 (direct)
Electron affinity χ (eV)	4.01	4	4.07
Electron effective mass m^* in units of the free-electron mass m (l = longitudinal, t = transversal)	$m_l^* = 0.98$ $m_t^* = 0.19$	$m_l^* = 1.64$ $m_t^* = 0.082$	$m^* = 0.063$
Hole effective mass m^*/m (lh = light hole, hh = heavy hole)	$m_{lh}^* = 0.16$ $m_{hh}^* = 0.49$	$m_{lh}^* = 0.044$ $m_{hh}^* = 0.28$	$m_{lh}^* = 0.09$ $m_{hh}^* = 0.48$
Number of equivalent conduction edge valleys	6	4	1
Effective density-of-states in the conduction band N_c (cm^{-3})	2.9×10^{19}	1.04×10^{19}	4.4×10^{17}
Effective density-of-states in the valence band N_v (cm^{-3})	1.1×10^{19}	6.1×10^{18}	8.2×10^{18}
Intrinsic carrier concentration n_i or p_i (cm^{-3})	1.02×10^{10}	2.4×10^{13}	5×10^6
Static dielectric constant	11.7	16	13.2
Carrier mobility (cm^2/V · s) μ_n (electrons) μ_p (holes)	1450 500	3800 1800	8500 400
Diffusion constant (cm^2/s) D_n (electrons) D_p (holes)	37.5 13	98 47	220 10
Linear thermal expansion coefficient $10^{-6}/°$C	2.6	5.7	5

we have also described the general methods to obtain the band energies $E_n(\mathbf{k})$ and the band wavefunctions $\psi_n(\mathbf{k}, \mathbf{r})$, exploiting the translational invariance of the crystal Hamiltonian.

A perfect periodic crystal, however, is an ideal entity. Real crystals contain a number of defects (substitutional or interstitial impurities, vacancies, dislocations, physical surfaces, etc.). The presence of defects implies, in general, a (total or partial) breaking of the translational symmetry of the crystal; in the allowed energy regions of the host crystals, the density-of-states function may be modified and resonances may appear; more important, in the forbidden energy gap of the host crystals, new energy levels can occur with wavefunctions localized around the impurity region.

The electronic structure of a crystal, in the presence of an impurity, can be studied by means of the Schrödinger equation

$$\left[\frac{\mathbf{p}^2}{2m} + V(\mathbf{r}) + V_I(\mathbf{r})\right]\phi(\mathbf{r}) = E\phi(\mathbf{r}),\tag{13.10}$$

where $V(\mathbf{r})$ is the periodic crystal potential, $V_I(\mathbf{r})$ is the modification due to the impurity, and the one-electron approximation is assumed to hold. The direct solution of Eq. (13.10) appears a quite demanding (if not impossible) job, since the break of periodicity due to $V_I(\mathbf{r})$ does not allow to use the basic Bloch theorem. However, in general, it is not necessary to solve "ex novo" Eq. (13.10); rather it is convenient to use the band wavefunctions as starting basis set on which to represent the wavefunctions of the impurity problem.

Within this line of approach, we expand the wavefunctions of the perfect crystal plus impurity system on the complete set of band wavefunctions of the perfect crystal in the form

$$\phi(\mathbf{r}) = \sum_{n'\mathbf{k}'} A_{n'}(\mathbf{k}')\psi_{n'}(\mathbf{k}', \mathbf{r}).\tag{13.11}$$

We insert Eq. (13.11) into Eq. (13.10), project on $\langle\psi_n(\mathbf{k}, \mathbf{r})|$ and use Eq. (13.9). We have that the coefficients $A_n(\mathbf{k})$ of the linear expansion (13.11) satisfy the set of coupled integral equations

$$[E_n(\mathbf{k}) - E]A_n(\mathbf{k}) + \sum_{n'\mathbf{k}'} U_{nn'}(\mathbf{k}, \mathbf{k}')A_{n'}(\mathbf{k}') = 0,\tag{13.12a}$$

where $\mathbf{k}$ and $\mathbf{k}'$ can be thought as continuous variables within the Brillouin zone, and

$$U_{nn'}(\mathbf{k}, \mathbf{k}') = \int \psi_n^*(\mathbf{k}, \mathbf{r})V_I(\mathbf{r})\psi_{n'}(\mathbf{k}', \mathbf{r})\,d\mathbf{r}.\tag{13.12b}$$

The general Eqs. (13.12) become of practical value when a small number of bulk bands are actually involved in the description of the impurity levels. Thus we consider appropriate simplifications of Eqs. (13.12) with the purpose to obtain (whenever possible) intuitive descriptions of impurity levels.

Single Conduction Band Model and Shallow Donor Levels

We consider here the particular case in which only *one band of the perfect crystal*, suppose the lowest conduction band, is of relevance for the description of the impurity electronic structure. We further assume that the conduction band of interest is parabolic, with a unique valley of effective mass $m_c^* > 0$ at $\mathbf{k} = 0$ and energy dispersion $E_c(\mathbf{k}) = E_c + \hbar^2k^2/2m_c^*$. With the above assumptions, Eqs. (13.12) become

$$\left[\frac{\hbar^2k^2}{2m_c^*} + E_c - E\right]A(\mathbf{k}) + \sum_{\mathbf{k}'} U(\mathbf{k}, \mathbf{k}')A(\mathbf{k}') = 0,\tag{13.13a}$$

with the kernel of the integral equation given by

$$U(\mathbf{k}, \mathbf{k}') = \int \psi_c^*(\mathbf{k}, \mathbf{r}) V_I(\mathbf{r}) \psi_c(\mathbf{k}', \mathbf{r}) \, d\mathbf{r}. \tag{13.13b}$$

In order to put in evidence relevant physical effects avoiding inessential details, we simplify the kernel $U(\mathbf{k}, \mathbf{k}')$ with considerations similar to what done in Section 7.1 for the integral equation of excitons. In Eq. (13.13b) we write the Bloch wavefunctions as the product of plane waves and periodic parts and obtain

$$U(\mathbf{k}, \mathbf{k}') = \int u_c^*(\mathbf{k}, \mathbf{r}) u_c(\mathbf{k}', \mathbf{r}) e^{-i(\mathbf{k}-\mathbf{k}')\cdot\mathbf{r}} V_I(\mathbf{r}) \, d\mathbf{r}.$$

In the case the impurity potential is reasonably smooth on the crystal unit cell, we can replace the product of periodic functions by the average value on the volume V of the crystal

$$u_c^*(\mathbf{k}, \mathbf{r}) u_c(\mathbf{k}', \mathbf{r}) \implies \frac{1}{V} \int u_c^*(\mathbf{k}, \mathbf{r}) u_c(\mathbf{k}', \mathbf{r}) \, d\mathbf{r} \implies \frac{1}{V} \quad \text{for} \quad \mathbf{k} \approx \mathbf{k}';$$

it follows

$$U(\mathbf{k}, \mathbf{k}') = \frac{1}{V} \int e^{-i(\mathbf{k}-\mathbf{k}')\cdot\mathbf{r}} V_I(\mathbf{r}) \, d\mathbf{r}; \tag{13.13c}$$

the above expression can be recognized as the Fourier transform of the impurity potential.

We now introduce the "envelope function" $F(\mathbf{r})$ defined as

$$F(\mathbf{r}) = \frac{1}{\sqrt{V}} \sum_{\mathbf{k}} A(\mathbf{k}) e^{i\mathbf{k}\cdot\mathbf{r}},$$

and normalized to unity. The integral eigenvalue equation (13.13a), with the kernel (13.13c), can be conveniently transformed into an ordinary differential eigenvalue equation for the envelope function

$$\left[-\frac{\hbar^2}{2m_c^*} \nabla^2 + V_I(\mathbf{r}) \right] F(\mathbf{r}) = (E - E_c) F(\mathbf{r}). \tag{13.14}$$

As an illustration, we consider here the case of great technological importance of *donor impurities* in silicon and germanium; donor impurities are formed by atoms of column V of the periodic table entering substitutionally into the silicon or germanium crystal. We can intuitively schematize the long-range part of the impurity potential in the form $V_I(\mathbf{r}) \approx -e^2/\varepsilon_s r$, where ε_s is the static dielectric constant of the host crystal (see Figure 13.4a). Notice that the short-range part (the so called "*central-cell-correction*") of the impurity potential would require a detailed (and demanding) analysis of the electronic structure before and after the substitution has taken place. Neglecting for simplicity central-cell-corrections (and also disregarding the actual multivalley structure of the conduction bands of Si and Ge), the Schrödinger equation (13.14) for the

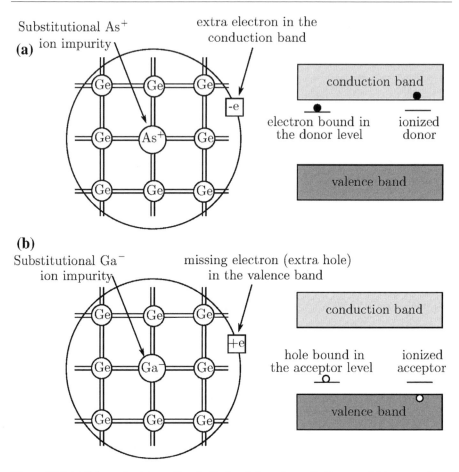

Figure 13.4 (a) Schematic picture of a substitutional donor impurity (i.e. arsenic) in germanium; the pentavalent impurity is embodied in the lattice as positive As^+ ion, with its four external electrons forming four covalent bonds with the nearest neighbor Ge atoms; each As^+ ion introduces a shallow donor level, below the bottom of the conduction band; the fifth electron, not incorporated in the bonding, is bound to the As^+ ion by the Coulomb attraction (at low temperature) and is donated to the conduction band (at sufficiently high temperature). (b) Schematic picture of a substitutional acceptor impurity (i.e gallium) in germanium; the trivalent impurity is embodied in the lattice as negative Ga^- ion, with its four external electrons forming four covalent bonds with the nearest neighbor Ge atoms; each Ga^- impurity introduces a shallow acceptor level, above the top of the valence band. The "missing electron" or "hole," created for the formation of the tetrahedral bonding, is bound to the Ga^- ion by Coulomb attraction (at low temperature) and is given to the valence band (at sufficiently high temperature).

donor impurity levels takes the form

$$\left[-\frac{\hbar^2}{2m_c^*}\nabla^2 - \frac{e^2}{\varepsilon_s r} \right] F(\mathbf{r}) = (E - E_c)F(\mathbf{r}), \tag{13.15}$$

which represents a hydrogenic atom characterized by ε_s and m_c^*.

The eigenvalue equation (13.15) supports the picture of a donor impurity as an extra electron of effective mass m_c^* bound to the defect by a screened Coulomb interaction. The binding energy $\varepsilon_d = E_c - E_d$ of the donor ground state is

$$\varepsilon_d = E_c - E_d = 13.606 \frac{m_c^*}{m} \frac{1}{\varepsilon_s^2} \text{ (eV)}, \tag{13.16a}$$

and the effective radius of the donor ground-state wavefunction is

$$a_d = a_B \frac{m}{m_c^*} \varepsilon_s. \tag{13.16b}$$

For instance in the case of Si, with $\varepsilon_s = 11.7$ and average effective mass $m_c^*/m \approx 0.3$, we expect $\varepsilon_d \approx 30$ meV and $a_d \approx 40\ a_B$. For Ge with $\varepsilon_s = 16$ and average effective mass $m_c^*/m \approx 0.2$, we expect $\varepsilon_d \approx 10$ meV and $a_d \approx 80\ a_B$. We thus see that the binding energy of donors in Si and Ge is rather small with respect to the energy gap of the host material, while the effective radius is rather large with respect to the lattice parameter (*shallow impurity level*). Further refinements are necessary to describe the short-range effects of the impurity, or the multivalley conduction band structure of the host semiconductors. In Figure 13.5, we report the experimental ionization energies for various donor impurities in silicon and germanium.

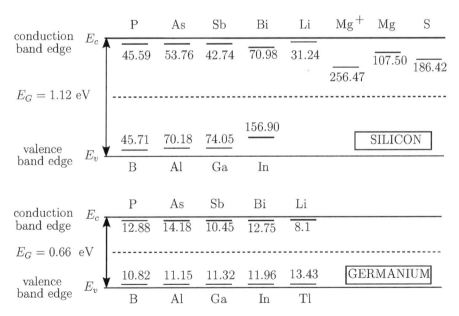

Figure 13.5 Experimental ionization energies (meV) for various donor and acceptor impurities (upper and lower part of the panels, respectively) in silicon and germanium. For donor impurities, the electron binding energy $\varepsilon_d = E_c - E_d$ is reported; for acceptor impurities the hole binding energy $\varepsilon_a = E_a - E_v$ is reported. The data are taken from A. K. Ramdas and S. Rodriguez, Rep. Progr. Phys. *44*, 1297 (1981).

Single Valence Band Model and Shallow Acceptor Levels

We consider now the particular case in which only *one valence band of the perfect crystal* is of relevance for the description of the electronic structure of the impurity. We also assume that the topmost valence band of interest is isotropic, with the maximum at $\mathbf{k} = 0$ and energy dispersion $E_v(\mathbf{k}) = E_v - \hbar^2 k^2/2m_v^*$ $(m_v^* > 0)$. The effect of a (reasonably smooth) impurity potential $V_I(\mathbf{r})$ on the valence band can be described within the envelope function formalism, in a way essentially similar to the above discussion of shallow donor levels. With appropriate relabeling of quantities entering in Eq. (13.14), we obtain that the envelope function for the impurity states associated to the valence band satisfies the Schrödinger equation

$$\left[+\frac{\hbar^2}{2m_v^*}\nabla^2 + V_I(\mathbf{r}) \right] F(\mathbf{r}) = (E - E_v)F(\mathbf{r}). \tag{13.17a}$$

As an illustration, we consider the case of *acceptor impurities* in silicon and germanium; these impurities are formed when atoms of column III of the periodic table enter substitutionally into the silicon or germanium crystal (see Figure 13.4b). We can intuitively schematize the impurity potential felt by an electron because of the negatively charged impurity in the form $V_I(\mathbf{r}) = +e^2/\varepsilon_s r$, where ε_s is the static dielectric constant of the host crystal. Eq. (13.17a), with both members multiplied by -1, gives

$$\left[-\frac{\hbar^2}{2m_v^*}\nabla^2 - \frac{e^2}{\varepsilon_s r} \right] F(\mathbf{r}) = -(E - E_v)F(\mathbf{r}). \tag{13.17b}$$

The binding energy of the ground acceptor level is

$$\varepsilon_a = E_a - E_v = 13.606 \frac{m_v^*}{m}\frac{1}{\varepsilon_s^2} \ (\text{eV}),$$

and the effective radius of the acceptor ground-state wavefunction is $a_B(m/m_v^*)\varepsilon_s$.

For more refined treatments of acceptor levels in Si and Ge, one has to describe accurately the short-range part of the impurity potential and the fact that the group of valence bands with the same energy at $\mathbf{k} = 0$ contribute on the same footing to the formation of the acceptor levels. In Figure 13.5, we report the experimental ionization energies for various acceptor impurities in silicon and germanium.

The concept of donor and acceptor impurities can be extended to III–V compounds (such as GaAs). A group VI impurity substituting a group V atom, or a group IV impurity substituting a group III atom, act as a donor; similarly, a group II impurity replacing a group III atom, or a group IV impurity replacing a group V atom, act as an acceptor.

Considerations on Deep Impurity Levels in Semiconductors

In most situations, the set of integral equations (13.12) cannot be decoupled in the form discussed above, and the analysis of impurity levels becomes more demanding, both from the point of view of the physical picture and from the technical point of view of an adequate solution. Consider for instance group II or group VI substitutional impurities in group IV elemental semiconductors silicon and germanium; in the case of doubly ionized impurities, the radius of the ground impurity orbit decreases, the dielectric screening is less effective at small distances and some of the assumptions inherent the envelope function formalism are at least doubtful. Levels tend to become deep in the forbidden gap, and their description in terms of linear combination of band wavefunctions may include a number of conduction and valence bands. Deep levels are relatively difficult to be ionized thermally, but are important as recombination-generation centers of electron and hole carriers.

Just to extract a simple model of deep impurity levels in semiconductors, consider the general integral equations (13.12) in the case the kernel $U_{nn'}(\mathbf{k}, \mathbf{k}')$ can be considered independent of $\mathbf{k}$ and $\mathbf{k}'$, and also independent of the band index for the group of bands under consideration; with the approximation $U_{nn'}(\mathbf{k}, \mathbf{k}') = U_0/N$ (where N is the number of unit cells of the crystal), Eq. (13.12a) reads

$$[E_n(\mathbf{k}) - E]A_n(\mathbf{k}) + \frac{U_0}{N} \sum_{n'\mathbf{k}'} A_{n'}(\mathbf{k}') = 0.$$

Dividing by $[E_n(\mathbf{k}) - E]$ (which is different from zero for E in the forbidden energy gap) and summing up over n and $\mathbf{k}$ one obtains the eigenvalue compatibility equation

$$\frac{1}{N} \sum_{n\mathbf{k}} \frac{1}{E - E_n(\mathbf{k})} = \frac{1}{U_0}. \tag{13.18}$$

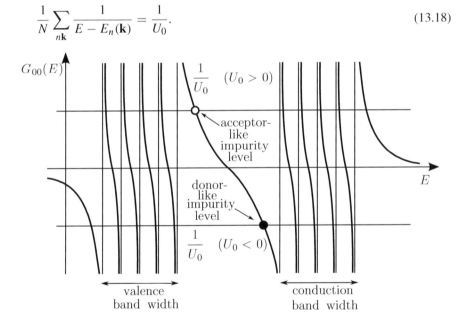

Figure 13.6 Schematic graphical solution of the compatibility equation $G_{00}(E) = 1/U_0$ for impurity levels in semiconductors. The vertical lines correspond to (arbitrarily discretized) energies within the allowed energy bands. The intersection of the Green's function $y = G_{00}(E)$ with the horizontal line $y = 1/U_0$ yields the energies of the impurity states (we focus here on the states within the energy gap). Small attractive values of U_0, produce a donor-like impurity level, split-off from the conduction band; small repulsive values of U_0 produce an acceptor-like impurity level, split-off from the valence band. Impurity states, deep in the forbidden energy gap, are controlled jointly by the conduction and valence band states.

The first member of Eq. (13.18) represents a Green's function matrix element, say $G_{00}(E)$; the structure of the equation $G_{00}(E) = 1/U_0$ has been discussed in Section 5.8.3 and already encountered in a number of problems; its graphical solution is schematically shown in Figure 13.6. The compatibility Eq. (13.18) summarizes the tight-binding Koster-Slater model of impurities in crystals [G. F. Koster and J. C. Slater, Phys. Rev. 95, 1167 (1954)]. The Green's function approach, with appropriate implementations, has been widely applied to investigate a number of defects, and in particular the substitutional deep traps in covalent semiconductors following the pioneering work of H. P. Hjalmarson, P. Vogl, D. J. Wolford and J. D. Dow, Phys. Rev. Lett. 44, 810 (1980). For further information on the vast theoretical and experimental investigations regarding the impurity levels in solids we refer to the literature.

13.3   Fermi Level and Carrier Density in Doped Semiconductors

Carrier Concentration in n-Type Semiconductors

We consider now extrinsic semiconductors, containing donor impurities, or acceptor impurities, or both, and we wish to study their influence on the Fermi level and the free carrier concentrations. We consider first the case of semiconductors in which only donor impurities are present (*n-type semiconductors*). The density N_d of donor impurities is supposed to be uniform in the sample, and the binding energy of the donor levels is ε_d. The schematic representation of the energy levels and occupancy (at $T = 0$) is given in Figure 13.7a.

In intrinsic semiconductors the Fermi level lies (basically) at the middle of the energy gap (see Eq. (13.6)). Doping with donors (or acceptor) levels is the most common method to change in a controlled way the position of the Fermi level within the energy gap. The presence of donor levels shifts the Fermi level from the middle of the energy gap toward the edge of the conduction band. Let us in fact define the temperature

$$k_B T_d \equiv \varepsilon_d,$$

where T_d can be considered as the "ionization temperature" of the donor levels. If $T \ll T_d$ we expect that practically all donor levels are occupied and thus the chemical potential must be located in the energy range $E_d < \mu(T) < E_c$. If T is comparable with T_d we expect that most donor levels are ionized and $\mu(T)$ lies somewhat below the donor energy E_d, but still very near to the conduction band edge. At temperatures so high that the intrinsic carriers are much larger than the concentration of donor impurities, doping becomes irrelevant and we expect that the chemical potential approaches the middle of the band gap. The chemical potential and the carrier concentration can be determined quantitatively from the knowledge of donor concentration, density-of-states of the bulk crystal, and appropriate Fermi-Dirac statistics for band levels and donor levels.

The impurity states within the energy gap are described by localized wavefunctions; a donor level can thus be empty, or occupied by one electron of either spin, but not by

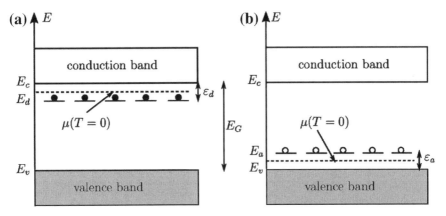

Figure 13.7 (a) Schematic representation of the energy levels of a homogeneously doped n-type semiconductor at $T = 0$ (in abscissa any arbitrary direction in the homogeneous material can be considered). Typical energy values are $E_G = E_c - E_v \approx 1$ eV and $\varepsilon_d = E_c - E_d \approx 10$ meV. The Fermi level at zero temperature lies at $(1/2)(E_d + E_c)$, which is the middle point between E_d and E_c. (b) Schematic representation of the energy levels of a homogeneously doped p-type semiconductor at $T = 0$; typical values of $\varepsilon_a = E_a - E_v$ are of the order of 10 meV. The Fermi level at zero temperature lies at $(1/2)(E_v + E_a)$, which is the middle point between E_v and E_a.

two electrons (of opposite spin) because of the penalty in the electrostatic repulsion energy. Due to this, the probability $P(E_d)$ that the level E_d is occupied by an electron of either spin is given by

$$P(E_d) = \frac{1}{(1/2)e^{(E_d-\mu)/k_BT} + 1};$$ (13.19)

the above expression has been derived in Appendix 3.C in the same way as the fundamental Fermi-Dirac statistics (13.1).

The chemical potential of the doped semiconductor is determined by enforcing the conservation of the total number of electrons as the temperature changes. In a semiconductor with N_d donor impurities per unit volume, the density $n_0(T)$ of electrons in the conduction band must satisfy the relation

$$\boxed{n_0(T) = N_d[1 - P(E_d)] + p_0(T)},$$ (13.20)

where n_0 and p_0 are given by expressions (13.2). Equation (13.20) is the straightforward generalization of Eq. (13.3); it states that the free electrons in the conduction bands are supplied by the thermal ionization of donor levels and by the thermal excitation of valence electrons. Equation (13.20) can also be interpreted as an overall *charge neutrality condition* in the sample: the concentration n_0 of negative charges equals the concentration of ionized donor impurities plus the concentration of holes.

Equation (13.20) can be solved (numerically) to obtain the Fermi level and hence the free carrier concentration. In the case the n-type semiconductor is non-degenerate (which is the ordinary situation, except for extremely high concentration of dopants), Eq. (13.20) can be simplified using Eqs. (13.5). We have:

$$N_c(T)e^{-(E_c-\mu)/k_BT} = N_d\frac{(1/2)e^{(E_d-\mu)/k_BT}}{(1/2)e^{(E_d-\mu)/k_BT}+1} + N_v(T)e^{-(\mu-E_v)/k_BT}. \quad (13.21)$$

This is a third order algebraic expression in $x = \exp(\mu/k_BT)$ that could be easily solved. We prefer to consider Eq. (13.21) in different regions of physical interest and handle it analytically.

(i) *Very low temperatures (or "freezing out region")*. Consider the semiconductor at very low temperatures $T \ll T_d$. In this temperature region we certainly have

$$E_d < \mu(T) < E_c.$$

Thus the second term in the right-hand side of Eq. (13.21) can safely be neglected; furthermore the denominator in the first term in the right-hand side of Eq. (13.21) can be taken as unity. It follows

$$N_c(T)e^{-(E_c-\mu)/k_BT} = \frac{1}{2}N_d\, e^{(E_d-\mu)/k_BT}; \quad (13.22a)$$

taking the logarithm of both members we obtain for the Fermi level

$$\mu(T) = \frac{1}{2}(E_d + E_c) + \frac{1}{2}k_BT\ln\frac{N_d}{2N_c(T)}. \quad (13.22b)$$

Insertion of Eq. (13.22b) into Eq. (13.22a) gives for the carrier density in the conduction band the expression

$$n_0(T) = N_c(T)e^{-(E_c-\mu)/k_BT} = \sqrt{N_c(T)\frac{N_d}{2}}\, e^{-\varepsilon_d/2k_BT}. \quad (13.23)$$

Thus, the temperature dependence of the free electron carriers in n-type semiconductors at temperatures $T \ll T_d$ has (approximately) the exponential form $\exp(-\Delta/k_BT)$, where Δ is *half the binding energy of the donor levels*. Notice that for high doping, Eq. (13.22b) shows a tendency of $\mu(T)$ to increase and possibly to invade the conduction band; in this situation one must consider directly the implicit Eq. (13.20) for the determination of the chemical potential.

(ii) *Saturation region*. Consider the semiconductor in the region $T_d < T \ll E_G/k_B$; we expect that (almost) all donor levels are ionized, while the thermal excitation of valence electrons is still negligible. We have

$$n_0(T) = N_c(T)e^{-(E_c-\mu)/k_BT} \cong N_d; \quad (13.24a)$$

from the logarithm of both members, we have for the chemical potential

$$\mu(T) = E_c + k_BT\ln\frac{N_d}{N_c(T)}. \quad (13.24b)$$

While the number $n_0(T)$ of majority carriers is essentially constant and equal N_d, the number of minority carriers is obtained by considering the mass-action law (13.7). In the *saturation region*, characterized by all donor levels ionized, and at temperatures

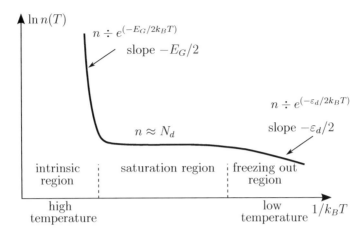

Figure 13.8 Schematic variation of the electron concentration as a function of $1/k_BT$ in an n-type semiconductor with N_d donor impurities per unit volume.

where $n_i(T) \ll N_d$, we have

$$n_0(T) \cong N_d \quad \text{and} \quad p_0(T) \cong \frac{n_i^2(T)}{N_d}. \tag{13.24c}$$

For instance, the intrinsic carrier concentration of silicon at room temperature is $n_i(T) \approx 10^{10}$ cm^{-3}. In n-type silicon with donor concentration $N_d \approx 10^{14}$ cm^{-3}, we have $n_0 \approx 10^{14}$ cm^{-3} and $p_0 \approx 10^6$ cm^{-3}; in the above situation there are eight orders of magnitude in the difference between the concentration of majority carriers and of minority carriers. Notice also that in silicon $N_c(T) \approx 10^{19}$ cm^{-3}; the chemical potential (13.24b) remains near the conduction band edge, but safely below it, so that the non-degeneracy conditions (13.4) are justified.

(iii) *Intrinsic region.* If we increase further the temperature, the thermal excitation of valence electrons into the conduction band increases, and eventually the intrinsic situation is recovered. The temperature dependence of the density of free electron carriers in an n-type semiconductor is schematically summarized in Figure 13.8.

Up to this point, impurities have been (tacitly) considered as isolated and independent; furthermore the doped semiconductor is assumed to remain non-degenerate, i.e. the Fermi level is several k_BT away from the band edges. As the concentration of dopants is increased new phenomena occur; for instance, the Fermi level may approach and invade the energy bands; the density-of-states of the semiconductor may be perturbed near the edges and a band gap narrowing may result; the impurity levels may interact forming an impurity band, with effects on the conductivity of the sample; here, we do not enter in these and other interesting consequences of heavy doping in semiconductors.

Carrier Concentration in p-Type Semiconductors

We consider now the case of semiconductors in which only acceptor impurities are present (*p-type semiconductors*). The density N_a of acceptor impurities is supposed to be uniform on the sample, and the binding energy of the acceptor levels is ε_a. The schematic representation of the energy levels and occupancy (at $T = 0$) is given in Figure 13.7b.

The chemical potential and the carrier concentration of p-type semiconductors can be determined from the knowledge of the acceptor concentration, density-of-states of the bulk crystal, and appropriate Fermi-Dirac statistics for band levels and acceptor levels. Concerning the statistics of acceptor levels, we notice that an acceptor level can be empty, or occupied by one hole of either spin, but not by two holes (of opposite spin) because of the penalty in the electrostatic repulsion energy. Due to this, the probability $P(E_a)$ that an acceptor level of energy E_a is occupied by a hole is

$$P(E_a) = \frac{1}{(1/2)e^{(\mu - E_a)/k_B T} + 1};$$

the above expression has been derived in Appendix 3.C in the same way as expression (13.19). It is seen by inspection that the carrier concentration in p-type semiconductors may be stated in terms of holes as the carrier concentration in n-type semiconductors is stated in terms of electrons; thus there is no need to repeat and report here the explicit results.

Up to now we have considered the cases of semiconductors doped either with donors or with acceptors. In actual semiconductors, beside the intentionally introduced donor (or acceptor) impurities, some concentration of uncontrolled acceptor (or donor) impurities may be present. A semiconductor, containing both donors with concentration N_d and acceptors with concentration N_a, is said to be *partially compensated* if $N_a \neq N_d$; the compensation is said to be ideal if $N_a = N_d$. The equilibrium carrier statistics of compensated semiconductors can be carried out with appropriate extension of the treatment developed so far. We do not enter in details, as the general equations of carriers balance can be easily established and solved in the actual situations of interest; here we only remark that it is essentially the heavier doping that dictates the p or n character of partially compensated semiconductors.

13.4   Non-Equilibrium Carrier Distributions

13.4.1   Drift and Diffusion Currents

In the previous sections, we have dealt with the distribution of electrons and holes at the thermodynamic equilibrium in uniformly doped (or undoped) semiconductors; in the present and the next section, still in uniformly doped materials, we consider non-equilibrium carrier distributions and transport phenomena of interest in the physics of semiconductors and devices.

In non-equilibrium and non-stationary conditions (produced for instance by applied fields, absorption of electromagnetic radiation of sufficiently high frequency, carrier injection or extraction, etc.), the electron and hole concentrations depend in general on space coordinate and time. In the following, we denote by $n(\mathbf{r}, t)$ and $p(\mathbf{r}, t)$ the actual non-equilibrium electron and hole concentrations in the semiconductor; the equilibrium electron and hole concentrations (at a given temperature T) are denoted as previously by n_0 and p_0; the differences $\Delta n(\mathbf{r}, t) = n(\mathbf{r}, t) - n_0$ and $\Delta p(\mathbf{r}, t) = p(\mathbf{r}, t) - p_0$, whether positive or not, are referred to as the "excess carrier concentrations" of electrons and holes, respectively.

We examine first the carrier transport in a semiconductor, with a uniform concentration of carriers, in the presence of an applied electric field. We consider the simplest possible model of semiconductor with two bands only: a parabolic conduction band of effective mass m_c^*, and a parabolic valence band of effective mass m_v^* (see Figure 13.3). A classical Drude-like picture of the dynamics and damping of carriers (or also a more rigorous approach based on the Boltzmann transport equation, see Chapter 11) gives for the electron current density

$$\mathbf{J}_n = \frac{ne^2 \tau_n}{m_c^*} \mathcal{E} \equiv ne\mu_n \mathcal{E}, \tag{13.25a}$$

where n is the density of electrons, e is the absolute value of the electronic charge, τ_n is the average interval between collisions suffered by an electron, and $\mu_n \equiv e\tau_n/m_c^*$ is the electron mobility, i.e. the average drift velocity of carriers divided by the applied electric field. Similarly for the hole current density we have

$$\mathbf{J}_p = \frac{pe^2 \tau_p}{m_v^*} \mathcal{E} \equiv pe\mu_p \mathcal{E}. \tag{13.25b}$$

Notice that $\mathbf{J}_n$ and $\mathbf{J}_p$ are proportional to the electric field (Ohm's law), and this is the expected behavior for sufficiently small electric fields.

In the following, we assume that the linear relationships (13.25) between current density and electric field hold. However we wish to mention that deviations from Ohm's law are expected at sufficiently high electric fields in semiconductors, due to several mechanisms. In metals, the high value of the conductivity (because of the high carrier density), and the consequent joule heating, strongly limit the values of applicable electric fields without fusion of the material; vice versa, in semiconductors high electric field effects may occur. As long as the applied electric field has low values ($\leq 10^3$ V/cm), the carriers lose energy mainly by emitting low energy acoustic phonons. As the field is increased, at around 10^4 V/cm, energy dissipation is through emission of optical phonons; in this region the drift velocity of the carriers is nearly independent of the electric field (*saturation velocity*). For still higher electric fields avalanche processes may become active. Another interesting source of deviations from Ohm's law occurs in semiconductors with secondary valleys in the conduction band structure at energies (slightly) higher than the principal minimum. In an *n*-type GaAs semiconductor, for instance, high frequency current oscillations (Gunn effect) are produced by the application of a constant electric field of the order of kvolt/cm. Electric fields of this order allow electrons in the lower energy band (of small mass and high mobility) to migrate into the higher valleys (of large mass and small mobility). Upon application of an electric field electrons move to higher valleys with a consequent decrease in current; vice versa the transition of the electrons to the lower valley leads to current increase; periodic current oscillations may thus result upon application of a constant potential.

We discuss now carrier transport in semiconductors in the presence both of an *electric field* and a *concentration gradient*. An intuitive classical picture that includes the concept of diffusion according to the Fick's law (or also the Boltzmann transport equation) gives for the electron current density the expression

$$\boxed{\mathbf{J}_n = en\mu_n\boldsymbol{\mathcal{E}} + eD_n\nabla n}\,, \tag{13.26}$$

where D_n is the diffusion coefficient. The current density $\mathbf{J}_n$ is the sum of two terms: the *drift term* proportional to the field (and already discussed), and the *diffusion term* proportional to the concentration gradient. The diffusion term takes into account the fact that the carriers tend to move from high concentration regions to lower concentration regions; for electrons the diffusion term and the concentration gradient have the same direction. In metals, n is so high to make the drift current in general dominant over the diffusion current; vice versa in semiconductors the two contributions may be of comparable size.

The mobility μ_n and the diffusion coefficient D_n are not independent but are related by the Einstein relation, already discussed in Section 11.6.2, and here reexamined for non-degenerate semiconductors on a more intuitive basis. Consider in fact a semiconductor, in stationary conditions and thermal equilibrium, with a superimposed potential $\phi(\mathbf{r})$; in an open circuit arrangement the current density vanishes, and Eq. (13.26) gives

$$\mathbf{J}_n(\mathbf{r}) \equiv 0 \quad \Longrightarrow \quad n(\mathbf{r})\mu_n[-\nabla\phi(\mathbf{r})] + D_n\nabla n(\mathbf{r}) = 0. \tag{13.27}$$

On the other hand, the bottom of the conduction band depends continuously on the space variable $\mathbf{r}$ via the relation

$$E_c(\mathbf{r}) = E_c + (-e)\phi(\mathbf{r}). \tag{13.28a}$$

From Eq. (13.5a), which applies to *non-degenerate* semiconductors, also the carrier density depends on $\mathbf{r}$ in the form

$$n(\mathbf{r}) = N_c(T)e^{-[E_c(\mathbf{r})-\mu]/k_BT}, \tag{13.28b}$$

where the chemical potential μ is independent of the position, since the system is at thermal equilibrium. From the gradient of both members of Eq. (13.28b) and the use of Eq. (13.28a), it follows

$$\nabla n(\mathbf{r}) = n(\mathbf{r})\frac{1}{k_BT}[-\nabla E_c(\mathbf{r})] = n(\mathbf{r})\frac{e}{k_BT}\nabla\phi(\mathbf{r})\,;$$

comparison with Eq. (13.27) yields

$$\boxed{eD_n = \mu_n k_BT}\,, \tag{13.29}$$

which is the Einstein relation for the diffusion coefficient.

A similar reasoning for the hole current density gives

$$\boxed{\mathbf{J}_p = ep\mu_p\boldsymbol{\mathcal{E}} - eD_p\nabla p}\,, \tag{13.30}$$

where the first term arises from the drift and the second term arises from the diffusion; notice that the diffusion term of the hole current is opposite to the direction of the hole concentration gradient. The diffusion coefficient and the mobility of holes are linked in non-degenerate semiconductors by the Einstein relation

$$\boxed{e D_p = \mu_p k_B T} . \tag{13.31}$$

The total current density $\mathbf{J}$ is then given by the equation

$$\mathbf{J} = e(n\mu_n + p\mu_p)\boldsymbol{\mathcal{E}} + e D_n \nabla n - e D_p \nabla p, \tag{13.32}$$

which is the sum of the drift and the diffusion currents of electrons and holes.

13.4.2   Continuity Equations and Generation Recombination Processes

In the theory of transport, the equation of continuity ensures that the carriers crossing a closed surface balance the change of the carriers enclosed therein. In semiconductors, we must express the continuity equation both for electrons and holes, taking into account that the electron and hole concentrations change not only because of drift and diffusion mechanisms, but also because of generation and recombination processes due to internal or external causes. We indicate by $G_n = G_n(\mathbf{r}, t)$ the number of electrons generated in the conduction band per unit time and unit volume, and $R_n = R_n(\mathbf{r}, t)$ the number of electrons subtracted from the conduction band per unit time and volume (space and time coordinates $\mathbf{r}$ and t are often left implicit, for simplicity). Similarly, G_p and R_p refer to the holes generated or subtracted in the valence band.

Generation and recombination processes may occur via a number of different mechanisms, and we mention in particular the following. (i) Electron and hole concentrations can change because of interband transitions from occupied valence states to empty conduction states or, vice versa, from occupied conduction states to empty valence states (band-to-band transitions); in this case we have evidently $G_n = G_p$ and $R_n = R_p$. (ii) Electron and hole concentrations can change because of transitions to or from impurity levels (band-to-impurity transitions); the impurity centers of relevance (if any) are those deep within the energy gap, since at temperatures of ordinary interest all the shallow impurity levels (donors and acceptors) are thermally ionized. (iii) Electron and hole concentrations can change because of carrier injection or extraction from "external sources" (for instance applied electrodes; illumination of the specimen with a beam of photons of energies higher than the energy gap; bombardment with charged particles that lose kinetic energy by creating electron-hole pairs; avalanche multiplication processes due to the presence of strong electric fields, direct tunneling of electrons from the valence band to the conduction band in strong electric fields, etc.).

In the case generation-recombination processes of relevance involve electron-hole pairs, electrons and holes are created or annihilated at the same rate. However, in general, the rates for electrons and holes may be different.

Let $n(\mathbf{r}, t)$ indicate the electron concentration in the semiconductor, and $\mathbf{J}_n$ the electron current density. In the presence of electron generation-recombination processes at a rate $G_n(\mathbf{r}, t)$ and $R_n(\mathbf{r}, t)$, the continuity equation for electrons takes the form

$$\frac{\partial n}{\partial t} = \frac{1}{e} \text{div } \mathbf{J}_n + G_n(\mathbf{r}, t) - R_n(\mathbf{r}, t);$$ (13.33a)

in fact, if we multiply both members of Eq. (13.33a) by the volume element dV and integrate over any given volume V enclosed by a surface S, we obtain a very intuitive balance between the change of the number of electrons in the volume V, and the electrons crossing the surface or being supplied by the generation-recombination processes.

In a completely similar way, the time and space dependent hole concentration $p(\mathbf{r}, t)$ satisfies the continuity equation

$$\frac{\partial p}{\partial t} = -\frac{1}{e} \text{div } \mathbf{J}_p + G_p(\mathbf{r}, t) - R_p(\mathbf{r}, t).$$ (13.33b)

We can also write a continuity equation for the total current density $\mathbf{J} = \mathbf{J}_n + \mathbf{J}_p$ and the total electron and hole charge density $\rho = (-e)n + ep$. From Eqs. (13.33) we obtain

$$\mathbf{J} = \mathbf{J}_n + \mathbf{J}_p \quad \Longrightarrow \quad \frac{\partial}{\partial t}(p - n) = -\frac{1}{e}\text{div } \mathbf{J} + G_p - R_p - G_n + R_n,$$

The continuity equation for the total current greatly simplifies when generation and recombination processes involve electron-hole pairs; in this case we have

$$R_n = R_p, \quad G_n = G_p \quad \Longrightarrow \quad \frac{\partial}{\partial t}(p - n) = -\frac{1}{e}\text{div } \mathbf{J}.$$ (13.34)

Differently from Eqs. (13.33), the continuity equation (13.34) for the total current density does not contain explicitly electron and hole recombination-generation rates, when the carriers are produced or annihilated in pairs.

The continuity equation (13.33a) for the electron concentration is more conveniently written using Eq. (13.26) and assuming, for simplicity, that the electric field and concentration gradients are in the x-direction. One obtains the basic equation

$$\boxed{\frac{\partial n}{\partial t} = D_n \frac{\partial^2 n}{\partial x^2} + \mu_n \frac{\partial(n\mathcal{E})}{\partial x} + G_n - R_n}.$$ (13.35a)

The interpretation of the right part of Eq. (13.35a) is very simple: the first term represents the change of concentration due to the diffusion, the second term the drift and diffusion in the presence of an electric field, and finally the last terms the generation and recombination processes. In a completely similar way, the continuity equation for holes reads

$$\boxed{\frac{\partial p}{\partial t} = D_p \frac{\partial^2 p}{\partial x^2} - \mu_p \frac{\partial(p\mathcal{E})}{\partial x} + G_p - R_p}.$$ (13.35b)

The two basic continuity equations (13.35) for electrons and holes are not independent; they are coupled through the presence of the electric field, related self-consistently to the total charge density via the Poisson equation.

13.5 Generation and Recombination of Electron-Hole Pairs in Doped Semiconductors

The solution of the basic continuity equations (13.35) requires the knowledge (or at least some minimal information) on the generation and recombination processes of electrons and holes in doped semiconductors. A microscopic description of all the possible mechanisms would be rather demanding, and is beyond of the scope of this text.

For simplicity, we confine our attention to a few qualitative considerations concerning semiconductors doped with shallow donor or acceptor levels, in the regime of full ionization of impurities. It is assumed that the kinetics of the carriers is mostly determined by radiative generation-recombination processes of electron-hole pairs, occurring with rates say R_{eh} and G_{eh}. In these processes the generation and recombination rates for electrons and holes are the same:

$$G_n = G_p = G_{eh}; \quad R_n = R_p = R_{eh}. \tag{13.36}$$

Band-to-band transitions are typical processes where electrons and holes participate in pairs. Electrons in the conduction band can make optical transitions to the valence band, with annihilation of an electron and a hole. Similarly, an electron from the valence band can be transferred to the conduction band with absorption of a photon.

The interband generation process of electron-hole pairs is controlled by the number of occupied electron states in the valence band, by the number of empty electronic states in the conduction band, and by the spectral density of photons, whose frequencies depend on the optical properties of the material and whose number is given by the Planck formula. The generation rate is practically independent of the possible presence of holes in the valence band and electrons in the conduction band (at their ordinary concentrations). It is thus reasonable to assume that the thermal generation process is determined by the intrinsic material properties, and takes the form

$$G_{eh} = G_{eh}(T).$$

The interband recombination process of electron-hole pairs is controlled by the number of available electrons in the conduction band and by the number of available holes in the valence band, besides the radiation field. We thus expect

$$R_{eh} = G_{eh}(n, p, T).$$

Taking into account that R_{eh} must vanish if either n or p vanish, it appears reasonable to assume R_{eh} proportional to the product np.

In summary, in the absence of external perturbations, the simplest phenomenological description of band-to-band carrier kinetics is provided by the bimolecular equation

$$R_{eh} - G_{eh} = B(np - n_0 p_0) \quad \left(n_0 p_0 = n_i^2 = p_i^2\right), \tag{13.37}$$

where $B(T)$ is the bimolecular recombination parameter for the material under investigation, n and p are the actual non-equilibrium electron and hole concentrations, n_0 and p_0 are the thermal equilibrium electron and hole concentrations of the semiconductors, n_i and p_i the intrinsic concentrations. In Eq. (13.37) the generation rate G_{eh} has been expressed, for convenience, in such a way to make evident that at thermal equilibrium $G_{eh} \equiv R_{eh}$, and the net effect of generation-recombination processes vanishes.

Expression (13.37) becomes particularly agile in the case of semiconductors with large difference in the density of majority and minority carriers. In fact, for n-type or p-type semiconductors the linearization of Eq. (13.37) puts in evidence the direct link between the phenomenological constant B and the lifetime of minority carriers, one of the most important concepts in the physics of semiconductors.

Consider first the case of an n-type semiconductor, where $n_0 \gg p_0$. We also suppose that the deviations Δn and Δp are much smaller than n_0, i.e. $\Delta n \approx \Delta p \ll n_0$ (this is called *low disturbance or low injection regime*). In an n-type semiconductor in the low disturbance regime, we can replace n with n_0 in the right-hand side of Eq. (13.37), and we have

$$R_{eh} - G_{eh} = Bn_0(p - p_0) \qquad n\text{-semiconductor.} \tag{13.38}$$

The quantity Bn_0 has the dimension of the inverse of a time. It is thus convenient to define the *minority carrier lifetime* $t_p = 1/Bn_0$. Notice that t_p can be interpreted as the lifetime of the minority carriers in the presence of a (hostile) concentration n_0 of majority carriers. Equation (13.38) can be rewritten as

$$R_{eh} - G_{eh} = \frac{p}{t_p} - \frac{p_0}{t_p} \qquad n\text{-semiconductor.} \tag{13.39}$$

The recombination-generation rate is proportional to the excess density of the minority carriers. The generation rate of electron-hole pairs p_0/t_p is just the equilibrium value of the minority carrier density divided by the minority carrier lifetime. Similarly the recombination rate is the actual value of minority carrier density divided by the lifetime.

We now specify Eq. (13.35b) in the case the electron-hole generation-recombination mechanism at work is described by Eq. (13.39); we obtain

$$\boxed{\frac{\partial p}{\partial t} = D_p \frac{\partial^2 p}{\partial x^2} - \mu_p \mathcal{E} \frac{\partial p}{\partial x} - \mu_p p \frac{\partial \mathcal{E}}{\partial x} + \frac{p_0 - p}{t_p}} \qquad n\text{-semiconductor.} \tag{13.40}$$

This is the basic continuity equation for holes minority carriers in an n-type semiconductor. In the case of a homogeneous semiconductor in the absence of electric fields and carrier gradients, we see that the minority carrier concentration (and hence also the majority carrier concentration) relaxes toward the equilibrium value with time constant t_p (typical values of t_p are often in the range $\approx 10^{-6} - 10^{-8}$ s).

Similar considerations can be done for p-type semiconductors, where holes are the majority carriers and electrons the minority ones. The partner equation of (13.40) is

$$\boxed{\frac{\partial n}{\partial t} = D_n \frac{\partial^2 n}{\partial x^2} + \mu_n \mathcal{E} \frac{\partial n}{\partial x} + \mu_n n \frac{\partial \mathcal{E}}{\partial x} + \frac{n_0 - n}{t_n}} \qquad p\text{-semiconductor.} \tag{13.41}$$

Equation (13.41) is the *basic continuity equation for electrons minority carriers in a p-type semiconductor.*

Equations (13.40) and (13.41) are reasonable manageable to be solved, and several analytic examples are provided in Appendix A. Equations (13.40) and (13.41) are the basic ingredient for the calculation of currents in p-n junctions and bipolar transistors, as discussed in Chapter 14.

Appendix A. Solutions of Typical Transport Equations in Uniformly Doped Semiconductors

Problem 1. Dielectric relaxation time and charge neutrality in uniformly doped semi-conductors.

Problem 2. Injection of minority carriers at one end of an n-type semiconductor in steady conditions.

Problem 3. Injection (or extraction) of carriers at both ends of an n-type semiconductor in steady conditions.

Problem 4. Injection of a narrow bunch of minority carriers and measurement of drift mobility (Haynes-Shockley experiment).

Problem 1. Dielectric Relaxation Time and Charge Neutrality in Uniformly Doped Semiconductors

Consider a n-type (or p-type) homogeneously doped semiconductor. Suppose that an unbalanced excess of carriers $\Delta p \neq \Delta n$ is produced in the sample. Show that the unbalance $\Delta p - \Delta n$ quickly goes to zero with time constant equal to the dielectric relaxation time $t_D = \varepsilon_s / \sigma_s$, where ε_s is the dielectric constant of the semiconductor and σ_s its conductivity.

Consider a homogeneously doped semiconductor (say n-type) with equilibrium carrier concentrations n_0 and p_0, and $p_0 \ll n_0$. At thermal equilibrium, the *charge neutrality condition* is satisfied over all the sample. This problem shows that also non-equilibrium carrier concentrations must satisfy a *charge quasi-neutrality condition* throughout the sample.

According to Eq. (13.34), the continuity equation for the total current density (in ordinary situations where the electron-hole generation-recombination processes occur in pairs) reads

$$\frac{\partial}{\partial t}(p - n) = -\frac{1}{e}\,\mathrm{div}\,\mathbf{J},$$

or equivalently

$$\frac{\partial}{\partial t}(\Delta p - \Delta n) = -\frac{1}{e}\,\mathrm{div}\,\mathbf{J}, \tag{A.1}$$

where $\Delta p = p - p_0$ is the excess hole concentration, and $\Delta n = n - n_0$ is the excess electron concentration. If $\Delta n \neq \Delta p$ a space charge $e(\Delta p - \Delta n)$ is set up in the material (assuming a regime of fully ionized impurity centers). Thus an electric field arises that satisfies the Poisson equation

$$\mathrm{div}\,\boldsymbol{\mathcal{E}} = \frac{e}{\varepsilon_s}(\Delta p - \Delta n), \tag{A.2}$$

where $\varepsilon_s = \varepsilon_r \varepsilon_0$ is the static dielectric constant of the semiconductor (we use here standard SI units). It is easy to infer that the unbalance $\Delta p - \Delta n$ can not survive in the uniformly doped semiconductor: the unbalance $\Delta p - \Delta n$ quickly goes to zero (or almost to zero) because of the drift current of the majority carriers in the electric field $\mathcal{E}$, created by the unbalance itself.

To link Eqs. (A.1) and (A.2) we assume for simplicity

$$\mathbf{J} \approx \sigma_s \mathcal{E} \qquad \sigma_s = en_0\mu_n + ep_0\mu_p \approx en_0\mu_n, \tag{A.3}$$

where σ_s denotes the conductivity of the n-type semiconductor, essentially due to the majority carriers (diffusion terms in the current density have been omitted in the low disturbance approximation). From Eqs.(A.1)–(A.3) one obtains

$$\frac{\partial}{\partial t}(\Delta p - \Delta n) = -\frac{\sigma_s}{\varepsilon_s}(\Delta p - \Delta n);$$

this equation shows that the space-charge decreases exponentially with time constant $t_D = \varepsilon_s/\sigma_s$ (*dielectric relaxation time*).

To estimate t_D, consider for instance an n-type GaAs semiconductor with $N_D = 10^{14}$ cm^{-3}. In the saturation regime $n_0 \approx N_D$. The electron mobility (see Table 13.1) is $\mu_n = 8500$ cm^2/V$\cdot$s; the dielectric constant of the semiconductor is $\varepsilon_s = \varepsilon_0\varepsilon_r$ with $\varepsilon_r = 13.2$ and $\varepsilon_0 \approx 8.85 \cdot 10^{-14}$ F/cm. We can estimate

$$t_D = \frac{\varepsilon_s}{\sigma_s} = \frac{\varepsilon_r\varepsilon_0}{en_0\mu_n} = \frac{13.2 \cdot 8.85 \cdot 10^{-14}}{1.6 \cdot 10^{-19} \cdot 10^{14} \cdot 8500} \approx 10^{-11} \text{ s.}$$

Notice that t_D is in general much smaller than typical lifetime of minority carriers.

The above considerations show that the space-charge neutrality condition $\Delta n \approx \Delta p$ is reached within the very short dielectric relaxation time t_D, because of the drift of the majority carriers: any deviation from the space-charge neutrality condition $\Delta n \approx \Delta p$ tends to be neutralized very quickly, within an interval of the order of the dielectric relaxation time. This estimate is confirmed by more detailed analyses, that show that charge quasi-neutrality holds in ordinary situations in homogeneously doped materials. In Chapter 14, in the study of semiconductors with non-uniform doping (p-n junction, for instance) or non-uniform energy band structure (heterostructures, for instance) we will see that space-charge regions with large built-in electric fields are formed in the non-homogeneity regions.

Problem 2. Injection of Minority Carriers at One End of an n-type Semiconductor in Steady Conditions

Consider a bar of homogeneously doped n-type semiconductor and suppose that an external source (for instance light irradiation or particle bombardment) maintains an excess of holes $\Delta p(x) = \Delta p(0)$, and an excess of electrons $\Delta n(x) = \Delta n(0) = \Delta p(0)$ for any $x \leq 0$. Carriers diffuse in the $x > 0$ part of the semiconductor. Determine the excess concentration $\Delta p(x)$ and $\Delta n(x)$ in stationary conditions; also determine the internal electric field, that automatically sets in and accompanies the diffusion process (even in the open circuit arrangement of Figure 13.9).

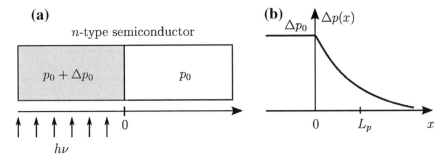

Figure 13.9 (a) Schematic representation of an n-type sample subject to irradiation. (b) Distribution of excess holes (and excess electrons) for $x > 0$.

The basic equation (13.40) controls the *dynamics of the minority carriers* in the region $x > 0$ (where only thermal and radiative electron-hole generation and recombination mechanisms are present). We ignore (momentarily) the internal electric field, which may accompany the diffusion processes, and also consider steady conditions. Equation (13.40) simplifies in the field-free steady-state continuity equation

$$D_p \frac{d^2 p}{dx^2} - \frac{p - p_0}{t_p} = 0 \quad n\text{-semiconductor} \quad (x > 0). \tag{A.4}$$

The boundary conditions for the carrier concentration $p(x)$ are given by

$$p(0) \equiv p_0 + \Delta p(0) \quad \text{and} \quad p(\infty) \equiv p_0. \tag{A.5}$$

The solution of Eq. (A.4), which fulfills the boundary conditions (A.5), is

$$p(x) - p_0 = \Delta p(0) e^{-x/L_p} \quad \text{with} \quad L_p^2 = D_p t_p. \tag{A.6}$$

The length L_p is called the *minority carrier diffusion length*. With $D_p \approx 10 \text{ cm}^2/\text{s}$ and $t_p \approx 10^{-5}$ s the diffusion length of minority carriers is $L_p \approx 0.01$ cm. The behavior of the excess of minority carriers concentration $\Delta p(x) = p(x) - p_0$ for $x > 0$ is plotted in Figure 13.9b. Notice that the mean square displacement of the excess particle distribution (A.6) is

$$\langle x^2 \rangle = \frac{\int_0^\infty x^2 \exp(-x/L_p) dx}{\int_0^\infty \exp(-x/L_p) dx} = 2L_p^2 = 2D_p t_p. \tag{A.7}$$

The hole diffusion current in the x-direction becomes

$$J_p^{(\text{diff})}(x) = -e D_p \frac{\partial p}{\partial x} = e D_p \frac{\Delta p(0)}{L_p} e^{-x/L_p}; \tag{A.8}$$

in particular for $x = 0$, we have

$$J_p^{(\text{diff})}(0) = e D_p \frac{\Delta p(0)}{L_p} = e L_p \frac{\Delta p(0)}{t_p}. \tag{A.9}$$

The expression of the hole diffusion current at $x = 0$ is particularly meaningful; in fact $L_p \Delta p(0)/t_p$ represents the number of holes that must be supplied per unit time through the plane $x = 0$ to keep approximately an excess of carriers $\Delta p(0)$ (with lifetime t_p) in the hostile region of width L_p in the n-type semiconductor.

Due to the local quasi-neutrality conditions discussed in Problem 1, the excess of majority carriers is given with good approximation

$$\Delta n(x) \approx \Delta p(x) \equiv \Delta p(0) e^{-x/L_p}.$$

The electron diffusion current in the x-direction becomes

$$J_n^{(\text{diff})}(x) = e D_n \frac{\partial n}{\partial x} = -e D_n \frac{\Delta p(0)}{L_p} e^{-x/L_p} = -b J_p^{(\text{diff})}(x), \tag{A.10}$$

where $b = D_n/D_p = \mu_n/\mu_p$.

From Eq. (A.10) we see that the sum of the diffusion currents due to electrons and holes is not zero. However, the *total current* (diffusion current plus drift current of electrons and holes) obeys equation (A.1), and so it is a constant in steady situation, and actually vanishes in the open circuit situation we are investigating. We thus must have at any x point

$$J_{\text{tot}}(x) = J_p^{(\text{diff})}(x) + J_n^{(\text{diff})}(x) + \sigma_n \mathcal{E}(x) \equiv 0, \tag{A.11}$$

where $\sigma_n = e n \mu_n$ is the conductivity of the electrons in the n-type material, and the much smaller conductivity of the minority carriers is neglected. From Eq. (A.11) we recover for the electric field

$$\mathcal{E}(x) = \frac{1}{\sigma_n} (b - 1) e D_p \frac{\Delta p(0)}{L_p} e^{-x/L_p}. \tag{A.12}$$

This shows that also the electric field decreases exponentially in the region $x > 0$; it is also evident that for $b = 1$ the electron and hole diffusion currents are equal and opposite, and the electric field is zero. In most ordinary conductivity situations, the electric field (A.12) is anyway so small that corrections to the field-free Eq. (A.4) are hardly necessary.

Problem 3. Injection (or Extraction) of Carriers at Both Ends of an n-type Semiconductor in Steady Conditions

Consider a homogeneously doped n-type semiconductor and suppose to maintain for $x \leq 0$ an excess (or a deficit) of holes $\Delta p(0)$ and for $x \geq w$ an excess (or deficit) of holes $\Delta p(w)$ (see Figure 13.10). Carriers diffuse in the $0 < x < w$ part of the semiconductor. Determine the excess carrier concentrations and currents in stationary conditions.

Similarly to Problem 2, the minority carrier distribution is described by the field-free steady-state continuity equation

$$(0 < x < w) \quad D_p \frac{d^2 p}{dx^2} - \frac{p - p_0}{t_p} = 0 \quad n\text{-semiconductor}. \tag{A.13}$$

The boundary conditions for the carrier concentration $p(x)$ are given by

$$p(0) \equiv p_0 + \Delta p(0) \quad \text{and} \quad p(w) \equiv p_0 + \Delta p(w). \tag{A.14}$$

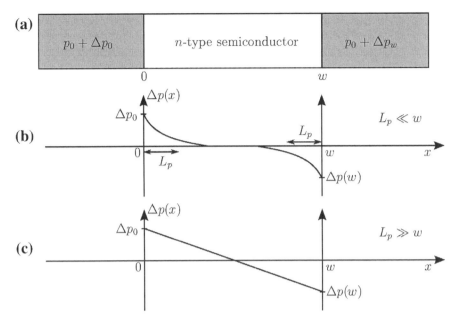

Figure 13.10 (a) Schematic representation of an n-type semiconductor, with hole concentration $p_0 + \Delta p(0)$ for $x < 0$ and $p_0 + \Delta p(w)$ for $x > w$; we assume $\Delta p(0) > 0$ and $\Delta p(w) < 0$. (b) Distribution of excess holes (and excess electrons) as a function of x in the case $L_p \ll w$. (c) Distribution of excess holes (and excess electrons) as a function of x in the case $L_p \gg w$.

The second order differential equation (A.13) has two linearly independent solutions $\exp(\pm x/L_p)$; the linear combination that fulfills the boundary conditions (A.14) is

$$\Delta p(x) = \Delta p(0)\frac{\sinh(w - x)/L_p}{\sinh(w/L_p)} + \Delta p(w)\frac{\sinh(x/L_p)}{\sinh(w/L_p)} \quad 0 \le x \le w. \quad (A.15)$$

The hole diffusion current associated to the above carrier distribution is

$$J_p^{(\text{diff})}(x) = -eD_p\frac{\partial \Delta p(x)}{\partial x} = eD_p\frac{\Delta p(0)}{L_p}\frac{\cosh[(w - x)/L_p]}{\sinh(w/L_p)}$$

$$- eD_p\frac{\Delta p(w)}{L_p}\frac{\cosh(x/L_p)}{\sinh(w/L_p)}. \quad (A.16)$$

We now discuss the concentration of minority carriers and diffusion currents in the two limiting cases $w \gg L_p$ and $w \ll L_p$, i.e. when the quasi-neutral central device is "long" or "short" with respect to the diffusion length.

In the case the width w is much larger than the diffusion length, Eqs. (A.15), (A.16) simplify, and we have in particular

$$\Delta p(x) = \Delta p(0)e^{-x/L_p} + \Delta p(w)e^{-(w-x)/L_p} \quad 0 \le x \le w \quad w \gg L_p$$

and

$$J_p^{(\text{diff})}(0) = eD_p \frac{\Delta p(0)}{L_p}, \quad J_p^{(\text{diff})}(w) = -eD_p \frac{\Delta p(w)}{L_p}. \tag{A.17}$$

In the long device configuration, the concentration of minority carriers, as well as diffusion currents penetrating in the semiconductors from the two ends, are completely decoupled (as shown in Figure 13.10b).

A much more interesting situation occurs in the case $L_p \gg w$. With appropriate series expansion of the terms at the right member of Eqs. (A.15) and (A.16), one obtains

$$\Delta p(x) = \Delta p(0) \frac{w - x}{w} + \Delta p(w) \frac{x}{w} + \cdots = \Delta p(0) + [\Delta p(w) - \Delta p(0)] \frac{x}{w} + \cdots$$

and

$$J_p^{(\text{diff})}(0) = J_p^{(\text{diff})}(w) = eD_p \frac{\Delta p(0) - \Delta p(w)}{w}. \tag{A.18}$$

Expressions (A.17) (long device configuration) and expression (A.18) (short device configuration) are rather similar, except for the use in the denominator of the smaller between the two lengths L_p and w. It is evident that in the short device configuration $w \ll L_p$ the diffusion current is greatly increased (a similar effect is crucial for the operation of the bipolar junction transistor, described in Chapter 14).

Due to the local quasi-neutrality condition the excess majority carrier concentration is given approximately by $\Delta n(x) \approx \Delta p(x)$; from the requirement that the total current must be zero, the (small) electric field in the sample can be obtained, with considerations similar to those worked out in Problem 2, and not repeated here.

Problem 4. Injection of a Narrow Bunch of Minority Carriers and Measurement of Drift Mobility (Haynes-Shockley Experiment)

Consider an n-type semiconductor and suppose that a narrow bunch of holes (and electrons) is produced by a source in the plane $x = 0$ at the time $t = 0$ (for instance by irradiation of light, or in practice by a voltage-pulse applied to an emitting contact). Determine the excess hole (and electron) concentration $\Delta p(x, t)$ at successive instants $t > 0$ in the semiconductor.

The equation that governs the dynamics of the holes minority carriers in the injection experiment is the basic Eq. (13.40); neglecting momentarily the terms containing the electric field, we have

$$\frac{\partial p}{\partial t} = D_p \frac{\partial^2 p}{\partial x^2} - \frac{p - p_0}{t_p} \quad n\text{-semiconductor.} \tag{A.19}$$

The solution of this equation with the boundary condition specified in the problem is

$$p(0, 0) = p_0 + N\delta(t)\delta(x) \Longrightarrow$$

$$\Delta p(x, t) = \frac{N}{\sqrt{4\pi D_p t}} \exp\left(-\frac{x^2}{4D_p t}\right) \exp\left(-\frac{t}{t_p}\right); \tag{A.20}$$

in fact it is easily seen by inspection that the above expression of $\Delta p(x, t)$ satisfies the differential equation (A.19), as well as the boundary conditions $\Delta p(0, 0) = N\delta(t)\delta(x)$.

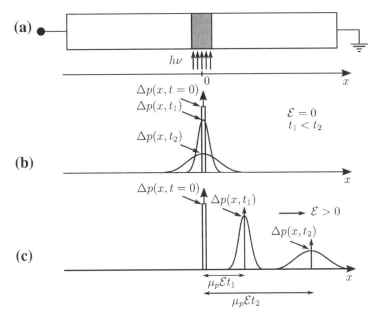

Figure 13.11 (a) Schematic representation of an n-type semiconductor with a narrow bunch of holes (and electrons) optically injected at $x = 0$ and $t = 0$. (b) Distribution of excess holes (and electrons) in the absence of applied electric fields. (c) Distribution of excess holes (and electrons) in the presence of a constant and uniform electric field.

The schematic description of the device and the behavior of $\Delta p(x, t)$ at different times in the semiconductor is illustrated in Figure 13.11.

It is of interest to evaluate the average square distance $\langle x^2(t) \rangle$ of the carriers at time t; we have

$$\langle x^2(t) \rangle = \int_0^\infty x^2 \Delta p(x, t) \, dx \Big/ \int_0^\infty \Delta p(x, t) \, dx = 2D_p t. \tag{A.21}$$

The average distance $\langle x^2 \rangle$ travelled by the carriers during their lifetime is then $\langle x^2 \rangle = 2D_p t_p$, in agreement with the previous expression in Eq. (A.7). The calculation of $\langle x^2(t) \rangle$ has been carried out using the following elementary integrals of the Gaussian function

$$\int_0^\infty e^{-ax^2} dx = \frac{1}{2}\sqrt{\frac{\pi}{a}} \quad \text{and} \quad \int_0^\infty x^2 e^{-ax^2} dx = \frac{1}{4a}\sqrt{\frac{\pi}{a}} \quad a > 0.$$

We discuss now the injection of a narrow bunch of holes in the presence of a uniform and constant electric field $\mathcal{E}$ (in the x-direction); instead of Eq. (A.19), we have to consider the partial differential equation of the form

$$\frac{\partial p}{\partial t} = D_p \frac{\partial^2 p}{\partial x^2} - \frac{p - p_0}{t_p} - \mu_p \mathcal{E} \frac{\partial p}{\partial x};$$

the solution with the boundary condition $\Delta p(0, 0) = N\delta(t)\delta(x)$ is

$$\Delta p(x, t; \mathcal{E}) = \frac{N}{\sqrt{4\pi D_p t}} \exp\left[-\frac{(x - \mu_p \mathcal{E} t)^2}{4D_p t}\right] \exp\left(-\frac{t}{t_p}\right).$$

The expression $\Delta p(x, t; \mathcal{E})$ has the same form as $\Delta p(x, t; \mathcal{E} \equiv 0)$ of Eq. (A.20), except that x is now replaced by $x - \mu_p \mathcal{E} t$; thus the excess carriers (both holes and electrons) follow the electric field with drift velocity $\mu_p \mathcal{E}$, as shown in Figure 13.11c.

The above results are the basis for the understanding of the Haynes-Shockley experiment for measuring the drift mobility of minority carriers. The method consists in the injection of a narrow bunch of minority carriers into a filament and sweeping the bunch along the filament by means of an electric field; if L is the distance between the emitter and collector contacts, from the measurement of the transit time $t = L/\mu_p \mathcal{E}$ the mobility of minority carriers can be obtained.

Further Reading

Bassani, F., Iadonisi, G., & Preziosi, B. (1974). Electronic impurity levels in semiconductors. *Reports on Progress in Physics, 37*, 1099.

Bhattacharya, P. (2004). *Semiconductor optoelectronic devices*. Englewood Cliffs, New Jersey: Prentice Hall.

Grundmann, M. (2006). *The physics of semiconductors*. Berlin: Springer.

Jaros, M. (1989). *Physics and applications of semiconductor microstructures*. Oxford: Clarendon Press.

Muller, R. S., Kamins, T. I., & Chan, M. (2002). *Device electronics for integrated circuits*. New York: Wiley.

Pantelides, S. T. (1978). The electronic structure of impurities and other point defects in semiconductors. *Reviews of Modern Physics, 50*, 797.

Sapoval, B., & Hermann, C. (1995). *Physics of semiconductors*. New York: Springer.

Seeger, K. (2004). *Semiconductor physics*. Berlin: Springer.

Smith, R. A. (1978). *Semiconductors*. Cambridge: Cambridge University Press.

Sze, S. M., & Lee, M.-K. (2012). *Semiconductor devices: Physics and technology*. New York: Wiley.

Wolfe, C. M., Holonyak, N., & Stillman, G. E. (1989). *Physical properties of semiconductors*. Englewood Cliffs, New Jersey: Prentice Hall.

14 Transport in Inhomogeneous Semiconductors

This chapter is the natural extension of the previous one and deals with semiconducting structures that are inhomogeneous, because of non-uniform doping or non-uniform band structure, or both. It would be impossible to describe all the electron devices that can be tailored in this way; thus we focus on simple selected models, which nevertheless should give a feeling of the art of band structure engineering. Particular attention is given to the bipolar junction transistor and to the metal-oxide-semiconductor field-effect-transistor, because of their prominent fundamental and technological importance.

We begin this chapter with a discussion of *p-n* homojunctions, obtained by doping different regions of the same semiconductor with acceptor and donor impurities. Among the devices, whose building blocks are *p-n* homojunctions, the main interest is devoted to the *p-n-p* (or *n-p-n*) bipolar structures, where the transistor action is achieved through the interaction of two back-to-back junctions. We then consider a variety of structures that can be envisaged when the flexibility in doping is accompanied also by the flexibility in the composing materials; this analysis includes semiconductor heterostructures, metal-semiconductor contacts, metal-oxide-semiconductor junctions. Finally we discuss the field-effect transistor action, achieved through modulation of the width of the conductivity channel in doped semiconductors.

14.1   Properties of the *p-n* Junction at Equilibrium

A *p-n* junction (or *p-n* diode) consists of a semiconducting crystal with spatially dependent concentration of shallow impurities; acceptor impurities are prevalent in the *p*-type region, while donor impurities are prevalent in the *n*-type region. The *p-n*

Solid State Physics, Second Edition. http://dx.doi.org/10.1016/B978-0-12-385030-0.00014-1

junction is perhaps the simplest electronic device manufactured with inhomogeneous doping of a semiconductor; the study of its physical properties is vital also for the comprehension of many other semiconducting devices.

In the following, just for the sake of simplicity, we consider an *ideal abrupt p-n junction*; it consists of a semiconducting single crystal with a uniform concentration N_a of acceptor impurities (at the left, say, of the $x = 0$ plane) and a uniform concentration N_d of donor impurities (at the right of the $x = 0$ plane). In reality the transition from the p-region to the n-region is not discontinuous, but gradual; furthermore a small number of donor impurities may be present in the p-type region (and vice versa); these details, however, are here disregarded because they do not change the essential aspects of the model.

Before considering the p-n junction at equilibrium, we represent schematically in Figure 14.1a the energy bands and the Fermi energies of the p-type semiconductor and n-type semiconductor, not yet in contact. As seen in Section 13.3, the Fermi energy E_{Fp} for the p-type semiconductor lies in the energy gap rather near to the top E_v of the valence band, while the Fermi energy E_{Fn} of the n-type semiconductor lies in the energy gap rather near to the bottom E_c of the conduction band. Both semiconductors are assumed non-degenerate. In the saturation regime (often considered for ready calculations) the hole concentration in the p-type region at thermodynamic equilibrium is $p_{p0} = N_a$; similarly, in the n-type region, $n_{n0} = N_d$ (the concentration of minority carriers in both regions is then determined by the mass-action law). In the n-type material, the majority carriers are mobile electrons, of density n_{n0}, embedded in the background of immobile positively ionized donors with the same density N_d; similarly, in the p-type material, the majority carriers are mobile holes, of density p_{p0}, embedded in the background of immobile negatively ionized acceptors with the same density N_a.

Now suppose that the p-type and n-type specimens are joined together. On formation of the junction, some electrons leave the n-type material and move toward the p-region, where the chemical potential is lower; the leaving electrons, once in the hostile p-region, recombine with holes. Similarly, holes from the p-type material move to the n-region and recombine with electrons. A region depleted of the majority carriers is thus formed near the junction (*depletion region*). In the depletion layer a *space-charge* is present: it is constituted by an amount of immobile non-neutralized positive donor centers in the n-side part of the semiconductor and by an equal amount of immobile non-neutralized negative acceptor centers in the p-side part. The space-charge produces a strong electric field so to oppose the flux of carriers; eventually, any net flux of carriers vanishes, when the Fermi level is spatially uniform throughout the system (see Figure 14.1b). At equilibrium, drift and diffusion currents of electrons and holes must balance exactly at any point of the junction, and we have:

$$J_n(x) = J_n^{(\text{drift})}(x) + J_n^{(\text{diff})}(x) = ne\mu_n \mathcal{E} + eD_n \frac{dn}{dx} \equiv 0, \tag{14.1a}$$

$$J_p(x) = J_p^{(\text{drift})}(x) + J_p^{(\text{diff})}(x) = pe\mu_p \mathcal{E} - eD_p \frac{dp}{dx} \equiv 0 \tag{14.1b}$$

(carrier gradients ∇n and ∇p, and electric field $\mathcal{E} = -\nabla \phi$, are assumed in the x-direction).

Let us indicate with $\phi(x)$ the electrostatic potential created by the internal space-charge region. As shown below, the potential $\phi(x)$ across the structure *varies slowly with respect to the lattice constant* (typical lengths of space-charge regions are in fact of the order of 10^3 Å or so). We can thus retain that, locally, the potential $\phi(x)$ rigidly shifts every energy level of the semiconductor by the x-dependent quantity $(-e)\phi(x)$; the x-dependence of the band structure energies is referred to as *energy band profile*. In particular, the band edges E_c and E_v (bottom of the conduction band and top of the valence band) are given by

$$E_c(x) = E_c + (-e)\phi(x), \qquad E_v(x) = E_v + (-e)\phi(x). \qquad (14.2a)$$

Notice that the built-in electric field $\mathcal{E} = -d\phi(x)/dx$ can be expressed as

$$\mathcal{E} = \frac{1}{e}\frac{dE_c(x)}{dx} \quad \text{or equivalently} \quad \mathcal{E} = \frac{1}{e}\frac{dE_v(x)}{dx}; \qquad (14.2b)$$

(a similar expression holds of course for any other level). Thus a flat-band profile is synonymous of vanishing electric field, while a steep band bending denotes a region of strong electric field.

We can now evaluate quantitatively the extent of the space-charge region, the potential profile and the electric field in the *p-n* junction. From Figure 14.1, it is seen by inspection that the total potential drop between the bulk part of the *n*-type semiconductor and the bulk part of the *p*-type semiconductor is given by the difference of the

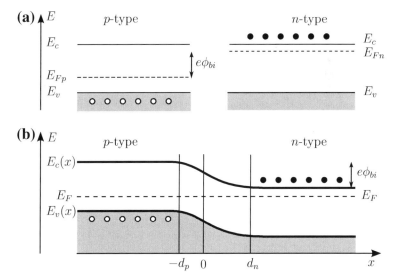

Figure 14.1 (a) Energy bands and *Fermi levels* in two physically separated *p*-type and *n*-type pieces of the same semiconductor. The majority carriers (electrons in *n*-type and holes in *p*-type regions) are indicated by full dots and empty dots, respectively, while minority carriers are not indicated for simplicity. (b) Energy band profile and *unique Fermi level* across the *p-n* junction at equilibrium.

two Fermi levels of the two separated materials

$$\boxed{e\phi_{bi} = E_{Fn} - E_{Fp}}.$$ (14.3)

The quantity ϕ_{bi} (> 0), defined by the difference of the Fermi levels of the two materials before contact, plays a key role in the study of transport properties, and is called *built-in potential* or *potential barrier*.

As discussed in Section 13.3, the Fermi levels for the separated n-type and p-type materials in the saturation regime are respectively given by

$$E_{Fn} = E_c + k_B T \ln \frac{N_d}{N_c(T)}, \qquad E_{Fp} = E_v + k_B T \ln \frac{N_v(T)}{N_a},$$

where $N_c(T)$ and $N_v(T)$ indicate the effective density-of-states in the conduction and valence bands, respectively. Thus

$$e\phi_{bi} = E_{Fn} - E_{Fp} = E_G + k_B T \ln \frac{N_a N_d}{N_v(T) N_c(T)},$$

and the potential barrier ϕ_{bi} is of the order of E_G/e.

In the p-n junction at equilibrium, the free electron carrier concentration $n(x)$ depends on the distance x from the junction via the relation

$$n(x) = N_c(T) e^{-[E_c(x) - E_F]/k_B T} \quad \text{with} \quad E_c(x) = E_c - e\phi(x),$$ (14.4a)

where $E_c(x)$ is the x-dependent bottom of the conduction band, given by Eq. (14.2a). Equation (14.4a) shows that electrons become numerous in regions where the bottom of the conduction band and the Fermi level are close. From Eq. (14.4a), we see that the carrier density $n(x_1)$ and $n(x_2)$, at any two arbitrary points of the semiconducting structure at thermodynamic equilibrium, are related by the expression

$$\boxed{n(x_1) = n(x_2) e^{e[\phi(x_1) - \phi(x_2)]/k_B T}} \quad \text{(junction at thermal equilibrium)}.$$ (14.4b)

This useful relation can also be obtained directly from Eq. (14.1a) by ready integration.

In a completely similar way we have for holes

$$p(x) = N_v(T) e^{-[E_F - E_v(x)]/k_B T} \quad \text{with} \quad E_v(x) = E_v - e\phi(x).$$ (14.5a)

Thus we see that holes are more numerous in regions where the top of the valence band and the Fermi level are closer. We have also

$$\boxed{p(x_1) = p(x_2) e^{-e[\phi(x_1) - \phi(x_2)]/k_B T}} \quad \text{(junction at thermal equilibrium)}.$$ (14.5b)

This useful relation can also be obtained directly from Eq. (14.1b) by ready integration.

It is of interest to apply Eq. (14.4b) in the case x_1 is well inside the p-type material and x_2 is well inside the n-type material. Then $n(x_1) = n_{p0}$ and $n(x_2) = n_{n0}$, where n_{p0}

and n_{n0} indicate the equilibrium concentration of electrons in the neutral bulk p-type material and in the neutral bulk n-type material, respectively. We have thus

$$n_{p0} = n_{n0}e^{-e\phi_{bi}/k_B T}. \tag{14.6a}$$

We can do a completely similar reasoning for holes and obtain

$$p_{p0} = p_{n0}e^{e\phi_{bi}/k_B T}, \tag{14.6b}$$

where p_{p0} and p_{n0} indicate the equilibrium concentration of holes in the p-side and in the n-side neutral bulk parts of the semiconductor, respectively.

Space-Charge Region and Internal Electric Field

We have indicated schematically in Figure 14.1b the energy band profile of a p-n junction at equilibrium; far from the junction, the n and p materials are neutral and have bulk equilibrium properties. For instance, well inside the bulk p-type material, the concentration of holes is constant and equal to N_a (in the saturation regime). As we approach the junction plane, the difference between the Fermi level E_F and $E_v(x)$ increases. Any increase exceeding a few $k_B T$ depresses severely the number of holes, as can be inferred from Eq. (14.5a); thus, on the p-side of the junction, a layer depleted of majority carriers is formed; in the depletion layer (of width d_p to be determined) the net charge density is practically $-eN_a$. Similarly, there is formation of a spatial region with charge density $+eN_d$ and extension d_n (to be determined) on the n-side of the junction. The space-charge region, shown in Figure 14.2, consists essentially of ionized donor impurities and ionized acceptor impurities in the regions of widths d_n and d_p from the junction, respectively. In the *full depletion approximation*, it is assumed that the depletion region is fully emptied of carriers while the adjacent regions remain perfectly neutral; with this approximation the space-charge $\rho(x)$ takes the form

$$\rho(x) = \begin{cases} -eN_a & -d_p < x < 0, \\ +eN_d & 0 < x < d_n, \\ 0 & \text{for } x < -d_p \text{ and } x > d_n. \end{cases} \tag{14.7}$$

The electrostatic potential $\phi(x)$ satisfies the Poisson equation

$$-\frac{d^2\phi}{dx^2} = \frac{\rho(x)}{\varepsilon} \tag{14.8}$$

($\varepsilon = \varepsilon_r \varepsilon_0$ is the static dielectric constant of the semiconductor; $\varepsilon_0 = 8.85 \times 10^{-14}$ F/cm).

We integrate Eq. (14.8) in the region $-d_p \leq x \leq 0$ with the boundary conditions $\phi(-d_p) = 0$ and $\phi'(-d_p) = 0$. We observe in fact that for $x \leq -d_p$, we have a homogeneous semiconductor in open circuit situation with null diffusion and drift current density, and hence null electric field; the potential $\phi(x)$ for $x \leq -d_p$ is thus a constant that can be set equal to zero. Similarly integration of Eq. (14.8) in the region $0 \leq x \leq d_n$ with the boundary conditions $\phi(d_n) = \phi_{bi}$ and $\phi'(d_n) = 0$, can be carried

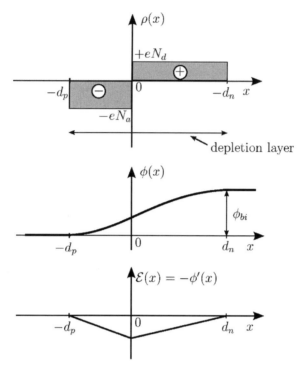

Figure 14.2 Charge density $\rho(x)$ at a p-n junction in the depletion layer approximation, electrostatic potential $\phi(x)$, and internal electric field $\mathcal{E}(x)$.

out and one obtains

$$
\phi(x) = \begin{cases}
\dfrac{eN_a}{2\varepsilon}(x + d_p)^2 & -d_p \le x \le 0, \\[2mm]
\phi_{bi} - \dfrac{eN_d}{2\varepsilon}(x - d_n)^2 & 0 \le x \le d_n.
\end{cases}
\tag{14.9}
$$

We now require the continuity of the potential $\phi(x)$ and of its derivative $\phi'(x)$ at $x = 0$; the two conditions determine the extensions d_n and d_p of the depletion layer. [The request of continuity of $\phi'(x) = -\mathcal{E}(x)$ at $x = 0$ follows from the continuity of the normal component of the displacement vector $\mathbf{D} = \varepsilon\mathcal{E}$ at the interface $x = 0$ between two media with the same dielectric constant.] The continuity of $\phi'(x)$ at $x = 0$ implies

$$
N_a d_p = N_d d_n,
\tag{14.10a}
$$

a condition which is synonymous of overall charge neutrality of the system. The continuity of the potential at $x = 0$ gives

$$
\phi_{bi} = \frac{e}{2\varepsilon}[N_a d_p^2 + N_d d_n^2].
\tag{14.10b}
$$

The solution of the two Eqs. (14.10) gives

$$d_p = \left[\frac{N_d}{N_a} \frac{1}{N_a + N_d} \frac{2\varepsilon\phi_{bi}}{e} \right]^{1/2} \quad \text{and} \quad d_n = \left[\frac{N_a}{N_d} \frac{1}{N_a + N_d} \frac{2\varepsilon\phi_{bi}}{e} \right]^{1/2}.$$

(14.11a)

The total width of the depletion layer is

$$w = d_p + d_n = \left[\frac{N_a + N_d}{N_a N_d} \frac{2\varepsilon\phi_{bi}}{e} \right]^{1/2}.$$

(14.11b)

As an example, consider for instance $\phi_{bi} = 1$ V, $N_a = N_d = 10^{16}$ cm^{-3}, $\varepsilon = \varepsilon_r \varepsilon_0$ with $\varepsilon_r = 10$ and $\varepsilon_0 = 8.85 \times 10^{-14}$ F/cm; one obtains $w \approx 5 \times 10^{-5}$ cm. The average value of the electric field ϕ_{bi}/w is approximately 2×10^4 V/cm.

Considerations on Other Types of Homojunctions

In the *p-n* junctions considered until now, we have (tacitly) assumed that the acceptor and donor doping levels N_a and N_d are moderate and comparable. In particular circumstances, it is convenient to consider junctions with very high concentrations of acceptors, or donors, or both; these junctions are indicated as p^+-n, p-n^+, and p^+-n^+ structures. These homojunctions, when treated in the full depletion layer approximation, are just particular cases of the formalism considered so far. For instance the one-sided abrupt junction p^+-n denotes a homojunction with $N_a \gg N_d$; in this case from Eqs. (14.11), one obtains

$$d_p \ll d_n \implies w \cong d_n = \left[\frac{1}{N_d} \frac{2\varepsilon\phi_{bi}}{e} \right]^{1/2}, \quad p^+\text{-}n \text{ junction};$$

(14.12)

the depletion layer extends prevalently on the low doped side of the p^+-n junction.

It is instructive to consider also the n^+-n homojunctions (or the p^+-p homojunctions); for our purpose, a few qualitative remarks are sufficient. In contrast to *p-n* junctions, in n^+-n junctions the electrons are the majority carriers in the whole structure; the concentration of electrons changes gradually from a very high value in the n^+-side to a high value on the *n*-side. Differently from the *p-n* junctions, the n^+-n junctions present an accumulation of majority carriers in the *n*-region rather than a depletion; thus while a *p-n* contact is a region of high resistivity with *rectifying properties* (see next section), an n^+-n contact is a region of low resistivity. Well designed n^+-n junctions are often used to realize *ohmic contacts* in electronic circuits (a contact is said to be "ohmic" when the potential drop across it can be safely disregarded with respect to the potential drops in other active parts of the electronic device).

14.2 Current-Voltage Characteristics of the *p-n* Junction

In a *p-n* diode at equilibrium, an internal potential barrier ϕ_{bi} is established so that no net flow of carriers occurs. In equilibrium situation, the drift and diffusion components of the current density both for electrons and for holes must cancel (see Eq. (14.1)). When a bias potential V is applied to the diode, the effective potential barrier is modified; drift and diffusion currents are no longer balanced, and current flows across the junction.

To determine the current-voltage characteristic of the device, and its rectifying properties, consider a p-n junction with a bias voltage V applied to it (see Figure 14.3). The external voltage V is counted positive if the potential barrier between p and n regions is decreased (forward bias) and thus the p-region tends to be positive with respect to the n-region; vice versa, the external potential V is considered negative, if the potential barrier between the regions is increased (reverse bias). In Figure 14.3 we illustrate schematically a direct-biased and a reverse-biased p-n diode.

We have already noticed that the space-charge region has high resistivity (compared with the bulk p and n regions) since the majority carriers are removed from it. We can thus assume that the external potential V applied to the diode (almost completely) drops across the depletion region; the total potential drop across the depletion layer in a biased diode is $\phi_{bi} - V$ (it is assumed $V < \phi_{bi}$). The width of the depletion layers in the presence of an external potential V can be obtained from Eqs. (14.11), replacing ϕ_{bi} with $\phi_{bi} - V$; we have

$$d_p = \left[\frac{N_d}{N_a} \frac{1}{N_a + N_d} \frac{2\varepsilon(\phi_{bi} - V)}{e} \right]^{1/2} , \quad d_n = \left[\frac{N_a}{N_d} \frac{1}{N_a + N_d} \frac{2\varepsilon(\phi_{bi} - V)}{e} \right]^{1/2} ,$$

$$(14.13a)$$

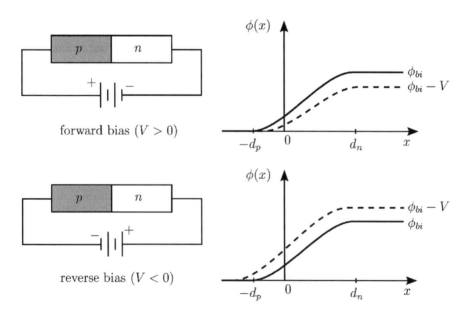

Figure 14.3 Forward and reverse bias for a p-n junction.

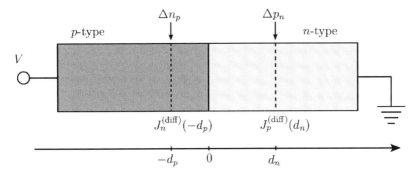

Figure 14.4 Schematic representation of a *p-n* homojunction with an applied potential V on the *p*-side (the *n*-side is grounded); Δn_p and Δp_n denote the change in the concentration of minority carriers at the boundaries between the space-charge region and the neutral regions. If V is positive, also Δn_p and Δp_n are positive (and vice versa). The electron diffusion current at $x = -d_p$, and the hole diffusion current at $x = d_n$ are indicated. The diode current is the sum $J_n^{(\text{diff})}(-d_p) + J_p^{(\text{diff})}(d_n)$ if recombination processes are negligible in the depletion region.

and

$$w = d_p + d_n = \left[\frac{N_a + N_d}{N_a N_d} \frac{2\varepsilon(\phi_{bi} - V)}{e} \right]^{1/2}, \qquad (14.13\text{b})$$

where $\phi_{bi} - V > 0$. Our following considerations are restricted to forward voltages smaller than the built-in potential ϕ_{bi}, while for reverse biases we have no restriction (except for the onset of avalanche breakdown or Zener breakdown).

We can examine quantitatively the change of minority carriers just at the boundaries of the depletion layer region of the *p-n* homojunction (see Figure 14.4), when the voltage V is applied to the device, so that the potential drop across the depletion layer becomes $\phi_{bi} - V$. We assume that in situations of quasi-equilibrium, the electron carrier densities and electrostatic potentials at the two boundaries of the depletion layer are still related by Eq. (14.4b) (strictly valid only at the thermodynamic equilibrium); with this assumption, the electron concentrations at $-d_p$ and d_n are related by

$$n_p(-d_p) = n_n(d_n)e^{-e(\phi_{bi} - V)/k_B T} \qquad (14.14)$$

(a similar treatment holds for holes). It is also reasonable to assume the *low injection condition*, and put $n_n(d_n) \approx n_{n0}$ in Eq. (14.14). Then

$$n_p(-d_p) = n_{n0}\, e^{-e(\phi_{bi} - V)/k_B T} = n_{p0}\, e^{eV/k_B T},$$

where use has been made of Eq. (14.6a).

In summary: the minority carriers at the edges of the depletion layer of a non-equilibrium *p-n* junction are given by the expression

$$\boxed{\begin{aligned}
n_p(-d_p) - n_{p0} &\equiv \Delta n_p(-d_p) = n_{p0}(e^{eV/k_B T} - 1) \\
p_n(+d_n) - p_{n0} &\equiv \Delta p_n(+d_n) = p_{n0}(e^{eV/k_B T} - 1)
\end{aligned}} \qquad (14.15)$$

The minority carrier concentrations at the edges of the depletion layer are controlled by the factor $\exp(eV/k_BT)$. In the forward bias condition we have injection of minority carriers, while extraction occurs in the reverse bias condition.

For $x < -d_p$ the semiconductor is (approximately) neutral; the excess (or deficit) of minority carriers $\Delta n_p(-d_p)$ gives rise to a diffusion electron current (as studied in Problem 2 of Appendix 13.A). The *electron diffusion current* at $x = -d_p$ is given by

$$J_n^{(\text{diff})}(-d_p) = eD_n \frac{\Delta n_p(-d_p)}{L_n} = eD_n \frac{n_{p0}}{L_n} \left(e^{eV/k_BT} - 1 \right).$$

The *total electron current* at the boundary $x = -d_p$ is the sum of the diffusion component and drift component; the latter is negligible, because so is the concentration of minority carriers, as well as the electric field in the quasi-neutral region $x < -d_p$. We have thus

$$J_n(-d_p) \approx J_n^{(\text{diff})}(-d_p) = eD_n \frac{n_{p0}}{L_n} \left(e^{eV/k_BT} - 1 \right). \tag{14.16a}$$

Similarly, for the hole current density on the right boundary $(x = d_n)$ of the space-charge region we have

$$J_p(d_n) \approx J_p^{(\text{diff})}(d_n) = eD_p \frac{p_{n0}}{L_p} \left(e^{eV/k_BT} - 1 \right). \tag{14.16b}$$

The space-charge region is so narrow that the current due to recombination processes in the depletion layer can be ignored (at least in a first approach). In this case, *electron and hole currents are constant throughout the depletion layer*, and the total current through the diode is then given by the sum of the contributions (14.16); it follows

$$\boxed{J = J_s \left(e^{eV/k_BT} - 1 \right)}, \tag{14.17a}$$

where

$$J_s = eD_p \frac{p_{n0}}{L_p} + eD_n \frac{n_{p0}}{L_n} = eL_p \frac{p_{n0}}{t_p} + eL_n \frac{n_{p0}}{t_n}; \tag{14.17b}$$

the characteristic J-V given by Eq. (14.17a) is called the *ideal diode equation*, and J_s is called the *reverse saturation density-current* of the diode. Notice that J_s is composed by the currents of the minority carriers across the depletion region. Holes are created thermally in the n-type region at a rate p_{n0}/t_p; those created within a distance L_p from the barrier region, reach the depletion layer by diffusion and are then swept by the internal electric field to the p-side of the semiconductor. Similar considerations hold for electrons thermally generated in the p-type region.

It is customary to define the "thermal potential" V_T from the relation

$$eV_T \equiv k_B T.$$

For the analysis of devices at room temperature, the value of V_T is usually set at $V_T = 25$ mV. From Eq. (14.17a) and proper account of the section A of the device, one obtains the ideal-diode I-V characteristics

$$I = I_s \left(e^{V/V_T} - 1 \right), \tag{14.17c}$$

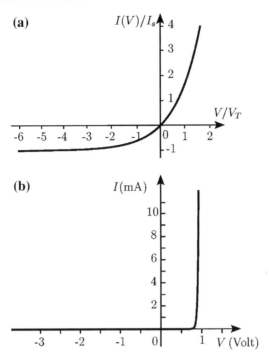

Figure 14.5 (a) Ideal characteristics $I(V) = I_s \left[\exp\left(V/V_T\right) - 1\right]$ of a *p-n* junction for values of V comparable with $V_T = 0.25$ mV. For diodes of small dimensions, typical values of I_s might be in the range around 10^{-15} A. (b) The $I(V)$ curve is plotted in an expanded scale, to make evident the rectifying characteristics of the diode.

where $I_s = J_s A$ is the reverse saturation current. In Figure 14.5a we report the I-V characteristics of a *p-n* junction for values of V comparable with $V_T = 25$ meV. In Figure 14.5b, for clarity, the same curve is reported for V extending in a much larger range of values up to one volt (or so), in order to point out the rectifying characteristics of the diode. Figure 14.5b shows that, from a practical point of view, the current through a diode can be considered negligible for V smaller than ≈ 0.5 V; also notice that a well conducting diode can sustain a potential drop in a small range of values around $\approx 0.7 \pm 0.1$ V.

The differential resistance (or impedance), defined as $R = (dI/dV)^{-1}$, is in general rather low in the direct bias configuration (typical values can be of the order of 10 Ω or so). Notice instead that the impedance is very high for reverse bias polarization, and can be of the order of 10^5 Ω or so. Notice also that, with appropriate degree of doping, it is possible to make electron current negligible with respect to hole current (or vice versa). For instance in a p^+-n junction, where the *p*-side is strongly doped with respect to the *n*-side, the hole current may represent almost all of the total current.

It is of interest to give some further consideration on the differential resistance of a diode in the direct bias configuration $0 < V < \phi_{bi}$. For simplicity, we focus on a p^+-n

junction, whose I-V characteristic (14.17) takes the simplified form

$$I(V) = A e D_p \frac{p_{n0}}{L_p} \left(e^{eV/k_B T} - 1 \right), \tag{14.18}$$

where A is the area of the junction. Using Eq. (14.18) the differential resistance of the diode becomes

$$\frac{1}{R_{\text{diode}}(V)} = \frac{\partial I}{\partial V} = A e D_p \frac{p_{n0}}{L_p} e^{eV/k_B T} \frac{e}{k_B T}.$$

It is seen by inspection that $R_{\text{diode}}(V)$ decreases rapidly as V increases. In particular for $V = 0$ we have

$$\frac{1}{R_{\text{diode}}(0)} = \frac{A}{L_p} e \, p_{n0} \, \mu_p, \tag{14.19a}$$

where use has been made of the Einstein relation $e D_p = \mu_p k_B T$. For $V = \phi_{bi}$ we have

$$\frac{1}{R_{\text{diode}}(\phi_{bi})} = \frac{A}{L_p} e \, p_{p0} \, \mu_p, \tag{14.19b}$$

where use has been made of Eq. (14.6b). The inverse of the second member of Eq. (14.19b) denotes the resistance R_s of a semiconductor of length L_p and dopant concentration $N_a = p_{p0}$. From Eqs. (14.19), it is seen that $R_{\text{diode}}(V)$ may indeed become rather low for V approaching ϕ_{bi}, but is always larger than the resistance R_s defined by Eq. (14.19b).

The ideal rectifier equation (14.17) has been obtained with a number of simplifying assumptions, that we briefly summarize as follows: abrupt depletion layer, carrier densities at the boundaries in quasi-equilibrium, low injection condition, no appreciable net generation-recombination processes in the space-charge region. We do not discuss the extensions of the theory necessary to overcome these limitations.

There are other mechanisms that can modify the I-V characteristic of a diode. When the reverse voltage exceeds a critical value, the inverse current begins to increase rapidly (breakdown effect). This can be due to direct tunneling of electrons from the valence band in the p-side semiconductor to the conduction band in the n-side, or to "avalanche" multiplication of carriers in strong electric fields. Also the forward I-V characteristics of a strongly doped p^+-n^+ junction may present a current component due to carriers tunneling through thin barriers, with the possibility of negative differential resistance.

Finally we should notice that the most important applications of p-n junctions regard not only electronic devices, but also photonic devices. Among them, we mention light-emitting diodes or diode lasers (with conversion of electrical energy into optical energy), solar cells and photovoltaic cells (with conversion of optical energy into electric energy), photodetectors (electronic detection of photons). In the following we focus on the use of p-n junctions as building blocks in transistors.

14.3 The Bipolar Junction Transistor

A bipolar transistor consists of a single crystal containing two back-to-back junctions in interaction; the device relies on the behavior of two types of carriers and is thus called bipolar. We consider the essential aspects of the p-n-p transistor (for the n-p-n transistor one can follow a completely specular treatment).

A p-n-p junction transistor is built by two p-type regions separated by a thin n-type layer of the same material. In Figure 14.6 the three regions (emitter, base, collector) are shown: the emitter is a strongly doped p^+ region, the base is a weakly doped n-type region, and the collector is a moderately doped p-region. The width of the base is much smaller than the diffusion length of minority carriers (holes) in the n-region.

We begin with a qualitative description of the operation of the p-n-p device, and we pass later to a semi-quantitative analysis. In the normal operating mode, the base-emitter junction is biased in the forward sense, and the base-collector junction is biased in the reverse sense. Taking the potential V_B of the base as reference zero potential, (i.e. $V_B \equiv 0$), we have $V_E > 0$ and $V_C \ll 0$. The forward bias of the emitter-base p^+-n junction determines a current due to the injection of holes in the base (the current due to the injection of the electrons in the emitter is relatively small, since the emitter is strongly doped, while the base is weakly doped). If the base width is small with respect to the diffusion length of minority carriers in the base, practically all the hole current from the emitter is collected by the collector. The *transistor* name has been coined to address this "tran(sfer)-(res)istor" action from the low-impedance forward-biased emitter-base junction to the high-impedance reverse-biased collector-base junction.

From what has been said, we expect that the collector current I_C is only slightly smaller than the emitter current I_E. It is customary to define the "current transfer" parameter α as the ratio of the collector and emitter current

$$\alpha = \frac{I_C}{I_E};$$
(14.20a)

for a well designed transistor, α is very close to 1, and a typical value is $\alpha \approx 0.99$. It is also convenient to define the "current gain" parameter β as the ratio between the

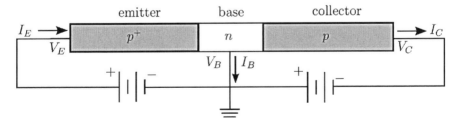

Figure 14.6 Schematic representation of a p-n-p junction transistor in the grounded base arrangement.

collector current I_C and the base current $I_B = I_E - I_C$; we have:

$$\beta = \frac{I_C}{I_B} = \frac{I_C}{I_E - I_C} = \frac{\alpha}{1 - \alpha}. \tag{14.20b}$$

We thus see that for $\alpha \approx 0.99$, the current gain factor β is typically 100. As we shall see by a more quantitative analysis, a large current gain requires a very thin base, and large diffusion length of the minority carriers of the base.

We analyze now quantitatively the currents flowing in the p-n-p transistor, when the emitter and the collector potentials are V_E and V_C, respectively, and the base is grounded. In Figure 14.7 we give a schematic representation of the structure of the p-n-p transistor; we indicate also the limits of the depletion layer regions, and the change in density of minority carriers in the emitter, base, and collector. We indicate by $\Delta n(x_E)$ the change of electron concentration at the boundary plane $x = x_E$ of the space-charge region within the p^+-side of the semiconductor; $\Delta p(x_{1B})$ denotes the change of hole concentration at the boundary plane $x = x_{1B}$ of the depletion layer at the emitter-base junction; similar notations are used for $\Delta p(x_{2B})$ and $\Delta n(x_C)$. Using the results obtained in Section 14.2, we have

$$\Delta n(x_E) = n_{E0}\left[e^{eV_E/k_B T} - 1\right], \tag{14.21a}$$

$$\Delta p(x_{1B}) = p_{B0}\left[e^{eV_E/k_B T} - 1\right], \quad \Delta p(x_{2B}) = p_{B0}\left[e^{eV_C/k_B T} - 1\right], \tag{14.21b}$$

$$\Delta n(x_C) = n_{C0}\left[e^{eV_C/k_B T} - 1\right] \tag{14.21c}$$

(n_{E0}, p_{B0}, and n_{C0} are the equilibrium concentrations of minority carriers in the emitter, in the base and in the collector, respectively).

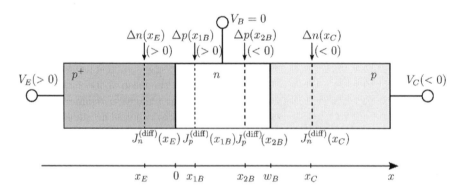

Figure 14.7 Effect of applied emitter and collector potentials on the concentration of minority carriers at the borders x_E, x_{1B}, x_{2B}, x_C of the space-charge regions; the signs of $\Delta n(x_E), \Delta p(x_{1B}), \Delta p(x_{2B}), \Delta n(x_C)$ correspond to $V_E > 0$ and $V_C < 0$. The electron diffusion current at $x = x_E$, the hole diffusion current at $x = x_{1B}$ and $x = x_{2B}$, and the electron diffusion current at $x = x_C$ are indicated. The emitter current is $J_E = J_n^{(\text{diff})}(x_E) + J_p^{(\text{diff})}(x_{1B})$; the collector current is $J_C = J_p^{(\text{diff})}(x_{2B}) + J_n^{(\text{diff})}(x_C)$; the base current is $J_B = J_E - J_C$.

The electron diffusion current, at the left boundary $x = x_E$ of the space-charge region between emitter and base is

$$J_n^{(\text{diff})}(x_E) = e D_E \frac{\Delta n(x_E)}{L_E} = e D_E \frac{n_{E0}}{L_E} \left[e^{eV_E/k_BT} - 1 \right], \tag{14.22a}$$

where $\Delta n(x_E)$ is the excess (or deficit) of electron concentration at the boundary of the neutral region, D_E is the diffusion constant, and L_E the diffusion length, of the minority carriers in the emitter. The hole diffusion current in the presence of hole concentration excess $\Delta p(x_{1B})$ at $x = x_{1B}$ and $\Delta p(x_{2B})$ at $x = x_{2B}$ has been calculated in Eq. (13.A16). With some obvious changes of notations we have

$$J_p^{(\text{diff})}(x_{1B}) = e D_B \frac{p_{B0}}{L_B} \frac{1}{\tanh(w_B/L_B)} \left[e^{eV_E/k_BT} - 1 \right]$$
$$- e D_B \frac{p_{B0}}{L_B} \frac{1}{\sinh(w_B/L_B)} \left[e^{eV_C/k_BT} - 1 \right], \tag{14.22b}$$

where $w_B = x_{2B} - x_{1B}$ is the width of the neutral region in the base (in ordinary situations, we can safely assume that the width w_B of the neutral region in the base and the width of the base are the same). The sum of the contributions (14.22a) and (14.22b) gives the emitter current density J_E; in a similar way, we can obtain the collector current density J_C. Their expressions can be written in the compact form

$$\boxed{\begin{aligned} J_E &= a_{11} \left(e^{eV_E/k_BT} - 1 \right) - a_{12} \left(e^{eV_C/k_BT} - 1 \right) \\ J_C &= a_{21} \left(e^{eV_E/k_BT} - 1 \right) - a_{22} \left(e^{eV_C/k_BT} - 1 \right) \end{aligned}} \tag{14.23}$$

where

$$\begin{cases} a_{11} &= e D_E \dfrac{n_{E0}}{L_E} + e D_B \dfrac{p_{B0}}{L_B} \coth \dfrac{w_B}{L_B}, \\[2mm] a_{12} = a_{21} &= e D_B \dfrac{p_{B0}}{L_B} \left[\sinh \dfrac{w_B}{L_B} \right]^{-1}, \\[2mm] a_{22} &= e D_C \dfrac{n_{C0}}{L_C} + e D_B \dfrac{p_{B0}}{L_B} \coth \dfrac{w_B}{L_B}. \end{cases} \tag{14.24}$$

In spite of the simplifications inherent Eqs. (14.23) and (14.24), these results are useful to give a rather intuitive discussion of the transistor action. Let us consider first the particular case in which $w_B \gg L_B$; the parameters (14.24) of the model become

$$a_{11} = e D_E \frac{n_{E0}}{L_E} + e D_B \frac{p_{B0}}{L_B}, \quad a_{22} = e D_C \frac{n_{C0}}{L_C} + e D_B \frac{p_{B0}}{L_B}, \quad a_{12} = a_{21} = 0. \tag{14.25}$$

We thus see that for $w_B \gg L_B$ the p-n-p device is nothing more than the sum of two independent back-to-back junctions. The achievement of the transistor action requires that *the base is much smaller than the diffusion length of minority carriers in the base*; in the case $w_B \ll L_B$ the parameters (14.24) become

$$a_{11} = eD_E \frac{n_{E0}}{L_E} + eD_B \frac{p_{B0}}{w_B}, \quad a_{22} = eD_C \frac{n_{C0}}{L_C} + eD_B \frac{p_{B0}}{w_B},$$

$$a_{12} = a_{21} = eD_B \frac{p_{B0}}{w_B}, \tag{14.26}$$

and the coupling parameter $a_{12} = a_{21}$ is now different from zero.

In order to show in a simple way the central features of Eqs. (14.23), let us consider the particular case of the forward mode operation, characterized by $eV_E \gg k_BT$ and $V_C \ll 0$. We can simplify expressions (14.23) in the form

$$J_E \approx a_{11}e^{eV_E/k_BT}, \quad J_C \approx a_{21}e^{eV_E/k_BT}, \tag{14.27}$$

and the current gain parameter of the transistor, defined in Eq. (14.20b), becomes

$$\beta = \frac{J_C}{J_E - J_C} = \frac{a_{21}}{a_{11} - a_{21}}. \tag{14.28}$$

In the case $w_B \gg L_B$, Eqs. (14.25) hold and the current gain of the p-n-p structure vanishes. In the case $w_B \ll L_B$, Eqs. (14.26) hold and we obtain

$$\beta = \frac{eD_B(p_{B0}/w_B)}{eD_E(n_{E0}/L_E)} = \frac{D_B}{D_E} \frac{N_{aE}}{N_{dB}} \frac{L_E}{w_B}, \tag{14.29}$$

where $N_{aE} \ (= n_i^2/n_{E0})$ is the acceptor impurity doping concentration in the emitter and, similarly, $N_{dB} \ (= n_i^2/p_{B0})$ is the donor impurity doping concentration in the base. When $\beta \gg 1$, we see that a small variation of the base current is associated with a much bigger variation of the collector current and the transistor operates as a current amplifier of gain β. From Eq. (14.29) it is evident that a large current gain β requires a heavy doping of the emitter region with respect to the base, and a base width w_B very small with respect to the diffusion length of holes in the base.

14.4 Semiconductor Heterojunctions

Until now we have considered homopolar junctions, formed by a single *host semiconductor* partly doped with donors and partly with acceptors. For particular applications it may be very useful to consider junctions between two different semiconductors, with different energy gaps. If the two semiconductors, although formed by different chemical elements, are similar for what concerns the nature of the crystalline binding and the lattice structure, it is possible to form ideal (or nearly ideal) "heterostructures." Molecular beam epitaxy (MBE) and metal-organic chemical-vapor deposition (MOCVD) have allowed the realization of several heterostructures for electronic and photonic devices.

The most studied heterojunction is the one between GaAs and the $Al_xGa_{1-x}As$ alloy, where x can vary from 0 to 1; the band gap passes from ≈ 1.42 eV in pure gallium

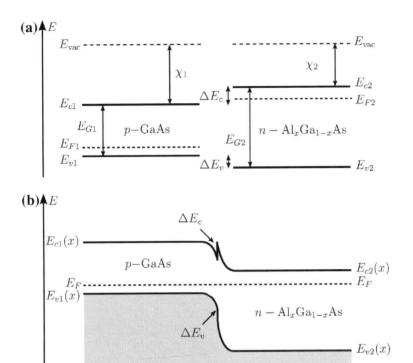

Figure 14.8 (a) Schematic energy bands and Fermi levels, referred to the vacuum level, in two separated semiconductors: p-type GaAs and n-type $Al_xGa_{1-x}As$. (b) Energy band profile and Fermi level across the heterostructure at thermal equilibrium.

arsenide to ≈ 2.17 eV in pure aluminum arsenide, while the lattice constants of the two solids are basically the same within less than 0.1%. The energy band diagrams (referred to the vacuum level) for the two isolated semiconductors are schematically shown in Figure 14.8a.

Of particular importance in the study of heterostructures, is the conduction band offset ΔE_c and the valence band offset ΔE_v, defined as

$$\Delta E_c = E_{c2} - E_{c1}, \quad \Delta E_v = E_{v1} - E_{v2}. \tag{14.30a}$$

These two quantities are not independent; in fact their sum equals the difference between the band gaps of the two semiconductors

$$\Delta E_c + \Delta E_v = E_{G2} - E_{G1}. \tag{14.30b}$$

In the case of GaAs-$Al_xGa_{1-x}As$ heterostructures, ΔE_c accounts for about 66% of the energy gap difference.

The conduction band offset can be expressed in the form

$$\Delta E_c = \chi_1 - \chi_2 + \delta_{int}, \tag{14.30c}$$

where χ_1 and χ_2 are the bulk electron affinities of the two solids (i.e. the energy difference between an electron at rest outside the solid and an electron at the bottom of the conduction band), and δ_{int} is a correction due to the specific microscopic properties of the interface under attention. The origin of δ_{int} is the quantum modification of wavefunctions at the interface that separates the two chemically different materials (this quantum modification is generally confined within a few lattice planes from the interface), and the charged double layer formed therein. For simplicity, we neglect the corrective term δ_{int} at the semiconductor-semiconductor interfaces (or also at the metal-semiconductor interfaces encountered in the following); often, the quantum correction δ_{int} is (tacitly) embodied with empirical or semi-empirical adjustments of the band offset.

We consider now equilibrium properties of semiconductor heterostructures, for instance between p-type GaAs and n-type $Al_x Ga_{1-x} As$. When the two semiconductors are brought into contact, carriers are exchanged between the two materials, until a unique Fermi level results in the whole structure. The potential barrier ϕ_{bi} at the interface is given by

$$\boxed{e\phi_{bi} = E_{F2} - E_{F1}},\qquad\qquad (14.31)$$

where E_{F1} and E_{F2} are the two Fermi levels before contact, referred to the vacuum level. Very much as in the p-n junction, a space-charge region is formed on both sides of the interface: in the n-type material we have a layer of width d_n of non-neutralized donor impurities (with charge density $+eN_d$); in the p-type material we have a layer of width d_p of non-neutralized acceptor impurities (with charge density $-eN_a$).

The electrostatic potential $\phi(x)$ across the heterostructure at equilibrium can be easily obtained following (with trivial modifications) the same procedure already applied in Section 14.1 for homojunctions. The potential $\phi(x)$, due to the space-charge region at the heterostructure, is given by a straightforward modification of Eq. (14.9); we have

$$\phi(x) = \begin{cases} (eN_a/2\varepsilon_1)(x + d_p)^2 & -d_p \le x \le 0, \\ \phi_{bi} - (eN_d/2\varepsilon_2)(x - d_n)^2 & 0 \le x \le d_n, \end{cases}$$

where ε_1 and ε_2 are the dielectric constants of the two materials. The two boundary conditions that allow to determine the widths d_p and d_n of the depletion layers are the continuity of the potential and the continuity of the normal component of the displacement vector $\mathbf{D} = \varepsilon \mathbf{\mathcal{E}}$ at the interface; namely:

$$\phi(0^-) = \phi(0^+), \quad \varepsilon_1 \phi'(0^-) = \varepsilon_2 \phi'(0^+).$$

A simple calculation gives

$$d_p = \left[\frac{N_d}{N_a} \frac{1}{\varepsilon_1 N_a + \varepsilon_2 N_d} \frac{2\varepsilon_1 \varepsilon_2 \phi_{bi}}{e} \right]^{1/2}, \quad d_n = \left[\frac{N_a}{N_d} \frac{1}{\varepsilon_1 N_a + \varepsilon_2 N_d} \frac{2\varepsilon_1 \varepsilon_2 \phi_{bi}}{e} \right]^{1/2}.$$

$$(14.32)$$

In the particular case $\varepsilon_1 = \varepsilon_2 = \varepsilon$, Eqs. (14.32) reduce to the corresponding Eqs. (14.11) of Section 14.1.

The profiles of the bands at equilibrium are easily obtained adding the contribution $-e\phi(x)$ to the band structure of the materials not yet in contact (see Figure 14.8); of course the discontinuities ΔE_c and ΔE_v at the interface must be preserved. This latter fact is of particular importance: indeed by appropriate tailoring the band offsets (for instance varying the concentration x in $Al_xGa_{1-x}As$) and appropriate choice of doping level, the conduction band of p-GaAs near the contact can be drawn close or below the Fermi level of the system. Under these circumstances, in the p-doped side of the interface, it is possible to realize a thin *inversion layer of electrons*, of fundamental importance for basic and applicative research. Notice that appropriate p-doping of the wide bandgap semiconductor and n-doping of the small bandgap semiconductor can allow the formation of an inversion layer of holes. The application of heterostructures in the physics of devices are numerous; we mention here their use for wide bandgap emitters in highly efficient bipolar transistors, the realization of heterostructure lasers, their use as window layers for detectors and photovoltaic cells (so that light can be absorbed in the active region of interest).

14.5   Metal-Semiconductor Contacts

Metal-Semiconductor Contacts at Thermal Equilibrium

In the previous sections, we have considered contacts formed by semiconductor homo-junctions or heterojunctions. We now study some properties of metal-semiconductor contacts (*Schottky contacts*). It is in fact important to recognize (at least in principle) when the contact has *blocking* or *ohmic* character; this is of major interest also in view of the necessity to weld metallic contacts on semiconducting elements in electronic circuits.

The properties of metal-semiconductor junctions can be obtained using and extending the models already adopted for homojunctions and heterojunctions. Additionally, we make the basic assumption that the *junction is ideal* (i.e. microscopically free of localized states, defects, strains, charged double layers, etc.), so that the basic transport properties of the contact are just determined by the difference of the Fermi levels in the two solids (before contact) and by the doping in the semiconductor. In general, however, a metal-semiconductor interface implies a strong local variation of the quantum mechanical potential, with possible disruption in local properties (these complications are not present in homojunctions, and hardly relevant in well lattice-matched hetero-junctions). Thus the results of the simplified models here adopted are to be intended only as qualitative; the actual realization of high performance contacts requires care and experience.

We consider specifically the contact between a metal and an n-type semiconductor (the contact between a metal and a p-type semiconductor could be done with similar procedures). We begin the discussion starting from the energy band diagrams and Fermi levels E_{Fm} and E_{Fs} of the two materials, the metal and the n-type semiconductor, not yet in contact (the energy bands of the two materials are referred to the vacuum level); as usual, the semiconductor is supposed to be non-degenerate, with the Fermi level E_{Fs} within the energy gap and separated by the band edges E_c and E_v by several k_BT.

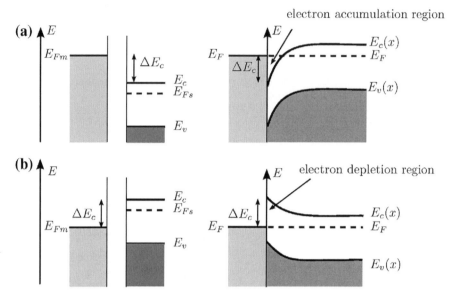

Figure 14.9 Schematic energy band diagrams of a metal and an n-doped semiconductor before contact (left panel) and after contact (right panel); the energy bands of the two materials before contact are referred to the vacuum level. Case (a) refers to $E_{Fm} > E_{Fs}$ in the initially separated metal and semiconductor, and corresponds to an ohmic contact. Case (b) refers to $E_v < E_{Fm} < E_{Fs}$ in the separated materials, and corresponds to a blocking contact. The energy difference between the Fermi level of the metal and the band edges of the semiconductor are indicated by ΔE_c and ΔE_v; these discontinuities are the same before and after contact.

In Figure 14.9a, we illustrate schematically the situation in which $E_{Fm} > E_{Fs}$ before the contact takes place. When the two solids are brought together, a unique Fermi level is established throughout the whole structure. In the n-doped semiconductor we have *accumulation of electrons near the junction*; a ready flow of electrons is possible and *the contact is said to be ohmic*. In Figure 14.9b, we illustrate the situation in which initially $E_v < E_{Fm} < E_{Fs}$; this means that, before contact, the Fermi level of the metal is lower than the Fermi level of the semiconductor and is placed within the forbidden energy band of the semiconductor. When the contact is established, a depletion layer in the semiconductor region adjacent the junction is formed; this region of high resistance gives rise to a *blocking contact*.

We consider now in more detail the electrostatic potential and the physical properties of a blocking contact at thermal equilibrium. Figure 14.10a illustrates the band diagram of the metal and the n-type semiconductor, not yet in contact. The Fermi level of the metal is taken to be in the interval $E_v < E_{Fm} < E_{Fs}$; W_m and W_s denote the work functions of the metal and the semiconductor respectively (i.e. the energy difference between the vacuum level and the Fermi level of the two solids); χ_s indicates the electron affinity of the semiconductor, and $E_G = E_c - E_v$ the energy gap.

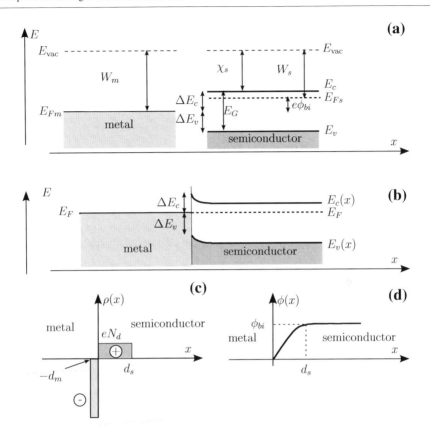

Figure 14.10 (a) Schematic energy bands of separated metal and n-type semiconductor. It is assumed that the Fermi energy of the metal is lower than the Fermi energy of the semiconductor and within the forbidden energy gap of the semiconductor. (b) Energy band profile for the metal and the n-type semiconductor in contact at equilibrium. The space-charge region $\rho(x)$ and the internal electrostatic potential barrier $\phi(x)$ are also shown.

In the study of the band profile at the contact, an essential parameter is the built-in potential barrier ϕ_{bi}, defined as usual by the difference of the Fermi levels (before contact)

$$\boxed{e\phi_{bi} = E_{Fs} - E_{Fm}} .$$
(14.33)

The discontinuities between the Fermi level of the metal and the bottom of the conduction band of the semiconductor, or the top of the valence band, are

$$\Delta E_c = E_c - E_{Fm}, \qquad \Delta E_v = E_{Fm} - E_v, \qquad \Delta E_c + \Delta E_v = E_G;$$

the discontinuities are preserved in the potential profile of the metal-semiconductor contact. On formation of the junction, some electrons leave the n-type semiconductor and move in the metal, where the chemical potential is lower; eventually, any net flux of carriers vanishes when the Fermi level is spatially uniform throughout the whole structure. A depletion layer of extension d_s is formed in the semiconductor region, with space-charge density $\rho(x) = +eN_d$ ($0 \le x \le d_s$) due to fixed ionized donors. On the metal side, just at the surface there is a corresponding negative charge (the free electron density in the metal is so high that the space-charge region within the metal is confined to the first few layers next to the surface). There is thus an electric field in the space-charge region, and a corresponding barrier potential $\phi(x)$.

The electrostatic potential $\phi(x)$ and the width of the depletion layer at the equilibrium can be evaluated from the Poisson equation

$$-\nabla^2 \phi(x) = \frac{eN_d}{\varepsilon}, \qquad 0 \le x \le d_s. \tag{14.34a}$$

We integrate this equation with the boundary condition $\phi(d_s) = \phi_{bi}$ and $\phi'(d_s) = 0$, and obtain

$$\phi(x) = \phi_{bi} - \frac{eN_d}{2\varepsilon}(x - d_s)^2, \qquad 0 \le x \le d_s; \tag{14.34b}$$

the boundary conditions $\phi(0) = 0$ gives for the width d_s of the depletion layer the expression

$$d_s = \left[\frac{1}{N_d} \frac{2\varepsilon\phi_{bi}}{e} \right]^{1/2}. \tag{14.34c}$$

The band structure profile, space-charge density and electrostatic potential are illustrated in Figure 14.10b–d respectively.

Non-Equilibrium Metal-Semiconductor Diode

The previous discussion refers to a metal-semiconductor junction at thermal equilibrium. If we apply a potential V to the metal side with respect to the semiconductor side, the total electrostatic potential across the contact becomes $\phi_{bi} - V$ (Figure 14.11). The potential V is counted as positive if the metal side is positively biased with respect to the semiconductor (forward bias); the potential V is counted as negative if the metal side is negatively biased with respect to the semiconductor (reverse bias). The thickness of the depletion layer, in the presence of an applied potential V, is given by

$$d_s = \left[\frac{1}{N_d} \frac{2\varepsilon(\phi_{bi} - V)}{e} \right]^{1/2}.$$

As in a *p-n* junction, the thickness of the barrier region decreases for forward bias, and increases for reverse bias.

The current through a metal-semiconductor junction can be described with various models, depending on the physical situation. In the case the depletion layer is long compared with the electron mean free path, the drift-diffusion model so far considered can be adapted and applied in the region of the depletion layer. In the opposite situation

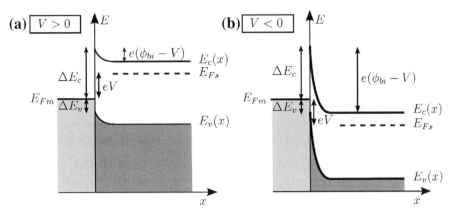

Figure 14.11 Schematic energy band profile of a junction between a metal and an *n*-type semiconductor under forward bias (a) and reverse bias (b).

of a high-mobility semiconductor, the actual shape of the barrier can be ignored, and the thermionic model becomes the appropriate tool for the description of the device. For sufficiently thin barriers, quantum mechanical tunneling may provide a significant contribution. In actual metal-semiconductor junctions, a combination of all these mentioned mechanisms may be possible. Without entering in all the details, we focus here on the thermionic mechanism, which assumes that sufficiently energetic electrons moving toward the interface overcome the barrier and determine the device current.

The thermionic emission from metals with a simple parabolic conduction band has been considered in Section 3.4, and it has been shown that the current density of escaping electrons is given by the Richardson expression

$$J_s = A\,e^{-W/k_B T}, \quad A = \frac{-emk_B^2}{2\pi^2\hbar^3}\,T^2, \tag{14.35}$$

where the work function W is the energy required to extract an electron from the Fermi level of the metal to the vacuum level, and A is the Richardson constant. It is readily seen by inspection that the same expression (14.35) holds also for semiconductors with a parabolic conduction band; in this case, the work function W is the energy difference between the vacuum level and the Fermi level of the semiconductor. Notice that the Fermi level of the semiconductor, and hence the work function, depend obviously on the doping type and impurity concentrations, but Eq. (14.35) holds in any case. The Richardson expression (14.35) applies to an ideally sharp energy potential barrier at the metal-vacuum surface (or semiconductor-vacuum surface), with the further assumption that all conduction electrons that arrive at the surface with an energy sufficient to overcome the surface barrier are actually transferred to the vacuum (reflection, image forces, space-charge accumulation effects are neglected).

We estimate the electronic current through the junction considering the metal and the semiconductor as two thermionic emitters facing each other. Within the thermionic model, we can analyze the total current into the component flowing from the metal to the semiconductor and vice versa. The thermionic current flowing from the metal to the semiconductor is given by an expression of the type

$$J^{(m \to s)} = A^* e^{-\Delta E_c / k_B T} = J_0, \tag{14.36a}$$

where ΔE_c is the barrier that an electron at the Fermi energy in the metal must overcome to pass to the semiconductor, and the order of magnitude of A^* is given by the Richardson constant. The current of Eq. (14.36a) is approximately independent of the applied potential V, as the barrier height ΔE_c is not changed by V (at least within the adopted simplifications).

The current that flows from the semiconductor to the metal at thermal equilibrium must equal $-J_0$. With an applied potential V, we see that an electron at the Fermi energy in the (bulk) semiconductor must overcome a barrier $\Delta E_c - eV$ to pass to the metal. Thus the thermionic current flowing from the semiconductor to the metal is

$$J^{(s \to m)} = A^* e^{-(\Delta E_c - eV)/k_B T} = J_0 \, e^{eV/k_B T}. \tag{14.36b}$$

Summing up algebraically the currents (14.36a) and (14.36b), one obtains the net thermionic emission of the metal-semiconductor diode

$$\boxed{J = J_0 \, (e^{eV/k_B T} - 1)}. \tag{14.37}$$

The current-voltage characteristic is controlled by an exponential factor, similarly to the p-n junction, and gives rise to rectifying behavior. [Also the drift-diffusion model of the metal-semiconductor junction gives an expression similar to Eq. (14.37), apart a different prefactor.] In ordinary situations, the thermionic contribution of the majority carriers, given by Eq. (14.37), is much more important than the contribution due to hole injection (or extraction) at the interface, and the metal-semiconductor diode is basically a unipolar device.

The above considerations refer to metal-semiconductor contacts with $E_v < E_{Fm} < E_{Fs}$ (see Figure 14.9b). In the case that $E_{Fm} > E_{Fs}$ (see Figure 14.9a), it is easily seen by inspection that an accumulation of electron carriers occurs at the interface; in close similarity with the n^+-n junction, the contact is now ohmic with low resistance. Another means to realize ohmic contacts is by strong doping, so that the passage through the barrier can also occur by tunneling effect.

14.6   Metal-Oxide-Semiconductor Structure

Properties of the Metal-Oxide-Semiconductor (MOS) Structure

As a natural development of the metal-semiconductor structure, we study now the metal-insulator-semiconductor structure, which is obtained when the metal and the semiconductor are separated by a thin insulating layer. For silicon structures, the insulator is a layer of silicon dioxide obtained by oxidation of the semiconductor itself (the band gap of SiO_2 is ≈ 10 eV). The metal-oxide-semiconductor structure is denoted

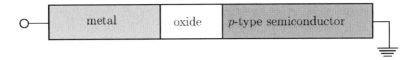

Figure 14.12 Schematic representation of a biased metal-oxide-semiconductor junction; no current flows in the structure, and carriers are at thermal equilibrium both in the metal and in the semiconductor.

with the MOS acronym. We assume that interfaces between the composing material are appropriately technologically fabricated, with negligible concentration of surface traps or other defects.

The peculiar advantage and flexibility of the MOS structure with respect to a standard Schottky contact is easily understood (see Figure 14.12). When a voltage is applied across a MOS structure, no current flows because of the presence of the insulator; carriers remain thus at the thermodynamic equilibrium both in the metal and in the semiconductor, although at two different Fermi levels; the relative position of the Fermi levels E_{Fm} and E_{Fs} in the metal and in the semiconductor can be tailored as desired by varying the sign and size of the applied potential. Accumulation or depletion of free carriers at the oxide-semiconductor junction, or even the appearance of an inversion layer, can thus be easily controlled.

We study now the carrier distribution within a MOS structure, in the case the semiconductor is p-doped (the reasoning for n-type semiconductor would follow a similar procedure). We assume for simplicity (but without loss of generality) that the Fermi level of the metal E_{Fm} and the Fermi level of the semiconductor E_{Fs}, not yet in contact, are the same (this assumption is called "ideal MOS structure"); thus when the metal and the semiconductor are connected with applied voltage $V \equiv 0$, no net flow of carriers occurs, no space-charge region is formed, and the energy bands *remain flat* (flat-band condition). In Figure 14.13a we report the Fermi level and the band diagram of an "ideal" MOS with zero applied voltage. [Notice that a "non-ideal" MOS reaches the flat-band condition when biased with a potential V_{FB} such that $eV_{FB} = E_{Fm} - E_{Fs}$. The quantity V_{FB} is called "flat-band voltage." The results for ideal MOS in applied voltage V also hold for non-ideal MOS in applied voltage $V - V_{FB}$; thus from now on we consider only ideal MOS structures.]

Suppose a potential V is applied to the metal side, while the semiconductor side is grounded (as indicated in Figure 14.12). In Figure 14.13 we show qualitatively the energy band profile of an ideal MOS structure under bias voltage V. For $V < 0$, there is an accumulation of holes in the semiconductor near the junction. If $V > 0$ a depletion of holes occurs in the semiconductor. Of the highest interest is the case $V \gg 0$, which leads to an inversion layer formation (see Figure 14.13d). All these situations can be realized and controlled continuously with the applied potential V.

We study now in detail the electrostatic potential and the energy band profile in the case the potential V applied to the metal is moderately positive, so that we have a depletion of holes, but not yet a strong accumulation of electrons (Figure 14.14). The space-charge density $\rho(x)$ is essentially formed by the ionized acceptor impurities

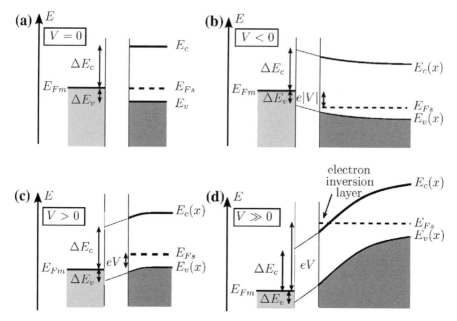

Figure 14.13 Schematic energy band structure of an "ideal" junction formed by metal, oxide, p-type semiconductor and energy band profiles for: $V = 0$ (flat-band situation), $V < 0$ (hole accumulation), $V > 0$ (hole depletion), $V \gg 0$ (inverse layer formation). The energy differences between the Fermi level of the metal and the band edges of the semiconductor are indicated by ΔE_c and ΔE_v; these discontinuities are independent of the applied electric field. For convenience, the curves $E_c(x)$ and $E_v(x)$ are continued linearly in the oxide region, regardless the absence of states in the oxide in the energy range considered.

(of charge $-e$) in the semiconductor region $0 \le x \le d_s$ (with d_s to be determined), and a corresponding number of ionized atoms of the metal, confined in proximity of the metal surface at $x = -d_{\text{ox}}$. We have

$$\rho(x) = \begin{cases} -eN_a & 0 \le x \le d_s, \\ Q_m \, \delta(x + d_{\text{ox}}) & Q_m = eN_a d_s, \end{cases} \tag{14.38}$$

where δ is the Dirac function.

The electrostatic potential within the semiconductor (of dielectric constant ε_s) is determined via the Poisson equation

$$-\frac{d^2\phi}{dx^2} = -\frac{eN_a}{\varepsilon_s}, \quad 0 \le x \le d_s.$$

Integrating twice the above equation with the boundary conditions that the electric field and the potential are zero at $x = d_s$ one obtains

$$\phi(x) = \frac{eN_a d_s^2}{2\varepsilon_s}\left(1 - \frac{x}{d_s}\right)^2, \quad 0 \le x \le d_s. \tag{14.39}$$

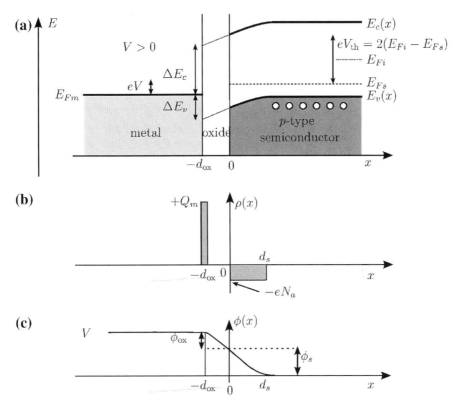

Figure 14.14 Schematic energy band profile, charge density and electrostatic potential of an ideal metal-oxide-semiconductor structure under (moderate) positive bias of the metal. The Fermi energy of the p-type semiconductor is indicated with E_{Fs}, while E_{Fi} denotes the intrinsic Fermi energy value. In the oxide region, in spite of the absence of energy levels, we have indicated the straight line connecting the energy level $E_{Fm} - \Delta E_v$ at the metal-oxide interface with $E_v(0)$; similarly, the other straight line connects the energy level $E_{Fm} + \Delta E_c$ in the metal to $E_c(0)$.

The potential ϕ_s at the surface of the semiconductor is thus

$$\boxed{\phi_s = \frac{eN_a d_s^2}{2\varepsilon_s}} \; ; \tag{14.40}$$

the surface potential ϕ_s represents the potential drop across the depletion layer formed within the semiconductor.

We can now calculate the potential drop ϕ_{ox} in the oxide region. From Eq. (14.39), we can obtain the electric field at the semiconductor surface: $\mathcal{E}_{surf} = -\phi'(0) = eN_a d_s/\varepsilon_s$. In the oxide region, the electric field is given by $\mathcal{E}_{ox} = eN_a d_s/\varepsilon_{ox}$, where ε_{ox} is the dielectric constant of the insulating material. In the oxide region the charge is zero, the electric field is constant and the potential varies linearly with the x-coordinate; the potential drop in the oxide $\phi_{ox} = \mathcal{E}_{ox} d_{ox}$ is given by

$$\phi_{ox} = \frac{d_{ox}}{\varepsilon_{ox}} e N_a d_s. \tag{14.41}$$

The potential V applied to the metal is evidently the sum of the potential drop ϕ_s in the semiconductor and the potential drop ϕ_{ox} in the oxide.

The partition of the applied voltage V between the semiconductor and the oxide is characterized by the ratio

$$r = \frac{\phi_{ox}}{\phi_s} = \frac{d_{ox}}{d_s} \cdot \frac{\varepsilon_s}{\varepsilon_{ox}}.$$

For an estimation of the parameter r, consider for instance a silicon MOS structure with $N_a = 10^{16}$ cm^{-3}, $d_{ox} = 20$ nm $= 2 \times 10^{-6}$ cm, $\varepsilon_{ox} = 3.9\,\varepsilon_0$, $\varepsilon_s = 11.7\,\varepsilon_0$, $\varepsilon_0 = 8.85 \times 10^{-14}$ F/cm, and $\phi_s = 1$ V. From Eq. (14.40) one has

$$d_s = \left(\frac{2\varepsilon_s \phi_s}{e N_a} \right)^{1/2} = \left(\frac{2 \cdot 11.7 \cdot 8.85 \cdot 10^{-14}}{1.6 \cdot 10^{-19} \cdot 10^{16}} \right)^{1/2} \approx 36 \times 10^{-6} \text{ cm,}$$

and

$$r = \frac{d_{ox}}{d_s} \cdot \frac{\varepsilon_s}{\varepsilon_{ox}} = \frac{2 \cdot 10^{-6} \cdot 11.7}{36 \cdot 10^{-6} \cdot 3.9} = 0.17.$$

The parameter r decreases for narrow oxide regions and for less doped semiconductors. In the following it is assumed, just for sake of simplicity, that the parameter r is reasonably small, so that the potential drop in the oxide region can be considered small and negligible with respect to the potential drop in the semiconductor. With this simplifying assumption, the potential V applied to the metal equals ϕ_s. [With a little of attention, the relations elaborated below for the case of small partition parameter can be generalized to the cases where r is not small; however, the overall features of the current-voltage characteristics are not changed in essential way.]

Onset of the Strong Inversion Condition

When the bias potential V applied to the metal increases, it is possible to reach a situation in which the concentration of minority carriers (the electrons) at the surface of the p-type semiconductor equals the concentration of holes in the bulk p-type semiconductor; the onset of this "strong inversion regime" can be estimated with the following argument.

Let us indicate with E_{Fi} and E_{Fs} the Fermi level of the intrinsic semiconductor and the Fermi level of the p-doped semiconductor, respectively (see Figure 14.14). We observe that

$$n_{p0} = n_{i0}\, e^{-(E_{Fi} - E_{Fs})/k_B T}, \quad p_{p0} = p_{i0}\, e^{(E_{Fi} - E_{Fs})/k_B T};$$

since $n_{i0} = p_{i0}$ it follows

$$n_{p0}\, e^{2(E_{Fi} - E_{Fs})/k_B T} = p_{p0}. \tag{14.42}$$

With an eye at Eq. (14.42), it is customary to define the so called threshold potential V_{th} for a p-type semiconductor as

$$\boxed{eV_{th} \equiv 2(E_{Fi} - E_{Fs})} .$$ (14.43)

From Eq. (14.42) and the definition (14.43), we see that *the concentration of electron carriers at the surface equals the equilibrium majority carrier concentration* when $\phi_s \equiv V_{th}$. Thus the potential V_{th} can be considered as the "threshold potential" for the onset of the strong inversion regime.

Let us indicate with d_s^* the length of the depletion layer in the p-type semiconductor, just at the beginning of the strong inversion condition. According to Eq. (14.40), the value of d_s^* reads

$$V_{th} \equiv \frac{eN_a d_s^{*2}}{2\varepsilon_s} \quad \Longrightarrow \quad d_s^* = \sqrt{\frac{2\varepsilon_s V_{th}}{eN_a}} .$$

If the potential V of the metal is further increased with respect to the threshold voltage, we can assume that d_s^* remains constant and an inversion electron layer begins to appear near the surface; the negative charge Q_{inv} per unit surface formed at the oxide-semiconductor interface is such that

$$-Q_{inv} \frac{d_{ox}}{\varepsilon_{ox}} = V - V_{th} \quad \text{for} \quad V > V_{th} ,$$

where the proportionality constant between the inversion charge and the applied voltage is the oxide layer capacitance per unit area $C_{ox} = \varepsilon_{ox}/d_{ox}$. From the last equation, we see that the surface charge density Q_{inv} of mobile electrons is related to the applied potential V ($> V_{th}$) by the expression

$$\boxed{Q_{inv} = -\frac{\varepsilon_{ox}}{d_{ox}}(V - V_{th})} \quad \text{if} \quad V > V_{th} .$$ (14.44)

Beyond threshold, the total charge in the semiconductor is the sum of the depletion layer charge $Q_d = -eN_a d_s^*$, due to the ionized acceptor impurities and the inversion layer charge Q_{inv} due to the mobile electrons.

14.7 Metal-Oxide-Semiconductor Field-Effect Transistor (MOSFET)

We have now all the elements for a semi-quantitative discussion of the I-V characteristics of an ideal metal-oxide-semiconductor field-effect transistor (MOSFET); a cross-section of the device is schematically given in Figure 14.15. A voltage is applied between the *source* and *drain* electrodes; the source is grounded ($V_s = 0$) and the potential of the drain is positive ($V_D > 0$). There is a third electrode, the *gate*, which is separated from the semiconductor by an oxide layer and held at a positive potential V_G.

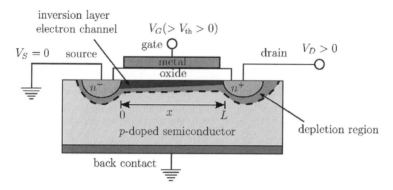

Figure 14.15 Schematic representation of a n-channel MOSFET.

The positive potential of the gate exceeds the threshold value required to convert the p-type semiconductor into an n-type material at the interface; when the n-type inversion layer is formed, electrons can flow between the two n-type source and drain regions, and the current through the channel can be controlled by the gate potential.

For the calculation of the current-voltage characteristics of a MOSFET, we begin to consider the situation in which the inversion layer channel extends from source to drain. Let us indicate with $V_c(x)$ the potential at a generic point x of the channel ($0 \leq x \leq L$); at the two ends of the channel we have $V_c(0) = 0$ and $V_c(L) = V_D$. If the channel potential is assumed to be a smoothly changing function of the position, the electron surface charge induced at a position x along the channel is given by Eq. (14.44) setting $V = V_G - V_c(x)$. In the gradual channel approximation, one has

$$Q_{\mathrm{inv}}(x) = -\frac{\varepsilon_{\mathrm{ox}}}{d_{\mathrm{ox}}}\left[V_G - V_{\mathrm{th}} - V_c(x)\right]. \tag{14.45}$$

Assuming a constant mobility in the channel, the drain current (taken as positive in the $-x$-direction) becomes

$$I_D = Q_{\mathrm{inv}}(x)W\mu_n \mathcal{E}(x) = Q_{\mathrm{inv}}(x)W\mu_n\left(-\frac{dV_c}{dx}\right), \tag{14.46}$$

where μ_n is the low-field electron mobility, W is the device width (orthogonal to the cross-section of Figure 14.15), $-dV_c(x)/dx$ is the component of the electric field parallel to the semiconductor-oxide interface. With separation of variables in Eq. (14.46) and use of Eq. (14.45), one obtains

$$I_D\, dx = W\mu_n\frac{\varepsilon_{\mathrm{ox}}}{d_{\mathrm{ox}}}\left[V_G - V_{\mathrm{th}} - V_c(x)\right]dV_c(x).$$

Integration with respect to x from the source to the drain (i.e. in the range $0 \leq x \leq L$) gives

$$\boxed{I_D = K\left[(V_G - V_{\mathrm{th}})V_D - \frac{1}{2}V_D^2\right]} \quad \text{with} \quad K = \frac{W}{L}\mu_n\frac{\varepsilon_{\mathrm{ox}}}{d_{\mathrm{ox}}}. \tag{14.47}$$

Equation (14.47) is applicable provided the n channel remains open up to the drain.
    For small drain-to-source voltages, Eq. (14.47) gives

$$I_D = K(V_G - V_{th})V_D \quad \text{for} \quad V_D \ll V_G - V_{th};$$

thus a linear region is expected in the I_D-V_D characteristics for small V_D, followed by
a quadratic region.
    Equation (14.47) is based on various assumptions, and in particular on the require-
ment that the mobility of carriers is field-independent, and that the conductivity channel
is open from source to drain. It is convenient to define the saturation potential $V_{D,\text{sat}}$ as
the potential for which $Q_{\text{inv}}(x)$ vanishes at the drain position $x = L$; from Eq. (14.45)
one has

$$V_G - V_{th} - V_{D,\text{sat}} \equiv 0 \implies V_{D,\text{sat}} = V_G - V_{th}. \tag{14.48}$$

The I_D-V_D relation (14.47) must be reconsidered with caution when V_D, although
smaller than $V_{D,\text{sat}}$, is very near to it. In this case the conductivity channel is almost
closed at the drain; according to Eq. (14.46), the flow of a finite drain current implies
that the electric field $\mathcal{E}(x)$ is becoming arbitrary large. Thus the assumption that the
mobility of electrons is constant is no more acceptable; it is in fact well known that the
average drift velocity of carriers in semiconductors is linear with the electric field for
small fields, but tends to saturate for strong electric fields. Without entering a detailed
analysis of transport processes in high electric fields, we can expect that for $V_D \approx V_{D,\text{sat}}$
the drain voltage loses control of the channel current and I_D remains (approximately)
constant as the drain voltage is further increased (saturation region). Using Eqs. (14.47)
and (14.48), the saturation current becomes

$$I_{D,\text{sat}} = K\left[(V_G - V_{th})V_{D,\text{sat}} - \frac{1}{2}V_{D,\text{sat}}^2\right] = \frac{1}{2}KV_{D,\text{sat}}^2 = \frac{1}{2}K(V_G - V_{th})^2. \tag{14.49}$$

The saturation current depends quadratically on the saturation potential.
    The results of Eq. (14.47) and the inherent discussion are schematically presented
in Figure 14.16. The I_D-V_D characteristics of a metal-semiconductor-insulator field-
effect transistor is reported in the case $V_{th} = 1$ V and $K = 2$ mA/V^2, for several
gate voltages. The locus of points $I_{D,\text{sat}} = (1/2)KV_{D,\text{sat}}^2$ is also indicated. From
Figure 14.16, we see that for a given V_G, the drain current first increases linearly with
the voltage (linear region), then gradually levels off to the saturation value (saturation
region); the saturation and non-saturation regimes are approximately separated by the
curve of Eq. (14.49). From the knowledge of the current-voltage characteristics of the
transistor, it is in principle possible to design MOSFET circuits for microelectronics.
    In the present and in the previous sections, we have considered *ideal* n-channel
MOSFETs (see Figure 14.13a), where the Fermi level of the metal and the Fermi level
of the p-type semiconductor not yet in contact, are the same; in realistic structures
the flat-band voltage must be taken in proper account, as discussed at the beginning
of Section 14.6. Realistic devices that have (or do not have) a conducting channel
between source and drain when no gate voltage is applied are called "normally on
depletion-mode devices" (or "normally off enhancement-mode devices"). Till now,

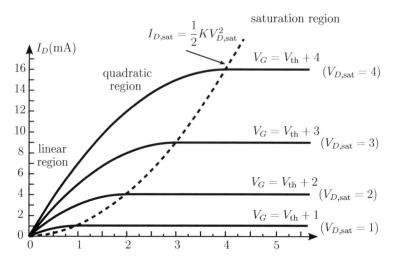

Figure 14.16 Idealized I_D-V_D characteristics of a MOSFET transistor. The parameters of the transistor are taken $V_{th} = 1$ V and $K = 2$ mA/V^2.

we have only outlined n-channel MOSFET structures, which have a p-type semiconductor substrate, a negative charge of ionized acceptors in the depletion layer, and a negative charge of mobile electrons in the inversion layer. Similar considerations could be done for p-channel MOSFET structures, which have a n-type semiconductor substrate, a positive charge of ionized donors in the depletion layer, and a positive charge of mobile holes in the inversion layer. Needless to say, of extreme importance are the complementary metal-oxide-semiconductors devices (CMOS) which require both n-type and p-type MOSFETS. The variety and flexibility of the architectures of these and similar structures are of dominant importance for the realization of logic circuits, high-density electronic memory devices, ultra large-scale integration in planar technology, and have fostered the continuous progressing toward the world of nanostructures and nanotechnology.

Further Reading

Ando, T., Fowler, A., & Stern, F. (1982). Electronic properties of two-dimensional systems. *Reviews of Modern Physics, 54*, 437.

Bhattacharya, P. (2004). *Semiconductor optoelectronic devices*. Englewood Cliffs, New Jersey: Prentice Hall.

Capasso, F., & Margaritondo, G. (Eds.). (1987). *Heterojunction band discontinuities. Physics and device applications*. Amsterdam: Elsevier.

Grundmann, M. (2006). *The physics of semiconductors*. Berlin: Springer.

Li, S. S. (2006). *Semiconductor physical electronics*. New York: Springer.

Muller, R. S., Kamins, T. I., & Chan, M. (2002). *Device electronics for integrated circuits*. New York: Wiley.

Sapoval, B., & Hermann, C. (1995). *Physics of semiconductors*. New York: Springer.

Sedra, A. S., & Smith, K. C. (2010). *Microelectronic circuits*. New York, USA: Oxford University Press.

Seeger, K. (2004). *Semiconductor physics*. Berlin: Springer.

Singh, J. (1994). *Semiconductor devices: An introduction*. New York: McGraw-Hill.

Smith, R. A. (1978). *Semiconductors*. Cambridge: Cambridge University Press.

Sze, S. M., & Lee, M.-K. (2012). *Semiconductor devices: Physics and technology*. New York: Wiley.

Wolfe, C. M., Holonyak, N., & Stillman, G. E. (1989). *Physical properties of semiconductors*. Englewood Cliffs, New Jersey: Prentice Hall.

15 Electron Gas in Magnetic Fields

In this chapter, we begin the study of the electronic structure of matter in the presence of magnetic fields, and we start with the treatment of the electron gas in magnetic fields. In the next chapter, we consider diamagnetism and paramagnetism of substances, such as ions, molecules, or impurities in solids; Chapter 17 concerns cooperative effects and magnetic ordering; the peculiar magnetic properties of superconductors (Meissner effect, magnetic flux quantization) are contained in Chapter 18.

When a magnetic field is applied to an otherwise freely moving electron, the classical motion is given by a helical path along the magnetic field direction. Quantistically, the cyclotron motion perpendicular to the field is quantized into discrete Landau levels; this has important implications on the physical properties of the electron gas and in particular on the magnetic susceptibility. We will see that the modification of the electron orbital motion due to the magnetic field leads to diamagnetism (Landau diamagnetism), while the spin of the electron gives rise to a paramagnetic contribution (spin or Pauli paramagnetism). Under very special conditions (strong magnetic fields on very pure specimens at low temperatures) the quantization of Landau levels manifests itself in characteristic oscillatory effects occurring in density-of-states, thermodynamic functions, and several physical properties; in particular the magnetic moment exhibits oscillations as a function of the magnetic field (de Haas-van Alphen effect), and precise information on the Fermi surface can be inferred.

Magnetic fields strongly influence the transport properties of ordinary (i.e. three-dimensional) metals or semiconductors, and we discuss some aspects of the rich

Solid State Physics, Second Edition. http://dx.doi.org/10.1016/B978-0-12-385030-0.00015-3

phenomenology related to the classical Hall effect and magnetoresistivity. We also describe the profound modifications that may occur in the transport properties of a two-dimensional electron gas and the integer and fractional quantum Hall effects. In the integer quantum Hall effect, in particular, the quantization due to the magnetic field manifests itself in well-defined plateaus in the Hall resistance, which allow to measure the fundamental resistance quantum h/e^2 with the astonishing accuracy of one part in billions (or so). In the fractional quantum Hall effect, a novel and challenging world of composite particles opens to the experimental and theoretical investigation.

15.1 Magnetization and Magnetic Susceptibility

Definitions and Generalities

In general, upon application of an external magnetic field H, a specimen acquires a magnetization M (*magnetic moment per unit volume*). When M is proportional to H, the *magnetic susceptibility* χ is defined as

$$\chi = \frac{M}{H};$$ (15.1a)

otherwise, the magnetic susceptibility χ takes the more general definition

$$\chi = \frac{\partial M}{\partial H}.$$ (15.1b)

If gaussian units are used χ is dimensionless.

The static magnetic susceptibility χ can be measured for instance with the method of the Gouy balance; precision up to 10^{-10} can be obtained. We can anticipate that the magnetic susceptibility is typically of the order of -10^{-6} for several diamagnetic substances, and of the order of 10^{-4} for several paramagnetic substances at ordinary temperatures. However, the range of possible values of χ in materials is extremely wide; for instance, in the case of perfect diamagnetism in superconductors, the susceptibility takes the value $\chi = -1/4\pi$, and the magnetic forces can outnumber the gravitational forces with typical levitation effects.

Throughout this chapter, we focus on electronic systems (such as the free-electron gas and free-electron-like metals) with extremely small magnetic susceptibility ($\chi \approx \pm 10^{-6}$ or so); in these systems we can safely assume negligible the difference between the external magnetic field H in the absence of the specimen and the induction magnetic field B in the presence of the specimen. In other words, in these systems one can adopt the simplifying assumption that the magnetic field in the sample and the magnetic field that has been applied to the sample are the same for all practical purposes. [Of course there are many other situations (for instance ferromagnetic materials, superconductors, etc.) for which the magnetization of the sample is not negligible compared to the applied magnetic field; in these situations, appropriate caution must be done in defining fields, in taking proper account of possible demagnetizing effects related to the geometry of the sample, and in measuring magnetic response. These and other warnings can be ignored in the special case of "weak magnetism" considered in this chapter.]

We consider now some definitions and properties, useful in the quantum treatment of the magnetic susceptibility of electronic systems. At $T = 0$, the *magnetization* of a homogeneous system of volume V in a *uniform magnetic field B* is defined as

$$M(T = 0, B) = -\frac{1}{V}\frac{\partial E_0(B)}{\partial B}, \tag{15.2a}$$

where $E_0(B)$ is the energy of the ground state of the sample in the presence of B. The magnetic susceptibility at zero temperature then becomes

$$\chi(T = 0, B) = -\frac{1}{V}\frac{\partial^2 E_0(B)}{\partial B^2}. \tag{15.2b}$$

The magnetic susceptibility at zero temperature is related to the second derivative of the ground-state energy $E_0(B)$ with respect to the magnetic field; if the curvature of $E_0(B)$ versus B is positive, χ is negative, and vice versa. In the (rather frequent) cases that $E_0(B)$ depends quadratically on the field, the magnetic susceptibility is independent of the strength of the field.

At a finite temperature T, if $F(T, B)$ is the free energy of the system, the magnetization is given by the thermodynamic relation

$$M(T, B) = -\frac{1}{V}\frac{\partial F(T, B)}{\partial B}. \tag{15.3a}$$

The meaning of this relation becomes obvious, if one considers the general expression of the free energy given in Eq. (3.A.13), and here rewritten for convenience

$$F = -k_B T \ln Z = -k_B T \ln \sum_n e^{-E_n(B)/k_B T},$$

where $E_n(B)$ are the energy values of the system in the presence of the magnetic field. Inserting the above expression into Eq. (15.3a), we obtain

$$M(T, B) = -\frac{1}{V}\frac{\sum_n [\partial E_n(B)/\partial B] e^{-E_n(B)/k_B T}}{\sum_n e^{-E_n(B)/k_B T}};$$

from the above expression, it is seen that the magnetization $M(T, B)$ is just the thermal equilibrium average of the magnetization of each pure state of the system. The magnetic susceptibility at temperature T becomes

$$\chi(T, B) = -\frac{1}{V}\frac{\partial^2 F(T, B)}{\partial B^2}. \tag{15.3b}$$

In the particular case that $T = 0$, one recovers Eq. (15.2b). In the case $F(T, B)$ changes quadratically with B, the magnetic susceptibility is independent of the strength of the applied field.

Before passing to the study of the magnetic susceptibility of specific quantum systems, we wish to comment briefly on the absence of magnetism from a purely classical point of view.

Absence of Magnetism from Purely Classical Arguments

From pure classical arguments, the magnetic susceptibility of any dynamical system is rigorously zero. This general result of classical statistical mechanics is known as *Bohr-van Leeuwen theorem*: it applies to any system of interacting (or non-interacting) particles, whose Hamiltonian depends exclusively on the positions and the momenta of the particles (hence excluding any other internal degree of freedom of particles, and spin in particular).

The theorem can be proved as follows. Consider a many-electron system with *N spinless electrons*, whose classical state is characterized by $6N$ coordinates and momenta $(q_1, q_2, \ldots, q_{3N}, p_1, p_2, \ldots, p_{3N})$. The classical partition function is given by

$$Z_{\text{classical}} = \int \exp[-H_0(q_i, p_i)/k_B T] \, dq_1 \, dq_2 \cdots dq_{3N} \, dp_1 \, dp_2 \cdots dp_{3N},$$

where $H_0(q_i, p_i)$ $(i = 1, \ldots, 3N)$ is the Hamiltonian of the electron system. In the presence of a magnetic field described by the vector potential $\mathbf{A}(\mathbf{r})$, each momentum $\mathbf{p}_i$ must be replaced by the generalized momentum $\mathbf{P}_i = \mathbf{p}_i + (e/c)\mathbf{A}(\mathbf{r}_i)$. This is the only way by which the magnetic field enters in the expression of the Hamiltonian (the electron spin, and hence the interaction energy of the magnetic field with each electron spin, is taken as a quantum phenomenon and disregarded in the present considerations of classical nature). The terms containing the vector potential can be eliminated by a simple shift in the momentum integration. Thus, the classical partition function does not depend on the magnetic field; hence the magnetization and the magnetic susceptibility vanish.

As an example, consider the partition function of a free electron, confined in a volume V, in the absence and in the presence of a uniform magnetic field B (supposed directed along the z-axis), and described (for instance) by the vector potential $\mathbf{A}(\mathbf{r}) = (-By, 0, 0)$. We have for the classical partition function

$$Z_{\text{classical}} = \int_V d\mathbf{r} \int d\mathbf{p} \exp\left[-\frac{(p_x - eBy/c)^2 + p_y^2 + p_z^2}{2mk_B T}\right] = V(2\pi m k_B T)^{3/2};$$

thus the classical partition function is not influenced by the presence of magnetic fields, and there is no thermal equilibrium magnetization in this context. Even for the study of the magnetic properties of a free-electron gas, a quantum theory is necessary from the very beginning.

15.2 Energy Levels and Density-of-States of a Free Electron Gas in Magnetic Fields

In this section we consider the effects produced by a uniform magnetic field on the dynamics of a free-electron gas. From a classical point of view, the dynamics of a free charged particle in a magnetic field is composed by a circular cyclotron motion in the plane perpendicular to the magnetic field and a free motion parallel to it. The quantum picture leads to quantization of the motion of the electron perpendicularly to the magnetic field, and converts the free-particle kinetic energy into a set of one-dimensional sub-bands, each corresponding to a given Landau level.

Before beginning with the study of magnetic fields, it is convenient to recall the energies and the density-of-states of free particles in the one-, two-, and three-dimensional cases (see Section 2.7). The dimension of the sample is supposed to be L_z for the one-dimensional case, $S = L_x L_y$ for the two-dimensional case, and $V = L_x L_y L_z$ for the three-dimensional case. The dispersion relations of the free-electron gas, and the corresponding density-of-states (*excluding spin degeneracy*), can be summarized as follows:

$$E(k_z) = E_0 + \frac{\hbar^2 k_z^2}{2m} \qquad D^{(1d)}(E) = \frac{L_z}{2\pi}\left(\frac{2m}{\hbar^2}\right)^{1/2}\frac{1}{\sqrt{E-E_0}}\Theta(E-E_0), \qquad (15.4a)$$

$$E(k_x, k_y) = E_0 + \frac{\hbar^2}{2m}\left(k_x^2 + k_y^2\right) \qquad D^{(2d)}(E) = \frac{S}{4\pi}\frac{2m}{\hbar^2}\Theta(E-E_0), \qquad (15.4b)$$

$$E(\mathbf{k}) = E_0 + \frac{\hbar^2 k^2}{2m} \qquad D^{(3d)}(E) = \frac{V}{4\pi^2}\left(\frac{2m}{\hbar^2}\right)^{3/2}\sqrt{E-E_0}\,\Theta(E-E_0), \qquad (15.4c)$$

where $\Theta(x)$ is the step function ($\Theta(x) = 0$ for $x < 0$, $\Theta(x) = 1$ for $x > 0$). When the spin degeneracy is included, the density-of-states for both spin directions is obtained multiplying by two the results of Eqs. (15.4).

15.2.1 Energy Levels of the Two-Dimensional Electron Gas in Magnetic Fields

Our analysis of the magnetic properties of matter starts considering a two-dimensional electron gas. As discussed in Chapter 14, appropriate semiconductor interfaces or metal-oxide-semiconductor structures provide the technical means to realize this system: carriers can be considered as freely moving parallel to the planar interfaces, while in the perpendicular direction they are described by wavefunctions strongly localized in a thin layer of atomic planes next to the interface.

The (effective) Hamiltonian of a two-dimensional electron gas (in xy plane) in a uniform magnetic field (parallel to z) can be written as

$$H = \frac{1}{2m}\left(p_x + \frac{e}{c}A_x\right)^2 + \frac{1}{2m}\left(p_y + \frac{e}{c}A_y\right)^2, \qquad (15.5)$$

where e is the absolute value of the electron charge, the effective mass m is taken equal to the electron mass (for simplicity), and $\mathbf{A}(\mathbf{r})$ is the vector potential of the magnetic field. The Hamiltonian of Eq. (15.5) focuses on the modification of the *orbital motion* of carriers due to the magnetic field, and neglects the interaction of the spin magnetic moment with the applied field (this interaction can be worked out separately, as discussed below in Section 15.4); spin degeneracy is taken into account introducing a factor two, whenever appropriate.

A possible choice of the vector potential, corresponding to a uniform magnetic field B in the z-direction, is given by the first Landau gauge

$$\mathbf{A}(\mathbf{r}) = (-By, 0, 0) \qquad \Longrightarrow \qquad \text{curl } \mathbf{A}(\mathbf{r}) = B(0, 0, 1). \qquad (15.6)$$

The vector potential is not uniquely determined; other frequently used gauges are the second Landau gauge $\mathbf{A}(\mathbf{r}) = (0, Bx, 0)$ and the symmetric gauge $\mathbf{A} = (B/2)(-y, x, 0)$.

The energies of the Hamiltonian H, in the absence of the external magnetic field, are given by

$$E(k_x, k_y) = \frac{\hbar^2}{2m}\left(k_x^2 + k_y^2\right).$$

Using Eq. (15.4b), we see that the corresponding two-dimensional density-of-states (excluding spin degeneracy) for a sample in a rectangular box of surface S is given by

$$\boxed{D(E, B = 0) = D_0 = \frac{m}{2\pi\hbar^2}S \quad \text{for} \quad E > 0}, \qquad (15.7)$$

thus the density of allowed states for a uniform two-dimensional electron gas is independent of energy.

In the presence of the magnetic field described within the first Landau gauge, the Hamiltonian of Eq. (15.5) becomes

$$H = \frac{1}{2m}\left(p_x - \frac{eB}{c}y\right)^2 + \frac{1}{2m}p_y^2. \qquad (15.8)$$

The above Hamiltonian does not contain x, and then p_x is a constant of motion with values $\hbar k_x$; the Hamiltonian of Eq. (15.8) can be rewritten in the form

$$H = \frac{1}{2m}p_y^2 + \frac{1}{2m}\left(\hbar k_x - \frac{eB}{c}y\right)^2 = \frac{1}{2m}p_y^2 + \frac{1}{2}m\omega_c^2(y - y_0)^2, \qquad (15.9a)$$

where

$$\omega_c = \frac{eB}{mc}, \quad y_0 = \frac{\hbar c}{eB}k_x, \quad \text{and} \quad l = \sqrt{\frac{\hbar c}{eB}} \qquad (15.9b)$$

are called, respectively, the *cyclotron resonance frequency*, the *centers of oscillations*, and the *magnetic length* of the system. Equation (15.9a) is the Hamiltonian of a displaced linear oscillator of frequency ω_c; its eigenvalues are

$$E_{nk_x} = \left(n + \frac{1}{2}\right)\hbar\omega_c \quad n = 0, 1, 2, \ldots, \qquad (15.10)$$

where the quantum number n indicates the *Landau level*. Thus the presence of a magnetic field modifies the continuous kinetic energy $E = (\hbar^2/2m)(k_x^2 + k_y^2)$ into discretized Landau levels, as indicated schematically in Figure 15.1. The energy separation

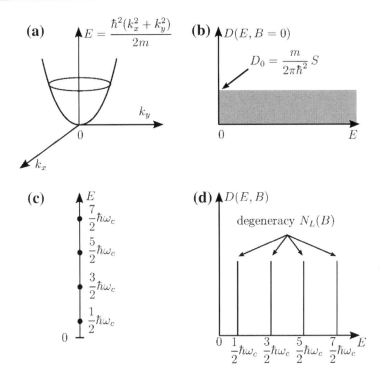

(a) $E = \dfrac{\hbar^2(k_x^2 + k_y^2)}{2m}$

(b) $D(E, B = 0)$

$D_0 = \dfrac{m}{2\pi\hbar^2} S$

(c)

(d) $D(E, B)$

degeneracy $N_L(B)$

Figure 15.1 Energy levels (a) and density-of-states excluding spin degeneracy (b) of a two-dimensional electron gas in the absence of a magnetic field. The energy levels and the density-of-states in the presence of a magnetic field are indicated in (c) and (d), respectively. The degeneracy of the Landau levels is $N_L(B) = (e/hc)BS = D_0\hbar\omega_c$.

between adjacent Landau levels is $\hbar\omega_c = (e\hbar/mc)B = 2\mu_B B$; the value of the Bohr magneton is $\mu_B = (e\hbar/2mc) = 0.05788$ meV/Tesla, and $\hbar\omega_c/B = 0.1158$ meV/Tesla (1 Tesla $= 10^4$ gauss).

The normalized eigenfunctions corresponding to the eigenvalues of Eq. (15.10) are

$$\phi_{nk_x}(\mathbf{r}) = \frac{1}{\sqrt{L_x}} e^{ik_x x} \left(\frac{m\omega_c}{\pi\hbar}\right) \frac{1}{2^{n/2}\sqrt{n!}} e^{-m\omega_c(y-y_0)^2/2\hbar} H_n\left[(y - y_0)\sqrt{\frac{m\omega_c}{\hbar}}\right],$$

where L_x is the length, in the x-direction, of the rectangular box where the electron is confined, and H_n are the Hermite polynomials for the harmonic oscillator. The wave-functions $\phi_{nk_x}(\mathbf{r})$ in the presence of a magnetic field (in the chosen Landau gauge) are plane waves in the x-direction, multiplied by harmonic oscillator wave functions in the y-direction, with typical extension $l_n = l_0\sqrt{2n+1}$, where $l_0 = (\hbar c/eB)^{1/2}$ is the classical cyclotron radius of an electron with kinetic energy $\hbar\omega_c/2$. For $B = 10^4$ gauss, with $hc/e = 4.136 \times 10^{-7}$ gauss $\cdot$ cm^2, the magnetic length l_0 is ≈ 257 Å.

The energy levels of Eq. (15.10) do not depend on k_x; thus their degeneracy is given by the number of allowed k_x values for the system. We can determine the orbital degeneracy $N_L(B)$ of a Landau level by imposing the condition

$$0 \le y_0 \equiv \frac{\hbar c}{eB} k_x < L_y \quad \text{or equivalently} \quad 0 \le k_x < \frac{eB}{\hbar c} L_y.$$

The number N_L of allowed k_x values in the above interval gives the number of harmonic oscillators, whose origin is confined within the sample; the degeneracy N_L of each Landau level is then

$$\boxed{N_L(B) = \frac{L_x}{2\pi} \frac{eB}{\hbar c} L_y \equiv \frac{e}{hc} BS}. \tag{15.11}$$

The orbital degeneracy $N_L(B)$ is *thus proportional to the magnetic field, whose effect is that of piling up a large number of states into discrete Landau levels.* Separation and degeneracy of Landau levels are both proportional to the applied field. The orbital degeneracy can be written in the effective form

$$N_L(B) = \frac{\Phi(B)}{\Phi_0},$$

where $\Phi(B) = BS$ is the flux of B through the sample, and

$$\Phi_0 = \frac{hc}{e} = 4.136 \times 10^{-7} \text{ gauss} \cdot \text{cm}^2$$

provides the natural unit of flux.

The degeneracy $N_L(B)$ of each Landau level can be written also in the following alternative form

$$N_L(B) = D_0 \hbar \omega_c \quad \text{where} \quad D_0 = \frac{m}{2\pi \hbar^2} S \tag{15.12}$$

indicates the constant two-dimensional density-of-states in the absence of the magnetic field (see Eq. 15.7). The quantity $D_0 \hbar \omega_c$ represents the number of states in the energy interval $\hbar \omega_c$ between two successive Landau levels; Eq. (15.12) explicitly shows that the overall number of states is unaffected by the magnetic field.

There is another instructive way to characterize a two-dimensional electron gas, with N free electrons (of both spin directions) in the surface S, and surface density $n_s = N/S$. As suggested by the expression of the Landau degeneracy (15.11), it is useful to introduce the filling factor ν defined as the number of electrons N divided by the number of magnetic flux quanta through the surface S

$$\nu = \frac{N}{\Phi/\Phi_0} = n_s \frac{hc}{eB}. \tag{15.13}$$

From Eq. (15.11), it is evident that integer values of ν denote that the lowest-energy spin-resolved ν Landau levels are fully occupied by electrons, while all the others are fully empty. Similarly, values of ν smaller than one, for instance $\nu = 1/2$, $\nu = 1/3$ etc. represent one-half, one-third, etc. occupancy of the lowest spin-resolved Landau level.

In the case ν is non-integer and greater than one, the part of ν exceeding the highest integer less than ν, represents the occupancy fraction of the partially filled Landau level.

Taking into account the value of Φ_0 the filling factor can be expressed as

$$\nu = 4.136 \frac{n_s}{B} 10^{-11} \qquad B \text{ in Tesla} \qquad n_s \text{ in } 10^{11} \text{ cm}^{-2}. \tag{15.14}$$

For typical values of the density $n_s \approx 2.5 \cdot 10^{11}$ cm^{-2}, the filling factor $\nu = 1$ is reached with applied magnetic fields $B \approx 10$ Tesla.

15.2.2 Energy Levels of the Three-Dimensional Electron Gas in Magnetic Fields

A free electron, in a uniform magnetic field of vector potential $\mathbf{A}(\mathbf{r}) = (-By, 0, 0)$, is described by the Hamiltonian

$$H = \frac{1}{2m} \left(\mathbf{p} + \frac{e}{c} \mathbf{A} \right)^2 = \frac{1}{2m} \left(p_x - \frac{eB}{c} y \right)^2 + \frac{1}{2m} p_y^2 + \frac{1}{2m} p_z^2 \, ; \tag{15.15}$$

(the interaction of the spin magnetic moment with the applied field is neglected at this stage, and is discussed in Section 15.4).

The eigenvalues of the Hamiltonian (15.15) are straightforwardly obtained from the eigenvalues (15.10) taking into account the free motion parallel to the direction z of the magnetic field; one has

$$E_{nk_z} = \left(n + \frac{1}{2} \right) \hbar \omega_c + \frac{\hbar^2 k_z^2}{2m} \, , \tag{15.16}$$

where each state of quantum numbers n and k_z has degeneracy $N_L(B) = (e/hc)BS$ (excluding spin degeneracy), and $S = L_x L_y$ is the section of the sample of volume $V = L_x L_y L_z$. The energy levels E_{nk_z} are schematically indicated in Figure 15.2; the magnetic field governs the spacing $\hbar \omega_c$ between the one-dimensional Landau sub-bands and their degeneracy $N_L(B)$.

For a three-dimensional electron gas the density-of-states for both spin directions in the absence of a magnetic field is obtained from Eq. (15.4c) (including a factor 2 due to spin degeneracy); we have

$$D(E, B = 0) = A \cdot E^{1/2} \qquad A = \frac{V}{2\pi^2} \left(\frac{2m}{\hbar^2} \right)^{3/2} \qquad E > 0. \tag{15.17}$$

The density-of-states corresponding to the n-th sub-band of Eq. (15.16) can be obtained using Eq. (15.4a) for the one-dimensional character of the dispersion along z, and taking into account the orbital degeneracy $N_L(B) = (e/hc)BS$ of the Landau levels and the spin degeneracy; we have

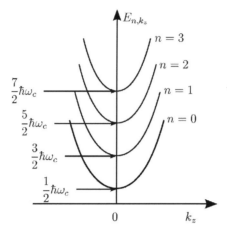

Figure 15.2 Schematic representation of the energy levels for a three-dimensional electron gas in the presence of a magnetic field in the z-direction.

$$D_n(E,B) = \frac{L_z}{2\pi}\left(\frac{2m}{\hbar^2}\right)^{1/2} \cdot N_L(B) \cdot 2\frac{1}{\sqrt{E - \left(n + \frac{1}{2}\right)\hbar\omega_c}} \quad \text{for} \quad E > \left(n + \frac{1}{2}\right)\hbar\omega_c.$$

The numerical factor in the right-hand side of the above equation equals $(1/2)\hbar\omega_c A$. Summing up the contributions from every Landau sub-band, we obtain for the total density-of-states (including spin degeneracy)

$$D(E,B) = \frac{1}{2}\hbar\omega_c A \sum_{n=0}^{\infty} \frac{1}{\sqrt{E - (n + \frac{1}{2})\hbar\omega_c}}\Theta\left[E - \left(n + \frac{1}{2}\right)\hbar\omega_c\right], \qquad (15.18)$$

where $\Theta(x)$ is the step function, which equals unity for $x > 0$ and vanishes otherwise. From Eq. (15.18), we see that the density of energy states in the presence of a magnetic field becomes infinite just above the edge of any Landau level and behaves as $E^{-1/2}$ in the neighborhood (see Figure 15.3). A further effect of the presence of the magnetic field is the increase ("blue shift") of $\hbar\omega_c/2$ of the energy spectrum threshold.

Until now we have considered the magnetic field effects on free electrons characterized by the dispersion relation $E(\mathbf{k}) = \hbar^2 k^2/2m$. In some materials (for instance simple metals, or a number of cubic semiconductors), the conduction band in the energy region of interest can be reasonably approximated with a parabolic energy band dispersion $E(\mathbf{k}) = E_c + \hbar^2 k^2/2m_c^*$, where m_c^* is the effective conduction mass and E_c is the bottom of the conduction band. In this case, upon application of a uniform magnetic field, the energy of the electrons at the bottom of the conduction band becomes quantized in the form

$$E_{nk_z} = E_c + \left(n + \frac{1}{2}\right)\hbar\omega_c^* + \frac{\hbar^2 k_z^2}{2m_c^*},$$

where the cyclotron frequency is $\omega_c^* = eB/m_c^* c$.

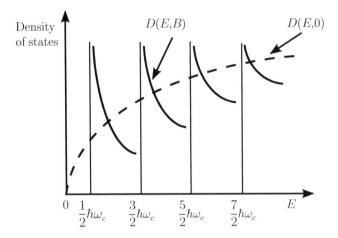

Figure 15.3 Density-of-states of a three-dimensional electron gas in the absence of a magnetic field (dashed line) and in the presence of a magnetic field (continuous lines).

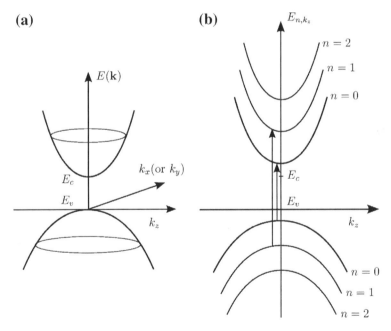

Figure 15.4 (a) Schematic representation of an isotropic two-band model semiconductor. (b) Schematic representation of quantization of electron states in the conduction band and in the valence band in the presence of a uniform magnetic field in the z-direction. Arrows indicate examples of optical transitions between valence and conduction states with the same values of n and k_z.

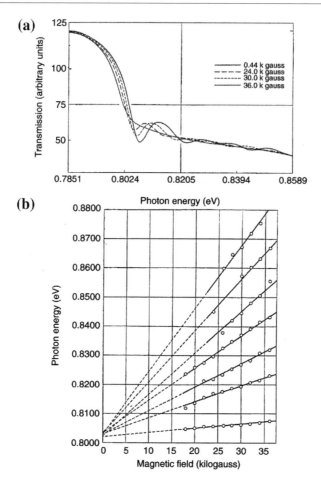

Figure 15.5 (a) Oscillatory magneto-absorption in germanium. (b) Energies of consecutive transmission minima as a function of the magnetic field. The lowest curve corresponds to transitions between Landau levels with $n = 0$; the successive curves correspond to transitions with higher quantum states of both conduction and valence bands [reprinted with permission from S. Zwerdling and B. Lax, Phys. Rev. *106*, 51 (1957); copyright 1957 by the American Physical Society].

The above considerations for the motion of electrons in a conduction band of effective mass m_c^* can be repeated for holes in semiconductors; in the case of a parabolic topmost valence band of effective mass m_v^*, the energy of holes becomes quantized in the form

$$E_{nk_z} = E_v - \left(n + \frac{1}{2}\right)\hbar\omega_v^* - \frac{\hbar^2 k_z^2}{2m_v^*}$$

($m_v^* > 0$; $\omega_v^* = eB/m_v^*c$). The energy levels around the fundamental gap for an isotropic two-band model semiconductor are shown in Figure 15.4. In small gap semiconductors, the effective masses of electrons and holes are in general rather small; this makes it easier to resolve in magneto-optic spectroscopy (or other experiments) several Landau levels, as their energy separation, at parity of magnetic field, increases for small effective masses. We wish also to mention that the effect of magnetic fields has been generalized to more complicated band structures, including the case of degenerate bands, anisotropic effects, and spin-orbit interaction.

In strong magnetic fields, the continuum of energy levels in the valence and conduction bands tend to pile up into strongly degenerate Landau sub-bands (see for instance Figures 15.3 and 15.4); were it not for the energy dispersion in the magnetic field direction, a full discretization of the energy levels would occur. The bunching effect of the energy levels, produced by the magnetic field, is at the origin of the oscillatory behavior of several physical properties; among them we mention the *oscillatory magneto-absorption effect* observed in semiconductors (pioneered by Zwerdling in late 1950s), and the oscillatory behavior of the magnetization observed in metals (the latter subject is discussed in the next section).

As an example of magneto-absorption measurements in semiconductors, we report in Figure 15.5a typical magneto-absorption curves concerning the direct transition at $\mathbf{k} = 0$ of germanium from the valence bands (heavy holes) to the conduction band; with increasing values of the magnetic field, pronounced minima in the transmission curve can be detected, together with a shift to higher energies of the direct absorption edge. Figure 15.5b shows the energy position of consecutive transmission minima (i.e. absorption maxima) as a function of the magnetic field; the figure shows a fan of straight lines, which extrapolate with high accuracy to the value of 0.803 eV at $B = 0$ and at room temperature; the so determined value of the direct gap of germanium exceeds by 0.14 eV the indirect gap of the material. An accurate analysis of the fine structure of the magneto-band effect provides invaluable information also on the light holes in germanium and on the detailed structure of the energy bands around the threshold of the direct transitions.

15.3   Landau Diamagnetism and de Haas-van Alphen Effect

In this section we discuss the orbital magnetic susceptibility of the three-dimensional electron gas; considerations on the spin magnetic susceptibility are reported in the next section.

The orbital contributions to the magnetic susceptibility of the free-electron gas in a uniform magnetic field consists of two main terms: (i) the Landau diamagnetism and (ii) the de Haas-van Alphen oscillations. The study for the electron gas is of interest in its own right, and furthermore provides guidelines for the interpretation of the response to magnetic fields of realistic metals.

The energy levels and the density-of-states of the free-electron gas in magnetic fields have been studied in the previous section. The density-of-states is all is needed in order to obtain the free energy and the magnetic response. However some relevant features can be better appreciated on the basis of more direct physical arguments, and in this line we begin the discussion of the Landau diamagnetism.

The Landau diamagnetism occurs because the quantization effect of the orbital motion perpendicular to the magnetic field is expected to increase the mean energy of the electrons. With some bookkeeping, one could attempt to estimate this increase in energy through the relationship

$$\Delta E(B) \approx D(E_F)\frac{1}{2}\hbar\omega_c\frac{1}{2}\hbar\omega_c, \tag{15.19}$$

where $D(E_F)$ is the density-of-states at the Fermi energy, and $D(E_F)\hbar\omega_c/2$ is an estimate of the number of states increasing their energy by the amount $\hbar\omega_c/2$. Notice that the naive estimate of Eq. (15.19) is not far from the exact value $D(E_F)\hbar^2\omega_c^2/24$ elaborated below. For energy change positive and proportional to B^2, the magnetic response is diamagnetic.

It is instructive to consider more closely the change in energy brought about by the application of a magnetic field to the free-electron gas. The ground state of the electron gas in the absence of magnetic fields, and zero temperature, can be represented in the reciprocal space by the Fermi sphere, with all the states occupied up to the Fermi energy. As illustrated in Figure 15.6a, in the presence of a magnetic field (say along the z-axis), it is convenient to consider a slice of states of the Fermi sphere with $k_z = $ const., since the k_z quantum number, and the energy contribution $\hbar^2 k_z^2/2m$ are unaffected by the magnetic field in the z-direction. The allowed energy levels of the electrons in the slice are

$$E_{nk_z} = \left(n + \frac{1}{2}\right)\hbar\omega_c + \frac{\hbar^2 k_z^2}{2m} \quad (n = 0, 1, 2, \dots). \tag{15.20}$$

We remind that the degeneracy of Landau levels (ignoring spin degeneracy) is

$$N_L(B) = D_0\hbar\omega_c \quad \text{with} \quad D_0 = \frac{m}{2\pi\hbar^2}S.$$

The quantity D_0, represents the density-of-states (independent of energy) of the two-dimensional electron gas in the absence of the magnetic field. We also remind that for the three-dimensional free-electron gas (see Chapter 3)

$$D(E_F) = \frac{3}{2}\frac{n}{E_F}V = \frac{3}{2}\frac{k_F^3}{3\pi^2}\frac{2m}{\hbar^2 k_F^2}V = \frac{mk_F}{\hbar^2\pi^2}V, \tag{15.21}$$

a relation that links the total density-of-states (included spin degeneracy) at the Fermi energy with the Fermi wavenumber.

The average change in energy of the two-dimensional electron gas of the slice k_z, due to the quantization, can be estimated as follows (see Figure 15.6b). In the absence of external magnetic field, electrons fill up the continuum of states from $\hbar^2 k_z^2/2m$ to E_F with density D_0. In the presence of the magnetic field, the allowed energy levels collapse into discrete Landau levels. Since the magnetic energy is $\hbar\omega_c$, in the energy interval $[E_F - \hbar\omega_c/2, E_F + \hbar\omega_c/2]$ there is just one Landau level; this very Landau level, the nearest in energy to E_F, is the only one responsible for the energy change in the section $k_z = $ const.

Suppose that, for the given slice under attention, the Landau level E_n nearest to the chemical potential differs from E_F by the quantity ε, with $0 < \varepsilon < \hbar\omega_c/2$ (as depicted

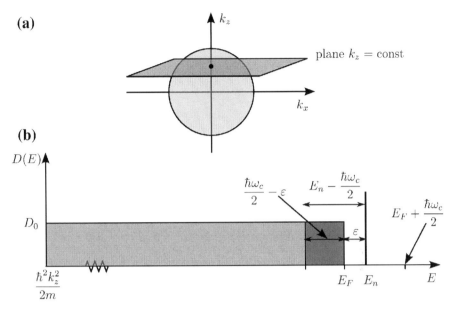

Figure 15.6 (a) Schematic representation of the ground state, at zero temperature, of the free-electron gas with a Fermi sphere of occupied states in the reciprocal space. The section of the Fermi sphere intersected by the plane with k_z = const. is also shown. (b) Schematic representation of the relevant features of the density of allowed states of the two-dimensional electron gas in the absence and in the presence of a uniform magnetic field. The whole energy region from $\hbar^2 k_z^2/2m$ to E_F represents the occupied states in the absence of magnetic field. In the presence of magnetic fields, the Landau level E_n nearest in energy to the chemical potential is indicated (in the figure it is assumed $E_n > E_F$). The electron states displaced to (or from) the Fermi energy (and then appropriately redistributed to other slices) are those indicated by the shaded energy region of width $(1/2)\hbar\omega_c - \varepsilon$.

in Figure 15.6b). The electrons belonging to the energy interval $[E_n - \hbar\omega_c/2, E_F]$ are shifted to the Fermi energy (and then redistributed to other slices). The number of states is $D_0 \cdot (-\varepsilon + \hbar\omega_c/2)$ and the average change of energy per particle is $(1/2)(-\varepsilon + \hbar\omega_c/2)$. The change of energy in the slice under attention is thus

$$\Delta U(\varepsilon) = D_0 \cdot \left(-\varepsilon + \frac{1}{2}\hbar\omega_c\right)\left(-\frac{1}{2}\varepsilon + \frac{1}{4}\hbar\omega_c\right) = \frac{1}{2}D_0 \cdot \left(\varepsilon^2 - \hbar\omega_c\varepsilon + \frac{\hbar^2\omega_c^2}{4}\right).$$

(15.22)

We average $\Delta U(\varepsilon)$, which is always positive, for the possible values of ε in the interval between 0 and $\hbar\omega_c/2$, gives

$$\langle\Delta U(\varepsilon)\rangle = \frac{1}{2}D_0 \frac{1}{\hbar\omega_c/2}\int_0^{\hbar\omega_c/2}\left(\varepsilon^2 - \hbar\omega_c\varepsilon + \frac{\hbar^2\omega_c^2}{4}\right)d\varepsilon$$

$$= \frac{D_0}{\hbar\omega_c}\left[\frac{1}{3}\varepsilon^3 - \frac{\hbar\omega_c}{2}\varepsilon^2 + \frac{\hbar^2\omega_c^2}{4}\varepsilon\right]_0^{\hbar\omega_c/2} = \frac{1}{24}D_0\hbar^2\omega_c^2.$$

(15.23)

It is seen by inspection that the same result $\langle \Delta U(\varepsilon) \rangle$ is obtained also in the case $-\hbar\omega_c/2 < \varepsilon < 0$. Furthermore, the excess of states transferred to the Fermi energy in the slices with $\varepsilon > 0$, are accommodated by an equal number of extra states available in the slices with $\varepsilon < 0$. Thus, for $\hbar\omega_c \ll E_F$, one can safely assume that the Fermi energy is independent of the presence or absence of magnetic fields.

To obtain the total change of energy for the system we sum the contributions from the various slices into which the Fermi sphere has been resolved; one obtains (including a factor 2 for spin degeneracy)

$$\Delta U_{\text{tot}} = 2 \frac{L_z}{2\pi} \int_{-k_F}^{+k_F} \frac{1}{24} D_0 \hbar^2 \omega_c^2 \, dk = \frac{D_0 L_z}{12} \frac{\hbar^2 \omega_c^2}{\pi} k_F = \frac{V}{24} \frac{m k_F}{\pi^2 \hbar^2} \hbar^2 \omega_c^2.$$

Using Eq. (15.21) we arrive thus at the final result

$$\boxed{\Delta U_{\text{tot}} = \frac{1}{24} D(E_F) \hbar^2 \omega_c^2}, \tag{15.24}$$

which shows an overall increase in energy with the magnetic field and a consequent diamagnetic susceptibility. The Landau contribution of Eq. (15.24), here derived with *ad hoc* procedures at low temperature, remains valid at any temperature: the demonstration requires some general elaborations. Momentarily, however, we continue the study of the magnetic properties of the free-electron gas with some qualitative considerations, before passing to a quantitative analysis.

Besides considering the overall increase in energy, we can consider in detail population and depopulation effects of the Landau levels crossing the Fermi energy; as the applied magnetic field is varied, we expect oscillations in the ground-state energy $E_0(B)$ and, as a consequence, in the magnetic moment. With reference to the electrons in a thin slice around $k_z = \text{const.} = 0$, we shall expect that the period of these oscillations is such that

$$\left(j + \frac{1}{2} \right) \hbar\omega_c = \left(j + \frac{1}{2} \right) \hbar \frac{eB}{mc} = E_F$$

with j integer, or equivalently

$$\frac{1}{B} = \left(j + \frac{1}{2} \right) \frac{\hbar e}{mc E_F}. \tag{15.25a}$$

Thus we expect a contribution to the magnetic susceptibility, which is *periodic in inverse magnetic field* $1/B$ with period $\hbar e/mc E_F$, called the de Haas-van Alphen period. The experimental determination of the de Haas-van Alphen oscillations requires the sharpness of the Fermi-Dirac distribution function, i.e. $k_B T \ll \hbar\omega_c$; it also requires the sharpness of Landau levels, i.e. $\omega_c \tau \gg 1$ (the relaxation time τ represents the mean time between successive collisions suffered by the electrons); in general, the de Haas-van Alphen effect can be detected for high magnetic fields (of the order of Tesla), in high pure samples, at a few Kelvin degrees.

It is possible to elaborate Eq. (15.25a), which holds for the free-electron gas and for its spherical Fermi surface, into a form suitable to more complicated Fermi surfaces. If we write $E_F = \hbar^2 k_F^2/2m$ in Eq. (15.25a), we obtain for the period of oscillations

$$\Delta\left(\frac{1}{B}\right) = \frac{e}{\hbar c} \frac{2\pi}{A_{\text{extremal}}}, \tag{15.25b}$$

where $A_{\text{extremal}} = \pi k_F^2$ is the extremal area determined by the intersection of the Fermi surface with planes perpendicular to the magnetic field. Eq. (15.25b) is valid also in real metals with more complicated Fermi surfaces and allows to determine extremal cross-sectional areas for any given direction of the external magnetic field.

The quantum mechanical treatment of the orbital magnetic susceptibility for the three-dimensional electron gas can be carried out analytically, with a little demanding but particularly instructive procedure. As shown in the appendix, according to Eq. (A.21), the free energy of the electron gas in a uniform magnetic field B at temperature T is given by the expression

$$F(T, B) = F(T, 0) + \frac{\hbar^2 \omega_c^2}{24} D(\mu) + \frac{\hbar^2 \omega_c^2}{4\pi^2} D(\mu) \left(\frac{\hbar\omega_c}{2\mu}\right)^{1/2} \cdot \sum_{s=1}^{\infty} \frac{(-1)^s}{s^{5/2}} \frac{2\pi^2 s k_B T}{\hbar\omega_c}$$

$$\times \frac{1}{\sinh\left(2\pi^2 s k_B T/\hbar\omega_c\right)} \cos\left(2\pi s \frac{B_F}{B} - \frac{\pi}{4}\right), \tag{15.26}$$

where $F(T, 0)$ is the free energy in the absence of the magnetic field, $D(\mu)$ is the density-of-states (for both spin directions) at the chemical potential in the absence of magnetic field, and the special field B_F is defined so that $\hbar(e B_F/mc) \equiv \mu$. The presence of oscillatory terms in the right-hand side of Eq. (15.26) explains the oscillatory behavior of the magnetic moment $M = -(1/V)\partial F/\partial B$ as a function of the magnetic field, in *the low temperature limit* $k_B T \ll \hbar\omega_c$; when the condition $k_B T \ll \hbar\omega_c$ is not satisfied, the oscillations are damped through a factor of the type $\exp\left(-2\pi^2 k_B T/\hbar\omega_c\right)$.

In the *high temperature limit* ($\hbar\omega_c \ll k_B T$) we can neglect the oscillatory terms in Eq. (15.26), and obtain

$$F(T, B) = F(T, 0) + \frac{1}{24}\hbar^2 \omega_c^2 D(\mu); \tag{15.27}$$

this result corroborates the physical arguments adopted to elaborate Eq. (15.24). The free energy (15.27) increases as B^2; the magnetic susceptibility $\chi = -(1/V)\partial^2 F/\partial B^2$ gives rise to the Landau diamagnetism

$$\boxed{\chi_L = -\frac{1}{12}\left(\frac{e\hbar}{mc}\right)^2 \frac{D(\mu)}{V} = -\frac{1}{3}\mu_B^2 \frac{D(\mu)}{V}}. \tag{15.28}$$

For the free-electron gas, we can also write (neglecting for simplicity the temperature dependence of the chemical potential)

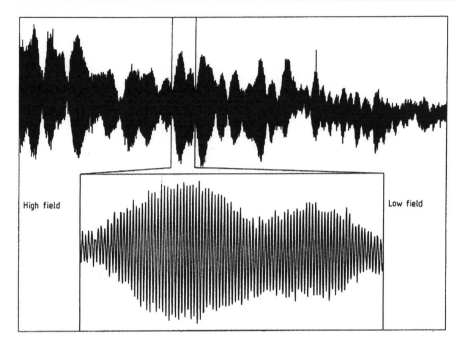

Figure 15.7 Experimental de Haas-van Alphen magnetization oscillations for a sample containing randomly oriented particles of lithium dispersed in paraffin wax. The full trace covers the field range 12.9–7.9 Tesla at a temperature of 20 mK. The portion of the trace between 11.1 and 10.8 T is shown expanded in the inset. [from M. B. Hunt, P. H. P. Reinders, and M. Springford, J. Phys. Condens. Matter *1*, 6589 (1989); copyright 1989 by the Institute of Physics.]

$$\chi_L = -\frac{1}{12}\frac{e^2\hbar^2}{m^2c^2}\frac{3}{2}\frac{n}{E_F} = -\frac{1}{12\pi^2}\frac{e^2}{mc^2}k_F \qquad (15.29)$$

(use has been made of the relations $E_F = \hbar^2 k_F^2/2m$ and $k_F^3 = 3\pi^2 n$). With $e^2/mc^2 = r_0 = 2.82 \times 10^{-13}$ cm and $k_F \approx (1/10^{-8})$ cm^{-1}, the order of magnitude of the susceptibility is $\chi_L \approx -10^{-6}$.

As an example of experimental de Haas-van Alphen oscillations, we report in Figure 15.7 the behavior of the magnetization of lithium grains as a function of the external field H. In lithium metal there is one conduction electron per atom and the Brillouin zone is large enough to contain the free-electron Fermi sphere; the Fermi surface of bcc lithium looks like a distorted spherical surface, pushed (slightly) outwards in the $\langle 1\,1\,0 \rangle$ directions and inwards in the $\langle 1\,0\,0 \rangle$ directions. In the free-electron model the Fermi energy of lithium is $E_F = 4.78$ eV, and the quantity B_F takes the value $B_F = 41250$ Tesla (taking $\hbar e/mc \approx 0.115$ meV/Tesla). In Figure 15.7 about 2000 de Haas-van Alphen oscillations are observed over the field range 12.9–7.9 Tesla; the oscillations show a complex beat structure, which is related to the anisotropy of the extremal cross-sectional areas of the randomly oriented particles.

15.4   Spin Paramagnetism of a Free-Electron Gas

Until now, in the discussion of the magnetic field effects on the free-electron gas, we have neglected the interaction of the spin magnetic moment with the applied field; when this interaction is taken into account, the Hamiltonian of an electron in a uniform magnetic field of vector potential $\mathbf{A}(\mathbf{r})$, becomes

$$H = \frac{1}{2m}\left(\mathbf{p} + \frac{e}{c}\mathbf{A}\right)^2 + g_0\mu_B\mathbf{s}\cdot\mathbf{B}, \tag{15.30}$$

where $\mathbf{s}$ is the electron spin operator (in units $\hbar$), and $g_0 \approx 2$ is the gyromagnetic factor for the electron spin. The eigenvalues of the Hamiltonian (15.30) are

$$E_{nk_z} = \left(n + \frac{1}{2}\right)\hbar\omega_c + \frac{\hbar^2 k_z^2}{2m} \pm \mu_B B;$$

they are obtained from Eq. (15.16) by adding the terms $\pm\mu_B B$ for spin up and spin down electrons.

The magnetic susceptibility of the free-electron gas including spin contribution, could be worked out following the same guidelines adopted for handling the closely related spectrum of Eq. (15.16). In a more simple way, the effect of the spin contribution to the magnetic susceptibility can be better illustrated with the following argument.

Consider a metal, with N free electrons (of both spin directions) in the volume V; the standard occupation of states for both spin directions in the absence of magnetic field is indicated in Figure 15.8a. In Figure 15.8b we indicate the situation occurring in the presence of a uniform field B (while keeping the electrons in their original states); there is an energy shift by $-\mu_B B$ (or $+\mu_B B$) for electrons with magnetic moment parallel (or antiparallel) to the applied magnetic field. In Figure 15.8c equilibrium is recovered, and an excess of electrons with parallel magnetic moment is established.

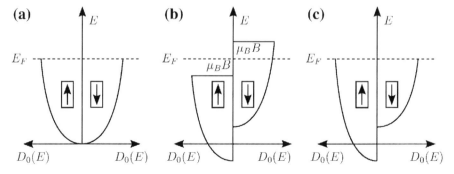

Figure 15.8 Schematic representation of the origin of the Pauli spin paramagnetism of a free-electron gas; arrows indicate the direction up or down of the electron spin magnetic moment; $D_0(E)$ denotes the density-of-states for one spin direction.

The number of electrons ΔN that flip their magnetic moment from $-\mu_B$ to μ_B are those contained in the energy interval $\mu_B B$ around the Fermi level:

$$\Delta N = \frac{1}{2}D(E_F)\mu_B B,$$

where $D(E_F)/2 = D_0(E_F)$ is the density-of-states at the Fermi level for one spin direction. The magnetic moment M per unit volume is then

$$M = \frac{1}{V}\Delta N 2\mu_B = \mu_B^2 B \frac{D(E_F)}{V};$$

the ratio M/B gives for the Pauli spin susceptibility the value

$$\boxed{\chi_P = \mu_B^2 \frac{D(E_F)}{V}}. \tag{15.31}$$

In the case of the free-electron gas, the net effect of the Pauli paramagnetic susceptibility (15.31) and the Landau diamagnetic susceptibility (15.28) is a paramagnetic behavior; in fact $\chi_L = -(1/3)\chi_P$. In actual materials, the "effective mass" for orbital motion, as well as the "effective gyromagnetic factor," can be rather different from the corresponding free-electron values, so changing the relative importance of the Pauli and Landau susceptibility; this explains why some metals may have a net diamagnetic behavior. For a more quantitative account, correlation, and exchange effects among electrons should be considered, because they may significantly influence the magnetic susceptibility.

15.5 Magnetoresistivity and Classical Hall Effect

General Considerations and Phenomenological Aspects

Transport effects in crystals in the presence of electric fields and temperature gradients have been considered in Chapter 11. In this section, we study some aspects of transport phenomena due to the simultaneous presence of electric and magnetic fields; the possible presence of thermal gradients adds further variety to the phenomenology, but here we confine our attention to samples at uniform temperature.

The study of magnetic field effects on the transport properties of metals and semiconductors has become a well-established and invaluable tool for the investigation of mobile carriers in crystals. In particular the Hall measurements, aimed at the determination of carrier concentration and charge sign, are routinely used for the characterization of materials. Also magnetoresistivity measurements, which determine the resistivity of materials in the presence of magnetic fields, offer a wide range of effects. In metals with closed Fermi surfaces (such as alkali metals), the magnetoresistivity does saturate for any crystal orientation (i.e. it approaches a constant value for sufficiently high magnetic fields, irrespective of orientation); the same occurs for n-type and p-type semiconductors. In metals with equal number of electrons and holes (such as Bi, Sb, and so on), the magnetoresistivity does not saturate for any crystal orientation and keeps on increasing as the magnetic field increases; the same occurs for semiconductors with equal numbers of electrons and holes. We also mention that in metals with open Fermi surfaces (such as Cu, Ag, Au, and so on), the magnetoresistivity saturates for most of the crystal orientations but does not saturate for others. Finally, and most importantly, for two-dimensional systems the magnetoresistance is quantized and the quantum Hall effect occurs. In this section we consider some aspects of the traditional magnetoresistivity and Hall phenomenology in three-dimensional crystals, while in the next section we consider the quantized Hall effect in two-dimensional systems.

In isotropic media the application of a (small) electric field drives a current density parallel and proportional to it, and the linear relationship holds

$$\mathbf{J} = \sigma \mathbf{E}, \tag{15.32a}$$

where the conductivity σ is a scalar quantity. In the presence of a magnetic field, carriers are deflected and in general the current density is no more parallel to the electric field; the conductivity becomes a tensor even for an isotropic material. Relation (15.32a) has to be replaced by the more general expression

$$J_i = \sum_j \sigma_{ij}(B) E_j, \tag{15.32b}$$

where $\sigma_{ij}(B)$ $(i, j = x, y, z)$ are the components of the *magnetoconductivity tensor* $\sigma(B)$. Similar considerations can be done for the resistivity of an isotropic medium in the presence of a magnetic field; the relationship between electric field and current density becomes

$$E_i = \sum_j \rho_{ij}(B) J_j, \tag{15.32c}$$

where $\rho_{ij}(B)$ $(i, j = x, y, z)$ are the components of the *magnetoresistivity tensor* $\rho(B)$. The magnetoconductivity tensor and magnetoresistivity tensor are the inverse of each other, and it holds

$$\rho_{ij}(B) = \left(\frac{1}{\sigma(B)} \right)_{ij}. \tag{15.32d}$$

The transport parameters $\rho_{ij}(B)$ are often determined experimentally *using the standard geometry* in which a magnetic field $\mathbf{B}$ is applied orthogonally to a long and thin current carrying conductor and the current flows, for instance, along the x-direction (see Figure 15.9); the x- and y-directions are often referred to as "longitudinal" and "transverse" directions, respectively.

In the standard geometry, in which transport is in the xy plane and furthermore $J_y \equiv 0$ (in stationary conditions), the density current $\mathbf{J}$, and the electric field $\mathbf{E}$ are related by

$$\begin{pmatrix} E_x \\ E_y \end{pmatrix} = \begin{pmatrix} \rho_{xx}(B) & \rho_{xy}(B) \\ \rho_{yx}(B) & \rho_{yy}(B) \end{pmatrix} \begin{pmatrix} J_x \\ J_y \equiv 0 \end{pmatrix}.$$

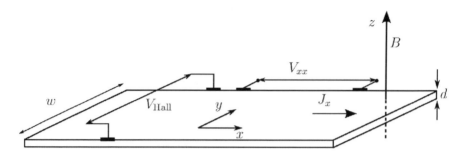

Figure 15.9 Standard geometry for Hall effect and magnetoresistivity measurements. V_{Hall} is the Hall potential, J_x is the current density in the flow direction, and w denotes the width of the strip.

The above matrix equation can be written explicitly in the form

$$E_x = \rho_{xx}(B)J_x, \tag{15.33a}$$
$$E_y = \rho_{yx}(B)J_x. \tag{15.33b}$$

Thus the diagonal element $\rho_{xx}(B)$ of the magnetoresistivity tensor is measured by the ratio between the longitudinal electric field E_x and the current density J_x in the x-direction. The off-diagonal component $\rho_{yx}(B)$ is measured by the ratio between the transverse electric field E_y and the current density J_x. It can also be inferred by inspection that $\rho_{xy}(B) = -\rho_{yx}(B)$ (as shown also in the models discussed below).

The transverse electric field E_y, also called *Hall field*, is produced by the space charges accumulated (in stationary conditions) at the borders of the conductor because of the deflection due to the magnetic field. One or the other of the off-diagonal magnetoresistivity components (i.e. ρ_{yx} or its opposite ρ_{xy}) are also known as *Hall resistivity*. Most often it is convenient to report the *Hall coefficient*, defined as

$$R_{Hall}(B) = \frac{1}{B}\rho_{yx}(B) = \frac{1}{B}\frac{E_y}{J_x}. \tag{15.34a}$$

Notice that the *Hall potential* is $V_{Hall} = E_y w$, where w is the transverse dimension of the sample; in the absence of magnetic field, both V_{Hall} and ρ_{yx} vanish. The current in the Hall bar is related to the current density J_x, and to the thickness d and width w of the strip by the relation $I = J_x w d$. We have thus

$$\rho_{yx}(B) = \frac{E_y}{J_x} = \frac{V_{Hall}\,d}{I} \quad \Longrightarrow \quad V_{Hall} = \rho_{yx}(B)\frac{I}{d}; \tag{15.34b}$$

thus at parity of other conditions the Hall potential is higher for bars of small thickness. The opportunity to use conductors of small thickness, was at the basis of the discovery by Hall of the effect that brings his name, after failing in observing the effect in massive metal samples. *"Owing probably to the fact that the metal disk used had considerable thickness, the experiment at that time failed to give any positive result. Prof. Rowland now advised me, in repeating this experiment, to use gold leaf mounted on a plate of glass as my metal strip. I did so, and, experimenting as indicated above, succeeded on the 28th of October in obtaining, as the effect of the magnet's action, a decided deflection of the galvanometer needle ..."* [Excerpt from the article by E. H. Hall, Amer. J. Math. **2**, 287 (1879)].

We pass now to study the Hall effect and the magnetoresistivity in a few simple models. We consider first the case of a single type of carriers in a parabolic band model, with a unique relaxation time. Next, we consider the case in which holes and electrons are present, both with isotropic masses. The models described below, provide an indicative picture of the transport phenomena in the presence of magnetic fields in somewhat idealized situations. We wish to remark that the description of magneto-transport effects in realistic materials is rather demanding and requires a proper account of several features (such as energy dependence of the relaxation time, deviations from parabolic bands, detailed shape of the Fermi surfaces especially in the presence of a complicated connectivity in the repeated zone scheme, accurate analysis of the Boltzmann transport equations). We cannot enter in these and other aspects, and we refer for more elaborated models and discussions to the classic book by R. A. Smith "Semiconductors" (Cambridge University Press, Cambridge, 1978)].

Model 1. Magnetoresistivity and Hall Effect in an Isotropic One-Band Model

We consider here the magnetoresistivity and the Hall effect in the case of a single type of carriers (electrons or holes) in a parabolic energy band. For simplicity we use a model approach to the motion of electrons (or holes); the treatment with the more rigorous Boltzmann equation would give in the present case the same results.

The classical equation of motion of an electron, in a dissipative medium, in the presence of an electric field $\mathbf{E}$, and a magnetic field $\mathbf{B}$ reads

$$m^* \frac{d\mathbf{v}}{dt} = (-e)\mathbf{E} + \frac{(-e)}{c} \mathbf{v} \times \mathbf{B} - \frac{m^*}{\tau} \mathbf{v}, \tag{15.35a}$$

where m^* is the effective mass of the electron, and a damping term with constant relaxation time τ has been included. In stationary conditions $d\mathbf{v}/dt = 0$, and Eq. (15.35a) becomes

$$\mathbf{v} = -\frac{e\tau}{m^*}\mathbf{E} - \frac{e\tau}{m^*c}\mathbf{v} \times \mathbf{B}. \tag{15.35b}$$

We specify the above equation in the geometry of Figure 15.9, with the electric field in the xy plane and the magnetic field in z-direction; Eq. (15.35b) becomes

$$\begin{cases} v_x = -\dfrac{e\tau}{m^*}E_x - \omega_c\tau v_y, \\[2mm] v_y = -\dfrac{e\tau}{m^*}E_y + \omega_c\tau v_x, \end{cases} \tag{15.36a}$$

where $\omega_c = eB/m^*c$ is the cyclotron frequency. From Eqs. (15.36a) we have

$$\begin{cases} v_x = -\dfrac{e\tau}{m^*}\dfrac{1}{1+\omega_c^2\tau^2}(E_x - \omega_c\tau E_y), \\[3mm] v_y = -\dfrac{e\tau}{m^*}\dfrac{1}{1+\omega_c^2\tau^2}(\omega_c\tau E_x + E_y). \end{cases} \tag{15.36b}$$

Thus the current density $\mathbf{J} = n(-e)\mathbf{v}$, (where n is the electron density) is related to the electric field via the magnetoconductivity tensor $\sigma(B)$ given by

$$\sigma(B) = \frac{ne^2\tau}{m^*} \frac{1}{1+\omega_c^2\tau^2} \begin{pmatrix} 1 & -\omega_c\tau \\ \omega_c\tau & 1 \end{pmatrix}. \tag{15.37}$$

Notice that $\sigma_{xy}(B) = -\sigma_{yx}(B)$, which is a particular case of the general Onsager relations.

Equation (15.37) provides the magnetoconductivity for a *parabolic band with constant (i.e. energy independent) relaxation time*. Inversion of the matrix (15.37) gives the magnetoresistivity tensor

$$\rho(B) = \frac{m^*}{ne^2\tau} \begin{pmatrix} 1 & +\omega_c\tau \\ -\omega_c\tau & 1 \end{pmatrix}. \tag{15.38}$$

From Eq. (15.38), we see that the diagonal (or parallel) magnetoresistivity $\rho_{xx}(B)$, the Hall magnetoresistivity $\rho_{yx}(B)$, and the Hall coefficient have the expressions

$$\rho_{xx}(B) = \frac{m^*}{ne^2\tau}, \quad \rho_{yx}(B) = -\frac{B}{nec}, \quad R_{\text{Hall}}(B) = -\frac{1}{nec}. \tag{15.39}$$

Thus in the parabolic one-band model with a single relaxation time, the diagonal magnetoresistivity turns out to be independent of B, and we have $\rho_{xx}(B) = \rho_{xx}(0) = m^*/(ne^2\tau)$. Even more important, the Hall coefficient is independent of the effective mass and of the relaxation time; it depends only on the carrier concentration and charge sign. Also notice that in the case of positive holes, the off-diagonal matrix elements in Eqs. (15.37)–(15.39) change sign.

The results summarized in Eq. (15.39), obtained in the rather idealized one-band model, are to be taken only as indicative, and cannot be used as they stand for quantitative descriptions of realistic conductors. It is important in fact to notice that a proper account of the energy dependence of the relaxation time, or of the anisotropy of the energy bands, modify the results of Eq. (15.39); in particular, a dependence of $\rho_{xx}(B)$ on B is actually always observed in experiments. For these reasons, we consider the slightly more sophisticated two-band model, representing two groups of carriers.

Model 2. Magnetoresistivity and Hall Effect in an Isotropic Two-Band Model

Interesting new features appear in the study of magnetoresistivity and Hall effect within the two-band model. For simplicity we suppose that the two bands are parabolic, with effective masses m_1 and m_2; we also assume that the relaxation times τ_1 and τ_2 are constant for each group of carriers. The two-band model is useful to provide insight of transport phenomena in crystals with two groups of carriers of the same type (but different masses or relaxation times) or for mixed type carriers; we consider specifically this last situation.

Consider a material with n electrons (per unit volume) of mass m_1 and relaxation time τ_1, and p holes of mass m_2 and relaxation time τ_2. The magnetoconductivity is just the sum of the contributions from each group of carriers. Using Eq. (15.37) for electrons, and the appropriate modified form for positive holes, we obtain

$$\sigma(B) = \begin{pmatrix} A_1 & -B_1 \\ B_1 & A_1 \end{pmatrix} + \begin{pmatrix} A_2 & B_2 \\ -B_2 & A_2 \end{pmatrix} = \begin{pmatrix} A_1 + A_2 & -B_1 + B_2 \\ B_1 - B_2 & A_1 + A_2 \end{pmatrix}, \quad (15.40)$$

where

$$A_1 = \frac{\sigma_1}{1 + \omega_1^2 \tau_1^2}, \quad B_1 = \frac{\sigma_1 \omega_1 \tau_1}{1 + \omega_1^2 \tau_1^2}, \quad \sigma_1 = \frac{ne^2 \tau_1}{m_1}, \quad (15.41a)$$

$$A_2 = \frac{\sigma_2}{1 + \omega_2^2 \tau_2^2}, \quad B_2 = \frac{\sigma_2 \omega_2 \tau_2}{1 + \omega_2^2 \tau_2^2}, \quad \sigma_2 = \frac{pe^2 \tau_2}{m_2}. \quad (15.41b)$$

The magnetoresistivity tensor is obtained by inverting the magnetoconductivity tensor (15.40); it holds

$$\rho(B) = \frac{1}{(A_1 + A_2)^2 + (B_1 - B_2)^2} \begin{pmatrix} A_1 + A_2 & B_1 - B_2 \\ -B_1 + B_2 & A_1 + A_2 \end{pmatrix}. \quad (15.42)$$

Consider first the parallel component $\rho_{xx}(B)$ of the magnetoresistivity tensor of Eq. (15.42); using expressions (15.41) one obtains

$$\rho_{xx}(B) = \frac{\sigma_1 + \sigma_2 + \sigma_1 \omega_2^2 \tau_2^2 + \sigma_2 \omega_1^2 \tau_1^2}{(\sigma_1 + \sigma_2)^2 + (\sigma_1 \omega_2 \tau_2 - \sigma_2 \omega_1 \tau_1)^2}. \quad (15.43)$$

It is straightforward to verify that $\rho_{xx}(B) > \rho_{xx}(0)$; thus $\rho_{xx}(B) - \rho_{xx}(0)$ is an essentially *positive quantity for any value of the magnetic field*.

For very high magnetic fields (such that $\omega_1 \tau_1 \gg 1$ and $\omega_2 \tau_2 \gg 1$), Eq. (15.43) shows that in general $\rho_{xx}(B \to \infty)$ is finite, and thus there is saturation of the magnetoresistivity; the only remarkable exception occurs when

$$\sigma_1 \omega_2 \tau_2 \equiv \sigma_2 \omega_1 \tau_1 \qquad \Longrightarrow \qquad n = p.$$

When the two groups of carriers of opposite type (electrons and holes) have the same concentration, then $\rho_{xx}(B \to \infty) = \infty$ and no saturation occurs.

Consider now the off-diagonal magnetoresistivity transport parameter $\rho_{yx}(B)$; from Eqs. (15.42) and (15.41) one obtains

$$\rho_{yx}(B) = \frac{-\sigma_1 \omega_1 \tau_1 (1 + \omega_2^2 \tau_2^2) + \sigma_2 \omega_2 \tau_2 (1 + \omega_1^2 \tau_1^2)}{(\sigma_1 + \sigma_2)^2 + (\sigma_1 \omega_2 \tau_2 - \sigma_2 \omega_1 \tau_1)^2}. \quad (15.44a)$$

For high magnetic fields (i.e. $\omega_i \tau_i \gg 1$), the above expression simplifies in the form

$$\rho_{yx}(B \to \infty) \approx -\frac{\omega_1 \tau_1 \omega_2 \tau_2}{\sigma_1 \omega_2 \tau_2 - \sigma_2 \omega_1 \tau_1} = -\frac{B}{(n-p)ec}. \quad (15.44b)$$

The Hall parameter becomes

$$R_{\text{Hall}}(B \to \infty) = -\frac{1}{(n-p)ec}, \quad (15.44c)$$

a result which is independent of relaxation time and is *governed by the difference of the density of electrons and holes*. It is easy to understand qualitatively the limiting result (15.44c); for high values of B, the deflection of carriers produced by the magnetic field increases. Since electrons and holes have opposite charges and move in opposite directions, they are deflected on *the same side;* thus the effective number of carriers entering in Eq. (15.44c) is given by the difference of the electron and hole concentrations.

15.6   Quantum Hall Effects

15.6.1   Integer Quantum Hall Effect

In the previous section we have considered some effects of magnetic fields on transport properties in three-dimensional materials. In this section we present some aspects of transport measurements under strong magnetic fields for the two-dimensional electron gas; the observed effects have opened new areas of investigation and brought major breakthroughs in the comprehension of the two-dimensional structures and new states of condensed matter.

The transport properties of *two-dimensional* conductors, when observed in high purity samples, at very low temperatures and strong magnetic fields, show striking departure from the classical behavior; in particular the Hall resistance $\rho_{xy}(B)$ versus B exhibits flat plateaus, from which one can obtain the universal constant h/e^2, now known as the von Klitzing constant. In the plateau regions the Hall resistance is given exactly by h/e^2 divided by an integer, and its experimental value is

$$R_K = \frac{h}{e^2} = 25812.807 \ \Omega. \tag{15.45}$$

Measurements, performed even on samples of different origin, turn out to be reproducible within the astonishing accuracy of one part per billion (or so), and for this reason the Hall effect has attained a special role in metrology as the standard of resistance [see for instance P. J. Mohr, B. N. Taylor and D. B. Newell "CODATA recommended values of the fundamental physical constants: 2006" Rev. Mod. Phys. *80*, 633 (2008)].

The quantum Hall effect was first reported for the two-dimensional electron gas in the inversion layer of a silicon metal-oxide-semiconductor field-effect-transistor at $T = 1.5$ K and $B = 18$ Tesla by K. von Klitzing, G. Dorda, and M. Pepper, Phys. Rev. Lett. *45*, 494 (1980). In the device, the density of surface electrons can be controlled and changed by varying the MOSFET gate voltage; the Hall resistance shows fixed values $(1/i)(h/e^2)$ (with i integer number) around experimentally well-defined surface carriers concentrations, while the longitudinal resistance is vanishingly small.

Degenerate two-dimensional electron systems can be also realized at the interface between GaAs and (n-doped) $Al_x Ga_{1-x} As$; nearly ideal semiconductor heterostructures are prepared by molecular beam epitaxy techniques. As already discussed in Section 14.4, the electrons at the interface are confined by the potential well originated from the conduction band offset; the motion perpendicular to the interface is quantized, and, even when all the carriers are trapped in the lowest ground state, the motion

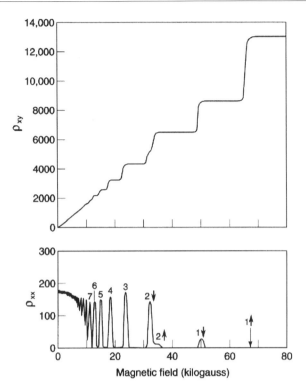

Figure 15.10 Hall resistance $\rho_{xy}(B)$ and longitudinal resistance $\rho_{xx}(B)$ (in ohm) of a two-dimensional electron gas at the GaAs-Al$_x$Ga$_{1-x}$As interface, at 50 mK. The quantum Hall effect is already evident at about 10 kilogauss. The numbers and the arrows above $\rho_{xx}(B)$ maxima refer to the Landau quantum number and the spin polarization of the levels; for the Landau levels $n \geq 3$ the spin splitting is not resolved and plateaus are observed at the corresponding even filling factors [reprinted with permission from M. A. Paalanen, D. C. Tsui, and A. C. Gossard, Phys. Rev. B **25**, 5566 (1982); copyright 1982 by the American Physical Society].

parallel to the interface is still free like. When a strong magnetic field B is applied perpendicularly to the two-dimensional electron gas, the quantum Hall effect is observed as a sequence of flat plateaus in the Hall resistance $\rho_{xy}(B)$ plotted as a function of B; in the same regions the parallel component of the electric resistance $\rho_{xx}(B)$ becomes vanishingly small, as shown in Figure 15.10. The effective mass of electrons in GaAs ($m^* \approx 0.07m_0$) is approximately three times lighter than that of electrons in silicon inversion layers, and thus the quantum Hall regime in Al$_x$Ga$_{1-x}$As heterostructures is reached at smaller values of magnetic fields (see Figure 15.10).

For a better appreciation of the striking features of the integer quantum Hall effect, it is convenient to start with a few qualitative remarks on what would be the expectations from a classical point of view.

Consider a two-dimensional electron system (in the xy plane) in the presence of a strong perpendicular magnetic field. Let n_s denote the *surface carrier density* (more precisely, n_s is the number of carriers, per unit surface, in the conductive layer of planes adjacent the surface). Similarly, let $\mathbf{J}^{(s)} = n_s(-e)\mathbf{v}$ denote the *surface current density* (notice that in the previous section $\mathbf{J}$ and n were bulk quantities). In the presence of a magnetic field, carriers are deflected and in general $\mathbf{J}^{(s)}$ is no more parallel to the electric field (in the xy plane); rather we have

$$J_i^{(s)} = \sum_j \sigma_{ij}(B)E_j, \tag{15.46a}$$

where $\sigma_{ij}(B)$ $(i, j = x, y)$ are the components of the conductance tensor $\sigma(B)$. Similarly, the relation between electric field and current density is

$$E_i = \sum_j \rho_{ij}(B)J_j^{(s)}, \tag{15.46b}$$

where $\rho_{ij}(B)$ $(i, j = x, y)$ are the components of the resistance tensor $\rho(B) = 1/\sigma(B)$. [It is common practice to use the same notations ρ_{xx} or ρ_{yx} to denote the components of the resistivity tensor in the three-dimensional case and the components of the resistance tensor in the two-dimensional case; notice however that in the former case the unit is $\Omega \cdot$ cm while in the latter is Ω].

In the standard geometry of Figure 15.9 we have

$$\rho_{yx}(B) = \frac{E_y}{J_x^{(s)}} = \frac{wE_y}{wJ_x^{(s)}} = \frac{V_{\text{Hall}}}{I},$$

where w is the width of the Hall bar in the y-direction. Thus we see that for the two-dimensional system the Hall resistance $\rho_{yx}(B)$ is given by the Hall voltage divided by the current flowing in the sample; $\rho_{yx}(B)$ can thus be measured with high accuracy, since the precise sample dimensions are not relevant and drop out [differently from the three-dimensional system as evident from comparison with Eq. (15.34b)].

The magneto-conductance and the magnetoresistance of a two-dimensional electron system in a model approach with constant relaxation time τ, effective mass m^*, and carrier surface density n_s is given by Eqs. (15.37) and (15.38), respectively, once n is replaced by n_s. In analogy with Eq. (15.38), the explicit expression of the magnetoresistance tensor reads

$$\boxed{\rho(B) = \frac{m^*}{n_s e^2 \tau} \begin{pmatrix} 1 & \omega_c \tau \\ -\omega_c \tau & 1 \end{pmatrix}} \qquad \omega_c = \frac{eB}{m^*c}. \tag{15.47}$$

Possible refinements in the classical estimation of the magnetoresistance do not change the essential facts of Eq. (15.47), whose contents can be summarized as follows:

i. The longitudinal component of the magnetoresistance in Eq. (15.47) reads

$$\rho_{xx}(B) = \rho_0 = \frac{m^*}{n_s e^2 \tau}. \tag{15.48a}$$

The longitudinal resistance ρ_{xx} is sensitive to the relaxation time of the system, but is independent (or weakly dependent) of the strength of the magnetic field.

ii. The transverse component of the magnetoresistance in Eq. (15.47) reads

$$\rho_{xy}(B) = \frac{m^*}{n_s e^2 \tau} \omega_c \tau \equiv \frac{1}{n_s ec} B. \tag{15.48b}$$

The transverse Hall resistance $\rho_{xy}(B)$ is independent (or weakly dependent) of the relaxation processes of the system, and is linear in B regardless of how weak or how strong is the applied magnetic field.

iii. It is possible to express $\rho_{xy}(B)$ as a function of the filling factor in the form

$$\rho_{xy}(\nu) = \frac{1}{\nu} \frac{h}{e^2}, \tag{15.48c}$$

where $\nu \equiv n_s hc/eB$ is the definition of the filling factor reported in Eq. (15.13). Of course no particular physical meaning (within the classical point of view we are considering) can be ascribed to ρ_{xy} in Eq. (15.48c) in the case ν takes integer values or fractional values (or any other real values). It is also to be noticed that the linear dependence of ρ_{xy} on B holds even in the hypothetical case of dissipationless regime (i.e. $\tau \to \infty$ and $\rho_{xx} \to 0$).

Consider now the experimental results of Figure 15.10. For small fields (up to 5 kilogauss, or so) it is seen that $\rho_{xx}(B)$ is (almost) constant and $\rho_{xy}(B)$ is linear with B. With increasing B, $\rho_{xx}(B)$ shows oscillations, which are a manifestation of the Shubnikov-de Haas effect: their origin is the quantization of the Landau levels and entailed effects on the density-of-states and scattering rates at the Fermi energy [we do not enter here in details, and simply notice the analogy with other oscillatory effects, such as the de Haas-van Alphen effect and the magneto-optical absorption, previously described]. With increasing B, we enter in the integer quantum Hall regime, with regions where $\rho_{xx}(B)$ approaches zero and simultaneously the resistance ρ_{xy} assumes quantized values. The dissipationless transport regime, obtained by the extreme quantum conditions realized experimentally, and the presence of so well-defined plateaus are the basic features that defy the classical treatment, and need a quantum treatment of transport from the very beginning.

15.6.2 Model Simulations of Quantum Transport in the Integer Hall Regime

The quantum Hall effects, besides opening novel areas of investigations, have greatly contributed to the cross-fertilization of interdisciplinary fields. These achievements have been recognized by the 1985 Nobel Prize in Physics awarded to von Klitzing for his discovery of the integer quantum Hall effect, and by the 1998 Nobel Prize to Laughlin, Stormer, and Tsui for the fractional quantum Hall effect.

In this subsection, we focus on some microscopic aspects of the electron transport in the integer Hall regime. We do not enter in all the multiform aspects of this area of research, and limit our discussion to reasonably simple but yet instructive models. A meaningful and useful model must be able to capture the basic features that are at the heart of the integer quantum Hall effect, and that can be summarized as follows:

(i) occurrence of quantized plateaus in the Hall resistance; (ii) dissipationless flow regime; (iii) role of edge states and backscattering suppression; and (iv) role of disorder. In constructing a reasonable simulation model of the integer quantum Hall effect at an acceptable level of technicality, we are primarily guided by the Bloch theorem for the description of magnetic energy bands, which give the dispersion curves of the electronic states in periodic systems in the presence of uniform magnetic fields. Among its invaluable merits already met throughout the various chapters, another jewel of the Bloch theorem is its capability to capture the key role of edge states in periodic systems, so providing the preliminary ingredients for understanding the experimental phenomenology of conductance quantization.

Magnetic Energy Levels of a Two-Dimensional Wire

In the initial chapters of this book we have considered several models of crystals, and in particular in Section 2.7 we have considered the case of "cubium," a simple cubic crystal with a single orbital per site and nearest neighbor interaction only. Reducing the dimension to two, we have a simple square lattice, which is often used as a starting framework of two-dimensional devices. The one-electron Hamiltonian of the simple square lattice can be written in the bra-ket notations as

$$H = \sum_{mn} E_0 |\phi_{mn}\rangle \langle \phi_{mn}| + t \sum_{mn} \left[|\phi_{m,n+1}\rangle \langle \phi_{m,n}| + |\phi_{m,n}\rangle \langle \phi_{m,n+1}| \right]$$
$$+ t \sum_{mn} \left[|\phi_{m,n}\rangle \langle \phi_{m+1,n}| + |\phi_{m+1,n}\rangle \langle \phi_{m,n}| \right]. \quad (15.49)$$

The Hamiltonian is schematically indicated in Figure 15.11.

By virtue of the Bloch theorem, the energy band of the Hamiltonian (15.49) is given by the expression

$$E(\mathbf{k}) = E_0 + 2t(\cos k_x a + \cos k_y a), \quad (15.50)$$

where E_0 is the on-site energy, a is the lattice parameter, and t is the hopping parameter supposed to be negative. The width of the energy band is $\Delta = 8|t|$; a typical bandwidth of 1 eV corresponds to $t = -0.125$ eV. Around the minimum at $\mathbf{k} = 0$ the series development of the energy reads

$$E(\mathbf{k}) = E_0 - 4|t| + \frac{\hbar^2 k^2}{2m^*} \quad \text{with} \quad \frac{m^*}{m} = \frac{1}{|t|} \frac{\hbar^2}{2ma^2}.$$

Having in mind carriers at the GaAs/AlGaAs heterointerface we set the effective mass $m^* = 0.068\, m$, which gives for the lattice parameter $a \approx 2$ nm.

We consider now the two-dimensional squared lattice embedded in a uniform magnetic field, oriented in the positive z-direction. In the tight-binding framework, the magnetic field is described through the appropriate Peierls-Berry phase factors. The Berry phase accumulated through a square path is determined by the ratio of the magnetic flux through the squared plaquette $\Phi_p(B)$ and the quantum of flux Φ_0, namely

$$\alpha = 2\pi \frac{\Phi_p(B)}{\Phi_0} = 2\pi \frac{a^2 B}{\Phi_0} = \frac{a^2}{l^2} \quad \text{where} \quad \Phi_0 = \frac{hc}{e} \quad \text{and} \quad l^2 = \frac{\hbar c}{eB}.$$

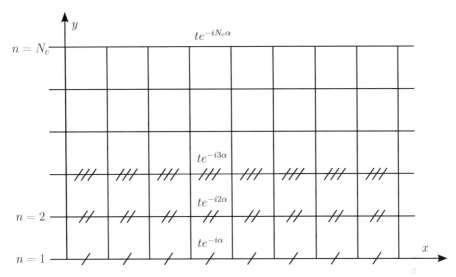

Figure 15.11 Pictorial representation of a strip of N_c longitudinal chains, and of the Hamiltonian (15.51), for the simulation of the two-dimensional Schrödinger electron gas in a perpendicular magnetic field. In the absence of magnetic field, the hopping parameters between adjacent sites are all taken equal to t. In the presence of magnetic fields, the hopping parameters between adjacent sites in the longitudinal direction are modified by the Peierls phase factors $\exp(-i\alpha n)$. The number of dashes on the bonds denote pictorially the value of n in the Peierls phase factors.

For instance, for typical values $B = 5$ Tesla, the magnetic length is $l \approx 11.5$ nm, and $\alpha \approx 0.03$. The value of the magnetic energy is $\hbar\omega_c^* = 8.51$ meV.

Taking into account the Peierls substitution, the tight-binding Hamiltonian (15.49) of the two-dimensional electron gas is modified as follows

$$H = \sum_{mn} E_0 |\phi_{mn}\rangle\langle\phi_{mn}| + t \sum_{mn} \left[|\phi_{m,n+1}\rangle\langle\phi_{m,n}| + |\phi_{m,n}\rangle\langle\phi_{m,n+1}| \right]$$

$$+ t \sum_{mn} \left[e^{-i\alpha n} |\phi_{m,n}\rangle\langle\phi_{m+1,n}| + e^{i\alpha n} |\phi_{m+1,n}\rangle\langle\phi_{m,n}| \right]. \quad (15.51)$$

In fact, according to the Peierls recipe, the off-diagonal matrix elements t_{mn} of a tight-binding Hamiltonian (in the absence of magnetic fields) are modified by the presence of a magnetic field of vector potential $\mathbf{A}(\mathbf{r})$ as follows

$$t_{mn} \Longrightarrow t_{mn} \exp\left[i\frac{e}{\hbar c} \int_{\mathbf{R}_m}^{\mathbf{R}_n} \mathbf{A}(\mathbf{r}) \cdot d\mathbf{l} \right],$$

where $\mathbf{R}_m$ and $\mathbf{R}_n$ are any pair of nearest neighbor lattice sites, and the integration path is the straight line connecting them (it is necessary to specify the integration path since curl $\mathbf{A}(\mathbf{r}) \neq 0$ in general). The phase factors introduced in the Hamiltonian (15.51) are those entailed by the use of the first Landau gauge $\mathbf{A}(\mathbf{r}) = (-By, 0, 0)$.

The Hamiltonian of Eq. (15.51) is translationally invariant in the longitudinal x-direction as also evident from Figure 15.11. We thus benefit of the Bloch theorem, and perform the one-dimensional Bloch sums

$$|\Phi_{k_x,n}\rangle = \frac{1}{\sqrt{N_x}} \sum_{t_m} e^{ik_x t_m} |\phi_{m,n}\rangle, \tag{15.52}$$

where $t_m = ma$. The matrix elements of the Hamiltonian (15.51) between the Bloch sums (15.52) of given k_x wavenumber are easily obtained. The diagonal elements (setting E_0 as the reference zero energy) are

$$\langle \Phi_{k_x,n} | H | \Phi_{k_x,n} \rangle = t e^{-i\alpha n} e^{ik_x a} + t e^{i\alpha n} e^{-ik_x a} = 2t \cos{(k_x a - n\alpha)};$$

the off-diagonal elements are different from zero, and equal to t, only between adjacent chains. In summary, on the basis functions (15.52) the Hamiltonian of the wire composed of N_c coupled linear chains (see Figure 15.11) takes the tridiagonal form

$$H(k_x, B) = \begin{bmatrix} 2tc_1 & t & 0 & 0 & 0 & \cdots \\ t & 2tc_2 & t & 0 & 0 & \cdots \\ 0 & t & 2tc_3 & t & 0 & \cdots \\ 0 & 0 & t & 2tc_4 & t & \cdots \\ \cdots & \cdots & \cdots & \cdots & \cdots & \cdots \end{bmatrix}_{N_c} \qquad c_n \equiv \cos{(k_x a - n\alpha)}. \tag{15.53}$$

The diagonalization of $H(k_x, B)$ gives the magnetic energy bands of the electrons in the quantum wire. The band structure produced by the diagonalization of the matrix (15.53) is clearly invariant if the arguments of all c_n are shifted by a same arbitrary quantity. We can use this gauge arbitrariness to require that $E(k_x, B) = E(-k_x, B)$. For this purpose we choose

$$c_n = \cos{(k_x a - n\alpha + Q\alpha)} \qquad \text{with} \qquad Q = \frac{N_c + 1}{2} \quad (n = 1, 2, \dots, N_c). \tag{15.54}$$

The magnetic energy bands, obtained by diagonalization of the matrix $H(k_x, B)$ with the choice (15.54) produce the same energy spectrum at $+k_x$ and $-k_x$.

As an exemplification, we report in Figure 15.12 the dispersion curves of the magnetic energy bands for a wire with $N_c = 101$ longitudinal chains and width 200 nm, effective mass of the carriers $m^* = 0.068\ m$, applied magnetic field $B = 5$ Tesla. The magnetic length and the magnetic energy corresponding to this field are $l \approx 11.5$ nm and $\hbar\omega_c^* = 8.1$ meV, respectively. As seen from Figure 15.12, the dispersion curves are rather flat in the bulk region and the group velocity $(1/\hbar)dE/dk_x$ is practically vanishing. The flat bands are seen to correspond to the Landau levels $(n + 1/2)\hbar\omega_c^*$. At the edges of the sample the confining potential (regardless of the fact that it is sharp as in this model, or smooth as in other simulations) produces an upward bending of the energy bands. From the study of the eigenvalues and eigenfunctions of the magnetic

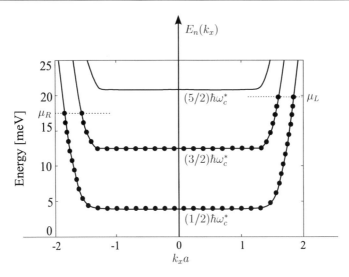

Figure 15.12 Magnetic energy bands for a 100-nm-wide wire threaded by a uniform magnetic field of 5 Tesla. The carrier effective mass is $m^* = 0.068\ m$. The bands are flat in the central part of the Brillouin zone, and correspond to Landau levels separated by the magnetic energy of $\hbar\omega_c^* = 8.1$ eV. The bands have chiral character near the border of the Brillouin zone, with states propagating along the boundaries of the wires. In the figure, it is also shown the occupancy of states in the case the device is in contact with two particle reservoirs, the left and right reservoirs, with chemical potentials μ_L and μ_R (and zero temperature, for simplicity) intersecting two energy bands.

bands, it is seen by inspection that the **k** states with negative group velocity are traveling in the lower edge of the sample, while the **k** states with positive group velocity are traveling in the upper edge of the sample. Thus the edges have a definite chiral character, i.e. they behave like "one-way roads," where only left moving electron trajectories are allowed in the lower boundary, while only right moving electron trajectories are allowed in the upper boundary (if the direction of the magnetic field is reversed, also the one-way roads are reversed). These quantum edge states correspond classically to the "skipping orbits" trajectories described by the electrons feeling electric and magnetic fields. It is in fact well known that a charged particle under the action of an electric field (entailed by the confining potential) and a perpendicular magnetic field, describes a cycloid. In summary: by virtue of the magnetic field and carrier confinement in the sample, there are extended quantum chiral states near the sample boundaries. At a given energy E the number of chiral states at each sample boundary equals the number of Landau levels of energy lower than E. All these conclusions are essentially based on the Bloch theorem, rather than the specific model under attention, and maintain thus a general validity beyond the model itself.

Charge Current in a Biased Two-Dimensional Wire in a Uniform Magnetic Field

We analyze now the charge transport current flowing through the wire. Consider first the equilibrium situation, with all the electronic states below a given equilibrium Fermi level are occupied, and all states above it are empty (for simplicity, we consider the zero temperature situation).

The magnetic energy bands of the wire satisfy the general property $E(k_x) \equiv E(-k_x)$; thus the corresponding group velocities are opposite and this assure that the total current through the ribbon at equilibrium is zero, as obviously expected. It is also apparent that, near the edges, "persistent currents" are flowing in opposite directions at the opposite sample edges, and balance perfectly to zero. [We can also add that in wires where time reversal symmetry is not broken by the presence of the magnetic field (and spin is ignored), the left moving states and the right moving states are the complex conjugate one of the other, occur in the same spatial region, and persistent currents are rigorously forbidden through any bond connecting the basis orbitals of the ribbon].

Consider now the wire in non-equilibrium conditions, i.e. suppose that the wire schematized in Figure 15.11 is in contact with two particle reservoirs (at the far left and right ends of the device) with chemical potentials μ_L and μ_R (and zero temperature). The electronic states moving from left to right are occupied for energies up to the chemical potential μ_L. Similarly carriers moving from right to left are occupied for energies up to the chemical potential μ_R.

The occupancy of the Bloch states in the case, for example, two channels are intersected by the chemical potentials, is illustrated schematically in Figure 15.12. The current carried in each channel can be calculated exactly as follows

$$I = 2(-e)\frac{1}{L}\sum_{k_x \text{occ.}} v(k_x) = \frac{2(-e)}{2\pi}\int_{-k_R}^{+k_L} v(k_x)\, dk_x \quad \text{with} \quad v(k_x) = \frac{1}{\hbar}\frac{dE(k_x)}{dk_x},$$

where the discrete sum over k_x has been replaced as usual by $L/2\pi$ times the corresponding integral, the factor 2 takes into account spin degeneracy, and $v(k_x)$ is the group velocity. Since the applied voltage V is related to the difference of chemical potentials by the equality $(-e)V = \mu_L - \mu_R$, one obtains *for each open channel*

$$I = \frac{2(-e)}{h}\int_{\mu_R}^{\mu_L} dE = \frac{2(-e)}{h}(\mu_L - \mu_R) = \frac{2e^2}{h}V \implies \boxed{R = \frac{h}{2e^2}}. \quad (15.55)$$

The conductance (or the resistance), that relates the total current to the voltage drop takes thus an universal quantized value for any channel, intersected by the Fermi level. Notice that the exact result (15.55) is completely independent of the actual shape of the energy bands (the shape could be modified by the superposition of the Hall potential or other self-consistent effects). Also notice that the voltage drop occurs conceptually at the contacts of the sample with the left and right reservoirs, and not along the sample in the chiral regime.

The good results provided so far are essentially an almost direct application of the Bloch theorem, and as such concern ideal wires with translational invariance. A real device contains countless impurities and disruptions, and so it is necessary to understand why the above results are so robust against disorder effects present in any realistic material.

At first sight, it could appear that any kind of disruption from periodicity in a system may lead to scattering between left moving states and right moving states, thus breaking in actual experiments the exact quantized relation (15.55) (whose accuracy in the quantum Hall regime is better than one part over billions). In the presence of magnetic fields, however, the left moving states and the right moving states (whose transverse width is of the order of the magnetic length) travel at the opposite edges of the sample, and the tunneling through the much larger bulk region is in general not possible. Thus the perfect quantization of the Hall resistance is insensitive to (moderate) disorder, and even to the unavoidable irregularities of the boundaries (once left moving and right moving trajectories remain far apart in real space).

While (moderate) disorder has no effect at all on the quantization of resistance (for energies far enough from the bulk Landau levels) it has a beneficial effect in explaining why the Fermi level can be pinned in a energy region between the bulk Landau levels, in spite of the fact that the edge states constitute in general a negligible fraction of the bulk states in realistic systems under investigation. [The ratio of edge states to bulk states is of the order of a/W, where a is the lattice parameter and W is the transverse width of the samples, typically in the range of micron, but possibly up to millimeters or so]. It is generally accepted that the presence of disorder broadens the orbitally degenerate states, within each Landau level, into a band of levels (see Figure 15.13). If the magnetic field

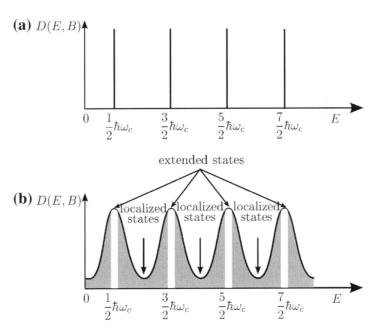

Figure 15.13 (a) Density-of-states $D(E, B)$ of the two-dimensional electron gas in a magnetic field. In a pure material, the density-of-states in the bulk is composed by a series of delta functions at the Landau levels. (b) In the presence of disorder, the Landau levels are broadened into bands, constituted by extended and localized states. The behavior of the sample in the bulk is metallic for energies at or near the Landau levels, and insulating for energies far from them.

is sufficiently strong, the energy overlap of the broadened δ-functions is minimal. While the states near the original Landau level make the bulk conducting, the states toward the tails become localized, according to the general concepts concerning the effect of disorder, and make the bulk insulating. Thus the Fermi level can be set at energies where the bulk electronic states are localized, while the edge conducting trajectories remain spatially separate on the sample boundaries. [For further details see for instance the book of Yoshioka (1998) and references quoted therein].

In conclusion, from the discussion based on the Bloch theorem of the general features of the magnetic energy bands in periodic quantum wires, with the emergence of chiral states and consequent backscattering suppression, the general framework of the integer quantum Hall regime is firmly established.

The tight-binding representation of electronic states, of which Eq. (15.51) is an elementary example, in conjunction with quantum transport theories, have greatly contributed to the understanding of mesoscopic devices and nanostructures. Typically, the tight-binding Hamiltonian of the system under investigation takes the general form

$$H = \sum_{mn} E_{mn} c_{mn}^{\dagger} c_{mn} + \sum_{mn \neq m'n'} t_{mn,m'n'} c_{mn}^{\dagger} c_{m'n'}, \tag{15.56}$$

where E_{mn} denote on site energies, and $t_{mn} \neq m'n'$ the hopping matrix elements between localized orbitals ϕ_{mn} and $\phi_{m'n'}$; the corresponding creation and annihilation operators are $c_{mn}^{\dagger}$ and $c_{m'n'}$. Local or extended imperfections, disordered regions, physical boundaries of given shape, and other types of scatterers, are represented by proper choice of diagonal and off-diagonal matrix elements of the Hamiltonian. In this framework magnetic fields are quite conveniently described by inserting appropriate Peierls phase factors in the off-diagonal matrix elements. It is also possible to enrich the localized basis set with spin degrees of freedom and handle spin-dependent terms in the Hamiltonian of spintronics devices. The quantum theories of transport, and in particular the nonequilibrium Keldysh formalism, provide principles and workable algorithms for the numerical evaluation of microscopic currents, and current profiles [see for instance A. Cresti, R. Farchioni, G. Grosso and G. Pastori Parravicini, Phys. Rev. B 68, 075306 (2003), and Eur. Phys. J. B 53, 537 (2006); B. K. Nikolić, L. B. Zârbo and S. Souma Phys. Rev. B 73, 075303 (2006) and references quoted therein].

The microscopic simulations of quantum transport, stirred by the quantum Hall effect, have greatly contributed to the progress of similar or related investigations in the wide field of nanoscience and nanotechnology. [For a thorough account see for instance the book of D. K. Ferry et al. (2009) cited in the bibliography].

15.6.3   Fractional Quantum Hall Effect

In the integer quantum Hall effect, the Hall resistance is quantized at $h/e^2 i$ where i is an integer number. For an overall account of the integer Hall effect a single-electron picture is sufficient; the effect is basically linked to the energy gaps resulting from the quantization of the kinetic energy of the free electrons into discrete Landau levels. Due to the presence of single particle energy gaps, electron-electron interaction does not modify in an essential way the states of the non-interacting particle picture.

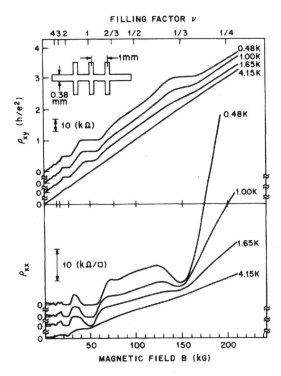

Figure 15.14 Hall resistance ρ_{xy} and longitudinal resistance ρ_{xx} versus B in a GaAs-AlGaAs sample with $n_s = 1.23 \times 10^{11}$ cm^{-2}, mobility $\mu = 90000$ cm^2 V^{-1} s^{-1} [reprinted with permission from D. C. Tsui, H. L. Stormer, and A. C. Gossard, Phys. Rev. Lett. *48*, 1559 (1982); copyright 1982 by the American Physical Society].

The situation becomes quite different when the electrons are all accommodated, and partially occupy the ground Landau level, and the occurrence of energy gaps is by necessity linked to the electron-electron interaction.

The discovery in 1980 of the integer quantum Hall effect was soon followed in 1982 by the experimental findings of D. C. Tsui, H. L. Stormer and A. C. Gossard, reproduced in Figure 15.14, that initiated the race to the challenging fractional quantum Hall effect. The results of Figure 15.14 show the clear tendency of the formation of a plateau at filling factor 1/3, accompanied by a minimum in the longitudinal resistance. This discovery made evident the importance of a proper account of the electron-electron interaction and the consequent need of a thorough many-body theory capable to describe the occurrence of gaps in strongly correlated electronic systems. The theoretical breakthrough into the physics of the fractional quantum Hall effect was provided by R. B. Laughlin [Phys. Rev. Lett *50*, 1395 (1983)], with his brilliant ansatz of the many-body ground-state wavefunction at filling factor $\nu = 1/3$. Since then, the experimental investigations on samples with higher mobilities and smaller temperatures put in evidence a rich structure of the Hall resistance, primarily at appropriate odd-denominator quantum numbers, but also at some even-denominators filling factors. The continued

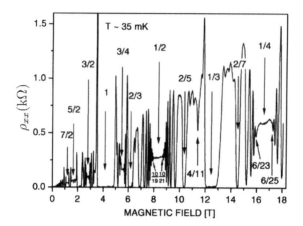

Figure 15.15 Longitudinal resistance ρ_{xx} versus B in a sample with $n_s = 10^{11}$ cm^{-2} and mobility $\mu = 10 \times 10^6$ cm^2 V^{-1} s^{-1} [reprinted with permission from W. Pan, H. L. Stormer, D. C. Tsui, L. N. Pfeiffer. K. W. Baldwin and K. W. West, Phys. Rev. Lett. *88*, 176802 (2002); copyright 2002 by the American Physical Society].

accumulation of experimental data have stirred and challenged the interpreting many-body theories, with the emerging scenario of interacting electrons condensed into novel quantum states, characterized by rational fractional filling factors and quasiparticles with exact fractional charge. Quasiparticles, originated by electrons (or holes) accompanied by magnetic flux quanta (composite fermions) are expected to have peculiar statistics properties, and can correlate on their turn into fractional states of composite particle themselves. We do not enter in all these topics, and in the following we confine ourselves to some elementary (and yet very instructive) aspects, that give a flavor of the ingenuity of the theoretical investigations in the field. Before beginning this intro-ductory discussion, we report in Figure 15.15 some experimental data on the fractional Hall effect, just to give an idea of the complexity and wealth of structures found in the experiments.

The study of a free electron in a magnetic field has been performed analytically in the initial sections of this chapter, where the Landau gauge was adopted (linked to a rectangular sample geometry). The problem can of course be considered in any other (convenient) gauge. In preparation of the ingredients needed for the study of the interacting electron system, it is useful to study analytically the quantum states of free electron in a magnetic field in the symmetric gauge (linked to a disk sample geometry). The Hamiltonian of a free electron in a magnetic field described within the symmetric gauge reads

$$H = \frac{1}{2m}\left(\mathbf{p} + \frac{e}{c}\mathbf{A}\right)^2 \quad \text{with} \quad \mathbf{A}(\mathbf{r}) = \frac{B}{2}(y, -x, 0), \tag{15.57}$$

where e is the modulus of the electronic charge, B is the modulus of the applied magnetic field, and the vector $\mathbf{B}$ points in the negative z-direction. [The choice of the negative direction for the magnetic field entails that the complex representation of the particle position encountered below can be defined as $z = x + iy$. The authors that prefer the magnetic field in the positive direction are used to define $z = x - iy$, to avoid to blur formulae with complex conjugate symbols].

The Hamiltonian (15.57) can be written in the form

$$
\begin{aligned}
H &= \frac{1}{2m}\left(-i\hbar\frac{\partial}{\partial x} + \frac{eB}{2c}y\right)^2 + \frac{1}{2m}\left(-i\hbar\frac{\partial}{\partial y} - \frac{eB}{2c}x\right)^2 \\
&= \frac{1}{2m}\frac{\hbar eB}{c}\left[-\frac{\hbar c}{eB}\frac{\partial^2}{\partial x^2} - \frac{\hbar c}{eB}\frac{\partial^2}{\partial y^2} + i\left(x\frac{\partial}{\partial y} - y\frac{\partial}{\partial x}\right) + \frac{1}{4}\frac{eB}{\hbar c}(x^2 + y^2)\right].
\end{aligned}
$$

(15.58)

We remember that $\hbar\omega_c = \hbar eB/mc$ is the magnetic energy, and $l^2 = \hbar c/eB$ is the square of the magnetic length. Using the magnetic length as unit of length, the Hamiltonian becomes

$$
H = \frac{1}{2}\hbar\omega_c\left[-\frac{\partial^2}{\partial x^2} - \frac{\partial^2}{\partial y^2} + i\left(x\frac{\partial}{\partial y} - y\frac{\partial}{\partial x}\right) + \frac{1}{4}(x^2 + y^2)\right]. \qquad (15.59)
$$

It is convenient to pass to cylindrical coordinates, and express the Hamiltonian operator in the new coordinates. We have

$$
x = r\cos\theta, \quad y = r\sin\theta, \quad l_z = -i\hbar\left(x\frac{\partial}{\partial y} - y\frac{\partial}{\partial x}\right) = -i\hbar\frac{\partial}{\partial\theta},
$$

$$
\nabla^2 = -\frac{\partial^2}{\partial x^2} - \frac{\partial^2}{\partial y^2} = -\frac{\partial^2}{\partial r^2} - \frac{1}{r}\frac{\partial}{\partial r} - \frac{1}{r^2}\frac{\partial^2}{\partial\theta^2}.
$$

The Hamiltonian (15.59) becomes

$$
H = \frac{1}{2}\hbar\omega_c\left[-\frac{\partial^2}{\partial r^2} - \frac{1}{r}\frac{\partial}{\partial r} - \frac{1}{r^2}\frac{\partial^2}{\partial\theta^2} + i\frac{\partial}{\partial\theta} + \frac{r^2}{4}\right]. \qquad (15.60)
$$

It is easily seen by inspection that the ground-state energy of H is the zero-point energy $(1/2)\hbar\omega_c$, and all the degenerate eigenfunctions of the ground state are given by

$$
\boxed{\psi_m(\mathbf{r}) = \psi_m(r, \theta) = \frac{1}{\sqrt{2\pi\,2^m m!}}\,r^m e^{im\theta}e^{-r^2/4}} \quad \text{with} \quad m = 0, 1, 2, \ldots \quad (15.61)
$$

In fact substitution of Eq. (15.61) into Eq. (15.60) gives

$$
\begin{aligned}
&\frac{\hbar\omega_c}{2}\left[-\frac{\partial^2}{\partial r^2} - \frac{1}{r}\frac{\partial}{\partial r} - \frac{1}{r^2}\frac{\partial^2}{\partial\theta^2} - i\frac{\partial}{\partial\theta} + \frac{r^2}{4}\right]r^m e^{im\theta}e^{-r^2/4} \\
&= \frac{\hbar\omega_c}{2}\left[-\frac{m(m-1)}{r^2} + \left(m+\frac{1}{2}\right) - \frac{r^2}{4} - \frac{m}{r^2} + \frac{1}{2} + \frac{m^2}{r^2} - m + \frac{r^2}{4}\right]r^m e^{im\theta}e^{-r^2/4} \\
&= \frac{\hbar\omega_c}{2}r^m e^{im\theta}e^{-r^2/4}.
\end{aligned}
$$

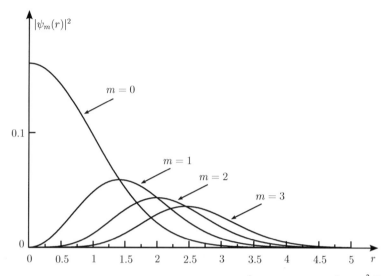

Figure 15.16 Representation of the wavefunctions $|\psi_m(\mathbf{r})|^2 = (1/2\pi 2^m m!)r^{2m}e^{-r^2/2}$, belonging to the lowest Landau level, for various values of the angular momentum. (The magnetic length is taken as unity).

It is possible to obtain analytically all the eigenfunctions of the Hamiltonian (15.57) for any higher energy Landau level, but we are here interested only in the ground-state wavefunctions.

Some comments on the properties of the wavefunctions (15.61) are useful. The wavefunction $\psi_m(\mathbf{r})$ has angular momentum m, and represents an electron localized on a circular orbit. The maximum of $|\psi_m(\mathbf{r})|^2$ occurs at $r_m = \sqrt{2m}l$ (where the magnetic length l has been restored), and the spread of the wavefunctions around the maximum equals the magnetic length; a few wavefunctions modulus square are shown in Figure 15.16. The flux of B through the disk of radius r_m is

$$\Phi(\mathbf{B}) = \pi r_m^2 B = 2\pi m B l^2 = m\Phi_0 \qquad \left[l^2 = \frac{\hbar c}{eB}; \Phi_0 = \frac{hc}{e} \right] \qquad (15.62)$$

and a state with angular momentum m encloses m flux quanta. The Berry phase of a state with angular dependence $\exp(im\theta)$ equals $m2\pi$. Loosely speaking, the wavefunction $\psi_m(\mathbf{r})$ of definite angular momentum m is envisaged to describe an electron dressed with m flux quanta, called vortices.

At this stage, it is convenient to rewrite the above wavefunctions (15.61) introducing the complex coordinate $z = x + iy$. The single particle states in the lowest Landau level are

$$\psi_m(\mathbf{r}) = \frac{1}{\sqrt{2\pi 2^m m!}} z^m e^{-|z|^2/4} \qquad \text{with} \qquad m = 0, 1, 2, \ldots \qquad (15.63a)$$

Consider the N single particle states of increasing radius

$$e^{-|z|^2/4}, ze^{-|z|^2/4}, z^2 e^{-|z|^2/4}, \ldots, z^{N-1}e^{-|z|^2/4}, \qquad (15.63b)$$

accommodating N electrons, one electron for each orbital (spin variables in this context of fully polarized Landau levels are ignored). The functions ψ_m ($m = 0, 1, \ldots, N-1$) represent rings of charge filling a disk up to the maximum radius

$$r_{\text{disk}} = \sqrt{2(N-1)}\, l \approx \sqrt{2N}\, l.$$

It is easily seen that the flux of the magnetic field through the disk surface is

$$\Phi(\mathbf{B}) = B\pi r_{\text{disk}}^2 \equiv N\Phi_0 \qquad \text{with} \qquad \Phi_0 = \frac{hc}{e}.$$

Thus the number of flux quanta through the disk and the number of electrons coincide: there is one flux quantum for each electron, and the filling factor is one.

In the independent particle approximation, the many-body wavefunction corresponding to the N single particle states (15.63b) occupied by electrons is given by the corresponding Slater determinant; factorizing apart the exponential part of the functions we have

$$\Psi(\mathbf{r}_1, \mathbf{r}_2, \ldots, \mathbf{r}_N) = \begin{vmatrix} 1 & 1 & \cdots & 1 \\ z_1 & z_2 & \cdots & z_N \\ z_1^2 & z_2^2 & \cdots & z_N^2 \\ \cdots & \cdots & \cdots & \cdots \\ z_1^{N-1} & z_2^{N-1} & \cdots & z_N^{N-1} \end{vmatrix} \cdot \exp\left[-\sum_{i=1}^{N} |z_i|^2/4 \right]$$

$$(15.64)$$

(normalization factor of the wavefunctions, when not indicated explicitly, must be taken in proper account whenever needed). It is easily seen that the above Slater determinant, formed with increasing powers of z_i ($i = 1, 2, \ldots, N$) has $(N-1)!$ zeroes of type $(z_i - z_j)$ for any of the $(N-1)!$ pairs of variables z_i, z_j. Then the zeroes must be simple, and the wavefunction (15.64) of the non-interacting electrons at the filling factor $\nu = 1$ can be cast in the form

$$\Psi(\mathbf{r}_1, \mathbf{r}_2, \ldots, \mathbf{r}_N) = \prod_{1 \le i < j \le N} (z_i - z_j) \cdot \exp\left[-\sum_{i=1}^{N} |z_i|^2/4 \right]. \qquad (15.65)$$

From Eq. (15.64) or the equivalent Eq. (15.65), it is evident that Ψ vanishes when two particles come close together, and the wavefunction changes sign if any two particles are interchanged, according to the routine properties required by many-body fermion statistics. Until now nothing new has been done, apart perhaps to notice the compact form originated by the complex representation of electronic positions, and we have exclusively confined our attention to the trivial systems of non-interacting particles.

At this stage, we introduce the Laughlin ansatz to describe the correlated motions of interacting electrons at odd fractional values at the filling factor $\nu = 1/q$. The Laughlin ansatz expresses the correlated ground state as a Jastrow-like function of the form

$$\Psi(\mathbf{r}_1, \mathbf{r}_2, \ldots, \mathbf{r}_N) = \prod_{1 \le i < j \le N} (z_i - z_j)^q \cdot \exp\left[-\sum_{i=1}^{N} |z_i|^2/4\right]. \tag{15.66}$$

The wavefunction Ψ must change sign under interchange of any two particles, so the exponent q must be odd integer (the first application was at $q = 3$, corresponding at $\nu = 1/3$). The considerations that lead to the proposal of the now called Laughlin wavefunctions (15.66) are a "condensate" of theoretical considerations (for instance the successful role of Jastrow-type wavefunctions for superfluid helium; estimation of total energy and excitation gaps borrowed from classic one-component plasma, etc.) and physical considerations (assumption that only the lowest Landau level states are of relevance, because of the relative smallness electron-electron interaction with respect to the magnetic energy; the physical picture of particles dressed by q vortices, etc.). It is not our purpose, nor it would be possible, to discuss in depth the far reaching implications of the Laughlin ansatz, and we limit our considerations to some basic physical concepts that emerge from its mathematical structure.

A first consideration we can do, is to show that the wavefunction (15.66) actually describes an electron system with filling factor $\nu = 1/q$. For this purpose consider the Jastrow wavefunction (15.66) in the case one coordinate (for instance z_1) is considered variable, while all the other coordinates are considered as fixed parameters, with values much smaller than z_1. Indicating momentarily z_1 by z, the wavefunction (15.66) in the stated conditions reads

$$\Psi(z) = Cz^{(N-1)q}e^{-|z|^2/4} \qquad \text{where} \qquad C = C(z_2, z_3, \ldots, z_{N-1})$$

is independent of z. In discussing the last function of Eq. (15.63b), we have seen that the maximum of $|z^{(N-1)} \exp(-|z|^2/4)|^2$ occurs at $\sqrt{2(N-1)}$ and entails the filling factor $\nu = 1$. Similarly it is seen that the maximum of $|z^{(N-1)q} \exp(-|z|^2/4)|^2$ occurs at $\sqrt{2(N-1)q}$ and entails the filling factor $\nu = 1/q$. Wavefunctions (15.66) describe the correlated motion of N electrons each carrying q flux quanta (or, in other words, each dressed with q vortices), at filling factor $\nu = 1/q$.

Suppose now to add one electron carrying q flux quanta to the Laughlin wavefunction (15.66). Indicating with z_0 the coordinate of the added electron with q flux quanta, the wavefunction of the correlated $N + 1$ particle system reads

$$\Psi(z_0, z_1, \ldots, z_N) = \left[\prod_{0 \le i < j \le N} (z_i - z_j)^q\right] \cdot \exp\left[-\sum_{i=0}^{N} |z_i|^2/4\right]$$

$$= \left[\prod_{j=1,N} (z_0 - z_j)^q\right] \cdot \left[\prod_{1 \le i < j \le N} (z_i - z_j)^q\right] \cdot \exp\left[-\sum_{i=0}^{N} |z_i|^2/4\right].$$

We can also write

$$\Psi(z_0, z_1, \ldots, z_N) = \left[\prod_{j=1,N} (z_0 - z_j) \right]^q \cdot \left[\prod_{1 \leq i < j \leq N} (z_i - z_j)^q \right]$$
$$\cdot \exp\left[-\sum_{i=0}^{N} |z_i|^2/4 \right]. \qquad (15.67)$$

The above equation shows that the repeated application for q times of the first operator in square brackets corresponds to adding an extra electron charge e together with q flux quanta to the correlated state under attention.

Consider now the trial excited wavefunction of the type

$$\widetilde{\Psi}(z_0, z_1, \ldots, z_N) = \left[\prod_{j=1,N} (z_0 - z_j) \right] \cdot \left[\prod_{1 \leq i < j \leq N} (z_i - z_j)^q \right]$$
$$\cdot \exp\left[-\sum_{i=0}^{N} |z_i|^2/4 \right] \qquad (15.68)$$

in which the first operator in square brackets is applied only once. To add a new particle to the condensate state with q quanta, we need to repeat q times the operation of adding a single flux quantum. We can thus argue that the trial excited state (15.68) represents the addition of a quasiparticle with fractional charge $e^* = e/q$ carrying a single vortex. Thorough theoretical elaborations confirm the description of quasiparticle with fractional electron charge, as well as the formation of energy gaps at fractional filling factors.

The Laughlin theory, with the discovery of fractionally charged excitations (dubbed "the electronic charge quarks"), constitutes one of the major triumphs of many-body formalism; and yet it is far from being complete. We have noticed that the exponent in the Laughlin wavefunctions (15.66) must be an odd integer; an even integer exponent would produce a boson wavefunction, while fractional exponents do not produce acceptable single valued wavefunctions. On the other hand, the observed fractions of the quantum Hall effect are not limited to $\nu = 1/q$ with odd q and to the particle-hole counterparts $1 - 1/q$. For instance the fractional states observed in Figure 15.15 include the values $\nu = 1/3$ and $\nu = 2/3$, but there are many other values and structures to be understood. Thus several generalized theories have been considered in the literature, including the idea that excitations with fractional charge can condense on their turn at appropriate fractional filling factors, in a hierarchical sequence of values. We cannot dwell further in these and other fascinating aspects, and we refer to the wide literature in the field for further information. In concluding, it can be said that the integer and fractional Hall effects have enriched the extraordinary baggage of condensed matter and have opened new horizons in the fundaments of microscopic interaction mechanisms among quasiparticles.

Appendix A. Solved Problems and Complements

Problem 1. Poisson summation formula and Fresnel integrals.
Problem 2. Ground-state energy of a system of independent fermions.
Problem 3. Free energy of a system of independent fermions.
Problem 4. Ground-state energy of the electron gas in a uniform magnetic field.
Problem 5. Special integrals involving exponential functions and the Fermi-Dirac
distribution function.
Problem 6. Special integrals involving trigonometric functions and the Fermi-Dirac
distribution function.
Problem 7. The free energy of an electron gas in a uniform magnetic field.

Problem 1. Poisson Summation Formula and Fresnel Integrals

*Discuss the essential features of the Poisson summation formula and of the Fresnel
integrals (in view of a rigorous evaluation of the free energy of the three-dimensional
electron gas in a uniform magnetic field).*

The forthcoming set of closely related problems is focused on the analytic evaluation
of the free energy of the three-dimensional electron gas in a uniform magnetic field.
This brief digression on the Poisson sum formula and Fresnel integrals is the starting
step in this direction.

Poisson sum formula. Let $f(x)$ be an arbitrary regular function for x in the interval
$[0, +\infty]$, and n be an integer number $n \geq 0$. For $n < x < n + 1$ we can write

$$f(x) = \sum_{s=-\infty}^{+\infty} a_s e^{-2\pi i s x} \qquad \text{with} \qquad a_s = \int_n^{n+1} f(x)\, e^{2\pi i s x}\, dx. \qquad (A.1)$$

From Eq. (A.1), we have in particular

$$f\left(n + \frac{1}{2}\right) = \sum_{s=-\infty}^{+\infty} (-1)^s \int_n^{n+1} f(x)\, e^{2\pi i s x}\, dx.$$

Summing up over n ($n = 0, 1, 2, \ldots$) provides the *Poisson sum formula*

$$\boxed{\sum_{n=0}^{+\infty} f\left(n + \frac{1}{2}\right) = \int_0^\infty f(x)\, dx + 2 \sum_{s=1}^\infty (-1)^s \int_0^\infty f(x) \cos(2\pi s x)\, dx}. \qquad (A.2)$$

The Poisson sum formula can be elaborated in a slightly different (although equivalent) form assuming that the function $f(x)$ and its derivative vanish at infinity, as it is
often the case. The integrals involving the cosine can be subjected to two integrations

by part:

$$\int_0^\infty f(x) \cos{(2\pi s x)} \, dx = \frac{1}{2\pi s} \int_0^\infty f(x) \, d \sin{(2\pi s x)}$$

$$= -\frac{1}{2\pi s} \int_0^\infty f'(x) \sin{(2\pi s x)} \, dx = \frac{1}{4\pi^2 s^2} \int_0^\infty f'(x) \, d \cos{(2\pi s x)}$$

$$= -\frac{1}{4\pi^2 s^2} f'(0) - \frac{1}{4\pi^2 s^2} \int_0^\infty f''(x) \cos{(2\pi s x)} \, dx.$$

Using the above result and the relationship

$$\sum_{s=1}^\infty \frac{(-1)^s}{s^2} = -\frac{\pi^2}{12},$$

the Poisson sum formula (A.2) can be cast into the form

$$\boxed{\begin{aligned} \sum_{n=0}^{+\infty} f\!\left(n + \frac{1}{2}\right) &= \int_0^\infty f(x) \, dx + \frac{1}{24} f'(0) \\ &\quad - \sum_{s=1}^\infty \frac{(-1)^s}{2\pi^2 s^2} \int_0^\infty f''(x) \cos{(2\pi s x)} \, dx \end{aligned}}$$

(A.3)

We can anticipate that the application of Eq. (A.3) to the free energy of the noninteracting electron gas in magnetic fields leads to three contributions: the first term in the right hand side of Eq. (A.3) is related to the free energy in the absence of the magnetic field, the second term is quadratic in the magnetic field, and represents the Landau diamagnetism, the third term is oscillating in the magnetic field and represents the de Haas-van Alphen effect (as thoroughly discussed in Problems 4 and 7).

Fresnel integrals. Integrals which can be brought into the form

$$\int_0^\infty \cos{(bx^2)} \, dx = \int_0^\infty \sin{(bx^2)} \, dx = \frac{1}{2} \sqrt{\frac{\pi}{2b}} \qquad (b > 0)$$

are called *Fresnel integrals*. The little more general form

$$\int_0^\infty \cos{(a \pm bx^2)} \, dx = \frac{1}{2} \sqrt{\frac{\pi}{b}} \cos{\left(a \pm \frac{\pi}{4}\right)} \qquad (b > 0) \tag{A.4}$$

is obtained exploiting the standard trigonometric formula

$$\cos{(a \pm bx^2)} = \cos a \cos bx^2 \mp \sin a \sin bx^2.$$

Problem 2. Ground-state Energy of a System of Independent Fermions

Consider a system of independent fermions with density-of-states $D(E)$. Denote by $P_1(E)$ the primitive of the density-of-states, and by $P_2(E)$ the primitive of the primitive.

Show that at zero temperature the ground-state energy is given by the expression

$$\boxed{E_0 = N\mu - P_2(\mu)}\,,\tag{A.5}$$

where N is the total number of fermions and μ is the chemical potential of the system.

At zero temperature, the ground-state energy of the non-interacting electron system becomes

$$E_0 = \int_0^\mu E D(E)\, dE \tag{A.6}$$

(for simplicity it is assumed that the density-of-states vanishes for $E < 0$; this situation can be always achieved with a shift in the reference energy, when necessary). In order to calculate Eq. (A.6), it is convenient to consider the primitive function of the density-of-states defined as

$$P_1(E) = \int_0^E D(E')\, dE' \quad \text{with} \quad P_1(0) = 0, \tag{A.7}$$

and similarly the next primitive function

$$P_2(E) = \int_0^E P_1(E')\, dE' = \int_0^E dE' \int_0^{E'} D(E'')\, dE'' \quad \text{with} \quad P_2(0) = 0\,. \tag{A.8}$$

It is evident that

$$P_1(\mu) = N,$$

where N is the total number of electrons (with both spin directions). An integration by parts in Eq. (A.6) gives

$$E_0 = \int_0^\mu E\, dP_1(E) = \mu P_1(\mu) - \int_0^\mu P_1(E)\, dE = N\mu - P_2(\mu);$$

this proves Eq. (A.5) and shows that the calculation of the ground-state energy at zero temperature of a system of non-interacting electrons (with any given density-of-states), is controlled by second primitive function of the density-of-states, evaluated at the chemical potential.

As an example consider the free-electron gas, whose density-of-states and second primitive are

$$D(E) = \frac{3}{2}\frac{N}{\mu}\left(\frac{E}{\mu}\right)^{1/2} \quad \text{and} \quad P_2(E) = \frac{2}{5}N\mu\left(\frac{E}{\mu}\right)^{5/2}.$$

From Eq. (A.5), one obtains $E_0 = (3/5)\,\mu N$, which is the standard expression of the ground-state energy of the free-electron gas.

Problem 3. Free Energy of a System of Independent Fermions

Consider a system of independent fermions with density-of-states $D(E)$. Denote by $P_1(E)$ and $P_2(E)$ the primitive and the next primitive of the density-of-states. Show that the free energy is given by the expression

$$\boxed{F(T) = N\mu - \int_0^\infty P_2(E)\left(-\frac{\partial f}{\partial E}\right)dE}, \tag{A.9}$$

where N is the total number of fermions, μ the chemical potential, and $f(E, T, \mu)$ the Fermi-Dirac distribution function.

The free energy of a system of independent fermions is given by Eq. (3.B7), here rewritten in the form

$$F(T) = N\mu - k_B T \sum_n \ln\left[1 + e^{(\mu - E_n)/k_B T}\right],$$

where E_n are all the possible one-electron energy states. Indicating by $D(E)$ the density-of-states, the free energy can be written as

$$F(T) = N\mu - k_B T \int_0^\infty \ln\left[1 + e^{(\mu - E)/k_B T}\right] D(E)\, dE. \tag{A.10}$$

In Eq. (A.10) we indicate $D(E)dE = dP_1(E)$, where $P_1(E)$ is the primitive of the density-of-states, and then we perform a first integration by parts; then we indicate with $P_2(E)$ the second primitive of the density-of-states and perform a second integration by parts; we also notice that

$$\frac{d}{dE}\ln\left[1 + e^{(\mu - E)/k_B T}\right] = \frac{-1}{k_B T}\frac{1}{e^{(E-\mu)/k_B T} + 1} = \frac{-1}{k_B T}f(E),$$

where $f(E)$ is the standard Fermi-Dirac function; we obtain

$$F(T) = N\mu - \int_0^\infty P_2(E)\left(-\frac{\partial f}{\partial E}\right)dE. \tag{A.11}$$

This form of the free energy can be easily recognized as the generalization of Eq. (A.5); in fact $(-\partial f/\partial E)$ is strongly peaked at $E = \mu$ and becomes a δ-function at $T = 0$.

Problem 4. Ground-State Energy of the Electron Gas in a Uniform Magnetic Field

Consider a three-dimensional electron gas of non-interacting particles, with parabolic dispersion curve $E(\mathbf{k}) = \hbar^2 k^2/2m$, at zero temperature and chemical potential (or Fermi energy) μ. A uniform magnetic field B is applied to the system. Show that the *ground-state energy of the system in the uniform magnetic field is given by the analytic expression*

$$
\begin{aligned}
E_0(B) = E_0(0) &+ \frac{\hbar^2 \omega_c^2}{24} D(\mu) \\
&+ \frac{\hbar^2 \omega_c^2}{4\pi^2} D(\mu) \left(\frac{\hbar \omega_c}{2\mu} \right)^{1/2} \sum_{s=1}^{\infty} \frac{(-1)^s}{s^{5/2}} \cos \left(\frac{2\pi s}{\hbar \omega_c} \mu - \frac{\pi}{4} \right),
\end{aligned}
\tag{A.12}
$$

where $E_0(0)$ is the ground-state energy of the free-electron gas at $T = 0$ in the absence of magnetic field, $D(\mu)$ is the density-of-states for both spin directions at the chemical potential in the absence of magnetic field, and $\hbar \omega_c = \hbar e B / mc$ is the magnetic energy.

The Poisson sum formula and the Fresnel integrals can be exploited to obtain the analytic expression (A.12) for the ground-state energy of the three-dimensional electron gas in a magnetic field, provided $\hbar \omega_c \ll \mu$, which is the standard situation.

For the three-dimensional electron gas the density-of-states for both spin directions in the absence of magnetic fields, and the second primitive are

$$
D(E) = A E^{1/2} \quad \text{and} \quad P_2(E) = \frac{4}{15} A E^{5/2}; \quad E > 0 \quad A = V \frac{1}{2\pi^2} \left(\frac{2m}{\hbar^2} \right)^{3/2} ;
$$

(following standard practice, the reference energy is here taken at the bottom of the conduction band).

The density-of-states (including spin degeneracy) of the free-electron gas in the presence of a magnetic field reads

$$
D(E, B) = \frac{1}{2} \hbar \omega_c A \sum_{n=0}^{\infty} \frac{1}{\sqrt{E - (n + 1/2)\hbar \omega_c}} \Theta \left[E - \left(n + \frac{1}{2} \right) \hbar \omega_c \right]
\tag{A.13}
$$

as shown in Eq. (15.18). The second primitive function of the density-of-states (A.13) is

$$
P_2(E, B) = \frac{2}{3} \hbar \omega_c A \sum_{n=0}^{\infty} \left[E - \left(n + \frac{1}{2} \right) \hbar \omega_c \right]^{3/2} \Theta \left[E - \left(n + \frac{1}{2} \right) \hbar \omega_c \right]
\tag{A.14}
$$

as can be easily seen by inspection, with double differentiation of Eq. (A.14) with respect of energy. Eq. (A.14) can be recast in the form

$$
P_2(E, B) = \frac{2}{3} \hbar \omega_c A \sum_{n=0}^{\infty} f \left(n + \frac{1}{2} \right) \quad \text{with} \quad f(x) = (E - x \hbar \omega_c)^{3/2} \Theta(E - x \hbar \omega_c).
$$

$$
\tag{A.15}
$$

From the Poisson summation formula of Eq. (A.3) we have

$$\sum_{n=0}^{+\infty} f\left(n + \frac{1}{2}\right) = \int_0^\infty f(x)\,dx + \frac{1}{24} f'(0)$$

$$-\sum_{s=1}^{\infty} \frac{(-1)^s}{2\pi^2 s^2} \int_0^\infty f''(x) \cos(2\pi s x)\,dx. \tag{A.16}$$

The first contribution in the right-hand side of Eq. (A.16) reads

$$\int_0^\infty f(x)\,dx = \int_0^{E/\hbar\omega_c} (E - x\hbar\omega_c)^{3/2}\,dx = \frac{2}{5}\frac{1}{\hbar\omega_c} E^{5/2}.$$

The second contribution gives

$$\frac{1}{24} f'(0) = \frac{1}{24}\frac{3}{2}(-\hbar\omega_c)E^{1/2} = -\frac{1}{16}\hbar\omega_c E^{1/2}.$$

The third contribution in Eq. (A.16) requires the evaluation of the integral

$$\int_0^\infty f''(x) \cos(2\pi s x)\,dx = \frac{3}{4}\hbar^2\omega_c^2 \int_0^{E/\hbar\omega_c} \frac{1}{\sqrt{E - x\hbar\omega_c}} \cos(2\pi s x)\,dx$$

$$\text{with the change of variable } \frac{E}{\hbar\omega_c} - x = t^2$$

$$= \frac{3}{4}\hbar^2\omega_c^2 \frac{2}{\sqrt{\hbar\omega_c}} \int_0^{\sqrt{E/\hbar\omega_c}} \cos\left(\frac{2\pi s}{\hbar\omega_c} E - 2\pi s t^2\right) dt$$

$$= \frac{3}{4}(\hbar\omega_c)^{3/2} \frac{1}{\sqrt{2s}} \cos\left(\frac{2\pi s}{\hbar\omega_c} E - \frac{\pi}{4}\right);$$

the last passage has been obtained assuming $E \gg \hbar\omega_c$ and applying Eq. (A.4) for Fresnel integrals. Inserting the above results into Eqs. (A.15) and (A.16), one obtains

$$P_2(E, B) = \frac{4}{15} A E^{5/2} - \frac{\hbar^2\omega_c^2}{24} A E^{1/2}$$

$$-\frac{A}{\pi^2} \left(\frac{\hbar\omega_c}{2}\right)^{5/2} \sum_{s=1}^{\infty} \frac{(-1)^s}{s^{5/2}} \cos\left(\frac{2\pi s}{\hbar\omega_c} E - \frac{\pi}{4}\right). \tag{A.17}$$

According to Problem 2, we can express the ground-state energy of the electron gas in the magnetic field in the form

$$E_0(B) = N\mu - P_2(\mu, B);$$

inserting Eq. (A.17) in the above expression provides the desired analytic expression (A.12).

Problem 5. Special Integrals Involving Exponential Functions and the Fermi-Dirac Distribution Function

Show that the Fourier transform of the derivative of the Fermi-Dirac distribution function is given by the analytic expression

$$\boxed{\int_{-\infty}^{+\infty} e^{ikE} \left(-\frac{\partial f}{\partial E} \right) dE = \pi \frac{k}{\beta} \frac{1}{\sinh{(\pi k/\beta)}} \qquad f(E) = \frac{1}{e^{\beta E} + 1} \qquad \beta = \frac{1}{k_B T}}.$$

(A.18)

In particular, consider the behavior of the Fourier transform in the limit of $k \ll \beta$ and $k \gg \beta$.

The Fermi-Dirac function can be expressed in the series form (indicating the energy variable E with z, in view of the extension to the complex plane)

$$f(z) = \frac{1}{e^{\beta z} + 1} \equiv \frac{1}{2} + \sum_{n=-\infty}^{+\infty} \frac{1}{(2n+1)\pi i - \beta z} = \frac{1}{2} + \frac{1}{\beta} \sum_{n=-\infty}^{+\infty} \frac{1}{i(\pi/\beta)(2n+1) - z}.$$

Derivation with respect to the z variable gives

$$\frac{\partial f}{\partial z} = \frac{1}{\beta} \sum_{n=-\infty}^{+\infty} \frac{1}{[z - i(\pi/\beta)(2n+1)]^2};$$

this shows that the derivative of the Fermi-Dirac distribution function has poles of the second order at $z_n = i(\pi/\beta)(2n+1)$, with n any integer number.

The integral under attention in Eq. (A.18) can be obtained closing the circuit in the upper (or in the lower) part of the complex plane, for positive (or negative) k respectively. In either case we obtain

$$\int_{-\infty}^{+\infty} e^{ikz} \left(-\frac{\partial f}{\partial z} \right) dz = -\frac{1}{\beta} 2\pi i \sum_{n=0}^{+\infty} ik e^{iki(\pi/\beta)(2n+1)}$$

$$= \frac{2\pi k}{\beta} e^{-\pi k/\beta} \sum_{n=0}^{+\infty} \left[e^{-2\pi k/\beta} \right]^n = \frac{\pi k}{\beta} \frac{1}{\sinh{(\pi k/\beta)}}.$$

In summary

$$\int_{-\infty}^{+\infty} e^{ikz} \left(-\frac{\partial f}{\partial z} \right) dz = \frac{\pi k}{\beta} \frac{1}{\sinh{(\pi k/\beta)}}$$

$$= \begin{cases} 1 & \text{if } k \ll \beta \\ (\pi k/\beta) \exp{(-\pi k/\beta)} \approx 0 & \text{if } k \gg \beta \end{cases}$$

and the behavior for large and small values of temperature is self-explanatory.

Problem 6. Special Integrals Involving Trigonometric Functions and the Fermi-Dirac Distribution Function

Consider an integral of the type

$$I_s = \int_0^\infty \cos\left(\frac{2\pi s}{\hbar\omega_c} E - \frac{\pi}{4}\right)\left(-\frac{\partial f}{\partial E}\right) dE, \tag{A.19}$$

where $f(E) = [\exp(E-\mu)/k_B T + 1]^{-1}$ *is the Fermi-Dirac function and* $\omega_c = eB/mc$ *is the cyclotron frequency. Show that the analytical expression of* I_s *is*

$$\boxed{I_s = \frac{2\pi^2 s k_B T}{\hbar\omega_c}\frac{1}{\sinh\left(2\pi^2 s k_B T/\hbar\omega_c\right)}\cos\left(\frac{2\pi s}{\hbar\omega_c}\mu - \frac{\pi}{4}\right).} \tag{A.20}$$

The integral I_s can be performed analytically, but much caution must be applied because of the rapid oscillations exhibited by the trigonometric functions; in particular the Sommerfeld expansion (see Section 3.2) cannot be applied in general, except for the case $k_B T \ll \hbar\omega_c$. In order to evaluate analytically the integral (A.19), we notice that $-\partial f/\partial E$ is strongly peaked at $E = \mu$ and we can thus replace the lower limit of integration with $-\infty$. From the identity

$$\cos\left(\frac{2\pi s}{\hbar\omega_c} E - \frac{\pi}{4}\right) \equiv \cos\left[\left(\frac{2\pi s}{\hbar\omega_c}\mu - \frac{\pi}{4}\right) + \frac{2\pi s}{\hbar\omega_c}(E - \mu)\right]$$

and from the fact that $-\partial f/\partial E$ is an even function of $E - \mu$ we have

$$I_s = \cos\left(\frac{2\pi s}{\hbar\omega_c}\mu - \frac{\pi}{4}\right)\int_{-\infty}^{+\infty}\cos\frac{2\pi s(E - \mu)}{\hbar\omega_c}\left(-\frac{\partial f}{\partial E}\right) dE.$$

We can add the odd function $i\sin(2\pi s(E - \mu)/\hbar\omega_c)$ to the cosine function under integral and perform a shift of energy equal to μ (i.e. the chemical potential is set equal to zero in the integrand). We have then

$$I_s = \cos\left(\frac{2\pi s}{\hbar\omega_c}\mu - \frac{\pi}{4}\right)\int_{-\infty}^{+\infty}\exp\left(i\frac{2\pi s}{\hbar\omega_c} E\right)\left(-\frac{\partial f}{\partial E}\right) dE$$

$$= \frac{2\pi^2 s k_B T}{\hbar\omega_c}\frac{1}{\sinh\left(2\pi^2 s k_B T/\hbar\omega_c\right)}\cos\left(\frac{2\pi s}{\hbar\omega_c}\mu - \frac{\pi}{4}\right),$$

where use has been made of Eq. (A.18). Eq. (A.20) is thus demonstrated.

Problem 7. The Free Energy of an Electron Gas in a Uniform Magnetic Field

Consider a three-dimensional electron gas of non-interacting particles, with parabolic dispersion curve $E(\mathbf{k}) = \hbar^2 k^2/2m$, *at temperature* T, *and chemical potential* μ. *A uniform magnetic field is applied to the system. Show that the free energy of the system in the presence of the magnetic field, is given by the analytic expression*

$$F(T, B) = F(T, 0) + \frac{\hbar^2 \omega_c^2}{24} D(\mu) + \frac{\hbar^2 \omega_c^2}{4\pi^2} D(\mu) \left(\frac{\hbar \omega_c}{2\mu} \right)^{1/2}$$

$$\times \sum_{s=1}^{\infty} \frac{(-1)^s}{s^{5/2}} \frac{2\pi^2 s k_B T}{\hbar \omega_c} \frac{1}{\sinh(2\pi^2 s k_B T / \hbar \omega_c)} \cos\left(\frac{2\pi s}{\hbar \omega_c} \mu - \frac{\pi}{4} \right), \quad (A.21)$$

where $F(T, 0)$ is the free energy of the electron gas in the absence of magnetic field, at temperature T, $D(\mu)$ is the density-of-states for both spin directions at the chemical potential in the absence of magnetic field, and $\omega_c = eB/mc$ is the cyclotron frequency.

The second primitive of the density of states in the presence of a uniform magnetic field is given by Eq. (A.17). According to Eq. (A.9) the free energy of the system under attention is

$$F(T, B) = N\mu - \int_0^\infty P_2(E, B) \left(-\frac{\partial f}{\partial E} \right) dE.$$

Exploiting the special integrals of Eq. (A.20), with straight elaboration we obtain for the free energy the expression reported in Eq. (A.21). In the low temperature limit $k_B T \ll \hbar \omega_c$ Eq. (A.21) coincides with Eq. (A.12). In the opposite limit that the thermal energy is much higher than the magnetic energy the oscillating terms in Eq. (A.21) can be neglected. This occurs because the hyperbolic sine term in Eq. (A.21) leads to a damping behavior with temperature, which depends on the cyclotron frequency.

At the end of the quantitative analysis of the magnetic susceptibility of the free electron gas a few final comments are worthwhile. The free-electron gas with "ideal" plane-wave-type wavefunctions has theoretically (almost) infinite conductivity (if not for impurities, lattice vibrations, and other defects) and small orbital diamagnetism ($\chi \approx -10^{-6}$). The free-electron gas is thus the prototype model of *materials with "supermetallic behavior," characterized by very high conductivity and very poor diamagnetism.* Similarly, ordinary metals with (almost) ideal Bloch-type electronic wavefunctions may exhibit extremely high conductivity and poor diamagnetism.

Materials characterized by perfect conductivity and perfect diamagnetism ($\chi = -1/4\pi$) are known as superconductors. For an introduction to superconductivity we refer to Chapter 18; here we only wish to notice that "extremely high conductivity" or even "ballistic transport regime" do not automatically entail high diamagnetic susceptibility. Only in superconductors, perfect conductivity and perfect diamagnetism are twin manifestations of a same macroscopic mechanism, discussed in Chapter 18.

Further Reading

Chakraborty, T., & Pietiläinen, P. (1995). *The quantum Hall effects.* Berlin: Springer; (1988). *The fractional quantum Hall effect.* Berlin: Springer.

Datta, S. (1995). *Electronic transport in mesoscopic systems.* Cambridge: Cambridge University Press.

Ezawa, Z. F. (2013). *Quantum Hall effects*. Singapore: World Scientific.

Ferry, D. K., Goodnick, S. M., & Bird, J. (2009). *Transport in nanostructures*. Cambridge: Cambridge University Press.

Harrison, W. A., & Webb, M. B. (Eds.). (1960). *The Fermi surface*. New York: Wiley.

Jeckelmann, B., & Jeanneret, B. (2001). The quantum Hall effect as an electrical resistance standard. *Reports on Progress in Physics, 64*, 1603.

Kelly, M. J. (1995). *Low dimensional semiconductors. Materials, physics, technology, devices*. Oxford: Clarendon Press.

Laughlin, R. B. (1999). Fractional quantization. *Reviews of Modern Physics, 71*, 863.

Morrish, A. H. (1965). *The physical principles of magnetism*. New York: Wiley.

Nolting, W., & Ramakanth, A. (2009). *Quantum theory of magnetism*. Berlin: Springer.

Pippard, A. B. (1965). *Dynamics of conduction electrons*. New York: Gordon and Breach.

Prange, R. E., & Girvin, S. M. (Eds.). (1990). *The quantum Hall effect*. New York: Springer.

Stone, M. (Ed.). (1992). *Quantum Hall effect*. Singapore: World Scientific.

Stormer, H. L. (1999). The fractional quantum Hall effect. *Reviews of Modern Physics, 71*, 875.

Tsui, D. C. (1999). Interplay of disorder and interaction in two-dimensional electron gas in intense magnetic fields. *Reviews of Modern Physics, 71*, 891.

von Klitzing, K. (1986). The quantized Hall effect. *Reviews of Modern Physics, 58*, 519.

White, R. M. (2007). *Quantum theory of magnetism*. Berlin: Springer.

Wilson, A. H. (1954). *Theory of metals*. Cambridge: Cambridge University Press.

Yoshioka, D. (1998). *The quantum Hall effect*. Berlin: Springer.

16 Magnetic Properties of Localized Systems and Kondo Impurities

Chapter Outline head

In the previous chapter we have considered the effects of magnetic fields on the *free-electron gas*. In this chapter we discuss the effects of magnetic fields on *localized electronic units*, such as atoms, ions, molecules, and in particular magnetic impurities embedded in solids. It is assumed that the units in the solids are sufficiently far apart, so that any mutual interaction among themselves can be neglected; the study of cooperative magnetic effects is postponed to the next chapter.

We begin with a preliminary study of magnetic field effects on atomic or molecular systems with no permanent magnetic moment. These systems are most often diamagnetic, but there are a few remarkable exceptions of paramagnetism (van Vleck paramagnetism); in both cases, the magnetic susceptibility is practically independent of temperature. We consider then the magnetic field effects in atoms, ions, or impurities, whose ground states have a permanent magnetic moment. These electronic systems exhibit in general a temperature-dependent paramagnetism (Curie paramagnetism), because of the balance between the orientation effect of the magnetic field on permanent magnetic dipoles and the opposite effect of temperature. The remaining part of this chapter is focused on the problem of localized magnetic impurities dissolved in normal metals, and on the rich phenomenology related to the Kondo problem.

Solid State Physics, Second Edition. http://dx.doi.org/10.1016/B978-0-12-385030-0.00016-5

16.1 Quantum Mechanical Treatment of Magnetic Susceptibility

From purely classical argument, the orbital magnetic susceptibility of any dynamical system is zero, according to the Bohr-van Leeuwen theorem (see Section 15.1). Since the response of an electronic system to a magnetic field includes in general a contribution due to the orbital motion, besides the spin contribution, a quantum mechanical treatment is necessary from the very beginning.

The magnetic moment of a single orbiting electron of angular momentum $\mathbf{l}$ (in units $\hbar$) and bare mass m, is

$$\boldsymbol{\mu}_l = -\mu_B \mathbf{l}, \quad \mathbf{l} = \frac{1}{\hbar} \mathbf{r} \times \mathbf{p}, \quad \mu_B = \frac{e\hbar}{2mc} = 0.05788 \text{ meV/Tesla},$$

where μ_B denotes the Bohr magneton. The magnetic moment associated to the electron spin is given by

$$\boldsymbol{\mu}_s = -g_0 \mu_B \mathbf{s} = -2\mu_B \mathbf{s},$$

where $2\mathbf{s} = \boldsymbol{\sigma}$ are the Pauli matrices for half-spin particles (explicitly reported in Section 16.5.2), and the gyromagnetic factor can be taken as $g_0 = 2$ (free electron without relativistic corrections).

Consider an atomic or polynuclear system (such as a molecule, a cluster, or a solid), composed by N_e electrons in interaction among themselves and with the nuclei (fixed in some configuration). The non-relativistic many-body electronic Hamiltonian can be written as

$$\mathcal{H}_0 = \sum_{i=1}^{N_e} \frac{\mathbf{p}_i^2}{2m} + \sum_{i=1}^{N_e} V(\mathbf{r}_i) + \frac{1}{2} \sum_{i \neq j}^{N_e} \frac{e^2}{|\mathbf{r}_i - \mathbf{r}_j|}. \tag{16.1}$$

The first term in the right-hand side of Eq. (16.1) is the kinetic energy of the electrons; the second term is the electronic-nuclear interaction energy, and the last one represents the electron-electron Coulomb repulsion. For simplicity in Eq. (16.1) the relativistic corrections, and in particular the spin-orbit interaction, are neglected; the spin-orbit operator couples together the spin and the orbital motion of the electrons, has far reaching consequences in the magnetic properties of systems with increasing atomic numbers, and will be discussed later.

We consider now a uniform magnetic field $\mathbf{H}$, described in the symmetric gauge by the vector potential

$$\mathbf{A}(\mathbf{r}) = \frac{1}{2} \mathbf{H} \times \mathbf{r} \quad \Longrightarrow \quad \text{curl } \mathbf{A} = \mathbf{H}; \quad \text{div } \mathbf{A} = 0; \quad \mathbf{A} \cdot \mathbf{p} = \mathbf{p} \cdot \mathbf{A}. \tag{16.2}$$

In the presence of a uniform magnetic field, the Hamiltonian (16.1) of the many-electron system is modified by the following terms. (i) In the expression of the electronic kinetic energy, the momentum operator $\mathbf{p}_i$ is replaced by the generalized momentum $\mathbf{P}_i = \mathbf{p}_i + (e/c)\mathbf{A}(\mathbf{r}_i)$, where e is the absolute value of the electronic charge. (ii) The interaction energy of relativistic origin $-\boldsymbol{\mu} \cdot \mathbf{H} = 2\mu_B \mathbf{s} \cdot \mathbf{H}$ of the spin magnetic moment of each electron with the magnetic field must be included. The Hamiltonian (16.1), with the modifications specified above, takes the form

$$\mathcal{H} = \sum_{i=1}^{N_e} \frac{1}{2m} \left[\mathbf{p}_i + \frac{e}{c} \mathbf{A}(\mathbf{r}_i) \right]^2 + \sum_{i=1}^{N_e} V(\mathbf{r}_i) + \frac{1}{2} \sum_{i \neq j}^{N_e} \frac{e^2}{|\mathbf{r}_i - \mathbf{r}_j|} + 2\mu_B \mathbf{H} \cdot \sum_{i=1}^{N_e} \mathbf{s}_i. \tag{16.3}$$

It is convenient to define the total spin operator (in units of $\hbar$) as

$$\mathbf{S} = \sum_{i=1}^{N_e} \mathbf{s}_i \tag{16.4a}$$

and the total orbital angular momentum operator (in units of $\hbar$) as

$$\mathbf{L} = \frac{1}{\hbar} \sum_{i=1}^{N_e} \mathbf{r}_i \times \mathbf{p}_i. \tag{16.4b}$$

In the presence of a uniform magnetic field (described in the symmetric gauge), the Hamiltonian (16.3) of the many-electron system becomes

$$\mathcal{H} = \mathcal{H}_0 + \mu_B \mathbf{H} \cdot (\mathbf{L} + 2\mathbf{S}) + \frac{e^2}{8mc^2} \sum_{i=1}^{N_e} (\mathbf{H} \times \mathbf{r}_i)^2, \tag{16.5}$$

where $\mathcal{H}_0$ is the unperturbed Hamiltonian of the electronic system.

The first term added to $\mathcal{H}_0$ in Eq. (16.5) is the Zeeman operator $\mathcal{H}_Z = -\boldsymbol{\mu} \cdot \mathbf{H}$, giving the energy of the magnetic dipole $\boldsymbol{\mu} = -\mu_B(\mathbf{L} + 2\mathbf{S})$ in the magnetic field; the second term is proportional to the square of the magnetic field strength; since both terms are in general small with respect to $\mathcal{H}_0$, we can treat them by perturbation theory. In adopting perturbation theory to determine the effect of a magnetic field on the ground state of $\mathcal{H}_0$, we have as usual to distinguish whether the ground state is degenerate or non-degenerate.

Suppose that the ground state Ψ_0 of $\mathcal{H}_0$ is *non-degenerate* (nor quasi-degenerate with other states; i.e. Ψ_0 is well separated in energy from the excited states Ψ_n). The expectation value of the Zeeman operator on Ψ_0 is zero, since $\langle \Psi_0 | \mathbf{L} | \Psi_0 \rangle = \langle \Psi_0 | \mathbf{S} | \Psi_0 \rangle \equiv 0$; this can be seen by inspection (the non-degenerate wavefunction Ψ_0 can be taken as real), or by parity considerations of Ψ_0, $\mathbf{L}$, and $\mathbf{S}$ under time-reversal operator. Thus corrections linear in the magnetic field do not occur, and the system is said to have *no permanent magnetic moment*. Corrections quadratic in the magnetic field involve the expectation value of $(\mathbf{H} \times \mathbf{r})^2$ on the state Ψ_0 and the Zeeman operator to second order.

The energy shift ΔE_0 of the non-degenerate ground state, using perturbation theory and consistently selecting terms up to second power in the magnetic field, is given by the expression

$$\Delta E_0 = \Delta E_D + \Delta E_P, \tag{16.6a}$$

where

$$\Delta E_D = \frac{e^2}{8mc^2} \langle \Psi_0 | \sum_{i=1}^{N_e} (\mathbf{H} \times \mathbf{r}_i)^2 | \Psi_0 \rangle \qquad (\text{notice } \Delta E_D > 0) \qquad (16.6b)$$

and

$$\Delta E_P = \sum_{n(\neq 0)} \frac{|\langle \Psi_n | \mu_B \mathbf{H} \cdot (\mathbf{L} + 2\mathbf{S}) | \Psi_0 \rangle|^2}{E_0 - E_n} \qquad (\text{notice } \Delta E_P < 0). \qquad (16.6c)$$

If we have N independent units in the volume V, each of them described by the Hamiltonian (16.1), the energy change of the ground state of the system due to the magnetic field is $N(\Delta E_D + \Delta E_P)$; from Eq. (15.2b) the magnetic susceptibility (when thermal excitations are negligible) becomes

$$\chi = -\frac{N}{V} \frac{\partial^2 \Delta E_D}{\partial H^2} - \frac{N}{V} \frac{\partial^2 \Delta E_P}{\partial H^2} = \chi_D + \chi_P \quad (\chi_D < 0; \ \chi_P > 0). \qquad (16.7)$$

Thus, the *positive term* ΔE_D contributes to the magnetic susceptibility with the *diamagnetic contribution* χ_D, while the *negative term* ΔE_P gives rise to the *paramagnetic contribution* χ_P; a net diamagnetic effect usually occurs. However for a number of molecules, whose lowest excitation energies $E_n - E_0$ are very small, paramagnetism may dominate over diamagnetism, and a net paramagnetic effect occurs (van Vleck paramagnetism). The van Vleck paramagnetism, as well as ordinary diamagnetism, is practically independent of temperature.

If the ground state of $\mathcal{H}_0$ is *degenerate*, then the diagonalization of the Zeeman operator in the degeneracy subspace removes the degeneracy of the ground state and leads in general to energy separations proportional to the magnetic field strength. The system is said to have a *permanent magnetic moment*, and presents in general a Curie-type paramagnetism; in this case, second-order perturbations are often of minor importance, but must sometimes be considered for better quantitative results. We pass now to illustrate the above concepts in the particular case of closed-shell spherically symmetric systems.

Diamagnetism of Closed-Shell Atoms or Ions

Consider a closed-shell atom or ion, and let Ψ_0 denote the non-degenerate ground-state wavefunction whose energy is in general well separated from the energy of the excited states. Assuming negligible spin-orbit interaction, Ψ_0 is also eigenfunction of the total spin operator and total angular momentum operator (with the nucleus site taken as origin) with zero eigenvalue. For the described spherically symmetric system, it holds

$$\mathbf{S}|\Psi_0\rangle \equiv 0; \quad \mathbf{L}|\Psi_0\rangle \equiv 0. \qquad (16.8)$$

Thus in Eqs. (16.6) the paramagnetic term is absent, and the ground-state energy shift becomes

$$\Delta E_0 = \frac{e^2}{8mc^2} \langle \Psi_0 | \sum_{i=1}^{N_e} (\mathbf{H} \times \mathbf{r}_i)^2 | \Psi_0 \rangle. \qquad (16.9)$$

Notice that, according to Eqs. (16.6), the determination of the energy shift of the ground state of a system in a magnetic field requires in general the knowledge both of the ground state and of the excited states; however, in the case of closed-shell atoms or ions only the ground state is needed, under the stated approximations.

From the spherical symmetry of the ground-state charge distribution of a closed-shell atom or ion it follows

$$\langle \Psi_0 | \sum_{i=1}^{N_e} (\mathbf{H} \times \mathbf{r}_i)^2 | \Psi_0 \rangle = \frac{2}{3} H^2 \langle \Psi_0 | \sum_{i=1}^{N_e} r_i^2 | \Psi_0 \rangle. \tag{16.10}$$

We can thus write Eq. (16.9) in the form

$$\Delta E_0 = \frac{e^2}{12mc^2} H^2 \langle r^2 \rangle \quad \text{where} \quad \langle r^2 \rangle = \langle \Psi_0 | \sum_{i=1}^{N_e} r_i^2 | \Psi_0 \rangle \tag{16.11}$$

represents the sum of the mean square of the radii of the electron orbits.

Consider now a system with N independent atoms (or ions) in the volume $V = N\Omega$. From Eq. (16.11), we obtain for the diamagnetic susceptibility the Larmor expression

$$\chi = -\frac{N}{V} \frac{\partial^2 \Delta E_0}{\partial H^2} = -\frac{N}{V} \frac{e^2}{6mc^2} \langle r^2 \rangle. \tag{16.12}$$

To estimate the order of magnitude of the susceptibility, consider for example a simple cubic lattice of parameter a_0, unit cell volume $\Omega = a_0^3$, with one atom per unit cell of atomic number Z_a, and average radius r_a. Equation (16.12) can be rewritten in the form

$$\chi = -\frac{1}{\Omega} \frac{e^2}{6mc^2} Z_a r_a^2 = -\frac{Z_a r_a^2}{6a_0^2} \frac{r_0}{a_0} \approx -\frac{r_0}{a_0} \approx -\frac{r_0}{a_B} \approx -10^{-6},$$

where $a_B = 0.529 \times 10^{-8}$ cm is the Bohr radius and $r_0 = e^2/mc^2 = 2.82 \times 10^{-13}$ cm is the classical electron radius. Thus, apart specific numerical factors, the Larmor diamagnetic susceptibility is to a large extent determined by the ratio r_0/a_B between the electron radius and the Bohr radius.

16.2 Permanent Magnetic Dipoles in Atoms or Ions with Partially Filled Shells

A permanent magnetic moment in atoms (or ions) can result only from incompletely filled shells. Consider a free atom (or ion) with an incomplete shell of orbital angular momentum l. Such a shell contains a total number $N = 2(2l + 1)$ of spin-orbitals (namely 2, 6, 10, 14 in the case of s, p, d, f shells, respectively). Let $n < N$ be the number of electrons in the shell; in the hypothetical case that the electrons do not interact, the electronic configuration would be $N!/n!(N - n)!$ times degenerate.

In free atoms the above degeneracy is removed (at least partially) by the electron-electron Coulomb interaction. In atoms with increasing atomic number, the role of spin-orbit interaction becomes of increasing importance, and must be properly taken into account. In the presence of an external magnetic field the Zeeman interaction has also to be considered. Of course, depending on the situation, the hierarchy of strength of the different contributions may vary, and additional terms (as for instance crystalline field effects if the ion is embedded in a crystalline matrix) may become of relevance. In this section we confine our considerations to a few elementary aspects, useful for the description of localized magnetic systems, while for a thorough discussion we refer for instance to A. Rigamonti and P. Carretta (2009) and bibliography quoted therein.

In removing the degeneracy of a partially filled electronic shell in atoms with not too large nuclear charge, the first relevant physical mechanism is the electron-electron Coulomb interaction. In general, this can be taken into account through the Russell-Saunders (or LS) coupling scheme: the spin angular momentum vectors s_i of the electrons combine to give a resultant vector S, and the orbital angular momentum vectors l_i of the electrons combine to give a resultant vector L. The energy levels so obtained are indicated with the notation ^{2S+1}L (where the value of L is indicated with the letters S, P, D, F, G, H, I for $L = 0, 1, 2, \ldots, 6$, respectively). A state LS, with total orbital angular momentum L and total spin S, is degenerate $(2L + 1)(2S + 1)$ times. If we include spin-orbit interaction in the form $\lambda L \cdot S$, a state LS is split into multiplets of definite total angular momentum J with $J = L + S, \ldots, |L - S|$; a multiplet is indicated with the notation $^{2S+1}L_J$.

It is well known that the degeneracy of a given multiplet is fully removed by a (weak) magnetic field that produces $2J + 1$ magnetic levels separated by $g\mu_B$; thus, the (effective) magnetic moment associated with a given multiplet is

$$\mu = -g\mu_B J, \tag{16.13}$$

where $g = g(JLS)$ is the Landé factor and μ_B is the Bohr magneton. For a given multiplet (JLS assigned), the Landé factor is expressed as

$$g = 1 + \frac{J(J + 1) + S(S + 1) - L(L + 1)}{2J(J + 1)}. \tag{16.14}$$

The determination of the states arising from a given electronic configuration is in general rather complicated; however in the study of magnetism we often need only the ground-state quantum numbers SLJ. Except for the heaviest ions, where spin-orbit coupling is rather strong and the Russell-Saunders coupling scheme may become inadequate, the lowest lying spectroscopic term can be obtained by applying Hund's rules. These give a simple prescription for obtaining the quantum numbers SLJ for the ground state of atoms (or ions) with incomplete filled shell.

The ground state for a system of equivalent electrons (electrons belonging to the same shell) is determined by the following three rules, essentially due to exchange, correlation, and relativistic effects:

i. *Hund's first rule:* The electron spins combine to give the maximum multiplicity $2S + 1$ consistent with the Pauli principle. This occurs because parallel spins imply electrons tendentially kept apart in real space, thus decreasing the Coulomb repulsion.

ii. *Hund's second rule:* The orbital momenta combine to give the maximum L consistent with the maximum spin multiplicity of rule (i) and with the exclusion principle. This is justified by the fact that, at parity of other conditions, the electron wavefunctions of higher L are more spread in real space, and electron-electron Coulomb interaction is minimized.

iii. *Hund's third rule:* The angular momenta S and L couple antiparallel giving $J = |L - S|$ if the shell is less than half-filled; S and L couple parallel giving $J = L + S$ for more than half-filled shells. The parallel or antiparallel coupling can be justified with the electron-hole symmetry of partially occupied or partially empty shells.

The physical picture behind the Hund's rules can be described as follows. The first step is the minimization of the *spin-spin* interaction energy, whose origin is the electrostatic potential in real space within the manifold of wavefunctions compatible with the Pauli principle. The second step is the minimization of the *orbit-orbit* interaction energy, whose origin is the electrostatic potential in real space, within the manifold of wavefunctions compatible with step one and with the Pauli principle. The third step is the minimization of the relativistic *spin-orbit* interaction energy, whose classical origin is related to the interaction of the spin magnetic moment with the magnetic field "seen" by the electrons, because of the apparent motion of the nuclear charge in the non-inertial frame, where the electrons are at rest. The classical picture, however, gives twice the correct relativistic value. Since the electron dynamics must be faster in the nuclei with higher atomic number Z, the spin-orbit coupling is of increasing importance for increasing Z. In the extreme case of actinides ($Z \geq 90$), the spin-orbit coupling for each electron is stronger than the spin-spin or orbit-orbit coupling between different electrons, and the $j\text{-}j$ coupling scheme applies, rather then the Russell-Saunders scheme.

Transition atoms or ions ($22 \leq Z \leq 28$) and rare earth atoms or ions ($58 \leq Z \leq 71$) follow the Russell-Saunders coupling with JLS good quantum numbers. As an example, consider the ion Fe^{2+}, whose external electron configuration is $3d^6$. The lowest spectroscopic term is obtained from the Hund's rules, as schematically shown in Figure 16.1. We indicate the five one-electron d-orbitals with $l_z = 2, 1, 0, -1, -2$, and accommodate the electrons following the Hund's rules. To obtain the highest value of S_z, we place as many electrons as possible with spin up. Since there are five d orbitals, we can place five electrons with spin up and the sixth electron with spin down in the $l_z = 2$ level. We have thus $S = 1/2 + 1/2 + 1/2 + 1/2 = 2$ and $L = 4 + 1 - 1 - 2 = 2$. Since the shell is more than half-filled, the value of the total angular momentum is $J = L + S = 4$; the ground state of Fe^{2+} is thus $^5D_4$.

As another example consider the ion Cr^{3+}, whose external electron configuration is $3d^3$. Then $S = 1/2 + 1/2 + 1/2 = 3/2$ and $L = 2 + 1 + 0 = 3$; since the shell is less

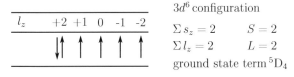

Figure 16.1 Application of the Hund's rules for the ground state of Fe^{2+} (configuration $3d^6$).

than half-filled we have $J = |L - S| = 3/2$. The ground state is thus $^4F_{3/2}$. With the Hund's rules, the quantum numbers of the ground states of atoms or ions can be easily worked out.

16.3 Paramagnetism of Localized Magnetic Moments

Consider an atom (or ion) whose ground state has angular momentum J and magnetic moment $\mu = -g\mu_B\mathbf{J}$. The interaction energy of the magnetic moment with an applied magnetic field (in the weak field regime that does not alter the LS coupling model) can be described by the Zeeman operator

$$\mathcal{H}_Z = -\mu \cdot \mathbf{H} = g\mu_B\mathbf{J} \cdot \mathbf{H}. \tag{16.15}$$

In the magnetic field, the ground state of the atom splits into $2J + 1$ Zeeman sublevels separated by the energy $g\mu_B H$.

We suppose that, at temperatures of practical interest, only the $2J + 1$ lowest states can be thermally excited. In this case the thermodynamical properties can be calculated starting from the partition function

$$Z = \sum_{i=-J}^{+J} e^{-g\mu_B Hi/k_B T}. \tag{16.16}$$

We confine first our attention to the case $J = 1/2$. With two terms only in the sum (16.16), the partition function becomes

$$Z = e^{g\mu_B HJ/k_B T} + e^{-g\mu_B HJ/k_B T} \qquad (J = 1/2).$$

The thermal average $\langle \mu_z \rangle$ at a given temperature is

$$\langle \mu_z \rangle = k_B T \frac{\partial \ln Z}{\partial H} \equiv g\mu_B J \frac{e^{g\mu_B HJ/k_B T} - e^{-g\mu_B HJ/k_B T}}{e^{g\mu_B HJ/k_B T} + e^{-g\mu_B HJ/k_B T}} = g\mu_B J \tanh \frac{g\mu_B HJ}{k_B T}.$$

The average magnetization $M = (N/V)\langle \mu_z \rangle$ in the field direction of a material with N (independent) magnetic atoms in the volume V takes the expression

$$\boxed{M = \frac{N}{V} g\mu_B J \tanh \frac{g\mu_B HJ}{k_B T}} \qquad (J = 1/2). \tag{16.17}$$

We consider now this expression in the high temperature limit (Curie law) and the high field limit (saturation).

In most ordinary conditions it holds $\mu_B H \ll k_B T$ (*paramagnetic region*); in this case the spacing of Zeeman sublevels is much smaller than the thermal energy $k_B T$, and $\tanh x$ can be approximated with x in Eq. (16.17). One obtains

$$\boxed{\chi = \frac{M}{H} = \frac{N}{V}\frac{\mu^2}{3k_BT}} \qquad \left[\mu^2 = g^2\mu_B^2 J(J+1); \ J = \frac{1}{2}\right]. \tag{16.18a}$$

The above equation can be rewritten in the form

$$\chi = \frac{C}{T} \quad \text{with} \quad C = \frac{1}{\Omega}\frac{\mu^2}{3k_B} = \frac{1}{a_0^3}\frac{\mu^2}{3k_B}, \tag{16.18b}$$

where $\Omega = a_0^3$ is the volume for each localized magnetic moment, and C is called Curie constant. The dependence $\chi \propto 1/T$ of the magnetic susceptibility on the temperature is known as Curie law. From measurements of χ as a function of $1/T$ one can infer the value of μ. The order of magnitude of the Curie constant can be estimated taking into account that $\mu_B^2/a_B^3 k_B \approx 4.21$ K; in typical situations at room temperature the susceptibility in Eq. (16.18a) is in the range $\chi \approx 10^{-4}$ or so.

The Curie law maintains its validity down to a few Kelvin degrees in typical magnetic fields. When $k_BT \ll \mu_B H$, $\tanh(\mu_B H/k_BT) \approx 1$, and Eq. (16.17) gives

$$M = \frac{N}{V}g\mu_B J. \tag{16.19}$$

This saturation value denotes that all magnetic moments are lined up along the applied field, while the susceptibility $\chi = \partial M/\partial H$ tends to zero at very low temperatures.

After the above considerations for the case $J = 1/2$, we observe now that higher values of J lead to quite similar physical results. The partition function (16.16) for a generic J, introducing the definition $\alpha = g\mu_B H/k_BT$, can be rewritten in the form

$$\begin{aligned}
Z &= e^{J\alpha}\left[1 + e^{-\alpha} + \cdots + e^{-2J\alpha}\right] = e^{J\alpha}\frac{1 - e^{-(2J+1)\alpha}}{1 - e^{-\alpha}} \\
&= \frac{e^{(2J+1)\alpha/2} - e^{-(2J+1)\alpha/2}}{e^{\alpha/2} - e^{-\alpha/2}} = \frac{\sinh[(2J+1)g\mu_B H/2k_BT]}{\sinh[g\mu_B H/2k_BT]}.
\end{aligned} \tag{16.20}$$

Using Eq. (16.20) we have for the thermal average of the magnetization in the direction of the applied field

$$\langle\mu_z\rangle = k_BT\frac{\partial \ln Z}{\partial H} \equiv g\mu_B J B_J(x), \qquad x = \frac{g\mu_B H J}{k_BT},$$

where

$$B_J(x) = \frac{2J+1}{2J}\coth\frac{(2J+1)x}{2J} - \frac{1}{2J}\coth\frac{x}{2J}$$

is called *Brillouin function*. The average magnetization of a system with N independent magnetic atoms of spin J in the volume V, becomes

$$\boxed{M = \frac{N}{V}g\mu_B J B_J\left(\frac{g\mu_B H J}{k_BT}\right)}. \tag{16.21}$$

Some straight properties of the Brillouin function are the following:

(i) $B_J(x) = \dfrac{J+1}{3J} x$ for $x \to 0$;   (ii) $B_J(x) = 1$ for $x \to \infty$;

(iii) $B_{1/2}(x) = 2 \coth 2x - \coth x \equiv \tanh x$;   (iv) $B_\infty(x) = \coth x - \dfrac{1}{x} \equiv L(x)$.

It is seen by inspection that the Curie law at high temperatures follows from the property (i). The saturation region result follows from (ii). In the case $J = 1/2$, the Brillouin function equals the hyperbolic tangent function, while for high values of J, it recovers the Langevin function $L(x)$.

Paramagnetism of Rare Earth Ions

Paramagnetism of atoms or ions requires the existence of partially filled electronic shells, and we consider here the rare earth group (incomplete $4f$ shell) and later the iron group (incomplete $3d$ shell). Other groups are the palladium group (incomplete $4d$ shell), the platinum group (incomplete $5d$ shell), and the uranium group or actinides (incomplete $5f$, $6d$ shells). In Table 16.1 we report the ground-state terms $^{2S+1}L_J$ of trivalent lanthanum group ions, obtained following the Hund's rules; we give also the calculated *effective number of Bohr magnetons* $p = g[J(J+1)]^{1/2}$. Finally, the values of p derived from measured values of χ and from the Curie law are also given for comparison; the experimental values refer to impurities in insulating crystals, and

Table 16.1 Ground states of rare earth ions with partially filled f shell, and effective Bohr magneton numbers.

Ion	Electronic configuration	Ground-state term	Calculated $p = g[J(J+1)]^{1/2}$	Measured p
La^{3+}	$4f^0 5s^2 5p^6$	$^1S_0$	0	Diamagnetic
Ce^{3+}	$4f^1 5s^2 5p^6$	$^2F_{5/2}$	2.54	2.4
Pr^{3+}	$4f^2 5s^2 5p^6$	$^3H_4$	3.58	3.5
Nd^{3+}	$4f^3 5s^2 5p^6$	$^4I_{9/2}$	3.62	3.5
Pm^{3+}	$4f^4 5s^2 5p^6$	$^5I_4$	2.68	–
Sm^{3+}	$4f^5 5s^2 5p^6$	$^6H_{5/2}$	0.84	1.5
Eu^{3+}	$4f^6 5s^2 5p^6$	$^7F_0$	0	3.4
Gd^{3+}	$4f^7 5s^2 5p^6$	$^8S_{7/2}$	7.94	8.0
Tb^{3+}	$4f^8 5s^2 5p^6$	$^7F_6$	9.72	9.5
Dy^{3+}	$4f^9 5s^2 5p^6$	$^6H_{15/2}$	10.65	10.6
Ho^{3+}	$4f^{10} 5s^2 5p^6$	$^5I_8$	10.61	10.4
Er^{3+}	$4f^{11} 5s^2 5p^6$	$^4I_{15/2}$	9.58	9.5
Tm^{3+}	$4f^{12} 5s^2 5p^6$	$^3H_6$	7.56	7.3
Yb^{3+}	$4f^{13} 5s^2 5p^6$	$^2F_{7/2}$	4.54	4.5
Lu^{3+}	$4f^{14} 5s^2 5p^6$	$^1S_0$	0	Diamagnetic

not to the free ion situation (see for further details the book of J. H. Van Vleck, or the book of S. Chikazumi, cited at the end of this chapter). A comparison is meaningful only if the crystalline potential has negligible effect on f-electrons; this is indeed a reasonable assumption because the paramagnetic electrons are in well localized inner shells and are screened by the outermost $5s^2 5p^6$ electrons [however, crystalline electric field effects may become important at low temperatures, as in the case of praseodymium and its magnetic excitations; see for instance J. Jensen and A. R. Mackintosh "Rare Earth Magnetism" (Clarendon, Oxford, 1991)].

It is seen from Table 16.1 that there is a good overall agreement between calculated and measured values of the effective number of Bohr magnetons, the only exceptions being Sm^{3+} and mainly Eu^{3+}; the discrepancy has been explained noticing that in this case excited levels mix to some extent with the ground state in the presence of a magnetic field, and the assumption of considering only the ground-state term becomes clearly inadequate.

Paramagnetism of Iron Group Ions

We give in Table 16.2 the measured and computed effective number of Bohr magnetons for some iron group ions; the measured values of p, derived from the experimental susceptibility and from the Curie law, refer to impurities in insulating crystals (see for further details the book of J. H. Van Vleck or the book of S. Chikazumi, cited at the end of this chapter). We see that in this case the effective number of Bohr magnetons cannot be estimated by the relation $p = g[J(J+1)]^{1/2}$; rather it is reasonably well described by the relation $p = 2[S(S+1)]^{1/2}$. Thus there is evidence that the LS coupling is broken because of the crystal field effects, and that the orbital angular momentum L is "quenched," leaving the electron spin as the only apparent source of the magnetic moment.

The origin of the quenching of the angular momentum, and the difference with respect to the rare earth ions, can be understood qualitatively with the following

Table 16.2 Ground states of iron group ions with partially filled d shell, and effective Bohr magneton numbers.

Ion		Electronic configuration	Ground-state term	p calculated values $g[J(J+1)]^{1/2}$	$2[S(S+1)]^{1/2}$	p measured values	
Ti^{3+}	V^{4+}	$3d^1$	$^2D_{3/2}$	1.55	1.73		1.8
V^{3+}		$3d^2$	$^3F_2$	1.63	2.83	2.8	
Cr^{3+}	V^{2+}	$3d^3$	$^4F_{3/2}$	0.77	3.87	3.7	3.8
Mn^{3+}	Cr^{2+}	$3d^4$	$^5D_0$	0	4.90	5.0	4.8
Fe^{3+}	Mn^{4+}	$3d^5$	$^6S_{5/2}$	5.92	5.92	5.9	5.9
Fe^{2+}		$3d^6$	$^5D_4$	6.70	4.90	5.4	
Co^{2+}		$3d^7$	$^4F_{9/2}$	7.54	3.87	4.8	
Ni^{2+}		$3d^8$	$^3F_4$	5.59	2.83	3.2	
Cu^{2+}		$3d^9$	$^2D_{5/2}$	3.55	1.73	1.9	

arguments. (i) The LS coupling in rare earth ions is larger because of the higher local-ization of f shells. (ii) The paramagnetic electrons are in an inner shell in rare earth ions, while they are in the external shell in transition group ions. Thus the crystalline potential is negligible on f states, because of screening of outermost electrons, while it is no more negligible for the less shielded d electrons with a consequent reduction, or quenching of the orbital angular momentum L. As a matter of fact, the effective value of the permanent magnetic moment of the same element or ion in different bulk or surface environments can change significantly, and the various competing microscopic interactions (spin-spin, orbit-orbit, spin-orbit, crystal field, external fields) and their hierarchic importance must be analyzed case by case.

Quenching of the Orbital Angular Momentum: An Example

We can understand qualitatively the tendency of the crystal field (or other similar perturbations) to quench the orbital angular momentum, considering the elementary example of an atom (or ion) with one electron in an incomplete d shell. When a magnetic field is applied to the free atom, the coupling of the orbital magnetic moment $-\mu_B \mathbf{l}$ of the electron under consideration with the external field is described by the Zeeman operator

$$\mathcal{H}_Z = \mu_B \mathbf{l} \cdot \mathbf{H}. \tag{16.22}$$

The Zeeman operator, diagonalized within the fivefold degenerate wavefunctions of the d level, produces the splitting into five levels of energies $\pm 2\mu_B H, \pm \mu_B H, 0$.

Suppose now that the atom is put in a cubic environment; the crystal field separates the fivefold degeneracy of the d orbitals into a triplet and a doublet, whose wavefunc-tions are

$$\begin{cases} \psi_1 \equiv d_{xz} = \sqrt{\dfrac{15}{4\pi}} \dfrac{xz}{r^2} = \dfrac{1}{\sqrt{2}}(-Y_{21} + Y_{2-1}), \\[2mm] \psi_2 \equiv d_{xy} = \sqrt{\dfrac{15}{4\pi}} \dfrac{xy}{r^2} = \dfrac{-i}{\sqrt{2}}(+Y_{22} - Y_{2-2}), \\[2mm] \psi_3 \equiv d_{yz} = \sqrt{\dfrac{15}{4\pi}} \dfrac{yz}{r^2} = \dfrac{i}{\sqrt{2}}(+Y_{21} + Y_{2-1}) \end{cases} \tag{16.23a}$$

and

$$\begin{cases} \psi_4 \equiv d_{3z^2-r^2} = \sqrt{\dfrac{5}{16\pi}} \dfrac{3z^2 - r^2}{r^2} = Y_{20}, \\[2mm] \psi_5 \equiv d_{x^2-y^2} = \sqrt{\dfrac{15}{16\pi}} \dfrac{x^2 - y^2}{r^2} = \dfrac{1}{\sqrt{2}}(Y_{22} + Y_{2-2}) \end{cases} \tag{16.23b}$$

(Y_{2m} with $m = -2, -1, \dots, +2$ are the spherical harmonics of order 2).

If we express the Zeeman operator (16.22) on the above basis (and take the magnetic field in the z-direction, and consider that $l_z Y_{lm} = m Y_{lm}$), we obtain for the matrix elements $M_{ij} = \langle \psi_i | \mu_B l_z H | \psi_j \rangle$ the expression

$$M = \mu_B H \begin{pmatrix} 0 & 0 & -i & 0 & 0 \\ 0 & 0 & 0 & 0 & 2i \\ i & 0 & 0 & 0 & 0 \\ 0 & 0 & 0 & 0 & 0 \\ 0 & -2i & 0 & 0 & 0 \end{pmatrix}. \tag{16.24}$$

It is immediate to verify that the eigenvalues of the above 5×5 matrix are $\pm 2\mu_B H$, $\pm \mu_B H$, 0 (as obviously expected). However if we add a cubic crystal field contribution, this produces a decoupling between the upper left 3×3 corner and the lower right 2×2 corner of the matrix given in Eq. (16.24). We have that the states of the doublet (with both null eigenvalues) behave as if a *total quenching* of the orbital momentum had occurred, while the states of the triplet behave as if a *partial quenching* of the angular momentum had occurred from $l = 2$ to $l = 1$. Although the treatment of realistic situations may be quite complicated, from the above simplified model the tendency of "suppression" of the orbital angular momentum by crystal field effects can be inferred. Also notice that if the degeneracy of the manifold is totally removed by the crystal field of a low symmetry environment, then the quenching of the orbital angular momentum could be complete since the diagonal matrix elements of l_z are all vanishing.

16.4 Localized Magnetic States in Normal Metals

The embedding of a magnetic atom in a normal metal leads to a variety of interesting effects. The foreign magnetic impurity, coupled to the (easily polarizable) Fermi sea of the host material, may lose or not its magnetic properties. For instance, iron group elements when dissolved in non-magnetic metals can lose their magnetic moments; on the contrary rare earth elements generally maintain their magnetic moments. Even more important for the entailed consequences, is the fact that *a preserved magnetic moment coupled to the Fermi sea can lead to a correlated non-magnetic ground state of the whole system*, constituted by the magnetic impurity plus the Fermi sea. Very peculiar properties then occur at temperatures sufficiently low that thermal excitations are negligible. In this section we study the problem of preservation or disappearance of magnetic moment for impurities embedded in normal metals, while in the next section we consider the occurrence of a correlated non-magnetic ground state.

The presence of a foreign magnetic atom in a normal metal gives rise to a complex situation, that can be qualitatively addressed following the Anderson reasoning on local magnetic moment preservation or loss. The system under consideration consists of a potentially magnetic impurity, embedded in a normal metal; in its essential aspects, the system can be described by means of a few phenomenological parameters through the following Anderson model:

$$\mathcal{H} = \sum_{\mathbf{k}\sigma} E_{\mathbf{k}} n_{\mathbf{k}\sigma} + E_d(n_{d\uparrow} + n_{d\downarrow})$$

$$+ \frac{1}{\sqrt{N}} \sum_{\mathbf{k}\sigma} \left[V_{\mathbf{k}d} c_{\mathbf{k}\sigma}^\dagger c_{d\sigma} + V_{\mathbf{k}d}^* c_{d\sigma}^\dagger c_{\mathbf{k}\sigma} \right] + U n_{d\uparrow} n_{d\downarrow}. \tag{16.25}$$

The first term in the right-hand side of Eq. (16.25) represents the conduction band of the host normal metal (for instance an s-like conduction band); $E_\mathbf{k}$ is the energy of the conduction band electronic state of wavevector $\mathbf{k}$ (and spin σ); $c_{\mathbf{k}\sigma}^\dagger$ and $c_{\mathbf{k}\sigma}$ are the corresponding creation and annihilation operators; $n_{\mathbf{k}\sigma} = c_{\mathbf{k}\sigma}^\dagger c_{\mathbf{k}\sigma}$ is the electron number operator. The second term in Eq. (16.25) represents a single localized non-degenerate impurity orbital of energy E_d (the possible degeneracy of the impurity orbital does not change the essential aspects of the treatment and is here neglected for simplicity); $n_{d\sigma} = c_{d\sigma}^\dagger c_{d\sigma}$ is the number operator for the impurity orbital. The third term in Eq. (16.25), called s-d mixing, represents the hybridization energy between the localized state and the conduction band wavefunctions of the metal; the interaction energy between the Bloch wavefunction of vector $\mathbf{k}$ and the localized d-orbital is denoted as $V_{\mathbf{k}d}/\sqrt{N}$ (where N is the number of unit cells of the metal); for simplicity, in the following the hybridization matrix is taken to be independent of $\mathbf{k}$ and is indicated as V_{sd}. The last term in Eq. (16.25) represents the Coulomb repulsive energy ($U > 0$) between electrons of opposite spins in the localized d orbital. It is evident that the Anderson Hamiltonian (16.25) contains two competing terms: the Coulomb repulsion between opposite spins tends to create unbalance of spin-up and spin-down occupancy of the localized state, while the hybridization tends to equalize them. This is indicated schematically in Figure 16.2.

The basic difficulty toward a rigorous solution of the Hamiltonian (16.25) is the presence of the last term, the intra-atomic correlation term, which is the product of four creation or annihilation operators. In order to find a workable (although approximate) way to handle it, we first drop it altogether and then we attempt some approximate account. We consider thus the Hamiltonian $\mathcal{H}_0$ obtained from Eq. (16.25) in the

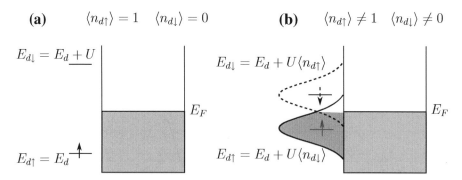

Figure 16.2 Schematic representation of the Anderson model. In (a) the hybridization of the localized level with the Fermi sea is neglected; it is assumed that $E_d < E_F$ and $E_d + U > E_F$, so that $\langle n_{d\uparrow} \rangle = 1$ and $\langle n_{d\downarrow} \rangle = 0$ (or vice versa); then $E_{d\uparrow} = E_d + U\langle n_{d\downarrow} \rangle = E_d$, while $E_{d\downarrow} = E_d + U$. In (b) the hybridization is switched on, and it tends to equalize spin-up and spin-down occupancies and energies. The occupied regions of the metal and of the localized resonances are shaded.

particular case $U \equiv 0$; we have

$$\mathcal{H}_0 = \sum_{\mathbf{k}\sigma} E_{\mathbf{k}} n_{\mathbf{k}\sigma} + E_d(n_{d\uparrow} + n_{d\downarrow}) + \frac{1}{\sqrt{N}} \sum_{\mathbf{k}\sigma} \left[V_{sd} c_{\mathbf{k}\sigma}^\dagger c_{d\sigma} + V_{sd}^* c_{d\sigma}^\dagger c_{\mathbf{k}\sigma} \right]. \quad (16.26)$$

The Hamiltonian (16.26) is an ordinary one-electron Hamiltonian, representing a local-ized state interacting with a continuum. It is evident that spin-up and spin-down occupancy of the localized d state are equal, and no net magnetic moment exists. Furthermore, following the same concepts and techniques adopted in the case of the Fano effect (see Section 12.6), we have that the localized state E_d develops into a resonance, because of the hybridization with the band continuum.

The shape of the resonance can be easily worked out from the diagonal matrix ele-ment of the Green's function on the localized state $|d\rangle$; by means of the renormalization approach (see Section 5.8.3) we obtain

$$G_{dd}^0(E) = \langle 0 | c_{d\sigma} \frac{1}{E - H_0} c_{d\sigma}^\dagger | 0 \rangle = \frac{1}{E - E_d - |V_{sd}|^2 \frac{1}{N} \sum_{\mathbf{k}} \frac{1}{E - E_{\mathbf{k}}}}, \quad (16.27)$$

where, as usual, the energy E is defined with the addition of a small imaginary part $i\varepsilon$ and the limit $\varepsilon \to 0^+$ is understood.

The sum over $\mathbf{k}$ appearing at the denominator of Eq. (16.27) can be split into a real part and an imaginary part; the real part represents an energy shift of the resonant level (and can be supposed to be embodied in E_d itself); for the imaginary part we have

$$\text{Im} \sum_{\mathbf{k}} \frac{1}{E + i\varepsilon - E_{\mathbf{k}}} = -\sum_{\mathbf{k}} \frac{\varepsilon}{(E - E_{\mathbf{k}})^2 + \varepsilon^2} = -\pi \sum_{\mathbf{k}} \delta(E - E_{\mathbf{k}}) = -\pi D_0(E),$$

where $D_0(E)$ denotes the density-of-states of the conduction band of the unperturbed metal, for one spin direction. In the following, for simplicity, we also assume that $D_0(E) \approx D_0(E_F)$ in the energy range of interest around the Fermi energy.

With the above approximations, the Green's function (16.27) can be written as

$$G_{dd}^0(E) = \frac{1}{E - E_d + i\Gamma}, \quad (16.28a)$$

where

$$\Gamma = \pi |V_{sd}|^2 \frac{1}{N} D_0(E_F) = \pi |V_{sd}|^2 n_0(E_F). \quad (16.28b)$$

The quantity $n_0(E_F) = D_0(E_F)/N$ denotes the electronic density-of-states at the Fermi level for one spin direction and per unit cell. The order of magnitude of $n_0(E_F)$ is $\approx 1/W_F$, where W_F is the energy width of the part of the conduction band occupied by electrons [W_F denotes the difference between the Fermi energy and the bottom of the conduction band; instead E_F denotes the actual position of the Fermi energy, often taken as the zero reference energy $E_F = 0$]. The order of magnitude of the parameter Γ is then $\approx |V_{sd}|^2/W_F$. The *local density-of-states at the impurity site* is

$$n_d^0(E) = -\frac{1}{\pi} \text{Im} \, G_{dd}^0(E) = \frac{1}{\pi} \frac{\Gamma}{(E - E_d)^2 + \Gamma^2}; \quad (16.28c)$$

the local density-of-states has thus a Lorentzian shape with half-width parameter Γ.

The integrated density-of-states up to the Fermi energy gives the average number of electrons at the impurity site with given spin direction, and can be easily calculated using the indefinite integral

$$\int \frac{b}{(x-a)^2 + b^2} \, dx = \text{arctg} \, \frac{x-a}{b}.$$

We have

$$\langle n_d^0 \rangle = \int_{-\infty}^{E_F} n_d^0(E) \, dE = \frac{1}{\pi} \int_{-\infty}^{E_F} \frac{\Gamma}{(E-E_d)^2 + \Gamma^2} \, dE = \frac{1}{2} + \frac{1}{\pi} \text{arctg} \, \frac{E_F - E_d}{\Gamma};$$

$$(16.29)$$

it is seen that $\langle n_d^0 \rangle$ changes monotonically from one, to one half and to zero, as the energy of the localized state is much less, equal or much larger than the Fermi level energy.

We have now to consider (at least in some approximate form) the intra-atomic correlation term so far neglected.

The simplest way to include approximately the effect of the Coulomb interaction $U n_{d\uparrow} n_{d\downarrow}$ is within the *mean field framework*: it is assumed that the average Coulomb field acting on the localized spin-up orbital is $U \langle n_{d\downarrow} \rangle$, where $\langle n_{d\downarrow} \rangle$ denotes the average occupancy of the localized spin-down orbital (and similar reasoning for the spin-down orbital). The *effective field assumption* thus entails the energy shifts

$$E_{d\uparrow} = E_d + U \langle n_{d\downarrow} \rangle; \quad E_{d\downarrow} = E_d + U \langle n_{d\uparrow} \rangle; \tag{16.30a}$$

these shifts are determined within the unrestricted Hartree-Fock approximation, that permits a different occupancy of the spin-up and spin-down impurity orbitals. Following the same arguments used above for obtaining Eq. (16.29), we arrive at the two coupled equations

$$\boxed{\begin{aligned} \langle n_{d\uparrow} \rangle &= \frac{1}{2} + \frac{1}{\pi} \text{arctg} \, \frac{E_F - E_d - U \langle n_{d\downarrow} \rangle}{\Gamma} \\ \langle n_{d\downarrow} \rangle &= \frac{1}{2} + \frac{1}{\pi} \text{arctg} \, \frac{E_F - E_d - U \langle n_{d\uparrow} \rangle}{\Gamma} \end{aligned}} \tag{16.30b}$$

These two equations must be solved self-consistently; it is seen by inspection that one can have solutions with $\langle n_{d\uparrow} \rangle \equiv \langle n_{d\downarrow} \rangle$ corresponding to non-magnetic behavior, and solutions $\langle n_{d\uparrow} \rangle \neq \langle n_{d\downarrow} \rangle$ corresponding to magnetic behavior.

Consider, for instance, the symmetrical case in which the energy of the d state is located at $U/2$ below the Fermi energy. In this case Eqs. (16.30) take the simplified form

$$\begin{cases} \langle n_{d\uparrow} \rangle = \frac{1}{2} + \frac{1}{\pi} \text{arctg} \left[\frac{U}{\Gamma} \left(\frac{1}{2} - \langle n_{d\downarrow} \rangle \right) \right], \\ \langle n_{d\downarrow} \rangle = \frac{1}{2} + \frac{1}{\pi} \text{arctg} \left[\frac{U}{\Gamma} \left(\frac{1}{2} - \langle n_{d\uparrow} \rangle \right) \right]. \end{cases} \tag{16.31}$$

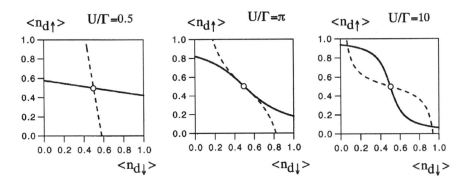

Figure 16.3 Graphical solution of the self-consistent equations (16.31) for different values of the dimensionless ratio U / Γ.

We can solve graphically the self-consistent equations (16.31), as illustrated in Figure 16.3. When $\Gamma \gg U$ we have *only* non-magnetic solutions of type $\langle n_{d\uparrow} \rangle \equiv \langle n_{d\downarrow} \rangle = 1/2$; if $\Gamma \ll U$ we have also the magnetic solution $\langle n_{d\uparrow} \rangle \approx 1$, $\langle n_{d\downarrow} \rangle \approx 0$ (or vice versa). The transition between the magnetic regime and the non-magnetic regime occurs when $U / \pi \Gamma = 1$, as can be established linearizing Eq. (16.31) when the parameters are near the phase transition curve. With similar procedures, we can analyze the occurrence or not of magnetic and non-magnetic solutions of Eqs. (16.30) at any values of the dimensionless parameters $\pi \Gamma / U$ and $(E_F - E_d)/U$, and obtain the phase diagram reported in Figure 16.4. The most favorable case for the formation of a magnetic moment is the symmetrical case $E_F - E_d = U/2$, while for $E_d \geq E_F$ or $E_d \leq E_F - U$ the magnetic moment is unfavored (or not possible) in the model.

As qualitative application of what has been said so far, consider the existence or non-existence of localized magnetic moments for transition element impurities dissolved in noble metals and aluminum. It is observed that the Cr impurity remains magnetic when inserted in metals such as Au, Cu, Ag, and is non-magnetic when inserted in Al, because of the higher value of the density-of-states at the Fermi level in the latter case. Similarly,

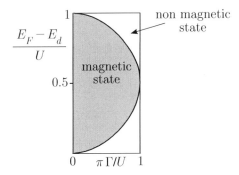

Figure 16.4 Phase diagram corresponding to the Hartree-Fock solution of the Anderson Hamiltonian for a magnetic impurity in a normal metal [from P. W. Anderson, Phys. Rev. *124*, 41 (1961)].

it is observed that Cr, Mn, and Fe impurities remain magnetic in Au, while Ti and Ni impurities are non-magnetic in Au; this trend can be understood considering that the orbital energy E_d decreases passing from Ti to Ni; at the beginning of the series (when E_d is higher or near the Fermi level of gold) and at the end of the series (when E_d is well below the Fermi level of gold) no magnetic moment is possible. It can also be noticed that, for iron group impurities, magnetism may or may not occur (estimates of U and Γ parameters are $U \approx 10$ eV and $\Gamma \approx 2\text{–}5$ eV). For rare earth impurities, where $U \approx 15$ eV and $\Gamma \approx 1$ eV or less, the magnetic regime is expected to occur very frequently.

16.5   Dilute Magnetic Alloys and the Resistance Minimum Phenomenon

16.5.1   Introduction and Some Phenomenological Aspects

In the previous section, we have seen that a magnetic impurity, when inserted in an ordinary metal, may lose or preserve its magnetic moment; which one of the two cases is likely to occur, has been discussed on the basis of the phenomenological Anderson model. In this section, we consider some transport properties of normal metals containing a small concentration of impurity atoms, which remain magnetic (*dilute magnetic alloys*). In experiments, typical concentrations are less than few hundred magnetic atoms per million host atoms (at this dilution, interaction between impurities are likely to be sufficiently small and are then neglected); typical host materials are copper, silver, gold, magnesium, zinc; typical magnetic impurities are chromium, manganese, iron, cobalt, nickel, vanadium, titanium.

It is a well-established fact that several thermal, electrical and magnetic properties of ordinary metals in the presence of dilute magnetic impurities appear to be "anomalous," with respect to the properties observed in the presence of ordinary (non-magnetic) impurities. For instance, the resistivity of an ordinary metal containing non-magnetic impurities is often reasonably described at low temperatures by a law of the type

$$\rho(T) = \rho_0[1 + AT^5 + \cdots], \tag{16.32}$$

where ρ_0 is the so-called residual resistivity due to the impurities, and the term with the fifth power of temperature is due to lattice vibrational effects. [Additional contributions due to disorder or electron-electron interactions are here disregarded; for them, we refer for instance to the article by P. A. Lee and T. V. Ramakrishnan, Rev. Mod. Phys. *57*, 287 (1985).]

Thus the resistivity (and hence the resistance, which is proportional to the resistivity in a given chosen sample geometry) decreases monotonically with decreasing temperature and becomes constant at very low temperatures. On the contrary, the resistivity of a normal metal in the presence of dilute magnetic impurities exhibits a rather shallow minimum (Kondo minimum) at very low temperatures T_K, typically of the order of a few Kelvin degrees; the resistivity eventually saturates (i.e. reaches a constant value) as the temperature is further decreased. In Figure 16.5 we report the "resistance minimum phenomenon" for dilute alloys of iron in copper, as an example.

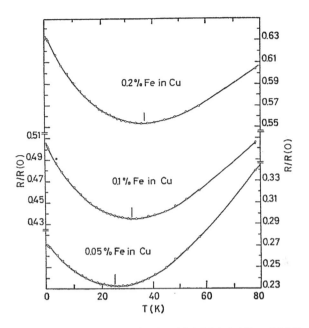

Figure 16.5 Resistance at low temperatures for Cu with 0.05%, 0.1%, and 0.2% of Fe impurities [from J. P. Franck, F. D. Manchester and D. L. Martin, Proc. Roy. Soc. (London) A *263*, 494 (1961); copyright 1961, The Royal Society].

As the temperature is decreased also the magnetic moment of the impurity, embedded in the ordinary metal, exhibits an "anomalous" behavior; in particular the magnetic susceptibility ceases to follow the Curie law, and rather saturates to a constant value at $T = 0$, thus indicating that the magnetic moment of the impurity is fully compensated and quenched by the conduction electrons of the host metal at very low temperatures. This is also confirmed by an extra contribution to the specific heat, corresponding to an entropy change of the order of $k_B \ln 2$, for each magnetic impurity.

Resistivity, magnetic susceptibility, specific heat, and several other properties such as magnetoresistance, thermoelectric effects, appear to be anomalous in ordinary metals with dilute magnetic impurities as the temperature is decreased in the neighborhood or below the temperature T_K. All these peculiar Kondo effects originate from the new physical situation that occurs at low temperatures.

For the description of the Kondo effects in non-magnetic metals, in general a full many-body theory is necessary. The many-body physics has originally considered the paradigmatic model of a spin 1/2 moment coupled with the channel of the itinerant electrons of a simple metal; later, more sophisticated models of higher spin and multi-channel variants have also been considered. Besides bulk and surface magnetism, the investigations are invaluable in several other branches of physics, including chemisorption, catalysis, and the variegated area of nanostructures and nanodevices.

It is not our purpose to discuss the wealth of experimental and theoretical investigations in this active field of research; rather we focus on semi-quantitative models, that can at least provide some guidelines in the complicated and challenging Kondo phenomenology. In this line, we limit our considerations just to a couple of particularly instructive issues, concerning a spin 1/2 moment in normal metal. In this section, we consider the resistance minimum phenomenon, which can be treated with perturbation theory (at least qualitatively in an appropriate temperature range). In the next section the focus is on the screening mechanism of the impurity magnetic moment by the delocalized electrons of the metal.

16.5.2   The Resistance Minimum Phenomenon

Model of the Magnetic Impurity Interaction with the Band Electrons

The study of a spatially localized magnetic impurity inside a normal non-magnetic metal is generally referred as the Kondo problem. When the effective exchange coupling between the magnetic impurity and the itinerant electrons of the metal is anti-ferromagnetic, various Kondo effects occur at low temperature. Among them, the observation of an apparent "divergence" of the electrical resistance opened historically this wide area of research.

An impurity, magnetic or non-magnetic, embedded in a normal metal interacts with the conduction electrons of the metal, thus giving a contribution to the resistivity. The reason why in dilute magnetic alloys the resistivity presents a broad minimum at low temperatures, while for non-magnetic alloys the resistivity drops monotonically to the residual resistivity, is related to the different kind of perturbation produced by impurities with or without magnetic moments. Such an effect is rather subtle, as the essential difference between the scattering amplitudes of spin-dependent and spin-independent interaction terms appears only at second (and higher orders) of perturbation theory, while at first order of perturbation theory there is no qualitative difference between them.

Let us consider the simplest possible picture of a normal non-magnetic impurity and of a magnetic impurity, embedded in an ordinary metal. For a normal impurity, the perturbation acting on the free-electron gas can be described in the one-electron picture by an impurity potential $V_{imp}(\mathbf{r})$, whose spatial range is of the order of $1/k_F$.

For a magnetic impurity of spin $\mathbf{S}$ (localized, say, at the origin), it is reasonable to assume an effective exchange coupling with the conduction electrons of the form

$$\mathcal{H}_K = -J\Omega\,\mathbf{S}\cdot\mathbf{s}\,\delta(\mathbf{r}), \tag{16.33}$$

where $\mathbf{s}$ is the spin of the electron in the conduction band, Ω is the volume of the unit cell, and J is a phenomenological parameter with the dimension of an energy [in the case Ω is omitted in Eq. (16.33), J becomes a phenomenological parameter with the dimension of energy times volume]. The contact type form (16.33) proposed by Kondo, is justified by more detailed microscopic models, which also show that in general $J < 0$ (anti-ferromagnetic spin-spin interaction); thus the interaction (16.33) shows the tendency of screening of the localized magnetic moments by the conduction band electron spins. Kondo first showed that the scattering by a spin-dependent potential of the type

(16.33) could explain the resistivity minimum which is observed in ordinary metals containing magnetic impurities. An elementary justification of the phenomenological spin-dependent interaction (16.33) is considered below in Section 16.5.3.

First-Order Calculation of Scattering Amplitude from a Magnetic Impurity

The spin of an electron is 1/2 (in units $\hbar$); the spin operators, and spin-up and spin-down wavefunctions of the operator s_z are represented in the form

$$s_x = \frac{1}{2}\begin{pmatrix} 0 & 1 \\ 1 & 0 \end{pmatrix}, \quad s_y = \frac{1}{2}\begin{pmatrix} 0 & -i \\ i & 0 \end{pmatrix}, \quad s_z = \frac{1}{2}\begin{pmatrix} 1 & 0 \\ 0 & -1 \end{pmatrix};$$

$$\alpha = \begin{pmatrix} 1 \\ 0 \end{pmatrix} = |\uparrow\rangle, \quad \beta = \begin{pmatrix} 0 \\ 1 \end{pmatrix} = |\downarrow\rangle.$$

From the above equalities it follows

$$\langle\alpha|\mathbf{s}\cdot\mathbf{S}|\alpha\rangle = \frac{1}{2}S_z, \qquad \langle\beta|\mathbf{s}\cdot\mathbf{S}|\alpha\rangle = \frac{1}{2}(S_x + iS_y) = \frac{1}{2}S_+,$$

$$\langle\beta|\mathbf{s}\cdot\mathbf{S}|\beta\rangle = -\frac{1}{2}S_z, \qquad \langle\alpha|\mathbf{s}\cdot\mathbf{S}|\beta\rangle = \frac{1}{2}(S_x - iS_y) = \frac{1}{2}S_-, \qquad (16.34)$$

where the operators S_+ and S_- raise and lower the z-component of the impurity spin. With the help of Eq. (16.34), we can now calculate the matrix elements and the scattering amplitudes of interest for the electrical resistivity in the first and in the second Born approximation, and we begin with the former.

At first order of perturbation theory, the spin-dependent operator $\mathcal{H}_K$ acting on the conduction states gives an *energy-independent scattering amplitude* (and hence an energy-independent scattering cross-section): nothing basically new appears with respect to an ordinary impurity potential. Consider for instance the scattering amplitude for an electron of the Fermi sea from an initial state $\mathbf{k}_i\alpha$ to the final state $\mathbf{k}_f\alpha$ (as usual, plane waves are normalized to one on the volume $V = N\Omega$ of the crystal); we have

$$t^{(1)}_{\mathbf{k}_f\alpha \leftarrow \mathbf{k}_i\alpha} = \langle\mathbf{k}_f\alpha|\mathcal{H}_K|\mathbf{k}_i\alpha\rangle = \langle\mathbf{k}_f\alpha|-J\Omega\mathbf{S}\cdot\mathbf{s}\,\delta(\mathbf{r})|\mathbf{k}_i\alpha\rangle$$

$$= \frac{1}{N\Omega}\int e^{-i\mathbf{k}_f\cdot\mathbf{r}}\delta(\mathbf{r})e^{+i\mathbf{k}_i\cdot\mathbf{r}}d\mathbf{r}\,(-J\Omega)\langle\alpha|\mathbf{S}\cdot\mathbf{s}|\alpha\rangle = -\frac{J}{N}\frac{1}{2}S_z. \qquad (16.35)$$

Similar expressions can be obtained for the scattering amplitude between spin-orbitals with different spins. Scattering amplitudes and hence transition probabilities (obtained with appropriate averages over the initial magnetic impurity states) are anyway energy independent to first-order perturbation theory.

Second-Order Calculation of Scattering Amplitude from a Magnetic Impurity

To second order of Born approximation, striking new features appear: it has been shown by Kondo in 1964 that the scattering amplitude of a conduction electron due to the contact exchange Hamiltonian diverges logarithmically at the Fermi energy. As a consequence, the resistivity due to a magnetic impurity is expected to increase with decreasing temperature. The Kondo discovery opened the path for the theoretical explanation of the resistivity minimum observed for dilute magnetic impurities in non-magnetic metals, a phenomenon well-known experimentally since about three decades before.

According to second-order perturbation theory, the scattering amplitude for an itinerant electron from an initial spin-orbital $\mathbf{k}_i\alpha$ to the final state $\mathbf{k}_f\alpha$ can be expressed in the form

$$t^{(2)}_{\mathbf{k}_f\alpha \leftarrow \mathbf{k}_i\alpha} = \sum_{\mathbf{q}\sigma} \frac{\langle \mathbf{k}_f\alpha|\mathcal{H}_K|\mathbf{q}\sigma\rangle\langle \mathbf{q}\sigma|\mathcal{H}_K|\mathbf{k}_i\alpha\rangle}{E(\mathbf{k}_i) - E(\mathbf{q})}(1 - f_\mathbf{q})$$
$$+ (-1)\sum_{\mathbf{q}\sigma} \frac{\langle \mathbf{q}\sigma|\mathcal{H}_K|\mathbf{k}_i\alpha\rangle\langle \mathbf{k}_f\alpha|\mathcal{H}_K|\mathbf{q}\sigma\rangle}{E(\mathbf{q}) - E(\mathbf{k}_f)}f_\mathbf{q}, \qquad (16.36)$$

where $f_\mathbf{q} = 1$ for $E(\mathbf{q}) < E_F$ and $f_\mathbf{q} = 0$ for $E(\mathbf{q}) > E_F$; the factors $1 - f_\mathbf{q}$ and $f_\mathbf{q}$ take into account the Pauli exclusion principle in the intermediate state; the intermediate state must be empty in direct processes (first term in the right-hand side of Eq. (16.36)) and full in the exchange processes (second term in the right-hand side of Eq. (16.36)). The two kinds of processes are schematically indicated in Figure 16.6. Similar expressions can be obtained for the scattering amplitude between spin-orbitals involving spin-flips. In elastic scattering $E(\mathbf{k}_i) = E(\mathbf{k}_f)$, and Eq. (16.36) becomes

$$t^{(2)}_{\mathbf{k}_f\alpha \leftarrow \mathbf{k}_i\alpha}$$
$$= \sum_{\mathbf{q}\sigma} \frac{\langle \mathbf{k}_f\alpha|\mathcal{H}_K|\mathbf{q}\sigma\rangle\langle \mathbf{q}\sigma|\mathcal{H}_K|\mathbf{k}_i\alpha\rangle(1 - f_\mathbf{q}) + \langle \mathbf{q}\sigma|\mathcal{H}_K|\mathbf{k}_i\alpha\rangle\langle \mathbf{k}_f\alpha|\mathcal{H}_K|\mathbf{q}\sigma\rangle f_\mathbf{q}}{E(\mathbf{k}_i) - E(\mathbf{q})}.$$
$$(16.37)$$

We now discuss the structure of the scattering elements in the case of an ordinary perturbation and in the case of a magnetic perturbation.

For an "ordinary" (spin-independent) perturbation, consider Eq. (16.37) with $\mathcal{H}_K$ replaced by $V_{\text{imp}}(\mathbf{r})$. In this case, the product of any two matrix elements at the numerator of Eq. (16.37) does not depend on their order, and the terms with $f_\mathbf{q}$ cancel out exactly

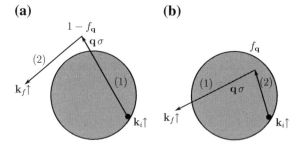

Figure 16.6 (a) Direct process in second-order Born approximation for scattering of an electron from the initial state $\mathbf{k}_i\uparrow$ to the final state $\mathbf{k}_f\uparrow$ via an empty intermediate state $\mathbf{q}\sigma$. (b) Exchange process contributing to second-order scattering amplitude: the filled state $\mathbf{q}\sigma$ is first scattered to the final state $\mathbf{k}_f\uparrow$ and is then occupied by the electron initially in the state $\mathbf{k}_i\uparrow$. In figure (a) and (b) the initial states of the Fermi gas can be represented by the same Slater determinant; it is then easily seen that the two final Slater determinants, at the end of the processes of (a) and (b), have a pair of electronic functions exchanged; hence the factor (-1) in Eq. (16.36).

(in other words, for an "ordinary" scattering potential, the Pauli principle in the inter-mediate states appears as non-operative, and can thus be neglected altogether); what is left in Eq. (16.37) represents in general a (small) correction to the first-order amplitude.

For magnetic impurities, consider now Eq. (16.37) as it stands with the spin-dependent operator $\mathcal{H}_K$: in this situation the order of the matrix elements in the numerator is vital and this gives an essentially new effect. Using Eq. (16.34), we can perform explicitly the sum over spin variables in Eq. (16.37) and obtain

$$t^{(2)}_{\mathbf{k}_f \alpha \leftarrow \mathbf{k}_i \alpha} = \left(\frac{J}{N}\right)^2 \frac{1}{4} \sum_{\mathbf{q}} \frac{(S_z^2 + S_- S_+)(1 - f_{\mathbf{q}}) + (S_z^2 + S_+ S_-) f_{\mathbf{q}}}{E(\mathbf{k}_i) - E(\mathbf{q})}.$$

From standard commutation rules, we have

$$S_+ S_- = S_- S_+ + 2 S_z$$

and thus

$$t^{(2)}_{\mathbf{k}_f \alpha \leftarrow \mathbf{k}_i \alpha} = \left(\frac{J}{N}\right)^2 \frac{1}{4} \sum_{\mathbf{q}} \frac{1}{E(\mathbf{k}_i) - E(\mathbf{q})} (S_z^2 + S_- S_+ + 2 S_z f_{\mathbf{q}}). \tag{16.38}$$

In Eq. (16.38) the key role is played by the Fermi function $f_{\mathbf{q}}$, whose presence can be traced back to the non-commutativity of the raising and lowering operators.

In order to estimate the sum over $\mathbf{q}$, suppose that the conduction band extends in the energy interval $(E_F - \Delta_c, E_F + \Delta_c)$, and here the density-of-states (per one spin direction) is constant, say $D_0(E_F)$; Eq. (16.38) becomes

$$t^{(2)}_{\mathbf{k}_f \alpha \leftarrow \mathbf{k}_i \alpha} = \left(\frac{J}{N}\right)^2 \frac{1}{4} (S_z^2 + S_- S_+) D_0(E_F) \int_{E_F - \Delta_c}^{E_F + \Delta_c} \frac{1}{E(\mathbf{k}_i) - E} dE$$
$$+ \left(\frac{J}{N}\right)^2 \frac{1}{2} S_z D_0(E_F) \int_{E_F - \Delta_c}^{E_F} \frac{1}{E(\mathbf{k}_i) - E} dE. \tag{16.39}$$

The integrations in Eq. (16.39) (to be understood in principal part) are straightforward, and it can be seen that for $E(\mathbf{k}_i) \approx E_F$ the first term in the right-hand side of Eq. (16.39) becomes negligibly small, while the second term in the right-hand side of Eq. (16.39) gives

$$t^{(2)}_{\mathbf{k}_f \alpha \leftarrow \mathbf{k}_i \alpha} = \left(\frac{J}{N}\right)^2 \frac{1}{2} S_z D_0(E_F) \ln \frac{\Delta_c}{|E(\mathbf{k}_i) - E_F|}; \tag{16.40}$$

thus $t^{(2)}$ is strongly energy dependent, and actually diverges when $E(\mathbf{k}_i) = E_F$.

At a given temperature T, we must average $t^{(2)}$ for $E(\mathbf{k}_i)$ in the thermal interval $k_B T$ around E_F; putting $E(\mathbf{k}_i) - E_F \approx k_B T$ in Eq. (16.40), and using Eq. (16.35), we obtain for the scattering amplitude up to second order

$$t = t^{(1)} + t^{(2)} = -\frac{J}{N} \frac{1}{2} S_z \left[1 - J n_0 \ln \frac{\Delta_c}{k_B T}\right], \tag{16.41}$$

where $n_0 = D_0(E_F)/N$ denotes the density-of-states at the Fermi level for one spin direction and per unit cell. The resistivity is related (after performing appropriate averages on initial spin directions and integrals on scattering angles) to the modulus square of the above expression; the temperature dependence is all embodied in the expression $[1 - Jn_0 \ln (\Delta_c/k_B T)]^2$, and the resistivity takes thus the form

$$\rho(T) = C\left[1 - 2Jn_0 \ln \frac{\Delta_c}{k_B T}\right] \qquad (J < 0), \qquad (16.42)$$

where C is approximately constant, and it is assumed $|J|n_0 \approx |J|/W_F \ll 1$ (W_F is the energy width of the part of the conduction band occupied by electrons).

From Eq. (16.42) we see that for $J < 0$, the contribution to the resistivity of the magnetic impurities increases logarithmically as the temperature approaches absolute zero. Since the contribution to the resistivity due to phonons decreases with decreasing temperature, we expect that the total resistivity goes through a minimum. Of course a more complete theory of the Kondo effect must take into account several other features, not embodied in the highly simplified model here considered; in particular, the increase in resistance on the low temperature side must eventually saturate, rather than diverge logarithmically. The strong energy dependence of the scattering relaxation time is also at the origin of the "giant" thermoelectric effect observed in some materials containing magnetic impurities.

16.5.3 Microscopic Origin of the Kondo Interaction: A Molecular Model

In the previous section we have seen that a phenomenological anti-ferromagnetic spin-dependent Hamiltonian of the type $\mathcal{H}_K = -J\Omega \mathbf{S} \cdot \mathbf{s}\, \delta(\mathbf{r})$, with $J < 0$, can explain the resistance minimum phenomenon of dilute magnetic alloys. Here we consider the simplest justification of the contact term $\mathcal{H}_K$ coupling the spin of the magnetic impurity and the spins of the conduction electrons, while for a somewhat more realistic approach we refer to Section 16.6.

To infer the interaction of a localized magnetic moment with the Fermi sea at the minimal level of details, we make here the drastic assumption to mimic the Fermi sea with just *one extended orbital* ψ_k (of energy E_k), which can be doubly occupied by electrons. The magnetic impurity is represented with a *single localized orbital* ψ_d (of energy E_d), where double occupancy is unfavored because of Coulomb repulsion effect, occurring when two electrons (of opposite spin) are forced to share the same orbital, strongly localized in space. [The localized orbital is typically a d-like or f-like atomic orbital, but for simplicity we ignore orbital degeneracy.] We consider thus the *two-orbital molecule* described by the Anderson-type Hamiltonian

$$\mathcal{H}_m = \sum_\sigma E_k n_{k\sigma} + E_d(n_{d\uparrow} + n_{d\downarrow}) + V_{sd} \sum_\sigma \left[c_{k\sigma}^\dagger c_{d\sigma} + c_{d\sigma}^\dagger c_{k\sigma}\right] + U n_{d\uparrow} n_{d\downarrow},$$

$$(16.43)$$

where the hybridization parameter V_{sd} is taken as real for simplicity. The physical meaning of the various terms has been already discussed for Eq. (16.25). The "molecular model" under attention is in fact the simplest version of the Anderson phenomeno-logical Hamiltonian (16.25), in which the conduction electrons of the normal metal are mimicked by just one itinerant orbital ψ_k at the Fermi energy $E_k \approx E_F$. In the following it is assumed that the energy of the magnetic level is well deep with respect to the Fermi energy, so that $|V_{sd}| \ll E_k - E_d \equiv \Delta$.

We examine the energy levels of two electrons accommodated in the described molecule first in the absence of correlation, $U = 0$ and then in the extreme correlated limit $U = \infty$. When $U = 0$, i.e. in the independent electron approximation, the Hamiltonian (16.43) trivially describes two orbitals of energy E_k and E_d, interacting via the off-diagonal matrix element V_{sd}. We are thus left with the determinantal equation

$$\begin{vmatrix} E_d - E & V_{sd} \\ V_{sd} & E_k - E \end{vmatrix} = 0,$$

which produces a bonding molecular state (below the energy E_d) and an antibonding molecular state (above the energy E_k). The eigenvalues are

$$E = \frac{1}{2}\left[E_d + E_k \mp \sqrt{(E_k - E_d)^2 + 4V_{sd}^2} \right].$$

With the assumption $|V_{sd}| \ll \Delta \equiv E_k - E_d$, well applicable in the present context, the bonding and antibonding eigenvalues take the expressions

$$E_b = E_d - \frac{V_{sd}^2}{\Delta}; \quad E_a = E_k + \frac{V_{sd}^2}{\Delta}.$$

The energy shift produced by the interorbital interaction V_{sd} is referred to as the *hybridization energy*. Typical values of the model parameters could be $\Delta \approx 1$ eV, hybridization parameter $|V_{sd}| \approx 10$–100 meV, and hybridization energy $V_{sd}^2/\Delta \approx 1$ meV.

Consider the case that two electrons occupy the molecule, when the correlation U is negligible. The two electrons of the molecule can be put with antiparallel spins in the bonding orbital (singlet state $S = 0$ with energy $2E_b$), or in the antibonding orbital (singlet state $S = 0$ with energy $2E_a$), or one in the bonding orbital and the other in the antibonding orbital ($S = 0, 1$, with energy $E_b + E_a = E_d + E_k$). The energy levels of the two-electron system when $U = 0$ *consists of three singlet states and one triplet*, and are reported in Figure 16.7a. It can be noticed that the ground state of the molecule is a singlet and the lowest excitation energy equals $\Delta + 2V_{sd}^2/\Delta \approx \Delta$; this excitation corresponds to the transfer of an electron from the d-like bonding state to the k-like antibonding state, and can be pictured as a "charge fluctuation."

Consider now two electrons on the molecule when the correlation effects are impor-tant, and actually U is so large to prevent double occupancy of the impurity orbital. For finite values of U, the number of possible two-electron molecular states is 6; in the limit $U \to +\infty$ double occupancy of the d state is avoided, and we can confine our considerations to the five states, reported for convenience in Figure 16.8.

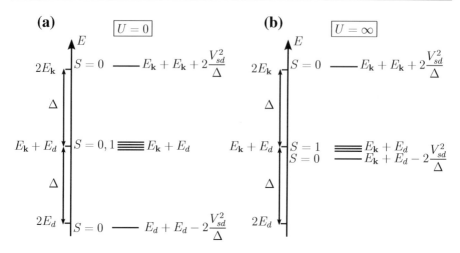

Figure 16.7 Energy levels of the two-electron two-orbital molecule, with one extended orbital of energy E_k and one localized orbital of energy E_d, in the case the Coulomb repulsive energy U between electrons of opposite spin on the localized orbital is zero or infinity.

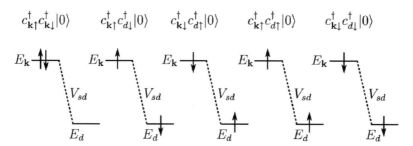

Figure 16.8 Possible molecular states for two electrons in a two-orbital molecule (double occupancy of the localized orbital ψ_d is not possible in the $U = \infty$ limit considered in the figure). The extended orbital ψ_k (of energy E_k) hybridizes with the localized orbital ψ_d (of energy E_d) via the matrix element V_{sd}, leading to the formation of the singlet molecular ground state.

From the five states listed in Figure 16.8 we can form *two singlet states and one triplet*. In the triplet state the U interaction plays no role, and the energy of the triplet is then $E_d + E_k$. The basis states for the singlets ($S = 0$) are

$$|\Psi_1^{(S=0)}\rangle = \frac{1}{\sqrt{2}} \left[c_{d\uparrow}^\dagger c_{k\downarrow}^\dagger - c_{d\downarrow}^\dagger c_{k\uparrow}^\dagger \right] |0\rangle \quad \text{and} \quad |\Psi_2^{(S=0)}\rangle = c_{k\uparrow}^\dagger c_{k\downarrow}^\dagger |0\rangle. \quad (16.44)$$

The representation of the Hamiltonian H_m on these states gives

$$H_m = \begin{bmatrix} E_d + E_k & \sqrt{2}\,V_{sd} \\ \sqrt{2}\,V_{sd} & 2E_k \end{bmatrix} \implies E = E_k + \frac{1}{2} \left[E_d + E_k \mp \sqrt{(E_k - E_d)^2 + 8V_{sd}^2} \right].$$

With the adopted assumption $|V_{sd}| \ll \Delta$, the expression of the ground state and upper state energies are

$$E_1 = E_d + E_k - \frac{2V_{sd}^2}{\Delta}; \quad E_2 = E_k + E_k + \frac{2V_{sd}^2}{\Delta}.$$

The energy levels of the correlated ($U = \infty$) molecular model, constructed by two orbitals and two electrons, are indicated in Figure 16.7b. It can be noticed that the ground state of the molecule is a singlet; *the singlet state is lower in energy than the triplet state*, as a result of the coupling between the singlet states of type (16.44) produced by the hybridization matrix element V_{sd}. The lowest excitation energy is rather weak $\approx 2V_{sd}^2/\Delta$ and corresponds to a spin-flip of one of the two electrons bound in the ground singlet state and can be pictured as a "spin fluctuation." A spin fluctuation requires to break a binding energy of the order of the hybridization energy, typically in the meV range. The molecular model, whatever rough it may appear, contains in germs what is an essential aspect of the Kondo phenomenology, namely: a localized magnetic impurity hybridizes with the surrounding extended orbitals and forms a weakly bound singlet ground state, so apparently losing its magnetic moment (for temperatures $k_B T$ smaller than $\approx 2V_{sd}^2/\Delta$). The quenching of the magnetic moment at sufficiently low temperatures is discussed in further detail in Section 16.6. The mixing of the singlet states of type (16.44) is also at the origin of the presence of states in the vicinity of the Fermi level when an impurity electron is added or removed from the ground singlet state.

Another message provided by the correlated molecular model is an intuitive insight on the origin of the spin-spin coupling Hamiltonian, assumed heuristically in Eq. (16.33). In fact, it is possible to express the energy difference between the triplet excited state and the singlet ground state (shown in Figure 16.7b) in such a way to put in evidence the effective coupling between the spins s_1 and s_2 of the two electrons. As far as we limit our attention to the ground singlet state and the triplet state, we can observe that the many-body Hamiltonian $\mathcal{H}_m$ (given by Eq. (16.43)) is equivalent to the much more manageable spin Hamiltonian

$$\mathcal{H} = -2J\mathbf{s}_1 \cdot \mathbf{s}_2 \quad \text{with} \quad J = -V_{sd}^2/\Delta. \tag{16.45}$$

In fact, from

$$(\mathbf{s}_1 + \mathbf{s}_2)^2 = \mathbf{s}_1^2 + \mathbf{s}_2^2 + 2\mathbf{s}_1 \cdot \mathbf{s}_2$$

we have

$$2\mathbf{s}_1 \cdot \mathbf{s}_2 = (\mathbf{s}_1 + \mathbf{s}_2)^2 - \mathbf{s}_1^2 - \mathbf{s}_2^2.$$

We notice that $\mathbf{s}_1^2 = \mathbf{s}_2^2 = s(s+1) = 3/4$; furthermore, $(\mathbf{s}_1 + \mathbf{s}_2)^2$ equals zero for singlet states and equals two for triplet states; then we have

$$\mathbf{s}_1 \cdot \mathbf{s}_2 = \begin{cases} -\dfrac{3}{4} & \text{for singlet states,} \\[2mm] +\dfrac{1}{4} & \text{for triplet states.} \end{cases} \tag{16.46}$$

Thus the many-body Hamiltonian (16.43) is "equivalent" (at least in an appropriate subspace of wavefunctions) to the Heisenberg-like spin-spin Hamiltonian (16.45) with antiferromagnetic interaction. Such equivalence, justified here in the extremely simplified one-impurity Anderson model of Eq. (16.43), is corroborated by the general formal relation between the Anderson and the Kondo Hamiltonians [see J. R. Schrieffer and P. A. Wolff, Phys. Rev. *149*, 491 (1966)].

16.6 Magnetic Impurity in Normal Metals at Very Low Temperatures

Singlet Ground State of a Magnetic Impurity Embedded in an Ordinary Electron Gas

In the previous sections we have considered some properties of the "electron gas plus magnetic impurity system" on the basis of a rather intuitive mean field model: the preserved magnetic impurity is described as a localized spin, weakly coupled through the contact interaction with the conduction electron spin density at the impurity. In this section we study the mechanism by which the *impurity forms a weakly bound singlet ground state with the surrounding extended orbitals*; as a consequence, at sufficiently low temperatures, the "electron gas plus magnetic impurity system" appears to collapse into a *singlet bound state*, constituted by the localized spin dressed and quenched by a cloud of conduction spin polarization. This quenching effect already appears in germs in the very idealized molecular model of Section 16.5.3. We consider now a more realistic (although still simplified) microscopic model that goes beyond the mean field approach previously elaborated. The localized magnetic level is typically a d-like or f-like atomic orbital (but for simplicity orbital degeneracy is ignored). Thus the magnetic impurity is schematized with a single orbital denoted ψ_d, of energy ε_d well below the Fermi energy, so that it can accept no more than one electron.

The model under attention starts, once again, from the by now familiar Anderson Hamiltonian of the type (16.25), here rewritten for convenience in slightly different notations

$$\mathcal{H} = \sum_{\mathbf{k}\sigma} \varepsilon_{\mathbf{k}} c_{\mathbf{k}\sigma}^{\dagger} c_{\mathbf{k}\sigma} + \varepsilon_d \left(c_{d\uparrow}^{\dagger} c_{d\uparrow} + c_{d\downarrow}^{\dagger} c_{d\downarrow} \right)$$

$$+ \frac{V_{sd}}{\sqrt{N}} \sum_{\mathbf{k}\sigma} \left[c_{\mathbf{k}\sigma}^{\dagger} c_{d\sigma} + c_{d\sigma}^{\dagger} c_{\mathbf{k}\sigma} \right] + U n_{d\uparrow} n_{d\downarrow}. \tag{16.47}$$

In Eq. (16.47), $c_{\mathbf{k}\sigma}^{\dagger}$ and $c_{\mathbf{k}\sigma}$ are the creation and annihilation operators corresponding to the conduction band wavefunctions of the metal, of wavevector $\mathbf{k}$, spin σ, and energy $\varepsilon_{\mathbf{k}}$ (for convenience we use the Fermi energy as the reference zero of energy). The operators $c_{d\sigma}^{\dagger}$ and $c_{d\sigma}$ are the creation and annihilation operators corresponding to the single impurity orbital, of energy ε_d (measured with respect to the Fermi energy). The Coulomb repulsion between the d electrons is assumed to be so high ($U \rightarrow \infty$) to allow occupancy not higher than one of the localized d orbital. The hybridization

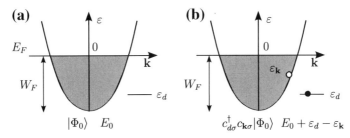

Figure 16.9 (a) Schematic representation of the model Hamiltonian describing a magnetic impurity embedded in a sea of conduction electrons; the magnetic impurity is schematized with a single impurity orbital, which can accept no more than one electron. The ground-state wavefunction and energy of the filled Fermi sea are denoted by $|\Phi_0\rangle$ and E_0, respectively; W_F is the energy width of the conduction band occupied by electrons, and the Fermi level E_F is taken to be zero. (b) Schematic representation of the basis states interacting with the filled Fermi sea $|\Phi_0\rangle$, and leading to the formation of the singlet state in the Kondo problem.

matrix elements between plane waves and the d orbital are considered constant and real ($V_{\mathbf{k}d} = V_{sd}$). The model, with its essential phenomenological parameters, is indicated schematically in Figure 16.9.

We consider the impurity in the magnetic regime (described in Section 16.4), and we suppose specifically that the parameter Γ, given by Eq. (16.28b), is such that

$$\Gamma = \pi V_{sd}^2 n_0(E_F) \ll |\varepsilon_d|; \tag{16.48}$$

the above condition specifies that the half-width Γ of the d-orbital is negligible with respect to $|\varepsilon_d|$, which represents the difference between the Fermi energy and the d-orbital energy. In this case, according to the mean field Hartree-Fock treatment of Section 16.4, a local magnetic moment is preserved with $\langle n_{d\uparrow}\rangle \approx 1$ and $\langle n_{d\downarrow}\rangle \approx 0$ (or vice versa). However, the effective field treatment is only approximate; it provides a spin-dependent energy shift of the impurity levels, but does not alter the essential features of the Fermi sea, where the impurity levels are embedded and broadened (see Eqs. (16.30)). In the following we focus on the mixing of the localized impurity levels and the delocalized Fermi sea electrons; it is seen that the impurity magnetic moment (at low temperatures) is quenched by the spin polarization of the conduction electrons, and the magnetic impurity and the electron gas are bound together into a singlet ground state.

The properties of the one-impurity Anderson Hamiltonian (16.47), and related versions, have been investigated in the literature with a number of approaches, including variational methods, renormalization group calculations, Bethe ansatz, various many-body techniques (for a very comprehensive description see for instance the monograph of Hewson cited in the bibliography). Here, to keep technicalities at the minimum, we adopt the simplest variational representation for the ground state of the Anderson model (16.47).

To describe the singlet ground state $|\Psi\rangle$ of the whole system, formed by the magnetic impurity in interaction with the Fermi sea, consider the subspace spanned by the singlet state $|\Phi_0\rangle$ (which is the filled Fermi sea of the conduction electrons with total energy E_0) and the singlet states $c_{d\sigma}^\dagger c_{\mathbf{k}\sigma}|\Phi_0\rangle$ (of energy $E_0 + \varepsilon_d - \varepsilon_k$), in which an electron from an occupied conduction state is transferred to the localized impurity state. In this subspace, $|\Psi\rangle$ can be expressed in the form

$$|\Psi\rangle = a_0|\Phi_0\rangle + \sum_{\mathbf{k}} a(\mathbf{k}) \left[c_{d\uparrow}^{\dagger} c_{\mathbf{k}\uparrow} + c_{d\downarrow}^{\dagger} c_{\mathbf{k}\downarrow} \right] |\Phi_0\rangle. \tag{16.49}$$

Notice that basis states with two or more holes below the Fermi energy and one or more electrons above it are not included in the expansion (16.49); this simplification is justified in the case of sufficiently small hybridization parameter V_{sd}.

The Anderson Hamiltonian (16.47), on the basis of the states $|\Phi_0\rangle$ and $c_{d\sigma}^{\dagger} c_{\mathbf{k}\sigma} |\Phi_0\rangle$ (with $k < k_F$, and $\sigma = \alpha$ or β) takes the form schematically indicated in Figure 16.10. It is immediate to obtain the Green's function on the state $|\Phi_0\rangle$ with the renormalization method (see Section 5.8.3); it is given by

$$G_{00}(E) = \frac{1}{E - E_0 - 2V_{sd}^2 \dfrac{1}{N} \sum_{\mathbf{k}}^{(occ)} \dfrac{1}{E - [E_0 + \varepsilon_d - \varepsilon_{\mathbf{k}}]}}, \tag{16.50a}$$

where the factor 2 takes into account spin degeneracy and the sum over $\mathbf{k}$ extends to the occupied states of the Fermi sea. The poles of the Green's function occur at the energies E such that

$$E - E_0 - 2V_{sd}^2 \frac{1}{N} \sum_{\mathbf{k}}^{(occ)} \frac{1}{E - E_0 - \varepsilon_d + \varepsilon_{\mathbf{k}}} = 0, \tag{16.50b}$$

and the equation can be conveniently solved with graphical methods.

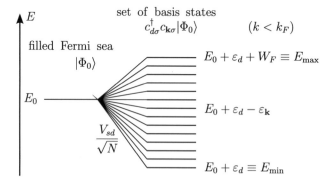

Figure 16.10 Representation of the Anderson Hamiltonian on the basis of states formed by $|\Phi_0\rangle$ and $c_{d\sigma}^{\dagger} c_{\mathbf{k}\sigma} |\Phi_0\rangle$ with $k < k_F$, connected by the hybridization interaction $V_{sd}/\sqrt{N}$. The Fermi energy of the normal metal is set to zero and $\varepsilon_d < 0$.

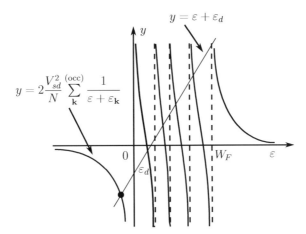

Figure 16.11 Graphical solution of the compatibility equation for the Kondo ground state (notice that $\varepsilon_d < 0$ and $0 \leq -\varepsilon_{\mathbf{k}} \leq W_F$).

For this purpose, it is convenient to measure energies from the minimum value $E_{\min} = E_0 + \varepsilon_d$ of the non-interacting system, and set

$$\varepsilon = E - E_0 - \varepsilon_d.$$

Then Eq. (16.50b) reads

$$\varepsilon + \varepsilon_d = 2V_{sd}^2 \frac{1}{N} \sum_{\mathbf{k}}^{(\text{occ})} \frac{1}{\varepsilon + \varepsilon_{\mathbf{k}}}. \tag{16.51}$$

The graphical solution of Eq. (16.51) is shown in Figure 16.11. The solution of Eq. (16.51) with negative ε individuates a collective state with peculiar properties (akin situations have been encountered in a number of problems, different from the physical point of view, but rather similar in the underlying mathematical aspects).

The numerical solution for the ground state, i.e. the collective state, of Eq. (16.51) can be elaborated as follows. We notice that for negative ε, the quantities $\varepsilon + \varepsilon_{\mathbf{k}}$ appearing in Eq. (16.51) are all different from zero since $-W_F \leq \varepsilon_{\mathbf{k}} \leq 0$. Thus we have

$$\sum_{\mathbf{k}}^{(\text{occ})} \frac{1}{\varepsilon + \varepsilon_{\mathbf{k}}} = D_0(E_F) \int_{-W_F}^{0} \frac{1}{\varepsilon + x}\, dx = D_0(E_F) \ln \left| \frac{\varepsilon}{\varepsilon - W_F} \right|, \tag{16.52}$$

where as usual we have changed the sum over $\mathbf{k}$ into a continuous integral over the energy; to avoid inessential details, we have supposed that the density-of-states of the conduction band (for one spin direction) is constant in the energy interval of interest around the Fermi level and equal to $D_0(E_F)$.

Inserting Eq. (16.52) into Eq. (16.51), and assuming $|\varepsilon| \ll W_F$ and $|\varepsilon| \ll |\varepsilon_d|$, we obtain

$$\varepsilon_d = 2V_{sd}^2 n_0 \ln \frac{|\varepsilon|}{W_F}, \tag{16.53}$$

where $n_0 = D_0(E_F)/N$ denotes the density-of-states at the Fermi level for one spin direction and per unit cell. The energy gain of the singlet ground state due to the hybridization is thus

$$
|\varepsilon| = W_F \exp\left[-\frac{|\varepsilon_d|}{2V_{sd}^2 n_0}\right].
\tag{16.54}
$$

The Kondo energy (16.54) describes the binding energy of the spin singlet, formed with the localized unpaired electron in interaction with the itinerant electrons.

It is convenient to define the Kondo temperature, so that

$$k_B T_K = |\varepsilon|;$$

for temperatures T much smaller than T_K, thermal excitations are negligible, and the system is essentially described by the non-magnetic singlet ground state; for temperatures comparable or higher than T_K the quenching of the impurity magnetic moment becomes eventually negligible.

We are now in the position to understand the finite, although large, paramagnetic susceptibility of the magnetic impurity for $T < T_K$. In the presence of a magnetic field H we suppose that the energy ε_d of the d state is split into $\varepsilon_d \pm \mu_B H$. We can proceed with a treatment similar to that used above, and we obtain that Eq. (16.53) is now replaced by

$$
\varepsilon_d = V_{sd}^2 n_0 \left[\ln\frac{|\varepsilon(H) + \mu_B H|}{W_F} + \ln\frac{|\varepsilon(H) - \mu_B H|}{W_F}\right],
\tag{16.55}
$$

where $\varepsilon(H)$ indicates the binding energy of the ground state of the system in the presence of the magnetic field. With $\mu_B H$ smaller than $|\varepsilon(H)|$, Eq. (16.55) reads

$$
\varepsilon_d = V_{sd}^2 n_0 \ln\frac{\varepsilon^2(H) - \mu_B^2 H^2}{W_F^2}.
\tag{16.56}
$$

Comparison of Eqs. (16.53) and (16.56) gives

$$\varepsilon^2(H) - \mu_B^2 H^2 = |\varepsilon|^2.$$

We have thus

$$|\varepsilon(H)| = \sqrt{|\varepsilon|^2 + \mu_B^2 H^2} = |\varepsilon| + \frac{1}{2}\frac{\mu_B^2 H^2}{|\varepsilon|}.$$

Since $\varepsilon(H)$ is negative, and $|\varepsilon| = k_B T_K$, the above equation can be written as

$$\varepsilon(H) = \varepsilon(H = 0) - \frac{1}{2}\frac{\mu_B^2 H^2}{k_B T_K}.$$

If we have a number N_I of isolated (i.e. non-interacting) impurities in the volume V, differentiating the above expression twice with respect to H yields the contribution to the magnetic susceptibility:

$$\chi_I = -\frac{N_I}{V}\frac{\partial^2\varepsilon(H)}{\partial H^2} = +\frac{N_I}{V}\frac{\mu_B^2}{k_B T_K}. \tag{16.57a}$$

It is interesting to compare the above expression with the Pauli paramagnetism of the Fermi gas $\chi_P = (1/V)D(E_F)\mu_B^2$ of Eq. (15.31); for a gas of N free electrons, we have

$$\chi_P = \frac{1}{V}\frac{3}{2}\frac{N}{E_F}\mu_B^2 = \frac{3}{2}\frac{N}{V}\frac{\mu_B^2}{k_B T_F}; \tag{16.57b}$$

thus the small Kondo temperature $T_K < T_F$ appearing in Eq. (16.57a) controls the contribution to the susceptibility from the magnetic impurities.

Before closing, a few more comments are worthwhile on the binding energy given by Eq. (16.54), after rewriting it in a slightly different form. We express $|\varepsilon| = k_B T_K$ and $W_F = k_B T_F$, and put $J_{\text{eff}} = 2J = -2V_{sd}^2/|\varepsilon_d|$ in evident analogy with the molecular model of Section 16.5.3. From Eq. (16.54), we obtain that the Kondo temperature T_K and the Fermi temperature T_F are linked by the equation

$$\boxed{T_K = T_F \exp[-1/|J_{\text{eff}}|n_0]}. \tag{16.58}$$

This relation is analogous in some ways to the expression of the Cooper pair binding energy encountered in superconductivity (see Section 18.2). It can be noticed that the exponential in the second member of Eq. (16.58) is *not* an analytic function of $|J_{\text{eff}}|$ for $|J_{\text{eff}}| \rightarrow 0$, and thus cannot be expanded in powers of $|J_{\text{eff}}|$; this explains the difficulties encountered attacking the Kondo problem with ordinary perturbation theory. Expression (16.58) was earlier inferred in the pioneer papers of Abrikosov and Suhl who applied the many body formalism to the treatment of the Kondo Hamiltonian $-J_{\text{eff}}\Omega\,\mathbf{S}\cdot\mathbf{s}\,\delta(\mathbf{r})$; for this the Kondo temperature T_K is also referred to in the literature as the Abrikosov-Suhl temperature.

Further Reading

Abragam, A., & Bleany, B. (1970). *Electron paramagnetic resonance of transition ions*. Oxford: Clarendon.

Chikazumi, S. (1997). *Physics of ferromagnetism*. Oxford: Clarendon Press.

Fulde, P. (1988). Introduction to the physics of heavy fermions. *Journal of Physics F: Metal Physics, 18*, 601.

Heeger, A. J. (1969). Localized moments and nonmoments in metals: The Kondo effect. *Solid State Physics, 23*, 283.

Hewson, A. C. (1993). *The Kondo problem to heavy fermions*. Cambridge: Cambridge University Press.

Kondo, J. (1969). Theory of dilute magnetic alloys. *Solid State Physics, 23*, 183.

Kubo, R., & Nagamiya, T. (Eds.) (1969). *Solid State Physics*. New York: McGraw-Hill.

Mattis, D. C. (2006). *The theory of magnetism made simple: An introduction to physical concepts and to some useful mathematical methods*. New Jersey: World Scientific.

Morrish, A. H. (1965). *The physical principles of magnetism*. New York: Wiley.

Nolting, W., & Ramakanth, A. (2009). *Quantum theory of magnetism*. Berlin: Springer.

Rado, G. T., & Suhl, H. (Eds.) (1973). *Magnetism. Magnetic properties of metallic alloys* (Vol. V). New York: Academic Press.

Rigamonti, A., & Carretta, P. (2009). *Structure of matter. An introductory course with problems and solutions*. Milano: Springer.

Van Vleck, J. H. (1952). *The theory of electric and magnetic susceptibilities*. Oxford: Oxford University Press.

White, R. M. (2007). *Quantum theory of magnetism*. Berlin: Springer.

17 Magnetic Ordering in Crystals

In the previous two chapters, among various topics on magnetism, we have discussed the occurrence of permanent magnetic moments in crystals and their paramagnetic susceptibility. In particular we have considered the situation of well-localized electronic wavefunctions with formation of localized magnetic moments (typified by the Curie paramagnetism), as well as the alternative situation of itinerant electronic wavefunctions (typified by the Pauli paramagnetism). It is well-known that many paramagnetic materials, below a critical temperature, present magnetic order even in the absence of applied magnetic fields. The most familiar order is the ferromagnetic one, with localized moments lined up in the same direction so that a spontaneous magnetization is apparent; but several other types of magnetic ordering (antiferromagnetic, ferrimagnetic, helical, etc.) are possible.

The occurrence of magnetic ordering in crystals implies some coupling mechanism between localized or delocalized magnetic moments. Among the microscopic models of coupling between the localized moments, we focus on the Heisenberg spin Hamiltonian, because of its formal simplicity and the clear insight on the electrostatic origin of the effective spin-spin interaction. The phase transition to magnetic ordering is first studied within mean field theories; this permits a justification of the Weiss molecular field assumptions, and a reasonably simple description of some aspects of the wide and rich phenomenology accompanying magnetic ordering and phase transitions. We also consider aspects beyond mean field theories. In particular, at low temperatures, we

Solid State Physics, Second Edition. http://dx.doi.org/10.1016/B978-0-12-385030-0.00017-7

study with the spin-wave theory the elementary excitations (magnons) of the Heisenberg ferromagnets; universal aspects at or near the critical temperature region are examined with the renormalization group theory, which establishes the common language for quantum critical phenomena.

In paramagnetic crystals characterized by itinerant electronic wavefunctions (band paramagnetism), the occurrence of magnetic ordering suggests the importance of correlation effects among the independent particle wavefunctions, as typified by the Stoner-Hubbard model of itinerant magnetism. We close this chapter with a discussion of spin-dependent transport in bulk and layered metallic ferromagnetic structures, with the extraordinary impact of giant magnetoresistance from both fundamental and technological points of view.

The vastness of the field of magnetism makes unavoidable drastic abridgments; interesting subjects, such as surface magnetism, molecular nanomagnets, magnetic tunneling, and others have been omitted. Although the topics we cover are so limited and simplified, they should give indicative guidelines in the subject of magnetic ordering in matter.

17.1   Ferromagnetism and the Weiss Molecular Field

Phenomenological Aspects of Ferromagnetism

We begin the study of cooperative effects in magnetic materials, by considering a system in which the microscopic magnetic moments have the same magnitude and tend to line up in the same direction. A ferromagnetic specimen, at sufficiently low temperatures, exhibits spontaneous magnetization. The spontaneous magnetization $M(T)$ depends on the temperature and vanishes above a *critical temperature T_c, called ferromagnetic Curie temperature*. The critical temperature T_c and *the saturation magnetization* $M(T = 0)$ for some substances are reported in Table 17.1 [For further data see for instance F. Keffer, Handbuch der Physik, Vol. 18 Part 2 (Springer, Berlin, 1966); for Fe and Ni see also T. Tanaka and K. Miyatani, J. Appl. Phys. *82*, 5658 (1997). For an extensive collection of experimental data on magnetism of bulk materials and nanostructures see for instance C. A. F. Vaz, J. A. C. Bland and G. Lauhoff "Magnetism in

Table 17.1 Ferromagnetic Curie temperature (in K) and saturation magnetization (in gauss) for some materials (above T_c, rare earth dysprosium makes a transition to helices).

Material	T_c	$M(T = 0)$
Fe	1043	1752
Co	1394	1446
Ni	630	510
Gd	293	1980
Dy	85	3000

ultrathin film structures" Rep. Prog. Phys. *71*, 056501 (2008) and the rich bibliography quoted therein]. Among the elements only a few are ferromagnetic; there is instead a relatively large number of ferromagnetic alloys and oxides.

For an estimate of the order of magnitude of the saturation magnetization, consider for example a cubic crystal with a localized magnetic moment $\mu_B = e\hbar/2mc$ in every unit cell of volume $\Omega = a^3$. Assuming parallel alignment of all the microscopic magnetic moments at zero temperature, the saturation magnetization becomes

$$M(0) = \frac{\mu_B}{\Omega} = \frac{a_B^3}{a^3}\frac{\mu_B}{a_B^3}, \tag{17.1a}$$

where μ_B is the Bohr magneton and a_B is the Bohr radius. For the evaluation of quantities of frequent interest in this chapter, the following relations may be useful

$$\mu_B = 0.05788\frac{\text{meV}}{\text{Tesla}}; \quad \frac{\mu_B^2}{a_B^3} = 0.363 \text{ meV} \Longrightarrow 4.21 \text{ K};$$

$$\frac{\mu_B}{a_B^3} = 6.27 \text{ Tesla} = 62700 \text{ gauss}. \tag{17.1b}$$

For typical values of the lattice parameter $a \approx 2 \text{ Å} \approx 4 - 5\,a_B$, the unit cell volume becomes $a^3 \approx 100\,a_B^3$. From Eqs. (17.1), it can be seen that the expected order of magnitude of the saturation magnetization is 1000 gauss, in the range of the observed values of Table 17.1.

The spontaneous magnetization of a ferromagnetic crystal can be explained only by some interaction mechanism, which favors parallel alignment of the microscopic magnetic moments. A little reflection on the order of magnitude of the critical temperatures given in Table 17.1, shows that the *magnetic dipolar coupling cannot be the origin of such interaction*. In fact two magnetic dipoles $\boldsymbol{\mu}_1$ and $\boldsymbol{\mu}_2$ at (sufficiently large) distance **R** interact with an energy

$$E_{\text{dip}} = \frac{1}{R^3}\left[\boldsymbol{\mu}_1 \cdot \boldsymbol{\mu}_2 - 3\frac{(\boldsymbol{\mu}_1 \cdot \mathbf{R})(\boldsymbol{\mu}_2 \cdot \mathbf{R})}{R^2}\right]; \tag{17.2}$$

from this, the dipolar coupling energy between dipoles μ_B at the lattice sites of a cubic crystal of parameter $a \approx 4 - 5\,a_B$, can be estimated as

$$E_{\text{dip}} \approx \frac{\mu_B^2}{a^3} = \frac{a_B^3}{a^3}\frac{\mu_B^2}{a_B^3} \approx 10^{-5} \text{ eV} \approx 0.1 \text{ K}. \tag{17.3}$$

Thus random thermal fluctuations would destroy alignment of magnetic moments at very low temperatures, of the order of tenths of kelvin degrees, much smaller than the observed critical temperatures, which are of the order of hundred or thousand kelvin degrees. The mechanism of spin alignment must be produced by an effective interaction among spins, which is larger than the dipolar magnetic interaction by a factor $10^3 - 10^4$ or so. The origin of such a "strong" field is basically electrostatics in nature and involves orbits in real space, and not spins. However, the global antisymmetry of the electronic wavefunctions entails a stringent connection between orbital and spin states: this connection can be envisaged as an effective spin-spin coupling in several situations of interest (as discussed in Section 17.2).

The Weiss Molecular Field

Consider a crystal of volume V, formed by N equal magnetic units (atoms or ions), each of angular momentum J and magnetic moment $\mu = -g\mu_B \mathbf{J}$, localized at the sites of a periodic lattice. In the previous chapter, we have discussed the magnetization of the sample in an applied field, under the assumption that the microscopic moments are independent. For instance, from Eq. (16.17), the magnetization of a paramagnetic substance, with $J = 1/2$ and gyromagnetic factor $g = 2$, is given by the thermodynamic average

$$M = \frac{N}{V}\mu_B \frac{e^{\mu_B H/k_B T} - e^{-\mu_B H/k_B T}}{e^{\mu_B H/k_B T} + e^{-\mu_B H/k_B T}} \quad \Longrightarrow \quad M = \frac{N}{V}\mu_B \tanh \frac{\mu_B H}{k_B T}, \quad (17.4a)$$

where H is the uniform magnetic field applied to the sample. In a paramagnetic substance, composed by independent magnetic dipoles, M is proportional to H for small H, and no spontaneous magnetization can occur in the absence of applied fields. With the approximation $\tanh x \approx x$ for small x in Eq. (17.4a), one finds that the magnetic susceptibility obeys the *Curie law*

$$\chi = \frac{M}{H} = \frac{C}{T} \quad \text{with Curie constant} \quad C = \frac{N}{V}\frac{\mu_B^2}{k_B} \quad \left(J = \frac{1}{2}; \ g = 2\right). \quad (17.4b)$$

In the case $J \neq 1/2$, expressions similar to Eqs. (17.4) hold with the hyperbolic tangent function replaced by the appropriate Brillouin function (and other minor changes if $g \neq 2$). For simplicity in this section we only consider paramagnetic substances with $J = 1/2$ and $g = 2$, since the generic Brillouin function $B_J(x)$ and the hyperbolic tangent function $B_{1/2}(x)$ have similar asymptotic behaviors. In the original work of Weiss the classical Langevin function $L(x) = B_\infty(x)$ was used.

The first phenomenological mechanism leading to magnetism was proposed by Weiss in 1907. It is based on the assumption that the effective magnetic field acting on a given dipole is given by

$$H_{\text{eff}} = H + \lambda M, \tag{17.5}$$

where H is the external magnetic field, λ is an appropriate constant, M is the magnetization, and the molecular field λM provides the cooperative effect. Originally the Weiss constant λ was considered as a phenomenological constant; the interpretation of λ in terms of microscopic quantum models appears later with the works of Heisenberg.

If we make the Weiss assumption that the effective field acting on a given dipole is $H + \lambda M$, we obtain for a ferromagnetic substance (with $J = 1/2$ and $g = 2$) the basic equation

$$\boxed{M = \frac{N}{V}\mu_B \tanh \frac{\mu_B(H + \lambda M)}{k_B T}.} \tag{17.6}$$

It is evident that the case $\lambda = 0$ corresponds to ordinary paramagnetism, while $\lambda > 0$ describes a ferromagnetic cooperative effect. The rest of this section is devoted to bring out the wealth of information contained in the transcendental equation (17.6), which represents historically the first mean field approach to phase transitions.

In zero field and for $M \to 0$, Eq. (17.6) is satisfied at the "critical" temperature T_c such that

$$M = \frac{N}{V} \mu_B \frac{\mu_B \lambda M}{k_B T_c} \quad \Longrightarrow \quad T_c = \lambda \frac{N}{V} \frac{\mu_B^2}{k_B} = \lambda C. \tag{17.7a}$$

Besides the critical temperature, it is also convenient to introduce the dimensionless magnetization m (expressed in term of saturation magnetization), so defined

$$m = \frac{V}{N} \frac{M}{\mu_B} = \frac{M}{M(0)}. \tag{17.7b}$$

Using the quantities defined in Eqs. (17.7), we can cast Eq. (17.6) in the dimensionless form

$$\boxed{m = \tanh\left(\frac{\mu_B H}{k_B T} + \frac{T_c}{T} m\right).} \tag{17.8}$$

The "normalized" or "reduced" magnetization m varies in the interval $0 \le m \le 1$. The case $m = 1$ corresponds to saturation magnetization with all the microscopic spins lined in the same direction, while the case $m = 0$ corresponds to random orientations of spins. We wish to establish the dependence of m from the temperature and from the applied magnetic field.

We begin to look for spontaneous magnetization, putting $H = 0$ in Eq. (17.8); then

$$m = \tanh\left(\frac{T_c}{T} m\right). \tag{17.9}$$

The solution of this equation is obtained plotting separately the first member (a straight line) and the second member as functions of m, and looking for the intersections between the curves; the graphical solution is shown in Figure 17.1. With the help of Figure 17.1, it is immediate to note that for $T \ge T_c$ we have only the trivial solution $m = 0$; however, below the critical temperature $T < T_c$ we have also a non-trivial solution with non-zero $m(T)$. The spontaneous magnetization $m(T)$ as a function of temperature can be obtained graphically, and its typical behavior is indicated in Figure 17.2.

Besides the graphical solution, it is instructive to obtain analytically the behavior of the magnetization for temperatures very high and very low with respect to T_c, and around the critical region.

High temperature limit. Consider first the magnetization of the ferromagnetic material for $T > T_c$, in the presence of an external magnetic field H. In the limit of sufficiently small H, also m is small, and linearization of Eq. (17.8) gives

$$m = \frac{\mu_B H}{k_B T} + \frac{T_c}{T} m \quad \Longrightarrow \quad \frac{m}{H} = \frac{\mu_B}{k_B (T - T_c)}; \tag{17.10a}$$

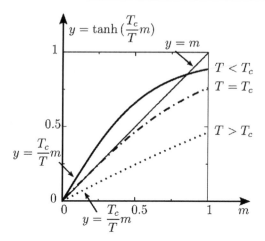

Figure 17.1 Graphical solution of the mean field equation $m = \tanh (mT_c/T)$ for the spontaneous magnetization of a ferromagnetic crystal.

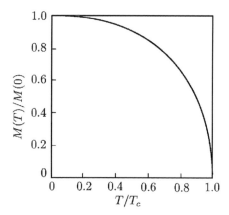

Figure 17.2 Spontaneous magnetization of a ferromagnetic crystal as function of the temperature in the mean field theory, with $J = 1/2$ in the Brillouin function.

this shows that the magnetic susceptibility follows the *Curie-Weiss law*

$$\chi = \frac{M}{H} = \frac{N}{V} \frac{\mu_B^2}{k_B(T - T_c)} = \frac{C}{T - T_c} \qquad (T > T_c). \qquad (17.10b)$$

Thus, in the Weiss molecular field theory, χ diverges as $(T - T_c)^{-1}$ as the temperature decreases toward the critical temperature. The tendency to parallel alignment of spins, enhances the magnetic susceptibility of the systems and leads to a phase transition below T_c.

Low temperature limit. For $T \ll T_c$ the argument of the hyperbolic tangent function in Eq. (17.9) becomes very large, and upon appropriate expansion, the spontaneous magnetization becomes for $T \to 0$

$$m = \tanh\left(\frac{T_c}{T}m\right) \quad \Longrightarrow \quad m = 1 - 2e^{-2(T_c/T)m} \approx 1 - 2\,e^{-2T_c/T}. \quad (17.11)$$

Thus, within the mean field theory, near zero temperature the spontaneous magnetization deviates from its saturation value by an exponential term; this prediction of mean field theory (as well as several others) must be appropriately corrected with treatments beyond the mean field approach.

Critical temperature region. Next, we examine the behavior near the critical temperature of the transcendental equation (17.8) in the case both H and m are small. It is convenient to elaborate Eq. (17.8) in such a way to handle functions of separate arguments H and m. For this purpose we write Eq. (17.8) in the form

$$m = \tanh(h + \tau m) \quad \text{with} \quad h = \frac{\mu_B H}{k_B T}, \quad \tau = \frac{T_c}{T};$$

then

$$m = \frac{\tanh h + \tanh(\tau m)}{1 + \tanh h \cdot \tanh(\tau m)} \quad \Longrightarrow \quad \frac{m - \tanh(\tau m)}{1 - m\tanh(\tau m)} = \tanh h.$$

The expansion of the first member of the above equation for small m gives

$$\frac{m - \tau m + \tau^3 m^3/3 + \cdots}{1 - \tau m^2 + \cdots} = (1-\tau)m + \left(\tau - \tau^2 + \frac{1}{3}\tau^3\right)m^3 \approx (1-\tau)m + \frac{1}{3}m^3$$

(the last passage takes into account that the coefficient of m^3 is finite and equals $1/3$ for $\tau \to 1$). In summary, near the critical temperature for sufficiently small values of h and m it holds

$$(1-\tau)m + \frac{1}{3}m^3 = h \quad \Longrightarrow \quad \boxed{\left(1 - \frac{T_c}{T}\right)m + \frac{1}{3}m^3 = \frac{\mu_B H}{k_B T}} \quad (17.12)$$

regardless if the temperature T, near T_c, is above or below it.

We consider now the behavior of the spontaneous magnetization for T near T_c. Setting $H = 0$ in Eq. (17.12), it is seen that for $T \geq T_c$ there is only the trivial solution $m = 0$. For T near T_c from below we have the non-trivial solution

$$m^2(T) = 3\frac{T_c - T}{T_c} \quad \Longrightarrow \quad m(T) \propto (T_c - T)^\beta \quad \text{with} \quad \beta = \frac{1}{2}, \quad T < T_c.$$
$$(17.13)$$

In the mean field theory, the spontaneous magnetization goes to zero when the temperature approaches T_c from below, according to a power law with critical exponent $\beta = 1/2$.

Another critical exponent can be obtained considering the behavior of the magnetization versus the applied field at the critical temperature. Setting $T = T_c$ in Eq. (17.12), it is seen that

$$m \propto H^{1/\delta} \quad \text{with} \quad \delta = 3. \tag{17.14}$$

Thus at the critical temperature the magnetization versus the applied field is described by a power law with critical exponent $\delta = 3$.

The critical exponent for the susceptibility can be found by differentiation of both sides of Eq. (17.12) with respect to H at constant temperature and in the zero-field limit:

$$\left(1 - \frac{T_c}{T}\right)\frac{\partial m}{\partial H} + m^2 \frac{\partial m}{\partial H} = \frac{\mu_B}{k_B T}.$$

For $T > T_c$ the magnetization at zero field vanishes, and the above equation gives

$$\frac{\partial m}{\partial H} = \frac{\mu_B}{k_B(T - T_c)} \sim (T - T_c)^{-\gamma} \quad \text{with} \quad \gamma = 1, \ T \to T_c^+.$$

The same exponent results for $T \to T_c^-$; in fact using the first of Eqs. (17.13) we have

$$\left(1 - \frac{T_c}{T}\right)\frac{\partial m}{\partial H} + 3\frac{T_c - T}{T}\frac{\partial m}{\partial H} = \frac{\mu_B}{k_B T} \quad \Longrightarrow \quad \frac{\partial m}{\partial H} = \frac{1}{2}\frac{\mu_B}{k_B(T_c - T)}.$$

The above results for the susceptibility critical exponent can be summarized as follows

$$\chi \sim |T - T_c|^{-\gamma} \quad \text{with} \quad \gamma = 1, \ T \to T_c. \tag{17.15}$$

The power law exponent for the magnetic susceptibility is the same for T near T_c from above or from below (although, the accompanying factors may be different in the two cases).

Another quantity of interest is the specific heat at zero field. The internal energy of the ferromagnet, contributed by the spontaneous magnetization, is $-(1/2)H_{\text{eff}} M = -(1/2)\lambda M^2$; using Eq. (17.13), it is seen that the specific heat, which is the derivative of the internal energy with respect to the temperature, is zero above the critical temperature and nearly constant just below it.

Near the critical temperature, the Weiss molecular field theory shows that several significant quantities, such as zero-field magnetization, magnetization at critical temperature, zero-field isothermal susceptibility, zero-field specific heat, follow simple power laws with appropriate exponents. Without entering in details, another significant quantity to be mentioned is the spin-spin correlation function, which is related to the probability to find the spin, at a given site, oriented in the same direction as the chosen reference spin. Near the critical point, and at large distances, the correlation function is expected to decay exponentially, with a correlation length going to infinity with a power-like dependence on temperature. At the critical temperature, it is found that the correlation function falls off much slower, in a power-like way, with increasing distance.

Table 17.2 Definition of some critical exponents for magnetic phase transitions. The values of the critical exponents in the mean field theory are given. The exact values of the two-dimensional Ising model are also reported. The approximate values for the three-dimensional Ising model and for the three-dimensional Heisenberg model, calculated with the renormalization group theory, are taken from the paper of J. C. Le Guillou and J. Zinn-Justin, Phys. Rev. B *21*, 3976 (1980). It can be noticed that the critical exponents satisfy the scaling laws relations $\alpha + 2\beta + \gamma = 2$; $\gamma = \nu(2 - \eta)$; $\gamma = \beta(\delta - 1)$; $\nu d = 2 - \alpha$ (d = lattice dimensionality).

		$(d=4)$ Mean field	$(d=2)$ Ising	$(d=3)$ Ising	$(d=3)$ Heisenberg
Zero-field specific heat	$C \approx \|T - T_c\|^{-\alpha}$	$\alpha = 0$ (Discontinuity)	$\alpha = 0$ (Logarithm)	$\alpha = 0.11$	$\alpha = -0.12$
Zero-field magnetization $(T < T_c)$	$M \approx (T_c - T)^\beta$	$\beta = 1/2$	$\beta = 1/8$	$\beta = 0.32$	$\beta = 0.37$
Zero-field isothermal susceptibility	$\chi \approx \|T - T_c\|^{-\gamma}$	$\gamma = 1$	$\gamma = 7/4$	$\gamma = 1.24$	$\gamma = 1.39$
Magnetization at $T = T_c$	$M \approx H^{1/\delta}$	$\delta = 3$	$\delta = 15$	$\delta = 4.82$	$\delta = 4.80$
Correlation length	$\xi \approx \|T - T_c\|^{-\nu}$	$\nu = 1/2$	$\nu = 1$	$\nu = 0.63$	$\nu = 0.71$
Pair correlation function at $T = T_c$	$\Gamma(r) \approx 1/r^{d-2+\eta}$	$\eta = 0$	$\eta = 1/4$	$\eta = 0.03$	$\eta = 0.03$

The definition and the values of the six critical exponents of more common use, mentioned in the above discussion, are summarized for convenience in Table 17.2. The critical exponents are not independent because of the scaling laws connecting them. In particular, from the knowledge of the "thermal" exponent α and of the "magnetic" exponent δ, all the other exponents can be obtained. For further details and relationships among critical point exponents see for instance the textbooks on statistical mechanics cited in the bibliography.

Inadequacies of the Weiss Molecular Field

The Weiss approach has the merit to shed light on some features of ferromagnetic behavior at the minimal technical level. However the Weiss molecular field, which is essentially a mean field theory, presents several inadequacies, that are summarized below; some aspects of the most relevant theories developed to overcome these inadequacies are outlined in the following of this chapter.

Origin of the Weiss molecular field. Historically, the more basic problem raised by the Weiss molecular field is the microscopic origin of the coupling between magnetic dipoles. We can estimate the order of magnitude of the Weiss parameter λ in terms of

T_c, using Eqs. (17.7) and the relations provided in Eq. (17.1b):

$$\lambda = \frac{V}{N} \frac{k_B T_c}{\mu_B^2} \approx \frac{a^3}{a_B^3} \frac{a_B^3}{\mu_B^2} k_B T_c \approx 10^4 - 10^5 \quad (a \approx 4 - 5a_B, \quad T_c \approx 1000 \text{ K}).$$

The phenomenological values of λ in the range $10^4 - 10^5$ *exclude* that the molecular field originates from magnetic dipole-dipole interactions; these dipolar interactions, once treated "mutatis mutandis" according to the Lorentz cavity field of Section 9.7.3, would give $H_{\text{eff}} = H + \lambda M$ with $\lambda \approx 4\pi/3$ (or so), a value thousand times smaller than the phenomenological values of λ. No other classical magnetic interaction can be envisaged as a source of the Weiss internal field: its origin requires a full quantum mechanical approach, as discussed in more detail in Section 17.2. The dipolar interactions, on the other hand, have a leading role in the tendency of a ferromagnetic specimen to break into "domains," so to reduce the magnetic dipole-dipole interaction energy. In the following, we will neglect this and other consequences of the dipolar interactions among spins; rather we will focus on the quantum origin of the effective coupling mechanism among the localized magnetic moments. This effective coupling can lead not only to ferromagnetism but also to other types of magnetic ordering (as discussed in Section 17.3).

Low temperature region. Near zero temperature the mean field theory predicts that spontaneous magnetization deviates from saturation by an exponential term. In contrast, experiments show a $T^{3/2}$ dependence in typical three-dimensional ferromagnets; a more accurate analysis in terms of the concepts of spin waves and magnons is needed (see Section 17.4). Another evident limit of the mean field theory is the fact that it is unable to correctly determine the effect of dimensionality on the phase transitions, as predicted for instance by the basic Mermin-Wagner theorem.

Critical temperature region. Near the critical temperature, the mean field theory shows that several significant quantities follow power-like laws. However the measured exponents of some ferromagnetic materials are rather different from the mean field result, although only minor differences seem to occur among the different materials, as seen from Table 17.3. A more detailed analysis of critical exponents is necessary in second order phase transitions; some aspects concerning this topic are discussed in Sections 17.5 and 17.6.

Table 17.3 Measured critical exponents for several ferromagnetic materials [see S. N. Kaul, J. Magn. Magn. Mater. *53*, 5 (1985) and references quoted therein].

Material	Critical exponent β	Critical exponent γ
Fe	0.39	1.33
Co	0.44	1.23
Ni	0.38	1.34
Gd	0.38	1.20

Itinerant band magnetism. The description of ferromagnetism, starting from a picture of magnetic moments localized at the lattice sites, may become inadequate for transition metals and other materials, in which the interplay between band formation and magnetization is relevant; some aspects are considered in Section 17.7.

17.2 Microscopic Origin of the Coupling Between Localized Magnetic Moments

17.2.1 The Spin Hamiltonian Approach to Magnetism

The spin Hamiltonian approach is a common and fruitful short-cut approach for the description of magnetism in several materials. We consider here, at a qualitative level and without efforts to be exhaustive, different types of microscopic mechanisms that can lead to an effective coupling between pairs of spin operators. We begin this discussion with some paradigmatic models, which produce an effective interaction of the form of *scalar product of spin operators*, usually written in the form $-2J\mathbf{S}_1 \cdot \mathbf{S}_2$, where J is the so called exchange parameter. Needless to say in realistic situations a great variety of different microscopic mechanisms can be at work in the energy range of interest, but not more than a cursory mention to some of them will be given for brevity. For an in-depth overview of fundamental aspects of magnetism, extending from traditional bulk materials to the rapidly expanding world of molecular nanomagnets, we refer for instance to the book of Gatteschi et al., cited in the bibliography.

The origin of the effective spin-spin interaction is in essence connected to the global antisymmetry of the electron wavefunctions. Alignment or not of spins in the *spin space* imposes requirements on the wavefunctions *in real space*, where the electrostatic interactions are at work. Under appropriate circumstances, an effective spin-spin Hamiltonian may result, whose energy spectrum mimics the one of the original Hamiltonian in an appropriate energy range; in typical situations, the interaction energy J can assume values of the order of $10 \approx 100$ meV (or so), which are several orders of magnitude higher than typical magnetic dipolar interactions, estimated in Eq. (17.3). The sign of the coupling can be either positive or negative, or even oscillatory with spin distance. In the case $J > 0$ the spins tend to line up in the same direction and the coupling is said to be ferromagnetic; in the case $J < 0$ the spins tend to line up in opposite directions and the coupling is said antiferromagnetic.

It is instructive to begin our qualitative analysis considering prototype two-electron systems where an effective coupling of scalar product between the electronic spins $\mathbf{s}_1$ and $\mathbf{s}_2$ appears to be at work.

A model of two-electron systems has been discussed in Section 16.5.3. In the introduction to the Kondo effect, we have considered in detail the energy levels of an idealized molecule with two electrons and two orbitals, one of them of extended nature (with the possibility of double occupancy by electrons of opposite spin) and the other of localized nature (with so large Coulomb repulsive energy between electrons of opposite spin to prevent double occupancy). The hybridization between the two orbitals makes the singlet state (with total spin $S = 0$) lower in energy than the triplet state (with total spin $S = 1$); in Section 16.5.3, it has also been shown that the energy difference and

multiplicity of the ground state and first excited molecular state can be mimicked by
the spin-spin Hamiltonian $-2J\,\mathbf{s}_1 \cdot \mathbf{s}_2$ with antiferromagnetic interaction $J < 0$.

We discuss now examples of two-electron systems with ferromagnetic spin-spin
interaction. Consider the case of two electrons in *incomplete shells* of ions or atoms
(for instance in the configurations of the type np^2 or nd^2 or nf^2). It is well-known
that the two external (or "optical") electrons line up their spins in the ground state, in
agreement with Hund's first rule; this means that the triplet state is lower in energy
than the singlet state, and an effective coupling of ferromagnetic type $-2J\mathbf{s}_1 \cdot \mathbf{s}_2$ (with
$J > 0$) appears to be at work.

To estimate the singlet-triplet separation avoiding inessential details, we consider
a system with *two electrons* and *two degenerate orbitals*, whose orthonormal wavefunctions are denoted by ϕ_a and ϕ_b. Accommodation of the two electrons on the four
available spin-orbitals ($\phi_a\alpha, \phi_a\beta, \phi_b\alpha, \phi_b\beta$) gives rise to six possible determinantal
states. For simplicity, we assume that *double occupancy of the same orbital is avoided*,
because of the strong Coulomb repulsive energy between electrons of opposite spin in
the same localized orbital. We are thus left with four determinantal states (illustrated in
Figure 17.3); from them, we can form the singlet and triplet states given respectively by

$$\Psi_S = \frac{1}{\sqrt{2}}[\phi_a(\mathbf{r}_1)\phi_b(\mathbf{r}_2) + \phi_a(\mathbf{r}_2)\phi_b(\mathbf{r}_1)]\frac{1}{\sqrt{2}}[\alpha(1)\beta(2) - \alpha(2)\beta(1)] \quad (17.16a)$$

$$\Psi_T = \frac{1}{\sqrt{2}}[\phi_a(\mathbf{r}_1)\phi_b(\mathbf{r}_2) - \phi_a(\mathbf{r}_2)\phi_b(\mathbf{r}_1)] \begin{cases} \alpha(1)\alpha(2), \\ [\alpha(1)\beta(2) + \alpha(2)\beta(1)](1/\sqrt{2}), \\ \beta(1)\beta(2). \end{cases}$$

$$(17.16b)$$

We now evaluate the energy difference between the singlet and the triplet states.

The Hamiltonian of the two-electron system has the general form

$$\mathcal{H} = \mathcal{H}_0(\mathbf{r}_1) + \mathcal{H}_0(\mathbf{r}_2) + \frac{e^2}{|\mathbf{r}_1 - \mathbf{r}_2|} \quad \text{with} \quad \mathcal{H}_0(\mathbf{r}) = \frac{\mathbf{p}^2}{2m} + V(\mathbf{r}),$$

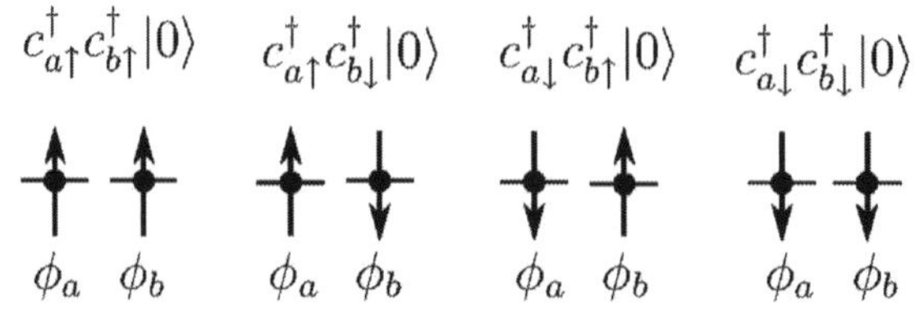

Figure 17.3 The four possible two-electron states of a system with two orbitals (here taken
as degenerate) and two electrons. It is assumed that double occupancy of the same orbital is
suppressed because of strong Coulomb repulsion.

where $\mathcal{H}_0(\mathbf{r})$ is a one-electron operator and e^2/r_{12} is the electron-electron Coulomb repulsion (the effect of core electrons is assumed to be embodied someway in the model "pseudopotential" $V(\mathbf{r})$). The one-electron operator needs not be further specified, since the energy splitting of the singlet and triplet states (17.16) is only due to the bielectronic operator e^2/r_{12}. A straightforward calculation gives for the difference

$$E_S - E_T = \langle \Psi_S | \mathcal{H} | \Psi_S \rangle - \langle \Psi_T | \mathcal{H} | \Psi_T \rangle = 2J, \qquad (17.17a)$$

where

$$J = \langle \phi_a \phi_b | \frac{e^2}{r_{12}} | \phi_b \phi_a \rangle \equiv \int \phi_a^*(\mathbf{r}_1) \phi_b^*(\mathbf{r}_2) \frac{e^2}{|\mathbf{r}_1 - \mathbf{r}_2|} \phi_b(\mathbf{r}_1) \phi_a(\mathbf{r}_2) \, d\mathbf{r}_1 \, d\mathbf{r}_2. \qquad (17.17b)$$

The bielectronic exchange integral J, defined by Eq. (17.17b), represents the self-energy of the (complex) charge distribution $\phi_a^*(\mathbf{r})\phi_b(\mathbf{r})$ and is thus a *positive definite quantity* (see Appendix 4.A); it follows that the ground state of the two-electron system is the (triplet) magnetic state. This is so because the Coulomb repulsion between electrons is lowered in the triplet states with respect to the singlet state; notice in fact that $\Psi_T \equiv 0$ if $\mathbf{r}_1 \equiv \mathbf{r}_2$. The direct exchange mechanism mimics a ferromagnetic interaction between the spins of the two external electrons.

A significant example of a system with two ("optical") electrons with ferromagnetic-like interaction is provided by the O_2 molecule; in the independent particle approximation, the (antibonding) π molecular orbitals are twofold degenerate, corresponding to the two possible orientations perpendicular to the axis of the molecule; on them, we can accommodate the two external electrons of the molecule, with their spins lined up to decrease the Coulomb repulsion.

In certain crystals, a spin-spin coupling may result due to a *superexchange interaction* with the following mechanism: two atoms (or ions) with magnetic moments interact with the polarization produced on a third non-magnetic atom (or ion). A typical case is that of transition-metal fluorides MnF_2, FeF_2, CoF_2; these materials are antiferromagnetic at low temperature and the coupling between cations is mediated through the halogen wavefunctions. This kind of superexchange (often antiferromagnetic) characterizes several magnetic insulators [for a thorough discussion of magnetic exchange interactions in insulators and semiconductors see P. W. Anderson, Solid State Physics *14*, 99 (1963)].

In some metals, such as rare earth metals, an effective interaction between localized spins may occur through the polarization of the free electron gas, where the localized spins are embedded. The magnetic moments of the incomplete f shells of rare earth atoms are so well localized that any direct interaction between f-electron wave-functions on different sites are safely negligible. The f electrons, however, are coupled to the conduction electrons, induce a spin polarization, and an effective coupling between localized f-electron spins can occur. This effective interaction is known as the Ruderman-Kittel-Kasuya-Yosida (RKKY) interaction; it is oscillatory with the distance between the localized spins, and thus can give rise both to ferromagnetic and antiferromagnetic kind of interaction, as discussed in Appendix A. Assuming spherical Fermi surface of energy E_F and wavevector k_F, the expression of the spin-spin interaction provided in Eq. (A.11), is here reported for convenience

$$E_{\mathrm{RKKY}}(R) = \frac{1}{4\pi^3} \frac{J^2}{E_F} (\Omega k_F^3)^2 F(2k_F R) \, \mathbf{S}_1 \cdot \mathbf{S}_2 \quad \text{with} \quad F(x) = \frac{x \cos x - \sin x}{x^4},$$

$$(17.18)$$

where Ω is the volume of the unit cell and J controls the strength of the interaction. Notice that $F(x) \approx -1/6x$ for small x, and $F(x) \approx \cos x / x^3$ for large x. Thus the interaction oscillates with distance and falls off with the inverse cube power law r^{-3} at large distances. [A similar long-range behavior has been found in Section 7.5 for charge density oscillations, or Friedel oscillations, produced by an ordinary impurity in a metal]. The Ruderman-Kittel-Kasuya-Yosida interaction (17.18) is ferromagnetic-like for small R; for large R it can be either ferromagnetic or antiferromagnetic depending on the distance of the two spins. The oscillatory behavior of $F(2k_F R)$ is responsible for the great variety of magnetic behavior occurring in rare earth metals and compounds, including ferro- and antiferromagnetism, incommensurate order, and spiral structures. It is also believed that a mechanism rather similar to RKKY is responsible of the oscillatory exchange coupling between the magnetic moments of the layered ferromagnets of the giant magnetoresistance devices (discussed in Section 17.7).

Depending on specific materials and microscopic environments, there is a variety of other effective spin-spin interactions, besides the scalar product of spin of previous models. In particular directional effects in spin space may develop as a consequence of spin-orbit coupling, or in the presence of external fields. Several other models of spin-spin interactions have been suggested in the literature to discuss and interpret the variegated subject of magnetism, but we finish here with examples and come to some concluding remarks.

In summary: in many-electron systems, real space and spin space are always interconnected quantum mechanically, because of the use of antisymmetric total wavefunctions. In a number of paradigmatic examples discussed above, it appears reasonable to get rid altogether of the real space variables, at the low cost of introducing an appropriate coupling among spin observables. More generally, the essential idea of the spin Hamiltonian approach to magnetism entails the elimination of the electronic space coordinates, assuming that their presence can be accounted for via effective spin interactions (at least in an appropriate energy range of interest).

17.2.2   The Heisenberg Spin Hamiltonian and the Mean Field Approximation

The localized picture of crystal magnetism assumes that microscopic magnetic moments are set at the lattice sites and interact cooperatively among themselves. In a simplified description, the contribution to the energy of a pair of spins in the m-th and n-th sites is taken in the isotropic form $W_{\mathrm{pair}} = -2J_{mn}\mathbf{S}_m \cdot \mathbf{S}_n$, where the exchange parameters can extend beyond nearest-neighbor sites. Under the stated approximation, the Hamiltonian of the localized moments on a regular lattice can be written in the Heisenberg form

$$\mathcal{H} = -2 \sum_{m<n} J_{mn} \mathbf{S}_m \cdot \mathbf{S}_n, \tag{17.19}$$

where $\mathbf{S}_m$ denotes the local angular momentum of the atom (or ion, or magnetic center) at the m-th site, J_{mn} denotes the exchange parameter between the spins at sites m and n, and the summation includes each pair of spins only once. For practical elaborations, it may be convenient to rewrite Eq. (17.19) in the equivalent form

$$\mathcal{H} = -\sum_{m \neq n} J_{mn} \mathbf{S}_m \cdot \mathbf{S}_n, \tag{17.20}$$

where the sum runs on all sites m and n with the exclusion of the self-interaction terms $m = n$. Notice that the summation in Eq. (17.20) includes each pair of spins twice; to avoid double counting of interactions, the factor 2 in front of J_{mn} is omitted. The Hamiltonian (17.20) can describe different kinds of magnetic order present in nature: ferromagnetism, antiferromagnetism, ferrimagnetism, helimagnetism, and other more complex structures, depending on the sign of nearest neighbor couplings, the presence of competing second neighbor or higher order interactions, the additional effect of external fields and on-site interactions.

Often spin-spin interactions are relevant only for nearest neighbor sites (mainly in insulators). In the case nearest neighbor interactions have the same parameter J, the Heisenberg Hamiltonian (17.20) can be recast in the form

$$\mathcal{H} = -J \sum_{m \neq n}^{\text{(n.n.)}} \mathbf{S}_m \cdot \mathbf{S}_n = -J \sum_{m \neq n}^{\text{(n.n.)}} (S_m^z S_n^z + S_m^x S_n^x + S_m^y S_n^y), \tag{17.21}$$

where the sum runs only on the sites m and n that are nearest neighbors (with the exclusion of the self-interaction terms $m = n$).

The Heisenberg exchange interaction is isotropic and depends only on the mutual direction of spins. When necessary, the isotropic spin-spin interaction can be modified to account for preferential orientations of the magnetic moments in the crystal, as crystal axes could be felt through the spin-orbit interaction. In some magnets, interactions among localized moments can be described by the anisotropic XY model, defined as

$$\mathcal{H}_{XY} = -J_1 \sum_{m \neq n}^{\text{(n.n.)}} S_m^z S_n^z - J_2 \sum_{m \neq n}^{\text{(n.n.)}} \left(S_m^x S_n^x + S_m^y S_n^y \right).$$

For $J_1 = J_2 = J$ we recover the isotropic Heisenberg model of Eq. (17.21). In the Ising model it is assumed that spins can only point up or down along a given direction, say the z-direction ($J_1 = J$, $J_2 = 0$); the Hamiltonian of the Ising model reads

$$\mathcal{H}_{\text{Ising}} = -J \sum_{m \neq n}^{\text{(n.n.)}} S_m^z S_n^z. \tag{17.22}$$

The above types of spin Hamiltonians have been of invaluable help for understanding the properties of a number of magnetic materials, and are the prototypes of several other related models.

Ferromagnetism in the Mean Field Approximation

We are now in the position to examine the microscopic origin of the Weiss molecular field. Consider a magnetic crystal composed by N spins $\mathbf{S}_i$ of magnetic moment $\boldsymbol{\mu}_i = -g\mu_B \mathbf{S}_i$ localized on the sites of a Bravais lattice, and interacting via a first neighbor spin-spin Hamiltonian. In the presence of an external magnetic field $\mathbf{H}$ (in the z-direction), the Hamiltonian of the system of spins reads

$$\mathcal{H} = -J \sum_{\substack{m \neq n}}^{(\text{n.n.})} \mathbf{S}_m \cdot \mathbf{S}_n + g\mu_B \mathbf{H} \cdot \sum_m \mathbf{S}_m, \tag{17.23}$$

where the first term is the Heisenberg spin-spin Hamiltonian, and the second term is the Zeeman energy. We treat the spin-spin coupling in the mean field approximation, and establish a relationship between the exchange parameter $J(>0)$ and the Weiss constant λ.

We define the magnetization of the sample as

$$\mathbf{M} = -\frac{1}{V} g\mu_B \sum_m \langle \mathbf{S}_m \rangle \quad \Longrightarrow \quad \mathbf{M} = -\frac{N}{V} g\mu_B \langle \mathbf{S} \rangle, \tag{17.24}$$

where the thermodynamic averages $\langle \mathbf{S}_m \rangle$ are site independent due to translational symmetry, and are then indicated simply as $\langle \mathbf{S} \rangle$.

We now express any spin operator as the sum of its average value and the fluctuation from the average value in the form

$$\mathbf{S}_m = \langle \mathbf{S} \rangle + \mathbf{S}_m - \langle \mathbf{S} \rangle$$

and consider the scalar product of two operators of the form

$$\mathbf{S}_m \cdot \mathbf{S}_n = [\langle \mathbf{S} \rangle + \mathbf{S}_m - \langle \mathbf{S} \rangle] \cdot [\langle \mathbf{S} \rangle + \mathbf{S}_n - \langle \mathbf{S} \rangle]$$

$$= \langle \mathbf{S} \rangle \cdot \langle \mathbf{S} \rangle + \langle \mathbf{S} \rangle \cdot [\mathbf{S}_n - \langle \mathbf{S} \rangle] + [\mathbf{S}_m - \langle \mathbf{S} \rangle] \cdot \langle \mathbf{S} \rangle + \overbrace{[\mathbf{S}_m - \langle \mathbf{S} \rangle] \cdot [\mathbf{S}_n - \langle \mathbf{S} \rangle]}.$$

The above expression is exact, but we now make the crucial approximation to neglect the last term, which is quadratic in the fluctuations. We obtain

$$\mathbf{S}_m \cdot \mathbf{S}_n \approx \langle \mathbf{S} \rangle \cdot \mathbf{S}_n + \mathbf{S}_m \cdot \langle \mathbf{S} \rangle - \langle \mathbf{S} \rangle \cdot \langle \mathbf{S} \rangle. \tag{17.25}$$

When fluctuations are neglected, inserting the basic simplification of Eq. (17.25) into the Heisenberg spin-spin Hamiltonian (17.21) gives

$$-J \sum_{\substack{m \neq n}}^{(\text{n.n.})} \mathbf{S}_m \cdot \mathbf{S}_n = -2n_c J \langle \mathbf{S} \rangle \cdot \sum_m \mathbf{S}_m - N n_c J \langle \mathbf{S} \rangle^2, \tag{17.26}$$

where n_c is the coordination number, i.e. the number of neighbors interacting with the chosen spin at a given lattice site.

It is convenient to write the first factor appearing in the right-hand side of Eq. (17.26) in the form

$$-2n_c J \langle \mathbf{S} \rangle \equiv g\mu_B \mathbf{H}_{\text{eff}},$$

where $\mathbf{H}_{\text{eff}}$ plays evidently the role of an effective magnetic field to be added to the external magnetic field in Eq. (17.23). We have

$$\mathbf{H}_{\text{eff}} = -\frac{2n_c J \langle \mathbf{S} \rangle}{g\mu_B} = \frac{V}{N} \frac{2n_c J}{g^2 \mu_B^2} \mathbf{M} \equiv \lambda \mathbf{M} \qquad \text{with} \qquad \boxed{\lambda = \frac{V}{N} \frac{2n_c J}{g^2 \mu_B^2}}. \quad (17.27)$$

Notice that the value of λ can be also obtained by enforcing the equality between the last term of Eq. (17.26) and the magnetization energy of the system, namely

$$-\frac{1}{2}\lambda M^2 V \equiv -N n_c J \langle \mathbf{S} \rangle^2 = -N n_c J \frac{V^2}{N^2} \frac{1}{g^2 \mu_B^2} M^2 \implies \lambda = \frac{V}{N} \frac{2n_c J}{g^2 \mu_B^2}.$$

The Weiss constant (17.27) can be estimated with the help of Eqs. (17.1), and taking typical values of the order

$$\frac{V}{N} = \Omega \approx 100\, a_B^3; \quad \frac{g^2 \mu_B^2}{a_B^3} \approx 1 \text{ meV}; \quad J \approx 10-100 \text{ meV} \implies \lambda \approx 10^4 - 10^5.$$

From the above estimate, we can safely conclude that the spin-spin model Hamiltonian provides a satisfactory microscopic explanation of the origin of the Weiss molecular field.

The mean field approach to phase transitions, although often useful as a preliminary step in three-dimensional systems, presents the major drawback to become less reliable in low-dimensional systems, where the effects of fluctuations are of increasing importance. The mean field approach does not account correctly of the effect of dimensionality, as can be inferred from the fact that the environment enters into play through the coordination number of the lattice and not through its dimensionality. Without entering in details, we add here a few considerations only on the isotropic Heisenberg model and the Ising model, both with finite range interactions, as described for instance by Eqs. (17.21) and (17.22). For a one-dimensional lattice of Ising spins there is no magnetic ordering at finite temperature; in two dimensions (and then in three dimensions) the Ising model presents a phase transition at finite temperature (as discussed in Section 17.5 with the transfer matrix method and in Section 17.6 with the renormalization theory). On the other hand a one-dimensional or two-dimensional isotropic Heisenberg model can be neither ferromagnetic nor antiferromagnetic at any non-zero temperature. This is rigorously proved in the theorem of N. D. Mermin and H. Wagner, Phys. Rev. Lett. *17*, 1133 (1966). In Section 17.4, the spin waves framework and heuristic considerations, corroborate the finding that a d-dimensional lattice of localized spins, whose exchange interactions are isotropic and of finite range, cannot have long range order for $d < 3$ at any finite temperature.

17.3   Antiferromagnetism in the Mean Field Approximation

In the Heisenberg model of ferromagnetism the coupling energy J between nearest neighbor spins is positive and parallel alignment of local spins is favorite. In the case the coupling energy J is negative, antiparallel orientation of neighbor spins is preferred, and a tendency of spins to order in a structure where up and down spins alternate is expected (Figure 17.4).

The simplest model of antiferromagnetism is the *two sublattice model*, in which the complete lattice of atoms is divided into two identical interpenetrating sublattices, say A and B, such that all nearest neighbors of A sites (with spin-up atoms) are B sites (with spin-down atoms), and viceversa.

The Weiss molecular field approach has been generalized by Néel to the treatment of antiferromagnetism, by assuming that the effective field acting on sites A (or B) contains a cooperative contribution, proportional and opposite to the magnetization on sublattice B (or A). In addition to the nearest neighbor AB (or BA) antiferromagnetic interaction, we also assume an antiferromagnetic interaction for second nearest neighbor AA and BB (this is not strictly necessary, but otherwise the model would be somewhat more restrictive).

At the absolute temperature $T = 0$, each of the two sublattices has a net magnetization (compatible with zero point fluctuations). Increasing the temperature the magnetization on each of the two sublattices decreases; however, differently from ferromagnetism the net magnetization of an antiferromagnet is zero at any temperature (from this the term "hidden magnetism" assigned sometimes to antiferromagnetism). Above a critical temperature T_N, called the *Néel temperature*, the magnetization on each sublattice vanishes, and the magnetic susceptibility decreases with a law of the type $\chi \propto 1/(T + \Theta)$, with Θ of the order of T_N. We show that all these facts can be accounted for within the mean field theory.

In the two sublattice model, the effective fields at A and B sites are given by

$$\mathbf{H}_a^{(\text{eff})} = \mathbf{H} - \lambda_1 \mathbf{M}_b - \lambda_2 \mathbf{M}_a, \tag{17.28a}$$

$$\mathbf{H}_b^{(\text{eff})} = \mathbf{H} - \lambda_1 \mathbf{M}_a - \lambda_2 \mathbf{M}_b, \tag{17.28b}$$

where $\mathbf{H}$ is the external field, $\mathbf{M}_a$ and $\mathbf{M}_b$ are the magnetization of A and B sublattices, λ_1 and λ_2 are *positive phenomenological constants* in the specific model of antiferromagnetism we are considering. [The mean field coefficient $\lambda_1 > 0$ must be positive in order to favor antiparallel alignment of the magnetic moments of the two sublattices; λ_2 could also be zero, or even negative, and in general $|\lambda_2| < \lambda_1$].

ferromagnet   antiferromagnet   ferrimagnet

Figure 17.4 Schematic illustration of ferromagnetic and antiferromagnetic order. The ferrimagnetic order, in which spins of different values alternate, is also shown.

With the effective fields provided by Eqs. (17.28), we can now obtain the magnetization of the two sublattices. We suppose that the local spins have $S = 1/2$ and $g = 2$. We indicate with z the direction of the applied field $\mathbf{H}$, and with M_a and M_b the z-components of the sublattices magnetization. In line with the procedure used to obtain Eq. (17.6), we have for the magnetization of the two sublattices

$$M_a = \frac{1}{2} \frac{N}{V} \mu_B \tanh \frac{\mu_B (H - \lambda_1 M_b - \lambda_2 M_a)}{k_B T}, \tag{17.29a}$$

$$M_b = \frac{1}{2} \frac{N}{V} \mu_B \tanh \frac{\mu_B (H - \lambda_1 M_a - \lambda_2 M_b)}{k_B T}, \tag{17.29b}$$

where N is the total number of sites of the crystal and V is the volume. We consider now the high temperature region $(T > T_N)$ and the low temperature region $(T < T_N)$.

Temperature region above the Néel temperature $(T > T_N)$. In the temperature region above the Néel temperature, we expect that both M_a and M_b are parallel to the applied field H and vanish when the applied field decreases to zero. For reasonably small applied fields, we can linearize Eqs. (17.29), and obtain

$$M_a = \frac{1}{2} \frac{N}{V} \frac{\mu_B^2}{k_B T} (H - \lambda_1 M_b - \lambda_2 M_a), \tag{17.30a}$$

$$M_b = \frac{1}{2} \frac{N}{V} \frac{\mu_B^2}{k_B T} (H - \lambda_1 M_a - \lambda_2 M_b). \tag{17.30b}$$

Summing up the two relations expressed by Eqs. (17.30), we have

$$M = M_a + M_b = \frac{C}{T} H - \frac{\Theta}{T} M$$

with

$$C = \frac{N}{V} \frac{\mu_B^2}{k_B} \quad \text{and} \quad \Theta = C \frac{\lambda_1 + \lambda_2}{2}. \tag{17.31}$$

The magnetic susceptibility of the antiferromagnet thus becomes

$$\boxed{\chi = \frac{M}{H} = \frac{C}{T + \Theta}}.$$

The susceptibility is reduced with respect to the non-interacting case $(\lambda_1 = \lambda_2 = 0 = \Theta)$ because the magnetic moments tend to arrange in an antiparallel way.

Temperature region below the Néel temperature $(T \leq T_N)$. In the temperature region below the Néel temperature, we expect a spontaneous magnetization on each sublattice, even in the absence of an external magnetic field. We furthermore expect that $M_a = -M_b$. Equation (17.29a) for the spontaneous magnetization of sublattice A becomes

$$M_a = \frac{1}{2} \frac{N}{V} \mu_B \tanh \frac{\mu_B (\lambda_1 - \lambda_2) M_a}{k_B T} \tag{17.32}$$

Table 17.4 Néel temperature T_N and Θ temperature (in K) for some antiferromagnetic materials [for further data see for instance F. Keffer, Handbuch der Physik, Vol. 18 Part 2 (Springer, Berlin, 1966)].

Material	T_N	Θ	Material	T_N	Θ
MnO	122	610	MnF$_2$	67	80
FeO	198	507	FeF$_2$	78	117
CoO	291	330	CoF$_2$	38	50
NiO	600	2000	KMnF$_3$	88	158

and a similar equation holds for M_b. Equation (17.32) for the spontaneous magnetization in each sublattice has the same structure as Eq. (17.6); in particular linearization of Eq. (17.32) gives the critical Néel temperature

$$M_a = \frac{1}{2}\frac{N}{V}\mu_B\frac{\mu_B(\lambda_1 - \lambda_2)M_a}{k_B T_N} \quad \Longrightarrow \quad T_N = C\frac{\lambda_1 - \lambda_2}{2}. \tag{17.33}$$

From Eqs. (17.31) and (17.33) we obtain

$$\frac{T_N}{\Theta} = \frac{\lambda_1 - \lambda_2}{\lambda_1 + \lambda_2}.$$

In the case the antiferromagnetic AA and BB interactions are negligible (i.e. $\lambda_2 = 0$) we have $T_N = \Theta$. With λ_2 positive, we have $T_N < \Theta$, as it is experimentally observed in a number of materials (see Table 17.4).

We do not dwell further on the analysis and application of the mean field theory to cooperatively interacting localized magnetic moments. The mean field theory, with the support of appropriate microscopic spin-spin Hamiltonians, can describe numerous other types of magnetic order present in the nature. Besides ferromagnetism and antiferromagnetism, we can mention ferrimagnetism, helimagnetism, anisotropic systems, modulated and incommensurate structures in the presence of competing interactions, and we refer to the literature for further information.

17.4 Spin Waves and Magnons in Ferromagnetic Crystals

So far, in the discussion of magnetic ordering, we have considered the Heisenberg Hamiltonian for a system of spins coupled by exchange interaction, and we have applied to it the Weiss (or Néel) molecular field approximation. In essence, in the molecular field treatment of the phase transitions, spin operators may be replaced by average values and spin fluctuations are neglected.

To improve the understanding of magnetism, one should know more about the eigenvalues and eigenstates of the spin Hamiltonian. For this purpose there is a variety of lines of attack; among them: (i) the spin-wave theory, which is particularly appropriate for magnetic systems in the low-temperature limit, as discussed below and (ii) the

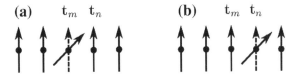

Figure 17.5 Schematic representation of trial excited states of the spin Hamiltonian. In (a) there is a spin deviation at the site $\mathbf{t}_m$, and in (b) there is a spin deviation at the adjacent site $\mathbf{t}_n$. In the spin-wave theory, the operators which couple spin deviations on different sites produce a wave-like Bloch state.

transfer matrix method and the renormalization group formalism, which focus on the physics of phase transitions near the critical points (see Sections 17.5 and 17.6).

It has been shown originally by Bloch that the low-lying energy states of a periodic system of localized spins, coupled by exchange interaction, are wavelike. The energy of a spin wave is quantized, and the unit of energy is called *magnon*. Spin waves have been studied for all types of magnetic ordering. For simplicity, we confine our attention to ferromagnetic materials (see Figure 17.5); for them the treatment of wavelike excitations is rather intuitive, compared with the more demanding cases of antiferromagnetism and other types of magnetic ordering.

Elementary Properties of Spin Operators

Before discussing the basic aspects of the spin-wave theory, it is useful to remind some elementary properties of spin operators. Consider the three component operators S^x, S^y, S^z of a given spin operator $\mathbf{S}$. The commutation rules are $[S^x, S^y] = i S^z$ and cyclic permutations of the indices x, y, z. The operators S^x, S^y, S^z are not independent, but are connected by the identity

$$\mathbf{S} \cdot \mathbf{S} = S_x^2 + S_y^2 + S_z^2 = S(S + 1).$$

The raising operator S^+ and the lowering operator S^- are defined as

$$S^+ = S^x + i S^y, \quad S^- = S^x - i S^y \quad \Longrightarrow \quad S^z = \frac{1}{2}[S^+, S^-]. \qquad (17.34)$$

The effect of S^+ or S^- on an eigenstate of S^z with eigenvalue M is

$$S^+|M\rangle = \sqrt{(S - M)(S + 1 + M)}\ |M + 1\rangle,$$
$$S^-|M\rangle = \sqrt{(S + M)(S + 1 - M)}\ |M - 1\rangle, \qquad (17.35)$$

where $-S \leq M \leq +S$.

It is convenient to define the spin deviation operator $\hat{n} = S - S^z$, which is diagonal on the representation where S^z is diagonal, and whose eigenvalues are integer numbers ranging from 0 to $2S$; $\hat{n}$ represents the spin deviations from the maximum value of spin S. Thus the states $|S\rangle, |S - 1\rangle, \ldots, |-S\rangle$ correspond to spin deviations $|0\rangle, |1\rangle, \ldots, |2S\rangle$, respectively. We now rewrite the operator S^z and the expressions

(17.35) in the newly relabeled basis and we have

$$
\begin{cases}
S^+|n\rangle = \sqrt{n}\,\sqrt{2S-(n-1)}\,|n-1\rangle, \\
S^-|n\rangle = \sqrt{2S-n}\,\sqrt{n+1}\,|n+1\rangle, \\
S^z|n\rangle = (S-n)|n\rangle,
\end{cases}
\tag{17.36}
$$

where the restriction $0 \le n \le 2S$ holds, and the arguments of the square root in Eqs. (17.36) are always non-negative. In formal analogy with the properties of annihilation and creation operators for the harmonic oscillator (Appendix 9.A), it is convenient to introduce the boson operators a and $a^\dagger$, which applied to the spin deviations $|n\rangle$ give

$$
a|n\rangle = \sqrt{n}|n-1\rangle, \quad a^\dagger|n\rangle = \sqrt{n+1}|n+1\rangle, \quad a^\dagger a|n\rangle = n|n\rangle.
$$

From the above relations and Eqs. (17.36), the spin operators can be expressed in the form

$$
\begin{cases}
S^+ = \sqrt{(2S-a^\dagger a)}\,a\,, \\
S^- = a^\dagger\sqrt{(2S-a^\dagger a)}\,, \\
S^z = S - a^\dagger a
\end{cases}
\tag{17.37}
$$

provided the subspace $0 \le n \le 2S$ is considered. Expressions (17.37) are known as the Holstein-Primakoff transformations.

The expressions (17.37) can be greatly simplified in situations in which the thermal average $\langle a^\dagger a \rangle$ is expected to be small with respect to $2S$. In this case we can use the Holstein-Primakoff relations in the simplified form

$$
\begin{cases}
S^+ = \sqrt{2S}\left[1 - \dfrac{a^\dagger a}{4S} - \dfrac{(a^\dagger a)^2}{32S^2}\cdots\right]a \approx \sqrt{2S}\,a, \\
S^- = a^\dagger\sqrt{2S}\left[1 - \dfrac{a^\dagger a}{4S} - \dfrac{(a^\dagger a)^2}{32S^2}\cdots\right] \approx \sqrt{2S}\,a^\dagger, \\
S^z = S - a^\dagger a.
\end{cases}
\tag{17.38}
$$

We will use the expressions (17.38) at the lowest order of approximation; whenever necessary higher order terms could be included.

Elementary Excitations in Ferromagnetic Crystals

Consider an ideal ferromagnetic crystal, of volume V, formed by N spins $\mathbf{S}_n$ localized at the points $\mathbf{t}_n$ of a simple Bravais lattice, with coordination number n_c. We assume a ferromagnetic interaction among nearest neighbor spins ($J > 0$), and consider the isotropic Heisenberg Hamiltonian

$$
\mathcal{H} = -J \sum_{\substack{(n.n.)\\ \mathbf{t}_m \ne \mathbf{t}_n}} \mathbf{S}_m \cdot \mathbf{S}_n = -J \sum_{\substack{(n.n.)\\ \mathbf{t}_m \ne \mathbf{t}_n}} \left(S_m^z S_n^z + \frac{1}{2}S_m^+ S_n^- + \frac{1}{2}S_m^- S_n^+ \right),
\tag{17.39}
$$

where the sum is over nearest neighbor lattice sites, and the scalar product between spin operators has been expressed with the help of the raising and lowering spin operators, defined by Eqs. (17.34).

The ground state $|0\rangle$ of the ferromagnetic crystal is that in which all spins are lined up in a given direction (say the z-direction), and can be written as

$$|0\rangle = |S\rangle_{\mathbf{t}_1} \cdot |S\rangle_{\mathbf{t}_2} \cdots |S\rangle_{\mathbf{t}_N}.$$

Applying the operator $\mathcal{H}$ to the state $|0\rangle$ one has

$$\mathcal{H}|0\rangle = E_0|0\rangle \quad \text{with} \quad E_0 = -N n_c J S^2 |0\rangle;$$

thus the state $|0\rangle$ is an eigenstate of $\mathcal{H}$ with energy E_0. The fact that we can individuate explicitly the ground state of a ferromagnet makes things much easier. We can now try to individuate the excited states. For this purpose it is convenient to perform on the periodic Hamiltonian (17.39) appropriate canonical transformations, suggested by the Bloch theorem.

We use the Holstein-Primakoff relations in the simplified form (17.38). The approximation made for S_m^+ and S_m^- is justified in the low temperature limit; in this limit, the number of excitations is small, the thermal average $\langle a_m^\dagger a_m \rangle$ is expected to be of the order $O(1/N)$ and can be safely neglected with respect to $2S$. Inserting Eqs. (17.38) into Eq. (17.39), gives

$$\mathcal{H} = -J \sum_{\mathbf{t}_m \neq \mathbf{t}_n}^{(n.n.)} \left[(S - a_m^\dagger a_m)(S - a_n^\dagger a_n) + S a_m a_n^\dagger + S a_m^\dagger a_n \right].$$

Keeping only the terms bilinear in the creation and annihilation operators, one obtains

$$\mathcal{H} = E_0 + 2 n_c J S \sum_{\mathbf{t}_m} a_m^\dagger a_m - 2 J S \sum_{\mathbf{t}_m \neq \mathbf{t}_n}^{(n.n.)} a_m^\dagger a_n, \tag{17.40}$$

where E_0 is the ground-state energy of the ferromagnet.

We now perform the standard canonical transformations suggested by the Bloch theorem

$$a_m = \frac{1}{\sqrt{N}} \sum_{\mathbf{q}} e^{i\mathbf{q}\cdot\mathbf{t}_m} a_{\mathbf{q}} \quad \text{and} \quad a_m^\dagger = \frac{1}{\sqrt{N}} \sum_{\mathbf{q}} e^{-i\mathbf{q}\cdot\mathbf{t}_m} a_{\mathbf{q}}^\dagger, \tag{17.41a}$$

or conversely

$$a_{\mathbf{q}} = \frac{1}{\sqrt{N}} \sum_{\mathbf{t}_m} e^{-i\mathbf{q}\cdot\mathbf{t}_m} a_m \quad \text{and} \quad a_{\mathbf{q}}^\dagger = \frac{1}{\sqrt{N}} \sum_{\mathbf{t}_m} e^{i\mathbf{q}\cdot\mathbf{t}_m} a_m^\dagger, \tag{17.41b}$$

where the allowed N vectors $\mathbf{q}$ are defined in the first Brillouin zone. In particular we have

$$\sum_{\mathbf{t}_m} a_m^\dagger a_m = \sum_{\mathbf{q}} a_{\mathbf{q}}^\dagger a_{\mathbf{q}}, \qquad \sum_{\mathbf{t}_m} a_m^\dagger a_{m+n} = \sum_{\mathbf{q}} e^{i\mathbf{q}\cdot\mathbf{t}_n} a_{\mathbf{q}}^\dagger a_{\mathbf{q}}, \tag{17.41c}$$

as can be seen by inspection (here $\mathbf{t}_n$ denotes a translation vector).

Using the relations (17.41), the Hamiltonian (17.40) can be recast in the form

$$\mathcal{H} = E_0 + 2n_c J S \sum_{\mathbf{q}} [1 - \gamma(\mathbf{q})] a_{\mathbf{q}}^\dagger a_{\mathbf{q}} \quad \text{where} \quad \gamma(\mathbf{q}) = \frac{1}{n_c} \sum_{\mathbf{t}_l} e^{i\mathbf{q}\cdot\mathbf{t}_l} \quad (17.42a)$$

and $\mathbf{t}_l$ ($I = 1, 2, \ldots, n_c$) indicate the translation vectors connecting a site with its first neighbors. We thus obtain an Hamiltonian which is equivalent to a set of harmonic oscillators with frequency

$$\hbar\omega(\mathbf{q}) = 2n_c J S [1 - \gamma(\mathbf{q})]. \qquad (17.42b)$$

The quanta of energy $\hbar\omega(\mathbf{q})$ are called *magnons*.

For a simple cubic lattice of parameter a, for instance, the dispersion relation (17.42b) reads

$$\hbar\omega(\mathbf{q}) = 12 J S \left[1 - \frac{1}{3} \cos q_x a - \frac{1}{3} \cos q_y a - \frac{1}{3} \cos q_z a \right].$$

For small values of q ($q \ll 2\pi/a$) we have for a ferromagnetic crystal the quadratic expression

$$\hbar\omega(\mathbf{q}) = 2 J S a^2 q^2 = D q^2 \quad \text{with} \quad D = 2 J S a^2, \qquad (17.43)$$

which suggests for the magnon an effective mass $m_{\text{mag}}^* = \hbar^2 / 4 J S a^2$; this amounts to about 20 electron masses (for $S = 1/2$, $a \approx 5 a_B$ and $J \approx 10^{-2}$ eV).

The energy-wavevector dispersion relation of the spin waves can be fruitfully studied by means of inelastic neutron scattering experiments. From measurements of the energy difference of the ingoing and outgoing neutron energies and wavevectors, the energies and the corresponding wavevectors of the spin wave can be obtained, and the spin-wave relationship $\hbar\omega(\mathbf{q})$ versus $\mathbf{q}$ can be worked out.

Low Temperature Magnetization in Ferromagnetic Crystals

We can now show how the spontaneous magnetization varies with temperature in a ferromagnet (at least in the low temperature limit, when the interaction among spin waves can be neglected and there is a small number of excitations). The average number of quasi particles, or magnons, in a state of wavevector $\mathbf{q}$ is given by the Bose-Einstein statistics

$$\langle n_{\mathbf{q}} \rangle = \frac{1}{e^{\hbar\omega(\mathbf{q})/k_B T} - 1}.$$

The spontaneous magnetization at $T = 0$ of a ferromagnetic crystal of volume V, at sufficiently low temperatures, where the spin-wave theory holds, is given by the expression

$$M(T) = \frac{1}{V} g\mu_B \sum_{\mathbf{t}_m} \langle S - a_m^\dagger a_m \rangle = \frac{1}{V} g\mu_B \left[NS - \sum_{\mathbf{t}_m} \langle a_m^\dagger a_m \rangle \right].$$

It follows

$$M(0) - M(T) = \frac{1}{V} g\mu_B \sum_{\mathbf{t}_m} \langle a_m^\dagger a_m \rangle \equiv \frac{1}{V} g\mu_B \sum_{\mathbf{q}} \langle a_\mathbf{q}^\dagger a_\mathbf{q} \rangle.$$

Converting, as usual, the sum over the vectors $\mathbf{q}$ into an integral over the three-dimensional Brillouin zone times $V/(2\pi)^3$, one finds

$$M(0) - M(T) = g\mu_B \frac{1}{(2\pi)^3} \int_{B.Z.} \frac{1}{\exp[\hbar\omega(\mathbf{q})/k_B T] - 1} \, d\mathbf{q}. \qquad (17.44a)$$

At small temperatures, only the occupation numbers of magnons of small q are significant; we thus use the quadratic form $\hbar\omega(\mathbf{q}) = Dq^2$ (for cubic lattices) and extend the integral over the whole reciprocal space (with negligible error); this gives

$$M(0) - M(T) = g\mu_B \frac{1}{(2\pi)^3} \int \frac{1}{\exp[Dq^2/k_B T] - 1} \, d\mathbf{q}. \qquad (17.44b)$$

With the change of variable

$$\frac{Dq^2}{k_B T} = x \quad \text{and} \quad d\mathbf{q} = 4\pi q^2 dq = 2\pi \left(\frac{k_B T}{D} \right)^{3/2} x^{1/2} \, dx$$

one obtains

$$M(0) - M(T) = g\mu_B \frac{1}{4\pi^2} \left(\frac{k_B T}{D} \right)^{3/2} \int_0^\infty \frac{x^{1/2}}{e^x - 1} \, dx. \qquad (17.44c)$$

Thus $M(0) - M(T)$ is proportional to $T^{3/2}$ a famous law due to Bloch, confirmed by experimental results. Notice that the mean field theory fails to give the correct power law $T^{3/2}$ of Eq. (17.44c), and rather gives the exponential behavior $\exp(-2T_c/T)$ of Eq. (17.11).

It is of interest to notice that the integral in Eq. (17.44b), at small wavevectors and finite temperature, takes the form $\int (1/q^2) \, d\mathbf{q}$; this integral converges at small wavevectors only if $d\mathbf{q}$ is three-dimensional. Thus if one calculates the spontaneous magnetization of the two- and the one-dimensional ferromagnets in analogy to the three-dimensional case, one finds an integral which diverges at small wavevectors. From this divergence at small q, it can be inferred that no spontaneous magnetization occurs in the one- and two-dimensional isotropic Heisenberg models, a conclusion in agreement with the exact theorem of Mermin and Wagner. Thus magnetic order is possible in the three-dimensional isotropic Heisenberg model, but not in lower dimensions. Notice, however, that this exact result applies to the case of isotropic exchange interaction, but not in the anisotropic case, where a gap in the spin-wave spectrum may occur, and ferromagnetic layers become possible.

Considerations on Spin Waves in Antiferromagnetic Crystals

The extension of the spin-wave formalism to antiferromagnetic crystals requires particular care and appropriate implementations, as it becomes apparent from the following
considerations. The simplest model of an antiferromagnetic crystal is that of two interpenetrating sublattices, say A and B, with spin-up and spin-down, respectively.

By analogy with the ferromagnetic state, one could attempt to write the ground state
of the antiferromagnet in the form

$$|S\rangle_{1A} \cdot |-S\rangle_{1B} \cdots |S\rangle_{mA} \cdots |-S\rangle_{nB} \cdots;$$

however such a state interacts ("resonates") with states in which the spin on a site of
sublattice A is decreased by 1 and the spin on a neighbor site of sublattice B is increased
by 1. Thus for antiferromagnets we cannot say easily what the ground state is; in fact the
ground state is not simply of the type $|\uparrow\downarrow\uparrow\downarrow\uparrow\downarrow \cdots\rangle$, but must be expressed as a super-
position of all spin configurations with total spin component S_z equal to zero (resonant
ground-state picture). The quantum fluctuations of the spins make the classical picture
and the quantum picture rather different; actually, even in three-dimensional antiferromagnets at $T = 0$, the full magnetization of a given sublattice is never achieved. We do
not dwell on this and other aspects of the spin-wave theory in antiferromagnetism and
other magnetic structures; for a thorough analysis, based on the Holstein-Primakoff
transformations, we refer to the literature [see for instance A. Altland and B. Simons
"Condensed Matter Field Theory" (Cambridge University Press, Cambridge, 2010)].

17.5   The Ising Model with the Transfer Matrix Method

In the previous sections, we have discussed some aspects of magnetism essentially on
the basis of the Heisenberg exchange Hamiltonian, of the type of Eq. (17.21). In particular we have considered the mean field theory and we have seen that its inadequacies
in the low temperature limit can be (partially) settled with the spin-wave theory. The
mean field theory presents serious limitations also in the study of the magnetic properties near the critical temperature. In order to study some elementary aspects of phase
transitions at criticality, we consider the Ising approximation (17.22) to the Heisenberg
Hamiltonian; the Ising model, in fact, permits a trivial simplification of the eigenvalue
problem and a consequent easier attention to the physics of phase transitions.

Transfer Matrix Method for the One-Dimensional Ising Model

The one-dimensional Ising model, with nearest neighbor spin interactions and in the
presence of an external magnetic field, can be described in its simplest formulation
as follows. We introduce on each site of a linear chain the spin variable s_i, which can
take only the values ± 1. The interaction energy between nearest neighbor spins and the
Zeeman interaction energy is described by the two terms of the following Hamiltonian

$$\mathcal{H} = -J \sum_{i=1}^{N} s_i s_{i+1} - h \sum_{i=1}^{N} s_i, \tag{17.45}$$

where the sum extends over a (very large) number N of sites, and periodic boundary conditions are adopted, so that $s_{N+1} \equiv s_1$ (in the thermodynamic limit $N \to \infty$ the choice of boundary conditions becomes irrelevant). We assume $J > 0$, which favors a parallel alignment of spins; h is the strength of the applied external magnetic field times the magnetic moment on each site (taken as unity for simplicity).

In the case of a one-dimensional Ising model, we can determine exactly the partition function and the thermodynamic functions, and then show that no spontaneous magnetization and no phase transition occur for a one-dimensional chain.

The partition function of the one-dimensional Ising chain is given by

$$Z = \sum_{\{s_i = \pm 1\}} \exp \left[\beta J (s_1 s_2 + s_2 s_3 + s_3 s_4 + \cdots) + \beta h (s_1 + s_2 + s_3 + \cdots) \right],$$

$$(17.46\text{a})$$

where $\beta = 1/k_B T$, and the sum runs over all the 2^N spin configurations, obtained by assigning the values $s_i = \pm 1$ to each of the N spins. To capture the fact that the multiple sum over the spin configurations is a matrix product, we rewrite the partition function (17.46a) in the form

$$Z = \sum_{\{s_i = \pm 1\}} \exp \left[\beta J s_1 s_2 + \beta h (s_1 + s_2)/2 \right] \cdot \exp \left[\beta J s_2 s_3 + \beta h (s_2 + s_3)/2 \right] \cdots$$

$$= \sum_{\{s_i = \pm 1\}} M(s_1, s_2) M(s_2, s_3) \cdots M(s_{N-1}, s_N), \qquad (17.46\text{b})$$

where the functions $M(s_i, s_{i+1})$ are defined as follow

$$M(s_i, s_{i+1}) = \exp \left[\beta J s_i s_{i+1} + \beta h (s_i + s_{i+1})/2 \right]. \qquad (17.46\text{c})$$

It is convenient to collect the four values of the function $M(s_i, s_j)$ in the form of a matrix M of rank 2. In a periodic lattice, these matrices are all equal, and the expression (17.46c) reads

$$M = \begin{pmatrix} e^{\beta J + \beta h} & e^{-\beta J} \\ e^{-\beta J} & e^{\beta J - \beta h} \end{pmatrix}. \qquad (17.46\text{d})$$

The above matrices are called *transfer matrices*, and borrow the name from the evident analogy with the transfer matrix technique describing wave propagation along a chain of atoms (as discussed in Chapter 1). The partition function (17.46b) thus takes the compact form

$$\boxed{Z = \text{Tr } M^N}, \qquad (17.47)$$

and the trace of the matrix M^N can be explicitly calculated as follows.

The two eigenvalues of the M matrix are real, positive and given by

$$\lambda_\pm = e^{\beta J} \left[\cosh \beta h \pm \sqrt{\sinh^2 \beta h + e^{-4\beta J}} \right].$$

The partition function (17.47) becomes

$$Z = \lambda_+^N + \lambda_-^N \approx \lambda_+^N;$$

since N is very large, the relevant contribution to the partition function is determined by the largest eigenvalue of the transfer matrix. For the free energy we have

$$F = -k_B T \ln \lambda_+^N = -NJ - Nk_B T \ln \left[\cosh \beta h + \sqrt{\sinh^2 \beta h + e^{-4\beta J}} \right].$$

The thermodynamic function $F(h, T)$ is a continuous function of h and T, without singularities; no phase transition is thus expected. In the presence of an external field, the magnetization of the Ising chain becomes

$$M = -\frac{1}{N} \frac{\partial F}{\partial h} = \frac{\sinh \beta h}{\sqrt{\sinh^2 \beta h + e^{-4\beta J}}}.$$

In particular, it is easily verified that the spontaneous magnetization, given by $-\partial F / \partial h$ in the limit $h \to 0$, vanishes at any finite temperature; thus a linear chain cannot be ferromagnetic.

We consider now the mean energy of the linear chain of interacting spins, and its heat capacity, in the absence of a magnetic field. We have

$$Z(h = 0, T) = [2\cosh \beta J]^N \tag{17.48a}$$

and

$$\langle E \rangle = k_B T^2 \frac{\partial \ln Z}{\partial T} = -NJ \tanh \beta J. \tag{17.48b}$$

The heat capacity becomes

$$C_V = \frac{\partial \langle E \rangle}{\partial T} = Nk_B \left(\frac{J}{k_B T} \right)^2 \frac{1}{\cosh^2 (J/k_B T)} \tag{17.48c}$$

and the behavior of C_V is shown in Figure 17.6.

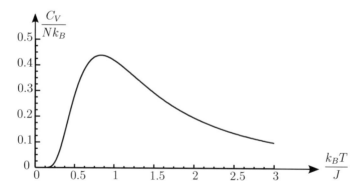

Figure 17.6 Specific heat of a linear chain of spins coupled by exchange interaction in the Ising model.

(a) $+ + + + + + + + + + + + +$

(b) $+ + + + + - - - - - - - -$

Figure 17.7 (a) Schematic representation of a linear Ising chain with all spins parallel to each other. (b) Suppression of long-range order in a linear chain, by a single break.

From the behavior of C_V, and from the fact that the spontaneous magnetization vanishes at any finite temperature, we can infer the following. (i) In a linear chain there is no long-range order (otherwise a spontaneous magnetization would occur) and (ii) In a linear chain, at low temperatures $k_B T \ll k_B T_J = J$, there is a short-range order; this short-range order is relaxed gradually as the temperature T increases at or above the temperature T_J.

We can understand qualitatively the results so far obtained, noticing that in a linear chain the long-range order is broken simply by a change of spin orientation at a single site, as schematized in Figure 17.7. The free energy of a linear chain of spins, with a single break, is given by

$$F = E_0 + 2J - k_B T \ln N,$$

where E_0 is the ground-state energy of a linear chain of fully ordered spins, $2J$ is the energy required to create a single break, $S = k_B \ln N$ is the entropy due to the fact that the break can occur at each of the N sites of the lattice. For any finite temperature T and for N sufficiently large, we can have a decrease of free energy (and thus a more stable situation) breaking the long-range order of the linear chain, rather than leaving it perfect.

Considerations on the Transfer Matrix Solution of the Two-Dimensional Ising Model

We have seen that the one-dimensional Ising model is a paramagnetic system, and no ferromagnetic phase transition can occur at finite temperature. The situation is entirely different in two- and three-dimensional Ising models, where a non-zero spontaneous magnetization can occur at finite temperatures, as can be inferred from the following qualitative remarks. Consider a two-dimensional lattice and two connected regions of opposite spins separated by a borderline involving N sites (see Figure 17.8). The free energy for the situation of Figure 17.8(b) is approximately

$$F \approx E_0 + 2NJ - k_B T \ln 3^N,$$

where E_0 is the energy of the fully ordered lattice of spins, $2NJ$ is (approximately) the energy required to create N breaks, and $S = k_B \ln 3^N$ is an estimate of the entropy; in fact at each point of the borderline we have approximately three choices of keeping on with the border line. Thus if

$$k_B T < \frac{2J}{\ln 3}$$

the ordered state is stable. [For three-dimensional crystals, the topological situation further favors long-range order].

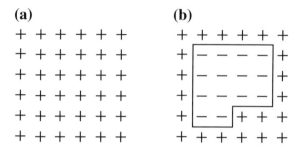

Figure 17.8 Break of long-range order in a two-dimensional Ising system.

The two-dimensional Ising model in zero magnetic field has been solved *exactly* by L. Onsager in his famous paper Phys. Rev. *65*, 117 (1944), using the transfer matrix technique: this formalism becomes extremely demanding from a technical point of view passing from one to two dimensions, although the entailed concepts are rather akin. The Onsager solution constitutes a remarkable piece of theoretical work. At the end of a tour-de-force treatment Onsager was able to find the analytic expression of the free energy, whose singularity shows that the infinite crystal has an order-disorder transition at a temperature T_c given by the condition

$$\boxed{\frac{J}{k_B T_c} = \frac{1}{2} \ln (1 + \sqrt{2}) = 0.4407}.$$

(17.49)

Later, within the transfer matrix formalism, the spontaneous magnetization of the two-dimensional Ising model (and the entailed critical exponents) have been calculated exactly by C. N. Yang, Phys. Rev. *85*, 808 (1952). The exact result of Eq. (17.49) for the critical temperature, and the exact results for the critical exponents (reported in Table 17.2) have been often used in the literature as the natural test for approximate methods. Although the analytic results appear surprisingly simple, no easy way has been found to reduce significantly the technical difficulties in the intermediate steps, or to obtain exact solutions in higher dimensions, or in more realistic situations [see for instance K. Huang "Statistical Mechanics", Wiley, New York, 1987, Chapter 15]. Due to the difficulty and exceptionality of exact analytic results, alternative approaches have been developed in the literature. The approaches include, among others, numerical methods, diagrammatic procedures, Monte Carlo simulations, renormalization group formalism; for brevity, we choose to confine our attention only on some relevant aspects of the latter. In next section, we will then consider at a minimal level of technicality the treatment with the renormalization group of the one-dimensional and two-dimensional Ising models: the motivation is not to learn more with respect to the exact results of the transfer matrix method, but rather to give a flavor of the innovative concepts of the renormalization formalism. [For an historical examination and philosophical reflection of the impact of the Wilson renormalization group approach in quantum field theory and in several other branches of the physics see for instance M. E. Fischer, Rev. Mod. Phys. *70*, 653 (1998)].

17.6   The Ising Model with the Renormalization Group Theory

The basic idea of the renormalization group theory to describe systems at the criticality is to define a transformation on the initial Hamiltonian so that the number of sites of the system (or more generally the number of degrees of freedom of the system) is thinned, say from N to $N/2$, the scale of lengths is increased by $2^{1/d}$ (where d is the dimensionality), and at the same time the partition function is not changed. It is apparent that the same transformation can be applied to the transformed system itself; the aim is to find the *fixed points* of this mapping procedure, since systems at criticality are invariant under scale change.

It is not our purpose to illustrate in full generality the renormalization group formalism and its procedures; rather we confine our attention to some practical and working aspects of the formalism using the paradigmatic one- and two-dimensional Ising models as examples. We should also notice that the renormalization-decimation procedure, described in Section 5.8.3 for the electronic properties of crystals, is aimed at keeping invariant the Green's function of the system (rather than the partition function); although quite different in technical aspects and purpose, the renormalization-decimation procedure of Section 5.8.3 owes much conceptually to the Wilson theory of critical phenomena.

Renormalization Group Theory for the One-Dimensional Ising Model

The one-dimensional Ising model offers the simplest opportunity to define a renormalization transformation. We have already seen, using the transfer matrix method, that the one-dimensional Ising model does not present a phase transition; it is however instructive to recover this same conclusion from the point of view of renormalization group theory; the key point in fact is that the renormalization theory can be also applied to Ising models in higher dimensions, differently from the already discussed transfer matrix method.

The Hamiltonian of the zero-field one-dimensional Ising model, with nearest neighbor interactions only, can be written in the form

$$\mathcal{H} = -J \sum_{i=1}^{N} s_i s_{i+1} \qquad (J > 0),$$

where the variables s_i can assume the values ± 1, and the sum extends over a (very large) number N of sites. In view of the calculation of the partition function, it is common practice to absorb the usual Boltzmann factor $\beta = 1/k_B T$ into the definition of the parameters entering in $\mathcal{H}$; this provides the so called "reduced" or "dimensionless" Hamiltonian

$$\mathcal{H} = -K \sum_{i=1}^{N} s_i s_{i+1} \qquad \left(K = \frac{J}{k_B T} > 0 \right), \tag{17.50}$$

where the dimensionless quantity K is the only coupling parameter of the present model.

The partition function corresponding to the Hamiltonian (17.50) reads

$$Z(K, N) = \mathrm{Tr}\, e^{-\mathcal{H}} = \sum_{\{s_i=\pm 1\}} \exp\left[K\,(s_1s_2 + s_2s_3 + s_3s_4 + s_4s_5 + \cdots)\right], \quad (17.51)$$

where the sum must be performed over the 2^N possible configurations $\{s_i\}$; a configuration is specified by the sequence of values ± 1 given to the variables $s_1, s_2, \ldots, s_N$. Notice that for $T \to 0$, the coupling parameter K goes to infinity and only the ground state is occupied; the limit $K = \infty$ implies perfect alignment of spins, and is called *ferromagnetic point*. Notice also that for $T \to \infty$ the coupling parameter K vanishes and $Z(0, N) = 2^N$; the value $K = 0$ is called *paramagnetic point*.

To perform the multiple sum in Eq. (17.51) it is convenient to cast it in the form

$$Z(K, N) = \sum_{\{s_i=\pm 1\}} \exp\left[K\,(s_1s_2 + s_2s_3)\right] \cdot \exp\left[K\,(s_3s_4 + s_4s_5)\right] \cdots, \quad (17.52a)$$

where we have collected in square brackets the terms containing the variables $s_2, s_4, \ldots$ (as our intention is to eliminate all even site variables). The first term in Eq. (17.52a), after summation over $s_2 = \pm 1$ can be written in the exponential form

$$e^{K(s_1+s_3)} + e^{-K(s_1+s_3)} \equiv A(K)e^{B(K)s_1s_3}. \tag{17.52b}$$

In fact, the requirement that Eq. (17.52b) holds for each of the four possibilities $s_1, s_3 = \pm 1$ gives

$$\begin{cases} s_1 = s_3 \;\; = +1 \; (\text{or} -1) & \Longrightarrow & 2\cosh 2K = A(K)e^{B(K)}, \\ s_1 = -s_3 = +1 \; (\text{or} -1) & \Longrightarrow & 2 = A(K)e^{-B(K)}. \end{cases}$$

From the above two equations we have

$$A(K) = 2\sqrt{\cosh 2K} \qquad \text{and} \qquad B(K) = \frac{1}{2}\ln\cosh 2K \equiv K^{(1)}. \tag{17.52c}$$

It can be noticed that

$$K^{(1)} = \frac{1}{2}\ln \frac{e^{2K} + e^{-2K}}{2} < \frac{1}{2}\ln e^{2K} = K; \tag{17.52d}$$

thus, in the present case the renormalized coupling is always smaller than the original one.

The elimination of all even sites in a single stroke allows us to express the partition function (17.51) in the form

$$Z(K, N) = [A(K)]^{N/2} \cdot Z(K^{(1)}, N/2), \tag{17.53}$$

where $Z(K^{(1)}, N/2)$ is the partition function corresponding to the renormalized Hamiltonian

$$\mathcal{H}^{(1)} = -K^{(1)} \sum_{i=1}^{N/2} s_i s_{i+1}. \tag{17.54}$$

In the renormalized Hamiltonian $\mathcal{H}^{(1)}$, the number of preserved sites is half of the original number, and the length scale is now $a^{(1)} = 2a$. According to Eq. (17.53), the original partition function $Z(K, N)$ is now expressed (apart a multiplicative factor) in terms of the partition functions of the new system with half the number of spins, and a change of the coupling constant. The free energy per spin of the original system and of the renormalized system are given by

$$f(K) = -\frac{1}{N} \ln Z(K, N) \quad \text{and} \quad f(K^{(1)}) = -\frac{1}{N/2} \ln Z(K^{(1)}, N/2).$$

Taking the logarithm of both members of Eq. (17.53), it is seen that the renormalization relation for the free energy per site reads

$$f(K^{(1)}) = 2f(K) + \ln A(K). \tag{17.55}$$

The renormalization procedure is schematically pictured in Figure 17.9.

The transformed Hamiltonian (17.54) in the present case has the same form as the original Hamiltonian (17.50), and is just ready as it stands for iterative procedures. The recursion transformation of the coupling constant is thus

$$K^{(1)} = \frac{1}{2} \ln \cosh 2K \tag{17.56a}$$

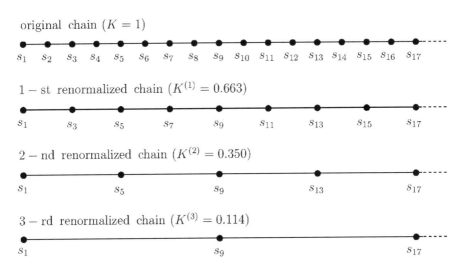

Figure 17.9 Schematic indication of the renormalization procedure for the one-dimensional Ising chain of spins with coupling constant K (taken for instance equal to 1). The sum over spin variables $s_2, s_4, s_6, \ldots$ is performed and this leaves a new chain that involves only the spin variables $s_1, s_3, s_5, \ldots$ interacting with coupling constant $K^{(1)} = (1/2) \ln \cosh 2K$. The procedure is iterated and this leaves a new chain that involves only the spin variables $s_1, s_5, s_9, \ldots$ interacting with coupling constant $K^{(2)} = (1/2) \ln \cosh 2K^{(1)}$. At the 20-th iteration, for instance, $2^{20} \equiv 1 \cdot 048 \cdot 576$, and the preserved sites are $s_1, s_{1 \cdot 048 \cdot 577}, s_{2 \cdot 097 \cdot 153}, \ldots$ and the coupling constant is $K^{(20)} < 10^{-6}$.

$$0 \qquad\qquad\qquad\qquad K \qquad\qquad\qquad\qquad \infty$$

Figure 17.10 Flow diagram of the recursive transformation $K^{(i+1)} = (1/2)\ln\cosh 2K^{(i)}$ for the coupling parameter of the one-dimensional Ising model. The two fixed points are at $K^* = 0$ and $K^* = \infty$. Arrows indicate the flow direction from the repulsive fixed point $K^* = \infty$ to the attractive fixed point $K^* = 0$ of the renormalization transformations.

and in general

$$K^{(i+1)} = \frac{1}{2}\ln\cosh 2K^{(i)}, \tag{17.56b}$$

(with $i = 1, 2, \ldots$). The fixed points of the mapping procedure (17.56) are defined as those points for which

$$\boxed{K^* = \frac{1}{2}\ln\cosh 2K^*}. \tag{17.57}$$

In the present mapping procedure, we have seen that the renormalized coupling is always smaller than the initial one; this means that the spins in the transformed system are therefore more weakly interacting. After repeating the procedure, we obtain a system with an even smaller coupling constant. After a sufficiently number of iterations, we arrive at a system of free spins, as indicated in Figure 17.9.

The fixed points of the mapping procedure (17.56) are well illustrated by the "flow diagram" graphical procedure. Suppose we specify some initial value of K; then we apply Eqs. (17.56) recursively. Since $K^{(1)}(K) < K$ for any K, we find that the coupling constant $K^{(i)}$, for increasing values of i, moves toward zero whatever is the chosen initial value of K. The flow diagram of Eqs. (17.56) is indicated in Figure 17.10. For the one-dimensional Ising problem, the flow diagram indicates that the only fixed points of the whole procedure are the trivial fixed points $K^* = 0$ (i.e. the paramagnetic one) and $K^* = \infty$ (i.e. the ferromagnetic one). The mapping procedure (17.56) does not have non-trivial fixed points, and thus no phase transition exists for the one-dimensional Ising model. The free energy obtained by iterative applications of Eq. (17.55), consists of a finite number of operations involving regular functions, and is thus a regular function itself.

Renormalization Group Theory for the Two-Dimensional Ising Model

Consider a square lattice, formed by N spins with only nearest neighbor interactions $J > 0$, as schematically indicated in Figure 17.11. The reduced zero-field Ising Hamiltonian of the square lattice can be written as

$$\mathcal{H} = -K\sum_{\langle ij\rangle} s_i s_j \qquad \left(K = \frac{J}{k_B T} > 0\right), \tag{17.58}$$

where $\langle ij\rangle$ denotes distinct nearest neighbor pairs. The partition function corresponding to the Hamiltonian (17.58) can be written as

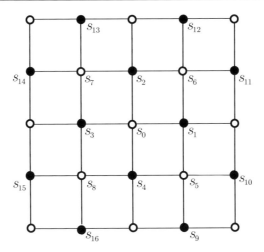

Figure 17.11 Two-dimensional square Ising lattice with nearest neighbor spin interactions; labels of sites are introduced for the application of the renormalization procedure.

$$Z = \sum_{\{s_i = \pm 1\}} \exp\left[K\left(s_0 s_1 + s_0 s_2 + s_0 s_3 + s_0 s_4\right)\right] \cdot \exp\left[K\left(s_5 s_1 + s_5 s_4 + s_5 s_9 + s_5 s_{10}\right)\right] \cdots,$$

where the first few sites near the origin have been labeled following Figure 17.11. The first term in the above expression, after summation over $s_0 = \pm 1$, can be written in the exponential for

$$\exp\left[K\left(s_1 + s_2 + s_3 + s_4\right)\right] + \exp\left[-K\left(s_1 + s_2 + s_3 + s_4\right)\right]$$
$$= A(K) \exp\left[B(K)\left(s_1 s_2 + s_2 s_3 + s_3 s_4 + s_4 s_1 + s_1 s_3 + s_2 s_4\right)\right] \exp\left[C(K) s_1 s_2 s_3 s_4\right].$$

In fact the requirement that the above equality holds for each of the sixteen distinct values of $s_1, s_2, s_3, s_4 = \pm 1$ gives the following three independent conditions

$$\begin{cases} s_1 = s_2 = s_3 = s_4 = +1 & \implies & 2\cosh 4K = A \exp\left(6B + C\right), \\ s_1 = s_2 = s_3 = -s_4 = +1 & \implies & 2\cosh 2K = A \exp\left(-C\right), \\ s_1 = s_2 = -s_3 = -s_4 = +1 & \implies & 2 = A \exp\left(-2B + C\right). \end{cases}$$

From the above expressions one obtains

$$A(K) = 2\cosh^{1/2} 2K \cosh^{1/8} 4K,$$
$$B(K) = (1/8) \ln \cosh 4K,$$
$$C(K) = (1/8) \ln \cosh 4K - (1/2) \ln \cosh 2K.$$

The elimination in a single stroke of all the "empty dots" in Figure 17.11 allows us to express the partition function in the form

$$Z(K, N) = [A(K)]^{N/2} \cdot Z(K_1, K_2, K_3, N/2), \tag{17.59}$$

where $Z(K_1, K_2, K_3, N/2)$ is the partition function corresponding to the renormalized Hamiltonian

$$\mathcal{H}^{(1)} = -K_1 \sum_{\langle ij \rangle} s_i s_j - K_2 \sum_{[ij]} s_i s_j - K_3 \sum_{\text{square}} s_i s_j s_k s_l,$$

$$K_1 = 2B(K), \qquad K_2 = B(K), \qquad K_3 = C(K) \tag{17.60}$$

(the factor 2 in the equality $K_1 = 2B(K)$ appears because each nearest neighbor coupling is generated by two different sums on adjacent empty dots). In the reduced Hamiltonian (17.60), the subscript $\langle ij \rangle$ denotes nearest neighbor pairs (of preserved sites) interacting with coupling parameter K_1, the subscript $[ij]$ denotes next nearest neighbor pairs interacting with parameter K_2, and the last term indicates spins on squares interacting with K_3. The number of preserved sites is half of the original number of sites, and the length scale is now $a^{(1)} = \sqrt{2}a$.

The transformed Hamiltonian (17.60) does not have the same form as the original one (17.58), because of the presence of the sum over second nearest neighbors and the sum over the sites of the square. We have thus to make some appropriate manipulations on the form of $\mathcal{H}^{(1)}$, before proceeding to a new renormalization, otherwise interactions of rapidly increasing complexity would be generated. This step is a key point of renormalization procedure, and can be solved (at least approximately) with the help of physical intuition and appropriate technical procedures. Among the various possible lines of attack, we consider here the simple and very instructive suggestion given by H. J. Maris and L. P. Kadanoff, Am. J. Phys. *46*, 652 (1978). [Another very effective and still simple procedure, based on first order cumulant approximation, is considered in Problems 4 and 5.] Needless to say, countless more accurate solutions are available in the literature, but our selection is motivated mainly by sake of technical simplicity, rather than sophistication.

An Approximate Solution of the Two-Dimensional Ising Model with Cutoff at Nearest Neighbor Interactions

Consider the transformed Hamiltonian (17.60) and the first two terms in its right-hand side: a given spin interacts with four neighbor spins with coupling parameter K_1; it also interacts with four next nearest neighbor spins with coupling constant K_2, which is smaller than K_1 by a factor two and has the same sign. It seems reasonable to mimic the aligning effect of second nearest neighbors on the reference spin by absorbing the coupling K_2 into K_1. Since the number of second and first neighbors in the square lattice is the same, we make the approximation $K^{(1)} = K_1 + K_2$, and obtain

$$\boxed{K^{(1)} = \frac{3}{8} \ln \cosh 4K}, \tag{17.61a}$$

and in general

$$\boxed{K^{(i+1)} = \frac{3}{8} \ln \cosh 4K^{(i)}} \quad (i = 1, 2, \dots). \tag{17.61b}$$

[The four-spin interaction K_3 in Eq. (17.60) is neglected; this can be justified *a posteriori* from its smallness at the critical value].

Suppose we specify some initial value of K; then we apply Eq. (17.61a) and iterate it recursively. In the limit of large number of recursions, we find three types of behaviors: (i) For $0 < K < K_c \equiv 0.507$ we find that $K^{(i)} \to 0$; this is evident for small K considering the series development of Eq. (17.61a), that gives $K^{(1)} = 3K^2$ (thus $K^{(1)}(K)$ is systematically smaller than K for small K). (ii) For $K > K_c$ we find $K^{(i)} \to \infty$; this is evident for large values of K considering the series development of Eq. (17.61a), that gives $K^{(1)} = (3/2)K$ (thus $K^{(1)}(K)$ is systematically larger than K for large K). (iii) There is just a unique value $K = K_c$ for which the effective coupling remains unchanged.

The fixed points of the mapping procedure (17.61) satisfy the equation

$$K^* = \frac{3}{8} \ln \cosh 4K^*;$$

the solutions are the trivial values $K^* = 0$, $K^* = \infty$, and the non-trivial value $K^* = 0.507 = K_c$. The flow diagram of Eqs.(17.61) has a non-trivial fixed point as shown in Figure 17.12. If K starts just at the right of K_c, it increases to infinity; if K starts just at the left of K_c, it decreases to zero; at the very value of K_c, the effective coupling between any two preserved nearest neighbor spins (no matter how far apart in real space they are) remains K_c; this fixed point is thus associated with the phase transition. The exact solution of the two-dimensional Ising problem provided by Onsager gives $K_c^{(\text{exact})} = 0.441$ (it is indeed remarkable that, with almost no mathematical or numerical labor, we have arrived so near to the exact result!).

We can now estimate the critical exponents as follows. Consider a renormalization transformation, where the number of spins is reduced from N to $N/2$. Similarly to Eq. (17.55), the renormalization relation for the free energy per spin reads

$$f(K^{(1)}) = 2f(K) + \ln A(K). \tag{17.62a}$$

Near criticality, i.e. $K \approx K_c$, the free energy per spin must contain a "non-analytic" or "singular" contribution, that we indicate by $f_s(K)$. From the observation that $A(K)$ is in any case a regular function of K, we have that the singular part of the free energy satisfies the relation

$$f_s(K^{(1)}) = 2f_s(K). \tag{17.62b}$$

As suggested by exactly soluble models, we assume a non-analyticity of f_s with power-type law

$$f_s(K) = c|K - K_c|^p, \tag{17.62c}$$

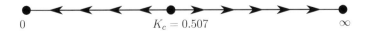

Figure 17.12 Flow diagram for the recursion relation $K^{(i+1)} = (3/8) \ln \cosh 4K^{(i)}$ in the approximate solution of the two-dimensional Ising model.

where c is a constant. We can obtain the value of p by requiring that Eq. (17.62b) is satisfied; we have

$$|K^{(1)} - K_c|^p \equiv 2|K - K_c|^p. \tag{17.62d}$$

The above equation is easily solved by "linearizing" the transformation function $K^{(1)}(K)$ near the critical point K_c. We have

$$K^{(1)}(K) = K_c + (K - K_c)\lambda_c, \tag{17.63a}$$

where

$$\lambda_c = \left(\frac{dK^{(1)}}{dK}\right)_{K=K_c} = \frac{3}{2}\tanh 4K_c = 1.449.$$

Notice that λ_c, which is larger than one, is a *relevant parameter*: unless $K \equiv K_c$ the system flows away from criticality, as evident from Eq. (17.63a). Inserting Eq. (17.63a) into Eq. (17.62d), we obtain

$$\lambda_c^p = 2 \quad \text{and} \quad p = \frac{\ln 2}{\ln \lambda_c} = 1.869. \tag{17.63b}$$

The specific heat, which is related to the second derivative of Eq. (17.62c), has the behavior

$$C_V \approx |K - K_c|^{p-2} \approx |T - T_c|^{p-2};$$

the critical exponent α (defined in Table 17.2) becomes

$$\alpha = 2 - p = 2 - \frac{\ln 2}{\ln \lambda_c} = 0.131. \tag{17.64a}$$

This exponent is just nearly equal to zero and should be compared with the Onsager exact result, which provides a logarithmic divergence in the specific heat, i.e. $\alpha = 0$.

We can do a similar reasoning to obtain the critical exponent for the correlation length, which is an estimate of the spatial extension of the spin-spin correlation function. Let us indicate with $\xi(K)$ the correlation length for the system near criticality, i.e. for $K \approx K_c$. After a renormalization transformation, where the number of spins is reduced from N to $N/2$, the new lattice retains its original topology with a change of scale of $\sqrt{2}$. Lengths, that are now measured in terms of the new lattice parameter, are reduced by a factor $1/\sqrt{2}$; thus we have

$$\xi(K^{(1)}) = \frac{1}{\sqrt{2}}\xi(K).$$

Again, by resorting to exactly soluble models, we assume that the correlation length obeys a power type law $\xi(K) \approx |K - K_c|^{-\nu}$; by substitution into the above equation and linearization, we obtain

$$\nu = \frac{\ln\sqrt{2}}{\ln\lambda_c} = 0.935, \tag{17.64b}$$

a value that compares favorably with the exact Onsager result $v = 1$. Notice that the critical exponents (17.64) satisfy the equality $vd = 2 - \alpha$ (with $d = 2$), which is one of the scaling relations mentioned in Table 17.2.

Before concluding this section, we can summarize some relevant features of the renormalization formalism as follows: the architecture of the renormalization group approach to critical phenomena is based on the calculation of the partition function of a system by a recursive sequence of steps, which successively decimate the preserved degrees of freedom. In each step, the initial system is transformed into a new system similar to the original one, but with different coupling constants. The fixed points of the mapping procedure, besides the trivial values zero and infinity (paramagnetic and ferromagnetic fixed points, respectively), may include non-trivial values. A system is subcritical or supercritical if the mapping procedure makes it flow to the trivial fixed points zero or infinity. The non-trivial fixed points of a mapping procedure are the fingerprints of phase transitions and also determine the critical exponents. The design of successful renormalization group transformations defies strict rules and requires a lot of ingenuity. "One cannot write a renormalization group cookbook" (K. G. Wilson, Adv. Math. *16*, 170 (1975)). Among the merits of the formalism stand the originality in finding novel analytic or numerical tools, and the advances in the art of bringing out universal aspects and common threads in various disciplines of physics.

17.7   Itinerant Magnetism

17.7.1   *The Stoner-Hubbard Model for Itinerant Magnetism*

The physical picture of magnetism, behind Heisenberg-type Hamiltonians, is that of localized spins at the lattice sites, cooperatively interacting among themselves. This phenomenological picture seems suitable for magnetic insulators and also for rare earth metals, whose f bands are quite narrow; the localized spin picture may become inadequate for transition metals with unfilled d bands, where the electrons participating in the magnetic state are itinerant. It would be desirable for these situations a physical picture of magnetism that starts from electrons described by Bloch functions, and then takes proper account of relevant correlation effects; the prototype model for the investigation of itinerant magnetism is constituted by the Stoner-Hubbard Hamiltonian, which embodies electronic correlations at a reasonably intuitive level.

Consider for simplicity a metal, described by a simple Bravais lattice, with a non-degenerate (s-like) conduction band partially filled by the available electrons; this minimal model is here adopted with the only purpose to bring out some qualitative considerations for the occurrence of band magnetism, and not to represent specific materials. In the one-electron approximation the conduction electrons are described by the Hamiltonian of the type

$$\mathcal{H}_0 = \sum_{\mathbf{k}} E(\mathbf{k}) \left[c_{\mathbf{k}\uparrow}^\dagger c_{\mathbf{k}\uparrow} + c_{\mathbf{k}\downarrow}^\dagger c_{\mathbf{k}\downarrow} \right],$$

where $E(\mathbf{k})$ are the conduction band energies, $c_{\mathbf{k}\sigma}^\dagger$ and $c_{\mathbf{k}\sigma}$ are creation and annihilation operators for electrons of spin σ in the conduction band state of vector $\mathbf{k}$. It is well known that the one-electron approximation is not satisfactory in keeping apart electrons of opposite spin. We thus introduce *ad hoc* a term that reflects the Coulomb repulsion of electrons with opposite spins on the same orbital, and arrive at the Stoner-Hubbard model Hamiltonian of the type

$$\mathcal{H} = \sum_{\mathbf{k}} E(\mathbf{k}) \left[c_{\mathbf{k}\uparrow}^{\dagger} c_{\mathbf{k}\uparrow} + c_{\mathbf{k}\downarrow}^{\dagger} c_{\mathbf{k}\downarrow} \right] + U \sum_{\mathbf{t}_m} c_{m\uparrow}^{\dagger} c_{m\uparrow} c_{m\downarrow}^{\dagger} c_{m\downarrow}, \tag{17.65}$$

where U is a phenomenological positive parameter that describes the Coulomb repulsion when two electrons of opposite spins are on the same orbital, and $c_{m\sigma}^{\dagger}$ and $c_{m\sigma}$ are creation and annihilation operators for electrons with spin σ in the orbital localized at the lattice site $\mathbf{t}_m$. The spin-orbital creation operators $c_{m\sigma}^{\dagger}$ (and $c_{m\sigma}$) can be expressed as linear combinations of the band operators $c_{\mathbf{k}\sigma}^{\dagger}$ (and $c_{\mathbf{k}\sigma}$) with coefficients given by the standard Bloch phase factors.

The Hamiltonian (17.65) can lead to macroscopic magnetic effects. Consider, for example, a material with N electrons to be accommodated in the conduction band with $2N$ states (the factor 2 takes into account spin degeneracy). If U is negligible with respect to the bandwidth, the ground state is non-magnetic with the lowest lying $N/2$ orbital Bloch states doubly occupied. In the opposite case in which the band dispersion is negligible, i.e. $E(\mathbf{k})$ is nearly independent of $\mathbf{k}$, then the ground state is magnetic with all states with spin-up occupied and all states with spin-down empty (or vice versa).

To give a semiquantitative analysis of intermediate situations, let us calculate the energy levels and the magnetic susceptibility of an electron system described by the Stoner-Hubbard Hamiltonian (17.65). This many-body system is quite complicated due to the correlation term; we adopt here the so-called "unrestricted Hartree-Fock approximation" and we replace $c_{m\uparrow}^{\dagger} c_{m\uparrow}$ (or $c_{m\downarrow}^{\dagger} c_{m\downarrow}$) with their expectation value on the ground state (we suppose to operate at $T=0$). We indicate by $N_{\uparrow}$ and $N_{\downarrow}$ the total number of conduction electrons with spin-up and spin-down, respectively, and by $n_{\uparrow} = N_{\uparrow}/N$ and $n_{\downarrow} = N_{\downarrow}/N$ the average number of electrons per site. We also consider an applied magnetic field H in the z-direction and the Zeeman energy $\pm \mu_B H$. The energy dispersion curves for electrons with spin-up and spin-down become

$$\sum_{\mathbf{k}} c_{\mathbf{k}\sigma}^{\dagger} c_{\mathbf{k}\sigma} = \sum_{\mathbf{t}_m} c_{m\sigma}^{\dagger} c_{m\sigma} \quad \Longrightarrow \quad \begin{cases} E_{\mathbf{k}\uparrow} = E(\mathbf{k}) + U n_{\downarrow} + \mu_B H, \\ E_{\mathbf{k}\downarrow} = E(\mathbf{k}) + U n_{\uparrow} - \mu_B H, \end{cases} \tag{17.66}$$

where use has been made of the equality indicated in the left part of Eq. (17.66).

For the crystal of volume V, let $D(E)$ denote the density-of-states for both spin directions corresponding to the dispersion curve $E = E(\mathbf{k})$. For the total number of electrons with spin-down we have

$$N_{\downarrow} = \int f(E + U n_{\uparrow} - \mu_B H) \frac{1}{2} D(E) \, dE = \int_{E_0}^{E_F - U n_{\uparrow} + \mu_B H} \frac{1}{2} D(E) \, dE, \tag{17.67}$$

where $f(E)$ is the Fermi-Dirac distribution function (at $T=0$), E_F is the Fermi energy, and E_0 is the bottom of the conduction band. Using Eq. (17.67) for $N_{\downarrow}$ and a similar expression for $N_\uparrow$ we have

$$N_\downarrow - N_\uparrow = \int_{E_0}^{E_F - U n_\uparrow + \mu_B H} \frac{1}{2} D(E)\, dE - \int_{E_0}^{E_F - U n_\downarrow - \mu_B H} \frac{1}{2} D(E)\, dE$$

$$= [U(n_\downarrow - n_\uparrow) + 2\mu_B H]\frac{1}{2} D(E_F) = \frac{1}{2} \frac{D(E_F)}{N} U[N_\downarrow - N_\uparrow] + \mu_B H D(E_F),$$

where $D(E)$ has been approximated by $D(E_F)$ around the Fermi energy; we thus obtain

$$N_\downarrow - N_\uparrow = \frac{\mu_B H D(E_F)}{1 - \dfrac{1}{2}\dfrac{D(E_F)}{N} U}.$$

The magnetization is given by $M = \mu_B(N_\downarrow - N_\uparrow)/V$, and the magnetic susceptibility becomes

$$\boxed{\chi = \frac{1}{V} \frac{\mu_B^2 D(E_F)}{1 - \dfrac{1}{2}\dfrac{D(E_F)}{N} U}.} \tag{17.68}$$

In the case $U = 0$, Eq. (17.68) regains the Pauli magnetic susceptibility of independent particles (see Eq. (15.31)). If the (essentially positive) quantity U is different from zero we see that the magnetic susceptibility is increased. In the case

$$\boxed{\frac{1}{2}\frac{D(E_F)}{N} U = 1} \tag{17.69}$$

the magnetic susceptibility diverges, and we expect a transition to ferromagnetism; relation (17.69) represents the Stoner criterion for onset of band magnetism.

Consider Eqs. (17.66) in the case the material is ferromagnetic (and $H = 0$); in the simplified picture we are considering, the spin-up and spin-down energy bands are split in energy by the amount $\Delta = U(n_\downarrow - n_\uparrow)$; the "*exchange splitting*" Δ is thus proportional to the magnetization and is independent from $\mathbf{k}$. In actual materials, majority and minority spin bands are not split rigidly over the whole Brillouin zone; nevertheless the trend for bands with opposite spins to have an almost constant splitting in the energy region around the Fermi level is confirmed by detailed calculations.

17.7.2   Transport in Ferromagnetic Structures and Giant Magnetoresistance

The Stoner criterion (17.69) for magnetism shows that the crystals, candidate to become ferromagnetic, should be characterized by narrow bands and large density-of-states at the Fermi level: this is just the situation of the transition metals Fe, Co, and Ni, the most common ferromagnetic materials.

The understanding of the band structure of ferromagnetic materials has received a great benefit from the extension of the density functional formalism (see Section 4.7)

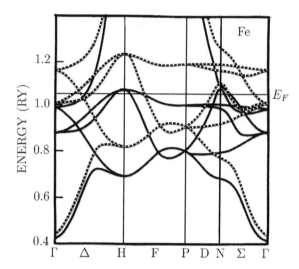

Figure 17.13 Energy bands in bulk ferromagnetic iron along high-symmetry directions. Solid lines are for majority-spin states and dashed lines are for minority-spin states. [reprinted with permission from T. Sasaki, A. M. Rappe and S. G. Louie, Phys. Rev. B *52*, 12760 (1995); copyright 1995 by the American Physical Society]

to spin polarized crystals. The "spin-density functional" formalism is based on the generalized theorem of Hohenberg and Kohn, which states that the ground-state energy of an inhomogeneous electron gas is a functional of the electron density and spin density. Ground-state properties, magnetic moment and spin polarized energy bands can be investigated with great accuracy. In Figure 17.13 we give as an example the band structure and spin splittings of ferromagnetic iron, efficiently described within the plane-wave pseudopotential approach; in Figure 17.14 the corresponding spin resolved density-of-states is reported.

The exchange splitting of the predominantly d-states is of the order of 1–2 eV, while the exchange splitting of predominantly s-like or p-like states is much smaller. Some portions of the d-bands are rather flat, and are responsible of the sharp peaks in the density-of-states. A few considerations on the conductivity of bulk ferromagnetic materials are worthwhile. Differently from ordinary metals, in ferromagnetic materials the conductivity due to charge carriers of the majority spin bands and of the minority spin bands are different, because the two types of bands are not filled equally. In particular electric transport is expected to be mainly sustained by carriers with predominant s-like or p-like character, with a scattering time strongly influenced by the states at the Fermi level with d-like character. Although rigorously speaking $s, p, d, \ldots$ orbitals are hybridized, it is a useful (approximate) picture to distinguish s-like (and p-like) electrons from d-like ones. The electrons with s-like character are spread in a large energy region, have in general small effective masses and high mobility; on the contrary d-like bands are in general rather narrow, with higher effective mass and smaller or negligible mobility. The large exchange splitting of d-bands with respect to the small exchange

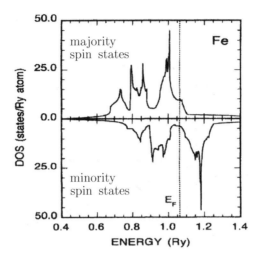

Figure 17.14 Density-of-states of iron for majority spin states (upper panel) and for minority spin states (lower panel). [reprinted with permission from T. Sasaki, A. M. Rappe and S. G. Louie, Phys. Rev. B *52*, 12760 (1995); copyright 1995 by the American Physical Society].

splitting of *s*-like or *p*-like bands produces a *spin-dependent scattering different for majority-spin carriers and minority-spin carriers*. These qualitative remarks justify the early suggestion by Mott of the *two-current model* for ferromagnetic metals, provided spin-flip processes are negligible. This subdivision of current needs of course appropriate refinements and elaborations, but already captures a most relevant peculiarity of transport in ferromagnetic materials.

A most important application of the above concepts occurs in the electric transport in magnetic multilayer systems. These systems are produced artificially and consist of alternating ferromagnetic and non-ferromagnetic layers of nanometer size. The structures are strongly influenced by the application of external magnetic fields, and show a very large magnetoresistance, known as giant magnetoresistance.

The first observations of giant magnetoresistance were done independently in 1988 by the Fert group on Fe/Cr superlattices, and by the Grünberg group on Fe/Cr/Fe trilayers. The investigated structures were designed so that the magnetization of adjacent Fe layers are coupled antiferromagnetically (the non magnetic Cr layer is a few nanometers thick). The phenomenon of "*interlayer exchange coupling*" between ferromagnetic layers has some features in common with the traditional RKKY coupling, and in particular it shows oscillations from positive to negative when the spacer thickness varies. The most remarkable discovery in the study of electronic transport is the huge change (up to the order of 50% or so) of the resistance, occurring when the magnetizations of the Fe layers, initially antiparallel, are aligned along the same direction by application of an external magnetic field. The name *giant magnetoresistance* has been coined to distinguish this type of mechanism from other ones (for instance anisotropic magnetoresistance originated by electron spin-orbit coupling), where the effects are rather small and have a different microscopic origin.

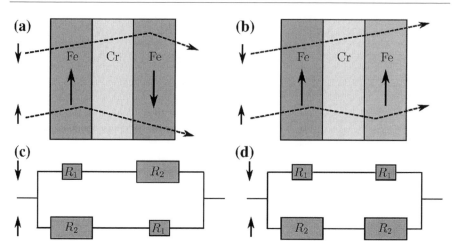

Figure 17.15 Schematic representation of a trilayer structure, formed by two ferromagnetic (Fe) layers and a spacer of non magnetic (Cr) material. (a) In the absence of magnetic fields the magnetic moments of the Fe layers are antiparallel due to the interlayer exchange coupling. (b) Applying a magnetic field the magnetic moments of the Fe layers align. Schematic paths of spin-up and spin-down currents through the device are indicated; a bend in the path indicates that the electron has been scattered. Panels (c) and (d) are the equivalent circuits for the evaluation of the resistances encountered by the two spin currents traveling through the structures of panel (a) and (b), respectively.

A schematic representation of the simplest giant magnetoresistance structure is illustrated in Figure 17.15(a) and (b). The structure consists of three layers: two ferro-magnetic layers (such as Fe) separated by a non-magnetic material (such as Cr). The left figure refers to the case without magnetic field, when the magnetizations of the two Fe layers are opposite; the right figure refers to the case when the magnetization of the Fe layers is aligned by means of an external magnetic field.

The resistance of the device, in the configurations of Figure 17.15(a) (antiparallel) and (b) (parallel) can be intuitively understood in terms of the two-current model, which considers separately the individual currents of majority-spin electrons and minority-spin electrons. When the spin of the carriers and the magnetization of the ferromagnetic layer have the same direction, we expect strong electron scattering and thus high resis-tance to the current flow. In fact from Figure 17.14 it is seen that the density-of-states at the Fermi level for majority spins is higher than the density-of-states for minority spin at the Fermi level, where a pronounced valley is apparent: thus in Fe the conductivity of minority spin carriers is larger than the conductivity of majority ones. [It should be noticed that in other ferromagnetic materials, for instance in Co, the conductivity is larger for majority electrons, due to the specific spin-split d-bands and density-of-states at the Fermi level in this material. The key point of the giant magnetoresistance effect is the asymmetric behavior of majority and minority spins, regardless of which of the two is favored or unfavored].

Let R_1 and R_2 ($R_1 \neq R_2$) denote the resistance felt by the spin electron current when crossing the magnetic layers, in the case its spin is parallel or antiparallel to the ferromagnetic layers magnetization. [For the sake of simplification, we assume current injection perpendicular to the layers and negligible spin flip processes, and do not enter in details for what concerns interface or impurity scattering, hybridization effects and other considerations]. Panels (c) and (d) of Figure 17.15 illustrate the equivalent resistor circuits adopted to interpret qualitatively the system resistance. In the antiparallel (AP) configuration of Figure 17.15(a) the electrons with spin down can travel almost freely through the first ferromagnetic layer, across the Cr region up to the ferromagnetic region with opposite magnetization, where they are strongly scattered; the reverse happens for spin-up electrons current (see Figure 17.15(c)). We thus expect a resistance of $R_1 + R_2$ for both spin polarizations, and a total resistance

$$\frac{1}{R_{AP}} = \frac{1}{R_1 + R_2} + \frac{1}{R_1 + R_2} \implies R_{AP} = \frac{R_1 + R_2}{2}.$$

In the parallel (P) configuration of Figure 17.15(b), spin-down electrons encounter negligible resistance across the structure while spin-up electrons are strongly scattered in both magnetic layers. We thus expect a resistance of $2R_1$ and $2R_2$ for the two spin polarizations, and a total resistance

$$\frac{1}{R_P} = \frac{1}{2R_1} + \frac{1}{2R_2} \implies R_P = \frac{2R_1 R_2}{R_1 + R_2}.$$

It is evident that $R_{AP} > R_P$, and the relative change of resistance, or magnetoresistance, becomes

$$\frac{\Delta R}{R} = \frac{R_{AP} - R_P}{R_{AP}} = \frac{(R_1 - R_2)^2}{(R_1 + R_2)^2}.$$

In essence a giant magnetoresistance material exploits the spin-dependent scattering of the majority-spin and minority-spin carriers, produced by the exchange splitting of the up and down electronic d-bands in ferromagnetic layers.

From an applicative point of view, the fundamental principles of giant magnetoresistance are of major interest in the technological developments for read heads in hard disks drivers, and the continuing advances on storage densities and magnetic sensors. Besides the applicative impact, a fundamental historical merit of giant magnetoresistance is to pioneer the theoretical and experimental investigations on the role of electron spin in transport phenomena, opening new approaches and perspectives in the worlds of electronics and information.

Appendix A. Solved Problems and Complements

Problem 1. Integrals of interest in the study of the indirect exchange interactions in simple metals.

Problem 2. Long-range magnetization oscillations of a localized magnetic impurity embedded in a simple metal.

Problem 3. Long-range exchange interactions between localized magnetic impurities embedded in a simple metal (RKKY interaction).

Problem 4. Ising model on a two-dimensional triangular lattice with the renormalization group.

Problem 5. Ising model on a two-dimensional square lattice with the renormalization group.

Problem 1. Integrals of Interest in the Study of the Indirect Exchange Interactions in Simple Metals

Evaluate the expression of the sum

$$I(k) = \frac{1}{V} \sum_{\mathbf{q}(q \neq k)} \frac{e^{i\mathbf{q}\cdot\mathbf{r}}}{k^2 - q^2} = -\frac{1}{4\pi r} \cos kr,$$ (A.1)

where V is the volume of the crystal, and the sum runs on the whole reciprocal space with the indicated exclusion.

The discrete sum in the reciprocal space is transformed as usual into $V/(2\pi)^3$ times the corresponding integral

$$I(k) = \frac{1}{(2\pi)^3} \mathcal{P} \int d\mathbf{q} \, \frac{e^{i\mathbf{q}\cdot\mathbf{r}}}{k^2 - q^2},$$

where $\mathcal{P}$ denotes "principal part of," and takes proper account of the restriction in the discrete sum. Passing to polar coordinates, and taking the polar axis along the **r** direction we have

$$I(k) = \frac{1}{(2\pi)^3} \mathcal{P} \, 2\pi \int_0^\infty q^2 dq \frac{1}{k^2 - q^2} \int_{-1}^{+1} e^{iqr\mu} \, d\mu \quad (\mu = \cos\theta)$$

$$= \frac{1}{4\pi^2} \frac{1}{ir} \mathcal{P} \int_0^\infty dq \frac{q}{k^2 - q^2} \left[e^{iqr} - e^{-iqr} \right].$$

The square bracket contains two terms; it is convenient to put $q = z$ in the first term and $q = -z$ in the second one; then the two terms can be collected in the compact form

$$I(k) = \frac{1}{4\pi^2} \frac{1}{ir} \mathcal{P} \int_{-\infty}^{+\infty} \frac{z e^{izr}}{k^2 - z^2} \, dz.$$

The integrand has two poles for $z = \pm k$, and near the poles we have the limits

$$R_1 = \lim_{z \to k} 2\pi i (z - k) \frac{z e^{izr}}{k^2 - z^2} = -\pi i e^{ikr};$$

$$R_2 = \lim_{z \to -k} 2\pi i (z + k) \frac{z e^{izr}}{k^2 - z^2} = -\pi i e^{-ikr}.$$

Closing the integral in the upper plane, we have

$$\oint \frac{z e^{izr}}{k^2 - z^2} \, dz = \mathcal{P} \int_{-\infty}^{+\infty} \frac{z e^{izr}}{k^2 - z^2} \, dz - \frac{1}{2} R_1 - \frac{1}{2} R_2 \equiv 0$$

and the result reported in Eq. (A.1) is recovered.

Another integral of interest (closely related to the previous one in the study of the indirect exchange interaction between magnetic impurities in metals) is the following

$$\boxed{\frac{1}{V} \sum_{\mathbf{k}}^{k < k_F} \frac{\cos kr}{4\pi r} e^{i\mathbf{k}\cdot\mathbf{r}} \equiv \frac{1}{V} \sum_{\mathbf{k}}^{k < k_F} \frac{\cos kr}{4\pi r} \cos \mathbf{k}\cdot\mathbf{r} = -\frac{k_F^4}{4\pi^3} F(2k_F r)}, \qquad (A.2)$$

where

$$\boxed{F(x) = \frac{x \cos x - \sin x}{x^4}}. \qquad (A.3)$$

The discrete sum under attention in Eq. (A.2) can be converted into the integral

$$\frac{1}{V} \sum_{\mathbf{k}}^{k < k_F} \frac{\cos kr}{4\pi r} \cos \mathbf{k}\cdot\mathbf{r} = \frac{1}{(2\pi)^3} \frac{1}{4\pi r} \int_{\substack{\text{Fermi} \\ \text{sphere}}} \cos kr \, \cos \mathbf{k}\cdot\mathbf{r} \, d\mathbf{k}$$

$$= \frac{1}{32\pi^4 r} 2\pi \int_0^{k_F} k^2 \, dk \cos kr \int_{-1}^{+1} \cos(kr\mu) \, d\mu = \frac{1}{16\pi^3 r^2} \int_0^{k_F} k \sin 2kr \, dk.$$

Using the indefinite integral

$$\int k \sin 2kr \, dk = \frac{1}{4r^2} \left[-2kr \cos 2kr + \sin 2kr \right]$$

the results summarized in Eqs. (A2) and (A3) are obtained. Notice that $F(x) \approx -1/6x$ for small x, and $F(x) \approx \cos x / x^3$ for large x.

Problem 2. Long-range Magnetization Oscillations of a Localized Magnetic Impurity Embedded in a Simple Metal

Consider a magnetic impurity of spin **S**, *embedded in a simple metal, and interacting with the spin* **s** *of the conduction band electrons via the contact coupling*

$$\mathcal{H}_{\text{int}} = -J\Omega \, \mathbf{S} \cdot \mathbf{s} \, \delta(\mathbf{r}), \qquad (A.4)$$

where J is a phenomenological constant with the dimension of an energy, and Ω is the volume of the unit cell of the crystal. Show that the contact interaction produces an oscillating spin polarization of the free electron gas, without charge polarization.

The conduction band wavefunctions of the metal are assumed to be plane waves of the form $W_{\mathbf{k}\sigma}(\mathbf{r}) = (1/\sqrt{V}) \exp(i\mathbf{k}\cdot\mathbf{r}) | \sigma \rangle$, normalized to one on the volume V of the crystal. At the lowest order in the perturbation $\mathcal{H}_{\text{int}}$, the conduction state $|W_{\mathbf{k}\sigma}\rangle$ is scattered into the state denoted with tilde superscript given by

$$
\begin{aligned}
|\widetilde{W}_{\mathbf{k}\sigma}\rangle &= |W_{\mathbf{k}\sigma}\rangle + \sum_{\mathbf{q}\sigma'(q\neq k)} |W_{\mathbf{q}\sigma'}\rangle \frac{\langle W_{\mathbf{q}\sigma'}|\mathcal{H}_{\text{int}}|W_{\mathbf{k}\sigma}\rangle}{E_{\mathbf{k}} - E_{\mathbf{q}}} \\
&= |W_{\mathbf{k}\sigma}\rangle + \sum_{\mathbf{q}(q\neq k)} |W_{\mathbf{q}}\rangle \frac{\langle W_{\mathbf{q}}|\delta(\mathbf{r})|W_{\mathbf{k}}\rangle}{E_{\mathbf{k}} - E_{\mathbf{q}}} (-J)\Omega\, \mathbf{S} \cdot \mathbf{s}\, |\sigma\rangle \\
&= |W_{\mathbf{k}\sigma}\rangle + \frac{1}{V} \sum_{\mathbf{q}(q\neq k)} \frac{e^{i\mathbf{q}\cdot\mathbf{r}}}{k^2 - q^2} \frac{1}{\sqrt{V}} \frac{2m}{\hbar^2} (-J)\Omega\, \mathbf{S} \cdot \mathbf{s}\, |\sigma\rangle,
\end{aligned} \tag{A.5}
$$

where the free particle energy dispersion $E(\mathbf{k}) = \hbar^2 k^2/2m$ is assumed, and the unit projection operator in the spin space is left implicit. Using the integral elaborated in Eq. (A.1), we have

$$
\widetilde{W}_{\mathbf{k}\sigma}(\mathbf{r}) = \frac{1}{\sqrt{V}} e^{i\mathbf{k}\cdot\mathbf{r}} |\sigma\rangle + \frac{1}{\sqrt{V}} \frac{2mJ\Omega}{\hbar^2} \frac{\cos kr}{4\pi r} \mathbf{S} \cdot \mathbf{s}\, |\sigma\rangle. \tag{A.6}
$$

For what concerns the spin operators we remind in particular

$$
\mathbf{S} \cdot \mathbf{s} = S_z s_z + S_x s_x + S_y s_y \quad \Longrightarrow \quad
\begin{aligned}
\mathbf{S} \cdot \mathbf{s}\, |\uparrow\rangle &= +\tfrac{1}{2} S_z |\uparrow\rangle + \tfrac{1}{2} S_+ |\downarrow\rangle, \\
\mathbf{S} \cdot \mathbf{s}\, |\downarrow\rangle &= -\tfrac{1}{2} S_z |\downarrow\rangle + \tfrac{1}{2} S_- |\uparrow\rangle
\end{aligned}
$$

as recovered from Eq. (16.34).

Setting $\sigma = \uparrow$ in Eq. (A.6), the density of electrons with spin-up corresponding to the wavefunction $\widetilde{W}_{\mathbf{k}\uparrow}(\mathbf{r})$ is given by the expression

$$
\begin{aligned}
n_{\mathbf{k}\uparrow}(\mathbf{r}) &= \left| \frac{1}{\sqrt{V}} e^{i\mathbf{k}\cdot\mathbf{r}} + \frac{1}{\sqrt{V}} \frac{2mJ\Omega}{\hbar^2} \frac{\cos kr}{4\pi r} S_z \frac{1}{2} \right|^2 \\
&= \frac{1}{V} \left[1 + \frac{2mJ\Omega}{\hbar^2} S_z \frac{\cos kr}{4\pi r} \cos \mathbf{k} \cdot \mathbf{r} \right] + O(J^2).
\end{aligned}
$$

Similarly $n_{\mathbf{k}\downarrow}(\mathbf{r}) = 0$ to $O(J^2)$. We sum now the above expression over the $\mathbf{k}$ vectors within the Fermi sphere; using Eq. (A.2) we obtain

$$
n_{\uparrow}(\mathbf{r}) = \frac{N_0}{V} - \frac{2mJ\Omega}{\hbar^2} S_z \frac{k_F^4}{4\pi^3} F(2k_F r) = n_0 \left[1 - \frac{3}{2\pi} \frac{J}{E_F} \Omega k_F^3 F(2k_F r) S_z \right],
$$

where N_0 is the number of electrons for each spin direction, $n_0 = N_0/V$ is the corresponding electron density for each spin direction, $F(x)$ is the range function defined in Eq. (A.3), $E_F = \hbar^2 k_F^2/2m$ is the Fermi energy, and the relation $k_F^3 = 3\pi^2 n = 6\pi^2 n_0$ has been used. It is seen by inspection that the amount of spin polarization is controlled by the ratio J/E_F, while the sign oscillations are controlled by the range function.

A completely similar calculation can be performed for the spin-down electron density; we obtain

$$n_\downarrow(\mathbf{r}) = n_0 \left[1 + \frac{3}{2\pi} \frac{J}{E_F} \Omega k_F^3 F(2k_F r) S_z \right].$$

From the above results, we see that there is *no net charge polarization* (or electric field gradient); in fact

$$n_\uparrow(\mathbf{r}) + n_\downarrow(\mathbf{r}) \equiv 2n_0;$$

however *there is a spin polarization*

$$n_\uparrow(\mathbf{r}) - n_\downarrow(\mathbf{r}) = -\frac{3}{\pi} \frac{J}{E_F} \Omega k_F^3 F(2k_F r) S_z . \qquad (A.7)$$

Since $F(x) \approx \cos x / x^3$ for large x, we see that the spin polarization is of long-range nature, oscillates with distance and falls off with the inverse cube power law r^{-3} at large distances; notice the analogies between the spin polarization oscillations produced by a magnetic impurity and the charge density oscillations (Friedel oscillations, Section 7.5) produced by an ordinary impurity in a metal.

Problem 3. Long-range Exchange Interactions between Localized Magnetic Impurities Embedded in a Simple Metal (RKKY Interaction)

Using second order perturbation theory, obtain the indirect exchange coupling between two localized spins, interacting with the Fermi sea electrons via a contact coupling, and discuss the results.

Suppose the two spins $\mathbf{S}_1$ and $\mathbf{S}_2$, are localized at $\mathbf{r} = 0$ and $\mathbf{R}$, respectively. The perturbation Hamiltonian of the two spins can be taken as

$$\mathcal{H}_{\text{int}} = \mathcal{H}_1 + \mathcal{H}_2 = -J\Omega\, \mathbf{S}_1 \cdot \mathbf{s}\, \delta(\mathbf{r}) - J\Omega\, \mathbf{S}_2 \cdot \mathbf{s}\, \delta(\mathbf{r} - \mathbf{R}). \qquad (A.8)$$

The microscopic mechanism of the indirect exchange interaction can be intuitively described as follows: the first spin $\mathbf{S}_1$ creates a spin polarization, and the other spin $\mathbf{S}_2$ feels this spin polarization via the contact interaction (or vice versa).

According to second order of perturbation theory, the energy change of a conduction electron initially described by the plane wave $W_{\mathbf{k}\sigma}$ becomes

$$\Delta E_{\mathbf{k}\sigma} = \sum_{\mathbf{q}\sigma'(q \neq k)} \frac{\langle W_{\mathbf{k}\sigma} | \mathcal{H}_1 + \mathcal{H}_2 | W_{\mathbf{q}\sigma'} \rangle \langle W_{\mathbf{q}\sigma'} | \mathcal{H}_1 + \mathcal{H}_2 | W_{\mathbf{k}\sigma} \rangle}{E_{\mathbf{k}} - E_{\mathbf{q}}}, \qquad (A.9)$$

and the parabolic dispersion curve $E_{\mathbf{k}} = \hbar^2 k^2 / 2m$ is assumed for the free electron energy.

We take advantage of the fact that $\mathcal{H}_{\text{int}}$ contains operators involving completely independent spin and space coordinates. We also select mixed terms in Eq. (A.9), namely: $\mathcal{H}_1$ in the first matrix element and $\mathcal{H}_2$ in the second, and vice versa. The mixed terms take proper account of the mutual influence of the two magnetic impurities (we thus ignore the self-interaction terms, which are irrelevant in the present context). In summary, we consider

$$
\Delta E_{\mathbf{k}\sigma}^{(\text{mixed})} = \frac{J^2 \Omega^2}{V^2} \frac{2m}{\hbar^2} \sum_{\mathbf{q}(q \neq k)} \frac{e^{i(\mathbf{k}-\mathbf{q})\cdot\mathbf{R}} + e^{-i(\mathbf{k}-\mathbf{q})\cdot\mathbf{R}}}{k^2 - q^2} \langle \sigma | \mathbf{S}_1 \cdot \mathbf{s}\, \mathbf{S}_2 \cdot \mathbf{s} | \sigma \rangle
$$

$$
= \frac{J^2 \Omega^2}{V} \frac{2m}{\hbar^2} \left[-\frac{1}{4\pi R} \cos kR \cdot 2 \cos \mathbf{k} \cdot \mathbf{r} \right] \langle \sigma | \mathbf{S}_1 \cdot \mathbf{s}\, \mathbf{S}_2 \cdot \mathbf{s} | \sigma \rangle;
$$

for the last passage Eq. (A.1) has been used. To obtain the total interaction energy, we carry out the sum

$$
E(R) = \sum_{\mathbf{k}\sigma}^{k<k_F} \Delta E_{\mathbf{k}\sigma}^{(\text{mixed})} = \frac{2m J^2 \Omega^2}{\hbar^2} \frac{k_F^4}{4\pi^3} F(2k_F R)\, \mathbf{S}_1 \cdot \mathbf{S}_2 \tag{A.10}
$$

exploiting Eq. (A.2). The isotropic scalar product $\mathbf{S}_1 \cdot \mathbf{S}_2$ appears because of the general property

$$
\langle \uparrow | \mathbf{S}_1 \cdot \mathbf{s}\, \mathbf{S}_2 \cdot \mathbf{s} | \uparrow \rangle + \langle \downarrow | \mathbf{S}_1 \cdot \mathbf{s}\, \mathbf{S}_2 \cdot \mathbf{s} | \downarrow \rangle = \frac{1}{2} \mathbf{S}_1 \cdot \mathbf{S}_2;
$$

the above identity can be easily worked out by inspection, using the matrix elements of Eq. (16.34), or more formally using general properties of spin operators. Equation (A.10) can also be written in the form

$$
\boxed{E(R) = \frac{1}{4\pi^3} \frac{J^2}{E_F} (\Omega k_F^3)^2 F(2k_F R)\, \mathbf{S}_1 \cdot \mathbf{S}_2}. \tag{A.11}
$$

The Ruderman-Kittel-Kasuya-Yosida interaction (A.11) is ferromagnetic-like for small R, while for large R it can be either ferromagnetic or antiferromagnetic depending on the distance of the two spins.

Problem 4. Ising Model on a Two-dimensional Triangular Lattice with the Renormalization Group

Consider the zero-field Ising model for a two-dimensional triangular lattice. Using blocks of three spins, establish an approximate form of the renormalization transformations, and evaluate the thermal-like critical parameters of the model.

The Ising model on a triangular lattice is schematically indicated in Figure 17.16. The zero-field Hamiltonian can be written as

$$
\mathcal{H} = -K \sum_{\langle ij \rangle} s_i s_j \qquad \left(K = \frac{J}{k_B T} > 0 \right), \tag{A.12}
$$

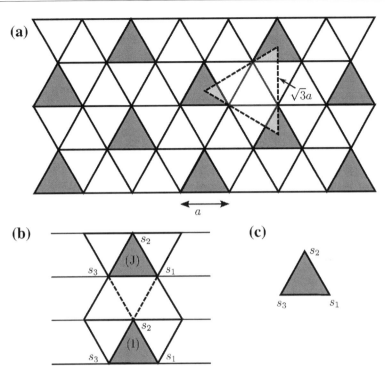

Figure 17.16 Ising model on a triangular lattice. (a) Blocks with three spins each are shaded, showing that block spins are again located on a triangular lattice. (b) Representation of two interacting nearest neighbor blocks, denoted as I, J. (c) The single building brick of the renormalization procedure.

where $J > 0$ is the ferromagnetic nearest-neighbor spin-spin coupling, and the symbol $\langle ij \rangle$ here denotes distinct nearest neighbor pairs of sites. The partition function is

$$Z = \sum_{\{s_i = \mp 1\}} \exp\left[-\mathcal{H}(s_1, s_2, \ldots)\right], \tag{A.13}$$

where the sum is extended over all possible spin configurations.

For the iterative calculation of the partition function, in line with Kadanoff block spin concepts, it is convenient to consider the Niemeijer-van Leeuwen transformation to a new spin variable, defined as the sum of three old spins. The geometrical procedure is illustrated in Figure 17.16(a): for every shaded triangle of three spins, we form a block spin; the resulting structure has the same topology as the initial one, with new lattice constant $a^{(1)} = \sqrt{3}a$.

Consider first the single block of three spins (where the interactions with the other blocks are momentarily neglected) represented in Figure 17.16(c). The Hamiltonian of this isolated cluster reads

$$\mathcal{H}_0 = -K(s_1 s_2 + s_1 s_3 + s_2 s_3). \tag{A.14}$$

The partition function corresponding to the operator $\mathcal{H}_0$ consists of the sum of $2^3 = 8$ terms, generated by the eight possible configurations $s_1 = \pm 1, s_2 = \pm 1, s_3 = \pm 1$. It is convenient to define a block spin $S(S = \mp 1)$ according to the majority rule

$$S = \text{sign}(s_1 + s_2 + s_3). \tag{A.15}$$

Out of the eight spin configurations, four have $S = 1$ and four have $S = -1$. We list below the four configurations with $S = 1$, and also give the corresponding contribution to the partition function.

$$
S = 1 \quad
\begin{array}{llll}
\{\sigma_1\} & \equiv (+1, +1, +1) & \implies & e^{-\mathcal{H}_0(\sigma_1)} = e^{+3K} \\
\{\sigma_2\} & \equiv (-1, +1, +1) & \implies & e^{-\mathcal{H}_0(\sigma_2)} = e^{-K} \\
\{\sigma_3\} & \equiv (+1, -1, +1) & \implies & e^{-\mathcal{H}_0(\sigma_3)} = e^{-K} \\
\{\sigma_4\} & \equiv (+1, +1, -1) & \implies & e^{-\mathcal{H}_0(\sigma_4)} = e^{-K}
\end{array}
\tag{A.16}
$$

The contribution to the partition function of the four configurations with $S = 1$ reads

$$Z_0(K) = \sum_{i=1,\dots,4} e^{-\mathcal{H}_0(\sigma_i)} = e^{+3K} + 3e^{-K}. \tag{A.17}$$

[The four configurations corresponding to the majority spin $S = -1$ have spin components exactly opposite to the ones of Eq. (A.16), but produce the same contribution $Z_0(K)$ to the partition function]. In the following we need the weighted average of the spin variables with the weight function $\exp[-\mathcal{H}_0(\sigma_i)]$ defined as usual

$$\langle s_m \rangle_0 = \frac{\sum_{i=1,\dots,4} s_m e^{-\mathcal{H}_0(\sigma_i)}}{\sum_{i=1,\dots,4} e^{-\mathcal{H}_0(\sigma_i)}} \quad (m = 1, 2, 3); \tag{A.18}$$

this gives

$$\langle s_1 \rangle_0 \equiv \langle s_2 \rangle_0 \equiv \langle s_3 \rangle_0 \equiv \frac{e^{+3K} + e^{-K}}{e^{+3K} + 3e^{-K}} S, \tag{A.19}$$

where $S = 1$ if the configurations of Eq. (A.16) are considered (and $S = -1$ for the opposite configurations).

The next step toward the decimation of the triangular lattice is to consider two blocks, indicated in Figure 17.16(b) and labeled with the indices I and J (in using the results elaborated for a single block we add the subscripts or superscripts I or J whenever necessary). The Hamiltonian of the two interacting blocks can be broken into two parts, describing the intra-block interactions and inter-block interactions, respectively:

$$\mathcal{H} = \mathcal{H}_0 + V,$$

$$\mathcal{H}_0 = -K(s_1^I s_2^I + s_1^I s_3^I + s_2^I s_3^I) - K(s_1^J s_2^J + s_1^J s_3^J + s_2^J s_3^J); \quad V = -K(s_2^I s_1^J + s_2^I s_3^J).$$

Using Eqs. (A.18) and (A.19), the weighted average of V becomes

$$\langle V \rangle_0 = -K \left[\langle s_2^I \rangle_0 \langle s_1^J \rangle_0 + \langle s_2^I \rangle_0 \langle s_3^J \rangle_0 \right] = -2K \left[\frac{e^{+3K} + e^{-K}}{e^{+3K} + 3e^{-K}} \right]^2 S_I S_J. \tag{A.20}$$

The same procedure is now applied to all the blocks of the infinite triangular lattice. Using Eq. (A.20), it is apparent that the weighted average $\langle V \rangle_0$ with all the inter-block interactions reads

$$\langle V \rangle_0 = -2K \left[\frac{e^{3K} + e^{-K}}{e^{3K} + 3e^{-K}} \right]^2 \sum_{\langle IJ \rangle} S_I S_J. \tag{A.21}$$

Equation (A.21) has the same structure as the original Hamiltonian (A.12) with the renormalized interaction:

$$\boxed{K^{(1)} = 2K \left[\frac{e^{4K} + 1}{e^{4K} + 3} \right]^2}. \tag{A.22}$$

The renormalization transformation, based on Eq. (A.21) and summarized by Eq. (A.22) is known as *first order cumulant approximation*, and more details on this approximation are given at the end of this problem.

The mapping procedure (A.22), besides the trivial points $K^* = 0$ and $K^* = \infty$, presents only one non-trivial fixed point given by the equation

$$\frac{e^{4K^*} + 1}{e^{4K^*} + 3} = \frac{1}{\sqrt{2}};$$

the solution is

$$e^{4K^*} = 1 + 2\sqrt{2} \quad \Longrightarrow \quad K^* = \frac{1}{4} \ln (1 + 2\sqrt{2}) \approx 0.336.$$

It is worthwhile to observe that, in spite of the various approximations adopted in the effort to keep technicality at the minimum possible level, so little labor has produced a so satisfactory result; in fact the exact theory of Onsager for triangular lattices gives $K^* = (1/4) \ln 3 \approx 0.275$ [see G. H. Wannier, Rev. Mod. Phys. *17*, 50 (1945)]. No wonder that, in the literature, an acceptable higher degree of sophistication has proved quite successful in the study of critical phenomena.

We can now linearize the transformation function near the critical point, and obtain

$$\lambda_c = \left(\frac{dK^{(1)}}{dK} \right)_{K=K^*} \approx 1.663.$$

Similarly to Eqs. (17.64) of the main text (taking into consideration that in the present problem the degrees of freedom are reduced from N to $N/3$ in each step), we have for the critical exponents α and ν the values

$$\alpha = 2 - \frac{\ln 3}{\ln \lambda_c} = -0.16, \qquad \nu = \frac{\ln \sqrt{3}}{\ln \lambda_c} = 1.08, \tag{A.23}$$

which again compare surprisingly well with the exact values $\alpha = 0$ and $\nu = 1$ of Onsager.

Before closing this problem, we give a few details on the first order cumulant approximation, adopted in this problem. The general Hamiltonian of the Ising block lattice can be split in the form

$$\mathcal{H} = \mathcal{H}_0 + V,$$

where $\mathcal{H}_0$ contains the interactions between the spins of the same blocks, while V contains the interactions between spins on adjacent blocks. The partition function reads

$$
\begin{aligned}
Z &= \sum \exp\left(-\mathcal{H}_0 - V\right) = \sum \exp\left(-\mathcal{H}_0\right) \exp\left(-V\right) \\
&= \sum \exp\left(-\mathcal{H}_0\right) \cdot \left[\frac{\sum \exp\left(-\mathcal{H}_0\right) \exp\left(-V\right)}{\sum \exp\left(-\mathcal{H}_0\right)} \right] \\
&\equiv \sum \exp\left(-\mathcal{H}_0\right) \cdot \langle \exp\left(-V\right) \rangle_0,
\end{aligned}
\tag{A.24}
$$

where the sum runs on all spin configurations, and the last factor in Eq. (A.24) is the standard way to indicate the average operation, with weight function $\exp\left(-\mathcal{H}_0\right)$. Equation (A.24) is still exact.

The first factor in Eq. (A.24) is related to the N non-interacting blocks, and is trivially given by the product of N partition functions of the isolated blocks. The second factor in Eq. (A.24) needs approximations. According to the lowest order cumulant approximation, it is assumed that the weighted average of the exponential of V can be reasonably replaced by exponential of the weighted average:

$$\langle \exp\left(-V\right) \rangle_0 \approx \exp\left(-\langle V \rangle_0\right). \tag{A.25}$$

Expansion of the exponentials in both members of Eq. (A25) justifies the equality at the lowest order of perturbation theory (and also shows how to proceed to higher order when necessary). Focusing our attention on Eq. (A21) is justified within the lowest order cumulant approximation. [Problem 4 is a simplified initial part of the instructive and highly accurate elaborations provided in the paper by Th. Niemeijer and J. M. J. van Leeuwen "Wilson theory for 2-dimensional Ising systems" Physica *71*, 17 (1974); in this paper magnetic-like critical exponents and higher order cumulant expansions are also considered. See also D. J. Wallace and R. K. P. Zia "The renormalization group approach to scaling in physics" Rep. Prog. Phys. *41*, 1 (1978)].

Problem 5. Ising Model on a Two-Dimensional Square Lattice with the Renormalization Group

Consider the zero-field Ising model for a two dimensional square lattice. Using blocks of five spins, establish an approximate form of the renormalization transformations, and evaluate the temperature-like critical parameters of the model.

This problem has several points in common with the previous one, so we can be rather schematic. The Ising model on a square lattice is illustrated in Figure 17.17, where we have also shown its construction with five-spin building blocks; the lattice of block spins is still a square with lattice constant $a^{(1)} = \sqrt{5}a$.

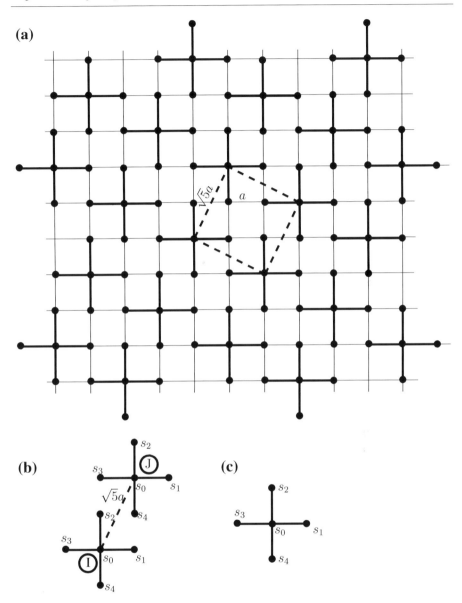

Figure 17.17 Ising model on a square lattice. (a) Blocks with five spins each are indicated, showing that block spins are again located on a square lattice. (b) Representation of two interacting nearest neighbor blocks, denoted as I, J. (c) The single building brick of the renormalized structure.

Consider first the single block of five spins represented in Figure 17.17(c). The Hamiltonian of this isolated cluster reads

$$\mathcal{H}_0 = -K(s_0 s_1 + s_0 s_2 + s_0 s_3 + s_0 s_4) = -K s_0 (s_1 + s_2 + s_3 + s_4). \quad (A.26)$$

The partition function corresponding to the operator $\mathcal{H}_0$ consists of the sum of $2^5 = 32$ terms, generated by the possible configurations of the five spins. It is convenient to define a block spin $S(S = \mp 1)$ according to the majority rule

$$S = \text{sign}(s_0 + s_1 + s_2 + s_3 + s_4). \quad (A.27)$$

The 16 spin configurations with $S = 1$ are listed below.

$$
\begin{array}{llll}
 & \{\sigma_1\} & \equiv (+1; +1, +1, +1, +1) & \Longrightarrow \quad e^{-\mathcal{H}_0(\sigma_1)} = e^{-4K} \\
 & \{\sigma_2\} & \equiv (-1; +1, +1, +1, +1) & \Longrightarrow \quad e^{-\mathcal{H}_0(\sigma_2)} = e^{+4K} \\[6pt]
 & \{\sigma_3\} & \equiv (+1; -1, +1, +1, +1), & \Longrightarrow \quad e^{-\mathcal{H}_0(\sigma_3)} = e^{-2K} \\
 & \{\sigma_4\} & \equiv (+1; +1, -1, +1, +1) & \Longrightarrow \quad e^{-\mathcal{H}_0(\sigma_4)} = e^{-2K} \\
 & \{\sigma_5\} & \equiv (+1; +1, +1, -1, +1) & \Longrightarrow \quad e^{-\mathcal{H}_0(\sigma_5)} = e^{-2K} \\
 & \{\sigma_6\} & \equiv (+1; +1, +1, +1, -1) & \Longrightarrow \quad e^{-\mathcal{H}_0(\sigma_6)} = e^{-2K} \\
 & \{\sigma_7\} & \equiv (-1; -1, +1, +1, +1) & \Longrightarrow \quad e^{-\mathcal{H}_0(\sigma_7)} = e^{+2K} \\
S = 1 & \{\sigma_8\} & \equiv (-1; +1, -1, +1, +1) & \Longrightarrow \quad e^{-\mathcal{H}_0(\sigma_8)} = e^{+2K} \\
 & \{\sigma_9\} & \equiv (-1; +1, +1, -1, +1) & \Longrightarrow \quad e^{-\mathcal{H}_0(\sigma_9)} = e^{+2K} \\
 & \{\sigma_{10}\} & \equiv (-1; +1, +1, +1, -1) & \Longrightarrow \quad e^{-\mathcal{H}_0(\sigma_{10})} = e^{+2K} \\[6pt]
 & \{\sigma_{11}\} & \equiv (+1; -1, -1, +1, +1) & \Longrightarrow \quad e^{-\mathcal{H}_0(\sigma_{11})} = 1 \\
 & \{\sigma_{12}\} & \equiv (+1; -1, +1, -1, +1) & \Longrightarrow \quad e^{-\mathcal{H}_0(\sigma_{12})} = 1 \\
 & \{\sigma_{13}\} & \equiv (+1; -1, +1, +1, -1) & \Longrightarrow \quad e^{-\mathcal{H}_0(\sigma_{13})} = 1 \\
 & \{\sigma_{14}\} & \equiv (+1; +1, -1, -1, +1) & \Longrightarrow \quad e^{-\mathcal{H}_0(\sigma_{14})} = 1 \\
 & \{\sigma_{15}\} & \equiv (+1; +1, -1, +1, -1) & \Longrightarrow \quad e^{-\mathcal{H}_0(\sigma_{15})} = 1 \\
 & \{\sigma_{16}\} & \equiv (+1; +1, +1, -1, -1) & \Longrightarrow \quad e^{-\mathcal{H}_0(\sigma_{16})} = 1
\end{array}
$$

The partition function corresponding to the 16 configurations is

$$Z_0(K) = \sum_{\sigma_i}^{i=1,\dots,16} \exp[-\mathcal{H}_0(\sigma_i)] = 2\cosh 4K + 8\cosh 2K + 6. \quad (A.28)$$

The averages of the spin variables s_1, s_2, s_3, s_4 with weight function $\exp[-\mathcal{H}_0(\sigma_i)]$ are given by

$$\langle s_1 \rangle_0 = \langle s_2 \rangle_0 = \langle s_3 \rangle_0 = \langle s_4 \rangle_0 = \frac{\cosh 4K + 2\cosh 2K}{\cosh 4K + 4\cosh 2K + 3}.$$

As indicated in Figure 17.17(b), the perturbation V between two blocks is

$$V = -K(s_1^I s_4^J + s_2^I s_3^J + s_2^I s_4^J).$$

The weighted average of V is then

$$\langle V \rangle_0 = -3K \left[\frac{\cosh 4K + 2 \cosh 2K}{\cosh 4K + 4 \cosh 2K + 3} \right]^2 = -3K \left[\frac{2 \cosh^2 2K + 2 \cosh 2K - 1}{2 \cosh^2 2K + 4 \cosh 2K + 2} \right]^2 .$$

The transformation function of the coupling parameter in the renormalization process is

$$K^{(1)} = 3K \left[\frac{2 \cosh^2 2K + 2 \cosh 2K - 1}{2 \cosh^2 2K + 4 \cosh 2K + 2} \right]^2 . \tag{A.29}$$

The non-trivial fixed point is determined by the relation

$$\frac{2 \cosh^2 2K^* + 2 \cosh 2K^* - 1}{2 \cosh^2 2K^* + 4 \cosh 2K^* + 2} = \frac{1}{\sqrt{3}} .$$

It holds

$$\cosh 2K^* = \frac{\sqrt{3} - 1 + \sqrt{24 + 10\sqrt{3}}}{4} \qquad \Longrightarrow \qquad K^* \approx 0.593,$$

in reasonable agreement with the exact value $(1/2) \ln (1 + \sqrt{2}) = 0.441$ of Onsager. Linearization of the transformation function at the fixed point gives

$$\lambda_c = \left(\frac{dK^{(1)}}{dK} \right)_{K=K^*} \approx 2.111. \tag{A.30}$$

For the critical exponents α and ν we have

$$\alpha = 2 - \frac{\ln 5}{\ln \lambda_c} \approx -0.154, \qquad \nu = \frac{\ln \sqrt{5}}{\ln \lambda_c} \approx 1.077$$

again very near to the exact Onsager results $\alpha = 0$ and $\nu = 1$, and also very close to the approximate values of the triangular lattice of the previous problem.

Further Reading

Blundell, S. (2001). *Magnetism in condensed matter*. Oxford: Oxford University Press.

Borsa, F., & Tognetti, V. (Eds.) (1988). *Magnetic properties of matter*. Singapore: World Scientific.

Cardy, J. (1996). *Scaling and renormalization in statistical physics*. Cambridge: Cambridge University Press.

Chikazumi, S. (1997). *Physics of ferromagnetism*. Oxford: Clarendon Press.

Coey, J. M. D. (2011). *Magnetism and magnetic materials*. Cambridge: Cambridge University Press.

Domb, C. & Green, M. S. (Eds.), (1972–1983). *Phase transitions and critical phenomena* (Vols. I-VIII). London: Academic Press.

Fert, A. (2008). Origin, development and future spintronics. *Rev. Mod. Phys. 80*, 1517; Grünberg, P. A. (2008). From spin waves to giant magnetoresistance and beyond. *Rev. Mod. Phys. 80*, 1531.

Gatteschi, D., Sessoli, R. & Villain, J. (2006). *Molecular nanomagnets*. Oxford: Oxford University Press.

Huang, K. (1987). *Statistical mechanics*. New York: Wiley.

Jensen, J. & Mackintosh, A. R. (1991). *Rare earth magnetism*. Oxford: Clarendon.

Majlis, N. (2007). *The quantum theory of magnetism*. Singapore: World Scientific.

Müller, K. A. & Rigamonti, A. (Eds.), (1976). *Local properties of phase transitions international school of physics "Enrico Fermi" Course LIX*. Bologna: Società Italiana di Fisica.

Nolting, W., & Ramakanth, A. (2009). *Quantum theory of magnetism*. Berlin: Springer.

Stanley, H. E. (1971). *Phase transitions and critical phenomena*. Oxford: Oxford University Press.

Tsymbal, E.Y., & Žutić, I. (Eds.), (2011). *Handbook of spin transport and magnetism*. Boca Raton, FL: CRC Press, Taylor and Francis.

Wilson, K. G. (1975). The renormalization group: critical phenomena and the Kondo problem. *Rev. Mod. Phys. 47*, 773; (1983). The renormalization group and critical phenomena. *Rev. Mod. Phys. 55*, 583.

White, R. M. (2006). *Quantum theory of magnetism*. New York: Springer.

18 Superconductivity

The history of superconductivity is full of fascinating surprises and challenging developments. The milestone work of Kamerlingh Onnes in 1911 on the electrical resistivity of mercury has opened a new world to the physical investigation, and the discovery of high-T_c superconductivity in barium-doped lanthanum cuprate, by Bednorz and Müller in 1986, has given a novel impetus to the subject.

The microscopic origin of superconductivity is linked to the possible occurrence of a (small) effective attractive interaction between conduction electrons (or valence holes in p-type conductors) and the consequent formation of electron pairs (or hole pairs), at sufficiently low temperatures. The mechanism of electron pairing is at the origin of perfect conductivity, perfect diamagnetism, anomalous specific heat and thermodynamical properties, magnetic flux quantization, coherent tunneling, and several other effects in superconductors. Empirical laws and semi-empirical models have accompanied the accumulation of the wide and rich phenomenology of superconductors. Eventually, the fundamental work of Bardeen, Cooper, and Schrieffer (1957) has transformed an

Solid State Physics, Second Edition. http://dx.doi.org/10.1016/B978-0-12-385030-0.00018-9

endless list of peculiar effects and conjectures into a logically consistent theoretical framework.

The subject of superconductivity is quite extensive, and by necessity in this chapter we can focus only on some relevant experimental and theoretical aspects. A special attention is devoted to the many-body microscopic theory of superconductivity of Bardeen, Cooper, and Schrieffer (BCS); without its concepts no serious discussion would be possible at all. Then our attention focuses on the order parameter approach to phase transitions and on the ingenious Ginzburg-Landau theory, which makes possible the study of various phenomena of superconductivity without using explicitly many-body wavefunctions (although the general concepts of the microscopic BCS theory provide the cultural background behind any approach to superconductivity).

After a survey of the relevant phenomenological aspects of superconductivity, this chapter describes the essential features of the of BCS microscopic theory and the Ginzburg-Landau macroscopic approach to phase transitions. The necessary elaborations are presented in a self-contained way, at an accessible technical level; with these tools, the reader is in a position to appreciate some remarkable pieces of theoretical work and to follow the intimate relation among the various phenomenological aspects. For the novice to the subject of superconductivity, we suggest the preliminary reading of the sections more descriptive in nature: these include the first two sections (phenomenological aspects of superconductors, and Cooper pair idea for the two-electron systems), and the last three sections (London model, macroscopic quantum phenomena, and Josephson effects). In a subsequent reading of the entire chapter, the framework that links together the phenomenological facts can be better appreciated.

18.1 Some Phenomenological Aspects of Superconductors

Old and New Superconducting Materials

Superconductivity was discovered in 1911 by Kamerlingh Onnes, soon after helium had been liquefied by him. In studying the electrical resistance of mercury at low temperatures, Kamerlingh Onnes found that, at about 4.2 K and in a range of 0.01 K, the electrical resistance sharply dropped by several orders of magnitude to non-measurable values; cooling the metal below this critical temperature apparently led to a new resistanceless state, referred to as the superconducting state. The resistanceless state is actually a state of *really zero resistivity* ρ (for continuous currents much smaller than a critical current value, temperatures much smaller than T_c and in zero external fields). When a current is started, for instance by magnetic induction, in a closed superconducting ring which is maintained at temperature well below T_c, it circulates for years without any detectable decay. The non-dissipative current flow in the superconducting ring, besides of course zero resistance, also entails that the magnetic flux threading the ring is quantized. It should be noticed that the non-dissipative current flow in a sample is not sufficient by itself to claim superconductivity. For instance, it is well known that the current flow in the quantum Hall regime is dissipationless. What defines unambiguously superconductivity is the combination of perfect conduction and perfect expulsion of (weak) magnetic fields from the bulk material (Meissner effect). In the specific case of current flowing in a superconducting ring, the quantization of the magnetic flux threading the ring is a manifestation of the Meissner effect, as discussed in detail in Section 18.7.3.

Since 1911, superconductivity has been found in more than 25 metallic elements and in more than one thousand alloys. The element with the highest transition temperature is niobium with $T_c = 9.25$ K. From 1972 to 1986 the alloy Nb_3Ge kept the record of highest critical temperature with $T_c = 23.3$ K. The run to the achievement of materials with the highest critical temperature has been sharply accelerated from 1986 after the discovery by Bednorz and Müller of the superconducting properties of La-based cuprates, with T_c in excess of 30 K [J. G. Bednorz and K. A. Müller, Z. Phys. B *64*, 189 (1986)]. Soon, other families of superconducting cuprate oxides were discovered, and the achievements of materials with critical temperatures above 77 K (the boiling point of liquid nitrogen) passed suddenly from dream to reality [M. K. Wu et al., Phys. Rev. Lett. *58*, 908 (1987)].

The copper oxide superconductors are characterized by crystal structures with one (or more) sheets of CuO_2, separated by insulating block layers. The first discovered high-T_c compounds belong to the family of La-based superconductors, and are obtained with appropriate doping of the insulating parent compound La_2CuO_4, whose body-centered-tetragonal structure is shown in Figure 18.1. The CuO_2 planes are $\approx 6.6\,Å$ apart and separated by two LaO planes; each copper ion is surrounded by an elongated octahedron of oxygens, with Cu-O distance $\approx 1.9\,Å$ within the plane and $\approx 2.4\,Å$ perpendicular to it.

The configurations of the outer electrons of the atoms forming the La_2CuO_4 compound are La: $5d^1 6s^2$; O: $2s^2 2p^4$; Cu: $3d^{10} 4s^1$. In the crystal, lanthanum loses three electrons and takes the closed-shell configuration La^{3+}; oxygen completes the $2p$ shell

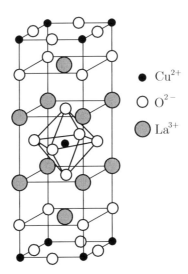

Figure 18.1 Crystal structure of the ceramic material La_2CuO_4. Appropriately doped, lanthanum-based cuprates opened the path to high-T_c superconductivity in 1986.

and becomes O^{2-}; to conserve charge neutrality, the copper atom loses the $4s$ electron and one d electron, and takes the ionic configuration $Cu^{2+} : 3d^9$. There is thus a hole in the d shell of the Cu^{2+} ion and, in an independent-electron model, the material should be a metal; in reality, hopping of holes between copper sites is prevented by the strong Coulomb repulsion, at work whenever two holes are on the same site; this correlation mechanism produces what is called a *Mott-Hubbard insulator*, since the $Cu^{2+}3d$ holes remain localized around their parent atoms. The magnetic moments of the Cu^{2+} ions in different unit cells are coupled antiferromagnetically, because of the super-exchange interaction through oxygens, typical of oxides (see Section 17.2); so La_2CuO_4 is an antiferromagnetic insulator, with Néel critical temperature ≈ 320 K.

Upon doping, La^{3+} ions are randomly replaced by Sr^{2+} or Ba^{2+} ions; thus fewer electrons are donated to the CuO_2 planes, and a number of mobile holes are produced (essentially) in the $2p$ orbitals of oxygen (hybridized with $3d_{x^2-y^2}$ orbitals of copper). Thus we can crudely schematize $La_{2-x}Sr_xCuO_4$ by means of a localized picture for the Cu^{2+} states and an itinerant picture of the oxygen $2p$ holes; a superconducting phase is found for $0.05 < x < 0.30$, with the optimum value at $x \approx 0.15$ yielding the maximum critical temperature $T_c \approx 38$ K.

The discovery of lanthanum-based superconductors paved the way to the discovery of other families of superconducting cuprate oxides based on yttrium, bismuth, thallium, and mercury. In spite of their apparent complexity, the basic structure of the cuprates can be described as an alternating sequence of electronically active metallic CuO_2 layers and other block layers, which act as "charge reservoirs" and provide the carriers (electrons or holes) to the active layers. In this category of lamellar copper oxides, critical temperatures in the range or higher than 100 K can be considered routine. In particular the families of Tl-bearing and Hg-bearing compounds appear promising materials in the race for higher critical temperatures.

In 1991 fullerene has entered the arena of challenging superconductor materials. Solid fullerene is constituted by molecules C_{60} (see Figure 2.12) arranged in a fcc lattice. The molecules C_{60} have a closed-shell ground state, and are weakly bound by van der Waals forces in the solid. When doped with alkali metals, excess electrons are accommodated in the lower unoccupied molecular orbital and solid C_{60} becomes conductive. An explosion of interest in fullerene has followed the discovery that the potassium-doped K_3C_{60} becomes superconductor with critical temperature at 18 K [A. F. Hebard et al., Nature *350*, 600 (1991)]; this critical temperature is striking high for a molecular superconductor, especially when compared with the critical temperature of 0.55 K in potassium intercalated graphite.

In the continued quest of novel superconducting materials, much interest has followed the discovery of the superconductivity of MgB_2 with critical temperature $T_c \approx 40$ K, a value which is remarkable high for a two-component system [J. Nagamatsu et al., Nature *410*, 63 (2001)]; the peculiar electronic structure of magnesium diboride has brought attention and stimulated research on the two-band (and multiband) manifestations of superconductivity. Another remarkable achievement [Y. Kamihara et al. J. Am. Chem. Soc. *130*, 3296 (2008)] concerns the discovery of superconductivity in the class of oxypnictides compounds of type $LaO_{1-x}F_xFeAs$ with critical temperatures in the range up to $T_c \approx 50$ K. [As and P are the pnictides elements, or group-V elements, frequently investigated]. This family of iron-based layered superconductors, with the great freedom allowed by chemical composition, offers new opportunities in the exploration of the microscopic mechanisms of high-T_c superconductivity.

Cuprate oxides compounds, together with the other families of high-T_c superconductors and other materials of lower critical temperature (such as heavy fermions, organic crystals), have contributed to expand the interest in the fundamental and applicative aspects of superconductivity; in Table 18.1 we report the critical temperature of some materials (just for indicative purpose and without any attempt of completeness or record update).

Table 18.1 Critical temperature of some selected superconductors, and zero-temperature critical field. For elemental materials, the thermodynamic critical field $H_c(0)$ is given in gauss. For the compounds, which are type-II superconductors, the upper critical field $H_{c2}(0)$ is given in Tesla (1 T $= 10^4$ G). The data for metallic elements and binary compounds of V and Nb are taken from G. Burns (1992). The data for MgB_2 and iron pnictide are taken from the references cited in the text, and refer to the two principal crystallographic axes. The data for the other compounds are taken from D. R. Harshman and A. P. Mills, Phys. Rev. B 45, 10684 (1992)]. A more extensive list of data can be found in the mentioned references.

Metallic elements	$T_c(K)$	$H_c(0)$ (gauss)
Al	1.17	105
Sn	3.72	305
Pb	7.19	803
Hg	4.15	411
Nb	9.25	2060
V	5.40	1410
Binary compounds	$T_c(K)$	$H_{c2}(0)$ (Tesla)
V_3Ga	16.5	27
V_3Si	17.1	25
Nb_3Al	20.3	34
Nb_3Ge	23.3	38
MgB_2	40	$\approx 5; \approx 20$
Other compounds	$T_c(K)$	$H_{c2}(0)$ (Tesla)
UPt_3 (heavy fermion)	0.53	2.1
$PbMo_6S_8$ (Chevrel phase)	12	55
$\kappa-[BEDT-TTF]_2Cu[NCS]_2$ (organic phase)	10.5	≈ 10
Rb_2CsC_{60} (fullerene)	31.3	≈ 30
$NdFeAsO_{0.7}F_{0.3}$ (iron pnictide)	47	$\approx 30; \approx 50$
Cuprate oxides	$T_c(K)$	$H_{c2}(0)$ (Tesla)
$La_{2-x}Sr_xCuO_4$ ($x \approx 0.15$)	38	≈ 45
$YBa_2Cu_3O_7$	92	≈ 140
$Bi_2Sr_2CaCu_2O_8$	89	≈ 107
$Tl_2Ba_2Ca_2Cu_3O_{10}$	125	≈ 75

General Considerations and Some Phenomenological Aspects

Superconductivity is related to the presence of an effective attractive interaction among conduction electrons (in n-type conductors) or valence holes (in p-type conductors) [from now on, for brevity, we use the terminology appropriate to the case that metallic conductivity is due to electron carriers; however, the considerations below are applicable also in the specular case that metallic conductivity is due to hole carriers]. In the so-called "conventional superconductivity" it is generally accepted that the small attractive interaction among electrons near the Fermi energy is mediated by the phonon field; in the more recent "non-conventional superconductors" other pairing mechanisms could play a role; in either cases the Bardeen-Cooper-Schrieffer theory provides the appropriate general framework to interpret experimental properties.

Above the critical temperature, the effect of the attractive interaction between electrons is in general of limited relevance (apart causing superconducting fluctuations and related phenomena). Below the critical temperature, the effect of the attractive interaction leads to the formation of highly correlated pairs of electrons (Cooper pairs) in an energy shell around the Fermi surface. As we shall see, many physical properties of this *correlated electron gas* are dramatically different from the corresponding properties of the *normal electron gas* of non-interacting electrons; in particular the pairing mechanism entails not only perfect conductance, but also perfect diamagnetism, anomalous specific heat and transport properties, energy gap in quasiparticle spectrum, dissipationless tunneling through non-conducting layers, and many other effects. Before discussing these effects in the light of the microscopic quantum theory, we need to present a few phenomenological facts on the magnetic properties of superconductors, because of their particular role in the historical development of superconductivity.

Besides resistanceless, a distinctive feature of superconductors is perfect diamagnetism: a magnetic field (smaller than a critical value) cannot penetrate into the bulk of a superconductor, regardless of its connectivity. This effect was first observed by Meissner and Ochsenfeld in 1933, and is usually referred to as *Meissner effect*. In Figure 18.2 we show schematically the Meissner-Ochsenfeld effect; the magnetic

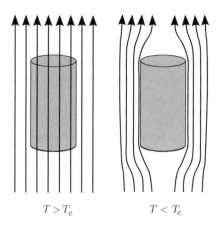

$$T > T_c \qquad\qquad T < T_c$$

Figure 18.2 Exclusion of a weak external magnetic field from the interior of a superconductor.

field is expelled from the bulk of the sample when it is cooled below the critical temperature. Contrary to the superconductor, a normal conductor would keep the magnetic field in which it is embedded, when it becomes perfectly conducting (we discuss in detail this point in Section 18.6). Thus *perfect conductivity does not imply perfect diamagnetism; it is the pairing mechanism, with the resulting modifications in the quasiparticle spectrum and wavefunctions, that links in a unique destiny perfect conductivity and perfect diamagnetism in superconducting materials.*

The absence of magnetic field in the interior of a superconductor has its origin in the "diamagnetic screening currents", which flow in a very thin surface layer of the order of 10^{-5} cm thick. It is worth noticing that even if it is well known that are these self-generated lossless "supercurrents" to make $\mathbf{B} = 0$ in the bulk superconductor, it is generally convenient to accept the diamagnetic description of the superconductor, attributing to it an internal magnetization $\mathbf{M}$ (magnetic moment per unit volume). Since

$$\mathbf{B} = \mathbf{H} + 4\pi\mathbf{M}$$

vanishes *inside* the superconductor, we have that the static magnetic susceptibility of a superconductor is $\chi = M/H = -1/4\pi$. In Chapter 15, in treating the orbital diamagnetism of a normal metal, we found $\chi \approx -10^{-6}$; the magnetic susceptibility of a superconductor is thus about *one million times larger* than the susceptibility of a normal metal (even if the normal metal were ideally conducting); a superconducting material is thus also a superdiamagnetic material. The energy per unit volume, associated with the presence of the screening currents holding the field out of the superconducting sample, is

$$-\int_0^H M \, dH = \frac{H^2}{8\pi}. \tag{18.1}$$

The strong diamagnetic behavior of superconductors is responsible of peculiar phenomena such as magnetic levitation. It is now a routine classroom experiment to show a disk of high-T_c material and a piece of permanent magnet levitating and freely floating above it, when the superconducting material is cooled below the critical temperature; with high-T_c materials, experiments can be easily performed at liquid nitrogen temperature [see for instance E. H. Brandt, Am. J. Phys. *58*, 43 (1990) and references quoted therein].

Sufficiently small magnetic fields are fully expelled by a superconductor; on the other hand sufficiently strong magnetic fields destroy the superconducting state. The modality of conversion to the normal state depends both on the intrinsic properties of the specimen and also on the geometry of the sample (the geometry determines appropriate demagnetization factors). In order to put in evidence genuine microscopic properties of the sample, we consider *exclusively* the structure of *long, thin, cylindrical rods with axis parallel to the applied magnetic field*. [This geometry assures that the field inside a normal specimen is spatially uniform and equal to the external applied field. In this "standard geometry", demagnetization factors are zero, and end effects are ignored. For the magnetic properties of samples with other shapes, for instance spherical, ellipsoidal, laminar, etc., we refer to the literature]. The behavior of superconductors in magnetic fields divides them into two classes; correspondingly they are called type-I (or soft) superconductors and type-II (or hard) superconductors.

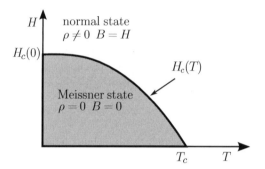

Figure 18.3 Schematic phase diagram illustrating normal and superconducting regions of a type-I superconductor.

Type-I superconductors. The behavior of type-I superconductors, at a given temperature T and in uniform external magnetic field H, can be described as follows. If H is smaller than a critical value $H_c(T)$, the superconductor completely expels the magnetic flux from its interior (Meissner effect); as the external field is increased above the critical value $H_c(T)$, the *entire specimen reverts from the superconducting to the normal state.*

The equilibrium curve $H_c(T)$ divides the $H-T$ plane in the "normal" region and in the "superconducting" (or "Meissner") region, as schematically shown in Figure 18.3. In a number of superconducting materials of type-I, the variation of the magnetic critical field with the temperature has approximately the parabolic form

$$H_c(T) = H_c(0) \left(1 - \frac{T^2}{T_c^2} \right), \tag{18.2}$$

where $H_c(0)$ is the critical field extrapolated at $T = 0$; for $T \to T_c$ from below, we have approximately the linear behavior $H_c(T) \propto (T_c - T)$.

A plot of the magnetization M versus the applied magnetic field H is shown in Figure 18.4. For $H < H_c(T)$ we have $B = H + 4\pi M \equiv 0$ and thus $-4\pi M = H$.

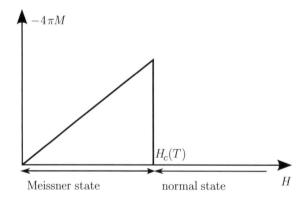

Figure 18.4 Magnetization versus applied field for type-I superconductors.

For $H > H_c(T)$ we have a normal metal; for the present considerations, the magnetic susceptibility of the normal metal (see Chapter 15) can be safely disregarded and hence $B = H$. Most of the elemental superconducting metals are type-I superconductors.

Type-II superconductors. The behavior of type-II superconductors, at a given temperature T and in a uniform external magnetic field H, is the following. If H is smaller than a critical value $H_{c1}(T)$, the magnetic flux is completely expelled from the sample (Meissner state). As the external magnetic field is increased above the lower critical value $H_{c1}(T)$ and below an upper critical value $H_{c2}(T)$, the flux partially penetrates the sample and subdivides into flux-bearing regions, arranged into a regular triangular lattice; each tube of flux is called *filament* or *vortex*, and the sample is said to be in a *mixed state* or *vortex state* or *Shubnikov phase*. As the external magnetic field increases, the density of vortices increases; finally, above $H_{c2}(T)$, the field penetrates uniformly and the whole material returns to the normal state.

The equilibrium curves $H_{c1}(T)$ and $H_{c2}(T)$ divide the $H-T$ plane in normal, mixed, and Meissner regions, as schematically shown in Figure 18.5. Notice that the upper critical field $H_{c2}(T)$ may be as high as several tens of Tesla; thus the mixed state can sustain high currents in high magnetic fields, still in a practically friction-free situation (provided appropriate material inhomogeneities, called "pinning centers", are present to hamper the motion of the flux tubes); for this reason type-II superconductors are of major interest for superconducting high field magnets and other technological applications. Also for the fields $H_{c1}(T)$ and $H_{c2}(T)$ approximate parabolic behavior of type (18.2) holds.

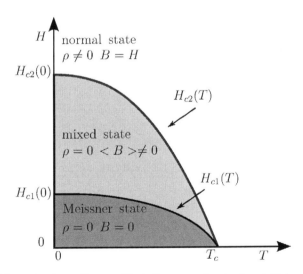

Figure 18.5 Schematic phase diagram illustrating normal, mixed and Meissner regions of a type-II superconductor (the vanishingly small resistivity of the mixed state occurs if flux lines are "pinned" by appropriate material defects); in the mixed state, $\langle B \rangle$ denotes the average magnetic field in the superconductor.

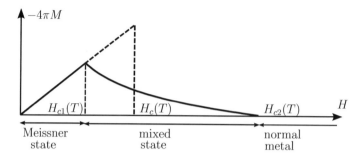

Figure 18.6 Magnetization versus applied field H for a type-II superconductor. The equivalent area construction of the thermodynamic field $H_c(T)$ is also illustrated.

A plot of the magnetization M versus the applied magnetic field is shown in Figure 18.6. Below $H_{c1}(T)$ the superconductor shows perfect diamagnetism within all the sample; above $H_{c2}(T)$ the magnetization is negligible; between these two critical fields, the region of mixed state appears, in which normal and superconducting regions coexist in the sample. In Figure 18.6 it is also illustrated the construction of the *thermodynamic critical field* $H_c(T)$: the low field magnetization is extrapolated linearly to the field $H_c(T)$ so to preserve the total area under the magnetization curve. In a type-II superconductor, we can thus distinguish three critical fields $H_{c1}(T) < H_c(T) < H_{c2}(T)$; a vortex state exists in the range between $H_{c1}(T)$ and $H_{c2}(T)$. In a type-I superconductor, the lower critical field, the thermodynamic critical field, and the upper critical field (which destroys superconductivity on the whole sample) coincide and are denoted as $H_c(T)$.

In the mixed state the magnetic flux penetrates the sample under the form of thin filaments, or vortices, carrying a quantum of magnetic flux. A quantized flux vortex is essentially formed by a core of normal metal (carrying a magnetic field) surrounded by a region of superconducting material (carrying the screening supercurrents which reduce the magnetic field to zero within the penetration depth). Each vortex carries a quantum of magnetic flux equal to

$$\Phi_0 = \frac{hc}{2e} = 2.0679 \times 10^{-7} \text{ gauss} \cdot \text{cm}^2. \tag{18.3}$$

As we will see in Section 18.7, the classification of superconductors into two classes is determined by the sign of the surface energy of the superconductor-normal metal interface: positive in type-I superconductors and negative in type-II superconductors.

The phenomenology concerning superconductors is particularly rich; it is not limited to perfect conductivity and magnetic properties, but includes a long list of peculiar effects as discussed in the suggested references. At this stage, however, it is convenient to describe the fundamental aspects of the microscopic quantum theory of superconductivity; with this tool in hands, we will continue more fruitfully the analysis of the main experimental facts of superconductivity.

18.2   The Cooper Pair Idea

The occurrence of an effective attractive interaction among conduction electrons of a metal, whatever striking it might appear at first sight, is the key point of the microscopic interpretation of superconductivity. A possible mechanism leading to mutual attraction is the indirect electron-electron interaction via phonons, suggested by Fröhlich. In reasonably simple metals, we are accustomed to describe the conduction electrons as a system of independent fermions, weakly interacting with lattice vibrations. Because of this coupling, one electron interacts with the lattice and polarizes it, and another electron interacts with the polarized lattice. This (second order) indirect electron-electron interaction can lead to an effective attractive interaction, as discussed in more detail in Appendix A. This coupling mechanism is operative among electrons lying near the Fermi energy E_F in an energy shell of the order of $\hbar\omega_D$, where ω_D is the Debye phonon frequency of the material.

Other mediator systems that could lead to effective pairing of carriers have been suggested in the literature, especially in connection with the families of high-temperature materials. Proposed microscopic mechanisms include modified phonon coupling (with attention to anharmonic vibrational modes or Jahn-Teller active modes), crucial role of strong correlation, coupling via spin fluctuations, or via bipolarons, or other mechanisms (see for instance the review articles by Lee et al. (2006), Scalapino (2012), and references therein). We do not enter here in the details and justifications of them: rather, *regardless the origin of the net attractive pairing interaction, we wish to analyze its consequences.*

A significant step toward the microscopic theory is due to the Cooper demonstration that the normal Fermi sea becomes unstable, if a whatever small attractive interaction is operative among electrons [L. N. Cooper, Phys. Rev. *104*, 1189 (1956)]. This can be understood along the following lines.

The ordinary ground state of a free-electron gas at $T = 0$ is obtained filling with electrons all the states with $k < k_F$, where k_F is the Fermi wavevector. Consider two "extra" electrons added to the normal Fermi sea, as schematically indicated in Figure 18.7. Let the two electrons interact with each other via a (small) *attractive* two-body potential $U(\mathbf{r}_1, \mathbf{r}_2)$ and feel the other non-interacting electrons of the Fermi sea only through the Pauli exclusion principle, i.e. they cannot occupy the states of the filled Fermi sea. The Schrödinger equation for the two "extra" electrons can be written as

$$\left[\frac{\mathbf{p}_1^2}{2m} + \frac{\mathbf{p}_2^2}{2m} + U(\mathbf{r}_1, \mathbf{r}_2) \right] \psi(\mathbf{r}_1\sigma_1, \mathbf{r}_2\sigma_2) = E\psi(\mathbf{r}_1\sigma_1, \mathbf{r}_2\sigma_2) \tag{18.4}$$

where $\mathbf{r}_1, \mathbf{r}_2, \sigma_1, \sigma_2$ are the space and spin coordinates of the two electrons. The Hamiltonian in Eq. (18.4) does not contain spin dependent terms; thus spin and spatial coordinates are separable and the wavefunction for the two electrons can be written as

$$\psi(\mathbf{r}_1\sigma_1, \mathbf{r}_2\sigma_2) = \phi(\mathbf{r}_1, \mathbf{r}_2)\chi(\sigma_1, \sigma_2). \tag{18.5a}$$

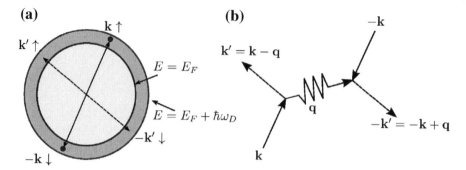

Figure 18.7 (a) Schematic representation of a *single* Cooper pair, added to the ground state of a free-electron gas. Two "extra electrons" in the pair state $(\mathbf{k}\uparrow, -\mathbf{k}\downarrow)$ scatter freely to the pair states $(\mathbf{k}'\uparrow, -\mathbf{k}'\downarrow)$, in the energy region $E_F < E_\mathbf{k}, E_{\mathbf{k}'} < E_F + \hbar\omega_D$, where the phonon-mediated attractive interaction is operative, and form a bound Cooper pair. (b) Schematic representation of the scattering of two electrons with wavevectors $(\mathbf{k}, -\mathbf{k})$ into the state $(\mathbf{k}', -\mathbf{k}')$ via the emission and subsequent absorption of a phonon of momentum $\hbar\mathbf{q}$.

For the spin part, we can consider the antisymmetric singlet state, corresponding to a state of total spin equal to 0

$$\chi^{(S=0)} = \frac{1}{\sqrt{2}}[\alpha(1)\beta(2) - \beta(1)\alpha(2)], \tag{18.5b}$$

where as usual α and β indicate spin-up and spin-down functions with respect to a given quantization axis. Alternatively we can consider the three partner functions

$$\chi^{(S=1)} = \begin{cases} \alpha(1)\alpha(2), \\ \dfrac{1}{\sqrt{2}}[\alpha(1)\beta(2) + \beta(1)\alpha(2)], \\ \beta(1)\beta(2), \end{cases} \tag{18.5c}$$

which constitute a symmetric triplet spin state where the total spin is equal to 1 with projection $+1, 0, -1$ on the quantization axis. The symmetry of the orbital part $\phi(\mathbf{r}_1, \mathbf{r}_2)$ is determined so that to respect the antisymmetry character of the total wavefunction (18.5a), imposed by the Pauli principle.

In order to solve Eq. (18.4), and in view of the smallness of $U(\mathbf{r}_1, \mathbf{r}_2)$, we express the orbital part of the total wavefunction on the basis set formed by the plane waves $W(\mathbf{k}, \mathbf{r}) = (1/\sqrt{V}) \exp(i\mathbf{k} \cdot \mathbf{r})$, normalized to unity in the volume V of the crystal.

In a homogeneous system, the two-body interaction potential entering Eq. (18.4) has the form $U(\mathbf{r}_1, \mathbf{r}_2) = U(\mathbf{r}_1 - \mathbf{r}_2)$, thus the total momentum $\hbar\mathbf{K}$, that describes the motion of the center of mass of the two electron system, is a constant of motion. In the present discussion we are interested in the lowest possible energy of the pair; thus we neglect the motion of the center of mass by setting $\mathbf{K} = 0$ so that $\mathbf{k}_1 = \mathbf{k}$ and $\mathbf{k}_2 = -\mathbf{k}$.

The most general pair wave function of total spin zero and total momentum zero can then be expressed in the form

$$\psi(\mathbf{r}_1\sigma_1,\mathbf{r}_2\sigma_2) = \sum_{\mathbf{k}} g(\mathbf{k})\frac{1}{V}e^{i\mathbf{k}\cdot(\mathbf{r}_1-\mathbf{r}_2)}\frac{1}{\sqrt{2}}[\alpha(1)\beta(2) - \beta(1)\alpha(2)] \quad g(\mathbf{k}) = g(-\mathbf{k}),$$

(18.6a)

where $g(\mathbf{k})$ is the probability amplitude for finding one electron in the plane wave $\mathbf{k}$ and the other electron in the plane wave $-\mathbf{k}$. In Eq. (18.6a), $g(\mathbf{k})$ is an even function of $\mathbf{k}$ to guarantee the global antisymmetry of $\psi(\mathbf{r}_1\sigma_1, \mathbf{r}_2\sigma_2)$ for exchange of the two fermions. Similarly, the most general pair wavefunction of total spin $S = 1$, definite spin component (say $S_z = 1$), and total momentum zero, has an expansion of type

$$\psi(\mathbf{r}_1\sigma_1, \mathbf{r}_2\sigma_2) = \sum_{\mathbf{k}} g(\mathbf{k})\frac{1}{V}e^{i\mathbf{k}\cdot(\mathbf{r}_1-\mathbf{r}_2)}\alpha(1)\alpha(2) \quad g(\mathbf{k}) = -g(-\mathbf{k}) \quad (18.6b)$$

where $g(\mathbf{k})$ is an odd function of $\mathbf{k}$.

We take into account the exclusion principle between the extra pair and the "passive" free-electron gas by requiring

$$g(\mathbf{k}) = 0 \quad \text{for} \quad E_{\mathbf{k}} = \frac{\hbar^2 k^2}{2m} < E_F,$$

where $E_{\mathbf{k}}$ is the unperturbed free-electron energy. Moreover, from the considerations on phonon-mediated electron-electron coupling of Appendix A, it is assumed that $U(\mathbf{r}_1 - \mathbf{r}_2)$ is attractive only among electrons in an energy shell of the order of $\hbar\omega_D$ around the Fermi level E_F; thus we take

$$g(\mathbf{k}) = 0 \quad \text{for} \quad E_{\mathbf{k}} > E_F + \hbar\omega_D.$$

The Debye energy $\hbar\omega_D$ is much smaller than the Fermi energy E_F (measured from the bottom of the conduction band) in most ordinary metals; thus we can assume that the condition $\hbar\omega_D \ll E_F$ is verified.

The matrix elements of the potential $U(\mathbf{r}_1 - \mathbf{r}_2)$ between an initial state with two electrons with wavevectors $(\mathbf{k}', -\mathbf{k}')$ and a final state with two electrons with wavevectors $(\mathbf{k}, -\mathbf{k})$ is

$$U_{\mathbf{k}\mathbf{k}'} = \iint \frac{1}{V}e^{-i\mathbf{k}\cdot(\mathbf{r}_1-\mathbf{r}_2)}U(\mathbf{r}_1-\mathbf{r}_2)\frac{1}{V}e^{i\mathbf{k}'\cdot(\mathbf{r}_1-\mathbf{r}_2)}d\mathbf{r}_1 d\mathbf{r}_2$$

$$= \frac{1}{N\Omega}\int e^{-i(\mathbf{k}-\mathbf{k}')\cdot\mathbf{r}}U(\mathbf{r})\,d\mathbf{r}$$

where $\mathbf{r} = \mathbf{r}_1 - \mathbf{r}_2$ is the relative motion coordinate and $V = N\Omega$ is the volume of the sample, formed by N unit cells of volume Ω.

We now transform the Schrödinger Eq. (18.4) into an integral equation for the coefficients $g(\mathbf{k})$ following the by now familiar procedure adopted in several other physical problems (excitons, plasmons, impurities, etc.). We insert expressions (18.6) (with $\mathbf{k}$ relabeled as $\mathbf{k}'$) into Eq. (18.4), premultiply both members of the resulting equation by $(1/V)\exp[-i\mathbf{k}\cdot(\mathbf{r}_1 - \mathbf{r}_2)]$ (times the singlet or triplet spin function), integrate over $\mathbf{r}_1$ and $\mathbf{r}_2$ and sum over spin variables, and obtain

$$(2E_{\mathbf{k}} - E)g(\mathbf{k}) + \sum_{\mathbf{k}'} U_{\mathbf{k}\mathbf{k}'}g(\mathbf{k}') = 0 \qquad E_F < E_{\mathbf{k}}, E_{\mathbf{k}'} < E_F + \hbar\omega_D \quad (18.7\text{a})$$

with $\mathbf{k}$ and $\mathbf{k}'$ thought of as continuous variables. Equation (18.7a) represents an integral equation, with kernel $U_{\mathbf{k}\mathbf{k}'}$, for the expansion coefficients $g(\mathbf{k})$ of the expressions (18.6).

To solve the integral Eq. (18.7a), we consider the particularly significant and simple case that the scattering amplitudes $U_{\mathbf{k}\mathbf{k}'}$ can be taken as constant in the (small) energy shell of width $\hbar\omega_D$ around the Fermi level. It is convenient to express $U_{\mathbf{k}\mathbf{k}'}$ in the form $U_{\mathbf{k}\mathbf{k}'} = -U_0/N$, where U_0 is a *positive energy*, independent of the number N of unit cells of the crystal. Equation (18.7a) then becomes

$$(2E_{\mathbf{k}} - E)g(\mathbf{k}) - U_0\frac{1}{N}\sum_{\mathbf{k}'} g(\mathbf{k}') = 0 \qquad E_F < E_{\mathbf{k}}, E_{\mathbf{k}'} < E_F + \hbar\omega_D; \quad U_0 > 0.$$

$$(18.7\text{b})$$

It is seen by inspection that the integral Eq. (18.7b) does not modify the energies of the triplet states of the electron pair; in fact, for triplet states $g(\mathbf{k}) = -g(-\mathbf{k})$, thus $\sum g(\mathbf{k}) \equiv 0$ and no effect of U_0 occurs. On the contrary, for singlet states the integral Eq. (18.7b) presents a bound state with energy lower than twice the Fermi energy (this hints at a possible instability of the Fermi sea). [In the model summarized by Eq. (18.7b), characterized by spherically symmetric band energy $E(\mathbf{k})$ and scattering interaction $U_{\mathbf{k}\mathbf{k}'}$ independent of the wavevector $\mathbf{k}$ and $\mathbf{k}'$ in an appropriate shell around the Fermi energy, only spin-singlet s-wave pairing is possible. However, in the presence of other types of interactions, different types of pairing could be favored. In particular, it is believed that in some high-temperature superconductors the formation of d-wave Cooper pairs is favored. We do not enter in detail, and continue our considerations in the case of s-wave pairing mechanism.]

To determine the energy E_{pair} of the Cooper pair in the bound singlet state ($E_{\text{pair}} < 2E_F$), let us put $E = E_{\text{pair}}$ in Eq. (18.7b), divide it by $(2E_{\mathbf{k}} - E_{\text{pair}})$ (which is always a non-vanishing and positive quantity), and sum up over $\mathbf{k}$. We obtain the integral compatibility equation

$$1 = U_0\frac{1}{N}\sum_{\mathbf{k}}\frac{1}{2E_{\mathbf{k}} - E_{\text{pair}}} \qquad E_F < E_{\mathbf{k}} < E_F + \hbar\omega_D.$$

The sum over the discrete wavevectors $\mathbf{k}$ can be converted into an energy integral in the usual form

$$1 = U_0\frac{1}{N}\int_{E_F}^{E_F+\hbar\omega_D} D_0(E)\frac{1}{2E - E_{\text{pair}}}\,dE, \qquad (18.7\text{c})$$

where $D_0(E)$ is the electronic density-of-states for one spin direction of the metal under consideration.

In Eq. (18.7c), in the small energy range of width $\hbar\omega_D$ around E_F, we can neglect the energy dependence of $D_0(E)$. It is convenient to denote by

$$n_0 = n_0(E_F) \equiv \frac{D_0(E_F)}{N}, \tag{18.8}$$

the density-of-states for one spin direction per unit cell at the Fermi energy. In particular for simple metals of valence Z, in the free-electron model one has $n_0 = (3/4)Z/E_F$. Equation (18.7c) becomes

$$1 = U_0\,n_0 \int_{E_F}^{E_F+\hbar\omega_D} \frac{1}{2E - E_{\text{pair}}}\, dE = \frac{1}{2} U_0\,n_0\, \ln \frac{2E_F + 2\hbar\omega_D - E_{\text{pair}}}{2E_F - E_{\text{pair}}}.$$

It follows

$$\frac{2E_F - E_{\text{pair}}}{2E_F + 2\hbar\omega_D - E_{\text{pair}}} = e^{-2/U_0 n_0};$$

indicating by $\Delta_b = 2E_F - E_{\text{pair}}$ the binding energy of the two electron pair, one obtains

$$\Delta_b = \hbar\omega_D \frac{e^{-1/U_0 n_0}}{\sinh[1/U_0 n_0]}.$$

In the *weak coupling limit*, i.e. when $U_0 n_0 \ll 1$ and then $\Delta_b \ll \hbar\omega_D$, we have

$$\Delta_b = 2\hbar\omega_D \exp[-2/U_0 n_0]. \tag{18.9}$$

The model considered until now, however simplified it might appear, is nevertheless useful to illustrate several key points. From Eq. (18.9), it can be seen that, no matter small U_0 is, the free-electron gas is unstable ($\Delta_b > 0$) and electrons are expected to group in "singlet *s*-wave Cooper pairs." We also notice that the binding energy Δ_b in Eq. (18.9) is not an analytic function of U_0 for $U_0 \to 0$ and thus cannot be expanded in powers of U_0. Conventional perturbation theory expressed in powers of the coupling constant is not valid due to the circumstance that the effect of U_0 is very significant near the Fermi energy, no matter small U_0 might be.

Another interesting message that can be inferred from Eq. (18.9) is that superconductivity is more likely to occur in materials with high values of U_0 (this entails high phonon contribution to electron scattering and electron resistivity) than in materials with low values of U_0. For instance a poor conductor such as lead becomes superconductor at 7.19 K, while excellent conductors (such as copper, silver, gold, or sodium, potassium, and the other alkali metals) do not appear to become superconductors even at the lowest temperatures. We notice that the singlet mechanism of coupling ($\mathbf{k}\uparrow, -\mathbf{k}\downarrow$) tends to be broken, for instance, by the presence of magnetic impurities; in fact, it is well established experimentally that even a small amount of magnetic impurities may severely depress conventional superconductivity.

A comparison with the integral equations we have encountered in the electronic structure calculations of crystals (for instance impurity, excitons and plasmons) shows that, in the case of the Cooper pair treatment, it is the *constant density-of-states* at the Fermi level responsible of the pair binding. In the Cooper model no threshold for the potential strength exists, differently for instance from what happens in the Slater-Koster model for impurities in three-dimensional crystals: in that case we are faced with a density-of-states of the form $n(E) \propto (E - E_G)^{1/2}$; a bound state of an impurity in the Slater-Koster model exists only for sufficiently *strong* attractive interaction.

Single Cooper Pair with the Second Quantization Formalism

Before concluding this section, it is instructive to describe the single Cooper pair by the second quantization formalism (see Appendix 4.B). The treatment of the single "two-electron" pair with the many-body apparatus represents a useful algebraic exercise, before considering the full many-body treatment of superconductivity of the next sections; this should allow the not yet expert reader to visualize better physical facts beyond the formalism.

We remark that the wavefunction (18.6a), using the property $g(\mathbf{k}) = g(-\mathbf{k})$, can be recast in the form

$$
\psi(\mathbf{r}_1\sigma_1, \mathbf{r}_2\sigma_2) = \sum_{\mathbf{k}} g(\mathbf{k}) \frac{1}{\sqrt{2}} \frac{1}{V} \left[e^{i\mathbf{k}\cdot(\mathbf{r}_1-\mathbf{r}_2)}\alpha(1)\beta(2) - e^{-i\mathbf{k}\cdot(\mathbf{r}_1-\mathbf{r}_2)}\beta(1)\alpha(2) \right]
$$

$$
\equiv \sum_{\mathbf{k}} g(\mathbf{k}) \, \mathcal{A} \left\{ W_{\mathbf{k}\uparrow} \, W_{-\mathbf{k}\downarrow} \right\} \equiv \sum_{\mathbf{k}} g(\mathbf{k}) \, c^{\dagger}_{\mathbf{k}\uparrow} c^{\dagger}_{-\mathbf{k}\downarrow} |0\rangle,
$$

where $\mathcal{A}$ denotes the determinantal state formed with the indicated spin-orbitals, $|0\rangle$ is the vacuum state, $c^{\dagger}_{\mathbf{k}\uparrow}$ creates an electron with momentum $\hbar\mathbf{k}$ and spin-up (and obvious meaning for $c^{\dagger}_{-\mathbf{k}\downarrow}$).

The normal ground-state $|\Psi_N\rangle$ of a free-electron metal can be represented as a Slater determinant, formed with spin-orbitals (product of plane waves and spin functions) with wavevectors $\mathbf{k}$ up to the Fermi wavevector k_F. In second quantization formalism, we can write $|\Psi_N\rangle$ in the form

$$
|\Psi_N\rangle = \prod_{\mathbf{k}}^{k<k_F} c^{\dagger}_{\mathbf{k}\uparrow} c^{\dagger}_{\mathbf{k}\downarrow} |0\rangle \quad \text{or equivalently} \quad |\Psi_N\rangle = \prod_{\mathbf{k}}^{k<k_F} c^{\dagger}_{\mathbf{k}\uparrow} c^{\dagger}_{-\mathbf{k}\downarrow} |0\rangle. \tag{18.10a}
$$

We choose to create electrons in pairs $(\mathbf{k}\uparrow, -\mathbf{k}\downarrow)$, in view of the forthcoming elaborations.

Let us consider now the wavefunction (18.6a) of the single Cooper pair with the restriction $g(\mathbf{k}) \equiv 0$ for $E_{\mathbf{k}} < E_F$; in the second quantization formalism it can be written in the equivalent form

$$
|\Psi_{\text{Cooper pair}}\rangle = \sum_{\mathbf{k}} g(\mathbf{k}) c^{\dagger}_{\mathbf{k}\uparrow} c^{\dagger}_{-\mathbf{k}\downarrow} |\Psi_N\rangle. \tag{18.10b}
$$

The above equation shows by inspection that the two extra electrons added to the normal metal ground state have total spin equal to zero and total momentum equal to zero. The Hamiltonian of the system of interacting fermions in the second quantized form becomes

$$H = \sum_{\mathbf{k}} E_{\mathbf{k}} \left(c_{\mathbf{k}\uparrow}^{\dagger} c_{\mathbf{k}\uparrow} + c_{-\mathbf{k}\downarrow}^{\dagger} c_{-\mathbf{k}\downarrow} \right) + \sum_{\mathbf{k}\mathbf{k}'} U_{\mathbf{k}\mathbf{k}'} \, c_{\mathbf{k}\uparrow}^{\dagger} c_{-\mathbf{k}\downarrow}^{\dagger} c_{-\mathbf{k}'\downarrow} c_{\mathbf{k}'\uparrow}. \qquad (18.10c)$$

The first term in H is the kinetic energy of a set of independent fermions of wavevectors $\mathbf{k}$ and energies $E_{\mathbf{k}} = \hbar^2 k^2 / 2m$; the second term is the pairing electron-electron interaction. The creation and annihilation operators obey the standard anticommutation rules

$$\left\{ c_{\mathbf{k}\sigma}, c_{\mathbf{k}'\sigma'} \right\} = \left\{ c_{\mathbf{k}\sigma}^{\dagger}, c_{\mathbf{k}'\sigma'}^{\dagger} \right\} = 0, \qquad \left\{ c_{\mathbf{k}\sigma}, c_{\mathbf{k}'\sigma'}^{\dagger} \right\} = \delta_{\mathbf{k}\mathbf{k}'}\delta_{\sigma\sigma'}. \qquad (18.11)$$

From Eqs. (18.10) and (18.11) we can re-obtain all the results of the isolated Cooper pair, and in particular the compatibility Eq. (18.7).

The next step of the theory is to consider not only a single pair of electrons added to the Fermi sea, but to allow pairing of all the electrons of the sea, on the same footing. This is done by the approach outlined below.

18.3 Ground State for a Superconductor in the BCS Theory at Zero Temperature

18.3.1 Variational Determination of the Ground-State Wavefunction

The Cooper model (1956) is only a rough (although for some aspects illuminating) starting point for a theory of superconductivity; it opened the way to the major breakthrough (1957) in the microscopic theory of superconductivity by Bardeen, Cooper, and Schrieffer (BCS). In the Cooper treatment we have seen that the normal electron gas in the presence of a (net) attractive electron-electron interaction is unstable, even with respect to the formation of a single Cooper pair. A realistic theory of superconductivity must be able to describe a *cooperative condensation process* in which many pairs of electrons of the normal Fermi sea are formed so to minimize the total energy of the system (at $T = 0$).

These demanding physical and formal requirements can be adequately described by the following BCS variational form of the ground-state wavefunction for superconductors

$$\boxed{|\Psi_S\rangle = \prod_{\mathbf{k}} \left(u_{\mathbf{k}} + v_{\mathbf{k}} c_{\mathbf{k}\uparrow}^{\dagger} c_{-\mathbf{k}\downarrow}^{\dagger} \right) |0\rangle}, \qquad (18.12)$$

where $|0\rangle$ is the vacuum state, $u_{\mathbf{k}}$ and $v_{\mathbf{k}}$ are real and even functions of $\mathbf{k}$, chosen in such a way to minimize the ground-state energy (the requirement of reality is made for simplicity, but could be relaxed when needed). Under the constraint $u_{\mathbf{k}}^2 + v_{\mathbf{k}}^2 = 1$ (which assures the normalization to unity of Ψ_S), $v_{\mathbf{k}}$ represents the probability amplitude that the pair $(\mathbf{k}\uparrow, -\mathbf{k}\downarrow)$ is occupied and $u_{\mathbf{k}}$ represents the probability amplitude that the pair $(\mathbf{k}\uparrow, -\mathbf{k}\downarrow)$ is unoccupied. In the superconducting ground state described by Ψ_S, electrons are involved only as pairs; in fact by carrying out explicitly the products in Eq.

(18.12) we see that there is a term without Cooper pairs and terms with one, two, and so on Cooper pairs. Notice also that the wavefunction (18.12) of the superconductor is reduced to the wavefunction (18.10a) of the normal metal in the particular case in which u_k and v_k are given by

$$\begin{cases} u_k = 0 & v_k = 1 \quad \text{for } k < k_F, \\ u_k = 1 & v_k = 0 \quad \text{for } k > k_F. \end{cases} \tag{18.13}$$

An aspect of the trial wavefunction (18.12) is that the total number of electrons is not well defined. Actually, it is not convenient to be tied up with a fixed number of electrons; in a manner which is usual in such cases, we proceed to minimize the expectation value $\langle \Psi_S | H | \Psi_S \rangle$ under the constraint $\langle \Psi_S | N_{op} | \Psi_S \rangle = N$ where $N_{op} = \sum_k \left(c_{k\uparrow}^\dagger c_{k\uparrow} + c_{-k\uparrow}^\dagger c_{-k\uparrow} \right)$ is the particle number operator and N is the number of electrons in the actual metal. The standard method to treat minimization plus one (or more) appropriate constraint consists in introducing one (or more) Lagrange multiplier μ, minimizing without restraints the quantity $\langle \Psi_S | H - \mu N_{op} | \Psi_S \rangle$ and finally determining μ through the condition $\langle \Psi_S | N_{op} | \Psi_S \rangle = N$; it turns out that μ is the Fermi energy, and it is the same for the normal or superconducting state.

In essence we have thus arrived at the minimization of the quantity

$$\boxed{W_S = \langle \Psi_S | H_{\text{BCS}} | \Psi_S \rangle}, \tag{18.14a}$$

where H_{BCS} is the Bardeen-Cooper-Schrieffer Hamiltonian given by

$$\boxed{H_{\text{BCS}} = \sum_k \varepsilon_k \left(c_{k\uparrow}^\dagger c_{k\uparrow} + c_{-k\uparrow}^\dagger c_{-k\uparrow} \right) + \sum_{kk'} U_{kk'} c_{k\uparrow}^\dagger c_{-k\downarrow}^\dagger c_{-k'\downarrow} c_{k'\uparrow}}, \tag{18.14b}$$

and $\varepsilon_k = E_k - \mu = (\hbar^2 k^2 / 2m) - \mu$ denotes the single particle energy measured with respect to the Fermi energy.

The expectation value (18.14a) can be easily evaluated keeping in mind the practical recipe of using the anticommutation rules (18.11) for systematically shifting the fermion annihilation operators to the right (or creation operators to the left) until they arrive to operate on the vacuum state (giving zero). Here are some matrix elements of interest. We first note that

$$\langle 0 | \left(u_k + v_k c_{-k\downarrow} c_{k\uparrow} \right) \left(u_k + v_k c_{k\uparrow}^\dagger c_{-k\downarrow}^\dagger \right) | 0 \rangle = u_k^2 + v_k^2 ; \tag{18.15a}$$

the normalization of the wavefunction $| \Psi_S \rangle$ is given by

$$\langle \Psi_S | \Psi_S \rangle = \prod_k \left(u_k^2 + v_k^2 \right) = 1 \quad \text{if} \quad u_k^2 + v_k^2 = 1 \quad \text{for every} \quad k. \tag{18.15b}$$

Similarly for the number operator we have

$$\langle \Psi_S | c_{k\uparrow}^\dagger c_{k\uparrow} | \Psi_S \rangle = \langle 0 | \left(u_k + v_k c_{-k\downarrow} c_{k\uparrow} \right) c_{k\uparrow}^\dagger c_{k\uparrow} \left(u_k + v_k c_{k\uparrow}^\dagger c_{-k\downarrow}^\dagger \right) | 0 \rangle = v_k^2. \tag{18.15c}$$

We have also

$$\langle \Psi_S | c_{\mathbf{k}\uparrow}^\dagger c_{-\mathbf{k}\downarrow}^\dagger | \Psi_S \rangle = \langle \Psi_S | c_{-\mathbf{k}\downarrow} c_{\mathbf{k}\uparrow} | \Psi_S \rangle = u_{\mathbf{k}} v_{\mathbf{k}}. \tag{18.15d}$$

Using the matrix elements (18.15) we obtain for the superconductor ground-state energy the following expression

$$W_S = \langle \Psi_S | H_{\mathrm{BCS}} | \Psi_S \rangle = 2 \sum_{\mathbf{k}} \varepsilon_{\mathbf{k}} v_{\mathbf{k}}^2 + \sum_{\mathbf{kk'}} U_{\mathbf{kk'}} u_{\mathbf{k}} v_{\mathbf{k}} u_{\mathbf{k'}} v_{\mathbf{k'}} ; \tag{18.16}$$

at this stage our problem has become a standard algebraic minimization problem of W_S as a function of the $u_{\mathbf{k}}$ and $v_{\mathbf{k}}$ (with $\mathbf{k}$ in the reciprocal space).

In order to minimize W_S under the constraint $u_{\mathbf{k}}^2 + v_{\mathbf{k}}^2 = 1$, let us represent the variables in polar form

$$\begin{cases} u_{\mathbf{k}} = \cos \theta_{\mathbf{k}} \\ v_{\mathbf{k}} = \sin \theta_{\mathbf{k}} \end{cases} \implies \quad \sin 2\theta_{\mathbf{k}} = 2 u_{\mathbf{k}} v_{\mathbf{k}}; \quad \cos 2\theta_{\mathbf{k}} = u_{\mathbf{k}}^2 - v_{\mathbf{k}}^2.$$

The superconductor ground-state energy (18.16) becomes

$$W_S = 2 \sum_{\mathbf{k}} \varepsilon_{\mathbf{k}} \sin^2 \theta_{\mathbf{k}} + \frac{1}{4} \sum_{\mathbf{kk'}} U_{\mathbf{kk'}} \sin 2\theta_{\mathbf{k}} \sin 2\theta_{\mathbf{k'}} .$$

The minimization condition gives

$$\frac{\partial W_S}{\partial \theta_{\mathbf{k}}} = 0 \implies 2\varepsilon_{\mathbf{k}} \sin 2\theta_{\mathbf{k}} + \sum_{\mathbf{k'}} U_{\mathbf{kk'}} \cos 2\theta_{\mathbf{k}} \sin 2\theta_{\mathbf{k'}} = 0$$

$$\implies 2\varepsilon_{\mathbf{k}} u_{\mathbf{k}} v_{\mathbf{k}} + \sum_{\mathbf{k'}} U_{\mathbf{kk'}} \left(u_{\mathbf{k}}^2 - v_{\mathbf{k}}^2 \right) u_{\mathbf{k'}} v_{\mathbf{k'}} = 0 \tag{18.17}$$

for every vector $\mathbf{k}$.

To solve this set of self-consistent equations for $u_{\mathbf{k}}$ and $v_{\mathbf{k}}$, let us define the *energy gap parameters* (the origin of the name will be clear soon):

$$\Delta_{\mathbf{k}} = - \sum_{\mathbf{k'}} U_{\mathbf{kk'}} u_{\mathbf{k'}} v_{\mathbf{k'}}. \tag{18.18}$$

The variational condition (18.17) thus becomes

$$\boxed{ 2\varepsilon_{\mathbf{k}} u_{\mathbf{k}} v_{\mathbf{k}} - \Delta_{\mathbf{k}} \left(u_{\mathbf{k}}^2 - v_{\mathbf{k}}^2 \right) = 0 }. \tag{18.19}$$

The solution of Eq. (18.19), together with the normalization condition $u_{\mathbf{k}}^2 + v_{\mathbf{k}}^2 = 1$ gives

$$u_{\mathbf{k}}^2 = \frac{1}{2} \left[1 + \frac{\varepsilon_{\mathbf{k}}}{\sqrt{\varepsilon_{\mathbf{k}}^2 + \Delta_{\mathbf{k}}^2}} \right] \quad \text{and} \quad v_{\mathbf{k}}^2 = \frac{1}{2} \left[1 - \frac{\varepsilon_{\mathbf{k}}}{\sqrt{\varepsilon_{\mathbf{k}}^2 + \Delta_{\mathbf{k}}^2}} \right]. \tag{18.20}$$

The choice of sign in Eqs. (18.20) has been determined by the requirement that, for vanishing $U_{\mathbf{kk}'}$ interaction and thus for vanishing $\Delta_{\mathbf{k}}$, the standard normal ground state described by Eqs. (18.13) is obtained. If we use Eqs. (18.20) to calculate $u_{\mathbf{k}'}v_{\mathbf{k}'}$, and insert the result into Eq. (18.18) we see that the energy gap parameters are determined by the self-consistent equations

$$\boxed{\Delta_{\mathbf{k}} = -\frac{1}{2}\sum_{\mathbf{k}'} U_{\mathbf{kk}'} \frac{\Delta_{\mathbf{k}'}}{\sqrt{\varepsilon_{\mathbf{k}'}^2 + \Delta_{\mathbf{k}'}^2}}} \qquad (18.21)$$

for every vector $\mathbf{k}$.

A trivial solution of Eq. (18.21) is $\Delta_{\mathbf{k}} = 0$ for every $\mathbf{k}$; this corresponds to the normal metal state with electrons filling the Fermi sphere up to k_F. In general the integral equation (18.21) is not easy to be solved. For simplicity, and in analogy with the procedure adopted in the treatment of a single s-wave Cooper pair, the matrix elements $U_{\mathbf{kk}'}$ are taken as constant in an appropriate energy shell around the Fermi level; in the so-called *average potential approximation*, one has

$$U_{\mathbf{kk}'} = \begin{cases} -U_0/N & \text{if } |\varepsilon_{\mathbf{k}}|, |\varepsilon_{\mathbf{k}'}| < \hbar\omega_D \quad (U_0 > 0), \\ 0 & \text{otherwise,} \end{cases} \qquad (18.22)$$

where U_0 is a *positive constant* independent of the number N of unit cells of the crystal. From the average potential approximation of Eq. (18.22), and from Eq. (18.18), it follows that

$$\Delta_{\mathbf{k}} = \begin{cases} \Delta_0 & \text{if } |\varepsilon_{\mathbf{k}}| < \hbar\omega_D, \\ 0 & \text{otherwise.} \end{cases}$$

The integral equation (18.21) simplifies as

$$1 = \frac{1}{2}U_0\frac{1}{N}\sum_{\mathbf{k}'} \frac{1}{\sqrt{\varepsilon_{\mathbf{k}'}^2 + \Delta_0^2}} \qquad \text{with} \qquad -\hbar\omega_D < \varepsilon_{\mathbf{k}'} < \hbar\omega_D.$$

As usual the sum over $\mathbf{k}'$ can be converted into an integral; assuming the electron density-of-states as constant in the small energy shell of interest around the Fermi level, one obtains

$$1 = \frac{1}{2}U_0 n_0 \int_{-\hbar\omega_D}^{\hbar\omega_D} \frac{d\varepsilon}{\sqrt{\varepsilon^2 + \Delta_0^2}}, \qquad (18.23a)$$

where $n_0 = n_0(E_F)$, defined in Eq. (18.8), denotes the density-of-states for one spin direction per unit cell at the Fermi energy.

The elaboration of Eq. (18.23a) can be carried out considering the indefinite integral

$$\int \frac{1}{\sqrt{x^2 + a^2}} dx = \sinh^{-1}\frac{x}{|a|} \quad \Longrightarrow \quad 1 = U_0 n_0 \sinh^{-1}\frac{\hbar\omega_D}{\Delta_0};$$

it follows

$$\Delta_0 = \frac{\hbar\omega_D}{\sinh\left(1/U_0 n_0\right)}.$$

In the weak coupling limit, when $U_0 n_0$ is rather small with respect to unity, one obtains

$$\boxed{\Delta_0 = 2\hbar\omega_D \exp[-1/U_0 n_0]} \tag{18.23b}$$

[the similarity in form of Eq. (18.23b) with Eq. (18.9) is fortuitous, as the weak coupling limit tends to produce exponentials]. For typical conventional superconductors, one has $E_F \approx 1\,\text{eV}$ (measured from the bottom of the conduction band) and $U_0 \approx 0.1$–$0.5\,\text{eV}$; the dimensionless coupling parameter $U_0 n_0 \approx U_0 / E_F$ is in the range 0.1–0.5; thus $\Delta_0 \approx 1\,\text{meV}$ is in general a small fraction (≈ 0.1–0.01) of $\hbar\omega_D$.

Until now, we have confined our discussion of the BCS theory of superconductivity within the *"weak coupling limit"* (i.e. $\Delta_0 \ll \hbar\omega_D$), and *"s-wave symmetry of the pairing state"*; also in the following we keep our considerations and guidelines within these simplifying approximations, that allow understanding of the underlying physical concepts at an acceptable technical level.

Needless to say, in the study of specific superconducting materials some assumptions must be revisited and when necessary appropriately modified. Along the lines of improvements concerning conventional low-T_c materials ($T_c \le 20\,\text{K}$) we can mention that the BCS theory has been generalized by various authors to describe strong-coupling situations [see W. L. McMillan, Phys. Rev. *167*, 331 (1968) and in particular the works of Eliashberg and other authors, quoted therein].

Another line of improvements concerns the superconducting properties of magnesium diboride, an intermediate T_c material ($T_c \approx 40\,\text{K}$) where the pairing mechanism remains the conventional electron-phonon coupling; the novelty of this material is the presence of two electronic bands and two superconducting gaps. This peculiarity is responsible of properties significantly apart from the single-band single-gap superconductors, considered in the original BCS theory, to which we confine our attention. [For more details, see for instance the review article by X. X. Xi "Two-band superconductor magnesium diboride" Rep. Progr. Phys. *71*, 116501 (2008) and references quoted therein.]

A major challenge to some key achievements of the original BCS formulation is constituted by high-T_c materials, as cuprates and iron pnictides. In the original BCS theory phonons are responsible of the formation of the Cooper pairs. In high-T_c superconductors the mechanism of Cooper pairs formation is much more complex. An active debate concerns the role of electronic correlation, and the possibility that unconventional pairing mechanisms may provide pairing states with non-zero angular momentum. Again we do not enter in details, and we refer to the literature for further information [see for instance D. C. Johnston, Advances in Physics *59*, 803 (2010) and bibliography cited therein].

18.3.2 Ground-State Energy and Isotope Effect

The *condensation energy of a superconductor* is defined as the energy difference between the superconductor ground-state energy W_S of Eq. (18.16) and the normal ground-state energy W_N; its expression is

$$W_S - W_N = 2 \sum_{\mathbf{k}} \varepsilon_{\mathbf{k}} v_{\mathbf{k}}^2 + \sum_{\mathbf{k}\mathbf{k}'} U_{\mathbf{k}\mathbf{k}'} u_{\mathbf{k}} v_{\mathbf{k}} u_{\mathbf{k}'} v_{\mathbf{k}'} - 2 \sum_{\mathbf{k}}^{k<k_F} \varepsilon_{\mathbf{k}}. \qquad (18.24)$$

Using Eq. (18.18), and then Eqs. (18.20), we obtain

$$W_S - W_N = \sum_{\mathbf{k}} \left[2\varepsilon_{\mathbf{k}} v_{\mathbf{k}}^2 - \Delta_{\mathbf{k}} u_{\mathbf{k}} v_{\mathbf{k}} \right] - \sum_{\mathbf{k}}^{k<k_F} 2\varepsilon_{\mathbf{k}}$$

$$= \sum_{\mathbf{k}} \left[\varepsilon_{\mathbf{k}} - \frac{\varepsilon_{\mathbf{k}}^2}{\sqrt{\varepsilon_{\mathbf{k}}^2 + \Delta_{\mathbf{k}}^2}} - \frac{1}{2} \frac{\Delta_{\mathbf{k}}^2}{\sqrt{\varepsilon_{\mathbf{k}}^2 + \Delta_{\mathbf{k}}^2}} \right] - \sum_{\mathbf{k}}^{k<k_F} 2\varepsilon_{\mathbf{k}};$$

the above expression is now evaluated in the case of average potential approximation and weak binding limit.

In the case of average potential approximation expressed by Eq. (18.22), $\Delta_{\mathbf{k}}$ is different from zero only for $|\varepsilon_{\mathbf{k}}| < \hbar\omega_D$. As usual, we convert the sum over $\mathbf{k}$ into an energy integral; assuming that the density-of-states in the energy interval of interest around the Fermi level is constant, we have

$$W_S - W_N = D_0(E_F) \int_{-\hbar\omega_D}^{\hbar\omega_D} \left(\varepsilon - \frac{2\varepsilon^2 + \Delta_0^2}{2\sqrt{\varepsilon^2 + \Delta_0^2}} \right) d\varepsilon - D_0(E_F) \int_{-\hbar\omega_D}^{0} 2\varepsilon \, d\varepsilon.$$

Using the indefinite integral

$$\int \frac{2x^2 + a^2}{\sqrt{x^2 + a^2}} \, dx = x\sqrt{x^2 + a^2}$$

$$\Longrightarrow W_S - W_N = D_0(E_F) \left[-\hbar\omega_D \sqrt{\hbar^2\omega_D^2 + \Delta_0^2} + \hbar^2\omega_D^2 \right].$$

In the weak binding limit $\Delta_0 \ll \hbar\omega_D$, with a series development of the square root, we obtain for the condensation energy the expression

$$\boxed{W_S - W_N = -\frac{1}{2} D_0(E_F) \Delta_0^2}. \qquad (18.25)$$

The condensation energy (18.25) of the superconductor can be interpreted as originated by the electrons $D_0(E_F)\Delta_0$, the ones in the energy shell Δ_0 around the Fermi energy, which decrease their energy by about Δ_0 because of the pairing mechanism (see Figure 18.8).

We can link the *thermodynamic magnetic field* $H_c(0)$ of the superconductor with the condensation energy per unit volume, via the relation

$$\frac{1}{V}(W_N - W_S) = \frac{H_c^2(0)}{8\pi} = \frac{1}{2} D_0(E_F) \Delta_0^2 \frac{1}{V} \Longrightarrow H_c^2(0) = 4\pi \, n_0 \, \Delta_0^2 \frac{1}{\Omega}, \qquad (18.26a)$$

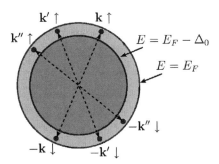

Figure 18.8 Schematic representation of the origin of the condensation energy in the BCS ground state. The electron pairs $(\mathbf{k}\uparrow, -\mathbf{k}\downarrow)$ in the blurred region of $\mathbf{k}$ space, within an energy shell Δ_0 around the Fermi level, contribute to the condensation energy of the superconductor (notice that, typically, $\Delta_0 \approx 1$ meV and $E_F \approx 1$ eV). The dashed lines are a remind of the binding energy of a Cooper pair.

where n_0 is the density-of-states at the Fermi level for one spin direction and for unit cell, given by Eq. (18.8), and Ω is the volume of the unit cell. To estimate the value of $H_c(0)$, we multiply the last equation by μ_B^2, set $n_0 \approx 1/E_F$ and $\Omega = (10\, a_B)^3$, and obtain

$$\mu_B^2 H_c^2(0) = 4\pi \, \frac{\Delta_0^2}{E_F} \, \frac{\mu_B^2}{(10\, a_B)^3}.$$

With the values

$$\mu_B = 5.788 \cdot 10^{-6} \, \frac{\text{meV}}{\text{gauss}}, \quad \Delta_0 \approx 1 \text{ meV}, \quad E_F = 10^3 \text{ meV}, \quad \frac{\mu_B^2}{a_B^3} = 0.363 \text{ meV},$$

we can estimate values of $H_c(0)$ of the order of several hundred or few thousand gauss.

It is interesting to notice that Eq. (18.26a) also contains the so-called *isotope effect*, which historically has been the hint of the phonon induced electron-electron coupling mechanism considered in the BCS theory. Compare in fact metals built with atoms of different isotopic masses M. It is reasonable to assume that the relevant effect of isotopic substitution is just to change the Debye frequency $\hbar\omega_D$, leaving essentially unchanged all the other parameters concerning the electronic part of the problem. In the Debye model for simple lattices $\hbar\omega_D$ is proportional to $1/\sqrt{M}$ (similarly to what expected for a simple oscillator). Using Eq. (18.23b) and (18.26a) we thus expect that

$$H_c(0) \cdot M^\alpha = \text{const} \tag{18.26b}$$

with $\alpha = 0.5$ (at least in the weak coupling limit). Experimental measurements show that the above relation is well verified for mercury and a few other non-transition metals. The experimental situation is however much richer; for instance for Mo, Re, and Os, α is lower than 0.5. For the conventional superconductors Ru and Zr the isotope effect is absent (or nearly absent). More complex appears the situation for several high-T_c superconductors. It should be remembered that the theoretical model we are presenting contains by necessity several simplifying assumptions, and the actual behavior and mediation mechanism in real materials must be discussed and analyzed case by case.

18.3.3 Momentum Distribution and Coherence Length

The probability of finding an electron in a state with momentum $\hbar\mathbf{k}$ and spin σ in the superconductor is given in terms of the single particle number operator by

$$\langle\Psi_S|c_{\mathbf{k}\sigma}^{\dagger}c_{\mathbf{k}\sigma}|\Psi_S\rangle = v_{\mathbf{k}}^2 = \frac{1}{2}\left[1 - \frac{\varepsilon_{\mathbf{k}}}{\sqrt{\varepsilon_{\mathbf{k}}^2 + \Delta_{\mathbf{k}}^2}}\right]; \tag{18.27a}$$

in the normal gas, the same probability is given by

$$\langle\Psi_N|c_{\mathbf{k}\sigma}^{\dagger}c_{\mathbf{k}\sigma}|\Psi_N\rangle = \begin{cases} 1 & \text{for } k < k_F, \\ 0 & \text{for } k > k_F. \end{cases} \tag{18.27b}$$

The two situations are schematically indicated in Figure 18.9.

We can do a similar analysis for the pair operator $c_{\mathbf{k}\uparrow}^{\dagger}c_{-\mathbf{k}\downarrow}^{\dagger}$ (or $c_{-\mathbf{k}\downarrow}c_{\mathbf{k}\uparrow}$). In the superconducting ground state we have

$$\langle\Psi_S|c_{\mathbf{k}\uparrow}^{\dagger}c_{-\mathbf{k}\downarrow}^{\dagger}|\Psi_S\rangle = u_{\mathbf{k}}v_{\mathbf{k}} = \frac{1}{2}\frac{\Delta_{\mathbf{k}}}{\sqrt{\varepsilon_{\mathbf{k}}^2 + \Delta_{\mathbf{k}}^2}}. \tag{18.28}$$

The above quantity is always zero in the normal state, but it is different from zero in a small shell around k_F in the superconductor state; some of the most typical effects of superconductors (for instance Meissner effect and Josephson tunneling) are related to this fact.

The region in reciprocal space where $u_{\mathbf{k}}v_{\mathbf{k}}$ is different from zero has an energy width Δ_0 around the Fermi energy and a momentum width δk around the Fermi wavevector such that

$$\frac{\hbar^2}{2m}\left[(k_F \pm \delta k)^2 - k_F^2\right] \approx \pm\Delta_0 \implies \frac{\hbar^2 k_F}{m}\delta k \approx \Delta_0 \implies \delta k \approx \frac{\Delta_0}{\hbar v_F},$$

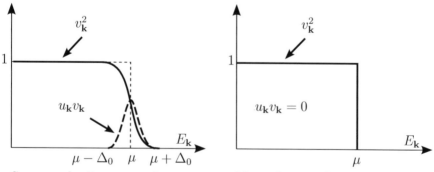

Superconducting ground state   Normal ground state

Figure 18.9 Behavior of the quantities $v_{\mathbf{k}}^2$ and $u_{\mathbf{k}}v_{\mathbf{k}}$ for the superconducting ground state and for the normal ground state.

where v_F is the Fermi velocity. From the uncertainty principle, we can estimate that the spatial extent $\xi_0 \approx 1/\delta k$ of the electrons in a Cooper pair is given by

$$\xi_0 = \frac{1}{\pi} \frac{\hbar v_F}{\Delta_0} = \frac{1}{\pi} \frac{\hbar^2 k_F}{m \Delta_0} \tag{18.29}$$

(the numerical factor π requires a more detailed analysis). The length ξ_0 is the so-called *BCS coherence length*, and represents the average distance in real space between the two electrons of the Cooper pair. The pair coherence length is proportional to the Fermi velocity and inversely proportional to the binding energy of the Cooper pair. Typical values of ξ_0 range from hundred or thousand Å in conventional superconductors, to some tens Å in high-T_c superconductors (where the traditional BCS framework becomes inadequate).

In conventional superconductors in general there is a large number of Cooper pairs within the spatial coherence length ξ_0. In fact the density of Cooper pairs is of the order of $\approx n \cdot (\Delta_0/E_F)$, where $n \equiv 1/\left[(4/3)\pi r_s^3 a_B^3\right]$ is the electron density of the metal, and Δ_0/E_F is the fraction of electrons that condensate in the ground superconducting state with an energy gain of $\approx 2\Delta_0$; we thus have that the average distance between the centers of mass of the pairs is related to the average distance $r_s a_B$ between electrons by the expression $d \approx r_s a_B (E_F/\Delta_0)^{1/3}$. For conventional superconductors, $E_F \approx 1$ eV, $\Delta_0 \approx 1$ meV, and $d \approx 10 \, r_s a_B \approx 10$ Å. With ξ_0 of the order of 1000 Å, there are million Cooper pairs within the coherence length; the individual pairs overlap strongly in space and the binding energy $2\Delta_0$ of any pair depends cooperatively on the presence of all the other pairs.

18.4   Excited States of Superconductors at Zero Temperature

18.4.1   *The Bogoliubov Canonical Transformation*

Until now we have studied the ground state of a superconductor *at zero temperature*. We could now study the excited states of a superconductor system (still remaining at $T = 0$) starting from the ground-state wavefunction (18.12), applying to it creation or annihilation operators and elaborating the trial excited states so obtained; this was the procedure originally followed by BCS. An equivalent although much more convenient procedure, based on the canonical transformations of Bogoliubov and Valatin, allows one to obtain in a single stroke the excitation spectrum of the superconductor [N. N. Bogoliubov, Nuovo Cimento 7, 794 (1958); J. G. Valatin, Nuovo Cimento 7, 843 (1958)].

Let us consider the superconductor BCS Hamiltonian of Eq. (18.14b), here rewritten for convenience

$$H_{\text{BCS}} = \sum_{\mathbf{k}\sigma} \varepsilon_{\mathbf{k}} c_{\mathbf{k}\sigma}^{\dagger} c_{\mathbf{k}\sigma} + \sum_{\mathbf{kk}'} U_{\mathbf{kk}'} c_{\mathbf{k}\uparrow}^{\dagger} c_{-\mathbf{k}\downarrow}^{\dagger} c_{-\mathbf{k}'\downarrow} c_{\mathbf{k}'\uparrow}. \tag{18.30}$$

We then write the ground-state wavefunction in the form

$$|\Psi_S\rangle = \prod_{\mathbf{k}} \left(u_{\mathbf{k}} + v_{\mathbf{k}} c_{\mathbf{k}\uparrow}^{\dagger} c_{-\mathbf{k}\downarrow}^{\dagger} \right) |0\rangle, \tag{18.31}$$

where $u_{\mathbf{k}}$ and $v_{\mathbf{k}}$ are not yet specified real quantities, satisfying $u_{\mathbf{k}}^2 + v_{\mathbf{k}}^2 = 1$ to assure the normalization of $|\Psi_S\rangle$. In the variational BCS procedure, $u_{\mathbf{k}}$ and $v_{\mathbf{k}}$ are determined so to minimize the energy of the superconductor ground state. In the Bogoliubov procedure, $u_{\mathbf{k}}$ and $v_{\mathbf{k}}$ are chosen as coefficients of a canonical transformation that gives diagonal form to the Hamiltonian (18.30) (once appropriately simplified). As we shall see, the two procedures are equivalent but the bonus of the Bogoliubov procedure in obtaining the one-particle excitation spectrum of the superconductor is invaluable.

In the BCS pairing theory of $(\mathbf{k}\uparrow, -\mathbf{k}\downarrow)$ spin-orbitals, we have considered for every pair of wavevectors $\pm\mathbf{k}$ the four Fermi operators (two creation and two annihilation operators) of the type

$$c_{\mathbf{k}\uparrow}, \ c_{\mathbf{k}\uparrow}^{\dagger}, \ c_{-\mathbf{k}\downarrow}, \ c_{-\mathbf{k}\downarrow}^{\dagger}.$$

The Bogoliubov procedures defines four new Fermi operators

$$\gamma_{\mathbf{k}\uparrow}, \ \gamma_{\mathbf{k}\uparrow}^{\dagger}, \ \gamma_{-\mathbf{k}\downarrow}, \ \gamma_{-\mathbf{k}\downarrow}^{\dagger}$$

via the linear canonical transformations

$$\begin{bmatrix} \gamma_{\mathbf{k}\uparrow} \\ \gamma_{-\mathbf{k}\downarrow}^{\dagger} \end{bmatrix} = \begin{bmatrix} u_{\mathbf{k}} & -v_{\mathbf{k}} \\ v_{\mathbf{k}} & u_{\mathbf{k}} \end{bmatrix} \begin{bmatrix} c_{\mathbf{k}\uparrow} \\ c_{-\mathbf{k}\downarrow}^{\dagger} \end{bmatrix} \quad \text{and} \quad \begin{bmatrix} \gamma_{\mathbf{k}\uparrow}^{\dagger} \\ \gamma_{-\mathbf{k}\downarrow} \end{bmatrix} = \begin{bmatrix} u_{\mathbf{k}} & -v_{\mathbf{k}} \\ v_{\mathbf{k}} & u_{\mathbf{k}} \end{bmatrix} \begin{bmatrix} c_{\mathbf{k}\uparrow}^{\dagger} \\ c_{-\mathbf{k}\downarrow} \end{bmatrix},$$

where $u_{\mathbf{k}}$ and $v_{\mathbf{k}}$ are (not yet specified) real and even functions of $\mathbf{k}$, with the only constraint $u_{\mathbf{k}}^2 + v_{\mathbf{k}}^2 = 1$. Keeping the spirit of Cooper pairing, the Bogoliubov transformations mix creation and annihilation operators for the spin-orbital $\mathbf{k}\uparrow$ with the ones for the spin-orbital $-\mathbf{k}\downarrow$.

We can write the Bogoliubov canonical transformation in the extended form

$$\begin{cases} \gamma_{\mathbf{k}\uparrow} = u_{\mathbf{k}}c_{\mathbf{k}\uparrow} - v_{\mathbf{k}}c_{-\mathbf{k}\downarrow}^{\dagger} \\ \gamma_{-\mathbf{k}\downarrow}^{\dagger} = v_{\mathbf{k}}c_{\mathbf{k}\uparrow} + u_{\mathbf{k}}c_{-\mathbf{k}\downarrow}^{\dagger} \end{cases} \quad \text{and} \quad \begin{cases} \gamma_{\mathbf{k}\uparrow}^{\dagger} = u_{\mathbf{k}}c_{\mathbf{k}\uparrow}^{\dagger} - v_{\mathbf{k}}c_{-\mathbf{k}\downarrow} \\ \gamma_{-\mathbf{k}\downarrow} = v_{\mathbf{k}}c_{\mathbf{k}\uparrow}^{\dagger} + u_{\mathbf{k}}c_{-\mathbf{k}\downarrow} \end{cases}, \quad (18.32)$$

while the inverse transformation reads

$$\begin{cases} c_{\mathbf{k}\uparrow} = u_{\mathbf{k}}\gamma_{\mathbf{k}\uparrow} + v_{\mathbf{k}}\gamma_{-\mathbf{k}\downarrow}^{\dagger} \\ c_{-\mathbf{k}\downarrow}^{\dagger} = u_{\mathbf{k}}\gamma_{-\mathbf{k}\downarrow}^{\dagger} - v_{\mathbf{k}}\gamma_{\mathbf{k}\uparrow} \end{cases} \quad \text{and} \quad \begin{cases} c_{\mathbf{k}\uparrow}^{\dagger} = u_{\mathbf{k}}\gamma_{\mathbf{k}\uparrow}^{\dagger} + v_{\mathbf{k}}\gamma_{-\mathbf{k}\downarrow} \\ c_{-\mathbf{k}\downarrow} = u_{\mathbf{k}}\gamma_{-\mathbf{k}\downarrow} - v_{\mathbf{k}}\gamma_{\mathbf{k}\uparrow}^{\dagger} \end{cases}. \quad (18.33)$$

The linear transformations in Eqs. (18.32) and (18.33) are canonical, since the fermion operators $\gamma_{\mathbf{k}\sigma}$ and $\gamma_{\mathbf{k}\sigma}^{\dagger}$ satisfy the same anticommutation rules as those between $c_{\mathbf{k}\sigma}$ and $c_{\mathbf{k}\sigma}^{\dagger}$, given by Eqs. (18.11). In the particular case of an *ordinary free-electron metal*, where Eqs. (18.13) apply, the operators $\gamma_{\mathbf{k}\sigma}$ and $\gamma_{\mathbf{k}\sigma}^{\dagger}$ represent annihilation and creation operators of *free-electrons and free-holes, of wavevector* $\mathbf{k}$ *and spin* σ. In the general case of a superconductor, we will see that the operators $\gamma_{\mathbf{k}\sigma}$ and $\gamma_{\mathbf{k}\sigma}^{\dagger}$ describe the annihilation and creation of *quasiparticles*, of wavevector $\mathbf{k}$ and spin σ, of the correlated system, i.e. of the interacting electron gas where an appropriate attractive two-body coupling is active among electrons.

To begin to appreciate the convenience of the Bogoliubov transformations, consider the application of any γ annihilation operator on the BCS wavefunction (18.31). For this purpose we observe that

$$\gamma_{\mathbf{k}\uparrow}\left(u_{\mathbf{k}} + v_{\mathbf{k}}c_{\mathbf{k}\uparrow}^{\dagger}c_{-\mathbf{k}\downarrow}^{\dagger}\right) = \left(u_{\mathbf{k}}c_{\mathbf{k}\uparrow} - v_{\mathbf{k}}c_{-\mathbf{k}\downarrow}^{\dagger}\right)\left(u_{\mathbf{k}} + v_{\mathbf{k}}c_{\mathbf{k}\uparrow}^{\dagger}c_{-\mathbf{k}\downarrow}^{\dagger}\right)$$

$$\equiv \left(u_{\mathbf{k}}^{2} + u_{\mathbf{k}}v_{\mathbf{k}}c_{\mathbf{k}\uparrow}^{\dagger}c_{-\mathbf{k}\downarrow}^{\dagger}\right)c_{\mathbf{k}\uparrow};$$

it follows

$$\gamma_{\mathbf{k}\uparrow}|\Psi_{S}\rangle = \gamma_{\mathbf{k}\downarrow}|\Psi_{S}\rangle \equiv 0,$$

i.e., the BCS wavefunction is the vacuum state for the Bogoliubov annihilation operators. Notice that this important result is valid with $u_{\mathbf{k}}$ and $v_{\mathbf{k}}$ not yet specified (except for the constraint $u_{\mathbf{k}}^{2} + v_{\mathbf{k}}^{2} = 1$). We now follow the Bogoliubov procedure to individuate the energies of the quasiparticles of the superconductor state.

We remark that in the operator H_{BCS} of Eq. (18.30), the source of difficulties is constituted by the presence of products of four fermion operators. With the aim to find an appropriate simplification, and in line with a routine procedure of mean field theory, we express the product of any two creation operators $c_{\mathbf{k}\uparrow}^{\dagger}c_{-\mathbf{k}\downarrow}^{\dagger}$ in the form

$$c_{\mathbf{k}\uparrow}^{\dagger}c_{-\mathbf{k}\downarrow}^{\dagger} = a_{\mathbf{k}} + \left(c_{\mathbf{k}\uparrow}^{\dagger}c_{-\mathbf{k}\downarrow}^{\dagger} - a_{\mathbf{k}}\right), \tag{18.34}$$

where $a_{\mathbf{k}}$ is the *expectation value of* $c_{\mathbf{k}\uparrow}^{\dagger}c_{-\mathbf{k}\downarrow}^{\dagger}$ *on the (not yet specified) ground state of the superconductor*

$$a_{\mathbf{k}} = \left\langle c_{\mathbf{k}\uparrow}^{\dagger}c_{-\mathbf{k}\downarrow}^{\dagger}\right\rangle, \tag{18.35}$$

and $(c_{\mathbf{k}\uparrow}^{\dagger}c_{-\mathbf{k}\downarrow}^{\dagger} - a_{\mathbf{k}})$ is called the *fluctuation operator*. A similar split is performed for the product of any two annihilation operators $c_{-\mathbf{k}\downarrow}c_{\mathbf{k}\uparrow}$. The split in Eq. (18.34) is useful whenever the fluctuation operators are small, so that it appears plausible to retain in the Hamiltonian (18.30) only terms up to first order in the fluctuations; in such a situation the resulting simplified model Hamiltonian can be easily diagonalized and the excitation spectrum of the superconductor explicitly worked out.

Let us follow in detail the outlined procedure. The Hamiltonian (18.30), using the operatorial identity (18.34) (and the hermitian conjugate of it), takes the form

$$H_{\text{BCS}} = \sum_{\mathbf{k}\sigma}\varepsilon_{\mathbf{k}}c_{\mathbf{k}\sigma}^{\dagger}c_{\mathbf{k}\sigma} + \sum_{\mathbf{k}\mathbf{k}'}U_{\mathbf{k}\mathbf{k}'}\left[a_{\mathbf{k}} + \left(c_{\mathbf{k}\uparrow}^{\dagger}c_{-\mathbf{k}\downarrow}^{\dagger} - a_{\mathbf{k}}\right)\right]\left[a_{\mathbf{k}'} + \left(c_{-\mathbf{k}'\downarrow}c_{\mathbf{k}'\uparrow} - a_{\mathbf{k}'}\right)\right]$$

(for simplicity we consider all $a_{\mathbf{k}}$ real, in close analogy to a similar assumption for the amplitudes $u_{\mathbf{k}}$ and $v_{\mathbf{k}}$ of the variational method). The above equation is still exact. Neglecting the presumably small terms that are second order in the fluctuations, we arrive at the simplified Bogoliubov Hamiltonian H_B given by

$$H_{B} = \sum_{\mathbf{k}\sigma}\varepsilon_{\mathbf{k}}c_{\mathbf{k}\sigma}^{\dagger}c_{\mathbf{k}\sigma} + \sum_{\mathbf{k}\mathbf{k}'}U_{\mathbf{k}\mathbf{k}'}\left[a_{\mathbf{k}}c_{-\mathbf{k}'\downarrow}c_{\mathbf{k}'\uparrow} + a_{\mathbf{k}'}c_{\mathbf{k}\uparrow}^{\dagger}c_{-\mathbf{k}\downarrow}^{\dagger} - a_{\mathbf{k}}a_{\mathbf{k}'}\right].$$

If we define the parameters

$$\Delta_{\mathbf{k}} = -\sum_{\mathbf{k}'} U_{\mathbf{k}\mathbf{k}'} a_{\mathbf{k}'}, \tag{18.36}$$

the model Hamiltonian H_B becomes

$$H_B = \sum_{\mathbf{k}\sigma} \varepsilon_{\mathbf{k}} c_{\mathbf{k}\sigma}^{\dagger} c_{\mathbf{k}\sigma} - \sum_{\mathbf{k}} \Delta_{\mathbf{k}} \left[c_{\mathbf{k}\uparrow}^{\dagger} c_{-\mathbf{k}\downarrow}^{\dagger} + c_{-\mathbf{k}\downarrow} c_{\mathbf{k}\uparrow} \right] + \sum_{\mathbf{k}} \Delta_{\mathbf{k}} a_{\mathbf{k}}. \tag{18.37}$$

We now express the Bogoliubov Hamiltonian (18.37) in terms of γ operators. For instance, using Eq. (18.33) and performing some straightforward commutation of fermion operators, we have

$$c_{\mathbf{k}\uparrow}^{\dagger} c_{\mathbf{k}\uparrow} + c_{-\mathbf{k}\downarrow}^{\dagger} c_{-\mathbf{k}\downarrow} = \left(u_{\mathbf{k}}^2 - v_{\mathbf{k}}^2 \right) \left[\gamma_{\mathbf{k}\uparrow}^{\dagger} \gamma_{\mathbf{k}\uparrow} + \gamma_{-\mathbf{k}\downarrow}^{\dagger} \gamma_{-\mathbf{k}\downarrow} \right]$$
$$+ 2 u_{\mathbf{k}} v_{\mathbf{k}} \left[\gamma_{\mathbf{k}\uparrow}^{\dagger} \gamma_{-\mathbf{k}\downarrow}^{\dagger} + \gamma_{-\mathbf{k}\downarrow} \gamma_{\mathbf{k}\uparrow} \right] + 2 v_{\mathbf{k}}^2 .$$

Similarly:

$$c_{\mathbf{k}\uparrow}^{\dagger} c_{-\mathbf{k}\downarrow}^{\dagger} + c_{-\mathbf{k}\downarrow} c_{\mathbf{k}\uparrow} = -2 u_{\mathbf{k}} v_{\mathbf{k}} \left[\gamma_{\mathbf{k}\uparrow}^{\dagger} \gamma_{\mathbf{k}\uparrow} + \gamma_{-\mathbf{k}\downarrow}^{\dagger} \gamma_{-\mathbf{k}\downarrow} \right]$$
$$+ \left(u_{\mathbf{k}}^2 - v_{\mathbf{k}}^2 \right) \left[\gamma_{\mathbf{k}\uparrow}^{\dagger} \gamma_{-\mathbf{k}\downarrow}^{\dagger} + \gamma_{-\mathbf{k}\downarrow} \gamma_{\mathbf{k}\uparrow} \right] + 2 u_{\mathbf{k}} v_{\mathbf{k}} .$$

The Bogoliubov Hamiltonian (18.37) thus becomes

$$H_B = \sum_{\mathbf{k}} \left[\varepsilon_{\mathbf{k}} \left(u_{\mathbf{k}}^2 - v_{\mathbf{k}}^2 \right) + 2 \Delta_{\mathbf{k}} u_{\mathbf{k}} v_{\mathbf{k}} \right] \left[\gamma_{\mathbf{k}\uparrow}^{\dagger} \gamma_{\mathbf{k}\uparrow} + \gamma_{-\mathbf{k}\downarrow}^{\dagger} \gamma_{-\mathbf{k}\downarrow} \right]$$
$$+ \sum_{\mathbf{k}} \left[2 \varepsilon_{\mathbf{k}} u_{\mathbf{k}} v_{\mathbf{k}} - \Delta_{\mathbf{k}} \left(u_{\mathbf{k}}^2 - v_{\mathbf{k}}^2 \right) \right] \left[\gamma_{\mathbf{k}\uparrow}^{\dagger} \gamma_{-\mathbf{k}\downarrow}^{\dagger} + \gamma_{-\mathbf{k}\downarrow} \gamma_{\mathbf{k}\uparrow} \right]$$
$$+ \sum_{\mathbf{k}} \left[2 \varepsilon_{\mathbf{k}} v_{\mathbf{k}}^2 - 2 \Delta_{\mathbf{k}} u_{\mathbf{k}} v_{\mathbf{k}} + \Delta_{\mathbf{k}} a_{\mathbf{k}} \right]. \tag{18.38}$$

A direct inspection of the model Hamiltonian (18.38) shows that it contains the diagonal particle number operators of the type $\gamma^{\dagger} \gamma$, but also the undesired terms of the type $\gamma^{\dagger} \gamma^{\dagger}$ and $\gamma \gamma$. At this stage we choose $u_{\mathbf{k}}$ and $v_{\mathbf{k}}$ so that the coefficient of $\gamma^{\dagger} \gamma^{\dagger}$ and $\gamma \gamma$ are zero and the model Hamiltonian (18.38) becomes diagonal in the particle number operators; for this aim we require that

$$\boxed{2 \varepsilon_{\mathbf{k}} u_{\mathbf{k}} v_{\mathbf{k}} - \Delta_{\mathbf{k}} \left(u_{\mathbf{k}}^2 - v_{\mathbf{k}}^2 \right) = 0}. \tag{18.39}$$

Thus we see that Eq. (18.39), which expresses the requirement of diagonalization of the model Hamiltonian (18.38), exactly coincides with the variational condition (18.19), which expresses energy minimization in the BCS theory. The solution of Eq. (18.39), together with the normalization condition $u_{\mathbf{k}}^2 + v_{\mathbf{k}}^2 = 1$, gives for $u_{\mathbf{k}}$ and $v_{\mathbf{k}}$ the explicit expressions already reported in Eqs. (18.20).

Further comparison of the present procedure with the variational procedure, makes transparent the equivalence of the variational BCS approach and the Bogoliubov approach, based on canonical transformations. Notice in particular that the constant term in the second member of Eq. (18.38) coincides with the expression of the super-conductor ground-state W_S, given by Eq. (18.16). Using Eqs. (18.20) for u_k and v_k, we easily verify that the quasiparticle energies in Eq. (18.38) become

$$\varepsilon_k \left(u_k^2 - v_k^2 \right) + 2\Delta_k u_k v_k = \sqrt{\varepsilon_k^2 + \Delta_k^2}.$$

The model Hamiltonian (18.38) can be recast in the form

$$\boxed{H_B = \sum_k w_k \left[\gamma_{k\uparrow}^\dagger \gamma_{k\uparrow} + \gamma_{-k\downarrow}^\dagger \gamma_{-k\downarrow} \right] + W_S} \quad \text{with} \quad w_k = \sqrt{\varepsilon_k^2 + \Delta_k^2} \ ;$$

$$(18.40)$$

it is apparent that w_k are the energies, above the ground state, of the *quasiparticles* created by the Fermi operators $\gamma_{k\uparrow}^\dagger$ and $\gamma_{-k\downarrow}^\dagger$.

In order to further clarify the meaning of the Hamiltonian (18.40) let us consider the *average gap approximation*. In this case, the energy gap parameters Δ_k are constant and equal to Δ_0 (in the energy shell $\pm \hbar\omega_D$ around the Fermi level). The quasiparticles excitations in the superconductor become

$$w_k = \sqrt{\varepsilon_k^2 + \Delta_0^2} \quad \text{(superconductor)}. \tag{18.41a}$$

In the limiting case of vanishing electron-electron interaction, i.e. for the normal metal, we would have $\Delta_0 = 0$ and hence

$$w_k = |\varepsilon_k| \quad \text{(normal metal)}. \tag{18.41b}$$

In Figure 18.10 we show the energies of the elementary excitations in a normal metal and in a superconductor. The quasiparticle spectrum of the superconductor exhibits

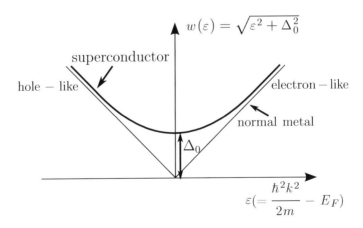

Figure 18.10 Quasiparticle energy spectrum for the normal metal and for the superconductor at zero temperature.

thus an energy gap given by Δ_0. In the superconductor, there are no electron-like states with energy in the interval $[E_F, E_F + \Delta_0]$, and no hole-like states in the energy interval $[E_F - \Delta_0, E_F]$.

The density of quasiparticle states in the superconductor can be obtained from the dispersion relation (18.41a) as follows. The constant energy surfaces in the reciprocal space are spheres, and the number of states $D_S(w)\, dw$ in the energy interval $[w, w+dw]$ are thus

$$D_S(w)\, dw = \frac{2V}{(2\pi)^3} 4\pi k^2 dk = \frac{V k_F^2}{\pi^2} dk, \tag{18.42a}$$

where V is the volume of the sample, and k has been replaced by k_F, since we consider energy intervals very close to the Fermi surface. From the dispersion relation (18.41a), we have

$$\frac{dw}{dk} = \frac{\varepsilon}{\sqrt{\varepsilon^2 + \Delta_0^2}} \frac{d\varepsilon}{dk} = \frac{\sqrt{w^2 - \Delta_0^2}}{w} \frac{\hbar^2 k_F}{m}, \tag{18.42b}$$

where $d\varepsilon/dk$ has been calculated at the Fermi energy. Inserting Eq. (18.42b) into Eq. (18.42a), gives for the density-of-states of the superconductor

$$D_S(w) = D(E_F) \frac{w}{\sqrt{w^2 - \Delta_0^2}}, \tag{18.42c}$$

where $D(E_F) = (m k_F / \pi^2 \hbar^2) V$ denotes the density-of-states of the normal metal at the Fermi energy.

It is convenient at this stage to measure the energy of quasiparticles on a fixed scale (instead of referring them to the Fermi energy). We obtain for the density of quasiparticles in the superconductor the expression

$$D_S(E) = D(E_F) \frac{|E - E_F|}{\sqrt{(E - E_F)^2 - \Delta_0^2}} \qquad \text{for} \qquad |E - E_F| > \Delta_0. \tag{18.42d}$$

From Eq. (18.42d) we see that the density of quasiparticle states in the superconductor presents a singularity at the energies $E_F \pm \Delta_0$; the schematic behavior of the density-of-states in a normal metal and in a superconductor is shown in Figure 18.11.

From the above considerations on quasiparticle excitations, we also infer that breaking a pair, giving rise to two quasiparticles, requires at least an energy $2\Delta_0$; the quantity $2\Delta_0$ can be interpreted as the binding energy of any one pair, due to the cooperative presence of many other pairs (all in the same quantum state of zero total spin and zero total momentum) in the superconductor ground state.

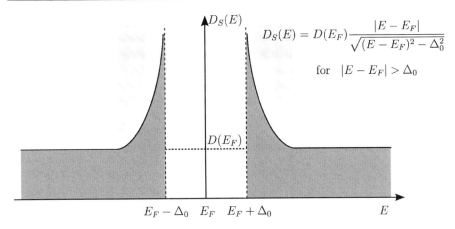

$$D_S(E) = D(E_F)\frac{|E - E_F|}{\sqrt{(E - E_F)^2 - \Delta_0^2}}$$

$$\text{for} \quad |E - E_F| > \Delta_0$$

Figure 18.11 Density-of-states for quasiparticles in a superconductor with energy gap $2\Delta_0$ around E_F. In the same energy region the density-of-states of the normal metal is (approximately) constant and equal to $D(E_F)$.

18.4.2   Persistent Currents in Superconductors

The existence of an energy gap at the Fermi level in the excitation spectrum of a superconductor has profound effects on the transport properties, that are sensitive to the modifications introduced at and near the Fermi level. These include, for instance, perfect conductance, diamagnetic properties, tunneling properties, specific heat, infrared photon absorption, Knight shift, nuclear-spin relaxation, ultrasonic attenuation, transport coefficients; here we discuss the zero electrical resistance of a superconductor.

Until now we have considered the ground state of a superconductor by pairing electrons $(\mathbf{k}\uparrow, -\mathbf{k}\downarrow)$ in states of total spin $S = 0$ and total momentum $\mathbf{K} = \mathbf{k} - \mathbf{k} = 0$. A current carrying state of a superconductor is obtained by pairing the electrons $(\mathbf{k} + \mathbf{Q}\uparrow; -\mathbf{k} + \mathbf{Q}\downarrow)$, and following for the rest the whole BCS microscopic formalism. In fact the total momentum $\hbar\mathbf{K}$, with $\mathbf{K} = (\mathbf{k} + \mathbf{Q}) + (-\mathbf{k} + \mathbf{Q}) = 2\mathbf{Q}$, of each Cooper pair is a constant of motion, and the kinetic energy associated with the center of mass motion of each Cooper pair is $E_{\text{kin}} = \hbar^2(2Q)^2/2(2m)$. For small values of $\mathbf{Q}$ (the ones of practical interest in transport), the kinetic energy E_{kin} is small with respect to the binding energy $2\Delta_0$ of any one pair, and is not sufficient to break the pair.

Suppose we attempt to degrade the current of the drifting superconducting state by changing with some scattering mechanism, the momentum of a number of drifting Cooper pairs $(\mathbf{k} + \mathbf{Q}\uparrow; -\mathbf{k} + \mathbf{Q}\downarrow)$. Due to the cooperative origin of the binding energy of the pairs, changing the momentum of a single pair with respect to the common value $2\mathbf{Q}$, requires a cost in energy equal to the binding energy $2\Delta_0$ (at least for small $\mathbf{Q}$). As soon as the number of pairs with random momenta increases, the energy penalty to sustain the situation becomes prohibitively large; furthermore, astray pairs tend naturally to "condensate" in the ground energy state, where all partner pairs share the same common momentum. The only process that could degrade current is the *simultaneous scattering of a huge number of drifting Cooper pairs* into a new coherent state, where the scattered pairs have the same final momentum so that the pairing energy gain is at work. It is not easy to imagine how such a scattering event could occur and therefore our arguments justify the persistent currents in superconductors.

18.5 Treatment of Superconductors at Finite Temperature and Heat Capacity

Self-Consistency at Finite Temperature

At finite temperature the quasiparticle states of the superconductor are thermally excited and a number of Cooper pairs are broken; this process, in turns, is accompanied by a decrease of the energy gap in quasiparticle excitations and eventually leads to the transition to the normal state.

In the BCS variational method, the treatment of superconductors at finite temperature can be done by minimizing the free energy at temperature T. Alternatively, and equivalently, we can extend at finite temperature the approach of Section 18.4.1; keeping with the spirit of the mean field theory of phase transitions, at finite temperature we express the product $c^{\dagger}_{\mathbf{k}\uparrow} c^{\dagger}_{-\mathbf{k}\downarrow}$ in the form

$$c^{\dagger}_{\mathbf{k}\uparrow} c^{\dagger}_{-\mathbf{k}\downarrow} = a_{\mathbf{k}} + \left(c^{\dagger}_{\mathbf{k}\uparrow} c^{\dagger}_{-\mathbf{k}\downarrow} - a_{\mathbf{k}} \right), \tag{18.43}$$

where $a_{\mathbf{k}}$ is the *thermal average expectation value*

$$a_{\mathbf{k}} = \left\langle c^{\dagger}_{\mathbf{k}\uparrow} c^{\dagger}_{-\mathbf{k}\downarrow} \right\rangle_{T} \tag{18.44}$$

and $(c^{\dagger}_{\mathbf{k}\uparrow} c^{\dagger}_{-\mathbf{k}\downarrow} - a_{\mathbf{k}})$ is the *fluctuation operator from the thermal average*. Equations (18.43) and (18.44) are the trivial generalization at finite temperature of Eqs. (18.34) and (18.35), and reduce to them at zero temperature. With these generalizations in mind, we now follow step-by-step the whole procedure of Section 18.4.1.

At finite temperature, the generalization of Eq. (18.36) takes the form

$$\Delta_{\mathbf{k}} = - \sum_{\mathbf{k}'} U_{\mathbf{k}\mathbf{k}'} \langle c_{-\mathbf{k}'\downarrow} c_{\mathbf{k}'\uparrow} \rangle_{T}. \tag{18.45}$$

It is convenient to express the product operator $c_{-\mathbf{k}\downarrow} c_{\mathbf{k}\uparrow}$ in terms of the γ operators by means of the canonical transformations (18.33); we have

$$c_{-\mathbf{k}\downarrow} c_{\mathbf{k}\uparrow} = \left(u_{\mathbf{k}} \gamma_{-\mathbf{k}\downarrow} - v_{\mathbf{k}} \gamma^{\dagger}_{\mathbf{k}\uparrow} \right) \left(u_{\mathbf{k}} \gamma_{\mathbf{k}\uparrow} + v_{\mathbf{k}} \gamma^{\dagger}_{-\mathbf{k}\downarrow} \right).$$

The thermal average of the above operators gives

$$\langle c_{-\mathbf{k}\downarrow} c_{\mathbf{k}\uparrow} \rangle_{T} = \left\langle u_{\mathbf{k}} v_{\mathbf{k}} \gamma_{-\mathbf{k}\downarrow} \gamma^{\dagger}_{-\mathbf{k}\downarrow} - u_{\mathbf{k}} v_{\mathbf{k}} \gamma^{\dagger}_{\mathbf{k}\uparrow} \gamma_{\mathbf{k}\uparrow} \right\rangle_{T}$$

$$= u_{\mathbf{k}} v_{\mathbf{k}} \left\langle 1 - \gamma^{\dagger}_{-\mathbf{k}\downarrow} \gamma_{-\mathbf{k}\downarrow} - \gamma^{\dagger}_{\mathbf{k}\uparrow} \gamma_{\mathbf{k}\uparrow} \right\rangle_{T} = u_{\mathbf{k}} v_{\mathbf{k}} [1 - 2 f(w_{\mathbf{k}})],$$

where $w_{\mathbf{k}}$ are the quasiparticle energies, and the Fermi-Dirac function $f(E)$ gives their excitation probability at thermal equilibrium.

At finite temperature the self-consistent Eq. (18.45) becomes

$$\Delta_{\mathbf{k}} = -\sum_{\mathbf{k}'} U_{\mathbf{k}\mathbf{k}'} u_{\mathbf{k}'} v_{\mathbf{k}'}[1 - 2f(w_{\mathbf{k}'})] \quad \text{with} \quad w_{\mathbf{k}} = \sqrt{\varepsilon_{\mathbf{k}}^2 + \Delta_{\mathbf{k}}^2}.$$

Using Eq. (18.20) and the equality

$$f(E) = \frac{1}{e^{\beta E} + 1} \implies 1 - 2f(E) = \tanh \frac{\beta E}{2},$$

we obtain the self-consistent equation

$$\Delta_{\mathbf{k}} = -\frac{1}{2} \sum_{\mathbf{k}'} U_{\mathbf{k}\mathbf{k}'} \frac{\Delta_{\mathbf{k}'}}{\sqrt{\varepsilon_{\mathbf{k}'}^2 + \Delta_{\mathbf{k}'}^2}} \tanh \frac{\beta \sqrt{\varepsilon_{\mathbf{k}'}^2 + \Delta_{\mathbf{k}'}^2}}{2}. \tag{18.46}$$

This is the desired generalization of Eq. (18.21), and reduces to it at zero temperature.

In the average potential approximation, Eq. (18.46) simplifies in the form

$$1 = \frac{1}{2} U_0 n_0 \int_{-\hbar\omega_D}^{\hbar\omega_D} \frac{d\varepsilon}{\sqrt{\varepsilon^2 + \Delta^2}} \tanh \frac{\beta \sqrt{\varepsilon^2 + \Delta^2}}{2}, \tag{18.47}$$

where n_0 denotes the density-of-states for one spin direction and per unit cell at the Fermi energy, as in Eq. (18.8). Equation (18.47) provides an implicit relation for the temperature dependence of the gap parameter $\Delta(T)$; for $T \to 0$ the hyperbolic tangent in the integrand equals unity and we recover the value Δ_0 given by Eq. (18.23). As T increases, the hyperbolic tangent is always less than 1 and the integral in the second member of Eq. (18.47) can preserve its constant value only if Δ decreases. Equation (18.47) thus puts in evidence that in the superconducting state a cooperative decrease of the energy gap parameter occurs as the temperature increases, leading eventually to the disappearance of the energy gap and transition to the normal state.

From Eq. (18.47) we see that the energy gap disappears at the critical temperature T_c implicitly determined by the relation

$$1 = U_0 n_0 \int_0^{\hbar\omega_D} \frac{1}{\varepsilon} \tanh \frac{\varepsilon}{2k_B T_c} d\varepsilon,$$

or equivalently, introducing the dimensionless variable $x = \varepsilon/2k_B T_c$ we have

$$\int_0^{\hbar\omega_D/2k_B T_c} \frac{1}{x} \tanh x \, dx = \frac{1}{U_0 n_0}.$$

To carry on, consider the integral of the type

$$I(a) = \int_0^a \frac{1}{x} \tanh x \, dx = [\tanh x \, \ln x]_0^a - \int_0^a \ln x \frac{1}{\cosh^2 x} dx$$

$$(\text{for } a \gg 1) \approx \ln a - \int_0^\infty \ln x \frac{1}{\cosh^2 x} dx = \ln a + \ln \frac{4\gamma}{\pi} \quad \gamma = 1.78107\ldots$$

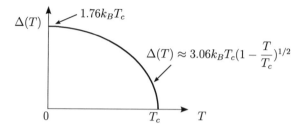

Figure 18.12 Behavior of the energy gap parameter $\Delta(T)$ for a superconductor in the BCS theory and in weak coupling limit.

where γ is the Euler constant [see I. S. Gradshteyn and I. M. Ryzhik "Table of Integrals, Series, and Products" Academic Press, New York, 1980, p. 580].

For large ratios $\hbar\omega_D/k_BT_c$ it holds

$$\int_0^{\hbar\omega_D/2k_BT_c} \frac{1}{x} \tanh x \, dx = \ln\left(\frac{2\gamma}{\pi}\frac{\hbar\omega_D}{k_BT_c}\right) \approx \ln\frac{1.13\hbar\omega_D}{k_BT_c} = \frac{1}{U_0n_0}.$$

Then, in the weak coupling limit $U_0n_0 \ll 1$ and $\hbar\omega_D/k_BT_c \gg 1$, we have

$$k_BT_c = 1.13\,\hbar\omega_D \exp[-1/U_0n_0].$$

Comparing this result with Eq. (18.23b) we have

$$\boxed{\Delta(0) = 1.76\,k_BT_c}. \tag{18.48a}$$

The behavior of $\Delta(T)$ obtained by numerical integration of Eq. (18.47) is shown in Figure 18.12. In particular for $T \to T_c$ (and $T < T_c$) the form of $\Delta(T)$ is found to be

$$\Delta(T) = 3.06\,k_BT_c\left(1 - \frac{T}{T_c}\right)^{1/2}, \tag{18.48b}$$

and the exponent 1/2 is a characteristic feature of the mean field theories.

Electronic Heat Capacity for a Superconductor

We can now discuss the electronic heat capacity of a superconductor. The transition to the superconducting state is accompanied by a quite drastic change of the electronic contribution to the heat capacity; the characteristic behavior is indicated in Figure 18.13. There is a sharp jump of the heat capacity of the superconductor at the critical temperature. For $T > T_c$ the electronic heat capacity of the normal material is linear with temperature (as studied in Section 3.3). For $T \approx 0$ the heat capacity decays exponentially to zero; the exponential behavior at low temperature is of the form expected when an energy gap exists in the energy spectrum of quasiparticles.

To calculate the electronic heat capacity of a superconductor, we start from the general expression of the entropy $S(T)$ of a system of fermionic quasiparticles (see Eq. (3.B8)); in the present case we have

$$S(T) = -2k_B \sum_{\mathbf{k}} [f_{\mathbf{k}} \ln f_{\mathbf{k}} + (1 - f_{\mathbf{k}}) \ln(1 - f_{\mathbf{k}})], \tag{18.49a}$$

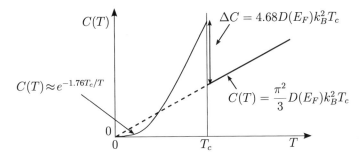

Figure 18.13 Characteristic behavior of the heat capacity of a normal metal and a superconductor, above and below the critical temperature.

where the factor 2 takes into account the two spin orientations, $f_\mathbf{k}$ is the Fermi-Dirac distribution function

$$f_\mathbf{k} = \frac{1}{\exp(\beta w_\mathbf{k}) + 1} \quad \text{with} \quad w_\mathbf{k} = \sqrt{\varepsilon_\mathbf{k}^2 + \Delta^2} \quad \text{and} \quad \Delta = \Delta(T);$$

the average potential approximation has been assumed for the quasiparticle excitations. The heat capacity is given by the expression

$$C(T) = T\frac{dS(T)}{dT} = \sum_\mathbf{k} \frac{\partial S}{\partial f_\mathbf{k}} \frac{\partial f_\mathbf{k}}{\partial T} \tag{18.49b}$$

according to Eq. (3.21b).

We notice that

$$\frac{\partial S}{\partial f_\mathbf{k}} = -2k_B \ln\frac{f_\mathbf{k}}{1 - f_\mathbf{k}} = 2\frac{1}{T}\sqrt{\varepsilon_\mathbf{k}^2 + \Delta^2}$$

$$\frac{\partial f_\mathbf{k}}{\partial T} = \frac{1}{k_B T^2}\frac{\exp(\beta w_\mathbf{k})}{[\exp(\beta w_\mathbf{k}) + 1]^2}\left[\sqrt{\varepsilon_\mathbf{k}^2 + \Delta^2} - T\frac{d}{dT}\sqrt{\varepsilon_\mathbf{k}^2 + \Delta^2}\right].$$

Eq. (18.49b) thus gives

$$C(T) = \frac{2}{k_B T^2} \sum_\mathbf{k} \frac{\exp(\beta w_\mathbf{k})}{[\exp(\beta w_\mathbf{k}) + 1]^2}\left[\varepsilon_\mathbf{k}^2 + \Delta^2 - \frac{T}{2}\frac{d}{dT}\Delta^2\right]. \tag{18.50a}$$

As usual, twice the sum over $\mathbf{k}$ in the reciprocal space can be replaced by the integral in the energy variable ε multiplied by the density-of-states $D(E_F)$ for both spin directions (the density-of-states is taken as constant in the small energy shell of interest around the Fermi level). Extending to infinity the limits of integration one obtains

$$C(T) = \frac{D(E_F)}{k_B T^2} \int_{-\infty}^{\infty} \frac{\exp\left(\beta\sqrt{\varepsilon^2 + \Delta^2}\right)}{\left[\exp\left(\beta\sqrt{\varepsilon^2 + \Delta^2}\right) + 1\right]^2}\left[\varepsilon^2 + \Delta^2 - \frac{T}{2}\frac{d}{dT}\Delta^2\right]d\varepsilon. \tag{18.50b}$$

From Eq. (18.50b) the behavior of the heat capacity as a function of temperature can be worked out. In particular for $T > T_c$ we have $\Delta(T) = 0$ and Eq. (18.50b) becomes

$$
\begin{aligned}
C(T) &= \frac{D(E_F)}{k_B T^2} 2 \int_0^\infty \frac{e^{\beta \varepsilon}}{(e^{\beta \varepsilon} + 1)^2} \varepsilon^2 d\varepsilon \\
&= 2k_B^2 T D(E_F) \int_0^\infty \frac{e^x}{(e^x + 1)^2} x^2 \, dx = \frac{\pi^2}{3} D(E_F) k_B^2 T,
\end{aligned}
$$

which is the standard result for a normal metal (see Eq. (3.23)).

For $T = T_c$ a discontinuity occurs in the heat capacity because of the term involving $d\Delta^2/dT$ which equals $-9.36 \, k_B^2 T_c$ (from Eq. 18.48b). From Eq. (18.50b) the value of this discontinuity is

$$
\Delta C(T_c) = 9.36 \, k_B D(E_F) \int_0^\infty \frac{e^{\beta \varepsilon}}{(e^{\beta \varepsilon} + 1)^2} \, d\varepsilon = 4.68 \, D(E_F) k_B^2 T_c.
$$

At the critical temperature, the heat capacity jump $C_s - C_n$ from the superconductor to the normal metal, divided by the normal metal heat capacity, gives the parameter free expression

$$
\frac{C_s - C_n}{C_n} = \frac{4.68 \cdot 3}{\pi^2} = 1.42.
$$

Finally, for $T \ll T_c$, Eq. (18.50b) exhibits an exponential decay for the superconductor heat capacity of the form $C(T) \approx \exp(-\beta \Delta(0)) = \exp(-1.76 T_c/T)$.

18.6   The Phenomenological London Model for Superconductors

Diamagnetism of Superconductors and Meissner Effect

Some phenomenological aspects concerning the behavior of superconductors in the presence of magnetic field have been presented in Section 18.1. We have seen that a distinctive feature of superconductivity is the Meissner effect: a (weak) magnetic field cannot penetrate in the bulk of the superconductor. The first phenomenological model to explain the Meissner effect was given by the brothers F. and H. London (1935); it is instructive to report from their model the aspects useful to clarify the distinctive transport and magnetic properties of *ordinary conductors, ideally perfect conductors* and *superconductors*.

Consider first an *ordinary conductor* with n electrons per unit volume with effective mass m. In an ordinary conductor, the electrons moving in the material can be scattered by impurities, phonons, and other defects. Assuming for simplicity that the carrier dynamics can be characterized with some finite relaxation time τ, the average drift velocity of the carriers in the presence of an electric field is $\mathbf{v}_d = (-e)\tau \mathbf{E}$, and the current density is given by the Ohm law $\mathbf{J} = \sigma \mathbf{E}$ with $\sigma = ne^2\tau/m$.

We consider now an ideally *pure metal*, with a parabolic conduction band, and n electrons per unit volume of effective mass m. In the presence of an electric field $\mathbf{E}$, the motion of each freely moving electron is described by the ballistic dynamic equation

$$m\frac{d\mathbf{v}}{dt} = (-e)\mathbf{E}.$$

The current density is then related to the electric field by the equation

$$\mathbf{J} = n(-e)\mathbf{v} \implies \boxed{\frac{\partial \mathbf{J}}{\partial t} = \frac{ne^2}{m}\mathbf{E}}. \tag{18.51}$$

This is known as the *first London equation* and it is written on account of ideal collision-less motion. Notice that in the case of a standard conductor, with a (finite) relaxation time τ, the first member of Eq. (18.51) is replaced by $\mathbf{J}/\tau$; thus the first London equation is simply a modification of the Ohm law needed to describe the dynamics of collisionless electrons.

We apply the curl-operator to both members of Eq. (18.51), exploit the appropriate Maxwell equation for the electric field, and obtain

$$\frac{\partial}{\partial t}\operatorname{curl}\mathbf{J} = \frac{ne^2}{m}\operatorname{curl}\mathbf{E} = -\frac{ne^2}{mc}\frac{\partial \mathbf{B}}{\partial t} \implies \frac{\partial}{\partial t}\left[\operatorname{curl}\mathbf{J} + \frac{ne^2}{mc}\mathbf{B}\right] = 0. \tag{18.52}$$

This is a general equation for any ideal conductor, but alone does not account for the Meissner effect. In fact Eq. (18.52) shows that the vectorial field $\mathbf{V} = \operatorname{curl}\mathbf{J} + (ne^2/mc)\mathbf{B}$ is constant in time, but this argument is not sufficient to establish whether fields and currents can penetrate or not in the interior of the sample.

The macroscopic London theory assumes that in the superconducting state of a metal, the electronic density is made of two contributions, one from normal electrons and one from super-electrons which condense into a macroscopic quantum state. It has been pointed out by the London brothers that, if the quantity in square brackets in Eq. (18.52), for super-electrons, is not only time-independent (as stated by Eq. 18.52), but actually *vanishes identically*, then the ideal perfect conductor also exhibits the Meissner effect, i.e. the property of excluding fields and currents from its interior. The equation

$$\boxed{\operatorname{curl}\mathbf{J}_s = -\frac{n_s e^2}{mc}\mathbf{B}} \tag{18.53}$$

is called the *second London equation*, and appeared originally in the literature as a conjecture of the London brothers for superconductors.

The first London equation (18.51) implies, in particular, that a perfect conductor in stationary conditions cannot sustain an electric field. It is easily seen that the second London equation (18.53) implies, in particular, that a superconductor in stationary conditions cannot sustain a magnetic field in its interior, except for a thin surface layer whose depth can be calculated as follows. In stationary conditions, the magnetic field is related to the current density by the Maxwell equation

$$\text{curl } \mathbf{B} = \frac{4\pi}{c} \mathbf{J}_s. \tag{18.54}$$

We apply the curl-operator to both sides of Eq. (18.54) and use Eq. (18.53); we obtain

$$\text{curl curl } \mathbf{B} = \frac{4\pi}{c} \text{curl } \mathbf{J}_s = -\frac{4\pi n_s e^2}{mc^2} \mathbf{B}.$$

We exploit the operator identity curl curl $=$ grad div $- \nabla^2$ and the Maxwell equation div $\mathbf{B} = 0$ obtaining

$$\nabla^2 \mathbf{B}(\mathbf{r}) = \frac{1}{\lambda_L^2} \mathbf{B}(\mathbf{r}), \tag{18.55}$$

where the *London penetration length* λ_L is given by

$$\lambda_L = \frac{c}{\omega_p} = \sqrt{\frac{mc^2}{4\pi n_s e^2}} \tag{18.56}$$

and ω_p is the plasma frequency corresponding to the electron density n_s. For electron concentrations typical of metals, the magnetic penetration length λ_L is of the order of $10^2 \approx 10^3$ Å, under the assumption that all electrons behave ballistically ($n = n_s$). [At $T = 0$ we have supposed that all the electrons are super-electrons. A more general theory is required to settle this point, in particular to establish the temperature dependence of the effective number of super-electrons].

We solve now the above Eq. (18.55) for a semi-infinite slab of superconductor, whose surface is in the xy plane, the region $z < 0$ being empty. We consider two simple and particularly significant situations.

(i) The magnetic field $\mathbf{B}$ is parallel to the z-axis and homogeneous in the xy-plane. This means that $\mathbf{B}$ can be written in the form $\mathbf{B} = (0, 0, B(z))$. The equation div $\mathbf{B} = 0$ gives $\partial B(z)/\partial z = 0$, and thus $B(z)$ is constant. Equation (18.55) gives $B(z) = 0$ for $z > 0$ and shows that it is impossible to have a magnetic field normal to the superconductor surface.

(ii) The magnetic field $\mathbf{B}$ is parallel to the x-axis and homogeneous in the xy-plane (see Figure 18.14). This means that $\mathbf{B}$ can be written in the form $\mathbf{B} = (B(z), 0, 0)$. The equation div $\mathbf{B} = 0$ is now automatically satisfied. From Eq. (18.55) we have

$$\frac{\partial^2 B(z)}{\partial z^2} = \frac{1}{\lambda_L^2} B(z),$$

whose solution is

$$B(z) = B(0)e^{-z/\lambda_L} \quad (z > 0).$$

Thus, except for a narrow surface region of width of the order of λ_L, the magnetic field parallel to the surface cannot penetrate into a superconductor (Meissner effect). A similar conclusion holds for the screening current, that is provided by Eq. (18.54), and is directed along the y-axis in the geometry of Figure 18.14.

Figure 18.14 Semi-infinite slab geometry for illustration of screening currents and penetration depth of the magnetic field parallel to the surface.

The London Gauge for Superconductors

Before concluding this section, we wish to consider some other consequences entailed by Eq. (18.53). We consider here stationary situation, in which case the continuity equation for the current density entails that its divergence is zero. For simplicity, we also assume that the superconducting sample has finite volume and simply connected geometry.

Equation (18.53) can be elaborated in a more convenient form, by expressing the magnetic induction $\mathbf{B}(\mathbf{r})$ in terms of the vector potential $\mathbf{A}(\mathbf{r})$ associated to it through the relation

$$\mathbf{B}(\mathbf{r}) = \operatorname{curl} \mathbf{A}(\mathbf{r}). \tag{18.57a}$$

It is well known that the above equation does not define the vector potential completely. The vector potential $\mathbf{A}(\mathbf{r})$ is determined within arbitrary gauge transformations of the type $\mathbf{A}'(\mathbf{r}) = \mathbf{A}(\mathbf{r}) + \nabla\phi(\mathbf{r})$, where $\phi(\mathbf{r})$ is any regular and single-valued function.

We now insert Eq. (18.57a) into Eq. (18.53), and obtain

$$\operatorname{curl} \left[\mathbf{J}_s(\mathbf{r}) + \frac{n_s e^2}{mc} \mathbf{A}(\mathbf{r}) \right] = 0 \quad (\mathbf{r} \text{ within the superconductor}). \tag{18.57b}$$

The vector in square brackets in Eq. (18.57b) is irrotational within the superconductor and can thus be expressed in the form

$$\mathbf{J}_s(\mathbf{r}) + (n_s e^2/mc)\mathbf{A}(\mathbf{r}) = -\nabla\chi(\mathbf{r}),$$

where $\chi(\mathbf{r})$ is a regular single-valued function for the simply connected geometry under attention. We can exploit the mentioned arbitrariness in $\mathbf{A}(\mathbf{r})$ to embody $\nabla\chi(\mathbf{r})$ into $\mathbf{A}(\mathbf{r})$ itself; thus we arrive at the London equation

$$\boxed{\mathbf{J}_s(\mathbf{r}) = -\frac{n_s e^2}{mc}\mathbf{A}(\mathbf{r})}. \tag{18.57c}$$

Equation (18.57c) shows that it is possible for connected superconductors to construct a gauge in which $\mathbf{J}_s(\mathbf{r})$ and $\mathbf{A}(\mathbf{r})$ are proportional (*this gauge is called the London gauge*).

Notice that Eq. (18.57c) is not gauge-invariant and holds exclusively in the London gauge; such a gauge is completely and uniquely specified by the following requirements

$$
\begin{cases}
\text{(i)} & \text{curl}\,\mathbf{A}(\mathbf{r}) = \mathbf{B}(\mathbf{r}) & \text{basic equation} \\
\text{(ii)} & \text{div}\,\mathbf{A}(\mathbf{r}) = 0 & \text{bulk supplementary condition} \\
\text{(iii)} & \mathbf{A}(\mathbf{r}_s) \cdot \mathbf{n} = -(mc/ne^2)\mathbf{J}_s(\mathbf{r}_s) \cdot \mathbf{n} & \text{surface supplementary condition.}
\end{cases}
$$

$$(18.58)$$

In the London gauge, the basic equation (i) $\text{curl}\,\mathbf{A}(\mathbf{r}) = \mathbf{B}(\mathbf{r})$ is supplemented by the familiar Coulomb condition (ii) $\text{div}\mathbf{A}(\mathbf{r}) = 0$ (compatible with $\text{div}\,\mathbf{J}_s = 0$ that holds in stationary conditions), and by the requirement that (iii) the normal component of $\mathbf{A}$ at the surface is assigned and equal to the normal component of $\mathbf{J}_s$ times $-(mc/n_s e^2)$ (in the case no current is fed into the superconductor, normal components of vector potential and current density at the surface are both zero). [It is evident that Eq. (18.58) determine uniquely the vector potential. This could be proved using exact mathematical theorems, or also with the following argument. Suppose two vector potentials $\mathbf{A}(\mathbf{r})$ and $\mathbf{A}'(\mathbf{r})$ satisfy the three conditions (18.58). Then the vector field $\mathbf{v}(\mathbf{r}) = \mathbf{A}(\mathbf{r}) - \mathbf{A}'(\mathbf{r})$ must satisfy (j) $\text{curl}\,\mathbf{v}(\mathbf{r}) = 0$, (jj) $\text{div}\,\mathbf{v}(\mathbf{r}) = 0$, (jjj) normal component of $\mathbf{v}(\mathbf{r})$ at the surface equal to zero. Because of (jj) and (jjj), the force lines of $\mathbf{v}(\mathbf{r})$ are closed within the connected region; then (j) entails that the line integral $\oint \mathbf{v}(\mathbf{r}) \cdot d\mathbf{l}$ vanishes for any closed line within the superconductor, force lines included, so that $\mathbf{v} \equiv 0$ and $\mathbf{A}(\mathbf{r}) \equiv \mathbf{A}'(\mathbf{r})$ is uniquely determined].

The London electrodynamics, summarized by Eq. (18.57c) is a *local one*: the current density at some point depends on the vector potential at the same point. The penetration depth of fields tangential to the surface occurs in a region given by the London length λ_L.

A detailed analysis based on the microscopic BCS theory shows that the London equation (18.57c) holds in the case $\mathbf{A}(\mathbf{r})$ is slowly varying on the length scale given by the coherence length ξ_0 of the Cooper pair. The assumption of local electrodynamics is valid for type-I superconductors, near the transition temperature where the effective penetration depth becomes large; it also holds, in general, for type-II superconductors (characterized by high values of the Ginzburg-Landau parameter κ, discussed in the next section). When the conditions of locality are not met, a nonlocal electrodynamics must be applied, following the guidelines introduced by Pippard. His approach is essentially based on the assumption (corroborated by experiments and in-depth theoretical analysis) that the current density is given by a weighted integral of the vector potential over a distance controlled appropriately by the correlation length of the Cooper pairs and by the mean free path of the electrons in the normal metal.

18.7 Macroscopic Quantum Phenomena

18.7.1 Order Parameter in Superconductors and the Ginzburg-Landau Equations

The BCS microscopic theory of superconductivity is very illuminating but also rather demanding in the formal and technical aspects, in spite of the fact that the systems considered up to now are bulk and clean homogeneous materials. In several problems of remarkable theoretical and technological interest, one has to consider non homogeneous systems with boundary effects, impurity effects, space-varying Cooper pairs density, magnetic fields etc. Important physical effects near the normal metal-superconductor phase transition can be well described and understood within the Ginzburg-Landau phenomenological theory of second order phase transitions. This approach is an ingenious body of assumptions, which are based on the phenomenology of superconductors and some findings of the microscopic theory, avoiding however the explicit use of many-body wavefunctions.

Let us specify some elements of the theory with reference to the transition to the superconducting phase. For the description of the superconducting phase, Ginzburg and Landau introduce a macroscopic complex function, the *order parameter* ψ, which characterizes the ordering which is reached passing from the disordered phase (for $T > T_c$) to the ordered one (for $T < T_c$). It is assumed that the complex order parameter $\psi(\mathbf{r})$ at a given temperature T in the superconducting phase is related to the local number of Cooper pairs by the relation

$$n_{\text{pairs}}(\mathbf{r}) = |\psi(\mathbf{r})|^2.$$

The basic assumption of the Ginzburg-Landau theory is that the superconductor free energy density near the transition temperature can be expanded in the form

$$f_S(T, \psi, \mathbf{A}) = f_N(T) + \alpha(T)|\psi|^2 + \frac{1}{2}\beta(T)|\psi|^4$$

$$+ \frac{1}{2m^*}\left|\left(\mathbf{p} - \frac{e^*}{c}\mathbf{A}\right)\psi(\mathbf{r})\right|^2 + \frac{\mathbf{B}^2}{8\pi}, \quad (18.59\text{a})$$

where $f_N(T)$ is the free energy density in the normal phase ($\psi = 0$). The total free energy in the volume V of the sample is given by the space integral

$$F_S(T) = \int_V f_S(T, \psi(\mathbf{r}), \mathbf{A}(\mathbf{r}))\, d\mathbf{r}. \quad (18.59\text{b})$$

The above assumptions, suggested by physical intuition, can be made plausible by the following considerations.

The terms in $|\psi|^2$ and $|\psi|^4$ in Eq. (18.59a) can be recognized as an expansion of the free energy up to second order in $\psi\psi^*$; neglecting higher order terms seems reasonable in the vicinity of the critical temperature, where $\psi\psi^*$ becomes small. The fourth term in the second member of Eq. (18.59a) is written as if $\psi(\mathbf{r})$ represents a true quantum mechanical wavefunction for a particle of charge e^* and mass m^* in the presence of spatial gradients of the order parameter and magnetic fields. From comparison with the BCS microscopic theory, we identify the charge e^* of the Cooper pair with $-2e$ (e is the absolute value of the electronic charge); ignoring effective mass effects, we also identify $m^* = 2m$. The last term corresponds to the electromagnetic energy density of the magnetic field $\mathbf{B}(\mathbf{r}) = \text{curl } \mathbf{A}(\mathbf{r})$.

The order parameter $\psi(\mathbf{r})$ is now obtained by minimization of the free energy $F_S(T)$ with respect to arbitrary variations of ψ^* (or ψ). The variational calculation shows that the order parameter function ψ satisfies the Ginzburg-Landau equation

$$\frac{1}{2m^*}\left(\mathbf{p} - \frac{e^*}{c}\mathbf{A}\right)^2 \psi(\mathbf{r}) + \beta|\psi(\mathbf{r})|^2\psi(\mathbf{r}) = -\alpha(T)\psi(\mathbf{r}) \;; \qquad (18.60)$$

notice that Eq. (18.60) is a nonlinear equation because of the presence of the term $|\psi(\mathbf{r})|^2$. This Ginzburg-Landau equation is complemented by the expression of the supercurrent density in terms of $\psi(\mathbf{r})$ (in the way usual in quantum mechanics)

$$\mathbf{J}_S(\mathbf{r}) = \frac{-i\hbar e^*}{2m^*}\left[\psi^*(\mathbf{r})\nabla\psi(\mathbf{r}) - \psi(\mathbf{r})\nabla\psi^*(\mathbf{r})\right] - \frac{e^{*2}}{m^*c}|\psi(\mathbf{r})|^2\mathbf{A}(\mathbf{r}) \;. \quad (18.61)$$

The above equation can also be obtained by minimization of the free energy $F_S(T)$ with respect to arbitrary variations of the vector potential $\mathbf{A}$.

Equation (18.61) assumes a particularly significant form if we write the complex macroscopic wavefunction in the polar form

$$\psi(\mathbf{r}) = |\psi(\mathbf{r})| \, e^{i\theta(\mathbf{r})}$$

where $|\psi(\mathbf{r})|$ represents the modulus of the wavefunction and the real quantity $\theta(\mathbf{r})$ represents the phase. Inserting the above expression into the Ginzburg-Landau equation (18.61), the current density becomes

$$\mathbf{J}_S(\mathbf{r}) = |\psi(\mathbf{r})|^2\left[\frac{e^*\hbar}{m^*}\nabla\theta(\mathbf{r}) - \frac{e^{*2}}{m^*c}\mathbf{A}(\mathbf{r})\right]. \qquad (18.62)$$

Intuitively, the order parameter $\psi(\mathbf{r})$ can be interpreted as the wavefunction which microscopically describes the motion of the center of mass of a Cooper pair (and of all the other Cooper pairs in the superconductor due to their condensation into the same "two-electron" quantum state): the magnitude $|\psi(\mathbf{r})|^2$ describes the local density of Cooper pairs, while the phase $\theta(\mathbf{r})$ describes locally the motion of the center of mass, as evident from Eq. (18.62) considered when fields are negligible.

Before closing this section we add a few details on the variational calculation that produces the Ginzburg-Landau equations (18.60) and (18.61) starting from Eqs. (18.59a) and (18.59b) of the local and total free energy. Consider an infinitesimal arbitrary variation of the function ψ^* (or independently of ψ) and the corresponding change of the total free energy given by Eq. (18.59b). We have

$$\psi^* \to \psi^* + \delta\psi^*$$

$$\delta F_S = \int_V \left[\alpha\,\delta|\psi|^2 + \frac{1}{2}\beta\,\delta|\psi|^4 + \frac{1}{2m}\delta\left|\left(\mathbf{p} - \frac{e^*}{c}\mathbf{A}\right)\psi\right|^2\right]d\mathbf{r} \qquad (18.63a)$$

It is trivial to see that for the first two terms of the integrand we have

$$\int_V \left[\alpha\,\delta(\psi^*\psi) + \frac{1}{2}\beta\,\delta\left(\psi^{*2}\psi^2\right)\right]d\mathbf{r} = \int_V \delta\psi^*(\mathbf{r})\left[\alpha\psi + \beta|\psi|^2\psi\right]d\mathbf{r}.$$

$$(18.63b)$$

The remaining term in Eq. (18.63a) can be elaborated with an integration by parts. [Some caution is required in setting up appropriate boundary conditions for a proper account of surface integrals encountered by application of the Green's theorem; we do not enter here in details concerning the "correct" boundary conditions at the sample surface, since often these conditions are better suggested by physical insight, situation by situation]. Integrating by parts the last term in Eq. (18.63a), and neglecting surface integrals, gives

$$\int \left[\left(i\hbar\nabla + \frac{e}{c}\mathbf{A} \right) \delta\psi^*(\mathbf{r}) \right] \cdot \left[\left(-i\hbar\nabla + \frac{e}{c}\mathbf{A} \right) \psi(\mathbf{r}) \right] d\mathbf{r}$$
$$= \int \delta\psi^*(\mathbf{r}) \left(-i\hbar\nabla + \frac{e}{c}\mathbf{A} \right)^2 \psi(\mathbf{r})\, d\mathbf{r}. \tag{18.63c}$$

From the above relations (18.63), the requirement that $\delta F_S \equiv 0$ for arbitrary variations $\delta\psi^*(\mathbf{r})$ provides the Ginzburg-Landau equation (18.60). With a similar procedure, the minimization of F_S as a functional of the vector potential, taking proper account of the Maxwell equation $\mathbf{J} = (c/4\pi)\mathrm{curl}\,\mathbf{B} = (c/4\pi)\mathrm{curl}\,\mathrm{curl}\,\mathbf{A}$, provides Eq. (18.61).

18.7.2   The Phenomenological Parameters of the Ginzburg-Landau Equations

The expression of the superconductor free energy introduces two phenomenological parameters, $\alpha(T)$ and $\beta(T)$, into the theory. To clarify their role, consider first a bulk superconductor in the absence of fields ($\mathbf{A}(\mathbf{r}) = 0$) and gradients ($\psi(\mathbf{r}) = $ const, and thus $\mathbf{p}\,\psi(\mathbf{r}) = -i\hbar\nabla\psi(\mathbf{r}) = 0$). Equation (18.59a) becomes in this case

$$f_S(T, \psi) - f_N(T) = \alpha(T)|\psi|^2 + \frac{1}{2}\beta(T)|\psi|^4, \tag{18.64a}$$

and represents the expansion of the free energy change up to second order in $\psi\psi^*$. The two phenomenological quantities α and β must satisfy $\beta(T) > 0$ and $\alpha(T) > 0$ for $T > T_c$, and $\beta(T) > 0$ and $\alpha(T) < 0$ for $T < T_c$; in fact only with these constraints, the free energy change of Eq. (18.64a) admits a negative minimum, at a finite value $|\psi|$, for $T < T_c$ (see Figure 18.15). The simplest possible temperature dependence of the parameters, compatible with the above requirements, is

$$\begin{cases} \alpha(T) = \alpha_0(T - T_c) & \text{with } \alpha_0 > 0 \text{ for } T \approx T_c, \\ \beta(T) = \beta_0 & \text{with } \beta_0 > 0 \text{ for } T \approx T_c. \end{cases} \tag{18.64b}$$

According to Eq. (18.64b), in a temperature range reasonably near to T_c, we consider $\alpha(T)$ linear in $T - T_c$, and disregard the temperature dependence of β.

We now examine more closely the structure of the Ginzburg-Landau equations, in order to link the phenomenological quantities α and β to empirically known quantities, such as the thermodynamic critical field $H_c(T)$ and the effective penetration depth $\lambda_{\mathrm{eff}}(T)$; a common procedure for isotropic systems is the following.

Thermodynamic Critical Field

Consider Eq. (18.60) in the bulk superconductor, in the absence of fields and gradients, i.e. $\mathbf{A}(\mathbf{r}) = 0$ and $\psi(\mathbf{r}) = $ const $= \psi_\infty$; for $T < T_c$ it holds $\beta|\psi(\mathbf{r})|^2 = -\alpha(T)$ which

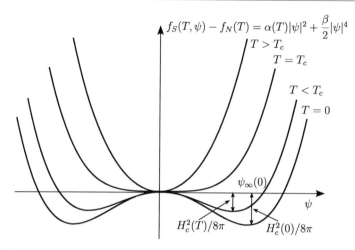

Figure 18.15 Schematic behavior of the Ginzburg-Landau free-energy change (in the absence of fields and gradients) for a typical second order phase transition. For graphical purposes ψ is taken real.

gives

$$|\psi_\infty|^2 = -\frac{\alpha(T)}{\beta} \qquad \text{for} \qquad T < T_c, \tag{18.65}$$

while $\psi_\infty = 0$ for $T > T_c$. Inserting Eq. (18.65) into Eq. (18.64a), the free energy density difference becomes

$$f_S(T, \psi_\infty) - f_N(T) = -\frac{\alpha^2(T)}{2\beta} \qquad \text{for} \qquad T < T_c.$$

The first member of this equation can be identified with the condensation energy per unit volume $-H_c^2(T)/8\pi$. From the empirical knowledge of the thermodynamic critical field $H_c(T)$ we obtain a first useful relationship

$$\frac{H_c^2(T)}{8\pi} = \frac{\alpha^2(T)}{2\beta} \qquad \Longrightarrow \qquad H_c(T) = 2\sqrt{\frac{\pi}{\beta}}\,|\alpha(T)|. \tag{18.66}$$

From the temperature dependence of $\alpha(T)$, specified in Eq. (18.64b), we have that $H_c(T) \rightarrow (T_c - T)$ for T approaching T_c from below.

Magnetic Penetration Depth

Consider now a superconductor and assume that $\psi(\mathbf{r}) = \psi_\infty$ not only in the bulk but everywhere up to the surface. Equation (18.61) becomes in this case

$$\mathbf{J}_S(\mathbf{r}) = -\frac{e^{*2}}{m^*c}|\psi_\infty|^2\mathbf{A}(\mathbf{r}) \qquad \Longrightarrow \qquad \frac{c}{4\pi}\text{curl }\mathbf{B} = -\frac{e^{*2}}{m^*c}|\psi_\infty|^2\mathbf{A}(\mathbf{r}),$$

where in the last relation $\mathbf{J}_S(\mathbf{r})$ has been expressed using the appropriate Maxwell equation. The curl-operator applied to both members of the last equation gives for the temperature-dependent effective penetration length

$$\frac{1}{\lambda_{\text{eff}}^2(T)} = \frac{4\pi}{c} \frac{e^{*2}}{m^*c} |\psi_\infty|^2 \quad \Longrightarrow \quad \lambda_{\text{eff}}^2(T) = \frac{m^*c^2\beta}{4\pi e^{*2}|\alpha(T)|}. \tag{18.67}$$

From the temperature dependence of $\alpha(T)$ we have that $\lambda_{\text{eff}}(T)$ diverges as $(T_c - T)^{-1/2}$ for T approaching T_c from below. By combining Eqs. (18.66) and (18.67) one can obtain the parameters α and β as a function of H_c and λ_{eff}.

Ginzburg-Landau Coherence Length

The Ginzburg-Landau equations, besides the magnetic penetration depth, contains another characteristic length, which represents the scale of spatial variations of the order parameter. Consider, for example, a semi-infinite superconductor in the $z > 0$ half-plane, and assume that the order parameter depends only on z. For this one-dimensional model, the Ginzburg-Landau equation (18.60), in the absence of magnetic fields, gives

$$-\frac{\hbar^2}{2m^*} \frac{d^2\psi(z)}{dz^2} + \beta|\psi(z)|^2\psi(z) = -\alpha(T)\psi(z).$$

For $T < T_c$, we divide both members of the above equation by $|\alpha(T)|$ and obtain

$$-\frac{\hbar^2}{2m^*|\alpha(T)|} \frac{d^2\psi(z)}{dz^2} + \frac{|\psi(z)|^2}{|\psi_\infty|^2}\psi(z) = \psi(z). \tag{18.68}$$

The coefficient of the first term of Eq. (18.68) allows to define naturally a new scale length $\xi_{GL}(T)$ given by

$$\xi_{GL}^2(T) = \frac{\hbar^2}{2m^*|\alpha(T)|}. \tag{18.69}$$

This length, referred to as the *Ginzburg-Landau coherence length*, represents the space scale on which the order parameter $\psi(\mathbf{r})$ varies when one introducessome inhomogeneity (for instance a surface). To prove this statement, we remark that the solution of Eq. (18.68), assuming boundary conditions $\psi(0) = 0$ and $\psi(\infty) = \psi_\infty$, is given by

$$\psi(z) = |\psi_\infty| \tanh \frac{z}{\sqrt{2}\,\xi_{GL}(T)},$$

as can be established by direct substitution into Eq. (18.68). Thus $\xi_{GL}(T)$ governs the way in which $\psi(z)$ approaches its bulk value ψ_∞ within the superconductor.

In superconductors, it is important to compare the Ginzburg-Landau spatial coherence length $\xi_{GL}(T)$ with the penetration depth $\lambda_{\text{eff}}(T)$; both lengths diverge as $(T_c - T)^{-1/2}$ for T approaching T_c from below. At a boundary between a normal region and a superconducting region, the coherence length governs the region where the condensation energy is partial, while the effective penetration depth governs the region where the Meissner effect is partial. It is thus useful to define the dimensionless temperature-independent parameter

$$\kappa = \frac{\lambda_{\text{eff}}(T)}{\xi_{GL}(T)} = \frac{m^*c}{|e^*|\hbar}\sqrt{\frac{\beta}{2\pi}}. \tag{18.70}$$

As we shall see in Section 18.7.4, the parameter κ determines whether the material is a type-I or a type-II superconductor. The Ginzburg-Landau equations have been particularly successful (at times beyond hope) in the description of numerous phenomena, including the basic distinction between type-I and type-II superconductors, boundary formation, thin films, impurity effects, etc. In the following we focus on some general consequences implied by the Ginzburg-Landau equations.

18.7.3   Magnetic Flux Quantization

Consider a multiply connected superconductor, for instance a superconducting ring, as illustrated in Figure 18.16. The magnetic flux, threading a loop lying in the bulk superconductor, cannot have arbitrary values; rather it has to be an integer multiple of the flux quantum $\phi_0 = hc/2e$. The origin of the magnetic flux quantization can be seen with the following arguments.

Consider the Ginzburg-Landau equation (18.62) for the current density, here rewritten for convenience in the form

$$\mathbf{J}_S(\mathbf{r}) = \frac{e^*}{m^*}|\psi(\mathbf{r})|^2\left[\hbar\nabla\theta(\mathbf{r}) - \frac{e^*}{c}\mathbf{A}(\mathbf{r})\right], \tag{18.71}$$

where $\theta(\mathbf{r})$ is the local phase of the order parameter. Let Γ be a loop completely inside the superconductor and deep enough from the surfaces of the superconducting ring that

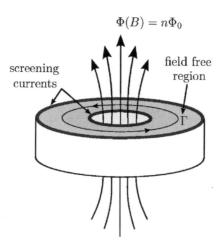

Figure 18.16 Quantization of the magnetic flux through a multiply connected superconductor with a ring shape.

magnetic field and screening current have dropped to zero (the circuit Γ must avoid surface regions of the order of the penetration depth, but for the rest is arbitrary). *Inside the superconductor in its Meissner state*, the current density vanishes; as a consequence in any point $\mathbf{r}$ of the circuit Γ it holds

$$\nabla\theta(\mathbf{r}) = -\frac{2e}{\hbar c}\mathbf{A}(\mathbf{r}) \quad \text{for} \quad \mathbf{r} \in \Gamma. \tag{18.72}$$

In the above equation e^* has been set equal to $-2e$ for a Cooper pair.

From Eq. (18.72), performing the line integral of both members around the closed circuit Γ within the superconductor gives

$$\oint \nabla\theta(\mathbf{r}) \cdot d\mathbf{l} = -\frac{2e}{\hbar c} \oint \mathbf{A}(\mathbf{r}) \cdot d\mathbf{l}.$$

Thus we have

$$\Delta\theta = -\frac{2e}{\hbar c}\Phi_\Gamma(\mathbf{B}), \tag{18.73a}$$

where $\Delta\theta$ gives the variation of the phase $\theta(\mathbf{r}) = \arg\psi(\mathbf{r})$ after going around Γ once, and $\Phi_\Gamma(\mathbf{B})$ is the flux of $\mathbf{B}$ through a surface that embraces the line Γ. Since $\psi(\mathbf{r})$ is a single-valued function we have that $\Delta\theta$ must be a multiple of 2π; this condition, inserted into Eq. (18.73a) gives the flux quantization result

$$\Phi_\Gamma(\mathbf{B}) = n\frac{\hbar c}{2e} = n\Phi_0, \tag{18.73b}$$

where n is an integer and Φ_0 is the elemental quantum of flux. Notice that in the case of simply connected superconductor n equals zero; but if the superconductor is multiply connected, or, equivalently if it contains normal regions, then n may be an integer number different from zero.

18.7.4 Type-I and Type-II Superconductors

One of the most important achievements of the Ginzburg-Landau theory is the natural classification between type-I and type-II superconductors. As discussed in Section 18.1, type-I superconductors are characterized by the fact that the entire superconducting specimen reverts to the normal state, when an external magnetic field higher than the critical value $H_c(T)$ is applied to the sample (in the standard geometry of a long, thin cylinder with the axis parallel to the external applied field). In type-II superconductors, for applied fields in the range $H_{c1}(T)$ and $H_{c2}(T)$, the flux begins to penetrate partially in the superconductors; this mixed state, referred to as the Shubnikov phase, reverts back to the ordinary metal state for applied magnetic fields higher than $H_{c2}(T)$. The Ginzburg-Landau theory, ingeniously extended by Abrikosov, allows to understand easily the occurrence of type-I and type-II behavior, the upper and lower critical fields, and the nature of the mixed state.

Consider the Ginzburg-Landau equation (18.60) for a superconductor of type-II, at temperature T, with a uniform applied magnetic field H extremely close to $H_{c2}(T)$.

Since we are extremely close to the superconductor-metal transition, the density of super-electrons $|\psi(\mathbf{r})|^2 \to 0$. We can thus linearize the Ginzburg-Landau equation (18.60) and write

$$\frac{1}{2m^*}\left(\mathbf{p} - \frac{e^*}{c}\mathbf{A}\right)^2 \psi(\mathbf{r}) = -\alpha(T)\psi(\mathbf{r}). \tag{18.74}$$

The operator in the first member of Eq. (18.74) is just the Hamiltonian of a free particle of mass m^* and charge e^* in a uniform magnetic field H; its lowest eigenvalue is

$$E_0 = \frac{1}{2}\hbar\omega_c = \frac{1}{2}\hbar\frac{|e^*|H}{m^*c}$$

(as seen in Section 15.2). Uniform magnetic field and beginning of superconductivity are compatible at the critical field $H_{c2}(T)$ such that:

$$\frac{1}{2}\hbar\frac{|e^*|H_{c2}(T)}{m^*c} = |\alpha(T)| \implies H_{c2}(T) = \frac{2m^*c}{|e^*|\hbar}|\alpha(T)|.$$

We have thus for the upper critical field the expression

$$\frac{H_{c2}(T)}{H_c(T)} \equiv \sqrt{2}\kappa, \tag{18.75}$$

where $H_c(T)$ is the thermodynamic critical field of Eq. (18.66), and κ is the dimensionless Ginzburg-Landau parameter of Eq. (18.70).

Thus we see that for $\kappa > 1/\sqrt{2}$ the critical field $H_{c2}(T) > H_c(T)$; the superconducting state appears at and below $H_{c2}(T)$. Thus the discrimination between type-I and type-II superconductors is just the value $\kappa = 1/\sqrt{2}$ of the Ginzburg-Landau parameter

$$\kappa = \frac{\lambda_{\text{eff}}(T)}{\xi_{\text{GL}}(T)}, \quad \begin{cases} \kappa < 1/\sqrt{2} & \text{type-I superconductor} \\ \kappa > 1/\sqrt{2} & \text{type-II superconductor}. \end{cases}$$

Most of the clean superconducting elemental metals are type-I superconductors. In contrast, superconducting alloys are of type-II, since disorder reduces meanfree path and correlation length, thus implying increase of κ. The high-T_c cuprates are extreme type-II superconductors with very large values $\kappa \approx 100$.

We can now give a few considerations on the mixed state that characterizes type-II superconductors. In type-II superconductors, starting from an applied magnetic field just higher than $H_{c1}(T)$, the flux begins to penetrate partially in the superconductor. It seems plausible to speculate that it enters in the form of cylindrical flux tubes, each carrying a quantum of flux $\Phi_0 = hc/2e$. In fact, considering an isolated flux tube, and a surrounding circuit Γ well *inside* the superconductor region, the magnetic flux quantization concept of Section 18.7.3 restricts the threading flux to integer values of the flux quantum Φ_0. A more detailed analysis shows that the single quantum flux formation is favored. The single-valuedness of the order parameter also imposes that $|\psi(\mathbf{r})|$ must be equal to zero along the axis of the tube. Thus a vortex can be visualized as a "core" of normal metal completely surrounded by the superconducting material:

superconductivity is destroyed (approximately) in a cylinder of radius $\approx \xi_{GL}$, while screening supercurrents extend to distance $\approx \lambda_{\text{eff}}$ from the axis of the vortex. These qualitative considerations are corroborated by detailed calculations.

We can also estimate the lower critical field $H_{c1}(T)$. The estimate is done by requiring that the flux within a cylinder of radius $\lambda_{\text{eff}}(T)$ is just a flux quantum

$$\pi \lambda_{\text{eff}}^2(T) H_{c1}(T) \approx \Phi_0 = \frac{hc}{2e}.$$

Thus we expect approximately

$$H_{c1}(T) \approx \frac{hc}{|e^*|} \frac{1}{\pi \lambda_{\text{eff}}^2(T)} = 2\sqrt{2} \frac{H_c(T)}{\kappa}, \tag{18.76}$$

where the last passage in Eq. (18.76) can be verified by inspection from Eqs. (18.66), (18.67) and (18.70). From Eqs. (18.75) and (18.76) it is seen that $H_c(T)$ is approximately the geometrical mean between $H_{c1}(T)$ and $H_{c2}(T)$. Many other aspects of vortices, regular array of flux tubes, flow and pinning in the presence of a current density and inhomogeneities could be studied, but we confine ourselves to the brief outline given so far.

18.8   Tunneling Effects

18.8.1   Electron Tunneling into Superconductors

A central feature of the BCS theory is the presence of an energy gap in the electron density-of-states of superconductors. A most direct evidence of the gap and of the electron structure of the superconductors is provided by the electron tunneling experiments, initiated by I. Giaever, Phys. Rev. Lett. 5, 147, 464 (1960).

Consider first a junction constituted by two normal metals, separated by a thin insulating film (typically $10 \approx 50$ Å). It is well known that, if a potential difference is applied across the junction, a current flows because of the capability of electrons to penetrate a thin barrier. For low fields, the tunneling current is proportional to the applied voltage V. In fact, in "quasi-equilibrium" conditions, the Fermi levels of the two metals are shifted by eV. The density-of-states in the two metals, as well as tunneling probabilities, are practically independent of energy in the few millielectronvolts of interest around E_F, and this leads to the ohmic behavior of the junction.

Let us now examine the electron tunneling (*Giaever tunneling*) across a junction formed by normal metal-insulator-superconductor (NIS junction).

In the ordinary metal (at zero temperature) all states below the Fermi energy E_F are filled, while all states above E_F are empty, with zero gap between occupied and empty states. In the superconductor the quasiparticle energies differ from the Fermi energy at least by the energy gap Δ_0 at $T = 0$. When a bias potential V is applied to a NIS junction, one-particle states are not available in the superconducting material for accepting or supplying electrons, unless the bias voltage exceeds Δ_0/e. When $V > \Delta_0/e$, the *I-V* characteristics and in particular the differential conductance $G =$

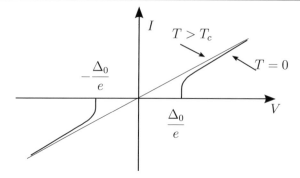

Figure 18.17 Schematic representation of the *I-V* characteristics of a normal metal-insulator-superconductor (NIS) junction at zero temperature and above the critical temperature; Δ_0 is the energy gap parameter of the superconductor. At finite temperature $0 < T < T_c$ the presence of thermally excited electrons contribute to smear out and shift the strongly nonlinear $T = 0$ current-voltage characteristics.

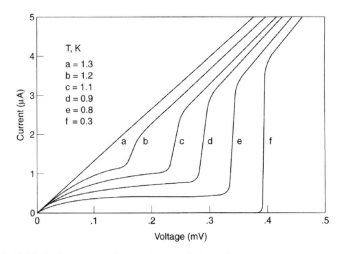

Figure 18.18 Typical current-voltage characteristics of an Al-Al$_2$O$_3$-Al superconductor-insulator-superconductor (SIS) junction (only quasiparticle tunneling current is considered). The onset of the sharp break in the curves for $T < T_c$ occurs when $eV = 2\Delta(T)$; the critical temperature is $T_c = 1.26$ K, and $2\Delta_0/e = 0.379$ mV [reprinted with permission from D. N. Langenberg, D. J. Scalapino and B. N. Taylor, Proc. IEEE *54*, 560 (1966); copyright 1966 IEEE].

dI/dV is related to the density-of-states of quasiparticles in the superconductor (since the density-of-states in the metal can be taken as constant in the few millielectronvolts of interest). Finally when $V \gg \Delta_0/e$ the ohmic behavior of the junction is recovered. These essential features of the *I-V* characteristics of the NIS junction are schematically indicated in Figure 18.17.

Let us now consider the tunneling between two equal superconductors, of energy gap parameter Δ_0, separated by a thin insulating barrier (SIS junction). At $T = 0$ we find that there is no quasiparticle tunneling until the bias voltage V exceeds $2\Delta_0/e$; in the case the two superconductors are different, the threshold voltage is $(\Delta_1 + \Delta_2)/e$. At the threshold voltage, we expect a discontinuous jump of the current, because of the singularity in the density-of-states of quasiparticles in the two superconductors (see Eq. (18.42)); eventually for higher bias voltages the ohmic behavior is recovered. All these features of the I-V characteristics of the SIS tunnel junction can be clearly seen in the specific example reported in Figure 18.18.

18.8.2 Cooper Pair Tunneling and Josephson Effects

We have considered above tunneling of "normal electrons" (or "quasiparticles") from a superconducting SIS junction, composed by the superconducting films (of the same material) separated by a very thin insulating layer. The highly nonlinear current-voltage I-V characteristic at zero temperature is schematically indicated in Figure 18.19; the onset of quasiparticle tunneling occurs for $V = 2\Delta_0/e$, where Δ_0 is the energy gap parameter of the superconductor.

It is important to notice that in the case of SIS junctions, besides the quasiparticle tunneling current discussed so far, we can have also a supercurrent tunneling due to Cooper pairs transfer between the two superconductors; this current, called Josephson current, can be observed in SIS junctions with extremely thin insulating layers (10–15 Å); in this situation, the coupling between the two superconductors is sufficiently strong that a definite phase relationship between pairs on opposite sides of the insulating barrier can be maintained (and controlled by electromagnetic fields).

For superconducting tunnel junctions with extremely thin insulating layers (in the range 10–15 Å), the electron pair correlations extend through the insulating barrier. In this situation, it has been predicted by Josephson that "paired electrons" can tunnel without dissipation from one superconductor to the other superconductor on the opposite side of the insulating layer [B. D. Josephson, Phys. Letters *1*, 251 (1962)]. The direct

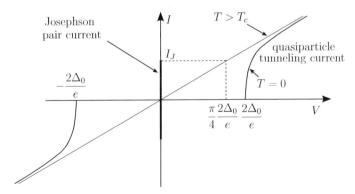

Figure 18.19 Schematic representation of the direct current I-V characteristics (at $T = 0$) of a superconductor-insulator-superconductor (SIS) junction displaying the Josephson current. In correspondence to $V \equiv 0$ a direct supercurrent can flow (up to a maximum value I_J). For $0 < V < 2\Delta_0/e$ an alternate supercurrent flows with frequency $\omega = 2eV/\hbar$, and no direct supercurrent is observed. For $V > 2\Delta_0/e$ the quasiparticle tunneling current is reported.

supercurrent of pairs, for currents less than a certain critical value I_J, flows with *zero voltage drop across the junction (dc Josephson effect)*, as illustrated in Figure 18.19. The width of the insulating barrier of the junction limits the maximum supercurrent that can flow across the junction, but introduces no resistance in the flow. Josephson also predicted that, in the case a *constant finite voltage V* is established across the junction, an alternating supercurrent $I_J \sin(\omega_J t + \phi_0)$ flows with frequency $\omega_J = 2eV/\hbar$ *(ac Josephson effect)*.

Weak Links and Josephson Effects

We discuss here only the most elementary aspects of this subject, remarkable both for its fundamental aspects and technological applications. Consider two superconductors (of the same or different materials) separated by a thin insulating barrier of width b, in the geometry indicated in Figure 18.20. In the case the insulating barrier is infinitely thick, the superconductor on the left side would be characterized by the order parameter ψ_1, that can be written as

$$\psi_1 = |\psi_1|e^{i\theta_1}, \tag{18.77a}$$

with $|\psi_1|$ and θ_1 uniform on the whole volume of superconductor 1. Similarly the superconductor on the right side would be characterized by the order parameter

$$\psi_2 = |\psi_2|e^{i\theta_2}, \tag{18.77b}$$

with $|\psi_2|$ and θ_2 space independent of the volume of superconductor 2. When the two superconductors are separated by a thin insulating barrier, we can expect that the superconducting order parameters ψ_1 and ψ_2 decay within the insulating region; we can guess that the order parameter $\psi(z)$ within the barrier can be expressed in the form

$$\psi(z) = \psi_1 e^{-\beta z} + \psi_2 e^{+\beta(z-b)} \qquad 0 < z < b, \tag{18.77c}$$

where β characterizes the damping within the barrier.

We insert Eq. (18.77c) into the Ginzburg-Landau equation (18.62) (assuming negligible magnetic fields) and obtain

$$J_S(z) = \frac{-i\hbar e^*}{2m^*}\left[\psi^*(z)\frac{d}{dz}\psi(z) - \psi(z)\frac{d}{dz}\psi^*(z)\right] = \frac{-i\hbar e^*}{m^*}\beta e^{-\beta b}(\psi_1^*\psi_2 - \psi_1\psi_2^*)$$

$$= \frac{2\hbar e^*}{m^*}\beta e^{-\beta b}|\psi_1||\psi_2|\sin(\theta_2 - \theta_1).$$

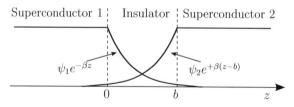

Superconductor 1 ⋮ Insulator ⋮ Superconductor 2

$\psi_1 e^{-\beta z}$ $\psi_2 e^{+\beta(z-b)}$

0   b   z

Figure 18.20 Schematic representation of a junction between two superconductors separated by a thin insulating barrier of width b. The behavior of the order parameter wavefunction for the superconductor-insulator-superconductor junction is indicated.

We thus see that the Josephson supercurrent, that flows between two superconductors separated by an insulating barrier, is related to the phase difference $\gamma = \theta_2 - \theta_1$ of the order parameters in the two superconductors by the relation

$$\boxed{I = I_J \sin \gamma}, \tag{18.78a}$$

where I_J depends on the geometrical and physical properties of the junction. Equation (18.78a) has been derived on the basis of a heuristic approach. A detailed analysis based on the microscopic BCS theory provides not only Eq. (18.78a), but also the maximum current $I = I_J$, which can flow across the junction without dissipation. The critical value I_J of an ideal SIS tunnel junction, made of the same material, is given by the current that would flow applying a voltage equal to $(\pi/4)(2\Delta_0/e)$ to the normal junction, in the same geometrical conditions. Notice finally that our treatment assumes that the magnetic field flux that threads the junction is negligible; in general, the dependence of I_J on the magnetic field (*diffractive pattern*) should be appropriately considered.

Consider now a superconducting junction biased with a (constant or time dependent) voltage V. If a potential difference is established between the two superconductors, the relative energy difference between Cooper pairs belonging to the superconductors is $2eV$. In perfect analogy with the quantum mechanical rate of change of phases of ordinary eigenfunctions (energy divided by $\hbar$), we expect for the time variation of the relative phase

$$\boxed{\frac{d\gamma}{dt} = \frac{2eV}{\hbar}}. \tag{18.78b}$$

It is now easy to see that the above Eqs. (18.78) describe both the dc Josephson effect as well as the ac Josephson effect.

If the potential V across the SIS junction is zero, we see from Eq. (18.78b) that γ is constant; from Eq. (18.78a) we have that any supercurrent with intensity ranging from $-I_J$ to $+I_J$ can flow through the junction (the actual value is determined by the external circuit); this is the origin of the zero resistance spike at $V = 0$ in the I-V characteristics for a Josephson junction shown in Figure 18.19.

A schematic illustration of the dc Josephson circuit is reported in Figure 18.21. A Josephson junction of maximum supercurrent I_J is placed in series with a resistance R and connected to a generator of potential difference V_0. In the case $I = V_0/R < I_J$ there is a stationary regime where the current I flows through the circuit with zero potential drop between the two superconductors and zero resistance of the insulating layer within them.

Suppose that a *constant potential* V ($0 < V < 2\Delta_0/e$) is established across the SIS junction (for simplicity we assume that an "ideal" voltage source maintains a constant voltage V across the SIS junction; this independently of the normal current, that could be flowing through the circuit). Integration of Eq. (18.78b) gives

$$\gamma(t) = \frac{2eV}{\hbar}t + \gamma_0.$$

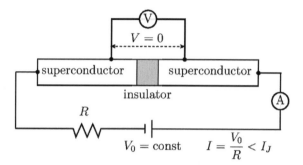

Figure 18.21 Schematic illustration of the dc Josephson effect; a direct supercurrent (up to a maximum value I_J) flows without dissipation through the insulating layer of the Josephson junction.

The phase difference is a linear function of time and the Josephson current is

$$I = I_J \sin\left(\frac{2eV}{\hbar}t + \gamma_0\right).$$

This pair current oscillates with angular frequency $\omega_J \equiv 2\pi \nu_J = (2e/\hbar)V$. The frequency ν_J can be expressed as

$$\nu_J = \frac{2e}{h}V \implies K_J = \frac{2e}{h} = 483597.9\,\frac{\text{GHz}}{V} = 483.5979\,\frac{\text{MHz}}{\mu V}\,;$$

the Josephson constant K_J is related only to the electronic charge e and to the Planck constant h, and it has a particular role in metrology.

As suggested by Josephson, experimental evidence of the ac current can be conveniently obtained by applying across the junction both a constant voltage V and a radiofrequency or microwave voltage, say $V_r \cos \omega_r t$. When the combined dc and ac voltage $V_{\text{tot}} = V + V_r \cos \omega_r t$ is applied, from Eq. (18.78b) we have

$$\frac{d\gamma}{dt} = \frac{2eV}{\hbar} + \frac{2eV_r}{\hbar}\cos \omega_r t. \tag{18.79a}$$

By integration, we obtain

$$\gamma(t) = \frac{2eV}{\hbar}t + \frac{2eV_r}{\hbar\omega_r}\sin \omega_r t + \gamma_0, \tag{18.79b}$$

where γ_0 is an appropriate integration constant. The Josephson current is thus

$$I(t) = I_J \sin\left[\frac{2eV}{\hbar}t + \gamma_0 + \frac{2eV_r}{\hbar\omega_r}\sin \omega_r t\right]. \tag{18.79c}$$

This is a frequency modulated current. To analyze it, we use the following mathematical expressions

$$\cos(a \sin x) = \sum_{n=-\infty}^{+\infty} J_n(a) \cos nx, \quad \sin(a \sin x) = \sum_{n=-\infty}^{+\infty} J_n(a) \sin nx,$$

and hence

$$\sin (b + a \sin x) = \sum_{n=-\infty}^{+\infty} J_n(a) \sin (b + nx),$$

where J_n are Bessel functions of first kind of order n. Equation (18.79c) becomes

$$I(t) = I_J \sum_{n=-\infty}^{+\infty} J_n \left(\frac{2eV_r}{\hbar \omega_r} \right) \sin \left[\frac{2eV}{\hbar} t + \gamma_0 + n\omega_r t \right].$$

We see from this expression that whenever V satisfies the relation

$$2eV \equiv n\hbar\omega_r \quad (n = 0, 1, 2, \dots), \tag{18.80}$$

a *zero frequency supercurrent* is obtained (this is called *inverse ac effect*). Thus well defined vertical spikes, known as Shapiro spikes, are expected to occur in the dc current voltage characteristics, whenever Eq. (18.80) is satisfied. The heights of the Shapiro spikes are $I_J J_n(2eV_r/\hbar\omega_r)$; in particular for the zero voltage spike, the critical dc Josephson current becomes

$$I_c = I_J J_0 \left(\frac{2eV_r}{\hbar \omega_r} \right),$$

and its dependence on the amplitude of the radiofrequency voltage can be used to measure it. The Shapiro spikes are so well defined that they are used for accurate determination of the ratio $2e/\hbar$. We do not pursue further the current-voltage characteristics of superconductor junctions; we can mention however their use as fast switches, and the interest for lossless computer elements.

Superconducting Quantum Interference Devices

Before concluding, we also briefly mention the superconducting quantum interference devices (SQUID). Consider a superconducting circuit with two Josephson junctions in parallel, as schematized in Figure 18.22. For simplicity we assume that both junctions have the same critical current I_J, that the magnetic flux threading the junctions is negligible, and the thickness of the superconducting samples is much larger than the magnetic penetration depth.

Let us indicate with γ_P and γ_Q the phase differences across the junctions P and Q, respectively. We have

$$\gamma_P = \theta(P_2) - \theta(P_1) \quad \text{and} \quad \gamma_Q = \theta(Q_2) - \theta(Q_1),$$

where P_1 and P_2 are points just at the left and at the right of the junction P (the same for Q_1 and Q_2 of junction Q), and θ is the phase of the superconducting wavefunction. The total current flowing in the circuit is

$$I = I_J(\sin \gamma_P + \sin \gamma_Q) = 2I_J \sin \frac{\gamma_P + \gamma_Q}{2} \cos \frac{\gamma_P - \gamma_Q}{2}. \tag{18.81}$$

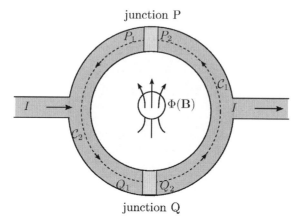

Figure 18.22 Representation of a dc superconducting quantum interference device for detecting an enclosed magnetic field of flux $\Phi(\mathbf{B})$.

In the interior of a superconductor in its Meissner state $\mathbf{J}_S = 0$; then from expression (18.62) we have

$$\nabla\theta(\mathbf{r}) = -\frac{2e}{\hbar c}\mathbf{A}(\mathbf{r}), \tag{18.82}$$

where $e^* = -2e$ for the Cooper pair. Line integration of expression (18.82) along the circuits $\mathcal{C}_1$ and $\mathcal{C}_2$ indicated in Figure 18.22 gives

$$\text{circuit } \mathcal{C}_1 \implies \theta(P_2) - \theta(Q_2) = -\frac{2e}{\hbar c}\int_{Q_2}^{P_2}\mathbf{A}(\mathbf{r})\cdot d\mathbf{l}$$

$$\text{circuit } \mathcal{C}_2 \implies \theta(Q_1) - \theta(P_1) = -\frac{2e}{\hbar c}\int_{P_1}^{Q_1}\mathbf{A}(\mathbf{r})\cdot d\mathbf{l}$$

Summing up the two contributions we obtain

$$\theta(P_2) - \theta(Q_2) + \theta(Q_1) - \theta(P_1) = -\frac{2e}{\hbar c}\oint\mathbf{A}(\mathbf{r})\cdot d\mathbf{l}\,;$$

it follows

$$\gamma_P - \gamma_Q = -\frac{2e}{\hbar c}\Phi(\mathbf{B}) = -2\pi\frac{\Phi(\mathbf{B})}{\Phi_0} \quad \text{with} \quad \Phi_0 = \frac{hc}{2e},$$

where Φ is the total flux through the loop. Equation (18.81) then gives for the current the expression

$$I = 2I_J\cos\pi\frac{\Phi(\mathbf{B})}{\Phi_0}\sin\frac{\gamma_P + \gamma_Q}{2}.$$

The above expression takes its maximum value

$$I_{\max} = 2I_J\left|\cos\pi\frac{\Phi(\mathbf{B})}{\Phi_0}\right|. \tag{18.83}$$

The maximum supercurrent is a periodic function of $\Phi(\mathbf{B})$, and changes from maxima to minima for a change of flux as small as $\Phi_0/2$; very small magnetic fields can thus be measured with high precision. The critical value $I_{\max}(\Phi)$ represents the maximum value of the current that can flow in the device without dissipation and zero potential voltage across the SQUID.

The sensitive superconducting quantum interference devices offer the opportunity of a number of interesting applications [see for instance D. Koelle, R. Kleiner, F. Ludwig, E. Dantsker and J. Clarke, Rev. Mod. Phys. *71*, 631 (1999)]. We mention, in particular, clinical applications such as magneto-encephalography and magneto-cardiography; these techniques use an appropriate array of SQUIDs to measure the tiny magnetic fields produced by the electric currents involved in the biomagnetism activity [see for instance M. Hämäläinen, R. Hari, R. J. Ilmoniemi, J. Knuutila, and O. V. Lounasmaa, Rev. Mod. Phys. *65*, 413 (1993); C. Del Gratta, V. Pizzella, F. Tecchio and G. L. Romani, Rep. Prog. Phys. *64*, 1759 (2001) and references quoted therein].

Appendix A. The Phonon-Induced Electron-Electron Interaction

The origin of the phenomena related to the superconductivity is a small attractive effective interaction between electrons. In the conventional superconductivity it is generally accepted that this indirect interaction occurs via the phonon field (Fröhlich attractive mechanism). While the quantitative analysis of effective interactions may constitute a quite demanding problem, from a qualitative point of view simple models can be considered to point out some basic aspects; in particular it is not difficult to show that one should expect an attractive interactions for electrons lying near the Fermi surface in an energy shell of the order of $\hbar\omega_D$, where ω_D is the Debye frequency. To make this result plausible, we exploit a well-known canonical transformation, and then we apply it to the indirect interaction via the phonon field.

A Canonical Transformation

Let us consider a Hamiltonian H; we can always perform a canonical transformation of the type

$$\widetilde{H} = e^{-S} H\, e^{S} = (1 - S + \frac{1}{2!}S^2 - \cdots)H(1 + S + \frac{1}{2!}S^2 + \cdots)$$

$$= H + [H, S] + \frac{1}{2!}[[H, S], S] + \frac{1}{3!}[[[H, S], S], S] + \cdots \quad (A.1)$$

with S any arbitrary antihermitian operator, i.e. $S^\dagger = -S$. The antihermiticity requirement guarantees that the transformation $U = \exp S$ is unitary, and so it preserves the norm of the states. It is evident that the eigenvalues of $\widetilde{H}$ and H coincide.

Suppose that the operator H is split in the form

$$H = H_0 + H_1. \quad (A.2)$$

Then Eq. (A.1) reads

$$\widetilde{H} = H_0 + H_1 + [H_0, S] + [H_1, S] + \frac{1}{2}[[H_0, S], S] + \cdots$$

We can exploit the arbitrariness of S to satisfy the relation

$$H_1 + [H_0, S] = 0. \tag{A.3}$$

With such a choice, $\widetilde{H}$ becomes

$$\widetilde{H} = H_0 + \frac{1}{2}[H_1, S] + \cdots$$

If we imagine to replace H_1 with λH_1 (with λ eventually put equal to 1) we see that the terms neglected in $\widetilde{H}$ are at least of order λ^3. Thus the eigenvalues of H and the eigenvalues of $\widetilde{H}$, where

$$\widetilde{H} = H_0 + H_{\text{indirect}}$$
$$H_{\text{indirect}} = \frac{1}{2}[H_1, S],$$

are equal to order λ^2 in the coupling parameter λ.

From Eq. (A.3), we see that the operator S in a representation in which H_0 is diagonal is

$$\langle n|S|m \rangle = \frac{\langle n|H_1|m \rangle}{E_m - E_n}$$

(to avoid singularities we assume that the diagonal matrix elements of H_1 in a representation in which H_0 is diagonal are zero; if not so, these terms could be included in H_0). The expression of the matrix elements of H_{indirect} on the eigenstates of H_0 is

$$\langle f|H_{\text{indirect}}|i \rangle = \frac{1}{2}\langle f|H_1 S - S H_1|i \rangle$$
$$= \frac{1}{2}\sum_\alpha [\langle f|H_1|\alpha \rangle\langle \alpha|S|i \rangle - \langle f|S|\alpha \rangle\langle \alpha|H_1|i \rangle]$$
$$= \frac{1}{2}\sum_\alpha \langle f|H_1|\alpha \rangle\langle \alpha|H_1|i \rangle \left[\frac{1}{E_i - E_\alpha} + \frac{1}{E_f - E_\alpha} \right]. \tag{A.4}$$

The meaning of H_{indirect} can be further clarified if we note that $\langle i|H_{\text{indirect}}|i \rangle$ is the standard second order correction given by perturbation theory. It can be noticed that the matrix elements $\langle f|H_{\text{indirect}}|i \rangle$ can also be obtained by the renormalization procedure of Section 5.8.3 (simply disregarding the coupling of the intermediate states $|\alpha \rangle$ with any other state but $|i \rangle$ and $|f \rangle$, and putting the energy E equal to E_i and E_f).

Indirect Electron-electron Interaction

The mechanism leading to the indirect electron-electron interaction via phonons is schematically represented in Figure 18.23. In the process pictured in Figure 18.23(a), one electron of wavevector $\mathbf{k}_1$ "emits" a phonon of wavevector $\mathbf{q}$ and is scattered into the state $\mathbf{k}_1' = \mathbf{k}_1 - \mathbf{q}$; the phonon of wavevector $\mathbf{q}$ is immediately "absorbed" by another electron of wavevector $\mathbf{k}_2$ that scatters into the state $\mathbf{k}_2' = \mathbf{k}_2' + \mathbf{q}$. A similar interpretation holds for Figure 18.23(b). The quantum mechanical analysis of these processes provides a net attractive interaction between pairs of electrons, whose energy difference is smaller than the phonon energy $\hbar\omega_\mathbf{q}$. To show this, we can apply the canonical transformation considered before to obtain the effective electron-electron interaction.

We consider final and initial electron states, coupled with the vacuum phonon state (the indirect interaction we obtain is, however, independent of the phonon occupation numbers). With reference to Figure 18.23(a), the states of interest and their unperturbed energy (neglecting electron-phonon interaction) are:

$$\begin{aligned} |i\rangle &= |\mathbf{k}_1, \mathbf{k}_2; 0\rangle & E_i &= E(\mathbf{k}_1) + E(\mathbf{k}_2), \\ |\alpha\rangle &= |\mathbf{k}_1', \mathbf{k}_2; 1\rangle & E_\alpha &= E(\mathbf{k}_1') + E(\mathbf{k}_2) + \hbar\omega_\mathbf{q}, \\ |f\rangle &= |\mathbf{k}_1', \mathbf{k}_2'; 0\rangle & E_f &= E(\mathbf{k}_1') + E(\mathbf{k}_2'). \end{aligned} \qquad (A.5)$$

Applying Eq. (A.4) with the states (A.5) and indicating with $M_\mathbf{q}$ and $M_\mathbf{q}^*$ the matrix elements for emission and absorption of a phonon of wavevector $\mathbf{q}$, we obtain

$$\langle f | H_{\text{indirect}} | i \rangle = \frac{1}{2} |M_\mathbf{q}|^2 \left[\frac{1}{E(\mathbf{k}_1) - E(\mathbf{k}_1') - \hbar\omega_\mathbf{q}} + \frac{1}{E(\mathbf{k}_2') - E(\mathbf{k}_2) - \hbar\omega_\mathbf{q}} \right]. \qquad (A.6)$$

By assuming $\hbar\omega_\mathbf{q} \approx \hbar\omega_D$, it is seen that for initial and final electron states in the energy shell of the order of $\hbar\omega_D$ around the Fermi level, the virtual process described by Eq. (A.6) leads to an attractive interaction.

It is customary to add to the term (A.6), corresponding to the diagram of Figure 18.23(a), the term corresponding to the diagram of Figure 18.23(b); the sum

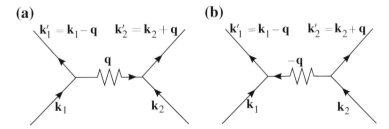

Figure 18.23 Origin of the indirect electron-electron interaction via phonons. In process (a) the electron $\mathbf{k}_1$ emits a phonon of wavevector $\mathbf{q}$, which is later absorbed by the electron $\mathbf{k}_2$. In process (b) the electron $\mathbf{k}_2$ emits a phonon of wavevector $-\mathbf{q}$, which is later absorbed by the electron $\mathbf{k}_1$.

gives

$$\langle f | H_{\text{indirect}} | i \rangle = |M_{\mathbf{q}}|^2 \left[\frac{\hbar \omega_{\mathbf{q}}}{[E(\mathbf{k}_1) - E(\mathbf{k}_1')]^2 - \hbar^2 \omega_{\mathbf{q}}^2} + \frac{\hbar \omega_{\mathbf{q}}}{[E(\mathbf{k}_2') - E(\mathbf{k}_2)]^2 - \hbar^2 \omega_{\mathbf{q}}^2} \right].$$
(A.7)

In the case of electron-phonon induced scattering $V_{\mathbf{kk'}}$ of paired electrons, initially in the state $(\mathbf{k}, -\mathbf{k})$ to the final state $(\mathbf{k'}, -\mathbf{k'})$, Eq. (A.7) gives

$$V_{\mathbf{kk'}} = |M_{\mathbf{q}}|^2 \frac{2\hbar \omega_{\mathbf{q}}}{[E(\mathbf{k}) - E(\mathbf{k'})]^2 - \hbar^2 \omega_{\mathbf{q}}^2}.$$
(A.8)

The phonon-induced electron-electron interaction (A.8) is attractive in the energy shell of width $\approx \hbar \omega_D$ around the Fermi level, and can overcome the appropriately screened repulsive Coulomb interaction between electrons.

Further Reading

Abrikosov, A. A. (1988). *Fundaments of the theory of metals*. North-Holland: Amsterdam.

Acquarone, M. (Ed.). (1996). *High temperature superconductivity*. Singapore: World Scientific.

Annett, J. F. (2004). *Superconductivity, superfluidity and condensates*. Oxford: Oxford University Press.

Bardeen, J., Cooper, L. N., & Schrieffer, J. R. (1957). Theory of Superconductivity. *Physical Review, 108*, 1175.

Barone, A., & Paternò, G. (1982). *Physics and applications of the Josephson effect*. New York: Wiley.

Bednorz, J. G., & Müller, K. A. (1986). Possible high-T_c superconductivity in the Ba-La-Cu-O system. *Zeitschrift fur Physik, 64*, 189.

Bednorz, J. G., & Müller, K. A. (1988). Perovskite-type oxides. The new approach to high-T_c superconductivity. *Reviews of Modern Physics, 60*, 585.

Bednorz, J. G., & Müller, K. A. (Eds.) (1990). *Early and recent aspects of superconductivity*. Berlin: Springer.

Burns, G. (1992). *High-temperature superconductivity*. New York: Academic Press.

Cyrot, M., & Pavuna, D. (1992). *Introduction to superconductivity and high-T_c materials*. Singapore: World Scientific.

de Gennes, P. G. (1966). *Superconductivity of metals and alloys*. New York: Benjamin; Reading, Massachusetts: Addison-Wesley, 1989.

Huebener, R. P. (1979). *Magnetic flux structure in superconductors*. Berlin: Springer.

Josephson, B. D. (1962). Possible new effects in superconducting tunneling. *Physics Letters, 1*, 251.

Larkin, A., & Varlamov, A. A. (2005). *Theory of fluctuations in superconductors*. Oxford: Clarendon Press.

Lee, P. A., Nagaosa, N., & Wen, X. -G. (2006). Doping a Mott insulator: Physics of high-temperature superconductivity. *Reviews of Modern Physics, 78*, 17.

London, F. (1950). *Superfluids*. New York: Wiley.

Parks, R. D. (Ed.). (1969). *Superconductivity* (Vols. 1 and 2). New York: Dekker.

Phillips, J. C. (1989). *Physics of high-T_c superconductors*. New York: Academic Press.

Poole, C. P., Farach, H. A., Creswick, R. J., & Prozorov, R. (2007). *Superconductivity*. Amsterdam: Academic Press.

Rao, C. N. R. (Ed.). (1991). *Chemistry of high T_c superconductors*. Singapore: World Scientific.

Rickayzen, G. (1965). *Theory of superconductivity*. New York: Wiley.

Rigamonti, A., & Varlamov, A. A. (2011). Superconductivity and applications after a century from the discovery (Vol. 5). *Scientifica Acta*, special issue. This review has the peculiar merit to capture the interest both of specialists and of beginners in the world of superconductivity.

Rogalla, H., & Kes, P. H. (Eds.). (2012). *100 years of superconductivity*. Boca Raton, FL: CRC Press, Taylor and Francis.

Rose-Innes, A. C., & Rhoderick, E. H. (1978). *Introduction to superconductivity*. Oxford: Pergamon Press.

Scalapino, D. J. (2012). A common thread: The pairing interaction for unconventional superconductors. *Reviews of Modern Physics, 84*, 1383.

Schrieffer, J. R. (1983). *Theory of superconductivity*. New York: Benjamin.

Tilley, D. R., & Tilley, J. (1990). *Superfluidity and superconductivity*. Bristol: Adam Hilger.

Tinkham, M. (1996). *Introduction to superconductivity*. New York: McGraw-Hill.

Waldram, J. R. (1996). *Superconductivity of metals and cuprates*. Bristol: Institute of Physics.

Index

Printed and bound by CPI Group (UK) Ltd, Croydon, CR0 4YY

13/10/2024

01773508-0001